Lecture ‖‖‖‖‖‖‖‖ D0347097 ‖ e   3102

Commenced Publication in 1973
Founding and Former Series Editors:
Gerhard Goos, Juris Hartmanis, and Jan van Leeuwen

Lecture Notes in Computer Science 3102

Editorial Board

Springer

Berlin
Heidelberg
New York
Hong Kong
London
Milan
Paris
Tokyo

Kalyanmoy Deb   Riccardo Poli
Wolfgang Banzhaf   Hans-Georg Beyer
Edmund Burke   Paul Darwen
Dipankar Dasgupta   Dario Floreano
James Foster   Mark Harman
Owen Holland   Pier Luca Lanzi
Lee Spector   Andrea Tettamanzi
Dirk Thierens   Andy Tyrrell (Eds.)

Genetic and Evolutionary Computation – GECCO 2004

Genetic and Evolutionary Computation Conference
Seattle, WA, USA, June 26-30, 2004
Proceedings, Part I

 Springer

Main Editor

Kalyanmoy Deb
Indian Institute of Technology Kanpur
Department of Mechanical Engineering
Kanpur, Pin 208 016, India
E-mail: deb@iitk.ac.in

CR Subject Classification (1998): F.1-2, D.1.3, C.1.2, I.2.6, I.2.8, I.2.11, J.3, G.2

ISSN 0302-9743
ISBN 3-540-22344-4 Springer-Verlag Berlin Heidelberg New York

Typesetting: Camera-ready by author, data conversion by PTP-Berlin, Protago-TeX-Production GmbH
Printed on acid-free paper   SPIN: 11017745   06/3142   5 4 3 2 1 0

Volume Editors

Riccardo Poli
Owen Holland
University of Essex
Department of Computer Science
Wivenhoe Park, Colchester, CO4 3SQ, UK
E-mail: {rpoli,owen}@essex.ac.uk

Wolfganz Banzhaf
Department of Computer Science
Memorial University of Newfoundland
St. John's, NL, A1B 3X5, Canada
E-mail: banzhaf@cs.mun.ca

Hans-Georg Beyer
University of Dortmund
Systems Analysis Research Group
Joseph-von-Fraunhoferstr. 20
44221 Dortmund, Germany
hans-georg.beyer@cs.uni-dortmund.de

Edmund Burke
University of Nottingham
School of Computer Science
and Information Technology
Jubilee Campus, Nottingham NG8 2BB, UK
E-mail: ekb@cs.nott.ac.uk

Paul Darwen
Anadare Pty Ltd
14 Annie Street, Brisbane
Queensland 4066, Australia
E-mail: darwen@ieee.org

Dipankar Dasgupta
University of Memphis
Division of Computer Science
Memphis, TN 38152, USA
E-mail: dasgupta@memphis.edu

Dario Floreano
Swiss Federal Institute of Technology
Autonomous Systems Laboratory
1015 Lausanne, Switzerland
E-mail: dario.floreano@epfl.ch

James Foster
University of Idaho
Department of Computer Science
P.O. Box 441010, Moscow
ID 83844-1010, USA
E-mail: foster@uidaho.edu

Mark Harman
Brunel University
Department of Information Systems
and Computing
Uxbridge, Middlesex, 11B8 3PH, UK
E-mail: Mark.Harman@brunel.ac.uk

Pier Luca Lanzi
Politecnico di Milano
Dipartimento di Elettronica e Informazione
Piazza Leonardo da Vinci, 32
20133 Milan, Italy
E-mail: lanzi@elet.polimi.it

Lee Spector
Hampshire College, Cognitive Science
Amherst, MA 01002, USA
E-mail: lspector@hampshire.edu

Andrea G.B. Tettamanzi
University of Milan
Department of Information Technology
Via Bramante 65, 26013 Crema, Italy
E-mail: andrea.tettamanzi@unimi.it

Dirk Thierens
Utrecht University
Department of Information
and Computing Sciences
P.O. Box 80.089, 3508 TB Utrecht
The Netherlands
E-mail: dirk.thierens@cs.uu.nl

Andrew M. Tyrrell
The University of York
The Department of Electronics
Heslington, York YO10 5DD, UK
E-mail: amt@ohm.york.ac.uk

Preface

These proceedings contain the papers presented at the sixth annual Genetic and Evolutionary Computation Conference (GECCO 2004). The conference was held in Seattle, during June 26–30, 2004.

A total of 460 papers were submitted to GECCO 2004. After a rigorous double-blind reviewing process, 230 papers were accepted for full publication and oral presentation at the conference, resulting in an acceptance rate of 50%. An additional 104 papers were accepted as posters with two-page extended abstracts included in these proceedings.

This year's GECCO constituted the union of the Ninth Annual Genetic Programming Conference (which has met annually since 1996) and the Thirteenth International Conference on Genetic Algorithms (which, with its first meeting in 1985, is the longest running conference in the field). Since 1999, these conferences have merged to produce a single large meeting that welcomes an increasingly wide array of topics related to genetic and evolutionary computation.

Since the fifth annual GECCO conference, the proceedings have been published by Springer-Verlag as part of their Lecture Notes in Computer Science (LNCS) series. This makes the proceedings available in many libraries as well as online, widening the dissemination of the research presented at the conference. In addition to these proceedings volumes, each participant of the GECCO 2004 conference received a CD containing electronic versions of the papers presented.

A new track entitled 'Biological Applications' was introduced this year to emphasize the use of evolutionary computing methods to various biological applications, such as bioinformatics and others.

In addition to the presentation of the papers contained in the proceedings, the conference included 16 workshops, 32 tutorials by leading specialists, the Evolutionary Computation in Industry special track, and presentation of late-breaking papers.

GECCO is sponsored by the International Society for Genetic and Evolutionary Computation (ISGEC). The ISGEC by-laws contain explicit guidance on the organization of the conference, including the following principles:

(i) GECCO should be a broad-based conference encompassing the whole field of genetic and evolutionary computation.

(ii) Papers will be published and presented as part of the main conference proceedings only after being peer reviewed. No invited papers shall be published (except for those of up to three invited plenary speakers).

(iii) The peer review process shall be conducted consistently with the principle of division of powers performed by a multiplicity of independent program committees, each with expertise in the area of the paper being reviewed.

(iv) The determination of the policy for the peer review process for each of the conference's independent program committees and the reviewing of papers for each program committee shall be performed by persons who occupy their positions by virtue of meeting objective and explicitly stated qualifications based on their previous research activity.

(v)   Emerging areas within the field of genetic and evolutionary computation shall be actively encouraged and incorporated in the activities of the conference by providing a semi-automatic method for their inclusion (with some procedural flexibility extended to such emerging new areas).

(vi) The percentage of submitted papers that are accepted as regular full-length papers (i.e., not posters) shall not exceed 50%.

These principles help ensure that GECCO maintains high quality across the diverse range of topics it includes.

Besides sponsoring the conference, the ISGEC supports the field in other ways. ISGEC sponsors the biennial "Foundations of Genetic Algorithms" workshop on theoretical aspects of all evolutionary algorithms. The journals *Evolutionary Computation* and *Genetic Programming and Evolvable Machines* are also supported by ISGEC. All ISGEC members (including students) receive subscriptions to these journals as part of their membership. ISGEC membership also includes discounts on GECCO registration rates as well as discounts on other journals. More details on ISGEC can be found online at http://www.isgec.org.

Many people volunteered their time and energy to make this conference a success. The following people in particular deserve the gratitude of the entire community for their outstanding contributions to GECCO 2004:

- Riccardo Poli, the General Chair of GECCO 2004 for his tireless efforts in organizing every aspect of the conference, which started well before GECCO 2003 took place in Chicago in July 2003
- David E. Goldberg, John Koza and Riccardo Poli, members of the Business Committee, for their guidance and financial oversight
- Stefano Cagnoni, for coordinating the workshops
- Maarten Keijzer, for editing the late breaking papers
- Past conference organizers, James Foster and Erick Cantú-Paz, for their constant help and advice
- John Koza, for his efforts as publicity chair
- Simon Lucas, for arranging competitions during GECCO 2004
- Mike Cattolico, for local arrangements
- Pat Cattolico, for her help in the local organization of the conference
- Carol Hamilton, Ann Stolberg, and the rest of the AAAI staff for their outstanding efforts administering the conference
- Thomas Preuss, for maintaining the ConfMaster Web-based paper review system
- Gerardo Valencia, for Web programming and design
- Jennifer Ballentine, Lee Ballentine and the staff of Professional Book Center, for assisting in the production of the proceedings
- Alfred Hofmann and Ursula Barth of Springer-Verlag for the production of the GECCO 2004 proceedings; and

the sponsors who made generous contributions to support student travel grants:

- Air Force Office of Scientific Research
- New Light Industries
- Philips Research
- Tiger Mountain Scientic
- Unilever.

The track chairs deserve special thanks. Their efforts in recruiting program committees, assigning papers to reviewers, and making difficult acceptance decisions in relatively short times, were critical to the success of the conference:

- Owen Holland, A-Life, Adaptive Behavior, Agents, and Ant Colony Optimization,
- Dipankar Dasgupta, Artificial Immune Systems
- James Foster and Wolfgang Banzhaf, Biological Applications
- Paul Darwen, Coevolution
- Hans-Georg Beyer, Evolution Strategies, Evolutionary Programming
- Dario Floreano, Evolutionary Robotics
- Edmund Burke, Evolutionary Scheduling and Routing
- Andy Tyrrell, Evolvable Hardware
- Dirk Thierens, Genetic Algorithms
- Lee Spector, Genetic Programming
- Pier Luca Lanzi, Learning Classifier Systems
- Andrea Tettamanzi, Real World Applications
- Mark Harman, Search Based Software Engineering

The conference was held in cooperation and/or affiliation with:

- The American Association for Artificial Intelligence (AAAI)
- The 2004 NASA/DoD Conference on Evolvable Hardware
- *Evolutionary Computation*
- *Genetic Programming and Evolvable Machines*
- *Journal of Scheduling*
- *Journal of Hydroinformatics*
- *Applied Soft Computing*

Above all, our special thanks are due to the numerous researchers and practitioners who submitted their best work to GECCO 2004, reviewed the work of others, presented tutorials, organized workshops, or volunteered their time in any other way. I am sure that the contributors to this proceedings will be proud of the results of their efforts and readers will get a glimpse of the current activities in the field of genetic and evolutionary computation.

April 2004

Kalyanmoy Deb
Editor-in-Chief, GECCO 2004

GECCO 2004 Conference Organization

Conference Committee

General Chair: Riccardo Poli
Proceedings Editor-in-Chief: Kalyanmoy Deb
Business Committee: David E. Goldberg, John Koza, Riccardo Poli
Chairs of Program Policy Committees:
    Owen Holland, A-Life, Adaptive Behavior, Agents, and Ant Colony Optimization
    Dipankar Dasgupta, Artificial Immune Systems
    James Foster and Wolfgang Banzhaf, Biological Applications
    Paul Darwen, Coevolution
    Hans-Georg Beyer, Evolution Strategies, Evolutionary Programming
    Dario Floreano, Evolutionary Robotics
    Edmund Burke, Evolutionary Scheduling and Routing
    Andy Tyrrell, Evolvable Hardware
    Dirk Thierens, Genetic Algorithms
    Lee Spector, Genetic Programming
    Pier Luca Lanzi, Learning Classifier Systems
    Andrea Tettamanzi, Real World Applications
    Mark Harman, Search Based Software Engineering

Late Breaking Papers Chair: Maarten Keijzer
Workshops Chair: Stefano Cagnoni

Workshop Organizers

E. Costa, F. Pereira , G. Raidl, Application of Hybrid Evolutionary Algorithms to Complex Optimization Problems
S.C. Upton and D.E. Goldberg, Military and Security Applications of Evolutionary Computation
H. Lipson, E. De Jong and J. Koza, Modularity, Regularity and Hierarchy in Open-Ended Evolutionary Computation
H. Suzuki and H. Sawai, Evolvability in Evolutionary Computation (EEC)
I. Parmee, Interactive Evolutionary Computing
M. Pelikan, K. Sastry and D. Thierens, Optimization by Building and Using Probabilistic Models (OBUPM 2004)
W. Stolzmann, P.L. Lanzi, S.W. Wilson, International Workshop on Learning Classifier Systems (IWLCS)
S. Mueller, S. Kern, N. Hansen and P. Koumoutsakos, Learning, Adaptation, and Approximation in EC, Jiri Ocenasek
M. O'Neill and C. Ryan, Grammatical Evolution (GEWS 2004)
T. Yu, Neutral Evolution in Evolutionary Computation
J.F. Miller, Regeneration and Learning in Developmental Systems (WORLDS)

I. Garibay, G. Holifield and A.S. Wu, Self-Organization on Representations for
Genetic and Evolutionary Algorithms
A. Wright and N. Richter, Evolutionary Computation Theory
Jason H. Moore and Marylyn D. Ritchie, Biological Applications of Genetic and
Evolutionary Computation (BioGEC 2004)
T. Riopka, Graduate Student Workshop
M.M. Meysenburg, Undergraduate Student Workshop

Tutorial Speakers

Erik Goodman, Genetic Algorithms
John Koza, Genetic Programming
Thomas Bäck, Evolution Strategies
Kenneth De Jong, A Unified Approach to EC
Tim Kovacs, Learning Classifier Systems
Martin Pelikan, Probabilistic Model-Building GAs
Russ Eberhart, Particle Swarm Optimization
Steffen Christensen and Mark Wineberg, Introductory Statistics for Evolution-
ary Computation
W.B. Langdon, Genetic Programming Theory
Jonathan Rowe, Genetic Algorithm Theory
J. Foster and W. Banzhaf, Biological Applications
Chris Stephens, Taxonomy and Coarse Graining in EC
Darrell Whitley, No Free Lunch
Kalyanmoy Deb, Multiobjective Optimization with EC
Ingo Wegener, Computational Complexity and EC
Julian Miller, Evolvable Physical Media
Tetsuya Higuchi, Evolvable Hardware Applications
Franz Rothlauf, Representations
Lee Altenberg, Theoretical Population Genetics
Ingo Rechenberg, Bionik: Building on Biological Evolution
Marco Tomassini, Spatially Structured EAs
Hideyuki Takagi, Interactive Evolutionary Computation
Garry Greenwood, Evolutionary Fault Tolerant Systems
Maarten Keijzer, GP for Symbolic Regression
Conor Ryan, Grammatical Evolution
Dario Floreano, Evolutionary Robotics
Al Biles, Evolutionary Music
Peter Ross, EAs for Combinatorial Optimization
Jürgen Branke, Optimization in Dynamic Environments
Ian Parmee, Evolutionary Algorithms for Design
Xin Yao, Evolving Neural Networks
Arthur Kordon, Guido Smits and Mark Kotanchek, Industrial Evolutionary
Computing

Keynote Speakers

Leroy Hood, President, Institute for Systems Biology, Seattle
François Baneyx, Professor of Chemical Engineering and adjunct Professor of
Bioengineering, Center for Nanotechnology at University of Washington, Seattle

Members of the Program Committee

Hussein Abbass	Wilker Bruce	Leandro de Castro
Andrew Adamatzky	Peter Brucker	Patrick De Causmaecker
Adam Adamopoulos	Anthony Bucci	Ivanoe De Falco
Alexandru Agapie	Bill P. Buckles	Hugo de Garis
Jose Aguilar	Dirk Bueche	Edwin de Jong
Jesus Aguilar-Ruiz	Larry Bull	David de la Fuente
Hernan Aguirre	Martin Butz	Anthony Deakin
Uwe Aickelin	Stefano Cagnoni	Kalyanmoy Deb
Javier Alcaraz Soria	Xiaoqiang Cai	Myriam Delgado
Lee Altenberg	Alexandre Caminada	Medha Dhurandhar
Giuliano Antoniol	Erick Cantú-Paz	Ezequiel Di Paolo
Shawki Areibi	Nachol Chaiyaratana	Jose Javier Dolado Cosin
Tughrul Arslan	Uday Chakraborty	Keith Downing
Dan Ashlock	Partha Chakroborty	Kath Dowsland
Anne Auger	Weng Tat Chan	Gerry Dozier
R. Muhammad Atif Azad	Alastair Channon	Rolf Drechsler
B.V. Babu	Kumar Chellapilla	Stefan Droste
Thomas Bäck	Shu-Heng Chen	Tim Edwards
Karthik Balakrishnan	Ying-ping Chen	Aniko Ekart
Gianluca Baldassarre	Prabhas	Mark Embrechts
Julio Banga	Chongstitvatana	Michael Emmerich
Ranieri Baraglia	John Clark	Maria Fasli
Alwyn Barry	Maurice Clerc	Francisco Fernandez
Thomas Bartz-Beielstein	André Coelho	Bogdan Filipic
Cem Baydar	Carlos Coello Coello	Peter Fleming
Theodore Belding	Myra Cohen	Stuart Flockton
Fevzi Bell	David Coley	Carlos Fonseca
Michael Bender	Philippe Collard	James Foster
Peter Bentley	Pierre Collet	Alex Freitas
Aviv Bergman	Clare Congdon	Clemens Frey
Ester Bernado-Mansilla	David Corne	Christian Gagné
Tim Blackwell	Luis Correia	Luca Gambardella
Jacek Blazewicz	Ernesto Costa	Josep Maria
Lashon Booker	Carlos Cotta	Garrell-Guiu
Peter Bosman	Peter Cowling	Michel Gendreau
Klaus Bothe	Bart Craenen	Pierre Gerard
Leonardo Bottaci	Keshav Dahal	Andreas Geyer-Schulz
Jürgen Branke	Rajarshi Das	Robert Ghanea-Hercock

Marco César Goldbarg
Faustino Gomez
Jonatan Gomez
Fabio Gonzalez
Tim Gosling
Jens Gottlieb
Buster Greene
Garrison Greenwood
Gary Greenwood
Michael Gribskov
Hans-Gerhard Gross
Steven Gustafson
Charlie Guthrie
Walter Gutjahr
Pauline Haddow
Hani Hagras
Hisashi Handa
Nikolaus Hansen
Dave Harris
Emma Hart
Inman Harvey
Jun He
Robert Heckendorn
Jeffrey Herrmann
Rob Hierons
David Hillis
Steven Hofmeyr
John Holmes
Jeffrey Horn
Daniel Howard
Jianjun Hu
Phil Husbands
Hitoshi Iba
Christian Igel
Auke Jan Ijspeert
Akio Ishiguro
Christian Jacob
Thomas Jansen
Yaochu Jin
Colin Johnson
Bryan Jones
Bryant Julstrom
Mahmoud Kaboudan
Sanza Kazadi
Maarten Keijzer
Douglas Kell
Graham Kendall

Mathias Kern
Didier Keymeulen
Joshua Knowles
Arthur Kordon
Bogdan Korel
Erkan Korkmaz
Petros Koumoutsakos
Tim Kovacs
Natalio Krasnogor
Krzysztof Krawiec
Kalmanje Krishnakumar
Renato Krohling
Gabriella Kůkai
Rajeev Kumar
Raymond Kwan
Sam Kwong
Han La Poutre
Shyong Lam
Gary Lamont
W. B. Langdon
Pedro Larranaga
Jesper Larsen
Claude Lattaud
Marco Laumanns
Claude Le Pape
Martin Lefley
Tom Lenaerts
K. S. Leung
Lukas Lichtensteiger
Anthony Liekens
Hod Lipson
Fernando Lobo
Jason Lohn
Michael Lones
Sushil Louis
Jose Lozano
Evelyne Lutton
Bob MacCallum
Nicholas Macias
Ana Madureira
Spiros Mancoridis
Vittorio Maniezzo
Elena Marchiori
Peter Martin
Andrew Martin
Alcherio Martinoli
Iwata Masaya

Shouichi Matsui
Dirk Mattfeld
Barry McCollum
Nic McPhee
Jörn Mehnen
Karlheinz Meier
Lawrence Merkle
Jean-Arcady Meyer
Christoph Michael
Zbigniew Michalewicz
Olivier Michel
Martin Middendorf
Stuart Middleton
Orazio Miglino
Julian Miller
Brian Mitchell
Chilukuri Mohan
Francesco Mondada
David Montana
Byung-Ro Moon
Frank Moore
Jason Moore
Alberto Moraglio
J. Manuel Moreno
Masaharu Munetomo
Hajime Murao
Kazuyuki Murase
Olfa Nasraoui
Bart Naudts
Norberto Eiji Nawa
Chrystopher Nehaniv
Miguel Nicolau
Fernando Nino
Stefano Nolfi
Peter Nordin
Bryan Norman
Cedric Notredame
Wim Nuijten
Una-May O'Reilly
Markus Olhofer
Sigaud Olivier
Michael O'Neill
Ender Ozcan
Anil Patel
Shail Patel
Martin Pelikan

Carlos-Andrés
  Pena-Reyes
Francisco Pereira
Sanja Petrovic
Hartmut Pohlheim
Daniel Polani
Marie-Claude Portmann
Jean-Yves Potvin
Alexander Pretschner
Thomas Preuss
Mike Preuss
Adam Prugel-Bennett
Joao Pujol
Günther Raidl
Khaled Rasheed
Al Rashid
Thomas Ray
Tapabrata Ray
Victor Rayward-Smith
Patrick Reed
Richard Reeve
Colin Reeves
Marek Reformat
Andreas Reinholz
Rick Riolo
Jose Riquelme Santos
Marc Roper
Franz Rothlauf
Rajkumar Roy
Guenter Rudolph
Kazuhiro Saitou
Arthur Sanderson
Eugene Santos
Kumara Sastry
Yuji Sato
Thorsten Schnier
Marc Schoenauer
Sonia Schulenburg

Alan Schultz
Hans-Paul Schwefel
Mikhail Semenov
Sandip Sen
Bernhard Sendhoff
Kisung Seo
Martin Shepperd
Alaa Sheta
Richard Skalsky
Jim Smith
Don Sofge
Terry Soule
Pieter Spronck
Peter Stadler
Kenneth Stanley
Chris Stephens
Harmen Sthamer
Christopher Stone
Matthew Streeter
Thomas Stuetzle
Raj Subbu
Keiki Takadama
Kiyoshi Tanaka
Uwe Tangen
Alexander Tarakanov
Gianluca Tempesti
Sam Thangiah
Scott Thayer
Lothar Thiele
Jonathan Thompson
Jonathan Timmis
Jon Timmis
Ashutosh Tiwari
Marco Tomassini
Jim Torresen
Paolo Toth
Edward Tsang
Shigeyoshi Tsutsui

Supiya Ujjin
Steven van Dijk
Jano van Hemert
Frederik Vandecasteele
Greet Vanden Berghe
Leonardo Vanneschi
Robert Vanyi
Oswaldo Velez-Langs
J. L. Verdegay
Fernando Von Zuben
Roger Wainwright
Matthew Wall
Harold Wareham
Jean-Paul Watson
Everett Weber
Ingo Wegener
Karsten Weicker
Peter Whigham
Shimon Whiteson
Darrell Whitley
R. Wiegand
Stewart Wilson
Mark Wineberg
Alden Wright
Annie Wu
Zheng Wu
Jinn-Moon Yang
Tina Yu
Hongnian Yu
Ricardo Zebulum
Andreas Zell
Byoung-Tak Zhang
Gengui Zhou
Fan Zhun
Tom Ziemke
Lyudmilla Zinchenko
Eckart Zitzler

A Word from the Chair of ISGEC

You may have just picked up your proceedings, in hard copy and CD-ROM, at GECCO 2004. We've chosen once again to work with Springer-Verlag, including our proceedings as part of their Lecture Notes in Computer Science (LNCS) series, which makes them available in many libraries, broadening the impact of the GECCO conference.

If you're now at GECCO 2004, we, the organizers, hope your experience is memorable and productive, and you will find the proceedings to be of continuing value. The opportunity for first-hand interaction among authors and other participants at GECCO is a big part of what makes it exciting, and we all hope you come away with many new insights and ideas.

If you were unable to come to GECCO 2004 in person, I hope you'll find many stimulating ideas from the world's leading researchers in evolutionary computation reported in the proceedings, and that you'll be able to participate in future GECCO conferences, for example, next year, in the Washington, DC area!

The International Society for Genetic and Evolutionary Computation, sponsoring organization of the annual GECCO conferences, is a young organization, formed through merger of the International Society for Genetic Algorithms (sponsor of the ICGA conferences) and the organization responsible for the annual Genetic Programming Conferences. It depends strongly on the voluntary efforts of many of its members. It is designed to promote not only exchange of ideas among innovators and practitioners of well-known methods such as genetic algorithms, genetic programming, evolution strategies, evolutionary programming, learning classifier systems, etc., but also the growth of newer areas such as artificial immune systems, evolvable hardware, agent-based search, and others. One of the founding principles is that ISGEC operates as a confederation of groups with related but distinct approaches and interests, and their mutual prosperity is assured by their representation in the program committees, editorial boards, etc., of the conferences and journals with which ISGEC is associated. This also insures that ISGEC and its functions continue to improve and evolve with the diversity of innovation that has characterized our field.

The ISGEC saw many changes last year, in addition to its growth in membership. We anticipate yet more advances in the next year. A second round of Fellows and Senior Fellows will be added to our society this year, after last year's inaugural group. GECCO continues to be subject to dynamic development – the many new tutorials, workshop topics, and tracks will evolve again next year, seeking to follow and encourage the developments of the many fields represented at GECCO. The best paper awards will be presented for the third time at this GECCO, and we hope many of you will participate in the balloting. This year, most presentations at GECCO will once again be made electronically, displayed with the LCD projectors that ISGEC purchased last year. Our journals, Evolutionary Computation and Genetic Programming and Evolvable Machines, continue to prosper, and we are exploring ways to make them even more widely available.

The ISGEC is your society, and we urge you to become involved or continue your involvement in its activities, to the mutual benefit of the whole evolutionary computation community. Three members were re-elected to five-year terms on the Executive Board at GECCO 2003 – Ken De Jong, David Goldberg, and Erik Goodman.

Since that time, the ISGEC has been active on many issues, through actions of the Board and our two Councils – the Council of Authors and the Council of Conferences. Last year, the Board voted to combine the Council of Authors and Council of Editors into a single body, the Council of Authors.

The organizers of GECCO 2004 are shown in this front matter, but special thanks are due to Riccardo Poli, General Chair, and Kalyanmoy Deb, Editor-in-Chief of the proceedings, as well as to John Koza and Dave Goldberg, the Business Committee. Each year has seen many new features in GECCO, and it is the outstanding efforts of this group that "make GECCO come together."

Of course, we all owe a great debt to those who chaired or served on the various Core and Special Program Committees that reviewed all of the papers for GECCO 2004. Without their effort, it would not be possible to put on a meeting of this quality.

Another group also deserves the thanks of GECCO participants and ISGEC members – the members of the ISGEC Executive Board and Councils, who are listed on the next page. I am particularly indebted to them for their thoughtful contributions to the organization and their continuing demonstrations of concern for the welfare of the ISGEC.

I invite you to communicate with me (goodman@egr.msu.edu) if you have questions or suggestions for ways ISGEC can be of greater service to its members, or if you would like to get more involved in ISGEC and its functions.

Don't forget about the eighth Foundations of Genetic Algorithms (FOGA) workshop, also sponsored by ISGEC, the biennial event that brings together the world's leading theorists on evolutionary computation. FOGA will be held January 5–9, 2005 at the University of Aizu, Japan, which will be a fascinating place to visit for those of us who haven't spent much time in Japan. I hope you'll join many of your fellow ISGEC members there!

Finally, I hope to see you at GECCO 2005 in the Washington, DC area. Get your ideas for new things for GECCO 2005 to Una-May O'Reilly, the General Chair of GECCO 2005, when you see her at GECCO 2004, and please check the ISGEC Web site, www.isgec.org, regularly for details as the planning for GECCO 2005 continues.

Erik D. Goodman
ISGEC Chair

ISGEC Executive Board

Council of Authors

David E. Goldberg, University of Illinois at Urbana-Champaign
Jens Gottlieb, SAP AG
Wolfgang A. Halang, FernUniversität, Hagen
John H. Holland, University of Michigan & Sante Fe Institute
Hitoshi Iba, University of Tokyo
Christian Jacob, University of Calgary
Francisco Herrera, University of Granada
Yaochu Jin, Honda Research Institute Europe
Robert E. Keller, University of Dortmund
Dimitri Knjazew, SAP AG
John R. Koza, Stanford University
Sam Kwong, City University of Hong Kong
W.B. Langdon, University College, London
Dirk C. Mattfeld, University of Bremen
Pinaki Mazumder, University of Michigan
Zbigniew Michalewicz, University of North Carolina at Charlotte
Eric Michielssen, University of Illinois at Urbana-Champaign
Melanie Mitchell, Oregon Health and Science University
Byung-Ro Moon, Seoul National University
Michael O'Neill, University of Limerick
Ian Parmee, University of North Carolina at Charlotte
Witold Pedrycz, University of Alberta
Frederick E. Petry, University of North Carolina at Charlotte
Riccardo Poli, University of Essex
Rajkumar Roy, Cranfield University
Elizabeth M. Rudnick, University of Illinois at Urbana-Champaign
Conor Ryan, University of Limerick
Marc Schoenauer, INRIA Futurs
Moshe Sipper, Swiss Federal Institute of Technology
James E. Smith, University of the West of England
Terence Soule, University of Idaho
William M. Spears, University of Wyoming
Lee Spector, Hampshire College
Wallace K.S. Tang, Swiss Federal Institute of Technology
Adrian Thompson, University of Sussex
Jose L. Verdegay, University of Granada
Michael D. Vose, University of Tennessee
Darrell Whitley, Colorado State University
Man Leung Wong, Lingnan University

Council of Conferences, Una-May O'Reilly (chair)

The purpose of the Council of Conferences is to provide information about the numerous conferences that are available to researchers in the field of Genetic and Evolutionary Computation, and to encourage them to coordinate their meetings to maximize our collective impact on science.

ACDM, Adaptive Computing in Design and Manufacture, Bristol, UK, April 2004, Ian Parmee, Ian.Parmee@uwe.ac.uk

EuroGP, European Conference on Genetic Programming, Coimbra, Portugal, April 2004, Ernesto Costa, ernesto@dei.uc.pt

EvoCOP, European Conference on Evolutionary Computation in Combinatorial Optimization, Coimbra, Portugal, April 2004, Günther Raidl, raidl@ads.tuwien.ac.at and Jens Gottlieb, jens.gottlieb@sap.com

EvoWorkshops, European Evolutionary Computing Workshops, Portugal, Coimbra, Portugal, April 2004, Stefano Cagnoni, cagnoni@ce.unipr.it

FOGA, Foundations of Genetic Algorithms Workshop, Fukushima, Japan, January 2005, Lothar M. Schmitt, info@foga05.org

GECCO 2004, Genetic and Evolutionary Computation Conference, Seattle, USA, June 2004, Riccardo Poli, rpoli@essex.ac.uk

PATAT 2004, 5th International Conference on the Practice and Theory of Automated Timetabling, Pittsburgh, USA, August 2004, Edmund Burke, ekb@cs.nott.ac.uk

PPSN-VIII, Parallel Problem Solving from Nature, Birmingham, UK, September 2004, Xin Yao, xin@cs.bham.ac.uk

SAB, 8th international conference on Simulation of Adaptive Behavior, Los Angeles, USA, July 2004, John Hallam, john@mip.sdu.dk and Jean-Arcady Meyer, jean-arcady.meyer@lip6.fr

EMO 2005, 3rd Evolutionary Multi-Criterion Optimization, Guanajuato, Mexico, March 2005, Carlos Coello Coello, coello@cs.cinvestav.mx

An up-to-date roster of the Council of Conferences is available online at http://www.isgec.org/conferences.html.

Papers Nominated for Best Paper Awards

In 2002, ISGEC created a best paper award for GECCO. As part of the double blind peer review, the reviewers were asked to nominate papers for best paper awards. The Chairs of Core and Special Program Committees selected the papers that received the most nominations for consideration by the conference. One winner for each program track was chosen by secret ballot of the GECCO attendees after the papers had been presented in Chicago. The titles and authors of all 32 papers nominated for the best paper award for GECCO 2004 are given below:

Robot Trajectory Planner Using Multi-objective Genetic Algorithm Optimization: E.J. Solteiro Pires, J.A. Tenreiro Machado, and P.B. de Moura Oliveira   I-615

Evolved Motor Primitives and Sequences in a Hierarchical Recurrent Neural Network: Rainer Paine and Jun Tani   I-603

Actuator Noise in Recombinant Evolution Strategies on General Quadratic Fitness Models: Hans-Georg Beyer   I-654

An Analysis of the $(\mu + 1)$ EA on Simple Pseudo-Boolean Functions: Carsten Witt   I-761

On the Choice of the Population Size: Tobias Storch   I-748

Table of Contents – Part I

A-Life, Adaptive Behavior, Agents, and Ant Colony Optimization – Posters

Artificial Immune Systems

Artificial Immune Systems – Posters

Biological Applications

Biological Applications – Posters

Coevolution

Coevolution – Posters

Evolutionary Robotics

Evolutionary Robotics – Poster

Evolution Strategies/Evolutionary Programming

Evolution Strategies/Evolutionary Programming – Posters

Evolvable Hardware

Genetic Algorithms

Table of Contents – Part II

Volume II

Genetic Algorithms (Continued)

Genetic Algorithms – Posters

Genetic Programming

Genetic Programming – Posters

Learning Classifier Systems

Learning Classifier Systems – Poster

Real World Applications

Real World Applications – Posters

Search-Based Software Engineering

Search-Based Software Engineering – Posters

Efficient Evaluation Functions for Multi-rover Systems

Adrian Agogino[1] and Kagan Tumer[2]

[1] University of California Santa Cruz, NASA Ames Research Center, Mailstop 269-3,
Moffett Field CA 94035, USA,
`adrian@email.arc.nasa.gov`
[2] NASA Ames Research Center, Mailstop 269-3, Moffett Field CA 94035, USA,
`kagan@email.arc.nasa.gov`

Abstract. Evolutionary computation can successfully create control policies for single-agent continuous control problems. This paper extends single-agent evolutionary computation to multi-agent systems, where a large collection of agents strives to maximize a global fitness evaluation function that rates the performance of the entire system. This problem is solved in a distributed manner, where each agent evolves its own population of neural networks that are used as the control policies for the agent. Each agent evolves its population using its own agent-specific fitness evaluation function. We propose to create these agent-specific evaluation functions using the theory of collectives to avoid the coordination problem where each agent evolves neural networks that maximize its own fitness function, yet the system as a whole achieves low values of the global fitness function. Instead we will ensure that each fitness evaluation function is both "aligned" with the global evaluation function and is "learnable," i.e., the agents can readily see how their behavior affects their evaluation function. We then show how these agent-specific evaluation functions outperform global evaluation methods by up to 600% in a domain where a collection of rovers attempts to maximize the amount of information observed while navigating through a simulated environment.

1 Introduction

Evolutionary computation combined with neural networks can be very effective in finding solutions to continuous single-agent control tasks, such as pole balancing, robot navigation and rocket control [10,7,8]. The single-agent task of these evolutionary computation methods is to produce a highly fit neural network, which is used as the controller for the agent. Applying these evolutionary computation methods to certain multi-agent problems such as controlling constellations of satellites, constructing distributed algorithms and routing over a data network offer a promising approach to solving difficult, distributed control problems. Unfortunately the single-agent methods cannot be extended directly to a large multi-agent environment due to the large state-space and possible communication limitations. Instead we use an alternative approach of having

K. Deb et al. (Eds.): GECCO 2004, LNCS 3102, pp. 1–11, 2004.

each agent use its own evolutionary algorithm and attempt to maximize its own fitness evaluation function. For such a system to produce good global solutions, two fundamental issues have to be addressed: (i) ensuring that, as far as the provided global evaluation function is concerned, the agents do not work at cross-purposes (i.e., making sure that the private goals of the agents and the global goal are "aligned"); and (ii) ensuring that the agents' fitness evaluation functions are "learnable" (i.e., making sure the agents can readily see how their behavior affects their evaluation function). This paper provides a solution that satisfies both criteria in the problem of coordinating a collection of planetary exploration rovers based on continuous sensor inputs.

Current evolutionary computation methods address multi-agent systems in a number of different ways [2,9,1]. In [2], the algorithm takes advantage of a large number of agents to speed up the evolution process in domains where agents do not have the problem of working at cross-purposes. In [9] beliefs about about other agents are update through global and hand-tailored fitness functions. In addition ant colony algorithms [6] solve the coordination problem by utilizing "ant trails," providing good results in path-finding domains. Instead this paper presents a framework based on the theory of collectives that directs the evolutionary process so that agents do not work at cross-purposes, but still evolve quickly. This process is performed by giving each agent its own fitness evaluation function that is both aligned with the global evaluation function and as easy as possible for the agent to maximize. These agents can then use these evaluation functions in conjunction with the system designer's choice of evolutionary computation method. New evolutionary computation methods can replace the one used here without changing the evaluation functions, allowing the latest advances in evolutionary computation to be leveraged, without modifying the design of the overall system.

This paper will first give a brief overview of the theory of collectives in Section 2, showing how to derive agent evaluation functions that are both learnable and aligned with the global evaluation function. In Section 3 we discuss the "Rover Problem" testbed where a collection of planetary rovers use neural networks to determine their movements based on a continuous-valued array of sensor inputs. In Section 4 we compare the effectiveness of three different evaluation functions. We first use results from the previous section to derive fitness evaluation functions for the agents in a simple version of the Rover Problem in a static environment. Then we show how these methods perform in a more realistic domain with a changing environment. Results show up to a 600% increase in the performance of agents using agent-specific evaluation functions.

2   Multi-agent System Evaluation Functions

This section summarizes how to derive good evaluation functions, using the theory of collectives described by Wolpert and Tumer [11]. We first assume that there is a **global evaluation function**, $G(z)$, which is a function of all of the environmental variables and the actions of all the agents, z. The goal of the multi-agent system is to maximize $G(z)$. However, the agents do not maximize $G(z)$ directly. Instead each agent, η, attempts to maximize its **private evaluation function** $g_\eta(z)$. The goal is to design $g(z)$s such that when all of the $g(z)$s are close to being maximized, $G(z)$ is also close to being maximized.

2.1   Factoredness and Learnability

For high values of the global evaluation function, G, to be achieved, the private evaluation functions need to have two properties, **factoredness** and **learnability**. First we want the private evaluation functions of each agent to be factored with respect to G, intuitively meaning that an action taken by an agent that improves its private evaluation function also improves the global evaluation function (i.e. G and g_η are aligned). Specifically when agent η takes an action that increases G then g_η should also increase. Formally an evaluation function g is **factored** when:

$$g_\eta(z) \geq g_\eta(z') \Leftrightarrow G(z) \geq G(z') \quad \forall z, z' \text{ s.t. } z_{-\eta} = z'_{-\eta} \ .$$

where $z_{-\eta}$ and $z'_{-\eta}$ contain the components of z and z' respectively, that are not influenced by agent η.

Second, we want the agents' private evaluation functions to have high **learnability**, intuitively meaning that an agent's evaluation function should be sensitive to its own actions and insensitive to actions of others. As a trivial example, any "team game" in which all the private functions equal G is factored [5]. However such systems often have low learnability, because in a large system an agent will have a difficult time discerning the effects of its actions on G. As a consequence, each η may have difficulty achieving high g_η in a team game. We call this signal/noise effect learnability:

$$\lambda_{\eta,g_\eta}(\zeta) \equiv \frac{\|\boldsymbol{\nabla}_{\zeta_\eta} g_\eta(\zeta)\|}{\|\boldsymbol{\nabla}_{\zeta_{-\eta}} g_\eta(\zeta)\|} \ . \tag{1}$$

Intuitively it shows the sensitivity of $g_\eta(z)$ to changes to η's actions, as opposed to changes to other agent's actions. So at a given state z, the higher the learnability, the more $g_\eta(z)$ depends on the move of agent η, i.e., the better the associated signal-to-noise ratio for η.

2.2   Difference Evaluation Functions

Consider **difference** evaluation functions, which are of the form:

$$D_\eta \equiv G(z) - G(z_{-\eta} + c_\eta) \tag{2}$$

where $z_{-\eta}$ contains all the variable not affected by agent η. All the components of z that are affected by agent η are replaced with the fixed constant c_η. Such difference evaluation functions are factored no matter what the choice of c_η, because the second term does not depend on η's actions [11]. Furthermore, they usually have far better learnability than does a team game, because of the second term of D, which removes a lot of the effect of other agents (i.e., noise) from η's evaluation function. In many situations it is possible to use a c_η that is equivalent to taking agent η out of the system. Intuitively this causes the second term of the difference evaluation function to evaluate the fitness of the system without η and therefore D evaluates the agent's contribution to the global evaluation.

3   Continuous Rover Problem

In this section, we show how evolutionary computation with the difference evaluation function can be used effectively in the Rover Problem. In this problem, there is a set of rovers on a two dimensional plane, which are trying to observe points of interests (POIs). A POI has a fixed position on the plane and has a value associated with it. The observation information from observing a POI is inversely related to the distance the rover is from the POI. In this paper the distance metric will be the squared Euclidean norm, bounded by a minimum observation distance, d:[1]

$$\delta(x,y) = min\{\|x - y\|^2, d^2\} .\tag{3}$$

While any rover can observe any POI, as far as the global evaluation function is concerned, only the closest observation counts[2]. The global evaluation function for a trial is given by:

$$G = \sum_t \sum_i \frac{V_i}{min_\eta \, \delta(L_i, L_{\eta,t})} ,\tag{4}$$

where V_i is the value of POI i, L_i is the location of POI i and $L_{\eta,t}$ is the location of rover η at time t.

At every time step, the rovers sense the world through eight continuous sensors. From a rover's point of view, the world is divided up into four quadrants relative to the rover's orientation, with two sensors per quadrant (see Figure 1). For each quadrant, the first sensor returns a function of the POIs in the quadrant at time t. Specifically the first sensor for quadrant q returns the sum of the values of the POIs in its quadrant divided by their squared distance to the rover:

$$s_{1,q,\eta,t} = \sum_{i \in I_q} \frac{V_i}{\delta(L_i, L_{\eta,t})}\tag{5}$$

---

[1] The square Euclidean norm is appropriate for many natural phenomenon, such as light and signal attenuation. However any other type of distance metric could also be used as required by the problem domain. The minimum distance is included to prevent singularities when a rover is very close to a POI

[2] Similar evaluation functions could also be made where there are many different levels of information gain depending on the position of the rover. For example 3-D imaging may utilize different images of the same object, taken by two different rovers.

where I_q is the set of observable POIs in quadrant q. The second sensor returns the sum of square distances from a rover to all the other rovers in the quadrant at time t:

$$s_{2,q,\eta,t} = \sum_{\eta' \in N_q} \frac{1}{\delta(L_{\eta'}, L_{\eta,t})} \qquad (6)$$

where N_q is the set of rovers in quadrant q.

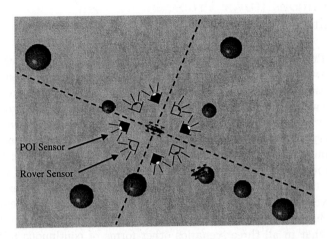

Fig. 1. Diagram of a Rover's Sensor Inputs. The world is broken up into four quadrants relative to rover's position. In each quadrant one sensor senses points of interests, while the other sensor senses other rovers.

With four quadrants and two sensors per quadrant, there are a total of eight continuous inputs. This eight dimensional sensor vector constitutes the state space for a rover. At each time step the rover uses its state to compute a two dimensional action. The action represents an x,y movement relative to the rover's location and orientation. The mapping from state to action is done with a multi-layer-perceptron (MLP), with 8 input units, 10 hidden units and 2 output units. The MLP uses a sigmoid activation function, therefore the outputs are limited to the range $(0, 1)$. The actions, dx and dy, are determined from substracing 0.5 from the output and multiplying by the maximum distance the rover can move in one time step: $dx = d(o_1 - 0.5)$ and $dy = d(o_2 - 0.5)$ where d is the maximum distance the rover can move in one time step, o_1 is the value of the first output unit, and o_2 is the value of the second output unit.

The MLP for a rover is chosen by a simple evolutionary algorithm. In this algorithm each rover has a population of MLPs. At the beginning of each trial, the rover selects the best MLP from its population 90% of the time and a random MLP from its population 10% of the time (ϵ-greedy selector). The selected MLP is then mutated by adding a value sampled from the Cauchy Distribution (with scale parameter equal to 0.3) to each weight, and is used for the entire trial. When the trial is complete, the MLP is evaluated by the rover's evaluation function and inserted into the population. The worst performing member of the population is then deleted. While this algorithm is not sophisticated, it is effective if the evaluation function used by the agents is factored with G and highly learnable. The purpose of this work is to show gains due to principled selection of evaluation functions in a multi-agent system. We expect more advanced algorithms from evolutionary computation, used in conjunction with these same evaluation functions, to perform even better.

4   Results

The Rover Problem was tested in three different scenarios. There were ten rovers in the first two scenarios and thirty rovers in the third scenario. In each scenario, a trial consisted of 15 time steps, and each rover had a population of MLPs of size 10. The world was 100 units long and 115 units wide. All of the rovers started the trial near the center (65 units from the left boundary and 50 units from the top boundary). The maximum distance the rovers could move in one direction during a time step, d, was set to 10. The rovers could not move beyond the bounds of the world. The minimum distance, d, used to compute δ was equal to 5. In the first two scenarios, the environment was reset at the beginning of every trial. However the third scenario showed how learning in changing environments could be achieved by having the environment change at the beginning of each trial. Note that in all three scenarios other forms of continuous reinforcement learners could have been used instead of the evolutionary neural networks. However neural networks are ideal for this domain given the continuous inputs and bounded continuous outputs.

4.1   Rover Evaluation Function

In each of the three scenarios three different evaluation functions were tested. The first evaluation function was the global evaluation function (G):

$$G = \sum_t \sum_i \frac{V_i}{\min_\eta \delta(L_i, L_{\eta,t})} \tag{7}$$

The second evaluation function was the "perfectly learnable" evaluation function (P):

$$P_\eta = \sum_t \sum_i \frac{V_i}{\delta(L_i, L_{\eta,t})} \tag{8}$$

Note that the P evaluation function is equivalent to the global evaluation funtion when there is only one rover. It also has infinite learnability in the way it is defined in Section 2, since the P evaluation function for a rover is not affected by the actions of the other rovers. However the P evaluation function is not factored. Intuitively P and G offer opposite benefits, since G is by definition factored, but has poor learnability. The final evaluation function is the difference evaluation function. It does not have as high learnability as P, but is still factored like G. For the rover problem, the difference evaluation function, D, is defined as:

$$D_\eta = \sum_t \left[\sum_i \frac{V_i}{\min_{\eta'} \delta(L_i, L_{\eta',t})} - \sum_i \frac{V_i}{\min_{\eta' \neq \eta} \delta(L_i, L_{\eta,t})} \right]$$

$$= \sum_t \sum_i I_{i,\eta,t}(z) \frac{V_i}{\delta(L_i, L_{\eta,t})}$$

where $I_{i,\eta,t}(z)$ is an indicator function, returning one if and only if η is the closest rover to L_i at time t. The second term of the D is equal to the value of all the information collected if rover η were not in the system. Note that for all time steps where η is not the closest rover to any POI, the subtraction leaves zero. The difference evaluation can be computed easily as long as η knows the position and distance of the closest rover to each POI it can see. If η cannot see a POI then it is not the closest rover to it. In the simplified form, this is a very intuitive evaluation function yet it was generated mechanically from the general form of the difference evaluation function [11]. In this simplified domain we could expect a hand-crafted evaluation function to be similar. However the difference evaluation function can still be used in more complex domains with a less tractable form of the global utility, even when it is difficult to generate and evaluate hand-crafted solution. Even in domains where an intuitive feel is lacking, the difference evaluation function will be provably factored and learnable.

4.2   Learning in Static Environment

The first experiment was performed using ten rovers and a set of POIs that remained fixed for all trials (see Figure 2). The POIs were placed in such a way as to create the potential of congestion problems around one highly valued POI, testing the cooperation level among the rovers. This was achieved by creating a grid of 15 POIs, with value 3.0, to the left of the rovers' starting location, and a high valued POI of value 10.0 to the right of the rovers' starting location. There was also a POI of value 10.0, which was ten units to the left of the rovers' starting location. System performance was measured by how well the rovers were able to maximize the global evaluation function (even though from a rover's point of view, it is trying to maximize its private evaluation function).

Results from Figure 3 (left) show that the rovers using D performed the best, by a wide margin. Early in training, rovers using P performed better than rovers using G. However since the learning curve of these rovers using P remained flat, while the ones using G increased, the rovers using G eventually overtook the ones using P. The error bars (smaller than the symbols) show that these results are statistically significant. In essence agents using P converge quickly to a poor solution, while agents using G move slowly towards a good solution, while agents using D converge rapidly to a good solution. This phenomenon is explained by factoredness and learnability. The P evaluation function is highly learnable since it is only affected by the moves of a single rover. However since

P is not factored, a rover that maximizes it occasionally takes actions that hurt its fitness with respect to the global evaluation. In contrast rovers using G learn slowly, since the global evaluation is effected by the actions of all the other rovers.

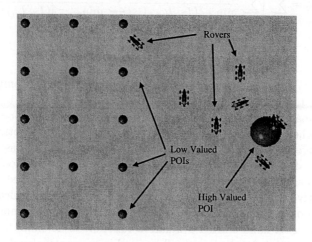

Fig. 2. Diagram of Static Environment. Points of interests are at fixed locations for every trial.

The second experiment is similar to the first one except that the value of a POI goes down each time it is observed by the closest agent. This reduction in value models a domain where an agent receives less useful new information with each successive observation. This is a harder problem than the previous one, since the values of the POIs change at every time step during a trial. Figure 3 (right) verifies that this domain is more difficult as rovers using all three evaluation function performed worse than in the pervious domain. However rovers using D, still performe significantly better than rovers using the other evaluation functions. Note also that rovers using G suffer the most in this domain since they were cut off at a steeper portion in their learning curve than the agents using the other evaluation function. By the time the trial had ended, the rovers using G had just begun to learn their domain because of G's low learnability.

4.3   Learning in Changing Environment

In the first two experiments, the environment was returned to its starting state at the beginning of each trial. Therefore the rovers learned specific control policies for a specific configuration of POIs. This type of learning is most useful when the rovers learn on a simulated environment and are then deployed to an environment closely matching the simulation. However it is often desirable for the rovers to be able to learn a control policy after they are deployed in their environment. In such a scenario the rovers generalize what they learned in the parts of the environment they were first deployed in, to other parts of the environment. The

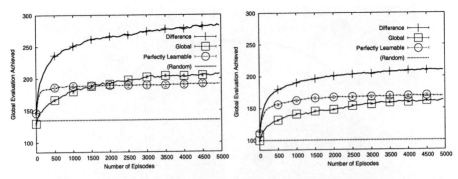

Fig. 3. Results for Three Different Evaluation Functions in Static Environment. Points of interests are at fixed locations for every trial. Right Figure: Results when POI values constant for duration of trial. Left Figure: Results when POI values decrease as they are observed. Difference evaluation is superior since it is both factored and learnable.

last experiment tests the rovers' ability to generalize what they learned in early environmental conditions to later environmental conditions.

In the last experiment there were thirty rovers and thirty POIs. POI locations were set randomly at the beginning of each trial using a uniform distribution within a 70 by 70 unit square centered on the rovers starting location (see Figure 4). The value of the POIs were set to random values uniformly chosen between one and ten. Changing locations at each trial forced the rovers to create a general solution, based on their sensor inputs, since each new trial was different from all of the trials they had previously seen. This type of problem is common in real world domains, where the rovers typically learn in a simulator and later have to apply their learning to the environment in which they are deployed. Note that learning in this scenario does not depend on the use of multiple trials as the rovers can continuously learn, and generalize from their past experience.

Figure 5 shows that rovers using D performed best in this scenario. Rovers using D were effective in generalizing the knowledge gained from exploring previous POI configurations and applying that knowledge to new POI configurations. In contrast, rovers using the P evaluation were especially ineffective in this scenario. We attribute this to the congested nature of the problem, where the rovers competed rather than cooperating with each other. Since a rover's P evaluation only returns the value of what that rover observes, a rover using the P evaluation tends to move towards the highest valued POI in its area. However all the other rovers in that vicinity are also moving towards the same high-valued POI, and thus many other POIs are not properly observed.

5 Conclusion

This paper has shown that even simple evolutionary algorithms can be used in complex multi-agent systems, if the proper evaluation functions are used. In

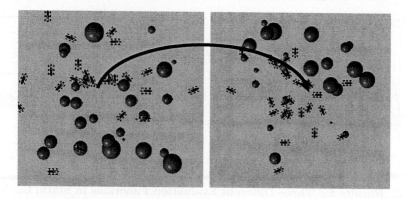

Fig. 4. Changing Environment. POIs are placed at random locations at the beginning of each trial. Rovers have to generalize their knowledge from one trial to the next.

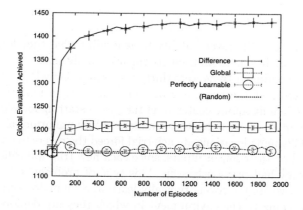

Fig. 5. Results in Changing Environment. Difference evaluation is superior since it is both factored and learnable.

simple continuous problems, the neural network based rovers using the difference evaluation function D, derived from the theory of collectives, were able to achieve high levels of performance because the evaluation function was both factored and highly learnable. These rovers performed 300% better (over random rovers) than rovers using the non-factored perfectly learnable utility and more than 200% better than rovers using the hard to learn global evaluations. These results were even more pronounced in a more difficult domain with a changing environment where rovers using the difference evaluation performed up to 600% better than rovers using global evaluations. These rovers were still able to learn quickly even though they had to generalize a solution learned in earlier environmental configurations to new environmental configurations. These results show the power of using factored and learnable fitness evaluation functions, which allow evolutionary computation methods to be successfully applied to large distributed systems.

References

1. Arvin Agah and George A. Bekey. A genetic algorithm-based controller for decentralized multi-agent robotic systems. In *In Proc. of the IEEE International Conference of Evolutionary Computing*, Nagoya, Japan, 1996.
2. A. Agogino, K. Stanley, and R. Miikkulainen. Online interactive neuro-evolution. *Neural Processing Letters*, 11:29–38, 2000.
3. T. Balch. Behavioral diversity as multiagent cooperation. In *Proc. of SPIE '99 Workshop on Multiagent Systems*, Boston, MA, 1999.
4. G. Baldassarre, S. Nolfi, and D. Parisi. Evolving mobile robots able to display collective behavior. *Artificial Life*, pages 9: 255–267, 2003.
5. R. H. Crites and A. G. Barto. Improving elevator performance using reinforcement learning. In D. S. Touretzky, M. C. Mozer, and M. E. Hasselmo, editors, *Advances in Neural Information Processing Systems - 8*, pages 1017–1023. MIT Press, 1996.
6. M. Dorigo and L. M. Gambardella. Ant colony systems: A cooperative learning approach to the travelling salesman problem. *IEEE Transactions on Evolutionary Computation*, 1(1):53–66, 1997.
7. D. Floreano and F. Mondada. Automatic creation of an autonomous agent: Genetic evolution of a neural-network driven robot. In *Proc. of Conf. on Simulation of Adaptive Behavior*, 1994.
8. F. Gomez and R. Miikkulainen. Active guidance for a finless rocket through neuroevolution. In *Proceedings of the Genetic and Evolutionary Computation Conference*, Chicago, Illinois, 2003.
9. E. Lamma, F. Riguzzi, and L.ÊM. Pereira. Belief revision by multi-agent genetic search. In *In Proc. of the 2nd International Workshop on Computational Logic for Multi-Agent Systems*, Paphos, Cyprus, December 2001.
10. K. Stanley and R. Miikkulainen. Efficient reinforcement learning through evolving neural network topologies. In *Proceedings of the Genetic and Evolutionary Computation Conference (GECCO-2002)*, San Francisco, CA, 2002.
11. D. H. Wolpert and K. Tumer. Optimal payoff functions for members of collectives. *Advances in Complex Systems*, 4(2/3):265–279, 2001.

A Particle Swarm Model of Organizational Adaptation

Anthony Brabazon[1], Arlindo Silva[2,3], Tiago Ferra de Sousa[3], Michael O'Neill[4], Robin Matthews[5], and Ernesto Costa[2]

[1] Faculty of Commerce, University College Dublin, Ireland.
anthony.Brabazon@ucd.ie
[2] Centro de Informatica e Sistemas da Universidade de Coimbra, Portugal.
ernesto@dei.uc.pt
[3] Escola Superior de Tecnologia, Instituto Politecnico de Castelo Branco, Portugal.
arlindo@est.ipcb.pt.
[4] Biocomputing & Development Systems, University of Limerick, Ireland.
Michael.ONeill@ul.ie
[5] Centre for International Business Policy, Kingston University, London.

Abstract. This study introduces the particle swarm metaphor to the domain of organizational adaptation. A simulation model (*OrgSwarm*) is constructed to examine the impact of strategic inertia, in the presence of errorful assessments of future payoffs to potential strategies, on the adaptation of the strategic fitness of a population of organizations. The results indicate that agent (organization) uncertainty as to the payoffs of potential strategies has the affect of lowering average payoffs obtained by a population of organizations. The results also indicate that a degree of strategic inertia, in the presence of an election mechanism, assists rather than hampers adaptive efforts in static and slowly changing strategic environments.

1 Introduction

The objective of this study is to investigate the affect of strategic inertia, in the presence of uncertainty as to future payoffs to potential strategies, on the rate of strategic adaptation of a population of organizations. Following a long-established metaphor of adaptation as search [14], strategic adaptation is considered in this study as an attempt to uncover peaks on a high-dimensional strategic landscape. Some strategic configurations produce high profits, others produce poor results. The search for good strategic configurations is difficult due to the vast number of configurations possible, uncertainty as to the nature of topology of the strategic landscape faced by an organization, and changes in the topology of this landscape over time. Despite these uncertainties, the search process for good strategies is not blind. Decision-makers receive feedback on the success of their current and historic strategies, and can assess the payoffs received by the strategies of their competitors [9]. Hence, certain areas of the strategic landscape are illuminated. In an organizational setting, a strategy can be conceptualized as being the choice of what activities an organization will per-

K. Deb et al. (Eds.): GECCO 2004, LNCS 3102, pp. 12–23, 2004.

form, and the subsequent choices as to how these activities will be performed [12]. These choices define the strategic configuration of the organization. Recent work by [10] and [13] has recognized that strategic configurations consist of interlinked individual elements (decisions), and have applied general models of interconnected systems such as Kauffman's NK model to examine the implications of this for processes of organizational adaptation. This study adopts a similar approach, and employs the NK framework to generate a series of strategic landscapes. The performance of a population of organizations in searching the landscapes for high payoff locations under differing search heuristics, is examined. A key characteristic of the framework which integrates the search heuristics examined in this study, is that organizations do not adapt in isolation, but interact with each other. Their efforts at strategic adaption are guided by 'social' as well as individual learning. The work of [5,8], drawing on a swarm metaphor, has emphasized similar learning mechanisms. We extend this work into the organizational domain, by constructing a simulation model (*OrgSwarm*) based on this metaphor to examine the affect of strategic inertia on the rate of strategic adaptation of a population of organizations.

2   Particle Swarm Algorithm

This section provides an introduction to the basic Particle Swarm algorithm (PSA).[1] A fuller description of this algorithm and the cultural model which inspired it is provided in [5,8]. Under the particle swarm metaphor, a swarm of particles (entities) are assumed to move (fly) through an n-dimensional space, typically looking for a function optimum. Each particle is assumed to have two associated properties, a current position and a velocity. Each particle also has a memory of the best location in the search space that it has found so far (*pbest*), and knows the location of the best location found to date by all the particles in the population (*gbest*). At each step of the algorithm, particles are displaced from their current position by applying a velocity vector to them. The size and direction of this velocity is influenced by the velocity in the previous iteration of the algorithm (simulates 'momentum'), and the current location of a particle relative to its *pbest* and *gbest*. Therefore, at each step, the size and direction of each particle's move is a function of its own history (experience), and the social influence of its peer group. A number of variants of the PSA exist. The following paragraphs provide a description of the basic continuous version described by [8]. Each particle i has an associated current position in search space x_i, a current velocity v_i, and a personal best position in search space y_i. During each iteration of the algorithm, the location and velocity of each particle is updated using equations (1-2). Assuming a function f is to be maximized, that the swarm consists of n particles, and that r_1, r_2 are drawn from a uniform distribution in the range (0,1), the velocity update is as follows:

---

[1] The term PSA is used in place of PSO (Particle Swarm Optimization) in this paper, as the object is not to develop a tool for 'optimizing', but to adapt and apply the swarm metaphor as a model of organizational adaptation.

$$v_i(t + 1) = \Upsilon(W v_i(t) + c_1 r_1 (y_i - x_i(t)) + c_2 r_2 (\hat{y} - x_i(t))) \tag{1}$$

where $\hat{y}$ is the location of the global-best solution found by all the particles. In every iteration of the algorithm, each particle's velocity is stochastically accelerated towards its previous best position and towards a neighborhood (global) best position. The weight-coefficients c_1 and c_2 control the relative impact of *pbest* and *gbest* locations on the velocity of a particle. The parameters r_1 and r_2 ensure that the algorithm is stochastic. A practical affect of the random coefficients r_1 and r_2, is that neither the individual nor the social learning terms are always dominant. Sometimes one or the other will dominate [8]. Although the velocity update has a stochastic component, the search process is not 'random'. It is guided by the memory of past 'good' solutions (corresponding to a psychological tendency for individuals to repeat strategies which have worked for them in the past [6]), and by the global best solution found by all particles thus far. W represents a momentum coefficient which controls the impact of a particle's prior-period velocity on its current velocity. Each component of a velocity vector v_i is restricted to a range $[-v_{max}, v_{max}]$ to ensure that individual particles do not leave the search space. The implementation of a v_{max} parameter can also be interpreted as simulating the incremental nature of most learning processes [6]. The value of v_{max} is usually chosen to be $k * x_{max}$, where $0 < k < 1$. Υ represents a *constriction coefficient* which reduces in value during iterations of the algorithm. This ensures that particles tend to converge over time, as the amplitude of their oscillations (caused by the velocity equation) decreases [8]. Once the velocity update for particle i is determined, its position is updated and pbest is updated (equations 3-4) if necessary.

$$x_i(t + 1) = x_i(t) + v_i(t + 1) \tag{2}$$

$$y_i(t + 1) = y_i(t) \text{ if, } f(x_i(t)) \leq f(y_i(t)), \tag{3}$$

$$y_i(t + 1) = x_i(t) \text{ if, } f(x_i(t)) > f(y_i(t)) \tag{4}$$

After all particles have been updated, a check is made to determine whether gbest needs to be updated (equation 5).

$$\hat{y} \in (y_0, y_1, ..., y_n) | f(\hat{y}) = \max (f(y_0), f(y_1), ..., f(y_n)) \tag{5}$$

Despite its simplicity, the algorithm is capable of capturing a surprising level of complexity, as individual particles are capable of both individual and social learning. Learning is 'distributed' and parallel. Communication (interactions) between agents (individuals) in a social system may be direct or indirect. An example of the former could arise when two organizations trade with one another. Examples of the latter include

i. The observation of the success (or otherwise) of a strategy being pursued by another organization, and

ii. 'Stigmergy' which arises when an organization modifies the environment, which in turn causes an alteration of the actions of another organization at a later time

The mechanisms of the Particle Swarm algorithm bear *prima facie* similarities to those of the domain of interest, organizational adaptation. It embeds concepts of a population of entities which are capable of individual and social learning. However, the model requires modification before it can employed as a plausible model of organizational adaptation. These modifications, along with a definition of the strategic landscape used in this study are discussed in the next section.

3   Simulation Model

The two key components of the simulation model are the landscape generator (environment), and the adaption of the basic Particle Swarm algorithm to incorporate the activities and interactions of the agents (organizations).

3.1   Strategic Landscape

In this study, the strategic landscape is defined using Kauffman's NK model [3, 4]. It is noted *ab initio* that application of the NK model to define a strategic landscape is not atypical and has support from existing literature in organizational science [10,13], [2]. The NK model considers the behavior of systems which are comprised of a configuration (string) of N individual elements. Each of these elements are in turn interconnected to K other of the N elements (K<N). In a general description of such systems, each of the N elements can assume a finite number of states. If the number of states for each element is constant (S), the space of all possible configurations has N dimensions, and contains a total of $\prod_{i=1}^{N} S_i$ possible configurations.

In Kauffman's operationalization of this general framework [4], the number of states for each element is restricted to two (0 or 1). Therefore the configuration of N elements can be represented as a binary string . The parameter K, determines the degree of fitness interconnectedness of each of the N elements and can vary in value from 0 to N-1. In one limiting case where K=0, the contribution of each of the N elements to the overall fitness value (or worth) of the configuration are independent of each other. As K increases, this mapping becomes more complex, until at the upper limit when K=N-1, the fitness contribution of any of the N elements depends both on its own state, and the simultaneous states of all the other N-1 elements, describing a fully-connected graph.

If we let s_i represent the state of an individual element i, the contribution of this element (f_i) to the overall fitness (F) of the entire configuration is given by $f_i(s_i)$ when K=0. When K>0, the contribution of an individual element to overall fitness, depends both on its state, and the states of K other elements to which it is linked ($f_i(s_i : s_{i1}, ..., s_{ik})$). A random fitness function ($U(0,1)$) is adopted, and the overall fitness of each configuration is calculated as the average of the fitness values of each of its individual elements. Therefore, if the fitness values of the individual elements are $f_1, ..., f_N$, overall fitness (F) is $F = \left[\frac{\sum_{i=1}^{N} f_i}{N}\right]$. Altering the value of K affects the ruggedness of the described landscape (graph), and consequently impacts on the difficulty of search on this landscape [3], [4]. As K increases, the landscape becomes more rugged, and the best peaks on the landscape become higher, but harder to find. The strength of the NK model in the context of this study is that by tuning the value of K it can be used to generate strategic landscapes (graphs) of differing degrees of local-fitness correlation (ruggedness). The strategy of an organization is characterized as consisting of N attributes [10]. Each of these attributes represents a strategic decision or policy choice, that an organization faces. Hence, a specific strategic configuration s, is represented as a vector $s_1, ..., s_N$ where each attribute can assume a value of 0 or 1 [13]. The vector of attributes represents an entire organizational form, hence it embeds a choice of markets, products, method of competing in a chosen market, and method of internally structuring the organization [13]. Good consistent sets of strategic decisions - configurations, correspond to peaks on the strategic landscape. The definition of an organization as a vector of strategic attributes finds resonance in the work of Porter [11,12], where organizations are conceptualized as a series of activities forming a value-chain. The choice of what activities to perform, and subsequent decisions as to how to perform these activities, defines the strategy of the organization. The individual attributes of an organization's strategy interact. For example, the value of an efficient manufacturing process is enhanced when combined with a high-quality sales force. Differing values for K correspond to varying degrees of payoff-interaction among elements of the organization's strategy [13].

3.2   Simulation Model

Five characteristics of the problem domain which impact on the design of a plausible simulation model are:

i. The environment is dynamic
ii. Organizations are prone to strategic anchoring (inertia)
iii. Organizations do not knowingly select poorer strategies than the one they already have (election operator)
iv. Organizations make errorful *ex-ante* assessments of fitness
v. Organizations co-evolve

In this study, our experiments consider the first four of these factors. Future work will include the fifth factor. We note that this model bears passing resemblance to the 'eleMentals' model of [7], which combined a swarm algorithm and an NK landscape, to investigate the development of culture and intelligence in a population of hypothetical beings called 'eleMentals'. However, the 'strategic' model developed in this study is differentiated from the eleMental model, not just on grounds of application domain, but because of the inclusion of an 'inertia' operator, and also through the investigation of both static and dynamic environments.

Dynamic environment. Organizations do not compete in a static environment. The environment may alter as a result of exogenous events, for example a 'regime change' such as the emergence of a new technology, or a change in customer preferences. This can be mimicked in the simulation by stochastically respecifying the strategic landscape during the course of a simulation run. These respecifications simulate a dynamic environment, and a change in the environment may at least partially negate the value of past learning (adaptation) by organizations. Minor respecifications are simulated by altering the fitness values associated with one of the N dimensions in the NK model, whereas in major changes, the fitness of the entire NK landscape is redefined.

Inertia. Organizations do not have complete freedom to alter their current strategy. Their adaptive processes are subject to 'conservatism' arising from inertia. Inertia springs from the organization's culture, history, and the mental models of its management. Inertia could be incorporated into the PSA in a variety of ways. We have chosen to incorporate it into the velocity update equation, so that the velocity and direction of the particle at each iteration is also a function of the location of its 'strategic anchor'. Therefore for the simulations, equation 1 is altered by adding an additional 'inertia' term:

$$v_i(t+1) = v_i(t) + R_1(y_i - x_i(t)) + R_2(\hat{y} - x_i(t) + R_3(a_i - x_i(t)) \qquad (6)$$

where a_i represents the position of the anchor for organization i (a full description of the other terms such as R_1 is provided in the pseudo-code below). The anchor can be fixed at the initial position of the particle at the start of the algorithm, or it can be allowed to 'drag', thereby being responsive to the recent adaptive history of the particle. Both the weight attached to the anchor parameter (relative to those attached to pbest and gbest), can be altered by the modeler. Two other alterations are made to the velocity update equation as originally stated in equation 1. The momentum term W and the constriction coefficient Υ are omitted on the grounds that these factors implicitly embed an inertia component. Including these terms could therefore bias the comparison of populations of organizations operating with/without an inertia heuristic.

Election operator. Real-world organizations do not usually intentionally move to 'poorer' strategies. Hence, an 'election' operator is implemented, whereby position updates which would worsen an organization's strategic fitness are discarded. In these cases, an organization remains at its current location. One economic interpretation of the election operator, is that strategists carry out a mental simulation or 'thought experiment'. If the expected fitness of the new strategy appears unattractive, the 'bad idea' is discarded. The simulation incorporates a *conditional update* or *ratchet* operator option, which if turned on, ensures that an organization only updates (alters) its strategy if the new strategy being considered is better than its current strategy. Unfortunately, such evaluations in the real-world, are subject to error. Strategists do not evaluate proposed strategies perfectly due to uncertainty and bounded rationality. The affect of errorful assessments is simulated by subjecting the assessments to 'noise' using the following formula:

$$\text{fitness estimate} = \text{actual fitness of the new strategy} * (1+ \text{'error'}) \quad (7)$$

where *error* is drawn from a normal distribution with a mean of zero and a modeler-defined standard deviation. Hence, despite the election operator, a strategist may sometimes choose a 'bad' strategy because of an incorrect assessment of its fitness.

Outline of algorithm. A number of further modifications to the basic PSA are required. As the strategic landscape is defined using a binary representation, the basic PSA is adapted for the binary case. The pseudocode for the algorithm is as follows:

```
For each dimension n
    v[n]=v[n]+R1*(p[n]-x[n])+R2*(l[n]-x[n])+R3*(a[n]-x[n])
    If(v[n]>Max) v[n]=Vmax
    If(v[n]<-Vmax) v[n]=-Vmax
    If(Pr<S(v[n]))t[n]=1
    Else t[n]=0
If(fitness(t)*err)>fitness(x))  //conditional move
    For each dimension n
        x[n]=t[n]
    UpdateAnchor(a)        //if iteratively update anchor
                          //option is selected
```

$R1$, $R2$ and $R3$ are random weights drawn from a uniform distribution ranging from 0 to $R1_{max}$, $R2_{max}$ and $R3_{max}$ respectively, and they weight the importance attached to gbest, lbest and the anchor in each iteration of the algorithm. $R1$, $R2$ and $R3$ are constrained to sum up to 4.0. Therefore changing the weight value of the anchor alters its significance in the adaptive process, relative to the importance of gbest and lbest. x is the particle's actual position, p is its past best position, l the local best and a is the position of the particle's anchor. V_{max} is set to 4.0. Pr is a probability value drawn from a uniform distribution ranging from 0 to 1, and S is the sigmoid function: $S(x) = \frac{1}{1+exp(-x)}$, which squashes v into a 0 to 1 range, in order to implement a binary PSA. t is a temporary record which is used in order to implement conditional moving. If the new strategy is accepted, t is copied into x, otherwise t is discarded and x remains unchanged. *err* is the error or noise, injected in the fitness evaluation, in order to mimic an errorful forecast of the payoff to a proposed strategy.

4   Results

All reported fitnesses are the average population fitnesses, and average environment best fitnesses, across 30 separate simulation runs at the conclusion of the 5,000 iterations. On each simulation run, the NK landscape is specified anew, and the positions and velocities of particles are randomly initialized at the start of each run. The simulations employ a population of 20 particles, with a circular neighborhood of size 18. 'Real-world' strategy vectors consist of a large array of strategic decisions. A value of N=96 is selected (arbitrary) in defining the landscapes in this simulation. A series of landscapes of differing K values (0,4 and 10), representing differing degrees of fitness interconnectivity, were used in the simulations.

Tables 1 and 2 provide the results for each of ten distinct PSA 'variants', at the end of 5,000 iterations, across three static and dynamic NK landscape 'scenarios'. In each scenario, the same series of simulations are undertaken. Initially, a basic PSA is employed, without an anchor or a conditional move operator. This simulates a population of organizations searching a strategic landscape, where members of the population have no strategic inertia, and where organizations do not utilize a ratchet (conditional move) operator in deciding whether to alter their position on the strategic landscape. The basic PSA is then supplemented by a series of strategic anchor formulations, ranging from a fixed position (fixed at a randomly chosen initial position) anchor which does not change position during the simulation, to one which adapts after a time-lag (moving anchor). In both the initial and moving anchor experiments, a weight value of 1 is attached to the inertia term in the velocity equation, and a time-lag of 20 periods is used for the moving anchor. In the experiments concerning the affect of error when assessing the future payoffs to potential strategies, three values of *error* are examined, 0, 0.05 (5%) and 0.20 (20%).

4.1   Static Landscape

Table 1 provides the results for the static NK landscape. Examining these results suggests that the basic PSA, without inertia or ratchet operators, performs poorly on a static landscape, even when there is no error in assessing the 'payoffs' to potential strategies. The average populational fitness (averaged over each population, across all 30 simulation runs) obtained after 5,000 iterations is not better than random search, suggesting that unfettered adaptive efforts, based on 'social communication' between organizations (gbest), and a memory of good past strategies (pbest) is not sufficient to achieve high levels of populational fitness, even when organizations can make error-free assessments of the 'payoff' of potential strategies. When a ratchet operator is added to the basic PSA (Ratchet PSA-No Anchor), a significant improvement (statistically significant at the 5% level) in both average populational, and average environment best fitness is obtained across landscapes of all K values, suggesting that the simple decision heuristic of *only abandon a current strategy for a better one* leads to notable increases in populational fitness.

Errorful Assessment of Strategic Fitness. In real-world organizations, assessments of the payoffs to potential strategies are not error-free. *A priori* we do not know whether this could impact positive or negatively on the evolution of populational fitness, as permitting errorful assessments of payoff could allow an organization to escape from a local optimum on the strategic landscape, and

Table 1. Average fitness after 5,000 iterations, static landscape.

Algorithm	Fitness		
	(N=96, K=0)	(N=96, K=4)	(N=96, K=10)
Basic PSA	0.4641	0.5002	0.4991
Ratchet PSA-No Anchor	0.5756	0.6896	0.6789
Ratchet-No Anchor, e=0.05	0.4860	0.6454	0.6701
Ratchet-No Anchor, e=0.20	0.4919	0.5744	0.5789
Ratchet-Initial Anchor, w=1	0.6067	0.6991	0.6884
Ratchet-Initial Anchor, w=1, e=0.05	0.5297	0.6630	0.6764
Ratchet-Initial Anchor, w=1, e=0.20	0.4914	0.5847	0.5911
Ratchet-Mov. Anchor (20,1)	0.6692	0.7211	0.6976
Ratchet-Mov. Anchor (20,1, e=0.05)	0.5567	0.6675	0.6770
Ratchet-Mov. Anchor (20,1, e=0.20)	0.4879	0.5757	0.5837

possibly therefore to uncover a new 'gbest'. In essence, an errorful assessment of payoff may allow a short-term 'wrong-way' move (one which temporarily reduces an organization's payoff), but which in the longer-term leads to higher payoffs. Conversely, it could lead to the loss of a promising but under-developed strategy, if an organization is led away from a promising part of the strategic landscape by an incorrect payoff assessment. To examine the impact of errorful payoff assessment, results are reported for the Ratchet PSA-No Anchor, for values of the error ratio of 0.05 and 0.20. Examining Table 1 shows that these produce lower results (statistically significant at 5%) than the error-free case. As the size of the error ratio increases, the average populational fitness declines, suggesting that the utility of the ratchet operator decreases as the level of error in assessing the payoff to potential strategies rises.

The experiments implementing strategic inertia (initial anchor with weight=1, and moving anchor on a 20-lag period with weight=1) for each of the three values of the error ratio generally indicate that the addition of strategic inertia enhances average populational fitness. Comparing the results for the two forms of strategic inertia indicates that a moving anchor performs better when organizations can make error-free assessments of the payoff to potential strategies, but when these payoffs are subject to error neither form of strategic inertia clearly dominates the other in terms of producing the higher average populational fitness. In summary, the results for the static landscape scenario do not support a hypothesis that errorful assessments of payoffs to potential strategies are beneficial for populations of organizations. In addition, the results broadly suggest that strategic inertia, when combined with an election operator, produces higher average populational fitness, but the benefits of this combination dissipates when the level of error in assessing *ex-ante* payoffs gets large.

4.2   Dynamic Landscapes

The real world is rarely static, and changes in the environment can trigger adaptive behavior by agents in a system [1]. When the strategic landscape is wholly or partially respecified, the benefits of past strategic learning by organizations is eroded. In this simulation, two specific scenarios are examined. Table 2 provides the results for the case where a single dimension of the NK landscape is respecified in each iteration of the algorithm with a probability of P=0.00025, and also the results for the case where the entire NK landscape is respecified with the same probability (Figures 1 and 2 provides a graphic of the adaptive trajectories of each search heuristic for K=4 and K=10, on both the static and dynamic 'full respecification' landscapes, and demonstrate that the simulation results are not qualitatively sensitive to the choice of end-point). Qualitatively, the results from both scenarios are similar to those obtained on the static landscape. The basic PSA does not perform any better than random search. Supplementing the basic PSA with the ratchet mechanism leads to a significant improvement in populational fitness, with a further improvement in fitness occurring when the ratchet is combined with an anchor. Adding errorful assessment of the payoffs to potential strategies leads to a deterioration in populational fitnesses as the error ratio increases, but as for the static landscape case, the addition of strategic inertia generally enhances average populational fitness for lower levels of error in assessing payoffs. Comparing the results for the two forms of strategic inertia indicates that a moving anchor performs better when organizations can make error-free assessments of the payoff to potential strategies, but when these payoffs are subject to error, neither form of strategic inertia dominates the other in terms of producing the higher average populational fitness.

Table 2.   Average fitness after 5,000 iterations, one dimension (entire landscape) respecified stochastically.

Algorithm	Fitness		
	(N=96, K=0)	(N=96, K=4)	(N=96, K=10)
Basic PSA	0.4667 (0.4761)	0.4987 (0.4886)	0.4955 (0.4961)
Ratchet PSA-No Anchor	0.5783 (0.5877)	0.6859 (0.6802)	0.6808 (0.6754)
R-No Anchor, e=0.05	0.4927 (0.5143)	0.6458 (0.6309)	0.6673 (0.6568)
R-No Anchor, e=0.20	0.4945 (0.5027)	0.5769 (0.5672)	0.5810 (0.5779)
R-Initial Anchor, w=1	0.6207 (0.6187)	0.6994 (0.6874)	0.6895 (0.6764)
R-Initial Anchor, w=1, e=0.05	0.5390 (0.5612)	0.6636 (0.6551)	0.6766 (0.6599)
R-Initial Anchor, w=1, e=0.20	0.4914 (0.5045)	0.5848 (0.5819)	0.5881 (0.5873)
R-Mov. Anchor (20,1)	0.6689 (0.6575)	0.7193 (0.7152)	0.6974 (0.6819)
R- Mov. Anchor (20,1, e=0.05)	0.5612 (0.5613)	0.6679 (0.6622)	0.6814 (0.6670)
R- Mov. Anchor (20,1, e=0.20)	0.4926 (0.5004)	0.5785 (0.5689)	0.5830 (0.5810)

5   Conclusions

In this paper, a novel synthesis of a strategic landscape defined using the NK model, and a Particle Swarm metaphor is used to model the strategic adaption of organizations. The results suggest that a degree of strategic inertia, in the presence of an election operator, can generally assist rather than hamper the adaptive efforts of populations of organizations in static and slowly changing strategic environments, when organizations can accurately assess payoffs to future strategies. The results also suggest that errorful assessments of the payoffs

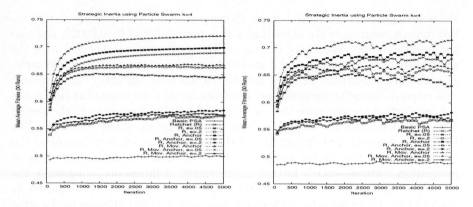

Fig. 1. Plot of the mean average fitness on the static (left) and dynamic (right) landscape where k=4.

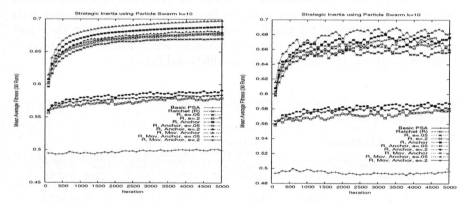

Fig. 2. Plot of the mean average fitness on the static (left) and dynamic (right) landscape where k=10.

to potential strategies leads to a deterioration in populational fitnesses as the error ratio increases. It is also noted that despite the claim for the importance of social learning in populations of agents, the results suggest that social learning is not always enough, unless learnt lessons can be maintained by means of an election mechanism.

No search heuristic will perform equally on all landscapes and across all scales of environmental change. Hence, we acknowledge that the results of this study will not generalize to all possible forms of landscape, and all rates of environmental change. The affect of gbest, pbest and inertia terms, is to 'pin' each organization to a region of the strategic landscape. To the extent that the entire population of organizations have converged to a relatively small region of the landscape, they may find it impossible to migrate to a new high-fitness region if that region is far away from their current location. This suggests that the benefits of an inertia heuristic for a population of organizations comes at a price, the risk of catastrophic failure of the entire population to adapt to a major change in the strategic landscape. In real-world environments, this is compensated for by the birth of new organizations. Finally, it is noted that the concept of inertia or 'anchoring' developed in this paper is not limited to organizations, but is plausibly a general feature of social systems. Hence, the extension of the social swarm model to incorporate inertia may prove useful beyond this study.

References

1. Blackwell, T. (2003). Swarms in Dynamic Environments, *Proceedings of GECCO 2003*, Lecture Notes in Computer Science (2723), Springer-Verlag, Berlin, pp. 1-12.
2. Gavetti, G. and Levinthal, D. (2000). Looking Forward and Looking Backward: Cognitive and Experiential Search, *Administrative Science Quarterly*, 45:113-137.
3. Kauffman, S. and Levin, S. (1987). Towards a General Theory of Adaptive Walks on Rugged Landscapes, *Journal of Theoretical Biology*, 128:11-45.
4. Kauffman, S. (1993). *The Origins of Order*, Oxford,England: Oxford University Press.
5. Kennedy, J. and Eberhart, R. (1995). Particle swarm optimization, *Proceedings of the IEEE International Conference on Neural Networks*, December 1995, pp.1942-1948.
6. Kennedy, J. (1997). The particle swarm: Social adaptation of knowledge, *Proceedings of the International Conference on Evolutionary Computation*, pp. 303-308: IEEE Press.
7. Kennedy, J. (1999). Minds and Cultures: Particle Swam Implications for Beings in Sociocognitive Space, *Adaptive Behavior*, 7(3/4):269-288.
8. Kennedy, J., Eberhart, R. and Shi, Y. (2001). *Swarm Intelligence*, San Mateo, California: Morgan Kauffman.
9. Kitts, B., Edvinsson, L. and Beding, T. (2001). Intellectual capital: from intangible assets to fitness landscapes, *Expert Systems with Applications*, 20:35-50.
10. Levinthal, D. (1997). Adaptation on Rugged Landscapes, *Management Science*, 43(7):934-950.
11. Porter, M. (1985). *Competitive Advantage:Creating and Sustaining Superior Performance*, New York: The Free Press.
12. Porter, M. (1996). What is Strategy?, *Harvard Business Review*, Nov-Dec, 61-78.
13. Rivkin, J. (2000). Imitation of Complex Strategies, *Management Science*, 46(6):824- 844.
14. Wright, S. (1932). The roles of mutation, inbreeding, crossbreeding and selection in evolution, *Proceedings of the Sixth International Congress on Genetics*, 1:356-366.

Finding Maximum Cliques with Distributed Ants

Thang N. Bui and Joseph R. Rizzo, Jr.

Department of Computer Science
The Pennsylvania State University at Harrisburg
Middletown, PA 17057
tbui@psu.edu, jrr200@cs.hbg.psu.edu

Abstract. In this paper we describe an ant system algorithm (ASMC) for the problem of finding the maximum clique in a given graph. In the algorithm each ant has only local knowledge of the graph. Working together the ants induce a candidate set of vertices from which a clique can be constructed. The algorithm was designed so that it can be easily implemented in a distributed system. One such implementation is also described in the paper. For 22 of the 30 graphs tested ASMC found the optimal solution. For the remaining graphs ASMC produced solutions that are within 16% of the optimal, with most being within 8% of the optimal. The performance of ASMC is comparable to existing algorithms.

1 Introduction

Let $G = (V, E)$ be a graph with vertex set V and edge set E. A *clique* in G is a complete subgraph, i.e., a subgraph of G in which there is an edge between any two vertices in the subgraph. The *size* of a clique is the number of vertices in the clique. The MAXCLIQUE problem is the problem of finding the largest clique in a given graph. MAXCLIQUE arises in a variety of problems such as finding good codes, identifying faulty processors in multiprocessor systems and finding counterexamples to Keller's conjecture in geometry [2][24][25][15][16]. However, it is well known that MAXCLIQUE is $\mathcal{NP}$-hard [11], hence it is not expected to have a polynomial time algorithm. The next best thing to have would be a good and efficient approximation algorithm. But it has been shown under various complexity assumptions that finding a good approximation to an instance of MAXCLIQUE is just as hard as finding an optimal solution. For example, it is known that unless $\mathcal{NP} = co - \mathcal{RP}$ no polynomial time algorithm can achieve an approximation factor of $n^{1-\epsilon}$ for MAXCLIQUE for arbitrarily small ϵ [13].

In practice heuristics are used to solve MAXCLIQUE. One of the simplest such heuristics is the greedy heuristic, a version of which is described in Section 3. Other heuristic approaches to MAXCLIQUE include tabu search, continuous-based heuristics, genetic algorithms and ant colony optimization [26][12][4][10]. A good review of MAXCLIQUE and its algorithms is given in [7].

In this paper we give an ant system algorithm, called ASMC, for MAX-CLIQUE. Our algorithm differs from ant colony optimization (ACO) algorithms

K. Deb et al. (Eds.): GECCO 2004, LNCS 3102, pp. 24–35, 2004.
© Springer-Verlag Berlin Heidelberg 2004

in that each ant in ASMC does not solve the entire problem, and each ant has only local knowledge of the graph. As in ACO, ants in ASMC use pheromone to help guide the search. Also included in ASMC is a local optimization step where a clique is constructed based on the positions of the ants in the graph. A major impetus for the development of ASMC was to facilitate a variety of distributed implementations. This paper describes one such implementation. The paper also gives a sequential implementation of ASMC. Experimental results on a set of test graphs from DIMACS [14] show that ASMC is comparable to existing algorithms for MAXCLIQUE.

The rest of the paper is organized as follows. In Section 2 we give some preliminaries. In Section 3 we describe our ant system algorithm for the MAXCLIQUE problem. Section 4 describes a distributed implementation of this algorithm. We compare the performance of our algorithm against some existing algorithms in Section 5. The conclusion is given in Section 6.

2   Preliminaries

Ant colony optimization (ACO) is a meta heuristic inspired by the behavior of foraging ants [9]. By using pheromone, ants are able to help each other find short paths to food sources. The main idea in ACO is to have a collection of ants each of which takes its turn solving the problem. For each solution found by an ant an amount of pheromone proportionate to the quality of the solution is placed in the appropriate places in the search space. The pheromone serves as a means of guiding later ants in their search for a solution. This technique has been successfully applied to a number of problems including MAXCLIQUE, e.g., see [9][10][21]. In an ACO algorithm, each individual ant has the full knowledge of the problem and the placement of pheromone is done after an ant has solved the problem, not while it is searching for the solution. In contrast, ants in our ant system have only local knowledge. Here the placement of pheromone is done by each ant as it is moving about the search space. Each individual ant follows the same set of rules and none solves the problem by itself. It is from their collective behavior that we obtain a solution to the problem. We can have more than one species of ants in the system. Ants in different species can behave differently or can follow the same sets of rules. The species may not interact directly except perhaps through the pheromone. Species can be either collaborative or competitive. This technique has been used successfully in solving other problems, e.g., see [5][6].

For MAXCLIQUE, each vertex in the input graph is considered as a location that ants can occupy. In principle, each vertex can hold an arbitrary number of ants. Ants in our system move from vertex to vertex along the edges of the graph. As an ant traverses an edge of the graph it also puts down a certain amount of pheromone on that edge. To allow for more exploration and possible escape from local optima, pheromone evaporates over time. In addition to using pheromone as a means of communication, ants in our system also use their positions to communicate. For example, a vertex that is occupied by more ants of the same species is more attractive to an ant of that species when it decides where to move. As in most search algorithms of this type there is a constant tug-of-war between exploration and exploitation of the search space. It is usually useful to have more exploration in the beginning and more exploitation nearer to the end so that the algorithm will not converge prematurely to a local optimum. To allow for this type of strategy, ants in our system have an adaptive behavior. For example, they rely less on pheromone and more on the structure of the graph in the beginning of the algorithm. As the algorithm progresses ants make more use of pheromone in determining their movement.

3   Ant System Algorithm for MAXCLIQUE (ASMC)

In this section we describe an ant system algorithm for the MAXCLIQUE problem. The main idea of the algorithm is as follows. Ants are distributed on the graph vertices. Each ant follows the same set of rules to move from vertex to vertex. The rules are designed so that ants are encouraged to aggregate on sets of vertices that are highly connected. These highly connected portions of the graph then serve as candidate sets of vertices from which we can construct cliques.

The algorithm starts by distributing ants of different species on to the vertices of the graph according to some predetermined configuration. It then goes through a number of stages. Each stage of the algorithm consists of a number of cycles. In each cycle a fraction of the ants are selected to move. Each ant that is selected to move will move with certain probability. Its destination is determined by various factors such as pheromone and the structure of the neighborhood around the ant. At the end of each stage, the algorithm constructs a clique based on the current configuration of the ants. The algorithm then shuffles the ants around to help them escape from local optima before moving on to the next stage. After finishing all stages, the algorithm returns the largest clique found in all stages. The algorithm is given in Figure 1. Details of the algorithm are given in the remainder of this section.

3.1   Initialization

The algorithm starts by instantiating $6n$ ants for each species, where n is the number of vertices in the graph. The number of species is a parameter to the algorithm. The description that follows applies to each species. The ants are then distributed to the vertices of the graph. To determine how the ants should be distributed, the algorithm first runs a simple greedy algorithm to find a clique. A large fraction (90%) of the ants are then distributed at random on vertices of the clique found by the greedy algorithm. The remaining ants are distributed randomly on the rest of the vertices. By distributing a majority of the ants to the vertices of a clique found by the greedy algorithm we help speed up the search process at the cost of a possible bias introduced by that clique. This potential bias is alleviated somewhat by the random distribution of the remaining ants and by the placement of different species.

```
Initialize ants
Distribute ants on vertices of the graph
for stage=1 to MaxStage
   for cycle=1 to MaxCycle
     Randomly activate 75% of all ants
       if an ant a is activated
           move(a)
   endfor
   Find cliques through local optimization
   Save solutions
   Shuffle ants
endfor
return the best solution found
```

Fig. 1. Ant System Algorithm for MAXCLIQUE

The greedy algorithm is given in Figure 2. This completes the initialization step. The algorithm then goes through a number of stages. Each stage in turn consists of a number of cycles and at the end of which a clique is constructed. In each cycle 75% of the ants are activated. An activated ant can decide to stay where it is or move to another vertex. In our experiments we found that it is sufficient to have about 25 stages and 10 cycles in each stage. The next subsection describes this process in more detail.

```
R ← ∅
for each v ∈ V
   mark v feasible
endfor
sort V into decreasing degree order
for each v ∈ V in the sorted order
   if v is feasible
       R ← R ∪ {v}
       for each w ∈ V that is not adjacent to v
         mark w infeasible
       endfor
endfor
return R
```

Fig. 2. A Greedy Algorithm for Finding a Clique

3.2   How an Ant Moves

When an ant is activated, i.e., selected to move, it first determines whether it will move or not. This choice is made probabilistically and ants that are older are less likely to move than younger ants. The age of an ant is the number of times that it has moved. This rule enables the ants to explore more in the beginning and move less later on.

If an ant decides to move then it can either move to a randomly selected vertex or to a vertex determined by the *vertex attractiveness (VA) heuristic*. The former choice is made with a fixed probability. The availability of this choice offers the ants a chance to escape from local optima. The latter choice, when selected, uses the VA heuristic to help determine the ant's destination. The VA heuristics computes, for each vertex adjacent to the vertex that the ant is currently on, the probability that the ant will move to that vertex. The ant then selects its destination based on this probability. More specifically, if an ant is currently on vertex i and j is a vertex adjacent to i then the probability that the ant will move to j, called $p_{i,j}$, is defined as follows.

$$p_{i,j} = \frac{\kappa \cdot \tau_{i,j}(t) + \lambda \cdot \nu_j(t) + \mu \cdot \sigma_j^d}{\sum_{j \in L(i)} [\kappa \cdot \tau_{i,j}(t) + \lambda \cdot \nu_j(t) + \mu \cdot \sigma_j^d]}$$

where κ, λ, and μ are nonnegative weights, $L(i)$ is a list of vertices that are adjacent to i, and $\tau_{i,j}(t), \nu_j(t), \sigma_j^d$ are the pheromone score, the population score and the connectivity score, respectively. The pheromone score, $\tau_{i,j}(t)$, is the total amount of pheromone on the edge (i,j) at time t. Each time an ant traverses an edge of the graph, it lays down a certain amount of pheromone on that edge. Under the assumption that frequently traversed edges most likely link groups of highly connected vertices, the pheromone score, acting as a form of memory that records the aggregate behavior of the ants, helps encourage ants to follow these edges. The population score, $\nu_j(t)$, is the number of same-species ants on vertex j at time t. This score helps to cluster ants of the same species. This score is helpful as the final clique is extracted from groups of vertices that are occupied by ants of the same species. Finally, the connectivity score, σ_j^d, measures how well the vertices in the neighborhood of j with radius d connect to each other. More formally, let $V_j^d = \{u \in V \mid u$ is reachable from j by a path of length at most $d\}$ and E_j^d be the set of edges of the graph that have both endpoints in V_j^d. Then,

$$\sigma_j^d = \frac{2|E_j^d|}{|V_j^d|(|V_j^d| - 1)}.$$

The inclusion of the connectivity score in the computation of $p_{i,j}$ helps ants to discover well-connected regions of the graph.

The nonnegative weights κ and λ are constants whereas μ decreases over time. In the beginning μ is set to a value that is higher than both κ and λ so that the structure of the graph plays a much larger role in the exploration of the ants. μ is decreased linearly over time to allow the aggregate behavior of the ants, expressed through pheromone scores and population scores, to help in guiding the movement of an ant. We stop decreasing the value of μ when its value is comparable to that of κ and λ.

The connectivity score depends on the neighborhood radius d, which increases as the ant's age increases. Initially, d is set to 1. Overall, as an ant gets older it moves less often but when it does it looks at a larger neighborhood before deciding where to move. This works well since, as time passes, more information are stored in the pheromone deposited on the edges and ant positions.

At the end of a stage, the algorithm extracts a clique from the current ant configuration. It also attempts to help the ants move away from a possible local optimum by randomly moving a small fraction of the ants to different locations on the graph.

3.3   Local Optimization

At the end of each stage, the algorithm takes the following actions for each species. It first extracts a candidate set of vertices C based on the positions of the ants in that species and the placement of pheromone on the graph. Specifically, each vertex v in the graph is given a *threshold score* by the following formula

$$\text{thresholdScore}(v) = \alpha n_v^S + \beta p_v,$$

where n_v^S is the number of ants of species S occupying vertex v, p_v is the total amount of pheromone on all edges incident to v, and α and β are weights that vary after each stage, from 10 to 5 and from 0.1 to 1, respectively. In this fashion, the threshold scores emphasize the number of ants in the earlier stages, when exploration is important, and emphasize the pheromone in the later stages when exploitation is more important. The candidate set C is obtained by taking those vertices whose threshold scores are in the top γ percent, where γ varies from 10% to 25% over the course of the algorithm. In the beginning stages it is not expected that ant configurations would reflect the highly connected regions of the graph very well as ants may not have enough time to explore the graph yet. In later stages, ant configurations would be more accurate and hence it is reasonable to increase γ.

The next step in the FindClique algorithm is to expand the size of the candidate set C. This is done by adding vertices to C until C has grown by δ%. The added vertices are selected in order of how well they are connected to C. That is, the more neighbors a vertex has in C the higher it is on the list of vertices to be selected. This step is done to ensure that any nearby local optima that the ant configuration missed are included in the candidate set. FindClique seems to work well when δ varies over time from 0% to 1.3%.

Finally, a clique is extracted from this candidate set in a greedy manner similar to the greedy algorithm described earlier. The full FindClique algorithm is given in Figure 3.

The clique R obtained by the FindClique algorithm is then improved in a simple manner by examining each vertex that is not in R and see if adding it to

```
FindClique (G = (V, E), S) // S: species name
// build candidate set
for each v ∈ V
   thresholdScore(v) = α · nᵥˢ + β · pᵥ
endfor
Let C be the set of vertices whose threshold
  scores are in the the top γ%
// expand candidate set
for each v ∈ V
   solutionDegree(v) = |{u ∈ C | (v, u) ∈ E}|
endfor
Select vertices, in decreasing order of solution
  degree, to add to C until C has grown by δ%
// identify clique
R ← ∅
while C ≠ ∅
   Update solutionDegree of vertices in C
   v ← highest solutionDegree vertex in C
   R ← R ∪ {v}
   C ← C − ({v} ∪ {u ∈ C | (u, v) ∉ E})
endwhile
return R
```

Fig. 3. The FindClique algorithm

R still gives us a clique. If so, the vertex is added to R. For each vertex that is added to R in this manner, the algorithm also adds more ants of the appropriate species to that vertex and adds pheromone to the edges incident to that vertex. This is done to enforce the fact that the new R is a clique which can be utilized by the ants in the next stage.

The last operation the algorithm performs before going to the next stage is to perturb the ant configuration enough to allow the ants a chance to break out of a possible local optimum. The perturbation is, however, not too large so that all previous information gathered by the ants are destroyed. Specifically, the algorithm selects 40% of the vertices at random. The ants on the selected vertices are swapped randomly among these vertices. Furthermore, the pheromone on 10% of the edges incident to each of these selected vertices is reduced by a small amount. Experimental results found that this perturbation produced desirable effects.

The algorithm is now ready to start another stage. When all stages have finished, the algorithm returns the largest clique found at the end of each stage.

4   The Distributed Implementation

ASMC is especially suited to distributed implementation since ants contribute partial solutions based on local information. Distribution makes it possible to optimize the algorithm for speed, due to parallel processing, or for space, due to partitioning ants and large graphs over several machines. Though the benefits gained depend on the specific implementation, ASMC itself does not preclude any of the benefits.

The distributed implementation built for this paper is a simple proof of concept and, as such, perfoms synchronous interprocess communication through a central server. Ants are distributed across four machines, each of which has a complete copy of the graph. The idea is for ants to move from processor to processor. One machine is designated the server and the others clients 1, 2 and 3. The server coordinates four synchronous transactions: starting, ant transfer, local optimization and ending.

- To **start**, the server first waits for each client to connect. It then partitions the vertices into four roughly equal sets and assigns each set to a processor, including itself. It sends each client a list of which vertices are "owned" by which processors.
- **Ant transfer** occurs at the end of each stage. The server asks each client for a list of ants that are moving to other processors. It sorts these ants according to their destinations and sends each client a list of incoming ants. Each processor instantiates the ants on its copy of the graph. It also updates cached data regarding vertices and edges connected to its "owned" vertices. The cached data makes it possible for ants to make informed decisions about whether to move to another processor in the future.
- **Local optimization** occurs after each ant transfer. Each client sends the server vertex and edge data, and the server performs the same operations as in the sequential implementation of ASMC. In principle, this step could be redesigned to be less centralized.
- The server **ends** by letting each client know that the run has ended. Synchronized starting and ending makes it simpler to invoke the program from a script.

Though this implementation is rather simple, other distributed ASMC designs are possible. For example, one could build a less centralized, peer-to-peer model. The ant transfers could occur asynchronously throughout each stage. The graph partitions could adapt over time to minimize the number of shared edges. The graph itself could be distributed so that each client is completely unaware of vertices it does not "own," thereby enabling runs on extremely large graphs in a reasonable time.

5   Experimental Results

In this section we describe the results obtained from testing the sequential implementation and the distributed implementation of ASMC.

Sequential Implementation. The algorithm was implemented in C++ and run on an Intel Pentium IV 2.4GHz machine. The algorithm was tested on a set of 30 graphs selected from the benchmark graphs of the Second DIMACS Implementation Challenge [14]. The graphs that we used have up to 1,500 vertices and over 500,000 edges. There are 9 classes of graphs. The C-fat graphs are from fault diagnosis problems [2], the Johnson and Hamming graphs are from problems in coding theory [24][25]. The Keller graphs are from tiling problems in geometry [15][16]. The remaining families of graphs: San, Sanr, Brock, P-hat and Mann are various types of random graphs with known optimal cliques. In our implementation of ASMC only one species was used.

For each graph, ASMC is run 100 times. Table 1 summarizes the results produced by ASMC. Overall, ASMC did very well for graphs in the classes C-fat, Keller, Johnson and Hamming. It always found the optimal solution. For all but two of the tested graphs in the San and Sanr families ASMC found the optimal solution. For the Mann graphs the solutions returned by ASMC were within 1% of the optimal solution. The most difficult families of graphs for ASMC were the Brock and P-hat graphs. For the P-hat graphs ASMC produced solutions that were within 10% of the optimal solution, with some being optimal. However, for the Brock graphs solutions given by ASMC were as bad as 18% from the optimal. These results seem to be consistent with those obtained by other algorithms. It should be noted that in developing ASMC, the parameters used were derived from experiments based only on three graphs from this set: keller4, san200_0.7_1 and p_hat300_1.

We compare the results given by ASMC against the following three algorithms for MAXCLIQUE: (i) GMCA – a hybrid genetic algorithm [4], (ii) CBH – a global optimization algorithm that uses a continuous formulation of MAX-CLIQUE [12], and (iii) IHN – a neural approximation algorithm based on discrete Hopfield networks [3]. These algorithms are chosen to represent different approaches to MAXCLIQUE as well as for the availability of their test results. Table 1 summarizes the results of ASMC together with these three algorithms. It is clear that ASMC is comparable to the other algorithms. We provided no running time comparisons since the algorithms were tested on different machines and there were not sufficient information for us to derive reasonable conversion factors for the running times. We believe, however, that ASMC might be slower than some of these algorithms.

There are other more recent algorithms for MAXCLIQUE [1][10][19]. Their results are not included in Table 1 since there are not enough overlapped test results [1][10] or the problem solved is slightly different from MAXCLIQUE (the size of the largest known clique is required as an input to the algorithm) [19]. Based on the available data ASMC is also comparable to the algorithms of [1] and [10].

Distributed Implementation. The distributed implementation of ASMC was run on four Sun Blade 100 450MHz workstations. Due to time limitations we were able to run the algorithm only on a subset of the 30 test graphs. Also, we ran the algorithm 100 times for each graph. The results are comparable to that of the sequential implementation. Table 2 summarizes the results.

Table 1. ASMC solution quality (sequential implementation)

Graph	Vertices	Edges	Opt	ASMC Best*	ASMC (StdDev) Avg*		Avg Time*†	GMCA Best	CBH Best	IHN Best
c-fat200-1	200	1534	12	12	12.00	(0.00)	0.46	12	12	12
c-fat500-1	500	4459	14	14	14.00	(0.00)	1.55	14	14	14
johnson16-2-4	120	5460	8	8	8.00	(0.00)	1.58	8	8	8
johnson32-2-4	496	107880	16	16	16.00	(0.00)	48.97	16	16	16
keller4	171	9435	11	11	10.44	(0.76)	3.16	11	10	–
keller5	776	225990	27	26	21.90	(1.20)	110.34	18	21	–
hamming10-2	1024	518656	512	512	512.00	(0.00)	281.04	512	512	512
hamming8-2	256	31616	128	128	128.00	(0.00)	11.20	128	128	128
san200_0.7_1	200	13930	30	30	18.81	(5.35)	4.85	30	15	30
san200_0.9_1	200	17910	70	70	47.72	(4.18)	5.75	–	–	70
san200_0.9_2	200	17910	60	60	40.80	(5.73)	5.76	–	–	41
san200_0.9_3	200	17910	44	37	32.72	(1.12)	6.02	–	–	–
san400_0.5_1	400	39900	13	13	8.35	(0.91)	16.61	7	8	–
san400_0.9_1	400	71820	100	100	55.74	(11.67)	29.83	50	50	–
sanr200_0.7	200	13868	18	18	15.32	(0.83)	4.80	17	18	17
sanr400_0.5	400	39984	13	13	10.58	(0.57)	16.56	12	12	12
san1000	1000	250500	15	15	9.66	(0.74)	135.42	8	8	10
brock200_1	200	14834	21	20	17.98	(0.93)	5.12	20	20	–
brock400_1	400	59723	27	25	20.14	(0.80)	25.23	20	23	–
brock800_1	800	207505	23	20	16.65	(0.77)	102.46	18	20	–
p_hat300_1	300	10933	8	8	7.17	(0.38)	4.26	8	8	8
p_hat300_2	300	21928	25	25	23.93	(0.81)	8.84	–	–	25
p_hat300_3	300	33390	36	36	31.82	(1.04)	12.83	–	–	36
p_hat500_1	500	31569	9	9	8.19	(0.47)	14.19	9	9	9
p_hat500_2	500	62946	36	36	32.03	(1.45)	29.40	–	–	36
p_hat700_1	700	60999	11	11	8.42	(0.54)	30.87	8	11	11
p_hat1000_1	1000	122253	10	10	8.73	(0.57)	68.00	8	10	10
p_hat1500_1	1500	284923	12	11	9.54	(0.66)	177.70	10	11	–
MANN_a27	378	70551	126	125	124.64	(0.52)	27.89	125	121	–
MANN_a45	1035	533115	345	341	338.93	(1.01)	282.15	337	336	–

*Results for 100 runs per graph.   †All times are in seconds.

6   Conclusion

This paper describes an ant system algorithm for MAXCLIQUE which seems to perfom comparably with the current best known algorithms for this problem. One possible improvement of ASMC is to increase the diversity of the initial ant configuration. This can be done by generating several cliques instead of just one using the same greedy algorithm. Another direction is to make the distributed implementation more scalable, more network efficient and less centralized.

Table 2. ASMC solution quality (distributed implementation)

Graph	Vertices	Edges	Opt	ASMC Best*	ASMC (StdDev) Avg*		Avg Time*†
c-fat200-1	200	1534	12	12	12.00	(0.00)	0.72
c-fat500-1	500	4459	14	14	14.00	(0.00)	2.61
johnson16-2-4	120	5460	8	8	8.00	(0.00)	1.37
keller4	171	9435	11	11	10.49	(0.64)	3.21
hamming8-2	256	31616	128	128	127.26	(2.78)	32.70
san200_0.7_1	200	13930	30	30	17.16	(1.93)	6.29
san200_0.9_1	200	17910	70	48	47.09	(0.40)	10.64
sanr200_0.7	200	13868	18	17	15.44	(0.69)	6.22
brock200_1	200	14834	21	20	18.49	(0.64)	7.21
p_hat300_1	300	10933	8	8	8.00	(0.00)	4.52
p_hat500_1	500	31569	9	9	8.11	(0.31)	33.60

*Results for 100 runs per graph.   †All times are in seconds.

Acknowledgements. The authors would like to thank Sue Rizzo and Linda Null for useful discussions.

References

1. R. Battiti and M. Protasi, "Reactive Local Search for Maximum Clique," Proceedings of the Workshop on Algorithm Engineering (WAE'97), Venice, G. F. Italiano and S. Orlando, Eds., Venice, Italy, 1997, pp. 74–82.

2. P. Berman and A. Pelc, "Distributed Fault Diagnosis For Multiprocessor Systems," Proceedings of the 20th Annual International Symposium on Fault-Tolerant Computing, pp. 340–346, Newcastle, UK, 1990.

3. A. Bertoni, P. Campadelli and G. Grossi, "A Discrete Neural Algorithm for the Maximum Clique Problem: Analysis and Circuit Implementation," Proceedings of the Workshop on Algorithm Engineering (WAE'97), Venice, G. F. Italiano and S. Orlando, Eds., Venice, Italy, 1997.

4. T. N. Bui and P. H. Eppley, "A Hybrid Genetic Algorithm for the Maximum Clique Problem," Proceedings of the Sixth International Conference on Genetic Algorithms (ICGA), L. Eshelman (Ed.), pp. 748–484, Morgan Kauffman Publishers, 1995.

5. T. N. Bui and C. M. Patel, "An Ant System Algorithm for Coloring Graphs," Computational Symposium on Graph Coloring and Generalizations (COLOR02), Ithaca, NY, September 2002.

6. T. N. Bui and L. C. Strite, "An Ant System Algorithm for Graph Bisection," GECCO 2002: Proceedings of the Genetic and Evolutionary Computation Conference, W. B. Langdon et al. (Eds.), pp. 43–51, Morgan Kauffman Publishers, 2002.

7. I. M. Bomze, M. Budinich, P. M. Pardalos, and M. Pelillo, "The Maximum Clique Problem," D.-Z. Du and P. M. Pardalos (Eds.), *Handbook of Combinatorial Optimization*, 4, Kluwer Academic Publishers, Boston, MA, 1999.

8. K. Corradi and S. Szabo, "A Combinatorial Approach for Keller's Conjecture," Periodica Mathematica Hungarica, 21, pp. 95–100, 1990.

9. M. Dorigo and G. Di Caro, "The Ant Colony Optimization Meta-Heuristic," *New Ideas In Optimization*, D.Corne, M. Dorigo and F. Glover (Eds.), McGraw-Hill, London, pp. 11–32, 1999.

10. S. Fenet and C. Solnon "Searching for Maximum Cliques with Ant Colony Optimization," in Applications of Evolutionary Computing, Proceedings of the EvoWorkshops 2003, April 2003 Lecture Notes in Computer Science, No. 2611, Springer-Verlag, pp. 236-245

11. M. Garey and D. Johnson, *Computers and Intractibility: A Guide to the Theory of NP-Completeness*, Freeman, San Francisco, 1979.

12. L. E. Gibbons, D. W. Hearn and P. M. Pardalos, "A Continuous Based Heuristic for the Maximum Clique Problem," in [14], pp. 103–124, 1996.

13. J. Hastad, "Clique Is Hard to Approximate within $n^{1-\epsilon}$," Acta Mathematica, 182, pp. 105–142, 1999.

14. D. S. Johnson and M. A. Trick (Editors), *Cliques, Coloring and Satisfiability – Second DIMACS Implementation Challenge 1993*, DIMACS Series in Discrete Mathematics and Theoretical Computer Science, American Mathematical Society, Volume 26 (1996).

15. O. H. Keller, "Über die lückenlose Erfüllung des Raumes mit Würfen," Journal für die reine und angewandte Mathematik, 163, pp. 231–238, 1930.

16. J. C. Lagarias and P. W. Shor, "Keller's Cube-Tiling Conjecture Is False In High Dimensions," Bulletin of the American Mathematical Society, 27(2), pp. 279–283, 1992.

17. J. MacWilliams and N. J. A. Sloane, *The Theory of Error Correcting Codes*, North-Holland, Amsterdam, 1979.

18. C. Mannino and A. Sassano, "Solving Hard Set Covering Problems," Operations Research Letters, 18, pp. 1–5, 1995.

19. E. Marchiori, "Genetic, Iterated and Multistart Local Search for the Maximum Clique Problem," Applications of Evolutionary Computing, Proceedings of the EvoWorkshops 2002: EvoCOP, EvoIASP, EvoSTIM/EvoPLAN, Kinsale, Ireland, S. Cagnoni, J. Gottlieb, E. Hart, M. Middendorf, G.R. Raidl (Eds.), pp. 112–121, April 3-4, 2002.

20. H. Minkowski, *Diophantische Approximationen*, Teubner, Leipzig, 1907.

21. G. Navarro Varela and M.C. Sinclair, "Ant Colony Optimisation for Virtual-Wavelength-Path Routing and Wavelength Allocation," Proceedings of the Congress on Evolutionary Computation (CEC'99), Washington DC, July 1999.

22. O. Perron, "Über lückenlose Ausfüllung des n-dimensionalen Raumes durch kongruente Würfel," Mathematische Zeitschrift, 46, pp. 1–26, 161–180, 1940.

23. L. Sanchis and A. Jagota, "Some Experimental and Theoretical Results on Test Case Generators for the Maximum Clique Problem," INFORMS Journal on Computing, 8(2), pp. 87–102, Spring 1996.

24. N. J. A. Sloane, "Unsolved Problems in Graph Theory Arising from the Study of Codes," Graph Theory Notes of New York, XVIII, pp. 11–20, 1989.

25. N. J. A. Sloane and F. J. MacWilliams, *The Theory of Correcting Codes,* North Holland, Amsterdam, 1979.

26. P. Soriano and M. Gendreau, "Tabu Search Algorithms for the Maximum Clique Problem," in [14], pp. 221–244, 1996.

Ant System for the k-Cardinality Tree Problem

Thang N. Bui and Gnanasekaran Sundarraj

Department of Computer Science
The Pennsylvania State University at Harrisburg
Middletown, PA 17057
{tbui,gxs241}@psu.edu

Abstract. This paper gives an algorithm for finding the minimum weight tree having k edges in an edge weighted graph. The algorithm combines a search and optimization technique based on pheromone with a weight based greedy local optimization. Experimental results on a large set of problem instances show that this algorithm matches or surpasses other algorithms including an ant colony optimization algorithm, a tabu search algorithm, an evolutionary algorithm and a greedy-based algorithm on all but one of the 138 tested instances.

1 Introduction

Let $G = (V, E)$ be a graph with vertex set V, edge set E and a weight function $w : E \to \mathbb{R}^+$ assigning a weight for each edge of G. A k-cardinality tree of G is a subgraph of G that is a tree having exactly k edges. The weight of a k-cardinality tree T is the sum of the weights of all the edges in T. The k-cardinality tree problem is defined as follows.

Input: Edge-weighted graph $G = (V, E)$ with a weight function $w : E \to \mathbb{R}^+$ and an integer k, where $1 \le k \le |V| - 1$.
Output: The minimum weight k-cardinality tree of G.

The k-cardinality tree problem was first described by Hamacher, Jornsten and Maffioli in [12] who also proved this problem to be strongly NP-hard. It remains NP-hard even if $w : E \to \{1, 2, 3\}$ [14]. However, it is solvable in polynomial time if the range of w has cardinality 2 [14]. This problem arises in various areas such as facility layout [9], graph partitioning [10], quorum cast routing [5], telecommunications [11], and matrix decomposition [4].

Several heuristic and meta-heuristic algorithms have been developed for this problem. Under the heuristic category, an integer programming approach is given in [8], and a Branch and Bound approach is given in [5]. These heuristic algorithms are based on the greedy and dual greedy strategies in addition to dynamic programming technique. Recent meta-heuristic approaches for this problem include ant colony optimization [2], evolutionary computation [3], tabu search [3], and variable neighborhood search [13].

In this paper, we present an ant system algorithm for the k-cardinality tree problem. We test the algorithm on a number of benchmark graphs and for a

K. Deb et al. (Eds.): GECCO 2004, LNCS 3102, pp. 36–47, 2004.

large number of k values. We compare our results against a number of existing heuristics including an ant colony optimization algorithm, an evolutionary algorithm, a tabu search algorithm and a greedy-based algorithm. The experimental results show that our algorithm performs very well against these algorithms, matching or surpassing all of them in all but one of the 138 problem instances. In fact, for more than half of the instances, our algorithm found solutions that are better than the previously best known solutions.

The rest of the paper is organized as follows. Section 2 describes the algorithm in detail. The experimental results comparing our algorithm against other known algorithms are given in Section 3 and the conclusion is given in Section 4.

2   Algorithm

The main idea for our algorithm is that a given edge in the input graph by itself may not have a low enough weight to be considered for the solution tree, but when combined with other neighboring edges without creating a cycle, this set of connected edges may have the lowest total weight compared to other equicardinality, acyclic, connected set of edges in its neighborhood. We use ants to discover such connected acyclic sets of edges in the graph that can be combined to obtain an optimal k-cardinality tree. In fact, it has been shown that the connectivity requirement is the crux of the difficulty in this problem [7].

The algorithm, called ASkCT and given in Figure 1, consists of two main phases. The first phase has two stages: discovery and construction. In the discovery stage ants are used to discover a potential set of edges from which a small weight k-cardinality tree can be constructed. In the construction stage a greedy algorithm similar to the Kruskal algorithm for finding the minimum cost spanning tree is used to construct a k-cardinality tree of small weight from the set of potential edges produced by the discovery stage. The greedy strategy used in this construction stage is based on the pheromone left on the edges by the ants in the discovery stage.

The second phase of the algorithm consists of a sequence of local optimization stages designed to reduce the weight of the k-cardinality tree produced by the first phase. In what follows, we describe each of the two phases in detail.

2.1   The Discovery Stage

In this stage, we let each ant, starting from a given vertex, to discover a path consisting of more than one edge. We then let the ant move to the other end vertex of the path while allowing the ant to deposit pheromone along the path. We call such a move a *step* and number of edges in that move the *step size*. The amount of pheromone deposited on each edge in the path is inversely proportional to the total weight of the edges in the path. The edges in these paths are stored in the ant's 'memory' so that they will not be rediscovered by the same ant when it is trying to find the next best path from the other end vertex of

```
ASkCT(G = (V, E),  w)    // w is the weight function
  Phase 1.
  for  s  =  minStepSize to maxStepSize    // s is the step size
      CandidateEdges ← Discover(G, w, s)           // Discovery Stage
      T ← ConstructTree(CandidateEdges, G, w)      // Construction Stage
      if T is better than the current best tree T* // Remember the best
          T* ← T
  end-step s

  Phase 2.
  T* ← Optimize(T*)
  return T*
```

Fig. 1. The ASkCT algorithm

```
Discover(G = (V, E), w, s)) //w is the weight function, s is the step size
  InitializePheromone()
  for i = 1 to 50
      distribute ants on the vertices
      for each ant a
          for step = 1 to numSteps
              find the best path of length s from the current vertex
                  without using the edges in a's memory
              apply pheromone on the s edges of this path
              add these edges to a's memory
              move a to the other end of the path
          end-step
      end-ant
      apply pheromone evaporation
      clear ants' memory
  end-for
```

Fig. 2. The Discover Algorithm

the already discovered path and also to make sure that no cycle is created. This stage is accomplished by the Discover algorithm given in Figure 2.

The Discover algorithm starts by applying an initial amount of pheromone on all the edges and distributing the ants randomly on the vertices. For our experiment this initial amount of pheromone is set to 0.5, the pheromone evaporation rate is set to 0.01, and the number of ants is set to 20% of the number of vertices in graph. The algorithm then runs through a number of iterations, which is set to 50 for our experiment. At the end of each iteration, the memory of each ant is cleared and each ant starts the next iteration from a randomly chosen vertex. In each iteration, an ant takes numSteps steps. We set numSteps to 2 in our current implementation.

2.2   The Tree Construction Stage

We use a modified version of the Kruskal algorithm [6] for constructing the tree. A k-cardinality tree is extracted from the candidate edges based on the pheromone left by the ants in the discovery stage.

First, the edges are sorted into order of decreasing pheromone values. The algorithm maintains a collection of disjoint sets as in the normal implementation of the Kruskal algorithm, each disjoint set is a tree. Starting from the edge that has the highest pheromone value, edges are added to the disjoint sets that contain one of their end vertices. For any edge, if there is no disjoint set that has one of its end vertices, a new disjoint set is created and the edge is added to that set. If the end vertices of an edge are in two different disjoint sets, the sets are merged into a single set and the edge is added to the merged set. If a disjoint set has both the end vertices of an edge, the algorithm checks the loop that is formed by the addition of this new edge. If the new edge is better in weight than any other edge in the loop, that edge is replaced by the new edge. Once an edge is added to a disjoint set, the sets are checked to see if any of them contain *k* or more edges. If there is one, a *k*-cardinality tree is constructed out of these edges. If the size is more than *k*, the leaf edges are trimmed off in a greedy manner until there are only *k* edges left, i.e., higher weight leaf edges are removed first. A *leaf edge* is an edge one of whose end point is a leaf. The newly constructed tree is compared with the current best tree, and the smaller of the two is kept as the current best tree. Once the tree is constructed from a disjoint set, the set is marked as processed. New edges will be added to the set that is marked as processed, but no new tree will be constructed from that disjoint set until it is merged with another disjoint set. The tree obtained at the end is considered as the best for that step-size. The algorithm is given in Figure 3.

```
ConstructTree(CandidateEdges, G = (V, E), w)
    Sort the edges in CandidateEdges into decreasing pheromone values
    for each edge e in the sorted order
        Find the disjoint sets that contain the end vertices of e
        if there is none
            create a new set, add e to it, mark the set unprocessed
        if there is only one set containing one of the end vertices of e
            add e to it
        if there is only one set containing both end vertices
            try to replace an edge in the loop formed; otherwise discard e
        if there are two different sets, each containing one end vertex of e
            merge the two sets
            add e to the merged set
            mark the merged set unprocessed

        for each disjoint set A
            if A has k or more edges
                extract a k-cardinality tree by trimming greedily if needed
                if the new tree has a lower weight than the best known so far
                    discard the previous best and take the new tree as the best
                mark A processed
        end-disjoint-set
    end-edge
    return the best k-cardinality tree
```

Fig. 3. The ConstructTree Algorithm

It can be noted from the ASkCT algorithm given in Figure 1 that the discovery and the tree construction stages are repeated for various step sizes, that is, the step size is varied from `minStepSize` to `maxStepSize`. For our experiment `minStepSize` is set to 2 and `maxStepSize` is set to 3. The algorithm then picks the best tree out of these step sizes to pass on to the next phase to be optimized.

2.3   The Local Optimization Phase

In the local optimization phase, we apply a sequence of greedy algorithms to reduce the weight of the k-cardinality tree produced by the previous phase. Unlike the previous phase, the strategy used here is based on the weight of the edges not the pheromone. The main idea here is to swap edges in and out of the tree so that the tree weight is reduced. Specifically, this phase consists of four stages: (i) the tree is grown first and then shrunk back, in a greedy manner, to its original size, (ii) the tree is shrunk and then grown back, in a greedy manner, to its original size, (iii) the tree is shrunk and then grown back, in a depth first manner, to its original size, and (iv) the tree is split into two by removing the highest weight edge, and the resulting trees are grown back until they meet or one of them has k edges. The algorithm is given in Figure 4.

In these stages, when a tree is grown by i edges in a greedy manner, we select i smallest weight edges among the edges that are connected to the tree by one endpoint only, and add them to the tree. Similarly, when a tree is trimmed by i leaf edges in a greedy manner, we select the i highest weight edges among the leaf edges of the tree and remove them. When a tree is grown by i edges in a depth-first manner, we select by using a depth-first search the lowest weight path of length i that has one end connected to the tree and add the edges in the path to the tree.

These stages were chosen so that the effect of the individual stages are complementary to each other. In Stages 1–3, we try to replace the leaf edges, whereas in Stage 4 we try to replace the non-leaf edges, beginning at the highest weight edge. In order to achieve a balance between the quality of the results and the overall running time of the algorithm, only 30% of the edges in the k-cardinality tree were considered for replacement in each of these stages except Stage 3.

We observed that the k-cardinality tree problem is usually more difficult when the cardinality k is small but not too small. The local optimization algorithm accounts for this in the third for loop with a variable number of iterations based on the size of k. Specifically, the function $\alpha(k)$ in the third for loop in Figure 4 is defined as follows.

$$\alpha(k) = \begin{cases} 6, & \text{if } 0.05|V| < k < 0.35|V|, \\ 3, & \text{otherwise.} \end{cases}$$

where V is the vertex set of the input graph.

```
LocalOptimize(T)      // T is a k-cardinality tree
Stage 1.
    T* ← T
    for i = 1 to ⌊3k/10⌋
        grow T by adding i edges in a greedy manner
        trim T by removing i leaf edges in a greedy manner
        if w(T) < w(T*)
            T* ← T
    end-for

Stage 2.
    T ← T*
    for i = 1 to ⌊3k/10⌋
        trim T by removing i leaf edges in a greedy manner
        grow T by adding i edges in a greedy manner
        if w(T) < w(T*)
            T* ← T
    end-for

Stage 3.
    T ← T*
    for i = α(k) downto 2
        trim T by removing i leaf edges in a greedy manner
        find the lowest weight path of length i with one end connected to T
        add the edges of this path to T
        if w(T) < w(T*)
            T* ← T
    end-for

Stage 4.
    T ← T*
    sort the edges of T into decreasing order of weight
    for i = 1 to ⌊3k/10⌋
        remove the ith highest weight edge from T to obtain two
            trees T₁ and T₂
        grow T₁ and T₂ independently in a greedy manner
            until they meet or one of them has k edges
        let T' be the resulting tree
        if T' has more than k edges,
            trim off the excess leaf edges in a greedy manner
        if w(T') < w(T*)
            T* ← T'
    end-for
    return T*
```

Fig. 4. The LocalOptimize algorithm

3   Experimental Results

In this section, we describe the results of running our algorithm on a collection of benchmark graphs for this problem, and compare them against the current best known results from four other algorithms: an ACO algorithm [2], an evolutionary algorithm [3], a tabu search algorithm [1], and a greedy based algorithm [7].

Our algorithm was implemented in C++ and run on a PC with Pentium IV 2.4GHz processor and 512MB of RAM. We tested our algorithm on three different classes of graphs: random graphs, grid graphs and Steiner graphs. There are four graphs in each class, for a total of twelve graphs. In order to be able to compare the performance of our algorithm against others, we selected these graphs from KCTLIB, a library for the edge-weighted k-cardinality tree problem, maintained by C. Blum and M. Blesa at http://iridia.ulb.ac.be/~cblum/kctlib/. We also used the same set of values for k in each graph and the same number of runs, which is 20, for each of these k values as used by the authors of KCTLIB. Since the configuration of the system used to run the other algorithms was not available, we could not compare the running time of our algorithm against others.

Of the 138 instances that we tested, our algorithm ASkCT matches the previous best known results in 61 cases. It provides better results than previously known in 76 cases. ASkCT did not match the best known result for the one remaining case. The difference in this case has an absolute difference of 7 or 0.46% of the best known value. The results shown in Tables 1 through 6 list the best (w_{best}), the average (w_{avg}), the standard deviation (σ), and the average running time in seconds (t_{avg}) for our algorithm. These tables also include the previous best-known values for each of the k values. It can be observed from these tables that the standard deviations are very small for the most part. In fact, the standard deviations for these results are no more than 3.5% of the best known value. The results shown in Tables 7 through 12 compare our best values against the best values from other algorithms. It should be noted that the previously best known results were not achieved by any one single algorithm alone. They were the bests obtained among the four algorithms. The data for these four algorithms were obtained from KCTLIB.

From our experiments, we observed that the influence of the local optimization phase on the final results were minimal. It improved the results by no more than 5-10%. For many k values, we were able to obtain the same results without the local optimization.

In our implementation, the values for minStepSize and maxStepSize in the ASkCT algorithm and the value for numSteps in the Discover algorithm were chosen based on our experiments on two graphs. The same is true of the values for $\alpha(k)$ in the LocalOptimize algorithm.

Table 1. ASkCT solution quality on random graphs with 400 vertices

g400-4-01 (400 vertices, 800 edges)					g400-4-05 (400 vertices, 800 edges)				
k	Best known	w_{best}	w_{avg}	σ	t_{avg} (sec)	k	Best known	w_{best}	w_{avg}	σ	t_{avg} (sec)
2	8	8	8.00	0.00	0.02	2	4	4	4.00	0.00	0.02
40	563	563	563.00	0.00	3.99	40	675	673	673.00	0.00	3.95
80	1304	1304	1304.85	0.36	6.35	80	1457	1445	1455.45	6.50	6.30
120	2137	2135	2139.45	1.40	11.00	120	2295	2293	2303.05	5.75	11.88
160	3066	3062	3065.95	1.91	5.18	160	3197	3193	3203.70	5.75	5.21
200	4105	4086	4086.00	0.00	7.12	200	4169	4156	4165.75	3.83	7.88
240	5238	5225	5228.80	0.87	15.87	240	5209	5202	5213.30	4.15	15.10
280	6499	6487	6488.10	0.83	24.63	280	6372	6350	6361.15	0.00	21.83
320	7888	7882	7882.00	0.00	22.52	320	7682	7682	7682.00	0.00	29.52
360	9471	9468	9468.00	0.00	38.90	360	9250	9249	9249.00	0.00	38.82
398	11433	11433	11433.00	0.00	35.20	398	11236	11236	11236.00	0.00	35.66

Table 2. ASkCT solution quality on random graphs with 1000 vertices

	g1000-4-01 (1000 vertices, 2000 edges)					g1000-4-05 (1000 vertices, 2000 edges)				
	Best				t_{avg}		Best				t_{avg}
k	known	w_{best}	w_{avg}	σ	(sec)	k	known	w_{best}	w_{avg}	σ	(sec)
2	6	6	6.00	0.00	0.07	2	7	7	7.00	0.00	0.07
100	1528	1523	1564.85	22.86	8.51	100	1654	1653	1665.00	5.92	5.06
200	3341	3329	3367.10	10.49	59.51	200	3639	3627	3665.30	10.18	26.13
300	5334	5333	5367.30	17.38	109.96	300	5842	5825	5836.90	5.38	122.01
400	7609	7581	7595.65	4.13	75.94	400	8302	8230	8233.65	1.06	79.76
500	10104	10052	10066.65	6.89	157.73	500	10893	10801	10810.85	5.30	187.49
600	12794	12708	12725.75	5.51	316.03	600	13725	13592	13606.75	4.80	305.38
700	15767	15675	15675.00	0.00	581.87	700	16803	16686	16688.15	0.78	569.16
800	19079	19037	19037.65	0.48	685.38	800	20128	20078	20078.00	0.00	756.91
900	22838	22830	22830.00	0.00	840.25	900	24035	24029	24029.00	0.00	866.67
998	27946	27946	27946.00	0.00	825.71	998	29182	29182	29182.00	0.00	834.94

Table 3. ASkCT solution quality on grid graphs with 225 vertices

	bb15x15_1 (225 vertices, 420 edges)					bb15x15_2 (225 vertices, 420 edges)				
	Best				t_{avg}		Best				t_{avg}
k	known	w_{best}	w_{avg}	σ	(sec)	k	known	w_{best}	w_{avg}	σ	(sec)
2	2	2	2.00	0.00	0.01	2	6	6	6.00	0.00	0.01
20	257	257	258.00	3.00	0.17	20	253	253	253.00	0.00	0.16
40	642	642	644.40	1.20	0.39	40	585	585	624.10	20.54	0.32
60	977	977	1005.50	9.38	0.75	60	927	927	986.05	30.61	0.61
80	1335	1335	1429.15	29.12	0.95	80	1290	1290	1348.35	19.19	1.26
100	1762	1762	1780.05	15.97	1.63	100	1686	1686	1726.25	11.88	0.90
120	2235	2235	2262.80	9.40	3.89	120	2120	2120	2143.55	8.46	2.81
140	2781	2781	2798.10	6.46	4.95	140	2634	2634	2639.60	4.07	3.76
160	3417	3417	3423.00	7.03	3.37	160	3260	3250	3272.75	9.59	3.33
180	4158	4158	4162.15	3.71	5.88	180	3915	3915	3915.00	0.00	4.94
200	5040	5040	5040.95	0.22	4.17	200	4718	4718	4718.00	0.00	6.15
220	6176	6176	6176.00	0.00	7.37	220	5862	5862	5862.00	0.00	7.17
223	6400	6400	6400.00	0.00	7.30	223	6101	6101	6101.00	0.00	7.25

Table 4. ASkCT solution quality on grid graphs with 1089 vertices

	bb33x33_1 (1089 vertices, 2112 edges)					bb33x33_2 (1089 vertices, 2112 edges)				
	Best				t_{avg}		Best				t_{avg}
k	known	w_{best}	w_{avg}	σ	(sec)	k	known	w_{best}	w_{avg}	σ	(sec)
2	3	3	3.00	0.00	0.08	2	3	3	3.00	0.00	0.08
100	1587	1587	1594.60	1.74	12.48	100	1524	1531	1568.65	15.43	6.27
200	3386	3366	3466.35	35.55	48.79	200	3378	3316	3530.60	94.65	35.63
300	5235	5235	5320.45	31.39	165.52	300	5289	5275	5360.10	28.81	116.51
400	7192	7166	7224.80	14.77	103.83	400	7366	7340	7582.05	98.69	75.56
500	9461	9256	9327.60	29.78	200.01	500	9626	9514	9624.70	55.98	173.90
600	11743	11579	11579.00	0.00	319.43	600	12113	11879	11889.30	2.45	319.74
700	14556	14309	14313.35	0.78	518.59	700	14664	14523	14523.00	0.00	552.12
800	17606	17399	17399.00	0.00	716.44	800	17667	17571	17571.00	0.00	723.64
900	21057	20921	20921.00	0.00	776.63	900	21037	21002	21002.00	0.00	774.41
1000	25235	25199	25199.00	0.00	1103.54	1000	25275	25274	25274.00	0.00	1145.04
1087	30417	30417	30417.00	0.00	1026.62	1087	30326	30326	30326.00	0.00	1044.47

4   Conclusion

In this paper, we gave an efficient ant system algorithm, called ASkCT, for the k-cardinality tree problem. The algorithm combines a search and optimization

Table 5. ASkCT solution quality on Steiner graphs with 500 vertices

	steinc5 (500 vertices, 625 edges)					steinc15 (500 vertices, 2500 edges)				
k	Best known	w_{best}	w_{avg}	σ	t_{avg} (sec)	k	Best known	w_{best}	w_{avg}	σ	t_{avg} (sec)
2	5	5	5.00	0.00	0.02	2	2	2	2.00	0.00	0.05
50	774	774	820.15	11.85	2.09	50	208	208	208.00	0.00	15.02
100	1712	1712	1734.65	6.92	5.44	100	481	481	488.45	1.77	27.86
150	2871	2865	2888.15	5.92	14.94	150	802	802	809.70	3.15	28.90
200	4279	4273	4273.00	0.00	8.29	200	1183	1182	1185.80	0.87	17.75
250	5965	5952	5955.20	4.26	15.67	250	1628	1628	1630.15	1.06	24.01
300	7986	7938	7938.00	0.00	24.67	300	2150	2148	2148.00	0.00	46.89
350	10292	10247	10248.20	0.98	39.97	350	2798	2796	2796.95	0.22	44.00
400	12992	12965	12965.00	0.00	59.48	400	3571	3571	3571.00	0.00	64.86
450	16321	16321	16321.00	0.00	32.62	450	4553	4553	4553.00	0.00	134.63
498	20485	20485	20485.00	0.00	70.20	498	5973	5973	5973.00	0.00	142.62

Table 6. ASkCT solution quality on Steiner graphs with 1000 vertices

	steind5 (1000 vertices, 1250 edges)					steind15 (1000 vertices, 5000 edges)				
k	Best known	w_{best}	w_{avg}	σ	t_{avg} (sec)	k	Best known	w_{best}	w_{avg}	σ	t_{avg} (sec)
2	3	3	3.00	0.00	0.04	2	2	2	2.00	0.00	0.16
100	1515	1503	1526.35	7.79	11.16	100	455	455	455.00	0.00	80.15
200	3469	3452	3456.50	1.50	51.92	200	1035	1029	1038.90	3.82	129.54
300	5897	5829	5873.45	16.03	139.64	300	1691	1680	1680.00	0.00	265.45
400	8886	8695	8716.15	7.48	85.97	400	2472	2451	2458.70	4.12	165.50
500	12172	12062	12085.70	13.09	184.10	500	3382	3366	3369.15	1.11	343.72
600	16091	15933	15933.00	0.00	281.90	600	4434	4423	4424.05	0.97	528.90
700	20646	20520	20539.45	10.20	449.47	700	5704	5686	5686.00	0.00	994.40
800	26103	26053	26053.00	0.00	769.75	800	7241	7236	7236.00	0.00	1530.38
900	32963	32963	32963.00	0.00	599.64	900	9256	9248	9248.00	0.00	1726.35
998	41572	41572	41572.00	0.00	631.28	998	12504	12504	12504.00	0.00	1658.98

Table 7. Results for all algorithms on random graphs with 400 vertices.

	g400-4-01 (400 vertices, 800 edges)						g400-4-05 (400 vertices, 800 edges)					
k	Best known	ASkCT Best	ACO Best	EC Best	TS Best	KCP Best	k	Best known	ASkCT Best	ACO Best	EC Best	TS Best	KCP Best
2	8	8	8	8	8	8	2	4	4	4	4	4	4
40	563	563	563	563	563	592	40	675	673	676	676	675	739
80	1304	1304	1304	1305	1307	1392	80	1457	1445	1457	1460	1466	1601
120	2137	2135	2137	2140	2140	2285	120	2295	2293	2295	2314	2318	2451
160	3066	3062	3066	3071	3070	3198	160	3197	3193	3197	3217	3217	3389
200	4105	4086	4105	4117	4112	4249	200	4169	4156	4169	4171	4171	4400
240	5238	5225	5247	5255	5238	5410	240	5209	5202	5209	5217	5216	5525
280	6499	6487	6509	6514	6499	6666	280	6372	6350	6383	6372	6378	6533
320	7888	7882	7903	7892	7888	8048	320	7682	7682	7713	7682	7682	7869
360	9471	9468	9494	9472	9471	9553	360	9250	9249	9295	9256	9250	9262
398	11433	11433	11454	11433	11433	11433	398	11236	11236	11278	11236	11236	11236

based on pheromone with a weight based optimization. Extensive experimental results show that ASkCT outperforms existing heuristics from different methodologies. We note that the weight based local optimization algorithm can also be used with other algorithms as an extra optimization. Possible future work includes improving the quality of ASkCT even further, particularly, for the cases when k is small compared to the number of vertices in the graph and for grid

Table 8. Results for all algorithms on random graphs with 1000 vertices

	g1000-4-01 (1000 vertices, 2000 edges)							g1000-4-05 (1000 vertices, 2000 edges)					
	Best	ASkCT	ACO	EC	TS	KCP		Best	ASkCT	ACO	EC	TS	KCP
k	known	Best	Best	Best	Best	Best	k	known	Best	Best	Best	Best	Best
2	6	**6**	6	6	6	6	2	7	**7**	7	7	7	7
100	1528	**1523**	1528	1558	1567	1684	100	1654	**1653**	1654	1657	1662	1782
200	3341	**3329**	3341	3445	3438	3652	200	3639	**3627**	3639	3680	3692	3994
300	5334	**5333**	5334	5500	5482	5828	300	5842	**5825**	5842	5875	5923	6320
400	7609	**7581**	7609	7749	7669	8329	400	8302	**8230**	8302	8320	8344	8875
500	10104	**10052**	10114	10104	10125	10837	500	10893	**10801**	10922	10893	10956	11492
600	12794	**12708**	12864	12794	12797	13584	600	13725	**13592**	13780	13743	13725	14392
700	15767	**15675**	15806	15772	15767	16223	700	16803	**16686**	16924	16803	16805	17399
800	19079	**19037**	19232	19090	19079	19494	800	20128	**20078**	20262	20134	20128	20576
900	22838	**22830**	23022	22839	22838	23076	900	24035	**24029**	24226	24045	24035	24272
998	27946	**27946**	28119	**27946**	**27946**	**27946**	998	29182	**29182**	29342	**29182**	**29182**	**29182**

Table 9. Results for all algorithms on grid graphs with 225 vertices

	bb15x15_1 (225 vertices, 420 edges)							bb15x15_2 (225 vertices, 420 edges)					
	Best	ASkCT	ACO	EC	TS	KCP		Best	ASkCT	ACO	EC	TS	KCP
k	known	Best	Best	Best	Best	Best	k	known	Best	Best	Best	Best	Best
2	2	**2**	2	2	2	2	2	6	**6**	6	6	6	6
20	257	**257**	257	257	257	267	20	253	**253**	253	253	253	253
40	642	**642**	642	642	642	650	40	585	**585**	585	585	592	620
60	977	**977**	977	988	977	1154	60	927	**927**	927	929	930	1075
80	1335	**1335**	1335	1359	1355	1518	80	1290	**1290**	1290	1315	1324	1471
100	1762	**1762**	1762	1764	1764	1998	100	1686	**1686**	1686	1725	1741	1907
120	2235	**2235**	2235	2235	2235	2554	120	2120	**2120**	2120	2127	2155	2342
140	2781	**2781**	2783	2781	2783	2956	140	2634	**2634**	2634	2638	2642	2773
160	3417	**3417**	3417	3417	3435	3475	160	3260	**3250**	3260	3278	3268	3289
180	4158	**4158**	4158	4158	4167	4216	180	3915	**3915**	3922	3922	3915	3968
200	5040	**5040**	5059	5040	5041	5104	200	4718	**4718**	4722	4718	4718	4867
220	6176	**6176**	6183	6176	6176	6176	220	5862	**5862**	5864	5862	5862	5862
223	6400	**6400**	6401	6400	6400	6400	223	6101	**6101**	6105	6101	6101	6101

Table 10. Results for all algorithms on grid graphs with 1089 vertices

	bb33x33_1 (1089 vertices, 2112 edges)							bb33x33_2 (1089 vertices, 2112 edges)					
	Best	ASkCT	ACO	EC	TS	KCP		Best	ASkCT	ACO	EC	TS	KCP
k	known	Best	Best	Best	Best	Best	k	known	Best	Best	Best	Best	Best
2	3	**3**	3	3	3	3	2	3	**3**	3	3	3	3
100	1587	**1587**	1587	1595	1615	1802	100	1524	1531	**1524**	1558	1569	1760
200	3386	**3366**	3386	3410	3537	3927	200	3378	**3316**	3378	3475	3431	3761
300	5235	**5235**	5235	5261	5433	6038	300	5289	**5275**	5289	5486	5510	5876
400	7192	**7166**	7192	7328	7415	8232	400	7366	**7340**	7366	7497	7562	8571
500	9461	**9256**	9461	9556	9584	11038	500	9626	**9514**	9626	9687	9964	10689
600	11743	**11579**	11743	11946	12056	13263	600	12113	**11879**	12113	12256	12260	13164
700	14556	**14309**	14556	14746	14775	15724	700	14664	**14523**	14664	14884	14979	15808
800	17606	**17399**	17636	17606	17618	18683	800	17667	**17571**	17780	17667	17798	18831
900	21057	**20921**	21266	21057	21072	21862	900	21037	**21002**	21252	21037	21055	22144
1000	25235	**25199**	25582	25235	25264	25882	1000	25275	**25274**	25620	25275	25299	25914
1087	30417	**30417**	30571	30417	30417	30417	1087	30326	**30326**	30467	30326	30326	30326

graphs. The grid graph bb33x33_2 with $k = 100$ is the only instance that ASkCT did not match or surpass the best known value among the tested graphs.

Table 11. Results for all algorithms on Steiner graphs with 500 vertices

	steinc5 (500 vertices, 625 edges)							steinc15 (500 vertices, 2500 edges)					
k	Best known	ASkCT Best	ACO Best	EC Best	TS Best	KCP Best	k	Best known	ASkCT Best	ACO Best	EC Best	TS Best	KCP Best
2	5	5	5	5	5	5	2	2	2	2	2	2	2
50	774	774	774	777	783	877	50	208	208	208	208	208	229
100	1712	1712	1712	1714	1712	1963	100	481	481	482	483	481	526
150	2871	2865	2871	2896	2892	3163	150	802	802	802	807	815	923
200	4279	4273	4279	4335	4318	4552	200	1183	1182	1183	1185	1190	1276
250	5965	5952	5983	5992	5965	6221	250	1628	1628	1628	1633	1633	1731
300	7986	7938	8014	8014	7986	8378	300	2150	2148	2150	2158	2153	2227
350	10292	10247	10315	10371	10292	10711	350	2798	2796	2802	2798	2799	2950
400	12992	12965	13060	13023	12992	13475	400	3571	3571	3578	3571	3571	3666
450	16321	16321	16359	16334	16321	16666	450	4553	4553	4575	4553	4555	4584
498	20485	20485	20495	20485	20485	20485	498	5973	5973	5986	5973	5973	5973

Table 12. Results for all algorithms on Steiner graphs with 1000 vertices

	steind5 (1000 vertices, 1250 edges)							steind15 (1000 vertices, 5000 edges)					
k	Best known	ASkCT Best	ACO Best	EC Best	TS Best	KCP Best	k	Best known	ASkCT Best	ACO Best	EC Best	TS Best	KCP Best
2	3	3	3	3	3	3	2	2	2	2	2	2	2
100	1515	1503	1515	1570	1587	1790	100	455	455	455	457	457	491
200	3469	3452	3469	3539	3596	4135	200	1035	1029	1037	1035	1041	1138
300	5897	5829	5897	6031	6027	6766	300	1691	1680	1696	1691	1708	1857
400	8886	8695	8886	8972	8941	9881	400	2472	2451	2552	2487	2472	2684
500	12172	12062	12267	12216	12172	13830	500	3382	3366	3599	3382	3396	3607
600	16091	15933	16091	16125	16139	17190	600	4434	4423	4797	4434	4437	4670
700	20646	20520	20700	20671	20646	21520	700	5704	5686	6034	5704	5707	5889
800	26103	26053	26227	26103	26147	27173	800	7241	7236	7837	7241	7245	7407
900	32963	32963	33119	33006	32963	33579	900	9256	9248	9771	9256	9276	9388
998	41572	41572	41637	41572	41572	41572	998	12504	12504	12759	12504	12504	12504

Acknowledgements. The authors would like to thank Christian Blum for helpful discussions and the anonymous reviewers for their valuable comments.

References

1. Blesa, M. J. and F. Xhafa, "A C++ Implementation of Tabu Search for k-Cardinality Tree Problem Based on Generic Programming and Component Reuse," Net.ObjectDsays 2000 Tagungsband, NetObjectDays Forum, Germany, 2000, pp. 648–652.

2. Blum, C., "Ant Colony Optimization for the Edge-Weighted k-Cardinality Tree Problem," Proceedings of the Genetic and Evolutionary Computation Conference, 2002, pp. 27–34.

3. Blum, C. and M. Ehrgott, "Local Search Algorithms for the k-cardinality Tree Problem," Technical report TR/IRIDIA/2001-12, IRIDIA, Université Libre de Bruxelles, Belgium.

4. Borndörfer, R., C. Ferreira and A. Martin, " Decomposing Matrices Into Blocks," SIAM Journal on Optimization, 9(1), 1998, pp. 236–269.

5. Cheung, S. Y. and A. Kumar, "Efficient Quorumcast Routing Algorithms," in Proceedings of INFOCOM'94, Los Alamitos, 1994.

6. Cormen, T. H., C. E. Leiserson, R. L. Rivest and C. Stein, *Introduction to Algorithms,* Second Edition, McGraw-Hill, 2001.
7. Ehrgott, M., J. Freitag, H. W. Hamacher and F. Maffioli, "Heuristics for the k-cardinality Tree and Subgraph Problem," Asia Pacific Journal of Operational Research, 14(1), 1997, pp. 87–114.
8. Fischetti, M., W. Hamacher, K. Jornsten and F. Maffioli, "Weighted k-Cardinality Trees: Complexity and Polyhedral Structure," Networks, 24, 1994, pp. 11–21.
9. Foulds, L. R. and H. W. Hamacher, "A New Integer Programming Approach to (Restricted) Facilities Layout Problems Allowing Flexible Facility Shapes," Technical Report 1992-3, University of Waikato, Department of Management Science, 1992.
10. Foulds, L. H. Hamacher and J. Wilson, "Integer Programming Approaches to Facilities Layout Models with Forbidden Areas," Annals of Operations Research, 81, 1998, pp. 405–417.
11. Garg, N. and D. Hochbaum, "An $O(\log k)$ Approximation Algorithm for the k Minimum Spanning Tree Problem in the Plane," Algorithmica, 18, 1997, pp. 111–121.
12. Hamacher, H. W., K. Jornsten and F. Maffioli, " Weighted K-Cardinality Trees," Technical Report 91.023, Dept. di Elettronica, Politecnico di Milano, 1991.
13. Mladenovic, N., "Variable Neighborhood Search for the k-Cardinality Tree Problem," Proc. of the Metaheuristics International Conference, MIC'2001, 2001.
14. Ravi, R., R. Sundaram, M. V. Marathe, D. J. Rosenkrantz and S. S. Ravi, "Spanning Trees – Short or Small," SIAM J. on Discrete Mathematics, 9(2), 1996, pp. 178–200.

A Hybrid Ant Colony Optimisation Technique for Dynamic Vehicle Routing

Darren M. Chitty and Marcel L. Hernandez

The Advanced Processing Centre, QinetiQ Ltd.,
St Andrews Road, Malvern,
Worcestershire, WR14 3PS, UK
{chitty, marcel}@signal.qinetiq.com

Abstract. This paper is concerned with a dynamic vehicle routing problem. The problem is dynamic in the sense that the time it will take to traverse each edge is uncertain. The problem is expressed as a bi-criterion optimisation with the mutually exclusive aims of minimising both the total mean transit time and the total variance in transit time. In this paper we introduce a hybrid dynamic programming - ant colony optimisation technique to solve this problem. The hybrid technique uses the principles of dynamic programming to first solve simple problems using ACO (routing from each adjacent node to the end node), and then builds on this to eventually provide solutions (i.e. Pareto fronts) for routing between each node in the network and the destination node. However, the hybrid technique updates the pheromone concentrations only along the first edge visited by each ant. As a result it is shown to provide the overall solution in quicker time than an established bi-criterion ACO technique, that is concerned only with routing between the start and destination nodes. Moreover, we show that the new technique both determines more routes on the Pareto front, and results in a 20% increase in solution quality for both the total mean transit time and total variance in transit time criteria. However the main advantage of the technique is that it provides solutions in routing between each node to the destination node. Hence it allows "instantaneous" re-routing subject to dynamic changes within the road network. [1]

1 Introduction

A requirement in the routing of a single vehicle through a road network from a starting depot to a destination depot, is the ability to manoeuver quickly to take into account events such as blocked roads or heavy traffic. Subsequently, the problem becomes dynamic because the road conditions are not known with certainty and are continuously changing. As a result, each road is characterised by two indices. The first of these is the mean transit time, averaged over different driving scenarios. The second is the variance in transit time on each road, which gives an indication of how the transit time will fluctuate about this mean value as the scenario changes.

---

[1] ©Copyright QinetiQ Ltd. 2004

K. Deb et al. (Eds.): GECCO 2004, LNCS 3102, pp. 48–59, 2004.
© Springer-Verlag Berlin Heidelberg 2004

In an ideal world, one would wish to find routes that have both low mean transit time and low variance in transit time. However, typically these priorities have conflicting objectives. Routes that have the shortest overall travel time may not have the smallest variance in travel time and vice-versa. In such circumstances we must then trade-off between these two conflicting aims.

This is the basis of bi-criterion optimisation (see [1]). Instead of attempting to find a solution that satisfies the minimisation of each objective, we seek out the set of non-dominated solutions that form the Pareto front in the two-dimensional objective function space. Evolutionary techniques, which simultaneously create and evaluate a set of possible solutions, are a natural approach to solving problems of this type (again, see [1]). Alternative techniques, such as linear programming, have also been used for multi-objective optimisation in which one objective is minimised with (worst acceptable) performance bounds placed on each of the other objectives [2].

In this paper, we build on a technique known as Ant Colony Optimisation [3], [4] to solve this bi-criterion routing problem. For a variety of reasons, Ant Colony Optimisation is a natural approach. Firstly, its foundations lie in the way in which real ant colonies find shortest path routes between different parts of their natural habitat [5]. As a result, the technique has been successfully proved to be particularly effective in solving networking problems.

Indeed, it has been successfully applied to the Travelling Salesman Problem (TSP) [3], [4]; the Graph Colouring Problem [6]; and the Vehicle Routing Problem [7], [8]. The Vehicle Routing Problem considered in [7] and [8] is different from the one analysed in this report in that these papers are concerned with the routing of a fleet of vehicles to satisfy a number of customer requests. The vehicles begin and end at a central depot, and once the customers are assigned to vehicles the Vehicle Routing Problem is reduced to several Travelling Salesman Problems.

In each of these cases, the model is deterministic, but Ant Colony Optimisation has also been applied to several dynamic problems. It has been used for routing in communication networks [9], [10] in which there is uncertain demand on each node, with requests forming a dynamic and uncertain sequence. Moreover, ant techniques have also been applied successfully to a dynamic Travelling Salesman Problem in which, at certain time instances, parts of the network are 'lost' and re-routing is necessary [11]. The ability of the technique to cope with problems of this type makes it a natural approach to both vehicle routing, and more generally, to solving problems that are subject to dynamic and uncertain change.

2   The Bi-criterion Optimisation Problem

Consider a road network repesented by $G = (N, E)$, where $N = (N_1, ... N_n)$ is the set of n nodes (i.e. junctions) and E is the set of (directed) edges (i.e. roads, where a direction of travel may be specified). The aim is to route vehicles so that they will reach their destination in the quickest time possible. However, the problem is subject to uncertainty. Traffic congestion may cause delays, other forms of disruption such as road works and/or driving accidents, may also affect transit times. As a result, we are unable to simply characterise each edge, E_{ij}, in terms of the time it will take to traverse.

However, we can instead characterise E_{ij} in terms of two indices; these being the average time: M_{ij}, and the variance in time: V_{ij}, it will take to traverse each edge. M_{ij}, is averaged over the different driving scenarios. The variance, V_{ij}, gives an indication of how the travel time will fluctuate about the mean value as the scenario changes.

We will specify each route as $R = (a_{ij})$ where:

$$a_{ij} = \begin{cases} 1 \text{ if node } j \text{ is visited after node } i \\ 0 \text{ otherwise} \end{cases} \tag{1}$$

for $i, j = 1, \ldots, n$.

It is assumed that the time taken to traverse edge E_{ij} is independent of the time taken to traverse each of the other edges. This assumption is not entirely true, and indeed delays on certain edges may be expected to have a knock-on effect on other edges within the network. However, provided the network is not densely congested, these effects are likely to be small and make this assumption a valid approximation. It then follows that the average total transit time, $T_m(R)$, and the variance in total transit time, $T_v(R)$, are given by:

$$T_m(R) = \sum_{i=1}^{n} \sum_{j=1}^{n} a_{ij} M_{ij} \qquad T_v(R) = \sum_{i=1}^{n} \sum_{j=1}^{n} a_{ij} V_{ij} \tag{2}$$

Clearly, if mean transit time is proportional to variance in transit time, i.e.

$$M_{ij} \propto V_{ij}^{k} \qquad \text{for some } k > 0 \tag{3}$$

then by determining a route, R, that minimises mean time (equation (2)) the variance in time (equation (3)) is also minimised. In this case we have effectively just a single objective function, and the optimum route can easily be found using a Dynamic Programming technique such as Dijkstra's algorithm [12].

However, in general we would not expect a relationship as simple as (4) to hold. Indeed, with small average transit times may typically represent urban routes, that can usually be traversed quickly, but can be easily disrupted by traffic congestion. Conversely, edges with large mean transit times may represent long motorway segments, which are designed to be less prone to traffic disruption.

With this in mind, a more suitable relationship between M_{ij} and V_{ij} is given by:

$$M_{ij} \propto \frac{1}{V_{ij}^{k}} \qquad \text{for some } k > 0 \tag{4}$$

This is the relationship that we will assume in later examples. Clearly now if we minimise $T_m(R)$ we will maximise $T_v(R)$ and, indeed the converse is also true. We are therefore faced with a bi-criterion optimisation problem (again, see [1]).

The general structure of the solution space is shown in figure 1. Instead of attempting to find a solution that simultaneously minimises each objective, which is clearly no longer possible, we seek the set of non-dominated solutions that form the Pareto front in the multi-dimensional objective function space. In figure 1 this Pareto front is represented by the solid thick black edge of the solution space. Any solution not on the Pareto front (i.e. within the red region) is 'dominated' by a solution on the Pareto front which has both lower mean and variance (in transit time) and is clearly better in every respect. Solutions on the Pareto front itself cannot dominate each other, and the solution that should be utilised depends on the scenario and the operational requirements (i.e. we may not accept a solution that has a variance $T_v(R) > V_{max}$, say).

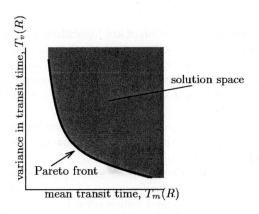

Fig. 1. The general structure of the solution space.

Techniques, such as Ant Colony Optimisation [3], [4], which simultaneously create and evaluate a set of possible solutions are a natural approach to solving problems of this type. Indeed, Ant Colony Optimisation has been used to solve several multi-criterion problems (see [13] for an overview) and, furthermore [13] introduces a general framework for solving bi-criterion optimisation problems that is utilised in this report.

3   Ant Colony Optimisation

Ant Colony Optimisation (ACO) [3], [4] is an evolutionary approach that is inspired by the way in which real ant colonies establish shortest path routes between their nest and feeding sources. Real ants establish such paths by depositing an aromatic essence known as pheromone along these paths [5]. The quantity of pheromone is proportional to the length of the path, or the quality of the feeding habitat. Ants are attracted to the pheromone and follow it. Hence, the pheromone concentrations along better paths will be further enhanced which will attract more ants. Eventually, the pheromone concentrations along better paths will become so great that all ants will uses these routes.

Like all evolutionary techniques, Ant Colony Optimisation provides a means of exploring promising areas of the solution space in order to find optimal/near optimal solutions. It does this by creating a trade-off between the conflicting aims of exploitation and exploration, where

- exploration is defined to be the search through the space of possible solutions in order to find the optimal/near optimal solution(s);

- exploitation is the fusion of information regarding the quality of solutions found so far, in order to focus on promising areas of the search space.

In addition to the natural application of ant algorithms to networking problems, they have been successfully applied to a host of other combinatorial optimisation problems (see [10] for an overview), including the Quadratic Assignment Problem [14] and the Job Shop Scheduling Problem [15]. Encouraging results have also been obtained when using Ant Colony Optimisation to build up classification rules for data mining [16].

A brief introduction to the technique follows, using an ant algorithm applied to a single objective problem. We will then show how the pheromone matrix can be adapted for a bi-criterion problem.

The basis of Ant Colony Optimisation is the pheromone matrix, $M = \tau_{ij}$. The probability, p_{ij} that $a_{ij} = 1$ (see equation (3-1)) is a function of the amount of pheromone τ_{ij} on 'edge' E_{ij} and information ν_{ij}, relating to the quality of this edge, i.e.

$$p_{ij} = \frac{\tau_{ij}^{\alpha} \nu_{ij}^{\beta}}{\sum\limits_{h \in S} \tau_{ih}^{\alpha} \nu_{ih}^{\beta}} \tag{5}$$

where α, β give the influence of the pheromone and heuristic information respectively; and S is the set of nodes not already visited.

Referring once again to a single objective vehicle routing problem, if the objective is to minimise the mean total transit time, then the heuristic information may be given as follows:

$$\nu_{ij} = 1/M_{ij} \qquad \text{for } i, j = 1, \ldots, n \tag{6}$$

i.e. the preferred edges are those that have the smallest mean transit time. The transition probabilities, p_{ij}, are then used to build candidate solutions, which are referred to as 'ants'. In turn, the quality of each solution determines the way in which the pheromone matrix is updated. 'Paths' that form part of high quality solutions have their pheromone levels reinforced and this enables promising areas of the search space to be identified and explored in subsequent iterations (generations). This is the analogy with real ant systems that reinforce pheromone concentrations along the better paths, leading the subsequent ants towards optimal or near optimal routes.

Pheromone update typically has an evaporation rate, ρ, and a deposit rate Δ_{ij}, i.e.

$$\tau_{ij} \rightarrow (1 - \rho)\tau_{ij} + \Delta_{ij} \tag{7}$$

The value of Δ_{ij} is dependent on whether the ant(s) used edge E_{ij}, i.e. whether $a_{ij} = 1$ or $a_{ji} = 1$ and how optimal the overall solution(s) were that used edge E_{ij}. Hence, in the case of attempting to minimise total mean time, if we define:

$$d_{ij}^r = \begin{cases} 1 \text{ if edge } E_{ij} \text{ is used by ant } r \text{ (i.e. } a_{ij}^r=1 \text{ or } a_{ji}^r=1) \\ 0 \text{ otherwise} \end{cases} \tag{8}$$

then we may take (as in [4]):

$$\Delta_{ij} = \sum_{r=1}^{M} \frac{d_{ij}^r}{T_m(r)} \qquad (= \Delta_{ji}) \tag{9}$$

where (to remind the reader), $T_m(r)$ is the total mean transit time of the route, a_{ij}^r taken by the ant r and M is the total number of ants in each generation.

Other pheromone update rules have also been proposed. For example, in some cases good results have been obtained by having an additional contribution from the highest quality solutions found so far (so called 'elitist') ants (see [8] and references therein).

The initial pheromone concentrations are usually set at some arbitrarily small value. The stopping criterion is typically either to terminate the ant algorithm if there has been no improvement in the best solution for a fixed number of generations; or we have reached the maximum number of generations permitted.

Clearly, the pheromone evaporation rate, p, provides one means by which the ant algorithm can control the trade-off between exploration and exploitation. If p has a large value (i.e. close to 1) then in each generation of ants, the pheromone matrix is highly dependent on the good solutions from the previous generation, which leads to high degree of search around these good solutions. The smaller the value of p, the greater the contribution of good solutions from all the previous generations and the greater the diversity of search through the solution space.

4   A Standard Bi-criterion Ant Model

We follow the approach of [13] and use two types of pheromone, one relating to the first objective of minimising the total mean transit time ($M_m = (\tau_{ij}^m)$) and the second matrix relating to the second objective of minimising the variance in the transit time, ($M_v = (\tau_{ij}^v)$). Each ant then uses the following transition probabilities:

$$p_{ij} = \frac{(\tau_{ij}^m)^\lambda (\tau_{ij}^v)^{(1-\lambda)}}{\sum_{h \in S} (\tau_{ih}^m)^\lambda (\tau_{ih}^v)^{(1-\lambda)}} \tag{10}$$

$\lambda \in [0, 1]$ is the importance of objective one in relation to objective two. If $\lambda = 1$, then the single objective is to minimise the mean transit time, if $\lambda = 0$, the single objective is to minimise the variance in transit time. For value of λ between these two extreme values we must trade off between these two conflicting objectives.

Pheromone update is again as follows:

$$\tau_{ij}^m \rightarrow (1 - \rho)\tau_{ij}^m + \sum_{r=1}^{M} \frac{d_{ij}^r}{T_m(r)} \qquad \tau_{ij}^v \rightarrow (1 - \rho)\tau_{ij}^v + \sum_{r=1}^{M} \frac{d_{ij}^r}{T_v(r)} \qquad (11)$$

where d_{ij}^r is given by equation (9). Only ants that reach the destination node update the pheromone matrix.

5   The Hybrid Ant Colony Optimisation Technique

5.1   Background

The standard ACO technique of the previous section can give good results for routing vehicles between a start node and a destination node. However, the time taken to produce a solution can be in excess of one minute[2], even for just a 100 node problem instance.

In addition, the technique determines a Pareto front of solutions only for the focal problem of routing from the start node to the destination node. If, for any unforeseen reason, a vehicle has to divert to another node not on the prescribed route (i.e. to avoid a road blocked as a result of an accident), it would no longer have any information that could be used to navigate it to its final destination. A new method is proposed which will lead to a quicker build up of the pheromone on edges that are important and will result in every node in the network having a Pareto front of non-dominated solutions. Hence, instantaneous re-routing can be performed if the target has to deviate from the original plan.

5.2   Features

The key features of this new technique are as follows:

- There are again two types of pheromone ($M_m = (\tau_{ij}^m)$ and $M_v = (\tau_{ij}^v)$).

- We use two kinds of ants, one for each criterion.

- Ants optimising the mean transit time criterion use the following transition probabilities:

$$p_{ij} = \frac{\tau_{ij}^m}{\sum_{h \in S} \tau_{ih}^m} \qquad (12)$$

with a similar expression giving the transition probabilities of ants optimising the variance in transit time criterion.

---

[2] When using C++ Version 6.0 run on a 600 MHz Athlon processor.

– Under each criterion, M ants travel to the destination node from each and
  every node in the network.

– Each ant travelling from node i updates a Pareto front, P_i of solutions (routes
  to the destination node) from that start location.

5.3   Pheromone Update

In each generation, pheromone evaporation occurs at a constant rate ρ. Hence,
for the two pheromone matrices:

$$\tau_{ij}^m \rightarrow (1 - \rho)\tau_{ij}^m \qquad \tau_{ij}^\nu \rightarrow (1 - \rho)\tau_{ij}^\nu \tag{13}$$

Pheromone update is performed in the following way.

– Ants only update their own type of pheromone.

– Only solutions on the global Pareto front, P_i (at each node, i) update the
  pheromone matrices.

– Pheromone update is performed only on the very first path taken by each
  ant.

Hence, for the rth $(r = 1, \ldots, M)$ ant starting from node i and optimising mean
transit time, pheromone update is then given by:

$$\tau_{ij}^m \rightarrow \tau_{ij}^m + \frac{\hat{d}_{ij}^r}{T_{m_{i_m}}(r)} \tag{14}$$

where

$$\hat{d}_{ij}^r = \begin{cases} 1 \text{ if:} & \begin{cases} \text{edge } E_{ij} \text{ is used by ant } r \\ \text{node } j \text{ is adjacent to node } i \\ (T_{m_{i_m}}(r), T_{\nu_{i_m}}(r)) \in P_i \end{cases} \\ \\ 0 \text{ otherwise} \end{cases}$$

$T_{m_{i_m}}(r)$ and $T_{\nu_{i_m}}(r))$ are the total mean transit time and total variance in transit
time respectively of the rth ant optimising mean transit time and travelling
from node i to the destination node. $(T_{m_{i_m}}(r), T_{\nu_{i_m}}(r)) \in P_i$ denotes that the
rth route (minimising mean transit time) from node i is on the Pareto front.
A similar equation (to (14)) then gives the variance in transit time pheromone
update.

5.4   Comment

The technique has a dynamic-programming basis [17] in which initially simple
problems are considered and through an iterative process complete solutions to

more challenging scenarios are built. The basis of the technique is to allow ants to travel from every single node in the network to the destination node, but only the first edge the ant takes being updated by its pheromone.

By building a Pareto front of solutions from each node in the network to the destination node it is the first decision of which node to go to that is of critical importance, because the solution space at/from that node is being simultaneously built. At nodes close to the destination node the number of potential routes decreases and more complete solution spaces (Pareto fronts) will exist. Iterating backwards, applying the fundamental principles of dynamic programming, we can build complete solution spaces for routing between each node in the network and the final destination.

We would expect the added complexity of this new (hybrid) technique to result in a huge increase in computation time/complexity when compared to the standard ant algorithm (of section 4). However, in updating the two pheromone matrices only the first edge selected by the ant is updated, and as a result it will be shown that the speed of the new technique far exceeds that of its predecessor.

6   Results

We shall now compare the capability of the two algorithms (the hybrid ant algorithm and the standard bi-criterion ant algorithm) to find high quality solutions. The two techniques were tested on randomly created networks of various sizes from 25 to 250 nodes with each node connected to its six nearest neighbors.

Figure 2(a) gives the lowest mean transit time determined under each algorithm for a range of networks of different sizes. Figure 2(b) gives the lowest variance in transit time in each case. These represent the two extreme edges of the Pareto front. We observe that in both cases the new technique outperforms the standard ant algorithm, the margin increasing with the size of the network.

These results may suggest that the new technique concentrates solely on finding the optimal mean transit time route and the optimal variance in transit time route, in which case an approach based solely on dynamic programming would be much better/quicker. Moreover, in comparing the two algorithms, it is insufficient simply to analyze each of the two objectives in isolation, because for the bi-criterion problem it is the combination of the mean and variance of each route that is important.

However, figure 3 clearly demonstrates that the new technique maintains a large number of non-dominated routes both at the start node (figure 3(a)) and at all nodes (figure 3(b)) across the network. We note that the average number of solutions on the Pareto front is lower when we average across the whole network because nodes near or adjacent to the destination node will obviously have few non-dominated solutions.

Furthermore, the reduced computational complexity of the new technique (hybrid) when compared to the standard technique is clearly demonstrated in figure 4(a). It can be seen that the hybrid technique typically finds the optimal solution in less than 15 seconds, whereas the standard technique did not find an optimal solution in less than 1 minute for any of the problem instances considered.

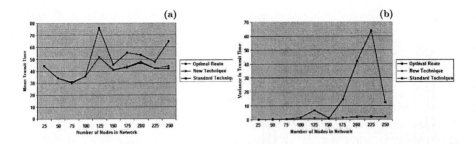

Fig. 2. **(a):** The lowest mean transit time, and **(b):** the lowest variance in transit time, for different network sizes.

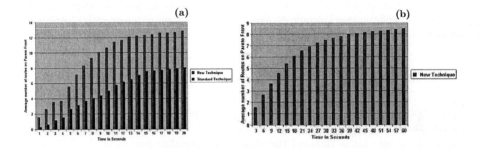

Fig. 3. **(a):** Number of non-dominated routes found by the algorithm at the start node, **(b):** number of non-dominated routes found, averaged over all the nodes in the network. In each case results are averaged over 150 runs on 100 node network problems.

With a set of non-dominated solutions held at every node the system is now also allowed to be dynamic. For example, consider a vehicle that is following the lowest mean transit route. If enroute it finds its path blocked it will have to divert. Using the new technique a new route is easily found by examining all the adjacent nodes for the route that most readily satisfies the operational requirements.

This is demonstrated by figure 4(b). The minimum mean transit time route (from the start node to the destination node) is shown by the green line. At several points the vehicle following the green route finds its path blocked by an obstacle (blue nodes) which was not there when the original route was determined. With this new technique, information is held at all the nodes in the network. Hence the adjacent nodes can be examined and the best route (here in terms of minimum mean transit time) across all the adjacent nodes is selected. This alternative route is shown by the red lines in figure 4(b).

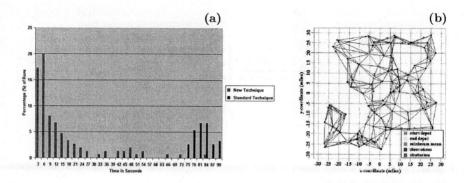

Fig. 4. (a): Time taken by the new ant algorithm to find the best routes in terms of mean transit time and variance in mean transit time. Results are averaged over 150 runs on 100 node network problems. **(b):** Route of minimum mean with diversions to avoid blocked vertices.

7   Conclusions

The hybrid ant algorithm presented in this report has been shown to be successful in solving bi-criterion Vehicle Routing Problems. The hybrid technique comprehensively outperformed the standard algorithm by:

– always finding the lowest mean transit and lowest variance transit time routes;

– finding considerably more routes along the Pareto front;

– running in much quicker time;

– finding a Pareto front of solutions at every single node, allowing instantaneous re-routing subsequent to network change.

Acknowledgments. This research was sponsored by the United Kingdom Ministry of Defence Corporate Research Programme CISP.

References

1. Deb, K. (1999),"Multi-Objective Optimization using Evolutionary Algorithms", Wiley.
2. Beasley, J. E. and Christofides, N. (1989), "An Algorithm for the Resource Constrained Shortest Path Problem" Networks, 19, pp 379-394.
3. Dorigo, M., Maniezzo, V. and Colorni, A. (1991), "The Ant System: An Autocatalytic Optimizing Process", Technical Report 91-016 (revised), Politecnico di Milano, Italy.

4. Colorni, A., Dorigo, M. and Maniezzo, V. (1991), "Distributed Optimization by Ant Colonies", Proceedings of the European Conference on Artificial Life, eds. Varela, F. and Bourgine, P. Elsevier.
5. Denebourg, J. L., Pasteels, J. M. and Verhaeghe, J. C. (1983), "Probabilistic Behaviour in Ants: a Strategy of Errors?", Journal of Theoretical Biology, 105, pp 259-271.
6. Costa, D. and Hertz, A. (1997), "Ants can colour graphs", Journal of the Operational Research Society, 48, pp 295.
7. Bullnheimer, B., Kotsis, G. and Strauss, C. (1997), "Applying the Ant System to the Vehicle Routing Problem", Proceedings of the Second International Conference on Metaheuristics.
8. Bullnheimer, B., Hartl, R. F. and Strauss, C. (1999), "An improved ant system algorithm for the vehicle routing problem", Annals of Operations Research, 89, pp 319-328.
9. Schoonderwoerd, R., Holland, O. and Bruten, J. (1996), "Ants for load balancing in telecommunications networks", Hewlett Packard Laboratory Technical Report.
10. Di Caro, G. and Dorigo, M. (1998), "AntNet: Distributed Stigmergetic Control for Communications Networks", Journal of Artificial Intelligence Research, 9, pp 317-365.
11. Guntsch, M., Middendorf, M. and Schmeck, H. (2001), "An Ant Colony Optimization Approach to Dynamic TSP", Proceeedings of the Genetic and Evolutionary Computation Conference, pp 860-867.
12. Jungnickel, D. (1998), "Graphs, Networks and Algorithms", Springer-Verlag.
13. Iredi, S., Merkle, D. and Middendorf, M. (2001), "Bi-Criterion Optimization with Multi Colony Ant Algorithms", Proceedings of the First International Conference on Evolutionary Multi-Criterion Optimization, ed. E. Zitzler, Springer-Verlag.
14. Maniezzo, V., Colorni, A. and Dorigo, M. (1994), "The Ant System Applied to the Quadratic Assignment Problem", Technical Report IRIDIA/94-28, Universite Libre de Bruxelles.
15. Colorni, A., Dorigo, M., Maniezzo, V. and Trubian, M. (1994), "Ant System for Job Shop Scheduling", Belgian Journal of Operations Research, Statistics, and Computer Science, 34, pp 39.
16. Parpinelli, R. S., Lopes, H. S. and Freitas, A. A. (2001), "An Ant Colony Based System for Data Mining: Applications to Medical Data", Proceeedings of the Genetic and Evolutionary Computation Conference, pp 791-797.
17. Bellman, R.E. (1957), "Dynamic Programming", Princeton University Press.

Cooperative Problem Solving Using an Agent-Based Market

David Cornforth and Michael Kirley

School of Environmental and Information Sciences
Charles Sturt University, Albury, NSW, Australia
dcornforth@csu.edu.au

Department of Computer Science and Software Engineering.
University of Melbourne, Melbourne, Vic, Australia
mkirley@cs.mu.oz.au

Abstract. A key problem in multi-agent systems research is identifying appropriate techniques to facilitate effective cooperation between agents. In this paper, we investigate the efficacy of a novel market-based aggregation technique in addressing this problem. An incremental transaction-based protocol is introduced where agents establish links by buying and selling from each other. Market transactions equate to agents coordinating their plans and sharing their resources to meet the global objective. An important contribution of this study is to clarify whether, in some circumstances, a market-based model leads to the effective formation of agent teams (or coalitions) and thus, solutions to the problem-solving task.

1   Introduction

An increasing number of computational systems may be viewed in terms of multiple, interacting autonomous agents. Interactions might include cooperation to achieve a joint goal, competition for resources, negotiation over a set of tasks to perform, or the buying or selling of resources [1]. If we adopt a game-theoretic perspective, the agents play a general-sum game in which particular coalitions or teams of agents working together have a higher utility (or relative fitness) than other agents in the population [2]. One way to view this coalition formation process is as a distributed search through the space of all possible configurations.

An alternative perspective favored by the Artificial Life community sees multi-agent systems as simulations based on metaphors inspired by ecological, economic or social communities. Here, the agents, their behavioral rules, and their mutual interactions define complex systems. Holland [3] suggests that the agents may be thought of as building blocks representing formal components of the model, built to understand the complex patterns of emergent behavior underlying the system. An inherent feature of these systems is the ability of the agent to group together to form composite entities, also known as modules, clusters, teams or coalitions, depending on the terminology of the respective discipline. However, it is still an open question as to

K. Deb et al. (Eds.): GECCO 2004, LNCS 3102, pp. 60–71, 2004.

how agents in a complex system form coalitions, and how these coalitions self-organize into hierarchies.

In this paper, we begin to address this question by focusing on cooperation in multi-agent systems. Specifically, we are interested in a situation such that given a specific goal, which cannot be satisfied by a single agent, a collective effort by several agents is required. In this instance, agents must coordinate their plans and share their resources to meet the global objective. We propose a novel economic market-based mechanism to facilitate cooperation in multi-agent systems. Using a highly idealized model, we illustrate how agents in a system can use a series of incremental transactions to form appropriate teams or coalitions for "solving" a given problem. Here, the multi-agent system may be viewed as a virtual market place populated by heterogeneous self-interested traders (agents) attempting to maximize their own utility. Buyers try to trade at the lowest possible price. Conversely, sellers try to trade at the highest price possible. Successful transactions represent steps in a bottom-up decentralized team formation protocol.

This study parallels (a) research into coalition formation protocols – where rational agents negotiate to join agent teams; (b) computational synthesis research – where low-level building blocks or features are combined to achieve given arbitrary high-level functionality in multi-agent systems, and (c) artificial symbiotic processes research – where alternative aggregation mechanisms based on a mutualism metaphor are used. An important contribution of this work is to clarify whether, in some circumstances, a market-based model leads to the effective formation of agent coalitions and thus, solutions to the problem solving task. To meet this objective, a number of simulations are presented focusing on the effectiveness of the aggregation process. In addition, we explore suitable mechanisms for fostering and maintaining diversity within the agent population.

The remaining sections of the paper are organized as follows. Related work and background material is presented in Section 2. In Section 3, the market-based model is described. Section 4 illustrates the functionality of the model using a pattern recognition task. In Section 5, we present the simulation results. We conclude with a discussion of the results and the implications of this work.

2 Background and Related Work

2.1 Coalitions and Cooperation

Cooperation is a key process in many multi-agent systems. Agents may cooperate to collectively solve some problem or perform some task, where a single agent could not succeed. In this scenario, each individual agent is able to carry out its tasks through interaction with a small number of *neighboring agents*. When interdependent problems arise, the agents in the systems must cooperate with one another to ensure that interdependencies are properly managed.

In computer science, cooperation has been studied extensively by Axelrod [4] and Huberman [5]. This work has been extended into the autonomous agent domain in the areas of auction theory [6], team formation [7] and coalition formation [8] [9] [10]

[11]. Much of this work has focused on how a group of agents make particular decisions and the associated utility values associated with the decisions. Sandholm and Lesser [12] present an interesting coalition formation process model for bounded-rational agents and a general classification of coalition games. They allow for varying coalition values, but provide the agents with heuristics that could be computed in polynomial time

The main question in every coalition formation application is how to determine which agents collaborate. While game theory is a useful analytical tool, Wooldridge and Jennings [13] suggest that it is not a good engineering tool primarily because of the type of representation employed by game theory. Wellman [14] suggest that agent interaction models employing market-based control mechanisms offer the possibility of fostering cooperation. It is this notion that provides some of the motivation for the model proposed in Section 3.

2.2 Artificial Symbiotic Models

Perhaps the most well known artificial symbiotic model is Potter and De Jong's [15] cooperative coevolutionary model. In this model, the artificial ecosystem consists of two (or more) species. Species (or modules) interact with one another within a shared domain model and have a cooperative relationship. Species evolve independently. However, the fitness value of an individual is directly related to how well that individual collaborated with representatives from each of the other species in solving the "super goal." Fundamentally, this model is a divide-and-conquer approach, where the system cycles between *decomposition – evolution – collaboration and evaluation*. In later work [16], the architecture was extended to include dynamic speciation and extinction. New species were added to the model based upon some measured stagnation in the evolutionary process. Other species were destroyed if they were no longer making significant contributions. This enhanced model was applied successfully to string covering problems and evolving cascade networks.

Watson and Pollack [17] have investigated how mechanisms based on abstract symbiotic processes affect adaptation in evolutionary systems. Specifically, they have developed algorithms where higher-level complexes are formed from simple "modules" or building blocks. This notion of building blocks is a fundamental principle of genetic algorithms. However, their "aggregation of modules" is directly related to coevolutionary interactions within the evolving population. The most important features of their model include: (a) techniques for combining modules based on symbiotic processes, (b) the introduction of explicit mechanisms ensuring that modules co-adapt to cover complementary parts of the problem domain, and (c) the use of appropriate techniques that can be used to determine the relative worth (fitness) of a module, including Pareto comparisons.

There are a number of similarities between each of the models described above. For instance, the flexibility inherent in Potter and De Jong's extended architecture and Watson and Pollack's hierarchical composition of modules have computational advantages. However, Daida and co-workers [18] have shown that many caveats exist either in adopting symbiosis as a computational heuristic, or in modeling symbiosis as an aspect of complex adaptive system behavior. They contended that in each case, symbiosis should be considered as a kind of operator instead of a state.

3 A Framework for Market-Based Problem Solving

The proposed market-based model is an incremental approach to problem-solving based on a bottom-up team (or coalition) formation protocol. The model consists of autonomous units – agents – interacting with each other as well as with an environment. At any time t, the system will contain a population of agents $\mathbf{A} = \{\ A_1,\ A_2, \ldots A_n\}$.

An agent refers to a localized entity with decision making capabilities. It can be a single individual or a coalition of individuals. In this instance, an agent represents a basic building block in the problem-solving task. Each agent has specific functionality and behaviors represented by the tuple $\mathbf{A_i} = <\ resource,\ P_{sell},\ G_{fit},\ P_{buy}\ L_{fit}>$ where: the *resource* represents the product encapsulated by the agent (which can be traded); P_{sell}, is the probability that the agent will offer its resource for sale; G_{fit} is the fitness gain – a weighting factor for the minimum selling price; P_{buy} is the probability that the agent will make a bid in the current market; and L_{fit} - the fitness loss – weighting factor for the maximum bid price.

We assume that all atomic agents are peers. That is, there is no default hierarchy among individual atomic agents. However, individual agents in the economy have heterogeneous beliefs concerning realization of possible outcomes.

The artificial market consists of a sequence of modified auctions. Randomly selected agents participate in a single auction and have the intention of buying the target resource from another agent. They maintain information about the resource they wish to purchase and their private valuation of this resource (the maximum amount that they are willing to pay for the desired item). A successful transaction, that is, the situation where agent A_i sells its resource to agent A_j establishes a trade-link or a coalition between the agents. Here, we use the term coalition to describe the team of agents drawn from A, who have worked together (traded-resources) to accomplish a task. In utility-theoretic terms, the utility (or fitness value) of each agent A_i is a function of G_{fit} and L_{fit} parameters of the individual agents.

To facilitate the functioning of the market, we have implemented a modified version of the Contract Net Protocol (CNET) [19]. CNET provides a general framework to describe negotiation processes between agents. Essentially, this protocol is based on a collection of agents, which cooperate in achieving sub-goals which, in turn collectively meet some high-level goal. CNET provides a means to find the "best" acquaintance for a given task. A key component of this protocol involves agents making decisions based on each agent's perspective of the current state of the world. The following steps encapsulate the basic functionality of the modified CNET protocol:

1. Task announcement and processing – corresponds to specifying the complete problem to be solved. On receipt of a task announcement, an agent decides if it is eligible for the task (or some part of the task). It does this by looking at the eligibility specifications contained in the announcement. If it is eligible (that is, the agent can solve some part of the problem), then details of the task are stored, and the agent will subsequently bid for the task.

2. Bid processing – agents who have responded to the task announcement bid to gain control of the selling agents resources. Details of the bid from would-be

contractors are stored by the would-be managers until the deadline for the task. The manager then awards the task to a single bidder.

3. Award processing – agents that bid for a task, but fail to be awarded it, simply delete details of the task. The successful bidder must attempt to expedite the task

It is important to note that the award processing phase may lead to different global utilities, depending upon who the successful agent/coalition is.

4 Problem Description

To illustrate the basic functionality of the trading model, a specific problem-solving task will be used – a string matching problem (pattern recognition of symbols). It is important to emphasize that we are interested in the general features of problem solving with agents that is applicable in a wide variety domains. However, we illustrate the phases of the market-based model using a concrete example.

A target string is drawn from a dictionary of English words longer than 16 letters. In this string-matching problem, the task of each agent is to synthesize the target string, but each agent is initialized with one letter (resource), and therefore must acquire other letters. All agents are able to buy or sell letters or word fragments to add to their collection. This task is not only a matter of permutation but also of the correct sequential composition, which in principle acts in parallel.

In Fig 1, we illustrate the outcome of repeated transactions within the market for the target string *acknowledgements*. The atomic agents – that is, agents encapsulating a single letter only – are shaded and can be found at the bottom of the hierarchy. When an agent purchases a letter or string fragment from another agent a coalition is formed. At the next level of the hierarchy the agent encapsulates the corresponding string.

As the model is iterated, agents trade with each other, buying and selling characters in exchange for a notional currency. After the model has been run, some agents will have

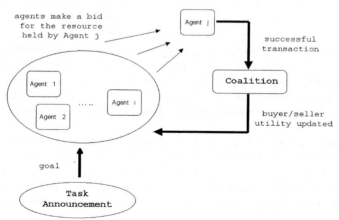

Fig. 1. An outline of the trading-model framework. The bidding-coalition formation phases are iterated until the goal is achieved (or a termination criterion is reached).

accumulated the correct letters to solve the problem , but they can still offer this complete collection for sale. Other agents will have no letters at all, having sold the letter they were initialized with. Other agents may have acquired partial solutions. All agents start with zero capital, but may gain or lose capital through trading. On each iteration of the model, one agent is selected at random to be the seller, and if its P_{sell} value is higher than a randomly generated number, the agent offers its letter or word fragment (coalition) for sale. If an agent has more than one letter, it cannot choose to offer some letters for sale and retain others: it offers all of its resources for sale. Other agents bid for these resources, as long as their P_{buy} value is higher than a randomly selected value. The price of any trade is determined by a simple formula that takes account of the number of letters that match the target word. Price setting is done using a tender system. The minimum selling price is determined by:

$$P_{min} = m^2 G_{fit}$$

(1)

where P_{min} is the minimum selling price, m is the number of matches, and G_{fit} is the fitness gain. The bid price for buyer is determined by:

$$P_{bid} = (m_n^2 + (m_n m_c))L_{fit}$$

(2)

where P_{bid} is the bid price, m_n is the number of new matches gained from the purchase, m_c is the number of current matches the agent has, and L_{fit} is the fitness loss. P_{bid} takes into account of the fact that if the current number of matches is zero, the bid should be greater than zero, but if the number of matches gained by the purchase is zero, the bid price should also be zero. The buyer agent has the problem of calculating how to join the two strings together to maximize the resulting number of matches. This is solved by sliding the new string over the old and assessing the total number of matches at each position. The agent then chooses a position to maximize the number of matches. The new string always replaces letters of the existing string. This is illustrated in Table 1, where G_{fit} and $L_{fit} = 0.5$. Notice that the letter 'l' is duplicated, so that the original letter 'l' is overwritten by the new string in the best combination (22 matches).

The agent that produces the highest bid purchase the resource (characters), as long as $P_{bid} > P_{min}$. The sellers' capital is increased by P_{bid} of the winning buyer, while the buyers' capital is decreased by the same amount. The problem is solved when a copy of the target string is held by one of the agents.

It is possible for more than one solution to be found if there are enough letters (resources) in circulation among the agents. The solution is collaborative in the sense that agents must be willing to trade in order for the letters to pass into the control of a single agent. This means that the agent parameters (P_{sell}, G_{fit}, P_{buy}, and L_{fit}) must have values conducive to trading. The agents that have contributed their letters to this solution can be considered to be part of a coalition, since they have received payment, and thus contribute to the fitness value of the agent controlling the solution. Agents that trade are deemed to belong to the same coalition, while agents that do not trade with each other belong to different coalitions, or have no membership. After a number of iterations, these coalitions form a hierarchy supporting a solution, as illustrated in Fig. 2. The data in this figure were produced during an actual run of our model.

Table 1. Sample utility calculations. The new string is positioned against the target string to maximise the number of matches obtained.

String	Resource	Calculations
Target	`acknowledgements`	
Offered for sale	`ledge`	$P_{min} = 5^2 * 0.5 = 12.5$
Belonging to potential buyer	`acknowl`	

New combined string	`acknowlledge`
	`acknowledge`
	`acknoledge`
	
	`etc`
		$P_{bid} =$
		$(0^2 + (0*5))*0.5 = 0$
		$(4^2 + (4*7))*0.5 = 22$
		$(4^2 + (4*6))*0.5 = 20$

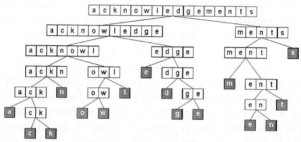

Fig. 2. Steps in the coalition formation process. Initially, all agents hold a single letter (resources). Agents then bid to gain control of additional letters to maximize their utility. In this diagram, we simply show the resource encapsulated by the agent at each level of the hierarchy for a given target word. The other agent parameters P_{Sell}, G_{fit}, P_{Buy}, and L_{fit} have been omitted for clarity.

5 Simulations

The market-based model has been designed as a continuous time, discrete-event simulation of a population of trading-agents. To examine the overall performance of model a number of simulation experiments were carried out. Specific parameters of interest include: the agent initialization states, behavioral rules (trading strategy), and the pattern of connectivity or possible interaction links.

5.1 Experiments

The aim of the first experiment was to establish the effectiveness and efficiency of the market-based model as a cooperative problem-solving tool. In particular, we are interested in determining how the agent/coalition capital levels (utility) are related to the number of correctly matching strings found.

The multi-agent system used in this simulation consisted of 1000 diverse agents. Here, diversity refers to the range of values that agent parameters are initialized with. The resource parameter of each agent was initialized with a randomly drawn letter from the English alphabet. Each of the other parameters – P_{sell}, G_{fit}, P_{buy}, and L_{fit} – was initialized with a random value from a uniform distribution between 0 and 1.0. These values do not change during a run of the model.

For a given target string, the market model was simulated for 5000 iterations. In this simulation, a total of 289 different target strings were considered – one run for each target string with 16 or more letters in the dictionary. As the agents were initialized with letters drawn from a uniform distribution, and as the distribution of letters in English words is far from uniform, the limit on the number of complete matches is the number of repeated letters in a single word. For example, it is common for the letter 'e' to be used 3 times in one word. This letter is expected to occur approximately 40 times (1000 / 26) in the initial population, so the maximum possible number of complete solutions is approximately 13 (40 / 3).

In the second experiment, we extend the simulation to include the possibility that some of the agents leave the market and new agents enter over the course of the run. This particular modification offers the possibility of: (a) promoting diversity across the agent populations and (b) mimicking real market places more closely. Determining which agents leave the market is not a straight forward problem. The utility of an agent (coalition) is a function of the number of matches obtained and its current capital. Agents are rewarded for matching characters in the solution, but punished for expending capital.

Consequently, we implement a form of "termination" based on an agents' utility. For example, if the termination threshold is set at 50%, then 50% of agents having zero matches are replaced, 50% of agents having one match are replaced, and so on until 50% of agents having a perfect match of the target string are replaced. When an agent is replaced, its string fragment is transferred to the new agent to minimize the loss of letters from the market. An exception is made for agents having a single letter that does not match the target string. These agents are replaced with a new agent having a letter drawn at random from a uniform distribution of letter in the alphabet. In either case, the new agents' capital is set to the average for agents with that number of matches. The remaining parameters are set in the same way as initialization.

Once again, this model configuration was executed for 5000 iterations for each of the target strings used in the first experiment.

5.2 Results

Fig. 3 plots the frequency distribution of correct string matches averaged over all agents across the 289 runs. The results of experiments 1 and 2 are compared. It is interesting to note the improved performance in Fig 3b. This difference may be attributed to the fact that agents who failed to satisfy a predefined performance criterion were removed at regular time intervals, and replaced by randomly initialized agents. The result of the selection process is that some poorly performing agents are eventually replaced by fitter agents.

In Fig 4, we plot the number of matches *vs.* capital possessed by agents at the end of each run. Again, we compare the results of experiments 1 and 2. For each of the corresponding match values, the average and standard deviation of the agent capital was calculated for all 289 target strings used. As the number of characters in each word was greater than 15, it can be seen that there are a number of complete solutions shown in the lower right hand part of the plots. A comparison of the two plots suggest that the diverse population has overcome, to some extent, the disadvantage of higher prices during trading, by eliminating some of the agents responsible for driving those higher prices.

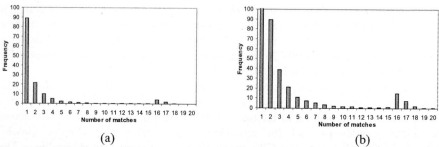

(a)   (b)

Fig. 3. The distribution of matches after 5,000 iterations: (a) Experiment 1 – using no termination; (b) Experiment 2 – using a 50% termination threshold. The average frequency of agents having 1 match is equal to 296, but the graph is shown with the same scale as (a) for purposes of comparison.

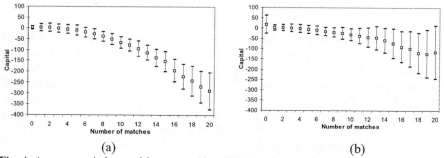

(a)   (b)

Fig. 4. Average capital owned by agents after 5000 iterations against number of matches: (a) Experiment 1 – using no termination; (b) Experiment 2 – using a 50% termination threshold.

6 Discussion and Conclusion

In this study, we have proposed a novel market-based cooperative problem solving mechanism. This incremental approach based on bottom-up team or coalition formation describes one possible mechanism for autonomous agents to coordinating their decisions without assuming *a priori* cooperation.

At the beginning of a run, agents were initialized with a random letter (resource). This particular resource may be attractive to other agents. As the model is iterated, other agents in the population may bid to gain control of the resource (resulting string fragments) in an attempt to satisfy the global objective. In this instance, the trading metaphor represents an effective communicating strategy, allowing agents to form coalitions using a bottom up methodology (Fig 2). Here, a coalition provides a framework for solving the given problem, which could not be solved by one agent working alone.

The preliminary results presented in this paper are very encouraging. The plots in Fig 3 illustrate that the agent populations is able to solve a given problem. In fact, a number of different successful coalitions have emerged, for the each of the target strings used in the simulations. The use of a notional currency preserves links between agents that otherwise yield their stake in the solution to the control of the buying agent. It is interesting to note that when some of the agents are removed from the population and replaced with new agents, an improvement in the number of matches found was noted. Fig 4 provides further empirical evidence supporting this notion based on the capital invested.

Coalition formation may be thought of generically as the process of devising a team of agents to work on a specific goal. In our model, agents continually interact with other agents and have to adapt to their environment. The repeated transactions between agents facilitate the developed of links (or trade networks). Our model is characterized by the nonlinear credit-assignment or utility function associated with the agent parameters. These parameters define the extent to which agents compete or cooperate. Coordination via this type of market mechanism is well suited for situations in which: (a) resources can be described easily or are commoditized, and (b) there are several agents offering the same (type) of resources and several agents that need them.

The cooperative problem solving model investigated in this study has focused on explicit subsystem interactions. As such, there are similarities between this work and the aggregation mechanisms inherent in Potter and De Jong's [15][16] cooperative coevolution model and the idea of combining together partial solutions into more complete solutions via sexual recombination (for example, the building-block hypothesis, Holland [3][20]). The simulation experiments described clearly illustrate two important characteristics of emergent properties in complex systems: (a) there must be a sufficiently large number of agents for the model to be effective, and (b) the model must include explicit self-reinforcing mechanisms.

This work also raises a number of questions in relation to the formation of coalitions or modules in complex systems. And, in particular how do these coalitions self-organize into hierarchies? Although the model described here is highly idealized, the underlying protocol may provide some insights into the characteristic interactions, which facilitates the transition from lower-level entities into new higher-level functional entities. Traditional practices in multi-agent systems rely on prepro-grammed interaction patterns, preventing adaptation to unexpected environmental changes. A market-based bottom-up protocol may offer an alternative means for self-assembled coalition/hierarchies to emerge.

Acknowledgements. This work was partially supported by a Communities of Scholars Award from the Faculty of Science and Agriculture, Charles Sturt University. We thank Leighton Weymouth for programming support.

References

1. Wooldridge, M.: *An Introduction to MultiAgent Systems*. John Wiley & Sons. UK. 2002.
2. Rosenschein, J.S. and Zlotkin, G. : *Rules of Encounters: Designing Conventions for Automated Negotiation among Computers*. MIT Press. Cambridge, MA. 1998.
3. Holland, J. H. : *Emergence: From chaos to order*. Reading, MA: Addison-Wesley. 1998.
4. Axelrod, R. : *The Evolution of Cooperation*. Basic Books, New York. 1984.
5. Huberman, B.A. : The performance of cooperative processes. In Emergent Computation – Special Issues. *Physics D*. 1991.
6. Wellman, M.P., Walsh, W.E., Wurman, P.R. and Mackie-Mason, J.K. : Auction protocols for decentralized scheduling. *Games and Economic Behaviour*. 35(1/2):271-303. 2001.
7. Tambe, M. : Towards flexible teamwork. *Journal of Artificial Intelligence Research*. 7:83-124. 1997.
8. Jennings, N. : Controlling cooperative problem solving in industrial multi-agent systems using joint interactions. *Artificial Intelligence Journal*. 75(2):1-46. 1995.
9. Shehory, O. and Kraus, S. : Formation of overlapping coalitions for precedence-ordered task-execution among autonomous agents. In *Proceedings of the Second International Conference on Multi-Agent Systems (ICMAS-96)* AAAI Press / MIT Press. pp 330-337. 1996.
10. Kraus, S., Shehory, O. and Taase, G. : Coalition Formation with Uncertain Heterogeneous Information. In J. S, Rosenschein et al., (eds). *Proceedings of the Second International Joint Conference on Autonomous Agents and Multiagent Systems*. ACM Press. pp 1-8. 2003.
11. Sims, M. Goldman, C.V. and Lesser, V. : Self-Organization through Bottom-Up Coalition Formation. In the *Proceedings of AAMAS'03*. ACM Press. pp 867-874. 2003.
12. Sandholm, T. and Lesser, V .: Coalition formation among bounded rational agents. *Artificial Intelligence Journal*. 94(1-2):99-137. 1997.
13. Wooldridge, M. and Jennings, N.R. : The Cooperative Problem Solving Process. *Journal of Logic & Computation*. 9(4):563-592. 1999.
14. Wellman, M .: Market Oriented Programming: Some Early Lessons, In S.H.Clearwater (ed). *Market-Based Control a Paradigm for Distributed Resource Allocation*. World Scientific Press. 1996.
15. Potter, M. and De Jong, K .: A Cooperative Coevolutionary Approach to Function Optimization. In *Parallel Problem Solving from Nature Conference PPSN III*. pp 249-257. Springer-Verlag. 1994.
16. Potter, M. and De Jong, K. : Cooperative Coevolution: An Architecture for Evolving Coadapted Subcomponents. *Evolutionary Computation* 8(1): 1-29. 2000.
17. Watson, R.A., and Pollack, J.B. : Symbiotic Combination as an Alternative to Sexual Recombination in Genetic Algorithms. *Parallel Problem Solving from Nature -- PPSN VI*, pp. 425-434. Springer. 2000.

18. Daida, J.M., Gasso, C.S., Stanhope, S.A. and Ross, S.J. : Symbioticism and Complex Adaptive Systems I: Implications of Having Symbiosis Occur in Nature. In *Evolutionary Programming V*. MIT Press. 1996.
19. Smith, R.G. : The contract net protocol. *IEEE Transactions on Computers*. C29(12). 1980.
20. Holland, J. : *Adaptation in Natural and Artificial Systems*. University of Michigan Press. 1975.

Cultural Evolution for Sequential Decision Tasks: Evolving Tic–Tac–Toe Players in Multi–agent Systems

Dara Curran and Colm O'Riordan

Dept. of Information Technology
National University of Ireland, Galway

Abstract. Sequential decision tasks represent a difficult class of problem where perfect solutions are often not available in advance. This paper presents a set of experiments involving populations of agents that evolve to play games of tic–tac–toe. The focus of the paper is to propose that cultural learning, i.e. the passing of information from one generation to the next by non–genetic means, is a better approach than population learning alone, i.e. the purely genetic evolution of agents. Population learning is implemented using genetic algorithms that evolve agents containing a neural network capable of playing games of tic–tac–toe. Cultural learning is introduced by allowing highly fit agents to teach the population, thus improving performance. We show via experimentation that agents employing cultural learning are better suited to solving a sequential decision task (in this case tic–tac–toe) than systems using population learning alone.

1 Introduction

Lifetime learning can take many forms - at its simplest it is a reaction to a particular stimulus and the adjustment of a world view that follows the reaction. Thus, very simple organisms are capable of learning to avoid harmful substances and are attracted to other beneficial ones. In computational terms, lifetime learning can be simulated using neural networks by employing an error reducing algorithm such as error back–propagation.

While this type of lifetime learning has been shown to be useful in the past, it relies on prior solution knowledge: in order to correctly train agents, the system must be aware of the solution to be attained. An alternative approach is the use of cultural learning, a subset of lifetime learning where agents are allowed to communicate information through a hidden verbal layer in each agent's neural network. Teacher agents are selected from the population and are assigned a number of pupils that follow the teacher as it performs its task.

Teachers and pupils exchange information each time a stimulus occurs and the pupils learn using back propagation to imitate the teacher's verbal output and behaviour. Since cultural learning does not require a priori solution knowledge, it is an ideal system for problems where perfect solutions are not available or are non–trivial to discover. Examples of such problems are sequential decision

K. Deb et al. (Eds.): GECCO 2004, LNCS 3102, pp. 72–80, 2004.

tasks, problems that can only be solved through repeated iterations, such as sorting problems and move–based games.

The focus of this paper is to examine the benefit of combining population and cultural learning over population learning alone for a simple sequential decision task problem: the game of tic–tac–toe. While the problem is not complex, it is of sufficient difficulty to illustrate the potential for cultural learning in domains of this kind.

The remainder of the paper is organised as follows. The next section describes some related work focusing on the types of learning that can be employed by multi–agent systems. Section 3 outlines the experimental setup, describing the artificial life simulator employed, the cultural learning framework and how these were adapted in order to learn the game of tic–tac–toe. Section 4 presents results of experiments where population learning was used to evolve tic–tac–toe players and where cultural learning was added. Section 5 concludes the paper and suggests avenues of future research.

2   Related Work

2.1   Evolving Game–Playing Agents

Many researchers have developed evolutionary techniques to generate game–playing agents for a variety of games [1,2,3,4]. However, little research to date has focused on the addition of cultural learning to such tasks. We feel that the game of tic–tac–toe, while a simple game, represents a good starting point for researching cultural learning in a game–playing domain.

2.2   Learning Models

A number of learning models can be identified from observation of nature. These can roughly be classified into two distinct groups: population and life-time learning. In this paper we consider another form of lifetime learning, cultural learning.

Population Learning. Population learning refers to the process whereby a population of organisms evolves, or learns, by genetic means through a Darwinian process of iterated selection and reproduction of fit individuals. In this model, the learning process is strictly confined to each organism's genetic material: the organism itself does not contribute to its survival through any learning or adaptation process.

Lifetime Learning. By contrast, there exist species in nature that are capable of learning, or adapting to environmental changes and novel situations at an individual level. Such learning, know as life-time learning, still employs population learning to a degree, but further enhances the population's fitness through its adaptability and resistance to change. Another phenomenon related to life-time learning, first reported by Baldwin [5], occurs when certain behaviours, first evolved through life-time learning, become imprinted onto an individual's genetic material through the evolutionary processes of crossover and mutation. This individual is born with an innate knowledge of such behaviour and, unlike the rest of the population, does not require time to acquire it through life-time learning. As a result, the individual's fitness will generally be higher than that of the population and the genetic mutation should become more widespread as the individual is repeatedly selected for reproduction.

Research has shown that the addition of life-time learning to a population of agents is capable of achieving much higher levels of population fitness than population learning alone [6,7]. Furthermore, population learning alone is not well suited to changing environment [8].

Cultural Learning. Culture can be succinctly described as a process of information transfer within a population that occurs without the use of genetic material. Culture can take many forms such as language, signals or artifactual materials. Such information exchange occurs during the lifetime of individuals in a population and can greatly enhance the behaviour of such species. Because these exchanges occur during an individual's lifetime, cultural learning can be considered a subset of lifetime learning.

Using genetic algorithms, the evolutionary approach inspired by Darwinian evolution, and the computing capacity of neural networks, artificial intelligence researchers have been able to achieve very interesting results.

Experiments conducted by Hutchins and Hazlehurst [9] simulate cultural evolution through the use of a hidden layer within an individual neural network in the population. This in effect, simulates the presence of a Language Acquisition Device (LAD), the physiological component of the brain necessary for language development, whose existence was first suggested by Chomsky [10]. The hidden layer acts as a verbal input/output layer and performs the task of feature extraction used to distinguish different physical inputs. It is responsible for both the perception and production of signals for the agent.

A number of approaches were considered for the implementation of cultural learning including fixed lexicons [11,12], indexed memory [13], cultural artifacts [14,15] and signal–situation tables [16]. The approach chosen was the increasingly popular teacher/pupil scenario [17,18,12] where a number of highly fit agents are selected from the population to act as teachers for the next generation of agents, labelled pupils. Pupils learn from teachers by observing the teacher's verbal output and attempting to mimic it using their own verbal apparatus. As a result of these interactions, a lexicon of symbols evolves to describe situations within the population's environment.

3   Experimental Setup

3.1   Simulator

The experiments outlined in this paper were performed using an artificial life simulator developed by Curran and O'Riordan [19,6,7]. The simulator allows populations of neural networks to evolve using a genetic algorithm and each network can also be trained during each generation of an experiment to simulate life–time learning.

Each member of the population is in possession of both a phenotype (a neural network) and a genotype (a gene code). The gene code is used to determine the individual's neural network structure and weights at birth. If the individual is selected for reproduction, the gene code is combined with that of another individual using the process of crossover and mutation to produce a genotype incorporating features from both parents.

In order for this mechanism to function correctly, a mapping of a neural network structure to a gene code is required. This is achieved using a modified version of marker based encoding which allows networks to develop any number of nodes and interconnecting links, giving a large number of possible neural network architecture permutations.

Marker based encoding represents neural network elements (nodes and links) in a binary string. Each element is separated by a marker to allow the decoding mechanism to distinguish between the different types of element and therefore deduce interconnections [20,21].

In this implementation, a marker is given for every node in a network. Following the node marker, the node's details are stored in sequential order on the bit string. This includes the node's label and its threshold value. Immediately following the node's details, is another marker which indicates the start of one or more node–weight pairs. Each of these pairs indicates a back connection from the node to other nodes in the network along with the connection's weight value. Once the last connection has been encoded, the scheme places an end marker to indicate the end of the node's encoding.

The networks undergo various stages throughout their lifetime. Firstly, the gene codes are decoded to create their neural network structure. Training is then performed using error back–propagation for a given number of iterations (training cycles). Each network is tested to determine its fitness and the population is ranked using linear based fitness ranking. Roulette wheel selection is employed to generate the intermediate population. Crossover and mutation operators are then applied to create the next generation.

3.2   Cultural Learning Framework

In order to perform experiments related to cultural evolution, it was necessary to adapt the existing simulator architecture to allow agents to communicate with one another. This was implemented using an extended version of the approach adopted by Hutchins and Hazlehurst. Their approach uses the last hidden layer of each agent's neural network as a verbal input/output layer (figure 1) and employs a fixed number of verbal input/output nodes. We have modified Hutchins and Hazlehurst's system to allow the number of verbal input/output nodes to evolve with the population, making the system more adaptable to potential changes in environment. In addition, this method does not make any assumptions as to the number of verbal nodes (and thus the complexity of the emerging lexicon) that is required to effectively communicate.

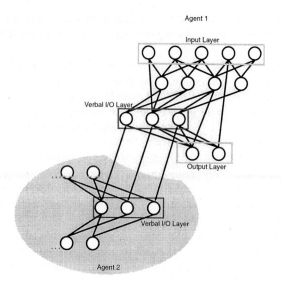

Fig. 1. Agent Communication Architecture

At the end of each generation, a percentage of the population's fittest networks are selected and are allowed to become teachers for the next generation. The teaching process takes place as follows: a teacher is stochastically assigned n pupils from the population where $n = \frac{N_{pop}}{N_{teachers}}$, where N_{pop} is the population size and $N_{teachers}$ is the number of teachers. Each pupil follows the teacher in its environment and observes the teacher's verbal output as it interacts with its environment. The pupil then attempts to emulate its teacher's verbal output using back-propagation. Once the teaching process has been completed, the teacher networks die and new teachers are selected from the new generation.

3.3 Tic Tac Toe

The sequential decision task chosen for this set of experiments is the game of tic-tac-toe. While this is a very simple game, we believe it serves to illustrate the benefit of cultural evolution for sequential decision tasks and can be used as a stepping stone to more difficult problems.

In order to evolve good players, it was decided that agents in the population would all compete against a perfect player rather than compete against each other. It was felt that populations of agents competing against each other would be likely to converge only to local maxima due to the lack of competitive pressure. To avoid over–fitting, the perfect player employs a modified minimax method to determine moves where the first move of the game is randomized so that agents play a variety of games rather than the same game at each iteration. Each agent plays four games in its lifetime: two where the agent moves first and the other two where the perfect player moves first.

In order to play tic-tac-toe, an agent's neural network structure must follow certain parameters. There are 18 input nodes, 2 for each board position where 01 is X, 10 is O and 11 is an empty square. Nine output nodes corresponding to each board position are used to indicate the agent's desired move where the node with the strongest response corresponding to a valid move is taken as the agent's choice. The simulator allows agents to evolve any number of hidden layers each with an unrestricted number of nodes, giving maximum flexibility to the evolutionary process. During the teaching process, a teacher agent plays alongside the pupil. At each move, both the pupil and teacher choose the next move and the pupil's verbal output is corrected with respect to the teacher's using error back-propagation.

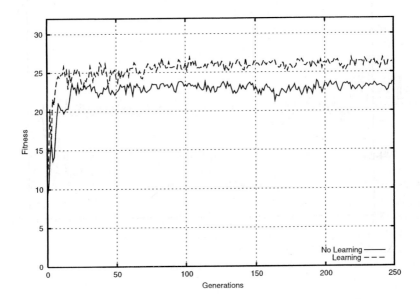

Fig. 2. Agent Fitness

Since the agents play against a perfect player, fitness is assigned according to how long each agent is capable of avoiding a loss situation. An agent's fitness is therefore correlated with the number of moves that each game lasts, rewarding agents capable of forcing the perfect player to as close to a draw as possible. The fitness function produces values in the range [0,32], where 32 is the maximum fitness (the situation where the agent draws all four games).

Populations of 100 agents were generated for these experiments and allowed to evolve for 250 generations. Crossover was set at 0.6 and mutation at 0.02. The teaching rate was set at five cycles and the value for n (the number of individuals selected to become teachers at each generation) was set to 10% of the population. In addition, a teaching mutation rate which modifies a teacher's output when training a pupil was incorporated and set at 0.02. The results presented are an average of 20 experiment runs.

4   Experimental Results

Two experiments were undertaken: one using only population learning to evolve players, and the other using population and cultural learning. Figure 2 shows the average fitness values for the two evolving populations. While both types of learning begin at similar levels of fitness, it is clear that agents employing cultural evolution are performing better as the experiment progresses.

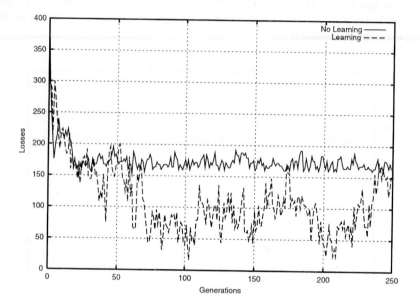

Fig. 3. Number of game losses

The number of game losses over time, illustrated in figure 3, show that while cultural evolution delivers less losses than pure population learning, it does not provide stability. While population learning losses seem to stabilise at above 150 losses, cultural learning losses seem to oscillate considerably, reaching a low of 24 losses and highs of 150. We posit that the inherent noisiness of the cultural learning approach causes agents to behave more erratically – the learning process may bring an output layer node from a dormant state to a sudden forefront position causing a dramatic change in the agent's playing pattern.

It is interesting to compare these results with those obtained by Angeline and Pollack [22] who used a competitive fitness function to evolve populations of neural network tic–tac–toe players. The population of evolving players was pitted against a number of 'expert' player strategies, including a perfect player. If we examine their results in terms of a draws/losses ratio, we find that their best evolved players (playing against a perfect player) obtain a ratio of 0.2405. By contrast, the cultural learning approach presented in this paper obtains highs of 0.94 and lows of 0.625.

5   Conclusions

The results of these experiments suggest that cultural learning is superior to population learning alone for simple sequential decision tasks. Future work will examine the effect of longer teaching cycles, varying teacher/population ratios and more complex tasks.

Acknowledgements. This research is funded by the Irish Research Council for Science, Engineering and Technology. The authors wish to thank the reviewers for their helpful feedback and comments.

References

1. D. Moriarty and R. Miikkulainen. Discovering complex othello strategies through evolutionary neural networks. *Connection Science*, 7(3–4):195–209, 1995.
2. L. Weaver and T. Bossomaier. Evolution of neural networks to play the game of dots-and-boxes. In *Artificial Life V: Poster Presentations*, pages 43–50, 1996.
3. K. Chellapilla and D. B. Fogel. Evolving neural networks to play checkers without relying on expert knowledge. In *IEEE Transactions on Neural Networks*, volume 10, pages 1382–1391, 1999.
4. Norman Richards, David Moriarty, Paul McQuesten, and Risto Miikkulainen. Evolving neural networks to play Go. In *Proceedings of the 7th International Conference on Genetic Algorithms*, East Lansing, MI, 1997.
5. J.M. Baldwin. A new factor in evolution. In *American Naturalist 30*, pages 441–451, 1896.
6. D. Curran and C. O'Riordan. On the design of an artificial life simulator. In V. Palade, R. J. Howlett, and L. C. Jain, editors, *Proceedings of the Seventh International Conference on Knowledge-Based Intelligent Information & Engineering Systems (KES 2003)*, University of Oxford, United Kingdom, 2003.
7. D. Curran and C. O'Riordan. Artificial life simulation using marker based encoding. In *Proceedings of the 2003 International Conference on Artificial Intelligence (IC-AI 2003)*, volume II, pages 665–668, Las Vegas, Nevada, USA, 2003.
8. D. Curran and C. O'Riordan. Lifetime learning in multi-agent systems: Robustness in changing environments. In *Proceedings of the 14th Irish Conference on Artificial Intelligence and Cognitive Science (AICS 2003)*, pages 46–50, Dublin, Ireland, 2003.
9. E. Hutchins and B. Hazlehurst. How to invent a lexicon: The development of shared symbols in interaction. In N. Gilbert and R. Conte, editors, *Artificial Societies: The Computer Simulation of Social Life*, pages 157–189. UCL Press: London, 1995.
10. N. Chomsky. On the nature of language. In *Origins and evolution of language and speech*, pages 46–57. Annals of the New York Academy of Science, New York. Vol 280, 1976.
11. H. Yanco and L. Stein. An adaptive communication protocol for cooperating mobile robots, 1993.
12. C. Angelo and D. Parisi. The emergence of a language in an evolving population of neural networks. *Technical Report NSAL–96004, National Research Council, Rome*, 1996.

13. L. Spector and S. Luke. Culture enhances the evolvability of cognition. In *Cognitive Science (CogSci) 1996 Conference Proceedings*, 1996.

14. E. Hutchins and B. Hazlehurst. Learning in the cultural process. In *Artificial Life II, ed. C. Langton et al.* MIT Press, 1991.

15. A. Cangelosi. Evolution of communication using combination of grounded symbols in populations of neural networks. In *Proceedings of IJCNN99 International Joint Conference on Neural Networks (vol. 6)*, pages 4365–4368, Washington, DC, 1999. IEEE Press.

16. B. MacLennan and G. Burghardt. Synthetic ethology and the evolution of cooperative communication. In *Adaptive Behavior 2(2)*, pages 161–188, 1993.

17. A. Billard and G. Hayes. Learning to communicate through imitation in autonomous robots. In *7th International Conference on Artificial Neural Networks*, pages 763–738, 1997.

18. D. Denaro and D. Parisi. Cultural evolution in a population of neural networks. In *M.Marinaro and R.Tagliaferri (eds), Neural Nets Wirn-96.New York: Springer*, pages 100–111, 1996.

19. D. Curran and C. O'Riordan. Learning in artificial life societies. In *Report number nuig-it-220202. Technical Report, Dept. of IT*. NUI, Galway, 2002.

20. H. Kitano. Designing neural networks using genetic algorithm with graph generation system. In *Complex Systems, 4, 461-476*, 1990.

21. G. F. Miller, P. M. Todd, and S. U. Hedge. Designing neural networks using genetic algorithms. In *Proceedings of the Third International Conference on Genetic Algorithms and Their Applications*, pages 379–384, 1989.

22. P. J. Angeline and J.B. Pollack. Competitive environments evolve better solutions for complex tasks. In S. Forrest, editor, *Proceedings of the Fifth International Conference on Genetic Algorithms*, pages 264–270, San Francisco, CA, 1993. Morgan Kaufmann.

Artificial Life and Natural Intelligence

Keith L. Downing

The Norwegian University of Science and Technology
Trondheim, Norway
keithd@idi.ntnu.no

Abstract. This paper reviews the neuroscience literature to sculpt a
view of intelligence from the artificial life (ALife) perspective. Three
key themes are used to motivate a journey down the *low road to cognition*. First, the origins of brain structures and dynamics exhibit considerable emergence at phylogenic, epigenetic, and ontogenetic levels. Second,
ALife complexity measures have interesting parallels in theoretical neuroscience. Finally, the cerebral internalization of sensory stimuli and motor
control explain, respectively, a) semantics in terms of differential complexity, and b) how neural evolution has overcome the limitations of
simple emergence.

1   Introduction

The vast majority of Artificial Life (ALife) research involves large populations
of extremely simple components whose collective behavior yields emergent sophistication. Often, these local units mimic simple biological organisms such as
bacteria, or serve as very high-level abstractions of complex organisms - e.g., simulated economic agents whose intelligence is restricted to a few simple buying
and selling activities.

The realm of high-level intelligence is traditionally unpopular in ALife for
several reasons. First, intelligence issues are often philosophical quagmires where
unwary visitors may disappear without a trace (of publishable work). Second, the
field of Artificial Intelligence (AI) once boldly marched into that dark swamp,
spouting claims of human-like computers just around the corner, thus fueling
rampant media hype. AI only barely escaped (from both the swamp and the
media) and has now retooled considerably to focus on intelligent systems without
significant anthropomorphic claims. ALife researchers are wary of a similar fate,
via either intelligence work of their own or merely association with AI. Finally,
the concept of complex localized controllers (i.e., brains) runs contrary to the
ALife philosophy of emergent global regulation from simple components.

Although intelligence is truly one of life's most perplexing riddles, contemporary neuroscience has made incredible strides in the past few decades, thus
burying many stale theories of mind that survived far too long due to lack of reasonable evidence. In a very strong sense, the neurophysiological evidence yields
intelligence even more foreboding than before. Now that we understand many of

K. Deb et al. (Eds.): GECCO 2004, LNCS 3102, pp. 81–92, 2004.

the local mechanisms such as neurotransmitter signaling, synaptic strengthening, neuronal migration and axonal growth, the distance from these primitives to cognition seems all the more ominous. The old theories of a CPU-like humunculous that ran the show were much easier to work with, but the von Neumann computer analogy was wrong, and, unfortunately, very misleading.

The relevance of intelligence for ALife is now obvious. Since no sophisticated central controller runs the brain, intelligence itself emerges from the interactions of billions of neurons distributed across many brain regions. Cognitive processes such as prediction, memory, learning, etc. should thus, ultimately, find explanations in the cooperative and competitive interactions among neural agents.

This paper examines neurophysiological evidence and theories in terms of two common ALife concepts, emergence and complexity, before showing how natural selection of brains combats the limitations of purely emergent behaviors at one level through hierarchical organization and ascending control.

2   Emergence, Adaptation, and Intelligence

Understanding intelligence from an emergent perspective involves both a) self-organization of structural and dynamic patterns within a given organism, and b) the crafting of sophisticated life forms through natural selection. Both mechanisms underlie life's 3 key adaptive mechanisms: a) development, b) learning, and c) evolution.

This paper reviews several basic mechanisms that are helpful in understanding the phylogenic (evolutionary), ontogenetic (developmental) and epigenetic (learning) aspects of emergent intelligence, and thus useful in tackling questions such as:

1. How can brains capable of intelligent behavior evolve?
2. How are neural topologies grown from genetic instructions?
3. How do neurons interact to facilitate intelligence?

2.1   Basic Mechanisms for Emergent Intelligence

Duplication and Differentiation. Something as complex as the brain cannot emerge in one evolutionary step, although this does not preclude a mixture of punctuated and gradual refinements over the millennia. Each incremental change, whether positive, negative or neutral with respect to intelligence, cannot compromise the species overall fitness. One simple means of guaranteeing a relatively monotonic progression is to duplicate existing genes and then allow the copies to gradually mutate until they achieve an intelligence-enhancing variant. While the copy explores function space, the copied gene continues to perform its normal role, thus providing protective fitness cover during exploration.

The classic illustration of this mechanism is the *homeobox*, a 180-base-pair DNA sequence found in the   *homeotic* genes. Similar sequences of homeotic genes appear in organisms as simple as hydra and fruit flies and as complex as chickens and humans [1]. The homeobox (along with other peripheral base pairs) has clearly been duplicated many times throughout evolution, with peripheral regions then differentiating to form more heterogeneous phenotypes. Incidentally, several of the homeotic genes are involved in the development of the hindbrain, the most primitive cerebral region.

During development, cellular duplication and differentiation are critical activities. After fertilization, the zygote undergoes rapid cleavage divisions to form the blastula, consisting of many identical copies of the original cell. Small asymmetries in the blastula eventually lead to differential gene expression among the cells, causing differences in inter-cellular chemical signaling, leading to further differentiation and an ensuing escalation in complexity.

The blastula transforms into the 3-layered gastrula, Within its neuroectoderm, all cells have the potential to become neuroblasts (neuron precursors). Random asymmetries lead to the formation of isolated neuroblasts, which then send chemical signals that inhibit nearby neuroblast formation and promote epidermal cells.

Migration and Extension. Neuroblasts migrate to the center of the gastrula to form the ventricular proliferation zone (VPZ), where they differentiate into either neurons or glial cells. Neurons undergo further differentiation to their final neural cell type before migrating back outward along the radial glial scaffolding. The brain thus forms from the inside out, as neurons migrate past their temporal predecessors to the periphery.

Since developmental timing effects in the VPZ can strongly influence eventual neural cell fates, a few simple (often genetically-controlled) timing effects can greatly alter the final brain anatomy in terms of the number and thickness of neural layers. Nonlinear competitive and cooperative interactions between the neurons of these layers can then lead to vastly different connection topologies as well. Essentially, neural development is a process poised on the edge of chaos, where small changes to initial (timing) conditions can have large-scale repercussions.

Once properly positioned, neurons sprout axonal and dendritic projections, with the former often growing many centimeters in length. The vast complexity of brain function stems largely from the ability of neurons to send direct, non-diffuse, signals to other particular (often distant) neurons. In contrast, a network in which neurons could merely broadcast signals within variable-sized radial neighborhoods could probably never achieve the same level of sophistication as the vertebrate nervous system.

As described in [19], 4 primary factors control axonal navigation: chemical attractants and repellents that either diffuse within intercellular spaces or cling to cell surfaces. Axons wind their way through fields of diffused attractants and repellents, while bouncing off repellent-laden cells and growing along certain surface attractants. Chemical signatures in the two neurons appear to determine the targets to which growing axons become synaptically coupled. In general, chemical signals direct axons into proper brain regions and to sub-populations of compatible dendrites, but finer tuning of interneural connections involves competitive and cooperative interactions based on correlated firing patterns.

Cooperation. In 1949, Donald Hebb [6] predicted that:

> When an axon of cell A is near enough to excite a cell B and repeatedly or persistently takes part in firing it, some growth process or metabolic change takes place in one or both cells, such that A's efficiency as one of the cells firing B, is increased.

In short, neurons that *fire together, wire together*. Neurophysiological research has revealed the underlying *metabolic changes* and largely confirmed Hebb's hypothesis [9]. This mechanism is central to both biological learning and associative memory formation in various artificial neural networks [7].

Hebb's rule embodies cooperation in at least two respects. First, and most obviously, by (nearly) simultaneously firing, neurons A and B work together to strengthen their intervening synapse. Second, and often overlooked in Hebb's quotation, is the clear implication that A is one of several neurons that actually stimulate B. This cooperation among presynaptic neurons underlies classical conditioning, wherein an initial C-B association essentially bootstraps an A-B association.

For example, assume a basketball player normally dribbles to the right (DR) when he sees the nearest defender moving to (the defender's) right (MR). With experience, he will notice that a defender quickly shifts his body weight to the left leg (SL) before moving to his right (MR). Eventually, this will lead to a hard-wired response wherein SL initiates DR: the dribbler moves right on seeing the weight shift and without waiting for MR.

As simplified in Figure 1, assume 3 neurons (more likely, 3 possibly-overlapping populations of neurons) that represent DR, MR and SL, respectively. Here, DR is the post-synaptic neuron, while SL and MR are pre-synaptic (i.e. they send signals across separate synapses to DR). Bootstrapping embodies cooperation in the following sense. MR is normally sufficient to fire DR. In situations where MR and SL fire almost simultaneously, MR still fires DR, but now SL and DR are also almost simultaneous. Hence, the SL-DR synapse strengthens such that, later, SL alone eventually suffices to fire DR. Essentially, MR has primed DR to enable the SL-DR association.

Dropping down a level, coincidence detectors are essential prerequisites for neural cooperation. In classical conditioning, NMDA receptors in the post-synaptic neuron's dendrites recognize the simultaneity of a) depolarization (firing) of the post-synaptic neuron (e.g. DR), and b) neurotransmitter release by the axon of the pre-synaptic neuron (e.g., SL). Only when both occur does the NMDA receptor open its associated calcium channel, setting off a chain of events that eventually enhance the SL-DR synapse [9].

In the pre-synaptic axon, adenyl cyclase (AC) detects the co-occurrence of a) pre-synaptic neuronal firing, and b) a general signal of excitement in the form of a neuromodulator (e.g. serotonin) broadcast to many parts of the brain by

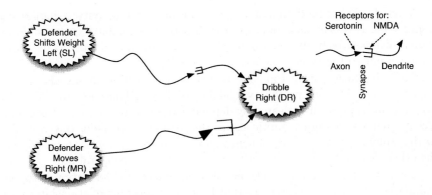

Fig. 1. Abstraction of neural circuitry involved in classical conditioning for learning to dribble right when the opponent shifts his body weight to his left side.

the limbic system during emotional arousal. This is a key mechanism behind operant conditioning, where proper actions are learned via punishment/reward. For example, as our basketball player is learning to shoot free throws, different sets of motor neurons correspond to particular shooting actions, since different sets trigger different muscle fibers. If a particular free-throw is good (due to the firings of the motor-neuron set M), then the emotional experience of success leads to serotonin release (by *broadcasting* neurons in the limbic system). The AC in the pre-synaptic axons of the M neurons will then detect the coincidence of serotonin and recent depolarization, thus leading to prolonged neurotransmitter release (via a complex, but well-understood [3] reaction sequence) at those same M-neuron synapses in the future. In short, AC's co-detection of success/excitement and M-neuron activity enhances the cooperative firing of the M neurons in the future.

This is one of many areas where a proposed ALife primitive is subsumed by other ubiquitous mechanisms. However, since coincidence detection forces a descent to the chemical level, where complexity quickly escalates, this paper will remain at a higher level of abstraction. Also, the general success of artificial neural networks (most of which abstract away all neural physics and chemistry) indicates that lower levels are not necessarily essential for artificial intelligence.

Competition. Whereas NMDA and AC underlie cooperation, chemical growth factors known as neurotrophins incite intense competition among neurons. Axons are dependent upon neurotrophins for growth and maintenance; without them, they wither away. During development, axons extend toward their targets, whose dendrites give off neurotrophins in inverse proportion to their activity. The limited supply of neurotrophins thus supports a restricted number of presynaptic axons, with the rest atrophying away. The proper match of axons to target dendrites (and thus a proper level of target activity) emerges from a simple negative-feedback loop: low target activity stimulates neurotrophin release, which positively affects presynaptic axonal growth, and more axons provide greater input to the target, increasing its activity.

Once the basic neural topology is established, neurotrophins also play an important role in learning, but via slightly different dynamics. In the mature brain, neurotrophins are released by depolarized dendrites and only taken up by recently-depolarized pre-synaptic axons. Hence, only those axons that contribute to the stimulation of the target neuron are rewarded with growth factor, while the others lose out in this competition for the fruits of cooperation.

Competition for neural growth factors may explain the formation of topological maps in the brain, i.e., regions that have an isomorphic relationship to some aspect of the environment. For example, neurons in layers 5 and 6 of the V1 area of the visual cortex respond maximally to lines at particular orientation angles in the visual field [8]. Most significantly, a) neighboring cells respond to similar angles, and b) horizontal transects represent a continuous sequence of monotonically changing preference angles. In short, the neuron space is isomorphic with orientation-angle space. A multitude of such maps exist in the brain, covering all types of sensory input. In fact, many of the initial levels of perceptual processing involve topological maps. Only in higher brain regions do the isomorphisms disappear, as sensory channels converge with one another and with top-down cognitive biases.

As detailed in [17], several different competition-based neural models suffice to generate topological maps of visual orientation angles, via self-organization. These include the classic Kohonen maps [11] in which *post-synaptic* neurons compete for pre-synaptic firing patterns. Our preliminary modelling efforts indicate that maps can also be generated by Kohonen-type networks based on a simple model of neurotrophin release and uptake.

2.2   The Emergent Integrated Hierarchy of Intelligence

The above primitives interact to form complex cognitive systems whose overall topology is convergent (i.e., high axonal feed-in from many regions), reentrant (i.e. loop-forming) and hierarchical (i.e., a series of layers whose functionalities vary along a spectrum from specific to general). For example, Figure 2 depicts some of the connections between sensory topological maps and the hippocampus, believed to be the center of long-term memory [9]. Many brain researchers agree that two key features of higher intelligences are convergence and reentrance/recurrence [13,5].

Convergence enables the integration of perceptual inputs. This helps provide the holistic awareness of a situated and embodied self, which many consider the basis of consciousness [4,5,14]. Many researchers point to the thalamus as the center of consciousness, since it combines multi-modal sensory inputs with memories and then feeds back to the sensory areas of the cerebral cortex, thus forming a memory/contextual bias on further perception. Consciousness is then

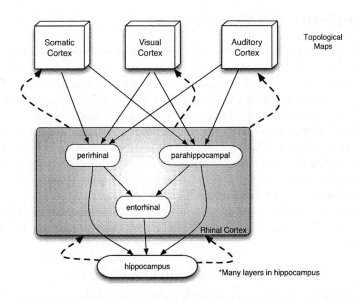

Fig. 2. Convergence and reentrance among the many hierarchical layers from the cortical maps to the hippocampus

more than just a simple awareness of present self and environment, but a partial integration of generalized-past and filtered-current experience.

Although convergence, recurrence and tight integration among components appear to be fundamental emergent properties of the brain, they are normally eschewed by engineers, who prefer systems of loosely-coupled, largely-independent, modules, since these are more amenable to top-down, divide-and-conquer design strategies. Hence, if considerable convergence and recurrence are necessary hallmarks of higher intelligence, then the design of truly artificial intelligences may actually require bottom-up selectionist techniques, such as evolving artificial neural networks.

In summary, the three adaptative mechanisms are essential prerequisites to higher intelligence, and each is grounded in (at least) the above basic mechanisms. Duplication and differentiation are necessary for genotypic and phenotypic complexification, since natural selection severely punishes most exploratory designs. They, along with migration and extension, are also fundamental tools of neural development. Also, beyond their obvious evolutionary influences, cooperation and competition are actively at work in development and learning.

Below this level lies a fascinating array of chemical devices, of which this paper only scratches the surface. Above it come emergent structures such as sensory maps, convergence zones and recurrent loops, whose integrated activity spawns abstract mental phenomena such as memory, awareness, decision-making and consciousness.

3   Complexity and Consciousness

Concepts such as intelligence and consciousness have successfully avoided precise formalization for centuries. However, Edelman and Tononi [5,18] provide a nice break from the philosophical confusion via a statistical quantification of neural complexity - and one which closely parallels classic ALife notions such as the edge of chaos [10,12] and self-organized criticality [2]. Also see [16].

The basic idea is quite simple: complex neural systems are those in which different regions can be in a large number of different states, but these states are highly correlated with the states of other regions. This motivates the expression: *differences that make a difference*, i.e., the (many) states of particular regions strongly influence the states of others.

Edelman and Tononi's complexity involves mutual information (MI), which is based on the comparative entropies (H) of two subsystems (A and B) and their union:

$$MI(A, B) = H(A) + H(B) - H(A \cup B) \qquad (1)$$

High entropy systems are those with many equiprobable states, so high mutual information involves subsystems with many equiprobable states but with few such states in the union, i.e. low $H(A \cup B)$, which directly reflects a high correlation between A and B.

Then, the neural complexity (CN) of an n-neuron system (S) involves the mutual information between every subsystem (s_j) and its complement ($S - s_j$), summing over the average MI values for each size class (k):

$$CN(S) = \sum_{k=1}^{n/2} \overline{MI(s_j^k, S - s_j^k)} \qquad (2)$$

A complex neural system is therefore one containing many subsystems that are both internally diverse and mutually integrated. Levels of diversity and integration show characteristic variations according to brain maturity [5], as shown in Figure 3.

1. Young brains exhibit high integration but low diversity, producing activity patterns in which many neurons in S synchronously change state, but local regions show few activation patterns. Hence, the system appears to run through low-period cycles of highly homogeneous states.
2. Old brains are weakly integrated but highly differentiated, leading to chaotic activation patterns in which each region behaves independently.
3. Mature brains portray both high integration and differentiation, thus producing long state cycles where each state has a structured (low entropy) appearance, but these patterns gradually change in a continuous manner, as a glider moves across a cellular-automata. They look alive!

In terms of the neural complexity metric:

$$CN(old) < CN(young) < CN(mature) \qquad (3)$$

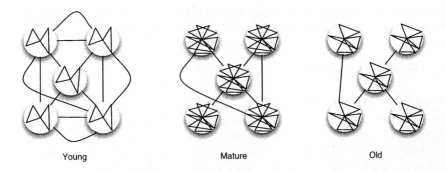

Young   Mature   Old

Fig. 3. Abstraction of neural topologies in young, mature and old brains, showing degree of connectivity both within and between different regions.

The parallels to ALife complexity definitions are striking. Old brains mirror a gaseous state, where chaos dominates. Young brains resemble a solid state, where state cycles are short. However, in contrast to a Kaufmann solid [10], where large regions are static, young brain cells do frequently change state, but individual regions show only a few local patterns: they are frozen into a small set of alternatives. Finally, the mature brain closely matches the liquid state, the edge of chaos, and self-organized criticality in that it has long state cycles and appears to exhibit both memory and information transfer as patterns gradually form and retain much of their shape as they move about.

Neural complexity links to consciousness via a few additional concepts [5]. First, a *functional cluster* is a neural region with a higher *cluster index* than any of its subset regions, where the cluster index (CI) is the ratio of a regions internal to external interactivity:

$$CI(S_i) = \frac{\sum_{j=1}^{m} H(s_{i,j}) - H(S_i)}{MI(S_i, S - S_i)} \qquad (4)$$

Here, the numerator denotes the integration ($I(S_i)$) of subsystem S_i: the amount of entropy that S_i's m subsystems ($s_{i,j}$) lose due to their mutual interactions. The denominator reflects S_i's contribution to S's complexity.

A functional cluster whose interface with its complement ($S - S_i$) may change on approximately a 1/10 second timescale is a *dynamic core* [5]. It is hypothesized that the neurons involved in consciousness at a particular time are members of a dynamic core that involves extensive reentrant looping between the cerebral cortex and thalamus, and has high neural complexity, i.e., many different states that make a difference outside the core.

Thus, consciousness is a process governed by the ever-changing neural constituency of a thalamocortically-centered dynamic core. As in ALife systems, a) the dynamics, not the substrate, define the phenomena, and b) the emerging global pattern exerts a strong influence upon the system, e.g., by biasing perceptual interpretations, the focus of attention, memory retrieval, etc.

4   Encephalization

Traditional ALife complexity analyses concern the effects of external perturbations upon internal dynamics, where the disturbance is small - such as the addition of one sand grain to a pile or the flipping of one bit in a quiesced boolean network - and the probability distribution over the sizes of the effects differentiates between the stable, chaotic and complex regimes.

However, living organisms interact with a complex environment whose perturbations are both intricate and extensive; and the resulting imprint on internal dynamics constitutes an adaptive response with real survival value. The view of brain and surroundings as coupled autopoietic (i.e. constantly self-maintaining) systems [15] captures this idea:

> The plastic splendor of the nervous system does not lie in its production of engrams or representations of things in the world; rather it lies in its continuous transformation in line with transformations of the environment ... The functioning organism, including its nervous system, selects the structural changes that permit it to continue operating, or it disintegrates (pg. 170).

Basically, salient aspects of the environment are internalized in the neural circuitry, or *encephalized*. To quantify this effect, Tononi et. al. [18] use CN to define *matching complexity*, CM: the complexity of the system due to sensory input, or, the degree to which internal correlations change due to external perturbations.

$$CM(S, P) = CN(S) - CN(S - P) - CN^E(S, P) \tag{5}$$

The matching complexity between a neural system S and its outer perceptual layer, P, is the total complexity of S reduced by both the complexity of S-P (the intrinsic complexity) and the complexity at the interface between P and S-P (the extrinsic complexity, $CN^E(S, P)$). Thus, CM measures the change in internal complexity due to the cascading effects of sensory input. This correlates well with stimulus familiarity, as seen in the experiments of [18], where ANNs trained on sample patterns and then exposed to similar new patterns show $CM > 0$, while novel test patterns yield $CM < 0$. Apparently, the familiar pattern calls a host of contextual information into play, producing much greater internal change than does a novel stimulus. Although a novel perturbation might cause great change to early levels of processing, its failure to link to previous experience quickly arrests any signaling cascades. Since CM seems to correlate with this memory/contextual factor, it implicitly measures the significance/meaning of a stimulus for the observer. Thus, the central ALife concept of complexity may provide insights into the philosophical conundrum of semantics.

In addition to sensory inputs, encephalization can also encompass motor outputs, as detailed by Llinas [14]. Here, motor activity patterns that were originally emergent from the direct electrical couplings between muscle cells have, through the course of evolution, become controlled by, first, spinal motor neurons, and later, neurons of higher brain regions. This higher-level control increases the potential complexity of the actions, since emergent oscillatory patterns - typically, spatial waves of muscle contractions - cannot approach the intricacy needed for walking a balance beam or playing the piano. But with a neural hierarchy, spatial activation waves at the higher levels can, via tangled top-down connections, differential propagation delays, etc., cause spatially diverse firing patterns at lower levels.

Llinas [14] believes that muscle oscillations in primitive animals and in developing vertebrate embryos have become internalized/encephalized to the 40 Hz activity of the thalamus. Since thalamic activity is a critical constituent of the dynamic core, these 40 Hz oscillations serve as a binding signal for mental activity. In a strong sense, this 40Hz signal is the heartbeat of the brain, and it arose via an evolutionary process that gradually translated emergent muscle-activation patterns into a high-level dynamic that coordinates all activity: perceptual, motor and cognitive. Thought is encephalized motricity.

Basically, evolutionary emergence combatted the limited motor complexity provided by simple emergence among locally-connected activators (i.e., muscle cells) by *designing* the nervous system, which permits intricate communication networks among non-adjacent cells. While primitive versions involve direct connections between sensory and motor apparatus, the brains of more intelligent organisms house many convergent and reentrant layers to realize high neural (and hence behavioral) complexity. This hierarchy manifests an encephalization of both environment and action, thus embedding reality for a selective survival advantage.

5   Conclusion

The potential synergies between ALife and neuroscience are abundant, albeit nonobvious. For instance, although a central controller is anathema to sciences of emergence, it becomes a crown jewel when its self-organizing processes are unveiled, as is now the case with many brain regions.

This paper has discussed several general mechanisms residing at an intermediate conceptual level between the electrochemical and the psychological. In all cases, these principles are quite well understood in terms of the lower levels, so they provide well-grounded intellectual scaffolding for bottom-up attacks on cognitive phenomena such as learning and memory.

Ideally, ALife systems could *begin* with these processes and derive intelligence. AI was driven by similar optimism, but primitives such as logic and best-first search had no obvious neural basis. Conversely, the staples of ALife (competition, cooperation, differentiation, etc.) do. In addition, ALife-related concepts such as neural complexity (CN) and matching complexity (CM) provide elegant metrics for cognitive emergence.

In general, the study of intelligence in terms of neural networks gives ALife researchers a host of interesting opportunities: 1) to understand complex phenomena such as memory, reasoning and semantics from an emergent perspective, 2) to recognize and formalize the limitations of behavioral emergence and the improvements accrued by a nervous system, 3) to quantitatively analyze the informational coupling between environments and brains as they coevolve, and 4) to analyze similar informational correlations between motor patterns and their proposed encephalizations in controlling neural circuitry.

References

1. J. ALLMAN, *Evolving Brains*, W.H. Freeman and Company, New York, NY, 1999.
2. P. BAK, C. TANG, AND K. WIESENFELD, *Self-organized criticality*, Physica Review A, 38 (1988).
3. T. J. CAREW, *Behavioral Neurobiology: The Cellular Organization of Natural Behavior*, Sinaur Associates, Sunderland, MA, 2000.
4. P. CHURCHLAND, *The Engine of Reason, the Seat of the Soul*, The MIT Press, Cambridge, MA, 1999.
5. G. EDELMAN AND G. TONONI, *A Universe of Consciousness*, Basic Books, New York, NY, 2000.
6. D. HEBB, *The Organization of Behavior*, John Wiley and Sons, New York, NY, 1949.
7. J. HOPFIELD, *Neural networks and physical systems with emergent collective computational abilities*, Proceedings of the National Academy of Sciences, 79 (1982), pp. 2554–2558.
8. D. HUBEL, *Eye, Brain, and Vision*, Scientific American Library, New York, NY, 1995.
9. E. KANDEL, J. SCHWARTZ, AND T. JESSELL, *Principles of Neural Science*, McGraw-Hill, New York, NY, 2000.
10. S. KAUFFMAN, *The Origins of Order*, Oxford University Press, New York, 1993.
11. T. KOHONEN, *Self-Organizing Maps*, Springer, Berlin, 2001.
12. C. LANGTON, *Studying artificial life with cellular automata*, Physica D, 22 (1986), pp. 120–149.
13. J. LEDOUX, *Synaptic Self: How Our Brains Become Who We Are*, Penguin Books, Middlesex, England, 2002.
14. R. R. LLINAS, *i of the vortex*, The MIT Press, Cambridge, MA, 2001.
15. H. MATURANA AND F. VARELA, *The Tree of Knowledge: The Biological Roots of Human Understanding*, Shambala Publishers, Boston, MA, 1998.
16. C. NEHANIV AND J. RHODES, *The evolution and understanding of hierarchical complexity in biology from an algebraic perspective*, Artificial Life, 6 (2000), pp. 45–67.
17. N. SWINDALE, *The development of topography in the visual cortex: A review of models*, Network: Computation in Neural Systems, 7 (1996), pp. 161–247.
18. G. TONONI, O. SPORNS, AND G. EDELMAN, *A complexity measure for selective matching of signals by the brain*, Proceedings National Academy of Sciences, 93 (1996), pp. 3422–3427.
19. L. WOLPERT, *Principles of Development*, Oxford University Press, New York, 2002.

Bluenome: A Novel Developmental Model of Artificial Morphogenesis

T. Kowaliw, P. Grogono, and N. Kharma

Departments of Computer Science and
Computer and Electrical Engineering, Concordia University,
1455 de Maisonneuve Blvd. Ouest, Montréal, QC, Canada, H3G 1M8
taras.kowaliw@utoronto.ca, grogono@cs.concordia.ca,
kharma@ece.concordia.ca

Abstract. The Bluenome Model of Development is introduced. The Bluenome model is a developmental model of Artificial Morphogenesis, inspired by biological development, instantiating a subset of two-dimensional Cellular Automata. The Bluenome model is cast as a general model, one which generates organizational topologies for finite sets of component types, assuming only local interactions between components. Its key feature is that there exists no relation between genotypic complexity and phenotypic complexity, implying its potential application in high-dimensional evolutionary problems. Additionally, genomes from the Bluenome Model are shown to be capable of re-development in differing environments, retaining many relevant phenotypic properties.

1 Introduction

Typical applications in Evolutionary Computation often involve a direct and simple relation between genotype and phenotype; Commonly, values from the genome are simply slotted into a fitness function in a bijective mapping. While this approach is sufficient for most practitioners, the dimensionality of the solution space (phenotypes) is directly translated into the dimensionality of the space of genotypes, potentially exceeding the size of space capable of being searched by a Genetic Algorithm. For larger and more complex problems direct relations between genotype and phenotype may be insufficient.

In a field inspired by Biology, it is often useful to re-examine the source: the human genome may be viewed as a tremendous compression of phenotypic complexity; The approximately 3 billion chemical base pairs of the genome map to approximately 100 trillion cells [8]. It is clear that the process of development plays a significant role in the addition of information to the phenotype; Indeed, models of biology often attempt to re-create the hierarchical structure inherently formed by the differentiation process, as in [5].

An emerging trend in Evolutionary Computation is to create a (relatively) small and simple genotype, and to increase the complexity of the phenotype through a developmental process. The field of Artificial Morphogenesis spans a wide array of

K. Deb et al. (Eds.): GECCO 2004, LNCS 3102, pp. 93–104, 2004.
© Springer-Verlag Berlin Heidelberg 2004

In this manner, a "good" genome will allow a single initial cell to grow to a robust agent. The process terminates when all telomeres are expended, or when there is no change from one developmental time step to the next. No other mechanisms are included – there are no special parameters for symmetry breaking, beyond that inherent in the directional bias.

One view of this process is as a subset of all 2-dimentional Cellular Automata with radius 3. The key differences between CAs and Bluenome are: (1) Bluenome begins with a single cell in the centre of a finite grid. Empty (white) cells cannot change their colour without a non-white neighbour[4]; (2) As the hormone collection process does not note direction, the rules instantiated by the Bluenome genome map to symmetrical patterns in a CA rule set; (3) Bluenome utilizes a measure to compute distance to a rule, unlike CAs, which are precise. This may be viewed as collapsing several similar rules into a single outcome; and (4) The lack of consideration of peripheral cells in the twelve-neighbourhood may be viewed as a further grouping of CA rules.

Fig. 2.1 shows the development of an interesting agent taken from Phase One, shown at various times in the development. An unfortunate point is that the majority of genotypes generate trivial agents – nearly 80% of random samples. However, it will be shown that selection quickly improves matters.

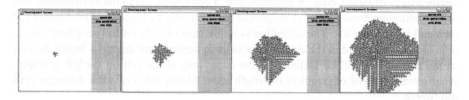

Fig. 2.1. Development of an interesting agent.

We can now make some estimates involving size: Firstly, we note that the maximum size of an agent with *numTel* telomeres will be $2*(numTel+1)^2$ – this is the size of a diamond with sides of length *(numTel+1)*. Hence, an agent with *numTel = 6* will have a maximum phenotypic size of 98 cells; 882 cells for an agent with *numTel = 20*. In contrast, and agent with *numColours = 5* and *numRules = 25* will have a genotypic size of 175, regardless of phenotypic size. The complexity of the developmental process is $O(numTel^3*numRules)$.

3   Phase One: Application-Neutral Experiments

The evolution of cellular automata is a notoriously difficult problem: the highly non-linear nature of the space of CAs, as well as the unpredictability of the forecasting problem [15], [13] makes the prospect of the evolution of complex patterns using GAs seem grim. As we have recognized the Bluenome model as a subset of the space of

---

[4] Similar to a Totalistic CA [15].

two-dimensional CAs, it is not obvious that evolution is possible in any reasonable measure of time. Phase One was a set of experiments utilizing an array of fitness functions to determine which high-level axis could potentially be evolved.

Phase one utilized a neutral view of phenotypes – an organism was treated as an image, fitness function being based on techniques from Image Processing. In all cases, grids of size 100x100 were used, and the agents were allowed to grow to fit the grid. A genetic algorithm was used to evolve phenotypes, typically running for 100 generations with a population size of 30.

Successful experiments showed promising results; Typically, highly fit population members were discovered prior to generation 100, with steady increase in mean fitness. These fitness functions included: selection for complexity, selection for similar but disconnected regions, selection for highly distinctive regions, and selection for a transport system.

An example of one such successful run was the attempt to generate images of increasing complexity. Here, fitness was a function of the size of the grown agent and the number of different cell types included in the phenotype. By generation 100 a robust agent utilizing all colours can be seen. Members of the population can be seen in Fig. 3.1. From generation 100 of this same experiment.

Fig. 3.1. Images from a Phase One experiment using complexity as a fitness. Members are shown from generations 0, 10, 40 and 100.

Fig. 3.2 shows a set of exemplar members chosen from the various experiments. In all cases, these members were found in less than 100 generations of evolution, utilizing population sizes of less than 30.

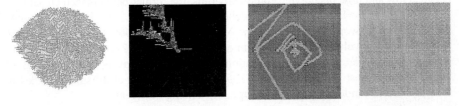

Fig. 3.2. Exemplar members from Phase One experiments, identified by generating fitness: (*far left*) maximal thin coverage; (*left*) disconnected regions of similarity; (*right*) disconnected regions of similarity; (*far right*) highly distinctive regions.

Another interesting result from Phase One was the attempt to re-grow a genotype in differing environments. This attempt is illustrated in Fig. 3.3, where again visual similarities may be seen most clearly, Additionally, under the fitness used in the experiment (the complexity function), fitness was nearly identical for each re-growth.

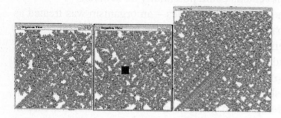

Fig. 3.3. The same genotype re-grown in differing environments. (*left*) original phenotype; (*centre*) environment with non-interacting foreign cells (*black*); (*right*) environment of double the size.

4   Phase Two: Application to an Artificial Problem

The goal of the experiments in Phase Two is the evolution of multi-cellular agents, capable of surviving as long as possible in an artificial world. The artificial agents are presented; An agent is a collection of cells, each with a defined behavior. These cells are laid out (connected) in a matrix of Grid Cells, and provided with an amount of food initially, a cell using one unit of food per discrete time step. Also in this environment are laid out patches of food; To survive longer, an agent must detect this food, move over top of it, absorb it, and distribute it to the remainder of cells in its body. All cells are capable of local interactions only – a cell communicates or passes food only in its local neighbourhood.

Additionally, a second model of development is presented, one in which the relation between genotype and phenotype is bijective. The purpose of this inclusion is to demonstrate that a developmental model may outperform a bijective model.

Worlds: A world is an infinite two-dimensional matrix of Grid Cells. Each world contains one agent at the centre, and a distribution of food. There are no collisions - instead, an agent will pass directly over top of food in the world, possibly absorbing it. Food is parceled in food pieces, each occupying one Grid Cell, having a value of $2*(numTel+1)^2$ food units. Food is distributed differently, depending on world type. Distances between the agent's starting point and the food batches varies between low and high phenotypic complexity runs, the former being placed closer.

Type 0 worlds contain eight batches of food, laid out in a circle surrounding the agent. *Type 1* worlds consist of a line of four patches of food, these patches being placed in successively longer distances in one direction. *Type 2* worlds consist of four patches of food placed in random locations, slightly farther away than the range of vision of the closest possible eye cell. *Type 3* worlds consist of 40 small batches of food distributed randomly in a donut shape surrounding the agent.

Agents: An agent is a collection of one or more cells, assumed to be connected. Each cell occupies one grid location. Agents behave as the sum of the behaviours of their cells. So long as one cell is declared "active", an agent is declared "active" - otherwise "inactive".

Cells may be viewed as independent agents of their own right - each maintains a food supply, and executes a particular program based on input and internal variables. Cells may communicate and pass food between adjacent cells (four or eight-neighbourhoods). A cell is "active" (coloured) if its food supply is greater than zero, otherwise "inactive" (black). An inactive cell will continue to occupy physical space, but will no longer be capable of processing input or output or absorbing food. All cells belong to one of the following classes: Eyes (Green), Nerves (Orange), Feet (Blue), Transports (Red) and Structure Cells (Gray).

Eyes: Eye cells can sense their external environment, and return a boolean value on the basis of the existence of food. However, the presence of other cells within its field of vision will block its ability to sense.

Nerves: Nerves are cells, which accept information from neighbouring eye or nerve cells, with up to four inputs, four outputs, or any combination thereof, determined by connections to eye cells. Nerves output the sum (*identity* nerves), the negative of the sum (*inverse* nerves), or the sum plus a random value from {-1, 0, 1} (*random* nerves).

Feet: Foot cells accept input from all neighbouring nerve cells. Following all other computation, an agent sums the motion of each foot, and moves accordingly (weighted by total size of agent). *Forward* foot cells move *forward* (backward for negative input), and *rotation* foot cells rotate counter-clockwise (clockwise).

Transports: Transport cells manage the collection and distribution of food. At each time step, a transport cell will: collect food from its environment, and pass food to all neighbours in the eight-neighbourhood.

An Agent in the World: An agent is initialized in the centre of a world, each cell containing 200 units of food, with time defined as zero[5]. The agent next executes the following process at every time step, considering *only* active cells:

1.   Replace any cells with no food with inactive cells (black, in the GUI)
2.   Each transport cell collects any food from the environment
3.   Each transport cell passes food to its neighbours
4.   Compute the depth of each nerve cell, where a nerve has depth 1 if it is connected to an eye cell, 2 if it is connected to a nerve connected to an eye, etc. Random nerve cells which are not connected to an eye cell are also labeled depth one.
5.   Eye cells are activated, returning 1 if food is within field of vision.
6.   All nerve cells of depth one collect input and compute output, continue for each successive depth
7.   Each foot cell collects input, adding output to the total
8.   The agent moves accordingly.

Fig. 4.1 is an illustration of perhaps the simplest agents capable of finding and absorbing food. As a curiosity, consider Fig. 4.2, an agent in a similar situation; This agent's actions would cancel each other out, leading to immobility.[6]

Fig. 4.1. One of the simplest agents capable of finding and absorbing food

---

[5] Hence, if an agent does not find food, all cells will die at time 200; Typically, most cells do, as most agents have imperfect methods for food distribution.

[6] Schopenhauer, eat your heart out.

Fig. 4.2. An immobile agent

Development: Two methods of development are used: the Bluenome method, as introduced in section 2, and a Bijective method. The bijective method of agent development is a simple model, in which there exists a one-to-one correspondence between elements in the genome, and cells present in the agent. The genome for an agent consists of an array of integer values, all between 0 and 9, inclusively. A bijective agent is developed by laying out the values of those integers one by one, in a spiral pattern, eventually forming a diamond of area $2*(numTel+1)^2$ – hence, an agent with $numTel = 6$ will have at most a genotypic and phenotypic complexity of 98; With $numTel = 20$, a complexity of 882. The genome values are mapped to cell types, where the 0 value is mapped to the empty cell. The spiral layout begins with the central point, and proceeds biased downwards and clock-wise.

Fitness: The base fitness of an agent in the world is a measure of its size and length of life, relative to a world w.

$$fitness_{base}(a)^W = \Sigma_t\, numCells(a,t) \qquad (1)$$

where $numCells(a,t)$ is the number of living cells in agent a at time t. Note: since the amount of food in a world is finite, so is $fitness_{base}$.

To help the Bluenome model overcome the development of trivial agents, we introduce a bonus to fitness (for both Bluenome and bijective versions). We define $numClasses \in [4]$ to be the number of those classes for which at least one cell exists in the fully developed agent.

$$fitness_{bonus}(a) = numClasses\,(a)^2*20*(numTel+1)^2 \qquad (2)$$

Finally, our fitness function is:

$$fitness(a)^W = fitness_{base}(a)^W + fitness_{bonus}\,(a) \qquad (3)$$

In any particular generation, an agent a will be subjected to two worlds, w_1 and w_2., chosen at random (our fitness is stochastic):

$$fitness(a) = fitness(a)^{W1} + fitness(a)^{W2} \qquad (4)$$

Experiments: Evaluation of the Bluenome system involves a series of experiments, differentiated by the model for growth (Bluenome versus Bijective, or bn versus bj), and also by value of $numTel$. The experiments consisted of one run of each the Bluenome and Bijective systems for $numTel \in \{6, 8, 10, 12\}$ (low phenotypic complexity). Additionally, there were three runs for each of the Bluenome and Bijective systems with $numTel = 20$; (high phenotypic complexity).

5   Data and Analysis

Data for the low phenotypic complexity runs ($numTel \in \{6, 8, 10, 12\}$) showed little variance between values of $numTel$; Hence, only data for the $numTel = 6$ runs are shown.

In the low phenotypic complexity runs, the bijective runs outperform substantially, as illustrated in Fig. 5.1. Also, Figure 5.2 shows a comparison between maximum time for the *numTel* = *6* runs – the bijective version typically outperforms the Bluenome version. Contrary to initial expectations, this is not a boon, but instead a drawback. The primary failing of the bijective method is its inability to generate an adequate transport system for distributing food throughout its body. The successes of the bijective model typically involve small groups of cells hoarding food, while no new food is found following time step 200.

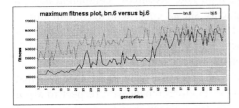

Fig. 5.1. Maximum fitness of the Bluenome versus the Bijective run for numTel = 6. The Bijective run clearly outperforms initially; While the Bluenome run catches up in later generations, it never reaches the same levels.

Fig. 5.2. Maximum time (of the most fit agent) plot for the Bluenome versus the Bijective run, with numTel = 6. The Bluenome version clearly shows a lower maximum time consistently.

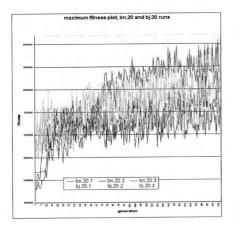

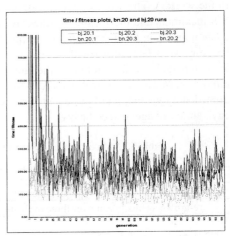

Fig. 5.3. Maximum Fitness plots for three of each of the Bluenome and Bijective runs, with numTel = 20. The Bijective runs *(light lines)* outperform initially, but two of the Bluenome runs *(dark lines)* catch up by generation 150. One run can be seen overtaking the Bijective runs, beginning with generation 70.

Fig. 5.4. Time / Fitness plots (of the most fit agent) of three Bluenome runs *(dark lines)* versus the Bijective runs *(light lines)*. The Bluenome runs are clearly higher consistently.

Fig. 5.3 show the fitness plots of the *numTel* = *20* runs. In these runs, a different trend is seen; Here, the bijective runs all follow a similar course. They begin with some visible evolution, until they reach maximum fitness values in a range of about 720 000 to 850 000 (in all cases prior to generation 100), where they appear to oscillate between values randomly; It appears that the complexity of the space involved exceeds the GAs ability to improve. The lowest of the Bluenome runs shows a similar course to the bijective runs, with some initial evolution and a seemingly random cycling of values following. However, the other two Bluenome plots show a continuous evolution proceeding up to generation 200, potentially continuing beyond this point. Additionally, the highest run quickly shows consistent maximum fitness values, which exceed the maximum, found in any of the bijective runs. Figure 5.4 clearly shows the continuation of the fitness / time trend – the Bluenome models clearly distribute food between body components more evenly.

An interesting and important property determined from Phase One was that of the Bluenome Model's resistance to changes in environment with respect to agent growth. In Phase Two, an experiment was undertaken which tested a similar situation – that of the re-use of an agent's genome in a differing setting. Populations of genomes were selected from a high-phenotypic complexity run (*numTel* = *20*) at a period late in evolution (generation 180). These genomes were re-developed, this time using a value of *numTel* = *8*, rather than 20. The developed agents were evaluated as normal in the *numTel* = *8* context (that is, using the lower distances for food locations in the worlds). The values obtained are comparable to the *numTel* = *8* run. In Table 5.1, maximum and mean fitness values are compared between the re-grown agents and a late population from the *numTel* = *8* run. While the original population outperforms the re-grown agents slightly, the mean and maximum fitnesses of the re-grown agents are comparable to those found in the later stages. Fig. 5.5 shows the first agent of the *numTel* = *20* run grown with *numTel* = *8, 20*; Visual similarities between the two are immediately visible, and both agents are members of the "Position-then-Rotate Strategy" family of agents (see below).

Table 5.1. Maximum and Mean fitness values from re-grown agents

	mean fitness	maximum fitness
original *numTel* = *8* pop., generation 94	75 252.67	126 364
re-grown agents, mean over three evaluations	73 211.06	119 344

Fig. 5.5. An agent from a run with *numTel* = *20*, generation 180 (left), re-grown having changes the value of *numTel* to 8 (below).

It has been noted that Phase Two presents an artificial problem for which human designers would experience difficulty; Three (of many) identified agent strategies are presented in the following figures: Fig. 5.6 illustrates a member of the *blind-back-and-forth* strategy: This strategy may be viewed as a local optimum which often dominates early generations. These agents typically do not include eye or rotation foot cells, relying instead solely on random nerves and forward cells, moving back and forth on the x-axis. This strategy works marginally well for worlds of type 0 and 3, but rarely for other worlds; Fig. 5.7 illustrates a member of the *rotate-then-forward* strategy, perhaps the most successful strategy found - The agent rotates randomly, until it sees food, then moves forward; Fig. 5.8 illustrates a member of the *position-then-rotate* strategy, another local optimum: initially the agent moves forward, until it is at the same distance as the first batch of food. Then, it begins to trace a constant circular path – this strategy works poorly on most worlds, except on type zero, where it may be a global optimum.

Fig. 5.6. Example of the *Blind Back-and-Forth* Strategy. The dark blue cells are Forward Foot cells, the mid-orange cells Random Nerve cells. The agent contains no Eye of Rotating Foot cells whatsoever, save a single Eye cell near the centre (its location guarantees it will never fire). The single Eye cell is included probably for the sole reason of maximizing fitness$_{bonus}$

Fig. 5.7. An example of the *Rotate-then-Forward* Strategy. There are two Eye cells on the periphery of the agent – centre left and upper right. These cells are somewhat buried, guaranteeing a narrow focus, and are connected through a large series of Identity Nerve cells to Forward Foot cells. In the centre of the agent are many Random Nerve cells, connected to Rotation Foot cells, providing the random rotation.

Fig. 5.8. Example of the *Position-then-Rotate* strategy. The agent has Eye cells connected to Forward Foot cells on both the left and right hand side, with more Forward foot cells on the left. Towards the centre of the agent, a series of Random Nerves connect to Rotation Foot cells.

6   Conclusions

In Phase One, several fitness functions were used to evolve images demonstrating suggestive principles. In nearly all cases, successful evolution was achieved quickly, generating images recovering some of the complexity of two-dimensional CAs.

In Phase Two, the Bluenome model was applied to a non-trivial artificial problem, one which involved the coordination of many non-linearly interacting components. In cases of low phenotypic complexity, the bijective methodology tended to outperform the Bluenome method, with a wide margin in initial generations, barely so in later generations. In cases of high phenotypic complexity, however, one of the Bluenome runs clearly outperformed all of the bijective runs, with a second matching with potential for further growth in later generations. The Bluenome methodology continued to develop in a high-dimensional space, while the bijective methodology stagnated early on.

In addition to this performance increase in high complexity runs, the Bluenome model showed an inherent ability to generate agents with better developed systems for the distribution of food throughout the body – this is no doubt a result of the inheritance of cell specialization creating a network of transport cells, a readily observed instance of the sorts of structural patterns inherent to the developmental process [5]. Finally, the resistance of the developmental process to changes in the environment was demonstrated. More intriguing still is the continuation of performance by the re-developed agents, both in terms of valuation by the fitness function in question, and in visual appearance.

A matter touched upon in the above discussions is the view of evolution as a mechanism for controlling complex processes; Indeed, this is an intriguing hypothesis, perhaps contributing to the success of the above system; If true, however, it begs an obvious question: by what mechanism? It is the hope of the authors that systems like Bluenome may serve as a test bed by which this claim may be evaluated and studied further.

References

1. Bentley, P., Kumar, S., The Ways to Grow Designs: A Comparison of Embryogenies for an Evolutionary Design Problem, in Proceedings of the Genetic and Evolutionary Computation Conference, GECCO-1999
2. Dellaert, F., Beer, R., A Developmental Model for the Evolution of Complete Autonomous Agents, in From Animals to Animats 4: Proceedings of the Fourth International Conference on Simulation of Adaptive Behavior, (1996)
3. Eggenberger, P., Evolving Morphologies of Simulated 3D Organisms Based on Differential Gene Expression., in Husbands, P., Harvey, I. (editors) Proceedings of the Fourth European Conference on Artificial Life (1997)
4. Fagotto, F., Gumbiner, B., Cell contact-dependent signaling, in Developmental Biology 180 (1996)
5. Furusawa, C., Kaneko, K., Emergence of Rules in Cell Society: Differentiation, Hierarchy, and Stability, in the Bulletin of Mathematical Biology (1998)
6. Hamahashi, S., Kitano, H. Simulation of Drosophilae Embryogenesis, in the Proceedings of the Sixth International Conference on Artificial Life (1998)
7. Hotz, P., Gomez, G., Pfiefer, R., Evolving the morphology of a neural network for controlling a foveating retina - and its test on a real robot, in Artificial Life VIII: The 8th International Conference on the Simulation and Synthesis of Living Systems (2003)
8. Karp, G., Cell and Molecular Biology, 3rd Ed., (John Wiley & Sons, Inc.; 2001)
9. Kowaliw, T., Bluenome: A Novel Developmental Model for the Evolution of Artificial Agents, Master's Thesis, Concordia University (2003)
10. Kvasnicka, V., Pospicjal, J., Emergence of Modularity in Genotype-Phenotype Mappings, in Artificial Life, v. 8, no. 4 (2002)
11. Olivera, G., Olivera, P. Omar, N. Definition and Application of a Five Parameter Characterization of One-Dimensional Cellular Automata Rule Space, in Artificial Life, Volume 7 Number 3, 2001
12. Stanley, K., Miikkulainen, R., A Taxonomy for Artificial Embryogeny, in Artificial Life, v.9, no. 2 (2003)
13. Wolfram, S.: A New Kind of Science. Wolfram Media New York (2002)

Adaptively Choosing Neighbourhood Bests Using Species in a Particle Swarm Optimizer for Multimodal Function Optimization

Xiaodong Li

School of Computer Science and Information Technology
RMIT University, VIC 3001, Melbourne, Australia
xiaodong@cs.rmit.edu.au
http://www.cs.rmit.edu.au/~xiaodong

Abstract. This paper proposes an improved particle swarm optimizer using the notion of species to determine its neighbourhood best values, for solving multimodal optimization problems. In the proposed species-based PSO (SPSO), the swarm population is divided into species sub-populations based on their similarity. Each species is grouped around a dominating particle called the species seed. At each iteration step, species seeds are identified from the entire population and then adopted as neighbourhood bests for these individual species groups separately. Species are formed adaptively at each step based on the feedback obtained from the multimodal fitness landscape. Over successive iterations, species are able to simultaneously optimize towards multiple optima, regardless of if they are global or local optima. Our experiments demonstrated that SPSO is very effective in dealing with multimodal optimization functions with lower dimensions.

1 Introduction

In recent years, Particle Swarm Optimization has been used increasingly as an effective technique for solving complex and difficult optimization problems [3, 6,7]. However, most of these problems handled by PSOs are often treated as a task of finding a single global optimum. In the initial PSO proposed by Eberhart and Kennedy [7], each particle in a swarm population adjusts its position in the search space based on the best position it has found so far, and the position of the known best-fit particle in the entire population (or neighbourhood). The principle behind PSO is to use these particles with best known positions to guide the swarm population to converge to a single optimum in the search space.

How to choose the best-fit particle to guide each particle in the swarm population is a critical issue. This becomes even more acute when the problem being dealt with has multiple optima, as the entire swarm population can be potentially misled to local optima. One approach to combat this problem is to allow the population to search for multiple optima (either global or local) simultaneously. Striving to locate multiple optima has two advantages. Firstly, by locating multiple optima, the likelihood of finding the global optimum is increased; sec-

K. Deb et al. (Eds.): GECCO 2004, LNCS 3102, pp. 105–116, 2004.

ondly, when dealing with real-world problems, for some practical reasons, it is often desirable for the designer to choose from a diverse set of good solutions, which may be equally good global optima or even second best optima.

The uniqueness of PSO's ability in adaptively adjusting particles' positions based on the dynamic interactions with other particles in the population makes it well suited for handling multimodal optimization problems. If suitable particles can be determined as the appropriate neighbourhood best particles to guide different portions of the swarm population moving towards different optima, then essentially we will be able to use a PSO to optimize over a multimodal fitness landscape. Ideally multiple optima will be found. Now the question is how to determine which particles would be suitable as neighbourhood bests; and how to assign them to the suitable particles in the population so that they will move towards different optima accordingly.

The paper is organized as follows: section 2 describes related work on multimodal optimization, and their relevance to the proposed species-based PSO (SPSO). Section 3 presents the classic PSO. Section 4 introduces the notion of species and its relation to multimodal optimization. Section 5 describes the proposed SPSO. Section 6 and 7 cover the performance measures and test functions respectively, followed by section 8 on experimental setup and then section 9 on results and discussion. Finally section 10 draws some conclusions and gives directions for future research.

2   Related Work

Although multimodal function optimization has been studied extensively by EA researchers, only few works have been done using particle swarm models. In [5], Kennedy proposed a PSO using a k-means clustering algorithm to identify the centers of different clusters of particles in the population, and then these cluster centers are used to substitute the personal bests or neighbourhood bests. However some serious limitations of this method can be identified:

1. In order to calculate the cluster centers, the method requires three iterations over all individuals in the population, which is very computationally expensive.
2. A cluster center identified is not necessarily the best-fit particle in that cluster. Consequently using these cluster centers as *lbest* is likely to lead to poor performance (see Fig. 1 of [5]).
3. The number of clusters must be pre-specified.

In [10] Parsopoulos and Vrahitis observed that when they applied the *gbest* method (i.e., the swarm population only uses a single global best) to a multimodal function, the swarm moved back and forth, failing to decide where to land. This behavior is largely caused by particles getting equally good information from those equally good global optima. To overcome this problem, they introduced a method in which a potentially good solution is isolated once it is found (if its fitness is below a threshold value ϵ), then the fitness landscape is "stretched" to keep other particles away from this area of the search space. The isolated particle is checked to see if it is a global optimum, and if it is below the desired accuracy, a small population is generated around this particle to allow a finer search in this area. The main swarm continues its search for the rest of the search space for other potential global optima. With this modification, their PSO was able to locate all the global optima of the test functions successfully.

Brits, et al. proposed a NichePSO [2], which has a number of improvements to Parsopoulos and Vrahitis's model. In NichePSO, multiple subswarms are produced from a main swarm population to locate multiple optimal solutions in the search space. Subswarms can merge together, or absorb particles from the main swarm. Instead of using the threshold ϵ as in Parsopoulos and Vrahitis's model, NichePSO monitors the fitness of a particle by tracking its variance over a number of iterations. If there is little change in a particle's fitness over a number of iterations, a subswarm is created with the particle's closest neighbour. The authors used a swarm of population size of 20-30, and NichePSO found all global optima of the test functions used within 2000 iterations.

Li, et al. introduced a species conserving genetic algorithm (SCGA) for multimodal optimization [9]. SCGA adopted a new technique for dividing the population based on the notion of species, which is added to the evolution process of a conventional genetic algorithm. Their results on multimodal optimization have shown to be substantially better than those found in literature.

The notion of species is very appealing. To some extent, it provides a way of addressing the three limitations we identified with the clustering approach used in the PSO proposed by Kennedy [5]. This paper proposes a species-based PSO (SPSO) incorporating the idea of species into PSO for solving the multimodal optimization problems. At each iteration step, SPSO aims to identify multiple species (each for a potential optimum) within a population and then determine a neighbourhood best for each species. These multiple adaptively formed species are then used to optimize towards multiple optima in parallel, without interference across different species.

3   Particle Swarm

The particle swarm algorithm is an optimization technique inspired by the metaphor of social interaction observed among insects or animals. The kind of social interaction modeled within a PSO is used to guide a population of individuals (so called particles) moving towards the most promising area of the search space. In a PSO algorithm, each particle is a candidate solution equivalent to a point in a d-dimensional space, so the i-th particle can be represented as $\mathbf{x_i} = (x_{i1}, x_{i2}, \dots, x_{id})$. Each particle "flies" through the search space, depending on two important factors, $\mathbf{p_i} = (p_{i1}, p_{i2}, \dots, p_{id})$, the best position the current particle has found so far; and $\mathbf{p_g} = (p_{g1}, p_{g2}, \dots, p_{gd})$, the global best position identified from the entire population (or within a neighbourhood). The rate of position change of the i-th particle is given by its velocity $\mathbf{v_i} = (v_{i1}, v_{i2}, \dots, v_{id})$. Equation (1) updates the velocity for each particle in the next iteration step, whereas equation (2) updates each particle's position in the search space [6]:

$$v_{id}(t) = \chi(v_{id}(t-1) + \varphi_1(\boldsymbol{p}_{id} - \boldsymbol{x}_{id}(t-1)) + \varphi_2(\boldsymbol{p}_{gd} - \boldsymbol{x}_{id}(t-1))) \quad (1)$$
$$\boldsymbol{x}_{id}(t) = \boldsymbol{x}_{id}(t-1) + \boldsymbol{v}_{id}(t) , \quad (2)$$

where

$$\chi = \frac{2}{|2 - \varphi - \sqrt{\varphi^2 - 4\varphi}|} \quad \text{and} \quad \varphi = \varphi_1 + \varphi_2, \ \varphi > 4.0. \quad (3)$$

Two common approaches of choosing $\boldsymbol{p}_g$ are known as *gbest* and *lbest* methods. In the *gbest* approach, the position of each particle in the search space is influenced by the best-fit particle in the entire population; whereas the *lbest* approach only allows each particle to be influenced by a fitter particle chosen from its neighbourhood. Kennedy and Mendes studied PSOs with various population topologies [8], and have shown that certain population structures could give superior performance over certain optimization functions.

4   Identifying Species

Central to the proposed SPSO in this paper is the notion of species. Goldberg and Richardson proposed a niching method based on speciation by fitness sharing [4], where a GA population is classified into groups according to their similarity measured by Euclidean distance. The smaller the Euclidean distance between two individuals, the more similar they are:

$$d(x_i, x_j) = \sqrt{\sum_{k=1}^{n} (x_{ik} - x_{jk})^2}, \quad (4)$$

where $\mathbf{x_i} = (x_{i1}, x_{i2}, \ldots, x_{in})$ and $\mathbf{x_j} = (x_{j1}, x_{j2}, \ldots, x_{jn})$ are vectors of real numbers representing two individuals i and j from the GA population.

The definition of a species also depends on another parameter r_s, which denotes the radius measured in Euclidean distance from the center of a species to its boundary. The center of a species, so called species seed, is always the best-fit individual in the species. All particles that fall within the r_s distance from the species seed are classified as the same species.

4.1   Determining Species Seeds from the Population

The algorithm for determining species seeds introduced by Li et al. is adopted here [9]. By applying this algorithm at each iteration step, different species seeds can be identified for multiple species and then used as the *lbest* for different species accordingly. Fig.1 summarizes the steps for determining the species seeds.

The algorithm (as given in Fig. 1) for determining the species seeds is performed at each iteration step. The algorithm takes as an input, L_{sorted}, a list

input   : L_{sorted} - containing all particles sorted in decreasing order fitness
output : S - containing dominating particles identified as species seeds

begin
  $S = \Phi$;
  while *not reaching the end of L_{sorted}* **do**
      found $\leftarrow$ FALSE;
      for *all $p \in S$* **do**
          if $d(s,p) \leq r_s$ **then**
              found $\leftarrow$ TRUE;
              break;
          end
      end
      if *(not found)* **then**
          let $S \leftarrow S \cup \{S\}$
      end
  end
end

Fig. 1. The algorithm for determining the species seeds.

containing all particles sorted in decreasing order of fitness. The species seed set S is initially set to Φ . All particles are checked in turn (from best to the least-fit) against the species seeds found so far. If a particle does not fall within the radius r_s of all the seeds of S, then this particle will become a new seed and be added to S. Fig. 2 provides an example to illustrate the working of this algorithm. In this case, applying the algorithm will identify s_1, s_2 and s_3 as the species seeds. Note that since a species seed is the best-fit particle in a species, other particles within the same species can be made to follow the species seed as the newly identified neighbourhood best (*lbest*). This allows particles within the same species to be attracted to positions that make them even fitter. Because species are formed around different optima in parallel, making species seeds the new neighbourhood bests will provide the right guidance for particles in different species to locate multiple optima.

The complexity of the above procedure can be estimated based on the number of evaluations of Euclidean distances between two particles that are required. Assuming there are N individuals sorted and stored on L_{sorted}, the **while** loop steps through L_{sorted} to see if each individual is within the radius r_s of the seeds on S. If S currently contains i number of seeds, then at best the **for** loop is executed only once when the particle considered is within r_s of the first seed compared; and at worst the **for** loop is executed i times when the particle falls outside of r_s of all the seeds on S. Therefore the number of Euclidean distance calculations required for the above procedure $T(N)$ can be obtained by the following [9]:

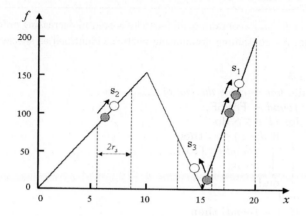

Fig. 2. An example of how to determine the species seeds from the population at each iteration step. s_1, s_2 and s_3 are chosen as the species seeds.

$$N \leq T(N) \leq \sum_{i=1}^{N}(i-1) = \frac{N(N-1)}{2}, \tag{5}$$

which gives the complexity of the procedure: $O(N^2)$.

5   The Species-Based PSO (SPSO)

Once the species seeds have been identified from the population, we can then allocate each seed to be the *lbest* to all the particles in the same species at each iteration step. The species-based PSO (SPSO) accommodating the above described algorithm for determining species seeds can be summarized in the following steps:

1. Generate an initial population with randomly generated particles;
2. Evaluate all particle individuals in the population;
3. Sort all particles in descending order of their fitness values (i.e., from the best-fit to least-fit ones);
4. Determine the species seeds for the current population (see Fig. 1);
5. Assign each species seed identified as the *lbest* to all individuals identified in the same species;
6. Adjusting particle positions according to equation (1) and (2);
7. Go back to step 2), unless the termination condition is met.

Considering the limitations of Kennedy's clustering-based PSO [5] (also discussed in section 2), SPSO improves in the following aspects:

1. SPSO only requires one iteration over all particles in the population in order to determine the species seeds, which are used as substitutes for neighbourhood bests (similar to the cluster centers in Kennedy's PSO).

2. In SPSO, an identified species seed is always the best-fit individual in that species.
3. There is no need to pre-specify the number of species seeds. They are automatically generated during a run.

6   Performance Measurements

The performance of SPSO in handling multimodal functions can be measured according to three criteria, *number of evaluations* required to locate the optima; *accuracy*, measuring the closeness to the optima, and *success rate*, i.e., the percentage of runs in which all global optima are successfully located.

To measure accuracy, we only need to check set S, which contains the species seeds identified so far. These species seeds are dominating individuals sufficiently different from each other, however they could be individuals with high as well as low fitness values (see Fig. 2). We can decide if a global optimum is found by checking each species seed in S to see if it is close enough to the known global optima (for all the test functions used in this study). A solution acceptance threshold ($0 < \epsilon \leq 1$) is defined to detect if the solution is close enough to a global optimum:

$$|f_{max} - f(x)| \leq \epsilon \tag{6}$$

where f_{max} is the known maximal (highest) fitness value for a test function (assuming maximization problems). If the number of global optima is greater than one, then all global optima will be checked for the required accuracy using equation (6) before a run is terminated.

7   Test Functions

The five test functions suggested by Beasley et al. [1] and the Rastrigin function (with different dimensions) were used to test SPSO's ability to locate a single or multiple maxima:

$$F1(x) = sin^6(5\pi x). \tag{7}$$

$$F2(x) = exp\left(-2log(2) \cdot \left(\frac{x - 0.1}{0.8}\right)^2\right) \cdot sin^6(5\pi x). \tag{8}$$

$$F3(x) = sin^6(5\pi(x^{3/4} - 0.05)). \tag{9}$$

$$F4(x) = exp\left(-2log(2) \cdot \left(\frac{x - 0.08}{0.854}\right)^2\right) \cdot sin^6(5\pi(x^{3/4} - 0.05)). \tag{10}$$

$$F5(x, y) = 200 - (x^2 + y - 11)^2 - (x + y^2 - 7)^2.$$ (11)

$$F6(\boldsymbol{x}) = \sum_{i=1}^{n} (x_i^2 - 10cos(2\pi x_i) + 10).$$ (12)

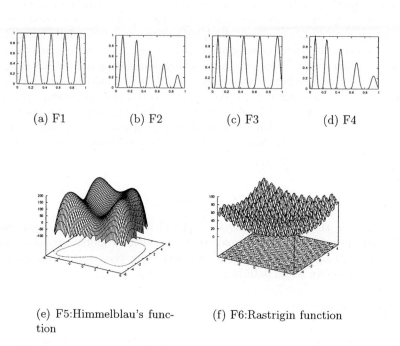

(a) F1   (b) F2   (c) F3   (d) F4

(e) F5:Himmelblau's func-   (f) F6:Rastrigin function
tion

Fig. 3. Test functions.

As shown in Fig. 3, F1 has 5 evenly spaced maxima with a function value of 1.0. F2 has 5 peaks decreasing exponentially in height, with only one peak as the global maximum. F3 and F4 are similar to F1 and F2 but the peaks are unevenly spaced. F5 Himmelblau's function has two variables x and y, where $-6 \leq x, y \leq +6$. This function has 4 global maxima at approximately (3.58,-1.86), (3.0,2.0), (-2.815,3.125), and (-3.78,-3.28). F6 Rastrigin function, where $-5.12 \leq x_i \leq 5.12, i = 1, \ldots, 30$, has one global minimum (which is (0,0) for dimension=2), and many local minima.[1] F6 with a dimension of 2, 3, 4, 5 and 6 variables were used to test SPSO's ability in dealing with functions with numerous local minima and of higher dimensions.

---

[1] Rastrigin function can be easily converted to a maximization function.

Table 1. Summary of performance results (averaged over 30 runs).

Function	Num. of global optima	ϵ	r_s	Num. of evals. (mean and std dev)	Success rate
F1	5	0.0001	0.05	1383.33 ± 242.95	100%
F2	1	0.0001	0.05	351.67 ± 202.35	100%
F3	5	0.0001	0.05	1248.33 ± 318.80	100%
F4	1	0.0001	0.05	503.33 ± 280.07	100%
F5	4	0.0001	2.0	3155 ± 402.22	100%

Table 2. Comparison of results on F1 and F5.

Function	Num. of global optima	Algorithm	Num. of evals. required	Success rate
F1	5	Sequential Niched GA (SNGA)	1900	99%
		Species Conservation GA (SCGA)	3310	100%
		SPSO	**1383.33**	100%
F5	4	Sequential Niched GA (SNGA)	5500	76%
		SPSO	**3155**	100%

8   Experimental Setups

A swarm population size of 50 was used for all the above test functions. SPSO was run 30 times, each run with a maximum of 1000 iteration steps. The accuracy threshold ϵ was set to 0.0001. A run is terminated if either the required accuracy for all the global optima or the maximum of 1000 iteration steps is reached. r_s was set normally to a value between 1/20 to 1/10 of the allowed variable range. Success rate is measured by the percentage of runs (out of 30) locating all the global optima within the 1000 iteration steps. The number of function evaluations required for finding all the global optima are averaged over 30 runs. Table 1 provides a summary of the results.

For PSO parameters in equation (1) and (2), φ_1 and φ_2 were both set to 2.05. The constriction factor χ was set to 0.729844 [8]. Using this χ value produces a damping effect on the amplitude of an individual particle's oscillations, and as a result, the particle will converge over time. V_{max} was set to be the lower and upper bounds of the allowed variable ranges.

9   Discussion of the Results

As shown in Table 1, for F1 - F5, SPSO has converged to the required accuracy of 0.0001 with 100% success rate. SPSO found all the global optima in all runs with less than 1000 iteration steps. In comparison, NichePSO [2] only obtained similar accuracy values on F1, F3 and F5 after 2000 iterations. Furthermore, on F2 and F4, SPSO got better accuracy values than the NichePSO.

Table 2 shows that SPSO has the best results comparing with the results of SNGA proposed by Beasley, et al. [1] and SCGA proposed by Li, et al. [9] on F1 (equal maxima function) and F5 (Himmelblau's function).

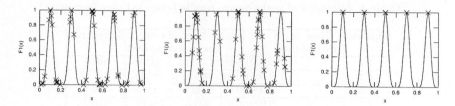

Fig. 4. A simulation run on F1(equal maxima), step 1, 4 and 74 from left to right.

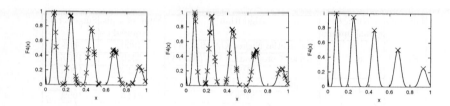

Fig. 5. A simulation run of SPSO on F4(uneven decreasing maxima) - step 1, 6 and 116 from left to right.

Fig. 4 shows that on F1, SPSO was able to locate all maxima, that is, at iteration step 74, all particles were able to converge to all 5 maxima. Fig. 5 shows that on F4, SPSO always found the highest peak first, then was also able to locate all other lower peaks in later iteration steps successfully, regardless of if they are global maxima or local optima. The results on F2 and F3 are similar to those of F1 and F4. Fig. 6 shows that on F5, many species seeds (based on the r_s value) were identified by SPSO initially as expected. Over the following iteration steps, these species were merged to form 4 groups around the 4 maxima. Eventually almost all particles converged to these 4 maxima at step 66.

Table 3 shows the results of SPSO on the Rastrigin function with dimension varying from 2 to 6. In this experiment the same parameter settings were used as the previous ones. It is interesting to note that on the Rastrigin function SPSO has increasing difficulty to converge to the required accuracy as the dimension is increased from 3 to 6. This is expected, as SPSO is designed to encourage forming species depending on the local feedback on the fitness landscape. The higher dimension and the presence of a large number of local minima of the Rastrigin function would demand SPSO to have a larger initial population in order to locate the global minimum. Further investigation on this will be carried out in future.

10   Conclusion

By using the concept of species, we have developed a PSO which allows the swarm population to be divided into different species adaptively, depending on the feedback obtained from the fitness landscape discovered at each iteration

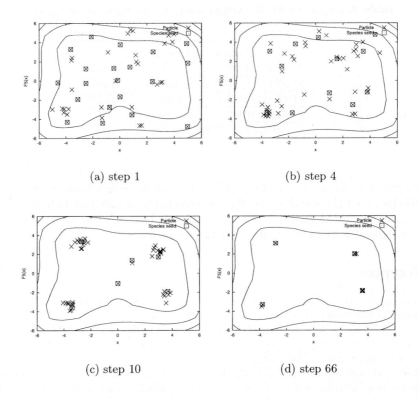

(a) step 1   (b) step 4

(c) step 10   (d) step 66

Fig. 6. A simulation run of SPSO on F5 - step 1, 4, 10 and 66.

step during a run. Particles from each identified species follow a chosen neighbourhood best to move towards a promising region of the search space. Multiple species are able to converge towards different optima in parallel, without interference across different species. In a classic GA algorithm, crossover carried out over two randomly chosen fit individuals often produces a very poor offspring (imagining the offspring are somewhere between two fitter individuals from two distant peaks). In contrast, SPSO seems to be able to alleviate this problem effectively.

Tests on a suite of widely used multimodal test functions have shown that SPSO can find all the global optima for the all test functions with one or two dimensions reliably (with 100% success rate), and with good accuracy (< 0.0001), although SPSO seemed to show increasing difficulty to converge as the dimension of the Rastrigin function was increased to more than three. Comparison of SPSO's results with other published works has demonstrated that SPSO is comparable or better than the existing evolutionary algorithms as well as another niche-based PSO for handling multimodal function optimization, not only with regard to success rate and accuracy, but also on computational cost. In future, we will apply SPSO to large and more complex real-world multimodal

Table 3. Results on **F6 Rastrigin** function (averaged over 30 runs).

Dimension	ϵ	r_s	Num. of evals. (mean and std dev)	Success rate
2	0.0001	2.0	3711.67 ± 911.87	100%
3	0.0001	2.0	9766.67 ± 4434.86	100%
4	0.0001	2.0	36606.67 ± 14662.38	33.3%
5	0.0001	2.0	44001.67 ± 10859.84	26.7%
6	0.0001	2.0	50000 ± 0.00	0%

optimization problems, especially problems that we have only little (or no) prior knowledge about the search space. We also need to investigate how to best choose the species radius, for example perhaps looking at how to adaptively choose the species radius based on the feedback obtained during the search.

References

1. Beasley, D., Bull, D.R., and Martin, R.R.: A Sequential Niche Technique for Multimodal Function Optimization. Evolutionary Computation, 1(2): 101–125 (1993)
2. Brits, R., Engelbrecht, A. P., and van den Bergh, F.: A Niching Particle Swarm Optimizer. In: Proceedings of the 4th Asia-Pacific Conference on Simulated Evolution and Learning 2002 (SEAL 2002) 692–696
3. Clerc, M. and Kennedy, J.: The Particle Swarm-Explosion, Stability, and Convergence in a Multidimensional Complex Space. In: IEEE Transactions on Evolutionary Computation, vol. 6, no. 1, (2002) 58–73
4. Goldberg, D.E. and Richardson, J.: Genetic algorithms with sharing for multimodal function optimzation. In: Grefenstette, J.J. (eds.): Genetic Algorithms and their Applications: Proceedings of the Second International Conference on Genetic Algorithms. Lawrence Earlbaum, Hillsidale, New Jersey (1987) 41–49
5. Kennedy, J.: Stereotyping: Improving Particle Swarm Performance with Cluster Analysis. In: Proceedings of the 2000 Congress on Evolutionary Computation. Piscataway, NJ: IEEE Service Center (2000) 1507–1512
6. Kennedy, J. and Eberhart, R.: Swarm Intelligence. Morgan Kaufmann Academic Press (2001)
7. Kennedy, J. and Eberhart, R.C.: Particle Swarm Optimization. In: Proceedings of the IEEE International Conference on Neural Networks, IV. Piscataway, NJ: IEEE Service Center (1995)1942–9148
8. Kennedy, J. and Mendes, R.: Population Structure and Particle Swarm Performance. In: Proceedings of the 2002 Congress on Evolutionary Computation. Piscatawat, NJ: IEEE service Center (2002) 1671–1675
9. Li, J.P., Balazs, M.E., Parks, G. and Clarkson, P.J.: A Species Conserving Genetic Algorithm for Multimodal Function Optimization. Evolutionary Computation 10(3): 207–234 (2002)
10. Parsopoulos, K. E. and Vrahatis, M. N.: Modification of the particle swarm optimizer for locating all the global minima. In: Proceedings of the International Conference on Artificial Neural Networks and Genetic Algorithms (ICANNGA 2001), Prague, Czech Republic. (2001) 324–327

Better Spread and Convergence: Particle Swarm Multiobjective Optimization Using the Maximin Fitness Function

Xiaodong Li

School of Computer Science and Information Technology
RMIT University, VIC 3001, Melbourne, Australia
xiaodong@cs.rmit.edu.au
http://www.cs.rmit.edu.au/~xiaodong

Abstract. Maximin strategy has its origin in game theory, but it can be adopted for effective multiobjective optimization. This paper proposes a particle swarm multiobjective optimiser, *maximinPSO*, which uses a fitness function derived from the maximin strategy to determine Pareto-domination. The maximin fitness function has some very desirable properties with regard to multiobjective optimization. One advantage is that no additional clustering or niching technique is needed, since the maximin fitness of a solution can tell us not only if a solution is dominated or not (with respect to the rest of the population), but also if it is clustered with other solutions, i.e., diversity information. This paper demonstrates that on the ZDT test function series, *maximinPSO* produces an almost perfect convergence and spread of solutions towards and along the Pareto-optimal front respectively, outperforming one of the state-of-art multiobjective EA algorithms, NSGA II, in all the performance measures used.

1 Introduction

Particle Swarm Optimization (PSO) has become increasingly popular as an efficient optimization method for single objective optimization, and more recently it has shown promising results for solving multiobjective optimization problems [3,4,5,6,7]. PSO is an optimization technique inspired by studies of the social behaviour of insects and animals [1][2]. The social behaviour is modelled in a PSO to guide a population of particles (or potential solutions) moving towards the most promising region of the search space. In PSO, each particle represents a candidate solution, $\mathbf{x_i} = (x_{i1}, x_{i2}, \ldots, x_{iD})$. D is the dimension of the search space. The i-th particle of the swarm population knows: **a)** its personal best position $\mathbf{p_i} = (p_{i1}, p_{i2}, \ldots, p_{iD})$, i.e., the best position this particle has visited so far that yields the highest fitness value; and **b)** the global best position, $\mathbf{p_g} = (p_{g1}, p_{g2}, \ldots, p_{gD})$, i.e., the position of the best particle that gives the best fitness value in the entire population; and **c)** its current velocity, $\mathbf{v_i} = (v_{i1}, v_{i2}, \ldots, v_{iD})$, which represents its position change. The following equation (1) uses the above information to calculate the new updated velocity

K. Deb et al. (Eds.): GECCO 2004, LNCS 3102, pp. 117–128, 2004.
© Springer-Verlag Berlin Heidelberg 2004

for each particle in the next iteration step. Equation (2) updates each particle's position in the search space.

$$v_{id} = wv_{id} + c_1 r_1 (p_{id} - x_{id}) + c_2 r_2 (p_{gd} - x_{id}) \tag{1}$$
$$x_{id} = x_{id} + v_{id} , \tag{2}$$

where $d = 1, 2, \ldots, D$; $i = 1, 2, \ldots, N$; N is the size of the swarm population; w is the inertia weight, which is often used as a parameter to control exploration/exploitation in the search space; c_1 and c_2 are two coefficients (positive constants); r_1 and r_2 are two random numbers within the range $[0, 1]$. There is also a V_{MAX}, which sets the upper and lower bound for velocity values.

Recently Balling in [8] proposed a very interesting multi-objective optimization technique based on fitness derived from using the maximin strategy [9]. In sharp contrast to almost all other existing multi-objective algorithms, Balling demonstrated that by using the maximin fitness, there is no need to use any additional niching technique, since using the maximin fitness by itself penalizes clustering of solutions.

In this paper, *maximinPSO*, a PSO model using the maximin fitness is proposed for multi-objective optimization. *maximinPSO* adopts a similar approach as NSPSO[7], except that it uses the maximin fitness to rank individuals in the population (rather than the non-dominated sorting procedure), and there is no niching method used. The paper is organized as follows: Section 2 first introduces the concept of dominance, from which the maximin fitness function is derived. Section 3 defines the maximin fitness function and describes its key properties in relation to the proposed *maximinPSO* algorithm. Section 4 introduces the *maximinPSO* algorithm formally. Section 5 presents test functions, performance measures, as well as the results and analysis of experiments carried out with the *maximinPSO* over the test functions. Finally Section 6 concludes the paper.

2 Multiobjective Optimization and the Notion of Dominance

Assuming minimization, multi-objective optimization strives to simultaneously minimize m objectives:

$$\textbf{Minimize } \mathbf{y} = f(\mathbf{x}) = (f_1(\mathbf{x}), f_2(\mathbf{x}), \ldots, f_m(\mathbf{x})) \tag{3}$$

where $\mathbf{x}$ is a n-dimensional decision variable vector, $\mathbf{x} = (x_1, \ldots, x_n) \in X$, and $\mathbf{y} = (y_1, \ldots, y_n) \in Y$. X is the decision variable space, whereas Y is the objective space. Each objective depends on the decision vector $\mathbf{x}$. A decision vector $\mathbf{u} \in X$ is said to strictly dominate another decision vector $\mathbf{v} \in X$ (denoted by $\mathbf{u} \prec \mathbf{v}$) if and only if

$$\forall i \in \{1, \ldots, m\} : f_i(\mathbf{u}) \leq f_i(\mathbf{v}) \quad \text{and} \quad \exists j \in \{1, \ldots, m\} : f_j(\mathbf{u}) < f_j(\mathbf{v}) \quad (4)$$

$\mathbf{u}$ weakly dominates $\mathbf{v}$ (denoted by $\mathbf{u} \preceq \mathbf{v}$) if and only if

$$\forall i \in \{1, \ldots, m\} : f_i(\mathbf{u}) \leq f_i(\mathbf{v}) \quad (5)$$

A decision vector $\mathbf{x} \in X$ is said to be Pareto-optimal with respect to X if and only if there is no other decision vector in X that dominates $\mathbf{x}$.

The set of all Pareto-optimal solutions in the decision variable space is called the *Pareto-optimal* set. The corresponding set of objective vectors is called the *Pareto-optimal front*. In this paper, for clarity, we denote the Pareto-optimal front as P^*, and the set of non-dominated solutions found as Q.

3   The Maximin Fitness Function

Maximin strategy has its origin in game theory [9]. Rawls in [10] used a nice example of the maximin strategy to illustrate his theory on principles of justice. Balling was the first to propose the use of the maximin fitness function for multiobjective optimization [8]. The maximin fitness for a decision vector $\mathbf{u}$ can be calculated through the following steps. First the **min** function is called to obtain the minimal value from set $\{f_i(\mathbf{u}) - f_i(\mathbf{v}) \mid \forall i \in \{1, \ldots, m\}$:

$$\mathbf{min}_{i=1,\ldots,m}\{f_i(\mathbf{u}) - f_i(\mathbf{v})\} \quad (6)$$

Then the **max** function is applied over the set of minimal values of all possible pairs of $\mathbf{u}$ and another decision vector (other than $\mathbf{u}$) in the population:

$$\mathbf{max}_{j=1,\ldots,N;u \neq v}\{\mathbf{min}_{i=1,\ldots,m}\{f_i(\mathbf{u}) - f_i(\mathbf{v})\}\} \quad (7)$$

In equation (7) two loops of comparison take place, with the **min** first stepping through all the objectives from 1 to m, and then the **max** looping through all candidate solutions in the population from 1 to N, except $\mathbf{u}$. To obtain all non-dominated solutions, another loop will be required to check each solution in the population, from 1 to N. As a result, the overall complexity is $O(mN^2)$.

The maximin fitness value for the decision vector $\mathbf{u}$ is defined as [8]:

$$f_{maximin} = \mathbf{max}_{j=1,\ldots,N;u \neq v}\{\mathbf{min}_{i=1,\ldots,m}\{f_i(\mathbf{u}) - f_i(\mathbf{v})\}\} \quad (8)$$

Given equation (8), it is obvious that for any solution (i.e., a decision vector) to be a non-dominated solution with respect to the current population, its maximin fitness value must be less than zero. Any solution with a maximin fitness equal to zero is a weakly-dominated solution. Any solution with a maximin fitness value greater than zero is a dominated solution.

One unique property that makes the maximin fitness function so appealing to multiobjective optimization is that the maximin fitness value can be used to reward diversity and penalize clustering of non-dominated solutions, therefore no additional diversity maintaining mechanism such as a niching technique is necessary. This can be illustrated by the two examples as shown in Fig. 1 [8]. Fig. 1 a) shows the maximin fitness values calculated for solution **A**, **B** and **C** respectively. Since the three solutions are non-dominated with each other, the maximin fitness values are negative (written in parentheses). Fig. 1 a) also shows that the maximin fitness is the same for the three equally-spaced solutions, however, Fig. 1 b) shows that the two closely-spaced solutions **B** and **C** are penalized by having a higher fitness than **A**. With the assumption of minimization, the smaller the fitness value is, the better the solution, so **A**(-1.5) is rewarded by getting an even smaller fitness than **A**(-1) in Fig. 1 a), whereas **B**(-0.5) and **C**(-0.5) are penalized by getting a higher fitness than **B**(-1) and **C**(-1) in Fig. 1 a). In the case when **B** and **C** are completely overlapped, the maximin fitness will be zero for both **B** and **C**.

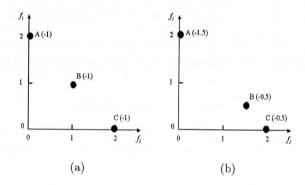

(a)   (b)

Fig. 1. a) Three non-dominated solutions with an equal fitness value; b) three non-dominated solutions with higher fitness values assigned to solutions that are close to each other.

Balling [8] also showed that maximin fitness favors the middle of a convex front, and the two extreme solutions of a concave front (as shown in Fig. 2). Fig. 2 might give you the impression that using maximin fitness will result in more solutions clustering in the middle of a convex front or two extreme ends of a concave front. This is not necessarily true when there is a sufficiently large number of solutions along the front, because maximin fitness works against clustering of the solutions. In fact, maximin fitness works against clustering of solutions regardless of if the front is convex or concave, as long as there are sufficient numbers of solutions along the front.

For a more detailed description of the properties of the maximin fitness function, the reader can be referred to [8].

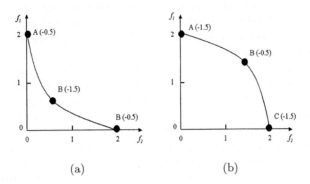

(a)   (b)

Fig. 2. a) The maximin fitness favors the middle of a convex front, whereas in b) the maximin fitness favors the two extreme solutions of a concave front.

Choosing p_g for each particle. Our objective is to propel particles in the population towards the current non-dominated front Q as well as the less crowded areas along Q. For each particle, we have decided to choose randomly its p_{gd} for each dimension $d = 1, \dots, D$ of the particle, from a pool of non-dominated particles with the smallest maximin fitness values (they should be negative too). Fig. 3 illustrates how this works. Note that this method allows the p_g of a particle to be composed of different p_{gd} from different non-dominated particles. This will have the effect of emphasizing the less crowded areas as a whole on the best known non-dominated front over iterations.

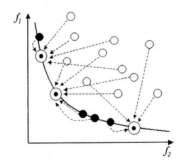

Fig. 3. A few "less crowed" non-dominated particles are used as a pool for choosing at random a p_{gd} (for each dimension $d = 1, \dots, D$) for a particle in the population.

4   The *maximinPSO* Algorithm

This paper proposes a PSO, *maximinPSO*, which makes use of the maximin fitnesses (according to equation (8)) of individuals in a swarm to facilitate dominance comparison and diversity maintenance for the purpose of multiobjective

(11). The number of evaluations taken for the run was also recorded. A set of $|P^*| = 500$ uniformly distributed Pareto-optimal solutions was used to calculate $\mathcal{M}_1^*$. For $\mathcal{M}_2^*$, the niche neighbourhood size σ^* was set to 0.01.

The results of $maximinPSO$ were compared with that of the real-coded NSGA II [12]. As in the $maximinPSO$, an initial population of 200 was used. NSGA II was run for 100 generations so it takes 20000 evaluations for each run, as NSGA II uses a constant population size. In contrast, $maximinPSO$ uses a population size that grows over time, which allows more non-dominated solutions to be discovered with relatively fewer evaluations. For NSGA II, we used the same parameter values as suggested by Deb et al. [13]. A crossover probability of 0.9 and a mutation probability of $1/n$ (n is the number of real-variables) were used. The SBX and real-parameter mutation operators, η_c and η_m were set to 20 respectively.

Both $maximinPSO$ and the real-coded NSGA II were run 30 times. The results were averaged and summarized in Table 1.

Table 1. $\mathcal{M}_1^*$, $\mathcal{M}_2^*$, $\mathcal{M}_3^*$, and the number of evaluations (averaged over 30 runs).

Metric	Algorithm	ZDT1	ZDT2	ZDT3	ZDT4	ZDT6
$\mathcal{M}_1^*$	$maximinPSO$	7.74E-04 ±1.72E-05	1.01E-02 ±5.13E-02	3.44E-03 ±1.09E-04	7.68E-04 ±1.50E-05	1.84E-03 ±6.06E-04
	real-coded NSGA II	1.14E-03 ±5.56E-05	8.26E-04 ±3.44E-05	4.90E-03 ±1.45E-04	5.77E-02 ±1.09E-01	1.04E-01 ±1.02E-02
$\mathcal{M}_2^*$	$maximinPSO$	2.65E+03 ±3.89E+02	2.51E+03 ±8.12E+02	2.15E+03 ±1.25E+02	2.59E+03 ±3.23E+02	2.35E+03 ±3.22E+02
	real-coded NSGA II	1.96E+02 ±6.62E-02	1.96E+02 ±5.90E-02	1.97E+02 ±1.60E-01	1.69E+02 ±5.50E+01	1.95E+02 ±2.74E-01
$\mathcal{M}_3^*$	$maximinPSO$	1.40E+00 ±9.75E-03	1.31E+00 ±3.58E-01	1.96E+00 ±5.26E-03	1.40E+00 ±7.13E-03	1.17E+00 ±3.22E-05
	real-coded NSGA II	1.41E+00 ±3.60E-03	1.41E+00 ±3.69E-03	1.93E+00 ±9.43E-02	1.29E+00 ±2.33E-01	1.13E+00 ±2.67E-02
Num. of Evals.	$maximinPSO$	5.56E+03 ±5.45E+02	5.65E+03 ±1.43E+03	1.13E+04 ±1.17E+03	5.26E+03 ±5.02E+02	5.30E+03 ±5.45E+02
	real-coded NSGA II	2.00E+04 ±0.00E+00	2.00E+04 ±0.00E+00	2.00E+04 ±0.00E+00	2.00E+04 ±0.00E+00	2.00E+04 ±0.00E+00

5.4   Discussion

From Table 1, comparing with the real-coded NSGA II, $maximinPSO$ consistently performed better on all test functions, except ZDT2. $maximinPSO$ outperformed NSGA II on all performance measures for ZDT3, ZDT4, and ZDT6. More specifically, $maximinPSO$ consistently converged better ($\mathcal{M}_1^*$), distributed solutions better along the non-dominated front ($\mathcal{M}_2^*$), had a better coverage ($\mathcal{M}_3^*$), and finally used fewer evaluations. For ZDT1, $maximinPSO$ outperformed NSGA II on all metrics except $\mathcal{M}_3^*$ (but just slightly). $maximinPSO$ has no difficulty in handling convex (ZDT1) and disconnected Pareto-fronts (ZDT3).

Handling non-convexity of Pareto-front. For ZDT2, we examined all 30 *maximinPSO* runs, and identified 3 very poor runs, which skewed the results for ZDT2 in the Table 1. Out of these 3 very poor runs, in two of which *maximinPSO* converged to a single end point of the Pareto-front, and in the 3rd run, *maximinPSO* only converged to a local front, but the spread of the solutions is still good. If the 3 poor runs were taken out, the remaining 27 runs are in fact equally good or better than NSGA II. This problem of converging to a single solution could be attributed to the fact that there is a higher probability that *maximinPSO* discovered non-dominated solutions on the end points of a concave Pareto-front too early. These non-dominated solutions around the end points subsequently attracted all other particles rather quickly before they even had a chance of going to other parts of the Pareto-front. To combat this problem, we increased the initial population size from 200 to 400 in order to encourage more particles to reach other parts of the Pareto-front. As expected, subsequently this problem was eliminated. The results of *maximinPSO* on ZDT2 using a population size of 400 are provided in Table 2 (averaged over 30 runs). Once again Table 2 shows that *maximinPSO* outperformed NSGA II on all performance metrics. It is also interesting to note that the number of evaluations used to get 2000 or more non-dominated solutions is not much greater than the *maximinPSO* with an initial population size of 200 (the last row in Table 1).

Table 2. Improved results of *maximinPSO* on ZDT2.

Algorithm	$\mathcal{M}^*1$	$\mathcal{M}^*2$	$\mathcal{M}^*3$	Num. of Evals.
maximinPSO	7.87E-04 ±9.70E-06	2.72E+03 ±4.55E+02	1.41E+00 ±1.46E-03	6.86E+03 ±6.58E+02
real-coded NSGA II	8.26E-04 ±3.44E-05	1.96E+02 ±5.90E-02	1.41E+00 ±3.69E-03	2.00E+04 ±0.00E+00

Fig. 4 shows the non-dominated solutions found for ZDT1, ZDT2, ZDT3, ZDT4 and ZDT6 in the final iteration step of a simulation run of *maximinPSO*. We achieved almost a perfect convergence, spread and coverage towards P^*.

Handling multiple local fronts. For ZDT4, where there is a large number of local fronts, *maximinPSO* was able to converge to P^* consistently, 30 out of 30 runs, while still maintaining a good spread and coverage of P^*. In contrast, all 30 NSGA II runs converged to a local front (see Fig. 4 d). These NSGA II runs also produced fewer non-dominated solutions, a poorer spread, and required a larger number of evaluations. Fig. 5 shows a few snapshots of a single *maximinPSO* run on ZDT4.

Handling non-uniform density of solutions. *maximinPSO* had no difficulty with ZDT6, where there is a non-uniform distribution of solutions (in decision variable space) corresponding to the Pareto-optimal front. *maximinPSO* was able to find a set of non-dominated solutions that corresponds to a set of smoothly distributed objective vectors in the objective space. In comparison, all

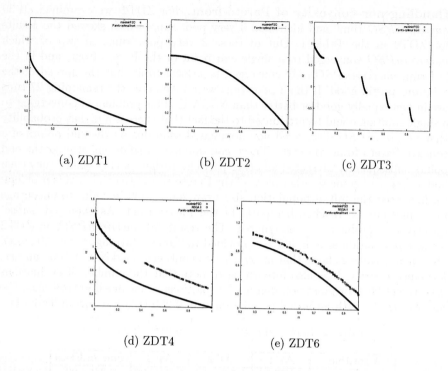

(a) ZDT1   (b) ZDT2   (c) ZDT3

(d) ZDT4   (e) ZDT6

Fig. 4. Non-dominated solutions found by *maximinPSO*. On ZTD4 and ZDT6, the results are also compared with solutions found by the real-coded NSGA II.

30 NSGA II runs only managed to converge to a local front, with a slightly worse spread (Fig. 4 e).

Making use of the discovered non-dominated solutions. One unique feature of *maximinPSO* is its ability to make use of the non-dominated solutions found so far to allow further improvement of the spread of solutions in future iteration steps. Since maximin fitnesses discourage clustering, those solutions appearing near the gaps in the current step would have lower maximin fitness values (see Fig. 6 step 8), hence more likely to be chosen to construct a new $\mathbf{p_g}$. As a result, other particles in the swarm will be more likely to be attracted to fill these gaps. Fig. 6 shows that during a run on ZDT6, *maximinPSO* was able to fill the gaps appearing on the distribution curve of the non-dominated solutions found over a number of iteration steps.

The importance of a larger population size. Population size plays a critical role in the performance of *maximinPSO*. Small population sizes do not work well when using *maximinPSO*. Especially for a problem with a concave front, the end points could become too dominated early on, hence attracting the rest of population to a single point. To avoid such problems, a reasonably large

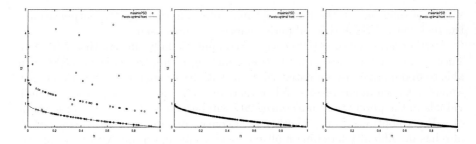

Fig. 5. Snapshots of a $maximinPSO$ run on ZDT4, showing all particles at step 8, 10 and 13 (from left to right).

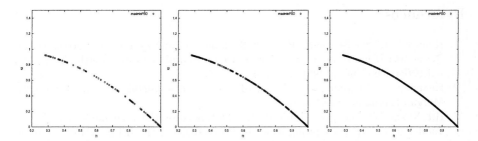

Fig. 6. On ZDT6, gaps among the distribution of solutions were filled over iteration step 8, 10 and 12 (from left to right).

population size is necessary to allow better sampling of the search space, thereby preventing certain individuals from becoming too dominated at the early stage of a run. As shown in the results on ZDT2 (Table 2), an initial population of 400 were needed to obtain a good convergence and spread of solutions consistently.

It may appear that using a larger population size would increase the amount of function evaluations, but for $maximinPSO$, at each iteration step we only evaluate the offspring produced. Parents are not evaluated again before being stored in $nextPopList$ for the next iteration step (see step 3 of the $maximinPSO$ algorithm). $maximinPSO$ generally converges very quickly, most of the time less than 20 iteration steps for all runs reported in Table 1, which is also why it used fewer function evaluations than the NSGA II counterpart. However, in order to determine if each particle is non-dominated with respect to the rest of population, the maximin fitnesses have to be calculated for all particles in $nextPopList$.

6   Conclusion

This paper has described a PSO, $maximinPSO$, using the maximin fitness function for multiobjective optimization. Our experiments on the ZDT test function series have shown that $maximinPSO$ is able to produce a convergence and spread of solutions on the Pareto-optimal front almost perfectly, outperforming the real-coded NSGA II in all performance measures used.

Furthermore, $maximinPSO$ is more computationally efficient than NSGA II, since it does not require any additional niching technique. $maximinPSO$ was able to find more non-dominated solutions with fewer numbers of function evaluations as shown in our results. When there are sufficiently large numbers of individuals in the population, $maximinPSO$ will become less sensitive to the shape of a Pareto-optimal front, whether it is convex or non-convex. $maximinPSO$ also has no difficulty handling multiple local fronts and fronts with non-uniform solutions. In future, it would be interesting to see how $maximinPSO$ performs on problems with constraints and those with more than two objectives.

References

1. Kennedy, J. and Eberhart, R.: Particle Swarm Optimization. In Proceedings of the Fourth IEEE International Conference on Neural Networks, Perth, Australia. IEEE Service Center(1995) 1942-1948.
2. Kennedy, J., Eberhart, R. C., and Shi, Y., *Swarm intelligence*, San Francisco: Morgan Kaufmann Publishers, 2001.
3. Hu, X. and Eberhart, R.: Multiobjective Optimization Using Dynamic Neighbourhood Particle Swarm Optimization. In Proceedings of the IEEE World Congress on Computational Intelligence, Hawaii, May 12-17, 2002. IEEE Press (2002).
4. Parsopoulos, K.E. and Vrahatis, M.N.: Particle Swarm Optimization Method in Multiobjective Problems, in Proceedings of the 2002 ACM Symposium on Applied Computing (SAC'2002) (2002) 603-607.
5. Fieldsend, E. and Singh, S.: A Multi-Objective Algorithm based upon Particle Swarm Optimisation, an Efficient Data Structure and Turbulence, Proceedings of the 2002 U.K. Workshop on Computational Intelligence, Birmingham, UK(2002) 37-44.
6. Coello, C.A.C. and Lechuga, M.S.: MOPSO: A Proposal for Multiple Objective Particle Swarm Optimization, in Proceedings of Congress on Evolutionary Computation (CEC'2002), Vol. 2, IEEE Press (2002) 1051-1056.
7. Li, X.: A Non-dominated Sorting Particle Swarm Optimizer for Multiobjective Optimization, in Erick Cantú-Paz et al. (editors), Genetic and Evolutionary Computation - GECCO 2003. Proceedings, Part I, Springer, Lecture Notes in Computer Science Vol. 2723, (2003) 37-48,.
8. Balling, R.: The Maximin Fitness Function; Multiobjective City and Regional Planning. In Proceedings of EMO 2003, (2003) 1-15.
9. Luce, R. D. and Raiffa, H.: Games and Decisions - Introduction and Critical Survey. John Wiley & Sons, Inc, New York (1957).
10. Rawls, J.: A Theory of Justice. Oxford University Press, London (1971).
11. Zitzler, E., Deb, K. and Thiele, L.: Comparison of multiobjective evolutionary algorithms: Empirical results. *Evolutionary Computation*, 8(2):173-195, April (2000).
12. Deb, K.: Multi-Objective Optimization using Evolutionary Algorithms, John Wiley & Sons, Chichester, UK (2001).
13. Deb, K., Agrawal,S., Pratap, A. and Meyarivan, T.: A fast and elitist multiobjective genetic algorithm: NSGA-II. IEEE Transactions on Evolutionary Computation 6(2): 182-197 (2002).

Evolving a Self-Repairing, Self-Regulating, French Flag Organism

Julian Francis Miller

Department of Electronics, University of York, Heslington, York, UK, YO10 5DD
jfm@ohm.york.ac.uk
http://www.elec.york.ac.uk/intsys/users/jfm7/

Abstract. A method for evolving programs that construct multicellular structures (organisms) is described. The paper concentrates on the difficult problem of evolving a cell program that constructs a fixed size French flag. We obtain and analyze an organism that shows a remarkable ability to repair itself when subjected to severe damage. Its behaviour resembles the regenerative power of some living organisms.

1 Introduction

The development of a fully formed adult from the zygote has to rank as one of the most remarkable feats of molecular engineering in the universe. From a single set of instructions inside one cell, an organism can grow to contain 10^{13} cells (for a human being) containing many hundreds of specialized cells performing distinct functions. How does nature achieve this feat of engineering? In his book biologist Frank M. Harold explains [9]:

> "Genes specify the cell's building blocks; they supply raw materials, help regulate their availability and grant the cell independence of its environment. But the higher levels of order, form and function are not spelled out in the genome. They arise by the collective self-organization of genetically determined elements, affected by cellular mechanisms that remain poorly understood."

The genotype-phenotype mapping employed by nature is highly complex and many-to-one. Despite this, in many branches of evolutionary algorithms genetic representations make no distinction between genotype and phenotype. This is a drawback if one is interested in problems that involve phenotypes of arbitrary size and complexity [1]. If higher level organisms were really colonies of cells with different genotypes it would have been much harder for evolution to evolve organisms of the complexity and sophistication of many living creatures. The poor scalability of directly encoded systems (i.e. a one-to-one mapping from genotype to phenotype) is particularly evident in the evolution of neural networks, where each link requires a floating-point weight that must be determined. The work presented in this paper builds on the author's previous work in evolving multicellular organisms [13]. This paper is devoted to a particularly difficult but interesting problem, that of the growth and regulation of a differentiated

K. Deb et al. (Eds.): GECCO 2004, LNCS 3102, pp. 129–139, 2004.
© Springer-Verlag Berlin Heidelberg 2004

multicellular organism that looks like a French flag. We examine in particular two evolved solutions that achieve growth regulation. One of the solutions appears to be static but shows a remarkable ability for self-repair that is reminiscent of the regenerative ability of some living organisms. The other solution shows interesting internal time dynamics. It is not the aim of this work to model closely natural developmental processes, but rather, to explore a simple idealization of biological development in the hope that it will exhibit some of the advantages of biological systems. The long term aim of this work is to investigate a new way of constructing software and hardware systems that are self-repairing and can achieve levels of complex and intelligent behaviour that top-down design methods are unable to attain.

The plan for the paper is as follows: A review of related work is given in section 2. Section 3 describes how the cells and their environment are represented, and the cell program's inputs and outputs. Section 4 describes the form of genetic programming used to evolve the cell program. Section 5 describes the experiments and the results obtained. In section 6, a solution is analyzed in detail. Section 7 examines the organism's behaviour under various kinds of damage. The paper concludes and discusses future work.

2 Related Work

Fleischer and Barr created a sophisticated multicellular developmental test bed and included realistic models of chemical diffusion, cell collision, adhesion and recognition [6]. Their purpose was to investigate cell pattern generation. They noted that the design of an artificial genotype that develops into a specific pattern is very difficult. They also noted that size regulation is critical and non-trivial and that developmental models tend to be robust to perturbations. Eggenberger suggests that the complex genotype-phenotype mappings typically employed in developmental models allow the reduction of genetic information without losing the complex behaviour. He stresses the importance of the fact that the genotype will not necessarily grow as the number of cells, thus he feels that developmental approaches will scale better on complex problems [5]. Bongard and Pfeifer have evolved genotypes that encode a gene expression method to develop the morphology and neural control of multi-articulated simulated agents [3]. Bentley and Kumar examined a number of genotype-phenotype mappings on a problem of creating a tessellating tile pattern [2]. They found that the indirect developmental mapping (that they refer to as an implicit embryogeny) could evolve the tiling patterns much quicker, and further, that they could be subsequently grown to (iterated) much larger sized patterns. One drawback that they reported was that the implicit embryogeny tended to produce the same types of patterns. Other researchers are more motivated by fundamental biological aspects of cell behaviour. Furusawa and Kaneko modeled cell internal dynamics and its relationship to the emergence of cell multicellularity[7]. Hogeweg has carried out impressive work in computer models of development and constructed a sophisticated model of cells (biotic) by modeling the internal dynamics by groups of cells in a cellular automaton that are subject to energy minimization [10][11]. The energy minimization leads to cell movement and sorting by differential cell adhesion. The cell genome was modeled as 24 node Boolean network that defined cell signaling and adhesion. She used a fitness criterion that was related to the difference in the gene expression in all the cells. She evolved organisms that exhibited many behaviours that are observed in living systems: cell migration and engulfing, budding and elongation, and cell death and re-differentiation. Streichert et al. have investigated the problems of growth regulation and self-repair in artificial embryos with a single cell type using Random Boolean Networks and S-Systems [15]. Recently, many of the research contributions in computational development have been presented in a single volume [12].

3 Cell and Chemical Representation

The cell's genotype is a representation of a feed-forward Boolean circuit (that implements the cell program). This maps the cell's input conditions to output behaviour. A cell is a square in a non-toroidal two-dimensional cellular automaton. The inputs to each live cell program are bits defining the cells states and the chemicals in the Moore neighbourhood. Using this information, the cell's program decides on the amounts of each chemical that it will produce (as binary bits), whether it will live, die, or change to a different cell type at the next time step, and how it will grow over the Moore neighbourhood. It also decides a single bit that represents whether it will obey the grow bits or not. This output was introduced to make it easier for a cell program to decide not to grow.

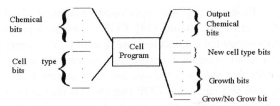

Fig. 1. The cell program's binary inputs and outputs

Unlike real biology, when a cell replicates itself, it is allowed to grow in any or all of the eight neighbouring cells simultaneously (this is done to speed up growth, mainly for reasons of efficiency). In all the experiments reported in this paper the amount of each chemical is represented by an eight-bit binary number. The cell types are represented by binary codes with zero reserved for the absence of a cell (or dead) and are synonymous with cell colour (1 - blue, 2-red, 3-white). In general, the user can decide how many cell types there are (a power of two), in the experiments reported here, only four cell types were required and also how many chemicals there are. Only live cells have their programs executed. Initially a single cell is placed in the grid (the zygote). If two or more cells decide to grow into the same location at the next time step, the last such cell in the scan path overwrites all previous growths. This was chosen as it greatly simplified the process of constructing the newly grown organism, though of course, it introduces a bias that isn't present in a truly parallel system. The process of constructing the new organism at time t+1 from the organism at time t is the following: Every live cell from the top-left to the bottom-right has its program run (all cells run the same program). A new map (initially empty) is created and filled with cells that have either grown, or not died, in the map at time t. After all the programs inside the living cells have been run, the map at time t+1 replaces the map at time t. For each chemical there is a rectangular array of the same dimensions and type as the cellular map. Chemicals obey a diffusion rule defined as follows: let N denote the neighbourhood with neighbouring position k,l, the chemical at position i,j at the new time step is given by (1).

$$(c_{ij})_{new} = 1/2(c_{ij})_{old} + \frac{1}{16}\sum_{k,l \in N}(c_{kl})_{old} \cdot \tag{1}$$

This ensures that over time, chemicals diffuse away from their point of origin. The rule was designed so that diffusing chemical would be conserved (apart from the loss when the level falls below a level of one). Note that, since cells can determine their own new level of chemical there is no strict conservation of chemical in the entire system (i.e. a cell with its chemicals overwrite those at the location that they grow into). The chemical map is scanned and updated in a similar manner to the cellular map. A depiction of the cell's inputs and outputs is shown in Fig. 1.

4   Cartesian Genetic Programming and the Cell Program

Cartesian Genetic Programming was developed from methods developed for the automatic evolution of digital circuits [14]. CGP represents a program or circuit as a list of integers that encode the connections and functions. The representation is readily understood from a small example. Consider the one bit binary adder circuit (Fig. 2). This has three inputs that represent the two bits to be summed and the carry-in bit. It has two outputs: sum and carry-out. CGP employs an indexed list of functions that represent in this example, various two input logic gates and three input multiplexers. Suppose that in a function lookup table AND is function 6, XOR is function 10 and MUX is function 16. The three inputs A, B, Cin are labeled 0, 1, 2. The output of the left (right) XOR gate is labeled 3 (6). The output of the MUX gate is labeled 5. The AND output is labeled 4. In Fig. 2, a genotype is shown and how it is decoded to a phenotype (the one-bit binary adder). The integers in italics represent the functions, and the others represent the connections between gates, however, if it happens to be a two input gate then the third input is ignored. It is assumed that the circuit outputs are taken from the last two nodes. The second group of four integers (shown in grey) represent an AND gate (with output 4) that is not part of the circuit phenotype. Since only feed-forward circuits are being considered, it is important to note that the connections to any gate can only refer to gates that appear on its left.

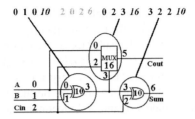

Fig. 2. The Cartesian genotype and corresponding phenotype for a one-bit adder circuit

Typically, CGP uses point mutation (that is constrained to respect the feed-forward nature of the circuit). Suppose that the first input of the MUX gate (0) was changed to 4. This would connect the AND gate into the circuit (defined by the four grey genes). Similarly, a point mutation might disconnect gates. Thus, CGP uses a many to one genotype-phenotype mapping, as redundant nodes may be changed in any way and the genotypes would still be decoded to the same phenotype. The (1+4)-ES evolutionary algorithm uses characteristics of this genotype-phenotype mapping to great advantage (i.e. genetic drift):

1. Generate 5 chromosomes randomly to form the population
2. Evaluate the fitness of all the chromosomes in the population
3. Determine the best chromosome (called it *current_best*)
4. Generate 4 more chromosomes (offspring) by mutating the *current_best*
5. The *current_best* and the four offspring become the new population
6. Unless stopping criterion reached return to 2

Step 3 is a crucial step in this algorithm: if more than one chromosome is equally good then the algorithm always chooses the chromosome that is not the *current_best* (i.e. equally fit but genetically different). This step allows a genetic drift process that turns out be of great benefit [16][18]. The mutation rate is defined to be the percentage of each chromosome that is mutated in step 4. In all the experiments described in this paper only four kinds of MUX logic gates were employed defined by the expression f(A,B,C)=AND(A, NOT(C)) OR AND(B, C). The four types correspond to cases where inputs A and B are either inverted or not. Program outputs are taken from consecutive nodes at the end of the phenotype with the leftmost of these being the grow/no grow output.

5 Evolutionary Experiments and Results

In the biological development of organisms, cells have to behave differently according to their position within the organism. Lewis Wolpert [17] proposed that this positional information arises from a combination of intercellular interactions and cellular responses to chemical gradients that form relative to organism boundaries. In the model, cells respond differently according to threshold concentrations of chemicals. He likened the problem to one of growing a French Flag; the developmental method of construction would be able to produce a recognizable flag of arbitrary size. This illustrates an important property of developmental systems in that they are scale free (i.e.

there is no relationship between the genotype size and the size of the phenotype). Wolpert's model was the inspiration for the task the maps of cells were to achieve. In this paper we describe experimental results for evolving organisms whose task is to grow from a single zygote into the 63 cell organism that looks like a French flag by a certain time (i.e. achieve maturity). This is a very difficult task as initially the cell program must replicate to grow to the desired size but must also somehow recognize that it has reached the appropriate size and no longer continue to replicate. It must also output the desired cell state signal (represented by colour) in the correct spatial region of the flag. The cell program (as with real embryos) is not given coordinates but must decide how to act by local interactions only. This is shown in Fig 3.

Fig. 3. Task definition: a single cell program beginning from a white state, at time 0, must replicate itself and differentiate itself into other cells, so that at iteration 6, it becomes a 63 cell French flag, and remains like that indefinitely. Initially, at the same location as the single start cell, there can be chemicals having values 0 or 255

The evolved organism and target organism were compared at iterations 6, 7, 8, 9, cell by cell, and a cumulative score of correctness (fitness) was calculated. By presenting exactly the same target at these four test points, we hoped to steer the evolution of the cell program towards growing into a fixed size French flag organism. The cell's Cartesian program was allowed 300 nodes and 20 runs of 30,000 generations were carried out (with 1% mutation) for five chemical scenarios: no chemicals up to four allowed chemicals. The amount of chemical initially present at the location of the zygote was initialized as follows: the first chemical is given 255, the second given 0, the third given 255, and the fourth given 0. The maximum fitness value is 1024, which occurs when all four 16 by 16 cell maps for the organism match perfectly with the target organism. A table of results for the five chemical scenarios is shown in Table 1. All solutions with fitness above 975 were iterated over 20 iterations to ascertain whether any of them stopped growing. Only two solutions were found with this property. The first was the solution with the highest overall fitness with two defined chemicals. The temporal behavior of this is shown in Fig. 4.

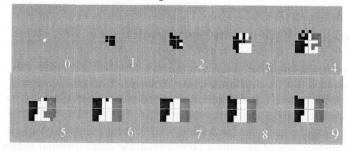

Fig. 4. Growth of fittest cell program from a white seed cell to a mature French flag (two chemicals)

The other solution that stopped growing occurred in the three chemical scenario, however, surprisingly it had fitness 988 (fourth best). This is shown in Fig. 5. Unlike the best overall solution this shows complex time dependent behaviour indefinitely (it was iterated over 50 iterations and did not grow). The results shown in Table 1. show that having chemicals makes it easier to find fitter organisms, and the results indicate that having either none or one chemical make it unlikely, if not impossible, to achieve solutions that grow and then stop growing that meet the target objective.

Table 1. Performance statistics for French flag problem. The maximum fitness is 1024

#chemicals	Average final best fitness (20 runs)	Standard deviation	Best fitness	Worst fitness	Average number of cell program nodes	#nodes in fittest program
0	875.85	16.41	909	855	155.05	153
1	924.80	30.48	975	862	153.90	165
2	938.65	31.00	1012	894	145.75	139
3	948.80	27.87	1008	907	146.45	144
4	941.50	20.35	988	918	142.95	139

The fourth fittest solution moves through a repeating cycle of activity indefinitely (as far as can be determined) always remaining bounded within a small region and with the majority of cells remaining the same.

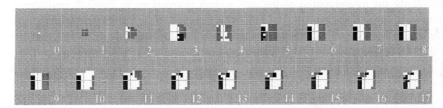

Fig. 5. Growth of fourth fittest cell program from a white seed cell (three chemicals)

6   Analysis

It is instructive to examine how the fittest French flag organism achieves apparent stasis at iteration 8 and subsequently. The cell program itself is too large and complicated to be shown in this paper: from Table 1 we see that it uses 139 binary IF statements (multiplexers). In addition the program cannot be understood without referring to the current state of the organism. However, in the compass of this paper we can illuminate aspects of the developmental program by showing (Fig. 6) the decisions being made by each cell in the organism. On the left, we see the organism itself at iteration 8 and on the right the cell growth and replication fate map. To clarify its interpretation, consider the line of four grey cells in the fate map that are live cells in the organism. These cells will obey the growth instructions at the next iteration (see Table 2) but will then die immediately afterward. Clearly as the organism remains unchanged, the cells around them grow over them, thus reconstructing the line of cells as if nothing had happened.

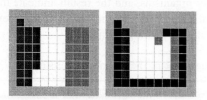

Key:

Black: cell will not grow
Grey: cell will grow and then die
Red/white/blue: cell will replicate according to colour

Fig. 6. Cell growth/replication fate map for the French flag organism at iteration 8.

Table 2. Cell growth instructions for actively growing cells in fittest French flag at iteration 8.

N	E	S	W	NE	SE	SW	NW
01101111	01001101	01001101	01001101	11011100			
01101111	01101100	01101100	01101100	01010100	01010101		11010101
01101111	01101100	01101100	01101100	01110100	01110101		11010101
01101111	01101100	01101100	01101100	01110100	01110101		11010101
01101101	01101100	01101100	01101100	01110100	01110101		11010101
01101101	01101100	01101100	01101100	01110100	01110101		11010101

To clarify the interpretation of the cell growth instructions in Table 2, consider the bottom right actively growing red cell, this has growth instructions 11010101. According to Fig. 6, it replicates its own colour in the directions: N, E, W, SE and NW. All the interior cells of the flag (within one cell inside) are actively growing and over-writing each other (according the top-left, bottom-right scan path), while most of the border cells will remain as they are. Note that none of the second column of blue cells in the flag overwrite their blue left neighbours, as they do not replicate to the west. The French flag is clearly being actively reconstructed even though it appears to be static.

7   Autonomous Behaviour after Damage

When the maturing French flag is damaged, it is often able to regenerate itself (or produce another similar French flag) in some cases, though sometimes it can be put into a state of continuous growth (further examples in Table 3). Fig. 7 shows the regeneration of a French flag cellular map from the original white central region of the original (fittest cell program). The cells replicate and then start to differentiate; eventually the growth of the organism slows and stops at iteration 20.

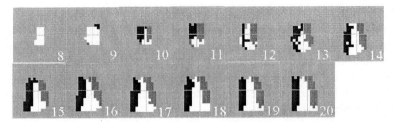

Fig. 7. Autonomous recovery of badly damaged French flag organism conditions (blue and red regions killed at iteration 8 - see Fig. 4). There is no further change after iteration 20

Fig. 8 shows what happens when the original cells of the fittest French flag are placed in a random but contiguous arrangement - chemical maps left intact - the phenotype grows a little at first and rapidly re-organizes over a period of time, eventually reaching stasis (by iteration 24). This behaviour is reminiscent of autonomous regeneration of the pond organism hydra, which can reform itself when its cells are dissociated and then re-aggregated in a centrifuge [8].

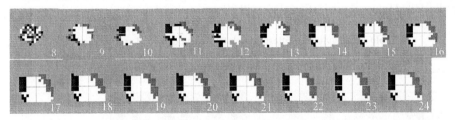

Fig. 8. Autonomous recovery of French flag from randomly rearranged cells (French flag at iteration 8 - see Fig. 4). There is no further change after iteration 24

In Table 3, we show the behaviour of the fittest French flag organism after it is disrupted or damaged at iteration 8. In many cases it achieves stability again fairly rapidly and recovers the approximate appearance of the French flag, however when the damage is too severe (as with a large hole or 25% cells disrupted randomly) the organism undergoes continuous growth and doesn't appear to stabilize (even when run for many more iterations). Such dynamic processes are hard to control in all circumstances and it illustrates the enormously difficult balancing act that living systems have to carry out.

8   Conclusions and Further Work

We have presented and investigated an idealized model of development and studied in detail the growth and regulation of an organism made of cells that can replicate, differentiate, and read and produce chemicals through local interactions. There are many avenues for further investigation such as, the roles of overwriting in self-repair and the diffusion law of chemicals, complexity of evolvable structures, and the evolvability of the representation. The software written also allows the possibility of cell movement (prior to growth) but as yet, this hasn't been investigated. There are many ways that the system can be made more sophisticated, for instance, by allowing cells to control the flow of chemicals and to add the possibility of cell adhesion. However, since the eventual aim of the work is toward technological applications it is important to try to keep the model as simple as possible. The great robustness of the evolved organisms to damage may be a consequence of the ability of cells to overwrite each other; this remains for further investigation.

Work is already underway in examining the possibility of using the developing organisms for robot control, thus giving the organism a function. It will be interesting to see if control programs can recover autonomously after damage. A detailed investigation also needs to be undertaken about the chemical information that is provided to the cell's program. It was discovered when the software was written to carry out this work, that the author inadvertently only provided the most significant bit of each chemical to the cell's program (i.e. cell's think chemicals are either high or low) and all the other bits were read as zeros. It has been found that providing all chemical bits or providing only the most significant bit made it very difficult, if not impossible, to solve the tasks presented. This is such an interesting finding that it warrants lengthy and detailed future investigation. This will be reported in due course.

Table 3. Behaviour of fittest French flag when damaged at iteration 8.

Initial condition	Final condition (iterations to stability below)

| all blue | all red | all white | 5 | 3 | 2 |

| large hole | small hole | continued growth | 9 |

| 25% | 12.5% | continued growth | 9 |

Random damage

References

1. Banzhaf, W., Miller J. F.: The Challenge of Complexity. In: Menon, A. (ed.): Frontiers of Evolutionary Computation. Kluwer Academic Publishers (2004)
2. Bentley, P. J., Kumar S.: Three Ways to Grow Designs: A Comparison of Embryogenies for an Evolutionary Design Problem. In: Proceedings of the Congress on Evolutionary Computation, IEEE Press (1999) 35-43

3. Bongard, J. C., Pfeifer R.: Repeated Structure and Dissociation of Genotypic and Pheno-typic Complexity in Artificial Ontogeny. In: Proceedings of the Genetic and Evolutionary Computation Conference, Morgan-Kaufmann, (2001) 829-836

4. Dellaert, F.: Toward a Biologically Defensible Model of Development, Masters thesis, Dept. of Computer Eng. and Science, Case Western Reserve University (1995)

5. Eggenberger, P.: Evolving morphologies of simulated 3D organisms based on differential gene expression, In: Proceedings of 4th European Conf. on Artificial Life (1997) 205-213

6. Fleischer, K., Barr, A. H.: A simulation testbed for the study of multicellular development: The multiple mechanisms of morphogenesis. In Langton C. G (ed.) Proceedings of the 3rd Workshop on Artificial Life, Addison-Wesley (1992) 389-416

7. Furusawa, C., Kaneko, K.: Emergence of Multicellular Organisms with Dynamic Differ-entiation and Spatial Pattern. In: Adami C. et al. (eds.) Proceedings of the 6th International Conference on Artificial Life, MIT Press (1998)

8. Gierer, A., Berking, S., Bode, H., David, C. N., Flick, K., Hansmann, G., Schaller, H., Trenkner E.: Regeneration of hydra from reaggregated cells, Nature New Biology, Vol. 239 (1972) 98-101

9. Harold, F. M.: The Way of The Cell. Oxford University Press (2001)

10. Hogeweg, P.: Evolving Mechanisms of Morphogenesis: on the Interplay between Differ-ential Adhesion and Cell Differentiation, J. Theor. Biol., Vol. 203 (2000) 317-333

11. Hogeweg, P.: Shapes in the Shadow: Evolutionary Dynamics of Morphogenesis, Artificial Life, Vol. 6 (2000) 85-101

12. Kumar, S., Bentley P. J. (eds.): On Growth, Form and Computers, Academic Press (2003)

13. Miller, J. F.: Evolving Developmental Programs for Adaptation, Morphogenesis and Self-Repair. In: Proceedings of the 7th European Conf. on Advances in Artificial Life. LNAI, Vol. 2801 (2003) 256-265

14. Miller, J. F., Thomson, P.: Cartesian Genetic Programming. In: Proceedings of the 3rd European Conf. on Genetic Programming. LNCS, Vol. 1802 (2000) 121-132

15. Streichert, F., Spieth, C., Ulmer, H., Zell, A.: Evolving the Ability of Limited Growth and Self-Repair for Artificial Embryos. In: Proceedings of the 7th European Conf. on Ad-vances in Artificial Life. LNAI, Vol. 2801 (2003) 289-298

16. Vassilev, V. K., Miller J. F.: The Advantages of Landscape Neutrality in Digital Circuit Evolution. In: Proceedings of 3rd Int. Conf. on Evolvable Systems: From Biology to Hardware, LNCS, Vol. 1801, Springer-Verlag (2000) 252-263

17. Wolpert, L.: Principles of Development. Oxford University Press (1998)

18. Yu, T., Miller, J. F.: Neutrality and the evolvability of Boolean function landscape. In: Proceedings of the 4th European Conference on Genetic Programming, Springer-Verlag (2001) 204-217

The Kalman Swarm

A New Approach to Particle Motion in Swarm Optimization

Christopher K. Monson and Kevin D. Seppi

Brigham Young University, Provo UT 84602, USA
{c,kseppi}@cs.byu.edu

Abstract. Particle Swarm Optimization is gaining momentum as a simple and effective optimization technique. We present a new approach to PSO that significantly reduces the number of iterations required to reach good solutions. In contrast with much recent research, the focus of this work is on fundamental particle motion, making use of the Kalman Filter to update particle positions. This enhances exploration without hurting the ability to converge rapidly to good solutions.

1 Introduction

Particle Swarm Optimization (PSO) is an optimization technique inspired by social behavior observable in nature, such as flocks of birds and schools of fish [1]. It is essentially a nonlinear programming technique suitable for optimizing functions with continuous domains (though some work has been done in discrete domains [2]), and has a number of desirable properties, including simplicity of implementation, scalability in dimension, and good empirical performance. It has been compared to evolutionary algorithms such as GAs (both in methodology and performance) and has performed favorably [3,4].

As an algorithm, it is an attractive choice for nonlinear programming because of the characteristics mentioned above. Even so, it is not without problems. PSO suffers from premature convergence, tending to get stuck in local minima [4,5, 6,7]. We have also found that it suffers from an ineffective exploration strategy, especially around local minima, and thus does not find good solutions as quickly as it could. Moreover, adjusting the tunable parameters of PSO to obtain good performance can be a difficult task [7,8].

Research addressing the shortcomings of PSO is ongoing and includes such changes as dynamic or exotic sociometries [6,9,10,11,12], spatially extended particles that bounce [13], increased particle diversity [4,5], evolutionary selection mechanisms [14], and of course tunable parameters in the velocity update equations [7,8,15]. Some work has been done that alters basic particle motion with some success, but the possibility for improvement in this area is still open [16].

This paper presents an approach to particle motion that significantly speeds the search for optima while simultaneously improving on the premature convergence problems that often plague PSO. The algorithm presented here, KSwarm, bases its particle motion on Kalman filtering and prediction.

K. Deb et al. (Eds.): GECCO 2004, LNCS 3102, pp. 140–150, 2004.

We compare the performance of KSwarm to that of the basic PSO model. In the next section, the basic PSO algorithm is reviewed, along with an instructive alternative formulation of PSO and a discussion of some of its shortcomings. Unless otherwise specified, "PSO" refers to the basic algorithm as presented in that section. Section 3 briefly describes Kalman Filters, and Section 4 describes KSwarm in detail. Experiments and their results are contained in Section 5. Finally, conclusions and future research are addressed in Section 6.

2   The Basic PSO Algorithm

PSO is an optimization strategy generally employed to find a global minimum. The basic PSO algorithm begins by scattering a number of "particles" in the function domain space. Each particle is essentially a data structure that keeps track of its current position x and its current velocity v. Additionally, each particle remembers the "best" (lowest valued) position it has obtained in the past, denoted p. The best of these values among all particles (the global best remembered position) is denoted g.

At each time step, a particle updates its position and velocity by the following equations:

$$v_{t+1} = \chi\big(v_t + \phi_1(p - x) + \phi_2(g - x)\big) \tag{1}$$

$$x_{t+1} = x_t + v_{t+1} . \tag{2}$$

The *constriction coefficient* $\chi = 0.729844$ is due to Clerc and Kennedy [15] and serves to keep velocities from exploding. The stochastic scalars ϕ_1 and ϕ_2 are drawn from a uniform distribution over $[0, 2.05)$ at each time step. Though other coefficients have been proposed in an effort to improve the algorithm [7,8], they will not be discussed here in detail.

2.1   An Alternative Motivation

Although the PSO update model initially evolved from simulated flocking and other natural social behaviors, it is instructive to consider an alternative motivation based on a randomized hill climbing search. A naive implementation may place a single particle in the function domain, then scatter a number of random sample points in the neighborhood, moving toward the best sample point at each new time step: $x_{t+1} = g_t$.

If the particle takes this step by first calculating a velocity, the position is still given by (2) and the velocity update is given by

$$v_{t+1} = g_t - x_t . \tag{3}$$

As this type of search rapidly becomes trapped in local minima, it is useful to randomly overshoot or undershoot the actual new location in order to do some directed exploration (after all, the value of the new location is already known). For similar reasons, it may be desirable to add momentum to the system, allowing particles to "roll out" of local minima. Choosing a suitable random scalar ϕ, this yields

$$v_{t+1} = v_t + \phi(g_t - x_t) \ . \tag{4}$$

The equation (4) is strikingly similar to (1). In fact, it is trivial to reformulate the PSO update equation to be of the same form as (1) [15,12].

The fundamental difference between this approach and PSO is the way that g is calculated. In PSO, g is taken from other particles already in the system. In the approach described in this section, g is taken from disposable samples scattered in the neighborhood of a single particle.

This suggests that the basic PSO is a hill climber that uses existing information to reduce function evaluations. It is set apart more by its social aspects than by its motion characteristics, an insight supported by Kennedy but for different reasons [16].

2.2   Particle Motion Issues

Given that PSO is closely related to an approach as simple as randomized hill climbing, it is no surprise that attempts to improve the velocity update equation with various scaling terms have met with marginal success. Instead, more fundamental changes such as increased swarm diversity, selection, and collision avoiding particles have shown the greatest promise [4,5,14].

Unfortunately these methods are not without problems either, as they generally fail to reduce the iterations required to reach suitable minima. They focus primarily on eliminating stagnation, eventually finding better answers than the basic PSO without finding them any faster.

It has been pointed out that nonlinear programming is subject to a fundamental tradeoff between convergence speed and final fitness [4], suggesting that it is not generally possible to improve one without hurting the other. Fortunately, this tradeoff point has not yet been reached in the context of particle swarm optimization, as it is still possible to find good solutions more quickly without damaging final solution fitness.

For example, the development of a PSO visualization tool served to expose a particularly interesting inefficiency in the basic PSO algorithm. As the particles close in on g they tend to lose their lateral momentum very quickly, each settling into a simple periodic linear motion as they repeatedly overshoot (and undershoot) the target. This exploration strategy around local minima is very inefficient, suggesting that a change to particle motion may speed the search by improving exploration.

Such a change should ideally preserve the existing desirable characteristics of the algorithm. PSO is essentially a *social* algorithm, which gives it useful emergent behavior. Additionally, PSO motion is stochastic, allowing for randomized *exploration*. Particles also have *momentum*, adding direction to the random search. The constriction coefficient indicates a need for *stability*. Alterations to particle motion should presumably maintain these properties, making the Kalman Filter a suitable choice.

3 The Kalman Filter

Kalman filters involve taking noisy observations over time and using model information to estimate the true state of the environment [17]. Kalman filtering is generally applied to motion tracking problems. It may also be used for prediction by applying the system transition model to the filtered estimate.

The Kalman Filter is limited to normal noise distributions and linear transition and sensor functions and is therefore completely described by several constant matrices and vectors. Specifically, given an observation column vector $\mathbf{z}_{t+1}$, the Kalman Filter is used to generate a normal distribution over a belief about the true state. The parameters $\mathbf{m}_{t+1}$ and $\mathbf{V}_{t+1}$ of this multivariate distribution are determined by the following equations [18]:

$$\mathbf{m}_{t+1} = \mathbf{F}\mathbf{m}_t + \mathbf{K}_{t+1}(\mathbf{z}_{t+1} - \mathbf{H}\mathbf{F}\mathbf{m}_t) \tag{5}$$

$$\mathbf{V}_{t+1} = (\mathbf{I} - \mathbf{K}_{t+1})(\mathbf{F}\mathbf{V}_t\mathbf{F}^\top + \mathbf{V}_\mathbf{x}) \tag{6}$$

$$\mathbf{K}_{t+1} = (\mathbf{F}\mathbf{V}_t\mathbf{F}^\top + \mathbf{V}_\mathbf{x})\mathbf{H}^\top \left(\mathbf{H}(\mathbf{F}\mathbf{V}_t\mathbf{F}^\top + \mathbf{V}_\mathbf{x})\mathbf{H}^\top + \mathbf{V}_\mathbf{z}\right)^{-1} . \tag{7}$$

In these equations, $\mathbf{F}$ and $\mathbf{V}_\mathbf{x}$ describe the system transition model while $\mathbf{H}$ and $\mathbf{V}_\mathbf{z}$ describe the sensor model. The equations require a starting point for the filtered belief, represented by a normal distribution with parameters $\mathbf{m}_0$ and $\mathbf{V}_0$, which must be provided.

The filtered or "true" state is then represented by a distribution:

$$\mathbf{x}_t \sim \text{Normal}(\mathbf{m}_t, \mathbf{V}_t) . \tag{8}$$

This distribution may be used in more than one way. In some applications, the mean $\mathbf{m}_t$ is assumed to be the true value. In others, the distribution is sampled once to obtain the value. In this work, the latter is done.

The above describes how to do Kalman *filtering*, yielding $\mathbf{m}_t$ from an observation $\mathbf{z}_t$. A simple form of *prediction* involves applying the transition model to obtain a belief about the next state $\mathbf{m}'_{t+1}$:

$$\mathbf{m}'_{t+1} = \mathbf{F}\mathbf{m}_t . \tag{9}$$

There are other forms of prediction, but this simple approach is sufficient for the introduction of the algorithm in the next section, and for its use in particle swarms.

4 The Kalman Swarm (KSwarm)

KSwarm defines particle motion entirely from Kalman prediction. Each particle keeps track of its own $\mathbf{m}_t$, $\mathbf{V}_t$, and $\mathbf{K}_t$. The particle then generates an observation for the Kalman filter with the following formulae:

$$z_v = \phi(g - x) \tag{10}$$

$$z_p = x + z_v . \tag{11}$$

Similar to PSO, ϕ is drawn uniformly from $[0, 2)$, and the results are row vectors. The full observation vector is given by making a column vector out of the concatenated position and velocity row vectors: $\mathbf{z} = (z_p, z_v)^\top$. This observation is then used to generate $\mathbf{m}_{t+1}$ and $\mathbf{V}_{t+1}$ using (5), (6), and (7)

Once the filtered value is obtained, a prediction $\mathbf{m}'_{t+2}$ is generated using (9). Together, $\mathbf{m}'_{t+2}$ and $\mathbf{V}_{t+1}$ parameterize a normal distribution. We say, then, that

$$\mathbf{x}_{t+1} \sim \text{Normal}(\mathbf{m}'_{t+2}, \mathbf{V}_{t+1}) . \tag{12}$$

The new state of the particle is obtained by sampling once from this distribution. The position of the particle may be obtained from the first half of $\mathbf{x}^\top_{t+1}$, and the velocity (found in the remaining half) is unused.

This method for generating new particle positions has at least one immediately obvious advantage over the original approach: there is no need for a constriction coefficient. Particle momentum comes from the state maintained by the Kalman Filter rather than from the transition model. In our experiments, this eliminated the need for any explicit consideration of velocity explosion.

5   Experiments

KSwarm was compared to PSO in five common test functions: Sphere, DejongF4, Rosenbrock, Griewank, and Rastrigin. The first three are unimodal while the last two are multimodal. In all experiments, the dimensionality $d = 30$. The definitions of the five functions are given here:

$$\text{Sphere}(\boldsymbol{x}) = \sum_{i=1}^{d} x_i^2 \tag{13}$$

$$\text{DeJongF4}(\boldsymbol{x}) = \sum_{i=1}^{d} i x_i^4 \tag{14}$$

$$\text{Rosenbrock}(\boldsymbol{x}) = \sum_{i=1}^{d-1} 100(x_{i+1} - x_i^2)^2 + (x_i - 1)^2 \tag{15}$$

$$\text{Rastrigin}(\boldsymbol{x}) = \sum_{i=1}^{d} x_i^2 + 10 - 10\cos(2\pi x_i) \tag{16}$$

$$\text{Griewank}(\boldsymbol{x}) = \frac{1}{4000} \sum_{i=1}^{d} x_i^2 - \prod_{i=1}^{d} \cos\left(\frac{x_i}{\sqrt{i}}\right) + 1 . \tag{17}$$

The domains of these functions are given in Table 1.

5.1   Experimental Parameters

In all experiments, a swarm size of 20 was used. Though various sociometries are available, the *star* (or *gbest* [1]) sociometry was used almost exclusively in

Table 1. Domains of Test Functions

Function	Domain
Sphere	$(-50, 50)^d$
DeJongF4	$(-20, 20)^d$
Rosenbrock	$(-100, 100)^d$
Griewank	$(-600, 600)^d$
Rastrigin	$(-5.12, 5.12)^d$

the experiments because it allows for maximum information flow [13]. Each experiment was run 50 times for 1000 iterations, and the results were averaged to account for stochastic differences. The parameters to (5), (6), and (7) are given below, and are dependent on the domain size of the function. The vector containing the size of the domain in each dimension is denoted w. The column vector $\mathbf{w} = (w, w)^\top$ is formed from two concatenated copies of w. In the following equations, $\mathbf{I}_n$ is an identity matrix with n rows.

$$\mathbf{m}_0 = 0 \qquad\qquad \mathbf{V}_0 = \theta \operatorname{diag}(\mathbf{w}) \qquad (18)$$

$$\mathbf{H} = \mathbf{I}_{2d} \qquad\qquad \mathbf{V}_\mathbf{z} = \theta \operatorname{diag}(\mathbf{w}) \qquad (19)$$

$$\mathbf{F} = \begin{pmatrix} \mathbf{I}_d & \mathbf{I}_d \\ \mathbf{0} & \mathbf{I}_d \end{pmatrix} \qquad\qquad \mathbf{V}_\mathbf{x} = \theta \operatorname{diag}(\mathbf{w}) \ . \qquad (20)$$

The initial mean $\mathbf{m}_0$ is a column vector of $2d$ zeros. The scalar θ indicates how large the variance should be in each dimension, and was set to 0.0001 for all experiments, as this produced a variance that seemed reasonable. The transition function simply increments position by velocity while leaving the velocity untouched.

All of the vectors used in the Kalman equations are of length $2d$ and all matrices are square and of size $2d$. This is the case because the model makes use of velocity as well as position, so extra dimensions are needed to maintain and calculate the velocity as part of the state. This implies that the sample obtained from (12) is *also* a vector of length $2d$, the first half of which contains position information. That position information is used to set the new position of the particle and the velocity information is unused except for the next iteration of the Kalman update equations.

5.2 Results

Table 2 shows the final values reached by each algorithm after 1000 iterations were performed. It is clear from the table that the KSwarm obtains values that are often several orders of magnitude better than the original PSO algorithm.

In addition to obtaining better values, the KSwarm tends to find good solutions in fewer iterations than the PSO, as evidenced by Figs. 1, 2, 3, 4, and 5. Note that each figure has a different scale.

Because the results obtained using the star sociometry were so striking, this experiment was also run using a sociometry where each particle had 5 neighbors. The corresponding results were so similar as to not warrant inclusion in this work.

Table 2. PSO vs. KSwarm Final Values

Function	PSO	KSwarm
Sphere	370.041	**4.723**
DejongF4	4346.714	**4.609**
Rosenbrock	2.61e7	**3.28e3**
Griewank	13.865	**0.996**
Rastrigin	106.550	**53.293**

These results represent a clear and substantial improvement over the basic PSO, not only in the final solutions, but in the speed with which they are found. It should be noted that much research has been done to improve PSO in other ways and that KSwarm performance in comparison to these methods has not been fully explored. The purpose of this work is to demonstrate a novel approach to particle motion that substantially improves the basic algorithm. The comparison and potential combination of KSwarm with other PSO improvements is part of ongoing research and will be a subject of future work.

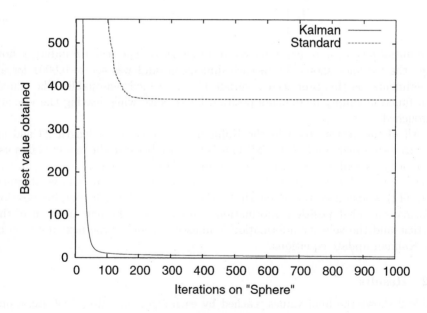

Fig. 1. Sphere

5.3   Notes on Complexity

It is worth noting that the Kalman motion update equations require more computational resources than the original particle motion equations. In fact, because

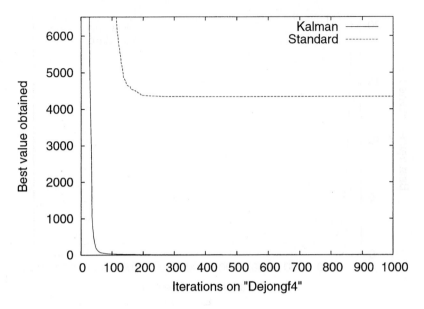

Fig. 2. DeJongF4

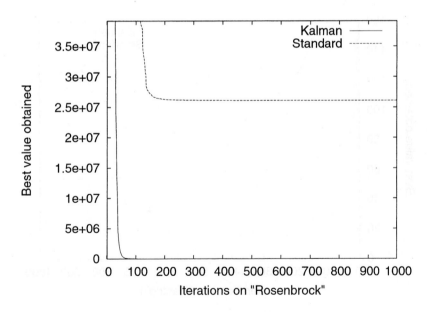

Fig. 3. Rosenbrock

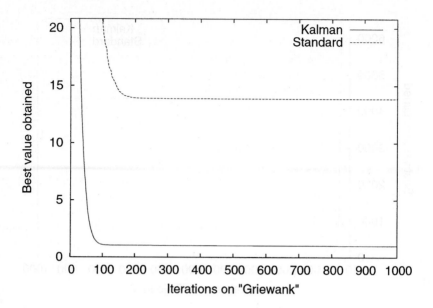

Fig. 4. Griewank

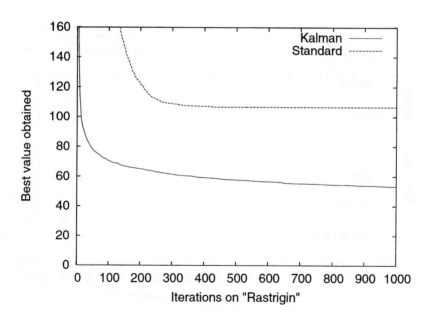

Fig. 5. Rastrigin

of the matrix operations, the complexity is $O(d^3)$ in the number of dimensions ($d = 30$ in our experiments). The importance of this increased complexity, however, appears to diminish when compared to the apparent exponential improvement in the number of iterations required by the algorithm. Additionally, the complexity can be drastically reduced by using matrices that are mostly diagonal or by approximating the essential characteristics of Kalman behavior in a simpler way.

6   Conclusions and Future Work

It remains to be seen how KSwarm performs against diversity-increasing approaches, but preliminary work indicates that it will do well in that arena, especially with regard to convergence speed. Since many methods which increase diversity do not fundamentally change particle motion update equations, combining this approach with those methods is simple. It can allow KSwarm to not only find solutions faster, but also to avoid the stagnation to which it is still prone.

Work remains to be done on alternative system transition matrices. The transition model chosen for the motion presented in this work is not the only possible model; other models may produce useful behaviors. Additionally, the complexity of the algorithm should be addressed. It is likely to be easy to improve by simple optimization of matrix manipulation, taking advantage of the simplicity of the model. More work remains to be done in this area.

KSwarm fundamentally changes particle motion as outlined in PSO while retaining its key properties of sociality, momentum, exploration, and stability. It represents a substantial improvement over the basic algorithm not only in the resulting solutions, but also in the speed with which they are found.

References

1. Kennedy, J., Eberhart, R.C.: Particle swarm optimization. In: International Conference on Neural Networks, IV (Perth, Australia), Piscataway, NJ, (IEEE Service Center) 1942–1948
2. Kennedy, J., Eberhart, R.C.: A discrete binary version of the particle swarm algorithm. In: Proceedings of the World Multiconference on Systemics, Cybernetics, and Informatics, Piscataway, New Jersey (1997) 4104–4109
3. Kennedy, J., Spears, W.: Matching algorithms to problems: An experimental test of the particle swarm and some genetic algorithms on the multimodal problem generator. In: Proceedings of the IEEE Congress on Evolutionary Computation (CEC 1998), Anchorage, Alaska (1998)
4. Riget, J., Vesterstroem, J.S.: A diversity-guided particle swarm optimizer - the ARPSO. Technical Report 2002-02, Department of Computer Science, University of Aarhus (2002)
5. Løvbjerg, M.: Improving particle swarm optimization by hybridization of stochastic search heuristics and self-organized criticality. Master's thesis, Department of Computer Science, University of Aarhus (2002)

6. Richards, M., Ventura, D.: Dynamic sociometry in particle swarm optimization. In: International Conference on Computational Intelligence and Natural Computing. (2003)

7. Vesterstroem, J.S., Riget, J., Krink, T.: Division of labor in particle swarm optimisation. In: Proceedings of the IEEE Congress on Evolutionary Computation (CEC 2002), Honolulu, Hawaii (2002)

8. Shi, Y., Eberhart, R.C.: Parameter selection in particle swarm optimization. In: Evolutionary Programming VII: Proceedings of the Seventh Annual Conference on Evolutionary Programming, New York (1998) 591–600

9. Kennedy, J., Mendes, R.: Population structure and particle swarm performance. In: Proceedings of the Congress on Evolutionary Computation (CEC 2002), Honolulu, Hawaii (2002)

10. Kennedy, J., Mendes, R.: Neighborhood topologies in fully-informed and best-of-neighborhood particle swarms. In: Proceedings of the 2003 IEEE SMC Workshop on Soft Computing in Industrial Applications (SMCia03), Binghamton, New York, IEEE Computer Society (2003)

11. Kennedy, J.: Small worlds and mega-minds: Effects of neighborhood topology on particle swarm performance. In Angeline, P.J., Michalewicz, Z., Schoenauer, M., Yao, X., Zalzala, Z., eds.: Proceedings of the Congress of Evolutionary Computation. Volume 3., IEEE Press (1999) 1931–1938

12. Mendes, R., Kennedy, J., Neves, J.: Watch thy neighbor or how the swarm can learn from its environment. In: Proceedings of the IEEE Swarm Intelligence Symposium 2003 (SIS 2003), Indianapolis, Indiana (2003) 88–94

13. Krink, T., Vestertroem, J.S., Riget, J.: Particle swarm optimisation with spatial particle extension. In: Proceedings of the IEEE Congress on Evolutionary Computation (CEC 2002), Honolulu, Hawaii (2002)

14. Angeline, P.J.: Using selection to improve particle swarm optimization. In: Proceedings of the IEEE Congress on Evolutionary Computation (CEC 1998), Anchorage, Alaska (1998)

15. Clerc, M., Kennedy, J.: The particle swarm: Explosion, stability, and convergence in a multidimensional complex space. IEEE Transactions on Evolutionary Computation **6** (2002) 58–73

16. Kennedy, J.: Bare bones particle swarms. In: Proceedings of the IEEE Swarm Intelligence Symposium 2003 (SIS 2003), Indianapolis, Indiana (2003) 80–87

17. Kalman, R.E.: A new approach to linear filtering and prediction problems. Transactions of the ASME–Journal of Basic Engineering **82** (1960) 35–45

18. Russel, S., Norvig, P.: Artificial Intelligence: A Modern Approach. Second edn. Prentice Hall, Englewood Cliffs, New Jersey (2003)

Adaptive and Evolvable Network Services

Tadashi Nakano and Tatsuya Suda

Department of Information and Computer Science
University of California, Irvine
Irvine, CA 92697-3425
{tnakano, suda}@ics.uci.edu

Abstract. This paper proposes an evolutionary framework where a network service is created from a group of autonomous agents that interact and evolve. Agents in our framework are capable of autonomous actions such as replication, migration, and death. An evolutionary mechanism is designed using genetic algorithms in order to evolve the agent's behavior over generations. A simulation study is carried out to demonstrate the ability of the evolutionary mechanism to improve the network service performance (e.g., response time) in a decentralized and self-organized manner. This paper describes the evolutionary mechanism, its design and implementation, and evaluates it through simulations.

1 Introduction

Swarm intelligence – the collective intelligence of groups of simple individuals often observed in social insects [1,8] has inspired many applications in a variety of fields such as optimization [5,9], clustering [4], communication networks [17] and robotics [10]. In these applications, individuals are capable of sensing the environment in which they operate, and act based only on partial information about the entire environment. Each individual is designed simply and does not have the ability to accomplish a goal. However, a collection of individuals exhibits intelligent behavior toward achieving a goal along with useful properties such as adaptability and scalability.

This paper proposes an evolutionary framework for developing distributed network services that require a large number of network components (e.g., data and software) to be replicated, moved and deleted in a decentralized manner. Such network services may include content distribution networks [2,7,14,15,16], content services networks [11] and peer-to-peer file sharing networks [13]. Analogous to those applications inspired by swarm intelligence, agents are simply designed using only local information without relying on global knowledge, and collectively provide adaptive and scalable network services.

In the proposed framework, a single network service is provided by a group of autonomous agents. Each autonomous agent implements an identical network service (e.g., a content hosting service or a web document), but can have different behavior in replication, migration and death (deletion of itself). An agent's behavior is governed by a set of genes embedded into each agent, and designed to evolve through a repro-

K. Deb et al. (Eds.): GECCO 2004, LNCS 3102, pp. 151–162, 2004.
© Springer-Verlag Berlin Heidelberg 2004

ductive process with genetic algorithms [12]. As opposed to generational genetic algorithms that rely on god-like central selection, agents in our framework are evaluated and selected by nearby agents in a decentralized and self-organized way.

A simulation study is carried out to present the ability of evolutionary adaptation to improve the service performance or fitness values (e.g., response time, bandwidth consumption, resource usage, etc.) It is shown through evolutionary adaptation processes that agents evolve their behavior over generations by which a network service becomes adapted to a variety of network environments.

The rest of the paper is organized as follows. Section 2 gives a broad overview of the proposed evolutionary framework, and then presents the design of agents and their evolutionary mechanisms. To evaluate the evolutionary mechanisms, preliminary simulations are run in Section 3 and an extensive set of simulation studies is carried out in Section 4. Section 5 concludes the paper with a brief summary.

2 Evolutionary Framework for Developing Network Applications

This section first provides an overview of the proposed framework for developing distributed network applications and then describes the evolutionary mechanisms designed for network applications to adapt to network environments.

2.1 Overview

Our framework assumes a fully distributed network environment in which a group of autonomous agents self-organizes a network service without centralized control. It is also assumed that communications and information are restricted to those locally available to agents (e.g., agents can communicate with each other only when residing on the same network platform or on adjacent ones.)

As illustrated in Figure 1, our network is modeled using three network components: agents, users and platforms, which exchange a common resource called *energy*.

Agents provide a service to users (i.e. end service consumers or an agent of an application) in exchange for energy. Agents use computing resources (e.g., CPU power, memory, and network bandwidth) provided by a hosting platform in exchange for energy. Agents also use energy to invoke behaviors such as replication, migration and death. Thus, the energy level of an agent is a measure of how efficiently the agent provides a service, uses computing resources and performs behavior.

The platforms are autonomous systems connected to each other. Platforms host agents and provide computing resources such as CPU, memory and bandwidth. Platforms provide an environment for agents where agents can migrate and replicate. Platforms periodically charge agents for energy, and expel agents who run out of energy. Thus, platforms perform natural selection and favor energy efficient agents.

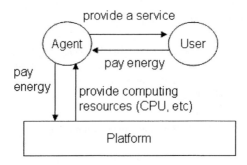

Fig. 1. Energy exchange

2.2   Agents and Their Behavior

Agents have an internal state that affects behaviors such as replication, migration and death. Example internal states include *energy intake* (the difference between acquired energy units and consumed energy units), *age* (time elapsed since birth) and *activeness* (the degree of willingness to invoke behavior). The internal state also includes an agent's performance directly related to a network service, such as *response time* (time taken for a user to receive a service from the agent).

Agents are capable of sensing the local environmental conditions of the platform that they reside on and also the environmental conditions of adjacent platforms. Environmental conditions include *request rate* (how often the platform that hosts an agent receives requests from a user), *request rate change* (how much the request rate increases or decreases), *population* (the number of agents on the platform), *resource cost* (the energy cost of resource at the platform), *behavior cost* (the energy cost of behavior at the platform).

According to an environmental condition or their internal state described above, agents autonomously replicate, reproduce, migrate or die. For instance, agents may replicate (make a copy of themselves) or reproduce (produce an agent with another agent) in response to an increased demand for the service; agents may migrate from one platform to another to perform a service in the vicinity of users; agents may die when the service they provide becomes outdated.

Figure 2 illustrates the behavior invocation mechanism embodied in each agent. In replication, reproduction and death, a set of input values (V_i) that includes internal states and environmental conditions described above is multiplied with a set of associated weights (W_i). If the weighted sum $(\Sigma V_i \times W_i)$ exceeds the threshold (θ), then a corresponding behavior is invoked. In migration, agents need to choose which platform to migrate to from the current platform. In this case, the equation shown in Figure 2 is examined for all the possible choices of platforms including the current platform. If multiple choices satisfy the equation, the behavior most likely triggered is the choice that produces the highest sum. If the current platform produces the highest sum, the agent remains on the current platform.

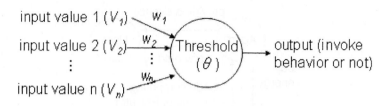

Fig. 2. Behavior invocation mechanism

2.3 Evolutionary Adaptation Mechanisms

In our framework, evolution occurs as a result of selection from diverse behavioral characteristics of individual agents. Agents have a set of genes (weights) that governs their behavior, and diverse behavioral agents arise from the genetic variation. When an agent reproduces, the agent selects a reproduction partner so as to improve one of the service performances (e.g., response time, bandwidth consumption, resource usage). Meanwhile, the diverse behavioral characteristics among agents are generated by genetic operators (mutation and crossover). Through successive generations, beneficial features are retained while detrimental behaviors become dormant, enabling the network service to adapt to the network environment. The following explains the evolutionary mechanism designed in this paper.

Representation. Each agent is represented as a vector of real values (a set of genes) with each real value corresponding to a weight shown in Figure 2.

Natural selection. Agents who run out of energy are eliminated by a platform. This mechanism is referred to as natural selection. Natural selection guarantees that inefficient behavioral agents, (which may be produced through the genetic operators such as crossover and mutation,) eventually become extinct.

Partner selection. In reproduction, an agent selects a partner agent on the same platform or adjacent platforms according to the following fitness assignment strategy that involves multiple fitness values including (1) *waiting time* (the average interval between the time that a service request arrives at a platform and the time that an agent provides the requested service), (2) *hop count* (the average number of platform hops that a service request travels from a user to the agent), and (3) *energy efficiency* (the fraction of the consumed energy and acquired energy). In the following, each of these fitness values is explained with more detail.

Waiting time is defined as the average interval between the time that a service request arrives at a platform and the time that an agent provides the requested service. Upon arrival at a platform, a service request is placed on a platform queue if all agents on the platform are busy processing other service requests. In order to improve this fitness value, agents are required to reproduce when there is insufficient number of agents to handle all service requests from users. Agents may further improve the waiting time by reproducing in advance when the number of service requests starts increasing.

Hop count is measured as the average number of platform hops that a service request travels from a user to the agent. It is assumed that a service request issued by a user is always directed to one of the agents nearest to the user. Agents can improve the hop count by staying around users or following users if they are moving around.

Energy efficiency is measured as the fraction of the consumed energy and acquired energy. The acquired energy is the total amount of energy that the agent obtains from its birth, and consumed energy is the total amount of energy that the agent consumes from its birth. The amount of acquired energy increases in proportion to the number of service requests that the agent processes. Consumed energy represents how efficiently the agent performs behavior and uses computing resources. For instance, inefficient behavior invocation (e.g., migrating too often) may incur a great energy loss. In order to improve energy efficiency, agents need to balance or even optimize its energy income relative to energy expenditure.

To select a reproduction partner, an agent probabilistically determines which fitness value needs improvement: waiting time, hop count or energy efficiency. The fitness value chosen for possible improvement is based on the agent's own fitness values and pre-defined desired fitness values (initially given to all agents) with respect to these three criteria.

Specifically, given the set of its own fitness values (F_i), the set of predefined required fitness values (R_i), the following equation defines the probability of the fitness value j being chosen: $(R_j/F_j)/\Sigma(R_i/F_i)$. Suppose that an agent is inefficient in satisfying a pre-defined level of energy efficiency, and then the agent is likely to choose the energy efficiency as the fitness value to improve in its reproduction. Similarly, the agent may choose the waiting time or hop count as the fitness value when it is inefficient in satisfying either of the pre-defined desired fitness values.

After selecting the fitness, candidate agents are ranked according to the selected fitness value through linear rank selection typically used in evolutionary computation [12].

Crossover. After an agent selects a reproduction partner, the two sets of weights from the two parent agents are crossed over to produce a new set of weights for a child agent. In crossover, a set of weights for a child agent is determined in such a way that more weights are inherited from the parent with a greater fitness value. The probability of a weight being chosen from a parent linearly increases in proportion to the number of its fitness values that are greater than those of the other parent.

Mutation. After two parental weights are crossed over, mutation may occur at each weight of a child agent with a probability called mutation rate. In mutation, each weight value is subject to random change within a certain range called mutation range.

The design of evolutionary mechanisms that were described in this section is an example, and alternative approaches developed in the evolutionary computation literature such as multiobjective optimization [3] and adaptive parameter control [6] are also applicable to implementing our agent-based evolutionary framework.

3 Simulations in Static Network Environments

In this section, the evolutionary mechanism designed in Section 2 is evaluated in relatively simple and homogenous network environments through a simulation study. Pseudo code of the simulator algorithm is shown in Figure 3. The various parameters used and simulation results are explained in the following.

3.1 Configurations

A simulated network is configured as an 8×8 mesh topology network with 64 nodes. Each node on the network hosts a single platform that charges each agent 1 energy unit per second for computing resources (one simulation cycle corresponds to 1 second). There are seven users in the network. Each user generates service requests at different rates ranging from 10 to 30 requests per second, totaling 150 requests per second on the entire network. A service request issued by a user is forwarded to one of the nearest platforms where agents exist. After forwarding, if no agents are available for processing the service request, the service request is placed on the platform queue and experiences a delay until an agent becomes available and processes it.

The simulations assume agents that are capable of reproduction, migration and death. In addition, each agent can process a maximum of 5 service requests per second. An agent receives 10 energy units from a user in exchange for each request processed. Every 15 seconds an agent makes a decision on whether to invoke death, reproduction, and migration. Agents consume 500 energy units for performing behavior such as reproduction and migration. Agents may invoke a single behavior (one out of

```
Initialize platforms, agents and users
While (not simulation last cycle)
        For each user do
                send service requests to one of the nearest agents according to
                configured service request rates.
        End For
        For each platform do
                charge agents platform cost
        End For
        For each agent do
                If received service requests do
                        Process the requests and receive energy
                End If
                make decision on reproduction, migration and death
                update average waiting time, hop count, energy efficiency
        End For
End While
```

Fig. 3. Pseudo code of simulation algorithm

Table 1. Three types of agents used in simulations: PR, PD and ES

Weight	PR (primary)	PD (productive)	ES (energy seeker)
R. Request Rate	0.1	0.3	0.1
R. Threshold	0.5	0.5	0.5
M. Request Rate	1.0	1.0	2.0
M. Resource Cost	0.5	0.5	0.0
M. Population	1.0	1.0	0.5
M. Activeness	0.5	0.5	0.5
M. Threshold	2.0	2.0	2.0

three) or multiple behaviors (reproduction and migration), or agents may decide not to invoke any behaviors according to a behavior invocation mechanism shown in Figure 2.

In reproduction, agents select a partner agent that is older than 30 simulated minutes. Agents replicate when there are no partner agents. In both reproduction and replication, a parent agent (the parent who decides to reproduce, not the parent selected as a reproduction partner) provides 3000 energy units to its child agent. Mutation occurs for each weight probabilistically according to the mutation rate of 0.2. In mutation, each weight is subject to random change within a range of mutation range of 0.1. Crossover and mutation are applied only to weights and not applied to thresholds. The predefined performance requirement used in partner selection is 1.0 seconds for the average waiting time, 0.5 hops for the average hop count, and 0.3 for the average energy efficiency.

In migration, agents are allowed to migrate only to adjacent nodes linked with the current node. Reproduction is allowed between two agents on the same platform or on adjacent platforms. Agents die when they exhaust their energy or reach a maximum age of 4 simulated hours.

3.2 Simulation Results

One of the three types of agents listed in Table 1 (See Section 2.2 for the detail of weight) is placed on the top left corner of the network, given 10000 energy units, and simulations are run for 10 simulated days.

Figures 4 through 6 depict simulation results for each of the four types of agents, comparing the performance of agents with the evolutionary mechanism against that of agents without the evolutionary mechanism. Figure 7 shows the dynamics of an agent population. In these figures, the horizontal axis indicates simulation time (in 1 hour increments), while the vertical axis indicates the total energy gain of all agents, average waiting time of all service requests, average hop count of all service requests, the number of agents.

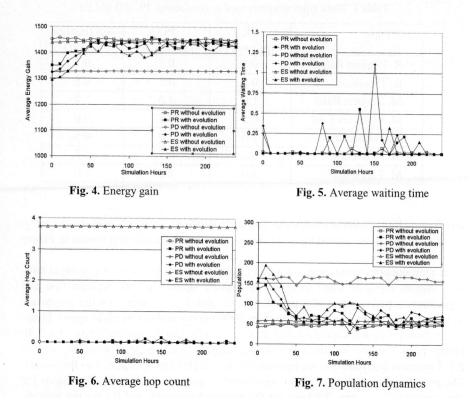

Fig. 4. Energy gain

Fig. 5. Average waiting time

Fig. 6. Average hop count

Fig. 7. Population dynamics

These simulation results are summarized as follows; PR doesn't evolve to improve any of the performance criteria considered, but rather degrades the performance; PD evolves toward invoking replication and reproduction less often, and improves the energy efficiency; ES evolves toward migrating to a user requesting a service and significantly improves the average hop count.

The simulation results explained above show that the evolutionary mechanism designed in this paper successfully improves performance, e.g., energy gain of PD, hop count of ES. However, the performance improvement comes with an overhead associated with the constantly repeating evolutionary process. In other words, when agents achieve a sufficiently low waiting time and hop count (e.g., the both averages reach nearly 0), they then try to minimize energy consumption. This means that in reproduction they select a partner agent based on energy efficiency (e.g., an agent who invokes reproduction or migration less often). This energy efficiency based selection is constantly performed, which creates a situation in which there are not enough agents to process all service requests on the network, resulting in occasional spikes in waiting time and hop count. A possible improvement would be brought about by applying an adaptive evolution technique traditionally suggested in the evolutionary computation literature [6], e.g., by reducing the mutation rate as solutions become closer to optima.

4   Evolution in a Variety of Network Environments

This section describes the extensive set of simulations performed in a variety of network environments and demonstrates the adaptability of the network application with evolutionary mechanisms.

4.1   Simulation Configurations

Evolutionary mechanisms are evaluated in the following three kinds of network environments. Note that each of these networks is a modified version of the static network used in the previous simulation and, unless otherwise stated, simulation configurations (e.g., the network topology, size, platform cost, etc) are the same as the previous ones. Also, only the PR agent (shown in Table 1) is used in the following simulations.

Network with varying resource cost: All platforms vary the resource cost depending on how many agents they are hosting. The following formula is experimentally used to determine the resource cost: *resource_cost = (the number of agents)×2/3*.Four users are placed on four different platforms. Each user has a different service request generation rate, 25, 50, 75 and 100 service requests per second.

Network with varying workload: A user requests either of two types of services: one service that takes agents 0.2 seconds to process, or another that takes 0.4 seconds. There are 10 users, who issue 5 service requests per second, randomly switching the types of services to request.

Network with platform failures: All platforms have a possibility of failure. Platform failures destroy all agents residing on the failed platform, and any service request in the platform queue is also discarded. The availability of all platforms starts with 1.0 and progressively decreases based on the following equation: *platform_availability = 1.0 − 0.025×simulation_hours*. Platforms probabilistically fail every single minute based on platform availability, and the failed platforms become available one second later. Five users are placed on five randomly selected platforms. The users stay on the selected platforms and generate 10 service requests per second throughout the simulations.

4.2   Simulation Results

Simulations are run with and without evolutionary mechanisms in each of the three kinds of networks. Results are shown in figures 8 through 11 and explained in the following.

In the network with varying resource cost, agents without the evolutionary mechanism suffer from a great energy loss leading to lower populations on a platform, and the inadequate number of agents leads to higher waiting time. Evolutionary adaptation allows agents to adapt to the locality of the network environment: those agents on expensive platforms evolve toward migrating away from the platforms, and those who

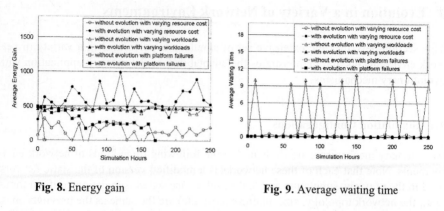

Fig. 8. Energy gain

Fig. 9. Average waiting time

Fig. 10. Average hop count

Fig. 11. Population dynamics

are on the affordable platform remain there. Consequently, the average hop count slightly increases, but the average energy gain significantly improves.

In the network with varying workload, agents without evolutionary adaptation severely increase the waiting time when the type of a service requested is changed. On the other hand, evolving agents appear to be responsive to the change: when experiencing a high waiting time, responsive agents that exhibit higher reproduction rate are likely to be selected as reproduction partners, which accelerates reproduction, resulting in a reasonable average waiting time.

In the network with platform failures, without the evolutionary mechanism, agents become extinct after consecutive platform failures and are unable to continue providing the network application. On the other hand, the evolutionary mechanism leads to population increase and dispersal, resulting in more available and robust network applications. This evolutionary adaptation occurs because agents that can distribute among platforms and that can produce more agents are more likely to survive in the face of platform failure.

5   Conclusions

This paper proposes an evolutionary framework where a network service is created from a group of simple agents that interact and evolve. This work has been done in order to improve the emergent properties such as scalability and adaptability of the Bio-Networking Architecture [18], which is a generic framework for building large-scale network services based on biological principles.

A design concept similar to the one applied to our framework is found in swarm intelligence [1,8], where simple individual agents collectively solve a problem. Swarm intelligence has been applied to some networking issues such as network routing and load-balancing [17], and we further apply the swarm intelligence to build adaptive and scalable network applications.

Although specific applications of our framework are not discussed in this paper, autonomous agents in our framework can represent any network objects (data and a service), and thus our framework can be applied, for instance, to replica placement of network objects for CDNs (Content Distribution Networks) and P2P networks. As opposed to replica placement algorithms proposed for CDNs and P2P networks [2,7,14,15,16], which are often complex, the behavior algorithm of agents in our framework is simple and easy to design, yet demonstrates through simulations that the service performance improves with evolutionary adaptation. An additional simulation study will be done focusing on specific network services or applications. In addition, the presented simulation study assumes a network that contains only one network service. Ongoing work is extending this model to include multiple network services that interact and coevolve.

Acknowledgements. This work was supported by the National Science Foundation through grants ANI-0083074 and ANI-9903427, by DARPA through Grant MDA972-99-1-0007, by Air Force Office of Scientific Research through Grant MURI F49620-00-1-0330, and by grants from the University of California MICRO Program, Hitachi, Hitachi America, Novell, Nippon Telegraph and Telephone Corporation (NTT), NTT Docomo, Fujitsu, and NS Solutions Cooperation.

References

1. E. Bonabeau, M. Dorigo and G. Theraulaz, "Swarm intelligence: from natural to artificial systems," Oxford University Press, 1999.
2. Y. Chen, R. H. Katz and J. D. Kubiatowicz, "Dynamic replica placement for scalable content delivery," in Proceedings of the First International Workshop on Peer-to-Peer Systems, pages 306-318, 2002.
3. C. A. Coello Coello, "A short tutorial on evolutionary multiobjective optimization," First International Conference on Evolutionary Multi-Criterion Optimization, Springer-Verlag, Lecture Notes in Computer Science, No. 1993, pages 21-40, 2001.
4. J. L. Denebourg, S. Goss, N. Franks, A. SendovaFranks, C. Detrain and L. Chretien, "The dynamics of collective sorting robot-like ants and ant-like robots," in Proceedings of the 1st Conference on Simulation of Adaptive Behavior: From Animal to Animats, MIT Press, pp. 356-365.

5. M. Dorigo and G. D. Caro, "Ant algorithms for discrete optimization", in Proceedings of the Congress on Evolutionary Computation, 1999.
6. A. E. Eiben, R. Hinterding and Z. Michalewicz, "Parameter control in evolutionary algorithms," IEEE Transactions on Evolutionary Computation," Vol. 3, No. 2, pages 124-141, 1999.
7. M. J. Kaiser, K. C. Tsui and J. Liu, "Adaptive distributed caching," in Proceedings of the IEEE Congress on Evolutionary Computation, pages 1810-1815, IEEE, 2002.
8. J. Kennedy and R. C. Eberhart, "Swarm intelligence," Morgan Kaufmann Publishers, 2001.
9. J. Kennedy and R.C. Eberhart, "Particle swarm optimization," in Proceedings of the IEEE International Conference on Neural Networks, Vol. 4, 1942-1948, 1995.
10. R. C. Kube and H. Zhang, "Collective robotics: from social insects to robots," Adaptive Behavior, Vol. 2, No. 2, pp.189-218, 1994.
11. W. Y. Ma, B. Shen and J. T. Brassil, "Content services networks: the architecture and protocol," in Proceedings of the 6th International Workshop on Web Caching and Content Distribution, 2001.
12. M. Mitchell, "An introduction to genetic algorithms," MIT Press, 1996.
13. A. Oram, "Peer-to-Peer: harnessing the power of disruptive technologies," O'Reilly & Associates, 2001.
14. G. Pierre, M. van Steen and A. Tanenbaum, "Dynamically selecting optimal distribution strategies for web documents," IEEE Transactions on Computers, 51(6), 2002.
15. L. Qiu, V. N. Padmanabhan and G. M. Voelker, "On the placement of web server replicas," in Proceedings of the IEEE INFOCOM 2001, pages1587-1596, 2001.
16. M. Rabinovich, I. Rabinovich, R. Rajaraman and A. Aggarwal, "A dynamic object replication and migration protocol for an Internet hosting service," in Proceedings of the International Conference on Distributed Computing Systems, pages 101-113, 1999.
17. R. Schoonderwoerd, O. E. Holland, J. L. Bruten and L. J. M. Rothkrantz, "Ant-based load balancing in telecommunications networks," Adaptive Behavior, Vol. 5, No. 2, MIT Press, pp.169-207, 1996.
18. T. Suda, T. Itao and M Matsuo, "The bio-networking architecture: the biologically inspired approach to the design of scalable, adaptive, and survivable/available network applications," in K. Park (ed.), The Internet as a Large-Scale Complex System, Oxford University Press, 2003.

Grammatical Swarm

Michael O'Neill[1] and Anthony Brabazon[2]

[1] Biocomputing and Developmental Systems Group
University of Limerick, Ireland
Michael.ONeill@ul.ie
[2] University College Dublin, Ireland
Anthony.Brabazon@ucd.ie

Abstract. This proof of concept study examines the possibility of specifying the construction of programs using a Particle Swarm algorithm, and represents a new form of automatic programming based on Social Learning, *Social Programming* or *Swarm Programming*. Each individual particle represents choices of program construction rules, where these rules are specified using a Backus-Naur Form grammar. The results demonstrate that it is possible to generate programs using the Grammatical Swarm technique.

1 Introduction

One model of social learning that has attracted interest in recent years is drawn from a swarm metaphor. Two popular variants of swarm models exist, those inspired by studies of social insects such as ant colonies, and those inspired by studies of the flocking behavior of birds and fish. This study focuses on the latter. The essence of these systems is that they exhibit flexibility, robustness and self-organization [1]. Although the systems can exhibit remarkable coordination of activities between individuals, this coordination does not stem from a 'center of control' or a 'directed' intelligence, rather it is self-organizing and emergent. Social 'swarm' researchers have emphasized the role of social learning processes in these models [2,3]. In essence, social behavior helps individuals to adapt to their environment, as it ensures that they obtain access to more information than that captured by their own senses.

This paper details an investigation examining the possibility of specifying the automated construction of a program using a Particle Swarm learning model. In the Grammatical Swarm (GS) approach, each particle or real-valued vector, represents choices of program construction rules specified as production rules of a Backus-Naur Form grammar.

This approach is grounded in the linear program representation adopted in Grammatical Evolution (GE) [4,5,6,7,8], which uses grammars to guide the construction of syntactically correct programs, specified by variable-length genotypic binary or integer strings. The search heuristic adopted with GE is thus a variable-length Genetic Algorithm. In the Grammatical Swarm technique presented here, a particle's real-valued vector is used in the same manner as the

K. Deb et al. (Eds.): GECCO 2004, LNCS 3102, pp. 163–174, 2004.
© Springer-Verlag Berlin Heidelberg 2004

genotypic binary string in GE. This results in a new form of automatic programming based on social learning, which we could dub *Social Programming*, or *Swarm Programming*. It is interesting to note that this approach is completely devoid of any crossover operator characteristic of Genetic Programming.

The remainder of the paper is structured as follows. Before describing the mechanism of Grammatical Swarm in section 4, introductions to the salient features of Particle Swarm Optimization (PSO) and Grammatical Evolution (GE) are provided in sections 2 and 3 respectively. Section 5 details the experimental approach adopted and results, section 6 provides some discussion of the results, and finally section 7 details conclusions and future work.

2   Particle Swarm Optimization

In the context of PSO, a swarm can be defined as '... a population of interacting elements that is able to optimize some global objective through collaborative search of a space.' [2](p. xxvii). The nature of the interacting elements (particles) depends on the problem domain, in this study they represent program construction rules. These particles move (fly) in an n-dimensional search space, in an attempt to uncover ever-better solutions to the problem of interest.

Each of the particles has two associated properties, a current position and a velocity. Each particle has a memory of the best location in the search space that it has found so far (*pbest*), and knows the location of the best location found to date by all the particles in the population (or in an alternative version of the algorithm, a neighborhood around each particle) (*gbest*). At each step of the algorithm, particles are displaced from their current position by applying a velocity vector to them. The velocity size / direction is influenced by the velocity in the previous iteration of the algorithm (simulates 'momentum'), and the location of a particle relative to its pbest and gbest. Therefore, at each step, the size and direction of each particle's move is a function of its own history (experience), and the social influence of its peer group.

A number of variants of the PSA exist. The following paragraphs provide a description of the basic *continuous* version described by [2].

   i. Initialize each particle in the population by randomly selecting values for its location and velocity vectors.

   ii. Calculate the fitness value of each particle. If the current fitness value for a particle is greater than the best fitness value found for the particle so far, then revise *pbest*.

  iii. Determine the location of the particle with the highest fitness and revise *gbest* if necessary.

  iv. For each particle, calculate its velocity according to equation 1.

   v. Update the location of each particle.

  vi. Repeat steps ii - v until stopping criteria are met.

The update algorithm for the velocity, v, of each dimension, i, of a vector is:

$$v_i^i = (w * v_i) + (c1 * R_1 * (p_{best} - p_i)) + (c2 * R_2 * (g_{best} - p_i)) \qquad (1)$$

where,
$$w = wmax - ((wmax - wmin)/itermax) * iter \tag{2}$$

$c1 = 1.0$ is the weight associated with the personal best dimension value, $c2 = 1.0$ the weight associated with the global best dimension value, R_1 and R_2 are a random real number between 0 and 1, p_{best} is the vector's best dimension value to date, p_i is the vector's current dimension value, g_{best} is the best dimension value globally, $wmax = 0.9$, $wmin = 0.4$, $itermax$ is the total number of iterations in the simulation, $iter$ is the current iteration value, and $vmax$ places bounds on the magnitude of the updated velocity value.

Once the velocity update for particle i is determined, its position is updated and pbest is updated if necessary.

$$x_i(t+1) = x_i(t) + v_i(t+1) \tag{3}$$

After all particles have been updated, a check is made to determine whether gbest needs to be updated.

$$\hat{y} \in (y_0, y_1, ..., y_n) | f(\hat{y}) = \max (f(y_0), f(y_1), ..., f(y_n)) \tag{4}$$

3 Grammatical Evolution

Grammatical Evolution (GE) is an evolutionary algorithm that can evolve computer programs in any language [4,5,6,7,8], and can be considered a form of grammar-based genetic programming. Rather than representing the programs as parse trees, as in GP [9,10,11,12,13], a linear genome representation is used. A genotype-phenotype mapping is employed such that each individual's variable length binary string, contains in its codons (groups of 8 bits) the information to select production rules from a Backus Naur Form (BNF) grammar. The grammar allows the generation of programs in an arbitrary language that are guaranteed to be syntactically correct, and as such it is used as a generative grammar, as opposed to the classical use of grammars in compilers to check syntactic correctness of sentences. The user can tailor the grammar to produce solutions that are purely syntactically constrained, or they may incorporate domain knowledge by biasing the grammar to produce very specific forms of sentences.

BNF is a notation that represents a language in the form of production rules. It is comprised of a set of non-terminals that can be mapped to elements of the set of terminals (the primitive symbols that can be used to construct the output program or sentence(s)), according to the production rules. A simple example BNF grammar is given below, where `<expr>` is the start symbol from which all programs are generated. These productions state that `<expr>` can be replaced with either one of `<expr><op><expr>` or `<var>`. An `<op>` can become either +, -, or *, and a `<var>` can become either x, or y.

```
<expr> ::= <expr><op><expr>  (0)
         | <var>             (1)
<op> ::= +                   (0)
       | -                   (1)
       | *                   (2)
```

```
<var> ::= x              (0)
        | y              (1)
```

The grammar is used in a developmental process to construct a program by applying production rules, selected by the genome, beginning from the start symbol of the grammar. In order to select a production rule in GE, the next codon value on the genome is read, interpreted, and placed in the following formula:

$$Rule = Codon\ Value\ \%\ Num.\ Rules$$

where % represents the modulus operator.

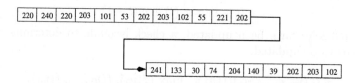

Fig. 1. An example GE individuals' genome represented as integers for ease of reading.

Given the example individuals' genome (where each 8-bit codon is represented as an integer for ease of reading) in Fig.1, the first codon integer value is 220, and given that we have 2 rules to select from for <expr> as in the above example, we get 220 % 2 = 0. <expr> will therefore be replaced with <expr><op><expr>. Beginning from the the left hand side of the genome, codon integer values are generated and used to select appropriate rules for the left-most non-terminal in the developing program from the BNF grammar, until one of the following situations arise: (a) A complete program is generated. This occurs when all the non-terminals in the expression being mapped are transformed into elements from the terminal set of the BNF grammar. (b) The end of the genome is reached, in which case the *wrapping* operator is invoked. This results in the return of the genome reading frame to the left hand side of the genome once again. The reading of codons will then continue unless an upper threshold representing the maximum number of wrapping events has occurred during this individuals mapping process. (c) In the event that a threshold on the number of wrapping events has occurred and the individual is still incompletely mapped, the mapping process is halted, and the individual assigned the lowest possible fitness value. Returning to the example individual, the left-most <expr> in <expr><op><expr> is mapped by reading the next codon integer value 240 and used in 240 % 2 = 0 to become another <expr><op><expr>. The developing program now looks like <expr><op><expr><op><expr>. Continuing to read subsequent codons and always mapping the left-most non-terminal the individual finally generates the expression y*x-x-x+x, leaving a number of unused codons at the end of the individual, which are deemed to be introns and simply ignored. Fig.2 draws an analogy between GE's mapping process and the molecular biological processes of transcription and translation. A full description of GE can be found in [4].

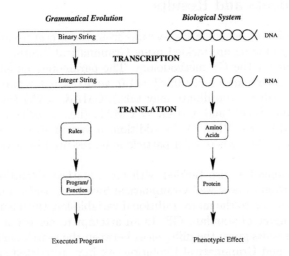

Fig. 2. A comparison between Grammatical Evolution and the molecular biological processes of transcription and translation. The binary string of GE is analogous to the double helix of DNA, each guiding the formation of the phenotype. In the case of GE, this occurs via the application of production rules to generate the terminals of the compilable program. In the biological case by directing the formation of the phenotypic protein by determining the order and type of protein subcomponents (amino acids) that are joined together.

4   Grammatical Swarm

Grammatical Swarm (GS) adopts a Particle Swarm learning algorithm coupled to a Grammatical Evolution (GE) genotype-phenotype mapping to generate programs in an arbitrary language. The update equations for swarm algorithm are as described earlier, with additional constraints placed on the velocity and dimension values, such that velocities are bound to VMAX=±255, and each dimension is bound to the range 0 to 255. Note that this is a continuous swarm algorithm with real-valued particle vectors. The standard GE mapping function is adopted with the real-values in the particle vectors being rounded up or down to the nearest integer value, for the mapping process. In the current implementation of GS, fixed-length vectors are adopted within which it is possible for a variable number of elements to be required during the program construction genotype-phenotype mapping process. A vector's values may be used more than once if the wrapping operator is used, and in the opposite case it is possible that not all elements will be used during the mapping process if a complete program comprised only of terminal symbols is generated before reaching the end of the vector. In this latter case, the extra element values are simply ignored and considered introns that may be switched on in subsequent iterations.

5   Experiments and Results

A diverse selection of benchmark programs from the literature on evolutionary automatic programming are tackled using Grammatical Swarm to demonstrate proof of concept for the GS methodology. The parameters adopted across the following experiments are c1 = 1.0, c2 = 1.0, wmax = 0.9, wmin = 0.4, CMIN = 0 (minimum value a coordinate may take), CMAX = 255 (maximum value a coordinate may take), and VMAX = CMAX (i.e., velocities are bound to the range +VMAX to -VMAX). In addition, a swarm size of 30 running for 1000 iterations is used, where each particle is represented by a vector with 100 elements.

The same problems are also tackled with Grammatical Evolution in order to get some indication of how well Grammatical Swarm is performing at program generation in relation to the more traditional variable-length Genetic Algorithm-driven search engine of standard GE. In an attempt to achieve a relatively fair comparison of results given the differences between the search engines of Grammatical Swarm and Grammatical Evolution, we have restricted each algorithm in the number of individuals they process, and using typical population sizes from the literature adopted for each method. Grammatical Swarm running for 1000 iterations with a swarm size of 30 processes 30,000 individuals, therefore, a standard population size of 500 running for 60 generations is adopted for Grammatical Evolution. The remaining parameters for Grammatical Evolution are roulette selection, steady state replacement, one-point crossover with probability of 0.9, and a bit mutation with probability of 0.01.

5.1   Santa Fe Ant Trail

The Santa Fe ant trail is a standard problem in the area of GP and can be considered a deceptive planning problem with many local and global optima [14]. The objective is to find a computer program to control an artificial ant so that it can find all 89 pieces of food located on a non-continuous trail within a specified number of time steps, the trail being located on a 32 by 32 toroidal grid. The ant can only turn left, right, move one square forward, and may also look ahead one square in the direction it is facing to determine if that square contains a piece of food. All actions, with the exception of looking ahead for food, take one time step to execute. The ant starts in the top left-hand corner of the grid facing the first piece of food on the trail. The grammar used in this problem is different to the ones used later for symbolic regression and the multiplexer problem in that we wish to produce a multi-line function in this case, as opposed to a single line expression. The grammar for the Santa Fe ant trail problem is given below.

```
<code> ::= <line> | <code> <line>
<line> ::= <condition> | <op>
<condition> ::= if(food_ahead()) { <line> } else { <line> }
<op> ::= left(); | right(); | move();
```

A plot of the mean best fitness and cumulative frequency of success for 30 runs can be seen in Fig.3. As can be seen, convergence towards the best fitness occurs, and a number of runs successfully obtain the correct solution.

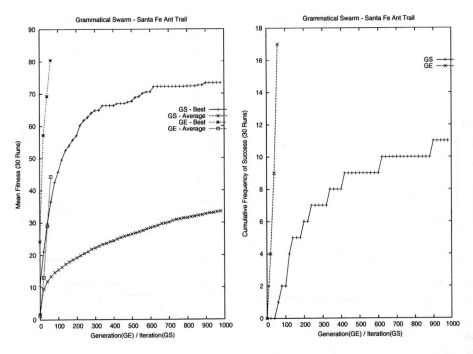

Fig. 3. Plot of the mean fitness on the Santa Fe Ant Trail problem instance (left), and the cumulative frequency of success (right).

5.2   Quartic Symbolic Regression

The target function is $f(a) = a + a^2 + a^3 + a^4$, and 100 randomly generated input-output vectors are created for each call to the target function, with values for the input variable drawn from the range [0,1]. The fitness for this problem is given by the reciprocal of the sum, taken over the 100 fitness cases, of the absolute error between the evolved and target functions. The grammar adopted for this problem is as follows:

```
<expr> ::= <expr> <op> <expr> | <var>
<op>   ::= + | - | * | /
<var>  ::= a
```

A plot of the cumulative frequency of success and the mean best fitness over 30 runs can be seen in Fig.4. As can be seen, a number of runs successfully find the correct solution to the problem, with convergence towards the best fitness occurring on average.

5.3   3 Multiplexer

An instance of a multiplexer problem is tackled in order to further verify that it is possible to generate programs using Grammatical Swarm. The aim with this

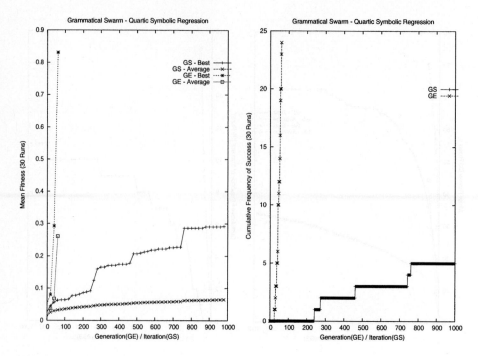

Fig. 4. Plot of the mean fitness on the quartic symbolic regression problem instance (left), and the cumulative frequency of success (right).

problem is to discover a boolean expression that behaves as a 3 Multiplexer. There are 8 fitness cases for this instance, representing all possible input-output pairs. Fitness is the number of input cases for which the evolved expression returns the correct output. The grammar adopted for this problem is as follows:

```
<mult>  ::= guess = <bexpr> ;
<bexpr> ::= ( <bexpr> <bilop> <bexpr> )
          | <ulop> ( <bexpr> )
          | <input>
<bilop> ::= and | or
<ulop>  ::= not
<input> ::= input0 | input1 | input2
```

A plot of the mean best fitness over 30 runs can be seen in Fig.5. As can be seen, convergence towards the best fitness occurs, and a number of runs successfully evolve correct solutions.

5.4   Mastermind

In this problem the code breaker attempts to guess the correct combination of colored pins in a solution. When an evolved solution to this problem (i.e. a combination of pins) is to be evaluated, it receives one point for each pin that

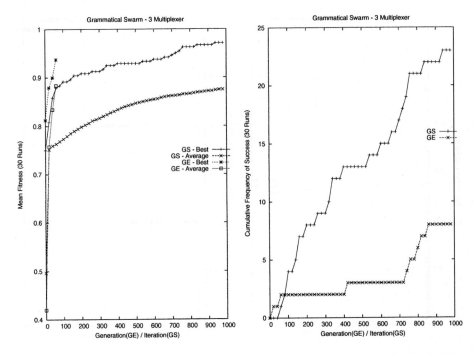

Fig. 5. Plot of the mean fitness on the 3 multiplexer problem instance (left), and the cumulative frequency of success (right).

has the correct color, regardless of its position. If all pins are in the correct order than an additional point is awarded to that solution. This means that ordering information is only presented when the correct order has been found for the whole string of pins.

A solution, therefore, is in a local optimum if it has all the correct color, but in the wrong positions. The difficulty of this problem is controlled by the number of pins and the number of colors in the target combination. The instance tackled here uses 4 colors and 8 pins with the following values 3 2 1 3 1 3 2 0.

Results are provided in Fig. 6 and the grammar adopted is as follows.

```
<pin> ::= <pin> <pin> | 0 | 1 | 2 | 3
```

6   Discussion

Table 1 provides a summary and comparison of the performance of Grammatical Swarm and Grammatical Evolution on each of the problem domains tackled. In two out of the four problems Grammatical Evolution outperforms Grammatical Swarm, Grammatical Swarm outperforms Grammatical Evolution on one problem instance, and there is a tie between the methods on the Mastermind problem. The key finding is that the results demonstrate proof of concept that Grammatical Swarm can successfully generate solutions to problems of interest.

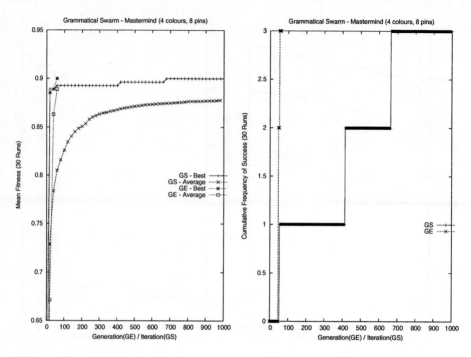

Fig. 6. Plot of the mean best and mean average fitness (left) and the cumulative frequency of success (right) on the Mastermind problem instance using 8 pins and 4 colors. Fitness is defined as the points score of a solution divided by the maximum possible points score.

In this initial study, we have not attempted parameter optimization for either algorithm, but results and observations of the particle swarm engine suggests that swarm diversity is open to improvement. We note that a number of strategies have been suggested in the swarm literature to improve diversity [15], and we suspect that a significant improvement in Grammatical Swarms' performance can be obtained with the adoption of these measures. Given the relative simplicity of the Swarm algorithm, the small population sizes involved, and the complete absence of a crossover operator synonymous with program evolution in GP, it is impressive that solutions to each of the benchmark problems have been obtained.

When analyzing the results presented one has to consider the fact that the Grammatical Evolution representation is variable-length with individuals' lengths restricted only by the machines physical storage limitations. In the current implementation of Grammatical Swarm fixed-length vectors are adopted in which a variable number of dimensions can be used, however, vectors have a hard length constraint of 100 elements. We intend to implement a variable-length version of Grammatical Swarm that will allow the number of dimensions of a particle to increase and decrease over simulation time to overcome this current limitation.

Table 1. A comparison of the results obtained for Grammatical Swarm and Grammatical Evolution across all the problems analyzed.

	Mean Best Fitness (Std.Dev.)	Mean Average Fitness (Std.Dev.)	Successful Runs
Santa Fe ant			
GS	73.3 (17.6)	33.6 (3.32)	11
GE	**80.4** (14.4)	**44.3** (5.7)	**17**
Multiplexer			
GS	**0.97** (0.05)	0.88 (0.01)	**23**
GE	0.94 (0.06)	0.88 (0.02)	15
Symbolic Regression			
GS	0.29 (0.35)	0.07 (0.02)	5
GE	**0.83** (0.33)	**0.26** (0.26)	**24**
Mastermind			
GS	**0.9** (0.03)	**0.88** (0.013)	**3**
GE	**0.9** (0.03)	**0.89** (0.001)	**3**

7   Conclusions and Future Work

This study demonstrates the feasibility of the generation of computer programs using Grammatical Swarm over four different problem domains. As such a new form of automatic programming based on social learning is introduced, which could be termed *Social Programming*, or *Swarm Programming*. While a performance comparison to Grammatical Evolution has shown that Grammatical Swarm is outperformed on two of the problems analyzed, the ability of Grammatical Swarm to generate solutions with such small populations, with a fixed-length vector representation, an absence of any crossover, no concept of selection or replacement, and without optimization of the algorithm's parameters is very encouraging for future development of the much simpler Grammatical Swarm, and other potential Social or Swarm Programming variants. Future work will involve developing a variable-length Particle Swarm algorithm to remove Grammatical Swarms length constraint, conducting an investigation into swarm diversity, the impact of a continuous encoding over a discrete encoding variant such as presented in [16], and considering the implications of a social learning approach to the automatic generation of programs.

References

1. Bonabeau, E., Dorigo, M. and Theraulaz, G. (1999). *Swarm Intelligence: From natural to artificial systems*, Oxford: Oxford University Press.
2. Kennedy, J., Eberhart, R. and Shi, Y. (2001). *Swarm Intelligence*, San Mateo, California: Morgan Kauffman.
3. Kennedy, J. and Eberhart, R. (1995). Particle swarm optimization, *Proc. of the IEEE International Conference on Neural Networks*, pp.1942-1948.
4. O'Neill, M., Ryan, C. (2003). *Grammatical Evolution: Evolutionary Automatic Programming in an Arbitrary Language*. Kluwer Academic Publishers.

5. O'Neill, M. (2001). *Automatic Programming in an Arbitrary Language: Evolving Programs in Grammatical Evolution.* PhD thesis, University of Limerick, 2001.
6. O'Neill, M., Ryan, C. (2001). Grammatical Evolution, *IEEE Trans. Evolutionary Computation.* Vol. 5, No.4, 2001.
7. O'Neill, M., Ryan, C., Keijzer M., Cattolico M. (2003). Crossover in Grammatical Evolution. *Genetic Programming and Evolvable Machines,* Vol. 4 No. 1. Kluwer Academic Publishers, 2003.
8. Ryan, C., Collins, J.J., O'Neill, M. (1998). Grammatical Evolution: Evolving Programs for an Arbitrary Language. *Proc. of the First European Workshop on GP,* 83-95, Springer-Verlag.
9. Koza, J.R. (1992). *Genetic Programming.* MIT Press.
10. Koza, J.R. (1994). *Genetic Programming II: Automatic Discovery of Reusable Programs.* MIT Press.
11. Banzhaf, W., Nordin, P., Keller, R.E., Francone, F.D. (1998). *Genetic Programming – An Introduction; On the Automatic Evolution of Computer Programs and its Applications.* Morgan Kaufmann.
12. Koza, J.R., Andre, D., Bennett III, F.H., Keane, M. (1999). *Genetic Programming 3: Darwinian Invention and Problem Solving.* Morgan Kaufmann.
13. Koza, J.R., Keane, M., Streeter, M.J., Mydlowec, W., Yu, J., Lanza, G. (2003). *Genetic Programming IV: Routine Human-Competitive Machine Intelligence.* Kluwer Academic Publishers.
14. Langdon, W.B., and Poli, R. (1998). Why Ants are Hard. In *Genetic Programming 1998: Proc. of the Third Annual Conference,* University of Wisconsin, Madison, Wisconsin, USA, pp. 193-201, Morgan Kaufmann.
15. Silva, A., Neves, A., Costa, E. (2002). An Empirical Comparison of Particle Swarm and Predator Prey Optimisation. In *LNAI 2464, Artificial Intelligence and Cognitive Science, the 13th Irish Conference AICS 2002,* pp. 103-110, Limerick, Ireland, Springer.
16. Kennedy, J., and Eberhart, R. (1997). A discrete binary version of the particle swarm algorithm. *Proc. of the 1997 Conference on Systems, Man, and Cybernetics,* pp. 4104-4109. Piscataway, NJ: IEEE Service Center.

A New Universal Cellular Automaton
Discovered by Evolutionary Algorithms

Emmanuel Sapin[1], Olivier Bailleux[1], Jean-Jacques Chabrier[1], and
Pierre Collet[2]

[1] Université de Bourgogne, 9 av. A. Savary, B.P. 47870, 21078 Dijon Cedex, France
emmanuelsapin@hotmail.com {olivier.bailleux,jjchab}@u-bourgogne.fr
[2] Laboratoire d'Informatique du Littoral, ULCO, Calais, France
Pierre.Collet@Univ-Littoral.Fr

Abstract. In *Twenty Problems in the Theory of Cellular Automata*,
Stephen Wolfram asks "how common computational universality and undecidability [are] in cellular automata." This papers provides elements of
answer, as it describes how another universal cellular automaton than the
Game of Life (*Life*) was sought *and found* using evolutionary algorithms.
This paper includes a demonstration that consists in showing that the
presented R automaton can both implement any logic circuit (logic universality) and a simulation of *Life* (universality in the Turing sense).
All the elements of the evolutionary algorithms that were used to find R
are provided for replicability, as well as the analytical description in R
of a cell of *Life*.

1 Introduction

Cellular automata are discrete systems [1] in which a population of cells evolves
from generation to generation on the basis of local transitions rules. They can
simulate simplified "forms of life" [2,3] or physical systems with discrete time,
space and local interactions [4,5,6].

In [7], Wolfram studies the space $\mathcal{I}$ of isotropic two states 2D automata
with a transition rule that takes into account the eight neighbours of a cell, to
determine the cell's state at the next generation. He talks of a special automaton
of $\mathcal{I}$ called *Game of Life* (hereafter referred to as *Life*) that was discovered by
Conway in 1970 and popularised by Gardner in [2]. In [8], Conway, Berlekamp
and Guy show that *Life* allows to compute any function calculable by a Turing
machine. Their demonstration of the universality of *Life* uses *gliders*, *glider guns*
and *eaters*. Gliders are patterns which, when evolving alone, periodically recover
their original shape after some shift in space. A glider gun emits a stream of
gliders that can be used to carry information. An eater absorbs gliders and,
along with stream collisions, allows to create and combine logic gates into logic
circuits. In [9], Rendell gives an explicit proof of the universality of *Life* by
showing a direct simulation of counter machines.

K. Deb et al. (Eds.): GECCO 2004, LNCS 3102, pp. 175–187, 2004.
© Springer-Verlag Berlin Heidelberg 2004

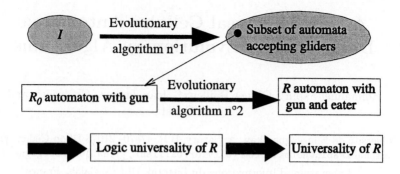

Fig. 1. Research approach used to find new universal automata.

In [10], Wolfram lists twenty problems on the theory of cellular automata. The sixteenth is introduced by the question *"How common are computational universality and undecidability in cellular automata ?"*

Up to now, *Life* was the only known dynamical universal automaton of $\mathcal{I}$: Margolus's cellular automaton billard-ball model [11] is universal but its transition rules take into account the parity of the generation number meaning that it is not in $\mathcal{I}$. Then, universal automata with more than two states [12,13] or more than two dimensions [14] are not in $\mathcal{I}$ either. Finally, Banks's 2D 2-state cellular automaton [15] is not dynamically universal because its wires are fixed.

This paper describes how another universal automaton of $\mathcal{I}$ was sought *and found* thanks to evolutionary algorithms. Section 2 describes the search process, that is developed in sections 3 and 4. Section 5 describes an analytical way to represent C.A.s, while sections 6 and 7 describe how patterns can be assembled into logic gates and how glider streams can be redirected and synchronised, in order to create logic circuits. Finally, section 8 presents a simulation of *Life* using the R automaton found in sections 3 and 4, and section 9 summarizes the presented results and discusses directions for future research.

2   Rationale

Figure 1 shows how new universal automata are sought. A first evolutionary algorithm is used to find in $\mathcal{I}$ a subset of automata accepting gliders.

In this subset, some automata accept glider guns. An automaton, called R_0, is chosen among them, on which attempts are made to demonstrate its universality.

In [8], Conway *et al.* used an eater to simulate a NAND gate. This observation is used as a starting point and an eater is sought among automata accepting the glider gun found in R_0, to be used as a building block to implement a NAND gate. This eater is found in an automaton called R, thanks to a second evolutionary algorithm.

The ability to create and assemble NAND gates into any logic circuit is then shown, which demonstrates the logic universality of R. Finally, *Life* is simulated in R as a proof that R is universal in the Turing sense, since *Life* was shown to be able to implement registers cf. [8] and counter machines cf. [9].

3   Finding an Automaton Accepting Gliders

The search space of the evolutionary algorithm is the set $\mathcal{I}$ of 2 states 2D automata described in the introduction. An automaton can be described by telling what will become of a cell in the next generation, depending on its neighbours.

If symmetric and rotated neighbourhoods are considered as having an identical effect on a cell (isotropic automata), there are "only" 102 possible different neighbourhoods for one cell (and 2^{102} different automata). Therefore, an individual can be coded as a bitstring of 102 booleans (cf. fig. 2).

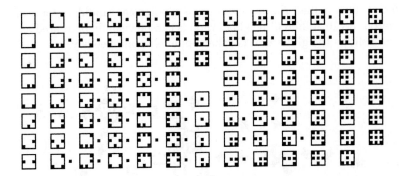

Fig. 2. A black cell on the right of the neighbourhood indicates a future central cell. The R automaton, can then be represented by 00000000011100011101111111 11111100100101111111100000000011101111101110000100010110100000000000000000000.

3.1   Description of the Evolutionary Algorithm That Searches Gliders

Fitness function. In order to find automata that accept gliders, the evolutionary algorithm attempts to maximise the number of gliders × the number of periodic patterns that appear during the evolution of a random configuration of cells by the tested automaton. (A more detailed description of the gliders and periodic patterns detector (inspired by Bays [16]) can be found in [17].)

Initialisation. The 102 bits of each individual are initialised at random.

Genetic operators. The mutation function simply consists in mutating one bit among 102, while the recombination is a single point crossover with a locus situated exactly on the middle of the genotype. This locus was chosen since the first 51 neighbourhoods determine the birth of cells, while the other 51 determine how they survive or die.

Evolution engine. It is very close to a $(\mu + \lambda)$ Evolution Strategy, although on a bitstring individual, and therefore without adaptive mutation: the population is made of 20 parents that are selected randomly to create 20 children by mutation only and 10 other by recombination. As in a straight ES+, the 20 best individuals among the 20 parents + 30 children are selected to create the next generation.

Stopping criterion. The algorithm stops after 200 generations, which, on a 800Mhz PC Athlon, takes around 20 minutes to complete.

3.2   The R_0 Automaton: An Experimental Result

The algorithm described above provided several automata accepting gliders. Among them, some would surprisingly generate glider guns for nearly every evolution of a random cell configuration.

One of them (called R_0) is picked up as a potential candidate for a universal automaton. Its 102 different neighbourhoods can be visually presented as in fig. 2 and fig. 3 shows the glider gun G_0 that appears spontaneously.

Fig. 3. An evolution of a random configuration of cell by R_0 showing G_0.

4   Looking for an "Eater" in Automata Accepting G_0

An eater is now needed in order to simulate a NAND logic gate, be it only because the glider gun of rule R_0 produces 3 too many glider streams that need to be suppressed to avoid interactions with other parts of a simulated logic circuit.

Ideal eaters are periodic patterns that, after the absorption of a glider, can resume their original shape and position, quickly enough to absorb another arriving glider (cf. fig. 4).

As no eater was found in R_0, a second evolutionary algorithm was elaborated to automatically find an eater in the space of all automata accepting the glider gun G_0. This space was determined, by finding deterministically which of the 102

Fig. 4. Eater of R in action, found by a second evolutionary algorithm.

neighbourhoods of automaton R_0 were needed for gun G_0 to operate normally. It turns out that G_0 uses 81 different neigbourhoods, meaning that the output of the 21 other neighbourhoods could be changed in order to find an automaton R that would both accept the glider gun G_0 *and* an eater (cf. [15]).

An eater being a periodic pattern, a collection of 10 small periodic patterns of R_0 appearing frequently *and using only* the 81 neighbourhoods necessary for G_0 were chosen. Those periodic patterns were therefore sure to appear in all of the 2^{21} automata implementing G_0.

Finally, in order to find an eater, one needs to perform on the established collection of periodic patterns what could be called a crash test: each periodic pattern is positioned in front of a stream of gliders, and its fitness is simply the number of crashes it survives.

The number of possibilities being quite large (10 patterns to be tested in different relative positions with reference to the stream of gliders among 2^{21} different automata), a second evolutionary algorithm was therefore elaborated, with the following characteristics:

Individual structure. An individual is made of:
1. a bitstring of 21 bits determining one automaton among 2^{21},
2. an integer between 1 and 10, describing one pattern among the 10 periodic patterns common to the 81 neighbourhoods needed by G_0,
3. the relative position of the pattern relatively to the stream of gliders, coded by two integers, x and y, varying between $[-8, 8]$ and $[0, 1]$.
   Individuals are initialised with R_0, a random integer between 1 and 10 and randomly within their interval for x and y.

Fitness function. Number of gliders stopped by an individual.

Genetic operators. The only operator is a mutator, since no really "intelligent" recombination function could be elaborated. The mutator is therefore called on all created offspring and can either choose any pattern among the 10 available, or mutate one bit in the bitstring, or move the position of the pattern by ± 1 within the defined boundaries for x and y.

Evolution Engine. It is this time closer to an Evolutionary Programming Engine, since it has no crossover, although the EP tournament was not implemented. 30 children are created by mutation of 20 parents selected uniformly. Among the 50 resulting individuals, the best 20 individuals are selected to create the next generation.

Stopping criterion. Discovery of an eater that would survive 50 000 collisions.

This algorithm allowed to discover the automaton R described in fig. 2, accepting *both* the glider gun G_0 and the eater shown in fig. 4.

Interestingly enough, other runs took between 1 and 20 minutes to complete, to always find the same eater pattern, although with different automata.

5   Analytical Description of a CA

Binary numbers can be implemented as a finite stream of gliders, where gliders represent 1s and missing gliders represent 0s.

The next step needed to prove the logic universality of R is to find a configuration of glider guns and eaters that can simulate a NAND gate on two streams of gliders representing two binary numbers.

Unfortunately, the cellular autamaton implementing a NAND gate would be too difficult to represent and explain by showing groups of cells on a grid, let alone a CA implementing a cell of *Life*. Therefore, a much clearer analytical description was needed, that should also allow replicability of the contents of this paper.

In order to simplify the representation of a CA, one can replace its building blocks by an analytical description, made of a letter, referring to the pattern, followed by three parameters (D, x, y) where D denotes a direction ($North$, $East$, $South$, $West$) and x, y the coordinates of a specific cell of the pattern (cf. [18]). An arrow is added in graphic descriptions to help visualising the CA.

Several patterns and their analytical representation are described in this section, namely a glider stream, an eater, a glider gun and a large glider gun:

Glider stream. Fig. 5 shows a glider stream $S(E, x, y)$, where x and y are the coordinates of the white cell whence an arrow is shooting.

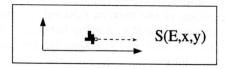

Fig. 5. A glider stream and its analytical representation $S(E, x, y)$.

Eater. The eater of fig. 4 can be identified as $E(N, x, y)$, where N denotes its northward orientation and x, y denote the position of the white cell.

Glider gun. The glider gun of fig. 3 is unfortunately not usable as is, because it shoots gliders spaced every nine cells only. The complex guns shown in fig. 6 shoot gliders spaced by 45 cells, which gives more slack to work on streams. Fig. 6 shows instances of this gun at generation 2, used later on in this paper, namely $Ga(S, x, y)$ and $Gb(S, x, y)$. The guns in other cardinal directions are obtained by rotation thanks to the isotropy of R.

Large glider gun. Another type of glider, called *large glider*, appears in R. A large gun (L) shooting a large glider every 45 cells, is made of two complex guns G, shooting their stream perpendicularly (cf. fig. 7).

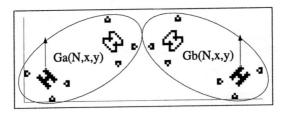

Fig. 6. Symmetric complex guns of R, viewed at generation 2. x, y are the coordinates of the white cell, whence an arrow is shooting.

When both streams collide, large gliders are created that sail on the direction of the stream of the top left gun. The large glider gun of fig. 7 will be referred to as $L(E, 16, 80)$, as it was decided that its coordinates would be those of the top left gun. The shape of a large glider is shown on the right of the gun.

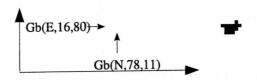

Fig. 7. Schematics of a large glider gun $L(E, 16, 80)$, made of two complex Gb guns.

6   Assembling Patterns into a Not Gate

[18] describes the simulation of an AND gate in R. Unfortunately, it is NAND gates that allow to build any logic circuits, and not AND gates. It is therefore very important to build a NOT gate for the R automaton to be logically universal. Before the description of the NOT gate begins, it is important to observe that:

1. The frontal collision of two identical large glider streams produces two orthogonal standard glider streams equal to the input streams. If only one output glider stream is needed, an eater can be placed in front of the second stream —cf. fig. 8 and $L(E, 16, 80)/L(W, 379, 80)/E(N, 217, 65)$ in fig. 9.
2. The collision between two standard gliders at a right angle, or between a standard glider and a large glider at a right angle destroys both gliders. Therefore, a glider gun (large or complex) G positioned so that it fires gliders at a right angle towards another stream of standard gliders A will complement the stream at a right angle: whenever a glider of stream A (symbolising 1) arrives at the collision point, the glider of stream A and the glider of G disappear (resulting in a 0). On the contrary, an absence of glider in stream A (symbolising 0) will let a glider of G get through the collision point (resulting in a 1) cf. second collision of fig. 8.

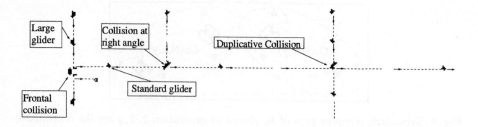

Fig. 8. This figure presents the three types of collisions used in this paper. 1's are represented by gliders, and 0's by an absence of gliders. The figure starts on the left with a frontal collision between two streams of large gliders, that creates a stream of standard gliders going towards the East. In this example, the stream then collides with a periodic stream of 101010... coming from the North at a 90 degrees angle, with the consequence that the information it carries is complemented and turned by 90 degrees into a horizontal eastward stream of 010101.... This stream finally hits another vertical stream made of ones coming from the North although slightly displaced, therefore creating a duplicating collision. The left to right 010101... stream goes through the duplicative collision unmodified, while a complemented stream 101010... is created and issued southwardly.

3. When the right angle stream of gliders is slightly displaced w.r.t. the initial stream, a collision of gliders does not result in a destruction of gliders. Instead, the initial glider goes through the collision, while the glider coming at a 90 degrees angle disappears, duplicating the stream into a complemented stream at a 90 degrees angle —cf. fig. 8 and $Gb(E, 179, 200)$ in fig. 9.

Thanks to these three observations, a NOT gate is built on fig. 9 below:

On this figure, the information stream A ($S(S, 210, 184)$) is shown as a dotted line. A complex glider gun $Gb(E, 179, 200)$ creates a complementary duplicate stream $\overline{A}$ towards the East. The two outputs are redirected by glider guns $Gb(W, 253, 142)$, $Ga(S, 159, 234)$ and $Ga(S, 273, 262)$ until they are vertical again. Then, they are complemented into large gliders by the two large glider guns $L(E, 16, 80)$ and $L(W, 379, 80)$. When the frontal collision between the two large glider streams occurs, a complementary stream $\overline{A}$ is created towards the same direction as the original A stream while the other one is "eaten" by $E(N, 217, 65)$.

7   Intersection and Synchronisation of Streams

In order to prove the logic universality of R, one needs to combine several NAND gates. This is possible if two streams in any position can be redirected in order to become the input streams of a NAND gate. This operation can be realised thanks to intersection and synchronisation patterns.

Intersection. Thanks to complex guns, gliders in a stream are separated by 45 cells. This means that it is possible to have two streams cross each other

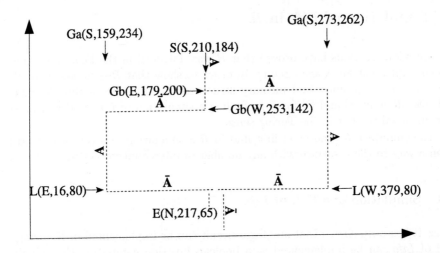

Fig. 9. Complementation of stream A: the analytical representation for the NOT gate is $\{L(E,16,80), Ga(S,159,234), Gb(E,179,200), S(S,210,184), E(N,217,65), Gb(W, 253,142), Ga(S,273,262), L(W,379,80)\}$.

without any interference. If, for a synchronisation reason, interference cannot be avoided, [19] shows how a stream intersection can be realised by a cunning combination of NAND gates.

Synchronisation. It is important to be able to delay one stream w.r.t another, in order to synchronise them properly just before they enter a logic gate, for instance. This can be done precisely by diverting four times the stream to be delayed with orthogonal guns (cf. fig. 10).

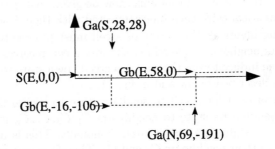

Fig. 10. Stream temporisation for synchronisation purposes.

Stream duplication, redirection, synchronisation and intersection allow to combine any number of NAND gates together, therefore proving the logic universality of R.

8   Simulation of *Life* in *R*

The previous sections have proved that R was universal in the logic sense (i.e.: it can implement any logic circuit). In order to show that R is universal *in the Turing sense*, one must find memory structures like registers within R. Since this was done in [8] and [9] for *Life*, finding a simulation of *Life* in R will prove the universality of R in the Turing sense.

To simulate *Life*, one must first find in R a simulation of a cell of *Life*, and then a way to tile a surface with any number of interconnected cells.

8.1   Simulation of a Cell of *Life*

Since it has been shown that any logic circuit can be implemented in R, a single cell of *Life* can be implemented as a boolean function computing the value of a cell S at generation $n + 1$ from the value of its eight neighbours $C_1 \ldots C_8$ at generation n.

The rules of *Life* are the following: a "living" cell dies at the next generation unless it has two or three neighbours. A dead cell comes alive at the next generation iff it has three neighbours in the current generation.

Supposing that the addition of $C_1 + \ldots + C_8$ gives a three bit number $n_2 n_1 n_0$, the rules of *Life* can be simply expressed by the formula $S_{n+1} = \overline{n_2}.n_1.(S_n + n_0)$, which can be translated into a combination of NAND gates. This function, implementing a cell of *Life*, is represented by the grey area of fig. 11 below, and analytically described in R in the appendix.

8.2   Interconnecting Cells of *Life*

In order to simulate *Life* in R, proof must now be given that it is possible to tile a surface with identical cells, each interconnected with their 8 neighbours.

All cells being identical, the inputs of a cell must physically correspond to the outputs of its neighbours. Therefore, the way a cell receives the state of its neighbours can be induced from the way it sends its own state to its neighbours, which is what is described below and in fig. 11.

It is straightforward for a cell to send its state to its cardinal neighbours C_2, C_4, C_5, C_7. Sending its state to neighbours C_1, C_3, C_6, C_8 is however more tricky, since those neighbours are situated diagonally. This is done by passing the information to their neighbours C_2 and C_7. Therefore, one can see in fig. 11 that the state S of the cell is sent three times to C_2 and three times to C_7, so that C_2 (resp. C_7) can keep one stream for its own use, and pass the two others to its horizontal neighbours, C_1 and C_3 (resp. C_6 and C_8).

S being itself a top neighbour of cell C_7, one sees how the state of C_7 is passed over to C_4 and C_5 in the same way that C_2 will pass over the information of the state of S to C_1 and C_3.

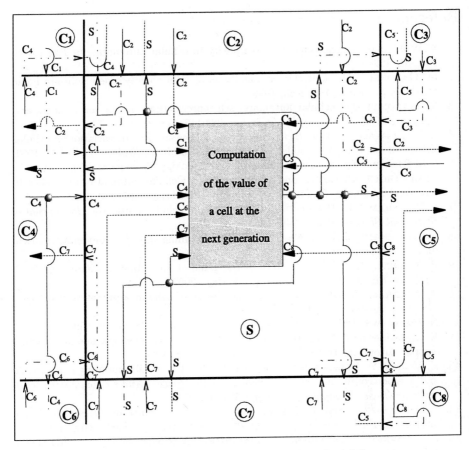

Fig. 11. Diagram of the simulation of a cell of *Life*.

9   Synthesis and Perspectives

An extensive bibliographic research seems to show that this paper actually presents the first proof that another 2D 2 state dynamical universal (in the Turing sense) automaton other than the famous *Life* exists in $\mathcal{I}$, therefore providing an element of answer to Wolfram's 16th problem.

Evolutionary algorithms played a key role in discovering gliders, and a rule R accepting a glider gun and eaters, in a very large search space.

Further goals are now to give a more complete answer to Wolfram's problem by finding whether other universal automata than *Life* and R exist, and how common they are. Then, another domain that seems worth exploring is how this approach could be extended to automata with more than 2 states.

Finally, the study of the construction of an automatic system of selection / discovery of such type of automata based on evolutionary algorithms is far more interesting, as it could lead to a new classification.

References

1. S. WOLFRAM. Universality and complexity in cellular automata. *In Physica D*, 10:1–35, 1984.
2. M. GARDNER. The fantastic combinaisons of john conway's new solitaire game "life". *In Scientific American*, 1970.
3. M. GARDNER. On cellular automata, self-reproduction, the garden of eden, and the game of life. *In Scientific American*, 224:112–118, 1971.
4. C. DYTHAM and B. SHORROCKS. Selection, patches and genetic variation: A cellular automata modeling drosophila populations. *Evolutionary Ecology*, 6:342–351, 1992.
5. I. R. EPSTEIN. Spiral waves in chemistry and biology. *In Science*, 252, 1991.
6. ERMENTROUT, G. LOTTI, and L. MARGARA . Cellular automata approaches to biological modeling. *In Journal of Theoretical Biology*, 60:97–133, 1993.
7. S. WOLFRAM N.H. PACKARD. Two-dimensional cellular automata. *In Journal of Statistical Physics*, 38:901–946, 1985.
8. E. BERLEKAMP, J.H CONWAY, and R.Guy. Winning ways for your mathematical plays. *Academic press, New York*, 1982.
9. P. RENDELL. Turing universaility of the game of life. *Andrew Adamatzky (ed.), Collision-Based Computing, Springer Verlag.*, 2002.
10. S. WOLFRAM. Twenty problems in the theory of cellular automata. *In Physica Scripta*, pages 170–183, 1985.
11. N. MARGOLUS. Physics-like models of computation. *In Physica D*, 10:81–95, 1984.
12. K. LINDGREN and M. NORDAHL. Universal computation in simple one dimensional cellular automata. *In Complex Systems*, 4:299–318, 1990.
13. K. MORITA Y. TOJIMA I. KATSUNOBO T. OGIRO. Universal computing in reversible and number-conserving two-dimensional cellular spaces. *Andrew Adamatzky (ed.), Collision-Based Computing, Springer Verlag.*, 2002.
14. A. ADAMATZKY. Universal dymical computation in multi-dimensional excitable lattices. *In International Journal of Theoretical Physics*, 37:3069–3108, 1998.
15. E. R. BANKS. *Information and transmission in cellular automata*. PhD thesis, MIT, 1971.
16. C. BAYS. Candidates for the game of life in three dimensions. *In Complex Systems*, 1:373–400, 1987.
17. E. SAPIN, O. BAILLEUX, and J.J. CHABRIER. Research of complex forms in the cellular automata by evolutionary algorithms. *In EA03.Lecture Notes in Computer Science*, 2936:373–400, 2004.
18. E. SAPIN, O. BAILLEUX, and J.J. CHABRIER. Research of a cellular automaton simulating logic gates by evolutionary algorithms. *In EuroGP03.Lecture Notes in Computer Science*, 2610:414–423, 2003.
19. A. DEWDNEY. The planiverse. *Poseidon Press*, 1984.

Appendix

This appendix contains an analytical description of a cell of the game of life in the R automaton, for replicability. Let us define X, Y and Z as:

$$X(S, 212, 65) = \{Gb(E, 18, 91), Gb(N, 80, 25), Gb(S, 212, 65), Gb(W, 241, 36), Gb(W, 253, 142), Gb(N, 370, 23), Gb(W, 433, 93), Ga(S, 159, 234), Gb(E, 179, 200), Ga(S, 273, 262), S(S, l, 210, 184), E(E, 292, 159), E(S, 219, 78)\},$$

$Y(E, 1, 1) = \{Gb(E, 1, 1), Ga(S, 99, 189), Ga(S, 173, 520), Gb(W, 184, 104), E(E, 183, 484)\}$

$Z(E, 1, 1) = \{Y(E, 1, 1), X(S, 425, 155), Gb(E, 721, 91), Gb(W, 843, 213), Ga(S, 749, 299),$
$Ga(S, 839, 630), Ga(S, 1019, 749), Gb(W, 1246, 261), E(E, 1229, 276), Gb(E, 990, 180), Gb$
$(E, 965, 277), Ga(S, 1071, 350), Ga(S, 1145, 457), E(E, 1161, 394), Ga(S, 423, 455), E(S,$
$427, -47)\}.$

A cell of the game of life can be described in R as:

$\text{LifeCell}(E, 1259, -436) = \{Z(E, 1, 1), Z(E, -759, 486), Z(E, -1519, 971), Z(E, -2279,$
$1456), Z(E, -3039, 1941), Z(E, -3799, 2426), Z(E, -4559, 2911), Y(E, -269, -89), Y(E,$
$-1028, 397), Y(E, -1298, 127), Y(E, -1787, 883), Y(E, -2057, 613), Y(E, -2327, 343),$
$Y(E, -2546, 1369), Y(E, -2816, 1099), Y(E, -3086, 829), Y(E, -3356, 559), Y(E, -3305,$
$1855), Y(E, -3575, 1585), Y(E, -3845, 1315), Y(E, -4115, 1045), Y(E, -4385, 775), Y$
$(E, -4064, 2341), Y(E, -4334, 2071), Y(E, -4604, 1801), Y(E, -4874, 1531), Y(E, -5144,$
$1261), Y(E, -5414, 991), Y(E, -4823, 2827), Y(E, -5093, 2557), Y(E, -5363, 2287), Y(E,$
$-5633, 2017), Y(E, -5903, 1747), Y(E, -6173, 1477), Y(E, -6443, 1207), Ga(S, 843, 33),$
$Ga(S, 933, 123), Ga(S, 1203, 213), X(S, 1205, -84), Gb(E, 663, -147), Y(E, 719, -437)\}.$

The input streams must arrive at coordinates –6227,988 for Sn, –5957,1258 for $C8$, –5687,1528 for $C7$, –5417,1798 for $C6$, –5147,2068 for $C5$, –4877,2338 for $C4$, –4607,2608 for $C3$, –4427,2788 for $C2$, –4337,2878 for $C1$.

The stream for $Sn + 1$ comes out at coordinates 1259,–436 after 21600 generations.

An Interactive Artificial Ant Approach to Non-photorealistic Rendering

Yann Semet[1], Una-May O'Reilly[2], and Frédo Durand[2]

[1] Optimization and Machine Learning Group, INRIA Futurs, Orsay, France
yann.semet@tremplin-utc.net,
[2] CSAIL, Massachusetts Institute of Technology
unamay@csail.mit.edu, fredo@mit.edu
http://graphics.csail.mit.edu/~semet/antsNPR

Abstract. We couple artificial ant and computer graphics techniques to create an approach to Non-Photorealistic Rendering (NPR). A user interactively takes turns with an artificial ant colony to transform a photograph into a stylized picture. In turn with the user specifying its control parameters, members of a colony of artificial ants scan the source image locally, following strong edges or wandering around flat zones, and draw marks on the canvas depending on their discoveries. Among a variety of obtained effects, two are painterly rendering and pencil sketching.

1   Introduction: Non-photorealistic Rendering and Ant Colonies

The goal of computer graphics has traditionally been to simulate the physics of light in order to produce photorealistic images. In contrast, the field of Non-Photorealistic Rendering (NPR), in recognition that realism is not always effective or superior, has the contradictory goal to develop traditional and novel pictorial rendering styles [1,2,3].

Our broad goal is to mediate creativity and artistry through decentralized activity and cooperation in a multiple-localized-agents metaphor. In this contribution we demonstrate how we have addressed one problem in the realm of NPR: creative transformation of a digital photograph into a natural and stylized picture. We have designed an artificial ant NPR system that is based on ants that navigate and sense the environment of a reference image. Ants deposit ink marks on an output picture according to where they are, what they sense and their short term memory which gives each its turn history for a few prior steps. The user interacts at a colony level to choose navigation and mark parameters. Then, the colony lives out its life on the image. These steps are repeated until the user is satisfied with the output picture. The approach launches the user and ant colony into an emergent, non-linear design process. The user interacts through the metaphor of an ant colony and, with acquired experience, it is possible to elicit a general artistic style. The system is able to provoke the user with specific and not entirely predictable applications of the directed style.

K. Deb et al. (Eds.): GECCO 2004, LNCS 3102, pp. 188–200, 2004.

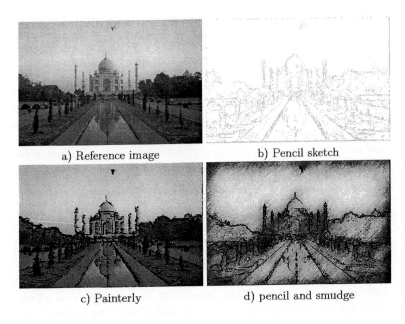

a) Reference image   b) Pencil sketch

c) Painterly   d) pencil and smudge

Fig. 1. We have designed a distributed agent system where 'ants' navigate and sense the environment of a reference image. Ants deposit ink marks on an output picture according to where they are, what they sense and their short term memory which gives them turn history for a few prior steps. The user interacts at a colony level to choose navigation and mark parameters. Then, the colony lives on the image. These steps are repeated until the user is satisfied with the output picture.

Our system is most precisely a distributed agent approach. Calling our agents 'ants' provides a convenient metaphor for its users to conceptualize agent navigation, image processing (as sensing) and ink deposition (as trail marking). Our system is capable of a range of styles. We have named three styles 'pencil sketching', 'painterly rendering' and 'sugar sculpture' . In Figure 1a) we present a reference image which is a photograph and not pictorially rendered. This is followed by example pictures, Figure 1b) through Figure 1d), that the system rendered during different user-interaction sequences while using the reference image.

We proceed as follows: in Section 2 we motivate our use of distributed agents (i.e. artificial ants). In Section 3 we describe how our system is designed. In Section 4 we describe how two pictorial styles can be generated and show a variety of pictorial outcomes. In Section 5 we try to 'put a finger' on what makes our system 'work' and compare it to related approaches. We conclude with our future vision for the system.

2   Our Rationale: Why Use Artificial Ant Colonies?

We considered three factors in our decision to address NPR using artificial ant colonies.

1. **Pictorial outcome:** One thread of NPR research involves transforming a digital photograph into a creative picture. With stroke approaches, randomization is used to achieve the inconsistent, uneven nature that strokes of a human artist exhibit [4,5,6]. Many systems draw strokes in random order or slightly randomize stroke length, the 'brush' texture or size. In addition, in general, ([7]'s importance map and individual stroke control such as [4] are exceptions) the output image is globally processed in a uniform manner each time a stroking style is selected and applied.

   With a distributed agent approach, the potential for non-global and semi-local treatment of the input image is possible. Ants are placed randomly and each ant's navigation is directed by the source image's varying local information. An ant's navigation history and memory capacity effects its ink deposition and navigation. Thus, conditions in different parts of the canvas have the opportunity to be treated individually. In addition, iterating colonies of ants with different behaviors for each colony allows another level on non-linearity and non-uniformity that one shot or multi-layers homogenous approaches can not achieve.

2. **Design initiative:** Our goal is for creative contributions to be made by both the user and the computational software. Ideally they are a design *team* with complementary roles and shared influence over the outcome. The ant approach has the user specify a process or behavior that will be executed, but not directly the outcome. This means of composing removes the onus on the user to be in complete control. Instead, the system is given more initiative in the design activity.

3. **User interaction:** Can a radical metaphor for mediating user interaction empower creativity? It is extremely eccentric for a user to work with an ant colony metaphor when attempting to create a pictorial artifact. However, perhaps this interface is so different it may inspire new creative process? There is an adage that creativity arises from constraints and in some respects the ant colony control metaphor is constraining. Additionally, creativity can also arise when existing approaches are exchanges for new resources and methods.

3   System Description

Our technique takes as input a digital photograph that is transformed into a non-photorealistic picture in the following way:

Step 1. The digital source image is preprocessed to create environmental maps.
Step 2. The user sets up parameters of a colony of artificial ants

Step 3. The ant colony navigates the source image by sensing it and the environmental maps. Some elements of navigation behavior are controlled by colony parameters. Each ant is able to deposit ink marks on an output image. Its decision to do so depends on what it senses and a short term memory of its turn angles. Ink marks emergently form the strokes of the picture. The colony is finished when each ant has exhausted its limit of navigation steps.

Step 4. Goto Step 2 unless user is satisfied with current output picture.

3.1 Preprocessing

Ants sense the digital input image by referencing its pixel values. In addition, they can sense the following pre-computed environmental maps:

Luminance Map. Assuming colour images, a *luminance* value is computed for every pixel through a weighted average of the Red (r), Green (g) and Blue(b) channels that reflects the global visual impact of the pixel as suggested in [8]: $L = 0.299r + 0.587g + 0.114b$.

Gradient Maps. The norm and orientation of the gradient are computed on the luminance map at every pixel using 3x3 Sobel filters.[9] A gradient map is shown in Figure 2.

Importance Map. A user can specify regions of the image that she deems particularly significant and that, as such, need particular attention. An example of an importance map is shown in Figure 2. Ants typically use this information to deposit finer brush strokes around important regions such as the eye of a portrait.

a)   b)

Fig. 2. a) A gradient map of the Taj Mahal image of Figure 1a). b) Close-up of a photograph containing a face and the corresponding importance map (important areas are in black). Marilyn's face and shoes as well as the artist's signature are preserved.

3.2 Setting Up a Colony

A colony is a collection of ants (typically around 500). Parameter values set for the colony apply to every ant (except the one that specifes colony size).

Table 1 shows all colony parameters. All but the last five listed are influential in navigation while the final five dictate how the ant draws its mark. In the course of describing ant navigation algorithms, the navigation parameters will individually be described. The five ink marking parameters set the mark's shape, length, angle, thickness and means of being colored.

The parameters of the colony are accessible and modifiable through our Graphical User Interface (Fig. 3). A keystroke quickly runs a colony with a given set of parameters. The elementary navigation behaviors of edge drawing, filling and hatching, and smudging have been combined into composite behaviors which are also executed with a keystroke. The process of getting to a final result is strongly time dependent so ready access to composite behaviors is an interface feature.

Table 1. User specified ant colony parameters.

Parameter Name and/or symbol	Type	Typical value
colony size	integer	$1 \ldots 1000$
marking behavior: a.k.a. navigation algorithm	edge drawing filling and hatching smudging	
step-size	pixels	$1 \ldots 200$
gradient start threshold (T_0)	[0.0...256.0]	80.0
gradient continues threshold (T_1)	[0.0...256.0]	20.0
gap threshold (T_{gap})	[0.0...256.0]	
crosshatch threshold (T_{hatch})	[0.0...256.0]	
memory size	integer	3
jump-radius	pixels	5
max-steps	integer	200
max-jumps	integer	5
sharp turn threshold	degrees	45
mark shape	circle, line, cross	
mark length	pixels	$1 \ldots 200$
mark angle (α)	degrees	
mark thickness	pixels	$1 \ldots 20$
color: underlying pixel in source image is used	copy RGB grey scale from luminance sketch effect	

3.3   Marking Behavior Is Directed by Ant Navigation Algorithms

An Ant Relies Solely on Local Information. Each ant is an agent situated on the source picture. It senses certain local, low level features of this picture (i.e. its environment) and acts solely upon this information. An ant is not subject to any centralized control that would direct it to coordinate with others to fulfill image-wide, high level intentions of the user. No global or high level features of the picture, such as strokes or shadows, are apparent to an ant. Each ant has an encapsulated individual state description in terms of current position, velocity and short term memory, see Table 2. Position is the x and y coordinates of the pixel the ant is on. Because an ant moves, it is characterized at each time step by a velocity vector that has two elements: angular orientation (normalized at the ant's level and expressed with x and y components), and step-size (i.e. magnitude) which does not vary over time and is set as a colony parameter. To avoid overly sharp turns, an ant stores its successive turning angles in memory and only makes a mark when a memory check confirms this will not occur. An ant also records the number of steps it has taken since its birth. This limits the maximum length of the strokes.

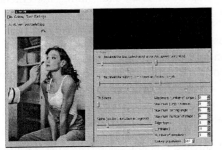

Fig. 3. Graphical User Interface

Table 2. An ant's local state information

Variable	Type	Representation
position (p_t)	(x, y)	floating point
orientation (θ_t)	(x, y)	floating point
velocity (v_t)	θ_t stepsize	integer
$memory[t]$	vector	floating points
steps counter	scalar	integer
jumps counter	scalar	integer

Values are time dependent except for counters. Stepsize is a global parameter but is shown here because it is an element of velocity. Using floating point representation for position and orientation is necessary to achieve smooth strokes.

The computational implementation of some of this local, non-feature-based information is crucial to our system's visual quality. In particular, storing position and the orientation value of velocity in floating point representation and rounding them to reference a pixel in the image yields smoother strokes.

Each ant is initially placed at a random position in the image. The system simulates one ant at the time, allowing it to move a maximum number of steps. Each ant, using its local information and environmental maps, executes a navigation algorithm for one of the three marking behaviors:

Edging Navigation. This is guided by the gradient of the original image. At a given time and position, ants base their trajectory calculation on the local gradient's norm and orientation. This is very similar to line integral convolution (LIC) [10] though LIC is a filtering approach with more rigourous normalization that is globally and uniformly applied across an image for the purposes of visualizing vector information. Both [5] and [6] also exploit gradient information (though for different purposes), the former to clip brush strokes and the latter for the placement of control points for anti-aliased cubic B-splines that model strokes.

Let us denote, G_{norm} to be the gradient map in terms of norm. Ants sense G_{norm} to follow edges, i.e. trails of high gradient points. The algorithm proceeds as follows for an ant originally situated on a pixel $p_0(x_0, y_0)$:

Edging Navigation:
1. If $G_{norm}(p) > T_0$, continue, else STOP
2. If jumps-counter < max-jumps :
   a) While *continuing conditions* hold do:
      i. Put a mark on the canvas.
      ii. Move from $p_0(x_0, y_0)$ to $p_1(x_1, y_1)$ with $p_1 = p_0 + v_t$
      iii. Increment steps counter.
      iv. Update short term angular memory $(memory[t] = \theta_t - \theta_{t-1})$.
   b) If gradient track vanishes, jump to a local maximum, increment jumps-counter.

Continuing conditions: The ant keeps on following the edge only if
1. It has not exceeded the maximum allowed number of steps.
2. It has not exceeded the sharp turn threshold as calculated from $memory[t]$.
3. It is still within the boundaries of the image.
4. The gradient still exists: G_{norm} at the current pixel $p_0(x_0, y_0)$ is over T_1.

Local jumps: When a gradient track vanishes, an ant scans the surrounding pixels within a given jump radius (a colony parameter) to look for other high gradient points greater than T_1. The ant then moves to the highest of them. To avoid sharp turns and backturns, the exploration of the neighbourhood is restricted to a portion of circle situated in front of the ant.

Filling and Hatching Navigation. This behavior macroscopically maps color information from areas of the source image to the output picture. It is macroscopic in two ways: it relies on multiple pixel information and it draws marks that are larger than a single pixel. If the ant starts at a pixel with a low gradient (i.e. $G_{norm} < T_0$), it copies the color of the source image pixel and uses it to draw a mark. This mark extends beyond the current pixel and is either a cross or a line (both angled using value α) or small circle. The ant then navigates ahead and keeps on depositing marks on the output canvas as long as the underlying pixel has a color that is sufficiently close to that of the pixel the ant started from. This comparison uses colony parameter, T_{gap}. In this way, the ant generates a group of identical marks that, referring to the source image, generalize one pixel to a group of similar, neighboring pixels.

The computation of this strategy is simpler (i.e. geometrically linear) than the edging algorithm. It requires extending the ant's state information to include local variables that record the 'gap' between the first pixel's and current pixel's color and the first pixel's color and position.

The details of drawing the mark depend on whether it is a circle, line or cross. A circle can be immediately drawn in a radius around the current pixel. A line is drawn in direction α then $-\alpha$. Hatches are drawn when luminance is less that T_{hatch}. Navigation is repeated from the first pixel while drawing a second line normal to the first.

Filll and Hatch algorithm:
1. Retrieve the underlying colour (C_0) or luminance (L_0) at P_0.
2. While *continuing conditions* hold do:
   a) Draw a mark. For a line, mark it in direction α.
   b) Move ahead
   c) Compute the gap: $gap = L_{local} - L_0$ or $gap = C_{local} - C_0$.
   d) Retrieve the local gradient norm $G_{norm}(p)$.
   e) Increment the steps counter.
3. Return directly to P_0.
4. Repeat Step 2 with $\alpha = -\alpha$.
5. If $L_0 < T_{hatch}$ do:
   a) Return to P_0.
   b) Repeat Steps 2, 3 with $\alpha = \alpha + \frac{\pi}{2}$

Continuing conditions:
The ant keeps on following the edge only if the three following conditions hold:
1. $gap < T_{gap}$: This ensures that an ant marks only where underlying pixels in the source image have sufficient color similarity.
2. $G_{norm}(p) < T_0$: Filling stops when the ant meets a high gradient point.
3. The maximum allowed number of steps has not been exceeded.

Smudging. For traditional media, smudging consists in rubbing one's finger or a small piece of paper over a portion of the picture to blur it or make it more uniform. To simulate this, an ant's smudge navigation behavior copies that of edging except that deposition of the ink mark is different. The ant uses the source image to compute the average luminance or colour within a square area of given radius around the pixel it is on and places its mark (circle, square, line or cross) using that value. A variation of smudging that led to an effect we call 'sugar' (see Section 4.3) actually introduced another dimension of non-linearity. An ant averages values on the current version of the output picture (not the source image).

4   Stylistic Results

By selecting a colony's parameters' values interactively each time before its ants navigate, a user is able to direct the style of the resulting picture. The picture itself is a time series of directly overlaying layers each of which can overwrite some or all pixels of the previous. From our personal experience with the system, an effective user design plan to explore outcomes in the vein of traditional styles is to progressively refine the picture detail each turn. In general, step-size controls the 'sketchiness' of emergent strokes. Coarseness is achieved by setting a high velocity for navigation. The step-size component of velocity (a colony parameter) controls the distance (in pixels) an ant travels between sensing steps. When gradient information farther apart is connected this effectively leads to strokes that are only roughly approximate to the photograph's edges and that are composed of long linear subsegments. Small step-size values will follow the photograph's edges more closely and have shorter linear subsegments.

Using knowledge of a technique of painting that conveys coarseness via long, wide, light brush strokes and detail, conversely, with short, fine, heavy brush strokes, velocity (i.e. step-size) can be complementarily teamed with appropriate mark parameter choices. Choosing the step size and mark parameters with this relationship will yield a style derived from conventional painting techniques. The user is also, however, free to invert traditional techniques or invent her own and explore the consequences of giving the colony this nature of direction. No two interaction sessions ever are the same due to both the random factor in ant placement and the user's inclination to explore something new, even if minor, each session.

Despite the developing nature of the system, it has already been able to produce some interesting results. Two traditional and easily identifiable effects that can be achieved are painterly rendering and pencil sketching. In addition, we observed a variety of intriguing, surprising and promising effects that might eventually form an element of a style (albeit not necessarily a rigorous pictorial one), one of which we call 'sugar sculpture'.

4.1   A Painterly Rendering Style

We obtained painterly effects by following Haeberli's [11] and Hertzmann's strategy [12] of producing a picture with multiple brush strokes of different sizes. See Figure 4 for an illustration of the process. Initially, each colony's navigation behavior is filling or hatching. The step-size component of each colony's velocity stays constant (it equals one in Figure 4 to provide maximum detail). By directing each ant of early colonies to deposit ink marks that are long and thick, (typically starting at 10 pixels wide and 15 pixels long), the canvas is first densely covered with long and thick brush strokes. The rendering is refined by gradually reducing the dimensions of the ink marks (down to 2 pixels wide and 4 pixels long in Figure 4) over successive colonies. Some emergent strokes may overwrite and overlap with earlier ones in a non-linear way as this reduction takes place. After filling and hatching, the smudging navigation behavior is used. The ants deposit tiny cross marks of average color and this results in fuzzy looking boundaries that roughly simulate the mixing of ink pigments.

4.2   Pencil Sketching

The key idea behind this effect relates to how an artist pencil sketches a portrait. The artist first draws the key edges very roughly using very long, linear strokes but with a very slight pressure. He then gradually refines his contours, making

Fig. 4. An interaction sequence in the *painterly rendering* style. The source image is at left. The middle two pictures are intermediate results between which multiple brush dimensions were successively decreased 5 times over 35 keystroke (i.e. colony) sequences of filling and hatching navigation. Afterwards, 10 colonies navigated while smudging with a 2X2 radius brush. The final result of the interaction is far right. The step-size was one and the colony size 200 throughout the session.

them both smoother, stronger and shorter. To achieve this technique in term of pencil pressure, the mark's luminance was set to be inversely proportional to the mark's length during edging navigation (with a minor luminance correction). Then, by using a series of decreasing step sizes that decrease the successive colonies' velocity, emergent strokes will more accurately follow the image's edges and be both darker and shorter. This results in a style like gradual sketching. This particular effect is illustrated in Figure 5 and Figure 6 where the images' contrast has been enhanced for display.

4.3   Sugar Sculpture Effect

In open exploration of the system's creative range, we obtained an interesting effect which we call the 'sugar sculpture': smudging is usually performed by averaging neighbouring values on the source image ; by averaging values on the target canvas instead and by replacing cross marks by circle marks, we obtained the hard to describe "sugar sculpture" look and feel as seen in Figure 6 which uses a digital photograph of the Eiffel Tower as its source image.

5   How Does It Manage to Work?

What is the source of our system's capabilities? Additionally, how does it compare to other NPR systems? Producing a stroke-based image is a major source of the system's compelling nature. Strokes are associated with traditional artistic style and media. Digital brush strokes convey media in addition to scene aspects. They contribute to a result's interpretation as an artistic rather than realistic rendering.

Using strokes to form the basis of a digital picture is widely practiced in computer graphics, for example, [13,14,15,16,17]. The basic concept of generating

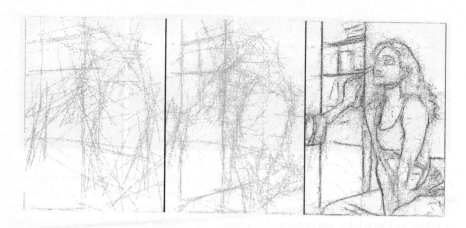

Fig. 5. A progression of gradual pencil sketching. Step-size is decreased in 7 non-uniform steps from 200 pixels to 7 pixels during the course of 35 colonies of 512 ants performing edge navigation. In between the final and previous to final sketch, the brush length is decreased from 100 to 7 in 5 steps over 20 colonies.

strokes is an approach we share with many graphics systems. In contrast to these examples, however, in our system a stroke is a one dimensional trajectory of ink marks determined by an ant's navigation behavior, its mark deposition behavior and its reference to memory of previous turn angles. This formulation creates strokes that are different in nature from the graphics systems. The use of gradient information creates strokes that are responsive to edge and varying aspects of content in the original image. Subtle, meaningful properties of the original image can be sensed and influence the outcome. This contrasts with the graphics systems where a stroke technique is uniformly applied and the only variation results from injected randomness. In addition, a stroke in our system

Fig. 6. Three pencil sketching examples (left) and a sugar sculpture effect at far right.

is not based on strictly local information, i.e. the information of a pixel and its immediate neighbors. It is created on *semi*-local information that is essentially gradient trails.

Another difference, this one obvious, is entirely external, i.e. in abstraction at the user's level: in the graphics systems, the user renders a picture by creating strokes. Stroke creation is an explicit task that user and system work on. By contrast, the user interface of our system does not provide access to a stroke. Ants give the user a unique control for influencing the rendering process. The user accesses ant-based navigation behavior, marking behavior and memory, instead. This radically different interface needs to be evaluated for its expressive capacity. Our limited, informal experience has shown it not to be taxing. However, it is more conducive to an explorative attitude rather than a functional one. It may better serve creative purposes rather than deliberate ones. The distributed and not wholly predictable nature of our ants balances creative exploration with purposeful intent.

The overall process of interaction between the user and the system during a 'design' session has novel consequences: the multiple, overlying layers result in unanticipated stroke interaction. One pixel may be overwritten multiple times - both by multiple ants in the same colony or by different colonies. In contrast, when a user controls strokes, it is less likely that many non-linear interactions can be simply explored. Or, in a system such as [6] which is automatic with access to only high level parameters, despite an iterative element to the algorithm's processing of increasing detail, an iteration is quite strictly defined: the user controls only brush size.

Our pencil sketch style results in drawings that are more pleasing than traditional computer vision edge detection. This is probably due to two factors. First, we successively use ants with different sets of parameters to build the drawing from coarse to fine, which results in a nice multi-scale creation of strokes that emphasize both the high-level appearance of the picture and the fine details. Second, the creation of strokes as trajectories and the use of a memory for the ants results in longer strokes that are also better shaped.

6   Conclusion and Future Work

This work demonstrates that an artificial ant system that capitalizes on NPR techniques from computer graphics can be teamed up with a user to accomplish visual representation in a novel way. We intend to incorporate interactive evolutionary computation to search for innovative navigation behaviors and stigmergic communication among ants plus conduct formal user evaluation studies.

Acknowledgements. We thank Martin C. Martin, Pierre Collet, Xavier Décoret and Stéphane Grabli.

References

1. Gooch, Gooch: Non-Photorealistic Rendering. AK-Peters (2001)
2. Strothotte, T., Schlechtweg, S.: Non-Photorealistic Computer Graphics. Modeling, Rendering, and Animation. Morgan Kaufmann, San Francisco (2002)
3. Durand, F.: An invitation to discuss computer depiction. In: Proc. of the ACM/Eurographics Symposium on Non-Photorealistic Animation and Rendering (NPAR). (2002)
4. Haeberli, P.: Paint By Numbers: Abstract Image Representation. In: SIGGRAPH 90 Conference Proceedings. (1990)
5. Litwinowicz, P.: Processing images and video for an impressionist effect. Proceedings of SIGGRAPH 97 (August 1997) 407–414
6. Hertzmann, A.: Painterly rendering with curved brush strokes of multiple sizes. Proceedings of SIGGRAPH 98 (July 1998) 453–460
7. Durand, F., Ostromoukhov, V., Miller, M., Duranleau, F., Dorsey, J.: Decoupling strokes and high-level attributes for interactive traditional drawing. Eurographics Workshop on Rendering (2001)
8. Foley, J., Van Dam, A., Feiner, S., Hughes, J.: Computer Graphics : Principles and Practice. 2nd edn. Addison Wesley (1997)
9. Jain, R., Kasturi, R., Schunck., B.: Machine Vision. New York, NY, McGraw-Hill. (1995)
10. Cabral, B., Leedom, L.C.: Imaging vector fields using line integral convolution. In: Proc. SIGGRAPH. (1993)
11. Haeberli, P.: Paint by numbers: Abstract image representations. Proce. SIGGRAPH (1990)
12. Hertzmann, A.: Painterly rendering with curved brush strokes of multiple sizes. Proc. SIGGRAPH (1998)
13. Salisbury, M.P., Anderson, S.E., Barzel, R., Salesin, D.H.: Interactive pen-and-ink illustration. Proceedings of SIGGRAPH 94 (July 1994) 101–108 ISBN 0-89791-667-0. Held in Orlando, Florida.
14. Sousa, M.C., Buchanan, J.W.: Observational models of graphite pencil materials. Computer Graphics Forum **19** (2000) 27–49 ISSN 1067-7055.
15. Durand, F., Ostromoukhov, V., Miller, M., Duranleau, F., Dorsey, J.: Decoupling strokes and high-level attributes for interactive traditional drawing. In: Eurographics Workshop on Rendering. (2001)
16. Ostromoukhov, V.: Digital facial engraving. Proc. SIGGRAPH (1999)
17. Curtis, C., Anderson, S., Seims, J., Fleischer, K., Salesin, D.: Computer-generated watercolor. Proc. SIGGRAPH (1997)

Automatic Creation of Team-Control Plans Using an Assignment Branch in Genetic Programming

Walter A. Talbott

Stanford Symbolic Systems Program
Stanford University
Stanford, California 94305
wtalbott@stanford.edu

Abstract. This paper is concerned with the introduction of a method for allowing genetic programming to automatically create team-control plans using an assignment branch. Team-control plans are representations of the composition and behavior of groups of agents and include the definition of one or more roles. Genetic programming is a general problem solving technique, and using genetic programming as a means for creating team plans would allow the user to specify very little, while producing effective results. I propose to show that the use of what I call an assignment branch in genetic programming provides a robust and general way to approach team composition and role definition problems.

1 Introduction

The creation of team-control plans is a problem that has become more pertinent as the power of computing has increased and as the sophistication of artificial intelligence methods has led to the improved ability to design situated agents. A team-control plan defines how a group of these agents act in combination with one another to solve problems and accomplish complex tasks. For most of these team plans, human designers must specifically define the way that the team will act. Because of the high level of human involvement, these teams are not as artificially intelligent as they could be. Genetic programming has recently emerged as a robust, domain-independent method for automatically generating the solution to difficult problems. It seems natural, therefore, to use the power of genetic programming for team organization and design problems.

A suitable problem for the exploration of genetic programming's aptitude for creation of team plans will ideally be impossible to solve without teamwork of some kind, and will require the creation of different roles. I define a role as a set of instructions that a subset of the team will carry out. If the team were intended to play soccer, for example, there would be roles for each position: goalkeeper, defense, and offense. Roles allow for much more elaborate team plans, and hence for the solution of much more elaborate problems, but are proportionately more difficult to create without much human intervention. Luke and Spector (1996), for example, apply genetic programming to multiple role teamwork problems, but require the number of roles to be selected beforehand. This paper will explore a method for automatically creating team plans, number of roles included, with genetic programming, using an artificial agent problem as an illustrative, if simple, problem domain.

K. Deb et al. (Eds.): GECCO 2004, LNCS 3102, pp. 201–212, 2004.
© Springer-Verlag Berlin Heidelberg 2004

1.1 The River Crossing Problem

The problem consists of a nine-by-nine toroidal world, in which there are a number of artificial agents that can move and interact with food pellets. Each agent can face in one of four directions, can be carrying either one or zero pieces of food at a time, and has an assigned role. The exact functions for control of the agents are described in Tables 1-3. The world has one strip that I define as the goal strip, to which the agents must move all the food pellets. The goal strip is surrounded on both sides by two rows of water, and this water is deadly to the agents. When an agent moves onto a square in the water, the agent dies, and can no longer gather food or move in any way. The dead agents, however, form a bridge for the remaining agents, such that if an agent moves onto a water square that has already been visited by another agent, the current visitor can continue to move as if it were on land. The agents are given a limited number of time steps to move the food, both so that the genetic programming run will eventually stop, and so that solutions are reasonably quick. Figure 1 shows an example world after an intermediate number of time steps.

The problem is designed to promote the need for teamwork, since no single agent could move even one pellet to the goal strip. Also, the problem is designed to facilitate a solution involving more than one role, but this facet of a solution is not strictly necessary. A solution, for instance, might benefit from designating a certain number of agents to the creation of a bridge, and the remaining agents to the collection of food. Because the creation of roles is not strictly necessary, I can also use this problem to compare solutions involving multiple roles to solutions with only one role. The success of the method in creating multiple roles is described further in the Results section below.

This paper is organized as follows. It first presents the methods used for representing the problem appropriately for genetic programming. Then, it presents the results gathered from four different experiments, and discusses these results. Finally, it suggests areas of possible future work and areas to which the results might be applied.

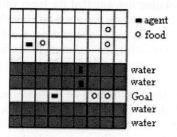

Fig. 1. Shows a world with four agents and five pieces of food. Two food pellets have successfully been placed on the goal strip.

2   Methods

Approaching the river-crossing problem and team control problems in general, with genetic programming methods offers several benefits. If biology is any example, it is clear that the basic principles of genetic programming have already been successful in nature; look at any pack of predators, or any school of fish. Genetic programming is an algorithm that uses the selection pressures found in natural systems to improve a randomly generated population of programs. These programs are built up from sets of terminals and functions into trees, as described by Koza (1992). Expressing the control of artificial agents with the population of tree-like programs for which genetic programming calls is relatively straightforward. Following the example of using zero-argument functions as terminals set by Koza's work with artificial ants (1992), I describe the movement of agents with the terminals outlined in Table 1.

Table 1. Description of zero-argument functions used as terminals.

Terminal Name	Description
MoveForward	Moves the agent one square in the direction that it is facing.
TurnLeft	Rotates the agent 90 degrees counter-clockwise.
TurnRight	Rotates the agent 90 degrees clockwise.
PickUpFood	Picks up food if the agent is currently on a square with food. Each agent can only carry one pellet at a time.
DropFood	Drops food on the current square if the agent is carrying food.
NoOp	The agent remains exactly how it is.

Table 2. Description of the functions used in the program tree

Function Name	Description
Prog2	Executes the two arguments in succession.
Prog3	Executes the three arguments in succession.
IfAtFood	Executes the first argument if the agent is currently on a square with food, and the second otherwise.
IfWaterAhead	Executes the first argument if the agent is facing a square with water and no agent bridge, and the second otherwise.
IfCarryingFood	Executes the first argument if the agent is carrying food, and the second otherwise.
IfGoalStrip	Executes the first argument if the agent is currently on the goal strip, and the second otherwise.
IfBridgeAhead	Executes the first argument if the agent is currently facing a square with an agent bridge, and the second otherwise.
ADF0	Executes an automatically defined function (Koza, 1992) that can use all previously listed functions.
ADF1	Executes an automatically defined function that can use all previously listed functions, including ADF0.
ADF2	Executes an automatically defined function that can use all previously listed functions, including ADF0 & ADF1.

With these terminals defined, the programs can use the functions in Table 2 to further control the action of the agents under certain conditions.

Using these terminals and functions, the experiments described in this paper define an individual in two different ways. One group of experiments uses a tree structure where only one role is allowed. The result-producing branch of the tree represents this role, and is executed once for each agent at every time step. The result-producing branch can call any of the automatically defined functions, which are each in turn represented by their own branch of the tree.

It seems, however, that at least the ability to support more than one role would expand the range of possible problems to which genetic programming could be applied, and make more powerful the solutions to the problems it can already create. Towards that end, the other group of experiments involves the use of an entirely separate branch, whose function set is entirely different than the result-producing branches. This is the assignment branch, and allows the genetic programming run to assign different numbers of agents to different roles. For the purposes of these experiments, up to three roles were allowed, and were represented by three result-producing branches. Proposed methods for automating the number of roles entirely are discussed in the Future Work section below. The functions of the assignment branch are defined in Table 3. During evaluation of the program, the assignment branch is run first, to set up the role of each agent in the world, and also how many total agents are present. These experiments cap the number of agents due to time constraints on each run. The assignment branch removes the necessity for team plans to rely on brute-force repetition of the same set of instructions for each agent. Simultaneously, it allows for efficiency of specification, since agents can be grouped together rather than defined independently and with a high chance of redundancy.

Table 3. The Functions and Terminal of the Assignment Branch

Function Name	Description
Role1	This function takes one integer argument, and inserts that number of agents into the world, all of which are assigned to role 1. If the maximum number of agents has already been assigned, this function does nothing.
Role2	This function takes one integer argument, and inserts that number of agents into the world, all of which are assigned to role 2. If the maximum number of agents has already been assigned, this function does nothing.
Role3	This function takes one integer argument, and inserts that number of agents into the world, all of which are assigned to role 3. If the maximum number of agents has already been assigned, this function does nothing.
Prog2	Executes the two arguments, and arbitrarily returns the value of the first argument.
Random Constant	The only terminal for the assignment branch, the random constant is an integer that is used to specify how many agents of each role are present.

2.1 Sample Branches

The following are illustrations of the structure of the branches used to represent each agent. They have been simplified greatly, and do not come close to solving the river-crossing problem, but serve to highlight the difference in structure between the result producing branches and the assignment branch. The automatically defined functions have been left out for simplicity, since they are nearly identical to a result producing branch.

Result-producing Branch:
```
(Prog2 (IfGoalStrip DropFood MoveForward)
       (Prog2 (IfAtFood PickUpFood MoveForward)
TurnLeft))
```

This branch represents an agent that, at each time step, will detect whether or not it is on the goal. If it is, it will attempt to drop food, the success of which depends on whether it is carrying food. If it is not, it moves one square forward. Then, regardless of the outcome of the goal test, the agent will test whether it is at food, pick it up if it is, and move forward otherwise. To end the time step, the agent will always turn left. The result producing branches are limited to a depth of 17, which did not seem to eliminate possible solutions. A depth of 17 allows a program with up to $3^{17} = 129,140,163$ nodes, which should be far more than sufficient for this problem.

Assignment Branch:
```
(Prog2 (Role1 (Role2 5)) (Role3 (Role3 2))
```

The assignment branch listed above assigns 5 agents to roles one and two, and 4 to role three, for a total of 14 agents. Each Role function returns the same integer that it takes as an argument, so the assignment branch chains backwards from the last node to the first, stopping when the maximum number of allowed agents has been reached. Note that, because constants are terminals, it is possible to have an expression such as:
```
(Prog2  4  5)
```
In such a case, no agents are assigned, and the individual would receive the worst possible fitness since none of the food can move itself. Because the assignment branch is only run once per individual, and because individuals such as this are quickly eliminated from the population, this quirk is acceptable.

2.2 Fitness Evaluation

Since it is desirable to generate a solution that is general enough to solve the river-crossing problem independent of where the food is located, the fitness is evaluated based on two fitness cases. The first case scatters the food across the squares of the board, and the second places all of the food in one square. The raw fitness of each program is just the number of food pellets that ended up on the goal strip, which is also the number of hits. A run terminates successfully when an individual has a raw fitness equal to the number of food pellets that exist in the world, over all fitness cases. Another experiment also evaluated fitness based on the number of steps it took to complete; raw fitness was the total number of food pellets present over all fitness cases minus the amount of food placed on the goal strip, plus the total number of steps it took. If a program did not get all the food pellets of a particular fitness case onto the goal strip, it could not complete that case in less than the maximum number of steps allowed, and so only those individuals who were relatively fit to begin with could gain from being faster. Because of this, the fitness evaluation with steps included can distinguish between individuals who would otherwise seem similar. This discrimination allows the comparison between successful individuals of the two types described earlier: those limited to one role, and those with the assignment branch.

2.3 Breeding and Run Parameters

These experiments use the breeding phases that were provided as default with lil-GP, the genetic programming software package in which the experiments were carried out. Each phase consists of tournament-selection for crossover 90% of the time and reproduction 10% of the time. Crossover could happen only between similar branches of the two parent trees. This is to isolate the evolution of the individual roles in the assignment branch individuals. Since they are somewhat structurally complex, inter-branch crossover might slow the progress of the evolution because it would introduce more noise into the process, which is not necessary for a problem of this magnitude. Mutation was not included, though it arguably should have been. Especially in the assignment branch, mutation could have helped fine-tune the random constants that were present, rather than forcing the program to rely on those constants generated at initialization.

Table 4. Tableau for the river crossing problem

Objective:	To produce a solution to the river crossing problem that uses the assignment branch, and compare with a solution that only uses one role.
Terminal set:	MoveForward, TurnLeft, TurnRight, PckUpFood, DropFood, NoOp, as defined earlier.
Function set:	One Role Individuals, defined in table 2. Assignment Branch Individuals, defined in table 3.
Fitness cases:	Two, one in which the food is scattered across the world, and one in which the food is stacked in one square
Raw fitness:	In some runs, raw fitness is the number of food pellets deposited on the goal strip. In others, it is the number of pellets on the goal strip weighted by the steps it took to get them there.
Hits:	Each pellet that ends up on the goal strip is counted as a hit.
Wrapper:	None.
Parameters:	M = 13,000 G = 95
Success predicate:	When an individual placed all food pellets on the goal strip, the run was terminated successfully, except on the runs where time mattered in the fitness, in which case the runs continued until all generations completed.

These runs were done with population size $M = 13,000$ and allowed to run until $G = 95$ generations. The population size, in some of the early runs, was set as high as 75,000, but each of these runs took up to about 48 hours. The reduced population size decreased each run to about 10 hours, and seemed sufficient for the problem. Because the structure of the programs is fairly cumbersome, with seven branches total in the individuals with the assignment branch, the number of generations has to be high enough to allow the single-node crossover operator to rearrange enough of the branches to form a good solution. In fact, 95 may be too low, but it was sufficient for these experiments. Table 4 presents the tableau summary of the method just described.

3 Results

The experiments lend themselves to a categorization into eight different groups. Each group will be presented in turn, and the results will be discussed in section 4 of the paper, below. The individuals limited to one role will be called simply one-role individuals, and those with the assignment branch will be called branch individuals. The fitness evaluation that takes only hits into account will be called basic fitness, whereas the evaluation that includes time steps will be called step fitness. The final difference between runs involved what happened when an agent that was carrying food hit the water. Originally, the agents would not be allowed to drop their food before they died and could no longer act. This required any fully successful plan to specify that no agent could ever hit the water if it was carrying food, which proved to be fairly difficult. And intuitively, this requirement seems to suggest the use of multiple roles, one for solely forming the bridge and leaving food alone, and one for gathering food. When this requirement was relaxed, agents that hit the water automatically dropped their food so that others could resume in their place. Runs where the requirement was in place will be called no-drop runs, and those without the requirement will be called drop runs. In every experiment, two fitness cases were used, with 25 food pellets each, which makes the highest possible hit score 50. In every experiment with a step score included, 150 steps were the maximum for each fitness case, so at most 300 is added to the difference between 50 and the number of hits to give the raw fitness score. A lower fitness score is better, with 0 being the best.

3.1 Basic Fitness with Drop

In all experiments conducted with basic fitness and drop, a solution was found before the 20th generation. Both one role and branch individuals solved the problem easily, and there were about as many single-role branch individuals as not. This group's results bear mentioning only for the sake of completeness.

3.2 Step Fitness with Drop

Given that the drop problem can be solved easily, I present the results of step fitness runs that allowed drowning agents to drop food in table 5.

Table 5. Results for Step Fitness Runs With Drop

	Generation of Best of Run	Nodes	Hits	Fitness
One Role	64.75	567	43.75	107.75
Branch	86	637	46.67	80.6

The numbers in the table are averages over all the runs conducted in this group. The generation of the best of run does not necessarily imply that a solution was found at the listed generation. The number of nodes is a good measure of structural complexity. It should be noted that in these runs, hits is not the same as fitness, because fitness is defined as the number of steps it takes the individual to complete its task. It is somewhat unfair to define solution as 50 hits in runs where steps contribute to the fitness function, because genetic programming is blind to hits, and only selects based on fitness. However, because individuals that get all the food to the goal strip in both fitness cases have such an advantage in the fitness measure, I are fairly safe defining solution as 50 hits. Out of the runs that did solve the problem, all but two resulted in fitness ratings below 50, and one of the one-role runs produced an individual with a fitness of 16. Of the branch individuals that solved the problem, 2 presented solutions with two distinct roles, and 2 presented solutions with only one.

3.3 Basic Fitness with No Drop

The problem changes drastically when no drowning agent is allowed to drop the food that it may be carrying. Of all the runs performed, only one produced a solution, though most came close. An assignment branch run generated the single solution, and the individual is partially presented below. All assignment branch runs resulted in the production of two roles. Table 6 outlines the results of these runs.

Table 6. Results for Basic Fitness Runs With No Drop

	Generation of Best of Run	Nodes	Hits	Fitness
One Role	35.25	354	41.5	8.5
Branch	55.75	424.5	43.5	6.5

This case is the most difficult for the process to solve, so the runs that produced the best results deserve closer inspection. Figure 2 and Figure 3 present a generation-by-generation breakdown of the hits from the best runs of the One Role and Branch tests. Figure 2 shows the highest number of hits achieved by any individual at each generation, and Figure 3 shows the mean number of hits over all individuals in each generation. Notice that the assignment branch run ended at generation 26, after which it terminated since the success predicate was achieved. The one role run stopped at generation 95, after failing to fulfill the success predicate. Both figures stop at generation 48, because the best individual from the entire one-role run was created in that generation.
Because only one individual managed to solve this problem, I will examine its assignment branch, printed in its entirety below.

```
ASSIGN:
  (role1 (prog2 (role1 (prog2 (role1 3) 4))
                (role1 (role3 (prog2 (role1 (role1 1))
  3)))))
```

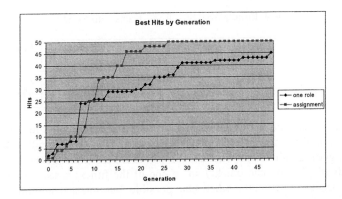

Fig. 2. Shows the highest hits achieved by an individual at each generation in the run.

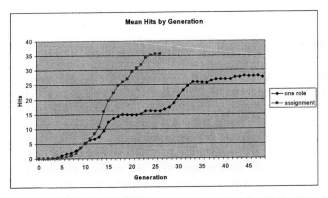

Fig. 3. Shows the mean hits achieved at each generation in the run. Notice that the assignment branch run stops at generation 26, when a solution was found.

The assignment branch has the effect of introducing 13 agents into the world, interestingly 7 less than the maximum of 20. 12 of these are role one agents, and 1 is a role three agent. Because the rest of the program is so complex, I will forego an analysis of the functionality of the program, letting it suffice to say that both role one and role three are substantive, and proscribe different courses of action for the agents they control. Because these roles are generated genetically, they do not define roles as humans might, with an explicit purpose for each, but instead piece together ideas that work to form something that is not intuitive to a human observer.

3.4 Step Fitness with No Drop

Though it is clear that a solution to the no drop restriction is very hard to produce, this next group defines its fitness by how quickly the agents can complete their task. Therefore, the hits are not as important as the overall fitness, since there is no selection pressure generated based on the number of hits. Although hits increase the fitness, they contribute only a small amount in comparison to the number of time steps taken. The results from these experiments are presented below in Table 7.

Table 7. Results from Step Fitness runs with No Drop

	Generation of Best of Run	Nodes	Hits	Fitness
One Role	59	406	35	270
Branch	56	342	29	251

Here I again run into the unfairness of defining a solution as 50 hits, because the runs that do not allow agents to drop food before they die are difficult to solve. This difficulty in getting individuals who can solve both fitness cases drives the population towards individuals who can quickly solve one of the fitness cases, and who do not necessarily perform well on the other. One possible, but untested solution to this is to alter the fitness function so that it only subtracts steps from the total fitness if all fifty hits have been achieved.

4 Discussion

These results show that, in all meaningful categories, assignment branch runs outperformed runs with only one role. The one category in which there was no improvement was not a difficult problem at all, and both solved it easily.

In general, the ratio of fitness of the assignment branch best-of-run individuals to one-role individuals was 1.2. In such a small problem, this increase is significant. Given that solutions can be found with only a single role, it is satisfying to note that an approach with less human involvement than the explicit creation of one role can provide such an increase in performance. Intuitively, note that solutions that allow more than one role are more likely to solve the problem since they can represent more complex tasks. The results presented in Figure 3 show the improvement over the one role method most clearly. The mean number of hits is much higher in the assignment branch runs. It seems reasonable to assume that had the run continued, the mean would have continued to increase. This is a good sign, because a higher mean suggests a more likely solution, and suggests that the assignment branch runs outperform one role runs. Perhaps it would have been better to observe the difference between a run that explicitly enforced the existence of two roles and the assignment branch runs. After running a few preliminary trials in explicitly enforcing more than one role, there still seems to be an improvement in fitness with the assignment branch runs over the hard-coded runs. This is most likely due to the fact that with the assignment branch, the genetic programming run can tailor a solution to the problem at hand, including both the number of agents in each role, and the total number of agents needed. Runs where the user must specify these parameters are confined to what is not necessarily the optimal configuration, or even a configuration that would allow a solution at all. The more general approach of the assignment branch is an attractive benefit. However, since the ultimate goal of this paper is to provide a method for applying genetic programming to automatically generate team plans, the increase in performance over even the simpler representation is a surprising bonus, and I would have been happy with comparability in performance.

It is interesting to observe that the generation at which the best of run individual was created was lower, with few exceptions, such as in the examples of figure 2 and 3, in individuals from the runs limited to one role. This suggests that one role individuals more quickly converge on their best answer. Runs with the assignment branch, since they are much more structurally complex, take longer for the crossover operation to shuffle meaningful parts of the tree around. This increase in the number of generations needed for the distribution of genetic information in the assignment branch runs does not seem a drastic setback to the assignment branch's utility in the creation of team plans, especially when weighed against the clean-hands approach and the benefit in fitness that the assignment branch affords. In fact, it could be the relative slowness of the assignment branch runs to converge that gives them an advantage in fitness. The one role runs might prematurely converge on a suboptimal result, and have no way to recover. However, most of the evidence seems to point to the fact that the assignment branch runs perform better because they allow for the automatic configuration of both the roles and the total number of agents present in each role.

5 Conclusions

I have shown that using an assignment branch to create team plans in the domain of artificial agents not only works as a method of removing responsibility from the human user of genetic programming, but also outperforms methods where the number of roles and number of agents in each role is set by the user before the run begins. My method leverages the innate power of genetic programming to tailor solutions closely to the problem, and though it may require longer runs to generate solutions, the tradeoff of having the computer automatically generate all aspects of the solution is a tempting one. Also, the assignment branch seems to have potential as a method for harnessing the power of genetic programming for the specific domain of multi-agent and teamwork problems.

6 Future Work

Though I have already shown some benefit to the assignment branch approach, it would be valuable to subject it to a more rigorous test. First, it would be beneficial to try the experiment using architecture altering operations (Koza, 1994) to allow the run to create the branches for each role as needed. This would require dynamic alteration of the function set of the assignment branch, and also an incorporation of mutation for at least the assignment branch so that the new role could have a chance to be incorporated into the solution. This would, of course, increase the time necessary for each run, since it might get stuck waiting for a mutation to occur, and would introduce difficulties if crossover occurred between individuals that had different roles defined. However, the benefit seems worth the potential difficulties.

Further work should be done in expanding this approach to more complex and demanding teamwork problems, such as the control of search and rescue teams, or of robot soccer teams. Also, introduction of support for genetically-defined communication between members of the team would be an interesting endeavor.

References

Koza, John R. 1992. *Genetic Programming: On the Programming of Computers by Means of Natural Selection.* Cambridge, MA: The MIT Press.

Koza, John R. 1994d. *Architecture-Altering Operations for Evolving the Architecture of a Multi-Part Program in Genetic Programming.* Stanford University Computer Science Department technical report STAN-CS-TR-94-1528. October 21, 1994.

Luke, Sean and L. Spector. 1996. "Evolving Teamwork and Coordination with Genetic Programming". In *Genetic Programming 1996: Proceedings of the First Annual Conference.* 141-149.

Implications of Epigenetic Learning Via Modification of Histones on Performance of Genetic Programming

Ivan Tanev[1] and Kikuo Yuta[1,2]

[1] ATR Network Informatics Laboratories, 2-2-2 Hikaridai, "Keihanna Science City",
Kyoto 619-0288, Japan
{i_tanev, kikuo}@atr.jp

[2] Department of System Science, Graduate School of Informatics, Kyoto University,
Kyoto 606-8502, Japan

Abstract. Extending the notion of inheritable genotype in genetic programming (GP) from the common model of DNA into chromatin (DNA and histones), we propose an approach of embedding in GP an explicitly controlled gene expression via modification of histones. Proposed double-cell representation of individuals features somatic cell and germ cell, both represented by their respective chromatin structures. Following biologically plausible concepts, we regard the plasticity of phenotype of somatic cell, achieved via controlled gene expression owing to modifications to histones (epigenetic learning, EL) as relevant for fitness evaluation, while the genotype of the germ cell – to reproduction of individual. Empirical results of evolution of social behavior of agents in predator-prey pursuit problem indicate that EL contributes to more than 2-fold improvement of computational effort of GP. We view the cause for that in the cumulative effect of polyphenism and epigenetic stability. The former allows for phenotypic diversity of genotypically similar individuals, while the latter robustly preserves the individuals from the destructive effects of crossover by silencing of certain genotypic fragments and explicitly activating them only when they are most likely to be expressed in corresponding beneficial phenotypic traits.

Keywords: epigenesis, histones, genetic programming, multi-agent system

1 Introduction

Until a few years ago, the role of histones (the family of proteins which DNA is wrapped around forming a super-coiled chromatin fiber) in molecular biology community was viewed as solely to help pack the long DNA into the tiny nucleus of eukaryotes' cells. However, as the results of resent research suggest, the histones play a significant role in regulating the synthesis, repair, recombination and transcription of DNA [8][15][18]. It is recognized that the regulation of DNA-transcription (and consequently, the overall gene expression) via histone code during cell division controls the specialization of the cells with the same DNA into variety of cell types. In addition, the histone code might control the variances in phenotypes (i.e. biochemistry, morphology, physiology and behavior) seen on different stages of life cycle of living organisms as developing, maturing and aging. Moreover, the onset of some genetically

K. Deb et al. (Eds.): GECCO 2004, LNCS 3102, pp. 213–224, 2004.

associated diseases (and even cancer) is viewed as a process triggered by both a sudden activation of the genes that "contribute" to the disease and/or sudden deactivation of the genes that "fight" the disease. Being an interface between the nurture and nature, the changeable histone code might be regarded as an integrating link in the information pathway of epigenesis of living organisms. As illustrated in Figure 1 the interaction between the phenotype and various environmental factors (such as food, viral infections, exposure to toxins, irradiation, light, UV, etc.) leads to corresponding variations in the histone code, which in turn result in modified (beneficial or detrimental) gene expression. Without touching the details of either the chromatin structure or the chemical processes in histones, we would like to generalize the recently emerged findings that transcription of the genes in DNA is controlled by the surrounding chemical structure of histones. The acetylation of histones correlates with transcriptional activity of the corresponding DNA gene, while the metylation - with transcriptional inactivity of the gene.

In our approach, extending the notion of inheritable genotype in GP from commonly considered model of DNA of simulated organisms (i.e. genetic programs) into chromatin (i.e. model of DNA with surrounding histone proteins) we attempt to mimic the naturally observed phenomenon of regulating gene expression via epigenetic modifications of histones (i.e. epimutations) into a software system. The system features epigenesis embedded in evolution (phylogenesis), simulated through GP. Because (i) we are interested in short-term (i.e. within the life cycle of the organisms), adaptive or developmental epimutations; and (ii) these epimutations are presumed to be beneficial to the performance of behavioral (rather than biochemical, morphological or physiological) aspect of the phenotype of simulated organisms, we consider our approach as a form of *epigenetic learning* (EL), incorporated in GP. The *objective* of our research is to explore the effects of EL on the performance, and namely – on the computational effort of evolution (phylogenesis) of emergent social behavior of autonomous software agents.

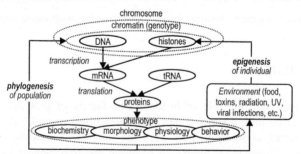

Fig. 1. Simplified Information pathways in phylogenesis and epigenesis in eukaryotic organisms. The inheritable genotype is illustrated as chromatin – a fiber of DNA wrapped around (balls of) histone proteins. The latter control the transcription of DNA by activating or silencing the corresponding nearby genes in DNA. Being an interface between the nature and nurture, histones are subject to modifications as a result of interaction between the phenotype and the environment during the lifetime development and adaptation of organisms

Our work can be viewed as related to the various aspects of approaches of employing heuristics [13], phenotype plasticity [4], Baldwin effect [5], and redundant code [14][16] in GP. In contrast to all of these approaches, the mechanism of EL does not imply direct manipulation on either the simulated DNA or the phenotype. Instead, the proposed EL is achieved through controllable and inheritable gene expression mechanism of simulated individuals. In our approach the genes, being silenced can still comprise the genotype without affecting the performance of individual's phenotype. In addition to being biologically more plausible, such an approach might offer (i) better phenotypic diversity of genotypically similar individuals in the populations and (ii) an efficient way to preserve the individuals from the destructive effects of crossover by explicit activation of the growing genetic combinations when they are most likely to be expressed as corresponding beneficial phenotypic traits.

The remainder of this document is organized as follows. Section 2 briefly introduces the task, which we use to test our hypothesis. The same section briefly explains the key properties of the algorithmic paradigm employed to evolve the functionality of agents. The proposed mechanism of EL is introduced in Section 3. The same section presents empirically obtained results of the implications of EL on the performance of evolution. The conclusion is drawn in Section 4.

2 Background

In this section, we briefly introduce the application and the main attributes of evolutionary algorithmic paradigm employed in our approach. The general, well defined and well studied yet difficult to solve predator-prey pursuit problem [2] is used to verify the implications of EL on the efficiency of evolution. The problem comprises four predator agents whose goals are to capture a prey by surrounding it on all sides in a world. In our work we consider an instance of the problem, which is more realistic and probably more difficult for predators than is commonly considered in the previous work [6][10]. The world is a simulated two-dimensional continuous torus and the moving abilities of four predator agents are continuous. We introduce a proximity perception model for predator agents in that they can see the prey and only the closest predator agent, and only when they are within the limited range of visibility of their simulated (covering an area of *360* degrees) sensors. The prey employs random wandering if there is no predator in sight and a priori handcrafted optimal escaping strategy as soon as predator(s) become "visible". The maximum speed of prey is higher than the maximum speed of predator (i.e. predator-agents feature inferior moving abilities). In order to allow predators to stalk and collectively approach the prey the range of visibility of predators is more than the range of visibility of the prey. We consider this case as a key prerequisite for creating inherently cooperative environment in that the mission of predators is nearly impossible unless they collaborate with each other.

The evolved social (surrounding) behavior of predator agents emerges from what we regard as Occam's razor in interactions between the predator agents: simple, local, implicit, proximity-defined, and therefore – robust and scalable interactions. Within the scope of this document, we consider the emergence as phenomena of local inter-action creating global properties [1][12]. Without providing explicit domain-specific knowledge about how to accomplish the task (e.g. how to surround the prey) the agents, evolved through GP behave "as if" they had such explicit knowledge, because the original source of problem-specific constrains is an integral part of the GP itself. The interaction between GP and the problem environment allows the appropriate knowledge about how to accomplish the task to emerge as a by-product. With the only difference in the representation of relevant domain-specific knowledge, from the out-side the evolved behavior is no different than one that allows the agents to accomplish the task by intentional, deliberate acts (of surrounding). Thus we consider the evolved behavior of agents as *emergent behavior*. Moreover, because in such seemingly inten-tional, deliberate acts each agent is acting "as if" it is well aware of objectives of other agents, shares these objectives, and anticipates the actions of other agents, the emerged (surrounding) behavior of agents is considered as a form of *social behavior*.

A set of stimulus-response rules is used as a natural way to model the reactive be-havior of predator agents [7], which in our approach is evolved using GP. GP is a domain-independent problem solving approach in which a population of computer programs (individuals) is evolved to solve problems [9]. The simulated evolution in GP is based on the Darwinian principle of reproduction and survival of the fittest. The strength of GP to automatically evolve a set of stimulus-response rules featuring arbi-trary complexity without the need to a priori specify the extent of such complexity might imply an enormous computational effort caused by the need to explore a huge search space while looking for the potential solution to the problem. The function set of GP comprises IF-THEN statement, arithmetical operations and comparison opera-tors. The terminal set features local, proximity defined sensory- and continuous mov-ing abilities. The representation of genetic programs is based on widely adopted document object model (DOM) and extensible markup language (XML) in a way as proposed in [17]: genetic programs are represented as a DOM-parsing trees featuring corresponding flat XML text. Both the genetic operations and the evaluation of indi-viduals are performed on their respective DOM-parsing trees using off-the shelf, plat-form- and language neutral DOM-parsers, and XML-text representation is employed as a flat format, feasible for migration of genetic programs among the computational nodes in the distributed implementation of GP. The genetic operations are binary tournament selection, random sub-tree and transposition mutations. The breeding strategy is homogeneous: the performance of a single genetic program, cloned to all the agents is evaluated. We consider such a strategy as adequate to the symmetrical nature of the world, which is unlikely to promote any behavioral specialization among predator agents. The fitness of the genetic program is evaluated as average of the fitness measured over 10 different, randomly created initial situations. The fitness measured during the trial starting with particular initial situation considers the per-formance of the team of agents [3] and accounts for (i) the average energy loss of the agents during the trial, (ii) the average distance of the agents to the prey by the end of the trial, and (iii) the elapsed time of the trial. The energy loss estimation takes into account both the basal metabolic rate of the agents and the energy loss for moving activities. Smaller values of fitness function correspond to better performing predator agents.

A trace of the entities in the world, where the team of predator agents is governed by sample best-of-run genetic program in one of the initial situations is shown in Figure 2. The prey, originally situated in the center of the world, is captured by time step 118. The emergence of following behavioral traits of predator agents are noticeable: (i) switch from greedy chase into surrounding approach (agent #2, time step 65, on the top, center of the world); (ii) zigzag move, which results in a lower chasing speed indicating "intention" to trap the prey (agent #1, after time step 40, center) and (iii) surrounding approach (agents #0 and #3, top; agent #2, bottom and top) demonstrated during the final stages of the trial.

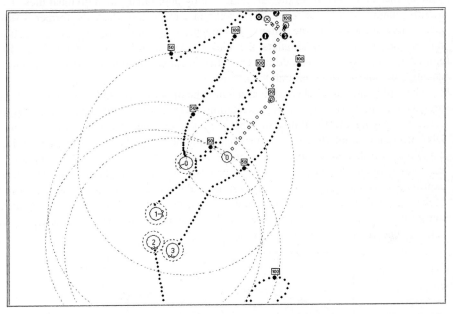

Fig. 2. Traces of the entities with predator agents governed by the sample best-of-run genetic program. The prey is captured in 118 simulated time steps (top). Large white and small black circles denote the predator agents in their initial and final position respectively. The small white circle indicates the prey, initially situated in the center of the world. The numbers in rectangles show the timestamp information

3   Embedding EL via Epimutations in GP

3.1   The Mechanism of EL

Chromatin Representation. In the proposed approach, we represent the predator agents as simulated individuals passing through the phases of birth, development and survival (reproduction) or death (Figure 3). At the phase of birth, the individual is represented as a single embryonic cell expressed by its respective chromatin. The simulated division of the embryonic cell into single germ cell and single somatic cell initiates the EL phase. Both cells are expressed by their respective chromatin structures. In contrast to the germ cell, the somatic cell is subject to EL via iterative epimutations. The completion of the learning phase of the organism is associated with fitness evaluation, based on the performance of the phenotype of modified somatic cell. Reproduction phase concludes the life cycle of the individual when, depending on the relative ranking of the individual in the population and the outcome of the following selection, either (i) a new individual (i.e. embryonic cell) is born by crossover with another individual from the surviving mating pool or (ii) the individual dies. The logical separation of the cells into germ cell and somatic cell where the former is subject to phylogenesis and the latter – to epigenesis and following fitness evaluation reflects our intention to simulate the biologically plausible presumption that the epimutations in somatic cells do not cross the so called Weissman barrier and consequently, are not inherited through the germline.

The representation of chromatin as a genotype of both germ and somatic cell of the organism in GP is based on the representation of DNA paired with isomorphic histone code. Employing the flexibility of XML, we implemented the chromatin in which DNA is organized into tree structure representing the evolved stimulus-response rules, combined with histones expressed as corresponding attributes of IF-THEN nodes of the rules (Figure 4a). The semantics of genotype is shown in Figure 4b. The gene expression mechanism, controlled by histone code implies that during the evaluation of the agent behavior only IF-THEN nodes with histone attributes equal to *"1"* feature transcriptional activity and therefore – only these nodes are parsed, considered as active phenotype and executed (Figure 4c).

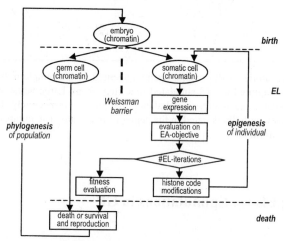

Fig. 3. Life cycle of simulated individual. Individual features double cell representation – germ cell and somatic cell where the former is subject to phylogenesis and the latter – to EL and following fitness evaluation. Only the best scoring (on EL objective) epigenetic changes to phenotype of somatic cell are evaluated for fitness

Algorithm of EL. We incorporated the EL into the fitness evaluation routine of GP as a special case of random local search embedded within GP (Figure 5). We consider the following relevant aspects of the algorithm of EL: (i) the way of selecting which histone should be modified, (ii) the algorithm of epimutations (histone modification), (iii) the objective of learning (i.e. the learning task), (iv) the learning interval and (v) the amount of EL-iterations.

Random histone is selected for modification and the modification algorithm is simply inverting the value of the selected histone. Both the selection and modification algorithms are implemented in the `Modify_Histones` function as illustrated in Figure 5, line 10.

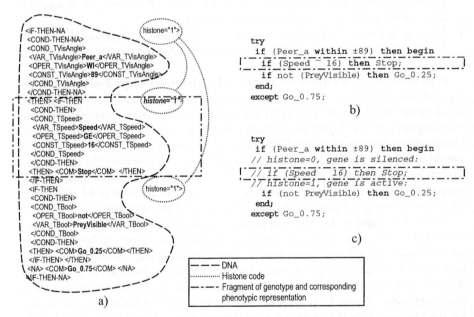

Fig. 4. Genetic representation. The genotype of somatic cell before the development phase of the individual is shown in (a). (b) and (c) illustrate the human readable semantics of the phenotype of somatic cell before the EL phase (b) and after EL (c) as a result of silencing an "`if-then`" statement due to change of the value of corresponding histone from "1" to "0".

```
 1.  Procedure EvalWithEL(GP: TChromatin; EL_Cycles: integer; [out] Fitness: TFitness);
 2.  var Dev_GP : TChromatin;
 3.      i, EL_Ability, Dev_EL_Ability : integer;
 4.  begin
 5.   Dev_GP:=GP; EL_Ability:=0;
 6.   for i:=1 to EA_Cycles do begin
 7.    Clone_GP_To_All_Agents(GP);
 8.    Eval_EL_Objective([out] Dev_EL_Ability);
 9.    if (Dev_EL_Ability better_than EL_Ability) then Dev_GP:=GP;
10.    if (i < EL_Cycles) then Modify_Histones(GP);
11.   end;
12.   Clone_GP_To_All_Agents(Dev_GP);
13.   Eval_Fitness ([out] Fitness);
14.  end;
```

Fig. 5. The algorithm of EL embedded in the routine of evaluation of genetic program. The output specifiers of function parameters are explicitly given for better readability

We considered three cases of EL featuring different learning objectives and epigenetic inheritance as follows:

- The learning objective is the same as the evolution attempts to achieve (EL case 1, EL1). The estimation of the learning ability (variable Dev_EL_Ability, Figure 5, assigned in line 8) is identical to fitness evaluation. The epimutations in case EL1, although inherited through the EL cycles of somatic cell of simulated individual, are not assumed to be inherited thought the germline,
- The learning objective aims to increase the amount of implicit interactions between predator agents (EL case 2, EL2). Learning ability accounts for the total number of references to any sensory variables related to perceiving the peer predator agents. The epimutations in case EL2 are not assumed to be inherited throught the germline of simulated individuals, and
- EL3: the learning objective is the same as in EL2 and the epimutations are assumed to be inherited thought the germline of simulated individuals.

In all three cases only the best-scoring (on EL objective) adapted individual is being evaluated for fitness. The considered cases of epigenetic inheritance through the development of the somatic cells (EL1 and EL2) is based on our intention to simulate the recognition that in Nature the epimutations are not inherited through the germline of individuals. The introduction of inheritance of epimutations through the germline of simulated individuals (EL3) is motivated by our interest in analyzing the implications of such inheritance on the efficiency of simulated evolution. Germline inheritance of epimutations implies that the histone code of germ cell is assumed to be identical to the mutated (through EL) histone code of somatic cell of the organism.

The case of learning objective, not identical to the fitness is introduced with objective to reduce the computational cost of learning (and consequently, to boost the overall performance of GP with embedded EL) by decreasing the learning interval. Because the qualitative value of the fitness heavily depends on the result of quantitative, discrete event of capturing the prey, the approach of noisy fitness evaluation [11] is hardly applicable for the EL1: the fitness values are simply undefined prior to capturing the prey or, ultimately, before the expiration of the allowed time interval of the trial. Exploring the feasibility of using simple, continuous function, evaluated over reduced learning interval for estimation of learning abilities of agents, we use the amount of implicit interactions among the agents. The benefits of the proposed approach are as follows: (i) it is biologically plausible – compared to the survival process, the learning acts in nature take place in a different time scale; and the learning objective usually features downgraded complexity, risk, and cost; (ii) being continuous, "anytime" indicator of the fitness, the EL objective function could be noisily evaluated over reduced learning interval; and (iii) due to the experimentally verified correlation of learning abilities with the fitness, the learning and evolution would synergistically influence each other.

The learning interval is 0.25 of the trial of fitness evaluation. The decision about the duration of learning interval is based on the anticipation that the degradation of computational effort due to noisy fitness evaluation would be insignificant (indeed, the experimentally obtained standard deviation of the noise in estimating the overall amount of interactions from the amount of interaction obtained during the considered learning interval is *0.35*) and this degradation is most likely to be overcompensated by the associated 4-fold improvement of computational performance.

The minimally possible amount of learning iterations is attempted: a single learning iteration in EL1, and *2* learning iterations in EL2 and EL3 (variable `EL_Cycles` in Figure 5 equals *1* and *2* respectively).

3.2 Effect of EL on the Performance of Evolution: Empirical Results

Values of Parameters. The values of parameters of GP used in our experiments are as follows: the population size is *400* genetic programs, the selection ratio is *0.1*, including *0.01* elitism, and the mutation ratio is *0.02*, equally divided between sub-tree mutation, transposition and histone modification. The latter is performed in identical way as in the EL. The termination criterion is defined as a disjunction of the following conditions: (i) fitness of the best genetic program in less than *300* and the amount of initial situations in which the prey is captured (successful situations) equals *10* (out of *10*), (ii) amount of elapsed generations is more than a *100*, and (iii) amount of recent generations without fitness improvement is more than 16. The raw fitness value of *300* roughly corresponds to successful team of predator agents, which in average (over all initial situations) capture the prey by the middle of the trial. The latter equals to *600* time steps, where each step is simulated by *500ms* of "real time" sampling interval. A superior sensory abilities of predators (range of visibility *400mm* vs. *200mm* for prey) and inferior moving abilities have been considered (*20mm/s* vs. *24mm/s*).

Computational Effort of GP Incorporating EL. The effect of EL on the computational effort, statistically estimated (in a way as suggested in [9]) from the probability of success p(t) of GP over 20 independent runs is shown in Figure 6. As figure illustrates, probability of success of 0.95 for GP (denoted as P, phylogenesis) is achieved for about 32,000 individuals, while for incorporated EL1 (denoted as P+EL1), EL2 (P+EL2) and EL3 (P+EL3) these values are 24,000 (reduced 1.3 times), 15,000 (2.1 times) and 14400 (2.2 times) individuals respectively. The reasons for improved computational effort of evolution with embedded EL we view in the cumulative effect of polyphenism and epigenetic stability introduced by histone code and exploited by proposed EL. The polyphenism allows for phenotypic diversity of genotypically similar individuals, while the epigenetic stability robustly preserves the individuals from the destructive effects of crossover by silencing of certain genotypic combinations and explicitly activating them only when they are most likely to be expressed in corresponding beneficial phenotypic traits.

The breakdown of computational effort is shown in Table 1. The total equivalent amount of evaluated individuals E_{TE}, depicted as an abscissa in Figure 9, which takes into account the reduced learning interval with respect to the trial interval of fitness evaluation, is calculated in accordance with the following rule:

$$E_{TE} = E_P + (T_{EA} / T_{FE}) \times E_{EL}$$

Where E_P and E_{EL} are the amount of fitness evaluations during the phylogenesis and EL (either EL1 or EL2) respectively, and T_{FE} and T_{EL} are the trial interval of fitness evaluation and learning interval respectively. The results, shown in Table 1 indicate that both P+EL2 and P+EL3 compared to P+EL1 not only reduce E_{TE} (which, in general might be due to the reduced learning interval alone). Moreover, EL2 and EL3 reduce E_P too, suggesting the presence of favorable effect of these cases of EL on the efficiency of phylogenesis. This effect overcompensates the expected deteriorative effect of the noisy evaluation of learning objective in EL2 and EL3 on the efficiency of evolution. The effect could be explained by beneficial dependencies between the amount of interactions and fitness. We assume that these dependencies are beyond the correlation (which alone should not result in decrease of E_P) as elaborated earlier in 3.1.2. Presumed statistical association of the climb in the learning landscape during EA with a move towards the proximity of optimal solutions in the fitness landscape could be viewed as a reason for the beneficial effect of considered cases of EL on the efficiency of evolution.

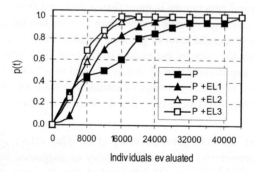

Fig. 6. Probability of success $p(t)$ for GP employing phylogenesis only (P), EL embedded in phylogenesis with learning objective the same as in the evolution (P+EL1), and learning objective of intensifying interactions (P+EL2 and P+EL3). The inheritance of epimutations in EL2 is through the EL cycles of somatic cell, while in EL3 the inheritance is both through the EL cycles of somatic cell and through the germline of simulated individual

Table 1. Breakdown of computational effort of GP

Case	EL objective	Epigenetic inheritance	EL interval, T_{EL}	EL cycles	Computational effort			Speedup
					E_P	E_{EL}	E_{TE}	
P	-	-	-	-	32000	-	32000	1
P+EL1	Same as fitness	In somatic cells only	T_{FE}	1	12000	12000	24000	1.3
P+EL2	Interactions	In somatic cells only	0.25^*T_{FE}	2	10000	20000	15000	2.1
P+EL3	Interactions	Both in somatic cells and through the germline	0.25^*T_{FE}	2	9600	19200	14400	2.2

4   Conclusion

We present the results of our work inspired by recently discovered findings in molecular biology suggesting that histones play a significant role in regulating the gene expression in eukaryotes. Extending the notion of inheritable genotype in GP from commonly considered model of DNA into chromatin, we propose an approach of epigenetic programming as way to incorporate the naturally observed phenomenon of regulated gene expression via modification of histones. Considering the individual as comprising of germ cell and somatic cell, both represented as a chromatin, we focus our attention on the development phase of the life cycle of simulated individuals. We mimic the biologically plausible hypothesis that the information contained in chromatin is inheritable both through the development of the somatic cells and through the germline, but that the epigenetic changes to somatic cells' histones are not believed to be inheritable through the germline. Thus, we regard the phenotype of the somatic cell (subject to beneficial EL via histone code modification) as relevant for fitness evaluation of the individual, while the genotype of the germ cell – as a genetic material involved in phylogenesis. The empirically obtained performance evaluation results indicate that epigenesis with biologically plausible epigenetic inheritance through the development the somatic cells contributes to 2.1-fold improvement of computational effort of genetic programming applied to evolve social behavior of predator agents in predator-prey pursuit problem. The simulated epigenetic inheritance through the germline of individuals yields a marginally better (2.2-fold) reduction of computational effort. We associate the benefits of embedding EL in evolution with the cumulative effect of polyphenism and epigenetic stability. The former allows for phenotypic diversity of genotypically similar individuals, while the latter robustly preserves the individuals from the destructive effects of crossover by silencing of certain genotypic combinations and explicitly activating them only when they are most likely to be expressed in corresponding beneficial phenotypic traits. In this context, our approach can be viewed as an attempt to co-evolve the most beneficial genotypic building blocks and the best possible combinations of their expressions.

In the near future we are planning to incorporate evolvable rather than handcrafted learning objective as used in our current approach in order to investigate whether it has even better effect on the performance of the phylogenesis. We are also interested in enhancing the currently used homogeneous breeding strategy into a strategy which allows for genotypically homogeneous team of agents to develop into phenotipically diverse one by means of EL (as in polyphenism, seen in social insects in the nature). In addition, we intend to investigate the feasibility of the proposed approach for modeling the morphogenesis of the somatic cells through EL in evolvable multi-cellular organisms. Finally, considering our approach as a domain-neutral we are planning to verify it on different tasks from various problem domains and to compare the obtained results with results of known approaches of incorporating other models of redundant genetic representations.

Acknowledgments. The authors thank Katsunori Shimohara for his immense support of this research. The research was conducted as part of "Research on Human

Communication" with funding from the Telecommunications Advancement Organization of Japan.

References

1. Angeline, P.J.: Genetic Programming and Emergent Intelligence. In Kinnear, K.E. Jr., editor, Advances in Genetic Programming, MIT Press (1994) 75-98
2. Benda, M., Jagannathan, B., Dodhiawala, R.: On Optimal Cooperation of Knowledge Sources. Technical Report BCS-G2010-28, Boeing AI Center, Boeing Computer Services, Bellevue,WA (1986)
3. Bull, L, Holland, O.: Evolutionary Computing in Multi-Agent Environments: Eusociality. In Koza, J.R., Deb, K., Dorigo, M., Fogel, D.B., Garzon, M., Iba, H., Riolo, R. (eds.), Proceedings of the Second Annual Conference on Genetic Programming, Morgan Kaufmann (1997) 347-352
4. Esparcia-Alcázar, A.I., Sharman, K.C.: Phenotype Plasticity in Genetic Programming: A Coparison of Darwinian and Lamarckian Inheritance Schemes, Genetic Programming, Procs. of the Second European Workshop (EuroGP'99), Lecture Notes in Computer Science, Springer (1999) 49-64
5. Floreano, D., Nolfi, S., Mondada, F.: Co-evolution and Ontogenetic Change in Competing Robots, In J.P.Mukesh, V.Honavar & K. Balakrishan (Eds.), Advances in Evolutionary Syntsis of Neural Networks. Cambridge, MA: MIT Press. (2001) 273-306.
6. Haynes, T., Sen, S.: Evolving Behavioral Strategies in Predators and Prey (1996)
7. Holand, J.H.: Emergence: From Chaos To Order, Cambridge, Perseus Books (1999)
8. Jenuwein, T., Allis, C.D.: Translating the Histone Code, Science, Vol.293 (2001) 1074-1080
9. Koza, J.R.: Genetic Programming: On the Programming of Computers by Means of Natural Selection, Cambridge, MA, MIT Press (1992)
10. Luke, S., Spector, L.: Evolving Teamwork and Coordination with Genetic Programming, Proceedings of the First Annual Conference on Genetic Programming (GP-96), MIT Press, Cambridge MA, (1996) 150-156
11. Miller, B.L., Goldberg, D.E.,: Genetic Algorithms, Tournament Selection, and the Efects of Noise, Illigal Report No. 95006, University of Illinois (1995).
12. Morowitz, H.J.: The Emergence of Everything: How the World Became Complex, Oxford University Press, New York (2002)
13. Moscato, P.: On Evolution, Search, Optimization, Genetic Algorithms and Martial Arts: Towards Memetic Algorithms, Tech. Rep. Caltech Concurrent Computation Program, Report. 826, California Institute of Technology, Pasadena, California, USA, 1989.
14. Nordin, P., Francone, F, Banzhaf, W.: Explicitly defined introns and destructive crossover in genetic programming. In J. P. Rosca, editor, Proceedings of the Workshop on Genetic Programming: From Theory to Real-World Applications, 9 July 1995, Tahoe City, California, USA (1995) 6-22
15. Phillips, H.: Master Code, New Scientist, Special Issue DNA: The Next 50 Years, 15 March (2003) 44-47
16. Rothlauf, F., Goldberg, D.E.: Redundant Representations in Evolutionary Computation, Evolutionary Computations, Vol.11, No.4 (2003) 381-415
17. Tanev, I.: DOM/XML-Based Portable Genetic Representation of Morphology, Behavior and Communication Abilities of Evolvable Agents, Proceedings of the 8th International Symposium on Artificial Life and Robotics (AROB'03), Beppu, Japan (2003), 185-188
18. True, H.L., Lindquist, S.L.: A Yeast Prion Provides a Mechanism For Genetic Variation And Phenotypic Diversity, Nature, Vol. 406 (Sep 2000) 477-483

Using Clustering Techniques to Improve the Performance of a Multi-objective Particle Swarm Optimizer

Gregorio Toscano Pulido and Carlos A. Coello Coello

CINVESTAV-IPN (Evolutionary Computation Group)
Depto. de Ing. Elect./Sección de Computación
Av. IPN No. 2508, Col. San Pedro Zacatenco, México, D.F. 07300, MEXICO
gtoscano@computacion.cs.cinvestav.mx, ccoello@cs.cinvestav.mx

Abstract. In this paper, we present an extension of the heuristic called "particle swarm optimization" (PSO) that is able to deal with multiobjective optimization problems. Our approach uses the concept of Pareto dominance to determine the flight direction of a particle and is based on the idea of having a set of sub-swarms instead of single particles. In each sub-swarm, a PSO algorithm is executed and, at some point, the different sub-swarms exchange information. Our proposed approach is validated using several test functions taken from the evolutionary multiobjective optimization literature. Our results indicate that the approach is highly competitive with respect to algorithms representative of the state-of-the-art in evolutionary multiobjective optimization.

1 Introduction

Particle swarm optimization (PSO) is a relatively recent heuristic inspired by the choreography of a bird flock which has been found to be quite successful in a wide variety of optimization tasks [1]. Its high speed of convergence and its relative simplicity make PSO a highly viable candidate to be used for solving not only problems with a single objective function, but also problems with several objectives (called "multiobjective optimization problems") [2]. In this paper, we present a proposal, called "another multi-objective particle swarm optimization" (AMOPSO), which extends PSO to deal with several objectives. The main novelty of the approach consists on using a clustering technique in order to divide the population of particles into several swarms in order to have a better distribution of solutions in decision variable space. The introduction of this mechanism significantly improves the quality of the Pareto fronts obtained, when comparing our results with respect to other multiobjective PSO previously reported in the literature and with respect to algorithms representative of the state-of-the-art in evolutionary multiobjective optimization.

2 Related Work

There have been several recent proposals to extend PSO to handle multiple objectives. We will review next the most important of them:

K. Deb et al. (Eds.): GECCO 2004, LNCS 3102, pp. 225–237, 2004.
© Springer-Verlag Berlin Heidelberg 2004

- **The Swarm Metaphor of Ray & Liew [3]**: This algorithm uses Pareto dominance and combines concepts of evolutionary techniques with the particle swarm. It uses crowding to maintain diversity and a multilevel sieve to handle constraints.
- **The algorithm of Parsopoulos & Vrahatis [4]**: This algorithm adopts different types of aggregating functions to solve multiobjective optimization problems.
- **Dynamic Neighborhood PSO proposed by Hu and Eberhart [5]**: In this algorithm, only one objective is optimized at a time using a scheme similar to lexicographic ordering. A revised version of this approach that uses a secondary population is presented in [8].
- **The Multi-objective Particle Swarm Optimizer (MOPSO) by Coello & Lechuga [6]**: This proposal is based on the idea of having a global repository in which every particle will deposit its flight experiences after each flight cycle. Additionally, the updates to the repository are performed considering a geographically-based system defined in terms of the objective function values of each individual; this repository is used by the particles to identify a leader that will guide the search.
- **The approach of Fieldsend & Singh [7]**: This approach incorporates an unconstrained elite archive (in which a special data structure called "dominated tree" is adopted) to store the nondominated individuals found along the search process. The archive interacts with the primary population in order to define local guides. This approach also uses a "turbulence" operator which is basically a mutation operator that acts on the velocity value used by PSO.
- **The algorithm of Mostaghim & Teich [9]:** This approach uses a sigma method in which the best local guides for each particle are adopted to improve the convergence and diversity of a PSO algorithm used for multiobjective optimization. They also use a "turbulence" operator, but applied on decision variable space. The use of the sigma values increases the selection pressure of PSO (which was already high). This may cause premature convergence in some cases.
- **The Nondominated Sorting PSO of Li [10]**: This approach incorporates the main mechanisms of the NSGA-II [11] into a PSO algorithm. The proposed approach showed a very competitive performance with respect to the NSGA-II (even outperforming it in some cases).

Our approach is based on the use of Pareto ranking and a subdivision of decision variable space into several sub-swarms (this is done using clustering techniques). Since independent PSOs are run into each swarm, our approach can be seen as a meta-MOPSO algorithm. After a certain (pre-defined) number of iterations, the leaders of each swarm are migrated to a different swarm in order to variate the selection pressure. This sort of scheme is a novel proposal to solve multiobjective optimization problems using PSO. Also, note that AMOPSO does not use an external population, since elitism in this case is an emergent process derived from the migration of leaders.

3   Description of the Proposed Approach

The analogy of particle swarm optimization with evolutionary algorithms makes evident the notion that using a Pareto ranking scheme could be the straightforward way to extend the approach to handle multiobjective optimization problems. However, merging a Pareto ranking scheme with the PSO algorithm will produce not one but a set of nondominated leaders and the selection of an "appropriate" leader becomes difficult (by definition, all nondominated solutions are equally good). Additionally, it is known that several difficult multiobjective optimization problems have a disconnected decision variable space. This issue is particularly important when using PSO, because it could be the case that a particle tries to follow a leader that resides in a disconnected region away from it. In this case, a lot of search effort would be wasted and the algorithm might not be able to converge to the true Pareto front of the problem. The use of neighborhoods may be useful in this case. However, we argue that the use of a neighborhood may delay convergence, because the selection pressure is significantly lowered when they are used, since in this case, particles spend most of their search effort following leaders that reside far away from the true Pareto front. Our proposal is to use several swarms (each with a fixed size). Each swarm will over-fly a specific region of the Pareto optimal set (i.e., decision variable space), and will have its own niche of particles and particle guides. The algorithm used to associate leaders into a swarm is the hierarchical single-connected clustering algorithm [13]. The appropriate selection of leaders is essential for the good performance of PSO when applied to multiobjective optimization problems. If the particle chooses an inappropriate leader (i.e., a leader who is too far away in the search space) then most of the flight will be fruitless because the particle will not be traversing promisory regions of search space. In this paper, we propose to use not one but several swarms to avoid this type of problem. However, even if we adopt a multi-swarm scheme, a good strategy to select a leader is still necessary. Some possible strategies for this sake are the following: (1) Randomly (a leader is randomly selected—no constraints are imposed on what sort of leader can a particle choose—, (2) The closest (a particle picks as a leader to the geographically closest leader), and (3) one at a time (a single leader is selected by all the particles at a time). In this paper, we adopted the first scheme (random selection of a leader). The way in which our algorithm works is shown next:

function AMOPSO Algorithm
Begin
    For each swarm
        1. Initialize its particles
        2. Initialize g_{leader} set (i.e., the set of global leaders)
    EndFor
    Do
        For each swarm
            Do
                For each particle
                    4. Select a leader
                    5. Perform the flight
                    6. Update values
                    If it is a leader **then** add to the g_{leader} set
                EndFor
        While maximum number of iterationsis not reached
        7. Store leaders in g_{leader} set in n_{swarms} groups

EndFor
8. Assign each leader group to a swarm
While maximum number of iterations is not reached
End.

The proposed algorithm requires the following parameters:

- $GMax$: it refers to the total number of generations that the algorithm will be executed.
- $n_{particles}$: it refers to the total number of particles that will be over-flying the search space.
- n_{swarms}: it refers to the number of particle groups. The swarm size is fixed because the total number of particles is a fixed value.
- $sgmax$: is the number of internal generations that the particles of each swarm will run before sharing their leaders.

The complete execution process of our algorithm can be divided in three stages: initialization, flight and generation of results. At the first stage, every swarm is initialized. Each swarm creates and initializes its own particles and generates the leaders set among the particle swarm set by using Pareto ranking. In the second stage is where the algorithm performs its strongest effort. First, it performs the execution of the flight of every swarm; next, it applies a clustering algorithm to group the guide particles. This is performed until reaching a total of $GMax$ iterations. The execution of the flight of each swarm can be seen as an entire PSO process (with the difference that it will only optimize an specific region of the search space). First, each particle will select a leader to which it will follow. At the same time, each particle will try to outperform its leader and to update its position. If the updated particle is not dominated by any member of the leaders set, then it will become a new leader. The execution of the swarm will start again until a total of $sgmax$ iterations is reached. Constraints are handled in AMOPSO when checking Pareto dominance. When we compare two individuals, we first check their feasibility. If one is feasible and the other is infeasible, the feasible individual wins. If both are infeasible, then the individual with the lowest amount of (total) constraint violation wins. If they both have the same amount of constraint violation (or if they are both feasible), then the comparison is done using Pareto dominance. Once all the swarms have finished theirs flights, a clustering algorithm takes the control by grouping the closest particle guides into n_{swarms} swarms. These particle guides will try to outperform each swarm in the next iteration. This is mainly done by grouping the leaders of all the swarms into a single set, and then splitting this set among n_{swarms} groups (clustering is done with respect to closeness in decision variable space). Each resulting group will be assigned to a different swarm. The third and final stage will present the results, i.e. it will report all the nondominated solutions found.

3.1 Clustering Algorithm

We use Johnson's algorithm to cluster the leaders in groups [13]. The pseudocode of this algorithm is shown next:

function Single-link clustering

Begin

    1.   Begin with the disjoint clustering having level $L(0) = 0$ and sequence number $m = 0$.

    Do

        2. Find the least dissimilar pair of clusters in the current clustering, say pair $(r), (s)$, according to $d[(r), (s)] = mind[(i), (j)]$ where the minimum is over all pairs of clusters in the current clustering

        3. Increment the sequence number: $m = m + 1$. Merge clusters (r) and (s) into a single cluster to form the next clustering m. Set the level of this clustering to $L(m) = d[(r), (s)]$

        4. Update the proximity matrix, D, by deleting the rows and columns corresponding to clusters (r) and (s) and adding a row and a column corresponding to the newly formed cluster. The proximity between the new cluster, denoted (r, s) and the old cluster (k) is defined as: $d[(k), (r, s)] = mind[(k), (r)], d[(k), (s)]$

    while objects are not in N clusters.

End.

The algorithm requires a proximity matrix as a parameter (in our case, we use a dissimilarity matrix, and the Euclidean distance in variable space to represent the dissimilarity).

4 Comparison of Results

Several test functions were taken from the specialized literature to compare our approach, but due to space limitations, only 3 are included in this paper. In order to allow a quantitative assessment of the performance of a multiobjective optimization algorithm, we adopted the following metrics:

1. **Error Ratio** (ER): This metric was proposed by Van Veldhuizen [14] to indicate the percentage of solutions (from the nondominated vectors found so far) that are not members of the true Pareto optimal set:

$$ER = \frac{\sum_{i=1}^{n} e_i}{n}, \tag{1}$$

where n is the number of vectors in the current set of nondominated vectors available; $e_i = 0$ if vector i is a member of the Pareto optimal set, and $e_i = 1$ otherwise. It should then be clear that $ER = 0$ indicates an ideal behavior, since it would mean that all the vectors generated by our algorithm belong to the Pareto optimal set of the problem.

2. **Generational Distance** (GD): This metric was proposed by Van Veldhuizen [14] as a way of estimating how far are the elements in the set of nondominated vectors found so far from those in the Pareto optimal set and is defined as:

$$GD = \frac{\sqrt{\sum_{i=1}^{n} d_i^2}}{n} \tag{2}$$

where n is the number of vectors in the set of nondominated solutions found so far and d_i is the Euclidean distance (measured in objective space) between each of these and the nearest member of the Pareto optimal set. It should be clear that a value of $GD = 0$ indicates that all the elements generated are in the Pareto optimal set.

3. **Spacing (SP):** Here, one desires to measure the spread (distribution) of vectors throughout the nondominated vectors found so far. Since the "beginning" and "end" of the current Pareto front found are known, a suitably defined metric judges how well the solutions in such front are distributed. Schott [15] proposed such a metric measuring the range (distance) variance of neighboring vectors in the nondominated vectors found so far. This metric is defined as:

$$ S \triangleq \sqrt{\frac{1}{n-1} \sum_{i=1}^{n} (\bar{d} - d_i)^2} , \tag{3} $$

where $d_i = \min_j (| f_1^i(\boldsymbol{x}) - f_1^j(\boldsymbol{x}) | + | f_2^i(\boldsymbol{x}) - f_2^j(\boldsymbol{x}) |)$, $i, j = 1, \ldots, n$, $\bar{d}$ is the mean of all d_i, and n is the number of nondominated vectors found so far. A value of zero for this metric indicates all members of the Pareto front currently available are equidistantly spaced.

In order to know how competitive was our approach, we decided to compare it against two multiobjective evolutionary algorithms that represent the state-of-the-art and with respect to a multi-objective particle optimizer (MOPSO) that is publicly available:

1. **Nondominated Sorting Genetic Algorithm II:** Proposed by Deb et al. [11], this algorithm is based on several layers of classifications of the individuals. It incorporates elitism (through the use of $(\mu + \lambda)$-selection), a crowded comparison operator and it keeps diversity without specifying any additional parameters. It remains as one of the most competitive multi-objective evolutionary algorithms known to date.

2. **Pareto Archived Evolution Strategy:** This algorithm was introduced by Knowles and Corne [16]. PAES consists of a (1+1) evolution strategy (i.e., a single parent that generates a single offspring) in combination with a historical archive that records some of the nondominated solutions previously found. This archive is used as a reference set against which each mutated individual is being compared. The archive is not only the elitist mechanism of PAES, but also incorporates an approach to maintain diversity (a crowding procedure that divides objective space in a recursive manner).

3. **Multiobjective Particle Swarm Optimization (MOPSO):** This approach was introduced in [6]. It uses the concept of Pareto dominance to determine the flight direction of a particle and it maintains previously found nondominated vectors in a global repository that is later used by other particles to guide their own flight. It also uses a mutation operator that acts both on the particles of the swarm, and on the range of each design variable of the problem to be solved.

The source code of the NSGA-II, PAES and MOPSO is available from the EMOO repository located at: http://delta.cs.cinvestav.mx/~ccoello/EMOO. The source code of AMOPSO is available upon request by email to the first author. In the following examples, the NSGA-II was run using a population size of 100, a crossover rate of 0.8 (uniform crossover was adopted), tournament selection, and a mutation rate of $1/L$, where L = chromosome length (binary representation was adopted). PAES was run using an adaptive grid with a depth of five, a size of the archive of 100, and a mutation rate of $1/L$, where L refers to the length of the chromosomic string that encodes the decision variables. MOPSO (with real-numbers representation) used a (main) population of 100 particles, a repository size of 100 particles, a mutation rate of 0.05, and 30 divisions for the adaptive grid. AMOPSO used 40 particles, a maximum number of generations of 20, a maximum number of generations per swarm of 5, and a total of 5 swarms. The values of all these parameters were empirically derived. The total number of fitness function evaluations was set to 4000 for all the algorithms compared in all the test functions shown next.

4.1 Test Function 1

For our first example, we used the following problem proposed in [17]: Maximize $F = (f_1(x, y), f_2(x, y))$, where

$$f_1(x, y) = -x^2 + y, \quad f_2(x, y) = \frac{1}{2}x + y + 1$$

subject to: $0 \geq \frac{1}{6}x + y - \frac{13}{2}, \quad 0 \geq \frac{1}{2}x + y - \frac{15}{2}, \quad 0 \geq 5x + y - 30$ and: $x, y \geq 0$.

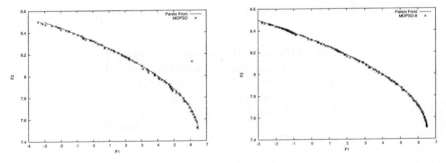

Fig. 1. Pareto fronts produced by MOPSO (left) and the AMOPSO (right) for the first test function.

Figures 1 and 2 show the graphical results produced by the PAES, the NSGA-II, MOPSO and our AMOPSO in the first test function chosen. The true Pareto front is shown as a continuous line. The solutions displayed correspond to the median result with respect to the generational distance metric for each of the algorithms compared. The true Pareto front of the problem is shown as a continuous line. Tables 1 and 2 show the comparison of results among the four algorithms considering the metrics previously described. It can be seen that the average performance of AMOPSO is the best with respect to the error ratio (by far), and with respect to generational distance. With respect

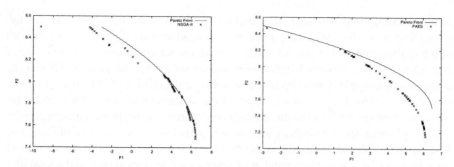

Fig. 2. Pareto fronts produced by the NSGA-II (left) and PAES (right) for the first test function.

Table 1. Results of the Error Ratio (ER) and the Generational Distance (GD) metrics for the first test function.

	Error Ratio				Generational Distance			
	MOPSO	AMOPSO	NSGA-II	PAES	MOPSO	AMOPSO	NSGA-II	PAES
Average	0.6542	**0.4926**	0.8955	1.0023	0.0150	**0.0038**	0.0518	0.1041
Best	0.5300	0.3763	0.7200	0.9600	0.0024	0.0017	0.0032	0.0114
Worst	0.8261	0.5814	1.0000	1.0156	0.0859	0.02109	0.3170	0.6380
Median	0.6500	0.4932	0.9100	1.0100	0.0078	0.0022	0.0092	0.0264
St. Dev.	0.06263	0.05164	0.07564	0.01630	0.01960	0.00447	0.08641	0.16033

Table 2. Results of the Spacing (SP) metric for the first test function.

SP	MOPSO	AMOPSO	NSGA-II	PAES
Average	0.109146	0.043361	**0.028961**	0.079803
Best	0.046508	0.028168	0.008999	0.021405
Worst	0.681124	0.125776	0.080856	0.230506
Median	0.059248	0.035124	0.025807	0.051504
St. Dev.	0.141827	0.023900	0.017047	0.060692

to spacing it places slightly below the NSGA-II, but with a lower standard deviation. By looking at the Pareto fronts of this test function, it can be easily seen that, except for MOPSO and our AMOPSO, none of the algorithms was able to cover the full Pareto front. It can also be seen that AMOPSO produced the best front. This is then an example in which a metric may be misleading, since the fact that the spacing metric provides a good value becomes meaningless if the nondominated vectors produced by the algorithm are not part of the true Pareto front of the problem.

5 Test Function 2

Our second test function was proposed in [18]:

$$\text{Min} f_1(\boldsymbol{x}) = \sum_{i=1}^{n-1} \left(-10 \exp\left(-0.2\sqrt{x_i^2 + x_{i+1}^2}\right)\right); \text{Min} f_2(\boldsymbol{x}) = \sum_{i=1}^{n} \left(|x_i|^{0.8} + 5 \sin(x_i)^3\right) \quad (4)$$

where: $-5 \leq x_1, x_2, x_3 \leq 5$

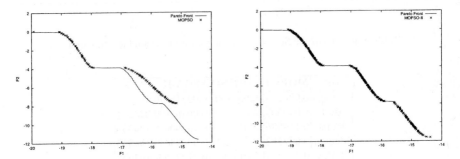

Fig. 3. Pareto fronts produced by MOPSO (left) and the AMOPSO (right) for the second test function.

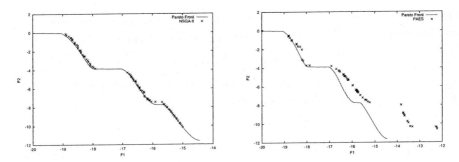

Fig. 4. Pareto fronts produced by the NSGA-II (left) and PAES (right) for the second test function.

Figures 3 and 4 show the graphical results produced by the PAES, the NSGA-II, MOPSO and our AMOPSO in the second test function chosen. Tables 3 and 4 show the comparison of results among the four algorithms considering the metrics previously described. It can be seen that the average performance of AMOPSO is the best with respect to all the metrics adopted. This result can be corroborated by looking at Figures 3 and 4. Except for our AMOPSO, all the algorithms missed important segments of the true Pareto front.

Table 3. Results of the Error Ratio (ER) and the Generational Distance (GD) metrics for the second test function.

	Error Ratio				Generational Distance			
	MOPSO	AMOPSO	NSGA-II	PAES	MOPSO	AMOPSO	NSGA-II	PAES
Average	0.7810	**0.3490**	0.7365	0.9320	0.0335	**0.0020**	0.0242	0.1730
Best	0.7300	0.2423	0.2300	0.3700	0.0318	0.0015	0.0030	0.0240
Worst	0.8500	0.4565	1.010	1.2500	0.0346	0.0035	0.1058	1.5743
Median	0.7800	0.3614	0.7950	1.0000	0.0336	0.0019	0.0072	0.0909
St. Dev.	0.03194	0.05107	0.24594	0.17914	0.00069	0.00050	0.02935	0.33300

Table 4. Results of the Spacing (SP) metric for the second test function.

SP	MOPSO	AMOPSO	NSGA-II	PAES
Average	0.086108	**0.037190**	0.038156	0.449358
Best	0.044509	0.019109	0.002839	0.094711
Worst	0.119586	0.055227	0.102906	5.124390
Median	0.092252	0.041811	0.035956	0.198565
St. Dev.	0.022214	0.014061	0.019729	1.102054

6   Test Function 3

Our third test function is to optimize a four-bar plane truss. The problem is the following [19]: Min $f_1(\mathbf{x}) = L(2x_1 + \sqrt{2}x_2 + \sqrt{x_3} + x_4)$, $f_2(\mathbf{x}) = \frac{FL}{E}\left(\frac{2}{x_2} + \frac{2\sqrt{2}}{x_2} - \frac{2\sqrt{(2)}}{x_3} + \frac{2}{x_4}\right)$ such that $(F/\sigma) \leq x_1 \leq 3 \times (F/\sigma)$, $\sqrt{2}(F/\sigma) \leq x_2 \leq 3 \times (F/\sigma)$, $\sqrt{2}(F/\sigma) \leq x_3 \leq 3 \times (F/\sigma)$, $(F/\sigma) \leq x4 \leq 3 \times (F/\sigma)$ where: $F = 10kN$, $E = (2)10^5 kN/cm^2$, $L = 200cm$, $\sigma = 10kN/cm^3$

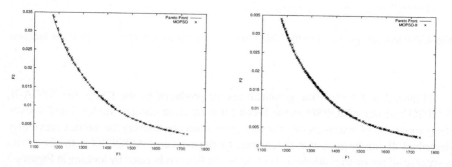

Fig. 5. Pareto fronts produced by MOPSO (left) and the AMOPSO (right) for the third test function.

Figures 5 and 6 show the graphical results produced by PAES, the NSGA-II, MOPSO and our AMOPSO in the third test function chosen. Tables 5 and 6 show the comparison of results among the four algorithms considering the metrics previously described. In this case, AMOPSO was the best with respect to the generational distance and spacing metrics, and it placed second (marginally) with respect to the error ratio metric (PAES was the best average performer with respect to this metric). Graphically, we can see that MOPSO and our AMOPSO were the only algorithms able to cover the entire Pareto front of this problem. Clearly, our AMOPSO produced the best front in this case (see Figures 5 and 6).

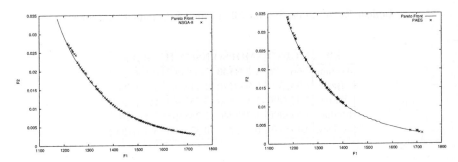

Fig. 6. Pareto fronts produced by the NSGA-II (left) and PAES (right) for the third test function.

Table 5. Results of the Error Ratio (ER) and Generational Distance (GD) metrics for the third test function.

	Error Ratio				Generational Distance			
	MOPSO	AMOPSO	NSGA-II	PAES	MOPSO	AMOPSO	NSGA-II	PAES
Average	0.447869	0.436210	0.447500	**0.386888**	0.452556	**0.215814**	0.364229	1.079089
Best	0.260000	0.293548	0.210000	0.190000	0.312648	0.152786	0.249701	0.168401
Worst	0.640000	0.558824	0.960000	0.640000	0.933802	0.348571	0.585769	14.222200
Median	0.425000	0.434563	0.355000	0.360000	0.380676	0.212191	0.354397	0.238549
St. Dev.	0.096677	0.061168	0.189150	0.115806	0.178124	0.048167	0.065211	3.126444

7   Conclusions and Future Work

We have presented a new proposal to extend particle swarm optimization to handle multiobjective problems using sub-swarms, Pareto ranking and clustering techniques. The proposed approach was validated using the standard methodology currently adopted in the evolutionary multiobjective optimization community. The results indicate that our approach is a viable alternative since it outperformed some of the best multiobjective evolutionary algorithms known to date. One aspect that we would like to explore in the future is the study of alternative mechanisms to handle constraints through the use of infeasible solutions that can act as leaders in a special swarm. We believe that this sort of mechanism could improve the performance of our AMOPSO, particularly when dealing with problems in which the Pareto front lies on the boundaries between the feasible and infeasible regions. We also want to perform an analysis of the impact of the mechanism adopted to select leaders in the performance of the approach. Finally, we aim to devise a way to eliminate the n_{swarm} parameter through the use of self-adaptation.

Table 6. Results of the Spacing (SP) metric for the third test function.

SP	MOPSO	AMOPSO	NSGA-II	PAES
Average	2.665576	**1.209140**	2.172087	4.180568
Best	1.932520	0.934140	1.115380	1.250790
Worst	3.925990	1.527800	2.557980	22.745200
Median	2.643000	1.257940	2.248595	2.314210
St. Dev.	0.495079	0.177295	0.375880	5.440785

Acknowledgements. The first author acknowledges support from CONACyT through a scholarship to pursue graduate studies at the Computer Science Section at CINVESTAV-IPN. The second author gratefully acknowledges support from CONACyT project 34201-A.

References

1. Kennedy, J., Eberhart, R.C.: Swarm Intelligence. Morgan Kaufmann Publishers, San Francisco, California (2001)
2. Coello Coello, C.A., Van Veldhuizen, D.A., Lamont, G.B.: Evolutionary Algorithms for Solving Multi-Objective Problems. Kluwer Academic Publishers, Boston (2002)
3. Ray, T., Liew, K.: A Swarm Metaphor for Multiobjective Design Optimization. Engineering Optimization **34** (2002) 141–153
4. Parsopoulos, K., Vrahatis, M.: Particle Swarm Optimization Method in Multiobjective Problems. In: Proceedings of the 2002 ACM Symposium on Applied Computing (SAC'2002), Madrid, Spain, ACM Press (2002) 603–607
5. Hu, X., Eberhart, R.: Multiobjective Optimization Using Dynamic Neighborhood Particle Swarm Optimization. In: Congress on Evolutionary Computation (CEC'2002). Volume 2., Piscataway, New Jersey, IEEE Service Center (2002) 1677–1681
6. Coello Coello, C.A., Salazar Lechuga, M.: MOPSO: A Proposal for Multiple Objective Particle Swarm Optimization. In: Congress on Evolutionary Computation (CEC'2002). Volume 1., Piscataway, New Jersey, IEEE Service Center (2002) 1051–1056

7. Fieldsend, J.E., Singh, S.: A Multi-Objective Algorithm based upon Particle Swarm Optimisation, an Efficient Data Structure and Turbulence. In: Proceedings of the 2002 U.K. Workshop on Computational Intelligence, Birmingham, UK (2002) 37–44

8. Hui, X., Eberhart, R.C., Shi, Y.: Particle Swarm with Extended Memory for Multiobjective Optimization. In: 2003 IEEE Swarm Intelligence Symposium Proceedings, Indianapolis, Indiana, USA, IEEE Service Center (2003) 193–197

9. Mostaghim, S., Teich, J.: Strategies for Finding Good Local Guides in Multi-objective Particle Swarm Optimization (MOPSO). In: 2003 IEEE Swarm Intelligence Symposium Proceedings, Indianapolis, Indiana, USA, IEEE Service Center (2003) 26–33

10. Li, X.: A Non-dominated Sorting Particle Swarm Optimizer for Multiobjective Optimization. In et al., E.C.P., ed.: Genetic and Evolutionary Computation—GECCO 2003. Proceedings, Part I, Springer. Lecture Notes in Computer Science Vol. 2723 (2003) 37–48

11. Deb, K., Pratap, A., Agarwal, S., Meyarivan, T.: A Fast and Elitist Multiobjective Genetic Algorithm: NSGA–II. IEEE Transactions on Evolutionary Computation **6** (2002) 182–197

12. Goldberg, D.E.: Genetic Algorithms in Search, Optimization and Machine Learning. Addison-Wesley Publishing Company, Reading, Massachusetts (1989)

13. Johnson, S.: Hierarchical Clustering Schemes. Psychometrika **32** (1967) 241–254

14. Veldhuizen, D.A.V.: Multiobjective Evolutionary Algorithms: Classifications, Analyses, and New Innovations. PhD thesis, Department of Electrical and Computer Engineering. Graduate School of Engineering. Air Force Institute of Technology, Wright-Patterson AFB, Ohio (1999)

15. Schott, J.R.: Fault Tolerant Design Using Single and Multicriteria Genetic Algorithm Optimization. Master's thesis, Department of Aeronautics and Astronautics, Massachusetts Institute of Technology, Cambridge, Massachusetts (1995)

16. Knowles, J.D., Corne, D.W.: Approximating the Nondominated Front Using the Pareto Archived Evolution Strategy. Evolutionary Computation **8** (2000) 149–172

17. Kita, H., Yabumoto, Y., Mori, N., Nishikawa, Y.: Multi-Objective Optimization by Means of the Thermodynamical Genetic Algorithm. In et al., H.M.V., ed.: Parallel Problem Solving from Nature—PPSN IV. Springer-Verlag, Berlin (1996) 504–512

18. Kursawe, F.: A Variant of Evolution Strategies for Vector Optimization. In Schwefel, H.P., Männer, R., eds.: Parallel Problem Solving from Nature. 1st Workshop, PPSN I. Volume 496 of Lecture Notes in Computer Science., Berlin, Germany, Springer-Verlag (1991) 193–197

19. Cheng, F., Li, X.: Generalized Center Method for Multiobjective Engineering Optimization. Engineering Optimization **31** (1999) 641–661

SWAF: Swarm Algorithm Framework for Numerical Optimization

Xiao-Feng Xie and Wen-Jun Zhang

Institute of Microelectronics, Tsinghua University, 100084 Beijing, China
xiexf@ieee.org, zwj@tsinghua.edu.cn

Abstract. A swarm algorithm framework (SWAF), realized by agent-based modeling, is presented to solve numerical optimization problems. Each agent is a bare bones cognitive architecture, which learns knowledge by appropriately deploying a set of simple rules in fast and frugal heuristics. Two essential categories of rules, the generate-and-test and the problem-formulation rules, are implemented, and both of the macro rules by simple combination and subsymbolic deploying of multiple rules among them are also studied. Experimental results on benchmark problems are presented, and performance comparison between SWAF and other existing algorithms indicates that it is efficiently.

1 Introduction

The general numerical optimization problems can be defined as:

$$\text{Minimize: } F(\vec{x}) \tag{1}$$

where $\vec{x} = (x_1, ..., x_d, ..., x_D) \in S \subseteq \mathbb{R}^D$ ($1 \le d \le D, d \in \mathbb{Z}$), and $x_d \in [l_d, u_d]$, l_d and u_d are lower and upper values respectively. $F(\vec{x})$ is the objective function. S is a D-dimensional *search space*. Suppose for a certain point $\vec{x}^*$, there exists $F(\vec{x}^*) \le F(\vec{x})$ for $\forall \vec{x} \in S$, then $\vec{x}^*$ and $F(\vec{x}^*)$ are separately the global optimum point and its value. The *solution space* is defined as $S_O = \{\vec{x} \mid F_\Delta(\vec{x}) = F(\vec{x}) - F(\vec{x}^*) \le \varepsilon_O\}$, where ε_O is a small positive value. In order to find $\vec{x} \in S_O$ with high probability, the typical challenges include: a) S_O/S is often very small; b) little *a priori* knowledge is available for the landscape; and c) calculation time is finite.

Many methods based on *generate-and-test* have been proposed, such as Taboo search (TS) [12], simulated annealing (SA) [14, 23], evolutionary algorithms (EAs) [3, 6, 21], and others algorithms [22, 32, 35], etc. If the set of problems that we feel interest in, called *FI*, is specified, it may be solved by using one or the combination of several ones of them. However, in practical applications, *FI* is generally varied, and it is difficult to find a universal algorithm to match all possible varieties of *FI* [36].

Autonomous cognitive entities are the products of biologic evolution while genes evolved to produce capabilities for learning [18]. Each entity, called agent [7, 16], is

K. Deb et al. (Eds.): GECCO 2004, LNCS 3102, pp. 238–250, 2004.
© Springer-Verlag Berlin Heidelberg 2004

an architecture of cognition executing production rules, which is featured by knowl-edge as the medium and the principle of rationality [9] as the law of behavior [27].

In this paper, due to the limited time and knowledge for numerical optimization, the *bounded rationality* [31] is used for guiding the behaviors of agents while not those *demon* rationalities [34] for existing cognitive architectures, such as SOAR [25] and ACT [1, 2]. Instead of a rule in generality, the specific rules based on *fast and frugal heuristics* [9, 34] that matching for full or a part of its *FI*, which avoiding the trap from specificity by their very simplicity, have been natural evolved for agents to adapt to environmental changes, typically that of situated in swarm systems, such as fish school, bird flock, primate society, etc., which each comprises a society of agents.

In swarm systems, individual agent can acquire phenotypic knowledge in two ways [20]: individual learning [33] and social learning [5, 18]. Both ways are not treated as independent processes. Rather, socially biased individual learning (SBIL) [19] is employed for *fast and frugal* problem-solving since it [8]: a) gains most of the advan-tages of both ways; b) allows cumulative improvement to the next learning cycles.

Rules deployment is necessarily when a set of rules for matching different parts of *FI* are available. The simple new macro rules by combining several rules [37] can be easily achieved. Moreover, it is significant to deploying adaptively, which the neural network [4, 9] instead of Bayesian inference [2] should be applied when no enough knowledge on prior odds and likelihood ratio for the rules available.

This paper studies a flexible swarm algorithm framework (SWAF) for numerical optimization problems. In SWAF, each point $\vec{x} \in S$ is defined as a *knowledge point*, which its goodness value is evaluated by the *goodness function* $F(\vec{x})$. In section 2, a multiagent framework is realized, which each agent is a bare bones cognitive archi-tecture with a set of fast and frugal rules. In section 3, the simple *generate-and-test* [15] rules in SBIL heuristics that matching to the social sharing environment are ex-tracted from two existing algorithms: particle swarm optimization (PSO) [11, 22] and differential evolution (DE) [32]. In section 4, the *problem-formulation* rules are then studied for forming and transforming the goodness landscape of the problems. In section 5, the deploying strategies for multiple rules are studied. In section 6, Experi-mental results on a set of problems [12, 26] are compared with some existing algo-rithms [12, 14, 17, 29]. In the last section, we conclude the paper.

2   Swarm Algorithm Framework (SWAF)

Formally, SWAF = <*E, Q, C*>. Here $Q = \{\Theta_i \mid 1 \le i \le N, i \in \mathbb{Z}\}$ comprise N agents (Θ). E is the environment that agents roam. C defines the communication mode.

2.1   Environment (*E*)

All the agents are roamed in an environment E [35]. It is capable of: a) evaluating each knowledge point ($\vec{x}$) via a functional form of the optimization problem; b) holding for social sharing information ($\underline{I}$) for agents.

2.2 Communication Mode (*C*)

The communication mode organizes information flows between *Q* and *E*, which determines the social sharing information (*I*) that available to agents. In SWAF, the simple *blackboard* mode is employed. Here the blackboard is a central data repository that contains the *I*. All the communication among the agents happens only through their actions that modifying the blackboard.

2.3 Agent (Θ)

Each agent (Θ) is a bare bones cognitive architecture in fast-and-frugal heuristics. Here it focuses on the essential model of numerical optimization. Many unconcerned details, such as the operations on goal stack in ACT [1, 2], are neglected. As shown in Fig. 1, it comprises two levels of description: a symbolic and a subsymbolic level.

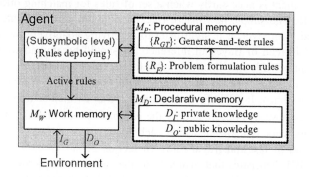

Fig. 1. Agent architecture in SWAF

Symbolic level provides the basic building blocks of cognition, which is interplayed between learning and memory. It includes one working memory and two long-term memories (LTM) [1, 28]: declarative memory and procedural memory.

Procedural memory (M_P) uses production rules [1] to represent procedural skill for the control of learning. Here we use two essential categories, which include generate-and-test rules $\{R_{GT}\}$ and problem-formulation rules $\{R_F\}$, solving problem as follows:

$$\text{Problem} \xrightarrow{\{R_F\}} F \xrightarrow{\{R_{GT}\}} \vec{x} \in S_O \qquad (2)$$

where each R_F forms the landscape F, and each R_{GT} generates the points in S_O.

Declarative memory (M_D) stores factual knowledge, such as knowledge points, which is divided into private and public knowledge (D_I&D_O). Only public knowledge (D_O) is updated to *I*. Instead of the infinite size in ACT [1, 2], the M_D employs an extremal forgetting mechanism: only the latest and/or the best several knowledge points are stored according to the pattern in a production rule.

As agent is activated, the most actively rules are sent into working memory (M_W).

The main topic of the subsymbolic level is adaptive deploying the active rule as there have more than one production rules in same class are available.

2.4   Working Process

The SWAF works in iterated learning cycles. If the maximum number of cycles is T, then at the tth ($1 \leq t \leq T, t \in \mathbb{Z}$) learning cycle, each agent in Q is activated in turn. The active rules, which deployed by the subsymbolic level, are pushed into M_W. As a frugal version, the ith agent generates and tests only one new knowledge point $\vec{x}_i^{(t+1)}$ by executing the active rules (The requirement on generating multiple points in one learning cycle can be achieved by multiple learning cycles), according to its own knowledge and information from the environment (E) determined by communication mode (C). For the convenience of discussion, the point with the best goodness value in $\{ \vec{x}_i^{(\tau)} \mid 1 \leq \tau \leq t, \tau \in \mathbb{Z} \}$ is defined as $\vec{p}_i^{(t)}$. The point with the best goodness value in $\{ \vec{p}_i^{(t)} \mid 1 \leq i \leq N, i \in \mathbb{Z} \}$ is defined as $\vec{g}^{(t)}$.

Each agent has same private goal, which is to find the best knowledge point $\vec{g}^{(t+1)}$ by the learning at the tth cycle. Then the public goal of SWAF consists with the collective of the private goals of all agents, which decreases $F(\vec{g}^{(t)}) \rightarrow F(\vec{x}^*)$ as $t \rightarrow T$.

3   {R_{GT}}: Generate-and-Test Rules

Each generate-and-test rule (R_{GT}) is the combination of a *generate* rule (R_G) and a *test* rule (R_T) [15], which is a process for acquiring declarative memory:

$$< M_D, I_G >^{(t)} \xrightarrow{R_G} \vec{x}^{(t+1)} \xrightarrow{R_T} < M_D, I_G >^{(t+1)} \tag{3}$$

Here we only discuss the {R_{GT}} matching to the sharing information, although the {R_{GT}} can be extract from some single starting point algorithms that without I, such pure random search (PRS), Taboo search (TS), simulated annealing (SA), etc.

The generate rule (R_G) generates a new knowledge point $\vec{x}^{(t+1)}$ based on *socially biased individual learning* (SBIL) [19] *heuristics*, i.e. a mix of reinforced practice of own experience in M_D and the selected information in I (especially for the successful point $\vec{g}^{(t)}$). Here the reinforced practice to an experience point $\vec{x}$ means to generate a point that is neighboring to $\vec{x}$. Then the test rule (R_T) updates $\vec{x}^{(t+1)}$ to M_D and I.

Both the R_G and the R_T call the problem-formulation rules to form the own goodness landscape of agent to evaluate $\vec{x}^{(t+1)}$ and information in M_D and I.

The {R_{GT}} in SBIL heuristics are extracted from two existing algorithms: particle swarm optimization (PSO) [11, 22] and differential evolution (DE) [32]. Both rules provide the bell-shaped variations with consensus on the diversity of points in I [37].

3.1   Particle Swarm (PS) Rule

Particle swarm rule uses three knowledge points in M_D, which $\vec{o}_{PS}^{(t)}$ and $\vec{x}_{PS}^{(t)}$ are situated in D_I, and $\vec{p}^{(t)}$ is situated in D_O, and the point $\vec{g}^{(t)}$ is in the I based on evaluation.

When PS rule is activated, its generate rule (R_G) generate one knowledge point $\vec{x}^{(t+1)}$ according to following equation, for the dth dimension [11, 22]:

$$x_d^{(t+1)} = x_d^{(t)} + CF \cdot (v_d^{(t)} + c_1 \cdot U_{\mathbb{R}}() \cdot (p_d^{(t)} - x_{PS,d}^{(t)}) + c_2 \cdot U_{\mathbb{R}}() \cdot (g_d^{(t)} - x_{PS,d}^{(t)})) \qquad (4)$$

where $CF = 2/(\sqrt{\varphi \cdot (\varphi - 4)} + \varphi - 2)$ [11], $\varphi = c_1 + c_2 > 4$, $\vec{v}^{(t)} = \vec{x}_{PS}^{(t)} - \vec{o}_{PS}^{(t)}$, $U_{\mathbb{R}}()$ is a random real value between 0 and 1. The default values for utilities are: $c_1 = c_2 = 2.05$.

The test rule (R_T) then set the $\vec{o}_{PS}^{(t+1)} := \vec{x}_{PS}^{(t)}$, $\vec{x}_{PS}^{(t+1)} := \vec{x}^{(t+1)}$, and if $F(\vec{x}^{(t+1)}) \leq F(\vec{p}^{(t)})$, then $\vec{p}^{(t+1)} := \vec{x}^{(t+1)}$. At last, the $\vec{p}^{(t+1)}$ is updated to $\underline{I}$.

3.2   Differential Evolution (DE) Rule

Differential evolution rule use one knowledge point in M_D, which $\vec{p}^{(t)}$ is situated in D_O, and one knowledge point $\vec{g}^{(t)}$ in the $\underline{I}$ based on evaluation.

When DE rule is activated, its R_G first sets $\vec{x}^{(t+1)} := \vec{p}^{(t)}$, and $DR = U_Z(1, D)$, where $U_Z(z_l, z_u)$ is a random integer value within [z_l, z_u]. For the dth dimension [32, 37]:

$$\text{IF } (U_{\mathbb{R}}() < CR \text{ OR } d == DR) \text{ THEN } x_d^{(t+1)} = g_d^{(t)} + SF \cdot \Delta_{N_V,d}^{(t)} \qquad (5)$$

where $0 \leq CR \leq 1$, DR ensures the variation at least in one dimension, $0 < SF < 1.2$. $\vec{\Delta}_{N_V}^{(t)} = \sum_1^{N_V} \vec{\Delta}_1^{(t)}$, where each *difference vector* $\vec{\Delta}_1^{(t)} = \vec{p}_{U_Z(1,N)}^{(t)} - \vec{p}_{U_Z(1,N)}^{(t)}$ is the difference of two knowledge points randomly selected from { $\vec{p}_i^{(t)} | 1 \leq i \leq N$ } that are available from the $\underline{I}$. The default values for utilities are: $N_V = 2$, $SF = 1/N_V = 0.5$.

The test rule (R_T) of DE rule updates $\vec{p}^{(t+1)}$ as same as PS rule.

4   {R_F}: Problem-Formulation Rules

The essentially role for {R_F} is forming the goodness landscape F. Moreover, it also takes the role for matching {R_{GT}} by transforming the landscape with extra knowledge.

4.1   Periodic Boundary Handling (PBH) Rule

It is essential to ensure the ultimate solution point belongs to S. In SWAF, such boundary constraints are handled by *Periodic* mode [37]. Each point $\vec{x} \notin S$ is not adjusted to S. However, $F(\vec{x}) = F(\vec{z})$, where $\vec{z} \in S$ is the *mapping point* of $\vec{x}$:

$$\tilde{M}_P(x_d \to z_d): \begin{cases} z_d = u_d - (l_d - x_d)\% s_d \text{ IF } x_d < l_d \\ z_d = l_d + (x_d - u_d)\% s_d \text{ IF } x_d > u_d \end{cases} \qquad (6)$$

where '%' is the modulus operator, $s_d = |u_d - l_d|$ is the parameter range of the dth dimension. The ultimate solution point $\vec{g}^* \in S$ is available by $\tilde{M}_P(\vec{g}^{(T)} \to \vec{g}^*)$.

4.2 Basic Constraint-Handling (BCH) Rule

For most real world problems, there have a set of constraints on the S:

$$\begin{cases} \text{Mimininze}: f(\vec{x}) \\ g_j(\vec{x}) \le 0 \quad (1 \le j \le m, j \in \mathbb{Z}) \end{cases} \tag{7}$$

where $g_j(\vec{x})$ are constraint functions. Moreover, it is usually to convert an equality constraint $h(\vec{x}) = 0$ into the form $g(\vec{x}) = |h(\vec{x})| - \varepsilon_h \le 0$ for a small value $\varepsilon_h > 0$ [13].

By defining the space that satisfies a g_j is $S_{F,g_j} = \{\vec{x} \in S \mid g_j(\vec{x}) \le 0\}$, the space that satisfies all the constraint functions is denoted as *feasible space* (S_F), which $S_F = S_{F,g_1} \cap ... \cap S_{F,g_m}$, and then $S_I = \overline{S}_F \cap S$ is defined as the *infeasible space*.

In SWAF, the basic goodness function is defined as $F(\vec{x}) = < F_{OBJ}(\vec{x}), F_{CON}(\vec{x}) >$, where $F_{OBJ}(\vec{x}) = f(\vec{x})$ and $F_{CON}(\vec{x}) = \sum_{j=1}^{m} r_j G_j(\vec{x})$ are the goodness functions for objective function and constraints, respectively, r_j are positive weight factors, which default value is 1, and $G_j(\vec{x}) = \max(0, g_j(\vec{x}))$. If $F_{CON}(\vec{x}) = 0$, then $\vec{x} \in S_F$.

To avoid adjusting penalty coefficient [29], and to follow criteria by Deb [13], the BCH rule for goodness evaluation is realized by comparing any two points $\vec{x}_A$, $\vec{x}_B$:

$$F(\vec{x}_A) \le F(\vec{x}_B), \text{ IF } \begin{cases} F_{CON}(\vec{x}_A) < F_{CON}(\vec{x}_B) \text{ OR} \\ F_{CON}(\vec{x}_A) = F_{CON}(\vec{x}_B) \text{ AND } F_{OBJ}(\vec{x}_A) \le F_{OBJ}(\vec{x}_B) \end{cases} \tag{8}$$

4.3 Adaptive Constraints Relaxing (ACR) Rule

The searching path of BCH rule is $S_I \to S_F \to S_O$. For discussion, the probability for changing $\vec{g}$ from space S_X to S_Y is defined as $P(S_X \to S_Y)$. The $P(S_F \to S_O)$ can be very small for current $\{R_{GT}\}$, especially for ridge function class with small *improvement intervals* [30], such as the S_F of problems with equality constraints [37].

"*If Mohammed will not go to the mountain, the mountain must come to Mohammed.*" Here extra knowledge for transforming the landscape is embedded for matching $\{R_{GT}\}$. The *quasi feasible space* is defined as $S_F' = \{F_{CON}(\vec{x}) \le \varepsilon_R\}$, where $\varepsilon_R \ge 0$ is threshold value, and the corresponding *quasi solution space* is defined as S_O', then an additional rule is applied on equation (8) in advance for relaxing constraints:

$$F_{CON}(\vec{x}) = \max(\varepsilon_R, F_{CON}(\vec{x})) \tag{9}$$

It has $S_F \subseteq S_F'$ after the relaxing, and the searching path becomes $S_I \to S_F' \to S_O'$. Compared with $P(S_F \to S_O)$, $P(S_F' \to S_O')$ can be increased dramatically due to the enlarged improvement intervals in the S_F', and then $P(S_I \to S_O') \geq P(S_I \to S_O)$.

Of course, S_O' is not always equal to S_O. However, the searching path can be built by decreasing ε_R so as to increasing $(S_O' \cap S_O)/S_O$. When $\varepsilon_R = 0$, $S_O' = S_O$.

The adjusting of $\varepsilon_R^{(t)}$ is referring to a set of points in I_G that are updated frequently, which is $\underline{P} = \{ \vec{p}_i^{(t)} \mid 1 \leq i \leq N, i \in \mathbb{Z} \}$ for both DE and PS rule. Then in $\underline{P}$, the number of elements with $F_{CON}(\vec{p}_i^{(t)}) > \varepsilon_R^{(t)}$ is defined as $N_\varepsilon^{(t)}$, and the minimum and maximum $F_{CON}(\vec{\kappa}_i^{(t)})$ values are defined as $\varepsilon_{R\min}^{(t)}$ and $\varepsilon_{R\max}^{(t)}$, respectively.

The adaptive constraints relaxing (ACR) rule is employed for ensuring $\varepsilon_R^{(T)} \to 0$. Initially, the $\varepsilon_R^{(0)}$ is set as $\varepsilon_{R\max}^{(0)}$. Then $\varepsilon_R^{(t+1)}$ is adjusted according following rule set:

$$\text{IF}(t \geq t_{th} \text{ AND } \varepsilon_{R\min}^{(t)} > 0) \text{ THEN } \varepsilon_R^{(t+1)} = \beta_f \cdot \varepsilon_R^{(t)} \quad \text{(Forcing sub-rule)} \tag{10}$$

$$\text{ELSE} \begin{cases} \text{IF}(N_\varepsilon^{(t)}/N_K \leq r_l) \text{ THEN } \varepsilon_R^{(t+1)} = \beta_l \cdot \varepsilon_R^{(t)} \\ \text{IF}(N_\varepsilon^{(t)}/N_K \geq r_u) \text{ THEN } \varepsilon_R^{(t+1)} = \beta_u \cdot \varepsilon_R^{(t)} \end{cases} \quad \text{(Basic sub-rules)}$$

where $0 \leq r_l < r_u \leq 1, 0 < \beta_l < 1 < \beta_u < 1/\beta_l$, $0 < \beta_f < 1$, and $0 \leq t_{th} \leq T$. The default values include: $r_l = 0.25$, $r_u = 0.75$, $\beta_b = \beta_l = 0.618$, $\beta_u = 1.312$, and $t_{th} = 0.5 \cdot T$.

The basic sub-rules try to keep a ratio between r_l and r_u for the points inside and outside the S_F'. The forcing sub-rule forces the $\varepsilon_R^{(t)} \mid_{t \to T} \to 0$ after $t \geq t_{th}$.

5 Deployment of Rules

Here we mainly discuss the deploying for $\{R_{GT}\}$. It is important to deploying multiple rules if an existing single rule cannot cover with the interested problems, which can be achieved from: a) macro rule at the symbolic level; and b) subsymbolic deploying.

5.1 Combined Macro Rule

A simple mode is the determinate combination (DC) of rules, which executing each rule in turn as t increasing. For instance, the DEPS macro rule [37] is the combination of a DE and a PS rule, which are sharing with the element $\vec{p}^{(t)}$ in M_D and $\{ \vec{p}_i^{(t)} \mid 1 \leq i \leq N \}$ in $\underline{I}$, performing complementally at odd and even t, respectively.

Another simple mode is the random combination (RC) of rules, which deploying each rule with specified probability at random.

5.2   Subsymbolic Deploying by Neural Network

To deploying rules adaptively, the neural network [4] instead of Bayesian inference [2] is applied since no enough knowledge for the rules available.

Considering a network with N_I input, N_J middle layer and N_K output neurons, as shown in figure 2. Each of the input neurons i ($1 \le i \le N_I$) is connected with each neuron in the middle layer j ($1 \le j \le N_J$) which, in turn, is connected with each output neuron k ($1 \le k \le N_K$) with synaptic strengths $w_s(j, i)$ and $w_s(k, j)$, respectively. Initially, all the synaptic strengths are set as $U_{\mathbb{R}}()$. The input neurons are associated with the available information, and the output neurons are associated to the rules.

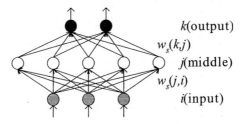

Fig. 2. Two-layer neural network

The deploying process goes as follows: a) Firstly, an input neuron i is chosen to be active at random, since no enough knowledge on the input information. Then the *extremal dynamics* [4] is employed, which only the neuron connected with the maximum w_s to the currently firing neuron is fired. It means that the neuron j_m with the maximum $w_s(j, i)$ is firing, and then the output neuron k_m with the maximum $w_s(k, j_m)$ is firing; b) The rule associated with the firing output neuron k_m is keep activating within an interval of learning cycles (T_l); c) Then a long-term depression (LTD) mechanism [28] is applied by punishing unsuccessful [9]: if the public knowledge of the agent is the worse ratio (R_W) part among all agents, $w_s(k_m, j_m)$ and $w_s(j_m, i)$ are both depressed by an amount $\xi = U_{\mathbb{R}}()$; d) Go to a), the process is repeated.

The process assures that the agent is capable of adapting to new situations, and yet readily recalls past successful experiences, in an ongoing dynamical process.

6   Experimental Results

Experiments were performed to demonstrate the performance. For SWAF, all the knowledge points at $t=0$ are initialized in the S at random, and the utilities of the rules are fixed as the default values if are not mentioned specially.

6.1   Unconstrained Examples

The SWAF was first applied for four unconstrained functions. They are Goldstein-Price (GP), Branin (BR), Hartman three-dimensional (H3), and Shubert (SH) functions [12]. The number of agents N=10, maximum learning cycles T=100. For $\{R_{GT}\}$, CR was fixed as 0.1 for DE rule. For $\{R_F\}$, only the PBH rule was employed since the problems have not constraint functions. 500 runs were done for each function.

Figure 3 gives the mean evaluation times T_E by simulated annealing (SA) [14], Taboo search (TS) [12] and the algorithms in SWAF by deploying different rules. The T_E is counted within 90% success runs (with the final result within 2% of the global optimum) as in [12]. It can be found that all the algorithms in SWAF perform faster than both SA and TS, especially for the functions H3 and SH.

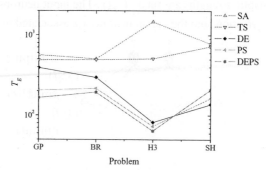

Fig. 3. Mean evaluation times T_E by different algorithms for unconstrained problems

6.2   Constrained Examples

The SWAF was then applied for 11 examples by Michalewicz et al [26]. N=70, T=2E3, then the evaluation times T_E=1.4E5. For $\{R_{GT}\}$, for DE rule, CR was fixed as 0.9, and for combined DEPS rule, CR were separately set as 0.1 and 0.9. For $\{R_F\}$, the PBH and the BCH rule are employed. 100 runs were done for each function. The results for algorithms in SWAF were compared with those for two previously published algorithms: a) (30, 200)-evolution strategy (ES) [29], T=1750, then T_E=3.5E5; and b) genetic algorithm (GA) [17], which N=70, T=2E4, then T_E=1.4E6.

Table 1. Mean results by different algorithms for problems with inequality constraints

F.	F*	ES [29]	GA [17]	DE	PS	DEPS (CR=0.1/0.9)	
G_1	-15	-15.000	-15.000	-14.672	-14.895	-15.000	-15.000
G_2	0.80362	0.7820	0.7901	0.6390	0.6347	0.7828	0.6433
G_4	-30665.5	-30665.5	-30665.2	-30665.5	-30665.5	-30665.5	-30665.5
G_6	-6961.81	-6875.94	-6961.8	-6961.81	-6961.81	-6961.8	-6961.81
G_7	24.306	24.374	26.580	24.352	25.118	24.490	24.306
G_8	0.095825	0.095825	0.095825	0.095825	0.095825	0.095825	0.095825
G_9	680.630	680.656	680.72	680.630	680.649	680.638	680.630
G_{10}	7049.248	7559.192	7627.89	7059.527	7444.366	7214.176	7049.501

Table 2. Comparison the results by SWAFs with existing results in worse/equal/better cases

W/E/B	DE	PS	DEPS (CR=0.1/0.9)	
ES [29]	2/2/4	3/2/3	1/3/4	1/3/4
GA [17]	2/2/4	2/2/4	1/3/4	1/3/4

Table 3. Mean results by different algorithm settings for problems with equality constraints

F	F^* (ε_h=1E-4)	ES [29]		GA [17]	{R_F}: BCH rule			{R_F}: ACR rule		
		P_f=0	P_f=0.45		DE	PS	DEPS	DE	PS	DEPS
G_3	1.0005	0.105	1.000	0.9999	0.35008	0.82137	0.96838	0.7060	1.0005	1.0005
G_5	5126.497	5348.683	5128.881	5432.080	5161.542	5361.89*	5192.810	5126.858	5131.842	5126.498
G_{11}	0.7499	0.937	0.750	0.750	0.75061	0.75566	0.7499	0.7499	0.7499	0.7499

* 16% runs were failed in entering S_F, and only successful runs were counted for the mean results

Table 1 gives the mean results by GA [17], ES [29], and algorithms in SWAF for eight examples with inequality constraints [26]. Table 2 gives the summary for comparing the results by the algorithms in SWAF with the existing results by GA and ES in worse/equal/better cases. For example, 1/3/4 for DEPS versus GA means that for the results of DEPS, 1 example was worse than, 3 examples were equal to, and 4 examples were better than that of GA. Here it can be found that the algorithms in SWAF were often performed better than GA and ES, especially for the combined DEPS rule. Moreover, for G_2, the results of DEPS (CR=0.1) was 0.7951, which was also better than GA [4], when T was increased to 5000 (i.e. T_E was increased to 3.5E5).

Table 3 summaries the mean results by GA [17], ES [29], and algorithms in SWAF for the rest three examples with equality constraints [26], which ε_h =1E-4. Here for ES, both the versions with (P_f=0.45) and without (P_f=0) *stochastic ranking* (SR) technique are listed. For the algorithms in SWAF, two {R_F} versions with: a) BCH rule; b) ACR rule are listed. For {R_{GT}}, CR was fixed as 0.9 for DE rule. For G_3, the learning cycles were set as T=4E3, and then $T_E(G_3)$=2.8E5.

The SWAF algorithms with BCH rule performed better than ES without SR technique, but worse than ES with SR technique and GA. However, with the ACR rule for transforming the landscape, the SWAF algorithms, especially for the combined DEPS, achieved better results than not only the SWAF with BCH rule, but also ES and GA.

6.3 Adaptive Deploying Example

The adaptive deployment was performed on a set of DE generate rules, which with eleven different $CR = 0.1 \cdot (k-1)$ ($1 \le k \le 11, k \in \mathbb{Z}$) in order to test the deploying for not only the rules, but also the utility values of a rule. Each rules were associated with an output neuron for a neural network with N_I=3, N_J=20, N_K=11. The interval learning cycles was set as $T_I = 100$, the worse ratio was set as $R_w = 20\%$. 100 runs were done.

Figure 4 gives the relative mean results for G_1 by comparing the adaptive deploying with the random combination, which each rules were selected in same probability. It can be found that the adaptive deploying performs better than the random combination.

7 Conclusions

This paper has presented a swarm algorithm framework that realized by a society of agents. Each agent is a bare bones cognitive architecture in fast and frugal heuristics, solving numerical optimization problems by deploying mainly two essential categories of rules: generate-and-test rules and problem-formulation rules. Both the simple combination and subsymbolic deploying of multiple rules are also studied.

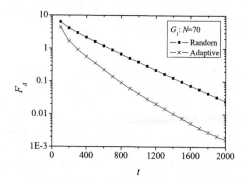

Fig. 4. Comparing the adaptive deploying with the random combination for F

The experiments on benchmark problems shows that the algorithms in SWAF, especially for the DEPS macro rule, cover with more problems than published results by some algorithms, such as TS, SA, GA, and ES, in much frugal evaluation time. Moreover, the $\{R_F\}$ improved the performance for problems that are hard for current $\{R_{GT}\}$ by transforming the landscape. It also showed that adaptive deploying by neural network performed better than random combination, at least for the tested example.

Comparing with the algorithms that can be situated in a single agent, such as TS and SA, it provides simple adjusting of utility values for generate-and-test rules. Comparing with the framework of EAs, it allows: a) evolving of new rules in arbitrary forms, which no longer restricted by genetic operations; b) frugal information utilizing by individual instead of population-based selection; c) subsymbolic deploying of rules.

By associating with the fields of optimization algorithms, agent-based modeling, and cognitive science, SWAF demonstrates the insight from swarm intelligence [7]: the complex individual behavior, including learning and adaptation, can emerge from agents following simple rules in a society. However, SWAF is still in its infant stage. Further works may focus on: a) finding new fast-and-frugal rules for matching new problems adaptively, which can be not only extracted from existing algorithms, but also evolved by genetic operations [24]; b) implementing the mechanism for discovering and incorporating the knowledge on the landscape of problems.

References

1. Anderson, J.R.: A theory of the origins of human knowledge. Artificial Intelligence, 40 (1989) 313-351
2. Anderson, J.R.: ACT: A simple theory of complex cognition. American Psychologist, 51 (1996) 355-365

3. Bäck, T., Hammel, U., Schwefel, H.-P.: Evolutionary computation: comments on the history and current state. IEEE Trans. Evolutionary Computation, 1 (1997) 3-17
4. Bak, P., Chialvo, D.R.: Adaptive learning by extremal dynamics and negative feedback. Physical Review E, 63, 031912 (2001) 1-12
5. Bandura, A.: Social Learning Theory. Prentice Hall, Englewood Cliffs, NJ (1977)
6. Beyer, H.-G., Schwefel, H.-P.: Evolution strategies: a comprehensive introduction. Natural Computing, 1 (2002) 35-52
7. Bonabeau, E.: Agent-based modeling: methods and techniques for simulating human systems. Proc. Natl. Acad. Sci. USA, 99 (2002) 7280–7287
8. Boyd, R., Richerson, P.J.: The Origin and Evolution of Cultures. Oxford Univ. Press (2004)
9. Chase, V.M., Hertwig, R., Gigerenzer, G.: Visions of rationality. Trends in Cognitive Sciences, 2 (1998) 206-214
10. Chialvo, D.R., Bak, P.: Learning from mistakes. Neuroscience, 90 (1999) 1137–1148
11. Clerc, M., Kennedy, J.: The particle swarm - explosion, stability, and convergence in a multidimensional complex space. IEEE Trans. Evolutionary Computation, 6 (2002) 58-73
12. Cvijovi , D., Klinowski, J.: Taboo search: an approach to the multiple minima problem. Science, 267 (1995) 664-666
13. Deb, K.: An efficient constraint handling method for genetic algorithms. Computer Methods in Applied Mechanics and Engineering, 186 (2000) 311-338
14. Dekkers, A., Aarts, E.: Global optimization and simulated annealing. Mathematical Programming, 50 (1991) 367-393
15. Dietterich, T.G.: Learning at the knowledge level. Machine Learning, 1 (1986) 287-316
16. Dorigo, M., Maniezzo, V., Colorni, A.: The ant system: optimization by a colony of cooperating agents. IEEE Trans. Systems, Man, and Cybernetics - Part B, 26 (1996) 1-13
17. Farmani, R., Wright, J.A.: Self-adaptive fitness formulation for constrained optimization. IEEE Trans. Evolutionary Computation, 7 (2003) 445-455
18. Flinn, M.V.: Culture and the evolution of social learning. Evolution and Human Behavior, 18 (1997) 23-67
19. Galef, B.G.: Why behaviour patterns that animals learn socially are locally adaptive. Animal Behaviour, 49 (1995) 1325-1334
20. Heyes, C.M.: Social learning in animals: categories and mechanisms. Biological Reviews, 69 (1994) 207-231
21. Holland, J.H.: Adaptation in natural and artificial systems. University of Michigan Press, Ann Arbor, MI (1975)
22. Kennedy, J., Eberhart, R.C.: Particle swarm optimization, IEEE Int. Conf. on Neural Networks, Perth, Australia (1995) 1942-1948
23. Kirkpatrick, S., Gelatt, C.D., Vecchi, M.P.: Optimization by simulated annealing. Science, 220 (1983) 671-680
24. Koza, J.R.: Genetic Programming: On the Programming of Computers by Means of Natural Selection. MIT Press, Cambridge, MA (1992)
25. Laird, J.E., Newell, A., Rosenbloom, P.S.: SOAR - an architecture for general intelligence. Artificial Intelligence, 33 (1987) 1-64
26. Michalewicz, Z., Schoenauer, M.: Evolutionary algorithms for constrained parameter optimization problems. Evolutionary Computation, 4 (1996) 1-32
27. Newell, A.: The knowledge level. Artificial Intelligence, 18 (1982) 87-127
28. Okano, H., Hirano, T., Balaban, E.: Learning and memory. Proc. Natl. Acad. Sci. USA, 97 (2000) 12403-12404
29. Runarsson, T.P., Yao, X.: Stochastic ranking for constrained evolutionary optimization. IEEE Trans. Evolutionary Computation, 4 (2000) 284-294

30. Salomon, R.: Re-evaluating genetic algorithm performance under coordinate rotation of benchmark functions. Biosystems, 39 (1996) 263-278
31. Simon, H.A.: Invariants of human behavior. Annual Review of Psychology, 41 (1990) 1-19
32. Storn, R., Price, K.V.: Differential evolution - a simple and efficient heuristic for global optimization over continuous spaces. J. Global Optimization, 11 (1997) 341-359
33. Thorndike, E.L.: Animal intelligence: An experimental study of the associative processes in animals. Psychological Review, Monograph Supplement, 2 (1898)
34. Todd, P.M., Gigerenzer, G.: Simple heuristics that make us smart. Behavioral and Brain Sciences, 23 (1999) 727-741
35. Tsui, K.C., Liu, J.M.: Multiagent diffusion and distributed optimization, Int. Joint Conf. on Autonomous Agents and Multiagent Systems, Melbourne, Australia (2003) 169-176
36. Wolpert, D.H., Macready, W.G.: No free lunch theorems for optimization. IEEE Trans. Evolutionary Computation, 1 (1997) 67-82
37. Zhang, W.J., Xie, X.F.: DEPSO: hybrid particle swarm with differential evolution operator, IEEE Int. Conf. on Systems, Man & Cybernetics, Washington D C, USA (2003) 3816-3821

Autonomous Agent for Multi-objective Optimization

Alain Berro and Stephane Sanchez

IRIT-UT1
Université Toulouse I Sciences Sociales,
1 place Anatole France 31042 Toulouse Cedex, France
{berro, sanchez}@univ-tlse1.fr

Abstract. In this article, we present an agent-based method associated with a local research inspired by strategies of evolution to solve multiobjective problems. In comparison with GA-based methods this method uses few parameters. Moreover a decision maker can easily understand the influence of these parameters on the result. The conception of this method led us to represent the Pareto optimal set with zones and not with points. This representation gives additional information which allows to choose between two non-dominated solutions.

1   Discussion

The implementation of the multiobjective optimization methods [2,3,4] requires in priority to control the process of a genetic algorithm and to understand the interaction of its various parameters. The relation between the value of a parameter and its action on the resolution of problem is thus very difficult to control without a perfect knowledge of used method. Moreover, much of these parameters have few meaning for a decision maker.

In the method that we carried out, we endeavoured to reduce these difficulties and to preserve qualities of convergence and maintain of diversity. We developed a method having a simple and understandable parameter setting for a non-initiated.

The majority of the methods describe the Pareto-optimal set by a whole of non-dominated points. But they do not provide any additional information which would allow the decision maker to choose a solution compared to another. For example, the figure on the right supposes that the triangle represents the theoretical Pareto-optimal set. This one will not be able to choose between the points A and B because they are both non-dominated. However we can note on the diagram that the point B is preferable at point A because B is very close to the limit of the optimal zone. By choosing the point B, the decision maker will have a safety margin larger than if he selects the point A.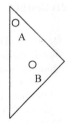

Our method allows to evaluate the smallest distance between a Pareto-optimal solution and the limit of the Pareto-optimal set. This characteristic comes from the fact that our process of optimization does not seek Pareto-optimal points but Pareto-optimal zones.

K. Deb et al. (Eds.): GECCO 2004, LNCS 3102, pp. 251–252, 2004.
© Springer-Verlag Berlin Heidelberg 2004

2 Our Research Method

Initially our method was carried out to seek the optima of a function in a dynamic environment [1] and has following characteristics :

- a self-adaptation of number of necessary agents to perform the research,
- only two parameters of adjustment are used,
- an evaluation of margin of error on the position of optimum.

We thus adapted this method for multiobjective optimization problems by using the concept of Pareto's optimum to compare various solutions and to lead agents towards the Pareto-optimal set. Our agent-based method uses a local research inspired of the strategies of evolution. The aim of an agent is to find a Pareto-optimal zone then to increase it. The original contributions of this method are the introduction to the concept of zone of influence and use of an accuracy factor as stop criterion of algorithm. During the design of this method, we also endeavoured to reduce the number of control parameters and to make these parameters understandable for decision makers.

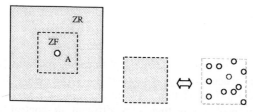

Fig. 1. On the left, the white circle represents the agent position A. The zone of influence *ZF* of an agent is an area around its position in which all solutions are non-dominated. The search zone *ZR* of an agent is a area with variable size in which the agent seeks non-dominated solutions. On the right, the zone of influence corresponds to a cluster of solutions.

References

1. Berro, A., Duthen, Y.: Search for optimum in dynamic environment : a efficient agent-based method. GECCO'2001 Workshop on Evolutionary Algorithms for Dynamic Optimization Problems, San Francisco, California, (2001) 51-54
2. Coello, C., A., C., Toscano, G., T.: Multiobjective Optimization using a Micro-genetic Algorithm. In Proceedings of the Genetic and Evolutionary Computation Conference (GECCO'2001), San Francisco, California (2001) 274-282
3. Corne D., W., and al.: PESA II : Region-based Selection in Evolutionary Multiobjective Optimization. In Proceedings of the Genetic and Evolutionary Computation Conference (GECCO'2001), San Francisco, California (2001) 283-290
4. Deb, K.: A Fast Elitist Non-Dominated Sorting Genetic Algorithm for Multiobjective Optimization : NSGA II. Parallel problem Solving form Nature – PPSN VI, Springer Lecture Notes in Computer Science (2000) 849-858

An Evolved Autonomous Controller for Satellite Task Scheduling

Darren M. Chitty

The Advanced Processing Centre, QinetiQ Ltd.,
St Andrews Road, Malvern,
Worcestershire, WR14 3PS, UK
chitty@signal.qinetiq.com

Abstract. A scheduling algorithm for satellites imaging tasks in a dynamic and uncertain environment. The environment is dynamic in the sense that imaging tasks will be added or removed from the given scenario and in addition, the parameters of individual tasks can change. The technique proposed develops an expert scheduling behaviour as opposed to a robust static schedule by using an evolutionary ALife methodology.[1]

1   Introduction

This paper is concerned with providing schedules for imaging satellites that make the most efficient use of the resources available. We employ a technique generally applied to robot collision avoidance systems to provide optimal schedules for imaging satellites which can be updated in real time and deal with uncertainty in the problem.

2   Principle

In this work we find near optimal schedules for imaging satellites using a technique known as a neural controller to form the decision making link between sensors and possible actions. A neural controller has actions and sensors that are fully connected to each other. The controller then performs the action that receives the greatest output from the sensors at any given time step. Each connection is weighted and the output for an action neuron is the accumulation of each connected sensory neuron multiplied by the weighting of the connection. This technique has been successfully used in industry to implement robot collision avoidance systems and in the development of walking robots [1]. Genetic algorithms were considered a good method to optimise the weights of a controller and have been successfully applied to the evolution of controllers for simple robots, such as Braitenberg vehicles [2] and Kerpera robots [3].

The neural controllers used to construct solutions to the problem have three actions that they can perform at each time step. Move the sensor 0.1 degrees, move the sensor -0.1 degrees or process (or continue processing) a task. The

---

[1] ©Copyright QinetiQ Ltd. 2004

K. Deb et al. (Eds.): GECCO 2004, LNCS 3102, pp. 253–254, 2004.

sensory inputs used by the neural controllers consist of 18 inputs which sense various attributes of a task such as the time it will take to image.

We represent each solution (set of weights) as a (genetic) string, each value in the string representing a weight. This string then represents the behavior of the neural controller. The fitness of each string is evaluated by running it in a simulation on multiple problems of tasks to be processed.

3   Conclusions

One neural controller was evolved with no uncertainty in its training set of problems, and another with some uncertainty. The controllers were evolved on problems that had time spans of up to 120 time steps with the number of tasks ranging from 20 to 50. The controllers were evolved using a large number of different problems. The evolved neural controllers are compared to a standard genetic algorithm approach that attempts to evolve a robust a plan for a single problem. The standard genetic algorithm evolves a plan over 500 generations with the fitness measure being the total task priority processed.

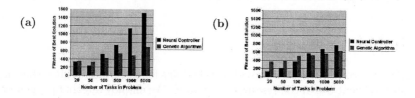

Fig. 1. **(a):** Neural controller technique compared with a standard GA technique on problems with no uncertainty. **(b):** with some uncertainty.

It can be seen in figure 1 that once the high cost of initially evolving a neural controller is completed it can produce good schedules instantly, this is a significant advantage over using a genetic algorithm. The neural controller can operate in real time and hence can respond immediately to any changes, which is a significant improvement over the genetic algorithm approach which would have to re-evolve a new plan to incorporate the changes.

Acknowledgments. This research was sponsored by the United Kingdom Ministry of Defence Corporate Research Programme CISP.

References

1. Pfeiffer, R. and Scheier, C. (1999) "Understanding Intelligence" The MIT Press.
2. Cliff, D., Husbands, P. and Harvey, I. (1992) "Evolving Visually Guided Robots" in J.-A. Meyer, H.L. Roitblat, and S.W. Wilson (eds), From Animals to Animats 2, pp. 374-383, MIT Press.
3. Nolfi, S., and Parisi, D. (1995) "Evolving non-trivial behaviours on real robots: An Autonomous robot that picks up objects" in proceedings of the Fouth Congress of the Italian Association for Artificial Intelligence, Firenze, Springer-Verlag.

Multi-agent Foreign Exchange Market Modelling Via GP

Stephen Dignum and Riccardo Poli

Department of Computer Science
University of Essex

Abstract. In this work genetic programming is used to express and evolve trading strategies for a foreign exchange currency market simulator.

1 Introduction

Multi-agent systems built using agents with simple behaviours can be used to demonstrate complicated, often counter intuitive, results seen in the real world [1]. A currency exchange market offers an ideal environment to put this idea into practice. Here, traders and their clients can be thought of as agents with the desire to exchange currencies to make a profit. A successful trader will pay no heed to past events, but simply react to current conditions quickly to take advantage of potential opportunities [2]. So, our approach is to use GP to evolve agents that react to trade requests from other agents, current trading rates, and news that could affect the value of currency.

2 Our Foreign Exchange Currency Market Simulator

A central location (the Currency Market) records and calculates trading rates between currencies to facilitate currency transactions. When a trade is transacted, a trading agent sends details of the agreed trade to the Currency Market, which updates the rate of the currency that has been purchased. To allow external influences upon the market, a component of the simulator provides current news to agents for each trading round. This information has been abstracted to a numerical form.

A *client agent* chooses a pair of currencies to trade, one to buy, another to sell, based on current market conditions. The choice is stochastic but is biased by current currency news. The client then selects a trader to transact with. This decision is based on previous transactions. If the trader refuses to trade another one is chosen.

Trading agents are more sophisticated. They have holdings in particular currencies and can increment and decrement those holdings by trading with both clients and other traders. Traders follow particular strategies in order to maximise their overall balance. To implement this functionality, trading agents have a mechanism to store their current holdings and to update those holdings when a transaction is successfully completed. The agent is also equipped with the ability to determine the value of its holdings. This value acts as a *fitness function* and determines whether the agent is allowed to continue trading i.e. whether it is still solvent.

K. Deb et al. (Eds.): GECCO 2004, LNCS 3102, pp. 255–256, 2004.
© Springer-Verlag Berlin Heidelberg 2004

On creation each trader is provided with a GP tree representing its trading strategy. This is used in two ways. First when a request to trade is received, if the amount returned from the expression is greater or equal to the amount requested in the trade, the trader accepts that trade. Secondly, when the trader is given the opportunity to trade on its own account, it chooses two currencies at random and selects the appropriate trade amount by evaluating the expression.

After initialisation, the simulator repeatedly executes trading rounds. A *trading round* is a session where all clients and traders have the opportunity to request a trade. A real world analogy for a trading round would be of a trading session within a financial market, the output of each round can be thought of as a set of 'close of business' valuations.

Not all trading rounds involve evolution of the trading expressions. So, an agent's performance can be judged over a number of rounds. In an *evolution round*, bankrupt traders and traders that have not traded by a specified number of rounds are removed from the main population. The simulator then creates new trading agents by using standard GP operators and an appropriate wealth-initialisation strategy.

3   Results and Conclusions

A variety of preliminary experiments have been performed with our GP-based trading simulator (see [3] for a fuller set of results). These have emphasized a number of interesting effects. For example, in many simulations we have seen a large number of occurrences of the sub-tree (*DIVIDE buyRate buyHolding*). This expression reduces an agent's desire to purchase a currency when his holding of the currency is high, which is a good trading strategy.

Modelling trader and client behaviours within a simulated trading environment can provide a valuable technique to analyse the dynamics of foreign exchange markets. Using a multi-agent approach may allow us to engineer and replicate, from simple behaviours, complicated, often emergent, aspects of financial markets.

Evolutionary methods allow agents to adapt to new market conditions, and provide a facility for the researcher to not only identify expressions and variables pertinent to certain market conditions, but also to discover robust trading strategies that are successful over a large range of trading circumstances. This paper has concentrated on GP as the primary technique for evolutionary change.

References

[1] J. Rauch, Seeing around Corners, *The Atlantic Monthly*, April 2002, pages 35-48 [http://www.theatlantic.com/issues/2002/04/rauch.htm], Last visited: 26/08/2003
[2] M. Lewis (1999), *Liars Poker: Two Cities True Greed*, Coronet Books, Hodder and Stoughton Ltd, 338 Euston Road, London NW1 3BH, 1999
[3] S. Dignum and R. Poli, Multi-agent Foreign Exchange Market Modelling via GP, Department of Computer Science, University of Essex, Tech. Rep. CSM-400, March 2004.

An Evolutionary Autonomous Agent with Visual Cortex and Recurrent Spiking Columnar Neural Network

Rich Drewes[1], James Maciokas[1], Sushil J. Louis[2], and Philip Goodman[1]

[1] Brain Computation Laboratory, http://brain.cs.unr.edu
[2] Evolutionary Computing Systems Lab, http://ecsl.cs.unr.edu
University of Nevada, Reno NV 89557, USA

Abstract. Spiking neural networks are computationally more powerful than conventional artificial neural networks [1]. Although this fact should make them especially desirable for use in evolutionary autonomous agent research, several factors have limited their application. This work demonstrates an evolutionary agent with a sizeable recurrent spiking neural network containing a biologically motivated columnar visual cortex. This model is instantiated in spiking neural network simulation software and challenged with a dynamic image recognition and memory task. We use a genetic algorithm to evolve generations of this brain model that instinctively perform progressively better on the task. This early work builds a foundation for determining which features of biological neural networks are important for evolving capable dynamic cognitive agents.

1 Introduction

We describe an evolutionary autonomous agent experiment designed to explore the computational power of certain biological features, such as *spiking* neural networks and *cortical-columnar* organization, in dynamic cognitive tasks. For these experiments, the agents are recurrent column-structured spiking neural networks with about 14000 total neurons and about ten times that many synapses. The model is divided into several areas that roughly mimic some of what is known of early mammalian visual processing, connected to "motor" output areas where the response of the model is interpreted as a rate-coded output. All learning occurs between generations, on an evolutionary time scale; each model is only given one chance to perform the challenge task in its own lifetime.

Because our long term goal is to replicate features of biological cognition, the task we have chosen to challenge our neural agents is modeled after a dynamic psychological recognition and memory test rather than a more typical artifical neural net mapping task such as static image recognition. Though important in other respects, we believe such mapping tasks are not interesting areas for the investigation of biological cognition in part because they are not dynamic (having no time constraints on response) and they are readily implemented in

K. Deb et al. (Eds.): GECCO 2004, LNCS 3102, pp. 257–258, 2004.

non-biological, non-spiking feedforward networks. In contrast, many explorations of human and animal visual working memory involve a delayed matching task [2]. In our variant an image is presented to the test subject momentarily and then removed. A short time later, a second test image is presented and the agent must decide if that image was the same as or different than the first image. To make the test more difficult, a "distractor" image is interposed between the test images. To succeed on the task, the evolved neural agents must actively *remember* some representation of the first image during the presentation of the distractor and then later *compare* this remembered representation with the second test image.

2   Results and Discussion

Little is known about the design and behavior of recurrent spiking neural networks, making prediction of their capabilities difficult and, we suspected, success unlikely. After considerable experimentation with model architecture and parameters—generally simplifications—we were able to consistently evolve agents to correctly perform our delayed matching task (figure 1). Any fitness value over 16 indicates that the agent got all four responses correct. Higher fitness values indicate an improved *ratio* of spikes in the correct vs. incorrect motor output regions. Generalization of the task and investigation of which model features are important for successful evolution of dynamic cognitive agents will follow.

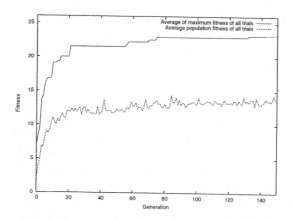

Fig. 1. Avg. of max. fitness, and avg. of pop. fitness, over 5 trials. Pop. size: 16

References

1. Maass, W.: Networks of Spiking Neurons: The Third Generation of Neural Network Models. Neural Networks 10(9), 1997, 1659–1671.
2. Miller, E. and Erickson, C., and Desimone, R.: Neural mechanisms of visual working memory in prefrontal cortex of the macaque. J Neurosci Aug 15, 1996, 5154–5167.

Arguments for ACO's Success

Osvaldo Gómez and Benjamín Barán

Centro Nacional de Computación
Universidad Nacional de Asunción - Paraguay
{ogomez, bbaran}@cnc.una.py
http://www.cnc.una.py

Abstract. Very little theory is available to explain the reasons underlying ACO's success. A population–based ACO (P-ACO) variant is used to explain the reasons of elitist ACO's success in the TSP, given a globally convex structure of the solution space.

1 Reasons Underlying ACO's Success

For this work a TSP tour is denoted as r_x, the optimal tour as r^* and a population of m tours as $P = \{P_i\}$. Distance $\delta(r_x, r_y)$ is defined as the number of cities n minus the number of common arcs between tours r_x and r_y. Inspired in [1], Fig. 1 (a) presents the length of a tour $l(r_x)$ as a function of its distance to the optimal solution $\delta(r_x, r^*)$ for the whole space S of a randomly chosen TSP with 8 cities. Fig. 1 (b) shows the length of $r_x \in S$ as a function of its mean distance to a population $\delta(P, r_x) = \frac{1}{m} \sum_{i=1}^{m} \delta(P_i, r_x)$ of randomly chosen good solutions for the same problem. As previously found for different TSP instances [1], a positive correlation is observed. Consequently, the TSP solution space has a globally convex structure for all tested instances [1].

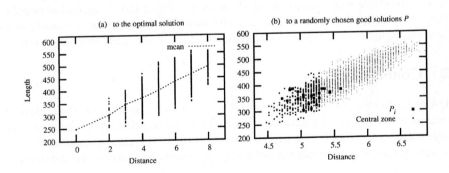

Fig. 1. Distance of the 2,520 solutions of the randomly chosen TSP with 8 cities

To understand the typical behavior of ACO, the n–dimensional TSP search space is simplified to two dimensions for a geometrical vision in Fig. 2. A population $P1 = \{P1_i\}$ of good solutions uniformly distributed is assumed in Fig. 2. Considering that the proposed variant of P-ACO [2] (called Omicron ACO or

K. Deb et al. (Eds.): GECCO 2004, LNCS 3102, pp. 259–260, 2004.

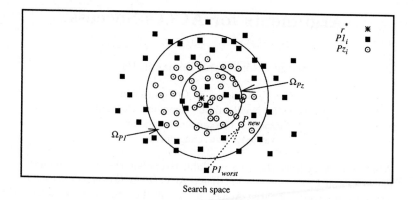

Fig. 2. Simplified vision of OA behavior

OA) gives more pheromones to the good solutions $P1_i$ already found, this can be seen as a search made close to each $P1_i$. Thus, OA concentrates the search of new solutions in a central zone of $P1$, denoted as Ω_{P1}, which is the zone close to all $P1_i$. Then OA typically replaces the worst solution of $P1$ ($P1_{worst}$) by a new solution P_{new} of smaller length. A new population $P2$ is created including P_{new}. This is shown in Fig. 2 with a dotted line arrow. As a consequence, it is expected that $\delta(P2, r^*) < \delta(P1, r^*)$ because there is a positive correlation between $l(r_x)$ and $\delta(r_x, r^*)$. Similarly, $\delta(P, P_{new}) < \delta(P, P_{worst})$ because there is a positive correlation between $l(r_x)$ and $\delta(P, r_x)$, therefore $\delta(P2) < \delta(P1)$ (where $\delta(P) = \frac{2}{m(m-1)} \sum_{i=1}^{m-1} \sum_{j=i+1}^{m} \delta(P_i, P_j)$ is the mean distance of P), i.e. it is expected that the subspace where the search of potential solutions is concentrated decreases. OA performs this procedure repeatedly to decrease the search zone where promising solutions are located, as seen in Fig. 1 (b). Considering population $Pz = \{Pz_i\}$ for $z >> 2$, Fig. 2 shows how Ω_{Pz} has decreased considerably as a consequence of the globally convex structure of the TSP solution space.

2   Conclusions

OA concentrates the search in a central zone Ω_P of its population P. In globally convex problems, good solutions are usually found in this region; therefore, OA concentrates its search in a promising subspace. Every time a good solution is found, it enters the population reducing the promising search zone iteratively.

References

1. Boese, K.D.: Cost Versus Distance in the Traveling Salesman Problem. Technical Report 950018, University of California, Computer Science Department (1995)
2. Guntsch, M., Middendorf, M.: Applying Population Based ACO to Dynamic Optimization Problems. In: Ant Algorithms, Proceedings of Third International Workshop ANTS 2002. Volume 2463 of LNCS. (2002) 111–122

Solving Engineering Design Problems by Social Cognitive Optimization

Xiao-Feng Xie and Wen-Jun Zhang

Institute of Microelectronics, Tsinghua University, 100084 Beijing, China
xiexf@ieee.org, zwj@tsinghua.edu.cn

Swarm systems are products of natural evolution. The complex collective behavior can emerge from a society of N autonomous cognitive entities [2], called as agents [5]. Each agent acquires knowledge in socially biased individual learning [4]. For human, the extrasomatic arbitrary symbols that manipulated by language allows for cognition on a grand scale [3], since agent can acquire social information that is no longer limited to direct observation to other agents. The individual learning then only plays secondary role due to the ubiquity and efficiency of social learning [1].

Social cognitive optimization (SCO) is based on human social cognition, which operates on a *symbolic space* (S) that encodes the problem. Each $\vec{x} \in S$ is a *knowledge point* that evaluated by the *goodness function* $F(\vec{x})$. The foundational entity of SCO is social cognitive (SC) agent, which includes a memory (M_D) and a set of action rules (R_A). Specially, the agent acquires *social sharing information* only from an external medium called *library* ($\underline{L}$), which stores N_L points. The cognition is realized by interplaying between learning and memory, which the M_D and the $\underline{L}$ are employed for storing knowledge for guiding future actions; and R_A are executed for acquiring memory.

Each SC agent is worked in iterated learning cycles. Suppose T is the number of maximum learning cycles. At the tth ($1 \le t \le T, t \in \mathbb{Z}$) learning cycle, its M_D stores the most recently point $\vec{x}^{(t)}$. The action rules (R_A) include: a) Selects a better point $\vec{\kappa}_B$ from $\underline{L}$ by a tournament size τ_B; b) Determines the model point X_M and the refer point X_R: if $F(\vec{\kappa}_B) \le F(\vec{x}^{(t)})$, then $X_M = \vec{\kappa}_B$, $X_R = \vec{x}^{(t)}$, else $X_R = \vec{\kappa}_B$, $X_M = \vec{x}^{(t)}$; c) Infers a new point $\vec{x}^{(t+1)}$: For the dth dimension: $x_d^{(t+1)} = U_{\mathbb{R}}(X_{R,d}, X_{B,d})$ is a random value between $X_{R,d}$ and $X_{B,d}$. Normally, $X_{B,d} = 2 \cdot X_{M,d} - X_{R,d}$ for the dth dimension. Besides, if it is required that $x_d^{(t+1)} \in [l_d, u_d]$, then the $X_{B,d}$ is replaced by both $\max(X_{B,d}, l_d)$ and $\min(X_{B,d}, u_d)$; d) Stores the $\vec{x}^{(t+1)}$ into M_D, and replaces a worse point $\vec{\kappa}_W$, which is selected from $\underline{L}$ by a tournament size τ_W, by its old $\vec{x}^{(t)}$. The default values of τ_B and τ_W are 2 and 4, respectively.

Fig 1 shows two SC models: a) Full sharing model (FSM): all agents shares with a common library ($\underline{L}$), which the evaluation times are $T_E = N_L + (1+T) \cdot N$; b) Partial sharing model (PSM): the *blackboard* ($\underline{B}$) serves as a central data repository, which the size is N_B. Each agent has own library ($\underline{L}$) and allows a specified *thinking time* (T_T). When $t = n_t \cdot T_T + 1$ ($0 \le n_t \le T/T_T, n_t \in \mathbb{Z}$), each agent updates its $\underline{L}$ by selecting N_L points from $\underline{B}$ at random. Then each agent works with its M_D and $\underline{L}$ as $n_t \cdot T_T + 1 \le t \le (n_t + 1) \cdot T_T$. After the $t = (n_t + 1) \cdot T_T$ cycle is finished, each agent only updates the point with best goodness in its $\underline{L}$ into $\underline{B}$. For PSM, $T_E = N_B + T \cdot N$.

K. Deb et al. (Eds.): GECCO 2004, LNCS 3102, pp. 261–262, 2004.
© Springer-Verlag Berlin Heidelberg 2004

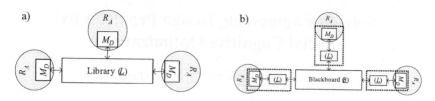

Fig. 1. SC models: a) full sharing model (FSM); b) partial sharing model (PSM)

Four engineering design problems [5] have been tested. They are speed reducer (*SR*), three-bar truss (*TB*), welded beam (*WB*), and tension spring (*TS*) problems. Specially, the *TS* is mixed-integer-continuous (MIC) problem. Table 1 summaries the global optimum F^*, the recently published results and the corresponding T_E by Ray et al. [5], and the results by SC models as T_E=20200, include FSM#1 (N=1, T=2E4, N_L=199), FSM#2 (N=40, T=500, N_L=160) and PSM (N=40, T=500, N_B=200, N_L=39, T_T=15) , which 500 runs were performed for each problem.

Table 1. Comparison of results between existing results [5] and SC models

F.	F^*	Ray et al. (T_E) [5]	FSM#1	FSM#2	PSM
SR	2994.471	2998.027 (110235)	2994.471	2994.471	2995.407
TB	263.8958	263.8989 (36113)	263.8972	263.8970	263.8963
WB	2.38113	2.96070 (64862)	2.47514	2.42723	2.39042
TS	0.012666	0.012923 (25167)	0.01365	0.01344	0.01283

It shows that even only one agent can get high quality solutions on three problems. Moreover, the models with N>1 are performed better than N=1 for most testing cases. For the MIC problem *TS*, the PSM shows better performance than FSM, which reduces the probability of the premature convergence by preventing some intermediate knowledge points from guiding the actions. At last, the PSM performs better than the published algorithm [5] for all problems in fewer evaluation times.

The SCO have few parameters that can be readily adjusting. The N can be set flexibly, even down to 1. Moreover, the trade-off between exploitation and exploration can be achieved by adjusting the size of library (N_L) while no significant impact on the T_E, which is mainly determined by the N and T.

References

1. Bandura, A.: Social Learning Theory. Prentice Hall, NJ (1977)
2. Bonabeau, E.: Agent-based modeling: methods and techniques for simulating human systems. Proc. Natl. Acad. Sci. USA, 99 (2002) 7280-7287
3. Flinn, M.V.: Culture and the evolution of social learning. Evolution and Human Behavior, 18 (1997) 23-67
4. Galef, B.G.: Why behaviour patterns that animals learn socially are locally adaptive. Animal Behaviour, 49 (1995) 1325-1334
5. Newell, A.: The knowledge level. Artificial Intelligence, 18 (1982) 87-127
6. Ray, T., Liew, K.M.: Society and civilization: an optimization algorithm based on the simulation of social behavior. IEEE Trans. Evolutionary Computation, 7 (2003) 386-396

Vulnerability Analysis of Immunity-Based Intrusion Detection Systems Using Evolutionary Hackers

Gerry Dozier[1], Douglas Brown[2], John Hurley[3], and Krystal Cain[2]

[1] Dept. of Computer Science & Software Engineering, Auburn University, AL
36849-5347 gvdozier@eng.auburn.edu
[2] Dept. of Computer Science Clark-Atlanta University, Atlanta, GA 30314
douglasbrown1982 KDJCain@aol.com
[3] Distributed Systems Integration The Boeing Company, Seattle, WA 98124
john.s.hurley@boeing.com

Abstract. Artificial Immune Systems (AISs) are biologically inspired problem solvers that have been used successfully as intrusion detection systems (IDSs). This paper describes how the design of AIS-based IDSs can be improved through the use of evolutionary hackers in the form of GENERTIA red teams (GRTs) to discover holes (in the form of type II errors) found in the immune system. GENERTIA is an interactive tool for the design and analysis of immunity-based intrusion detection systems. Although the research presented in this paper focuses on AIS-based IDSs, the concept of GENERTIA and red teams can be applied to any IDS that uses machine learning techniques to develop models of normal and abnormal network traffic. In this paper we compare a genetic hacker with six evolutionary hackers based on particle swarm optimization (PSO). Our results show that genetic and swarm search are effective and complementary methods for vulnerability analysis. Our results also suggest that red teams based on genetic/PSO hybrids (which we refer to Genetic Swarms) may hold some promise.

1 Introduction

Intrusion detection [11,12,13,14,18,19,21,22] can be viewed as the problem of classifying network traffic as normal (self) or abnormal (non-self). Researchers in this area have developed a variety of intrusion detection systems (IDSs) based on: statistical methods [18,19], neural networks [3], decision trees [2], and artificial immune systems (AISs) [1,6,7,11,12,13,15,16,23,26]. One of the primary objectives of machine learning is to develop a hypothesis that has low error and generalizes well to unseen instances [20]. There are two types of error associated with the hypotheses developed by any learning algorithm [19,20]: false positives, known as type I errors, and false negatives, referred to as type II errors. In the context of network security, type II errors represent 'holes' [12] in an IDS.

Since type II errors do exist in IDSs, a dilemma associated with IDS design, development, and deployment is, "Does one try to identify and/or patch holes

K. Deb et al. (Eds.): GECCO 2004, LNCS 3102, pp. 263–274, 2004.

in advance?" Or "Does one allow the *hackers* to identify the holes and only then try to patch them?" In this paper, we demonstrate how GENERTIA red teams (GRTs), in the form of genetic and particle swarm search [5,17], can be used to discover holes in IDSs. This information can then be used by the designers of IDSs to develop patches or it can be used by an IDS to 'heal' itself. This research is motivated by the fact that cyber-terrorists are now turning towards automated agent-based warfare [14,19,24]. The GRT can be seen as a 'white-hat' hacker agent.

The GRTs presented in this paper are part of a larger system named GEN- ERTIA which contains two sub-systems based on evolutionary algorithms [5]: a GENERTIA blue team (GBT) and a GRT. The objective of the GBT is to design AIS-based IDSs that have high attack detection rates, low error rates, and use a minimal number of detectors. The objective of the GRT is to analyze IDSs and provide feedback to the GBT. Figure 1 shows the architecture of GENERTIA for a host-based IDS. The GBT is used to design an IDS based on input from the network manager. After a preliminary host-based IDS has been designed, the GRT performs a strength and vulnerability analysis of the IDS, based on the input specifications of the network manager. The GRT analysis results in infor- mation concerning the relative strength of detectors (labeled as RSD in Figure 1) comprising the IDS as well as a list of vulnerabilities (holes in the IDS). This information is then given to the GBT to be used to re-design the IDS.

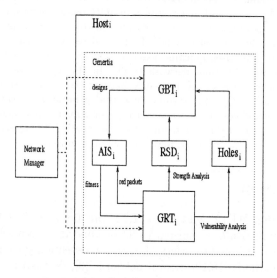

Fig. 1. Architecture of GENERTIA for a Host-Based IDS

2   AIS-Based Intrusion Detection Systems

A number of researchers [1,6,12,13] have developed artificial immune systems (AISs) for networks in an effort to protect them from malicious attacks. A typical AIS-based IDS attempts to classify network traffic as either self or non-self by allowing each host to maintain and evolve a population of detectors. The evolutionary process of these detectors is as follows.

Initially, for each host, a randomly generated set of immature detectors is created. These detectors are exposed to normal network traffic (self) for a user-specified amount of time (where time is measured in packets), $t_{immature}$. If an immature detector matches a self packet then the detector dies and is removed[1]. This process of removing immature detectors that match self packets is referred to as negative selection [11,12].

All immature detectors that survive the process of negative selection become mature detectors. Thus, mature detectors will typically match non-self packets. Mature detectors are also given a user-specified amount of time, t_{mature}, to match m_{mature} non-self packets. This time represents the learning phase of a detector [11,12]. Mature detectors that fail to match m_{mature} non-self packets during their learning phase die and are removed from the detector population. If a mature detector matches m_{mature} non-self packets during its learning phase, a primary response (alarm) is invoked.

Once a mature detector sounds an alarm, it awaits a response from the network administrator to verify that the detector has identified a valid attack on the system. This response is referred to as co-stimulation [11,12]. If co-stimulation does not occur within a prescribed amount of time the mature detector will die and be removed from the detector population. Those detectors that receive co-stimulation are promoted to being memory detectors and are assigned a longer life time of t_{memory}. Memory detectors invoke stronger (secondary) responses when they match at least m_{memory} non-self packets (where usually $m_{mature} > m_{memory}$).

2.1 Advantages Offered by AIS-Based IDSs

There are a number of advantages to using AIS-based intrusion detection. One advantage is that they provide a form of passively proactive protection via negative selection. This enables an AIS-based IDS to detect novel attacks. Another advantage is that AISs are systems that are capable of adapting to dynamically changing environments. As the characteristics of self traffic change over time, a properly tuned AIS will be able to effectively adapt to the dynamically changing definition of self. Finally, the detectors evolved by AIS-based IDSs can easily be converted into Snort or tcpdump filters. This is especially true when constraint-based detectors are used [15,16,26]. These constraint-based detectors are easy to understand and can provide network administrators with important forensic information when new attacks are detected.

---

[1] Any time a detector is removed from the detector population it is automatically replaced with a randomly generated immature detector. This keeps the size of the detector population constant.

2.2   A Disadvantage of Using AIS-Based IDSs

A disadvantage with the use of AIS-based IDSs[2] is that, at present, there is no way to know exactly what types of attacks will pass through undetected. However, a number of techniques have been developed to reduce the size of potential holes [1,7,12] but none of these can be used to alert the designer or users as to the types of attacks that will go undetected by the AIS. In the next section we demonstrate how quickly a GRT can discover holes in an AIS-based IDS.

3   The GENERTIA AIS

As stated earlier, the purpose of the GBT is to develop an AIS in the form of a set of detectors to catch new, previously unseen attacks. The purpose of the GRT is to discover holes in the AIS. As shown in Figure 1, the AIS communicates with the GRT by receiving 'red' packets in the form of attacks from the GRT and returning the percentage of the detector set that failed to detect the 'red' packet, referred to as the fitness of the 'red' packet. The preliminary results presented in this paper are based on a single host-based IDS. However, these results can be generalized to a network where each host has an instance of GENERTIA running on it.

3.1   Representation of Packets

For our AIS, packets are represented as triples (which we will refer to as data triples) of the form (**ip_address, port, src**), where **ip_address** represents the IP address of the remote host, **port** represents the port number the receiving host, and **src** is assigned to 0 if the packet is incoming or 1 if the packet is outgoing.

3.2   The Representation and Behavior of Detectors

The AIS maintains a population of constraint-based detectors of the form: $(lb_0 .. ub_0, lb_1..ub_1, lb_2..ub_2, lb_3..ub_3, lb_{port}..ub_{port}, src)^3$, where the first 4 intervals represent a set of IP addresses, the fifth interval represents the lower and upper bounds on the port, and where src is 0 to denote that the detector should be used on incoming traffic or 1 to denote that the detector should be used on outgoing traffic.

---

[2] Actually, this is the case with all IDSs that operate by building a model of self/non-self.

[3] Notice that detectors based on the above representation can be viewed as constraints. Thus, a detector population can be viewed as a population of constraints where self corresponds to a set of all solutions (data triples) that satisfies the constraints and where non-self corresponds to the set of all solutions that violate at least one constraint. Therefore the intrusion detection problem can be viewed as a distributed constraint satisfaction problem [25].

An *any-r intervals* matching rule [15,16,26] is used to determine a match between a data triple and a detector. That is, if any r numbers representing a data triple fall within the corresponding r intervals of a detector then that detector is said to match the data triple. For the experiments presented in this paper, $r = 3$.

If an immature detector matches a self packet then it is first relaxed by splitting it into two parts based on where the self packet intersects an interval and randomly removing either the lower or the upper part. This relaxation method is based on the the split-detector method [26].

In our experiments, if an immature detector fails to match a self packet after being exposed to $t_{immature} = 200$ self packets then it becomes a mature detector and is given a life time, t_{mature}, that insures its survival for the remainder of an experiment. If a mature detector matches a self packet it is relaxed using the split detector method. For our experiments, the values for m_{mature} and m_{memory} were set to 1.

4   Genetic and Swarm-Based Red Teams

The GRTs compared in this paper come in the form of a steady-state GA and six variants of particle swarm optimization [17]. Each of the seven GRTs evolved a population of 300 data triples. The fitness of a GRT data triple was simply the percentage of detectors that failed to detect it. For each cycle of the GA-based GRT, the worst fit data triple was replaced by an offspring if the offspring had a better fitness. An offspring was created by selecting two parents from the GRT population using binary tournament selection [10] and mating the parents using the BLX-0.5 crossover operator [8]. By using a steady-state GRT, the 300 best 'red' data triples will always remain in the population.

The swarm-based GRTs evolved a swarm of 300 particles where each particle contained: (a) a p-vector that recorded the best 'red' data triple that it has ever encountered, (b) a p-fitness value which represents the fitness of the p-vector, i.e. the percentage of the detectors of an AIS that failed to detect it, (c) an x-vector that recorded the current data triple that particle was visiting, (d) an x-fitness that recorded the fitness of the x-vector, and (e) a v-vector (velocity vector) which when added to the x-vector results in a new candidate 'red' data triple. The swarm-based GRTs were instances of the canonical PSO [4] where offspring were created as follows:
$$v_{id} = v_{id} + \eta_c\varphi_c(p_{id} - x_{id}) + \eta_s\varphi_s(p_{gd} - x_{id})$$
$$x_{id} = x_{id} + v_{id}$$
Where v_{id} represents the dth component of the ith particle's velocity vector, η_c and η_s represent the learning rates for the cognition and social components, φ_c and φ_s represent random numbers within $[0..1]$ for the cognition and social components, and g represents the index of the particle with the best p-fitness in the neighborhood of particle i. Based on the suggestions of [4] the four swarm-based GRTs use asynchronous updating. Thus, each time a particle is updated the newly form x-vector is submitted to the AIS-based IDS where its x-fitness

is assigned a value. The values of η_c and η_s were set to 2.3 and 1.8 according to [4].

The distinctions of the six swarms (denoted SW0, SW0+, SW1, SW2, SW3, and SW4) are based on: neighborhood (local, global), whether particles terminate their search when they discover a vulnerability, which we referred to as particle termination (PT), and whether the best particle, g, used for updating a particular particle is randomly selected or is the best particle with the lowest index (this only applies to those swarms that use a global neighborhood). This distinction is denoted, RB, for random best. The particles are arranged in a ring topology with a local neighborhood for a particle consisting of the particles adjacent to it. Table 1 shows the distinctions in terms of neighbohood, PT, and RB.

Table 1. The Distinctions of the Swarm-Based GRTs in Terms of Neighborhood, PT, and RB

Alg	Neighborhood	PT	RB
SW0	local	no	no
SW0+	local	yes	no
SW1	global	no	no
SW2	global	no	yes
SW3	global	yes	no
SW4	global	yes	yes

5   Training and Test Sets

Our training and test sets were obtained from the 1998 MIT Lincoln Lab data. The Lincoln Lab data represents 35 days of simulated network traffic for a Class B network. To obtain a host-based set, we extracted packets involving host 172.16.112.50 only. From this data, we converted each packet into data triples and filtered out all duplicates and data triples involving port 80. The ports were mapped into 70 distinct ports according to [11]. We then extracted the normal traffic to form the training set. Our final training set consisted of 112 self data triples. Our AISs were trained on approximately 80% of the training set, 89 self data triples. The other 23 self data triples were used to test for false positives (type I errors). Our test set consisted of all attacks launched during the 35 day period. This test set consisted of a total of 1604 data triples.

6   Experiment

Using the above training and test sets we conducted a comparison of the seven GRTs. Initially, an AIS was developed using a detector population size of 400.

The training of the AIS consisted of selecting a data triple from the training set (self set) and exposing the detector population to it. This process was repeated 400 times. After an AIS was trained, it was exposed to the test set. After the AIS was developed, each GRT, evolving a population of 300 malicious ('red') data triples, was allowed to interact with the AIS in order to discover holes in the system. A total of 5000 data triples were evaluated. This process was repeated ten times.

7   Results and Conclusions

Table 2 shows the average performance of the GRTs described above in terms of the number of total number of holes discovered, the number of duplicates, and the number of distinct holes. The ten AIS-based IDSs had an average detection rate of 0.747 with an average false positive rate of 0.4.

In Table 2, one can see that the GA outperforms all of the swarms with respect to number of holes and distinct number of holes discovered. Of the six swarms, SW0+ has the next overall best performance. This shows that terminating the motion of a particle once it has discovered a hole improves the performance of swarm search. When a particle is terminated it allows other particles in the swarm to have an increased number of trials. One can see in Table 2 that the swarms that used PT found a greater number of holes than those that did not; however, these swarms also had the greatest number of duplicates as well. Our results also show that those swarms that used a local neighborhood outperformed those that used a global neighborhood. In this study, the use of RB did not seem to provide a performance improvement.

Table 2. Comparison of the Seven GRTs on the 10 AIS-Based IDSs with an Average Detection Rate of 0.747 and an Average False Positive Rate of 0.4

Alg.	Holes	Duplicates	Distinct
GA	300.0	4.9	295.1
SW0	271.7	8.4	263.3
SW0+	298.4	21.0	277.4
SW1	297.8	51.7	246.1
SW2	292.5	50.2	242.3
SW3	299.1	62.7	236.4
SW4	300.0	62.4	237.6

Figure 2 and 3 show the convergence rates in terms of the average fitness of the population and the best fitness within the population of the GA, SW0, and SW0+ GRTs. In Figure 2, the average fitness of the populations increases rapidly from 30% to 95% within 500 evaluations (of red packets). After this point, the algorithms slowly converge on an average fitness that is close to 100%.

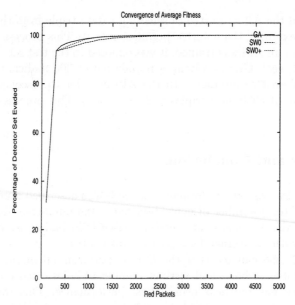

Fig. 2. Visualization of the Convergence Behavior in Terms of Average Fitness of the GA, SW0, and SW0+ GRTs Evolving a Population Size of 300 Data Triples

Figure 3 shows the convergence rates of the GA, SW0, and SW0+ GRTs with respect to the best fitness within a population. One can see that all of the algorithms start with an individual in the population that is capable of evading at least 98% of the 400 detectors. Each of the three algorithms finds their first hole in less than 1000 evaluations.

Figures 4-6 show a 3D visualization of the holes where the x-axis reprents the network, the y-axis represents the host and the z-axis represents the port. In this visualization we assume that the red packets are coming from a Class B network. In Figure 4-6, one can see that the GA and the swarms have a completely different search behavior. The GA tends to find solutions in clusters while the swarms seem to discover holes at the boundaries of the search space. This is interesting because based on our representation of a detector (constraint based intervals) the extreme values of source and host IP addresses and extreme values of port numbers are the hardest to cover. Based on the Figures 5 and 6 one can conclude that a better form of detector would one that uses wrap around intervals. This is a direction for future research. The results seen in Figures 4-6 suggest the possibility that a hybrid GA/PSO algorithm, one which we refer to as a genetic swarm, may be more effective than either GA or PSO alone. This is also a direction for future research.

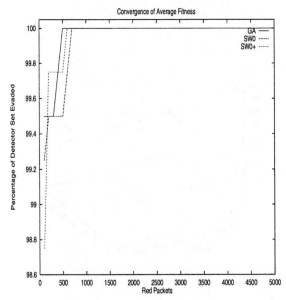

Fig. 3. Visualization of the Convergence Behavior in Terms of Average Fitness of the GA, SW0, and SW0+ GRTs Evolving a Population Size of 300 Data Triples

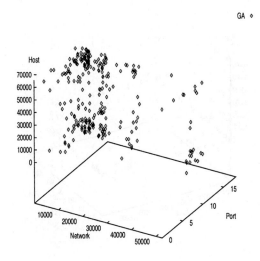

Fig. 4. Visualization of the Holes Evolved by the GA-Based GRT

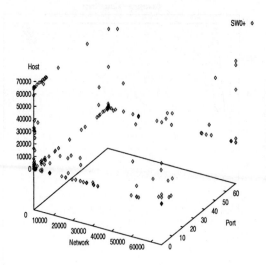

Fig. 5. Visualization of the Holes Evolved by the SW0+-Based GRT

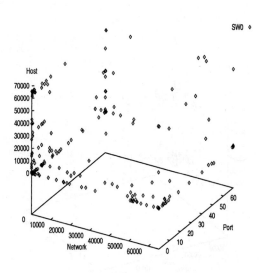

Fig. 6. Visualization of the Holes Evolved by the SW0-Based GRT

The number of non-# symbols in a schema H is called the order *O(H)*, and the distance between the outermost two non-# symbols is called the defining length *L(H)* of the schema. The length of a schema is important because it determines the likelihood of a member of the schema being disrupted by crossover--the further apart its genomes are, the less likely they are to stay together. Suppose the length of individuals be *l*, accordingly the number of variable genes in a chromosome is *l-O(H)*. In general, GAs employ binary encoding schemes, so the total search space is 2^l, where *l* denotes the length of chromosomes. If there exists one gene which is important and keeps invariable at the same position of all the chromosomes, the total search space is reduced to the half, and is further reduced to 2^{l-k} if there are *k* such important genes. In the process of evolution, a new schema is likely to be created by mutation rather than by crossover. Holland's schema theorem explains how schemata are expected to propagate from generation to generation by selection, crossover and mutation. Schemata theorem is also used to explain why GAs realise an optimum search strategy[14].

Generally there is more than one schema in the searching space. Each schema represents a common pattern hidden in a group of individuals. And one individual may be covered by multiple schemata. For example, considering in the three dimensional space of $2^{l=3}$ (see Figure 1), there are six surface hyperplains, each of which has one fixed gene at the same position of the four chromosomes such that a schema can be derived. The top hyperplain has the schema: #1# and the front hyperplain contains the schema: ##1. The two diagonal hyperplains do not form schemata.

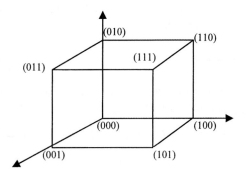

Fig. 1. Individuals and schemata in three dimensional space

Finding common schemata is also inspired from the natural immune system. Bacteria are inherently different from human cells, and many bacteria have cell walls made from polymers that do not occur in humans, so we can expect the immune system to recognize bacteria partially on the basis of the existence of these unusual molecules[15]. Common schemata represent generic properties of the antigen population. With this as motivation, we can construct antibodies which are complementary to the antigens in the bits of schema so that one antibody can recognise multiple antigens. This may explain to some extent why the natural immune system is able to recognize an enormous number of foreign pathogens with relatively limited resources.

2.2 Representation of Schemata

In this section, we first introduce the notation of anomaly detection, and then we present the representations of detectors and schemata which characterise both self and non-self data space.

A problem space $S=x_1 \times x_2 \times \cdots \times x_n$ is a n-dimensional vector space, where x_j is either a categorical feature or a numeric feature. $s_1^*, s_2^*, \cdots, s_n^*$ are the definite states in the space S, where $* \in \{normal, abnormal\}$. The normal state subspace is called self-state set denoted by $SS \subset S$, and the non-self state subspace NS is defined as the complement space of SS and thus $NS = S - SS$. The two subspaces are described as follows:

$$Self\text{-} state\ space:\ SS = \{\ s_i^*\ |\ *= normal,\ i=1,2,,,p\}$$
$$Non\text{-}self\ space:\ NS = \{\ s_j^*\ |\ *= abnormal,\ j=1,2,,,q\}$$

where $SS \cup NS = S$ and $SS \cap NS = \varnothing$.

The characteristic function χ_{self} for differentiating self and non-self is defined as follows:

$$\chi_{self}(s_j) = \begin{cases} 1, & if\ s_j \in SS \\ 0, & if\ s_j \in NS \end{cases}$$

As we mentioned above, a detector can be represented as a detection rule, the structure of which is described as follows:

$$x_1 \in [val_1^1, val_2^1] \wedge \ldots \wedge x_j = d_j \wedge x_m \in [val_1^m, val_2^m] \rightarrow abnormal$$

$$x_1 \in [val_3^1, val_4^1] \wedge \ldots \wedge x_j = d_j \wedge x_m \in [val_3^m, val_4^m] \rightarrow abnormal$$

where $x_i \in [val_1^i, val_2^i]$ represents that feature x_i is a real-valued feature and $x_j = d_j$ indicates that feature x_j is a categorical feature. Each rule(self or non-self) defines a hypercube which covers some states in the descriptor space, The detection rules cover non-self data space, while self rules characterise self-data space (see Figure 2).

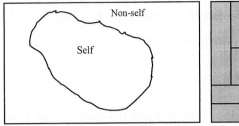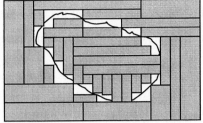

Fig. 2. A two-dimension space characterised by schemata

In reality, both self data space and non-self data space contain schemata. Generally, abnormal events of the same kind may have common characteristics which can be considered as common schemata. These common schema ta represent the dimension-reduced subspace to which the abnormal events belong. The same situation exists in self data space, that is, there may be some characteristic subspaces, each of which may contain a schema. For example, we get three rules which characterize normal events as follows:

$$x_1 \in [val_1^1, val_2^1] \wedge x_2 \in [val_1^2, val_2^2] \wedge x_3 \in [val_1^3, val_2^3] \wedge x_4 \in [val_1^4, val_2^4] \rightarrow normal$$

$$x_1 \in [val_1^1, val_2^1] \wedge x_2 \in [val_1^2, val_2^2] \wedge x_3 \in [val_1^3, val_2^3] \wedge x_5 \in [val_1^5, val_2^5] \rightarrow normal$$

$$x_1 \in [val_1^1, val_2^1] \wedge x_2 \in [val_1^2, val_2^2] \wedge x_4 \in [val_1^4, val_2^4] \wedge x_5 \in [val_1^5, val_2^5] \rightarrow normal$$

The common schema induced from the above three rules is:

$$x_1 \in [val_1^1, val_2^1] \wedge x_2 \in [val_1^2, val_2^2]$$

Definition 1. A real-valued based representation of a schema r is defined as the conjunction of feature-interval pairs as follows:

$$r = x_1 \in [val_1^1, val_2^1] \wedge ... \wedge x_k \in [val_1^k, val_2^k]$$

Definition 2. The coverage of a schema r is defined as the ratio of the number of self states contained in the hypercube determined by r to the total number of self states in self-state space SS:

$$Coverage(r) = \frac{|\{s_i \in SS | s_i \in r\}|}{|SS|} \qquad (1)$$

Definition 3. The volume of the hypercube that a schema $r = x_1 \in [val_1^1, val_2^1]$ and ... and $x_l \in [val_1^l, val_2^l]$ determines is represented as:

$$Volume(r) = \prod_{i=1}^{l} (val_2^i - val_1^i) \qquad (2)$$

Since each numeric variable is normalized between 0 and 1, the volume that a schema determines is less than 1, and the longer a schema is, the smaller of the volume it determines. Thus the fitness of a schema r is represented as:

$$fitness(r) = \alpha \, Coverage(r) + \beta \, Volume(r) \qquad (3)$$

where α, β are the weights and $\alpha + \beta = 1$.

2.3 Finding Common Schemata by Coevolutionary GA

We use coevolutionary genetic algorithm to evolve a number of schemata that are randomly initialised and will finally cover the entire self-data space.

In [15], Stephanie Forrest et al. proposed and solved a problem that whether or not the conventional GAs can find multiple common schemata. As we know that the conventional GAs have the property of convergence, we face the problem that how conventional GAs can maintain multiple sub-population, or how conventional GAs can find multiple peaks in the meantime. Their experiments showed if the population of chromosomes are initialised with schemata of the correct answers, GAs can maintain the correct schemata after some generations. But the right problem to be solved is "Can the conventional GAs find all the schemata in different parts of the space with a randomly initialised population?" and if can, then "what are the key parameters?". They found that the conventional GAs can discover and maintain multiple schemata, and also found the population size is an important parameter to maintain the diversity, that is, a bigger size of population can find more schemata. Further more, the mutation rate plays an important role in maintaining the diversity of the population.

In this paper, we exploit coevolutionary genetic algorithm proposed by Potter and De Jong in [16]. The population consists of a number of non-interbreeding subpopulation of species. Each species represents only a partial solution to the problem, and there is neither cooperation nor competition among subpopulations. Although nothing explicitly prevent multiple subpopulations from containing the identical schema, in practice, each subpopulation tends to be dominated by each species. In our work, each subpopulation is randomly initialised with a species which will converge on a specific schema after some generations of evolution. All the schemata form a schema space of self data, and the detectors are produceed in the schema complementary space. This approach has been proven to use a variety of settings and has been applied to concept learning[16] and Web document classification[6]. The chromosome is designed to be composed of 4 genes as follows.

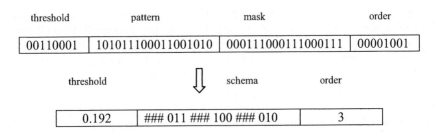

Fig. 3.   Binary scheme for evolving common schemata

The first 8-bit gene is the threshold. Its real value is calculated first by converting the gene to a decimal integer and then dividing this integer by 255. A match between a schema and an instance is considered to occur when the binding strength is greater than the threshold. The pattern gene and the mask gene are the same length and are combined to form a schema. To employ binary coding scheme, each numeric feature is discretized into several different intervals, maximally 8 in our approach so that it can be represented by a 3-bit binary string(000~111). Therefore, both the pattern and the mask gene have 3 times bits as many as the number of the feature vector. The mask gene is used to overlaid the pattern gene, that is, a three-mask bits of "111" keeps the corresponding bit in the pattern gene unchanged, and a three-mask bits of "000" generates a "don't care " schema value. The advantage that is captured by this representation is the ability of a schema matching a wider range of instances. In this way the schema is formed by copying the pattern gene and is modified by mutating the mask gene. The last gene in the genome represents the order of the potential schema. Its value is the total number of non-# in the schema. As we mentioned above, each feature in our approach is represented by 3 binary bits, so the actual order of the schema is the integer divided by 3. The fitness of each individual is calculated by formula (3), and the algorithm for evolving schemata is described as follows:

Algorithm: Coevolutionary genetic algorithm for evolving schemata
    Input :   A feature vector table, a group of parameters
    Output: a group of schemata
    1   discretize numeric feature vector and encode in a binary string;
    2   create the first species;
    3   while the number of species is less than a given number
    4   For each species
    5   Bind each individual to all the feature vectors;
    6   Calculate each individual's fitness;
    7   Do selection, crossover, and mutation;
    8   endfor
    9   Calculate the total fitness of the population;
    10   if the total fitness fails to increase for a few consecutive generations
    11   add a new species to the population;
    12   remove the individuals that do not contribute to the total fitness;
    13   endwhile
    14 decode each species into a common schema;

The algorithm starts with initializing a single species which represents a potential common schema, and new species are added into the population till the total number of species reaches the specified value. In each generation, the fitness of a single individual is determined by its binding ability, and the fittest individual in each species is generated and then the total fitness of the population is calculated. Child species are created by selecting two parents from the same species using fitness-proportionate selection with balanced linear scaling, and then by using uniform crossover and bit flipping mutation. As we mentioned above, the population tends to maintain the correct schema if it has initially been given a correct schema, and also has the capability to evolve out a new schema in the case of being randomly initialized.

3 Generating Detectors

We propose an extended negative selection algorithm in which detection rules are not completely randomly generated. First, it prefers choosing the features that appear more frequently in the set of schemata to the features that do not appear at all or appear less frequently. The reason for this is that a feature occurring more frequently in the schemata must be more important in the problem space and thus has a higher likelihood to be included in the detection rules. That is, a frequently-occurred feature with its infrequently-occurred interval-values is more likely to be selected into a detection rule. Second, when a detection rule is generated, it matches against the common schemata (see Figure 4). If a rule does not contain any common schema, it is considered as a detector and is stored in the detection rule set, otherwise it is rejected. This process is repeated until an appropriate number of detection rules are obtained. In the monitoring phase, each instance is matched against the detection rules. A change is considered to be detected if any match occurs.

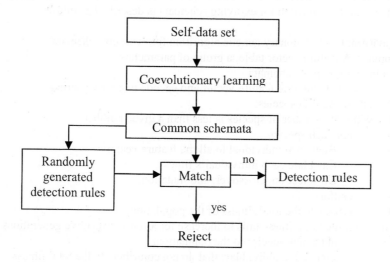

Fig. 4. Diagram of generating detection rules

4 Experimental Results

We perform this experiment with the published data set *iris* which consists of 5 numeric features and 150 instances. The last feature is the class label of three classes. We use the 100 instances of class 1 and class 3 as self data, and take the 50 instances of class 2 away as non-self data. Each of the four conditional numeric features is discreted as maximally 8 different intervals and encoded by 3-bit binary strings. The data set in coevolutionary phase includes the four conditional features of class 1 and class 3. The experimental parameters are set up as: Species number=9, Species size= 100, crossover rate=0.6, mutation=0.3, threshold=1.0, $\alpha = 0.7$ and $\beta = 0.3$.

Table 1. Evolved common schemata covering class 1 and class 3

No.	Schema	Class	Coverage	Fitness
1	PL∈[1.00, 1.90] and PW∈[0.10, 0.40]	1	0.49	0.371
2	PL∈ [5.00,5.60] and SL∈ [5.70, 6.30]	3	0.12	0.181
3	PL∈[5.00, 5.60] and SL∈ [6.40, 6.90]	3	0.13	0.182
4	PL∈[5.00, 5.60] and PW∈[1.80, 2.10]	3	0.14	0.182
5	PL∈[5.00, 5.60] and PW∈[2.20, 2.50]	3	0.08	0.110
6	PL∈[5.70, 6.90] and PW∈[1.80, 2.10]	3	0.09	0.114
7	PL∈[5.70, 6.90] and PW∈[2.20, 2.50]	3	0.09	0.114
8	PL∈[5.00, 5.60] and SW∈[2.40, 3.20]	3	0.22	0.298
9	PW∈[2.20, 2.50] and SL∈[6.40, 6.90]	3	0.09	0.113

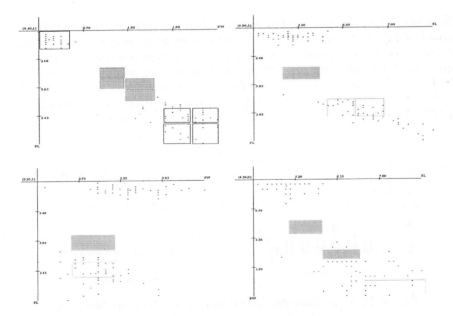

Fig. 5. Distributions of schemata and detectors in the two dimensional space

It is interesting to find that 49 of 50 instances in class 1 are covered by the schema: PL∈[1.00, 1.90] and PW∈[0.10, 0.40], and a group of schemata for class 3 are produced after some generations of evolution. Note that the binding threshold is setup to 1.0, which means an exact matching between a schema and an instance is required. In the second phase, the schemata guide the generation of the detection rules by getting rid of any detector which matches any of the above schemata. Eight generated detection rules are shown in Table 2, and the schemata and detectors are distributed in the two dimensional space showed in Figure 5, from which we can see that some small schemata and detectors can be merged into big ones. The transparent rectangles are the schemata in the two dimensional self data space, while the gray-filled rectangles are the detection rules.

Table 2. Generated detection rules

No	Detectors
1	PL∈[3.00, 3.60] and PW∈[0.90, 1.20] → class 2
2	PL∈[3.70, 4.30] and PW∈[0.90, 1.20] → class 2
3	PL∈[3.70, 4.30] and PW∈[1.30, 1.50] → class 2
4	PL∈[4.40, 4.90] and PW∈[1.30, 1.50] → class 2
5	PL∈ [3.70,4.30] and SW∈[2.40, 3.20] and SL∈[5.00, 5.60]→ class 2
6	SL∈[5.00, 5.60] and PW∈[0.90, 1.20] → class 2
7	PL∈[3.00, 3.60] and SL∈ [5.00, 5.60] and PW∈[0.90,1.20]→ class 2
8	SL∈[5.70,6.30] and PW∈[1.60, 1.70] → class 2

The number of instances detected by each detection rule is shown in the left chart of figure 6 and the total number of instances detected by all the detection rules exceeds 50, which means some instances are detected by more than one detectors. The right curve shows the detection rate and the false alarm rate, which are computed at different numbers of detectors. The first false positive error begins to occur when the number of detectors exceeds 7, which means a self data is detected as an anomaly. The false alarm rate increases as the number of detection rules grows.

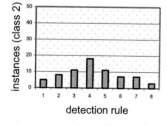
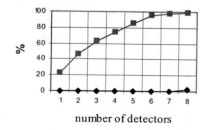

Fig. 6. Coverage of each detection rule(left), detection rate and false alarm

5   Conclusions and Discussions

To overcome the drawback of inefficiency that is created by the approach generate-and-test, we propose an extended negative selection algorithm for anomaly detection, which first evolves a number of common schemata through coevolutionary genetic algorithm in self-data space, and then constructs detectors in the schema complementary space. These common schemata characterize self-data space and thus guide the creation of detection rules. We use the published data set *iris* to test the effectiveness of our approach. The preliminary conclusions are obtained as follows:

- Converting data space into schema space and then constructing detectors in the schema complementary space is a novel and effective method. It eliminates the exponentially computational cost that is created by generate-and-test.
- Some methodologies of computational intelligence can be exploited to learn schemata from self data set. In this paper, we use coevolutionary genetic algorithm to evolve a group of schemata in the self data space.
- The discreteness of numeric features affects the number of schemata evolved and thus affects the number of the final detection rules. Generally, the more intervals a numeric feature is discretized, the more schemata and thus the more detection rules are generated.
- Our approach does not rely on the structured representation of the data and thereby is applied to the problem of general anomaly detection.

References

1. Steven A. Hofmeyr, and S. Forrest, "Architecture for an artificial immune system", IEEE transaction on Evolutionary Computation, 8(4) (2000) 443-473.
2. Dipankar Dasgupta and Fabio Gonzalez "An immunity-based Technique to Characterize Intrusions in Computer Networks", IEEE transaction on evolutionary computation 6(3),pages 1081-1088 June 2002.
3. Paul K. harmer, Paul D. Williams, Gregg H.Gunch and Gary B.Lamont, "An Artificial Immune System Architure for Computer Security Application" IEEE transaction on evolutionary computer, vol.6. No.3   June 2002.
4. Dipankar Dasgupta and Stephanie Forrest, "Artificial immune system in industrial application", In the proceeding of *International conference on Intelligent Processing and Manufacturing Material (IPMM). Honolulu, HI (July 10-14, 1999)*.
5. Dipankar Dasgupta and Stephanie Forrest,   "Novelty Detection in Time Series data using ideas from Immunology", *In the proceedings of the 5th International Conference on Intelligent Systems, Reno, June 19-21, 1996*

6. Jamie Twycross and Steve Cayzer, "An Immune-based approach to document classification", http://citeseer.nj.nec.com/558965.html.
7. S.Forrest, A.Oerelson, L.Allen, and R.cherukuri. "Slef-nonself discrimination in a computer", *In the proceedings of IEEE symposium on research in security and privacy, 1994.*
8. Fabio A.Gonzalez and Dipankar Dasgupta, "An Immunogenetic Technique to detect animalies in network traffic", *In the proceeding of GECCO 2002: 1081-1088*
9. Jonatan Gomez, Fabio Gonzalez and Dipankar Dasgupta, "An Immune-Fuzzy Approach to Anomaly detection", *In Proceedings of The IEEE International Conference on Fuzzy Systems, St. Louis, MO, May 2003.*
10. Fabio Gonzalez, Dipankar Dasgupta and Luis Fernando Nino, "A Randomized Real-Value Negative Selection Algorithm", ICARIS-2003.
11. M. Ayara, J. Timmis, R. de Lemos, L. deCastro and R. Duncan, "Negative Selection: How to Generate Detectors", 1st ICARIS, 2002.
12. D. Dasgupta, Z.Ji and F.Gonzalez, "Artificial Immune System Research in the last five years". *In the proceedings of the international conference on Evolutionary Computation Conference (CEC), Canbara, Australia, December 8-12, 2003.*
13. Jungwon Kim and Peter Bentley, "Negative selection and Niching by an artificial immune system for network intrusion detection", *In the proceeding of Genetic and Evolutionary Computation Conference (GECCO '99), Orlando, Florida, July 13-17.*
14. Riccardo Poli and William B. Langdon. *Schema theory for genetic programming with onepoint crossover and point mutation.* Evolutionary Computation, 6(3):231-252, 1998.
15. Stephanie Forrest, Brena Javornik, Robert E.Smith and Alan S.Perelson, "Using genetic algorithm to explore pattern recognition in the immune system", *Evolutionary Computation, 1(3) (1993) 191-211*
16. Mitchell. A. Potter and Kenneth A.De. Jong, The Coevolution of Antibodies for concept Learning. In the proceeding of the *fifth international conference on parallel problem solving form nature*, September 1998,Amsterdam, The Netherlands.
17. M. A. Potter, K. A. De Jong, and J. J. Grefenstette. *A coevolutionary approach to learning sequential decision rules.* In Larry J. Eshelman, editor, Proceedings of the 6th International Conference on Genetic Algorithms (ICGA95), pages 366--372. Morgan Kaufmann Publishers, 1995.

Real-Valued Negative Selection Algorithm with Variable-Sized Detectors

Zhou Ji[1], Dipankar Dasgupta[2]

[1] St. Jude Children's Research Hospital
Memphis, TN 38105
zhou.ji@stjude.org
[2] The University of Memphis
Memphis, TN 38152
ddasgupt@memphis.edu

Abstract. A new scheme of detector generation and matching mechanism for negative selection algorithm is introduced featuring detectors with variable properties. While detectors can be variable in different ways using this concept, the paper describes an algorithm when the variable parameter is the size of the detectors in real-valued space. The algorithm is tested using synthetic and real-world datasets, including time series data that are transformed into multiple-dimensional data during the preprocessing phase. Preliminary results demonstrate that the new approach enhances the negative selection algorithm in efficiency and reliability without significant increase in complexity.

1 Introduction

Soft computing is an increasingly active research area in computational intelligence. Artificial Immune Systems are soft computing techniques that are based on metaphor of the biological immune system. [1][2][3][4]. The immune system shows computational strength from different aspects in problem solving. Most existing AIS algorithms imitate one of the following mechanisms of the immune system: negative selection, immune network, or clonal selection. Negative selection-based algorithm [1][2] has potential applications in various areas, in particular anomaly detection. The inspiration of negative selection comes from the T cell maturation process in the immune system: if a T cell in thymus recognizes any self cell, it is eliminated before deploying for immune functionality. In a similar manner, the negative selection algorithm generates detector set by eliminating any detector candidate that match elements from a collection of self samples. These detectors subsequently recognize non-self data by using the same matching rule. In this way, it is used as an anomaly detection algorithm that only requires normal data to train [5].

Most works in negative selection used the problem in binary representation [6][7]. There are at least two obvious reasons of this choice: first, binary representation provides a finite problem space that is easier to analyze; second, binary presentation is straightforward to use for categorized data. However, many applications are natural to be described in real-valued space. Furthermore, these problems can hardly be processed properly using negative selection algorithm in binary representation [8]. On

K. Deb et al. (Eds.): GECCO 2004, LNCS 3102, pp. 287–298, 2004.
© Springer-Verlag Berlin Heidelberg 2004

the other hand, this work and some other works [9][10] demonstrated that despite the intrinsic difficulty of real-valued representation, it can also provide unique opportunity in dealing with higher dimensionality.

Matching rule is one of the most important components in a negative or positive pattern detection algorithm [6][7][8][11][12]. For binary representation, there exist several matching rules like rcb (*r*-contiguous bit), *r*-chunks, and Hamming distance [6][8]. For real-valued representation, however, the Euclidean distance is primarily used [8][9][10][13]. Matching is determined when the distance between a data point and some detector is within a certain threshold. In some cases, variations of Euclidean distance are used, for example, a Euclidean distance defined in a lower dimensional space projected from the original higher dimensional problem space [13].

Independent of the type of matching rule, the detectors usually have some basic characteristics, e.g., the number of bits, r, in binary representation, or the distance threshold, Δ, to decide a matching in real-valued representation, that are constant through out the entire detector set. However, the detector features can reasonably be extended to overcome this limitation. The algorithm introduced in this paper demonstrates that allowing the detectors to have some variable properties will enhance the performance of negative detectors. We call this idea and the algorithm based on it as *V-detector*. In the case of real-valued negative selection algorithm, the detectors are in fact hyper-sphere-shaped. The threshold used by Euclidean distance matching rule defines the radius of the detectors. The radius is an obvious choice to make variable considering that the non-self regions to be covered by detectors are very likely to be in different scales. The flexibility provided by the variable radius is easy to realize. However, variable radius is not the only possibility provided by *V-detector*. Detector variability can also be achieved by other ways, such as different detector shapes, variable matching rules, etc.

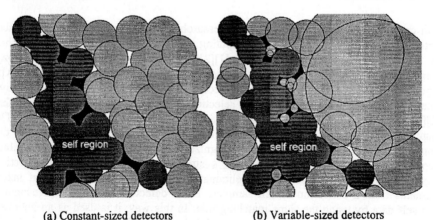

(a) Constant-sized detectors   (b) Variable-sized detectors

Fig. 1. Main concept of Negative Selection and *V-detector*

Figure 1 illustrates the core idea of variable-sized detectors in 2-dimensional space. The dark grey area represents the actual self region, which is usually given through the training data (self samples). The light grey circles are the possible detectors covering the non-self region. Figure 1(a) shows the case where the detectors are of constant size. In this case, a large number of detectors are needed to cover the large area of non-self space. The well-known issues of "holes" are illustrated in black. In figure 1 (b), using variable-sized detectors, the larger area of non-self space can be covered by fewer detectors, and at the same time, smaller detectors can cover the holes. Since the total number of detectors is controlled by using the large detectors, it becomes more feasible to use smaller detectors when necessary.

Another advantage of this new method is that estimated coverage, instead of the number of detectors, can be used as a control parameter. The algorithm can evaluate the estimated coverage automatically when the detector set is generated. On the other hand, we need to set the number of detectors in (advance) when constant sized detectors are used. This will be discussed in more details in the following sections.

2   Algorithm and Analysis

Detector - Set(S, m, r_s)

S : set of self samples

m : number of detectors

r_s : self radius

1 : $D \leftarrow \varnothing$

2 : Repeat

3      $x \leftarrow$ random sample from $[1, 0]^n$

4   Repeat for every s_i in $S = \{s_i, i = 1, 2, ...\}$

5          $d \leftarrow$ Euclidean distance between s_i and x

6 :   if $d \leq r_s$, go to 2

7 : $D \leftarrow D \cup \{x\}$

8 : Until $|D| = m$

9 : return D

Fig. 2. Detector generation algorithm in negative detection using constant-sized detectors

A negative selection algorithm basically consists of two phases. First, the detector set is generated in the training or generation phase. Then, the new sample is examined using the detector set during the detection phase. To highlight the feature of *V-detector*, let us first describe the real-valued negative detection algorithm using constant-sized detectors, where candidate detectors are generated randomly. Those that match any self samples (training data) using Euclidean distance matching rule are eliminated. The generation phase finishes when a preset number of detectors are obtained. The generation phase of this algorithm is shown in the figure 2. The time complexity of this algorithm is $O(m|S|)$, where m is the preset number of detectors and $|S|$ is the size of training set (self samples). Self radius r_s in this case is the same as the detector radius, which represents the allowed variability of the self points [10].

V - Detector - Set(S, T_{max}, r_s, c_o)

S : set of self samples

T_{max} : maximum number of detector

r_s : self radius

c_0 : estimated coverage

1 : $D \leftarrow \varnothing$

2 : Repeat

  $\leftarrow 0$

  $T \leftarrow 0$

5    $r \leftarrow$ inifinite

  $x \leftarrow$ random sample from $[1, 0]^n$

7   Repeat for every d_i in $D = \{d_i . i = 1, 2, ...\}$

8      $d_d \leftarrow$ Euclidean distance between d_i and x

9   if $d_d \leq r(d_i)$ then, where $r(d_i)$ is the radius of d

10        $t \leftarrow t +$

      if $t \geq 1/(1 - c_0)$ then return D

12   go to 4 :

13   Repeat for every s_i in S

14      $d \leftarrow$ Euclidean distance between s_i and x

15   if $d - r_s \leq r$ then $r \leftarrow d - r_s$:

16   if $r > r_s$ then $D \leftarrow D \cup \{< x, r >\}$, where $< x, r >$ is a detector

17   else $T \leftarrow T + 1$

18 :   if $T > 1/(1 - $ maximum self coverage $)$ exit

19 : Until $| D | = T_{max}$

20 : return D

Fig. 3. Detector generation algorithm of *V-detector*

V-detector algorithm also generates candidate detectors randomly. However, when we check the matching rule of Euclidean distance, we keep the distance in record and assign a variable radius based on the minimum distance to each detector that is going to be retained. The detector's detector generation phase is described in figure 3. Comparing with the version of constant-sized detectors, the most important differences lie in steps 13 through 15. Now that we let each detector has its own radius in addition to the location, the radius is basically decided by the closest self sample. Self radius still specifies the variability represented by the training data, but it is not used as detector radius anymore.

The algorithms of detection phase are similar for constant and variable detectors except that matching threshold for each variable-sized detector is difference. In the experiments of this paper, matching is decided by the closest detector.

The control parameters of *V-detector* are mainly self radius r_s and estimated coverage c_0. Maximum number of detectors, shown as T_{max} in figure 3, is preset to be the maximum allowable in practice, which does not need much further discussion. Self radius is an important mechanism to balance between detection rate and false alarm rate, in the other words, the sensitivity and accuracy of the system.

Estimated coverage is a by-product of variable detectors. If we sample m points in the considered space and only one point is not covered, the estimated coverage would be $1-1/m$. Therefore, when we randomly try m times without finding an uncovered point, we can conclude that the estimated coverage is at least $\alpha = 1-1/m$. Thus, the necessary number of tries to ensure estimated coverage α is

$$m = 1/(1-\alpha) \tag{1}$$

Despite the enhancement, complexity of *V-detector* is not increased comparing to basic negative selection. The computation of radius has linear complexity to the number of the training set size. Steps 13 through 15 has complexity O $(|S|)$, where $|S|$ is the set size of training data, just the same as steps 4 through 6 in figure 2's basic negative selection. Furthermore, not only are the complexities of the same order of n, but the times to actually compute the distance, which is potentially a costly step, is the same as well. If the final number of detectors is m, the total complexity if $O(m|S|)$. If m has the same order of magnitude as the preset number in the algorithm using constant-sized detectors, the complexity doesn't change; if m is reduced significantly, the complexity is further improved.

Similarly, the complexity of detection algorithms is $O(m)$ although m has different interpretation in the two methods. The difference in space complexity also only lies in the possible different m, which can always be limited in *V-detector*.

The *V-detector* algorithm normally converges in one of the two ways. Type 1 convergence is when the estimated coverage is reached in step 11 of figure 3. This is the scenario that *V-detector* shows more of its strength in controlling detector number. Type 2 convergence is when the limit of detector number is reached in step 19. Even in this case, the algorithm still has the potential to cover holes better then basic algorithm. There is another possibility that the algorithm will halt, which we are not going to discuss further. If the training data cover almost all space, say 99.99%, the algorithm terminates as a special case at step 18. It may happen when self-radius is set to be too big so that the whole space is all "normal".

The small "holes" are easier to be covered not by just using smaller detectors, rather by using the automatic decision of how small the detectors need to be. The total numbers of detectors, on the other hand, are dealt with by using larger detectors whenever possible.

3   Experiments and Results

3.1   *V-detector*'s Basic Property on Synthetic Data

A synthetic 2-dimensional datasets are used to demonstrate the properties of *V-detector* algorithm. Figures 4(a) shows a cross-shaped self region over the entire space (unit square) $[0, 1]^2$. The training set is 100 random picked points in the self region and the test set is 1000 random distributed points over the entire space. The shaded area in figures 4(b) and 4(c) shows the coverage achieved by detector set generated using different self radius. Comparing (b) and (c), it is easy to see the effect of self radius on the results. The smaller self radius would result in high detection rate but high false alarm rate too, so it is suitable for the scenario when detecting all or most anomalies is very important. On the other hand, larger self radius would result in low detection rate and low false alarm rate, thus suitable when we need to try the best to avoid false alarm.

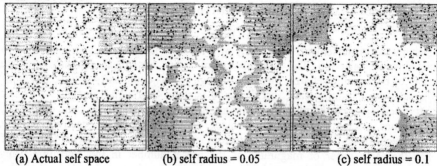

(a) Actual self space   (b) self radius = 0.05   (c) self radius = 0.1

Fig. 4. Cross-shaped self space

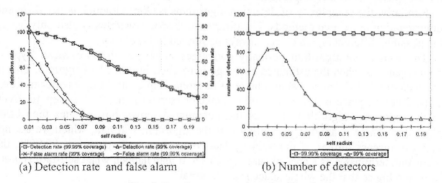

(a) Detection rate and false alarm   (b) Number of detectors

Fig. 5. Results on cross-shaped self region

Figure 5 shows the complete trend of self radius' affect on the results for self radius from 0.01 up to 0.2. The results using two different values of estimated coverage, 99% and 99.99%, are presented together to show that parameter's influence. All the results shown in this figure are average of 100 repeated experiments. Detection rate and false alarm rate are defined as

$$DR = TP/(TP+FN), \tag{2}$$

$$FA = FP/(FP+TN), \tag{3}$$

respectively, where TP, FN, FP, TN are the counts of true positive, false negative, false positive, and true negative. As shown in these results, high detection rate and low false alarm rate are the two goals between which we need to balance according to specific application. While *V-detector* algorithm uses much fewer detectors for both cases, more detectors are needed to obtain the estimated coverage when the self radius is small. The shape of self region also have direct effect on the detector number.

3.2 Comparison with Similar Methods on Real-World Data

To study the property and possible advantages of *V-detector*, experiments were also carried out to compare with the results obtained using other anomaly detection methods that only use normal data to train as *V-detector*. Two such AIS methods were reported in [13], namely MILA (*M*ultilevel *I*mmune *L*earning *A*lgorithm) and NSA (*N*egative *S*election *A*lgorithm, single level to compare with MILA). MILA is a multilevel model of combined negative detection and positive detection [13][14]. It provides a very flexible yet complex mechanism for anomaly detection. Single Level NSA to be compared is to some extent similar to the negative selection using constant-sized detectors described earlier in this paper. However, both MILA and Single Level NSA use a subset of all the dimensions of the problem space to calculate the Euclidean distance for the matching rule. While MILA's model involves multiple ways to choose the subset, Single Level NSA can be seen a extended version of rcb (*r*-contiguous bits) – *r* contiguous dimensions out of all the dimensions [13][14]. Nevertheless, the detectors are constant sized both in MILA and in Single Level NSA.

Table 1 shows the comparison using the famous benchmark Fisher's Iris Data (self raduis 0.1, estimated coverage 99%). The results shown are the summary of 100 repeated tests for each method and parameter setting. One of the three types of iris is considered as normal data, while the other two are considered abnormal. The normal data are either completely or partially used to train the system. Although the partial training set may seem small in this case, it is necessary to demonstrate the system's capability to recognize unknown normal data. As we have seen, self radius is an important control parameter of *V-detector* to balance its performance. The results in this table are obtained using self radius $r_s = 0.1$ considering that the Single Level NSA and MILA results were from threshold 0.1. However, we have to note that the threshold used in Single Level NSA or MILA is not strictly comparable to the self radius in *V-detector*. In the results cited here, MILA or Single Level NSA uses a sliding window of size 2, so the distance is defined in 2-dimensional space, not the original 4-dimensional space. The maximum detector set size is set to be 1000 for the reason of comparison too. *V-detector* has comparable detection rate but lower false alarm rate in most cases, especially when fewer training data were used. *V-detector*'s another obvious advantage is the potentially smaller number of detectors. Table 1 also shows that *V-detector* can obtain similar or better results using much smaller detector number in all cases.

The main control parameters, self radius and estimated coverage, can be used to balance between high detection rate and low false alarm rate. In the experiments we just described, false alarm did not really become a problem when all available training data are used, so the issue is more readily illustrated when only partial data are used to train.

Table 1. Comparison between *V-detector* and other methods using Fisher's Iris Data

Training Data	Algorithm	Detection Rate		False Alarm rate		Number of Detectors	
		Mean	SD	Mean	SD	Mean	SD
Setosa 100%	MILA	95.16	1.79	0	0	1000*	0
	NSA	100	0	0	0	1000	0
	V-detector	99.98	0.14	0	0	20	7.87
Setosa 50%	MILA	94.02	2.44	8.42	1.56	1000*	0
	NSA	100	0	11.18	2.17	1000	0
	V-detector	99.97	0.17	1.32	0.95	16.44	5.63
Versicolor 100%	MILA	84.37	2.79	0	0	1000*	0
	NSA	95.67	0.69	0	0	1000	0
	V-detector	85.95	2.44	0	0	153.24	38.8
Versicolor 50%	MILA	84.46	2.70	19.60	2.00	1000*	0
	NSA	96	0.45	22.2	1.25	1000	0
	V-detector	88.3	2.77	8.42	2.12	110.08	22.61
Virginica 100%	MILA	75.75	2.01	0	0	1000*	0
	NSA	92.51	0.74	0	0	1000	0
	V-detector	81.87	2.78	0	0	218.36	66.11
Virginica 50%	MILA	88.96	2.04	24.98	2.56	1000*	0
	NSA	97.18	0.71	33.26	0.96	1000	0
	V-detector	93.58	2.33	13.18	3.24	108.12	30.74

* MILA has actually 1000 T-cell detectors and 1000 groups of B-cell detector.

Similar comparison was done for a biomedical dataset, which is blood measurement of a group of 209 patients [15]. Each patient has four different types of blood measurements. These blood measures were used to screen a rare genetic disorder. 134 of the patients are normal; 75 patients are carrier of the disease, the "anomalies" to be detected. Table 2 compares the results from MILA and Single Level NSA, and the results from *V-detector* using self radius 0.1 and self radius 0.05. When all the available normal data were used to train the system, the false alarm didn't occur. However, the detection rate is lower than the cases trained by only part of the normal data. Considering the balance between detection rate and false alarm, and the much less number of detectors used, *V-detector*'s results are comparable. Figure 6 shows the balance over a whole range of self radius. The results by Single Level NSA and MILA are plotted as individual points at a comparable self radius on the graph. *V-detector*'s results appears better if we consider both detection performance and false alarm issue. It further confirms *V-detector*'s advantage in balancing the goals.

Table 2. Comparison between *V-detector* and other methods using biomedical data

Training Data	Algorithm	Detection Rate		False Alarm rate		Number of Detectors	
		Mean	SD	Mean	SD	Mean	SD
100% training	MILA	59.07	3.85	0	0	1000*	0
	NSA	69.36	2.67	0	0	1000	0
	r=0.1	30.61	3.04	0	0	21.52	7.29
	r=0.05	40.51	3.92	0	0	14.84	5.14
50% training	MILA	61.61	3.82	2.43	0.43	1000*	0
	NSA	72.29	2.63	2.94	0.21	1000	0
	r = 0.1	32.92	2.35	0.61	0.31	15.51	4.85
	r=0.05	42.89	3.83	1.07	0.49	12.28	4
25% training	MILA	80.47	2.80	14.93	2.08	1000*	0
	NSA	86.96	2.72	19.50	2.05	1000	0
	r=0.1	43.68	4.25	1.24	0.5	12.24	3.97
	r=0.05	57.97	5.86	2.63	0.77	8.94	2.57

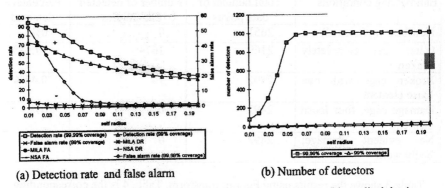

(a) Detection rate and false alarm   (b) Number of detectors

Fig. 6. Balance between detection rate and false alarm rate (biomedical data)

3.3 Application on Time Series Data

V-detector algorithm is used to detect ball bearing fault. The raw data are the time series of measured acceleration of ball bearings [16]. As preprocessing, the time series is first transformed into multiple-dimensional data using two common methods of signal analysis. The first method is basically DFF (Discrete Fourier Transform). It takes overlapped segments of 64 points from the raw time series. The step between the segments is chosen to be 8. Fast Fourier Transform (FFT) is performed with Hanning windowing to each segment [17][18]. Half of the Fourier transform coefficients are taken as data points to be detected. The data is thus 32-dimensional. The second method uses statistical moments to represent the property of each segment of 128 points [19]. The moments of first (mean), second (variance), third, four, and fifth order are used, so the resulted data points become 5-dimensional.

Table 3. Detection Results on Fast Fourier Transform of Different Ball Bearings

Ball bearing conditions	Total number of data points	Number of detected anomalies	Percentage detected
New bearing (normal)	2739	0	0%
Outer race completely broken	2241	2182	97.37%
Broken cage with one loose element	2988	577	19.31%
Damage cage, four loose element	2988	337	11.28%
No evident damage; badly worn	2988	209	6.99%

Table 4. Detection Results on Statistical Moments of Different Ball Bearings

Ball bearing conditions	Total number of data points	Number of detected anomalies	Percentage detected
New bearing (normal)	2651	0	0%
Outer race completely broken	2169	1674	77.18%
Broken cage with one loose element	2892	14	0.48%
Damage cage, four loose element	2892	0	0%
No evident damage; badly worn	2892	0	0%

Table 3 shows the results using Fourier transform. Table 4 is the corresponding results using statistical moments. Throughout all the different conditions of ball bearing, Fourier transform seems to be more sensitive to detect any anomaly than statistical moments. Both methods detect better when the damage is more severe.

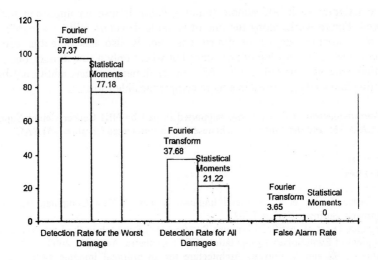

Fig. 7. Summary of Detection Results on Ball Bearing Data

Figure 7 summarizes the performance in terms of detection rate and false alarm. Detection rates are evaluated on two different assumptions: first, only the worst damage (broken race) is considered as fault to be detected; second, all the three types of damages or breakings are regarded as real fault. For complete new ball bearing, there is no false alarm. Assuming the last type of condition (no evident damage) does not count as fault to be detected, false alarm rate is also plotted in figure 7.

4   Conclusion

The paper proposed an extension of real-valued negative selection algorithm with a variable coverage detector generation scheme. Experimental results demonstrated that *V-detector* scheme is more effective in using smaller number of detectors because of their variable sizes. Moreover, it provides a more concise representation of the anomaly detectors derived from the normal data. The detector set generated by *V-detector* is more reliable because the estimated coverage instead of the arbitrary number of detectors is obtained by the algorithm at the end of the run.

The following are some advantages of *V-detector* algorithm:

- Time to generate detectors and to examine new samples is saved by using smaller number of detectors. It also requires less space to store them.
- Holes can be better covered. The smaller detectors are more acceptable because fewer detectors are used to cover the large non-self region.
- The coverage estimate is very useful to provide prediction of the algorithmic performance even if the detection rate (for some specific cases) is not very high due to incomplete or noisy data.

The influence of estimated coverage as a control parameter needs further study, including more experiments and formal analysis. The implication of self radius, or how to interpret each self sample (training data), is also an important topic to be explored. Future works along the line of variable detectors will be variable shape of detectors, variable number of dimensions, etc. It also has potential use for the problems that have very high dimensions but where only a few dimensions affect the detection process. Limited number of detector dimensions has additional benefit of extracting knowledge or rules in a more comprehensible form.

Acknowledgement: This work was supported in part by NIH Cancer Center Support Core Grant CA-21765 and the American Lebanese Syrian Associated Charities (ALSAC).

References

1. de Castro, L. N., et al, Artificial Immune System: A New Computational Intelligence Approach, Springer-Verlag. 2002
2. Dasgupta, D., et al, Artificial Immune System (AIS) Research in the Last Five Years, IEEE Congress of Evolutionary Computation (CEC), Canberra, Australia, 2003
3. Hofmeyr, S., and S. Forrest, Architecture for an artificial immune system, Evolutional Computation Journal, vol. 8, no.4, 2000
4. de Castro, L., N., J. I. Timmis, Artificial Immune Systems as a Novel Soft Computing Paradigm, Soft Computing Journal, vol. 7, issue 7, 2003
5. Dasgupta, D., et al, An Anomaly Detection Algorithm Inspired by the Immune System, in *Artificial Immune System and Their Application*, ed. by D. Dasgupta et al, 1999
6. Esponda, F., S. Forrest, P. Helman, A Formal Framework for Positive and Negative Detection Scheme, IEEE Transaction on Systems, Man, and Cybernetics, 2003
7. Ayara, M., J. Timmis, R. de Lemos, L. de Castro, and R. Duncan, Negative Selection: How to Generate Detectors, 1st International Conference on Artificial Immune System (ICARIS), UK, 2002
8. Gonzalez, F., D. Dasgupta, J. Gomez, The Effect of Binary Matching Rules in Negative Selection, Genetic and Evolutionary Computation Conference (GECCO), Chicago, 2003
9. Gonzalez, F., D. Dasgupta, L. F. Nino, A Randomized Rea-Valued Negative Selection Algorithm, 2nd International Conference on Artificial Immune System (ICARIS), UK. 2003
10. Gonzalez, F., D. Dasgupta, Anomaly Detection Using Real-Valued Negative Selection, Genetic Programming and Evolvable Machine, vol. 4. pp. 383-403, 2003
11. Ceong, H. T., et al, Complementary Dual Detectors for Effective Classification, 2nd International Conference on Artificial Immune System (ICARIS), UK, 2003
12. Kim, J., et al, An evaluation of negative selection in an artificial immune system for network intrusion detection, in Proceedings Genetic and Evolutionary Computation Conference (GECCO), San Francisco, 2001
13. Dasgupta, D., et al, MILA – Multilevel Immune Learning Algorithm, Genetic and Evolutionary Computation Conference (GECCO), Chicago, 2003
14. Ji, Z., Multilevel Negative/Positive Selection in Real-Valued Space, Research Report, The University of Memphis, December 21st, 2003
15. StatLib – Datasets Archive, {http://lib.stat.cmu.edu//dataset/}
16. Structural Integrity and Damage Assessment Network, Public Datasets, {www.brunel.ac.uk/research/cnca/sida/html/data.html}
17. Paul Bourke, Analysis {http://astronomy.swin.edu.au/~pbourke/analysis/}
18. Interstellar Research, FFT Windowing {http://www.daqarta.com/ww00wndo.htm}
19. Institute for Communications Engineering, Higher-order Statistical Moments, {http://speedy.et.unibw-muenchen.de/forsch/ut/moment/}

An Investigation of R-Chunk Detector Generation on Higher Alphabets

Thomas Stibor[1], Kpatscha M. Bayarou[2], and Claudia Eckert[1]

[1] Department of Computer Science
Darmstadt University of Technology
{stibor,eckert}@sec.informatik.tu-darmstadt.de
[2] Fraunhofer-Institute Secure Telecooperation (SIT)
bayarou@sit.fraunhofer.de

Abstract. We propose an algorithm for generating all possible generatable r-chunk detectors, which do not cover any elements in self set S. In addition, the algorithm data structure is used to estimate the average number of generatable detectors, dependent on set size S, r-chunk length r and alphabet size Σ. We show that higher alphabets influence the number of generatable detectors in a negative manner.

1 Introduction

The biological immune system is responsible for protecting organisms against disease caused by pathogens and is able to detect and eliminate most pathogens. The immune system consists of certain types of white blood cells, called lymphocytes, that cooperate to detect pathogens and assist in the destruction of those pathogens. These lymphocytes can be thought of as detectors which recognize pathogens and destroy them, provided a binding threshold between lymphocyte and pathogen is reached. Detectors can also recognize self molecules[1], which results in autoimmune disease and can lead to death. To avoid this reaction, the immune system eliminates through a process called *negative selection*, those lymphocytes (detectors) which bind to self molecules.

From the view of computer scientists, the immune system provides a rich set of methods, ideas, principles and properties to solve computational problems [1]. A short list of immune system properties that are highly appealing from a computational perspective are :

- *Pattern recognition:* the immune system is capable of recognizing and distinguishing between self and foreign molecules.
- *Distributed detection:* the detectors used by the immune system are small, efficient, highly distributed and not subject to centralized control or coordination.
- *Anomaly detection:* the immune system can detect and react to pathogens that the body has never before encountered.

---

[1] produced naturally in the body

K. Deb et al. (Eds.): GECCO 2004, LNCS 3102, pp. 299–307, 2004.
© Springer-Verlag Berlin Heidelberg 2004

Artificial immune systems (AIS) abstract these biological methods and principles and apply them to problem oriented computational paradigms. Application of artificial immune systems are manifold (computer network security, data analysis, machine learning, search and optimization methods). In this paper we focus on anomaly detection, especially the analysis of detector generation.

2   Negative Selection Algorithm

Forrest et al. [3] developed the *negative selection algorithm* based on the *negative selection* immune system process. The algorithm operates on a representation space U, self set S and non-self set N with

$$U = S \cup N \quad and \quad S \cap N = \emptyset.$$

and returns a detector set D which recognizes elements from $U \setminus S$. The *negative selection algorithm* is summarized in the following steps.

1. Define self as a set S of elements of length l in representation space U.
2. Generate a set D of detectors, such that each fails to match any element in S.
3. Monitor S for changes by continually matching the detectors in D against S.

This principle is adaptable to nearly all computer systems, where normal system behavior (self) is appropriately mapped in self set S. A deviate from S can be recognized through the generated detectors and classified as anomaly (non-self). The problem is to generate the smallest possible detector set, which recognizes a maximum part of N. More precisely, a perfect detector set $D_{perfect}$ contains a minimal number of detectors which recognize *all* elements in $U \setminus S$. D'haesleer [4] has shown, that $D_{perfect}$ must be approximately the same size (in bits) as the self set S. Another problem arises in commonly occurring distinct self elements, which induces so called *holes*. Holes are elements from N, for which no detectors can be generated and, therefore can not be recognized and classified as non-self elements.

2.1   R-Chunk Matching

The *r-chunk* matching rule was first proposed by Balthrop et al. [5] and is an improved variant of the r-contiguous matching rule developed by Percus et al. [6]. The r-contiguous matching rule was one of the earliest rules which focused on the biological immune system as a model and abstracts the binding between antibody and antigen [6]. Informally, two elements, with the same length, match under r-contiguous rule, if at least r contiguous characters are identical. This matching rule was theoretically and practically investigated in [4,7,8]. Gonzalez et al. [8] has compared the matching performance most of the well-known (in AIS literature) matching rules. He experienced that the r-chunk matching rule achieves the highest matching performance compared with the other matching rules over the binary alphabet. Since matching performance is strongly influenced by the number of generatable detectors, we focus on the number of generatable r-chunk detectors over arbitrary alphabet sizes.

Given a space U_l^Σ, which contains all elements of length l over an alphabet Σ and a detector set $D \subset U_l^\Sigma$.

Definition 1. *An element $e \in U_l^\Sigma$ with $e = e_1 e_2 \ldots e_l$ and detector $d \in D$ with $d = (p, d_1 d_2 \ldots d_r)$, with $r \leq l$, $p \leq l - r + 1$ match with r-chunk rule iff $e_i = d_i$ for $i = p, \ldots, p + r - 1$.*

Informally, element and detector match if, at position p, there is a sequence of length r where all the characters are identical.

3   Detector Generation Algorithms

We propose an algorithm called *BUILD-RCHUNK-DETECTORS* which generates all possible r-chunk detectors, which do not cover any element in S. This algorithm uses a hashtable $\mathcal{H}$ data structure to insert, delete and search efficiently for boolean values which are indexed with a key composite of r-chunk string concatenated with detector position p. Since the algorithm needs all keys from $[0..|\Sigma|^r]$ concatenated with p, $\mathcal{H}$ contains $p|\Sigma|^r$ elements, where r is the $r - chunk$ length. In addition, the hashtable is an appropriate data structure to analyse randomized operations. Figure 1 shows the hashtable with generated detectors for alphabet $\Sigma = \{0, 1\}$. The symbol | means concatenation of two elements, symbol || means logical value *true* or *false*.

BUILD-RCHUNK-DETECTORS(r, l, S, Σ)
```
 1   for i ← 0 to |Σ|^r − 1            / * init phase * /
 2       do for p ← 0 to l − r
 3           do H.put(i|p, true)
 4   for each s in S                   / * label phase * /
 5       do c ← 0
 6           while r + c < length[s]
 7               do rchunk ← substring[s, r + c]
 8                  H.put(rchunk|c, false)
 9                  c ← c + 1
10   for i ← 0 to |Σ|^r − 1            / * find phase * /
11       do for p ← 0 to l − r
12           do C ← H.get(i|p)
13               if C = true
14                   then
15                       D[k] ← H.get(i|p)
16                       k ← k + 1
17   return D
```

The *BUILD-RCHUNK-DETECTORS* algorithm needs four input parameters, r-chunk length r, self set S, element length l, alphabet Σ and outputs an array of all possible detectors, which do not match self elements. The algorithm is divided into three phases. The initial phase (line 1 to 3) initializes all keys with the boolean value *true*. The label phase (line 4 to 9) iterates with a length r sliding window[2] over all self elements s and replaces the hashtable boolean value, whose key matched the r-chunk with *false*. The last phase named *find* (line 10 to 17), iterates over all hashtable elements and extracts those, which have a boolean value of *true*. The returned array D contains all possible detectors which do not cover any self element.

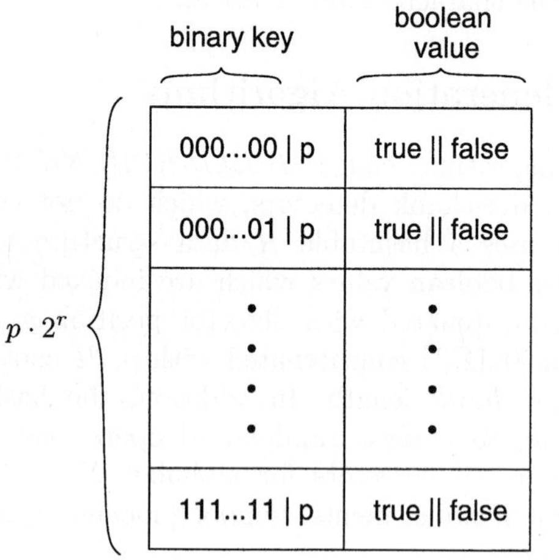

binary key	boolean value
000...00 | p	true || false
000...01 | p	true || false
. . .	. . .
111...11 | p	true || false

$p \cdot 2^r$

Fig. 1. Hashtable $\mathcal{H}$ configuration with $\Sigma = \{0, 1\}$

The space complexity is determinated by r and position p, where p is negligible. The hashtable uses keys of length r and, therefore the total space size results in $O(|\Sigma|^r)$. The runtime complexity is determinated by r, S and the self elements length l. The three phases need $O((l - r) \cdot |\Sigma|^r) + O(|S| \cdot (l - r + 1)) + O(|\Sigma|^r)$ time to generate all possible detectors. The total runtime complexity results in $O(|\Sigma|^r)$ which runs exponential in the length r. D'haesleer et al. [7] has proposed an algorithm to generate r-contiguous detectors with nearly equal runtime complexity as our proposed algorithm. He chooses r such that $|S| = O(|\Sigma|^r)$ and estimated the total runtime complexity as linear in $|S|$. This estimation is only acceptable, if $|S|$ and r are suitable chosen.

---

[2] substring operation

4   Detector Generation Analysis

In this section we estimate the average number of generatable r-chunk detectors. This number is determined by the cardinality $|S|$, the element length l of S and the r-chunk length r. We use the hashtable $\mathcal{H}$ which was defined in the algorithm to estimate the average number of generatable detectors.

Proposition 1. *Given a universe U_l^{Σ} which contains all elements of length l over the alphabet Σ, r-chunks length r and a self set S randomly drawn from U_l^{Σ}, the average number of detectors which do not cover any element in S is*

$$\left(1 - \frac{1}{(l-r+1)\cdot|\Sigma|^r}\right)^{|S|\cdot(l-r+1)} \cdot (l-r+1)\cdot|\Sigma|^r$$

Proof. The hashtable $\mathcal{H}$ contains $p|\Sigma|^r$ elements. We draw $n = |S|\cdot(l-r+1)$ elements and want to find zero labeled *false* elements. The probability distribution therefore is $P(k) = \binom{n}{k}q^k\cdot(1-q)^{n-k}$. For $k=0$ and $q = \left((l-r+1)\cdot|\Sigma|^r\right)^{-1}$ this results in

$$\left(1 - \frac{1}{(l-r+1)\cdot|\Sigma|^r}\right)^{|S|\cdot(l-r+1)}$$

and the total average number results in

$$|D| = \left(1 - \frac{1}{(l-r+1)\cdot|\Sigma|^r}\right)^{|S|\cdot(l-r+1)} \cdot (l-r+1)\cdot|\Sigma|^r \qquad (1)$$

$\square$

As it can be seen, term (1) strongly depends on $|S|$ and $|\Sigma|^r$. Increasing $|S|$, implies a decreasing detector set size, where $|\Sigma|^r$ also influences the amount of generatable detectors.

4.1   Number of Generatable Detectors

We investigate the parameter dependencies of $|\Sigma|, |S|, r$ and their effects on the number of generatable detectors. Therefore, we plot term (1) with small computable parameters, since term (1) increases exponentially. We choose $l = 16, r = 8, \ldots, l - 1, |\Sigma| = 2, 3, 4, 5$ and select S randomly from U_l^{Σ} with a percentage proportion of $|S|/|U_l^{\Sigma}| = 0\%, \ldots, 25\%$ of total universe U_l^{Σ}. In the plots the ordinate depicts the amount of generatable detectors in proportion to the total universe. Since the universe and the amount of detectors increased with higher alphabets we represent the relative number between $|D|$ and $|U_l^{\Sigma}|$. As it can be seen in figure 2(a) to 2(c), detectors can be only generated for $|\Sigma| = 2$ and $|S|/|U_l^{\Sigma}| < 5\%$. For higher alphabets it is not possible to generate

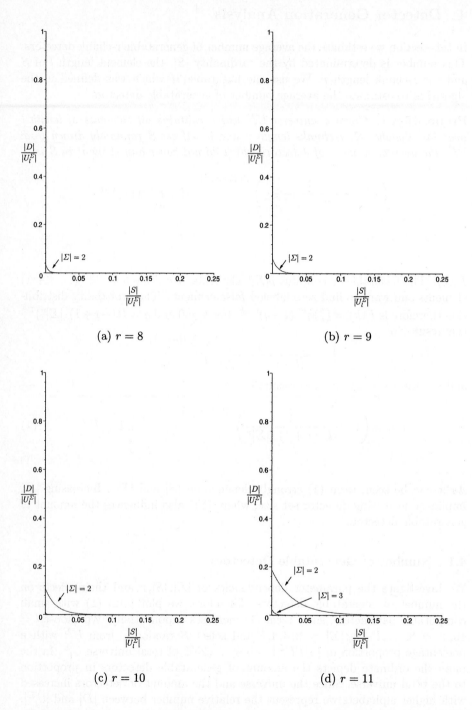

(a) $r = 8$

(b) $r = 9$

(c) $r = 10$

(d) $r = 11$

Fig. 2. (a-d) Plots of term (1), with $l = 16$, $r = \{8, 9, 10, 11\}$ and $|\Sigma| = \{2, 3, 4, 5\}$

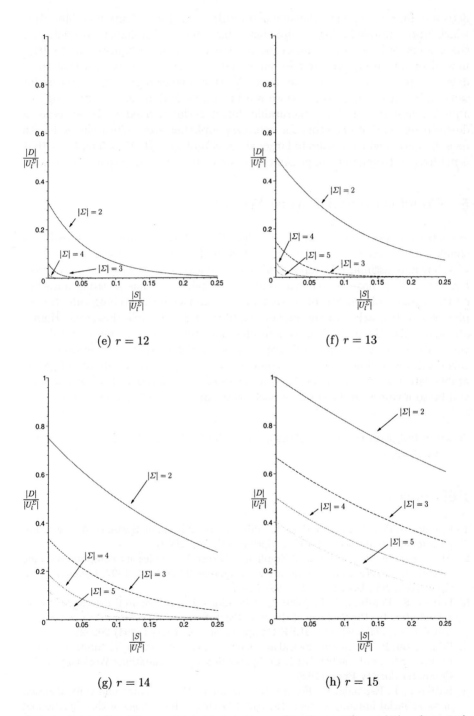

Fig. 2. (e-h) Plots of term (1), with $l = 16$, $r = \{12, 13, 14, 15\}$ and $|\Sigma| = \{2, 3, 4, 5\}$

detectors for $r \leq 10$. This phenomenon results from the r-chunk matching rule, which is not suitable for higher alphabets, since increased alphabet sizes influence the amount of generatable detectors. As it can be seen in figures 2(a) to 2(h), most detectors are generatable for an alphabet of size two. It is clear that these detectors recognize less elements from U_l^Σ than higher alphabets detectors for same value of l and r. But, as we see in figures 2(a) to 2(h), increasing the alphabet size implies less generatable detectors for a fixed r. To generate a destined amount of detectors for arbitrary alphabet sizes, the r-chunk length must lie near l, which results in larger space complexity. If $|\Sigma| > 5$ and $r > 16$ it is not feasible to generate all possible detectors, due to the large space complexity.

5   Conclusion and Future Work

We have proposed and analyzed an algorithm which generates all possible r-chunk detectors over arbitrary alphabet sizes. In addition, an average analysis was shown, to estimate the number of generatable detectors, by given parameters l, r, S and Σ. It has been shown, that the alphabet size has a strong influence on the number of generatable detectors. For the r-chunk matching rule, the alphabet size two achieves the highest number of generatable detectors. Higher alphabets bias the amount of generatable detectors in proportion to total universe. To generate a sufficiently large number of detectors r must lie near l and this results in an infeasible space complexity. So far, the total number of generatable detectors was not considered in terms of non-self cover. The further work will be to investigate the total non-self cover on arbitrary alphabet sizes.

Acknowledgments. The authors are grateful to Bob Lindsey for his comments and annotations.

References

1. Leandro N. de Castro, Jonathan Timmis: Artificial Immune Systems: A New Computational Intelligence Approach. Springer-Verlag (2002)
2. Anil Somayaji, Steven Hofmeyr, Stephanie Forrest: Principles of a computer immune system. In: Meeting on New Security Paradigms, 23-26 Sept. 1997, Langdale, UK, (New York, NY, USA : ACM, 1998) 75–82
3. Forrest, S., Perelson, A.S., Allen, L., Cherukuri, R.: Self-nonself discrimination in a computer. In: Proceedings of the 1994 IEEE Symposium on Research in Security and Privacy, Oakland, CA, IEEE Computer Society Press (1994) 202–212
4. D'haeseleer, P.: An immunological approach to change detection: Theoretical results. In: Proceedings of the 9th IEEE Computer Security Foundations Workshop, IEEE Computer Society Press (1996)
5. Balthrop, J., Esponda, F., Forrest, S., Glickman, M.: Coverage and generalization in an artificial immune system. In: GECCO 2002: Proceedings of the Genetic and Evolutionary Computation Conference, New York, Morgan Kaufmann Publishers (2002) 3–10

6. Jerome K. Percus, Ora E. Percus, Alan S. Perelson: Predicting the Size of the T-Cell Receptor and Antibody Combining Region from Consideration of Efficient Self-Nonself Discrimination. Proceedings of National Academy of Sciences USA **90** (1993) 1691–1695
7. P. D'haeseleer, S. Forrest, P. Helman: An immunological approach to change detection: algorithms, analysis, and implications, Proceedings of the 1996 IEEE Symposium on Computer Security and Privacy (1996)
8. F. Gonzalez, D. Dasgupta, J. Gomez: The effect of binary matching rules in negative selection. In: Genetic and Evolutionary Computation Conference. (2003)

A Comment on Opt-AiNET: An Immune Network Algorithm for Optimisation

Jon Timmis and Camilla Edmonds

Computing Laboratory, University of Kent. Canterbury. Kent. CT2 7NF. UK
j.timmis@kent.ac.uk, camillaedmonds@hotmail.com

Abstract. Verifying the published results of algorithms is part of the usual research process. This helps to both validate the existing literature, but also quite often allows for new insights and augmentations of current systems in a methodological manner. This is very pertinent in emerging new areas such as Artificial Immune Systems, where it is essential that any algorithm is well understood and investigated. The work presented in this paper results from an investigation into the opt-aiNET algorithm, a well-known immune inspired algorithm for function optimisation. Using the original source code developed for opt-aiNET, this paper identifies two minor errors within the code, propose a slight augmentation of the algorithm to automate the process of peak identification: all of which affect the performance of the algorithm. Results are presented for testing of the existing algorithm and in addition, for a slightly modified version, which takes into account some of the issues discovered during the investigations.

1 Introduction

This paper investigates the now popular aiNET algorithm, first proposed in [1]. The initial system was initially designed for data clustering, but later adapted for optimisation [2,6]. This investigation forms part of a larger on-going project into the usefulness and viability of the AIS approach: the investigations call for a detailed re-implementation and testing of existing algorithms. To this end, a detailed testing of an algorithm called opt-aiNET was undertaken. This investigation employed both the existing code written by the authors of [2] (obtained with their permission) and a reimplementation of the system in Java by the authors of this paper. Through the process of revisiting the algorithm, testing the original system and undertaking reimplementation, existing results can be verified and augmentations and improvements can be proposed. It was found that results from [2] were not exactly reproducible, whilst this is to be expected to some degree (as this is a stochastic algorithm) we observed that when averaged over a reasonable number of independent runs (in this case 50), the results were slightly different than first reported in [2]. This is not to say that results in [2] were inaccurate, merely incomplete, as results for multiple independent results were not reported. A reimplementation of opt-aiNET was then undertaken, to further investigate the performance of the algorithm and try to

K. Deb et al. (Eds.): GECCO 2004, LNCS 3102, pp. 308–317, 2004.

identify reasons for these differences. Reimplementation is a useful tool and this is especially true in new paradigms such as AIS [3]. Indeed, work as already shown problems with algorithms such as AINE [4] another immune network approach. Work by [5] demonstrated premature convergence of the algorithm, which was not observed in the original work. This in itself, acted to some degree to motivate this work and validate a similar style immune network algorithm. It is felt that only through rigorous investigations can the field of AIS hope to grow and be taken as a serious competitor and viable alternative to other techniques.

This paper presents the original opt-aiNET, and identifies a number of minor issues relating to the actual implementation of that algorithm. The paper then proposes slight modifications to the algorithm and results testing both the original and re-implemented versions of opt-aiNET. This paper assumes knowledge of the AIS area in general, if the reader is not familiar with the area, they are directed to [3] for further information.

2   Artificial Immune Systems in Optimisation

There is a natural parallel between the immune system and optimisation. Whilst the immune system is not specifically an optimiser, the process of the production of antibodies in response to an antigen is evolutionary in nature: hence the comparison with optimisaiton, the location of better solutions. The process of clonal selection (a theory widely held by many immunologists) describes how the production of antibodies occurs in response to an antigen, and also explains how a memory of past infections is maintained. This process of clonal selection has proved to be a source of inspiration for many people in AIS and there have been a number of algorithms developed for optimisation inspired by this process [3,6,12,13].

One algorithm that has received much attention is aiNET [1,6]. AiNET has the capability to not only perform unimodal search, but also multimodal, without the need for any enhancements, unlike hybrid genetic algorithms [2]. The rest of this paper focuses on the algorithm proposed in [1], then extended in [2]. The reason for this is two fold: (1) It is becoming more widely used, so it is important to ascertain that the algorithm behaves the way it is reported and (2) it allows for greater confidence in the area of AIS if results are verified.

2.1   AiNET

The aiNET algorithm is a discrete immune network algorithm that was developed for data compression and clustering [1], and was also extended slightly and applied to optimization to create the algorithm opt-aiNET [2]. This has subsequently been developed further and applied to areas such as bioinformatics [7] and even modeling of simple immune responses [8].

Opt-aiNET, proposed in [2], evolves a population, which consists of a network of antibodies (considered as candidate solutions to the function being optimised). These undergo a process of evaluation against the objective function, clonal expansion, mutation, selection and interaction between themselves. Opt-AiNET creates a memory set of antibodies that represent (over time) the best candidate solutions to the objective function. Opt-aiNET is capable of either unimodal or multimodal optimisation and can be characterised by five main features:

- The population size is dynamically adjustable;
- It demonstrates exploitation and exploration of the search space;
- It determines the locations of multiple optima;
- It has the capability of maintaining many optima solutions;
- It has defined stopping criteria.

2.1.1 The Algorithm
Assuming the following terminology:

Network Cell	Individual of the population. Opt-aiNET does not employ any mechanism of encoding – real values are used in a Euclidean shape space.
Fitness	Measure of how good a particular cell is performing in relation to the objective function
Affinity	The Euclidean distance between two network cells
Clone	Offspring cells that are identical copies of their parent cell.
Mutated Clone	A clone that has undergone somatic hypermutation

1.   Randomly initialise population
2.   While (stopping criteria is not met) do

   I.   Determine fitness of each network cell against objective function

   II.   Generate Nc clones for each network cell

   III.   Each clone undergoes somatic hypermutation in proportion to the fitness of the parent cell (see Eq. 1)

   IV.   Determine the fitness of all network cells (including new clones and mutated clones)

   V.   For each clone select the most fit and remove the others

   VI.   Determine average error (distance from solution), if different from previous iteration, repeat from step 2

3.   Determine highest affinity network cells and perform network suppression.

4.   Introduce a percentage d of randomly generated network cells.

In step 1, the initial population consists of N network cells (randomly created). Each cell is a real value vector, which represents a candidate solution. During steps I-V each network cell undergoes a process of clonal expansion (N x Nc) and affinity maturation. Clones of each cell are mutated according to the affinity of the parent cell. The fitness represents the value of the function for the specific candidate solution. The affinity proportion mutation is performed according to the following equation:

$$C' = c + \alpha N (0.1) \tag{1}$$

$$\alpha = (1/\beta) \exp (-f^*)$$

where α is the amount of mutation, c is the parent cell, c' is the mutated clone of c, N (0,1) is a Gaussian random variable of zero mean and standard deviation of 1, β is a parameter that controls the decay of the inverse exponential function and f^* is the fitness of c normalised in the interval [0..1]. As c' represents a candidate solution, it must be within the range of the functions specified domain. If c' exceeds that, then it is rejected and removed from the population.

The fitness of each clone (and parent cell) is evaluated, then the fittest individual being selected to become a memory cell and the algorithm adopts an elitist approach to achieve this by always selecting the most fit. This is an iterative process that continues unto the average error value (distance from objective function) stabilises (this must be less than 0.0001). Once stablilisation occurs, the algorithm then proceeds to steps 3 and 4. Network suppression removes any similar or non-stimulated antibodies and antibodies that fall below the pre-determined suppression threshold σ. By removing similar cells, opt-aiNET prevents antibodies clustering on a single peak. This reduces the amount of cells maintained in the memory set,

It should be noted at this point, that the network interactions within opt-aiNET are only suppressive in nature and they do not contribute to the stimulation of the cells in any way. This algorithm is not faithful to the traditional immune network theory by Jerne [9], which proposes interactions of suppression and stimulation between B-cells.

2.2 Observations on Opt-AiNET

A detailed investigation of the code for opt-aiNET both employing the existing code (implemented by the authors of [2]) and a reimplementation of the algorithm by authors of this paper. These investigations uncovered a number of minor issues within the code implanted in [2]. The implication of this was that if one were to implement a version directly from the paper, then the results reported there would not be obtained. Uncovering this minor points, whilst not dramatically altering the algorithms performance, does have some impact, however. This paper proposes the small fixes required to be implemented, which we feel more accurately represents the algorithm as it was intended. This work also extends the current system by employing an automated mechanism for the location of peaks in search space, a mechanism which was lacking from the original. Discrepancies between the original results and results obtained using the same code were also identified. It is suggested that the original results reported were not in any way meant to be misleading, but the discrepancy is more likely to be caused by a lack of independent runs being reported in [2]. A technique which has been employed in this paper.

The observations we would like to make are:

1. Population fitness is calculated assuming that the overall fitness has increased and both values are positive.

Every 5 iterations, the degree of similarity between the present average fitness (*Avfit*) and the previous average fitness, is evaluated (this number is left out from any literature and is hard coded into the algorithm. Experimental evidence also suggested tat this could play an important role in the convergence of the algorithm). If the similarity is less than 0.0001 then the algorithm the algorithm can proceed to the suppression function.

The average fitness delta is calculated using the following equation:

If I - *avefitold/avefit* < 0.0001;   (2)

Calculating the value in this manner assumes that Avfit > avefitold are positive and that *Avfit* is greater than *avefitold*. This creates a problem if both values are negative, as the absolute value of avefitold is greater than the absolute value of *Avfit* (as in the existing code, it is the absolute value that is taken). This would result in the criteria for similarity being met even if there is a large difference between the two values. In order to fix this problem, these assumptions have been lifted.

2. The selection function within opt-aiNET was not elitist.

In the original system, the suppress function, calls the DIST function in MATLAB. This calculates the Euclidean distance between two vectors. If the distance is less than the suppression threshold, then the first vector is suppressed. No attempt is made to evaluate which of the two possess the greater fitness. This of course can result in the deletion of a potential optimum solution. The greater the threshold, the greater the chance of this occurring.

In order to overcome this, a simple ordering was placed on the network cells (they were sorted by fitness), to ensure that the least fit was always removed, as opposed to another candidate solution that had a higher fitness.

3. The calculation of peaks was done manually

This is very important for the reliability of the results. Upon inspection of the code taken from [2], it was clear that there was no reliable mechanism for the analysis of peaks within the system. Therefore, a more reliable automated approach was created that counted the number of optima found within the population, so as to accurately record the number. This is done simply by comparing the fitness of the candidate solution with the fitness value of its neighbors (determined by stimulation threshold). The candidate solution with the highest fitness is determined to be the highest point in that neighborhood.

3   Results

3.1   Experimental Protocol

In order to assess the opt-aiNET, we were obliged to follow as far as possible the same experimental protocol as the authors of [2]. Therefore, we used the same three functions presented by those authors, namely:

Multi Function: Range [-2,2]

$$G(x,y) = x.sin(4\Pi x) - y.sin(4\Pi y + \Pi) + 1 \tag{3}$$

This function has a single global optimum solution, with many local optima distributed non-uniformly

Roots Function: Range [-2,2]

$$g(z) = 1/1 + |z^6 - 1| \tag{4}$$

Where z is a complex number $z = x + iy$
This function has six maxima, which are located on slender peaks that rise to a height of 1 ($g(z) == 1$) from a plateau of 0.5 centered on (0,0). These maxima are at the six roots of the unity of the complex plane.

Schaffer's Function: Range [-10,10]

$$g(z) = 0.5 + \frac{\sin^2\left(\sqrt{x^2 + y^2}\right) - 0.5}{\left(1 + 0.001(x^2 + y^2)\right)} \tag{5}$$

This function has as single global optima and an infinite number of local optima, which form concentric rings expanding out from the optima. The global optimum is hard to locate due to the similarity with the best local optima.

The parameters employed by the original authors were:

Parameter	Value
Suppression threshold	0.2
Initial population size	20
Number of clones generated	10
Percentage of random new cells each iteration	40%
Scale of affinity proportion selection	100
Maximum iteration number	500

The results in the following section record the number of peaks located (Peaks), ItG is the number of iterations required to reach the global optima and ItC is the number of iterations required for convergence. Each experiment was performed 50 times, and the standard deviation is presented. Results are presented for the published version of the results (taken from [2]) identified as **published**, testing of the original code, identified as **recorded** and taken from the reimplementation, identified as **new**.

3.2 Multi Function

Analysis the results in detail revealed some interesting results. First, there is an issue of extreme convergence, which affects the average value of convergence (ItG) and also the number of peaks found. The empirical evidence would suggest that on this function, opt-aiNET failed to converge at least 50% of the time, but premature convergence was less frequent. When there is a failure to converge, it was observed that the suppress function was never activated which lead to a large number of candidate solutions to be located on a single peak. The authors of [2] comment that opt-aiNET avoids a "waste of resources" by positioning only a single individual on a peak. This assumes that it converges before reaching the maximum number of iterations allowed and as peaks were counted manually, that process was very prone to error.

Table 1. Results for opt-aiNet

Opt-aiNET (Published)			Opt-aiNET (Recorded)		
Peaks	ItG	ItC	Peaks	ItG	ItC
56.10±4.36	53.50±47.19	278.50±70.09	57.2±13.83	58.8±45.18	410±145.36

	Opt-aiNET (New)	
Peaks	ItG	ItC
56.5±17	212.8±140	384.6±143

Secondly, the global optima solution was found to be very unstable. Although in each instance the algorithm did reach a point that was considered the global maximum, as many as 90% of the time, this optimum solution was lost (due to the problems highlighted earlier).

Overall, the obtained and previously published results are very similar, with the exception of the number of iterations. As the work in [2] does not report how many independent runs were undertaken, one can only assume this figure is somehow affected in the work presented here, due to the number of experiments undertaken.

However, when one observes the results using the version with the enhancements (which were highlighted above), then the results appear different. The ability to find peaks has not been affected greatly, but the time taken to find the first optima solution, has. In addition, the overall convergence rate increases. This is most likely explained by the elitist mechanism now implemented in the new version of the algorithm. This may well lead to an overall quicker convergence rate being achieved.

It is also worth pointing out that the standard deviations are rather large, even for a stochastic search algorithm, this might indicate a lack of reliability and robustness in the system.

3.3 Roots Function

Overall, with this function, the results that were obtained were consistent with those published, but did vary with the reimplementation of the algorithm. The time it takes to first locate the optima solution is significantly longer than the original versions and indeed, average convergence is also longer. In some ways this is surprising, as the new version operates with an elitist approach. Clearly the average fitness within the network is staying lower for longer, hence the increase in time to convergence. Also, the ability to identify peaks has been diminished slightly Whilst not overly significant, we fell that it is worthy of note at least as the standard deviation is quite high and often the maximum number of peaks was not obtainable.

Table 2. Results comparing the Roots Function

	Opt-aiNET (Published)			Opt-aiNET (Recorded)	
Peaks	*ItG*	*ItC*	*Peaks*	*ItG*	*ItC*
6.0	86.89±34.31	295±129.74	5.9±0.32	93.2±31.99	308.8±112.62

	Opt-aiNET (New)	
Peaks	*ItG*	*ItC*
5.2±0.79	149.2±87.1	344±99

3.4 Schaffer's Function

For this function, things were a little more complicated. According to the published results [2], the algorithm should not converge within the 500 iterations. However, this was found not to be the case, where on average, the number of iterations for convergence was 219.2±199.97 It is proposed that the cause for this is the unstable average fitness level within the population. The instability is such that convergence occurs frequently.

After further investigations, it was discovered that the reason that convergence is occurring so frequently is that with a suppression threshold of 0.2 (as taken from the already published work) it is highly probable that no network cells will undergo the suppression. Upon examination of the actual original code, if this does occur, than the algorithm will terminate. A simple remedy to this problem is increasing the suppression threshold, but this will reduce the number of peaks that are located by the search. An alternative to that would be to increase the size of the initial population. This would hope to reduce the chances of no suppression taking place (as there will be more members and a higher probability that some will be close enough to be below the suppression threshold). However, if this approach is adopted that there will be an adverse affect on the running time, as a larger population increases the execution time.

4   Conclusions

This paper has revisited an immune inspired algorithm called opt-aiNET. A number of minor modifications to the original system have been proposed, which, it is felt more accurately reflect the intended and previously described system.

An identical test protocol based on the original work was established, with systematic tests using the existing code and newly developed code, being undertaken. It was observed that the results obtained from testing the system were not quite the same as reported in [2], but this may well be due to more tests being produced. Minor problems with existing code were identified, and these corrected in a new implementation. Whilst not affecting the performance significantly, the algorithm does behave differently, and is subject (independent to the alterations) to a very high standard deviation when experiments are performed multiple times. We are confident that the system that has been developed as part of this investigation is faithful to the original intended system and the code is available from the authors upon request.

References

[1] De Castro, L.N and Von Zuben, F. (2001). "aiNET: An Artificial Immune Network for Data Analysis", in Data Mining: A Heuristic Approach. Abbas, H, Sarker, R and Newton, C (Eds). Idea Group Publishing.

[2] De Castro, L.N and Timmis, J. (2002) An Artificial Immune Network for Multimodal Function Optimisation. Proc. Of IEEE World Congress on Evolutionary Computation. Pp. 669-674

[3] De Castro, L.N and Timmis, J. (2002) Artificial Immune Systems: A New Computational Intelligence Approach. Springer-Verlag.

[4] Timmis, J and Neal, M. (2001). A Resource Limited Artificial Immune System for Data Analysis. Knowledge Based Systems, 14(3-4):121-130.

[5] Knight, T and Timmis, J (2001). AINE: An Immunological Approach to Data Mining. In Cercone, N, Lin, T and Wu X. (Eds) IEEE International Conference on Data Mining. Pp. 297-304, San Jose. CA.

[6] De Castro, L. N. & Von Zuben, F. J. (2002), Learning and Optimisation Using the Clonal Selection Principle. IEEE Transactions on Evolutionary Computation, Special Issue on Artificial Immune Systems, 6(3), pp. 239-251.

[8] Jerne, N (1975). Towards a Network theory for the Immune System. Annals of Immunology., Inst. Pasture.

[9] (name removed for blind review) Artificial Immune Networks and Multimodal Optimisation. MSc Thesis. (place removed for blind review)

[10] Bezerra, B and De Castro, L.N. (2003) Bioinformatics data analysis using an artificial immune network. Proceedings of ICARIS 2003, eds. Timmis, J., Bentley, P. and Hart, E. Lecture Notes in Computer Science 2787, pp. Springer-Verlag, 2003.

[11] De Castro, L.N. (2003). The Immune response of an Artificial Immune Network (aiNET). Congress on Evolutionary Computation (CEC) pp. 1273-1280. IEEE press.
[12] Walker, J and Garrett, G (2003). Dynamic Function Optimisation: Comparing the Performance of Clonal Selection and Evolutionary Strategies. LNCS 2787. 273-284. Timmis, J, Bentley, P and hart, E. (Eds.)
[13] Kelsey, J, Timmis, J and Hone, A. (2003). Chasing Chaos. Proceedings of the Congress on Evolutionary Computation (CEC). Pages 413-219. IEEE.

A Novel Immune Feedback Control Algorithm and Its Applications

Zhen-qiang Qi[1], Shen-min Song[1], Zhao-hua Yang[2], Guang-da Hu[1], and Fu-en Zhang[1]

[1] Department of Control Science and Engineering, Harbin Institute of Technology, Harbin 150001, China;
qizq@hit.edu.cn, songsm19981998@yahoo.com.cn
[2] Department of Automation Measurement & Control Engineering, Harbin Institute of Technology, Harbin 150001, China
xryzh@sina.com

Abstract. This paper first analyzes the feedback principle of nature immune system and then the immune process is imitated by virtue of nonlinear molecular dynamics. Then the mathematic model of immune system is founded. The model implicates two important processes in immune system. One is that antibodies (Ab) and killer T cells (Tkill) rapidly respond to the change of the number of antigens (Ag). Another is that suppressor T cells (Tsup) inhibit and adjust the number of Ab and Tkill. The model manifests the good performance that immune system can respond to foreign materials rapidly and stabilize itself simultaneously. In this paper, foreign disturbances, input errors and measurement noises are regarded as Ag. The process in which creature presents immune response, produces antibodies and removes antigens is regarded as the control process of disturbances eliminating and differences adjusting. So, we designed a novel control algorithm based on immune feedback principle (IFC). The results of simulation show that the IFC algorithm can make the system respond quickly and get to steady state rapidly, and has good noise-proof feature. Its performance is superior to that of ordinary controllers. It also shows that the IFC algorithm suits for controlling the large time-delay system especially.

Keywords: immune; feedback; nonlinear; control algorithm

1 Natural Immune Feedback Principle

The natural immune feedback process is shown in figure 1. We view the process of antigen defending and removing by immune system as the main feedback process, and the process of balance restoration of the system by suppressor T cells as the inhibition feedback process. By interplay of main feedback and inhibition feedback, immune system can rapidly respond to foreign material and stabilize itself simultaneously.

K. Deb et al. (Eds.): GECCO 2004, LNCS 3102, pp. 318–320, 2004.

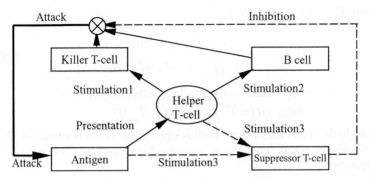

Fig. 1. Scheme of the whole immunity Circle

2 Modeling of Immune Process

Helper T cell :

$$
\begin{cases}
\dfrac{ds}{dt} = \sigma(Ag - s) \\[2mm]
\dfrac{dT}{dt} = \alpha + (\dfrac{s}{1+T} - 1)T
\end{cases}
\tag{1}
$$

Killer T cells:

$$
T_{kill}(t) = k_c \cdot T(t)
\tag{2}
$$

B cell :

$$
\begin{cases}
\dfrac{dB}{dt} = k_1 - k_2 \cdot B \cdot T \cdot H(Ag; \theta_B) - d_1 \cdot B \\[2mm]
\dfrac{dB_p}{dt} = k_2 \cdot B \cdot T \cdot H(Ag; \theta_B) - d_2 \cdot B_p \\[2mm]
H(Ag; \theta_B) = \dfrac{Ag^2}{\theta_B^2 + Ag^2}
\end{cases}
\tag{3}
$$

Antibody:

$$
Ab(t) = k_B \cdot B_p(t)
\tag{4}
$$

Suppressor T:

$$T_{sup}(t) = k_2 \left\{ \begin{array}{l} [T_{kill}(t-d) + Ab(t-d)] \\ -[T_{kill}(t-d-1) + Ab(t-d-1)] \end{array} \right\}^2 \cdot Ag(t) \tag{5}$$

The overall number of immune cells (attacking cells in figure 1) is

$$Attack(t) = T_{kill}(t) + Ab(t) - T_{sup}(t) \tag{6}$$

Equation (6) describes the immune feedback principle, which is a function about Ag and t. If we regard Ag as input signal, equation (6) describes a nonlinear feedback control algorithm.

3 Conclusions

We apply the IFC algorithm to control a second-order large time delay plant and the outlet temperature of the heat exchange unit in the thermal station of centralized heating system. According to simulation results, we can draw the following conclusions:

1) The nonlinear modeling method from the viewpoint of molecular dynamics is feasible.
2) The feedback principle of nature immune system is able to be led into the domain of the control engineering. The IFC algorithm also ensures the real-time performance of the control system.
3) The immune feedback control algorithm can make the system respond rapidly and stabilize quickly, which is just like the nature immune system. It improves the swiftness and stabilization of the system simultaneously. The all-round property of the immune feedback controller is superior to that of the ordinary PID controller.
4) Moreover, the IFC algorithm possesses good noise-proof feature and robustness.

Computer-Aided Peptide Evolution for Virtual Drug Design

Ignasi Belda[1], Xavier Llorà[2], Marc Martinell[1], Teresa Tarragó[1], and
Ernest Giralt[1,3]

[1] Institut de Recerca Biomèdica de Barcelona, Parc Científic de Barcelona,
Universitat de Barcelona, E 08028 Barcelona, Spain.
{ibelda,mmartinell,ttarrago,egiralt}@pcb.ub.es
[2] Illinois Genetic Algorithms Laboratory, National Center for Supercomputing
Application, University of Illinois at Urbana-Champaign, Urbana, IL 61801.
xllora@illigal.ge.uiuc.edu
[3] Departament de Química Orgànica. Universitat de Barcelona,
E 08028 Barcelona, Spain.

Abstract. One of the goals of computational chemistry is the automated *de novo* design of bioactive molecules. Despite significant progress
in computational approaches to ligand design and efficient evaluation of
binding energy, novel procedures for ligand design are required. Evolutionary computation provides a new approach to this design issue. A
reliable framework for obtaining ligands via evolutionary algorithms has
been implemented. It provides an automatic tool for peptide *de novo*
design, based on protein *surface patches* defined by user. A special emphasis has been given to the evaluation of the proposed peptides. Hence,
we have devised two different evaluation heuristics to carry out this task.
Then, we have tested the proposed framework in the design of ligands
for the protein Prolyl oligopetidase, p53, and DNA Gyrase.

1 Introduction

Since one of the goals of computational chemistry—as well as general drug
design—is the automated *de novo* design of bioactive molecules, significant
progress in computational approaches to ligand design and efficient evaluation
of binding energy have been done [1]. Nevertheless, novel procedures for ligand
design are required. This is motivated by the always increasing number of protein targets for drug design, being functionally and structurally characterized.
This situation is the result of major advances in both experimental methods for
structure determination [2] and high-throughput modeling [3].

Peptides are emerging as promising future drugs for several illnesses. Passing
from promising drugs to real drugs was difficult in the past because of their bad
ADME (absorption, distribution, metabolism, excretion) properties. However,
this handicap is soon to disappear thanks to the development of modern methods
of drug delivery, and the use of derivatives—D amino acids—metabolically more
stable.

K. Deb et al. (Eds.): GECCO 2004, LNCS 3102, pp. 321–332, 2004.
© Springer-Verlag Berlin Heidelberg 2004

Nowadays, several research projects are trying to develop new methodologies for designing drug-oriented peptides. These methodologies can be classified in several ways. One of them is structure-based drug design, where the design process is seen as an engineering problem. Another well-known approach is combinatorial chemistry, where a huge number of compounds are screened against the target. Our approach hybridizes both methodologies. We proposed a new methodology to screen large quantities of drug candidates. However these peptides are designed following a semi-rational process, in our case, evolutionary computation. From now on, we will use *in silico* to refer to virtual structure-based drug design. In our approach ligands are built from scratch, which is usually termed as *de novo* design in the literature. The main advantage of this approach is that novel structures, not contained in any database, can be devised.

To achieve this goal, algorithms must address two main tasks. First, a competent search method must be provided to explore this high-dimensional chemical space. Second, the search space (the set of all algorithmically treatable molecules) must be structured into regions of higher and lower quality to allow the prediction of desired properties. In order to perform the search task, we implemented and tested four different evolutionary algorithms: Darwinist genetic algorithm (GA) [4], Lamarckian genetic algorithm (LGA) [5], population-based incremental learning (PBIL) [6], and Bayesian optimization algorithm (BOA) [7]. In this evaluation, we have also approached the second task, to structure the search space into regions of higher and lower quality, by implementing two different heuristics to calculate the fitness (binding energy) of each individual (peptide) proposed by the evolutionary algorithms. The first heuristics was built to be computationally affordable, whereas the second focused on achieving and accurate binding energy estimate—throughout referred as docking energy. It must be pointed out that we used AutoDock 3.0.5 [8] for the docking calculations required by the heuristics.

The main goal of this work is to design peptide drugs which serve as effective ligands to the target protein area defined by the user. This area is also known as surface patch. One application of such peptide drugs could be to act as inhibitors of some pathological functionalities of the target protein [9]. Finally, we tested the developed methodology in several specific cases. In particular the proteins prolyl oligopetidase, p53, and DNA gyrase (still under study and for this reason the results are not summarized in this paper.) The results obtained are encouraging. We have compared the proposed peptides with some others designed using a purely chemical-knowledge based approach by the peptide design experts from our research group. In all the tested cases, the peptides designed in silico present better docking energies than their counterparts designed by chemical experts. We are now currently synthesizing these peptides in order to do *in vitro* comparisons.

2   Related Work

Several approaches to *de novo* ligand design have been previously reported. However, only one of these approaches is quite similar to that implemented in this work. ADAPT [10] tries to design from scratch small organic molecules. The underlying mechanism uses docking as part of the fitness measure. The main difference between ADAPT and our approach is that our work is very specifically optimized for the synthesis of sequential compounds—peptides—whereas ADAPT is focused in small organic molecules. Therefore, ADAPT would have to be modified in order to design peptides to act as protein binders.

Other groups are using different approaches, such as growing [11,12], linking [13,14], and physico-chemical properties [15]. The growing strategy is a clear example of a different approach. The building-up process start from a seed structure pre-placed in the surface patch. The user can assign certain growing sites on the seed structure and then the program will try to replace each growing site by a candidate fragment. The new structure will serve as the seed structure for the next growing cycle.

The building-up process in a linking strategy also starts from a pre-placed seed structure. However, the seed structure consists of several separate pieces that have been positioned to maximize the interactions with the target protein. The pieces grow simultaneously, while the linking program tries to link these pieces in an feasible way. This process continues until all the pieces are integrated into a single molecule.

Finally, in the physico-chemical approach, the main goal is to obtain an optimal set of physico-chemical properties that the peptide must exhibit in order to be a good ligand. Thus, a peptide is designed having in mind the obtained set of physico-chemical properties.

Previous approaches are aiming the design of new natural peptide ligands. Such peptides may present a poor bioavailability and, hence, become ineffective for their usage as drugs. Proteases catalyze the splitting of peptides into amino acids by a process known as proteolysis. Overcoming such constraints is another motivation of the methodology presented in this paper. Although we are currently working with natural amino acids (L amino acids), the framework is able to design new peptide ligands using D amino acids. The D amino acids are artificial amino acids which are the mirror image of the L amino acids. Therefore proteases do not recognize and split them [16], making them ideal candidates for drug design.

3   Computer-Aided Peptide Design

Evolutionary computation is being widely applied to different areas of bioinformatics [17]. Evolutionary computation techniques have recently been applied in many aspects related to drug and compound library design [14,18,19,20,21,22].

Evolutionary algorithms are perfect candidates for applications were deterministic or analytic methods fail, for instance, problems where the underlying mathematical model is ill-defined or the search space is too big. In this work both aspects are present. The underlying mathematical model being optimized is given by AutoDock 3.0.5. However, is important to note that docking algorithms are not a perfect simulation of the reality yet [23]. Also, the search space is too big to be systematically explored and each evaluation using the fast heuristics developed takes more than 30 minutes of calculations on a Pentium IV 1.60 GHz. Hence, these circumstances lead us to use evolutionary algorithms to steer the search for *de novo* design. These assumptions have been previously used by other researchers [11,12,14].

3.1   The Problem

Our goal is to design good peptide ligands to a user-defined surface patch of a protein. To reach this goal we have implemented and tested several evolutionary algorithms and we have developed two different fitness functions—throughout called *heuristics*.

Each individual of the evolutionary algorithms (peptides) is defined by a 6 genes chromosome. Each gene can adopt 7 different amino acid values. We have only used 7 different amino acids among the 20 natural amino acids. We have only selected those amino acids representative of different physico-chemical properties. They are *alanine, arginine, glutamic acid, serine, isoleucine, triptophan* and *proline*.

The algorithms used to steer the search are four evolutionary algorithms. They are: Darwinist genetic algorithm [4], Lamarckian genetic algorithm [5] where for the local search we have used evolutionary strategies $(1 + 1)$ [24], population-based incremental learning [6] and Bayesian Optimization Algorithm [7]. Details on how we have used these algorithms can be found in [25], as well as in the previous listed references.

3.2   Heuristics

The evolutionary algorithms need a *fitness function* to guide the search space into regions of higher and lower quality evaluating the peptides proposed. For this purpose, we developed two heuristics to dock the peptides to the surface patch defined by the user, giving us the docking energy. The first one is a quick one, whereas the second is an accurate one, hence computationally expensive. The user can choose the heuristics that better adapts to his necessities and constraints.

Docking. Docking algorithms are *in silico* tools that try to find the best mode of interaction between a small, possibly flexible, ligand and a large, usually rigid, macromolecular receptor. This is done by minimizing the energy of interaction, which is a complex function of variables describing the position, orientation, and conformation of the ligand. In our work, the small flexible ligands are the peptide proposed by the evolutionary algorithms, and the macromolecular receptor is the surface patch of the protein defined by user, *i.e.:* a site on a protein to which a peptide ligand is desired.

To perform the docking experiments we have used the program AutoDock 3.0.5 [8]. At the beginning the heuristics prepare all the input files needed by AutoDock, and at the end they can process output files in order to extract the information needed. The AutoDock parameters are tuned as suggested in [8]. AutoDock uses a LGA as the algorithm for minimize the energy of interaction with 50 individuals per generation, 10000 generations, 300 steps of local optimization. And all this has been run ten times with different initializations for each docking experiment.

Quick Heuristics. The first heuristics developed is a quick one. We say that it is quick because it only has one docking experiment while the accurate heuristics has five. This heuristics has five stages: (1) three-dimensional reconstruction, (2) energetic minimization, (3) flexible angles definition, (4) docking, and (5) Boltzmann averaged binding energy.

Three-dimensional reconstruction. The internal representation of the peptides in the evolutionary algorithms is not a three-dimensional structure. It is a sequence of amino acids. But, the docking experiments need a structure, therefore we made a program using the NAB language [26] which takes at its input a sequence of letters representing amino acids, and gives us at the output a PDBQ file (PDB with charges) [27] with the extended structure of that peptide.

Energetic minimization. The next step is to do a short energetic minimization over the extended structure obtained in the previous step. To perform this, we have implemented the energetic minimization inside the program developed for the three-dimensional reconstruction [26]. We used a conjugate gradient minimization until the root-mean-square of the components of the gradient is less than 1.0.

Flexible angles definition. Before starting the docking calculations we redefined our ligand (peptide) to be flexible. We fixed the backbone but we give flexibility to the side chains. To perform this task we use AutoTors, which is an auxiliary script of the AutoDock.

Docking. In this stage we perform ten flexible docking experiments using the ligand built in the previous stages over the surface patch defined by the user.

Boltzmann averaged binding energy. Finally, we compute a Boltzmann averaged binding energy of each of the most stable structures found in each of the ten runs of the docking algorithm. The value obtained in this step is the fitness value for the peptide being evaluated in this moment.

Accurate Heuristics. The second heuristics developed implies a more accurate evaluation than the first heuristics. The main idea of this evaluation is to make five different docks, each one with a different probable peptide backbone structure. Once the docks have been carried out, we keep the more stable structure from those obtained. This heuristics has five steps: Secondary structure prediction, rotamers construction, flexible angles definition, docking, and Boltzmann averaged binding energy.

Secondary structure prediction. First of all we use the Chou-Fasman method for secondary structure prediction of peptides attending at their amino acid sequence. This method assigns at each conformation (α-helix, β-strand or random coil) a number that inform us about the "probability" of adopting that conformation.

Rotamers construction. Attending to the number obtained using the Chou-Fasman prediction, we build five rotamers of the same peptide. In each rotamer, all the ϕ and ψ angles of the peptide are chosen randomly from the area of Ramachandran map which represents the secondary structure being built. Angle ω is fixed to 180°.

Flexible angles definition. Flexibility is assigned to the five peptide structures in the side chains.

Docking. One flexible docking calculation is carried out for each rotamer built. Each of the docking calculations is carried out in the same manner than the docking calculation of the first heuristics.

Boltzmann averaged binding energy. Finally, we analyze which is the more stable rotamer docked, we keep it and we take as the individual fitness the Boltzmann averaged binding energy of each of the more stable structures found in each of the ten runs.

4   Results

We have tested the proposed methodology for peptide design using three proteins with high therapeutical interest. Making an extended comparison of the performance of each one of the evolutionary algorithms developed is not within of the scope of this paper. However, we explored the performance of the algorithms PBIL, LGA and BOA in three different problems. GA has not been used in these results because we decided to explore its extended version: LGA. No special criteria guided the choice of algorithms used in each domain.

The three systems where we have tested our *in silico* methodology are proteins: POP, p53 and DNA gyrase (since it is currently under study, the results are not reported in this paper.) These three proteins have been selected because each problem presents interesting features. The surface patch of the POP is inside a large tunnel, the surface patch of the p53 is a large surface, and the surface patch of the DNA gyrase is an interaction area with DNA. For a proper comparison we have carried out two experiments for each of the studied systems. The first one is to make some docking experiments with the peptides designed by the design experts of our research group or some well-known natural ligands reported in the literature. The second followed the methodology reported in this paper. In this manner we compared rational- and evolutionary-designed peptides.

4.1 Prolyl Oligopeptidase

Introduction. Prolyl oligopeptidase (POP, EC 3.4.21.26) is a cytosolic serine peptidase characterized by oligopeptidase activity. The three-dimensional structure of POP revealed a two-domain organization: (1) the catalytic one, and (2) the structural one. In the catalytic domain there are the residues Ser554, His 680 and Asp641 which form the catalytic triad of the protease. Those residues are essential for the catalytic activity of the enzyme and are located in a large cavity at the interface of the two domains [28]. Based on this information available from the crystal structure we have defined a surface patch around the residues of the catalytic center Ser554, His 680 and Asp641. The surface patch also comprises the cavity formed by the structural domain which corresponds to the interior of the β-propeller domain. Inside this cavity the substrate interacts with many residues of the enzyme and is directed to the active site. The box is big enough to accommodate peptides of six residues that are tested in this study.

Table 1. Docking energy comparison for prolyl oligopeptidase peptides.

Docking energy of fragments of POP inhibitors described at the literature.				Docking energy of the best five individuals found by the algorithm	
Peptide	*Docking Energy*	*Peptide*	*Docking Energy*	*Peptide*	*Docking Energy*
GKPPIG	-7.186	GKPPVG	-7.067	WWPWPP	-13.737
GVEIPE	-5.122	GYPIPF	-5.843	WWPSWA	-13.216
HLPPPV	-8.542	KPRRPY	-3.562	WSPSWP	-13.032
LLSPFW	-6.314	LSPFWN	-9.538	PWPEWA	-12.991
MPPPLP	-8.862	MTPPLP	-7.436	WWPWSP	-12.936
QNCPLG	-7.977	QNCPRG	-8.874		
RPKPQQ	-7.546	SPFWNI	-7.457		
TPPLPA	-7.913				

POP is involved in the maturation and degradation of proline-containing neuropeptides related with learning and memory [29]. Physiological studies with human plasma have shown that POP activity is increased in mania, schizophrenia, post-traumatic stress and anxiety, while POP activity is decreased in depression, anorexia and bulimia nervosa [30]. Modulation of the POP activity with specific peptide inhibitors can be useful for the treatment of those diseases.

Ligand Peptides Found in the Literature. As POP hydrolyzes the peptide bond at the C-terminal side of prolyl residues, for the docking experiments we have chosen fragments of POP substrates and POP inhibitors containing 6 residues. The criterion followed to choose the fragments has been that at least one of those residues should be Proline. In table 1 are shown these peptides with their docking energies

In Silico **Designed Peptides.** In this stage we have run a LGA. In table 1 are shown the energies of the 5 best individuals found by the algorithm. All of them are proline rich, the same as the natural ligands. In figure 1 there is a representation of the best individual found docked to the target protein.

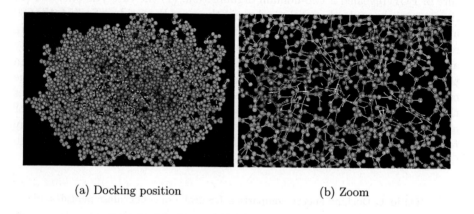

(a) Docking position   (b) Zoom

Fig. 1. Peptide proposed `WWPWPP` docked in the user-defined surface patch of the protein POP. High resolution color plates and 3D models of the figures displayed above can be found at `http://www-illigal.ge.uiuc.edu/~xllora/Pool/Papers/GECCO-2004-Figures/`.

4.2   p53

p53 protein is a transcription factor involved in different cellular functions, such as cell cycle control, programed cell death (or apoptosis) and differentiation. Actually the p53 gene was the first tumor-suppressor gene to be identified, and it was found that p53 protein does not function correctly in most of human cancer—around a 50% of cancer patients present patological mutations affecting the protein functionality.. The p53 can be divided in four domains: an N-terminus transactivation domain, a DNA-binding domain (DBD), a tetramerization domain (TD) and a C-terminus regulatory domain. The function of p53 is mediated by protein-DNA and by protein-protein interactions. The tetramerization domain has an essential role in its function, because only the tetrameric structure is active [31]. Peptides able to recognize this tetramerization domain and stabilize its native conformation could be of great interest in cancer research.

Expert Designed Peptides. In our group, we have studied the recognition of p53 tetramerization domain using a tetraguanidinium compound [32]. Based on this molecule, we have designed and synthesized a peptide which is also able to interact with the domain. In order to improve the affinity, we designed a peptide library based on the substitution of the arginines for other residues. These substitutions are mainly based on modifications of the chain length and/or the functional group. In table 2 are shown the docking energies of the five best peptides found in the 47 peptide library. More details of the design and evaluation of those peptides can be found in [33].

Table 2. Docking energy comparison for p53 ligands.

Docking energy of the top-five peptides designed		Docking energy of the best five individuals found by the algorithm	
Peptide identification	*Docking Energy*	*Peptide*	*Docking Energy*
Rpa3R_46	-11.067	WWPWWW	-13.294
Rpa4_47	-10.870	WWWWWW	-13.260
Oxaa_33	-10.180	AWWWWW	-13.182
Rab4_51	-9.707	WWWWWA	-13.113
Dapa3R_20	-9.545	AWRWWW	-12.827

In Silico **Designed Peptides.** In this stage we have run a PBIL. In table 2 are shown the energies of the 5 best individuals found by the algorithm. In figure 2 there is a representation of the best individual found docked to the target protein.

5   Discussion

Recent advances in bioinformatics have provided a new *in silico* approach for the design of peptide ligands. This approach offers interesting complementary advantages to laboratory methods. The time invested in performing each run or experiment implies three issues: preparation, performance and evaluation. In each of them, *in silico* methods are faster than laboratory ones, although less accurate. However, such virtual screening can help to guide laboratory search. Another important issue is the cost, which is directly related, among other things, to the number of workers, equipment, time and reagents needed. Even if it does not seem obvious, the costs of performing an experiment might exceed enormously the expenses of executing a run. On the other hand, the available computational resources are the major constraint to *in silico* design accuracy. However, computational horsepower increases each month at the same time that it become cheaper.

We have observed that if the docking calculations were extremely good, the evolutionary algorithms developed would be a good tool for exploring the huge

(a) Docking possition   (b) Zoom

Fig. 2. Peptide proposed WWPWWW docked in the user-defined surface patch of the protein p53. High resolution color plates and 3D models of the figures displayed above can be found at http://www-illigal.ge.uiuc.edu/~xllora/Pool/Papers/GECCO-2004-Figures/.

search space that this combinatorial search implies. Unfortunately, this assumption is not true. The scientific community is still working to improve docking techniques, as well as the experimental techniques in chemistry. Such cooperative endeavor may lead in the future to better virtual tools. However, methodologies such as the one proposed in this paper are ready to be used to recommend compounds to be synthesize, thus drug research can narrow the focus of their research to those regions suggested by the virtual screening tools.

Another conclusion of the work presented in this paper is that the peptides found by evolutionary algorithms have better docking energies. This fact has been shown comparing them to those designed using a rational chemistry approach in our research group, and those found in the related literature. Recently, we selected some evolved peptides, starting their synthesis *in vitro*. Once these new compounds become available, besides the comparison based on the docking energy, we will be be able to compare their behavior *in vitro*.

Acknowledgments. This work was partially supported by grants from Fundación BBVA, Fundació Marató TV3, Ministerio de Ciencia y Tecnología FEDER (BIO 2002-2301 and EET2001-4813), the Air Force Office of Scientific Research, Air Force Materiel Command, USAF (F49620-03-1-0129), and by the Technology Research, Education, and Commercialization Center (TRECC), at University of Illinois at Urbana-Champaign, administered by the National Center for Supercomputing Applications (NCSA) and funded by the Office of Naval Research (N00014-01-1-0175). The US Government is authorized to reproduce and distribute reprints for Government purposes notwithstanding any copyright notation thereon.

The views and conclusions contained herein are those of the authors and should not be interpreted as necessarily representing the official policies or endorsements, either expressed or implied, of the Air Force Office of Scientific Research, the Technology Research, Education, and Commercialization Center, the Office of Naval Research, or the U.S. Government.

References

1. Böhm, H.J.: Prediction of binding constants of protein ligands: A fast method for the priorization of hits obtained from de novo design or 3D database search programs. Journal of Computer-Aided Molecular Design **12** (1998) 309–323
2. Codina, A., Gairí, M., Tarragó, T., Vigueras, A.R., Feliz, M., Ludevid, D., Giralt, E.: 1h(n), 15n, 13co, 13ca, 13b assignement and secondary structure of a 20 kda a-l-fucosidade from pea using TROSY. Jornal of Biomolecular NMR **22** (2002) 295–296
3. Thormann, M., Pons, M.: Massive docking of flexible ligands using environmental niches in parallelized genetic algorithms. Journal of Computational Chemistry **22** (2001) 1971–1982
4. Holland, J.: Adaptation in Natural and Artificial Systems. MIT Press (1975)
5. Krasnogor, N.: Studies on the Theory and Design Space of Memetic Algorithms. PhD thesis, University of the West England, Bristol (2002)
6. Baluja, S., Caruana, R.: Removing the genetics from standard genetic algorithm. In Prieditis, A., Russell, S., eds.: Proceedings of the International Conference on Machine Learning, Morgan Kaufmann (1995) 112–128
7. Pelikan, M., Goldberg, D., Cantú-Paz, E.: BOA: The bayesian optimization algorithm. In: Proceedings of the Genetic and Evolutionary Computation Conference GECCO-99. Volume 1., Morgan Kaufmann (1999)
8. Morris, G., Goodsell, D., Halliday, R., Huey, R., Belew, R., Olson, A.: Automated docking using a lamarckian genetic algorithm and and empirical binding free energy function. Journal of Computational Chemistry **19** (1998) 1639–1662
9. Zeng, J.: Mini-review: Computational structure-based design of inhibitors that target proteins surfaces. Combinatorial Chemistry & High Throughput Screening **3** (2000) 355–362
10. Pegg, S., Haresco, J., Kuntz, I.: A genetic algorithm for structure-based de novo design. Journal of Computer-Aided Molecular Design **15** (2001) 911–933
11. Douglet, D., Thoreau, E., Grassy, G.: A genetic algorithm for the automated generation of small organic molecules: drug design using an evolutionary algorithm. Journal of computer-aided molecular design **14** (2000) 449–466
12. Budin, N., Majeux, N., Tenette, C., Caflisch, A.: Structure-based ligand design by a build-up approach and genetic algorithm search in conformational space. Journal of Computational Chemistry **22** (2001) 1956–1970
13. Böhm, H.J.: Computational tools for structure-based ligand design. Program biophysical molecular biology **3** (1996) 197–210
14. Wang, R., Gao, Y., Lai, L.: Ligbuilder: A multi-purpose pogram for structure-based drug design. Journal of molecular modeling **6** (2000) 498–516
15. Mandell, A., Selz, K., Shlesinger, M.: Algorithmic design of peptides for binding and/or modulation of the funcions of receptors and/or other proteins (2002)
16. Haack, T., González, M., Sánchez, Y., Giralt, E.: D-Amino acids in protein de novo design. II. Protein-diastereomerism versus protein-enantiomerism. Letters in Peptide Science **4** (1997) 377–386

17. Fogel, G.B., Corne, D.W., eds.: Evolutionary Computation in Bioinformatics. Elsevier Science (2002)
18. Teixido, M., Belda, I., Rosello, X., Gonzalez, S., Fabre, M., Llorà, X., Bacardit, J., Garrell, J.M., Vilaro, S., Albericio, F., Giralt, E.: Development of a genetic algorithm to design and identify peptides that can cross the blood-brain barrier. QSAR and Combinatorial Sciences **22** (2003) 745–753
19. Patel, S., Stott, I., Bhakoo, M., Elliott, P. In: Patenting Evolved Bactericidal Peptides. In Creative Evolutionary Systems. Morgan Kaufmann Publishers (2001) 525–545
20. Kamphausen, S., Höltgen, N., Wirsching, F., Morys-Wortmann, C., Riester, D., Goetz, R., Thürk, M., Schwienhorst, A.: Genetic algorithm for the design of molecules with desired properties. Journal of Computer-Aided Molecular Design **16** (2002) 551–567
21. Michaud, S., Zydallis, J., Lamont, G., Pachter, R.: Detecting secondary peptide structures by scaling a genetic algorithm. In: Technical Proceedings of the 2001 International Conference on Computational Nanoscience and Nanotechnology. (2001) 29–32
22. Goh, G.K.M., Foster, J.A.: Evolving molecules for drug design using genetic algorithms via molecular trees. In: Proceedings of the Genetic and Evolutionary Computation Conference (GECCO-2000), Morgan Kaufmann (2000) 27–33
23. Shoichet, B., McGovern, S., Wei, B., Irwin, J.: Lead discovery using molecular docking. Current Opinion in Chemical Biology **6** (2002) 439–446
24. Back, T.: Evolutionary Algorithms in Theory and Practice. Oxford University Press (1997)
25. Belda, I., Llorà, X., Piqueras, M.G., Teixido, M., Nicolas, E., Giralt, E.: Evolutionary algorithms and de novo peptide design. Technical Report 2003005, Illinois Genetic Algorithms Laboratory (2003)
26. Macke, T., Case, D.: NAB User's Manual. Departament of Molecular Biology, The Scripps Research Institute, University of California, La Jolla, California. (1999)
27. Berman, H.M., Westbrook, J., Feng, Z., Gilliland, G., Bhat, T.N., Weissig, H., Shindyalov, I.N., Bourne, P.E.: The protein data bank. Nucleic Acids Research **28** (2000) 235–242
28. Fülöp, V., Bocskei, Z., Polgár, L.: Prolyl oligopeptidase: an unusual b-propeller domain regulates proteolysis. Cell **94** (1998) 161–170
29. Mentlein, R.: Proline residues in the maturation and degradation of peptide hormones and neuropeptides. FEBS Letters **234** (1988) 251–256
30. Maes, M., Goossens, F., Scharpé, S., Calabrese, J., Desnyder, R., Meltzer, H.: Alterations in plasma prolyl endopeptidase activity in depression, mania, and schizophrenia: Effects of antidepressants, mood stabilizers, and antipsychotic drugs. Psychiatry Research **58** (1995) 217–225
31. Chene, P.: The role of tetramerization in p53 function. Oncogene **20** (2001) 2611–2617
32. Salvatella, X., Martinell, M., Gairí, M., Mateu, M.G., Feliz, M., Hamilton, A.D., de Mendoza, J., Giralt, E.: A tetraguanidinium ligand binds to the surface of the tetramerization domain of protein p53. Angewantde Chemie International Edition **43** (2004) 196–198
33. Martinell, M.: Disseny, síntesi i estudi de lligands peptídics capaços de reconèixer la superfície de la p53. PhD thesis, Universitat de Barcelona (2004) (in preparation).

Automating Genetic Network Inference with Minimal Physical Experimentation Using Coevolution

Josh C. Bongard and Hod Lipson

Computational Synthesis Laboratory
Sibley School of Mechanical and Aerospace Engineering
Cornell University, Ithaca, New York 14850
{JB382,HL274}@cornell.edu

Abstract. A major challenge in system biology is the automatic inference of gene regulation network topology—an instance of reverse engineering—based on limited local data whose collection is costly and slow. Reverse engineering implies the reconstruction of a hidden system based only on input and output data sets generated by the target system. Here we present a generalized evolutionary algorithm that can reverse engineer a hidden network based solely on input supplied to the network and the output obtained, using a minimal number of tests of the physical system. The algorithm has two stages: the first stage evolves a system hypothesis, and the second stage evolves a new experiment that should be carried out on the target system in order to extract the most information. We present the general algorithm, which we call the *estimation-exploration algorithm*, and demonstrate it both for the inference of gene regulatory networks without the need to perform expensive and disruptive knockout studies, and the inference of morphological properties of a robot without extensive physical testing.

Keywords: Bioinformatics, System Identification, Evolutionary Robotics

1 Introduction

System biology is concerned with the synthesis of large amounts of biological detail in order to infer the structure of complex structures with many interacting parts. In this way, system biology can be viewed as an example of reverse engineering.

Figure 1 depicts the basic cycle underlying any attempt to reverse engineer a target system: obtain output from the target system using some input, update the system hypothesis, generate some new input, and repeat. The cycle continues until enough data has been obtained from the target system to ensure a sufficiently accurate reconstruction. This paper presents a new approach to the automation of reverse engineering using an evolutionary algorithm that is both independent of the applied problem domain and requires minimal testing of the physical system.

Evolutionary algorithms can be used in two different ways: to evolve a completely new system, or evolve a system that approximates some target system. Examples of the former approach involve the evolution of robot morphology/controller pairs [25] [19] [4] [10] and the use of genetic programming to evolve agent behaviors (eg. [17]). The latter approach involves the use of evolutionary algorithms for reverse engineering: the system hypothesis is evolved based on input/output data sets. Examples of the evolution of reverse engineering include symbolic regression (eg. [17] and [7]), evolution of artificial

K. Deb et al. (Eds.): GECCO 2004, LNCS 3102, pp. 333–345, 2004.
© Springer-Verlag Berlin Heidelberg 2004

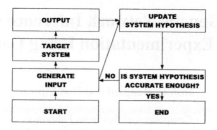

Fig. 1. Flow of a generalized reverse engineering algorithm.

neural networks that reproduce specific input/output vectors (useful for learning) [28], and electronic [21] [22], metabolic [16] or genetic circuit [11] inference.

The field of gene network inference is a rapidly burgeoning subfield in system biology [15], and is concerned with inferring genetic regulatory networks based on the results of a set of tests performed on the network in question. Many different models of the underlying genetic network have been used, usually classified based on the amount of biological detail inherent in the model (see [8] and [12] for an overview).

One of the most popular models is the Random Boolean Network (RBN) proposed by Kauffman [13], which is a discrete system in which a network of n genes, each of which is discretely regulated (turned on or off) by k other genes. The RBN's popularity as a model stems from its simplicity and general nature: it contains the minimum of biological detail. More detailed gene network models have been used (eg. [2]), as well as modeling the networks as differential equations [24] [11] or weight matrices [27]. In many of these models, the underlying gene network topology can be or is represented as a graph: nodes indicate genes, and directed labeled edges indicate gene regulation. In this paper we employ a graph-based method for representing gene networks.

In addition to the type of model, several methods have been used to infer genetic networks, including clustering algorithms (see [8] for an overview), correlation metrics [1], linear algebra [6], simulated annealing [23] and genetic algorithms [26] [11].

A number of input and output data pairs are required in order to obtain enough information about the target network to infer its structure correctly. As pointed out in [8], it is desirable to minimize the number of input/output pairs required, so that a minimum of experiments have to be conducted. Also, the *type* of experiment required should be as cheap as possible in terms of experimental difficulty and accuracy of acquired output data. Iba & Mimura [11] showed that by using a multi-population evolutionary algorithm to not only infer the hidden network, but also to propose additional experiments that would most help to refine the current best evolved network hypothesis. However their model requires the experimenter to perform costly knockout or lesion experiments in order to supply the algorithm with an actual subset of the regulatory network.

Knockout studies in genetics (eg. [18]), lesion studies in neuroscience (eg. [29]) and ablation studies in embryology (eg. [9]) are a related set of time-honored tools in those fields, but they have three major drawbacks: such experiments are often difficult to perform, they are destructive to the object of study, and often provide misleading data about the relationships between parts of the system under study. For example, in higher model organisms such as mice, one or more generations must be raised in order to accurately measure the phenotypic effect of a disabled gene.

By carefully selecting the input to be processed by a target network, output data rich enough in information to infer topology can be obtained such that more costly knockout studies are not required. Here we present a coevolutionary algorithm that not only evolves the hidden network, but also evolves these desirable input data sets. The next section describes the algorithm and the model of genetic networks we use; section 3 presents the results of our algorithm; section 4 provides discussion regarding the power and generality of this algorithm; and the final section offers some concluding remarks.

2   Methods

This paper presents an algorithm for reverse engineering networks. The networks can be interpreted as either biological or electronic networks, or can be interpreted as a representative of any coupled, non-linear system that takes inputs and produces outputs.

We first describe a graph-based model for representing genetic networks, and then describe the application of the estimation-exploration algorithm for inferring hidden instances of such networks based on sets of input and output gene product concentrations.

2.1   The Network

Many models of genetic regulatory networks employ a graph-based representation, in which nodes represent genes, and directed labeled edges from gene i to gene j indicate that gene i somehow influences the expression of gene j. For our purposes we have chosen to represent regulatory networks of n genes using an $n \times n$ matrix $\mathbf{R}$ with entries r_{ij}. If $r_{ij} > 0$, gene i contributes to the enhancement of gene j; if $r_{ij} < 0$, gene i contributes to the inhibition of gene j; if $r_{ij} = 0$, gene i does not directly regulate gene j. Now let an input data vector of gene product concentrations be represented by $\mathbf{g}^{(t+1)} = \{g_1^{(t)}, g_2^{(t)}, ...g_n^{(t)}\}$, in which $g_i^{(t)}$ indicates the gene product concentration at the beginning of some experiment. We can then calculate new gene product concentrations after some time period has elapsed using

$$g_j^{(t+1)} = \min(\ \max(\ 0\ ,\ g_j^{(t)} + \sum_{i=1}^{n} r_{ij} g_i^{(t)}\ )\ ,\ 1\ ) \tag{1}$$

Since the g_j variables indicate concentration, the *min* and *max* functions bound the value between 0 (for no concentration) and 1 (for concentration saturation). The vector $\mathbf{g}^{(t+1)} = \{g_1^{(t+1)}, g_2^{(t+1)}, ...g_n^{(t+1)}\} = [(\mathbf{R}+\mathbf{I})\mathbf{g}^{(t)}]_0^1$ then represents the bounded output data vector obtained from the hidden regulatory network $\mathbf{R}$, given input $\mathbf{g}^{(t)}$. All values of the input vector $\mathbf{g}^{(t)}$ and output vector $\mathbf{g}^{(t+1)}$ lie in the range $[0, 1]$. The values of $\mathbf{R}$ lie in the range $[-1, 1]$.

This discrete map approximates the differential equation model of regulation:

$$\frac{dg_i}{dt} = f_i(\mathbf{g}), \quad 1 \leq i \leq n \tag{2}$$

in which the product concentration of gene i changes as a function of the product concentrations of the other genes (possibly including gene i). In our formulation, f_i is the thresholded multiplication of row i in $\mathbf{R}$ by the column of initial concentrations $\mathbf{g}^{(t)}$.

2.2 The Estimation-Exploration Algorithm

In this paper we present a general algorithm that allows for the reverse engineering of a hidden network based solely on input/output data: in this case, the hidden network is the connection matrix $\mathbf{R}$. Our algorithm has two stages: the estimation and exploration phase, each of which has an associated genetic algorithm. The first stage evolves a plausible connection matrix based on input/output data sets (estimation), and the second stage evolves a new input vector that should produce an output vector rich in information when processed by the actual target connection matrix (exploration). In this way the algorithm performs two functions: it reverse engineers the hidden network, and it proposes *useful* experiments that will accelerate the inference process.

In the application of the estimation-exploration algorithm to genetic networks, first a random input vector is selected, and the corresponding output vector is computed using the actual target connection matrix. Then the resulting single input/output vector pair is passed into the algorithm.

The estimation genetic algorithm begins with a population of genomes, each of which is comprised of an $n \times n$ connection matrix $\mathbf{R}'$: the values are generated randomly over $[-1, 1]$ using a uniform distribution. Each genome is evaluated as follows. Each of the input vectors applied to the actual target network so far (which during the first iteration of the estimation phase is only one vector) is used to calculate a corresponding output vector based on the genome's connection matrix $\mathbf{R}'$. The *subjective error* associated with the genome is set to

$$\text{error}_{\text{subj}}(\mathbf{R}') = \frac{\sum_{k=1}^{x} \sum_{i=1}^{n} |g_{ik}^{(\text{tar})(t+1)} - g_{ik}^{(\text{guess})(t+1)}|}{xn} \qquad (3)$$

where x is the total number of experiments performed on the target network so far, $g_{ik}^{(\text{tar})(t+1)}$ is the resulting concentration of gene i when experiment k is performed on the target network, and $g_{ik}^{(\text{guess})(t+1)}$ is the resulting concentration of gene i when experiment k is performed on $\mathbf{R}'$.

Subjective error is then an indirect measure of how well $\mathbf{R}'$ approximates the hidden network $\mathbf{R}$, based how well $\mathbf{R}'$ can reproduce the experimental results produced by $\mathbf{R}$. The *absolute error* of the network hypothesis $\mathbf{R}'$ is then defined as

$$\text{error}_{\text{abs}}(\mathbf{R}') = \frac{\sum_{i=1}^{n} \sum_{j=1}^{n} |r_{ij} - r'_{ij}|}{n^2}, \qquad (4)$$

which is a direct indication of how good the approximation is. Note that this error is not available to the algorithm, but can be used to measure the efficacy of the algorithm itself.

Once all of the genomes have been evaluated, pairs of genomes are selected, and the connection matrix of the genome with higher subjective error is replaced by the connection matrix of the genome with the lower subjective error. The copied matrix is then mutated: one randomly selected element of the matrix is replaced either by a new random value in $[-1, 1]$ (50% probability), or nudged up or down by 0.0001 (50% probability). Crossover is currently not used, but may be implemented in future improvements to the algorithm. For the work reported here, a population size of 1000 is used, and a total of 750 replacements are performed after each generation. Note that a given connection matrix may undergo more than one mutation if it is selected, copied, mutated and then selected again. Once a set of selections, replacements and mutations have occurred, all of the new genomes in the population are evaluated. This process is continued for 30 generations. When the estimation phase terminates at the end of the 30 generations, the $\mathbf{R}'$ with the least subjective error is passed on to the exploration phase.

The exploration phase also begins with a set of randomly-generated genomes, but the EA for this phase maintains genomes that encode input vectors instead of connection matrices. The vectors are initialized with random floating-point values in $[0, 1]$ chosen using a uniform distribution. Each genome is then evaluated as follows. The encoded input vector is applied to the connection matrix $\mathbf{R}'$ obtained from the estimation phase, and an output vector is obtained. The *information* associated with the genome is set to

$$i = 1.0 - \frac{n_0 + n_s}{n} \tag{5}$$

where n_0 is the number of genes in the output vector that have zero concentration, and n_s is the number of genes that have a saturation concentration of 1. The same method of selection and replacement is then applied as described for the estimation phase: the output vectors of genomes with higher information replace the output vectors of genomes with lower information. The new genomes (input vectors) are mutated as follows: a single gene is chosen at random, and replaced with a new concentration chosen from $[0, 1]$ with a uniform distribution. This method of selecting genomes based on their expected information content is an attempt to try to evolve input vectors that, when supplied to the target network, will produce output vectors that contain high information: gene concentrations in the output vector that have either zero or saturation concentration levels indicate less about the state of their regulating genes' concentrations than concentrations between these extrema.

The exploration phase is also executed for 30 generations. When it terminates, the best input vector is supplied to the target network $\mathbf{R}$. This input vector is used by $\mathbf{R}$ to compute a new output vector, which is then passed back into the next iteration of the estimation phase, along with the $j - 1$ previously-generated input/output vector pairs. When the estimation phase is run again, the initial random population is seeded with the best connection matrix found so far in order to accelerate evolution. The entire process—calculation of the output vector from $\mathbf{R}$, execution of the estimation and then the exploration phase, and experiment suggestion—iterates for 100 cycles, leading to 100 experiments performed on the target network.

This same algorithm can be applied for the non-destructive inference of other physical systems. For example we have applied it in the domain of evolutionary robotics. In that case the algorithm was contained within a robot simulator: the algorithm evolved controllers for an actual robot (exploration phase), and also evolved hypotheses regarding possible damage suffered by the actual robot (estimation phase). Based on the sensory feedback from the actual robot, the simulator could both refine its hypothesis regarding what damage had occurred, and modify the controller so that the actual robot could regain as much functionality as possible. Figure 2 compares these algorithms, as well as providing a general framework for how this co-evolutionary algorithm can be applied such that a bidirectional dialogue is maintained between the physical nonlinear system (such as a robot or biological network) and the algorithm.

The next section presents some results generated using this algorithm.

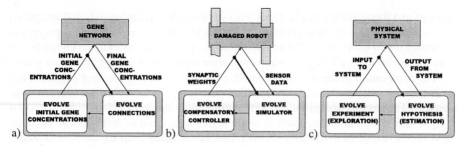

Fig. 2. Instantiations of the estimation-exploration algorithm. a: In order to reverse engineer a genetic network, the estimation phase evolves a connection matrix that describes the wiring of the network. Based on the best evolved connection matrix hypothesis, an input vector is evolved by the exploration phase that should return an information-rich output vector when applied to the hidden network. **b:** A neural network controller is evolved for an actual robot (exploration phase) that allows it to move. If it suffers damage, the estimation phase evolves a robot simulator that explains the damage. **c:** The general framework for applying the algorithm to a nonlinear system.

3   Results

In order to test the algorithm, a set of 40 target networks were generated for various numbers of genes (n) and number of incoming regulatory connections (k). The first 10 networks contained 5 genes ($n = 5$) and each gene was regulated by 2 genes ($k = 2$). The second 10 networks were generated using $n = 5$ and $k = 5$; the third 10 runs using $n = 10$ and $k = 2$; and the final 10 runs using $n = 10$ and $k = 10$.

The algorithm was applied to each of the 40 networks, and the cycle described in the above section was iterated 100 times; thus 100 experiments were performed on the hidden target network. Each pass through either the estimation or exploration phase involved the evolution of a population of 1000 genomes for 30 generations.

Two control algorithms were also applied to the same 40 networks as the proposed algorithm. The first control algorithm was random search: the algorithm performed as before, but instead of replacing genomes of low fitness with genomes of high fitness (for both phases), the genome with low fitness was replaced with a randomly generated genome. In the second control algorithm, the exploration phase is disabled: this phase simply returns a randomly generated input vector. Both control algorithms were executed for the same number of iterations (100), the same population size (1000), and the same number of generations (30) as the proposed algorithm.

Ten independent runs of the algorithm were conducted for each of the 40 hidden networks, and 10 independent runs of both control algorithms were also conducted for each network, leading to a total of $3 \times 10 \times 40 = 1200$ independent runs. Figure 3 shows the evolutionary progress of a typical run of both the proposed algorithm and the second control algorithm (in which the exploration phase is disabled) on a hidden network with $n = 10$ and $k = 10$. Figure 4 reports the resulting output vectors from the 100 experiments suggested by each of these two runs. Figure 5 shows the average performance for the proposed algorithm, compared to the two control algorithms, for the four different types of hidden target networks.

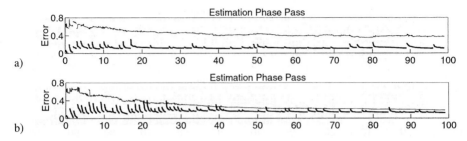

Fig. 3. Sample evolutionary progress for the second control and proposed algorithm. The thin line indicates the absolute error (see equation 4) between the actual connection matrix $\mathbf{R}$ and the best connection matrix evolved during that generation of the estimation phase, $\mathbf{R}'$. Note that this information is not available to the algorithm. The thick line indicates the subjective error of the best genome in the population at that generation according to equation 3. **a**: A sample run of the control algorithm with the exploration phase disabled on one of the hidden networks with $n = 10$ and $k = 10$. **b**: The progress of the proposed algorithm on the same hidden network. The evolutionary progress of the passes through the exploration phase for **b** are not shown.

4   Discussion

As can been seen from Figure 3, the downward-sloping thick curves indicate that during each pass through the estimation phase, the subjective error of the best genome in the population decreases. However after a new experiment has been performed on the hidden target system, the subjective error of the best hypothesis so far (which is used to seed the first generation of the next pass through the estimation phase) tends to increase: this is indicated by the successive curves seen during the first 500 generations of the second control algorithm (Figure 3a) and the first 1000 generations of the proposed algorithm (Figure 3b). This indicates that the new experiment exposed some previously hidden information about the target system that the best hypothesis so far did not account for.

   Importantly, the proposed algorithm tends to generate such information-rich experiments far beyond the 40th experiment, as compared to the control algorithm, in which the experiments lose their explanatory value (the curved subjective error trajectories become flat) before the 20th experiment. Although both algorithms eventually evolve a hypothesis that explains most of the input/output data pairs (indicated by the equally low subjective error curves at the end of the two runs), the proposed algorithm explains more *informative* input/output data pairs, thus leading to a better approximation of the hidden network (the absolute error (thin line) eventually becomes lower for the proposed algorithm, compared to the control algorithm.)

   Figure 4 shows why the proposed algorithm is able to outperform the control algorithm. Between the 20th and 40th experiments, the proposed algorithm has enough informative input/output data pairs to evolve a good approximation of the hidden network. Using this approximation, it can evolve input vectors that produce a greater fraction of informative gene product concentrations: concentrations that fall between the two extremal concentrations of 0 and 1. This is indicated by the greater density of such concentrations (the black squares) in the output vectors seen in Figure 4b, compared to those in the output vectors obtained by the control experiment (Figure 4a). However given a random network, it is more difficult to obtain intermediate concentrations for some genes compared to others: both the proposed and control algorithm obtained only

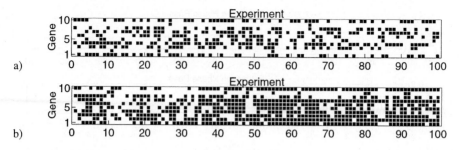

Fig. 4. Sample experimental results derived from the second control and proposed algorithm. Each column represents an output vector obtained when an input vector suggested by either the second control algorithm (**a**) or the proposed algorithm (**b**) was applied to the hidden target network. Blank squares indicate gene product concentrations that were either 0 or 1; dark squares indicate concentrations that fell between these extrema.

sporadic intermediate concentrations for gene 9 for this particular target network. Further improvements to the algorithm will entail changes in equation 5 in order to maximize intermediate concentrations for all genes.

Figure 5d indicates that the proposed algorithm consistently outperforms both the first control algorithm (random search) and the second control algorithm, in which experiments are suggested randomly. This indicates that evolving informative experiments in step with model hypotheses does improve the discovery of the hidden networks. Not surprisingly, both algorithms consistently outperform random search, for all four network types. However, interestingly, evolving informative tests is only beneficial for networks that are comprised of many genes. On further inspection this is not so surprising, because networks with many genes and high connectivity have a higher likelihood of reaching either of the extremal concentrations than smaller, less dense networks.

Figure 6 supports this claim: 1000 random networks were generated using values of n selected from $[2, 30]$ with a uniform distribution, and values of k selected from $[1, n' - 1]$, where n' is an already randomly selected value for n. For each of the 1000 random networks, 10 input vectors were randomly constructed, and the corresponding 10 output vectors were calculated using 1. The average fraction of output concentrations that were either 0 or 1 were computed for each network, and are plotted in Figure6a as a function of the number of genes in the network (n), and in Figure6b as a function of connectivity (k/n). Clearly, the fraction of non-informative output elements increases both with the number of genes, and with connectivity. Thus it becomes increasingly valuable not only to evolve hypotheses, but also to evolve informative experiments for the inference of larger and more complex gene networks.

Most importantly though, unlike the algorithm proposed by Iba [11], the experiments performed here on the target system do not require any internal or disruptive perturbation such as a knockout study: we simply supply a carefully evolved new set of initial conditions. Introducing gene product concentrations into a cell or subjecting the cell to external chemicals, rather than invasive knockout studies, may be a more attractive option for certain model organisms. However more study is required to determine how our input/output concentration data may be translated into actual biological experiments. A second advantage of our algorithm over Iba's algorithm is that it does not require direct comparison between proposed regulatory networks: both hypothesis

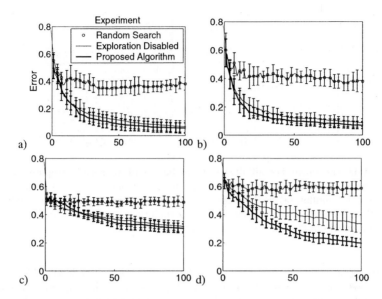

Fig. 5. Average performance of the three algorithms for the four different target network types. Comparative performance of the three algorithms, each averaged over 10 different hidden networks with $n = 5$ and $k = 2$ (**a**). Comparative performance for 10 different networks with $n = 5$ and $k = 5$ (**b**); for 10 networks with $n = 10$ and $k = 2$ (**c**); and for 10 networks with $n = 10$ and $k = 10$ (**d**).

and experiment quality is determined based on experimental output data, rather than on the internal topology of a given network hypothesis.

Human experimenters usually prefer to modify some initial set of conditions only slightly, and then measure the effect on the system in order to infer something about the internal structure of the system. However since the experiment is proposed automatically by the exploration phase, and the results of the experiment are analyzed automatically by the estimation phase, the algorithm is free to suggest input vectors that are very different from those tested on the network before. It is believed that this will greatly speed the inference of the actual system, but remains to be tested rigorously.

4.1   Multiple and Diverse Applications

This mutual dialogue was found to hold in the application of the estimation-exploration algorithm to a completely different problem domain: evolutionary robotics. Instead of evolving networks (estimation) and experiments (exploration) as the algorithm does here, the evolutionary robotics application evolved neural network-based controllers for an actual robot (the exploration phase). When the actual robot suffered some unknown damage and returned sensory data, the estimation phase attempts to evolve a robot simulator that describes the damage. Using this hypothesis the exploration phase then tries to re-evolve a compensatory controller that allows the robot to regain functionality despite its handicap. Figure 7 shows a sample run. For more details, refer to [3].

Just like the gene network inference application shown here, data returned by the physical system (the robot) enhances the estimation phase's ability to predict the state of

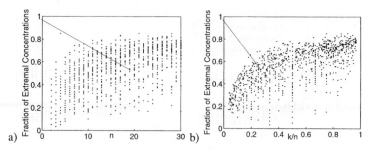

Fig. 6. Sample experimental results derived from the second control and proposed algorithm. Each column represents an output vector obtained when an input vector suggested by either the second control algorithm (**a**) or the proposed algorithm (**b**) was applied to the hidden target network. Blank squares indicate gene product concentrations that were either 0 or 1; blank squares indicate concentrations that fell between these extrema.

the system. The model of the system output by the estimation phase (for the robot application, a hypothesis about the damage suffered) aids the exploration phase by increasing its ability to evolve a controller for the damaged robot.

Also, the application of the estimation-exploration algorithm to robotics allows for recovery after minimal physical testing. Previously, evolutionary algorithms have been used to repair (but not diagnose) damaged physical systems, but they rely on extensive hardware testing (on the order of thousands of evaluations), which is prohibitive for most physical systems [5] [20].

5   Conclusions

In this paper we have presented a co-evolutionary algorithm composed of two phases that automates the process of reverse engineering: the estimation phase evolves a model of the hidden system under study (in this case a genetic network), and the exploration phase evolves an experiment to be performed on the hidden system to gain more information about it. Moreover, as the estimation phase refines its description of the physical system, the exploration phase is able to propose more valuable experiments because its internal model of the physical system is more accurate. In this way the algorithm serves two functions: it infers the internal structure of some hidden system, and proposes increasingly useful experiments to be performed on it. Indeed the idea of automating scientific inquiry has received a lot of interest as of late [14].

Our model has three benefits for gene network inference. First, it does not require invasive, expensive, slow and disruptive experiments such as knockout or lesion studies. Rather, the exploration phase carefully evolves a low-cost experiment (a change in the initial gene product concentrations) that yields a large amount of information about the physical system. Second, the number of experiments performed on the hidden system is minimized, because each proposed experiment is carefully chosen. Finally, the careful selection of experiments becomes increasingly valuable as the hidden networks become larger and more densely interconnected, because large, dense networks often produce information-poor output data. These three points suggest that our algorithm may prove very useful for genetic network inference in particular, and for system biology research in general.

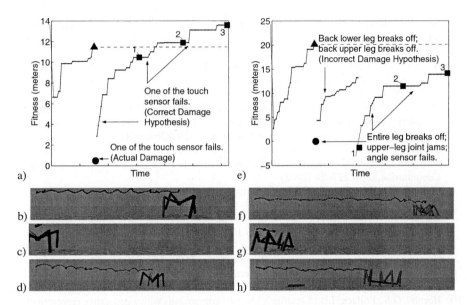

Fig. 7. Two typical robot damage recoveries. a: The evolutionary progress of four passes through the exploratory phase for a quadrupedal robot when it undergoes a failure of one of its touch sensors. The hypotheses generated by the three passes through the estimation phase (all of which are correct) are included. The small circles indicate the fitness (maximum forward locomotion) of the best controller after each generation of the estimation phase. The triangle shows the fitness of the first evolved controller on the physical robot (the behavior of the 'physical' robot (which is also simulated for now) with this controller is shown in **b**); the large circle shows the fitness of the robot after the failure occurs (the behavior is shown in **c**); the squares indicate the fitness of the physical robot for each of the three hardware trials (the behavior of the 'physical' robot during the third trial is shown in **d**). **e-h** The recovery of a hexapedal robot when it experiences severe, compound damage. The trajectories in **b-d** and **f-h** show the change in the robot's center of mass over time (the trajectories are displaced upwards for the sake of clarity).

We have also shown how general this approach is by describing and showing how it can be applied to a completely different nonlinear physical system: a robot. Future prospects for this work will include improving the selection of experiments so that even more information is extracted from the system, generalizing the model still further by applying it to various other nonlinear physical systems, and formulating a rigorous set of guidelines for how to apply the algorithm to a large class of physical systems.

Acknowledgment. This work was supported by the National Academies Keck Futures Grant for interdisciplinary research, number NAKFI/SIG07.

References

1. Arkin, A., Shen, P., Ross, J.: A test case for correlation metric construction of a reaction pathway from measurements. In: *Science* **277**: 1275–1279 (1997)

2. Arkin, A., Ross, J., McAdams, H.H.: Stochastic kinetic analysis of developmental pathway bifurcation in phage lambda-infected Escherichia coli cells. In: *Genetics* **149**: 1633–1648 (1998)
3. Bongard, J.C., Lipson, H.: Automated damage diagnosis and recovery for remote robotics. To appear in: *Proceedings of the 2004 International Conference on Robotics and Automation (ICRA)*, New Orleans, USA (2004)
4. Bongard, J.C., Pfeifer, R.: Repeated structure and dissociation of genotypic and phenotypic complexity in Artificial Ontogeny. In: Spector, L., Goodman, E.D. (eds.): *Proceedings of The Genetic and Evolutionary Computation Conference*: 829–836 (2001)
5. Bradley, D.W. and Tyrrell, A.M.: Immunotronics: novel finite-state-machine architectures with built-in self-test Using self-nonself differentiation. In *IEEE Transactions on Evolutionary Computation*, 6(3): 227–38 (2002)
6. Chen, T., He, H.L., Church, G.M.: Modeling gene expression with differential equations. In: *Pacific Symposium on Biocomputing* **4**: 29–40 (1999)
7. Davidson, J.W., Savic, D.A., Walters, G.A.: Symbolic and numerical regression: Experiments and applications. In: *Information Sciences* **150**(1-2): 95–117 (2003)
8. D'haeseleer, P., Liang, S., Somogyi, R.: Genetic network inference: From co-expression clustering to reverse engineering. In: *Bioinformatics* **16**(8): 707–726 (2000)
9. Hill, R.J., Sternberg, P.W.: Cell fate patterning during *C. elegans* vulval development. In: *Development* **suppl.** 9–18 (1993)
10. Hornby, G.S., Pollack, J.B.: Creating high-level components with a generative representation for body-brain evolution. In: *Artificial Life* **8**(3): 223–246 (2002)
11. Iba, H., Mimura, A.: Inference of a gene regulatory network by means of interactive evolutionary computing. In: Information Sciences **145**: 225–236 (2002)
12. de Jong, H.: Modeling and simulation of genetic regulatory systems: A literature review. In: *J. Comput. Biol.* **9**(1): 69–105 (2002)
13. Kauffman, S.A.: *The Origins of Order*, Oxford University Press. Oxford, UK. (1993)
14. King, R. D., Whelan, K. E., Jones, F. M., Reiser, P. G. K., Bryant, C. H., Muggleton, S. H., Kell, D. B., Oliver, S. G.: Functional genomic hypothesis generation and experimentation by a robot scientist. In: *Nature* **427**: 247–252 (2004)
15. Kitano, H.: *Foundations of Systems Biology*, MIT Press. Cambridge, MA. (2001)
16. Koza, J.R., Mydlowec, W., Lanza, G., Yu, J., Keane, M.A.: Reverse engineering of metabolic pathways from observed data using genetic programming. In: Altman, R. B. et al (eds.): *Pacific Symposium on Biocomputing*: 434–445 (2001)
17. Koza, J.R.: *Genetic Programming: On the Programming of Computers by Natural Selection*, MIT Press. Cambridge, MA. (1992)
18. Lewis, E.B.: Clusters of master control genes regulate the development of higher organisms. In: *Journal of the American Medical Association* **267**: 1524–1531 (1992)
19. Lipson, H. and Pollack, J.B.: Automatic design and manufacture of artificial lifeforms. In *Nature*, 406: 974–978 (2000)
20. Mahdavi, S.H. and Bentley, P.J.: An evolutionary approach to damage recovery of robot motion with muscles. In *Seventh European Conference on Artificial Life (ECAL03)*: 248—255 (2003)
21. Miller, J.F., Job, D., Vassilev, V.K.: Principles in the evolutionary design of digital circuits–Part I. In: *Journal of Genetic Programming and Evolvable Machines* **1**(1): 8–35 (2000)
22. Miller, J.F., Job, D., Vassilev, V.K.: Principles in the evolutionary design of digital circuits–Part II. In: *Journal of Genetic Programming and Evolvable Machines* **3**(2): 259–288 (2000)
23. Mjolsness, E., Sharp, D.H., Reinitz, J.: A connectionist model of development. In: *J. Theor. Biol.* **152**: 429–454 (1991)
24. Sakamoto, E., Iba, H.: Identifying gene regulatory network as differential equation by genetic programming. In: *Genome Informatics*: 281–283 (2000)
25. Sims, K.: Evolving 3D morphology and behaviour by competition. In: *Artificial Life IV*: 28–39 (1994)

26. Tominaga, D., Okamoto, M., Kami, Y., Watanabe, S., Eguchi, Y.: Nonlinear numerical optimization technique based on a genetic algorithm. http://www.bioinfo.de/isb/gcb99/talks/tominaga
27. Weaver, D. C.: Modeling regulatory networks with weight matrices. In: *Proc. Pacific Symp. Bioinformatics* **5**: 251–258 (2000)
28. Yao, X.: Evolving artificial neural networks. In: *Proceedings of the IEEE* **87**(9): 1423–1447 (1999)
29. Young, R.M.: *Mind, Brain and Adaptation in the Nineteenth Century. Cerebral Localization and its Biological Context from Gall to Ferrier*, Clarendon Press. Oxford, UK. (1970)

A Genetic Approach for Gene Selection on Microarray Expression Data

Yong-Hyuk Kim[1], Su-Yeon Lee[2], and Byung-Ro Moon[1]

[1] School of Computer Science & Engineering, Seoul National University
Shillim-dong, Kwanak-gu, Seoul, 151-742 Korea
{yhdfly, moon}@soar.snu.ac.kr
[2] Program in Bioinformatics, Seoul National University
Shillim-dong, Kwanak-gu, Seoul, 151-742 Korea
suylee@soar.snu.ac.kr

Abstract. Microarrays allow simultaneous measurement of the expression levels of thousands of genes in cells under different physiological or disease states. Because the number of genes exceeds the number of samples, class prediction on microarray expression data leads to an extreme "curse of dimensionality" problem. A principal goal of these studies is to identify a subset of informative genes for class prediction to reduce the curse of dimensionality. We propose a novel genetic approach that selects a subset of predictive genes for classification on the basis of gene expression data. Our genetic algorithm maximizes correlation between genes and classes and minimizes intercorrelation among genes. We tested the genetic algorithm on leukemia data sets and obtained improved results over previous results.

1 Introduction

With the development of microarray technology, scientists can now examine multiple genome-wide gene expression patterns at the same time. Microarrays have been powerful experimental tools for extracting functional information from genome [5] [15]. As well as the diagnosis of disease, the classification of disease types is one of the most useful applications of microarrays. Recently, microarrays were used to profile the global gene expression patterns of normal and transformed human cells in several tumors, such as leukemia [11]. These researches may shed light on identifying biomarkers for cancer classification (molecular diagnosis). A wide-spread technique for microarray data analysis is clustering analysis [1] [3] [4] [10] [9] [13]. Clustering analysis groups genes that have correlated patterns of expression which can provide insight into gene-to-gene interactions and gene functions.

While microarrays have been extensively used in the gene expression profiling of tumor cells or tissues, successful applications of the microarray technology in cancer classification rely on data mining tools. This is because, among a lot of genes examined, only a fraction present distinct profiles for different classes of samples. Thus, it is critical to have computational tools that are capable of

K. Deb et al. (Eds.): GECCO 2004, LNCS 3102, pp. 346–355, 2004.
© Springer-Verlag Berlin Heidelberg 2004

identifying a subset of informative genes embedded in a large dataset that is contaminated with high-dimensional noise.

Microarray data consist of a large number of genes (parameters) and relatively a small number of samples. It makes a "curse of dimensionality" problem; i.e., too many parameters for the data points. To reduce this problem, we try to identify a small subset of relevant genes. The major topic of this paper is to introduce an approach for gene selection with the help of a genetic algorithm.

Since typical microarray data consist of a large number of genes, many subsets of genes that distinguish between different classes of samples may exist. Our strategy is to find many such subsets and then evaluate the relative importance of genes for sample classification by examining inter-correlations of gene pairs in the subset. When selected genes were used for sample classification using a test set, samples were classified with accuracy. Other computational methods that select a subset of genes for sample classification were also developed [11] [2] [14] [12] [16]. The patterns of gene selection and the classification reliability of the selected genes using an independent test set are analyzed. We examine the sensitivity of gene selection results to the assignment of samples to the training set. We do this by dividing the dataset into a training set and test set in different ways, resulting in different training and test sets for the same data. Each training set is used to select a subset of genes.

In this paper, leukemia dataset is used as a benchmark dataset. We report the detailed analysis of the leukemia data using a genetic approach to find a subset of genes that can discriminate between acute lymphoblastic leukemia (ALL) and acute myeloid leukemia (AML). The results are compared with previous works.

The remainder of this paper is organized as follows. In Section 2, we summarize dataset and class predictor used in this paper. We propose a genetic approach for gene selection in Section 3. In Section 4, we present experimental results. Finally, we make our conclusions in Section 5.

2   Preliminaries

Recently, Golub *et al.* [11] proposed a method for selecting a subset of discriminative genes for sample classification. They successfully applied neighborhood analysis to identify a subset of genes that discriminates between AML and ALL, using a separation measure. The 50 genes that best distinguish AML from ALL in 38 training set samples were chosen as a class predictor that correctly classified 36 of the 38 training set samples. When these genes were subsequently used to predict the class of the test samples, 29 of the 34 samples were correctly classified with high confidence. In our implementation, four of the five samples were not classified (undecided) and one of the five samples was misclassified.

2.1   Dataset

The original leukemia dataset was downloaded from the web site[1]. The data contain the expression levels of 6,817 genes across 72 samples, of which 47 was classified as ALL and 25 as AML [11]. We divided the dataset into a training set (first 38 samples) and a test set (34 samples) following Golub *et al.* [11]. The training set was used to obtain a subset of genes that can discriminate between AML and ALL. The 50 most informative genes obtained using the training set were subsequently used in validation, to predict the classification of the test samples.

2.2   Class Predictor

Golub *et al.* [11] developed a procedure that uses a fixed subset of informative genes and makes a prediction based on the expression level of these genes in a new sample. Figure 1 shows the structure of their class predictor. Each informative gene casts a weighted vote for one of the classes, with the magnitude of each vote dependent on the expression level in the new sample and the degree of that gene's correlation with the class distinction. The votes are summed to determine the winning class as well as a prediction strength (PS), which is a measure that ranges from -1 to 1. The sample is assigned to the winning class if PS exceeds a predetermined threshold, and is considered undecided otherwise. We used the threshold of 0.3 following [11].

3   A Genetic Algorithm

We propose a genetic algorithm (GA) for gene selection to choose a good subset of genes. It selects genes based on the training set. It conducts a search for a good subset of genes using a correlation-based evaluation function. The search space with n genes has $2^n - 1$ elements if all nonempty subsets are considered. If the number of genes to be selected is predetermined, the optimal subset of size k can be found by enumerating and testing all possibilities, which requires $\binom{n}{k}$ tests. Then, this makes the problem intractable. Our GA provides an alternative search method to find a good subset with a predetermined size.

The dataset consists of 6,817 genes. If all the genes are considered as a candidate of informative genes, the problem size becomes intractable. So, we used the gene set filtered by the correlation ρ'. We considered three cases: $|\rho'| > 0.8$ (136 genes), $|\rho'| > 0.7$ (299 genes), and $|\rho'| > 0.5$ (980 genes).

The dataset is divided into two independent sets: the training set and the test set. Our GA runs on the training set until a termination criterion is satisfied and selects a predefined number of genes (50 genes in our experiments[2]). After our GA selects a subset of genes, the predictive model is tested on the test set.

---

[1] http://www.genome.wi.mit.edu/MPR

[2] To compare with the previous work [11] under the same condition, we fixed the number of genes to 50.

```
ClassPredictor(sample x = (x₁, x₂, ..., x_{#genes}))
{
```

$// x_i$: expression level of gene i

$V_{AML} \leftarrow 0, V_{ALL} \leftarrow 0;$

for each informative gene g,

$\mu_{AML}(g) \leftarrow$ mean expression levels of g for the samples in AML;

$\mu_{ALL}(g) \leftarrow$ mean expression levels of g for the samples in ALL;

$\sigma_{AML}(g) \leftarrow$ SD expression levels of g for the samples in AML;

$\sigma_{ALL}(g) \leftarrow$ SD expression levels of g for the samples in ALL;

$\rho'(g, C) \leftarrow (\mu_{AML}(g) - \mu_{ALL}(g))/(\sigma_{AML}(g) + \sigma_{ALL}(g));$

$v_g \leftarrow \rho'(g, C) \cdot (x_i - (\mu_{AML}(g) + \mu_{ALL}(g))/2);$

if $(v_g > 0)$ $V_{AML} \leftarrow V_{AML} + v_g;$

else $V_{ALL} \leftarrow V_{ALL} - v_g;$

PS $\leftarrow (V_{AML} - V_{ALL})/(V_{AML} + V_{ALL});$

if $(|PS| < threshold)$ **return** *undecided*;

else if $(PS > 0)$ **return** *AML*;

else return *ALL*;

```
}
```

Fig. 1. The structure of class predictor [11]

3.1   Genetic Operators

The general structure of steady-state genetic algorithms is used in our GA.

- *Encoding*: In this problem, a chromosome is represented by binary encoding. A gene has value one if the gene belongs to the informative gene subset; otherwise, it has value zero.
- *Initialization*: We first create p subsets at random. The only constraint on a chromosome is that the number of 1's should be 50. We set the population size p to be 100.
- *Selection*: We assign to each chromosome in the population a fitness value calculated from its object value. We use the roulette-wheel-based *proportional selection* scheme.
- *Crossover and Mutation Operators*: A crossover operator creates a new offspring by combining parts of the two parents. In our experiments, we use one-point crossover and use element-swap mutation that swaps the values of a random pair of genes. After the crossover, an offspring may not satisfy the constraint. It then selects random points on the chromosome and changes the required number of 1's to 0's (or 0's to 1's). This adjustment also produces some mutation effect.
- *Replacement*: After generating an offspring and applying a local optimization on it, we replace a member of the population with the offspring. We use the replacement scheme of [6]. The offspring tries to first replace the more similar parent, measured in bitwise difference; if it fails, then it tries to replace the other parent (replacement is done only when the offspring is better than one

Table 2. Data Comparison in Bootstrap Samples

Method	Training data		Independent data	
	Undecided Ave($\sigma/\sqrt{n}$)	Error Ave($\sigma/\sqrt{n}$)	Undecided Ave($\sigma/\sqrt{n}$)	Error Ave($\sigma/\sqrt{n}$)
Random	2.10(0.12)	0.12(0.03)	4.69(0.22)	0.61(0.09)
Golub *et al.* [11]	1.70(0.10)	0.04(0.02)	3.08(0.16)	0.48(0.06)
Greedy	1.77(0.11)	0.12(0.03)	3.08(0.14)	0.67(0.07)
DGA	0.93(0.09)	0.00(0.00)	3.60(0.19)	**0.29**(0.05)

Sampling in Random and DGA: $|\rho'| > 0.7$.
Average # of candidate genes = 174.69.
# of training samples = 38.
# of independent samples = 34.
Average over 100 datasets.

5   Discussion

As more genes were included, leading to the curse-of-dimensionality problem, the number of misclassified samples increased. This emphasizes that not all expression data are relevant to the distinction between ALL and AML. It is evident that not all genes are relevant to sample classification. Thus, the identification of informative genes is essential. The important issue is that microarray data consist of a large number of genes and a small number of samples, and, as a result, a great number of distinct and effective classifiers may exist for the same training set. Most of current literature methods seek a single subset of discriminative genes. Often, the informative genes identified for a given dataset vary from method to method. In conclusion, a number of methods have been developed for sample classification based on gene expression data. Our algorithm selected a good subset of genes and improved the predictive quality of the existing prediction model. As the quantitative aspect of the microarray technology is improved and computational methods that mine the resulting large dataset are further developed, this study will have a notable impact on biology and related areas.

Acknowledgments. This study was supported by a grant of the International Mobile Telecommunications 2000 R&D Project, Ministry of Information & Communication, Republic of Korea. This was also partly supported by grant No. (R01-2003-000-10879-0) from the Basic Research Program of the Korea Science and Engineering Foundation, and by Brain Korea 21 Project. The ICT at Seoul National University provided research facilities for this study.

References

1. U. Alon, N. Barkai, D. A. Notterman, K. Gish, S. Ybarra, D. Mack, and A. J. Levine. Broad patterns of gene expression revealed by clustering analysis of tumor and normal colon tissues probed by oligonucleotide arrays. *Proc. Natl Acad. Sci. USA*, 96:6745–6750, 1999.
2. A. Ben-Dor, L. Bruhn, N. Friedman, I. Nachman, M. Schummer, and Z. Yakhini. Tissue classification with gene expression profiles. In *The Fourth International Conference on Computational Molecular Biology (RECOMB2000)*. ACM Press, New York, 2000.
3. A. Ben-Dor, R. Shamir, and Z. Yakhini. Clustering gene expression patterns. *J. Comput. Biol.*, 6:281–297, 1999.
4. M. Bittner, P. Meltzer, and J. Trent. Data analysis and integration: of steps and arrows. *Nature Genetics*, 22:213–215, 1999.
5. P. O. Brown and Botstein D. Exploring the new world of the genome with DNA microarrays. *Nature Genetics*, 21:33–37, 1999.
6. T. N. Bui and B. R. Moon. Genetic algorithm and graph partitioning. *IEEE Trans. on Computers*, 45(7):841–855, 1996.
7. B. Efron. *The jacknife, the bootstrap, and other resampling plans*. Society for Industrial and Applied Methematics, 1982.
8. B. Efron and R. Tibshirani. *Cross-validation and the bootstrap: Estimating the error rate of a prediction rule*. Dept. of Statistics, Stanford University, 1995.
9. G. Getz, E. Levine, and E. Domany. Coupled two-way clustering analysis of gene microarray data. *Proc. Natl Acad. Sci. USA*, 97:12079–12084, 2000.
10. G. Getz, E. Levine, E. Domany, and M. Q. Zhang. Superparamagnatic clustering of yeast gene expression profiles. *Physica A*, 279:457–464, 2000.
11. T. R. Golub, D. K. Slonim, P. Tamayo, C. Huard, M. Gaasenbeek, J. P. Mesirov, H. Coller, M. L. Loh, J. R. Downing, M. A. Caligiuri, C. D. Bloomfield, and E. S. Lander. Molecular classification of cancer: class discovery and class prediction by gene expression monitoring. *Science*, 286:531–537, 1999.
12. I. Guyon, J. Weston, S. Barnhill, and V. Vapnik. Gene selection for cancer classification using support vector machines. *Machine Learning*, 46(1-3):389–422, 2002.
13. E. Hartuv, A. O. Schmitt, J. Lange, S. Meier-Ewert, H. Lehrach, and R. Shamir. An algorithm for clustering cDNA fingerprints. *Genomics*, 66:249–256, 2000.
14. L. Li, T. A. Darden, C. R. Weinberg, and L. G. Pedersen. Gene assessment and sample classification for gene expression data using a genetic algorithm/k-nearest neighbor method. *Combinatorial Chemistry & High Throughput Screening*, 4:727–739, 2001.
15. D. J. Lockhart and E. A. Winzeler. Genomics, gene expression and DNA arrays. *Nature*, 405:827–836, 2000.
16. H. Iba S. Ando. Artificial immune system for classification of gene expression data. In *Genetic and Evolutionary Compatation Conference*, pages 1926–1937, 2003.
17. J. W. Sammon, Jr. A non-linear mapping for data structure analysis. *IEEE Transactions on Computers*, 18:401–409, 1969.
18. D. Whitley and J. Kauth. Genitor: A different genetic algorithm. In *Rocky Mountain Conference on Artificial Intelligence*, pages 118–130, 1988.

Fuzzy Dominance Based Multi-objective GA-Simplex Hybrid Algorithms Applied to Gene Network Models

Praveen Koduru[1], Sanjoy Das[1], Stephen Welch[2], and Judith L. Roe[3]

[1] Electrical and Computer Engineering
[2] Department of Agronomy
[3] Division of Biology
Kansas State University
Manhattan, KS 66506

Abstract. Hybrid algorithms that combine genetic algorithms with the Nelder-Mead simplex algorithm have been effective in solving certain optimization problems. In this article, we apply a similar technique to estimate the parameters of a gene regulatory network for flowering time control in rice. The algorithm minimizes the difference between the model behavior and real world data. Because of the nature of the data, a multi-objective approach is necessary. The concept of fuzzy dominance is introduced, and a multi-objective simplex algorithm based on this concept is proposed as a part of the hybrid approach. Results suggest that the proposed method performs well in estimating the model parameters.

1 Gene Regulatory Network Models

Molecular geneticists are rapidly deciphering the genomes of an increasing number of organisms. As of November 2003, 166 organisms had completely sequenced genomes with another 775 in progress [1]. The current challenge is to understand how the genes in each organism interact with each other and the environment to determine the characteristics (*i.e.*, the *phenotype*) of the organism. In the agricultural contexts familiar to the authors, this is called the "genotype to phenotype" or "GP" problem. In [2] it has been stated that this problem is the most significant issue confronting crop improvement efforts today.

For over 40 years plant physiologists, systems engineers, and computer scientists have been employing "top-down" analysis methods to predict plant phenotypes based on varietal and environmental inputs [3, 4]. Recently, a "bottom" up approach has been applied [5, 6, 7] that models gene interactions directly at the expression level. Small groups of one to four interacting genes can synthesize a wide variety of signal processing functions including Boolean logic gates, linear arithmetic units, delays, differentiators, integrators, oscillators, coincidence detectors, and bi-stable devices [8]. This is consistent with the apparent small-scale modularity of gene networks [9]. Models of this type extrapolate phenotypes by explicitly tracking the status of key genetic developmental switches, accumulators, *etc.*

K. Deb et al. (Eds.): GECCO 2004, LNCS 3102, pp. 356–367, 2004.

Models of this type require efficient, multi-dimensional, multi-objective, derivative-free, global methods for parameter estimation. The problem is characterized by high dimensionality due to the large numbers of genes. Multi-objective optimization methods are appropriate because (1) multiple data types (continuous, discrete, and/or categorical) for both dependent and independent variables make the design of a single objective function problematic, (2) individual data sets come from different sources and may contain within- or between-set inconsistencies not apparent in the metadata, and (3) the models are incomplete and, therefore, may not be equally consistent with every data set. Because actual biophysical systems cannot harbor internal inconsistencies, the Pareto fronts associated with these problems are ideally single points. However, when data and/or model inconsistency exists, the size of the front is a useful measure of its magnitude. Finally, nonlinearities and data discontinuities can lead to exceptionally rough, multi-modal response surfaces (e.g., [7]) that mandate global, derivative free methods.

The following sections of this paper present a new algorithm that posses these features. The algorithm is described and then the algorithm is tested with the following single-gene model that demonstrates the features just described. In [25] the levels of messenger RNA was measured every 3h under short-days (SD, 9h) and long-days (LD, 15h) for *HEADING DATE 1 (Hd1)*, an important flowering time control gene in rice (Oryza sativa). In [8] the authors modeled this data with the equation(s):

$$\frac{d}{dt}(Hd1) = \left.\begin{matrix} R_D \\ R_L \end{matrix}\right\} g_{NN}(C(t)) - (Hd1) \left\{\begin{matrix} \lambda_D \\ \lambda_L \end{matrix}\right. \tag{1, 2}$$

where R's and λ's are constants and L and D denote light and dark periods. The clock input is C(t) = A*Sin($2\pi/p$ + θ) + μ, where A is amplitude, p is period, θ is phase angle, μ is a phase factor and $g_{NN} = \dfrac{1}{1+\exp(-c)}$ [5]. The state variable, Hd1, is dimensionless as expression levels are routinely normalized. The parameters have to be found such that model satisfies both SD and LD data with minimal MSE error with experimental data. So the approach of multi-objective optimization is used to find the possible solutions. Agronomic research on this point is underway. Thus, possible objective functions are the MSE between the model predicted SD and LD time series data with actual data obtained experimentally.

2 The Multi-objective Evolutionary Approach

Evolutionary algorithms have emerged as one of the most popular approaches for the complex optimization problems [10]. They draw upon Darwinian paradigms of evolution to search through the solution space (the set of all possible solutions). Starting with a set (or population) of solutions, in each generation of the algorithm, new solutions are created from older ones by means of two operations, mutation and crossover. Mutation is accomplished by imparting a small, usually random perturbation to the solution. In a manner similar to the Darwinian paradigm of survival of the fittest, only the better solutions are allowed to remain in a population, the degree of optimality of the solution being assessed through a measure called fitness.

When dealing with optimization problems with multiple objectives, the conventional concept of optimality does not hold good [11, 12, 13, 14]. Hence, the concepts of dominance and Pareto-optimality are applied. Without a loss of generality, if we assume that the optimization problem involves minimizing each objective $e_i(.)$, $i = 1...M$, a solution u is said to dominate over another solution v iff $\forall i \in \{1, 2, ..., M\}$, $e_i(u) \le e_i(v)$ with at least one of the inequalities being strict, i.e. for each objective, u is better than or equal to v and better in at least one objective. This relationship is represented as $u \prec v$. In a population of solution vectors, the set of all non-dominating solutions is called the Pareto front. In other words, if S is the population, the Pareto Front Γ is given by,

$$\Gamma = \{u \in S \mid \forall v \in S, \neg(v \succ u)\} \tag{3}$$

The simplistic approach of aggregating multiple objectives into a single one often fails to produce good results. It produces only a single solution. Multi-objective optimization on the other hand involves extracting the entire Pareto front from the solution space. In recent years, many evolutionary algorithms for multi-objective optimization have been proposed [14, 15, 16, 17].

We propose a hybrid algorithm that combines genetic algorithms (GAs), an evolutionary algorithm, with a well-known approach for function optimization known as the simplex algorithm [18]. While several GA-simplex algorithms have been proposed, our version is the only one that is equipped to carry out multi-objective optimization. This is accomplished by means of a concept, that we introduce, called fuzzy dominance.

2.1 Fuzzy Dominance

We first introduce the concept of fuzzy dominance. We assume a minimization problem involving M objective functions $e_i(\cdot)$, $i = 1...M$. The solution space, the set of all possible solution vectors, will be denoted as $\Psi \subset \Re^n$, where n is the dimensionality of the multi-objective problem.

Definition 1 *Fuzzy i-dominance by a solution*
Given a monotonically non-decreasing function $\mu_i^{dom} : \Psi \to [0,1]$, $i \in \{1, 2, ..., n\}$ such that $\mu_i^{dom}(0) = 0$, solution $u \in \Psi$ is said to i-dominate solution $v \in \Psi$, if and only if $e_i(u) < e_i(v)$. This relationship will be denoted as $u \prec_i^F v$. If $u \prec_i^F v$, the degree of fuzzy i-dominance is equal to $\mu_i^{dom}(e_i(v) - e_i(u)) \equiv \mu_i^{dom}(u \prec_i^F v)$. Fuzzy dominance can be regarded as a fuzzy relationship $u \prec_i^F v$ between u and v [19].

Definition 2 *Fuzzy dominance by a solution*
Solution $u \in \Psi$ is said to fuzzy dominate solution $v \in \Psi$ if and only if $\forall i \in \{1, 2, ..., M\}, u \prec_i^F v$. This relationship will be denoted as $u \prec^F v$. The degree of

fuzzy dominance can be defined by invoking the concept of fuzzy intersection [19]. If $u \prec^F v$, the degree of fuzzy dominance $\mu^{dom}\left(u \prec^F v\right)$ is obtained by computing the intersection of the fuzzy relationships $u \prec_i^F v$ for each i. The fuzzy intersection operation is carried out using a family of functions called t-norms, denoted with a $*$. Hence,

$$\mu^{dom}\left(u \prec^F v\right) = \overset{M}{\underset{i=1}{*}} \mu_i^{dom}(u \prec_i^F v).$$
(4)

Definition 3 *Fuzzy dominance in a population*

Given a population of solutions $S \subset \Psi$, a solution $v \in S$ is said to be fuzzy dominated in S iff it is fuzzy dominated by any other solution $u \in S$. In this case, the degree of fuzzy dominance can be computed by performing a union operation over every possible $\mu^{dom}\left(u \prec^F v\right)$, carried out using t-co norms, that are denoted with a $\oplus$. Hence the degree of fuzzy dominance of a solution $v \in S$ in the set S is given by,

$$\mu^{dom}(S \prec^F v) = \underset{u \in S}{\oplus} \mu^{dom}(u \prec^F v).$$
(5)

Using the above definitions, one can redefine the Pareto front as the set of all solutions in S that are not dominated in S. In other words,

$$\Gamma = \left\{u \in S \mid \neg(S \prec^F u)\right\}.$$
(6)

2.2 The Simplex Algorithm

A simplex in n-dimensions consists of $n+1$ solutions u_k, $k = \{1,2,\ldots n+1\}$ [18]. In a plane, this corresponds to a triangle as shown in Figure 1. The solutions are evaluated in each step and the worst solution w is identified. The centroid of the simplex is then evaluated, excluding the worst solution and the worst point is reflected along the centroid. If c is centroid such that $nc = \sum_k u_k - w$, the reflected solution is

$$r = c + (c - w)$$
(7)

Usually, the worst point w is replaced with the reflected point r in the simplex, but if the r is better than any solution in the simplex, the simplex is further expanded as,

$$r_e = c + \eta(c - w)$$
(8)

where η is called the expansion coefficient. However, if the reflected solution r is worse than w, the simplex is contracted and the reflected solution is placed on the same side of the centroid. When solution r is not worse than w, but worse than any other solution in the simplex, the simplex is still contracted, but the reflection is allowed to remain on the other side of the simplex. Reflection is carried out as follows,

$$r_c = c \pm \kappa(c - w) \tag{9}$$

In the above equation, κ is called the contraction coefficient. Solution w is replaced with the new one, r, r_e, or r_c in the next step. The simplex algorithm is allowed to run for multiple steps before it converges.

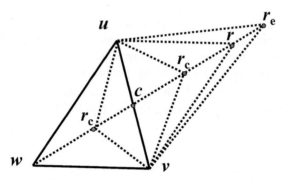

Fig. 1. A simplex in 2 dimensions

2.3 The GA-Simplex Hybrid Algorithm

Since genetic algorithms use a population of individuals, they are capable of performing an exploratory search over the entire solution space. In many complex optimization problems, they are hybridized with local search operations. The local algorithms improve single solutions by exploiting local information around the vicinity of the solutions. Hybrid algorithms combine the advantages of exploration and exploitation forms a new area of research. Hybrid algorithms that use the simplex algorithm discussed earlier, for local search are popular in continuous optimization problems [20, 21, 22, 23].

One of the hybrid approaches proposed uses the simplex algorithm as a post-processor for improving the solutions obtained by a GA [23]. The simplex approach has been used as an operator to improve the solutions obtained from the genetic operations of mutation and crossover in accordance with Lamarckian theory [20]. In our approach, the simplex has been used as an operation within each iteration of the genetic algorithm as in [20, 22]. But only a fraction of the next generation is obtained by carrying out crossover and mutation with the solutions in the present population. The rest of the population is established by using the simplex algorithm. Our approach is similar to [21]. However unlike the approach of [21] where only the elite individuals are used by the simplex algorithm, our approach uses solutions that are chosen from the entire population. Figure 2 is a schematic that shows the approach used in the present research.

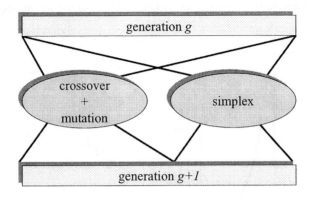

Fig. 2. A schematic of the hybrid approach

In order to apply the simplex algorithm, a total of $n+1$ solutions must be selected from the population S. This is done by first picking at random the $n+1$ solutions from S and computing their centroid C. Any solution vector u at a distance $\|c-u\| > \rho_{simplex}$ is rejected and replaced with another one drawn at random, where $\rho_{simplex}$ is the radius parameter of the simplex approach and $\|\cdot\|$ is the Euclidean norm. This process is repeated until either all the sample solutions fit within the radius $\rho_{simplex}$, or the total replacements exceed r_{max}. After selecting the initial vectors, the simplex algorithm is run for a total of α times. The best $n+1$ solutions are selected to be inserted into the population in the next generation. The genetic algorithm is applied by selecting individuals based on the fuzzy dominance and assigning performing standard crossover and mutation operations that are discussed in the next section.

3   Implementation

3.1   Fuzzy Dominance

In order to calculate the fuzzy dominance relationship between two solution vectors, trapezoidal membership functions were used. Therefore,

$$\mu_i^{dom}\left(u \prec_i^F v\right) = \begin{cases} 0 & \text{if } e_i(v) - e_i(u) < 0, \\ (e_i(v) - e_i(u))/p_i & \text{if } 0 \leq e_i(v) - e_i(u) < p_i, \\ 1 & \text{otherwise.} \end{cases} \qquad (10)$$

In the above equation, the parameter p_i determines the length of the linear region of the trapezoid for the objective function $e_i(\cdot)$. The t-norm and t-co norms were defined as $x * y = xy$ and $x \oplus y = x + y - xy$. Both are standard forms of operators [19]. This choice has an interesting property that makes it attractive for our application. While solutions located away from the Pareto front in any population S are always fuzzy dominated in S, those that are more towards the periphery are less dominated. Figure 3 explains this clearly.

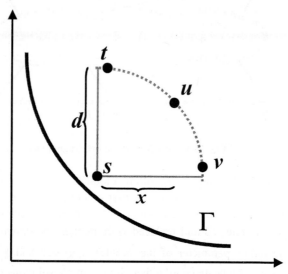

Fig. 3. Effect of fuzzy dominance

Figure 3 shows three solutions t, u and v that are all located at the same distance d from another solution s. Assume that this distance lies within the linear region of the trapezoidal membership function and that the distances of u from s along the horizontal and vertical directions are x and $\sqrt{d^2 - x^2}$ respectively. Hence, $\mu^{dom}\left(u \prec^F v\right) = x\sqrt{d^2 - x^2}$. This will be maximized when $x = \frac{1}{\sqrt{2}}d$, i.e. when u is as far away from the Pareto front as possible. In other words, t and v will be less dominated than u. This property can be extended to more than two objectives and aids the genetic algorithm that uses fuzzy dominance as a measure of inverse fitness during the selection. Solutions that are more peripherally located along the front are preferred to those closer to the center. Hence it assists the GA in maintaining a diverse Pareto front that is as spread as possible. The simplex algorithm works efficiently also, since it identifies worse solutions based on this measure as well and hence, when the simplex is 'flipped' along the centroid, the movement is kept approximately orthogonal to the Pareto front.

The fuzzy dominances of all solutions in the population are calculated at the beginning of each iteration and stored as a two dimensional array, each entry of which is a fuzzy dominance relationship between two solution vectors. However, in order to simplify the calculations, within the simplex algorithm, the fuzzy dominances are considered among the $n+1$ solutions only that are selected by the simplex algorithm.

3.2 Genetic Operators

A tournament selection was implemented in the GA that selected λ individuals at random from the population with replacement, and picked the one with the least fuzzy dominance. An offspring t, was computed from two parents u and v in the following manner,

$$t = \zeta u + (1 - \zeta)v \tag{11}$$

where ζ is a uniformly distributed random number in [0, 1].

Solutions were mutated with a probability of β, by adding a random number with zero mean, that followed a Gaussian distribution with a spread σ, according to,

$$u = u + N(0, \sigma) \tag{12}$$

Elitism was implemented by selecting the non-dominated points in a population and copying them to an elite-set. Selection is done on a union of elite-set and current population.

4 Results

We have applied the proposed method to estimate the parameters of the genetic network model in (1,2). The equations involve a total of 8 parameters. Additionally, two initial conditions for both a LD and SD periods exist, which makes up a total of 10 parameters to be computed. The network is simulated and the predicted Hd1 values for the SD as well as LD periods are compared with corresponding experimental data. The mean squared error (MSE) in each of the LD and SD periods predictions with the experimental data is computed, and the goal is to find a set of parameters that simultaneously minimize the MSE for the two periods.

In order to compare the algorithm performance with a standard multi-objective algorithm we applied SPEA as explained in [24] to the same problem. SPEA was chosen over other algorithms since it is one of the most recently proposed algorithms for multi-objective optimization that is also fairly easy to implement. In our test runs we have used a population size of 100 for the proposed method. The mutation rate was set at 0.4 and crossover rate at 0.7. These were found to be optimal for the parameter estimation problem after multiple trails. One simplex was implemented in each generation and the parameters used are $\alpha = 10$, $\eta = 1.5$ and $\kappa = .5$. SPEA was implemented as explained in [24] with a population size of 70. In SPEA clustering was invoked to reduce Pareto fronts whose size exceeded 30 individuals [24]. Tournament selection was used in each algorithm. Both the algorithms were run for a total of 30,000 function evaluations. Figure 4 shows the Pareto front obtained by both algorithms. It is clear that, unlike the proposed algorithm, SPEA was unable to converge to the Pareto front. SPEA was observed to be slower in convergence than our algorithm. Multiple runs were done with different data sets on both algorithms and it was evident that SPEA was unable to reach the Pareto front in the given number of function evaluations. However, in case of our algorithm with the help of simplex and fuzzy dominance a significantly better front was obtained in the same number of function evaluations. We believe that when estimating parameters of genetic network the fitness landscape contains good minima with basins large enough for the simplex algorithm to converge to with little effort. Figure 5 shows the time series simulation of one of the solutions in the Pareto front along with the experimental data. Figures 6, 7 show the convergence plots of minimum of each of the objectives vs. function evaluations for different mutation rates of 0.1, 0.4 and 0.7. A high mutation rate of 0.4 produced the best possible results. Further research is necessary to explain this phenomenon.

5 Conclusions

In this paper, we have proposed an effective approach to perform parameter estimation of gene regulatory network models. The algorithm, which hybridizes a multi-objective version of the simplex algorithm, based on a newly introduced concept of fuzzy dominance, with a standard genetic algorithm, is shown to converge well for a genetic network model of flowering time control in *Oryza sativa*. Our algorithm consistently outperformed a standard multi-objective optimization approach, SPEA.

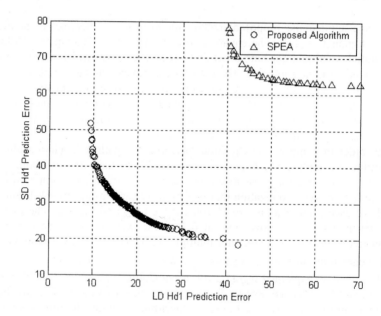

Fig. 4. Pareto front obtained by the proposed algorithm on Heading date prediction and its comparison with SPEA for same initial population.

The proposed algorithm combines the exploratory nature of genetic algorithms with the exploitative behavior of simplex search to carry out parameter estimation of gene regulatory models effectively. We believe that this approach can be applied to similar multi-objective optimization problems as well. Future work will be directed in testing the proposed method for parameter estimation problems with similar network model with additional objectives and more extensive comparison with more multi-objective methods.

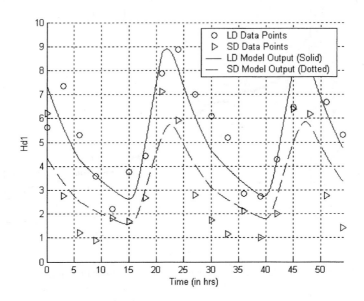

Fig. 5. Time series plot of Hd1 for one of the solutions in Pareto front from proposed algorithm with experimental data

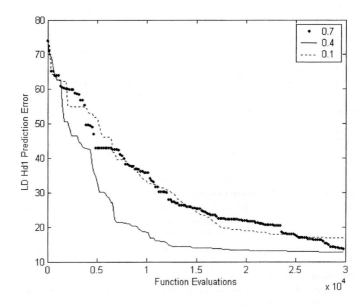

Fig. 6. Convergence of LD Hd1 Prediction error vs. Function evaluations for different mutation rates

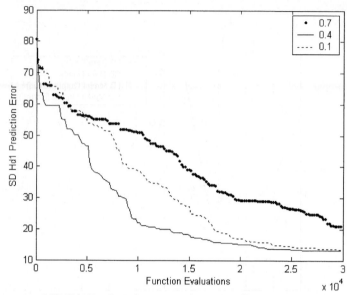

Fig. 7. Convergence of SD Hd1 Prediction error vs. Function evaluations for different mutation rates

Acknowledgment. This research was supported in part by USDA grant 2003-35304-13217 to Kansas State University.

References

1. http://ergo.integratedgenomics.com/GOLD/
2. Cooper, M., Chapman, S.C., Podlich, D.W. & Hammer, G.L. *In Silico Biol.* 2 (2002), 151-164.
3. Sinclar, T.R. and Seligman, N.G.: Crop modelling: From infancy to maturity. Agron. J. 88 (1966) 698-704.
4. Hammer, G. T., Sinclair, S. Chapman, E. van Oostererom.: On systems thinking, systems biology and the *in silico* plant. Plant Physiology. Scientific Correspondence (2004) (*in press*).
5. Welch, S.M., Roe, J.L., and Dong, Z.: A genetic neural network model of flowering time control in Arabidopsis thaliana. Agron. J. (2003) 95, 71-81.
6. Welch, S.M., Dong, Z., and Roe, J.L.: Modelling gene networks controlling transition to flowering in *Arabidopsis*. Proceedings of the 4th International Crop Science Congress, Brisbane, Au. Sep 26 – Oct 1, 2004. (*Under review*).
7. Dong, Z. Incorporation of genomic information into the simulation of flowering time in *Arabidopsis thaliana*. Ph.D. dissertation, Kansas State University (2003).
8. Welch, S.M., Roe, J.L., Das, S., Dong, Z., R. He, M.B. Kirkham.: Merging genomic control networks with soil-plant-atmosphere-continuum (SPAC) models. Agricultural Systems (2004b) (*submitted*).
9. Ravasz, E., Somera, A., Mongru, D., Oltvai, Z., and Baraba´si, A.L.:. Hierarchical organization of modularity in metabolic networks. Science 297 (2002) 1551-1555.

10. D. E. Goldberg, Genetic Algorithms in Search, Optimization, and Machine Learning, Addison Wesley, Reading, MA (1989).
11. Fonseca, C.M., Fleming, P.J.: An overview of evolutionary algorithms in multiobjective optimization, Evolutionary Computation, Vol. 3, no. 1 (1995) 1-16, Spring.
12. Coello Coello, C.A. : A comprehensive survey of evolutionary-based multiobjective optimization techniques, Knowledge and Information Systems, Vol. 1, no. 3 (Aug 1999) 269-308.
13. Van Veldhuizen, D.A., Lamont, G.B.: Multiobjective evolutionary algorithms: Analyzing the state-of-the-art, Evolutionary Computation, Vol. 8, no. 2, (2000) 125-147.
14. Jaszkiewicz, A.: Do multiple-objective metaheuristics deliver on their promises? A computational experiment on the set-covering problem, IEEE Transactions on Evolutionary Computation, Vol. 7, no. 2 (Apr. 2003) 133-143.
15. Haiming, L., Yen,G.G.: Rank-density-based multiobjective genetic algorithm and benchmark test function study, IEEE Transactions on Evolutionary Computation, Vol. 7, no. 4 (Aug. 2003).
16. Knowles, J., Corne, D.: Properties of an adaptive archiving algorithm for storing nondominated vectors, IEEE Transactions on Evolutionary Computation, Vol. 7, no. 2 (Apr. 2003) 100- 116.
17. Zitzler, E., Thiele, L., Laumanns, M., Fonseca, C.M., da Fonseca, V.G.: Performance assessment of multiobjective optimizers: An analysis and review, IEEE Transactions on Evolutionary Computation, Vol. 7, no. 2 (Apr. 2003) 117-132.
18. Nelder, J.A., Mead, R., A simplex method for function minimization, Computer Journal, Vol 7, no. 4 (1965) 308-313.
19. Mendel, J.M.: Fuzzy logic systems for engineering: A tutorial, Proceedings of the IEEE, Vol 83, No. 3 (March 1995) 345-377.
20. Renders, J.M., Flasse, S.P.: Hybrid methods using genetic algorithms for global optimization, IEEE Transactions on Systems, Man and Cybernetics Part-B, , Vol. 28, no. 2 (Apr. 1998) 73-91.
21. Yen, J., Liao, J.C., Lee, B., Randolph, D.: A hybrid approach to modeling metabolic systems using a genetic algorithm and simplex method, IEEE Transactions on Systems, Man and Cybernetics Part-B, Vol. 7, no. 1 (Feb. 2003) 243-258.
22. Bersini, H.: The immune and chemical crossovers, IEEE Transactions on Evolutionary Computation, Vol. 6, no. 3 (June 2002) 306-313.
23. "Simulation and evolutionary optimization of electron-beam lithography with genetic and simplex-downhill algorithms", IEEE Transactions on Evolutionary Computation, pp. 69-82, Vol. 7, no. 1, Feb. 2003.
24. Zitzler, E., Thiele, L.: Multiobjective Evolutionary Algorithms: A comparative case study and the strength Pareto approach, IEEE Transactions on Evolutionary Computation, Vol. 3, no. 4 (Nov. 1999) 257-271.

2.4   The Genetic Operators

The genetic reproduction operators adopted are listed below:

- The two-point crossover (2-X) operator generates two offspring by exchanging the genes between two randomly chosen cut points in the parents chromosomes.
- The simulated binary crossover (SBX), which assigns more probability for offspring to remain closer to their parents than away from them, generates two offspring as described in [23].
- The non-uniform mutation (NUM) operator [24], when applied to an individual x_i at generation gen, mutates a randomly chosen variable x_i^j according to

$$x_i^j \leftarrow \begin{cases} x_i^j + \Delta(gen, b^j - x_i^j) \text{ if } \tau = 0 \\ x_i^j - \Delta(gen, x_i^j - a^j) \text{ if } \tau = 1 \end{cases} \quad (4)$$

where a^j and b^j are respectively the lower and upper bounds for the variable x^j, τ is randomly chosen as 0 or 1, and the function $\Delta(gen, y)$ is defined as

$$\Delta(gen, y) = y(1 - r^{(1 - \frac{gen}{maxgen})^\beta}) \quad (5)$$

with r randomly chosen in $[0, 1]$ and the parameter β set to 2. It is clear that this operator reduces the amplitude of the perturbations as the number of generations increases.

2.5   Selection and Insertion Schemes

Due to the high modality of the fitness landscape for the docking problem, a critical issue is the maintenance of useful population diversity in order to permit the investigation of several high fitness regions in parallel and reduce the chances of convergence to low quality local optima. Among the techniques proposed to deal with high modality landscapes [25], *fitness sharing*, introduced by Holland [18] and enhanced by Goldberg & Richardson [26], has the drawback of requiring knowledge about the search space (such as distance between optima) in order to set the dissimilarity threshold. *Crowding*, introduced by De Jong [27] and enhanced by Mahfoud [28] insert new offspring in the population replacing similar ones. We are particularly interested in the idea of restricted tournament selection (RTS) proposed by Harik [17] which nicely blends with our SSGA.

In this work, we have tested three selection-insertion schemes: (i) rank-based selection [15] of parents with replacement of the worst individual in the population, (ii) restricted tournament selection (RTS) [17], and (iii) a new modified RTS scheme.

In the RTS scheme, parent individuals are selected randomly from the population and the new offspring generated is placed in the population replacing the closest existing individual found in a tournament of size w, provided that the new individual is better than the winner of the tournament. The metric used was euclidean norm weighted so that all genes have the same influence in spite of their different ranges. It is important to point out that $w = 1000$, does not mean that all individuals in the population will take part in the tournament. As the selection is random, one individual can be drafted more than once or not be drafted at all. We have also implemented a new modified RTS scheme where two tournaments are made. In the first (resp. second) tournament $w1$ (resp. $w2$) individuals that are better (resp. worse) than the new offspring are drafted. The winner of the first tournament, $CBetter$, is the closest individual (in the genotype space) to the new offspring, among the $w1$ individuals drafted in the first tournament. The winner of the second tournament, $CWorse$, is the closest individual (in the genotype space) to the new offspring, among the $w2$ individuals drafted in the second tournament.

The offspring is then inserted in the population in the following way:

- If the new offspring is closer to $CWorse$ than $CBetter$, then $CWorse$ is replaced by the newly generated offspring
- Else, If the RMSD between the new offspring and $CBetter$ is greater than 2.0 Å, then $CWorse$ is replaced by the new offspring. Otherwise, the new offspring is discarded.

In both cases the new offspring replaces $CWorse$. The modified RTS scheme uses information both from genotype space (chromosome) and the phenotype space (RMSD of all atoms coordinates). The criterion RMSD ≤ 2.0 Å is used to avoid an offspring insertion when a very similar and better individual already exists in a particular region of the search space. In this work we used $w1=w2$. If $w1 = w2 = 100\%$ is used, it means that the tournament size $w1$ is equal to the number of individuals that are better than the offspring, and the tournament size $w2$ is equal to the number the individuals that are worse than it.

3   Results

We have tested the SSGA on four HIV1 protease-ligand complexes. The experimental structures were obtained from the Protein Data Bank (PDB). The number of dihedral angles/torsions, total number of degrees of freedom (dimension) and the PDB file code, for each ligand molecule, are shown in Table 1. The structures and the dihedral angles of the four ligands tested are shown in Figure 1.   The grid is centered in the protein active site, with 23 Å of dimension

Table 1. HIV-1 protease ligands complexes tested

Ligand	Torsions	Dimension	PDB ID
NELFINAVIR	12	19	1ohr
INDINAVIR	14	21	1hsg
SAQUINAVIR	15	22	1hxb
RITONAVIR	20	27	1hxw

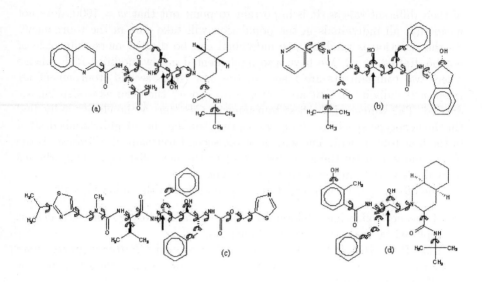

Fig. 1. Structural formulæ of HIV1 protease ligands and dihedral angles considered:
(a) Saquinavir; (b) Indinavir; (c) Ritonavir; (d) Nelfinavir. Arrow: reference atom

in each direction, and a spacing of 0.25 Å. We are interested in the performance of the SSGA in identifying the experimental binding mode of the ligand
molecule in the protein active site. To each ligand 30 independent runs were
performed. The CPU time for each SSGA run varied from 10 to 13 minutes on
a 2.0 GHz Pentium 4 with 256 MB of RAM. The algorithm success is measured
by the RMSD between the crystallographic structure (from the corresponding
PDB file) and the structure found by the algorithm. A structure with a RMSD
≤ 2.0 Å is classified as docked and that is considered a good result. A structure
with a RMSD ≤ 2.5 Å is classified as partially docked, but for large ligands,
with more than 15 dihedral angles, that is still considered a good result. The
success ratio is the number of structures found with RMSD ≤ 2.0 Å in 30 runs.
For the three selection-insertion schemes tested, we used a population of 1000
individuals, 1.000.000 energy evaluations, and probability of 0.15 for two-point
crossover, 0.15 for SBX crossover, and 0.7 for non-uniform mutation. The flexible
ligand docking results using linear ranking selection are shown in Table 2. We
tested the standard RTS using a tournament size $w = 500$ and $w = 1000$. The
docking results using standard RTS are shown in Table 3. The modified RTS
was tested with a tournament size $w1 = w2 = 50\%$ and $w1 = w2 = 100\%$. The
docking results using the modified RTS with 50% and 100% tournament sizes
are shown in Table 4.

Table 2. Docking results using linear rank selection

Ligand	Lowest Energy[a]	Mean Energy[a]	Mean RMSD (Å)	Success Ratio (%)
NELFINAVIR	−57.53	−2.43	5.776	6.6
INDINAVIR	−62.97	32.96	6.049	3.3
SAQUINAVIR	−65.12	−25.17	4.764	13.3
RITONAVIR	−87.69	−7.99	5.305	10.0

[a] kcal/mol

Table 3. Docking results using standard RTS and two tournament sizes (w)

Ligand	w	Lowest Energy[a]	Mean Energy[a]	Mean RMSD (Å)	Success Ratio (%)
NELFINAVIR	500	−58.13	−52.98	1.101	83.3
	1000	−58.12	−53.02	1.427	73.3
INDINAVIR	500	−62.78	−51.16	2.573	63.3
	1000	−62.87	−51.78	1.765	76.7
SAQUINAVIR	500	−65.75	−58.26	1.225	83.3
	1000	−65.67	−62.11	0.726	86.7
RITONAVIR	500	−107.27	−84.26	2.583	50.0
	1000	−105.69	−71.91	3.137	6.7

[a] kcal/mol

Table 4. Docking results using the modified RTS and two tournament sizes (w)

Ligand	w	Lowest Energy[a]	Mean Energy[a]	Mean RMSD (Å)	Success Ratio (%)
NELFINAVIR	50%	−58.12	−50.97	2.009	63.3
	100%	−58.12	−55.55	0.394	93.3
INDINAVIR	50%	−62.87	−54.13	2.147	73.3
	100%	−62.79	−53.58	2.204	76.7
SAQUINAVIR	50%	−65.69	−61.64	1.031	90.0
	100%	−65.72	−61.44	0.824	93.3
RITONAVIR	50%	−107.11	−88.21	2.291	50.0
	100%	−107.34	−89.56	2.167	53.3

[a] kcal/mol

3.1   RMSD Analysis

In the results shown in Tables 3 and 4, the success ratio was the number of times that the lowest energy structure found by the algorithm corresponds to the respective crystallographic structure (RMSD ≤ 2.0 Å). In many cases, we found structures with a higher energy, but with a lower RMSD than the best (minimum energy) solution. The results shown in the RMSD analysis use the same final population employed in the previous energy analysis. In the RMSD analysis the best solution in each run is the solution with lowest RMSD relative to the experimental structure, and not the structure with the lowest energy, as done in the previous energy analysis. The RMSD analysis of docking results using the standard RTS are shown in Table 5. The RMSD analysis of docking results using the modified RTS are shown in Table 6.

Table 5. RMSD analysis of docking results using the standard RTS and two tournament sizes (w)

Ligand	w	Lowest Energy[a]	Mean Energy[a]	Mean RMSD (Å)	SR[b] (%) ≤ 2.0 Å	SR[b] (%) (2.0,2.5] Å
NELFINAVIR	500	−58.10	−45.85	0.693	90.0	0.0
	1000	−58.09	−48.68	0.657	86.7	6.7
INDINAVIR	500	−62.61	−10.99	1.103	83.3	10.0
	1000	−62.24	−41.63	0.734	93.3	6.7
SAQUINAVIR	500	−65.28	−50.22	0.649	100.0	0.0
	1000	−65.24	−56.86	0.483	100.0	0.0
RITONAVIR	500	−106.95	−54.35	1.447	80.0	17.0
	1000	−102.29	−33.88	1.809	66.7	23.3

[a]kcal/mol

[b]Success Ratio

Table 6. RMSD analysis of docking results using the modified RTS and two tournament sizes (w)

Ligand	w	Lowest Energy[a]	Mean Energy[a]	Mean RMSD (Å)	SR[b] (%) ≤ 2.0 Å	SR[b] (%) (2.0,2.5] Å
NELFINAVIR	50%	−58.11	−49.57	1.126	73.3	6.7
	100%	−58.08	−54.96	0.296	96.7	0.0
INDINAVIR	50%	−62.71	−50.83	1.111	83.3	0.0
	100%	−62.42	−47.48	0.840	90.0	3.3
SAQUINAVIR	50%	−65.32	−60.31	0.877	93.3	3.3
	100%	−65.28	−58.45	0.574	96.7	3.3
RITONAVIR	50%	−106.84	−83.66	1.530	73.3	10.0
	100%	−106.71	−81.48	1.265	76.7	20.0

[a]kcal/mol

[b]Success Ratio

4   Discussion

The results show that with the implementation of the RTS technique we obtained a substantial increase in the algorithm performance. Comparing the results obtained using the linear rank selection (LRS) and using the standard RTS (energy analyses), we found that the mean success ratio increased from 8.4% (Table 2) to 70.1% (tournament size of 500 individuals, Table 3) and to 60.9% (tournament size of 1000 individuals, Table 3). These results indicate the standard RTS technique as a promising methodology for flexible ligand docking problems. The modified RTS (with an insertion criterion based on the RMSD between the ligand conformations) also shows a good performance to find structures close to the experimental structure with the lowest energy. The success ratio obtained using the modified RTS (tournament size of 50%, Table 4) ranges from 50% to 90% with a mean success ratio of 69.2%, and (using a tournament size of 100%, Table 4) from 53% to 100% with a a mean success ratio of 79.2%. Using the standard RTS the results show that a smaller tournament size produces a better result, while in the modified RTS the use of a tournament size of 100% showed to be the best choice. The modified RTS shows a slightly better performance than the standard RTS to find solutions closer to the experimental one and with better mean RMSD.

Considering all solutions in the final population (see RMSD analysis section), for both standard and modified RTS, we observe an increase in the success ratio regarding the experimental structure. The mean success ratio (including all ligands) obtained using the standard RTS are 88.4% and 86.7%, using a tournament size of 500 individuals (Table 5) and a tournament size of 1000 individuals (Table 5), respectively. Using the modified RTS, the mean success ratios obtained are 80.9% and 90.0%, using a tournament size of 50 % (Table 6) and a tournament size of 100% (Table 6), respectively. Once more the best results were obtained using a tournament size of 500 individuals for the standard RTS, and a tournament size of 100% for the modified RTS. Analyzing the results shown in Tables 5 and 6 we observe that RITONAVIR (the largest and most flexible ligand tested) shows the greater increase in performance when we consider the RMSD $\leq$ 2.5 Å criterion for computing the success ratio. In fact, for larger and highly flexible ligand molecules a RMSD $\leq$ 2.5 Å from the experimental structure can be considered a good result. Applying this criterion for all ligands, we observe that the mean success ratios are 95% and 96.7% for the standard RTS (tournament size of 50%, Table 5) and for the modified RTS (tournament size of 100%, Table 6) respectively. For all ligands a success ratio greater than 90% was obtained.

The results obtained in this work showed that the implementation of a multisolution RTS technique can be a very valuable approach in the flexible docking problem when dealing with highly flexible ligand molecules which are usually associated with a very complex energy hypersurface. Moreover, there are important advantages in applying a multisolution strategy in this type of problem. Usually the ligand-docking strategies approximate the real problem in the following points: (i) the absence of important factors associated with the ligand-receptor energy function (*e.g.*, entropic and solvatation effects); (ii) the absence of explicit water molecules which can intermediate ligand-receptor hydrogen bonds and; (iii) the receptor is usually considered rigid or partially rigid. Following a multisolution docking strategy, the best distinct solutions can be used as starting points in more sophisticated and computationally expensive strategies (*e.g*, explicit solvent molecular dynamics simulations). Secondly, in real world drug design research projects, the ligand to be docked is only a drug prototype which will be probably modified several times in order to account for several chemical and pharmacological properties (*e.g*, toxicity, metabolic stability, synthetic tractability, etc.). In this sense, finding and analyzing several ligand-receptor binding modes can increase the possibilities of successful improvements in a drug prototype molecule. A more specific analysis considering the relation between the final population diversity (number and quality of solutions) and the SSGA/RTS (standard and modified) docking parameters is under progress and will be reported elsewhere.

Acknowledgements. The authors thank the Carcará Project at LNCC for computational resources, CNPq (grants no. 302299/2003-3 and 402003/3003-9), MCT/LNCC/PRONEX, and FAPERJ (grant no. E26/171.401/01).

The authors would also like to thank the reviewers for the corrections and suggestions which helped improve the quality of the paper.

References

1. Gane P.J., Dean P.M.: Recent Advances in Structure-Based Rational Drug Design. Current Opinion in Structural Biology **10** (2000) 401–404
2. Marrone T.J., Briggs J.M., McCammon J.A.: Structure-based Drug Design. Annu Rev Pharmacol Toxicol **37** (1997) 71–90
3. Wang R., Lu Y., Wang S.: Comparative Evaluation of 11 Scoring Functions for Molecular Docking. J. Med. Chem. **46** (2003) 2287–2303
4. Diller D.J., Verlinde C.L.M.J.: A Critical Evaluation of Several Global Optimization Algorithms for the Purpose of Molecular Docking. J Comp Chem 20 **16** (1999) 1740–1751
5. McConkey B.J., Sobolev V., Edelman M.: The Performance of Current Methods in Ligand-Protein Docking. Current Science 83 **7** (2002) 845–855
6. Brooijmans N., Kuntz I.D.: Molecular Recognition and Docking Algorithms. Annu. Rev. Biophys. Biomol. Struct. **32** (2003) 335–373
7. Ewing T.J.A., Kuntz I.D.: Critical Evaluation of Search Algorithms for Automated Molecular Docking and Database Screening. J Comp Chem **18** (1997) 1175–1189
8. Jones G., Willett P., Glen R.C., Leach A.R., Taylor R.: Development and Validation of a Genetic Algorithm for Flexible Docking. J Mol Biol **267** (1997) 727–748
9. Morris G.M., Goodsell D.S., Halliday R.S., Huey R., Hart W.E., Belew R.K., Olson A.J.: Automated Docking Using a Lamarckian Genetic Algorithm and an Empirical Binding Free Energy Function. J. Comp. Chem. 19 **14** (1998) 1639–1662
10. Carlson H.A., McCammon J.A.: Accommodating Protein Flexibility in Computational Drug Design. Molecular Pharmacology **57** (2000) 213–218

11. Wong C.F., McCammon J.A.: Protein Flexibility and Computer-Aided Drug Design. Annu. Rev. Pharmacol. Toxicol. **43** (2003) 31–45

12. Osterberg F., Morris G.M., Sanner M.F., Olson A.J., and Goodsell D.S.: Automated Docking to Multiple Target Structures: Incorporation of Protein Mobility and Structural Water Heterogeneity in AutoDock. PROTEINS: Structure, Function and Genetics **46** (2002) 34–40

13. Hart W.E., Rosin C., Belew R.K., Morris G.M.: Improved Evolutionary Hybrids for Flexible Ligand Docking in AutoDock. Optimization in Computational Chemistry and Molecular Biology. (eds.) C.A. Floudas & P.M. Pardalos (2000) 209–229

14. Thomsen R.: Flexible Ligand Docking Using Evolutionary Algorithms: Investigating the Effects of Variation Operators and Local Search Hybrids. BioSystems **72** (2003) 57–73

15. Whitley D.: The GENITOR Algorithm and Selective Pressure. Proc. of the Third Int. Conf. on Genetic Algorithms and their Applications. (eds) J.D. Schaffer, Morgan Kaufmann, San Mateo, CA (1989)

16. van Gunsteren W.F., Berendsen H.J.C.: Groningen Molecular Simulation (GROMOS) Library Manual. Biomos, Groningen (1987)

17. Harik G.R.: Finding Multimodal Solutions Using Restricted Tournament Selection. Proc. of the Sixth Int. Conf. on Genetic Algorithms. (eds.):Larry Eshelman, Morgan Kaufmann, San Francisco, CA (1995) 24–31

18. Holland J.H.: Adaptation in Natural and Artificial Systems. University of Michigan Press, Ann Arbor, MI (1975)

19. Goldberg D.E.: Genetic Algorithms in Search, Optimization and Machine Learning. Addison-Wesley, New York (1989)

20. Maillot P.G.: In Graphics Gems. A. S. Glassner, (ed.): Academic Press, London, (1990) 498

21. Pascutti P.G., Mundim K.C., Ito A.S., Bisch P.M. Polarization Effects on Peptide Conformation at Water-membrane Interface by Molecular Dynamics Simulation. J. Comp. Chem. 20 **9** (1999) 971–982

22. Arora, N., Jayaram B.: Strength of Hydrogen Bonds in α-helices. J. Comp. Chem. 18 **9** (1997) 1245–1252

23. Deb K., Beyer H.: Self-adaptive Genetic Algorithms with Simulated Binary Crossover. Technical report no. CI-61/99, University of Dortmund, Department of Computer Science/LS11 (1999)

24. Michalewicz Z.: Genetic Algorithms + Data Structures = Evolution Programs. Springer-Verlag, New York (1992)

25. Sareni B., Krähenbühl" L.: Fitness Sharing and Niching Methods Revisited. IEEE Transactions on Evolutionary Computation 2 **3** (1998) 97–106

26. Goldberg D., Richardson J.: Genetic Algorithms with Sharing for Multimodal Function Optimization. Proc. of the Second Int. Conf. on Genetic Algorithms (eds.): J. J. Grefenstette, Lawrence Erlbaum Associates, Hillsdale, New Jersey (1987) 41–49

27. De Jong K.: Analysis of Behavior of a Class of Genetic Adaptive Systems. PhD Thesis, The University of Michigan (1975)

28. Mahfoud S.W.: Crowding and Preselection Revisited, Parallel Problem Solving from Nature 2. (eds.): Reinhard Männer and Bernard Manderick, North-Holland, Amsterdam (1992) 27–36

A GA Approach to the Definition of Regulatory Signals in Genomic Sequences

Giancarlo Mauri, Roberto Mosca, and Giulio Pavesi

Bioinformatics and Natural Computing Group
University of Milano–Bicocca
mauri,roberto.mosca,pavesi@disco.unimib.it

Abstract. One of the main challenges in modern biology and genome research is to understand the complex mechanisms that regulate gene expression. Being able to tell when, why, and how one or more genes are activated could provide information of inestimable value for the understanding of the mechanisms of life. The wealth of genomic data now available opens new opportunities to researchers. We present how a method based on genetic algorithms has been applied to the characterization of two regulatory signals in DNA sequences, that help the cellular apparatus to locate the beginning of a gene along the genome, and to start its transcription. The signals have been derived from the analysis of a large number of genomic sequences. Comparisons with related work show that our method presents different improvements, both from the computational viewpoint, and in the biological relevance of the results obtained.

1 Introduction

One of the main challenges in modern biology in general, and in the analysis of genome data in particular, is to understand the complex mechanisms that regulate the expression (i.e. the activation) of the genes of a given organism. The expression of a gene starts when the corresponding region in the double–stranded DNA sequence is *transcribed* into a single stranded RNA sequence, that later on is translated into the protein encoded by the gene (see Fig. 1). At any given time, not all the genes present in the genome of a given organism are expressed, but only a subset of them: this accounts for example for cell differentiation, that is, the genes that are active in a neural cell are different from those active, say, in a muscle cell. Moreover, genetic diseases are often caused by alterations occurring not within the genes themselves, but in the apparatus governing their activation, thus leading to anomalous expression levels. Transcription is initiated when one or more dedicated molecules called *transcription factors* (TFs) (that are proteins in turn encoded by some genes in the genome) bind to the DNA region adjacent to the gene (region called *promoter* of the gene), causing the double–strand to open and thus allowing the transcription of the gene. In some cases, the binding of a TF has the opposite effect, blocking transcription. Each TF recognizes a set of specific targets along the sequence, that is, short nucleotide fragments it can

K. Deb et al. (Eds.): GECCO 2004, LNCS 3102, pp. 380–391, 2004.

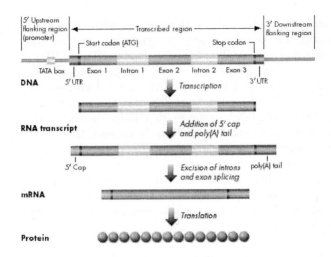

Fig. 1. Gene expression: from DNA to RNA to protein.

bind to, called *binding sites*. Binding sites thus function as regulatory signals in the genome.

Since the experimental characterization and identification of the binding sites of a given TF is a long and painstaking work, the huge amount of genomic data now available to researchers provides a invaluable source of information for shedding further light on this process. If we identify in the promoter of a gene known TF binding sites, we may better understand by whom, and when a gene is activated. Unfortunately, also the computational description and discovery of the binding sites of a given TF is far from being an easy task. The main difficulty lies in the very fact that each TF does not recognize a single binding site, but a set of them, that, although similar, differ in their nucleotide composition. This set is usually referred to as *signal* or *motif*.

1.1   TATA Box and CAP Binding Sites

Usually, each TF influences a relatively small number of organism–specific genes. However, it has been experimentally observed that in virtually every eukaryotic organism a large number of genes present two characteristic signals located in the proximity of the point of transcription initiation (*transcription start site*, TSS). The first one, called *TATA box*, is located along the DNA double helix about 25-30 pairs of nucleotides (base pairs, bp) before (upstream of) the TSS. Its name derives from the fact that when it was discovered the majority of its instances contained the stretch of nucleotides TATA. The TATA box is bound by a large complex of some 50 different proteins, including Transcription Factor IID (TFIID) – a complex of the TATA-binding protein (TBP, which is the part of the molecule that binds the TATA box) and 14 other proteins – and Transcription Factor IIB (TFIIB). The CAP signal (also called Initiator, or Inr) instead straddles the TSS, and often cooperates with the TATA box in starting the transcription of the gene. There is significant experimental evidence that also this signal is bound by Transcription Factor IID (TFIID) [1]. While most of the regulatory signals are spread at different positions in genome regions adjacent to genes, both the TATA and CAP motifs have instead a very precise location, perhaps with the purpose of directing the transcription apparatus to the right spot along the DNA sequence, signalling where exactly the gene begins and transcription has to start. While largely present, these two signals are not ubiquitous: in every organism, some genes contain both the signals, some either one, and some neither of them.

Giving a precise characterization of these two signals, as well as a classification of genes according to which signal they are regulated by, could thus provide substantial information on a basic mechanism for gene expression, present virtually in every organism.

2   Describing Binding Sites

The computational characterization of regulatory signals in genomic sequences is usually composed by two different steps. First, we need a method to describe a signal, that is, to represent the set of valid binding sites for a given TF. One possible way is to describe them with a *frequency matrix*, that is built as follows. Let B_S be a set of DNA fragments (all having the same length m) that a transcription factor is known to bind. Since all the fragments have the same length we can align them obtaining m columns. Let us consider the first column, containing the first nucleotide of each fragment. We can count the number of times each nucleotide is present in the column, and compute its frequency. The same operation can be performed for the other columns. In this way, a $4 \times m$ matrix P can be built, where the element $P(i,j)$ is the frequency of nucleotide i ($i \in \{A, C, G, T\}$) in the j-th column of the alignment:

$$P(i,j) = \frac{n_{i,j}}{N} \tag{1}$$

where $n_{i,j}$ is the number of times nucleotide i is found in the j-th position of the fragments considered, and N is the overall number of binding sites used. Given a set of promoter sequences of genes known (or suspected) to be regulated by the same transcription factor, the problem of finding its binding sites can be defined as the problem of finding the best frequency matrix. The basic idea is to find the matrix whose frequencies differ most from those that would be obtained by putting together random fragments from the sequences. Different measures for this task have been proposed so far, with some success [2,3].

Then, the frequency matrix can be used to determine whether a given DNA fragment can be considered a candidate binding site for the corresponding TF. In fact, given a frequency matrix P describing a signal, and a generic fragment S^f of length m the probability of the fragment to be a binding site for the corresponding TF can be estimated by

$$P_{\text{match}}(S^f) = \prod_{j=1}^{m} P(S_j^f, j) \tag{2}$$

where S_j^f is the j-th nucleotide of S^f. Fragment S^f is a said to be a binding site when $P_{match}(S^f)$ is greater than the probability with which S^f appears in the genome:

$$P_{bg}(S^f) = \prod_{j=1}^{m} b_{S_j^f} \tag{3}$$

where b_i is the frequency with which nucleotide i appears in the genome. The b_i values can be estimated for example by considering the frequency of each nucleotide in the sequence examined, or, since these data are now available, in the whole genome the sequence considered belongs to.

Stated in another way, a fragment S^f can be suspected to be a binding site if

$$\frac{P_{match}(S^f)}{P_{bg}(S^f)} = \prod_{j=1}^{m} \frac{P(S_j^f, j)}{b_{S_j^f}} > 1 \Leftrightarrow \log \frac{P_{match}(S^f)}{P_{bg}(S^f)} = \sum_{j=1}^{m} \log \frac{P(S_j^f, j)}{b_{S_j^f}} > 0 \tag{4}$$

so to have negative score for fragments that are not bound by the TF. That is, we evaluate the probability of a sequence fragment to be bound by the TF described by matrix P by checking whether it fits the description of the matrix $(P_{match}(S^f))$, and by comparing this result with the expected frequency $(P_{bg}(S^f))$ of the fragment in the genome.

Given a sequence S of arbitrary length $L > m$ we can say that the signal characterized by the frequency matrix P occurs at position t in the sequence if the fragment of length m starting at position t yields a positive value in the above equation.

Nearly all the methods proposed so far for the discovery of signals in genomic sequences are position–independent, that is, do not make any assumption on the location of binding sites along the input sequences. Moreover, they require all (or most of) the sequences studied to contain an instance of a binding site. In our case, instead, the position of the binding sites is known in advance, making the problem somewhat easier. This advantage, however, is balanced by the fact that while general case methods work on a set of pre–selected sequences, most of which supposedly share binding sites for the same (unknown) TF, in the case of the TATA and CAP signals we cannot choose the sequences to examine beforehand, but we have to work on virtually every gene promoter sequence available. Thus, the problem can be recast as choosing a subset of the promoters, and using fragments located in their TATA box and CAP positions to build a frequency matrix. Also, we have to define a suitable function to evaluate the best matrix, that is, the one providing the best partition between sequences containing the signals and those that do not contain it.

Matrices describing the TATA and CAP signals have been first characterized by Bucher in a seminal work in the late '80s [4], and are still used by researchers today. The method used started from two initial matrices, that were optimized using a local search technique. The datasets analyzed, however, composed by the sequences available at the time, were relatively small. Moreover, the fact that the method started from an initial matrix derived from human inspection of the data, inevitably skewed the results according to the starting point (basically, it had been observed that the TATA box contained the fragment TATAA, while the CAP motif started with CA). Thus, in this work our aim was to see whether the results obtained with a much larger dataset were consistent with the previous ones, and at the same time avoiding any preliminary constraint for the matrix.

In the following, we first introduce a formalization of the problem. Then, we present the genetic algorithm we used for the optimization of the scoring function used to evaluate candidate matrices. Finally, we compare our results with previous characterizations of the same signals.

3   The Problem

First of all, few variables must be introduced for the formalization of the problem:

1. $D_S = \{S^1, S^2, \ldots, S^n\}$ is the dataset composed by n sequences S^i of length L. Every sequence starts at position p_0 and ends at positon p_f, measured with respect to the TSS. Usually $p_0 < 0$, and $p_f > 0$, that is, each sequence encompasses the TSS of a different gene.
2. $\bar{s}$ is the position of the signal with respect to the TSS (for example -2 for the CAP signal).

Since we already know the position $\bar{s}$ of the binding site under investigation, the problem can be defined as finding a *partition* of the sequences in two subsets. Let us suppose that a particular signal of length m starts at position $\bar{s}$. Consider the set of fragments $B = \{S^i_{\bar{s}} S^i_{\bar{s}+1} \ldots S^i_{\bar{s}+m-1} | S^i \in D_S\}$ consisting of all the fragments of length m starting at position $\bar{s}$ in all the sequences belonging to the dataset. In general not all of them will be binding sites. Let us suppose that just a subset $\bar{S}$ of the sequences in D_S contains the signal. Then, considering $B_{\bar{S}} = \{S^i_{\bar{s}} S^i_{\bar{s}+1} \ldots S^i_{\bar{s}+m-1} \in B | S^i \in \bar{S}\}$ it is possible to build a frequency matrix from the subset $B_{\bar{S}}$ of fragments. The fragments used for this purpose are all the substrings of length m starting at position $\bar{s}$ in the sequences belonging to the subset $\bar{S}$.

Once a matrix has been built we need to define a score value for it. Different approaches have been introduced so far, including information content, MAP (Maximum A posteriori Probability) score and other approaches related to the sensitivity and specificity of the matrix [2,3,5,6,7]. Here, instead, we defined the scoring function in order to reflect also the fact that the signals we are trying to describe are position specific. Thus, they should not appear elsewhere along the sequence, so not to confuse the TF that binds them. In other words, we want the probability of the fragments in the signal position to be a binding site to be higher than the probability associated with all the fragments of the sequences in other positions. We associate positive scores to the fragments appearing in the correct position in the sequences selected:

$$POS(\bar{S}) = \frac{\sum_{S^i \in D_S} S_P(S^i, \bar{s})}{n} \tag{5}$$

where $S_P(S^i, \bar{s})$ is the score of the fragment of length m starting at position $\bar{s}$ in sequence S^i defined in (4) as

$$S_P(S^i, \bar{s}) = \sum_{j=1}^{m} \log \frac{P(S^i_{\bar{s}+j-1}, j)}{b_{S^i_{\bar{s}+j-1}}} \ . \tag{6}$$

Conversely, we associate a negative score with the fragments appearing in the other positions, that is, we want as less instances of the signal as possible to appear in wrong positions in the sequences selected:

$$NEG(\bar{S}) = \frac{\sum_{S^i \in D_S} \sum_{j=p_0, j \neq \bar{s}}^{p_0 + L - m} S_P(S^i, j)}{n\,(L - m)} \ . \tag{7}$$

Thus, the score associated with subset $\bar{S}$ and the corresponding frequency matrix is given by the difference between (5) and (7)

$$S_P(\bar{S}) = POS(\bar{S}) - NEG(\bar{S}) \ . \tag{8}$$

According to this score measure, the best matrix will be the one providing the largest difference between the occurrences of the signal in the position selected ($\bar{s}$) and its occurrence elsewhere in the sequences. The greater is the difference between the two terms and the more selective we can consider the matrix describing the binding site in the position under consideration. The goal is to find the subset S^* leading to the maximum score.

$$S^* = \arg \max_{\bar{S} \in \mathcal{P}(D_S)} \left\{ S_{P_{\bar{s}}}(\bar{S}) \right\} . \tag{9}$$

where $P_{\bar{s}}$ is the frequency matrix calculated from the subset $\bar{S}$ and $\mathcal{P}(D_S)$ is the power-set of D_S.

4   The Genetic Algorithm

For the solution of the problem we employed a genetic algorithm. For this purpose, two things must be provided: a method to encode an instance of the problem and a fitness function for the genome itself. A very simple solution is to use a binary string genome whose length equals the number n of sequences in the dataset. Given a genome string g the variable g_i indicates the i-th bit in the string. If $g_i = 1$ then the i-th sequence S^i in the dataset is included in the subset $\bar{S}$ of the positive sequences, otherwise it is not included. The fitness evaluation for each genome is done in the following way:

1 Given a genome g, derive the frequency matrix $P_{\bar{s}}$ from the subset of sequences $\bar{S} = \left\{ S^i \in D_S | g_i = 1 \right\}$.
2 Compute the score $S_{P_{\bar{s}}}(\bar{S})$. Negative scores are truncated to 0.

For the implementation of the GA we used the Galib library [8]. We employed one point crossovers with probability p_c to every couple of individuals during the evolution step. The mutation operator flips one bit in a genome with a probability p_m. Parent genomes are selected with a roulette wheel scheme, then mating and mutation are applied. The offspring genomes completely replace the parent population. The fitness of each individual was obtained from the score value with a linear scaling system as described in [9]. Different numbers of evolution steps and termination criteria have been tried. Usually no improvements on the fitness of the best individual were obtained after 20000 steps.

On the best genome output by the algorithm we also applied a local optimization procedure. For each sequence in the dataset, this procedure tries to perform either of the following actions:

1. if the sequence was selected by the GA for the matrix computation, it tries to exclude it;
2. if the sequence was not selected by the GA, it tries to include it.

If the change increases the score, then it is accepted, otherwise it is rejected and a new sequence is processed. This step is repeated until no further improvement can be made to the score.

Bucher's approach, even with some differences in the scoring function, performed, instead, just this procedure, starting from fragments that began with CA for the CAP signal and TATAA for the TATA box.

5   Results

Our method has been applied to three different datasets retrieved from two databases of sequences freely accessible on the web. The first one is the Eukaryotic Promoter Database (http://www.epd.isb-sib.ch/, [10]), a database of promoter sequences belonging to eukaryotic organisms. Each of these sequences belongs to a different gene, encompassing the exact point of transcription initiation. From the release 74 of this database we retrieved sequences belonging to 2199 genes of vertebrate organisms (EPD Vertebrates) 1796 of which were human sequences (EPD Homo Sapiens). It can be clearly seen how this is just a small subset of all the thousands and thousands of genes now available. However, in most of the cases the exact location along the genome of the TSS is not known. The EPD contains promoter sequences of genes where at least one TSS has been determined experimentally. An analogous species–specific database is the Drosophila Core Promoter Database (DCPD, http://www-biology.ucsd.edui /labs/Kadonaga/DCPD.htm, [11]) which contains several promoters belonging to the genome of the fruit fly (*Drosophila melanogaster*). Sequences taken from EPD were 101 nucleotides long starting from position -50 with respect to the TSS. From DCPD we retrieved 205 gene sequences, starting at position -47 and 92 nucleotides long. We performed our analysis on all the EPD promoters of vertebrates, and then on human promoters only. The fruit fly dataset has been used to validate the results, since on these sequences the occurrences of CAP and TATA box in each have been verified experimentally. In all the cases, some sequences contained both signals, some either of them, and some none. On each dataset we ran the genetic algorithm in order to obtain the best matrix for the

Table 1. Parameters of the genetic algorithm used in the experiments.

Parameter	Value
Population	500
Generations	20000
Crossover prob. (p_c)	0.9
Mutation prob. (p_m)	0.04
$\bar{s}$	-2 (CAP)
	-30 (TATA-box)
Matrix length m	6 (CAP)
	8 (TATA-box)

CAP signal (choosing position -2 with length 6) and the TATA box. While the TATA box is usually found in a random position between -36 and -24, we fixed position -30 (matrix of length 8), which has the highest frequency of occurrence. The parameters used in the GA are shown in Table 1. The application of the local optimization procedure after different runs of the GA on the various datasets converged to virtually the same matrix in each case. The signals obtained can be thus trusted to be strong (perhaps global) optima for the scoring function employed.

5.1 The CAP Signal

The CAP signal matrices computed by the genetic algorithm on the sequences of the first dataset (EPD Vertebrates and the Homo Sapiens subset) are shown in Table 2. As shown in the table, the local optimization procedure is able to "clean" the frequencies in the matrix, getting a more conserved distribution of nucleotides. Indeed, after the local optimization we can observe that the CAP signal defined over the EPD datasets has either a C or a T in position -1 (just before the TSS), and either an A or a G in position 0. No A is present in position -2 and no G in position 2. The former characterization on eukaryotes of the CAP signal of Bucher [4], where the matrix length was fixed to 8 nucleotides, is shown in Table 3. It is the result of a refinement of an initial matrix built using fragments containing a C in position -1 and an A in position 0. This matrix is still reported in EPD as the reference matrix for the description of this signal, and for this reason we compared the matrix obtained with our method with this one.

The final matrix maintains this strict constraint of a CA dinucleotide at position -1 (clearly determined by the initial choice), while our matrix shows also the presence of possible CG, TA, TG dinucleotides. The biological feasibility of this result is supported by the fact that, in the case of fruit fly sequences (see Table 4), the matrix shows the possible presence of CA and TA dinucleotides at position -1, a fact that is consistent with the results obtained experimentally by Kutach and Kadonaga in [11].

In Table 5 we show a comparison between the scores of the matrices obtained with our technique on the three different datasets and score of the matrix obtained by Bucher. For every matrix the score is calculated with the method

Table 2. CAP signal frequency matrix obtained with the genetic algorithm on the EPD datasets.

	-2	-1	0	1	2	3	-2	-1	0	1	2	3
	EPD Vertebrates						*EPD Homo Sapiens*					
	Non optimized						*Non optimized*					
A	0.069	0.000	**0.608**	0.139	0.226	0.184	0.017	0.000	**0.623**	0.136	0.259	0.174
C	0.345	**0.685**	0.129	0.223	0.240	0.252	0.356	**0.712**	0.092	0.189	0.234	0.229
G	0.294	0.009	**0.263**	0.301	0.151	0.278	0.327	0.000	**0.285**	0.341	0.121	0.293
T	0.293	**0.305**	0.000	0.337	0.383	0.286	0.300	**0.288**	0.000	0.334	0.387	0.303
	Optimized						*Optimized*					
A	0.001	0.000	**0.738**	0.144	0.304	0.177	0.000	0.000	**0.714**	0.119	0.328	0.167
C	0.366	**0.728**	0.000	0.247	0.253	0.231	0.374	**0.725**	0.000	0.227	0.225	0.229
G	0.310	0.000	**0.262**	0.360	0.000	0.323	0.311	0.000	**0.286**	0.385	0.000	0.332
T	0.323	**0.272**	0.000	0.279	**0.444**	0.270	0.315	**0.275**	0.000	0.270	**0.447**	0.271

Table 3. CAP signal obtained by Bucher in [4].

	-2	-1	0	1	2	3	4	5
A	0.162	0.000	**0.950**	0.086	0.254	0.221	0.149	0.165
C	0.158	**1.000**	0.000	0.267	0.314	0.281	0.281	0.317
G	0.228	0.000	0.000	0.383	0.000	0.241	0.241	0.185
T	0.452	0.000	0.050	0.264	**0.432**	0.330	0.330	0.333

described in section 2. Even if the latter is 2 nt longer than ours, its score is always lower. Moreover, in our description the signal appears in a greater number of sequences, at the same time maintaining a higher specificity for the position considered.

5.2   The TATA-Box Signal

The TATA-box signal has been searched in position −30 with length 8. In fact, this signal usually appears in the range between −36 and −24, but shows a high preference for position −30. In this case the score function has been modified, using for the positive term of (5), the average score over the range $[-36, -24]$ instead of position −30 alone. The negative term was taken to be the average

Table 4. CAP signal obtained with the genetic algorithm on the DCPD dataset.

	-2	-1	0	1	2	3
A	0.185	0.000	**1.000**	0.000	0.000	0.228
C	0.000	**0.772**	0.000	0.130	0.185	0.326
G	0.130	0.000	0.000	**0.565**	0.000	0.000
T	**0.685**	0.228	0.000	0.304	**0.815**	0.446

Table 5. Comparison between the score of the frequency matrix obtained by Bucher and the matrices computed using genetic algorithms on different datasets. Column *Seqs.* shows the number of sequences containing the signal.

Signal	Dataset	# of seqs.	Bucher Score	Bucher Seqs.	GA with opt. Score	GA with opt. Seqs
CAP	EPD Vertebrates	2199	3.364	581	5.001	917
	EPD Homo Sapiens	1796	3.469	489	5.447	790
	DCPD	205	7.497	114	9.106	102
TATA-box	EPD Vertebrates	2199	2.044	708	3.111	772
	EPD Homo Sapiens	1796	1.427	406	2.048	495
	DCPD	205	1.724	96	1.996	126

score in all the other positions. The results are shown in Table 6. A comparison of the scores obtained by our matrices and the one defined by Bucher (see Table 7) can be found in Table 5. Here the score is always computed on the range $[-36, -24]$. As the table shows, the matrix we obtained has always the greater score and describes a signal present in a higher number of sequences. By looking at the nucleotide frequencies, we can see that the matrices describe a sequence of A and T rich positions, with no definite preference for either nucleotide as in previous characterizations. Even if the absence of the usual TATAA sequence might look surprising, this result is consistent with the finding that the TBP (the TF part that recognizes the TATA box) recognizes the minor groove of DNA, where protein-DNA interactions are typically influenced by A/T-content, but not by the specific nucleotide sequence [12,13]. To our knowledge, this is the first time where a computational method was able to reproduce this result, without the canonical TATAA stretch in the signal. Anyway, further experimental investigation is needed, in order to establish whether the blurring of the TATAA motif depends on overlapping of occurrences of TATAA fragments in the surrounding positions, or actually describes an effective binding site for the TBP that is not strictly related to the usual consensus sequence.

6   Conclusions

In this paper we presented a method for the characterization of regulatory signals in genomic sequences, and we have shown its application to two important examples, the TATA box and CAP signals. The signals found have proved to be consistent with those described experimentally, as we have shown in the case of fruit fly signals. While more general than descriptions proposed in the past, our frequency matrices seem however to be able to characterize with better specificity the respective signals. The matrices obtained can be used to further investigate the mechanism of transcription regulation. For example, most of the genes usually present more than one possible point of transcription initiation. While on the dataset used for its construction (where the experimentally mapped TSSs had in many cases alternatives in the same sequence) the signal was present in less than half of the sequences, when we applied our CAP matrix to a selected set of gene sequences experimentally known to have a strong preference for a single

Table 6. TATA-box signal obtained with the genetic algorithm on the EPD and DCPD datasets after the local optimization procedure.

	0	1	2	3	4	5	6	7
				EPD Vertebrates				
A	0.372	0.471	0.538	0.640	0.794	0.643	0.551	0.225
C	0.000	0.000	0.000	0.000	0.000	0.000	0.000	0.163
G	0.135	0.000	0.000	0.000	0.000	0.151	0.274	0.465
T	0.492	0.529	0.462	0.360	0.206	0.206	0.175	0.148
				EPD Homo Sapiens				
A	0.362	0.436	0.521	0.612	0.793	0.649	0.553	0.239
C	0.000	0.000	0.000	0.000	0.000	0.000	0.000	0.133
G	0.154	0.000	0.000	0.000	0.000	0.149	0.287	0.489
T	0.484	0.564	0.479	0.388	0.207	0.202	0.160	0.138
				DCPD				
A	0.600	0.425	0.725	0.725	0.775	0.550	0.325	0.250
C	0.000	0.000	0.000	0.000	0.000	0.000	0.250	0.400
G	0.000	0.000	0.000	0.000	0.000	0.325	0.425	0.350
T	0.400	0.575	0.275	0.275	0.225	0.125	0.000	0.000

Table 7. TATA-box signal obtained by Bucher in [4].

	-3	-2	-1	0	1	2	3	4	5	6	7	8	9	10	11
A	0.157	0.041	0.905	0.008	0.910	0.689	0.925	0.571	0.398	0.144	0.213	0.211	0.211	0.175	0.198
C	0.373	0.118	0.000	0.026	0.000	0.000	0.008	0.005	0.113	0.347	0.378	0.326	0.303	0.275	0.260
G	0.391	0.046	0.005	0.005	0.013	0.000	0.051	0.113	0.404	0.386	0.329	0.329	0.329	0.357	0.360
T	0.080	0.794	0.090	0.961	0.077	0.311	0.015	0.311	0.085	0.123	0.080	0.134	0.157	0.193	0.183

TSS, the percentage of sequences having the signal in the correct position rose to about 70%. Nowadays genomic data are often flanked by transcriptome analysis projects describing for each gene how many TSSs have been detected, the frequency with which each is used, as well as their precise location [14]. Therefore, an interesting study would be to further investigate possible correlations between the presence of TATA and CAP signals and the most frequently used TSSs of a gene. From the computational point of view, the main advantage of this method is the fact that, differently from previous approaches to the same problem, it does not make any prior assumption about the signal to be characterized, and also takes advantage of the specific localization of the signals considered. Moreover, in order to apply our method it is not necessary to select in advance a set of sequences containing the signal. In fact, the method finds the best partition of the dataset between sequences containing and non containing the signal. This distinction is done by computing the frequency matrix that gives the best score in the predefined signal position while penalizing all the other positions.

References

1. Bellorini, M., Dantonel, J.C., Yoon, J.B., Roeder, R.G., Tora, L., Mantovani, R.: The major histocompatibility complex class II EA promoter requires TFIID binding to an initiator sequence. Mol Cell Biol. **16** (1996) 503–512

2. Stormo, G.D.: DNA binding sites: representation and discovery. Bioinformatics **16** (2000) 16–23

3. Pavesi, G., Mauri, G., Pesole, G.: Methods for pattern discovery in unaligned biological sequences. Briefings in Bioinformatics **2** (2001) 417–430

4. Bucher, P.: Weight matrix descriptions of four eukaryotic RNA polymerase II promoter elements derived from 502 unrelated promoter sequences. J. Mol. Biol. **212** (1990) 563–78

5. Hertz, G.Z., Stormo, G.D.: Identifying DNA and protein patterns with statistically significant alignments of multiple sequences. Bioinformatics **15** (1999) 563–77

6. Bailey, T., Elkan, C.: Unsupervised learning of multiple motifs in biopolymers using expectation maximization. Machine Learning **21** (1995) 51–80

7. Lawrence, C., Altschul, S., Boguski, M., Liu, J., Neuwald, A., Wooton, J.: Detecting subtle sequence signals: a Gibbs sampling strategy for multiple alignment. Science **262** (1993) 208–214

8. Wall, M.: (http://lancet.mit.edu/ga/)

9. Goldberg, D.E.: Genetic Algorithms in Search, Optimization, and Machine Learning. Addison-Wesley, Massachusetts (1989)

10. Praz, V., Périer, R., Bonnard, C., Bucher, P.: The eukaryotic promoter database, EPD: new entry types and links to gene expression data. Nucleic Acids Res. **30** (2002) 322–324

11. Kutach, A., Kadonaga, J.: The downstream promoter element DPE appears to be as widely used as the TATA box in Drosophila core promoters. Mol. Cell Biol. **20** (2000) 4754–64

12. Kim, J.L., Nikolov, D.B., Burley, S.I.: Co-crystal structure of TBP recognizing the minor groove of a TATA element. Nature **365** (1993) 520–527

13. Lo, K., Smale, S.T.: Generality of a functional initiator consensus sequence. Gene **182** (1996) 13–22

14. Okazaki, Y., Furuno, M., Kasukawa, T., et al: Analysis of the mouse transcriptome based on functional annotation of 60,770 full-length cDNAs. Nature **420** (2000) 563–573

Systems Biology Modeling in Human Genetics Using Petri Nets and Grammatical Evolution

Jason H. Moore and Lance W. Hahn

Center for Human Genetics Research, Department of Molecular Physiology and Biophysics,
519 Light Hall, Vanderbilt University, Nashville, TN, USA 37232-0700
{Moore, Hahn}@chgr.mc.Vanderbilt.edu

Abstract. Understanding the hierarchical relationships among biochemical, metabolic, and physiological systems in the mapping between genotype and phenotype is expected to improve the diagnosis, prevention, and treatment of common, complex human diseases. We previously developed a systems biology approach based on Petri nets for carrying out thought experiments for the generation of hypotheses about biological network models that are consistent with genetic models of disease susceptibility. Our systems biology strategy uses grammatical evolution for symbolic manipulation and optimization of Petri net models. We previously demonstrated that this approach routinely identifies biological systems models that are consistent with a variety of complex genetic models in which disease susceptibility is determined by nonlinear interactions between two DNA sequence variations. However, the modeling strategy was generally not successful when extended to modeling nonlinear interactions between three DNA sequence variations. In the present study, we develop a new grammar that uniformly generates Petri net models across the entire search space. The results indicate that choice of grammar plays an important role in the success of grammatical evolution searches in this bioinformatics modeling domain.

1 Introduction

Understanding how interindividual differences in DNA sequences map onto interindividual differences in phenotypes is a central focus of human genetics. Genotypes contribute to the expression of phenotypes through a hierarchical network of biochemical, metabolic, and physiological systems. The availability of biological information at all levels in the hierarchical mapping between genotype and phenotype has given rise to a new field called systems biology. One goal of systems biology is to develop a bioinformatics framework for integrating multiple levels of biological information through the development of theory and tools that can be used for mathematical modeling and simulation [1]. The promise of both human genetics and systems biology is improved human health through the improvement of disease diagnosis, prevention, and treatment.

K. Deb et al. (Eds.): GECCO 2004, LNCS 3102, pp. 392–401, 2004.
© Springer-Verlag Berlin Heidelberg 2004

Central to any study of common human disease is the realization that the mapping relationship between genotype and phenotype is extremely complex. Part of this complexity is due to epistasis or nonlinear gene-gene interactions. The historical biological definition of epistasis is one gene masking or standing upon the effects of another gene [2] while the statistical definition is a deviation from additivity in a linear model [3]. Today, epistasis is believed to be a ubiquitous component of the genetic architecture of common human diseases [4]. As a result, the identification of genes with genotypes that confer an increased susceptibility to a common disease will require a research strategy that embraces, rather than ignores, the complexity of these diseases.

We have developed a computational systems biology strategy for carrying out thought experiments about the complexity of biological systems that are consistent with a given genetic model [5-8]. With this approach, discrete dynamic systems modeling is implemented using Petri nets. Symbolic manipulation and optimization of Petri net models is carried out using grammatical evolution. This approach routinely generated Petri net models that were consistent with a variety of genetic models in which disease susceptibility is dependent on nonlinear interactions between two DNA sequence variations [5,6]. However, when applied to higher-order genetic models, the strategy did not consistently yield perfect Petri nets [7]. In fact, only one perfect model was discovered out of 100 independent runs. The goal of the present study is to develop a strategy that is able to discover Petri net models of biological systems that are consistent with nonlinear interactions between three DNA sequence variations.

A.   Model 1

	CC			Cc			cc		
	BB	Bb	bb	BB	Bb	bb	BB	Bb	bb
AA	.06	.06	.01	.04	0	.10	.02	.08	0
Aa	.02	.06	0	.09	.01	.06	.03	.07	.01
aa	.01	.07	.04	.01	.08	.02	.01	0	.10

B.   Model 2

	CC			Cc			cc		
	BB	Bb	bb	BB	Bb	bb	BB	Bb	bb
AA	.05	.06	.03	.09	.02	.06	.06	.03	.01
Aa	.02	.08	.04	.06	.02	.08	.03	.08	0
aa	.07	0	.02	.01	.10	.02	.09	.01	.10

Fig. 1. Penetrance functions for nonlinear gene-gene interaction models 1 (A) and 2 (B). High-risk genotype combinations are shaded while low-risk combinations are unshaded.

2 The Nonlinear Gene-Gene Interaction Models

Our two high-order, nonlinear, gene-gene interaction models are based on penetrance functions. Penetrance functions represent one approach to modeling the relationship between genetic variations and risk of disease. Penetrance is simply the probability (P) of disease (D) given a particular combination of genotypes (G) that was inherited (i.e. P[D|G]). A single genotype is determined by one allele (i.e. a specific DNA sequence state) inherited from the mother and one allele inherited from the father. For most genetic variations, only two alleles (encoded by A or a) exist in the biological popula-tion. Therefore, because the ordering of the alleles is unimportant, a genotype can have one of three values: AA, Aa or aa. Figures 1A and 1B (above) illustrate the penetrance functions used for Models 1 and 2, respectively. Each model was discov-ered using a modified version of the software of Moore et al. [9]. What makes these models interesting is that disease risk is dependent on each particular combination of three genotypes. Each single-locus genotype has effectively no main effect on disease risk when the allele frequencies are equal and the genotypes are consistent with Hardy-Weinberg proportions.

3 An Introduction to Petri Nets for Modeling Discrete Dynamic Systems

Petri nets are a type of directed graph that can be used to model discrete dynamical systems [10]. Goss and Peccoud [11] demonstrated that Petri nets could be used to model molecular interactions in biochemical systems. The core Petri net consists of two different types of nodes: places and transitions. Using the biochemical systems analogy of Goss and Peccoud [11], places represent molecular species. Each place has a certain number of tokens that represent the number of molecules for that particular molecular specie. A transition is analogous to a molecular or chemical reaction and is said to fire when it acquires tokens from a source place and, after a possible delay, deposits tokens in a destination place. Tokens travel from a place to a transition or from a transition to a place via arcs with specific weights or bandwidths. While the number of tokens transferred from place to transition to place is determined by the arc weights (or bandwidths), the rate at which the tokens are transferred is determined by the delay associated with the transition. Transition behavior is also constrained by the weights of the source and destination arcs. A transition will only fire if two precondi-tions are met: 1) if the source place can completely supply the capacity of the source arc and, 2) if the destination place has the capacity available to store the number of tokens provided by the full weight of the destination arc. Transitions without an input arc, act as if they are connected to a limitless supply of tokens. Similarly, transitions without an output arc can consume a limitless supply of tokens. The firing rate of the transition can be immediate, delayed deterministically or delayed stochastically, de-pending on the complexity needed. The fundamental behavior of a Petri net can be controlled by varying the maximum number of tokens a place can hold, the weight of each arc, and the firing rates of the transitions.

4 Our Petri Net Modeling Strategy

The goal of identifying Petri net models of biochemical systems that are consistent with observed population-level gene-gene interactions is accomplished by developing Petri nets that are dependent on specific genotypes from two or more genetic variations. Here, we make transition firing rates and/or arc weights genotype-dependent yielding different Petri net behavior. Each Petri net model is related to the genetic model using a discrete version of the threshold model from population genetics [12]. With a classic threshold or liability model, it is the concentration of a biochemical or environmental substance that is related to the risk of disease, under the hypothesis that risk of disease is greatly increased once a particular substance exceeds some threshold concentration. Conversely, the risk of disease may increase in the absence of a particular factor or with any significant deviation from a reference level. In such cases, high or low levels are associated with high risk while an intermediate level is associated with low risk. Here, we use a discrete version of this model for our deterministic Petri nets. For each model, the number of tokens at a particular place is recorded and if they exceed a certain threshold, the appropriate risk assignment is made. If the number of tokens does not exceed the threshold, the alternative risk assignment is made. The high-risk and low-risk assignments made by the discrete threshold from the output of the Petri net can then be compared to the high-risk and low-risk genotypes from the genetic model. A perfect match indicates the Petri net model is consistent with the gene-gene interactions observed in the genetic model. The Petri net then becomes a model that relates the genetic variations to risk of disease through an intermediate biochemical network.

5 The Grammatical Evolution Algorithm

5.1 Overview of Grammatical Evolution

Evolutionary computation arose from early work on evolutionary programming [13,14] and evolution strategies [15,16] that used simulated evolution for artificial intelligence. The focus on representations at the genotypic level lead to the development of genetic algorithms by Holland [17,18] and others. Genetic algorithms have become a popular machine intelligence strategy because they can be effective for implementing parallel searches of rugged fitness landscapes [19]. Briefly, this is accomplished by generating a random population of models or solutions, evaluating their ability to solve the problem at hand, selecting the best models or solutions, and generating variability in these models by exchanging model components between different models. The process of selecting models and introducing variability is iterated until an optimal model is identified or some termination criteria are satisfied. Koza [20] developed an alternative parameterization of genetic algorithms called genetic programming where the models or solutions are represented by binary expression trees. Koza [21] and others [22] have applied genetic programming to modeling metabolic networks.

Grammatical evolution (GE) has been described by O'Neill and Ryan [23, 24] as a variation on genetic programming. Here, a Backus-Naur Form (BNF) grammar is specified that allows a computer program or model to be constructed by a simple genetic algorithm operating on an array of bits. The GE approach is appealing because only a text file specifying the grammar needs to be altered for different applications. There is no need to modify and recompile source code during development once the fitness function is specified.

5.2   A Grammar for Petri Net Models in Backus-Naur Form

Moore and Hahn [5-8] developed a grammar for Petri nets in BNF. Backus-Naur Form is a formal notation for describing the syntax of a context-free grammar as a set of production rules that consist of terminals and nonterminals [25]. Nonterminals form the left-hand side of production rules while both terminals and nonterminals can form the right-hand side. A terminal is essentially a model element while a nonterminal is the name of a possibly recursive production rule. For the Petri net models, the terminal set includes, for example, the basic building blocks of a Petri net: places, arcs, and transitions. The nonterminal set includes the names of production rules that construct the Petri net. For example, a nonterminal might name a production rule for determining whether an arc has weights that are fixed or genotype-dependent. We show below the production rule that was executed to begin the model building process for the study by Moore and Hahn [7] and then describe the modifications evaluated in the present study.

<root> ::= <pick_a_gene> <pick_a_gene> <pick_a_gene> <net_iterations>
<expr> <transition> <transition> <place_noarc>

When the initial <root> production rule is executed, a single Petri net place with no entering or exiting arc (i.e. <place_noarc>) is selected and a transition leading into or out of that place is selected. The arc connecting the transition and place can be dependent on the genotypes of the genes selected by <pick_a_gene>. The nonterminal <expr> is a function that allows the Petri net to grow. The production rule for <expr> is shown below.

<expr> ::= <expr> <expr>   0
          | <arc>   1
          | <transition>   2
          | <place>   3

Here, the selection of one of the four nonterminals (0, 1, 2, or 3) on the right-hand side of the production rule is determined by a combination of bits in the genetic algorithm chromosome.

The base or minimum Petri net that is constructed using the <root> production rule consists of a single place, two transitions, and an arc that connects each transition to the place. Multiple calls to the production rule <expr> by the genetic algorithm chromosome can build any connected Petri net. In addition, the number of times the Petri net is to be iterated is selected with the nonterminal <net_iterations>. Many other production rules define the arc weights, the genotype-dependent arcs and transitions, the number of initial tokens in a place, the place capacity, etc. All decisions made in the building of the Petri net model are made by each subsequent bit or combination of bits in the genetic algorithm chromosome. The complete grammar is too large for detailed presentation here but can be obtained from the authors upon request.

While the grammar described above was successful for modeling interactions among two DNA sequence variations [5, 6], it was not successful for modeling interactions among three DNA sequence variations [7]. A potential concern is that the grammar starts with very a simple Petri net in the root production rule. Through the use of <expr>, the grammar is theoretically capable of generating larger, more complex Petri nets. However, there is clearly a bias towards smaller, simpler Petri nets. For example, Moore and Hahn [6] discovered Petri nets that almost always consisted of one place, two transitions, and two arcs. In the present study, we have modified the grammar to uniformly generate Petri nets from a defined search space. The new grammar builds a Petri net with one to five places, one to 20 transitions, and four to 24 arcs. Each architecture within these limits is equally probable. We also require that at least one conditional element be present for each of the DNA sequence variations in the genetic model.

5.3 The Fitness Function

Once a Petri net model is constructed using the BNF grammar, as instructed by the genetic algorithm chromosome, the model fitness is determined. As described in detail by Moore and Hahn [6, 7], this is carried out by executing the Petri net model for each combination of genotypes in the genetic dataset and comparing the final token counts at a defined place to a threshold constant to determine the risk assignments. The optimal threshold is determined by systematically evaluating different threshold constants. Fitness of the Petri net model is determined by comparing the high risk and low risk assignments made by the Petri net to those from the given nonlinear gene-gene interaction model (e.g. see Figure 1). Fewer inconsistencies are associated with a better fitness. Ideally, the risk assignments made by the Petri net are perfectly consistent with those in the genetic model.

5.4 The Genetic Algorithm Parameters

Details of the genetic algorithm are given by Moore and Hahn [7]. Briefly, each genetic algorithm chromosome consisted of 14 32-bit bytes. In implementing the grammar, it is possible to reach the end of a chromosome with an incomplete instance of the grammar. To complete the instance, chromosome wrap-around was used [23, 24].

In other words, the instance of the grammar was completed by reusing the chromosome as many times as was necessary.

We ran the genetic algorithm a total of 100 times with different random seeds for each gene-gene interaction model. Each run consisted of a maximum of 800 generations. The genetic algorithm was stopped when a model with a classification error of zero was discovered. We used a parallel search strategy of 10 demes, each with a population size of 2000, for a total population size of 20,000. A best chromosome migrated from each deme to all other demes every 25 generations. The recombination probability was 0.6 while the mutation probability was 0.02.

6   Results

The grammatical evolution algorithm was run a total of 100 times for each of the two high-order, nonlinear gene-gene interaction models. For genetic model one, the modified grammatical evolution strategy yielded 32 perfect Petri net models. For genetic model two, 36 perfect Petri net models were discovered. These results are in contrast to the grammar of Moore and Hahn [7] that yielded no perfect Petri nets for the first genetic model and only one perfect Petri net for the second genetic model.

Table 1 below summarizes the mode (i.e. most common) and range of the number of places, arcs, transitions, and conditionals that define the genotype-dependencies of the elements in the best Petri net models found across the 100 runs for each model. For each gene-gene interaction model, most Petri net models discovered consisted of one place and a minimum of 11 transitions, seven arcs, and seven elements that are conditional on genotype. In general, these Petri net models were much larger than those described by Moore and Hahn [7] due to the improved grammar that samples the search space in a uniform manner. Figure 2 illustrates example Petri net architectures that were discovered by the grammatical evolution algorithm for each genetic model. All generated Petri net models are available upon request.

Table 1. Summary of the distribution (mode and range) of the number of different Petri net elements identified across 100 grammatical evolution runs for the two nonlinear gene-gene interaction models.

	Mode (range) number of Petri net elements	
Petri net element	Model 1	Model 2
Place	1 (1-5)	1 (1-4)
Arc	7 (5-30)	12 (5-24)
Transition	19 (3-25)	11 (3-22)
Conditional	7,10 (5-29)	10 (5-30)

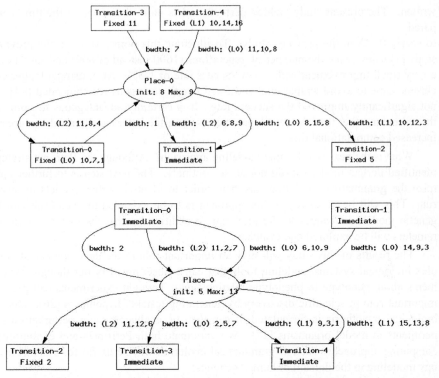

Fig. 2. Examples of Petri nets discovered for genetic model 1 (top) and 2 (bottom). Each place has an initial (init) and maximum (max) token count. Transitions have either a fixed firing rate of one token every one (immediate) to 16 time steps or a firing rate that is dependent on the genotype (e.g. 10, 14, 16 for *AA, Aa, aa*, respectively) at a particular DNA sequence variation (e.g. L0, L1, or L2). Similarly, arcs have a bandwidth (bwdth) that is either fixed (e.g. 7) or is dependent on the genotype (e.g. 11, 10, 8 for *AA, Aa, aa*, respectively) at a particular DNA sequence variation.

7   Discussion

The main conclusion of this study is that a grammar that generates Petri net models in a uniform fashion across the search space of all possible models significantly outperforms a grammar that relies on <expr> in the production rule to build Petri net complexity from an initial simple starting point. This study points to the importance of grammar selection on the ultimate performance of the search. In fact, choice of grammar may sometimes be more important than parameters such as the number of generations used in the evolutionary computing search. For example, the same parameter settings (described in Section 5.4 above) were used in the present study and the study by Moore and Hahn [7]. The only difference was the how the grammar was

written. The present study yielded perfect models more than 30% of the time compared
to nearly 0-1% in the previous study. To illustrate this point, we reran the previous study [7] using twice the number of generations (1600 instead of 800) and found only a very small improvement with a new success rate of 1-5%. Also, using a longer GA chromosome to avoid wraparound and fixed length codons as recommended [24] did not significantly improved the success rate. It is certainly advantageous to optimize the grammar rather than the search parameters since the latter usually results in greatly increased computational time.

What is the next step for this modeling approach? Although the Petri net strategy identified perfect models, it did not do so routinely. The next step is to further optimize the grammatical evolution search in order to identify perfect models on every run. This will be necessary if this approach is to be extended to even higher-order genetic models. Changes to the grammar in addition to changes to the search parameters will be explored and evaluated.

The results of this study will play an important role in the development of complex biological systems modeling tools that can be used to carry out thought experiments about genotype to phenotype relationships. Thought experiments can play an important role in scientific discovery by generating testable hypotheses [26]. Once a biological hypothesis is formulated, it can then be evaluated using perturbation experiments in model organisms [27]. We anticipate both Petri nets and evolutionary computing approaches such as grammatical evolution will be useful for systems biology modeling in the domain of human genetics.

Acknowledgements. This work was supported by National Institutes of Health grants HL65234, HL65962, GM31304, AG19085, and AG20135.

References

1. Ideker, T., Galitski, T., Hood, L.: A new approach to decoding life: systems biology. Annual Review of Genomics and Human Genetics 2 (2001) 343-72
2. Bateson, W.: Mendel's Principles of Heredity. Cambridge University Press, Cambridge (1909)
3. Fisher, R.A.: The correlation between relatives on the supposition of Mendelian inheritance. Transactions of the Royal Society of Edinburgh 52 (1918) 399-433
4. Moore, J.H. The ubiquitous nature of epistasis in determining susceptibility to common human diseases. Human Heredity 56 (2003) 73-82
5. Moore, J.H., Hahn, L.W. Grammatical evolution for the discovery of Petri net models of complex genetic systems. In: Cantu-Paz et al., editors, Genetic and Evolutionary Computation – GECCO 2003. Lecture Notes in Computer Science 2724 (2003) 2412-13
6. Moore, J.H., Hahn, L.W. Evaluation of a discrete dynamic systems approach for modeling the hierarchical relationship between genes, biochemistry, and disease susceptibility. Discrete and Continuous Dynamical Systems: Series B 4 (2004) 275-87
7. Moore, J.H., Hahn, L.W. Petri net modeling of high-order genetic systems using grammatical evolution. BioSystems 72 (2003) 177-86

8. Moore, J.H., Hahn, L.W. An improved grammatical evolution strategy for Petri net modeling of complex genetic systems. In: G.R. Raidl et al., editors, Applications of Evolutionary Computing, Lecture Notes in Computer Science 3005 (2004) 62-71

9. Moore, J.H., Hahn, L.W., Ritchie, M.D., Thornton, T.A., White, B.C: Application of genetic algorithms to the discovery of complex genetic models for simulation studies in human genetics. In: W.B.Langdon et al., editors, Proceedings of the Genetic and Evolutionary Computation Conference. Morgan Kaufmann Publishers, San Francisco (2002)

10. Desel, J., Juhas, G.: What is a Petri net? Informal answers for the informed reader. In H. Ehrig, G. Juhas, J. Padberg, and G. Rozenberg, editors, Unifying Petri Nets, Lecture Notes in Computer Science 2128, pp 1-27. Springer (2001)

11. Goss, P.J., Peccoud, J.: Quantitative modeling of stochastic systems in molecular biology by using stochastic Petri nets. Proceedings of the National Academy of Sciences USA 95 (1998) 6750-5

12. Falconer, D.S., Mackay, T.F.C: Introduction to Quantitative Genetics, 4th edition, Longman, Essex (1996)

13. Fogel, L.J. Autonomous automata. Industrial Research 4 (1962) 14-19

14. Fogel, L.J., Owens, A.J., Walsh, M.J. Artificial Intelligence through Simulated Evolution. John Wiley, New York (1966)

15. Rechenberg, I. Cybernetic solution path of an experimental problem. Royal Aircraft Establishment, Farnborough, U.K., Library Translation No. 1122 (1965)

16. Schwefel, H.-P. Kybernetische Evolution als Strategie der experimentellen Forschung in der Stromungstechnik. Diploma Thesis, Technical University of Berlin (1965)

17. Holland, J.H. Adaptive plans optimal for payoff-only environments. In: Proceedings of the 2nd Hawaii International Conference on Systems Sciences. University of Hawaii, Honolulu (1969) 917-920

18. Holland, J.H. Adaptation in Natural and Artificial Systems. University of Michigan Press, Ann Arbor (1975)

19. Goldberg, D.E.: Genetic Algorithms in Search, Optimization, and Machine Learning. Reading: Addison-Wesley (1989)

20. Koza, J.R.: Genetic Programming: On the Programming of Computers by Means of Natural Selection. The MIT Press, Cambridge London (1992)

21. Koza, J.R., Mydlowec, W., Lanza, G., Yu, J., Keane, M.A. Reverse engineering of metabolic pathways from observed data using genetic programming. Pacific Symposium on Biocomputing 6 (2001) 434-45

22. Kitagawa, J., Iba, H. Identifying metabolic pathways and gene regulation networks with evolutionary algorithms. In: Evolutionary Computation and Bioinformatics. Fogel, G.B., Corne, D.W., editors. Morgan Kaufmann Publishers, San Francisco (2003) 255-278

23. O'Neill, M., Ryan, C.: Grammatical evolution. IEEE Transactions on Evolutionary Computation 5 (2001) 349-358

24. O'Neill, M., Ryan, C.: Grammatical evolution: Evolutionary Automatic Programming in an Arbitrary Language. Kluwer Academic Publishers, Boston (2003)

25. Marcotty, M., Ledgard, H.: The World of Programming Languages. Springer-Verlag, Berlin (1986)

26. Di Paolo, E.A., Noble, J., Bullock, S.: Simulation models as opaque thought experiments. In: Proceedings of the Seventh International Conference on Artificial Life. Dedau, M.A. et al, editors. The MIT Press, Cambridge (2000)

27. Jansen, R.C.: Studying complex biological systems using multifactorial perturbation. Nature Reviews Genetics 4 (2003) 145-51

Evolutionary Computation Techniques for Optimizing Fuzzy Cognitive Maps in Radiation Therapy Systems

K.E. Parsopoulos[1,3], E.I. Papageorgiou[2,3], P.P. Groumpos[2,3], and
M.N. Vrahatis[1,3]

[1] Department of Mathematics, University of Patras, GR–26110 Patras, Greece,
{kostasp,vrahatis}@math.upatras.gr
[2] Department of Electrical and Computer Engineering, University of Patras,
GR–26500 Patras, Greece,
{epapageo,groumpos}@ee.upatras.gr
[3] University of Patras Artificial Intelligence Research Center (UPAIRC),
University of Patras, GR–26110 Patras, Greece

Abstract. The optimization of a Fuzzy Cognitive Map model for the supervision and monitoring of the radiotherapy process is proposed. This is
performed through the minimization of the corresponding objective function by using the Particle Swarm Optimization and the Differential Evolution algorithms. The proposed approach determines the cause–effect
relationships among the concepts of the supervisor–Fuzzy Cognitive Map
by computing its optimal weight matrix, through extensive experiments.
Results are reported and discussed.

1 Introduction

Several pathological illness cases can be addressed by eliminating the infected
cells through the application of ionizing radiation to the patient. This procedure
is widely known as *radiotherapy*. In the case of cancer cells, the radiation consists
mainly of photons or electrons. Healthy cells are also affected by the radiation.
Clearly, the determination of the dosage distribution of radiation, as well as
information regarding the affection of the tumor by irradiation and the affection
of the healthy tissues, are of major importance [1].

Radiotherapists–doctors must take into consideration many different (complementary, similar or conflicting) factors that influence the selection of the radiation dose and, consequently, the final result of the therapy. All these factors
are usually incorporated in an optimization process, where the main objectives
are to minimize the total amount of radiation at which the patient is exposed,
maximize the minimum final radiation dose received by the tumor, minimize the
radiation to critical structure(s) and healthy tissues, and produce acceptable
dosage distributions with the smallest computational effort [1].

Several algorithms have been proposed and used for the optimization of radiation therapy treatment plans [2,3]. Dose calculation algorithms [4,5], dose–
volume feasibility search algorithms [6], and biological objective algorithms [7]

K. Deb et al. (Eds.): GECCO 2004, LNCS 3102, pp. 402–413, 2004.

have been employed for the determination of dosage distributions in treatment planning systems under multiple criteria and dose–volume constraints [3]. Different algorithms have been proposed for the optimization of beam weights and beam directions [8]. Gradient–descent methods have been used to optimize the objective functions as well as the intensity distributions [9]. Moreover, methods related to knowledge–based expert systems and neural networks, have been proposed for the optimization of treatment variables and the support of decisions during radiotherapy planning [10,11].

The kind, the nature, as well as, the number of the parameters–factors that are taken into consideration for the determination of the radiation therapy treatment, give rise to a highly complex, uncertain and fuzzy overall model. Fuzzy Cognitive Maps (FCMs) have been applied for the modeling of the decision–making process of radiation therapy, with promising results [12]. FCMs can model complex systems that involve different factors, states, variables, and events, integrating the influence of several controversial factors in a decision–making process [13]. In FCMs, the causal effects among different factors are taken into consideration in the calculation of the values of all causal concepts that determine the radiation dose, so as to keep the dose at a minimum level, while destroying the tumor and inflicting the minimum injuries to healthy tissues and organs [1].

In this paper, two different algorithms, Particle Swarm Optimization (PSO) and Differential Evolution (DE), coming from the fields of Swarm Intelligence and Evolutionary Computation, respectively, are employed for the optimization of the supervisor–FCM used in an established radiation therapy treatment planning system. Both methods have proved to be very efficient in a plethora of applications in science and engineering. Also, PSO has recently proved to be very efficient algorithm for FCMs learning in an industrial problem [14].

The rest of this article is organized as follows: the PSO and DE algorithms are briefly presented in Sections 2 and 3, respectively. A review of the basic concepts and notion of FCMs, as well as a description of the FCM model for the supervision of the radiation therapy process, are given in Section 4. The proposed approach and experimental results are analyzed in Section 5. The paper concludes in Section 6.

2   The Particle Swarm Optimization Algorithm

Particle Swarm Optimization (PSO) is a stochastic optimization algorithm. It belongs to the class of *Swarm Intelligence* algorithms, which are inspired from, and based on the social dynamics and emergent behavior in socially organized colonies [15,16,17]. PSO is a population based algorithm, i.e. it exploits a population of individuals to synchronously probe promising regions of the search space. In this context, the population is called a *swarm* and the individuals (i.e. the search points) are called *particles*. Each particle moves with an adaptable velocity within the search space, and retains a memory of the best position it ever encountered. In the *global* variant of PSO, the best position ever attained by all individuals of the swarm is communicated to all the particles at each iteration. In the *local* variant, each particle is assigned to a neighborhood consisting of a prespecified number of particles. In this case, the best position ever attained by the particles that comprise a neighborhood is communicated among them [16, 18].

Assume a D–dimensional search space, $S \subset \mathbb{R}^D$, and a swarm consisting of N particles. Let $X_i = (x_{i1}, x_{i2}, \ldots, x_{iD})^\top \in S$, be the i–th particle and $V_i = (v_{i1}, v_{i2}, \ldots, v_{iD})^\top \in S$, be its velocity. Let also the best previous position encountered by the i–th particle in S be denoted by $P_i = (p_{i1}, p_{i2}, \ldots, p_{iD})^\top$. Assume g_i to be the index of the particle that attained the best previous position among all the particles in the neighborhood of the i–th particle, and G to be the iteration counter. Then, the swarm is manipulated by the equations [19]:

$$V_i(G+1) = \chi \left[V_i(G) + c_1 r_1 (P_i(G) - X_i(G)) + c_2 r_2 (P_{g_i}(G) - X_i(G)) \right], (1)$$

$$X_i(G+1) = X_i(G) + V_i(G+1), \tag{2}$$

where $i = 1, \ldots, N$; χ is a parameter called *constriction factor*; c_1 and c_2 are two parameters called *cognitive* and *social* parameters, respectively; and r_1, r_2, are random numbers uniformly distributed within $[0, 1]$.

Alternatively, a different version of the velocity's update equation, which incorporates a parameter called *inertia weight*, has been proposed [20,21]:

$$V_i(G+1) = w V_i(G) + c_1 r_1 (P_i(G) - X_i(G)) + c_2 r_2 (P_{g_i}(G) - X_i(G)), \tag{3}$$

where w is the *inertia weight*.

Both the constriction factor and the inertia weight are mechanisms for controlling the magnitude of velocities. However, there are some major differences regarding the way these two are computed and applied. The constriction factor is derived analytically through the formula [19],

$$\chi = \frac{2\kappa}{|2 - \phi - \sqrt{\phi^2 - 4\phi}|}, \tag{4}$$

for $\phi > 4$, where $\phi = c_1 + c_2$, and $\kappa = 1$. Different configurations of χ, as well as a thorough theoretical analysis of the derivation of (4), can be found in [19,22]. On the other hand, experimental results suggest that it is preferable to initialize the inertia weight w to a large value, giving priority to global exploration of the search space, and gradually decrease it, so as to obtain refined solutions [20,21]. This finding is intuitively very appealing. In conclusion, an initial value of w around 1.0 and a gradual decline towards 0 is considered a proper choice for w.

Regarding the social and cognitive parameter, although the default values $c_1 = c_2 = 2$ have been proposed and usually used, experimental results indicate that alternative configurations, depending on the problem at hand, may produce superior performance [17,19,23]. The initialization of the swarm and the velocities, is usually performed randomly and uniformly in the search space, although more sophisticated initialization techniques can enhance the overall performance of the algorithm [24].

3   The Differential Evolution Algorithm

The Differential Evolution (DE) algorithm has been developed by Storn and Price [25]. It utilizes N, D–dimensional vectors $X_i = (x_{i1}, x_{i2}, \dots, x_{iD})^\top$, $i = 1, \dots, N$, as a population for each iteration (generation), G, of the algorithm. The initial population is taken to be uniformly distributed in the search space. At each generation, the *mutation* and *crossover* (recombination) operators are applied on the individuals, and a new population arises. Then, the selection phase starts, where the two populations compete each other, and the next generation is formed [25].

According to the *mutation* operator, for each vector $X_i(G)$, $i = 1, \dots, N$, a *mutant vector*, $V_i(G + 1) = (v_{i1}, v_{i2}, \dots, v_{iD})^\top$, is determined through the equation:

$$V_i(G + 1) = X_{r_1}(G) + F\left(X_{r_2}(G) - X_{r_3}(G)\right), \qquad (5)$$

where $r_1, r_2, r_3 \in \{1, \dots, N\}$, are mutually different random indexes, and, $F \in (0, 2]$. The indexes r_1, r_2, r_3, also need to differ from the current index, i. Consequently, N must be greater than or equal to 4, in order to apply mutation.

Following the mutation phase, the *crossover* operator is applied on the population. Thus, a *trial vector*, $U_i(G + 1) = (u_{i1}, u_{i2}, \dots, u_{iD})^\top$, is generated, where,

$$u_{ij} = \begin{cases} v_{ij}, & \text{if } (\text{randb}(j) \leqslant CR) \text{ or } j = \text{rnbr}(i), \\ x_{ij}, & \text{if } (\text{randb}(j) > CR) \text{ and } j \neq \text{rnbr}(i), \end{cases} \qquad (6)$$

where, $j = 1, 2, \dots, D$; $\text{randb}(j)$, is the j-th evaluation of a uniform random number generator in the range $[0, 1]$; CR is the (user specified) crossover constant in the range $[0, 1]$; and, $\text{rnbr}(i)$ is a randomly chosen index from the set $\{1, 2, \dots, D\}$.

To decide whether or not the vector $U_i(G + 1)$ should be a member of the population comprising the next generation, it is compared to the initial vector $X_i(G)$. Thus,

$$X_i(G + 1) = \begin{cases} U_i(G + 1), & \text{if } f\left(U_i(G + 1)\right) < f\left(X_i(G)\right), \\ X_i(G), & \text{otherwise.} \end{cases}$$

The procedure described above is considered as the standard variant of the DE algorithm. Different mutation and crossover operators have been applied with promising results [25]. In order to describe the different variants, the scheme $DE/x/y/z$, is used, where x specifies the mutated vector ("rand" for randomly selected individual or "best" for selection of the best individual); y is the number of difference vectors used; and, z denotes the crossover scheme (the scheme described here is due to independent binomial experiments, and thus, it is denoted as "bin") [25]. According to this description scheme, the DE variant described above is denoted as $DE/rand/1/bin$. One highly beneficial scheme that deserves special attention is the $DE/best/2/bin$ scheme, where,

$$V_i(G + 1) = X_{\text{best}}(G) + F\left(X_{r_1}(G) + X_{r_2}(G) - X_{r_3}(G) - X_{r_4}(G)\right). \qquad (7)$$

The usage of two difference vectors seems to improve the diversity of the population, if N is high enough. A parallel implementation of DE is reported in [26].

4   A Fuzzy Cognitive Map Model for Supervision of the Radiation Therapy Process

Fuzzy Cognitive Maps (FCMs) have been introduced by Kosko [27] to describe a cognitive map model with the following characteristics:

1. Causal relationships between nodes are fuzzified, i.e. instead of using only signs to indicate positive or negative causality, a number is associated with each relationship to express the degree of causality between two concepts.

2. The system is dynamic and has feedback, i.e. the effect of change in a concept node also affects other nodes, which in turn can affect the node initiating the change. The presence of feedback introduces temporality to the operation of FCMs.

The concepts reflect attributes, characteristics, and qualities of the system. The interconnections among the concepts signify the cause and effect relationships among the concepts. Let us denote by C_i, $i = 1, \ldots, M$, the nodes–concepts of an FCM. Each concept represents one of the key–factors of the system, and it takes a value $A_i \in [0, 1]$, $i = 1, \ldots, M$. Each interconnection between two concepts C_i and C_j, has a weight $w_{ij} \in [-1, 1]$, which is analogous to the strength of the causal link between C_i and C_j. The sign of w_{ij} indicates whether the relation between the two concepts is direct or inverse. There are three types of causal relationships among concepts: positive causality $(w_{ij} > 0)$, negative causality $(w_{ij} < 0)$, and no relation $(w_{ij} = 0)$. So the FCM provides qualitative as well as quantitative information regarding the relationships among concepts [11].

In general, the value of each concept is calculated by aggregating the influence of the other concepts to the specific one [10], by applying the following rule:

$$
A_i^{(t)} = f \left(A_i^{(t-1)} + \sum_{\substack{j=1 \\ j \neq i}}^{M} w_{ji} A_j^{(t)} \right),
\tag{8}
$$

where $A_i^{(t)}$ is the value of C_i at time t, and f is a sigmoid threshold function.

The methodology for developing FCMs primarily draws on a group of experts who are asked to define the concepts and describe the relationships among them. IF–THEN rules are used to describe the cause and effect relationships among the concepts, and infer a linguistic weight for each interconnection [10]. Each expert describes independently every interconnection with a fuzzy rule; the inference of the rule is a linguistic variable, which describes the relationship and determines the grade of causality between the corresponding concepts. Subsequently, the inferred fuzzy weights suggested by the experts, are aggregated to a single linguistic weight, which is transformed to a numerical weight, $w_{ij} \in [-1, 1]$, using the Center of Area (CoA) defuzzification method [13]. This weight represents the aggregated suggestion of the whole experts' group. Thus, an initial weight matrix, $W^{\text{initial}} = [w_{ij}]$, with $w_{ii} = 0$, $i = 1, \ldots, M$, is obtained. Using the initial concept values, A_i, which are also provided by the experts, the matrix W^{initial} is used for the determination of the steady state of the FCM, through the application of the rule defined in (8).

The most significant weaknesses of FCMs are the critical dependence on the experts' opinions, and the potential convergence to undesired steady states. Learning procedures constitute means to increase the efficiency and robustness of FCMs, by updating the weight matrix so as to avoid convergence to undesired steady states. The desired steady state is characterized by values of the FCM's output concepts that are accepted by the experts [14].

Radiation therapy is a complex process involving a large number of treatment variables. The objective of radiotherapy is to deliver the highest radiation dose to the smallest possible volume that encloses the tumor, while retaining at a minimum the exposure of healthy tissues and critical organs to radiation. Treatment planning is another complex process that takes place before the final treatment execution. The performance criteria for this process include the maximization of the final dose received by the target volume (tumor), the maximization of the dose derived from the treatment planning within the target region, and dose minimization for the surrounding critical organs and normal tissues. To achieve these goals, several factors need to be taken into consideration [12,13].

In [12], an FCM with 33 concepts (factor–concepts, selector–concepts and output–concepts) has been developed, to model the aforementioned treatment planning and determine the dose distribution for the target volume, the healthy tissues and the critical organs. A different, more abstract FCM model is needed to supervise the whole radiotherapy process. This model must consist of more abstract concepts that represent the final parameters before the treatment execution, simulating, thus, the doctor's decision–making. In the proposed model [12], the supervision process is modeled with another FCM (supervisor–FCM) that models, monitors, and evaluates the whole process of radiation therapy. The supervisor–FCM is based on the knowledge of experts that supervise the actual process, and it consists of the following six concepts:

1. C_1–*Tumor localization*: It depends on the patient's contour, sensitive critical organs and tumor volume. It embodies the values and influences among these factor–concepts.

2. C_2–*Dose prescribed from the treatment planning*: This concept describes the prescribed dose and it depends on the concepts of the delivered dose to the target volume, normal tissues and critical organs, which are determined by the treatment planning model of the first level's FCM.

3. C_3–*Machine factors*: This concept describes the equipment characteristics.

4. C_4–*Human factors*: This is a general concept describing the experience and knowledge of the medical staff.

5. C_5–*Patient positioning and immobilization*: This concept describes the cooperation of the patient with the doctors and his willingness to follow their instructions.

6. C_6–*Final dose received by the target volume*: A measurement of the radiation dose received by the target tumor.

The supervisor–FCM has been developed following the methodology described previously in this section. Three oncologists were independently asked to describe the relationships among the concepts $C_1, \ldots, C_6$, using IF–THEN rules, and infer a linguistic weight for each interconnection [28]. The degree of influence is represented by a member of the fuzzy set,

{ positive very high, positive high, positive medium, positive weak,

zero, negative weak, negative medium, negative low, negative very low }.

The following connections among the concepts of the supervisor–FCM were suggested:

1. *Linkage 1*: Connects C_1 with C_6. It relates the tumor localization with the delivered final dose.
2. *Linkage 2*: Relates C_2 with C_1; when the dose derived from treatment planning is high, the value of tumor localization increases by a small amount.
3. *Linkage 3*: Connects C_2 with C_6; when the dose from treatment planning is high, the final dose given to the patient will be also high.
4. *Linkage 4*: Relates C_3 with C_2; when the machine parameters increase, the dose from treatment planning decreases.
5. *Linkage 5*: Connects C_3 with C_6; any change to machine parameters influences negatively the final dose given to the target volume.
6. *Linkage 6*: Relates C_4 with C_6; the human factors cause decrease in the final dose.
7. *Linkage 7*: Connects C_4 with C_5; the presence of human factors causes a decrease in the patient's positioning.
8. *Linkage 8*: Relates C_5 with C_4; any change on the patient positioning influences negatively the factors related to humans.
9. *Linkage 9*: Connects C_5 with C_6; when the patient positioning increases, the final dose also increases.
10. *Linkage 10*: Connects C_6 with C_5; when the final dose reaches an upper value, the patient positioning is influenced positively.
11. *Linkage 11*: Connects C_6 with C_1; any change in final dose causes change in tumor localization.
12. *Linkage 12*: Connects C_6 with C_2; when the final dose increases to an acceptable value, the dose from treatment planning also increases to a desired value.

After the determination of the linkages among concepts, experts suggested fuzzy values for the weights of the linkages. The fuzzy values were defuzzified and transformed in numerical weights, resulting in the following weight matrix:

$$
W^{\text{supervisor}} = \begin{pmatrix}
0.00 & 0.00 & 0.00 & 0.00 & 0.00 & 0.43 \\
0.28 & 0.00 & 0.00 & 0.00 & 0.00 & 0.57 \\
0.00 & -0.30 & 0.00 & 0.00 & 0.00 & -0.39 \\
0.00 & 0.00 & 0.00 & 0.00 & -0.32 & -0.43 \\
0.00 & 0.00 & 0.00 & -0.37 & 0.00 & 0.68 \\
0.22 & 0.67 & 0.00 & 0.00 & 0.54 & 0.00
\end{pmatrix}.
$$

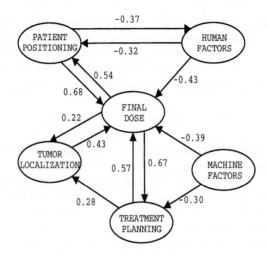

Fig. 1. The supervisor–FCM.

The final obtained supervisor–FCM is illustrated in Fig. 1.

The control objectives for the supervisor–FCM are to keep the amount of the final dose, FD, which is delivered to the patient, as well as the dose, D, prescribed by the treatment planning, within prespecified ranges:

$$FD_{\min} \leqslant FD \leqslant FD_{\max}, \tag{9}$$

$$D_{\min} \leqslant D \leqslant D_{\max}. \tag{10}$$

These objectives are defined by the related AAPM and ICRP protocols [1,10, 11], for the determination of accepted dose levels for each organ and part of the human body. The supervisor–FCM evaluates the success or failure of the treatment by monitoring the value of the "Final Dose" concept. Successful treatment corresponds to values of the final dose that lie within the desired bounds. The value of final dose identifies the supervisor model [29]. The supervisor–FCM has been incorporated to an integrated two–level hierarchical decision making system for the description and determination of the specific treatment outcome and for scheduling the treatment process before its treatment execution [12]. Thus, optimizing the supervisor–FCM, i.e. detecting the weights that correspond to the maximum values of the concepts FD and D, within their prespecified ranges, results in an enhanced control system which models the radiotherapy procedure more accurately and makes decision–making more reliable.

5   The Proposed Approach and Results

It has already been mentioned that the optimization of the supervisor–FCM described in Section 4, enhances the simulation ability of the system, resulting in more reliable decision–making. For this purpose, the PSO and DE algorithms have been used for the optimization of the supervisor–FCM, through the minimization of an appropriate objective function. The selection of these algorithms was based solely on their superior performance on a diverse field of applications, as well as due to the minimum effort required for their implementation.

The objective function can be straightforwardly defined as:

$$f(W) = -FD(W) - D(W), \tag{11}$$

where $FD(W)$ and $D(W)$ are the values of the final dose and the dose prescribed from the treatment planning, respectively, that correspond to the weight matrix W. The minus signs are used to transform the maximization problem to its equivalent minimization problem. Thus, the main optimization problem under consideration is the minimization of the objective function f, such that the constraints (9) and (10) hold.

The weight matrix W can, in general, be represented by a vector which consists of the rows of W in turn, excluding the elements of its main diagonal, $w_{11}, w_{22}, \ldots, w_{MM}$, which are by definition equal to zero. In the supervisor–FCM, the experts determined only 12 linkages, as described in Section 4, and thus, the corresponding minimization problem is 12–dimensional. Moreover, the bounds determined for the parameters FD and D, have been determined by the three radiotherapy oncologists (experts) in the following ranges:

$$0.90 \leqslant FD \leqslant 0.95, \tag{12}$$
$$0.80 \leqslant D \leqslant 0.95. \tag{13}$$

Taking into consideration the fuzzy linguistic variables that describe the cause–effect relationships among the concepts, as they were suggested by the three experts, the following ranges for the weight values were determined:

$$
\begin{array}{lll}
0.3 \leqslant w_{16} \leqslant 0.5, & -0.4 \leqslant w_{36} \leqslant -0.1, & 0.5 \leqslant w_{56} \leqslant 0.8, \\
0.2 \leqslant w_{21} \leqslant 0.4, & -0.5 \leqslant w_{45} \leqslant -0.2, & 0.2 \leqslant w_{61} \leqslant 0.4, \\
0.5 \leqslant w_{26} \leqslant 0.7, & -0.5 \leqslant w_{46} \leqslant -0.2, & 0.6 \leqslant w_{62} \leqslant 0.9, \\
-0.4 \leqslant w_{32} \leqslant -0.2, & -0.6 \leqslant w_{54} \leqslant -0.1, & 0.5 \leqslant w_{65} \leqslant 0.9.
\end{array}
\tag{14}
$$

These ranges were incorporated as constraints on the parameter vector in the experiments conducted.

The two most common variants of PSO and DE were used in the experiments. Specifically, the local versions of both the constriction factor and the inertia weight PSO variant, as well as the DE/rand/1/bin and DE/best/2/bin DE variants were used. Default values of the PSO parameters were used: $\chi = 0.729$, $c_1 = c_2 = 2.05$, and w decreasing from 1.2 to 0.1 [19,22]. Regarding the DE parameters, the values $F = 0.5$ and $CR = 0.5$ were selected after a trial and error process. The swarm (or population) size was always equal to 50. For each algorithm variant, 100 independent experiments were performed, to enforce the reliability of the results. Each algorithm was allowed to perform 1000 iterations (generations) per experiment.

The same solution was obtained by all algorithms in every experiment:

$$
W^* = \begin{pmatrix}
0.0 & 0.0 & 0.0 & 0.0 & 0.0 & 0.5 \\
0.4 & 0.0 & 0.0 & 0.0 & 0.0 & 0.7 \\
0.0 & -0.2 & 0.0 & 0.0 & 0.0 & -0.1 \\
0.0 & 0.0 & 0.0 & 0.0 & -0.2 & -0.2 \\
0.0 & 0.0 & 0.0 & -0.6 & 0.0 & 0.8 \\
0.4 & 0.9 & 0.0 & 0.0 & 0.9 & 0.0
\end{pmatrix},
$$

which corresponds to the following final state of the concepts after the convergence of the FCM:

$$
C_1^* = 0.819643, \quad C_2^* = 0.819398, \quad C_3^* = 0.659046,
$$

$$
C_4^* = 0.501709, \quad C_5^* = 0.824788, \quad C_6^* = 0.916315.
$$

Obviously, the values of C_2 and C_6 lie within the desired regions defined by the relations (12) and (13), while the weights fulfill the constraints (14) posed by the experts. This result supports the claim that the obtained solution seems to be the true optimal solution.

An interesting remark is the high positive influence of the concept C_6 (final dose) to the concept C_2 (dose prescribed from the treatment planning) as well as to the concept C_5 (patient positioning). This means that if we succeed to deliver the maximum dose to the target volume, then the initial calculated dose from treatment planning is the desired and the same happens with patient positioning. Another interesting fact is that the estimated weights assume their optimum values at the edges of the suggested fuzzy sets. This behavior has been also identified by other researchers [30,31].

The optimal values of "Final Dose" and "Dose Prescribed from the Treatment Planning" are acceptable according to the ICRU protocols [32,33], optimizing the whole treatment process. This supports the claim that the proposed approach is efficient and useful for the FCM–controlled radiation therapy process.

6   Conclusions

A Fuzzy Cognitive Map model, which supervises and monitors the radiotherapy process, resulting in a sophisticated decision support system, is optimized using the Particle Swarm Optimization and the Differential Evolution algorithms. The objective of the radiation treatment procedure is to give the acceptable–optimum amount of delivered dose to the target volume. The proposed methods determine the cause–effect relationships among concepts, determining the optimal weight matrix for the supervisor–FCM model.

Extensive experiments were performed, using different variants of the two stochastic optimization algorithms, always resulting in the same solution, which satisfies all optimality criteria and constraints imposed by the experts. This numerical evidence supports the claim that the obtained solution can be considered as an optimal by the user. The results contribute towards the direction of more reliable decision support system for radiation therapy.

Future work will focus on the optimization of the generic hierarchical system in all levels of radiation therapy, taking also into consideration the treatment planning (low–level) model of the system.

Acknowledgment. We thank the anonymous referees for their useful comments and suggestions. Also, we acknowledge the partial support by the "Pythagoras" research grant, awarded by the Greek Ministry of Education and Religious Affairs, as well as, the European Union.

References

1. Khan, F.: The Physics of Radiation Therapy. Williams & Wilkins, Baltimore (1994)
2. Brahme, A.: Optimization of radiation therapy and the development of multileaf collimation. Int. J. Radiat. Oncol. Biol. Phys. **25** (1993) 373–375
3. Brahme, A.: Optimization of radiation therapy. Int. J. Radiat. Oncol. Biol. Phys. **28** (1994) 785–787
4. Gibbons, J.P., Mihailidis, D.N., Alkhatib, H.A.: A novel method for treatment plan optimisation. In: Proc. 22–nd Ann. Int. Conf. IEEE Engin. in Med. and Biol. Soc. Volume 4. (2000) 3093–3095
5. Mageras, G.S., Mohan, R.: Application of fast simulated annealing to optimization of conformal radiation treatments. Med. Phys. **20** (1993) 639–647
6. Starkschall, G., Pollack, A., Stevens, C.W.: Treatment planning using dose–volume feasibility search algorithm. Int. J. Radiat. Oncol. Biol. Phys. **49** (2001) 1419–1427
7. Brahme, A.: Treatment optimization using physical and biological objective functions. In Smith, A., ed.: Radiation Therapy Physics. Springer, Berlin (1995) 209–246
8. Rowbottom, G., Khoo, V.S., Webb, S.: Simultaneous optimization of beam orientations and beam weights in conformal radiotherapy. Med. Phys. **28** (2001) 1696–1702
9. Soderstrom, S.: Radiobiologically Based Optimization of External Beam Radiotherapy Techniques Using a Small Number of Fields. PhD thesis, Stockholm University, Stockholm, Sweden (1995)
10. Wells, D., Niederer, J.: A medical expert system approach using artificial neural networks for standardized treatment planning. Int. J. Radiat. Oncol. Biol. Phys. **41** (1998) 173–182
11. Willoughby, T., Starkschall, G., Janjan, N., Rosen, I.: Evaluation and scoring of radiotherapy treatment plans using an artificial neural network. Int. J. Radiat. Oncol. Biol. Phys. **34** (1996) 923–930
12. Papageorgiou, E.I., Stylios, C.D., Groumpos, P.P.: An integrating two–level hierarchical system for decision making in radiation therapy using fuzzy cognitive maps. IEEE Transactions on Biomedical Engineering (2003) accepted for publication.
13. Papageorgiou, E.I., Stylios, C.D., Groumpos, P.P.: Decision making in external beam radiation therapy based on fuzzy cognitive maps. In: Proc. 1st Int. IEEE Symp. Intelligent Systems, Varna, Bulgaria (2002)
14. Parsopoulos, K.E., Papageorgiou, E.I., Groumpos, P.P., Vrahatis, M.N.: A first study of fuzzy cognitive maps learning using particle swarm optimization. In: Proceedings of the IEEE 2003 Congress on Evolutionary Computation, Canberra, Australia, IEEE Press (2003) 1440–1447

15. Kennedy, J., Eberhart, R.C.: Particle swarm optimization. In: Proceedings IEEE International Conference on Neural Networks. Volume IV., Piscataway, NJ, IEEE Service Center (1995) 1942–1948
16. Kennedy, J., Eberhart, R.C.: Swarm Intelligence. Morgan Kaufmann Publishers (2001)
17. Parsopoulos, K.E., Vrahatis, M.N.: Recent approaches to global optimization problems through particle swarm optimization. Natural Computing **1** (2002) 235–306
18. Eberhart, R.C., Simpson, P., Dobbins, R.: Computational Intelligence PC Tools. Academic Press (1996)
19. Clerc, M., Kennedy, J.: The particle swarm–explosion, stability, and convergence in a multidimensional complex space. IEEE Trans. Evol. Comput. **6** (2002) 58–73
20. Shi, Y., Eberhart, R.C.: A modified particle swarm optimizer. In: Proceedings IEEE Conference on Evolutionary Computation, Anchorage, AK, IEEE Service Center (1998) 69–73
21. Shi, Y., Eberhart, R.C.: Parameter selection in particle swarm optimization. In Porto, V.W., Saravanan, N., Waagen, D., Eiben, A.E., eds.: Evolutionary Programming. Volume VII. Springer (1998) 591–600
22. Trelea, I.C.: The particle swarm optimization algorithm: Convergence analysis and parameter selection. Information Processing Letters **85** (2003) 317–325
23. Parsopoulos, K.E., Plagianakos, V.P., Magoulas, G.D., Vrahatis, M.N.: Objective function "stretching" to alleviate convergence to local minima. Nonlinear Analysis, Theory, Methods & Applications **47** (2001) 3419–3424
24. Parsopoulos, K.E., Vrahatis, M.N.: Initializing the particle swarm optimizer using the nonlinear simplex method. In Grmela, A., Mastorakis, N., eds.: Advances in Intelligent Systems, Fuzzy Systems, Evolutionary Computation. WSEAS Press (2002) 216–221
25. Storn, R., Price, K.: Differential evolution–a simple and efficient heuristic for global optimization over continuous spaces. J. Global Optimization **11** (1997) 341–359
26. Plagianakos, V.P., Vrahatis, M.N.: Parallel evolutionary training algorithms for "hardware–friendly" neural networks. Natural Computing **1** (2002) 307–322
27. Kosko, B.: Fuzzy cognitive maps. Int. J. Man–Machine Studies **24** (1986) 65–75
28. Jang, J.S., Sun, C.T., Mizutani, E.: Neuro–Fuzzy and Soft Computing. Prentice Hall, Upper Saddle River, NJ (1997)
29. Dechlich, F., Fumasoni, K., Mangili, P., Cattaneo, G.M., Iori, M.: Dosimetric evaluation of a commercial 3–d treatment planning system using report 55 by aapm task group 23. Radiotherap. Oncol. **52** (1999) 69–77
30. Tsadiras, A.K., Margaritis, K.G.: Cognitive mapping and certainty neuron fuzzy cognitive maps. Information Sciences **101** (1997) 109–130
31. Tsadiras, A.K., Margaritis, K.G.: An experimental study of the dynamics of the certainty neuron fuzzy cognitive maps. Neurocomputing **24** (1999) 95–116
32. Determination of absorbed dose in a patient irradiated by beams of x or gamma rays in radiotherapy procedures. Technical Report 24, International Commission on Radiation Units and Measurements, Washington, USA (1976)
33. Prescribing, recording and reporting photon beam therapy. Technical Report 50, International Commission on Radiation Units and Measurements, Washington, USA (1993)

Identification of Informative Genes for Molecular Classification Using Probabilistic Model Building Genetic Algorithm

Topon Kumar Paul and Hitoshi Iba

Graduate School of Frontier Sciences, The University of Tokyo
Kashiwanoha 5-1-5, Kashiwa-shi, Chiba 277-8561, Japan
{topon,iba}@iba.k.u-tokyo.ac.jp

Abstract. DNA microarray allows the monitoring and measurement of the expression levels of thousands of genes simultaneously in an organism. A systematic and computational analysis of this vast amount of data provides understanding and insight into many aspects of biological processes. Recently, there has been a growing interest in classification of patient samples based on these gene expressions. The main challenge here is the overwhelming number of genes relative to the number of available training samples in the data set, and many of these genes are irrelevant for classification and have negative effect on the accuracy of the classifier. The choice of genes affects several aspects of classification: accuracy, required learning time, cost, and number of training samples needed. In this paper, we propose a new Probabilistic Model Building Genetic Algorithm (PMBGA) for the identification of informative genes for molecular classification and present our unbiased experimental results on three bench-mark data sets.

1 Introduction

The central dogma of molecular biology states that information is stored in DNA, transcribed to mRNA and then translated into proteins. The process by which mRNA and eventually protein is synthesized from the DNA template of each gene is called gene expression. Gene expression level indicates the amount of mRNA produced in a cell during protein synthesis; and is thought to be correlated with the amount of corresponding protein made. Expression levels are affected by a number of environmental factors, including temperature, stress, light, and other signals, that lead to change in the level of hormones and other signaling substances. Gene expression analysis provides information about dynamical changes in functional state of living beings. The hypothesis that many or all human diseases may be accompanied by specific changes in gene expressions has generated much interest among the Bioinformatics community in classification of patient samples based on gene expressions for disease diagnosis and treatment.

Classification based on microarray data faces with many challenges. The main challenge is the overwhelming number of genes compared to the number

K. Deb et al. (Eds.): GECCO 2004, LNCS 3102, pp. 414–425, 2004.

of available training samples, and many of these genes are not relevant to the distinction of samples. These irrelevant genes have negative effect on the accuracy of the classifier, and increase data acquisition cost as well as learning time. Moreover, different combination of genes may provide similar classification accuracy. Another challenge is that DNA array data contain technical and biological noises. So, development of a reliable classifier based on gene expression levels is getting more attention.

The main target of gene identification task is to maximize the classification accuracy and minimize the number of selected genes. For a given classifier and a training set, the optimality of a gene identification algorithm can be ensured by an exhaustive search over all possible gene subsets. For a data set with n genes, there are 2^n gene subsets. So, it is impractical to search whole space exhaustively, unless n is small. There are two approaches [19]: filter and wrapper approaches for gene subset selection. In filter approach, the data are preprocessed and some top rank genes are selected independently of the classifier. Although filter approaches tend to be much faster, their major drawback is that an optimal subset of genes may not be independent of the representational biases of the classifier that will be used during the learning phase [19].

In wrapper approach, the gene subset selection algorithm conducts the search for a good subset by using the classifier itself as a part of evaluation function. The classification algorithm is run on the training set, partitioned into internal training and holdout sets, with different gene subsets. The internal training set is used to estimate the parameters of a classifier, and the holdout set is used to estimate the fitness of a gene subset with that classifier. The gene subset with highest estimated fitness is chosen as the final set on which the classifier is run. Usually in the final step, the classifier is built using the whole training set and the final gene subset, and then accuracy is estimated on the test set. When number of samples in training data set is smaller, cross-validation technique is used. In k-fold cross-validation, the data D is randomly partitioned into k mutually exclusive subsets, $D_1, D_2, \ldots, D_k$ of approximately equal size. The classifier is trained and tested k times; each time $i(i = 1, 2, \ldots, k)$, it is trained with $D \backslash D_i$ and tested on D_i. When k is equal to the number of samples in the data set, it is called Leave-One-Out-Cross-Validation (LOOCV)[6]. The cross-validation accuracy is the overall number of correctly classified samples, divided by the number of samples in the data. When a classifier is stable for a given data set under k-fold cross-validation, the variance of the estimated accuracy would be approximately equal to $\frac{a(1-a)}{N}$ [6], where a is the accuracy and N is the number of samples in the data set. A major disadvantage of the wrapper approach is that it requires much computation time.

Numerous search algorithms have been used to find an optimal gene subset. In this paper, we propose a new method based on Probabilistic Model Building Genetic Algorithm (PMBGA) [16], which generates offspring by sampling the probability distribution calculated from the selected individuals under an assumption about the structure of the problem, as a gene selection algorithm. For classification, we use separately Naive-Bayes (NB) classifier [3] and the weighted voting classifier [5,18]. The experiments have been done with three well-known data sets. The experimental results show that our proposed algorithm is able to provide better accuracy with selection of smaller number of informative genes as compared to Multiobjective Evolutionary Algorithm (MOEA) [10].

2 Related Works in Molecular Classification Using Evolutionary Algorithms

Previously, Non-dominated Sorting Genetic Algorithm-II (NSGA-II) [4], Multi-objective Evolutionary Algorithm(MOEA) [10] and Parallel Genetic Algorithm [9] with weighted voting classifier have been used for the selection of informative genes responsible for the classification of the DNA microarray data.

In the optimization using NSGA-II, three objectives have been identified. One objective is to minimize the size of gene subset, the other two are the minimization of mismatches in the training and test samples, respectively. The number of mismatches in the training set is calculated using LOOCV procedure, and that in the test set is calculated by first building a classifier with the training data and the gene subset and then predicting the class of the test samples using that classifier. Due to inclusion of the third objective, the test set is, in reality, has been used as a part of training process and is not independent. Thus the reported 100% classification accuracy for the cancer data sets is not generalized accuracy, rather a biased accuracy on available data. In supervised learning, the final classifier should be evaluated on an independent test set that has not been used in any way in training or in model selection [7,17].

In the work using MOEA, also three objectives have been used; the first and the second objectives are the same as above, the third object is the difference in error rate among classes, and it has been used to avoid bias due to unbalanced test patterns in different classes. For decision making, these three objectives have been aggregated. The final accuracy presented is the accuracy on the training set (probably on the whole data) using LOOCV procedure. It is not clear how the available samples are partitioned into training and test sets, and why no accuracy on the test set has been reported.

In the gene subset selection using parallel genetic algorithm, the first two objectives of the above are used and combined into a single one by weighted sum, and the accuracy on the training and test sets (if available) have been reported. In our work, we follow this kind of fitness calculation.

3 Classifiers and Accuracy Estimation

3.1 Naive-Bayes Classifier

Naive-Bayes classifier uses probabilistic approach to assign the class to a sample. That is, it computes the conditional probabilities of different classes given the values of the genes and predicts the class with highest conditional probability. During calculation of conditional probability, it assumes the conditional independence of genes. Let C denote a class from the set of m classes, $\{c_1, c_2, \ldots, c_m\}$, $\mathbf{X}$ is a sample described by a vector of n genes, i.e., $\mathbf{X} = < X_1, X_2, \ldots, X_n >$;

the values of the genes are denoted by the vector $\mathbf{x} =< x_1, x_2, \ldots, x_n >$. Naive-Bayes classifier tries to compute the conditional probability $P(C = c_i|\mathbf{X} = \mathbf{x})$ (or in short $P(c_i|\mathbf{x})$) for all c_i and predicts the class for which this probability is the highest. The conditional probability takes the following form:

$$P(c_i|\mathbf{x}) \propto P(x_1|c_i)P(x_2|c_i)\cdots P(x_n|c_i)P(c_i) . \tag{1}$$

Taking logarithm we get,

$$\ln P(c_i|\mathbf{x}) \propto \ln P(x_1|c_i) + \cdots + \ln P(x_n|c_i) + \ln P(c_i) . \tag{2}$$

For a continuous gene, the conditional density is defined as

$$P(x_j|c_i) = \frac{1}{\sqrt{2\pi}\sigma_{ji}} e^{-\frac{(x_j - \mu_{ji})^2}{2\sigma_{ji}^2}} \tag{3}$$

where μ_{ji} and σ_{ji} are the expected value and standard deviation of gene X_j in class c_i. Taking logarithm of equation (3) we get,

$$\ln P(x_j|c_i) = -\frac{1}{2}\ln(2\pi) - \ln \sigma_{ji} - \frac{1}{2}\left(\frac{x_j - \mu_{ji}}{\sigma_{ji}}\right)^2 . \tag{4}$$

Since the first term in (4) is constant, it can be neglected during calculation of $\ln P(c_i|\mathbf{x})$. The advantage of the NB classifier is that it is simple and can be applied to multi-class classification problems.

3.2   Classifier Based on Weighted Voting

Classifier based on weighted voting has been proposed in [5,18]. We will use the term *Weighted Voting Classifier (WVC)* to mean this classifier. To determine the class of a sample, weighted voting scheme has been used. The vote of each gene is weighted by the correlation of that gene with the classes. The weight of a gene g is the correlation metric defined as

$$W(g) = \frac{\mu_1^g - \mu_2^g}{\sigma_1^g + \sigma_2^g} \tag{5}$$

where μ_1^g, σ_1^g and μ_2^g, σ_2^g are the mean and standard deviation of the values of gene g in class 1 and 2, respectively. The weighted vote of a gene g for an unknown sample x is

$$V(g) = W(g)\left(x_g - \frac{\mu_1^g + \mu_2^g}{2}\right) \tag{6}$$

where x_g is the value of gene g in that unknown sample. Then, the class of the sample x is

$$class(x) = sign\left\{\sum_{g\in G} V(g)\right\} \tag{7}$$

where G is the set of selected genes. If the computed value is positive, the sample x belongs to class 1; negative value means x belongs to class 2. But this kind of prediction does not make classification with reasonable confidence [5,18]. For more confident classification, we need to consider prediction strength. If we define V_+ and V_- are the absolute values of sum of all positive $V(g)$ and negative $V(g)$, respectively, the prediction strength,

$$ps = \left| \frac{V_+ - V_-}{V_+ + V_-} \right| . \tag{8}$$

The classification according to (7) is accepted if $ps > \theta$ (θ is the prefixed prediction strength threshold), else the sample is classified as undetermined. In our experiment, we consider undetermined samples as misclassified samples. This classifier is applicable to two-class classification tasks.

3.3 Accuracy Estimation

We use LOOCV procedure during the gene selection phase to estimate the accuracy of the classifier for a given gene subset and a training set. In LOOCV, one sample from the training set is excluded, and rest of the training samples are used to build the classifier. Then the classifier is used to predict the class of the left out one, and this is repeated for each sample in the training set. The LOOCV estimate of accuracy is the overall number of correct classifications, divided by the number of samples in the training set. Thereafter, a classifier is built using all the training samples, and it is used to predict the class of all test samples one by one. Final accuracy on the test set is the number of test samples correctly classified by the classifier, divided by the number of test samples. Overall accuracy is estimated by first building the classifier with all training data and the final gene subset, and then predicting the class of all samples (in both training and test sets) one by one. Overall accuracy is the number of samples correctly classified, divided by total number of samples. This kind of accuracy estimation on test set and overall data is unbiased because we have excluded test set during the search for the best gene subset.

4 Gene Selection Method

A new method based on Probabilistic Model Building Genetic Algorithm (PMBGA) [16] has been used as a gene selection method. PMBGA replaces the crossover and mutation operators of traditional evolutionary computations; instead, it uses probabilistic model building and sampling techniques to generate offspring. It explicitly takes into account the problem specific interactions among the variables. In evolutionary computations, the interactions are kept implicitly in mind; whereas in a PMBGA, the interrelations are expressed explicitly through the joint probability distribution associated with the individuals of variables, selected at each generation. The probability distribution is calculated from a database of selected candidate solutions of previous generation. Then, sampling this probability distribution generates offspring. The flow chart

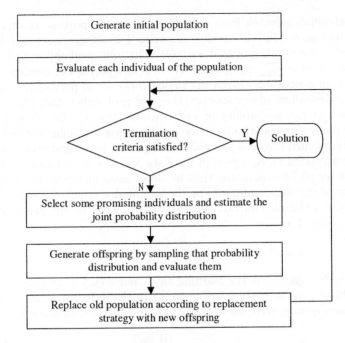

Fig. 1. Flowchart of a PMBGA

of a PMBGA is shown in figure 1. Since a PMBGA tries to capture the structure of the problem, it is thought to be more efficient than the traditional genetic algorithm. The other name of PMBGA is Estimation of Distribution Algorithm (EDA), which was first introduced in the field of evolutionary computations by Mühlenbein in 1996 [11]. A PMBGA has the follow components: encoding of candidate solutions, objective function, selection of parents, building of a structure, generation of offspring, selection mechanism, and algorithm parameters like population size, number of parents to be selected, etc.

The important steps of the PMBGA are the estimation of probability distribution, and generation of offspring by sampling that distribution. Different kinds of algorithms have been proposed on PMBGA. Some assume the variables in a problem are independent of one another, some consider bivariate dependency, and some multivariate. If the assumption is that variables are independent, the estimation of probability distribution as well as generation of offspring becomes easier. Reviews on PMBGA can be found in [8,12,13,14,15]. For our experiments, we propose another one which is described in the next subsection.

4.1   Proposed Method

Before we describe our proposed method, we need to give some notations. Let $X = \{X_1, X_2, \ldots, X_n\}$ be the set of n binary variables corresponding to n genes in the data set, and $x = \{x_1, x_2, \ldots, x_n\}$ be the set of values of those variables with $x_i \in \{0, 1\} (i = 1, \ldots, n)$ being the value of the variable X_i. N is the number of individuals selected from a population for the purpose of reproduction. $p(x_i, t)$ is the probability of variable X_i being 1 in generation t and $M(x_i, t)$ is the marginal distribution of that variable. The joint probability distribution is defined as $p(x, t) = \prod_{i=1}^{n} p(x_i, t | pa_i)$ where $p(x_i, t | pa_i)$ is the conditional probability of X_i in generation t given the values of the set of parents pa_i. If the variables are independent of one another, the joint probability distribution becomes the product of the probability of each variable $p(x_i, t)$. To select informative genes for molecular classification, we consider that variables are independent. We use binary encoding and probabilistic approach to generate the value of each variable corresponding to a gene in the data set. The initial probability of each variable is set to zero assuming that we don't need any gene for classification. Then, that probability is updated by the weighted average of marginal distribution and the probability of previous generation. That is, the probability of X_i has been updated as

$$p(x_i, t + 1) = \alpha p(x_i, t) + (1 - \alpha) M(x_i, t) \overline{w}(g_i) \tag{9}$$

where $\alpha \in [0, 1]$ is called the learning rate, and $\overline{w}(g_i) \in [0, 1]$ is the normalized weight of gene g_i corresponding to X_i in the data set. This weight is the correlation of gene g_i with the classes. This is calculated as follows:

$$\overline{w}(g_i) = \frac{|W(g_i)|}{MAX\{|W(g_1)|, |W(g_2)|, \ldots |, W(g_n)|\}} \tag{10}$$

where each $W(g_i)$ is calculated according to (5). The marginal distribution of X_i is calculated as follows:

$$M(x_i, t) = \frac{\sum_{j=1}^{N} \delta_j^i}{N} \tag{11}$$

where $\delta_j^i \in \{0, 1\}$ is value of variable X_i in the selected j^{th} individual. By sampling $p(x_i, t + 1)$, the value of X_i is generated for the next generation. Let us give an example of generating an offspring. Suppose, there are 4 genes in the data set; the normalized weight vector of genes and the probability vector of the corresponding variables at generation t are $\overline{w}(g) = (0.01, 1.0, 0.75, 0.05)$ and $p(x, t) = (0.001, 0.8, 0.4, 0.5)$, respectively. The selected individuals from a population at generation t are $(0, 1, 1, 0), (0, 1, 0, 1)$ and $(1, 1, 1, 0)$. Then, the corresponding marginal probabilities of variables will be $M(x, t) = (0.33, 1.0, 0.67, 0.33)$. If $\alpha = 0.1$, the updated probability according to (9) will be $p(x, t+1) = (0.00307, 0.98, 0.49225, 0.06485)$. Now generate 4 random numbers from uniform distribution and suppose they are $(0.001, 0.45, 0.6, 0.07)$. Comparing each random value with corresponding $p(x_i, t)$, we get an offspring $(1, 1, 0, 0)$.

4.2 Encoding and Fitness Calculation

In our experiments, the individuals in a population are binary-encoded with each bit for each gene. If a bit is '1', it means that the gene is selected in the gene subset; '0' means its absence. The fitness of an individual has been assigned as the weighted sum of the accuracy and dimensionality of the gene subset corresponding to that individual. It is

$$fitness(X) = w_1 * a(X) + w_2 * (1 - d(X)/n) \tag{12}$$

where w_1 and w_2 are weights from $[0, 1]$, $a(X)$ is the accuracy of X on training data, $d(X)$ the number of genes selected in X, and n is the total number of genes. This kind of fitness calculation was used in [9].

5   Experiments

5.1   Data Sets

We evaluate our method on three cancer data sets: Leukemia, Lymphoma and Colon. Leukemia and Lymphoma data sets need some preprocessing because the first one has some negative values while the second one has some missing values; we used the preprocessed data from
http://www.iitk.ac.in/kangal/bioinformatics.

Leukemia Data Set. This is a collection of gene expressions of 7129 genes of 72 leukemia samples reported by Golub et al. [5]. The data set is divided into an initial training set of 27 samples of Acute Lymphoblastic Leukemia (ALL) and 11 samples of Acute Myeloblastic Leukemia (AML), and an independent test set of 20 ALL and 14 AML samples. The data sets can be downloaded from http://www.genome.wi.mit.edu/MPR. These data sets contain many negative values which are meaningless for gene expressions, and need to be preprocessed. The negative values have been replaced by setting the threshold and maximum value of gene expression to 20 and 16000, respectively. Then genes that have $max(g) - min(g) > 500$ and $max(g)/min(g) > 5$ are excluded, leaving a total of 3859 genes. This type of preprocessing has been used in [4]. Then the data has been normalized after taking logarithm of the values.

Lymphoma Data Set. The Diffused Large B-Cell Lymphoma (DLBCL) data set [1] contains gene expression measurements of 96 normal and malignant lymphocyte samples, each measured using a specialized cDNA microarray, containing 4026 genes expressed in lymphoid cells or which are of known immunological or oncological importance. The expression data in raw format are available at http://llmpp.nih.gov/lymphoma/data/figure1/figure1.cdt. It contains 42 samples of DLBCL and 54 samples of other types. There are some missing gene expression values which have been replaced by applying k-nearest neighbor algorithm in [4]. Then the expression values have been normalized, and the data set is randomly divided into mutually exclusive training and test sets of equal size.

Colon Data Set. This data set, a collection of expression values of 62 colon biopsy samples measured using high density oligonucleotide microarrays containing 2000 genes, is reported by Alon et al. [2]. It contains 22 normal and 40 colon cancer samples. It is available at http://microarray.princeton.edu/oncology. These gene expression values have been log transformed, and then normalized. We divide the data randomly into training and test sets of equal size. The samples in one set are exclusive of the other set.

5.2   Experimental Results

For each data set and each experiment, the initial population is generated with each individual having 10 to 60 random bit positions set to '1'. This has been done to reduce the run time. For calculation of marginal distribution, we select best half of the population (truncation selection, $\tau = 0.5$). The settings of other parameters are: population size=500, maximum generation=10, elite=10%, α=0.1, w_1=0.75 and w_2=0.25. Elitism replacement has been used to prevent the so far found best individual of previous generations from being lost. We use both Naive-Bayes and weighted voting classifiers separately to predict the class of a sample. The algorithm terminates when either there is no improvement of the fitness value of the best individual in 5 consecutive generations or maximum number of generations has passed.

For all data sets, the average classification accuracy returned and number of genes selected by our algorithm in 50 independent runs are provided. For comparison, we provide only the experimental results of MOEA by Liu and Iba [10]. Though it is stated in the paper that the accuracy presented is on training set, it is actually the accuracy on all data (since all data have been used as training set) with prediction strength threshold 0. In the presented results, each value of the form $x \pm y$ indicates the average value x with the standard deviation y. The experimental results are shown in tables 1, 2 and 3. In the tables, WVC stands for weighted voting classifier and θ is the prediction strength threshold.

From the experimental results, we find that our algorithm using either Naive-Bayes or weighted voting classifier with $\theta = 0$ returns the same average accuracy on all three training data, but using weighted voting classifier (θ=0) provides better accuracy on test data as compared to using Naive-Bayes classifier. Weighted voting classifier with $\theta = 0.30$, provides 100% and 90%, 97% and 92%, 92% and 74% average accuracy on training and test sets of Leukemia, Lymphoma and Colon data, respectively. Our algorithm performs badly on colon data as compared to on other two data sets. According to our knowledge, there have been reported no algorithms and no classifiers that return 100% accuracy on this data set. The overall average accuracy returned by our method on each data set using both classifiers is superior to the accuracy returned by MOEA.

During the experiments, we found that our algorithm was selecting in each independent run only 2 genes using both Naive-Bayes and weighted voting classifier with prediction strength threshold 0 in the case of Leukemia data set. Weighted voting classifier with prediction strength threshold of 0.30 selects 2.02, 2.04 and 2.48 genes on the average for the three data sets: Leukemia, Lymphoma and Colon, respectively. In the case of Lymphoma and Colon data sets, Naive-

Bayes classifier selects higher number of genes with higher standard deviations than weighted voting classifier. The number of genes selected by applying MOEA are larger than those selected by our algorithm.

Table 1. The average accuracy returned by our algorithm on training and test data using weighted voting and Naive-Bayes classifiers

Data Set	WVC(θ=0)		WVC(θ=0.30)		Naive-Bayes Classifier	
	Train Set	Test set	Train Set	Test set	Train Set	Test Set
Leukemia	1.0 ± 0.0	0.92 ± 0.05	1.0 ± 0.0	0.90 ± 0.03	1.0 ± 0.0	0.91 ± 0.09
Lymphoma	0.99 ± 0.01	0.92 ± 0.04	0.97 ± 0.02	0.92 ± 0.05	0.99 ± 0.01	0.91 ± 0.05
Colon	0.96 ± 0.03	0.80 ± 0.07	0.92 ± 0.04	0.74 ± 0.09	0.96 ± 0.04	0.79 ± 0.06

Table 2. The average number of genes selected by our algorithm using weighted voting and Naive-Bayes classifiers

Data Set	WVC (θ=0)	WVC (θ=0.30)	NB Classifier	MOEA
Leukemia	2.0 ± 0.0	2.02 ± 0.14	2.0 ± 0.0	15.20 ± 4.54
Lymphoma	2.66 ± 2.22	2.04 ± 0.20	3.96 ± 4.70	12.90 ± 4.40
Colon	2.30 ± 0.61	2.48 ± 0.81	3.11 ± 1.32	11.4 ± 4.27

Table 3. The overall average accuracy reported by our algorithm on three data sets using weighted voting and Naive-Bayes classifiers

Data Set	WVC (θ=0)	WVC (θ=0.30)	NB Classifier	MOEA
Leukemia	0.96 ± 0.02	0.95 ± 0.02	0.95 ± 0.05	0.90 ± 0.07
Lymphoma	0.95 ± 0.02	0.95 ± 0.02	0.95 ± 0.02	0.90 ± 0.03
Colon	0.88 ± 0.03	0.83 ± 0.05	0.88 ± 0.03	0.80 ± 0.08

6   Discussion

Identification of the most useful genes for classification of available samples into two or more classes is a multi-objective optimization problem. There are many challenges for this classification task. Unlike other functional optimizations which use the values of the functions as fitness, this problem needs something beyond these values. It may be the case that you get 100% accuracy on training data but 0% accuracy on test data. So, the selection of proper training and test sets, and design of a reliable search method are very important. This problem has been solved in the past using both supervised and unsupervised methods. In this paper, we propose a new PMBGA for the selection of the gene subsets. Our method outperforms MOEA by selecting the most useful gene subset resulting in better classification accuracy.

In microarray data, overfitting is a major problem because the number of training samples given is very small compared to the number of genes. To avoid it, many researchers use all the data available to guide the search and report the accuracy that was used during the gene selection phase as the final accuracy. This kind of estimation is biased towards the available data, and may predict poorly when used to classify unseen samples. But our accuracy estimation is unbiased because we have isolated the test data from training data during gene selection phase. Whenever a training set is given, we have used that one only for the selection of genes, and the accuracy on the independent test set is presented using the final gene subset; whenever the data is not divided, we randomly partition it into two exclusive sets: training and test sets, and provide accuracy as described before.

Our algorithm finds smaller numbers of genes but results in more accurate classifications. This is consistent with the hypothesis that for a smaller training set, it may be better to select a smaller number of genes to reduce the algorithm's variance; and when more training samples are available, more genes should be chosen to reduce the algorithm's bias [7].

7   Summary and Future Work

In this paper, we have proposed a novel PMBGA for selection of informative genes aimed at maximizing classification accuracy for classification of DNA microarray data using either Naive-Bayes or weighted voting classifier. In our algorithm, the normalized weight of a gene is incorporated into the equation of updating the probability of the corresponding variable. The two objectives of the problem have been scalarized into one objective. We used the Leave-One-Out-Cross-validation technique to calculate the accuracy of an individual (a gene subset) on training data. By performing experiments, we found that the the accuracy was notably improved and the number of gene selected was smaller as compared to MOEA.

However, we have not attempted to identify the accession numbers of the selected genes and to maintain population diversity during the experiments; which are very important for cancer diagnosis and multimodal optimization. In the future, we would like to take care of these issues during experiments. We also plan to extend our algorithm for noisy DNA microarray data. As Naive-Bayes classifier is applicable to multi-class classification, we would like to apply our algorithm using this classifier on larger multi-class cancer data sets.

References

1. Alizadeh, A. A., Eisen, M. B., et al.: Distinct types of diffuse large B-cell lymphoma identified by gene expression profiling. Nature **403**(2000), 503–511.
2. Alon, U., Barkai, N., et al.: Broad patterns of gene expression revealed by clustering analysis of tumor and normal colon tissues probed by oligonucleotide arrays. Proceedings of National Academy of Science, Cell Biology, vol. 96, 1999, 6745–6750.
3. Cestnik, B.: Estimating probabilities: a crucial task in machine learning. Proceedings of the European Conference on Artificial Intelligence, 1990, 147–149.
4. Deb, K. and Reddy, A.R.: Reliable classification of two-class cancer data using evolutionary algorithms. BioSystems **72**(2003),111–129.
5. Golub, G.R., et al.: Molecular classification of cancer: class discovery and class prediction by gene expression monitoring. Science **286**(15)(1999), 531–537.

6. Kohavi, R.: A study of cross-validation and bootstrap for accuracy estimation and model selection. Proceedings of the International Joint Conference on Artificial Intelligence, 1995.

7. Kohavi, R. and John, G. H.: Wrappers for feature subset selection. Artificial Intelligence **97**(1-2)(1997), 273–324.

8. Larrañaga, P. and Lozano, J.A: Estimation of Distribution Algorithms: A New Tool for Evolutionary Computation. Kluwer Academic Publishers, Boston, USA, 2001.

9. Liu, J. and Iba, H.: Selecting Informative Genes with Parallel Genetic Algorithms in Tissue Classification. Genome Informatics **12**(2001), 14–23.

10. Liu, J. and Iba, H.: Selecting Informative Genes Using a Multiobjective Evolutionary Algorithm. Proceedings of the World Congress on Computation Intelligence(WCCI-2002), 2002, 297–302.

11. Mühlenbein, H. and Paaß, G. : From Recombination of Genes to the Estimation of Distribution I. Binary parameters. Parallel Problem Solving from Nature-PPSN *IV*, Lecture Notes in Computer Science (LNCS) 1411, Springer-Verlag, Berlin, Germany, 1996, 178–187.

12. Paul, T. K. and Iba, H.: Linear and Combinatorial Optimizations by Estimation of Distribution Algorithms. Proceedings of the 9th MPS Symposium on Evolutionary Computation, IPSJ, Japan, 2002, 99–106.

13. Paul, T. K. and Iba, H.: Reinforcement Learning Estimation of Distribution Algorithm. Proceedings of the Genetic and Evolutionary Computation Conference 2003 (GECCO2003), Lecture Notes in Computer Science (LNCS) 2724, Springer-Verlag, 2003, 1259–1270.

14. Paul, T. K. and Iba, H.:Optimization in Continuous Domain by Real-coded Estimation of Distribution Algorithm. Design and Application of Hybrid Intelligent Systems, IOS Press, 2003, pp. 262–271.

15. Pelikan, M., Goldberg, D.E. and Cantú-paz, E.: Linkage Problem, Distribution Estimation and Bayesian Networks. Evolutionary Computation **8**(3)(2000), 311–340.

16. Pelikan, M., Goldberg, D.E. and Lobo, F.G.: A Survey of Optimizations by Building and Using Probabilistic Models. Technical Report, Illigal Report no. 99018, University of Illinois at Urbana-Champaign, USA (1999).

17. Rowland, J.J: Generalization and Model Selection in Supervised Learning with Evolutionary Computation. EvoWorkshops 2003, LNCS 2611, Springer, 2003, pp. 119–130.

18. Slonim, D. K., Tamayo, P., et al.: Class Prediction and Discovery Using Gene Expression Data. Proceedings of the 4th Annual International Conference on Computational Molecular Biology, 2000, 263–272.

19. Yang, J. and Honavar, V.: Feature subset selection using a genetic algorithm. Feature extraction, construction and selection, Kluwer Academic Publishers, 1998, pp. 118–135.

GA-Facilitated Knowledge Discovery and Pattern Recognition Optimization Applied to the Biochemistry of Protein Solvation

Michael R. Peterson, Travis E. Doom, and Michael L. Raymer

Department of Computer Science and Engineering, Wright State University, Dayton, OH 45345 {mpeterso,doom,mraymer}@cs.wright.edu

Abstract. The authors present a GA optimization technique for cosine-based k-nearest neighbors classification that improves predictive accuracy in a class-balanced manner while simultaneously enabling knowledge discovery. The GA performs feature selection and extraction by searching for feature weights and offsets maximizing cosine classifier performance. GA-selected feature weights determine the relevance of each feature to the classification task. This hybrid GA/classifier provides insight to a notoriously difficult problem in molecular biology, the correct treatment of water molecules mediating ligand binding to proteins. In distinguishing patterns of water conservation and displacement, this method achieves higher accuracy than previous techniques. The data mining capabilities of the hybrid system improve the understanding of the physical and chemical determinants governing favored protein-water binding.

1   Introduction

Computational pattern recognition has proven to be a valuable tool in the analysis of biological data. Generally, objects are gouped into classes (such as diseased and healthy cells), and then characterized according to a variety of features. Feature selection facilitates classification by removing non-salient features. Even features providing some useful information may reduce accuracy when there are a limited number of training points available [1]. This "curse of dimensionality", along with the expense of measuring additional features, motivates feature dimensionality reduction. Though no known deterministic algorithm finds the optimal feature set for a classification task, a wide range of feature selection algorithms may find near-optimal feature sets [2].

The accuracy of some types of classification rules, such as k-nearest neighbors, improves by multiplying the value of each feature by a value proportional to its usefulness in classification. The assignment of weights to each feature as a form of feature extraction improves classifier accuracy over the knn classifier alone, and aids in the analysis of large datasets by isolating combinations of salient features [3]. Through use of a bit-masking feature vector, GAs have successfully performed feature selection in combination with a knn classifier [4]. This approach has been expanded for feature extraction [3,5] by searching for an ideal

K. Deb et al. (Eds.): GECCO 2004, LNCS 3102, pp. 426–437, 2004.

set of feature weights. Prior to classification, each feature value is multiplied by normalized values of GA-identified weights. The hybrid GA/$k$nn classifier described in [6] combines feature masking and feature weighting to simultaneously perform feature selection and extraction. The GA employs a weight vector for extraction and a mask vector for selection, allowing the GA to test the effect of completely eliminating a feature from consideration without reducing its associated weight completely to zero. The GA fitness function rewards smaller feature sets, leading to a tendency to mask features prematurely and not reintroduce them when appropriate.

Here, we present a novel hybrid GA/$k$nn system that eliminates the mask vector and instead employs a population-adaptive mutation technique allowing for improved simultaneous feature selection and extraction on the weight vector. Additionally, a cosine similarity measure replaces the traditional Euclidian distance metric for $k$nn classification. Cosine similarity is an effective similarity measure for diverse applications, including document classification [7] and gene expression profiling [8]. As with Euclidian distance, $k$nn classifiers employing cosine similarity may achieve improved classification through careful adjustment of feature weights [9]. Furthermore, the cosine similarity measure allows for a novel form of GA optimization by searching for an optimal set of feature offsets. These offsets affect the cosine of the angles between various data points considered by the $k$nn classifier, and thus may be optimized. In some cases, cosine similarity may be less prone to errors in attribute measurements. Euclidian distance is highly dependent upon the magnitude of measured attributes, since fluctuations in magnitude will directly affect the calculated distance between points. In contrast, the cosine measure depends more on the overall shape of the data distribution than on feature magnitude. Thus, in cases where the magnitude of features measured across experiments can vary, as in many biological experiments, cosine similarity is less susceptible to noise-induced error than Euclidian distance metrics [8].

This hybrid GA/$k$nn system provides new insight into the role of water molecules during the binding of drugs or other ligands to the protein surface. Protein surface-bound water molecules often form hydrogen bonds to a docking drug or other ligand, and are an essential part of the protein surface with respect to ligand screening, docking, and design [10]. It is thus important to identify the areas of the protein surface where water molecules will not be displaced upon ligand binding. However, the identification of favorable protein surface sites for solvent binding has proven difficult, in part because the majority of protein surface residues are hydrophilic.

Among the various attempts to treat water molecules during ligand binding, the *Consolv* system [10] employs a (GA/$k$nn) classifier to distinguish water molecules bound in the protein's ligand-binding site from those displaced upon ligand binding. *Consolv* improved on previous solvation techniques including *Auto-Sol* [11] and *AquariusII* [12]. The hybrid GA/classifier described here improves upon *Consolv's* reported accuracy while mining feature weights to aid in understanding the properties governing protein-water interactions.

2   Methods

2.1   Cosine-Based Knn Classification

The hybrid GA/classifier described here employs k-nearest neighbors classification. Unlike many common learning algorithms, $k$nn techniques do not construct an explicit description of the target function when a training set is provided. These algorithms only generalize beyond the training points when a query point is presented. Based upon the assumption that the classification of an unseen data point will be most similar to that of training points that share similar attributes, $k$nn approximates the target function over a small neighborhood of training points most similar to each test point.

When selecting the most similar neighbors, it is important to employ an appropriate similarity measure. There are several available for $k$nn classifiers; the most common of which is the Euclidean distance between two points within d-dimensional space, where d is the number of measured attributes. Another is the cosine of the angle between two vectors, each representing a data point within d-dimensional space. In addition to these measures, any other distance metric, such as Mahalanobis distance, may also be employed.

The $k$nn classifier described in this work employs cosine similarity as defined below. If $\boldsymbol{x}_i$ and $\boldsymbol{x}_j$ are attribute vectors representing two data points, then the cosine of the angle between them is defined as

$$cos(\boldsymbol{x}_i, \boldsymbol{x}_j) = \frac{\boldsymbol{x}_i \cdot \boldsymbol{x}_j}{\|\boldsymbol{x}_i\|\|\boldsymbol{x}_j\|} \tag{1}$$

where "$\cdot$" represents the dot product between the two vectors, and $\|\boldsymbol{x}_j\|$ represents vector length. Larger cosine values represent a greater degree of similarity between vectors. When taking the cosine similarity between a query point and all training points, the k points with the largest similarity values are the nearest neighbors.

After neighbor identification, a class is assigned to the query point. Unlike traditional $k$nn classification, the cosine-based $k$nn classifier described here does not use a simple voting scheme for class assignment. Classification occurs using a weighted scheme based on how similar each neighbor is to the query point. If the data contains only two classes, then the query point $\boldsymbol{x}$ is classified by the value of the measure q[13]:

$$q = \sum_{i=1}^{n} cos(\boldsymbol{x}_i, \boldsymbol{x})c(\boldsymbol{x}_i) \tag{2}$$

where

$$c(\boldsymbol{x}_i) = \begin{cases} 1 & : \quad \text{if } \boldsymbol{x}_i \in \text{ the positive class} \\ -1 & : \quad \text{otherwise} \end{cases}$$

If q is positive, then the query point is assigned to the positive class, otherwise it is assigned to the negative class. For problems with more than two classes, a seperate q function can be applied for each class, with the largest resulting function representing the class label applied to the query point.

In addition to feature weight evolution, the accuracy of the cosine classifier can be further improved by transformation of the coordinate space. Typically, the angle between two vectors is determined relative to the origin within the feature space. If the data is shifted by different amounts in each feature dimension, then the relative angle between any two given points changes. By shifting the origin within the feature space, a GA can improve the classification of new data. As demonstrated in Figure 1, this shifting may change the assigned class label for a test point. Figure 1(a) illustrates the behavior of an unshifted $k = 5$ nearest neighbor classifier in two-dimensional feature space. Here, no offset is applied. Among the points with the highest angular similarity to the test pattern, three belong to class 1, and 2 belong to class 2. The test pattern is labeled as belonging to class 1 since the sum of the cosine of angles to class 1 points is larger than that of class 2 points. In (b), the origin is shifted, thus changing the point of reference. Now, all of the nearest neighbors in terms of cosine similarity belong to class 2, so the test point is labeled as belonging to class 2. The hybrid GA/classifier system described here optimizes cosine $k$nn classifiers with respect to feature weights, feature offsets, and the k-value.

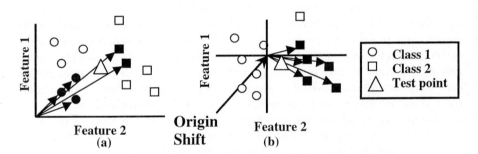

Fig. 1. Effect of the origin position on cosine-based $k$nn classification

2.2 Use of the Genetic Algorithm

Our goal is to provide a pattern recognition technique that improves classification accuracy while enabling knowledge discovery. To that end, an optimization technique should preserve the independence of features; that is, it must not define new features as combinations of the originals. In this manner, the influence of each feature upon the performance of the optimized classifier may easily be examined, thereby providing useful insight into the nature of the given problem.

The GA used here employs the chromosome shown in Figure 2. For an n-dimensional dataset, the first n genes on the chromosome represent the unnormalized real-valued weights of each feature. Weights evolve on the interval $[0.0 \ldots 100.0]$. The next n genes represent real-valued feature offsets for each feature. Prior to classification, each feature in a dataset is normalized by sum to the interval $[1.0 \ldots 10.0]$. Typically, offsets are permitted to evolve on the interval $[-15.0 \ldots 30.0]$ so that the classifier may view a feature either from within or from outside of its range. During the training process, the GA learns appropriate reference points by shifting offsets. For cases where classifiers are trained only through weight and k-value optimization, the offsets are simply ommited from the chromosome. The final gene is an integer representing the k-value for the $k$nn classifier. Typically, k may take any value over $[1 \ldots 100]$, though for small datasets a more limited range may be specified.

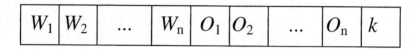

Fig. 2. Structure of the GA chromosome

In previous work involving GA optimization of $k$nn classifiers, dimensionality reduction is accomplished by employing feature masks using bit sets on the chromosome to explicitly perform feature selection [6]. The seperation of feature extraction and selection on the chromosome introduces the possibility that a partially relevant feature may be masked prematurely. If the GA gives preference to chromosomes masking a larger number of features, then a prematurely masked feature may never be unmasked. To avoid this, dimensionality reduction is accomplished using population-adaptive mutation. In the absence of an explicit feature mask, the weight of a feature must be reduced to zero in order to remove the feature from consideration. Population-adaptive mutation increases the likelihood that the weights of spurious features will be reduced to zero. When a gene is selected for mutation, the value of the gene is shifted randomly within a range depending upon the current variance of the gene across the GA population. The new value is randomly chosen from a Gaussian distribution with a mean equal to the gene's current value and a standard deviation based on the feature variance across the population. Under population-adaptive mutation, a gene may mutate either above or below its permitted range. In such cases, the minimum or maximum value is set as the new gene value. Thus, mutation may easily cause a feature weight to be set to zero due to this boundary effect, effectively removing that feature from consideration. As a feature weight gradually decreases across the population, the probability that the feature will be masked increases. Conversely, features with generally high weights are unlikely to become masked under population-adaptive mutation, thus preventing premature masking. Because population-adaptive mutation easily performs both feature selection and extraction, it is especially useful for mining relevant meaning from the remaining feature weights, which indicate the relevance of selected features for a given classification task.

Selection operations are implemented using tournament selection with a tournament of size 2. Recombination is implemented using uniform crossover with a crossover probability of 0.5 per gene. Because population-adaptive mutation largely drives dimensionality reduction, the GA uses a fairly high mutation rate of 0.1 mutations per gene. The population size is typically 50 or 100 chromosomes. The GA runs either to convergence or for 200 generations, whichever occurs first. During training, chromosomes are evaluated by applying their weight and offset vectors, as well as the k-value, to the feature set by performing classification on a set of patterns of known class using a cosine-based knn classifier. The fitness function contains components measuring overall classification accuracy, the balance of accuracy among classes, and the number of features employed. The fitness function gives preference to chromosomes providing high, balanced accuracy using as few features as possible. The GA-minimized cost function is:

$$cost(\boldsymbol{w}, k) = C_{pred} \times \% \text{ of incorrect predictions}$$
$$+ C_{mask} \times \# \text{ of unmasked features}$$
$$+ C_{bal} \times \text{class accuracy difference}$$

where C_{pred}, C_{mask}, and C_{bal}, are the cost function coefficients. For the authors' experiments the empirically derived values: $C_{pred} = 25.0$, $C_{bal} = 10.0$, and $C_{mask} = 1.0$ are used. The function gives highest preference to maintaining high overall accuracy, with balanced accuracy as a secondary goal. The number of features employed recieves a relatively small coefficient in order to prevent the GA from prematurely removing features from consideration.

For each experiment, data patterns are randomly split into class-balanced training and test sets, with the remaining points withheld for bootstrap validation upon the completion of GA training. At the completion of each GA experiment, the quality of the optimized classifier is assesed using a variant of the bootstrap test method [14] in order to obtain an unbiased accuracy measurement as well as a simple measure of this measurement's variance. The bootstrap test helps ensure that reported accuracies are not the result of GA overfitting of the test data.

2.3   Experiments on Biological Datasets

In order to demonstrate the utility of offset optimization, experiments are performed using a dataset containing diabetes diagnostic information for Native Americas of Pima heritage [15]. The dataset consists of diagnostic information for 768 adult women, 500 of whom tested negative for diabetes and 268 of whom tested positive. Six of the eight features representing clinical measurements are quantitative and continuous. The remaining two features are quantative and discrete. There are no missing feature instances. This dataset is suitable for testing the ability of a GA/classifier system to simultaneously extract features and boost accuracy due to its moderate dimensionality, its completeness, and its unbalanced class representation. This dataset is available from the UCI machine learning repository [16]. For comparison purposes, results for this dataset using a number of well-known classification techniques, as implemented in the Waikato Environment for Knowledge Analysis (WEKA) data mining software package [17] are also presented. Results from WEKA classifiers reflect accuracy after 10-fold cross-validation.

To demonstrate the knowledge discovery abilities of the hybrid GA/classifier, experiments are performed on two datasets describing protein-water interactions, created during development of the *Consolv* system. The first dataset describes the environments of water molecules bound to protein surfaces. Water molecules in this set belong to one of two classes: those displaced from the protein surface when a molecule (such as a drug) binds to the protein, and those conserved. When a ligand binds to a protein, it may displace water molecules at some locations and bind directly to the surface. In other locations, the ligand forms a hydrogen bond to a water molecule, which in turn forms a hydrogen bond to the protein surface, as illustrated in Figure 3. By accounting for water molecules involved in protein-ligand binding, accurate prediction of water conservation or displacement facilitates the design of ligands with higher complimentarity to the protein surface. Eight features are provided to characterize the local environment of water molecules in 30 independently solved, unrelated protein structures [6]. The chemical and physical features describing each water molecule include the number of protein atoms surrounding the water molecule (ADN), the frequency with which the types of atoms surrounding the water molecule are found to bind water molecules in another database of proteins (AHP), a measure of the thermal mobility (crystallographic temperature factor) of the water molecule (BVAL), the number of hydrogen bonds between the water molecule and the protein (HBDP), the number of hydrogen bonds to other water molecules (HBDW), and three additional normalizations on temperature factor of either the water molecule (MOB) or of its neighboring atoms (ABVAL and NBVAL). The goal of the GA/knn classifier is to distinguish conserved from displaced water molecules using a minimal set of features. Examination of the selected features and their corresponding weights found by the GA/knn classifier leads to a better understanding of the underlying physical and chemical properties salient to water binding interactions. The dataset describing water conservation and displacement consists of 5542 water molecules; 3405 conserved and 2137 displaced.

The second dataset consists of a set of all surface water molecules from the same 30 proteins, and an equal number of non-solvated probe sites, for a total of 11,084 samples. For each water molecule and probe site, all features (except for BVAL and MOB) from the first dataset are used. For this dataset, the goal of the GA/knn classifier is to distinguish solvation sites from non-sites with high accuracy using a minimal feature set. As before, the selected feature weights provide insight into the properties governing the interactions between water molecules and the protein surface.

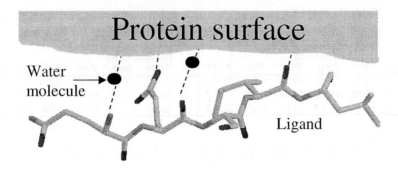

Fig. 3. Ligand-binding interface

For each GA run, the first dataset is split into class-balanced training and test datasets consisting of 1068 waters each, with all remaining waters reserved for bootstrap validation. The second dataset is split into balanced training and test sets consisting of 2128 waters each, with remaining patterns reserved for bootstrap validation. As with the diabetes dataset, results using WEKA classifiers are presented for comparison purposes. The two water datasets will be made publicly available at birg.cs.wright.edu/water/.

3   Results

3.1   Offset Inclusion During Optimization

Table 1 presents the best three optimizations obtained for the Pima Diabetes dataset including offset optimization (left) and using only weight and k-value optimization (right). Results reflect bootstrap validation. When offsets are included on the GA chromosome, the classifier achieves between 75 and 77% accuracy with an accuracy balance of approximately 7% between the classes. In contrast, the GA-trained classifier is unable to achieve better than 70% accuracy with a 10% class imbalance when offsets are not optimized. For this dataset, cosine-based $k$nn classifiers clearly benifit from offset optimization. Table 2 presents the cross-validated performance of the best 3 of 18 WEKA classifiers according to both accuracy and class accuracy balance. Logistic is a regression-based classifier, SMO is a support vector machine, DecisionStump and j48 are both decision tree-based classifiers, and NeuralNetwork is a backpropagation-based neural classifier. The most accurate WEKA classifier, Logistic, achieves higher accuracy than the GA-trained classifier, but it exhibits a high classification bias toward negative labelling, with an imbalance of 31.86%. Even the most balanced WEKA classifier, DecisionStump, exhibits a high imbalance of 16.18%. The GA is able to train a cosine-based $k$nn classifier to achieve an accuracy competitive with the best WEKA classifiers without resorting to an unbalanced bias toward one class or the other.

Table 1. Pima Diabetes results, with and without offset optimization

	With Offsets Accuracy (%)						Weights Only Accuracy (%)					
ID	Total	Neg	Pos	Bal	K	#F	ID	Total	Neg	Pos	Bal	K	#F
1	76.72	75.0	78.38	8.50	12	7	1	69.88	66.54	73.13	10.71	25	6
2	76.08	67.27	84.66	17.39	12	7	2	69.28	63.57	74.84	13.23	4	6
3	75.16	71.59	79.63	10.24	37	6	3	68.99	66.38	71.53	9.58	10	7

Table 2. Pima Diabetes WEKA classification results

Top 3 by Accuracy(%)					Top 3 by Balance(%)				
Classifier	Total	Neg	Pos	Bal	Classifier	Total	Neg	Pos	Bal
Logistic	77.08	88.20	56.34	31.86	DecisionStump	71.35	77.00	60.82	16.18
SMO	76.43	89.00	52.99	36.01	j48.J48	74.35	81.20	61.57	19.63
NaiveBayesSimple	75.91	83.80	61.19	22.61	NeuralNetwork	74.48	81.40	61.57	19.83

3.2   Ligand-Binding Water Conservation

The primary goal for experimental research on protein-bound water molecules is to classify whether specific water molecules on the protein surface are conserved or displaced upon ligand binding. This goal is manifested by achieving a high classification accuracy for this dataset during GA training. A secondary goal remains elucidation of the determinants of water conservation. This goal is met by examining the final relative feature weights of the various features evolved during classifier optimization. The inclusion of feature weights on the GA chromosome and the use of population-adaptive mutation for feature selection and extraction successfully yields combinations of features that provide improved distinction between conserved and displaced water molecules.

The left side of Table 3 presents the bootstrap results of the three best optimizations for the water conservation dataset. For comparison, the three best WEKA classifier results in terms of both accuracy and balance are presented on the right side of the table. While the most accurate WEKA classifiers achieve slightly higher accuracy, they all exhibit a notable bias towards the conserved class, indicating that they are unable to distinguish meaningful information in the dataset and thus resort to a preference toward the more frequently occuring class. In contrast, the GA-trained cosine $k$nn classifier achieves similar accuracy without significant bias toward either class, using as few as 4 of the 8 available features. The utility of using a GA favoring class-balanced results during optimization is clear. Section 4 discusses the biological implications of the optimized feature weights obtained for this dataset.

3.3   Solvation Site Prediction

As with the previous dataset, the goals for experiments investigating the physical and chemical determinants of solvation sites on the protein surface are two-fold.

Table 3. Water conservation results, GA (left) and WEKA classifiers (right)

Bootstrap Accuracy (%)					Feature Weights, Offsets			Top 5 WEKA Classifiers by Accuracy				
Total	Disp	Cons	Avg Bal	K	ADN	AHP	BVAL	Classifier	Total (%)	Disp	Cons	Bal
65.29	66.57	64.00	4.10	48	-	-	.426, 4.70	NeuralNetwork	66.62	44.17	80.70	36.53
64.76	63.91	65.61	3.89	29	-	.096, 4.321	.281, 2.29	j48.J48	66.02	37.06	84.20	47.14
64.31	63.52	65.10	3.61	26	-	-	.207, 1.08	ADTree	65.97	44.27	79.59	35.32

	Feature Weights, Offsets					Top 5 WEKA Classifiers by Balance				
Total	HBDP	NBVAL	HBDW	MOB	ABVAL	Classifier	Total (%)	Disp	Cons	Bal
65.286	-	.101, 3.68	-	.406, -.51	.067, -12.94	IB1	61.35	48.62	69.34	20.72
64.762	-	.078, 2.63	.115, 4.85	.238, 2.88	.192, -13.67	KernelDensity	61.30	47.73	69.81	22.08
64.309	.193, -14.96	-	.281, -9.48	.227, -1.95	.093, -12.51	NaiveBayesSimple	64.06	49.93	72.92	22.99

The first goal is to train a classifier to accurately identify favored solvation sites given the properties of a protein surface at varying localities. The second goal to determine the relative importance of the various chemical and physical factors governing solvation. Examination of the selected features and their evolved weights within a trained classifier leads to biological insights into the properties governing protein solvation.

The left side of Table 4 presents the best three results obtained for the solvation dataset, while the right side presents the best three WEKA classifiers in terms of both classification accuracy and class balance. The best GA-trained classifier achieves a mean bootstrap accuracy of 69.91% using five of the six available features. In contrast, the best WEKA classifiers achieve similar though slightly lower accuracy than the best optimized cosine knn classifier while maintaining a similar level of prediction balance. The main benefit of employing a hybrid GA/classifier system for the solvation dataset is the ability to elucidate the biological relevance of each feature through feature selection and extraction in order to form a more complete understanding of protein-water interactions.

Table 4. Water solvation results, GA (left) and WEKA classifiers (right)

Bootstrap Accuracy (%)					Weights, Offsets		Top 5 WEKA Classifiers by Accuracy				
Total	non	site	Avg Bal	K	ADN	AHP	Classifier	Total(%)	Non	Site	Bal
69.91	67.78	72.04	4.36	80	.252, -11.84	.177, -7.70	Logistic	69.33	65.50	73.16	7.67
69.48	65.79	73.16	7.38	67	.271, -8.42	.253, -15.00	NeuralNetwork	69.29	66.00	72.58	6.58
69.42	64.38	74.47	10.08	83	.242, -5.36	.222, 9.70	VotedPerceptron	69.25	66.75	71.74	4.98

	Feature Weights, Offsets				Top 5 WEKA Classifiers by Balance				
Total	HBDP	HBDW	ABVAL	NBVAL	Classifier	Total(%)	Non	Site	Bal
69.91	.352, -11.77	.105, -1.31	-	.113, 2.75	IB1	63.56	63.45	63.68	0.23
69.48	.154, -14.90	.113, -15.0	-	.209, -12.89	j48.J48	68.99	68.75	69.24	0.49
69.31	.205, -5.86	.165, 15.00	-	.166, 9.64	j48.PART	68.03	68.56	67.49	1.06

4   Discussion

Results obtained for the Pima diabetes dataset demonstrate the utility of optimizing offsets in addition to feature weights for a cosine-based knn classifier.

Evolution of both feature weights and offsets provides the GA an opportunity for classifier optimization that cannot be leveraged in Euclidian distance-based $k$nn classifiers. GA optimization of cosine-based classifiers may significantly boost the performance of a pattern recognition system. When compared to WEKA classifiers, results indicate that the hybrid GA/classifier system described here outperforms all other tested methods in terms of simultaneously increasing classification accuracy while maintaining class balance.

The features selected by the GA and their corresponding weights provide an opportunity to mine biologically relevant information from the optimized classifier systems. Consider the resulting features obtained by the GA during the best run on the conserved water dataset. All four selected features (BVAL, MOB, ABVAL, and NBVAL) relate to the thermal mobility of a given water molecule or of its surrounding atoms. In previous research, the *Consolv* system often favored BVAL and MOB, but almost always removed the other two features from consideration. Other features, such as AHP, HBDP, and HBDW, each relating to a water molecule's surrounding atomic environment, are more frequently considered [6]. In that work, 64.2% bootstrap accuracy is the highest reported result. Here, the GA-optimized cosine classifier increases accuracy over previously published results using only variations on thermal mobility. These results suggest that most of the information necessary to determine water molecule conservation upon ligand binding may be extracted from the thermal mobility and occupancy values of the water molecule and its neighbors. While other features may be related to conservation, they may be correlated with the temperature factor in such a way that they bring no additional information to the classification problem.

For the solvation dataset, the trained classifier consistently employs all measured features execpt for ABVAL. Features such as ADN, AHP, and ABDP typically recieve higher weights. These three features each depend upon the amount and type of atoms neighboring a probe site, suggesting that the local atomic environment of a probe site is more relevant than the thermal mobility of atoms surrounding the site in determining the favorability for solvation at the given site.

While maintaining a balanced accuracy level competitive with contemporary classification techniques, the hybrid GA/cosine classifier system described here provides the ability to mine insight into the relative importance of the various features provided for a given problem. This property permits the GA/cosine classifier system to be employed in cases where traditional techniques that do not maintain feature independence would not be well-suited for knowledge discovery.

References

[1] G. V. Trunk, "A problem of dimensionality: A simple example," *IEEE Transactions on Pattern Analysis and Machine Intelligence*, vol. 1, pp. 306–307, 1979.

[2] H. Liu and H. Motodata, *Feature Selection for Knowledge Discovery and Data Mining*, pp. 73–95. Boston, MA: Kulwer Academic Publishers, 1998.

[3] J. D. Kelly and L. Davis, "Hybridizing the genetic algorithm and the k nearest neighbors classification algorithm," in *Proceedings of the Fourth International Conference on Genetic Algorithms and their Applications*, pp. 377–383, 1991.

[4] W. Siedlecki and J. Sklansky, "A note on genetic algorithms for large-scale feature selection," *Pattern Recognition Letters*, vol. 10, pp. 335–347, 1989.

[5] W. F. Punch, E. D. Goodman, M. Pei, L. Chia-Shun, P. Hovland, and R. Enbody, "Further research on feature selection and classification using genetic algorithms," in *Proc. International Conference on Genetic Algorithms 93*, pp. 557–564, 1993.

[6] M. L. Raymer, W. F. Punch, E. D. Goodman, L. A. Kuhn, and A. K. Jain, "Dimensionality reduction using genetic algorithms," *IEEE Transactions on Evolutionary Computation*, vol. 4, no. 5, pp. 164–171, 2000.

[7] E. Han and G. Karypis, "Centroid-based document classification: Analysis & results," in *Principles of Data Mining and Knowledge Discovery: fourth European Conference*, pp. 424–431, 2000.

[8] M. P. S. Brown, W. N. Grundy, D. Lin, N. Cristianini, C. W. Sugnet, T. S. Furey, M. A. Jr., and D. Haussler, "Knowledge-based analysis of microarray gene expression data by using support vector machines," *Proceedings of the National Academy of Science*, vol. 97, pp. 262–267, 2000.

[9] E. Han, G. Karypis, and V. Kumar, "Text categorization using weight adjusted *k*-nearest neighbor classification," in *Advances in Knowledge Discovery and Data Mining: fifth Pacific-Asia Conference*, pp. 53–65, 2001.

[10] M. L. Raymer, P. C. Sanschagrin, W. F. Punch, S. Venkataraman, E. D. Goodman, and L. A. Kuhn, "Predicting conserved water-mediated and polar ligand interactions in proteins using a k-nearest-neighbors genetic algorithm," *J. Mol. Biol.*, vol. 265, pp. 445–464, 1997.

[11] A. Vedani and D. W. Huhta, "An algorithm for the systematic solvation of proteins based on the directionality of hydrogen bonds," *J. Am. Chem. Soc.*, vol. 113, pp. 5860–5862, 1991.

[12] W. R. Pitt, J. Murray-Rust, and J. M. Goodfellow, "AQUARIUS2: Knowledge-based modeling of solvent sites around proteins," *J. Comp. Chem.*, vol. 14, no. 9, pp. 1007–1018, 1993.

[13] M. Kuramochi and G. Karypis, "Gene classification using expression profiles: a feasibility study," in *Proceedings of the Second Annual IEEE International Symposium on Bioinformatics and Bioengineering*, pp. 191–200, 2001.

[14] A. K. Jain, R. C. Dubes, and C. C. Chen, "Bootstrap techniques for error estimation," *IEEE Transactions on Pattern Analysis and Machine Intelligence*, vol. 9, pp. 628–633, Sept. 1987.

[15] J. W. Smith, J. E. Everhart, W. C. Dickson, W. C. Knowler, and R. S. Johannes, "Using the ADAP learning algorithm to forecast the onset of diabetes mellitus," in *Proceedings of the Symposium on Computer Applications and Medical Care*, pp. 261–265, IEEE Computer Society Press, 1988.

[16] C. L. Blake and C. J. Merz, "UCI repository of machine learning databases." University of California, Irvine, Dept. of Information and Computer Sciences, 1998. http://www.ics.uci.edu/~mlearn/MLRepository.html.

[17] I. H. Witten and E. Frank, *Data Mining - Practical Machine Learning Tools and Techniques with Java Implementations*, pp. 265–319. San Francisco, GA: Morgan Kaufmann, 2000.

Genetic Programming Neural Networks as a Bioinformatics Tool for Human Genetics

Marylyn D. Ritchie[1], Christopher S. Coffey[2], and Jason H. Moore[1]

[1]Center for Human Genetics Research, Department of Molecular Physiology and Biophysics, Vanderbilt University, 519 Light Hall, Nashville, TN 37232
{ritchie, moore @chgr.mc.vanderbilt.edu}
[2]Department of Biostatistics, University of Alabama at Birmingham, Ryals Public Health Bldg., Rm. 327M, Birmingham, AL, 35294
{CCoffey@ms.soph.uab.edu}

Abstract. The identification of genes that influence the risk of common, complex diseases primarily through interactions with other genes and environmental factors remains a statistical and computational challenge in genetic epidemiology. This challenge is partly due to the limitations of parametric statistical methods for detecting genetic effects that are dependent solely or partially on interactions. We have previously introduced a genetic programming neural network (GPNN) as a method for optimizing the architecture of a neural network to improve the identification of gene combinations associated with disease risk. Previous empirical studies suggest GPNN has excellent power for identifying gene-gene interactions. The goal of this study was to compare the power of GPNN and stepwise logistic regression (SLR) for identifying gene-gene interactions. Using simulated data, we show that GPNN has higher power to identify gene-gene interactions than SLR. These results indicate that GPNN may be a useful pattern recognition approach for detecting gene-gene interactions.

1 Introduction

One goal of genetic epidemiology is to identify genes associated with common, complex multifactorial diseases. Success in achieving this goal will depend on a research strategy that recognizes and addresses the importance of interactions among multiple genetic and environmental factors in the etiology of diseases such as essential hypertension [1, 2]. One traditional approach to modeling the relationship between discrete predictors such as genotypes and discrete clinical outcomes is logistic regression [3]. Logistic regression is a parametric statistical approach for relating one or more independent or explanatory variables (e.g. genotypes) to a dependent or outcome variable (e.g. disease status) that follows a binomial distribution. However, as reviewed by Moore and Williams [2], the number of possible interaction terms grows exponentially as each additional main effect is included in the logistic regression model. Thus, logistic regression is limited in its ability to deal with interactions involving many factors. Having too many

K. Deb et al. (Eds.): GECCO 2004, LNCS 3102, pp. 438–448, 2004.

independent variables in relation to the number of observed outcome events is a well-recognized problem [4, 5] and is an example of the curse of dimensionality [6].

In response to this limitation, Ritchie et al. [7] developed a genetic programming optimized neural network (GPNN). Neural networks (NN) have been utilized in genetic epidemiology, however, with little success. A potential weakness in the previous NN applications is the improper selection of NN architecture. GPNN was developed in an attempt to improve upon the trial-and-error process of choosing an optimal architecture for a pure feed-forward back propagation neural network. The GPNN optimizes the inputs from a larger pool of variables, the weights, and the connectivity of the network including the number of hidden layers and the number of nodes in the hidden layer. Thus, the algorithm attempts to generate optimal neural network architecture for a given data set. This is an advantage over the traditional back propagation NN in which the inputs and architecture are pre-specified and only the weights are optimized.

Although previous empirical studies suggest GPNN has excellent power for identifying gene-gene interactions, a comparison of GPNN with a traditional statistical method has not yet been performed. The goal of the present study was to compare the power of GPNN and stepwise logistic regression (SLR) for identifying gene-gene interactions using data simulated from a variety of gene-gene interaction models. This study is motivated by the number of studies in human genetics where SLR has been applied. We wanted to determine if GPNN is more powerful than the status quo in the field. We find that GPNN has higher power to detect gene-gene interactions than stepwise logistic regression. These results demonstrate that GPNN may be an important pattern recognition tool for studies in genetic epidemiology.

2   Methods

2.1   A Genetic Programming Neural Network Approach

GPNN was developed to improve upon the trial-and-error process of choosing an optimal architecture for a pure feed-forward back propagation neural network (NN) [7]. Optimization of NN architecture using genetic programming (GP) was first proposed by Koza and Rice [8]. The goal of this approach is to use the evolutionary features of genetic programming to evolve the architecture of a NN. The use of binary expression trees allow for the flexibility of the GP to evolve a tree-like structure that adheres to the components of a NN. Figure 1 shows an example of a binary expression tree representation of a NN generated by GPNN. The GP is constrained such that it uses standard GP operators but retains the typical structure of a feed-forward NN. While GP could be implemented without constraints, the goal was to evolve NN since they were being explored as a tool for genetic epidemiology.

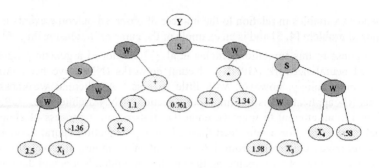

Fig. 1. An example of a NN evolved by GPNN. The Y is the output node, S indicates the activation function, W indicates a weight, and X_1-X_4 are the NN inputs.

Thus, we wanted to make an improvement to a method already being used. A set of rules is defined prior to network evolution to ensure that the GP tree maintains a structure that represents a NN. The rules used for this GPNN implementation are consistent with those described by Koza and Rice [8]. The flexibility of the GPNN allows optimal network architectures to be generated that consist of the appropriate inputs, connections, and weights for a given data set.

The GPNN method has been described in detail [7]. The steps of the GPNN method are shown in Figure 2 and described in brief as follows. First, GPNN has a set of parameters that must be initialized before beginning the evolution of NN models. These include an independent variable input set, a list of mathematical functions, a fitness function, and finally the operating parameters of the GP. These operating parameters include number of demes (or populations), population size, number of generations, reproduction rate, crossover rate, mutation rate, and migration [7]. Second, the data are divided into 10 equal parts for 10-fold cross-validation. Here, we will train the GPNN on 9/10 of the data to develop a NN model. Later, we will test this model on the 1/10 of the data left out to evaluate the predictive ability of the model.

Third, training of the GPNN begins by generating an initial population of random solutions. Each solution is a binary expression tree representation of a NN, similar to that shown in Figure 1. Fourth, each GPNN is evaluated on the training set and its fitness recorded. Fifth, the best solutions are selected for crossover and reproduction using a fitness-proportionate selection technique, called roulette wheel selection, based on the classification error of the training data [9]. Classification error is defined as the proportion of individuals where the disease status was incorrectly specified. A predefined proportion of the best solutions will be directly copied (reproduced) into the new generation. Another proportion of the solutions will be used for crossover with other best solutions. The new generation, which is equal in size to the original population, begins the cycle again. This continues until some criterion is met at which point the GPNN stops. This criterion is either a classification error of zero or the

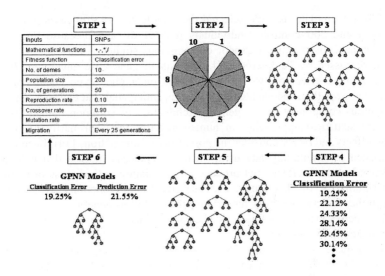

Fig. 2. The Steps of the GPNN algorithm

maximum number of generations having been reached. A "best-so-far" solution is chosen after each generation. At the end of the GPNN evolution, the one "best-so-far" solution is selected as the optimal NN. Sixth, this best GPNN model is tested on the 1/10 of the data left out to estimate the prediction error of the model. Prediction error is a measure of the ability to predict disease status in the 1/10 of the data. Steps two through six are performed ten times with the same parameters settings, each time using a different 9/10 of the data for training and 1/10 of the data for testing.

The results of a GPNN analysis include 10 GPNN models, one for each split of the data. In addition, a classification error and prediction error is recorded for each of the models. A cross-validation consistency can be measured to determine those variables which have a strong signal in the gene-gene interaction model [7, 10, 11, 12]. Cross-validation consistency is the number of times a particular combination of variables are present in the GPNN model out of the ten cross-validation data splits. Thus a high cross-validation consistency, ~10, would indicate a strong signal, whereas a low cross-validation consistency, ~1, would indicate a weak signal and a potentially false positive result.

2.2   Data Simulation

The goal of the simulation was to generate data sets that exhibit gene-gene interactions for the purpose of evaluating the power of GPNN in comparison to the power of SLR. We simulated a collection of models varying several conditions including number of interacting genes, allele frequency, and heritability. Heritability is defined in the broad sense as the proportion of phenotypic variation that is attributed to genetic factors. Loosely, this means the strength of the genetic effect. Thus a higher heritability will be a larger effect and easier to detect. Heritability is calculated using equations described in [13]. Additionally, we used a constant sample size for all simulations. We selected the sample size of 200 cases (individuals with disease) and 200 controls (individuals without disease) because this is a typical sample that is used in many genetic epidemiology studies.

As discussed by Templeton [14], epistasis, or gene-gene interaction, occurs when the combined effect of two or more genes on a phenotype could not have been predicted from their independent effects. It is anticipated that epistasis is likely to be a ubiquitous component of the genetic architecture of common human diseases [15]. Current statistical approaches in human genetics focus primarily on detecting the main effects and rarely consider the possibility of interactions [14]. In contrast, we are interested in simulating data using different epistasis models that exhibit minimal independent main effects, but produce an association with disease primarily through interactions.   In this study, we use penetrance functions as genetic models. Penetrance functions model the relationship between genetic variations and disease risk.   Penetrance is defined as the probability of disease given a particular combination of genotypes.

To evaluate the power of GPNN and SLR for detecting gene-gene interactions, we simulated case-control data using a variety of epistasis models in which the functional genes are single-nucleotide polymorphisms (SNPs). We selected models that exhibit interaction effects in the absence of any main effects. Interactions without main effects are desirable because they provide a high degree of complexity to challenge the ability of a method to identify gene-gene interactions. If main effects were present, it could be difficult to evaluate whether particular genes were detected due to the main effects or the interactions or both. In addition, it is likely that a method that can detect interacting genes in the absence of main effects will be able to detect main effect genes as well.

To generate a variety of epistasis models for this study, we selected three criteria for variation. First, we selected epistasis models with a varying number of interacting genes: either two or three. Previous studies had only investigated the power of GPNN using two-gene models [7]. We speculate that common diseases will be comprised of complex interactions among many genes. The number of interacting genes simulated here may still be too few to be biologically relevant. However, few, if any complex gene-gene interaction models are known at this time. Next, we selected two different allele frequencies. An allele frequency of 0.2/0.8 was selected so that we could evaluate the ability of GPNN in situations where there is a relatively rare allele. In addition, the frequency of 0.4/0.6 was selected to allow for the situation where both alleles are relatively common. Finally, we selected a range of heritability values including 3%, 2%, 1.5%, 1%, and 0.5%. These heritability values fall into the realm of very small genetic effects. In comparison, the heritability of many common diseases is much higher. For example, Alzheimer's disease is estimated to have heritability exceeding 60% [16] while breast, colorectal, and prostate cancers are 27%, 35%, and 42% respectively [17]. We chose to simulate data using epistasis models with such small heritability values to test the lower limits of GPNN. Based on previous studies, GPNN has over 80% power when the heritability is between 2%-5% [7]. For this particular study, we wanted to explore even smaller genetic effects to identify the point at which GPNN loses power.

We generated models using software described by Moore et al. [18]. We selected models from all possible combinations of number of interacting genes, allele frequency, and heritability, resulting in 20 total models. The penetrance tables for combinations of two SNPs are shown in Tables 1-10. The penetrance tables for the three SNP models are available from the authors by request. All 20 models were selected because they exhibit interaction effects in the absence of any main effects when genotypes are generated using the Hardy-Weinberg equation. Although the biological plausibility of these models is unknown, they represent the worst-case scenario for a disease-detection method because they have minimal main effects. If a method works well with minimal main effects, presumably the method will continue to work well in the presence of main effects.

Table 1. Model 1 – Two SNPs, allele frequency 0.2/0.8, $h^2 = 0.030$

	AA	Aa	aa
BB	0.0998	0.0984	0.0022
Bb	0.0933	0.0996	0.0002
bb	0.0028	0.0000	0.0574

Table 1 is an example of a penetrance function for a two-gene epistasis model with no main effects. Each gene is a single SNP with two alleles and three genotypes. In this example, the alleles each have a biological population frequency of $p = 0.2$ $q = 0.8$ with genotype frequencies of p^2 for AA and BB, $2pq$ for Aa and Bb, and q^2 for aa and bb, consistent with Hardy-Weinberg equilibrium. Thus, assuming the frequency of the AA genotype is 0.16, the frequency of Aa is 0.32, and the frequency of aa is 0.64, then the marginal penetrance of BB (i.e. the effect of just the BB genotype on disease risk) can be calculated as (0.04 * 0.0998) + (0.32 * 0.0984) + (0.64 * 0.0022) = 0.03. This means that the probability of disease given the BB genotype is 0.03, regardless of the genotype at the other genetic variation. Similarly, the marginal penetrance of Bb can be calculated as (0.04 * 0.0933) + (0.32 *0.0996) + (0.64 * 0.0002) = 0.03. Note that for this model, all of the marginal penetrance values (i.e. the probability of disease given a single genotype, independent of the others) are equal, which indicates the absence of main effects (i.e. the genetic variations do not independently affect disease risk). This is true despite the table penetrance values not being equal. Here, risk of disease is greatly increased by inheriting one of the following high-risk genotype combinations: AABB, AABb, AaBB, AaBb, and slightly increased by inheriting genotype combination aaBb.

Each data set consisted of 200 cases and 200 controls. We simulated 100 data sets of each model consisting of the functional SNPs and either seven or eight non-functional SNPs for a total of ten SNPs. This resulted in 2000 total datasets. We used a dummy variable encoding for the genotypes where n-1 dummy variables are used for n levels (or genotypes) [19]. Based on the dummy coding, these data would have 20 input variables.

Table 2. Model 2 - Two SNPs, allele frequency 0.2/0.8, $h^2 = 0.020$

	AA	Aa	aa
BB	0.0786	0.0003	0.0967
Bb	0.0010	0.0013	0.1001
bb	0.0948	0.0998	0.0428

Table 3. Model 3 - Two SNPs, allele frequency 0.2/0.8, $h^2 = 0.015$

	AA	Aa	aa
BB	0.0276	0.0942	0.0287
Bb	0.0941	0.0996	0.0226
bb	0.0277	0.0198	0.0657

Table 4. Model 4 - Two SNPs, allele frequency 0.2/0.8, $h^2 = 0.010$

	AA	Aa	aa
BB	0.0884	0.0894	0.0307
Bb	0.0710	0.0036	0.0737
bb	0.0368	0.0711	0.0404

Table 5. Model 5 - Two SNPs, allele frequency 0.2/0.8, $h^2 = 0.005$

	AA	Aa	aa
BB	0.0539	0.0732	0.0416
Bb	0.007	0.0207	0.0685
bb	0.0732	0.066	0.044

Table 6. Model 6 - Two SNPs, allele frequency 0.4/0.6, $h^2 = 0.030$

	AA	Aa	aa
BB	0.0848	0.0754	0.0053
Bb	0.0705	0.0135	0.0967
bb	0.0118	0.0937	0.0131

Table 7. Model 7 - Two SNPs, allele frequency 0.4/0.6, $h^2 = 0.020$

	AA	Aa	aa
BB	0.0093	0.0281	0.0902
Bb	0.0491	0.0763	0.0063
bb	0.0625	0.0161	0.0824

Table 8. Model 8 - Two SNPs, allele frequency 0.4/0.6, $h^2 = 0.015$

	AA	Aa	aa
BB	0.0381	0.0151	0.073
Bb	0.0485	0.0618	0.0067
bb	0.0288	0.0209	0.0693

Table 9. Model 9 - Two SNPs, allele frequency 0.4/0.6, $h^2 = 0.010$

	AA	Aa	aa
BB	0.0465	0.0368	0.0706
Bb	0.0666	0.0691	0.02
bb	0.0314	0.0329	0.0818

Table 10. Model 10 - Two SNPs, allele frequency 0.4/0.6, $h^2 = 0.005$

	AA	Aa	aa
BB	0.0161	0.0514	0.0573
Bb	0.0287	0.0442	0.0614
bb	0.0867	0.0511	0.0253

2.3   Data Analysis

Next, we used GPNN and SLR to analyze 100 data sets for each of the epistasis models. The GP parameter settings for GPNN included 10 demes, population size of 200 per deme, 50 generations, reproduction rate of 0.10, crossover rate of 0.90, mutation rate of 0.0, and migration every 25 generations. GPNN is not required to use all the variables as inputs. Here, GPNN performed random variable selection in the initial population of solutions. Through evolution, GPNN selects those variables that are most relevant. We calculated a cross-validation consistency for each SNP in each data set. This measure is defined as the number of times each SNP is in the GPNN model across the ten cross validation intervals. Thus, one would expect a strong signal to be consistent across all ten or most of the data splits, where a false positive signal may be present in only one or a few of the cross validation intervals. We estimated the power of GPNN as the number of times the correct functional SNPs had a cross-validation consistency that was higher than all other SNPs in the dataset, divided by the total number of datasets for each epistasis model. Either one or both of the dummy variables could be selected to consider a gene present in the model.

SLR is based on a statistical algorithm that determines the importance of variables and either includes them or excludes them from the model. The importance is determined by the statistical significance of the variable based on a chi-squared test [3]. Here, we used a p-Value of 0.20 to enter the model, and a p-Value of 0.10 to remain in the model. This type of model building procedure can also be referred to as hierarchical model building because to consider interactions among the variables, each variable must remain in the model due to its statistical significance on its own. Thus, using this approach, one can only detect interactions in the presence of main effects of each of the interacting variables. We performed this SLR procedure on each data set. We estimated power of SLR as the number of times the interaction term for the correct functional SNPs was selected in the final SLR model.

3   Results

The results of this study are shown in Table 11. Here, we list the 20 epistasis models sorted by number of genes, allele frequency, and heritability along the vertical axis. SLR has no power to detect the functional genes in any of the models studied. GPNN, on the other hand, has higher power than SLR for all of the epistasis models. The power of GPNN is higher for the models with two functional genes, and similarly for the models with higher heritability values.

4   Discussion

Identifying disease susceptibility genes associated with common complex, multifactorial diseases is a major challenge for genetic epidemiology. One of the dominating factors in this challenge is the difficulty in detecting gene-gene interactions with currently available statistical approaches. To deal with this issue, new statistical approaches have been developed such as the GPNN. GPNN has been shown to have higher power than a back propagation NN using simulated data generated under five two-gene epistasis models [7]. The goal of the current study was to compare the power of GPNN and SLR for detecting gene-gene interactions using data simulated from a variety of epistasis models. Computationally, GPNN is more burdensome than SLR. However, in human genetics the goal is to identify disease susceptibility genes. If one method is more powerful, even if it is more computationally expensive, it may be money well spent. Based on the results shown in Table 11, SLR had no power to detect a statistically significant interaction term. In comparison, GPNN had high power for most of the models examined. These results led to some skepticism that logistic regression (LR) may not be able to model the interactions that we had simulated. To be certain that LR was able to model these nonlinear interactions, we performed a forward selection LR analysis using only the two or three functional SNPs and their corresponding interaction term (Table 12). We estimated the power of LR using the number of data sets where the interaction term was statistically significant. In this study, LR had between 5-100% and 0-25% power for the two and three gene models respectively. Thus, LR was theoretically able to model these
interactions. We conclude that LR may be a successful procedure when the selection of variables has been conducted prior to the modeling process. However, when variable selection and modeling is taking place simultaneously, GPNN may provide higher power to detect such gene-gene interaction effects.

While these results demonstrate the lower limits of GPNN's power to detect gene-gene interactions, there are still many more questions to be addressed. First, it will be important to extend the simulation studies to include more interacting genes, larger sample sizes and a larger range of higher heritability values. In addition, a larger set of epistasis models including those with a small degree of main effect would provide further evidence of the power of GPNN. Finally, it would be interesting to use a different model validation procedure, such as the three-way data split [20], instead of ten-fold cross validation.

Table 11. Power comparison of GPNN and Stepwise Logistic Regression (SLR)

# Genes	Model Allele frequency	h^2	GPNN	SLR
			Power (%)	
2	0.2/0.8	0.030	100	0
2	0.2/0.8	0.020	94	0
2	0.2/0.8	0.015	97	0
2	0.2/0.8	0.010	81	0
2	0.2/0.8	0.005	24	0
2	0.4/0.6	0.030	100	0
2	0.4/0.6	0.020	99	0
2	0.4/0.6	0.015	99	0
2	0.4/0.6	0.010	77	0
2	0.4/0.6	0.005	16	0
3	0.2/0.8	0.030	99	0
3	0.2/0.8	0.020	94	0
3	0.2/0.8	0.015	22	0
3	0.2/0.8	0.010	4	0
3	0.2/0.8	0.005	3	0
3	0.4/0.6	0.030	75	0
3	0.4/0.6	0.020	35	0
3	0.4/0.6	0.015	20	0
3	0.4/0.6	0.010	3	0
3	0.4/0.6	0.005	1	0

The results of this study show that GPNN has higher power than SLR to detect gene-gene interactions in models with very small heritability values. Since most common diseases have overall heritability estimates greater than 20%, and GPNN was shown to have 100% power for heritability of 5% due to the genes examined [7], GPNN should have high power for detecting interactions in most common diseases. GPNN is likely to be a powerful pattern recognition approach for the detection of gene-gene interactions in future studies of common human disease.

Table 12. Power of Explicit Logistic Regression (LR)

# Genes	Model Allele frequency	h^2	LR
			Power (%)
2	0.2/0.8	0.030	92
2	0.2/0.8	0.020	61
2	0.2/0.8	0.015	100
2	0.2/0.8	0.010	60
2	0.2/0.8	0.005	64
2	0.4/0.6	0.030	7
2	0.4/0.6	0.020	5
2	0.4/0.6	0.015	29
2	0.4/0.6	0.010	44
2	0.4/0.6	0.005	89
3	0.2/0.8	0.030	25
3	0.2/0.8	0.020	16
3	0.2/0.8	0.015	2
3	0.2/0.8	0.010	5
3	0.2/0.8	0.005	2
3	0.4/0.6	0.030	0
3	0.4/0.6	0.020	7
3	0.4/0.6	0.015	0
3	0.4/0.6	0.010	12
3	0.4/0.6	0.005	2

Acknowledgements. This work was supported by National Institutes of Health grants HL65234, HL65962, GM31304, AG19085, AG20135, and LM007450.

References

1. Kardia S.L.R.: Context-dependent genetic effects in hypertension. Curr. Hypertens. Reports. 2 (2000) 32-38
2. Moore J.H. and Williams S.M.: New strategies for identifying gene-gene interactions in hypertension. Ann. Med. 34 (2002) 88-95
3. Hosmer D.W. and Lemeshow S.: Applied Logistic Regression. John Wiley & Sons Inc., New York (2000)
4. Concato J., Feinstein A.R., Holford T.R.: The risk of determining risk with multivariable models. Ann. Int. Med. 118 (1996) 201-210
5. Peduzzi P., Concato J., Kemper E., Holford T.R., Feinstein A.R.: A simulation study of the number of events per variable in logistic regression analysis. J. Clin. Epidemiol. 49 (1996) 1373-1379
6. Bellman R.: Adaptive Control Processes. Princeton University Press, Princeton (1961)
7. Ritchie M.D., White B.C., Parker J.S., Hahn L.W., Moore J.H.: Optimization of neural network architecture using genetic programming improves detection of gene-gene interactions in studies of human diseases. BMC Bioinformatics, 4 (2003) 28
8. Koza J.R. and Rice J.P.: Genetic generation of both the weights and architecture for a neural network. IEEE Press Vol II (1991) 397-404
9. Mitchell M.: An Introduction to Genetic Algorithms. MIT Press, Cambridge (1996)
10. Moore J.H.: Cross validation consistency for the assessment of genetic programming results in microarray studies. In: Lecture Notes in Computer Science Vol 2611 ed. by: Raidl, G, et al. Springer-Verlag, Berlin (2003) 99-106
11. Moore J.H., Parker J.S., Olsen N.J., Aune T.S.: Symbolic discriminant analysis of microarray data in autoimmune disease. Genet Epidemiol 23 (2002) 57-69
12. Ritchie M.D., Hahn, L.W., Roodi N., Bailey L.R., Dupont W.D., Parl F.F., Moore J.H.: Multifactor dimensionality reduction reveals high-order interactions among estrogen metabolism genes in sporadic breast cancer. Am. J. Hum. Genet. 69 (2001) 138-147
13. Culverhouse R., Suarez B.K., Lin J., Reich T.: A Perspective on Epistasis: Limits of Models Displaying No Main Effect. Am J Hum Genet 70 (2002) 461-471
14. Templeton A.R.: Epistasis and complex traits. In: Epistasis and Evolutionary Process. ed. by: Wolf J., Brodie III B., Wade M. Oxford University Press, Oxford (2000)
15. Moore J.H.: The ubiquitous nature of epistasis in determining susceptibility to common human diseases. Hum Hered 56 (2003) 73-82
16. Ashford J.W. and Mortimer J.A.: Non-familial Alzheimer's disease is mainly due to genetic factors. J Alzheimers Dis. 4 (2002) 169-77
17. Hemminki K. and Mutanen P.: Genetic epidemiology of multistage carcinogenesis. Mutat. Res. 473 (2001) 11-21
18. Moore, J.H., Hahn L.W., Ritchie M.D., Thornton T.A., White B.C.: Application of genetic algorithms to the discovery of complex genetic models for simulations studies in human genetics. In: Proceedings of the Genetic and Evolutionary Algorithm Conference ed. by W.B. Langdon, E. Cantu-Paz, K. Mathias, R. Roy, D. Davis, R. Poli, K. Balakrishnan, V. Honavar, G. Rudolph, J. Wegener, L. Bull, M.A. Potter, A.C. Schultz, J.F. Miller, E. Burke, and N. Jonoska. Morgan Kaufman Publishers San Francisco (2002) 1150-1155
19. Ott J.: Neural networks and disease association. Am. J. Med. Genet. 105 (2001) 60-61
20. Rowland J.J. Generalisation and model selection in supervised learning with evolutionary computation. In: Lecture Notes in Computer Science Vol 2611 ed. by: Raidl, G, et al. Springer-Verlag, Berlin (2003) 119-130

Evolving Better Multiple Sequence Alignments

Luke Sheneman and James A. Foster

Initiative for Bioinformatics and Evolutionary Studies (IBEST)
Department of Computer Science
University of Idaho, Moscow, ID 83844-1010, USA
+1 208.885.7062
{sheneman,foster}@cs.uidaho.edu

Abstract. Aligning multiple DNA or protein sequences is a fundamental step in the analyses of phylogeny, homology and molecular structure. Heuristic algorithms are applied because optimal multiple sequence alignment is prohibitively expensive. Heuristic alignment algorithms represent a practical trade-off between speed and accuracy, but they can be improved. We present EVALYN (EVolved ALYNments), a novel approach to multiple sequence alignment in which sequences are progressively aligned based on a *guide tree* optimized by a genetic algorithm. We hypothesize that a genetic algorithm can find better guide trees than traditional, deterministic clustering algorithms. We compare our novel evolutionary approach to CLUSTAL W and find that EVALYN performs consistently and significantly better as measured by a common alignment scoring technique. Additionally, we hypothesize that evolutionary guide tree optimization is inherently efficient and has less time complexity than the commonly-used neighbor-joining algorithm. We present a compelling analysis in support of this scalability hypothesis.

1 Introduction

Aligning multiple DNA or amino acid sequences is an extremely important task in modern biology. Researchers apply multiple sequence alignment (MSA) to a diverse set of problems. MSA is used to find positional similarity across distinct biological sequences as a first step in inferring sequence homology and the evolutionary relationships between organisms. MSA is used in gene identification and discovery and in identifying similarity in molecular structure and function. Among other practical applications, MSA plays a critical role in the diagnoses of genetic disease and the development of modern pharmaceuticals.

A sequence alignment is composed of two or more biological sequences which are arranged such that similar positions within the sequences are grouped (aligned). Alignments are often represented as a two-dimensional matrix where rows are sequences and columns are sequence positions. A good alignment is one which maximizes the positional similarity across all columns and all sequences. Alignments are constructed by inserting or deleting sequence segments in order to group similar characters into columns. Since it is impossible to know whether positions have been in

K. Deb et al. (Eds.): GECCO 2004, LNCS 3102, pp. 449–460, 2004.

serted or deleted relative to one another, these insertions/deletions are simply called *indels*. Indels can be represented as *gaps* in a sequence. Placing a single gap in a sequence causes the remainder of the sequence to shift by one position. Gaps placed in the optimal positions will result in an alignment where positional similarity is maximized as shown in Fig. 1.

```
A-CTTCAACTAAGT-ATTG-AATAAA-CT-GCTTAGATATATCTCCAAATTATTAGCTATCGCTTAT-GGATTATATTAC
ACCTTTA--TAAGTCATTG-ACT-AAGCTCGCCTAGAT--------AATTACCCGCTATCG---ATATCC-CCTATTAC
-CC-TCAACTAAGT-ATTG-AATAAAG---GCTTAGATATATCTCCAAATTACTAGCTAT----TATATCCTCATAT---
```

Fig. 1. Here is an example of a multiple sequence alignment of three DNA sequences in which gaps are denoted by the dash (-) character

Before the advent of alignment algorithms, researchers laboriously aligned multiple sequences by hand. This task was both error-prone and time-consuming. In the 1970's, researchers developed simple pairwise alignment algorithms based on dynamic programming (DP) and proved that they produce optimal alignments with respect to any given scoring system. [1, 2]. Although these algorithms extend easily to the simultaneous and optimal alignment of multiple sequences, they are NP-Hard [3] and have a time-complexity of $O(L^N)$ where L is the average length of the sequences being aligned and N is the number of sequences being aligned. Using DP, the simultaneous optimal alignment of more than a handful of sequences is prohibitively expensive. As a result, heuristic approaches trade quality for speed.

Progressive multiple sequence alignment is the most common heuristic [4] and is depicted in Fig. 2. In traditional progressive MSA, a distance matrix is formed by using DP to compute the optimal edit distance between all possible combinations of sequence pairs. A clustering algorithm such as neighbor-joining [5] takes a distance matrix as input and deterministically constructs a *guide tree* based on these distances, grouping closely related sequences prior to more divergent sequences. Once a guide tree has been constructed, sequences are progressively pairwise aligned in the order dictated by the guide tree. Closely related sequences are aligned prior to more distant sequences. Progressive MSA avoids the computationally intractable problem of the simultaneous alignment of multiple sequences by instead performing incremental pairwise alignments.

However, traditional progressive MSA has fundamental problems. Most importantly, after sequences are pairwise aligned, any inserted gaps in that pairwise alignment become immutable, and subsequent alignments with other sequences cannot retroactively add additional information to improve previously aligned sequences. This is a form of error propagation, also known as "*once a gap, always a gap*". The guide tree has a direct qualitative impact on this error propagation, as the amount of error is heavily dependent on the order in which sequences are progressively aligned. Since guide tree construction algorithms such as neighbor-joining are greedy and starting-point dependent, they are easily trapped in local optima, often resulting in suboptimal multiple alignments. We hypothesize that an evolutionary algorithm is better able to avoid entrapment in local optima and will therefore perform better than neighbor-joining in constructing good guide trees. Better guide trees result in better multiple sequence alignments.

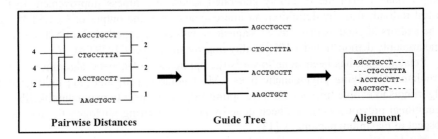

Fig. 2. In traditional progressive MSA, clustering algorithms use computed pairwise edit distances to construct a guide tree which clusters similar sequences prior to divergent sequences. A guide tree specifies an ordering of pairwise alignment operations which construct a complete multiple sequence alignment

Additionally, for large datasets, neighbor-joining has a prohibitive time complexity of $O(N^3)$, where N is the number of input sequences. As researchers apply MSA to larger and larger datasets, neighbor-joining scales poorly as N grows large. It is our hypothesis that an evolutionary computational approach to guide tree construction is more scalable than neighbor-joining.

2   Previous Work

Notredame and Higgins performed the seminal work in applying a genetic algorithm (GA) to MSA with a tool known as *Sequence Alignment by Genetic Algorithm* (SAGA) [6]. SAGA evolves a population of alignments using a complex set of 22 crossover and mutation operators in an attempt to gradually improve the fitness of the alignments in the population. Providing meaningful scores for sequence   alignments can be somewhat problematic, and SAGA relies on a weighted sum-of-pairs approach [7] in which each pair of sequences in an alignment is compared and scored and then the scores from all of the pairwise alignments are summed to produce a representative score for the entire alignment.

Although SAGA produces high quality results which are comparable (or sometimes better) than other popular heuristic techniques, SAGA scales poorly [6] when aligning more than 20 sequences. SAGA applies a large and overly-complex litany of crossover and mutation operators, which are dynamically scheduled via a sophisticated adaptive, self-tuning mechanism. By contrast, the GA approach outlined herein uses only one form of crossover and one mutation operator, thus simplifying the implementation and analysis of the algorithm.

Thomsen et al. [8] developed an alignment post-processing program that uses a genetic algorithm to *improve* alignments constructed by algorithms such as CLUSTAL

V [9]. A population of alignments is initialized by randomly distributing gaps throughout the individual alignments yet seeding the population with a single alignment produced by CLUSTAL V. Assuming that this CLUSTAL-derived seed was of higher quality than the randomly generated seeds, any fitness improvement in the fittest individual is, by definition, an improvement over the output of CLUSTAL V. The authors aligned as many as 71 sequences with an average length of 100 residues and arguably demonstrated a 10% quality improvement over CLUSTAL V.

Related work has been done towards the application of genetic algorithms to the problem of evolving phylogenetic trees. Most notably, [10] and [11] used genetic algorithms to evolve trees which were optimized with respect to maximum likelihood. Additional previous work has been done by [12] in inferring phylogenetic trees with respect to maximum parsimony [13].

Notably, the manipulation of tree-based data structures with genetic algorithms has been widely explored in the *genetic programming* literature [14, 15].

3   Algorithm Implementation

We present EVALYN, a novel progressive multiple sequence alignment (MSA) program that utilizes a genetic algorithm (GA) to optimize guide trees. EVALYN starts with a steady-state population of randomly constructed binary trees and iteratively optimizes this population using a combination of selection, crossover, and mutation. Guide trees are rooted binary trees. Each node in the guide tree contains an alignment which in turn contains at least one sequence. Leaf nodes have no children and contain the original input sequences.

3.1   Crossover and Mutation Operators

In each iteration of the genetic algorithm, EVALYN selects two unique parents for crossover based on an exponential distribution of relative rank. This ensures that highly fit trees are selected for crossover far more often than unfit trees, yet all trees are viable crossover candidates. Similarly, EVALYN selects a single unfit guide tree to be replaced by the offspring of a crossover operation.

EVALYN's crossover operator is depicted in Fig. 3. We implement tree crossover in a way similar to that described in GAML [10], a genetic algorithm for phylogenetic inference. Both selected parents are copied and a randomly chosen *crossover point* (internal node) is selected in the first parent. The first parent is re-rooted at the crossover point, and the remainder of the tree above the crossover point is discarded. All leaf nodes which exist in this new, smaller tree are removed from the second parent, and the second parent is *collapsed* into a typical bifurcating tree. This collapsed second parent is then *attached* to the first parent at a randomly chosen *insertion point*.

With some small probability, EVALYN mutates the child tree by performing a same-tree branch swap. Conveniently, this is implemented by performing a crossover operation on two copies of the *same* child guide tree, effectively swapping branches within the same tree.

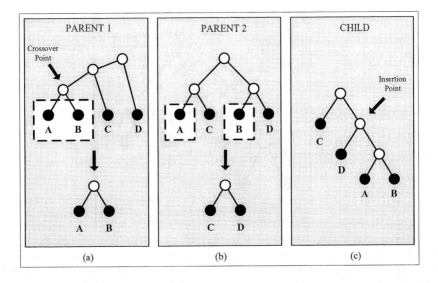

Fig. 3. Crossover is a three-step process. First, a copy of *PARENT 1* is rooted at a randomly selected crossover point and all nodes above this new root are discarded as shown in *(a)*. Next, all leaves are removed from a copy of *PARENT 2* which exist in the newly-rooted tree from *(a)*. As shown in *(b)*, leaves *A* and *B* are removed from *PARENT 2*, and the tree is collapsed to form a new bifurcating tree containing only leaves *C* and *D*. In *(c)*, the final child tree is constructed by combining the sub-trees from *(a)* and *(b)* at a randomly chosen insertion point

3.2 Measuring Guide Tree and Alignment Fitness

As shown in Fig. 4, the fitness of a particular guide tree is measured by performing progressive MSA in the order dictated by the guide tree and then scoring the resulting alignment.

We use the common sum-of-pairs score (SPS) as our scoring method as outlined in Fig. 5. In the case of protein, the evolutionary distance between every pair of residues in each column of the alignment is computed using a probabilistic residue substitution model such as PAM [16] or BLOSUM [17]. DNA is similarly handled using simple nucleotide substitution models which properly weight transitions and transversions. The SPS for the entire alignment is simply the sum of the SPS for each column in the alignment. Gaps are typically assigned large penalties, while substitutions are assigned smaller negative penalties or positive rewards. For our purposes, a higher SPS indicates a better alignment. Therefore, there is selective pressure favoring alignments with higher sum-of-pairs scores.

As in other MSA implementations, EVALYN implements *affine gap penalties*, in which the leading gap in a subsequence of contiguous gaps invokes significantly higher penalties than non-leading gaps. Affine gap penalties result in dense, contiguous gapped regions instead of sparsely distributed, isolated gaps. This affine gap model results in more biologically realistic alignments.

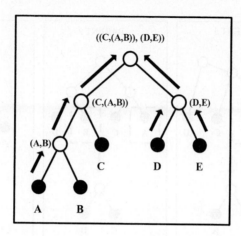

Fig. 4. Evaluating the fitness of a guide tree is accomplished by performing the progressive sequence alignment in the order dictated by the guide tree. EVALYN performs a depth-first traversal of the guide tree and aligns *A* and *B* first. Sequence *C* is then aligned to the alignment of *A* and *B* to form a 3-sequence alignment of sequences *A*, *B*, and *C*. The complete multiple sequence alignment of all sequences is performed by aligning the two alignments on either side of the root node of the guide tree. The sum-of-pairs score of the final alignment is the fitness of the alignment

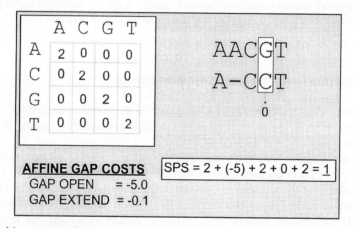

Fig. 5. In this toy example, a sum-of-pairs score (SPS) is computed for a simple pairwise alignment. A substitution matrix assigns points for nucleotide matches/mismatches and affine gap penalties are applied

4 Algorithm Analysis

We hypothesize that EVALYN has less computational complexity than neighbor-joining, and is therefore more scalable than CLUSTAL W as a function of the number of input sequences. We briefly analyze the time complexity of EVALYN and compare it to the time complexity of the neighbor-joining algorithm used in CLUSTAL W to demonstrate support for this hypothesis. We show that with only regard to the number of input sequences, EVALYN is an $O(N)$ algorithm. By contrast, CLUSTAL W's neighbor-joining algorithm is $O(N^3)$.

EVALYN evaluates guide tree fitness by performing the progressive MSA in the order dictated by the guide tree and computing the sum-of-pairs score for the resulting alignment. At each step in the progressive alignment, we compute an optimal global pairwise alignment [1]. Each pairwise alignment has an $O(L^2)$ complexity, where L is the average length of the sequences or partial alignments being aligned. There are N-1 such pairwise alignments for every evaluation of a guide tree, resulting in an $O(N \times L^2)$ complexity, where N is the number of input sequences being aligned, and L is the average length of all sequences. This fitness evaluation happens once per iteration. With I iterations, EVALYN becomes an $O(I \times N \times L^2)$ algorithm. Finally, when evaluating the fitness of the initial, randomly generated population of size P, the fitness of each guide tree must be computed prior to iteration, resulting in an initial cost of $O(P \times N \times L^2)$. In typical usage, $I >> P$ and we can simplify our analysis to $O(I \times N \times L^2)$. Although EVALYN is an iterative algorithm and has large amounts of constant-time overhead in the form of the multiple I, it does have a linear time complexity with respect to the N input sequences.

Although CLUSTAL W is fast in the typical usage scenario, it performs very poorly as N grows very large (thousands of input sequences). CLUSTAL W uses neighbor-joining to construct guide trees, and neighbor-joining has been shown to possess a $O(N^3)$ time complexity [18].

The *practical* question remains as to whether or not EVALYN is capable of finding comparable or better guide trees in less time than CLUSTAL W when aligning extremely large numbers of sequences. For example, if N is very large, it may be the case that I must be similarly large in order for EVALYN to converge on guide trees which score better than those constructed via neighbor-joining. We've shown that EVALYN is $O(N)$, but how fast does I (or P) need to grow as a function of N in order to get good alignments? Future experiments will focus on characterizing this behavior.

5   Experimental Setup and Results

Our central hypothesis is that a genetic algorithm (GA) is capable of finding better guide trees than those which are constructed using traditional deterministic clustering algorithms such as neighbor-joining. To test this hypothesis, we compare EVALYN to the popular CLUSTAL W progressive MSA tool [19]. CLUSTAL W uses neighbor-joining to construct guide trees based on a computed pairwise distance matrix. Both EVALYN and CLUSTAL W compute the sum-of-pairs score for the final multiple sequence alignment, and this is used as an objective metric of alignment quality.

First, we simulated DNA sequences according to the Jukes-Cantor model [20] of sequence evolution in which transitions and transversions are equally probable. We simulated the DNA sequences by producing a random template sequence of the desired length and then used this template sequence to generate related sequences with no more than 50% sequence divergence. In this way, we generated 10 independent sets of 50 sequences, all of which were 100 nucleotides in length. We performed 10 experimental runs of EVALYN on each of the 10 input datasets and averaged the results.

After generating our input sequences, CLUSTAL W was run with *default* parameters on each of the 10 input datasets and we recorded the final sum-of-pairs score of the output alignment. In 10 independent trials, EVALYN used the same 10 inputs and saved the best guide tree after 2500 iterations. In all cases, EVALYN used a population size of 500 guide trees, a mutation rate of 0.01, and ran for 2500 iterations as shown later in Table 1.

Where possible, EVALYN was parameterized identically to CLUSTAL W with respect to gap penalties and nucleotide substitution costs. The results from this experiment are shown in Fig. 6.

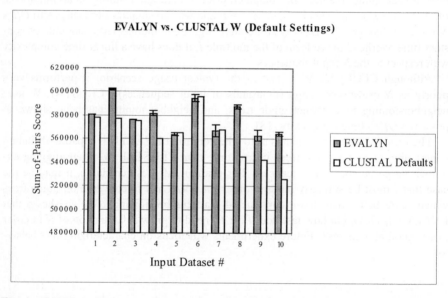

Fig. 6. CLUSTAL W run with default settings. In 70% of the runs, EVALYN produced better scoring alignments. The mean SPS and standard deviation across repeated trials is shown

We then invoked CLUSTAL W *again* for each dataset, but instead of generating its own guide trees, CLUSTAL W instead used the best guide trees produced *by EVALYN*. This novel technique removed any experimental error due to possible inconsistencies in alignment scoring between CLUSTAL W and EVALYN. We computed the mean and standard deviation of the sum-of-pairs scores across all 10 trials for each of the 10 inputs. We also calculated the standard deviation across EVALYN runs to assess the error and statistical significance of our results. Results indicate that EVALYN typically (70% of the time) outperforms CLUSTAL W when using CLUSTAL W with its default settings.

Additionally, EVALYN outperforms CLUSTAL W significantly and consistently when the two programs have identical parameterization as shown in Fig. 7. Tests of the statistical significance of all results were performed using a non-parametric Wilcoxon signed-rank test and showed that these results are statistically significant under that test.

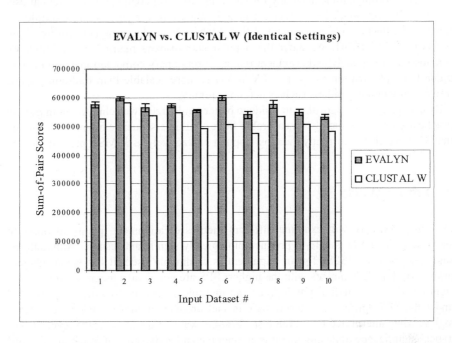

Fig. 7. CLUSTAL W and EVALYN were run with nearly identical parameterization (same substitution matrix, same gap penalties, etc.) Across all 10 input sets, EVALYN produced better guide trees which resulted in better alignments with higher sum-of-pairs scores

Table 1. Experimental configuration

Population Type	Steady-state
Population Size	500 guide trees
Crossover Rate	100% (steady-state population)
Mutation Rate	0.01
Iterations	2500
Selection Type	Rank-Based
Substitution Matrix	Matches = +1.9, Mismatches = 0
Gap Open Penalty	-15.0
Gap Extension Penalty	-6.66

By taking optimized guide trees produced by EVALYN and providing them as input to CLUSTAL W, we have found strong evidence to support our main hypothesis that guide trees evolved via a genetic algorithm produce better multiple sequence alignments than guide trees constructed using neighbor-joining.

6 Conclusions

Aligning multiple sequences of biological data is a critical aspect of modern biology. Since constructing optimal alignments is prohibitively time-intensive, popular heuristic MSA algorithms trade accuracy for speed in order to maximize their practical usefulness. We introduced EVALYN, a genetic algorithm for performing multiple sequence alignment. We demonstrated that EVALYN produces higher-scoring alignments than CLUSTAL W under the popular sum-of-pairs metric. In addition, we provided a strong analytical argument that an evolutionary computational approach to guide tree optimization as used in EVALYN is more scalable than traditional guide tree construction algorithms such as neighbor-joining.

By using a genetic algorithm to produce better guide trees, we achieved an important practical goal of producing a measurably better multiple sequence alignment program than CLUSTAL W, currently the most popular and actively used MSA program today.

7 Future Work

In future work, we will examine different metrics of alignment quality in order to show that EVALYN is able to produce biologically significant results. Although the sum-of-pairs score (SPS) for an alignment is commonly used, it has some inherent problems. First, alignments with the highest SPS are not always the most biologically significant or meaningful. Guide trees and multiple sequence alignments constructed under the SPS optimality criterion may in fact produce alignments which are meaningless when interpreted by a biologist. Finding ways of quantifying the level of biological significance of an alignment is an ongoing and active area of research. Toward this end, we intend to test EVALYN against the Benchmark Alignment dataBASE (BAliBASE) [21], which is a carefully designed set of protein alignments that were aligned and verified with elucidated or inferred structural and functional information. The BAliBASE alignments are composed of real protein sequences, and are intended to be a kind of "gold standard" by which alignment algorithms can be tested for their ability to recover biologically significant alignments. In addition to testing EVALYN against BAliBASE, we will develop new fitness functions to maximize meaningful biological signal in alignments. For example, future fitness functions may take into account predicted or elucidated secondary structure.

All phylogenetic analysis begins with multiple sequence alignment, which establishes the positional homology of the sequence data that is used for the basis of constructing and optimizing phylogenetic trees. Phylogenetic analysis is extremely dependent on the assumptions, biases, and accuracy of the initial multiple sequence alignment. State-of-the art work in phylogenetic inferencing attempts to address this by simultaneously optimizing *both* alignment *and* phylogeny. As guide trees serve as rough estimations of sequence phylogeny, EVALYN also simultaneously optimizes phylogeny and alignment. In effect, EVALYN currently performs phylogenetic tree optimization by using alignment sum-of-pairs scores as the optimality criterion. Along these lines, we will explore additional measures of guide tree fitness such as parsimony [13] and the more statistically rigorous approach of maximum-likelihood [22].

Finally, empirically characterizing the scalability of EVALYN across different numbers, lengths, and types of input sequences is a key focus of future work. In this paper, we analyzed EVALYN to show that it has a linear time complexity with respect to the number of input sequences. However, the rate of solution convergence as a function of the number of input sequences is not yet well understood. Future experimentation will explore the relationship between population size, GA convergence properties, sequence divergence effects, and alignment quality.

Acknowledgments. The project described was supported by NIH Grant Number P20 RR16454 from the BRIN Program of the National Center for Research Resources. Experiments were run on the IBEST Beowulf cluster, which is funded in part by NSF EPS 00809035, NIH NCRR 1P20 RR16448 and NIH NCRR 1P20 RR16454. Foster was partially funded for this research by NIH NCRR 1P20 RR16448.

References

1.   Needleman, S.B., Wunsch, C.D., *A general method applicable to the search for similarities in the amino acid sequences of two proteins.* Journal of Molecular Biology, 1970. **48**(3): p. 443-53.
2.   Smith, T.F., Waterman, M.S., *Identification of Common Molecular Subsequences.* Journal of Molecular Biology, 1981. **48**: p. 443-453.
3.   Just, W., *Computational Complexity of Multiple Sequence Alignment with SP-Score.* Journal of Computational Biology, 2001. **8**(6): p. 615-623.
4.   Notredame, C., *Recent progresses in multiple sequence alignment: a survey.* Pharmacogenomics, 2002. **3**(1).
5.   Saitou, N., Nei, M., *The neighbor-joining method: A new method for reconstructing phylogenetic trees.* Molecular Biology Evolution, 1987. **4**: p. 406-425.
6.   Notredame, C., Higgins D.G., *SAGA: sequence alignment by genetic algorithm.* Nucleic Acids Research, 1996. **24**(8): p. 1515-1524.
7.   Carillo, H., Lipman, D., *The multiple sequence alignment problem in biology.* SIAM Journal on Applied Mathematics, 1988. **48**(5): p. 1073-1082.
8.   Thomsen, R., Fogel, G.B., Krink, T. *A Clustal alignment improver using evolutionary algorithms.* in *Congress on Evolutionary Computation (CEC).* 2002. Honolulu, Hawaii.
9.   Higgins, D.G., Bleasby A.J., Fuchs, R., *CLUSTAL V: improved software for multiple sequence alignment.* Comput Appl Biosci, 1992. **8**(2): p. 189-191.
10. Lewis, P.O., *A genetic algorithm for maximum-likelihood phylogeny inference using nucleotide sequence data.* Molecular Biology Evolution, 1998. **15**(3): p. 277-283.
11. Matsuda, H. *Protein phylogenetic inference using maximum likelihood with a genetic algorithm.* in *Pacific Symposium on Biocomputing.* 1996: World Scientific, London.
12. Congdon, C.B. *Gaphyl: An Evolutionary Algorithms Approach for the Study of Natural Evolution.* in *Genetic Evolutionary Computation Confernece (GECCO).* 2002: Morgan Kaufmann.
13. Fitch, W.M., *Toward defining the course of evolution: Minimum change for a specific tree topology.* Systematic Zoology, 1971. **20**: p. 406-416.

14. Koza, J., *Genetic Programming: On the Programming of Computers by Means of Natural Selection.* 1992: MIT Press.

15. Soule, T., Foster, J.A., *Effects of code growth and parsimony pressure on populations in genetic programming.* Evolutionary Computation, 1998. **6**(4): p. 293-309.

16. Dayhoff, M.O., Schwartz, R.M., and Orcutt, B.C., *A model of evolutionary change in proteins.*, in *Atlas of Protein Sequence and Structure.* 1978.

17. Henikoff, S., Henikoff, J.G. *Amino acid substitution matrices from protein blocks.* in *National Academy of Sciences of the USA.* 1992.

18. Howe, K., Bateman, A., Durbin, R., *QuickTree: building huge Neighbor-Joining trees of protein sequences.* Bioinformatics, 2002. **18**(11): p. 1546-1547.

19. Thompson, J.D., Higgins, D.G., Gibson, T.J., *CLUSTAL W: improving the sensitivity of progressive multiple sequence alignment through sequence weighting, position-specific gap penalties and weight matrix choice.* Nucleic Acids Research, 1994. **22**(22): p. 4673-4680.

20. Jukes, T.H., Cantor, C.R., *Evolution of protein molecules*, in *Mammalian Protein Metabolism*, H.N. Munro, Editor. 1969, Academic Press. p. 21-132.

21. Thompson, J.D., Plewniak, F., Poch, O., *BAliBASE: A benchmark alignments database for the evaluation of multiple sequence alignment programs.* Nucleic Acids Research, 1999. **27**(13): p. 2682-2690.

22. Felsenstein, J., *Evolutionary trees from DNA sequences: A maximum likelihood approach.* Journal of Molecular Evolution, 1981. **17**: p. 368-376.

Optimizing Topology and Parameters of Gene Regulatory Network Models from Time-Series Experiments

Christian Spieth, Felix Streichert, Nora Speer, and Andreas Zell

Centre for Bioinformatics Tübingen (ZBIT), University of Tübingen,
Sand 1, D-72076 Tübingen, Germany,
spieth@informatik.uni-tuebingen.de,
http://www-ra.informatik.uni-tuebingen.de

Abstract. In this paper we address the problem of finding gene regulatory networks from experimental DNA microarray data. Different approaches to infer the dependencies of gene regulatory networks by identifying parameters of mathematical models like complex S-systems or simple Random Boolean Networks can be found in literature. Due to the complexity of the inference problem some researchers suggested Evolutionary Algorithms for this purpose. We introduce enhancements to the Evolutionary Algorithm optimization process to infer the parameters of the non-linear system given by the observed data more reliably and precisely. Due to the limited number of available data the inferring problem is under-determined and ambiguous. Further on, the problem often is multi-modal and therefore appropriate optimization strategies become necessary. We propose a new method, which evolves the topology as well as the parameters of the mathematical model to find the correct network.

1 Introduction

The inference of regulatory dependencies between genes from time series data has become one of the most challenging tasks in the field of functional genomics. With new experimental methods like DNA microarrays, which have become one of the key techniques in the area of gene expression analysis in the past few years, it is possible today to monitor thousands of genes in parallel. Therefore, these techniques can be used as a powerful tool to explore the regulatory mechanisms of gene expression in a cell. However, due to the huge number of components within the regulatory system, a large amount of experimental data is needed to infer genome-wide networks. This requirement is impracticable to meet today, because of the high costs of these experiments and due to the combinatorial nature of gene interaction.

The earliest models to simulate regulatory systems found in the literature are Boolean or Random Boolean Networks (RBN) [6]. In Boolean Networks gene expression levels can be in one of two states: either 1 (on) or 0 (off). The

K. Deb et al. (Eds.): GECCO 2004, LNCS 3102, pp. 461–470, 2004.

quantitative level of expression is not considered. Two examples for inferring Boolean Networks are given by Akutsu *et al.* [1] and the REVEAL algorithm [10] by Liang *et al.* These models have the advantage that they can be solved with only small computational effort. But they suffer from the disadvantage of being tied to discrete system states. In contrast, qualitative network models allow for multiple levels of gene regulation. An example for this kind of approach is given by Thieffry and Thomas in [16]. But these models use only qualitative dependencies and therefore only a small part of the information hidden in the time series data. Quantitative models based on linear models for gene regulatory networks like the weighted matrix model by Weaver *et al.* [18] or the singular value decomposition method by Yeung *et al.* [19] consider the continuous level of gene expression. Other approaches to infer regulatory systems from time series data by using Artificial Neural Networks [7] or Bayesian Networks [4] have been recently published, but face some drawbacks as well. Bayesian networks, for example, do not allow for cyclic networks. More general examples for mathematical non-linear models like S-Systems to infer regulatory mechanisms have been examined by Maki *et al.* [11] or Kiguchi *et al.* [8].

In our method we try to use the advantages of flexible mathematical models like S-Systems. We introduce a method, which separates the inference problem into two subproblems. The first task is to find the topology or structure of the network with a Genetic Algorithm. In the second task the parameters of a mathematical model are optimized for the given topology with an Evolution Strategy. The second problem can be seen as a local search phase of a Memetic Algorithm (MA).

The remainder of this paper is structured as follows. Section 2 describes the proposed algorithm and the mathematical model used in the optimization process. Applications and results are listed in section 3 and the conclusions and an outlook are given in section 4.

2   Inference Method

The following section gives an overview over the proposed algorithm.

2.1   Memetic Algorithm

Evolutionary Algorithms have proven to be a powerful tool for solving complex optimization problems. Three main types of Evolutionary Algorithms have evolved during the last 30 years: Genetic Algorithms (GA), mainly developed by J.H. Holland [3], Evolution Strategies (ES), developed by I. Rechenberg [12] and H.-P. Schwefel [14] and Genetic Programming (GP) by J.R. Koza [9]. Each of these uses different representations of the data and different main operators working on them. They are, however, inspired by the same principles of natural evolution. Evolutionary Algorithms are a member of a family of stochastic search techniques that mimic the natural evolution of repeated mutation and selection as proposed by Charles Darwin.

In the current implementation we used a combination of a Genetic Algorithm for optimizing the topology together with an Evolution Strategy to locally find the best parameters for the given topology. The general principle is outlined in Fig. 1.

```
begin                                      eval(GApop) {
  initGApop()                                for each topology
                                               initESpop()
  eval(GApop)                                  eval(ESpop)
  while (termination criteria not met)         while (termination criteria not met)
    selectGAparentPop()                          selectESparentPop()
                                                 createESoffsprings()
    createGAoffsprings()                         eval(ESpop)
    eval(GApop)                                  selectNewESpop()
                                               do
    selectNewGApop()                           setFitness(GAind, bestESfitness)
  do                                         do
end                                        }
```

Fig. 1. Pseudo-code describing the general principle of the Memetic Algorithm

2.2   Global Genetic Algorithm

In our implementation the Genetic Algorithm evolves populations of structures of possible networks. These structures are encoded as bitsets where each bit represents the existence or absence of an interaction between genes and therefore of non-zero parameters in the mathematical model. The evaluation of the fitness of each individual within the GA population uses a local search described below.

2.3   Local Evolution Strategy

For evaluation of each structure suggested by the global optimizer an Evolution Strategy is used, which is suited for optimizing problems based on real values. The ES optimizes the parameters of the mathematical model used for representation of the regulatory network.

Fitness. For assessing the quality of the locally obtained results we used the following equation for calculation of the fitness values for the ES optimization process:

$$f = \sum_{i=1}^{N} \sum_{k=1}^{T} \left(\frac{\hat{x}_i(t_k) - x_i(t_k)}{x_i(t_k)} \right)^2 \tag{1}$$

where N is the total number of genes in the regulatory system, T is the number of sampling points taken from the time series and $\hat{x}$ and x distinguish between estimated data and experimental data. The overall problem is to minimize the fitness value f.

Mathematical Model. On an abstract level, the behavior of a cell is represented by a gene regulatory network of N genes. Each gene g_i produces a certain amount of mRNA x_i, when expressed, and therefore changes the concentration of this mRNA over time: $x_i(t+1) = h(\boldsymbol{x}(t))$ with $\boldsymbol{x}(t) = (x_1, \cdots, x_n)$, where h describes the changing of each RNA level depending on all or only on some RNA concentrations at the previous time step.

To model and to simulate regulatory networks we decided to use S-Systems since we think they are flexible enough to model important gene regulatory dependencies like feed back loops, etc. But there are alternatives as listed in section 1, which will be the subject of research in future applications.

S-Systems. S-Systems are a type of power-law formalism, which has been suggested by Irvine and Savageau [5,13] and can be described by a set of nonlinear differential equations:

$$\frac{dx_i(t)}{dt} = \alpha_i \prod_{j=1}^{N} x_j(t)^{\mathcal{G}_{i,j}} - \beta_i \prod_{j=1}^{N} x_j(t)^{\mathcal{H}_{i,j}} \tag{2}$$

where $\mathcal{G}_{i,j}$ and $\mathcal{H}_{i,j}$ are kinetic exponents, α_i and β_i are positive rate constants and N is the number of equations in the system. The equations in (2) can be seen as divided into two components: an excitatory and an inhibitory component.

The kinetic exponents $\mathcal{G}_{i,j}$ and $\mathcal{H}_{i,j}$ determine the structure of the regulatory network. In the case $\mathcal{G}_{i,j} > 0$ gene g_j induces the synthesis of gene g_i. If $\mathcal{G}_{i,j} < 0$ gene g_j inhibits the synthesis of gene g_i. Analogously, a positive (negative) value of $\mathcal{H}_{i,j}$ indicates that gene g_j induces (suppresses) the degradation of the mRNA level of gene g_i.

The S-System formalism has a major disadvantage in that it includes a large number of parameters that have to be estimated. The total number of parameters in S-Systems is $2N(N+1)$, with N the number of state variables x_i (genes). This causes problems with increasing number of participating genes due to the quadratically increasing number of parameters to infer. The parameters of the S-System $\boldsymbol{\alpha}$, $\boldsymbol{\beta}$, $\mathcal{G}$, and $\mathcal{H}$ are optimized with Evolutionary Algorithms described in the previous paragraphs.

3   Results

To verify the concepts of our idea we first compare two network inference examples where the first inference process is initialized without any prior knowledge of the network structure. In the second case we incorporate the correct topology of the dependencies of each gene together with the experimental data to validate the theoretical ability of our approach to find the correct model. After this verification step, we use the proposed method to infer gene regulatory systems from artificial microarray expression data.

3.1   Preliminary Experiments

For validation purposes we examined a small example of gene regulatory networks described in the literature, which has been studied by a variety of researchers in the past. It was first introduced by Savageau [13] and was subject of several attempts to re-engineer networks: Tominaga *et al.* [17] tried to infer only selected genes in their work and Kiguchi *et al.* [8] and Maki *et al.* [11] proposed new methods to reverse engineer the complete system but changed the parameters of the original system for unknown reasons.

Fig. 2 shows the dependencies of the regulatory network as used in all of the publications listed above.

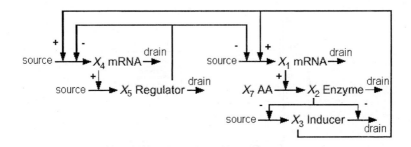

Fig. 2. 5-dimensional gene regulatory network [Savageau [13]]

The total number of parameters to be optimized with the Evolution Strategy in this example was $N = 60$ if modelled with an S-System. Fig. 3 shows the time courses for each mRNA level of the regulatory system. The optimization process was repeated $m = 20$ times to gain averaged fitness courses.

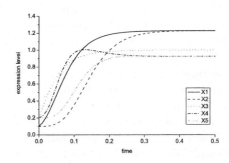

Fig. 3. Time dynamics of the 5-dim regulatory system

Without topological information. In the case of the inference without any additional knowledge a (μ,λ)-ES with $\mu = 5$ parents and $\lambda = 35$ offspring is used together with a Covariance Matrix Adaptation (CMA) mutation operator and no recombination to evolve individuals in the optimization process. The CMA operator is one of the most powerful self adaption mechanisms today available for ES. For further details see [2].

Fig. 4 shows the averaged fitness course of the gene regulatory network (GRN) model optimized with a standard ES with no topological information provided. As can be seen in this graph, the ES converges prematurely after approximately 2,000 generations into a local optimum and the fitness remains static until the end of the optimization process.

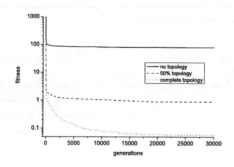

Fig. 4. Fitness of an ES optimization with and without topological information

With partial topological information. The second trial incorporated partial information about the topology. In this test case, 50% of the correct interactions of the regulatory network graph was used during optimizing the mathematical model. To use the correct topological information 50% of the values of $\mathcal{G}$ and $\mathcal{H}$ representing no dependencies in the network graph, i.e. $\mathcal{G}_{ij} = 0.0$ or $\mathcal{H}_{ij} = 0.0$, were excluded from the optimization process and therefore fixed to 0.0. This was implemented by a reduced vector of decision variables for the ES. The fitness of this test case is given Fig. 4.

With complete topological information. As a third test case we incorporated the correct information about the topology to verify the idea of our method, i.e. solving the overall problem by first finding the correct topology and then identifying the corresponding parameters. Fig. 4 shows the fitness of the second case, optimized by an ES with the optimization settings as given in the previous section. The results of the third test case with the additional information about the structure of the network yields far better fitness values compared to the first test case and better results than the second case as can be seen in Fig. 4.

3.2   Artificial Regulatory Network

To test the method on larger systems we created two artificial microarray data sets, which were to be reverse engineered by our algorithm. The first data set represented the time dynamics of an artificial 10-dimensional regulatory system, i.e. the relationships between 10 genes, which were randomly assigned and simulated. The second example was an artificially created 20-dimensional GRN, which consists of 20 components. All settings for the Evolutionary Algorithms were determined in preliminary experiments.

10-dimensional network. Due to the fact that GRNs in nature are sparse systems, we created regulatory networks randomly with a maximum cardinality of $k \leq 3$, i.e. each of the $N = 10$ genes depends on three or less other genes within the network. The dynamics of the example can be seen in Fig. 5.

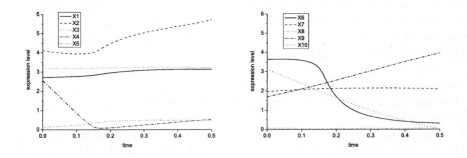

Fig. 5. Time dynamics of the 10-dim regulatory system

The optimization process was performed using a (μ,λ)-ES with $\mu = 10$ parents and $\lambda = 50$ offsprings together with a Covariance Matrix Adaptation (CMA) mutation operator without recombination. This optimization was repeated $m = 20$ times with different starting populations. After evolving the models for $200{,}000$ generations (total number of $1{,}000{,}000$ fitness evaluations), the m best fitness values found were averaged, as shown in Fig. 6.

As illustrated by the fitness plot the standard ES was not able to find a solution for the optimization problem, because it got stuck in local optima. The proposed method on the other hand found solutions with very good fitness values.

Unfortunately, our MA found only twice the correct target system with respect to the topology and parameter values. In the remaining 18 optimization runs, systems were found, which fitted the experimental data, but showed different relationships between the component genes. We address this problem in the discussion in sect. 4.

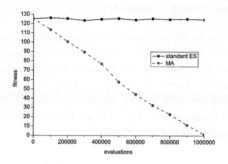

Fig. 6. Fitness graph of the 10-dim regulatory system (standard ES vs. MA)

20-dimensional network. The second GRN inferred with the proposed method is an artificial 20-dimensional system. As in the example before, we created the dependencies of the network randomly with a cardinality $k \leq 3$. The simulated time courses are not given here because the number of components of the system make the graph unclear.

The optimization was performed with the same parameter settings as described in sect. 3.2. Due to the larger number of system components, we increased the total number of fitness evaluations to $1,500,000$, thus increasing computing time as well.

Fig. 7 shows the fitness course averaged over the 20 repeated optimization runs. Again, the standard ES did not find a solution whereas the MA converged to optima with good fitness values.

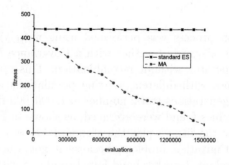

Fig. 7. Fitness graph of the 20-dim regulatory system (standard ES vs. MA)

As before in the 10-dimensional example the problem of finding the correct target system emerged again. The resulting time courses fitted the experiment data (yielding good fitness values) but showed different dependencies and inter-actions between the participating genes of the regulatory network.

4   Discussion

In comparison to a standard ES with CMA the proposed method yielded far better fitness values. In both test cases the ES was not even able to find models that fit the given data at all.

Additionally, our algorithm proved to work even for middle-sized examples. Most examples found in literature are artificial and very small, i.e. with a total number of ten genes or lower, while in biological networks even small systems have at least 50–100 components. We showed that our method is able to handle sparse systems ($k \leq 3$) with 20 genes, restricted currently only by computational performance. Because we use a bitset representation of the topology the algo-rithm reduces the total number of parameters and makes it therefore possible to infer larger systems. Future experiments on high performance computers will address large-scale systems with a minimum number of 100 genes.

Further on, the solutions found by the MA are sparse due to the preceding structure optimization. Because in nature GRNs are sparse systems the solutions of the MA represent better resemblance to biological systems than the standard ES, which resulted always in complete and thus dense matrices. Therefore, the proposed method shows promising results to be suitable to infer gene regulatory systems.

Due to the large number of model parameters and the small number of data sets available, the system of equations is highly under-determined. Therefore, multiple solutions exist, which fit the given data, but show only little resemblance with the original target system. This problem is known in literature but there are currently only few publications reflecting on this issue. Recently, Spieth *et al.* published a new method to incorporate data sets obtained by additional experiments [15]. In future enhancements of our algorithm we plan to incorporate additional methods to identify the correct network.

In future work we also plan to include a-priori information into the inference process like partially known pathways or information about co-regulated genes, which can be found in literature. For better coverage of the solution space of the optimizer we plan to use a cluster-based niching algorithm, which was developed in our group. Additional models for gene regulatory networks will be examined for simulation of the non-linear interaction system as listed in sect. 1 to overcome the problems with a quadratic number of model parameters of the S-System.

Further on, we will continue to test our method with real microarray data in close collaboration with biological researchers at our university.

References

1. T. Akutsu, S. Miyano, and S. Kuhura. Identification of genetic networks from a small number of gene expression patterns under the boolean network model. In *Proceedings of the Pacific Symposium on Biocomputing*, pages 17–28, 1999.

2. N. Hansen and A. Ostermeier. Adapting arbitrary normal mutation distributions in evolution strategies: the covariance matrix adaptation. In *Proceedings of the 1996 IEEE Int. Conf. on Evolutionary Computation*, pages 312–317, Piscataway, NJ, 1996. IEEE Service Center.

3. J. H. Holland. *Adaption in Natural and Artificial Systems: An Introductory Analysis with Applications to Biology, Control and Artificial Systems*. The University Press of Michigan Press, Ann Arbor, 1975.

4. S. Imoto, T. Higuchi, T. Goto, K. Tashiro, S. Kuhara, and S. Miyano. Combining microarrays and biological knowledge for estimating gene networks via bayesian networks. In *Proceedings of the IEEE Computer Society Bioinformatics Conference (CSB 03)*, pages 104 –113. IEEE, 2003.

5. D. H. Irvine and M. A. Savageau. Efficient solution of nonlinear ordinary differential equations expressed in S-systems canonical form. *SIAM Journal of Numerical Analysis*, 27(3):704–735, 1990.

6. S. A. Kauffman. *The Origins of Order*. Oxford University Press, New York, 1993.

7. E. Keedwell, A. Narayanan, and D. Savic. Modelling gene regulatory data using artificial neural networks. In *Proceedings of the International Joint Conference on Neural Networks (IJCNN 02)*, volume 1, pages 183–188, 2002.

8. S. Kikuchi, D. Tominaga, M. Arita, K. Takahashi, and M. Tomita. Dynamic modeling of genetic netowrks using genetic algorithm and s-sytem. *Bioinformatics*, 19(5):643–650, 2003.

9. J. R. Koza. *Genetic Programming: On the Programming of Computers by Means of Natural Selection*. MIT Press, Cambridge, MA, USA, 1992.

10. S. Liang, S. Fuhrman, and R. Somogyi. REVEAL, a general reverse engineering algorithm for inference of genetic network architectures. In *Proceedings of the Pacific Symposium on Biocomputing*, volume 3, pages 18–29, 1998.

11. Y. Maki, D. Tominaga, M. Okamoto, S. Watanabe, and Y. Eguchi. Development of a system for the inference of large scale genetic networks. In *Proceedings of the Pacific Symposium on Biocomputing*, volume 6, pages 446–458, 2001.

12. I. Rechenberg. *Evolutionsstrategie - Optimierung technischer Systeme nach Prinzipien der biologischen Evolution*. Frommann-Holzboog, Stuttgart, 1973.

13. M. A. Savageau. 20 years of S-systems. In E. Voit, editor, *Canonical Nonlinear Modeling. S-systems Approach to Understand Complexity*, pages 1–44, New York, 1991. Van Nostrand Reinhold.

14. H.-P. Schwefel. *Numerical optimization of computer models*. John Wiley and Sons Ltd, 1981.

15. C. Spieth, F. Streichert, N. Speer, and A. Zell. Iteratively inferring gene regulatory networks with virtual knockout experiments. In R. et al., editor, *Proceedings of the 2nd European Workshop on Evolutionary Bioinformatics (EvoWorkshops 2004)*, volume 3005 of *LNCS*, pages 102–111, 2004.

16. D. Thieffry and R. Thomas. Qualitative analysis of gene networks. In *Proceedings of the Pacific Symposium on Biocomputing*, pages 77–87, 1998.

17. D. Tominaga, N. Kog, and M. Okamoto. Efficient numeral optimization technique based on genetic algorithm for inverse problem. In *Proceedings of German Conference on Bioinformatics*, pages 127–140, 1999.

18. D. Weaver, C. Workman, and G. Stormo. Modeling regulatory networks with weight matrices. In *Proceedings of the Pacific Symposium on Biocomputing*, volume 4, pages 112–123, 1999.

19. M. K. S. Yeung, J. Tegner, and J. J. Collins. Reverse engineering gene networks using singular value decomposition and robust regression. In *Proceedings of the National Academy of Science USA*, volume 99, pages 6163–6168, 2002.

Comparing Genetic Programming and Evolution Strategies on Inferring Gene Regulatory Networks

Felix Streichert, Hannes Planatscher, Christian Spieth, Holger Ulmer, and
Andreas Zell

Centre for Bioinformatics Tübingen (ZBIT), University of Tübingen,
Sand 1, 72076 Tübingen, Germany,
streiche@informatik.uni-tuebingen.de
http://www-ra.informatik.uni-tuebingen.de/

Abstract. In recent years several strategies for inferring gene regulatory networks from observed time series data of gene expression have
been suggested based on Evolutionary Algorithms. But often only few
problem instances are investigated and the proposed strategies are rarely
compared to alternative strategies. In this paper we compare Evolution
Strategies and Genetic Programming with respect to their performance
on multiple problem instances with varying parameters. We show that
single problem instances are not sufficient to prove the effectiveness of a
given strategy and that the Genetic Programming approach is less prone
to varying instances than the Evolution Strategy.

1 Introduction

In recent years modern technologies like microarrays allowed scientists to measure large numbers of gene expression data for thousands of genes at the same
time. With this technique at hand scientists are also able to measure gene activities through time. Such time series nourish the idea that it could be possible
to reconstruct or infer the underlying gene regulatory networks. This problem
of inferring the real gene regulatory networks from time series data has recently
become one of the major topics in bioinformatics.

The strategies for inferring regulatory networks depend on the mathematical model used to represent the behavior of the real gene regulatory network.
Currently, both discrete and continuous models are used to model regulatory
networks, but to represent the activity of real regulatory networks continuous
models are believed to be the most suitable.

For discrete models like boolean or random boolean networks [17], several
efficient heuristics have been suggested [1]. More realistic models are given by
qualitative networks, which use several levels of activation rather than just 'on'
or 'off' [12]. For qualitative networks, Akutsu et al. have suggested a special
heuristic for inferring such networks from time series data [2].

Quantitative networks on the other hand consider the continuous level of gene
expression and are therefore more realistic. A parametrized model with discrete

K. Deb et al. (Eds.): GECCO 2004, LNCS 3102, pp. 471–480, 2004.
© Springer-Verlag Berlin Heidelberg 2004

time and a linear relationship between the genes is given by weight matrices [16]. The parameters for such weight matrices have been reverse engineered by means of Genetic Algorithms (GA) [15]. Other researchers use linear differential equations to model regulatory networks and use special heuristics to find the necessary parameters [3]. Another parameterized model based on differential equations is given by S-systems (*synergistic* and *saturable* systems) [9]. To infer the unknown parameters of this model Tominaga et al. applied a GA with special operators biased towards few non-zero parameters leading to sparsely connected regulatory networks [14].

An example for non-parameterized quantitative networks are arbitrary systems of differential equations, which are more powerful and flexible to describe the relations between genes, since the real structure underlying the observed data is unknown. The most prominent method that is able to optimize the structure and the parameters of differential equations to fit a given time series is Genetic Programming (GP) [5]. Sakamoto et al. applied a GP augmented with a least mean square method for parameter optimization for inferring differential equations for regulatory networks [8].

In this paper we compare two inferring strategies for quantitative networks based on Evolutionary Algorithms (EA). On the one hand to fix the network model *a priori* and reduce the inferring problem to a parameter optimization problem, which can be solved by means of Evolution Strategies (ES). We decided to use S-systems as a parameterized quantitative network, since they derive from a Taylor approximation of a general ordinary differential equation and are rather flexible. And on the other hand to leave the choice of the network structure to the inferring algorithm. This non-parameterized network model requires GP for inferring a suitable structure to met the target.

We compare both approaches on several examples generated from artificial data to determine, which approach is more suitable for inferring gene regulatory networks. We also try to identify, which are the most important properties of a regulatory network that make it difficult to reconstruct from time series data. Therefore, we vary multiple parameters of the artificial network and examine how the changes impact the performance.

In sec. 2 we give details on the experimental settings and our implementation of the optimization algorithms used. ES and GP are then compared in sec. 3 on several examples generated from artificial regulatory networks. Conclusions and an outlook of future research are given in sec. 4.

2   Experimental Settings

Since there are only few publicly available time series for gene expression and the correct or best model is not known for those regulatory networks, a comparison of inferring strategies cannot rely on real data. Therefore, it is necessary to create time series from artificial regulatory networks as benchmark problems. In this way one is in control of all properties of the target network and can arbitrarily vary the dimension of the problem and the connectivity of the network. One may also validate the results obtained by comparing it to the artificial target.

Unfortunately, it is unknown, which kind of model is most suitable to represent the same dynamical properties as real regulatory networks. Also there are currently artificial benchmark problems for inferring regulatory network that researchers commonly agree on. Only recently Mendes et al. proposed a set of benchmark networks [6], which were published after our experiments were conducted, but will be included in our future studies.

In this paper we examine S-systems as an example for parameterized quantitative networks as tentative benchmark problem, since they were derived from a simplified Taylor expansion of a general ordinary differential equation and should be rather flexible. An S-system for n artificial genes is given by a parameterized set of nonlinear differential equations:

$$\frac{dx_i(t)}{dt} = \alpha_i \prod_{j=1}^{n} x_j(t)^{\mathcal{G}_{i,j}} - \beta_i \prod_{j=1}^{n} x_j(t)^{\mathcal{H}_{i,j}} \tag{1}$$

where x_i is the state variable of the measured expression level of gene i. With $\alpha_i \geq 0$ and $\beta_i \geq 0$ the first product describes all synthesizing influences and the second product all degrading influences. Depending on the values of $\mathcal{G}_{i,j}$ and $\mathcal{H}_{i,j}$ the influence may be inhibitory, if the value in the matrix is smaller than zero, or excitatory, if greater than zero.

With this model for an artificial gene regulatory network we can generate example problems by choosing random values for α_i, β_i, $\mathcal{G}_{ij}$ and $\mathcal{H}_{ij}$ while checking whether they are stable or not. We can increase the dimensionality of the regulatory network by increasing n and we can change the level of interdependence between the genes by adding or removing zero-valued parameters.

In our experiments we generate a problem instance by simulating the target artificial regulatory network and store the calculated expression values at certain time points. This corresponds to real experiments where the number of microarrays and therefore the number of actually measured time points is limited.

We compare the performance of an EA approach based on a parameterized S-system using Evolution Strategies (ES) to identify suitable parameters to fit the measured time series to a Genetic Programming (GP) approach to search for the proper right hand side of a system of ordinary differential equations. For both EA methods the fitness f of an individual a is given by the Relative Standard Error (RSE) of the resulting estimation of gene expression $\hat{x}$ to the measured gene expression x:

$$f = \sum_{i=1}^{N} \sum_{k=1}^{T} \left(\frac{\hat{x}_i(t_k) - x_i(t_k)}{x_i(t_k)} \right)^2 \tag{2}$$

over all measured time points t_j over all genes n.

This fitness function is very much straight forward and commonly used in this area of research. But it suffers from a serious problem: a single measured time series is not sufficient to identify a unique solution. It is only one path in a phase diagram and from such a single path no general conclusions of the overall behavior of the dynamic system can be drawn. An extreme example is given in the phase diagram of a simple two dimensional example in fig. 2. The two genes

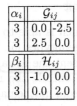

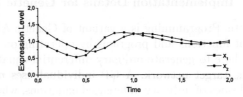

α_i	$\mathcal{G}_{ij}$	
3	0.0	-2.5
3	2.5	0.0

β_i	$\mathcal{H}_{ij}$	
3	-1.0	0.0
3	0.0	2.0

Fig. 2. Parameters and dynamics of the 2D example given in [13]

3.1   Two-Dimensional Examples: Increase of Connectivity

The parameters and the dynamics of the original two-dimensional S-system are given in fig. 2. To examine the impact of different levels of interdependence between the genes (connectivity) we varied the number of non-zero parameters from 0 to 8. We expected that the inference problem would become more difficult with increasing connectivity.

Fig. 3 shows that both methods perform well on all problems in each example, the RSE drops below 0.02 or even 0.01 in case of the S-system based ES. The ES performs slightly better, but when taking into account the overall low level of RSE the difference becomes marginal. And it has to be noted that the function set of the GP is insufficient while the ES has the same structure as the target.

Unfortunately, the performance of the inferring strategies seems to be independent of the connectivity level of the target network. This can be explained by taking into account that the inferred solutions are not necessarily similar to the target systems. Even if the target system has a low connectivity (sparse matrices $\mathcal{G}_{ij}$ and $\mathcal{H}_{ij}$) the solution is usually not sparse. Therefore, the problem is not really easier for targets with low connectivity.

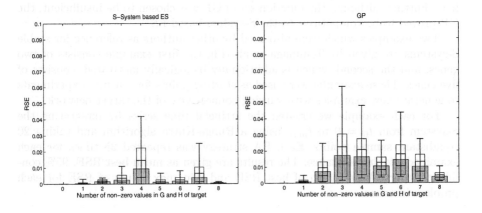

Fig. 3. Comparing ES and GP on two-dimensional examples, $t_{max} = 2.0$

α_i	$\mathcal{G}_{ij}$				
15	0.0	0.0	1.0	0.0	-0.1
10	2.0	0.0	0.0	0.0	0.0
10	0.0	-0.1	0.0	0.0	0.0
8	0.0	0.0	2.0	0.0	-1.0
10	0.0	0.0	0.0	2.0	0.0
β_i	$\mathcal{H}_{ij}$				
10	2.0	0.0	1.0	0.0	0.0
10	0.0	2.0	0.0	0.0	0.0
10	0.0	-0.1	2.0	0.0	0.0
10	0.0	0.0	0.0	2.0	0.0
10	0.0	0.0	0.0	0.0	2.0

Fig. 4. Parameters and dynamics of the 5D example given in [13]

3.2 Five-Dimensional Examples: Increase of Connectivity

In the second experiment we increased the problem dimension to five and again varied the connectivity of the artificial network. The parameters and the dynamics of the original five-dimensional S-system are given in fig. 4.

First, it has to be noted that the overall performance dropped considerably with the increased problem dimension, see fig. 5. But the performance did not suffer from increasing the connectivity of the target network. Instead the ES showed the worst results on the examples with 23-38 non-zero parameters, but performed well again on the examples with 43 and 48 non-zero parameters.

Comparing the ES to the GP, the GP performed better and more reliable than the ES on all five-dimensional examples regarding the mean RSE and the standard deviation. Only the best results of the S-system based ES are better than the best results of the GP in nearly all examples. The high variation of the

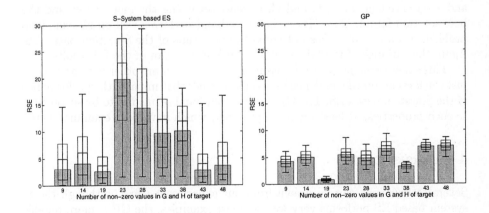

Fig. 5. Comparing ES and GP on five-dimensional examples, $t_{max} = 1.0$

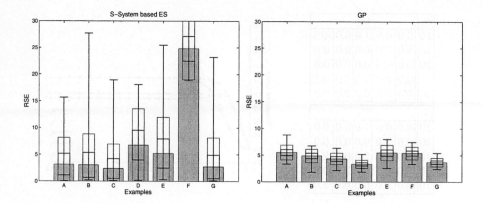

Fig. 6. Comparing ES and GP on five-dimensional examples of same connectivity, $t_{max} = 1.0$

ES could be accounted for by the multi-modal search space of the ES, but on the other hand the search space of the GP is even rougher and also multi-modal.

Interestingly enough the performance of the S-system based ES depends heavily on the instance of the problem regardless of the dimension of the problem, see also example 4 in fig. 3. Therefore, we decided to conduct another experiment with multiple problem instances with the same dimension and the same level of interdependence between the genes.

3.3   Five-Dimensional Examples: Constant Connectivity

We created seven artificial regulatory networks of the same dimension and with same number of non-zero values in $\mathcal{G}_{ij}$ and $\mathcal{H}_{ij}$. Again the performance of the ES depends on the problem instance, see fig. 6. For example the ES fails for problem instance F, while it performs quite well on all other problem instances and even excellent on A, C and G, at least regarding the mean value and the best solution found. The GP on the other hand performs well on all examined problem instances, but does not equal the best runs of the S-system based ES. Again the variance of the GP results is much lower than of the ES results.

This experiment supports the assumption that there are other properties than just the level of interdependence between the genes that impacts the performance of the S-system based ES. The GP on the other hand seems not to be susceptible to such properties, at least for those problems which have been examined here.

4   Discussion

Several conclusions can be drawn from our experiments: first, although the S-system based ES performs very well on most examples, the GP is more reliable and more versatile. Secondly, good RSE values do not necessarily indicate similarity to the target system, neither for the S-system based ES nor for the GP strategy. Third, and most important, the performance on a single example is not sufficient to evaluate a strategy as it is currently the case in most publications.

Regarding the primary objective of this paper we were able to show that the GP proved to be competitive to the S-system based ES regarding the RSE values reached, although the function set of the GP was insufficient, while the ES utilized the very same model that was used to generate the artificial data. While the ES had the best average results on at least half the examples examined, there were several problem instances where the ES performed much worse than the GP. Also, on all examples the worst results of single runs were produced by the ES. The GP on the other hand performed not as well as the ES regarding the quality of the best solutions found, but the results were more reliable. Further, the GP showed a smaller standard deviation of the results on each example. The GP also proved to be more robust on the different problem instances than the ES. Taking into account that it is unknown whether or not S-systems or similar mathematical models are suitable to represent true gene regulatory networks, GP seems to be the method of choice, since it requires no a priori assumptions about the structure of the gene regulatory network. Regarding the secondary objective we could, show that both approaches suffer from increased problem dimension, while the level of connectivity seems not to be of major relevance. The GP performed slightly better with increased problem dimension than the ES. This could be accounted to the quadratic increase of parameters in case of the S-system based ES compared to the linear increase of complexity for GP.

Secondly, although good RSE values in the experiments might suggest that the target system had been correctly identified, this was not the case in most examples. Especially the S-system based ES produced parameter sets that where often neither sparse like the target system nor were the parameters of the same magnitude as in the target system. The same holds true for GP, although most GP solutions could be considered 'sparse', due to the limited tree depth.

Finally, the experiments showed that the performance of the S-system based ES was heavily depending on the problem instance. This suggests that multiple problem instances are necessary to reliably specify the performance of a given inference strategy, instead of testing the strategy on just one or two examples. Therefore, it is necessary to find a whole set of artificial benchmark regulatory networks based on multiple mathematical models to evaluate inferring strategies.

To actually infer gene regulatory networks from real microarray time series data two issues need to be addressed in future work. First, the problem of ambiguity needs to be resolved. Either by utilizing additional experimental data to remove ambiguity or by introducing biologically motivated constraints to the fitness function like for example partially known gene interactions, preferring sparse networks over fully connected networks or favoring 'robust' networks regarding disturbance over networks with instable dynamics. Second, to tackle problems with higher dimension than the typical five to ten dimensions used in most papers, we need to explore separation strategies or develop new problem specific strategies to escape the curse of dimensionality. Otherwise we need to limit ourself to simpler models like RBN or weight matrices, where more efficient heuristics than EA can be applied.

References

1. T. Akutsu, S. Miyano, and S. Kuhara. Identification of genetic networks from a small number of gene expression patterns under the boolean network model. *Pacific Symposium on Biocomputing*, 4:17–28, 1999.
2. T. Akutsu, S. Miyano, and S. Kuhara. Algorithms for identifying boolean networks and related biological networks based on matrix multiplication and fingerprint function. *Journal of Computational Biology*, 7(3):331–343, 2000.
3. T. Chen, H. L. He, and G. M. Church. Modeling gene expression with differential equations. *Pacific Symposium on Biocomputing*, 4:29–40, 1999.
4. N. Hansen and A. Ostermeier. Adapting arbitrary normal mutation distributions in evolution strategies: The covariance matrix adaption. In *Proceedings of the 1996 IEEE International Conference on Evolutionary Computation*, pages 312–317, 1996.
5. J. R. Koza, David Andre, F. H. Bennett III, and M. Keane. *Genetic Programming 3: Darwinian Invention and Problem Solving*. Morgan Kaufman, Apr. 1999.
6. P. Mendes, W. Sha, and K. Ye. Artificial gene networks for objective comparison of analysis algorithms. *Bioinformatics*, 19(2):122–129, 2003.
7. R. Morishita, H. Imade, I. Ono, N. Ono, and M. Okamoto. Finding multiple solutions based on an evolutionary algorithm for inference of genetic networks by s-system. In *Congress on Evolutionary Computation*, pages 615–622, 2003.
8. E. Sakamoto and H. Iba. Inferring a system of differential equations for a gene regulatory network by using genetic programming. In *Proceedings of Congress on Evolutionary Computation*, pages 720–726. IEEE Press, 27-30 2001.
9. M. Savageau. 20 years of S-systems. In E. Voit, editor, *Canonical Nonlinear Modeling. S-systems Approach to Understand Complexity*, pages 1–44, New York, 1991. Van Nostrand Reinhold.
10. H.-P. Schwefel. *Evolution and Optimum Seeking*. John Wiley & Sons, New York, 1995.
11. C. Spieth, F. Streichert, N. Speer, and A. Zell. Iteratively inferring gene regulatory networks with virtual knockout experiments. In *Applications of Evolutionary Computing, EvoWorkshops2004: EvoBIO, EvoCOMNET, EvoHOT, EvoIASP, EvoMUSART, EvoSTOC*, volume 3005 of *LNCS*, pages 102–111, Coimbra, Portugal, 5-7 April 2004. Springer Verlag.
12. D. Thieffry and R. Thomas. Qualitative analysis of gene networks. *Pacific Symposium on Biocomputing*, 3:77–88, 1998.
13. D. Tominaga, N. Kog, and M. Okamoto. Efficient numerical optimization technique based on genetic algorithm for inverse problem. In *Proceedings of German Conference on Bioinformatics*, pages 127–140, 1999.
14. D. Tominaga, M. Okamoto, Y. Maki, S. Watanabe, and Y. Eguchi. Nonlinear numerical optimzation technique based on genetic algorithm for inverse problem: Towards the inference of genetic networks. In *Proceedings of Genetic and Evolutionary Computation Conference*, pages 251–258. Morgan Kaufmann, 2000.
15. M. Wahde and J. Hertz. Coarse-grained reverse engineering of genetic regulatory networks. *Biosystems*, 55:129–136, 2000.
16. D. C. Weaver, C. T. Workman, and G. D. Stormo. Modeling regulatory networks with weight matrices. In *Pacific Symposium on Biocomputing*, volume 4, pages 112–123, Singapore, 1999. World Scientific Press.
17. A. Wuensche. Genomic regulation modeled as a network with basins of attraction. In R. Altman, A. Dunker, L. Hunter, and T. Klein, editors, *Pacific Symposium on Biocomputing*, volume 3, pages 89–102, Singapore, 1998. World Scientific Press.

An Evolutionary Approach with Pharmacophore-Based Scoring Functions for Virtual Database Screening

Jinn-Moon Yang, Tsai-Wei Shen, Yen-Fu Chen, and Yi-Yuan Chiu

Department of Biological Science and Technology & Institute of Bioinformatics
National Chiao Tung University, Hsinchu, 30050, Taiwan
moon@cc.nctu.edu.tw

Abstract. We have developed a new tool for virtual database screening. This tool, referred to as the Generic Evolutionary Method for molecular DOCKing (GEM-DOCK), combines an evolutionary approach and a new pharmacophore-based scoring function. The former integrates discrete and continuous global search strategies with local search strategies to speed up convergence. The latter simultaneously serves as the scoring function of both molecular docking and post-docking analysis to improve the number of the true positives. We accessed the accuracy of our approach on HSV-1 thymidine kinase using a ligand database on which competing tools were evaluated. The accuracies of our predictions were 0.54 for the GH score and 1.62% for the false positive rate when the true positive rate was 100%. We found that our pharmacophore-based scoring function indeed is able to reduce the number of the false positives. These results suggest that GEMDOCK is robust and can be a useful tool for virtual database screening.

1 Introduction

Virtual screening of compound databases has emerged as one of the most powerful and inexpensive approach to discover novel rational lead compounds for drug development [1,2]. It is based on high-throughput molecular docking methods and the crystal structures of the target protein. Virtual screening is increasingly used for a number of drivers: explosions of high-resolution crystal protein structures, advent of the structural proteomics technologies, enriching the hit rate of high-throughput screening [2], and reducing cost of drug discover. Virtual screening encompasses four phases, including target protein modeling, compound database preparation, molecular docking, and post-docking analysis. In general, a computational method for virtual screening involves two basic critical elements that are molecular docking and a good scoring method.

A molecular docking method for virtual screening should be able to screen a large number of potential ligands with reasonable accuracy and speed. Many molecular docking approaches have been developed and can be roughly divided into rigid docking [3], flexible ligand docking [4,5], and protein flexible docking methods. Recently, the flexible docking tools were mostly used for virtual screening, such as incremental and fragment-based approaches (DOCK [6] and FlexX [5]) and genetic algorithms (GOLD [4], AutoDock [7], and GEMDOCK [8]).

Scoring methods for virtual screening should encompass two basic features: effectively discriminating between correct binding states and non-native docked conforma-

K. Deb et al. (Eds.): GECCO 2004, LNCS 3102, pp. 481–492, 2004.

tions during the molecular docking phase, and discriminating a small number of active compounds from hundreds of thousands of non-active compounds during the post-docking analysis phase. Various scoring functions have been developed for calculating binding free energy, including knowledge-based [9], physic-based [10], and empirical [11] scoring functions. In general the performance of these scoring functions is often inconsistent across different systems [12,13]. The inaccuracy, inadequately predicting the true binding affinity of a ligand for a receptor, of the scoring methods is probably major weakness for virtual screening. Combining multiple scoring functions, called consensus scoring, is a popular strategy and has been shown to improve the enrichment of true positive [12,13].

In this paper, we proposed a tool, GEMDOCK (Generic Evolutionary Method for DOCKing molecules) modified from our previous studies [8,14], for virtual screening. Our tool used a pharmacophore-based scoring function and an evolutionary approach. The former is able to simultaneously serve as the scoring function of both molecular docking and post-docking analysis. In order to balance exploration and exploitation, the core idea of our evolutionary approach, an efficient flexible docking tool, is to design multiple operators cooperating with each other by using the family competition which is similar to a local search procedure.

Our new pharmacophore-based scoring function is able to reduce the number of false positives for screening large database. This scoring function integrates a simple empirical scoring function and a pharmacophore-based scoring function. The former is used to quickly recognize potential ligands for the target receptor. It consists of electrostatic, steric, and hydrogen-bonding potentials with a linear model. The latter encompasses ligand preferences and the pharmacophore preferences that exploit knowledge from existing ligands to aid the docking process. The electrostatic and hydrophilic constrains were considered for ligand preferences. The pharmacophore-based preferences were assigned according to the binding-site preferences of protein-ligand interactions, such as hydrogen bonding and stacking force.

To evaluate the strengths and limitations of GEMDOCK and to compare with several widely used methods (e.g. DOCK, GOLD, and FlexX), we first tested our program on docking 10 active ligands, obtained from Protein Data Bank (PDB), back the respective complexes with experimentally x-ray structures. Second, we tested GEMDOCK on HSV-1 thymidine kinase, proposed by Bissabtz et al. [12], to evaluate GEMDOCK's screening utility. The docking accuracy of GEMDOCK was comparable with the best available methods and the screening performance of GEMDOCK was better than that of competing methods on these test cases.

2   Method

GEMDOCK is a nearly automatic tool, which was enhanced and modified from our original technique [8,15], for virtual screening. GEMDOCK consists of four computational phases, including target protein and ligand database preparation, molecular docking and post-docking analysis. First we specified the coordinates of target protein atoms from the PDB, the ligand binding area, atom formal charge, and atom types (Table 1). When we prepared the target protein and ligand database, GEMDOCK filters out some impossible compounds and pharmacological preferences by exploiting knowledge from existing ligands to improve screening speed. After GEMDOCK prepares the ligand database and the target protein, GEMDOCK sequentially reads the atom coordinates of a ligand from the database and executes flexible docking for each ligand. Finally GEMDOCK re-ranks all docked ligand conformations for the post-docking analysis according to the scoring values of our pharmacophore-based scoring function.

Here, we briefly presented our approach for flexible docking. Please refer our previous studies [8,16] for the details. First our method randomly generates a starting population with N solutions by initializing the orientation and conformation of the ligand relating to the center of the receptor. Each solution is represented as a set of three n-dimensional vectors (x^i, σ^i, ψ^i), where n is the number of adjustable variables of a docking system and $i = 1, \ldots, N$ where N is the population size. The vector x represents the adjustable variables to be optimized in which x_1, x_2, and x_3 are the 3-dimensional location of the ligand; x_4, x_5, and x_6 are the rotational angles; and from x_7 to x_n are the twisting angles of the rotatable bonds inside the ligand. σ and ψ are the step-size vectors of decreasing-based Gaussian mutation and self-adaptive Cauchy mutation. In other words, each solution x is associated with some parameters for step-size control. The initial values of x_1, x_2, and x_3 are randomly chosen from the feasible box, and the others, from x_4 to x_n, are randomly chosen from 0 to 2π in radians. The initial step sizes σ is 0.8 and ψ is 0.2. After GEMDOCK initializes the solutions, it enters the main evolutionary loop which consists of two stages in every iteration: decreasing-based Gaussian mutation and self-adaptive Cauchy mutation. Each stage is realized by generating a new quasi-population (with N solutions) as the parent of the next stage. These stages apply a general procedure "FC_adaptive" with only different working population and the mutation operator.

The FC_adaptive procedure employs two parameters, namely, the working population (P, with N solutions) and mutation operator (M), to generate a new quasi-population. The main work of FC_adaptive is to produce offspring and then conduct the family competition. Each individual in the population sequentially becomes the "family father." With a probability p_c, this family father and another solution that is randomly chosen from the rest of the parent population are used as parents for a recombination operation. Then the new offspring or the family father (if the recombination is not conducted) is operated by the rotamer mutation or by differential evolution to generate a quasi offspring. Finally, the working mutation operates on the quasi offspring to generate a new offspring. For each family father, such a procedure is repeated L times called the family competition length. Among these L offspring and the family father, only the one with the lowest scoring function value survives. Since we create L children from one "family father" and perform a selection, this is a family competition strategy. This method avoids the population prematureness but also keeps the spirit of local searches. Finally, the FC_adaptive procedure generates N solutions because it forces each solution of the working population to have one final offspring.

2.1 Recombination Operators

GEMDOCK implemented modified discrete recombination and intermediate recombination [17]. A recombination operator selected the "family father (a)" and another solution (b) randomly selected from the working population. The former generates a child as follows:

$$x_j^c = \begin{cases} x_j^a \text{ with probability } 0.8 \\ x_j^b \text{ with probability } 0.2. \end{cases} \tag{1}$$

The generated child inherits genes from the "family father" with a higher probability 0.8. Intermediate recombination works as:

$$w_j^c = w_j^a + \beta(w_j^b - w_j^a)/2, \tag{2}$$

where w is σ or ψ based on the mutation operator applied in the FC_adaptive procedure. The intermediate recombination only operated on step-size vectors and the modified discrete recombination was used for adjustable vectors (x).

2.2 Mutation Operators

After the recombination, a mutation operator, the main operator of GEMDOCK, is applied to mutate adjustable variables (x).

Gaussian and Cauchy Mutations: Gaussian and Cauchy Mutations are accomplished by first mutating the step size (w) and then mutating the adjustable variable x:

$$w_j' = w_j' A(\cdot), \tag{3}$$
$$x_j' = x_j + w_j' D(\cdot), \tag{4}$$

where w_j and x_j are the ith component of w and x, respectively, and w_j is the respective step size of the x_j where w is σ or ψ. If the mutation is a self-adaptive mutation, $A(\cdot)$ is evaluated as $\exp[\tau' N(0,1) + \tau N_j(0,1)]$ where $N(0,1)$ is the standard normal distribution, $N_j(0,1)$ is a new value with distribution $N(0,1)$ that must be regenerated for each index j. When the mutation is a decreasing-based mutation $A(\cdot)$ is defined as a fixed decreasing rate $\gamma = 0.95$. $D(\cdot)$ is evaluated as $N(0,1)$ or $C(1)$ if the mutation is, respectively, Gaussian mutation or Cauchy mutation. For example, the self-adaptive Cauchy mutation is defined as

$$\psi_j^c = \psi_j^a \exp[\tau' N(0,1) + \tau N_j(0,1)], \tag{5}$$
$$x_j^c = x_j^a + \psi_j^c C_j(t). \tag{6}$$

We set τ and τ' to $(\sqrt{2n})^{-1}$ and $(\sqrt{2\sqrt{n}})^{-1}$, respectively, according to the suggestion of evolution strategies [17]. A random variable is said to have the Cauchy distribution ($C(t)$) if it has the density function: $f(y;t) = \frac{t/\pi}{t^2+y^2}$, $-\infty < y < \infty$. In this paper t is set to 1. Our decreasing-based Gaussian mutation uses the step-size vector σ with a fixed decreasing rate $\gamma = 0.95$ and works as

$$\sigma^c = \gamma \sigma^a, \tag{7}$$
$$x_j^c = x_j^a + \sigma^c N_j(0,1). \tag{8}$$

Table 1. Atom types of GEMDOCK

Atom type	Heavy atom name
Donor	primary and secondary amines, sulfur, and metal atoms
Acceptor	oxygen and nitrogen with no bound hydrogen
Both	structural water and hydroxyl groups
Nonpolar	other atoms (such as carbon and phosphorus)

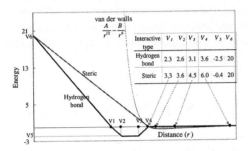

Fig. 1. The linear energy function of the pair-wise atoms for the steric interactions and hydrogen bonds in GEMDOCK (bold line) with a standard Lennard-Jones potential (light line).

2.3 Scoring Function

In this work, we have developed a new scoring function which was able to simultaneously serve as the scoring function of both molecular docking and post-docking analysis. It consisted of a simple empirical scoring function and a pharmacophore-based scoring function to reduce the number of false positives. The energy function can be dissected into the following terms:

$$E_{tot} = E_{bind} + E_{phama} + E_{ligpre}, \tag{9}$$

where E_{bind} is the empirical binding energy used during the molecular docking; E_{phama} is the energy of binding-site pharmacophores; E_{ligpre} is a penalty value if the ligand unsatisfied the ligand preferences. E_{phama} and E_{ligpre} were used to improve the number of true positives by discriminating active compounds from hundreds of thousands of non-active compounds. The empirical binding energy (E_{bing}) is given as

$$E_{bind} = E_{inter} + E_{intra} + E_{penal}, \tag{10}$$

where E_{inter} and E_{intra} are the intermolecular and intramolecular energy, respectively, and E_{penal} is a large penalty value if the ligand is out of range of the search box. In this paper, E_{penal} is set to 10000. The intermolecular energy is defined as

$$E_{inter} = \sum_{i=1}^{lig} \sum_{j=1}^{pro} \left[F(r_{ij}^{B_{ij}}) + 332.0 \frac{q_i q_j}{4 r_{ij}} \right], \tag{11}$$

$r_{ij}^{B_{ij}}$ is the distance between the atoms i and j with the interaction type B_{ij} forming by the pair-wise heavy atoms between ligands and proteins; B_{ij} is either a hydrogen bond or a steric state; q_i and q_j are the formal charges and 332.0 is a factor that converts the electrostatic energy into kilocalories per mole. The lig and pro denote the numbers of the heavy atoms in the ligand and receptor, respectively. $F(r_{ij}^{B_{ij}})$ is a simple atomic pair-wise potential function (Figure 1) modified from previous works [8,11]. In this atomic pair-wise model, the interactive types are only hydrogen binding and steric potential which have the same function form but with different parameters, $V_1, \ldots, V_6$ (defined in Figure 1). The energy value of hydrogen binding should be larger than the one of steric potential. In this model, the atom is divided into four different atom types (Table 1) : donor, acceptor, both, and nonplar. The hydrogen binding can be formed by the following pair atom types: donor-acceptor, donor-both, acceptor-both, and both-both. Other pair-atom combinations form the steric state.

The intramolecular energy of a ligand is

$$E_{intra} = \sum_{i=1}^{lig} \sum_{j=i+2}^{lig} F(r_{ij}^{B_{ij}}) + \sum_{k=1}^{dihed} A[1 - \cos(m\theta_k - \theta_0)], \tag{12}$$

where $F(r_{ij}^{B_{ij}})$ is defined as Equation 11 except the value is set to 1000 when $r_{ij}^{B_{ij}} <$ 2.0 Å and $dihed$ is the number of rotatable bonds. We followed the work of Gehlhaar et al. (1995) to set the values of A, m, and θ_0. For the $sp^3 - sp^3$ bond A, m, and θ_0 are set to 3.0, 3, and π; and $A = 1.5$, $m = 6$, and $\theta_0 = 0$ for the $sp^3 - sp^2$ bond.

The pharmacophore-based interaction (E_{phama}) between the ligand and the protein is calculated by summing up all hot-spot atoms:

$$E_{phama} = \sum_{i=1}^{lig} \sum_{j=1}^{hs} f(w_j, B_{ij})F(r_{ij}^{B_{ij}}), \tag{13}$$

where w_j is the pharmacophore weight of the hot-spot atom j, $F(r_{ij}^{B_{ij}})$ is defined as Equation 11, lig is number of the heavy atoms in the ligand, and hs is the number of hot-spot atoms in the receptor. The value of $f(w_j, B_{ij})$ is w_j or 0. $f(w_j, B_{ij})$ is w_j if the interaction type (B_{ij}) equals to the type of hot spots found between the target receptor and ligands.

In this paper the ligand preferences include electrostatic (i.e., the number of electrostatic atoms) and hydrophilic characteristic (i.e., the atom numbers of hydrogen donor and acceptor). The E_{ligpre} is a penalty value for a ligand which is unable to satisfy the ligand preferences and is defined as

$$E_{ligpre} = WP_{elec} + WP_{hb} \tag{14}$$

where WP_{elec} and WP_{hb} are the penalties for the electrostatic and hydrophilic preferences, respectively. In this paper WP_{elec} and WP_{hb} are set to 20.

3   Results

3.1   Parameters of GEMDOCK

Table 2 indicates the setting of GEMDOCK parameters, such as initial step sizes, family competition length ($L = 2$), population size ($N = 200$), and recombination probability ($p_c = 0.3$) in this work. The GEMDOCK optimization stops when either the convergence is below certain threshold value or the iterations exceed a maximal preset value which was set to 60. Therefore, GEMDOCK generated 800 solutions in one generation and terminated after it exhausted 48000 solutions in the worse case. These parameters were decided after experiments conducted to recognize complexes of test docking systems with various values. On average, GEMDOCK took 135 seconds for a docking run on a Pentium 1.4 GHz personal computer with a single processor.

Table 2. Parameters of GEMDOCK

Parameter	Value of parameters
Initial step sizes	$\sigma = 0.8, v = \psi = 0.2$ (in radius)
Family competition length	$L = 2$
Population size	$N = 200$
Recombination rate	$p_c = 0.3$
# of the maximum generation	60

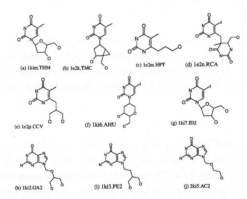

Fig. 2. Ten HSV-1 thymidine kinase ligands used as active compounds in evaluating docking accuracy and in screening performance. Each ligand systematically using four characters followed by three characters. For example, in the ligand "1kim.THM", "1kim" denotes the PDB code and "THM" is the ligand name in the PDB.

3.2   Target and Database Preparations

In order to evaluate GEMDOCK and to compare GEMDOCK with several widely used methods, we tested GEMDOCK on docking 10 active ligands (Figure 2) of HSV-1 thymidine kinase [18] back the complexes with experimentally x-ray structures from PDB. Each ligand systematically using four characters followed by three characters. For example, in the ligand "1kim.THM", "1kim" denotes the PDB code and "THM" is the ligand name in the PDB. When we evaluated the accuracy of GEMDOCK for molecular docking, the crystal coordinates of the ligand and protein atoms were taken from PDB, and were separated into different files. Our program then assigned the atom formal charge and atom type (i.e., donor, acceptor, both, or nonpolar) for each atom of both the ligand and protein. The bond type ($sp^3 - sp^3$, $sp^3 - sp^2$, or others) of a rotatable bond inside a ligand was also assigned.

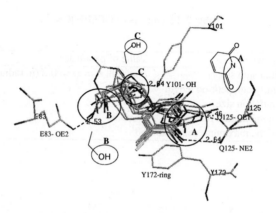

Fig. 3. Binding-site pharmacophores identified by superimposing ten crystal structures of HSV-1 thymidine kinase shown in Figure 2. Three pharmacological preferences and interactions are identified and circled as A (an amide binding site), B (a hydroxyl binding site), and C (a hydroxyl binding site). A stack force binding area (Y172-ring) is also indicated. The dash lines indicate the hydrogen binding.

To evaluate GEMDOCK's screening utility, we used HSV-1 thymidine kinase (TK), proposed by Bissabtz et al. [12] as the target protein with a ligand database, including 10 known active ligands (Figure 2) of TK and 990 randomly chosen non-active compounds from the ACD. When preparing the target protein, the atom coordinates for virtual screening were taken from the crystal structure of the TK complex (PDB entry 1kim). The atom coordinates of each ligand were sequentially taken from the database. Our program automatically decided the formal charge and atom type of each ligand atom. The ligand characteristics (i.e., the numbers of electrostatic atoms, hydrogen donor, and hydrogen acceptor) and the bond types of single bonds inside a ligand were also calculated. These variables were used in Equation 9 to calculate the scoring value of a docked conformation. Finally GEMDOCK re-ranked all docked ligand conformations for the post-analysis.

Figure 3 shows the binding-site pharmacophores and ligand preferences that were identified by superimposing ten crystal structures of TK shown in Figure 2. Three binding-site pharmacological preferences and interactions were identified and circled as A, B, and C. A stack force binding area (Y172-ring) was also indicated. The dash lines indicate the hydrogen binding. According to these observations, we added following pharmacological weights: Q125-OE1 and Q125-NE2 are hydrogen bonds with weighted value 4.0; Y101-OH and E83-OE2 are hydrogen bonds with weighted value 2.5; and six C atoms of Y172-ring form stacking force with weighted value 1.5. These weights were used in Equation 13 for calculating the value E_{phama}. For TK ligand preferences, the number of electrostatic atoms was set to 2 because all active ligands (Figure 2) have no charged atoms. The hydrophilic preference was not assigned in this target. In Equation 14, therefore, WP_{hb} is zero and WP_{elec} is 20 if the number of charged atoms inside a ligand was more than 2.

Table 3. Comparison GEMDOCK with three docking methods on docking 10 thymidine kinase ligands into the binding site of the target protein 1kim

Ligand[a]	No. of polar atoms[b]	No. of hydrogen bonds[c]	GEMDOCK pharmacophore weight (yes)	pharmacophore weight (no)	GOLD[d]	FlexX[d]	DOCK[d]
1e2k.TMC	6	5	0.75	0.79	1.19	7.56	1.11
1e2m.HPT	5	6	0.41	0.37	0.49	1.02	4.18
1e2n.RCA	9	6	1.54	1.41	2.33	9.62	13.3
1e2p.CCV	6	8	0.58	0.53	0.93	2.02	3.65
1ki2.GA2	9	4	3.56	2.15	3.11	3.01	6.07
1ki3.PE2	8	5	3.34	3.29	3.01	4.10	5.96
1ki6.AHU	7	6	0.43	0.39	0.63	1.16	0.88
1ki7.ID2	7	6	0.45	0.56	0.77	9.33	1.03
1kim.THM	7	4	0.47	0.48	0.72	0.82	0.78
2ki5.AC2	8	5	2.94	2.95	2.74	3.08	2.71

[a] The four characters and three characters separated by a period denote the PDB code and the ligand name in the Protein Data Bank, respectively.
[b] The number of the atoms that may form a hydrogen bond; i.e., the atom type is either both, donor, or acceptor.
[c] The number of hydrogen bonds formed between the ligand and the protein was derived from the native crystal conformations based on our scoring function (Equation 10).
[d] These results were directly taken from [12].

3.3 Molecular Docking Results on Ten TK Complexes

GEMDOCK executed 3 independent runs for each complex. The solution with lowest scoring function was then compared with the observed ligand crystal structure. First GEMDOCK docked each ligand of 10 TK ligands (Figure 2) back into its respective complex. We based the results on root mean square deviation (RMSD) error in ligand heavy atoms between the docked conformation and the crystal structure. The RMSD values of all ten docked conformations are less than 1.0 Å. Second we docked all ten TK ligands into the reference protein (1kim) and the results were shown in Table 3. During flexible docking GEMDOCK obtained similar results whether the pharmacophore preferences (i.e., E_{phama} and E_{ligpre}) were considered or not. The docked conformations with RMSD values less than 1.5 Å for seven pyrimidine-based ligands. On the other hand, three purine-based ligands (i.e., 1ki2.GA2, 1ki3.PE2, and 2ki5.AC2) could not be successfully docked into the reference protein because the side-chain conformation of GLN125 in the reference protein 1kim differs from the ones of these purine-based complexes, i.e., 1ki2, 1ki3, and 2ki5. GEMDOCK was the best among these four competing methods (GEMDOCK, GOLD, FlexX, and DOCK) on this test set.

Table 4. Comparison of GEMDOCK with four methods on screening 1000 compounds with false positive rates

True Positive(%)	GEMDOCK	Surflex[a]	DOCK[a]	FlexX[a]	GOLD[a]
80	0.8[b] (8/990)	0.9	23.4	8.8	8.3
90	1.0 (10/990)	2.8	25.5	13.3	9.1
100	1.6 (16/990)	3.2	27.0	19.4	9.3

[a] These results were directly taken from [12] and [19].
[b] the false positive rate from 990 random ligands (%).

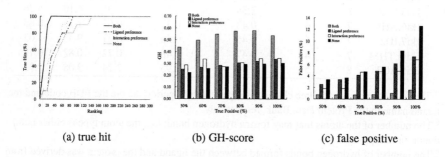

(a) true hit   (b) GH-score   (c) false positive

Fig. 4. GEMDOCK results for (a) true hit, (b) GH-score, and (c) the false positive rate for different true positive rates. GEMDOCK yielded good performance when it used both ligand and receptor pharmacological preferences.

3.4   Virtual Screening of TK Substrates

Figure 4 shows the overall accuracy of GEMDOCK using different combinations of pharmacophore preferences in screening the substrates of HSV-1 thymidine kinase (TK) from a data set with 1000 compounds. This data set, including 10 active and 990 random ligands proposed by Bissantz [12], was used to evaluate the performance of three docking tools (DOCK, FlexX, and GOLD) with different combinations of seven scoring functions[12]. The results of the comparison are also shown in Table 4.

Four common metrics were used to evaluate the screening quality, including true hit (the percentage of active ligands retrieved from database), yield (the percentage of active ligands in the hit list), goodness-of-hit (GH), and false positive rate. The GH score is defined as

$$GH = (\frac{A_h(3A + T_h)}{4T_hA})/(1 - \frac{T_h - A_h}{T - A}), \tag{15}$$

where A_h is the number of active ligands in the hit list, T_h is the total number of compounds in the hit list, A is total number of active ligands in the database, and T is the total number of compounds in the database. The yield (hit rate) can be given as $100\frac{A_h}{T_h}\%$. The false positive (FP) rate is given as $100\frac{T_h - A_h}{T}\%$. In the TK case A and T are 10 and 1000, respectively.

The main objective of this study was to evaluate whether the new scoring function was applicable to both molecular docking and ligand scoring in virtual screening. Figure 4 shows these results of GEMDOCK using different combinations of pharmacophore preferences that are ligand preferences (E_{ligpre}) and binding-site pharmacophore (E_{phama}). GEMDOCK generally improves the screening quality by considering both ligand preferences and binding-site pharmacophore weights although we did not attempt to refine any parameters of these combinations. The binding-site pharmacophores seem more important than ligand preferences. As shown in Figure 4(a), the hit rates of GEMDOCK for different combinations are 38% (both), 12% (ligand preferences), 13% (binding-site pharmacophore) and 7% (none) when the TP rate is 100%. If GEMDOCK applied binding-site and ligand preferences, the GH score is 0.54 (Figure 4(b)) and the FP rate is 1.62% (Figure 4(c)) when the TP rate is 100%.

Table 4 compares GEMDOCK with four docking methods (Surflex, DOCK, FlexX, and GOLD) on the same target protein and screening database at true positive rates ranging from 80% to 100%. For GEMDOCK on the target TK, the ranks of the ten active ligands were 3, 7-9, 12-14, 16, 19, and 26. For the true positive rate of 100%, the FP rate for GEMDOCK is 1.6%. In contrast, the FP rates for competing methods are 3.2% (Surflex), 27% (DOCK), 19.4% (FlexX), and GOLD (9.3%). DOCK is the worst and GEMDOCK is the best among these five approaches on this data set.

4 Conclusions

In summary, we have developed an automatic tool with a novel scoring function for virtual screening by applying numerous enhancements and modifications to our original techniques. By integrating a number of genetic operators, each having a unique search mechanism, GEMDOCK seamlessly blends the local and global searches so that they work cooperatively. Our new scoring function is able to be applied to both flexible docking and post-docking analysis for reducing the number of false positives. Experiments verify that the proposed approach is robust and adaptable to virtual screening.

References

1. P. D. Lyne. Structure-based virtual screening: an overview. *Drug Discovery Today*, 7:1047–1055, 2002.

2. B. K. Shoichet, S. L. McGovern, B. Wei, and J. Irwin. Lead discovery using molecular docking. *Current Opinion in Chemical Biology*, 6:439–446, 2002.

3. I. D. Kuntz, J. M. Blaney, S. J. Oatley, R. Langridge, and T. E. Ferrin. A geometric approach to macromolecular-ligand interactions. *Journal of Molecular Biology*, 161:269–288, 1982.

4. G. Jones, P. Willett, R. C. Glen, A. R. Leach, and R. Taylor. Development and validation of a genetic algorithm for flexible docking. *Journal of Molecular Biology*, 267:727–748, 1997.

5. B. Kramer, M. Rarey, and T. Lengauer. Evaluation of the flexX incremental construction algorithm for protein-ligand docking. *Proteins: Structure, Function, and Genetics*, 37:228–241, 1999.

6. T. J. Ewing, S. Makino, A. G. Skillman, and I. D. Kuntz. Dock 4.0: search strategies for automated molecular docking of flexible molecule databases. *Journal of Computer-Aided Molecular Design*, 15:411–428, 2001.

7. G. M. Morris, D. S. Goodsell, R. S. Halliday, R. Huey, W. E. Hart, R. K. Belew, and A. J. Olson. Automated docking using a lamarckian genetic algorithm and empirical binding free energy function. *Journal of Computational Chemistry*, 19:1639–1662, 1998.

8. J.-M. Yang. Development and evaluation of a generic evolutionary method for protein-ligand docking. *Journal of Computational Chemistry*, 25:843–857, 2004.

9. H. Gohlke, M. Hendlich, and G. Klebe. Knowledge-based scoring function to predict protein-ligand interactions. *Journal of Molecular Biology*, 295:337–356, 2000.

10. S. J. Weiner, P. A. Kollman, D. A. Case, U. C. Singh, C. Ghio, G. Alagona, S. Profeta, Jr., and P. Weiner. A new force field for molecular mechanical simulation of nucleic acids and proteins. *Journal of the American Chemical Society*, 106:765–784, 1984.

11. D. K. Gehlhaar, G. M. Verkhivker, P. Rejto, C. J. Sherman, D. B. Fogel, L. J. Fogel, and S. T. Freer. Molecular recognition of the inhibitor AG-1343 by HIV-1 protease: conformationally flexible docking by evolutionary programming. *Chemistry and Biology*, 2(5):317–324, 1995.

12. C. Bissantz, G. Folkers, and D. Rognan. Protein-based virtual screening of chemical databases. 1. evaluation of different docking/scoring combinations. *Journal of Medicinal Chemistry*, 43:4759–4767, 2000.

13. M. Stahl and M. Rarey. Detailed analysis of scoring functions for virtual screening. *Journal of Medicinal Chemistry*, 44:1035¡V1042, 2001.

14. J.-M. Yang and C.-Y. Kao. A robust evolutionary algorithm for training neural networks. *Neural Computing and Application*, 10(3):214–230, 2001.

15. J.-M. Yang and C.-C. Chen. Gemdock: A generic evolutionary method for molecular docking. *Proteins: Structure, Function, and Bioinformatics*, 55:288–304, 2004.

16. J.-M. Yang, C.-H. Tsai, M.-J. Hwang, H.-K. Tsai, J.-K. Hwang, and C.-Y. Kao. GEM: A gaussian evolutionary method for predicting protein side-chain conformations. *Protein Science*, 11:1897–1907, 2002.

17. T. Bäck. *Evolutionary Algorithms in Theory and Practice*. Oxford University Press, New York, USA, 1996.

18. J. N. Champness, M. S. Bennett, F. Wien, R. Visse, C. W. Summers, P. Herdewijn, E. de Clerq, T. Ostrowski, R. L. Jarvest, and M. R. Sanderson. Exploring the active site of herpes simplex virus type-1 thymidine kinase by x-ray crystallography of complexes with aciclovir and other ligands. *Proteins*, 32:350–361, 1998.

19. A. N. Jain. Surflex: Fully automatic flexible molecular docking using a molecular similarity-based search engine. *Journal of Medicinal Chemistry*, 46:499–511, 2003.

Statistical Test-Based Evolutionary Segmentation of Yeast Genome

Jesus S. Aguilar–Ruiz, Daniel Mateos, Raul Giraldez, and Jose C. Riquelme

Dept. of Computer Science, University of Seville, Spain
{aguilar,mateos,giraldez,riquelme}@lsi.us.es

Segmentation algorithms emerge observing fluctuations of DNA sequences in alternative homogeneous domains, which are named segments [1]. The key idea is that two genes that are controlled by a single regulatory system should have similar expression patterns in any data set. In this work, we present a new approach based on Evolutionary Algorithms (EAs) that differentiate segments of genes, which are represented by its level of meiotic recombination[1]. We have tested the algorithm with the yeast genome [2][3] because this organism is very interesting for the research community, as it preserves many biological properties from more complex organisms and it is simple enough to run experiments. We have a file with about 6100 genes, divided into sixteen yeast chromosomes (N). Each gene is a row of the file. Each column of file represents a genomic characteristic under specific conditions (in this case, only the activity of meiotic recombination). The goal is to group consecutive genes properly differentiated from adjacent segments. Each group will be a segment of genes, as it will maintain the physical location within the genome. To measure the relevance of segments the Mann–Whitney statistical test has been used.

Each individual of the population is an array of natural numbers with size C, and it represents a collection of cutpoints within the yeast genome. Fifteen of these cutpoints correspond to the boundaries of the sixteen chromosomes of the yeast genome, and they are permanent. The sixteen cutpoints corresponding to centromeres also are permanent, so we have 31 constant cutpoints. The centromere is approximately in the middle of a chromosome and separates it in two branches (L and R). Although these fixed cutpoints (FC=31) cannot be moved, they have been included in all of the individuals, making easier the computational process. For example, if a cutpoint array includes the values 34, 57, 7, 25 and 80, it means that there is a cutpoint between the 34th and the 35th genes, between the 57th and the 58th genes, between the 7th and the 8th genes, etc. We have chosen the Mann-Whitney test as the fitness function. The Mann–Whitney test, also known as the Wilcoxon rank sum test, is a non–parametric test used to test for difference between the medians of two independent groups. This test is the non–parametric equivalent of the two–sample t–test. No distributional assumptions are required for this test, so the test does not assume that the populations follow Gaussian distributions. The choice of this method is due to the necessity of differentiating adjacent segments clearly. If we choose the mean as

---

[1] Meiotic recombination is the exchange of chromosomal segments between the paternal and maternal homologs during meiosis

K. Deb et al. (Eds.): GECCO 2004, LNCS 3102, pp. 493–494, 2004.
© Springer-Verlag Berlin Heidelberg 2004

494   J.S. Aguilar–Ruiz et al.

representative statistical value for a segment, we can know when the mean of two adjacent segments is significantly different with the Mann–Whitney test. In order to verify the quality of the fitness function, we run the algorithm with the original data, and with randomized versions. We can understand that a fitness function is correct if the results obtained with the random data are lower in quality than those obtained from the original data. Otherwise, we can say that we have an "artifact"[2]. The fitness function returns the sum of all tests (for all the cutpoints of an individual).

Due to the intrinsic characteristics of the problem, a variant of the uniform crossover has been chosen. That is, a new individual is built by randomly choosing cutpoints from both parents. Also, we tested other well–known operators (one–point and two–points crossovers), but they did not provide better results. This operator has to maintain the diversity, controlling values different than those assigned to the boundaries of chromosomes and centromeres.

The mutation operator alters each cutpoint according to two probabilities: p_1 and p_2. The probability p_1 controls if a cutpoint is going to be modified; and the probability p_2 controls if the mutation will result in a random cutpoint within the range, or in a slight variation (currently, 5 genes to the left or to the right) of the cutpoint. Basically, these two options are: $indiv[i] := random(N)$ and $indiv[i] := indiv[i] + (-1)^{random(2)} * random(5)$, respectively. The choice of value 5 is not critical, other values around 5 can be used as well. However, if that value is high, consecutive populations will present greater diversity than its ancestors. Logically, the genes at the boundaries of chromosomes and centromeres are not mutated.

Experiments show that the genomic distribution in yeast genome is not random under the perspective of the activity of meiotic recombination. The Evolutionary Algorithm has a very satisfactory performance from the biologist point of view, as it can find a high percentage of valid adjacent segments, which can add knowledge to the biological research of functional properties of groups of genes. The results reported in this work are not comparable, as we have not found any other system that addresses the segmentation problem by using numerical information. We are now designing a bench mark test based on dynamic programming to avoid the computationally unapproachable exhaustive search.

References

1. Elton, R.: Theoretical models for heterogeneity of base composition in DNA. Journal of Theoretical Biology **45** (1974) 533–553
2. Goffeau, A., et al.: The yeast genome directory. Nature **387** (1997) 5–105
3. http://www.yeastgenome.org/

---

[2] An apparent experimental result that is not actually real but is due to the experimental methods

Equilibrium and Extinction in a Trisexual Diploid Mating System: An Investigation

Erik C. Buehler[1], Sanjoy Das[1], and Jack F. Cully, Jr.[2]

[1]Electrical and Computer Engineering Department,
Kansas State University, Manhattan, KS 66506
mrbuehler@mailcircuit.com
sdas@ksu.edu

[2]Kansas Cooperative Fish and Wildlife Research Unit, Division of Biology,
Kansas State University, Manhattan, KS 66504
bcully@ksu.edu

In order to study the dynamics of a three-sex (trisexual) mating system, we have chosen to extend the heterogametic sex-determining mechanism, used in many species, to include three sexes: XX, XY and YY. In this model, non-like types may mate, but like-types may not mate. Yeasts and fungi are known to have multiple mating types (sometimes numbering in the thousands), but the mechanics of these sex-determining systems are markedly different from the heterogametic system we are interested in studying [5]. Our motivation for using this scheme stems from the knowledge that in some species, such as most fish, XX is female, and XY is male [1]. Under certain conditions, a YY individual may be produced, and in the case of fish, this usually develops into a male. Our goal is to discover the emergent behavior of a hypothetical "diploid trisexual mating system" (DTMS) where the YY type is its own distinct "mating type", in order to shed light on why such a system is not observed in nature today.

We first constructed a stochastic computer model to simulate a DTMS under Hardy-Weinberg conditions, and found that the mating-type frequencies F_{xx}, F_{xy} and F_{yy}, tend to hover near a Hardy-Weinberg-like equilibrium point. However, the system inevitably converged to a two-sex system containing either ½ XX and ½ XY or ½ YY and ½ XY.

We derived the formulas describing the theoretical equilibrium points for the mating type frequencies, which agree very well with our simulation results. These formulas were then used to compute the starting points for subsequent simulations, to measure frequency changes from one generation to the next.

To give credence to the possibility that a DTMS may exist in nature, we explored the idea of mating efficiency. Because we are assuming random mating, any individual has an equal probability of attempting to mate with any other individual in the population. Therefore, in a two-sex system, 50% of pairings are viable, but in the proposed DTMS, ~57.7% of pairings are viable (assuming the system is balanced at the central equilibrium point, previously derived). As the system falls away from the central equilibrium point, it approaches a two-sex system, and the pairing efficiency approaches 50%. This hints at an advantage for a species to maintain a three-sex mating system.

K. Deb et al. (Eds.): GECCO 2004, LNCS 3102, pp. 495–496, 2004.
© Springer-Verlag Berlin Heidelberg 2004

Next, we carried out an "equilibrium point sweep" on the simulation, in order to uncover evidence for a bias against the less-frequent homozygote. Beginning with an F_{xx} of 0, we computed the corresponding F_{xy} and F_{yy}, and then ran the simulation from this starting point enough times to reliably compute the average frequency change of F_{xx} in the subsequent generation. We incremented F_{xx} slightly, and repeated the process until $F_{xx} = \frac{1}{2}$. This analysis showed no evidence of any bias for or against the less-frequent homozygote.

Finally, we used the "equilibrium point sweep" technique to measure the variance of the next-generation frequency change, and found that this variance is linearly proportional to the frequency its self. The decreased next-generation frequency change variance results in a lesser ability to recover a frequency loss in the subsequent generation, and inevitably, one of the homozygotes eventually goes extinct. We also found that this tends to occur in roughly $\frac{1}{2}N$ generations (N being the population size).

We approximated the next-generation frequency change distributions using the binomial distribution, constructed a simulation which used the binomial distribution to model the next-generation frequency change, and found its behavior to be nearly identical to that of our original DTMS simulation model. We therefore conclude that this linearly proportional relationship between frequency and next-generation frequency change variance is the root cause of the inevitable two-sex convergence.

Future work is needed to derive the average time to two-sex convergence, and to derive the actual statistical distributions of the next-generation mating-type frequency changes.

References

1. J.J. Bull, Sex determining mechanisms: An evolutionary perspective. [In] *The Evolution of Sex and Its Consequences*, S.C. Stearns, 1987. 93-114. Birkhäuser Verlag Basel, 1987
2. P. Coker and C. Winter, N-Sex reproduction in Dynamic Environments. *Proceedings of the Fourth European Conference on Artificial Life*, 1997
3. R.A. Fisher, The genetic theory of natural selection. Oxford: Clarendon Press, 1930
4. K. Jaffe, The dynamics of the evolution of sex: Why the sexes are, in fact, always two? *Interciencia*, 21(6): 259-267, 1996
5. E. Kothe, Tetrapolar fungal mating types: sexes by the thousands. *FEMS Microbiology Review*, 18: 65-87, 1996
6. C. Stern, The Hardy-Weinberg Law. *Science*, 97: 137-138, 1943

On Parameterizing Models of Antigen-Antibody Binding Dynamics on Surfaces – A Genetic Algorithm Approach and the Need for Speed

Daniel J. Burns[1] and Kevin T. May[2]

[1] Air Force Research Laboratory, Information Directorate, Rome, NY 13440 USA
burnds@rl.af.mil
[2] Department of Computer Engineering, Clarkson University, Potsdam, NY, USA 47907
mayk@clarkson.edu

Abstract. This paper discusses the performance of a simple GA for parameterizing a particular biomodel consisting of a set of coupled non-linear ordinary differential equations. Comments are offered on the need for speed that motivates choice of language and processing platform for solving scaled problems.

1 Introduction

Ag-Ab (Antigen-Antibody) binding dynamics at surfaces is of interest to biologists because of the critical need for biosensors and diagnostic tests for the presence of targeted substances in clinical, biological, or environmental samples [Zheng 1]. Accurate models of binding dynamics are needed to support the design and performance optimization of biosensor systems. Developing accurate models requires parameterizing them to fit experimental data. This is a hard optimization problem that can easily become analytically intractable for complex nonlinear bio-models. This motivates reduced order modeling that involves techniques such as variable replacement by exogenous functions, temporal windowing, sequential parameter fitting, etc. [Rundell 2]. Even with reduced order modeling, the computing time required for parameterizations can be many minutes. The present work evaluated whether a simple GA approach run on a full, unreduced model could be more efficient. This work served as a test case for developing working C codes that could be ported to parallel and embedded computer platforms to achieve extreme speed-ups known to be needed for certain hard problems.

2 Summary of the Work and Results

We wrote codes for simulating the Model and for parameterizing the Model with a simple GA in Labview and in MatLab, and translated them to C to achieve reasonable run times (100-10,000 X faster). Our GA used binary representations of the floating point Model parameters. Values were searched over adjustable width ranges that were different for each parameter. Adjustable GA parameters included # bits in each gene,

K. Deb et al. (Eds.): GECCO 2004, LNCS 3102, pp. 497–498, 2004.

chromosomes, # genes in each chromosome, # in initial population, # in running population, % of population selected to produce children, % of gene bits mutated at each generation, maximum # generations, ranking method for selection (probability based on fitness or rank), and termination criteria (based on total fitting error, or maximum single point error). We used uniform probability crossover for all genes, and elitism (keeping the current best individual).

The Model equations predict the# of Ag particles bound to an Ab functionalized surface, with one equation describing each attachment *valency* (i.e. the # of epitope sites binding an Ag particle). The Model has 6 parameters (initial association rate, initial dissociation rate, transport coefficient due to diffusion, transport coefficient due to gravity, crosslinking association rate, and crosslinking dissociation rate). A *complete* raw data set from the Model consisted of about 7 concentration vs time curves tracing out a binding and release experiment over 200 time points. Experimentally it is only possible to measure the *total* # of bound Ag, not the *valency* of a bound Ag. Therefore, we summed the curves to produce a single *total bound* concentration vs time curve. We measured the performance of the GA parameterization tool working with both *complete* data sets and reduced *total bound* data sets. The fitness function calculated the sum of the least squares differences of data sets from a *known* individual (with a set of preselected parameter values) and individuals in the population. Various supervisory programs were written to gather statistical performance results over typically 20 or 30 fitting runs, and to display results. Also, a number of metrics were defined to characterize the performance of the GA fitter (speed, accuracy, convergence) using different sets of GA parameters, and for determining how the metrics behaved with scaling, e.g. as a function of population sizes, maximum generations, etc.

We observed that fitting all six parameters to about 2% accuracy using *complete* data sets was easy using populations of about 200 run for 1000 generations. This involved about 200,000 function evaluations and took about 30 seconds. Times for fitting 5 parameters (without kg which hardly effects the data) were much better, typically about 2 seconds. We also determined that varying the ranges over which parameter values were fit (from 2 to 4 to 10 times the parameter values) changed the average number of generations required to converge from 500 to 800 to 1250.

Fitting all 6 Model parameters simultaneously from *total bound* data sets proved to be much harder, with typically only 10% of runs achieving <10% fitting errors. We are currently experimenting with a progressive GA (PGA) parameter fitting strategy that mimics the progressive fitting strategy described in [2] to improve competence.

References

1. Zheng, Y.; Rundell, A., "Biosensor Immuno-surface Engineering Inspired by B-cell Membrane Bound Antibodies: Modeling and Analysis of Multivalent Antigen Capture by Immobilized Antibodies, IEEE Transactions on NanoBioscience, 2(1):14-25, 2003.
2. Rundell, A.; DeCarlo, R.; Doerschuk, P.; HogenEsch, H.; "Parameter Identification for an Autonomous 11th Order Nonlinear Model of a Physiological Process", Proceedings of the 1998 American Control Conference, 6: 3585-3589, 1998.

Is the Predicted ESS in the Sequential Assessment Game Evolvable?

Winfried Just and Xiaolu Sun

Department of Mathematics, Ohio University, Athens, OH 45701, U.S.A.

Abstract. The Sequential Assessment Game model of animal contests predicts an evolutionarily stable strategy (ESS) that is a sequence of thresholds for giving up. Simulated evolution experiments reveal that the selection pressure on higher-numbered thresholds is most likely too low to allow for the theoretically predicted ESS to evolve in nature.

1 Simulating the Sequential Assessment Game

In game-theoretic models of animal contests, contestants are treated as players whose objective is to maximize their expected number of offspring. A strategy is a prescription of how a player should behave in every possible situation. A strategy S is evolutionarily stable, or an ESS, if a population of players that all follow S cannot be invaded by a mutant strategy S' [2]. Being an ESS is only a necessary condition for a strategy to actually evolve. Simulated evolution can be used to investigate whether theoretically predicted ESS's are indeed evolvable.

The objective in the classical Sequential Assessment Game of Enquist and Leimar [1] is to gain access to a resource. The game proceeds in succesive bouts. After each bout, each player makes a decision to either give up or continue. The decision is based on the player's currend estimate of relative fighting ability. These estimates are made by statistical sampling and become more reliable after each bout. The resource is obtained by the player who persists longer in the game. Both players incur fitness costs for each bout of the game.

A strategy in this game can be conceptualized as a sequence of thresholds $S = \{S_1, S_2, \ldots, S_k, \ldots\}$ for giving up: Let X_i denotes a player's estimate of relative fighting ability after the i-th bout. A player who follows Strategy S gives up after k bouts if, and only if, $X_1 > S_1, \ldots, X_{k-1} > S_{k-1}, X_k \leq S_k$, unless his opponent gives up earlier. In [1], an apparent ESS for the Sequential Assessment Game was found that corresponds to an increasing sequence of thresholds.

We tested the model by simulated evolution of finite populations of competing strategies. The original model does not put an upper limit on the number of bouts in this game. In order to represent strategies in the computer, we limited fights to T bouts, and each strategy was coded as the sequence of thresholds for giving up after each one of these T bouts. If none of the players gave up after T or fewer bouts, then the program randomly picked a winner and subtracted an extra punishment (equivalent, on average, to continuing for E additional bouts) from the fitness of both the winner and the loser. For more details see [3].

K. Deb et al. (Eds.): GECCO 2004, LNCS 3102, pp. 499–500, 2004.

In a number of trial runs with this setup and for a variety of parameter settings we did not observe a pattern of evolution towards a strategy that would resemble the predicted ESS. While means for the first two thresholds were usually highly consistent between different runs of the simulation, the means for the higher-numbered thresholds evolved towards very different numbers, depending on the seed. In order to understand this phenomenon, we ran simulations that monitored the number of fights in the last season that were terminated after any given bout. Table 1 gives some representative numbers for the last season in simulations with $T = E = 5$, out of 30,000 total encounters.

Table 1. Total number of encounters that were terminated after bout t

Seed	$t = 1$	$t = 2$	$t = 3$	$t = 4$	$t = 5$
1	28,405	743	118	734	0
2	27,582	68	2,350	0	0
3	27,920	1,563	309	208	0

2 Discussion

Our results strongly suggest that the reason for the observed lack of a clear pattern in the evolution of higher-numbered thresholds is the absence of a suffi-ciently strong selection pressure on them. We obtained three independent pieces of evidence for this: The absence of clear patterns of evolution in our simulations, the low occurrence of higher-numbered bouts, and the fact that a simple change in the rules that created artificial selction pressure on higher-numbered bouts lead to fairly consistent outcomes that fit the predictions of [1] remarkably well. The latter results were reported in [3].

We believe that our results show a serious weakness of the classical Sequential Assessment model. It is not clear how the predicted ESS could evolve in nature. We suggest that the problem of selection pressure on individual thresholds needs to be investigated by researchers in mathematical biology.

Acknowledgement. This research was partially supported by NSF grant DBI-9904799 to W.J.

References

1. Enquist, M., Leimar, O.: Evolution of Fighting Behaviour: Decision Rules and the Assessment of Relative Strength. J. Theor. Biol. **102** (1983), 387–410.
2. Maynard Smith, J., Price, G. R.: The logic of animal conflict. Nature **246** (1973) 15–18.
3. Sun, X., Just, W.: Evolution of Strategies in Modified Sequential Assessment Games. To appear in Proceedings of CEC 2004.

Automated Extraction of Problem Structure

Anthony Bucci[1], Jordan B. Pollack[1], and Edwin de Jong[2]

[1] DEMO Lab, Brandeis University, Waltham MA 02454, USA
{abucci***,pollack}@cs.brandeis.edu
http://demo.cs.brandeis.edu/
[2] Decision Support Systems Group, Universiteit Utrecht
dejong@cs.uu.nl

Abstract. Most problems studied in artificial intelligence possess some form of structure, but a precise way to define such structure is so far lacking. We investigate how the notion of problem structure can be made precise, and propose a formal definition of problem structure. The definition is applicable to problems in which the quality of candidate solutions is evaluated by means of a series of tests. This specifies a wide range of problems: tests can be examples in classification, test sequences for a sorting network, or opponents for board games. Based on our definition of problem structure, we provide an automatic procedure for problem structure extraction, and results of proof-of-concept experiments. The definition of problem structure assigns a precise meaning to the notion of the underlying objectives of a problem, a concept which has been used to explain how one can evaluate individuals in a coevolutionary setting. The ability to analyze and represent problem structure may yield new insight into existing problems, and benefit the design of algorithms for learning and search.

1 Introduction

Most problems studied in artificial intelligence possess some form of *structure*. Taking chess as an example, different players can be compared with regard to strategy, tactics, and other aspects of their play. Thus, there are several *dimensions* along which the behavior of players can be compared. Precise knowledge of such problem structure would benefit both our insight into problems and the design of algorithms. It has so far been unclear however how such dimensions might be defined precisely, and how informative dimensions might be determined. We investigate how these notions can be made precise, and propose a formal definition of problem structure. Based on this, we describe an automatic mechanism to explore and represent problem structure.

We consider problems where the performance of a candidate solution, or *candidate* for short, is determined by the outcomes of *tests*. We consider problems where the performance of a candidate is determined by the outcomes of *tests*. For example, a classifier may be evaluated on the errors it makes in classifying

---

*** Corresponding author

K. Deb et al. (Eds.): GECCO 2004, LNCS 3102, pp. 501–512, 2004.

test examples; an evolved checkers player may be evaluated on its scores against some set of opponents; and a sorting network can be evaluated on its ability to sort test sequences. This class of *test-based problems* defines a broad range of problems.

The structure of a problem consists of a space and a mapping of the candidates and tests into this space. The structure space is such that the outcome of a candidate on any test can be uniquely determined given only the coordinates of the candidate. Furthermore, a structure space is of minimum dimension given this constraint. The structure space captures essential information about a problem in an efficient manner.

Since the quality of a candidate is determined by its outcomes on tests, tests may be viewed as objectives in the sense of Multi-Objective Optimization (MOO; see [1], e.g.). In this view, the structure space may be seen as a projection of the tests onto a smaller set of dimensions or objectives, such that a one-to-one mapping exists between the candidate objective vectors for the two spaces. This resulting set of objectives will typically be unknown at first, but is fundamental in the sense that it represents all relevant relations between candidates and tests in an optimally compact way. The axes spanning the structure space may therefore be called the *underlying objectives* of a problem. The term *underlying objectives* was first introduced in work on coevolution [2], where it was observed that the tests in a coevolutionary algorithm tended to identify the objectives that governed the evaluation of learners. A simpler version of the same idea was presented in the form of the *ideal test set* and *test dimension* of [3].

In the realm of machine learning of game strategies, Arthur Samuel notes that terms for the evaluation polynomial of his checkers player should ideally be generated by the learning program itself. Samuel mentions the idea of an orthogonal set of terms to be used in this evaluation polynomial [4]. Along similar lines, Susan Epstein has argued that to be optimized, a game player should experience a variety of opponents with varying skill levels [5]. We feel the present work offers a precise way to discuss concepts like "orthogonal set of terms," and to clarify which variety of opponents a game player requires to be optimized.

The structure of this paper is as follows. In section 2, we present a mathematical definition of a *coordinate system* for a test-based problem. We give an example from geometry to motivate our choice of definitions, and explore some of the properties and implications of the definitions. In section 3, we present a polynomial-time dimension-extraction algorithm which, given a problem, constructs a coordinate system for it. The coordinate system need not be minimal, but it is guaranteed to *span* the problem in a certain sense and satisfy an independence criterion. Finally, in section 4, we present some experimental validation of the formal and algorithmic ideas. We run the dimension-extraction algorithm on the population in a coevolutionary simulation run on a game with known dimension; we see that the algorithm correctly deduces the dimension or overestimates it, depending on the game.

2   Geometrical Problem Structure

Let $p : S \times T \to 2$ be any function, where S and T are *finite* sets[1] and 2 is the partially ordered set $0 < 1$. Here the set S is interpreted as the set of candidate solutions; T is the set of tests or test cases, and 2 is the outcome of applying a test to a candidate. The function p encodes the interaction between a test and a candidate; intuitively, we can think of it as a payoff function. Such functions appear often in optimization and learning problems. For example:

Example 1 (Function approximation). Let $f : T \to \mathbb{R}$ be a target function defined over a set T, and let S be a set of model functions $T \to \mathbb{R}$. The problem is to find a function in S that matches f as closely as possible. Notice that if $h \in S$ is some candidate function, then a point $t \in T$ can serve as a test of h. For example, we can define $p : S \times T \to 2$ by $p(h, t) = \delta(f(t), h(t))$, δ the Kronecker delta function.

Example 2 (Chess). Let $S = T = \{\text{deterministic chess-playing strategies}\}$. For any two strategies $s_1, s_2 \in S$, define $p(s_1, s_2) = 1$ if s_1 beats s_2, 0 otherwise. Then p is of form $S \times T \to 2$.

Example 3 (Multi-objective Optimization). Let S be a set of candidate solutions, and for each $0 \leq i \leq n-1$, let $f_i : S \to 2$ be an objective. The optimization task is to find (an approximation of) the non-dominated front of these n objectives. Let $T = \{f_0, \ldots, f_{n-1}\}$, and define $p : S \times T \to 2$ by $p(s, f_i) = f_i(s)$ for any $s \in S$, $f_i \in T$.

In this section, we will use such a function p to define an abstract coordinate system on the set S. This coordinate system will give a precise meaning to the notion of *underlying objectives*. In all of our examples, S will be finite. At first glance it is not obvious what a coordinate system on a finite set might look like. One of the major contributions in this paper is forwarding an idea about how we might do that.

2.1   Motivation

As a motivating example for the definitions to follow, let us consider the 2-dimensional Euclidean space E_2, namely the set $\mathbb{R} \times \mathbb{R}$ with its canonical coordinate system and pointwise order. Write $x : E_2 \to \mathbb{R}$ and $y : E_2 \to \mathbb{R}$ for the two coordinate axes; for any point in E_2, the function x returns the point's x coordinate and the function y returns its y coordinate. $p \leq q$ holds for two points $p, q \in E_2$ exactly when $x(p) \leq x(q)$ and $y(p) \leq y(q)$ both hold. Now consider these two families of subsets of E_2. For each $r, s \in \mathbb{R}$:

$$X_r = \{p \in E_2 | x(p) \geq r\} \tag{1}$$
$$Y_s = \{p \in E_2 | y(p) \geq s\} \tag{2}$$

---

[1] The finiteness assumption is not strictly necessary, but it greatly eases the exposition.

Geometrically, X_r is the half plane consisting of the vertical line $x = r$ and all points to the right of it. Y_s is the half plane consisting of the horizontal line $y = s$ and all points above it. Figure 1 illustrates these two families.

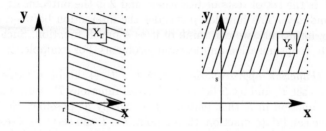

Fig. 1. Typical members of the families $\mathcal{X}$ and $\mathcal{Y}$; see text for details.

For brevity, let us write $\mathcal{X}$ for the family $(X_r)_{r\in\mathbb{R}}$ and $\mathcal{Y}$ for $(Y_s)_{s\in\mathbb{R}}$. In other words, an element of the family $\mathcal{X}$ is one of the sets X_r, and an element of $\mathcal{Y}$ is one of the sets Y_s. We would like to show that $\mathcal{X}$ and $\mathcal{Y}$ can act as stand-ins for the coordinate functions x and y. In particular, $\mathcal{X}$ and $\mathcal{Y}$ satisfy the following three properties:

1. **Linearity**: For all $r, s \in \mathbb{R}$, $X_r \subset X_s$ or $X_s \subset X_r$. Furthermore, $X_r = X_s$ implies $r = s$. Similarly, $Y_r \subset Y_s$ or $Y_s \subset Y_r$ and $Y_r = Y_s$ implies $r = s$.
2. **Independence**: There exist $r, s \in \mathbb{R}$ such that X_r and Y_s are incomparable; that is, neither is a subset of the other.
3. **Spanning**: For all $p \in E_2$ define $f(p) = \inf_r\{p \in X_r\}$ and $g(p) = \inf_r\{p \in Y_r\}$. Then f and g are well-defined functions from E_2 to $\mathbb{R}$, and $p \le q$ in E_2 exactly when $f(p) \le f(q)$ and $g(p) \le g(q)$ both hold.[2]

Property 1 states that the family $\mathcal{X}$ is linearly ordered by $\subset$; $\mathcal{Y}$ is as well. Property 2 states that the two families $\mathcal{X}$ and $\mathcal{Y}$ give independent information about E_2. Finally, property 2 states that $\mathcal{X}$ and $\mathcal{Y}$ can together be used to recover the order on E_2; this is the sense in which they span the space.

Properties 1-2 make no reference to the special qualities of E_2. In fact, they require only the family $\mathcal{X} \cup \mathcal{Y}$ of subsets of E_2. Since we can define families of subsets in any set, particularly finite ones, these three properties are a suitable abstract notion of coordinate system which can be fruitfully extended to finite sets.

2.2   Terminology

We will require some terminology from discrete math, which we review next.

Recall that a *preorder* on a set S is a reflexive, transitive, binary relation on S. Unless we state otherwise, the symbol $\le$ will be used for preorders; we will also

---

[2] In this example $f = x$ and $g = y$. This property is the definition of the order on E_2 in disguise.

write $s_1 \geq s_2$ to mean $s_2 \leq s_1$. The *reflexive* property means that for any $s \in S$, $s \leq s$ holds. The *transitivity* property means that for any three $s_1, s_2, s_3 \in S$, $s_1 \leq s_2$ and $s_2 \leq s_3$ together imply $s_1 \leq s_3$. A preorder is similar to a partial order. *Partial orders* are also *antisymmetric*, meaning: whenever $s_1 \leq s_2$ and $s_2 \leq s_1$ both hold, it must be that $s_1 = s_2$. In a preorder, antisymmetry may fail: both these relations may hold, but it may still be that $s_1 \neq s_2$. Preorders commonly arise from functions into sets that are already ordered. For instance, if $f : S \to \mathbb{R}$, then we can compare two $s_1, s_2 \in S$ using f. Namely, there is a preorder $\leq_f$ on S defined: $s_1 \leq_f s_2$ exactly when $f(s_1) \leq f(s_2)$. Antisymmetry of $\leq_f$ is then equivalent to f being injective.

A *linear order* is a partial order which satisfies the *trichotomy law*: for any two s_1, s_2, either $s_1 \leq s_2$, $s_2 \leq s_1$, or $s_1 = s_2$ must hold. A partial order need not satisfy this property. In other words, a partial order can have *incomparable* elements, meaning two $s_1, s_2 \in S$ such that neither is $\leq$ the other. The canonical example of a partial order is the power set of a set. The power set is partially ordered by inclusion: given any two subsets of a set, it need not be true that one is a subset of the other. We refer the reader to a discrete mathematics text such as [6] for more details and discussion of these concepts.

2.3   Coordinate Systems

Before defining a coordinate system on S, we will need some preliminary definitions to simplify notation.

Let $p : S \times T \to 2$ be any function on the finite sets S and T. For each $t \in T$, define the set $V_t = \{s \in S | p(s,t) = 0\}$. The set V_t is therefore the subset of all candidates which do poorly against the test t. We can use these sets to define a preordering on T. Namely, define $t_1 \leq t_2$ if $V_{t_1} \subset V_{t_2}$. Observe that in general this will be a preorder: there is no guarantee that $V_{t_1} = V_{t_2}$ implies $t_1 = t_2$. However, reflexivity and transitivity hold. It will be convenient to define two formal elements $t_{-\infty}$ and t_∞ and extend the order $\leq$ from T to $\overline{T} = T \cup \{t_{-\infty}, t_\infty\}$ by defining $t_{-\infty} < t < t_\infty$ for all $t \in T$. That is, $t_{-\infty}$ and t_∞ are respectively the minimum and maximum of $\leq$ extended to $\overline{T}$. For any subset $U \subset T$, we will write $\overline{U}$ for $U \cup \{t_{-\infty}, t_\infty\}$. Under the mapping $t \mapsto V_t$, $t_{-\infty}$ corresponds to $\emptyset$ and t_∞ corresponds to S. This formal device will make certain arguments easier. In particular, for any $s \in S$ and any $U \subset T$, there will always be $t_1, t_2 \in \overline{U}$ such that $p(s, t_1) = 0$ and $p(s, t_2) = 1$. $\overline{U}$ will always have a minimum and a maximum.

[3] argues that a function like p induces a natural ordering on the set S which is related to the idea of Pareto dominance in multi-objective optimization. We argue that this ordering captures important information about how two candidate solutions in S compare to one another in an optimization problem defined by p. Let us write $\preceq$ for this ordering; then for any $s_1, s_2 \in S$, $s_1 \preceq s_2$ holds if $p(s_1, t) \leq p(s_2, t)$ for all $t \in T$. For instance, in the multi-objective optimization example, $s_1 \preceq s_2$ exactly when $f_i(s_1) \leq f_i(s_2)$ for all objectives $f_i \in T$. In the multi-objective optimization literature the latter condition means s_2 *covers* s_1.

With these preliminaries, we can define a coordinate system on S. The sets V_t will play a role analogous to the X_r and Y_r above. The ordering $\preceq$ on S is the one we wish to span.

Definition 1 (Coordinate System). *A family $\mathcal{T} = (T_i)_{i \in I}$ of subsets of T is a coordinate system for S (with axes T_i) if it satisfies the following two properties:*

1. **Linearity:** *Each T_i is linearly ordered by $\leq$; in other words, for $t_1, t_2 \in T_i$, either $V_{t_1} \subset V_{t_2}$ or $V_{t_2} \subset V_{t_1}$.*
2. **Spanning:** *For each $i \in I$, define $x_i : S \to \overline{T_i}$ by: $x_i(s) = \min_{t \in \overline{T_i}}\{s \in V_t\} = \min_{t \in \overline{T_i}}\{p(s,t) = 0\}$, where the minimum is taken with respect to the linear ordering on $\overline{T_i}$. Then, for all $s_1, s_2 \in S$, $s_1 \preceq s_2$ if and only if $\forall i \in I, x_i(s_1) \leq x_i(s_2)$.*

The definition of $x_i(s)$ as the minimal $t \in \overline{T_i}$ such that $p(s,t) = 0$ implies that $p(s,t) = 1$ for all $t < x_i(s)$. The requirement that T_i be linearly ordered guarantees that if $s \in V_{t_1}$ and $t_1 < t_2$, then $s \in V_{t_2}$ as well. It follows that if $t > x_i(s)$, then $s \in V_t$; i.e., $p(s,t) = 0$. Consequently, if $T_i = \{t_0 < t_1 < \cdots < t_{k_i}\}$ is an axis and $x_i(s) = t_j$, we can picture s's placement on the axis like this:

$$
\begin{array}{ccccccccc}
p(s,t) & 1 & & 1 & \cdots & 1 & 0 & \cdots & 0 \\
T_i & t_0 & \longrightarrow & t_1 & \longrightarrow \cdots \longrightarrow & t_{j-1} & \longrightarrow t_j & \longrightarrow \cdots \longrightarrow & t_{k_i}
\end{array}
$$

This picture is the crux of what we mean by "axis." For any candidate s, the above picture holds. s's coordinate on a particular axis is exactly that place where it begins to fail against the tests of the axis. Intuitively, we can think of an axis as representing a dimension of skill at the task, while s's coordinate represents how advanced it is in that skill.

We have not assumed independence because we would like to consider coordinate systems that might have dependent axes. Much as in the theory of vector spaces, we can show that a coordinate system of minimal size must be independent. However, as we will see shortly, in this discrete case there is more than one notion of independence which we must consider.

Definition 2 (Dimension). *The dimension of S, written $\dim(S)$, is the minimum of $|\mathcal{T}|$ taken over all coordinate systems $\mathcal{T}$ for S.*

Remark 1. Because S and T are finite, $\dim(S)$ will be well-defined if we can show at least one coordinate system for S exists. We will do so in section 2.4.

In the meantime, let us assume coordinate systems exist and explore some of their properties.

Definition 3 (Weak Independence). *A coordinate system $\mathcal{T}$ for S is weakly independent if, for all $T_i, T_j \in \mathcal{T}$, there exist $t \in T_i$, $u \in T_j$ such that V_t and V_u are incomparable, meaning neither is a subset of the other.*

Then we have a theorem reminiscent of linear algebra:

Theorem 1. *Let $\mathcal{T}$ be a coordinate system for S such that $|\mathcal{T}| = \dim(S)$. Then $\mathcal{T}$ is weakly independent.*

Sketch of Proof. Suppose $\mathcal{T}$ is not weakly independent. Then there are two axes, call them T_i and T_j, such that all tests in T_i are comparable to all tests in T_j. Consequently, we can create a new coordinate system $\mathcal{T}'$ as follows. First, $\mathcal{T}'$ has all the axes as $\mathcal{T}$ except T_i and T_j. Create a new axis T_k by forming $T_i \cup T_j$ and then arbitrarily removing duplicates (which are t, u such that $V_t = V_u$). The resulting T_k is then linearly ordered, and so can be an axis. Put T_k in $\mathcal{T}'$. Then, $\mathcal{T}'$ is also a coordinate system for S, but $|\mathcal{T}'|$ is one less than $|\mathcal{T}|$, contradicting the fact that $\mathcal{T}$ was minimal. Thus, $\mathcal{T}$ must be independent.   □

2.4   Existence of a Coordinate System

In this section we prove that any function $p : S \times T \to 2$ with S and T finite gives rise to a coordinate system on S. Simply put, the set of all chains in T satisfies definition 1. Once we can show one such coordinate system exists, we know that a minimal one exists and there is a reasonable notion of the dimension of S.

Definition 4. *A chain in T is a subset $C \subset T$ such that, for all $t_1, t_2 \in C$, either $V_{t_1} \subset V_{t_2}$ or $V_{t_2} \subset V_{t_1}$; further, $V_{t_1} = V_{t_2}$ implies $t_1 = t_2$.*

Let $\mathcal{C}$ be the set of all chains in T. Then:

Theorem 2. *$\mathcal{C}$ is a coordinate system for S.*

Proof. Write $\mathcal{C} = (C_i)_{i \in I}$. By definition, each C_i is linear. Thus we need only check that this family spans $\preceq$.

($\Rightarrow$) Assume $s_1 \preceq s_2$. We want to show $\forall i, x_i(s_1) \le x_i(s_2)$. Consider a $C_i \in \mathcal{C}$ and imagine $C_i = \{t_0 < t_1 < \cdots < t_{k_i}\}$. If $x_i(s_1) \not\le x_i(s_2)$, i.e. $x_i(s_1) > x_i(s_2)$, we must have the following situation:

$$
\begin{array}{ccccccccccc}
p(s_1,t) & 1 & 1 & \cdots & 1 & 1 & \cdots & 1 & 0 & \cdots & 0 \\
C_i & t_0 \to & t_1 \to & \cdots \to & t_{j_2-1} \to & t_{j_2} \to & \cdots \to & t_{j_1-1} \to & t_{j_1} \to & \cdots \to & t_{k_i}
\end{array}
$$

$$
\begin{array}{ccccccccccc}
p(s_2,t) & 1 & 1 & \cdots & 1 & 0 & \cdots & 0 & 0 & \cdots & 0 \\
C_i & t_0 \to & t_1 \to & \cdots \to & t_{j_2-1} \to & t_{j_2} \to & \cdots \to & t_{j_1-1} \to & t_{j_1} \to & \cdots \to & t_{k_i}
\end{array}
$$

where $x_i(s_1) = t_{j_1}$ and $x_i(s_2) = t_{j_2}$. However, then $p(s_1, t_{i_2}) > p(s_2, t_{i_2})$, which contradicts the assumption that $s_1 \preceq s_2$. Thus, $x_i(s_1) \le x_i(s_2)$. This argument holds for any C_i and any $s_1, s_2 \in S$; therefore we have our result.

($\Leftarrow$) Assume $\forall i \in I, x_i(s_1) \le x_i(s_2)$. We have the following for each $C_i \in \mathcal{C}$:

$$p(s_1,t) \quad 1 \quad 1 \quad \ldots \quad 1 \quad 0 \quad \ldots \quad 0 \quad 0 \quad \ldots \quad 0$$
$$C_i \qquad t_0 \to t_1 \to \ldots \to t_{j_1-1} \to t_{j_1} \to \ldots \to t_{j_2-1} \to t_{j_2} \to \ldots \to t_{k_i}$$

$$p(s_2,t) \quad 1 \quad 1 \quad \ldots \quad 1 \quad 1 \quad \ldots \quad 1 \quad 0 \quad \ldots \quad 0$$
$$C_i \qquad t_0 \to t_1 \to \ldots \to t_{j_1-1} \to t_{j_1} \to \ldots \to t_{j_2-1} \to t_{j_2} \to \ldots \to t_{k_i}$$

where $x_i(s_1) = t_{j_1}$ and $x_i(s_2) = t_{j_2}$. It is clear from the diagram that for all $t \in C_i, p(s_1,t) \le p(s_2,t)$. This fact holds for any C_i. That is, we have for all $t \in \bigcup_{i \in I} C_i, p(s_1,t) \le p(s_2,t)$. However, $\bigcup_{i \in I} C_i = T$, meaning we have $s_1 \preceq s_2$.

Combining the above two implications, we have shown that $s_1 \preceq s_2$ if and only if $\forall i \in I, x_i(s_1) \le x_i(s_2)$, for any $s_1, s_2 \in S$. Hence, $\mathcal{C}$ is a coordinate system for S, as we set out to show. $\qquad \square$

3   Dimension-Extraction Algorithm

In this section we give a polynomial-time algorithm that finds a weakly-independent coordinate system for a set of candidates. The algorithm accepts as input a set of candidates, a set of tests, and the outcome of each candidate for each test. Given this input, the goal is to construct a coordinate system such that (i) the position of a candidate in the constructed space uniquely identifies which tests it passes and fails, and (ii) the dimension of this coordinate system is minimal. Since an efficient optimal algorithm is not available, an algorithm will be presented that satisfies (i) but uses heuristics to minimize the dimension, and is therefore not guaranteed to satisfy (ii).

The main idea of the algorithm is as follows. We start out with an empty coordinate system, containing no axes. Next, tests are placed in the coordinate system one by one, constructing new axes where necessary. A new axis is required when no axis is present yet, or when a test is *inconsistent* with tests on all existing axes. Two tests t, u are inconsistent if V_t and V_u are incomparable. We now discuss two aspects of coordinate systems that inform our algorithm.

In a valid coordinate system, the tests on each axis are ordered by strictness; any test must at least fail the candidates failed by its predecessors on the axis. This knowledge informs our heuristic for choosing the order in which to consider tests: the first step of the algorithm is to sort the tests based on the number of candidates they fail.

A second aspect of coordinate systems is that a test whose set of failed candidates is the union of the sets of candidates failed by two other tests can be viewed as the combination of those tests. For example, if a test A on the first axis fails candidates 1 and 3 and a test B on the second axis fails candidates 2 and 5, then a test located at position (A,B) in the coordinate system must fail the union of the candidate sets: candidates 1,2,3, and 5. Since such a composite test provides no additional information about which tests a candidate will pass of fail, it can be safely discarded. Therefore, the second step of the algorithm is to remove any tests that can be written as the combination of two other tests.

Once the tests have been sorted and superfluous tests removed, the procedure is straightforward; tests are processed in order and are either placed on an existing axis if possible, or on a new axis if necessary. The pseudocode of the algorithm is as follows:

Input:
List $candidates, tests$
boolean **play**$(cand, test)$
boolean **consistentWith**$(test1, test2)$
Test **and**$(test1, test2)$

Output:
Tree $dimensions$

Algorithm:
$sort\ tests\ by\ number\ of\ fails$
for each $test1, test2, test3 \in tests$ $(with\ test1 \neq test2 \neq test3)$
  if $test3 = $ **and**$(test1, test2)$
    $remove\ test3\ from\ tests$
  end
end

for each $test \in tests$
  for each $leaf \in dimensions$
    if $consistentWith(test, leaf)$
      $add\ test\ as\ child\ to\ leaf$
    end
    if $test\ was\ not\ added\ to\ a\ leaf$
      $add\ test\ as\ new\ leaf\ to\ root\ of\ dimensions$
    end
  end
end

Fig. 2. Algorithm for coordinate system construction. The algorithm accepts sets of candidates and tests and their outcomes, and constructs a coordinate system that reflects the structure of the problem. Axes in this coordinate system consist of tests, and the location of a candidate in this induced space uniquely identifies which tests it will fail or pass.

4   Experiments

As a validation of the ideas presented in the previous sections, we applied our dimension-extraction algorithm to the populations of a coevolutionary simulation. Here we report the procedure we used and the results of the experiments.

Naturally, the question arises whether this algorithm will really extract useful coordinate systems from a problem. This question clearly bears much further empirical study. Here we are content to address the simpler question of whether the dimension extraction algorithm will give meaningful answers for particular problems in which we know what the underlying objectives are.

4.1   Method

The algorithm of fig. 2 was applied to the populations in a variant of the Population Pareto Hill Climber (P-PHC) algorithm presented in [7]. Briefly, a population of candidates and a separate population of tests is maintained by the algorithm. At each time step, the tests are treated as objectives that the candidates are trying to maximize. Each candidate is given a single offspring, and the parent is replaced if the offspring does at least as well as the parent on each test.

Tests are incented to find distinctions between candidates. If a and b are two candidates, a test t makes a distinction between them $t(a) \neq t(b)$. Each test is given one offspring; an offspring replaces its parent if it makes a distinction the parent does not make. It is possible for an offspring to lose distinctions which the parent also makes; we are not concerned with this possibility in this algorithm. Except for this variation in test selection, all other algorithm details are the same as those reported in [7].

Two numbers games were used as test problems [8]. The first domain was the COMPARE-ON-ONE game presented in [2]. In this game, candidates and tests are both n-tuples of numbers. c and t are compared on the single coordinate where t is maximal. c "wins" the game if it is larger than t on that coordinate. This game has been shown to induce a pathology known as "focusing" or "overspecialization;" in conventional coevolutionary algorithms; see [7] or [2] for details.

The second domain was the TRANSITIVE game. Again, candidates and tests are n-tuples of numbers. This time, when a candidate c interacts with a test t, c wins if it is at least as large as t on all dimensions.

Observe that the coevolutionary algorithm does not have access to the fact that individuals are tuples of numbers. The games are given as black boxes to the P-PHC algorithm and it must make best use of this win/loss information. Consequently, when we run our dimension-extraction algorithm on the P-PHC populations, we are hoping to see the algorithm discover the number n which is the true dimension of the game.

We used the following procedure to estimate the number of dimensions. 10 independent copies of P-PHC were run for 2,000 time steps. At each time step, the estimated number of dimensions in the current population was output according to the dimension-extraction algorithm. This value was averaged across the 10 runs to obtain a single "average run." Then the following statistics were calculated across all 2,000 time steps: the 10th and 90th percentiles; the upper and lower quantiles; and the median. The number of true dimensions of the underlying problem was varied from 1 to 16 and statistics were gathered for each number of dimensions.

4.2   Results

Our results are presented in figure 3. These figures are box plots of the estimated number of dimensions versus the true number of dimensions. The boxes span the lower and upper quartiles of the dimension estimates; the whiskers give the 10th and 90th percentiles. The plus marks the median of the dimension estimates. The dotted line gives the expected answer.

The figure on the left gives the results for COMPARE-ON-ONE. There is good agreement between the estimated value of the number of dimensions and theoretical value for dimension ranging from 1 to 16. Further, the variance in the estimates is generally quite small.

The figure on the right gives the results for TRANSITIVE. In this case the algorithm consistently overestimates the number of dimensions of the problem. There is a larger amount of variance in the estimate as well when compared with COMPARE-ON-ONE. We only display up to 10 dimensions in the figure, enough to see the trend.

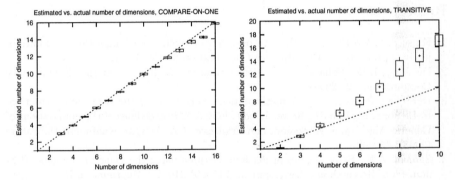

Fig. 3. Estimated number of dimensions in two numbers games, applying the algorithm in fig. 2 to the populations of a coevolutionary algorithm; see text for details. The left figure is the estimate for the COMPARE-ON-ONE game; note the tight correspondence with the theoretical number of dimensions. On the right is the estimate for the TRANSITIVE game; here the algorithm consistently overestimates.

5   Conclusions

A notion of *problem structure* with application to a broad class of problems in artificial intelligence, including learning and search, has been proposed. Problem structure here takes the form of a *coordinate system* whose axes consist of tests, and knowledge of the position of a test uniquely specifies the behavior of that test.

The structure of a problem is an intrinsic property. Thus, any existing problems for which candidates are evaluated using tests must have an associated coordinate system of the kind defined in this paper. For most problems, the question of what the underlying objectives are is new, and it is given a precise meaning by the definition of problem structure presented here. The definition, and the preliminary algorithm for extracting problem structure, may therefore yield new insight into existing problems. While computationally challenging, this permits asking intriguing questions such as: what is the dimension of chess, and what are the underlying dimensions of chess?

The formal definition of problem structure that has been presented directly suggests ways of extracting problem structure automatically. A preliminary algorithm for coordinate system construction has been provided, and demonstrated on example problems. It is our hope that the notion of problem structure that has been proposed may incite the study of problem structure as a general property of problems; if efficient algorithms for problem structure extraction can be identified, it may become possible to better understand existing problems of interest by the algorithmic analysis of their structure, thereby providing new insight into existing problems in an automatic manner.

References

1. Fonseca, C.M., Fleming, P.J.: An overview of evolutionary algorithms in multiobjective optimization. Evolutionary Computation **3** (1995) 1–16
2. De Jong, E.D., Pollack, J.B.: Ideal evaluation from coevolution. Evolutionary Computation **12** (2004)
3. Bucci, A., Pollack, J.B.: A mathematical framework for the study of coevolution. In De Jong, K., Poli, R., Rowe, J., eds.: FOGA 7: Proceedings of the Foundations of Genetic Algorithms Workshop, San Francisco, CA, Morgan Kaufmann Publishers (2003) 221–235
4. Samuel, A.L.: Some studies in machine learning using the game of checkers. IBM Journal of Research and Development **3** (1959) 210–229 Reprinted in E. A. Feigenbaum and J. Feldman (Eds.) 1963, *Computers and Thought*, McGraw-Hill, New York.
5. Epstein, S.L.: Toward an ideal trainer. Machine Learning **15** (1994) 251–277
6. Scheinerman, E.R.: Mathematics: A Discrete Introduction. 1st edn. Brooks/Cole, Pacific Grove, CA (2000)
7. Bucci, A., Pollack, J.B.: Focusing versus intransitivity: Geometrical aspects of coevolution. In Erick Cantú-Paz et al., ed.: Genetic and Evolutionary Computation - GECCO 2003. Volume 2723 of Lecture Notes in Computer Science., Springer (2003) 250–261
8. Watson, R., Pollack, J.B.: Coevolutionary dynamics in a minimal substrate. In L. Spector et al., ed.: Proceedings of the Genetic and Evolutionary Computation Conference, GECCO-2001, San Francisco, CA, Morgan Kaufmann Publishers (2001)

Modeling Coevolutionary Genetic Algorithms on Two-Bit Landscapes: Random Partnering

Ming Chang[1], Kazuhiro Ohkura[2], Kanji Ueda[3], and Masaharu Sugiyama[1]

[1] Gifu Prefecture Research Institute of Manufacturing Information Technology,
19-179-4 Sue, Kakamigahara city, Gifu 509-0108, Japan
{chang,sugi}@gifu-irtc.go.jp
[2] Faculty of Engineering, Kobe University, 1-1 Rokkoda-cho, Nada-ku, Kobe
657-8501, Japan. ohkura@mech.kobe-u.ac.jp
[3] Research Into Artifacts Center for Engineering, The University of Tokyo 4-6-1,
Komaba, Meguro, Tokyo 153-8904 Japan. ueda@race.u-tokyo.ac.jp

Abstract. A model of coevolutioinary genetic algorithms (COGA) consisting of two populations coevolving on two-bit landscapes is investigated in terms of the effects of random partnering strategy, different population updating schemes, and changes in mutation rate and evolution rate. The analytical and numerical approaches showed that even in such a simple model, the dynamics can change dramatically with different evolutionary scenarios in such an extent that deserves our attention from the point of view of algorithm design.

1 Introduction

The theory of natural selection is the only acceptable explanation for the origin and maintenance of adaptation among organisms. The Darwin's and Wallace's original idea about organic evolution has gone beyond biology as far as epistemology, psychology and economics *etc.* In computer science, the term evolutionary computation (EC) stands for a family of algorithms based on the belief that modeling the process of natural selection could help us to solve difficult real-world problems. While a mathematical theory of EC is indispensable by all means to constructing effective and efficient evolutionary algorithms (EAs), such a general and coherent theory is still far beyond our grasp.

To develop a mathematical theory of coevolutionary EAs (COEAs) maybe even more difficult. Different from standard EAs, where individuals are evaluated separately from each other according to predefined objective function(s), one main characteristic of COEAs is that the evaluation procedure involves more than one individuals, and the fitness of an individual is depending on its interaction with its partners. It is intuitively comprehensible that the implemented partner selection strategies can have significant influences on the algorithm's dynamics and optimization performance. In this paper, the effects of random partnering strategy, population updating scheme, mutation rate and evolution rate are investigated through a model of coevolutionary genetic algorithms (COGA) that consists of two populations coevolving on two-bit landscapes.

K. Deb et al. (Eds.): GECCO 2004, LNCS 3102, pp. 513–524, 2004.

There have been a lot of applications of modeling coevolutionary process to problem-solving [6,10], as well as analytic and empirical results about the dynamics of coevolutionary algorithms [2,3,9]. Our purpose is to give a relatively extensive investigation about the dynamics of COGA. This paper is organized as follows. After a brief review of the Schema Theorem, the basic model is introduced. Section 4 shows analytic and computer simulation results of different coevolutionary scenarios and Section 5 examines the model from an evolutionary game-theoretical point of view. Finally, Section 6 concludes with a brief of our findings and a few remarks about future work.

2   The Schema Theorem

GAs can be described using the well-known Schema Theorem [8]:

$$m(H,t+1) \geq m(H,t)\frac{f(H)}{\bar{f}}\left[1 - p_c\frac{\delta(H)}{l-1} - o(H)p_m\right] \qquad (1)$$

where $m(H,t)$ is the number of instances of schema H at time t, $f(H)$ is the average fitness of individuals in the population representing H at time t, $\bar{f}$ is the average fitness of individuals in the population, $\delta(H)$ is the defining length of H, $o(H)$ is the order of H, strings are of length l, p_c is the probability of crossover, and p_m the probability of mutation. The Schema Theorem shows that short, low-order, and highly fit schemata (referred to as building blocks) are given exponentially increasing numbers of instances. The building block hypothesis assumes that instead of building high-performance strings by trying every conceivable combination, GAs work with building blocks and construct better and better strings from the best partial solutions of past samplings [5, p.41]. But, as pointed out by Goldberg, due to the coding and the objective function itself, building blocks sometimes may be misleading, and make it difficult if not impossible to find the optimal solutions. The simplest case is a two-bit problem, which known as minimal deceptive problem (MDP) [5].

Grefenstette [4] however, showed that *deception* is neither necessary nor sufficient for problems to be GA-hard. The reason, he pointed out, lies in the fact that while the notion of deception [5] is defined in terms of the *static* average fitness of hyperplanes, what really important to GAs' ability of finding better solutions is their dynamic behaviour as described by the Schema Theorem. For example, consider the fitness landscape shown in Fig. 1, which can be considered both a two-bit landscape and a landscape of order-two schemata. The fitnesses are given as $f(01) = 1.0$, $f(11) = 0.0$, $f(00) = 0.0$ and $f(10) = 2.0$, 10 is the maximum. The static building block hypothesis assumes no deception existing here since $f(1*) = \frac{f(10)+f(11)}{2} = 1.0 > f(0*) = \frac{f(00)+f(01)}{2} = 0.5$ and $f(*0) = \frac{f(00)+f(10)}{2} = 1.0 > f(*1) = \frac{f(01)+f(11)}{2} = 0.5$. However, consider a GA population consisting of individuals as $p(00) = 16\%$, $p(01) = 64\%$, $p(10) = 4\%$ and $p(11) = 16\%$, respectively. Then the *observed* fitness [4] of the schema $1*$ is $f(1*) = \frac{p(10)f(10)+p(11)f(11)}{p(10)+p(11)} = 0.4$, and in the same sense $f(0*) = 0.8$,

$f(*0) = 0.4$, $f(*1) = 0.8$. Since $f(0*) > f(1*)$ and $f(*1) > f(*0)$, according to the Schema Theorem, the population might also converge to 01, the local optimum, if $p_c = 1.0$ and $p_m = 0.0$. This indicates that whether a problem is GA-hard is searcher-depended, and the characterization of GA-hard problems "must take into account the basic features of the GA, especially its dynamic, biased sampling strategy" [4].

3   A Model of Coevolutionary GA

Bull [3] extended the Schema Theorem to coevolutionary systems. In a $(1+S)$-population model with random partnering and asynchronous reproduction, suppose scheme $H = \{H_k, H_{c_1}, H_{c_2}, \ldots, H_{c_S}\}$, where H_k is sub-schema in current population and H_{c_i} sub-schema in the other S populations, the Schema Theorem reads

$$m(H, t+1) \geq m(H_k, t)\frac{f(H_k)}{\bar{f}} \cdot \left[1 - p_c\frac{\delta(H_k)}{l_k - 1} - o(H_k)p_m\right] \cdot \prod_{i=1}^{S} \frac{m(H_{c_i}, t)}{P_{op}} \quad (2)$$

where P_{op} stands for population size and

$$m(H_{c_i}, t) \geq m(H_{c_i}, t-1)\frac{f(H_{c_i})}{\bar{f}} \cdot \left[1 - p_c\frac{\delta(H_{c_i})}{l_{c_i} - 1} - o(H_{c_i})p_m\right]. \quad (3)$$

As shown in Eqn. (2), the basic idea of COGA is "divide-and-conquer", namely divide the problem at hand into a number of subproblems and challenge each of them separately, then construct the complete solutions from the subproblems' solutions. This kind of COGA has been named as cooperative coevolutionary GA [10]. To our purpose of investigating coevolutionary behaviour of GA, a model that consists of two populations coevolving on two-bit landscapes is constructed, where finding the maximum is divided into two subproblems, and each is challenged by a Simple GA (SGA). Individuals in the two population A and B represent the first and second bit respectively, and to evaluate individuals in A, individuals from B are required and *vice versa*. Unless otherwise stated, of the two-bit strings used in this paper, the first bit corresponds to individuals of A, the second to that of B.

4   Random Partnering Strategy

Let P_a (P_b) be the proportion of "0" individuals of A (B), f_a^0 (f_a^1) the fitness of "0" ("1") individuals of A, f_b^0 (f_b^1) that of B, respectively. By using random partnering strategy, an individual's partners are picked up from the other population at random, and the expectation of the individual's fitness reads

$$\begin{cases} E(f_a^0) = P_b \cdot f(00) + (1 - P_b) \cdot f(01) = 1 - P_b \\ E(f_a^1) = P_b \cdot f(10) + (1 - P_b) \cdot f(11) = 2P_b \\ E(f_b^0) = P_a \cdot f(00) + (1 - P_a) \cdot f(10) = 2 - 2P_a \\ E(f_b^1) = P_a \cdot f(01) + (1 - P_a) \cdot f(11) = P_a. \end{cases} \quad (4)$$

Those values can also be considered the fitness obtained when using *complete mixing strategy*, in which an individual interacts in a pairwise way with all individuals in the other population. Because complete mixing strategy is computational expensive, it is seldom used in COGA but is popular in literature of evolutionary game theory. By setting $E(f_a^0) = f_a^0$, $E(f_a^1) = f_a^1$, $E(f_b^0) = f_b^0$ and $E(f_b^0) = f_b^0$, the resulting behaviour of COGA with complete mixing strategy can also be interpreted as the "expected" behaviour of COGA with random partnering strategy. Let F_a (F_b) be the average fitness of individuals in A (B), it holds

$$\begin{cases} F_a = P_a \cdot f_a^0 + (1 - P_a) \cdot f_a^1 = P_a + 2P_b - 3P_aP_b \\ F_b = P_b \cdot f_b^0 + (1 - P_b) \cdot f_b^1 = P_a + 2P_b - 3P_aP_b. \end{cases} \tag{5}$$

Denote the changes of P_a and P_b due to selection by ΔP_a and ΔP_b, then the proportion of "0" in A and B after selection will be

$$\begin{cases} P_a' = P_a + \Delta P_a \\ P_b' = P_b + \Delta P_b \end{cases} \tag{6}$$

and the implemented roulette-wheel selection scheme in SGA leads

$$\begin{cases} \Delta P_a = K_a P_a \left(f_a^0/F_a - 1 \right) \\ \Delta P_b = K_b P_b \left(f_b^0/F_b - 1 \right) \end{cases} \tag{7}$$

where K_a and K_b are coefficients that scale the rate of evolutionary change [1]. Let M_a and M_b be mutation rate of A and B, then the proportion of "0" in the next generation can be expressed as

$$\begin{cases} P_a'' = P_a'(1 - M_a) + (1 - P_a')M_a \\ P_b'' = P_b'(1 - M_b) + (1 - P_b')M_b. \end{cases} \tag{8}$$

4.1   Without Mutation

The behaviour of the above model can be characterized by its fixed points and their stability properties. At the beginning, suppose $M_a = M_b = 0$, then the fixed points of Eqn. (8) must satisfy $\Delta P_a = \Delta P_b = 0$ and this leads

$$\begin{cases} P_a(1 - P_a)(1 - 3P_b) = 0 \\ P_b(1 - P_b)(2 - 3P_a) = 0. \end{cases} \tag{9}$$

In consequence, there are 5 fixed points: $(0,0)$, $(1,0)$, $(0,1)$, $(1,1)$ and $(\frac{2}{3}, \frac{1}{3})$. Since the sign of ΔP_a and ΔP_b in Eqn. (7) depends on f_a^0, F_a, f_b^0 and F_b, by comparing f_a^0 and f_a^1, f_b^0 and f_b^1 in Eqn. (4), we obtain

$$\begin{cases} f_a^0 > f_a^1 \text{ if } 0 < P_b < \frac{1}{3} \\ f_a^0 < f_a^1 \text{ if } \frac{1}{3} < P_b < 1 \end{cases} \text{ and } \begin{cases} f_b^0 > f_b^1 \text{ if } 0 < P_a < \frac{2}{3} \\ f_b^0 < f_b^1 \text{ if } \frac{2}{3} < P_a < 1. \end{cases}$$

As a result, the phase space is divided into four regions I, II, III and IV (Fig. 2), all orbits in region II converge to $(0,1)$ and all orbits in region IV to $(1,0)$.

Since any perturbations that made the system apart from (1,0) and (0,1) but still remain in region II and IV respectively will not change the system's ultimately ending up to them, those two fixed points are stable. By contrast, any perturbation that causes the system to apart from (0,0) and (1,1) will make the system ultimately converge to (1,0) and (1,0) respectively. So, (0,0) and (1,1) are unstable fixed point. For $(\frac{2}{3},\frac{1}{3})$, the Jacobian of Eqn. (6) at this point is

$$A = \begin{bmatrix} \frac{\partial P_a'}{\partial P_a} & \frac{\partial P_a'}{\partial P_b} \\ \frac{\partial P_b'}{\partial P_a} & \frac{\partial P_b'}{\partial P_b} \end{bmatrix} = \begin{bmatrix} 1 & -K_a \\ -K_b & 1 \end{bmatrix} \tag{10}$$

where the two eigenvalues are $1 \pm \sqrt{K_a K_b}$. As $0 < (1 - \sqrt{K_a K_b}) < 1 < (1 + \sqrt{K_a K_b})$, $(\frac{2}{3},\frac{1}{3})$ is a saddle, and there are two orbits converging to it, one of them in region I and the other in region III. Those two orbits consist the separatrix that divides the phase space into two basins of attraction, all orbits above the separatrix converge to (0,1), and all orbits below it converge to (1,0).

Fig. 3 shows a set of evolutionary trajectories where A and B evolving at the same rate $K_a = K_b = 0.1$. It can be seen that almost all trajectories end at one of the two stable fixed points, (1,0) and (0,1), which correspond to the global and local maximum respectively. The eventual end-point is completely determined by the initial starting positions, all orbits beginning at points on the same side of the separatrix lead to the same fixed point.

A point is called a *sink* (*source*), if very orbit converges to it when time runs toward $t \to +\infty$ ($t \to -\infty$). In the model, by setting $t \to -\infty$, the original problem can be converted to a problem of searching the minimum on the landscape, where the two minima have the same value 0.0. The time-reversed evolution processes can be realized by rewriting Eqn. (7) as

$$\begin{cases} \Delta P_a = -K_a P_a (f_a^0/F_a - 1) \\ \Delta P_b = -K_b P_b (f_b^0/F_b - 1) \end{cases} \tag{11}$$

Fig. 4 shows a set of evolutionary trajectories of this case. All trajectories start from the points satisfying $P_a + P_b = 1$ will end up to the saddle point $(\frac{2}{3},\frac{1}{3})$, and the seperatrix $P_a + P_b = 1$ divides the phase space into two basins of attraction of the same size since the two minima have the same fitness value. All orbits above the separatrix converge to (1,1), all orbits below it converge to (0,0).

To show how changing of the evolution rate K_a and K_b affects dynamics, Fig. 5 gives trajectories where B evolves 5 times faster than A. It can be seen that although the number and positions of the fixed points do not change, the shape of the separatrix is bent. As shown in Fig. 2, since orbits that beginning in region II and region IV end up to (0,1) and (1,0) respectively, the ultimate end-up point of orbits that beginning in region I and III depends on which region, namely II or IV, is reached first. If it is region II reached first, the orbits will end up to (1,0), otherwise end up to (0,1). The evolution rate K_a and K_b in this sense, represent velocity of the orbits moving along x axis and y axis. B evolving 5 times faster than A means that the velocity along y axis is 5 times faster than

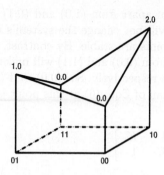

Fig. 1. A two-bit deceptive fitness landscape.

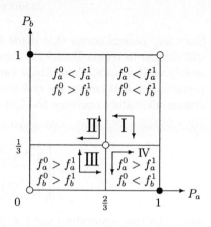

Fig. 2. Phase portrait of the model with random partnering strategy.

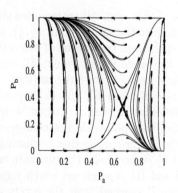

Fig. 3. $K_a = K_b = 0.1$, $M_a = M_b = 0$.

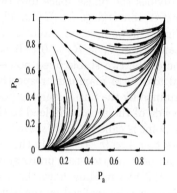

Fig. 4. $t \to -\infty$, $K_a = K_b = 0.1$, $M_a = M_b = 0$.

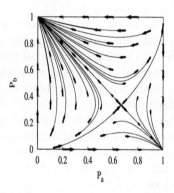

Fig. 5. B evolves fast than A. $K_a = 0.02$, $K_b = 0.1$, $M_a = M_b = 0$.

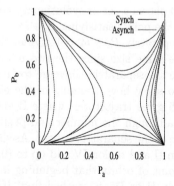

Fig. 6. Updating A first then B. $K_a = K_b = 0.1$, $M_a = M_b = 0$.

that along x axis. In region III, since $f_b^0 > f_b^1$, increasing K_b will increase the velocity translating positively along y axis, and make more orbits beginning in region III reach region II first and end up to $(1,0)$. On the other hand, In region I, since $f_b^0 < f_b^1$, increasing K_b will increase the velocity translating negatively along y axis, and make more orbits beginning in region I reach region IV first and end up to $(1,0)$.

Next, the effects of population updating scheme are investigated. Fig. 6 shows eight pairs orbits of which each pair orbits start from the same position but using synchronous and asynchronous updating strategies, respectively. In asynchronous updating, A is updated first, then B. This fact means that all orbits will move along x axis one-step-ahead than y axis, although the direction can be both positive and negative, depending on the value of f_a^0, f_a^1, f_b^0 and f_b^1. As it is shown in Fig. 2, in region III, since $f_a^0 > f_a^1$, updating A first made orbits beginning in this region move positively along x axis one-step-ahead. In consequence, there are more orbits reach region IV first and end up to $(0,1)$. On the other hand, in region I, because $f_a^0 < f_a^1$, updating A first made orbits beginning in this region move negatively along x axis one-step-ahead, and consequently, there are more orbits reach region II first and end up to $(1,0)$.

4.2 With Mutation

Firstly, consider what happens if there were no selection pressure imposed on both A and B. At the equilibrium, where $P_a'' = P_a' = P_a$ and $P_b'' = P_b' = P_b$, we have

$$\begin{cases} P_a = P_a(1 - M_a) + (1 - P_a)M_a \Rightarrow p_a = 0.5 \\ P_b = P_b(1 - M_b) + (1 - P_b)M_b \Rightarrow P_b = 0.5 \end{cases} \tag{12}$$

This means that when no selection pressure imposed, mutation makes all orbits converge to the single fixed point $(0.5,0.5)$.

Now, let us see what happens when both mutation and selection are assumed in A while only selection in B. In this case, from Eqn. (6)(7)(8), we have

$$\begin{cases} \Delta P_a = M_a(1 - 2P_a) + (1 - 2M_a)K_a P_a(f_a^0/F_a - 1) \\ \Delta P_b = K_b P_b(f_b^0/F_b - 1) \end{cases} \tag{13}$$

and at the equilibrium, where $\Delta P_a = \Delta P_b = 0$, we obtain three fixed points:

$$\begin{cases} p_1 = \left(\frac{M_a(1-2K_a)+K_a}{2M_a(1-K_a)+K_a}, 0 \right) \\ p_2 = \left(\frac{M_a}{2M_a(1-K_a)+K_a}, 1 \right) \\ p_3 = \left(\frac{2}{3}, \frac{1}{3} - \frac{M_a}{K_a(1-2M_a)} \right). \end{cases} \tag{14}$$

It can be seen from Eqn. (14) that by keeping M_a constant, increasing K_a will cause p_1 moving along x axis positively ($+x$ direction) toward $(1,0)$, p_2 along $y = 1$ negatively ($-x$ direction) toward $(0,1)$, p_3 along $x = \frac{2}{3}$ with $+y$ direction toward $(\frac{2}{3}, \frac{1}{3})$. On the other hand, taking M_a as constant, increasing M_a will treated as *objectives* in the sense of Evolutionary Multi-Objective Optimization. The resulting solution concept is the Pareto-front, containing all learners that are non-dominated as determined by their test outcomes.

Broadly, there are two approaches to the aim of guaranteed progress in co-evolution. The first approach is to strive towards accurate evaluation; if this can be achieved, then progress can be guaranteed simply by using an elitist selection mechanism based on the coevolutionary evaluation function. The second is to maintain an archive whose quality increases monotonically according to some performance criterion. The first approach is taken e.g. in [7] with the DELPHI algorithm; this approach will be discussed briefly below. Here, we will be concerned with the second, archive-based approach.

1.1 Reliable Progress by Means of Accurate Evaluation

One approach to reliable progress in coevolution is to consider how tests can be evolved that provide accurate evaluation. Several authors have investigated the accuracy of coevolutionary evaluation [16,12,9,3]. Based on Ficici's notion of *distinctions* [9], it has been shown that coevolution can in principle provide ideal evaluation [6,7]. The DELPHI algorithm is based on this principle, and will be used in comparison experiments here. For a discussion of the algorithm, its motivation, and experimental results, the reader is referred to [7].

1.2 Archive-Based Methods for Monotonic Improvement

A common technique in coevolution aimed at improving reliability is the use of an archive. Several archive mechanisms exist that are intended to improve the reliability of coevolutionary algorithms but do not provide any specific guarantees. Here, since our aim is to study how progress may be guaranteed, we will be concerned solely with archives that provide some form of progress guarantee.

The process that supplies new individuals to the archive will be referred to as the *generator*. The generator will typically use the archive for testing purposes, but an archive may also be valuable as a basis for generating new individuals. Any progress guarantee relies on the ability of the generator to produce new individuals. The aim for an archive is therefore to guarantee that regress is avoided; if this is guaranteed, then any changes in performance must represent progress in some aspect, as will be made precise.

A central requirement for a coevolution archive is that is should guarantee monotonic progress. Apart from this requirement, there are at least three other characteristics that determine the practical value of the archive:

- **Generality.** *Generality* reflects the scope of the archive method; an archive that guarantees progress for all forms of coevolution would be maximally general.
- **Sensitivity.** An archive is *sensitive* if it is able to detect small improvements in the quality of learners. This property applies to both learners and tests. If an archive is sensitive in accepting learners, it can accept many of the

learners that represent improvement, which positively affects both the generation of new learners and the evaluation of future tests. If an archive is sensitive in accepting tests, it will subsequently be more likely to detect improvements made by learners. The property of sensitivity necessarily depends on the solution concept.

- **Efficiency.** An archive is *efficient* if it consumes a limited amount of resources, notably computation time and storage capacity.

The majority of archives employed in the coevolution literature can be described as *best-of-generation* models, where the archive contains the fittest learners of the m past generations, and a sample of the archive is used for testing the current learners [10]. In such setups, tests are selected based on their quality as learners, rather than on their ability to provide informative evaluation. The maintenance of individuals performing well against a sample of previous learners is not by itself sufficient to guarantee progress in coevolution. In the following, we will discuss methods that do provide a progress guarantee.

Rosin describes the *covering competitive algorithm* [16], which alternates between finding a first-player strategy that beats all second-player strategies in the archive and *vice versa*. Under the assumption of an unbounded archive, the covering competitive algorithm guarantees monotonic progress. The algorithm assumes the existence of a first-player strategy that defeats *all* second-player strategies. For many test-based problems however, no learner can simultaneously achieve the highest attainable score on all possible tests, as there can be trade-offs between the different tests.

If the covering competitive algorithm is to be used for a problem featuring multiple underlying objectives and thus possibly more than one Pareto-optimal learner, every such learner on the Pareto-front would form a local optimum; whenever the method finds one learner on the Pareto-front and the tests it solves, no further progress can be made, and the method will stall. A similar argument holds for the *dominance tournament* [18], which was proposed as a method for tracking progress in coevolution but can also be used as a coevolutionary archive [10].

Schmitt [17] presents a stochastic model intended to demonstrate that coevolution can converge to a global optimum if for at least some species strictly dominant individuals exist that maximize performance over all possible evaluation environments. Here, the aim will be to guarantee progress under broader, less strict conditions.

A recent archive mechanism providing a progress guarantee is the *Nash memory* [10], which employs the Nash equilibrium as a solution concept. A mixed strategy Nash equilibrium is a combination of mixed strategies such that no player can profitably deviate given the strategies of the other players. An attractive feature of the Nash equilibrium as a solution concept is that the set of learners it represents can be relatively small compared to the Pareto-front, which is a valuable property for coevolutionary search. A disadvantage is that learners, and assign the outcomes such that each learner i is dominated by its successor $i + 1$; this can be done by letting each learner solve all tests solved by its antecessor plus one extra test. The learner update procedure will now result in the removal of the first learner, as it resides in layer $n + 1 > n$. All of the remaining learners solve the test solved by learner 1, making this test superfluous. Next, repeat the following procedure n times: add a new learner that solves all tests solved by the existing learners, and a new test solved only by the newly added learner. After each addition, the update procedure causes the most ancient learner and test to be removed from the archive. After n cycles, all learners solving the first test will have been discarded. Since the remaining learners were added after the first test was removed and hence do not solve it, the ability to solve the first test has been lost, and regress has thus occurred. This proof sketch demonstrates that no number of layers n is sufficient to guarantee the avoidance of regress.

4   The Incremental Pareto-Coevolution Archive (IPCA)

In this section, we describe an archive-based algorithm that guarantees monotonic progress for Pareto-Coevolution without simply keeping all learners or tests. The algorithm is called the *Incremental Pareto-Coevolution Archive* (IPCA). IPCA consists of a learner archive and a test archive. The algorithm provides procedures to decide which newly generated learners and tests will enter the archive. The learner archive is periodically updated to maintain non-dominated learners only.

The algorithm operates as follows. A newly generated learner is *useful* with respect to a set of learners LS and a set of tests TS if it is not dominated by any learner in LS, and if there is no learner in LS which has equal outcomes for all tests in TS:

$$useful(L, LS, TS) =$$

$$\nexists L' \in LS : L' \overset{TS}{\succ} L \ \land$$

$$\nexists L' \in LS : \forall T \in TS : G(L, T) = G(L', T)$$

where $G(L, T)$ is the outcome of learner L against test T, and $\succ$ represents Pareto-dominance. A related function called $useful-tests(TG, T^t, LG, L^t)$ identifies tests in a new generation of tests TG that are required in addition to T^t in order to determine that certain learners in LG are useful with respect to L^t. Specifically, if a learner is not useful based on T^t but is useful based on $T^t \cup T_1, T_2, \ldots T_k \in TG$, then some or all of $T_1, T_2, \ldots T_k$ are useful tests and a subset of TG with this same property will be returned by $useful - tests$. Additionally, if for any learner $L \in LG$, there is a test $T \in TG$ that defeats the learner and the learner is not defeated by any test in T^t, then L and T are marked as *useful*.

Using these functions, the IPCA algorithm can be described as follows as follows.

$L^0 := \emptyset$
$T^0 := \emptyset$
$t := 0$
while $\neg done$
  $L^t := non - dominated(L^t, T^t)$
  $L^{t+1} := L^t$
  $T^{t+1} := T^t$
  $LG := generate - learners(L^t)$
  $TG := generate - tests(T^t)$
  $TS := useful - tests(TG, T^t, LG, L^t)$
  $T^{t+1} := T^{t+1} \cup TS$
  for $i = 1 : |LG|$
    if $useful(L_i, L^{t+1}, T^{t+1})$
      $L^{t+1} := L^{t+1} \cup L_i$
  end
  if $L^{t+1} \neq L^t$
    $t := t + 1$
end

Fig. 1. The Incremental Pareto-Coevolution Archive (IPCA). Monotonic progress can be guaranteed for this archive-based Pareto-Coevolution algorithm.

4.1   Monotonicity and Convergence

The above algorithm is called the Incremental Pareto-Coevolution Archive (IPCA). The operation of any archive inevitably depends on the new individuals provided by the generator. The criterion required of a coevolution archive is therefore that progress can be guaranteed given the arrival of new individuals which occasionally represent progress. This can be guaranteed for example by generating every possible individual with a non-zero probability.

A proof that the algorithm guarantees monotonic progress as defined by Definition 1 is provided in the Appendix. If the number of different learners and tests is finite and all learners and tests are generated with non-zero probability, then the property of monotonic improvement implies convergence to the global optimum of the Pareto-front over all possible tests.

5   Experimental Results

To demonstrate the operation of the IPCA, we now investigate its performance on test problem that requires exploration, and compare performance with the DELPHI algorithm.

In COMPARE-ON-ONE [7], learners and tests are n-dimensional real-valued vectors. A tests assesses a learner on the dimension in which the test itself is highest. It assigns a score of 1 if the learner is at least as high as the test in this dimension, and -1 otherwise.

In the discretized COMPARE-ON-ONE problem, the value in each dimension of the learner and test is rounded to the nearest multiple of $\delta = 0.25$ below it before evaluation, without affecting the genotype. Thus, $[0.23, 0.30, 0.47]$ is mapped to $[0, 0.25, 0.25]$. This procedure greatly reduces the amount of gradient present.

To further increase the difficulty of the test problem, a mutation bias of 0.025 on a mutation range of 0.25 is used, meaning mutation adds a value randomly chosen from $[-0.15, 0.1]$. This bias towards regress is intended to model the situation in problems of practical interest, where the variation operator is typically more likely to produce regress than progress.

The generator that supplies candidate learners and tests to IPCA produces offspring using crossover (50%) and mutation (50%). With probability 0.1, it uses an archive member as a parent. The generator maintains a learner and test populations, both of size 10. The objectives for learners are their outcomes against tests and the distinctions between tests. The learner objectives are based on the union of the current population and new generation of tests. The objectives for the tests are analogous, namely their outcomes against and distinctions between individuals in the current population and new generation of learners, resulting in a symmetric setup.

For each objective achieved by an individual, a score is assigned that equals one over the number of other individuals that achieve the objective, as in *competitive fitness sharing* [16]. The sum of an individual's scores on the n outcome objectives and on the n^2 distinction objectives, where n is the size of the population plus the new generation, are added to yield a single total score for the individual.

The highest scoring individuals of the new generation are lined up with the lowest scoring individuals of the current population. Then k is determined as the highest number for which the summed scores of the first k generation members is still at least as high as that of the first k population members. The lowest scoring k population members are discarded and replaced by k *randomly* selected individuals from the new generation, thus yielding an explorative generator.

The performance criterion is the lowest value among all dimensions of an individual; if this value increases, progress is made on all dimensions. Performance is plotted as a function of the number of actual generations, and averaged over 50 runs.

Figure 2 shows the behavior of the DELPHI algorithm on both the standard and discretized COMPARE-ON-ONE problem with mutation bias. While DELPHI achieves stable progress on the standard COMPARE-ON-ONE problem, it fails on the discretized version of the problem. This is expected given the operation of the method; since new individuals can only be accepted into the population if they dominate an existing individual, the method cannot make progress on problems where exploration is required before such improvements can be identified.

Figure 2 shows the behavior of the IPCA on the same two problems. While IPCA improves slower than DELPHI on the continuous problem, it does make reliable progress, as expected. Moreover, IPCA makes substantial and reliable progress on the discretized problem as well. The main limitation of IPCA is

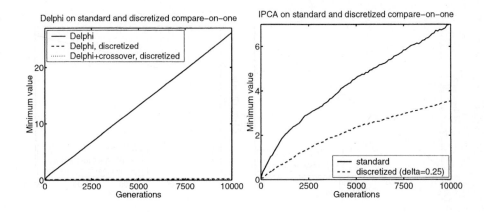

Fig. 2. Left: Performance of the DELPHI algorithm on the standard and discretized COMPARE-ON-ONE problem. DELPHI works well on the standard version but fails on the discretized version, also when using crossover in 50% of the cases. Right: The Incremental Pareto-Coevolution Archive (IPCA) makes consistent progress on both versions of the problem by virtue of its monotonic progress guarantee.

the size of the test archive; as Figure 3 (left) shows, the learner archive, which consists of the current approximation of the Pareto-front, is stable and small in size, while the test archive grows steadily over time as no individuals are pruned.

As a control experiment we apply two more standard coevolution algorithms to the problem. The first method is the generator used with IPCA without the archive itself. The second is a symmetric competition coevolutionary algorithm where the tests use their outcomes against the learners as objectives, *vice versa*. Both methods result in quick regress rather than progress; see Figure 3 (right). Apparently, the methods are insufficiently selective to cope with the mutation bias, which makes regress likely unless the replacement of individuals is highly selective and based on an informative set of tests.

An interesting question is whether and how the test archive may be pruned while retaining reliable progress. In a follow-up paper, we investigate a layered variant of IPCA called LAPCA [5]. While a layered approach cannot guarantee monotonic progress, as discussed in section 3, the method can produce sustained progress on the discretized COMPARE-ON-ONE PROBLEM with small and stable learner and test archives. This method may be of some practical interest, but the question of how archive sizes may be limited at a minimal reduction of reliability remains an important open issue. The analysis of the underlying objectives or structure of a test-based problem [7,4] may bring insight into this matter.

6   Conclusions

The Incremental Pareto-Coevolution Archive (IPCA) has been presented. The archive consists of a learner archive maintaining non-dominated individuals, in

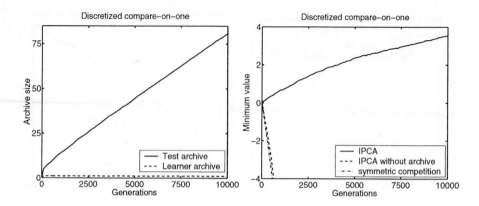

Fig. 3. Left: Sizes of the learner and test archives for IPCA. Since no tests are pruned, the test archive grows over time. Right: Comparison of IPCA with two control methods (see text). Due to the mutation bias, unreliable coevolution methods can regress, and only reliable methods can make sustained progress on the problem.

combination with an incrementally informative test archive. IPCA guarantees monotonic progress for Pareto-Coevolution.

IPCA is both general, in that asymmetric problems involving learners and tests can be addressed, and sensitive, as any nondominated learner and any test revealing new qualities of learners are accepted into the archive.

We have presented experiments based on the discretized three-dimensional COMPARE-ON-ONE problem with mutation bias. This problem requires exploration, and its mutation operator is biased towards regress. The Incremental Pareto-Coevolution Archive was found to produce sustained progress on this challenging test problem. An important remaining open question is how archive sizes may be limited at a minimal reduction of reliability.

Acknowledgements. The author wishes to thank the reviewers for detailed and thoughtful comments, and the Decision Support Systems Group at Utrecht University for a pleasant and fruitful research environment.

References

1. Robert Axelrod. The evolution of strategies in the iterated prisoner's dilemma. In Lawrence Davis, editor, *Genetic Algorithms and Simulated Annealing*, Research Notes in Artificial Intelligence, pages 32–41, London, 1987. Pitman Publishing.
2. Nils Aall Barricelli. Numerical testing of evolution theories. Part I: Theoretical introduction and basic tests. *Acta Biotheoretica*, 16(1–2):69–98, 1962.
3. Anthony Bucci and Jordan B. Pollack. A mathematical framework for the study of coevolution. In *Foundations of Genetic Algorithms (FOGA-2002)*, San Francisco, CA, 2003. Morgan Kaufmann.

4. Anthony Bucci, Jordan B. Pollack, and Edwin D. De Jong. Automated extraction of problem structure. In *Proceedings of the Genetic and Evolutionary Computation Conference, GECCO-04*, 2004.
5. Edwin D. De Jong. Towards a bounded Pareto-Coevolution archive. In *Proceedings of the Congress on Evolutionary Computation, CEC-04*, 2004.
6. Edwin D. De Jong and Jordan B. Pollack. Learning the ideal evaluation function. In E. Cantú-Paz et al., editor, *Proceedings of the Genetic and Evolutionary Computation Conference, GECCO-03*, pages 274–285, Berlin, 2003. Springer.
7. Edwin D. De Jong and Jordan B. Pollack. Ideal evaluation from coevolution. *Evolutionary Computation*, 12(2), 2004.
8. Sevan G. Ficici and Jordan B. Pollack. A game-theoretic approach to the simple coevolutionary algorithm. In M. Schoenauer et al., editor, *Parallel Problem Solving from Nature, PPSN-VI*, volume 1917 of *LNCS*, Berlin, 2000. Springer.
9. Sevan G. Ficici and Jordan B. Pollack. Pareto optimality in coevolutionary learning. In Jozef Kelemen, editor, *Sixth European Conference on Artificial Life*, Berlin, 2001. Springer.
10. Sevan G. Ficici and Jordan B. Pollack. A game-theoretic memory mechanism for coevolution. In E. Cantú-Paz et al., editor, *Genetic and Evolutionary Computation – GECCO-2003*, volume 2723 of *LNCS*, pages 286–297, Chicago, 12-16 July 2003. Springer-Verlag.
11. D. W. Hillis. Co-evolving parasites improve simulated evolution in an optimization procedure. *Physica D*, 42:228–234, 1990.
12. Hugues Juillé. *Methods for Statistical Inference: Extending the Evolutionary Computation Paradigm*. PhD thesis, Brandeis University, 1999.
13. Ludo Pagie and Paulien Hogeweg. Evolutionary consequences of coevolving targets. *Evolutionary Computation*, 5(4):401–418, 1998.
14. Jan Paredis. Coevolutionary computation. *Artificial Life*, 2(4), 1996.
15. Mitchell A. Potter and Kenneth A. De Jong. Cooperative coevolution: An architecture for evolving coadapted subcomponents. *Evolutionary Computation*, 8(1):1–29, 2000.
16. Christopher D. Rosin. *Coevolutionary Search among Adversaries*. PhD thesis, University of California, San Diego, CA, 1997.
17. Lothar M. Schmitt. Theory of coevolutionary genetic algorithms. In Minyi Guo and Laurence Tianruo Yang, editors, *Parallel and Distributed Processing and Applications, International Symposium, ISPA 2003*, pages 285–293, Berlin, 2003. Springer.
18. Kenneth O. Stanley and Risto Miikkulainen. The dominance tournament method of monitoring progress in coevolution. In Alwyn M. Barry, editor, *GECCO 2002: Proceedings of the Bird of a Feather Workshops, Genetic and Evolutionary Computation Conference*, pages 242–248, New York, 8 July 2002. AAAI.
19. Richard A. Watson. *Compositional Evolution: Interdisciplinary Investigations in Evolvability, Modularity, and Symbiosis*. PhD thesis, Brandeis University, 2002.
20. Richard A. Watson and Jordan B. Pollack. Symbiotic combination as an alternative to sexual recombination in genetic algorithms. In M. Schoenauer et al., editor, *Parallel Problem Solving from Nature, PPSN-VI*, volume 1917 of *LNCS*, Berlin, 2000. Springer.
21. R. Paul Wiegand. *An Analysis of Cooperative Coevolutionary Algorithms*. PhD thesis, George Mason University, Fairfax, Virginia, 2003.

Appendix: Proof of Monotonic Progress

In the following, we prove that the Incremental Pareto-Coevolution Archive (IPCA) algorithm described in Section 4 guarantees monotonic progress as defined by Definition 1. The definition specifies two requirements for any t and $t' > t$:

$$1.\ \forall TS \subseteq T^t : \quad [\exists L \in L^t : solves(L, TS) \implies$$
$$\exists L' \in L^{t'} : solves(L', TS)]$$
$$2.\ \exists TS \subseteq T^{t'} : \quad [\nexists L \in L^t : solves(L, TS) \wedge$$
$$\exists L' \in L^{t'} : solves(L', TS)]$$

To show that monotonic progress is made over time, it is sufficient to prove that monotonic progress is made from one time-step to the next, i.e. $t' = t + 1$. This will now be shown.

Ad 1). We must show that given an $L \in L^t$ that solves TS, there must be some $L' \in L^{t+1}$ that solves TS. We distinguish between two cases: (A) L is retained, or (B) L is removed from the archive. In the first case, the requirement is satisfied by L itself. In the second case, the only situation in which a learner can be removed from the archive is if it is dominated by another learner. Let us denote this latter learner by L'. Then:

$$\forall T \in T^{t+1} : G(L', T) \geq G(L, T)$$

Since $TS \subseteq T^t \subseteq T^{t+1}$, this shows that the requirement also holds for the second case.

Ad 2). The algorithm only makes a transition from time-step t to $t + 1$ if learners have actually been added to the archive. A learner L' is only added to the archive if it satisfies the *useful* relation. Thus, there must be some $L' \in L^{t+1}$ for which

$$\nexists L \in L^{t+1} : L \overset{T^{t+1}}{\succ} L' \quad \wedge$$
$$\nexists L \neq L' \in L^{t+1} : \forall T \in T^{t+1} : G(L', T) = G(L, T)$$

Let TS be the set of tests solved by L': $TS = \{T \in T^{t+1} | solves(L', T)\}$. Assume $\exists L \in L^t$ such that $\forall T \in TS : solves(L, T)$. Given the second clause of the above relation, we know that there must be some $T \in TS$ for which $G(L, T) \neq G(L', t)$. Since tests are binary and L' by definition solves all tests in TS, this implies L does not solve T. This contradicts our assumption, and therefore $\nexists L \in L^t$: $\forall T \in TS : solves(L, T)$, which completes our proof. ∎

A Cooperative Coevolutionary Multiobjective Algorithm Using Non-dominated Sorting

Antony W. Iorio and Xiaodong Li

School of Computer Science and Information Technology,
Royal Melbourne Institute of Technology University,
Melbourne, Vic. 3001, Australia
{iantony, xiaodong}@cs.rmit.edu.au
http://goanna.cs.rmit.edu.au/~xiaodong/ecml/

Abstract. The following paper describes a cooperative coevolutionary algorithm which incorporates a novel collaboration formation mechanism. It encourages rewarding of components participating in successful collaborations from each sub-population. The successfulness of the collaboration is measured by a non-dominated sorting procedure. The algorithm has demonstrated it can perform comparably with the NSGA-II on some multiobjective function optimization problems.

1 Introduction

In nature, coevolution is the process of reciprocal genetic change in one species, or group, in response to another. This process can also be utilised within evolutionary algorithms, and recently there has been a growing interest in the application of coevolution within multiobjective evolutionary algorithms. The reciprocal change observed in coevolution can be considered either as a competitive arms race, such as the coevolution of test cases for a problem (predators) with the solutions (prey) [1], or more recently, cooperative approaches where separate sub-populations evolve components of the solution [2,3].

The cooperative coevolutionary algorithm (CCA) [3], which was utilised in this work, separates the components of a problem solution into sub-populations, where each sub-population is subject to an evolutionary process. Solution components are rewarded in terms of their participation within good candidate solutions. In principle, the process of rewarding components of candidate solutions should also improve convergence towards the global Pareto front on particular multiobjective problems.

The CCA has already demonstrated benefits in a number of single objective optimization problem domains such as inventory control [4] and learning sequential decision rules [5]. Typically in this kind of algorithm, one picks the best collaborators from each sub-population to form a candidate solution, however this greedy approach can sometimes lead to premature convergence. We have proposed a new collaborator selection mechanism for the multiobjective domain; components of a solution can have the same rank if they belong to the same non-domination level. For example, a solution component of non-domination level 1

K. Deb et al. (Eds.): GECCO 2004, LNCS 3102, pp. 537–548, 2004.

Lohn *et. al.* have also devised a competitive algorithm with solutions in one population and another population containing a population of target objective vectors (TOVs). These vectors contain targets for the trial population to overcome [15, 16].

4 The Non-dominated Sorting Cooperative Coevolutionary Genetic Algorithm (NSCCGA)

In the following section we will describe the NSCCGA. The algorithm begins with a random initialization of all individuals across all sub-populations. Each sub-population is responsible for a particular parameter, x_i, from the decision space. If there are n decision variables there will be n sub-populations according to a natural decomposition of the problem.

4.1 Method

In the first generation random collaborations are formed and evaluated. This step is similar to the procedure outlined in Figure 1 step 1, except that we select random individuals from the complete set of components in each of the other sub-populations. Once these collaborations are formed, they are evaluated on the objectives, and the results from the evaluation are assigned back to the individual undergoing evaluation. In the first generation there will be only the current sub-populations, and the non-dominated sorting will only be over entirely random collaborations formed by this first generation of sub-populations. This is the only difference between collaboration formation in the first generation and following generations.

After the first generation, the resulting child sub-populations Q_1 to Q_n are evaluated (where Q_i is the child sub-population dealing with variable x_i) by forming collaborations with randomly selected components from the 'best' no-domination levels in the previous generation's sub-populations, P_1 to P_n (Figure 1 step 1). This collaboration formation is explained in more detail in section 4.2.

Following collaborator formation and evaluation, we perform a fast non-dominated sorting procedure [9] in step 2, over all collaborations from Q_1 to Q_n and the parent sub-populations P_1 to P_n (The parents have been assigned evaluations from their participation in collaborations in the previous generation). The sorting occurs on the values resulting from the evaluation on the objective functions. This will assign a front membership (non-domination level) F, to each of the individuals from the child sub-populations and parent sub-populations. F_1 contains the best candidates, F_2 the next best, and so on.

In step 2 the crowding distance [9] is calculated for each of the collaborations as well, just as it is for the NSGA-II. If two solutions are of the same non-domination level, the crowding distance sort determines which is better. Solutions with a higher crowding distance are preferred because they contribute to a more uniform non-dominated front. Both the non-dominational level and crowding distance are assigned back to the individual undergoing evaluation.

Step 3 applies the elitism operator which removes a number of the worst individuals from each sub-population which is in proportion to the number of children that were added. This allows the sub-population sizes to remain constant, while preserving good components. Good candidate components from the previous generation will have an opportunity to continue participating as well. The resulting sub-populations are the new parent sub-population P_1^{t+1} to P_n^{t+1}. The elitism operator is the same as NSGA-II except it operates on sub-populations.

Using the tournament selection operator, individuals are selected for mating from each sub-population and inserted into their respective mating pools. Crossover and mutation are then performed for each mating pool producing children for each of the sub-populations. The tournament operator rule we use for selection [9] states that a solution i wins a tournament against solution j if any of the following conditions are true:

- If solution i has a better non-domination level.
- If they have the same non-domination level but solution i has a better crowding distance than solution j.

After generating the new child sub-populations, the algorithm iterates until some termination condition is met. For further details regarding the elitism and non-dominated sorting procedures, including mechanisms for maintaining a diverse set of solutions, the reader is referred to the following paper on the NSGA-II [9]. The NSCCGA is implemented as a real coded GA, therefore the simulated binary crossover [17,18] and mutation operators [19] were used for recombination. These operators were also used in the NSGA-II.

4.2   A Novel Collaboration Formation Mechanism

Previous work with the CCA on single objective problems has primarily selected the current 'best' components from each sub-population to merge into a collaboration, or performed a tournament with candidate solutions formed with the current 'best' and randomly selected components. In a multiobjective scenario there may be a number of individuals in each sub-population which are parts of overall solutions that are non-dominated in relation to each other, so we cannot favour one over the other. In this case, we have proposed a novel collaboration formation mechanism where we select collaborators randomly from the 'best' non-domination level in each sub-population (Step 1 in Figure 1). The non-dominated sorting procedure sorts the collaborations into a number of separate non-dominated fronts. Individuals from the best non-dominated front, F_1, are given a non-domination level of 1, followed by 2 for the second front, F_2, and so on. By selecting randomly from the collaborators with the 'best' non-domination level in each sub-population we increase the chances of finding more diverse solutions. Early on in the search, there may be only a few individuals from the 'best' non-domination level, so the random selection of collaborators is from a in performance with a higher mutation rate at the sub-population level, by exploring more of the search space without significant disruption to the exploitation process. Therefore, we also conducted experiments with relatively high mutation rates with the NSCCGA, and a mutation rate of 0.6 was found to demonstrate generally good performance across the ZDT test problems. We have not explored the optimality of control parameters further because this paper is primarily concerned with demonstrating the utility of a new collaborator selection mechanism.

A population size of 100 was used for the NSGA-II, where 100 children were added each generation. The NSCCGA used a population size of 200 individuals for the sub-population evolving x_1, and added 200 children to this population each generation. For each of the other populations 40 individuals were used, with 40 new individuals added to each population each generation. A constant sub-population size is maintained through an elitism operator which culls a proportion of the least fit individuals from each sub-population. A crossover rate of 0.9 was used for both the NSCCGA and NSGA-II. For the $\mathcal{M}_2^*$ metric, σ^* was set to 0.01.

7   Results and Discussion

Figure 2 shows the typical non-dominated fronts, and Table 1 tabulates the results acquired using the performance metrics $\mathcal{M}_1^*$, $\mathcal{M}_2^*$, and $\mathcal{M}_3^*$. From this table it is apparent that the NSCCGA is comparable in performance to the NSGA-II upon the ZDT test functions. The NSCCGA also has the advantage of being able to acquire a large number of diverse non-dominated solutions for an equivalent number of evaluations of the NSGA-II (Metric $\mathcal{M}_2^*$ in Table 1) when the mutation rate is sufficiently high. This is by virtue of the collaboration formation mechanism within the NSCCGA, where n populations with m individuals can potentially form nm solutions through collaboration.

The $\mathcal{M}_1^*$ metrics demonstrate comparable performance between the algorithms on all but the ZDT6 function, where the NSCCGA was able to converge closer to the Pareto-optimal front than the NSGA-II. It is also apparent that the NSCCGA was able to achieve a slightly better measure of spread across all the functions from the $\mathcal{M}_3^*$ metric.

This can be understood in terms of the much larger number of collaborations which are formed within the NSCCGA, resulting in a greater likelyhood of good coverage across the front, including the extreme end points of the front.

7.1   Rotated Problems

Rotated problems introduce significant parameter interactions [21]. On the rotated problem, we have observed that the NSCCGA performed much worse than the NSGA-II. This is understandable because the NSCCGA assumes a problem that is decomposable, and can be solved by breaking it down into components. The CCA this work is based on has difficulties converging to good solutions with

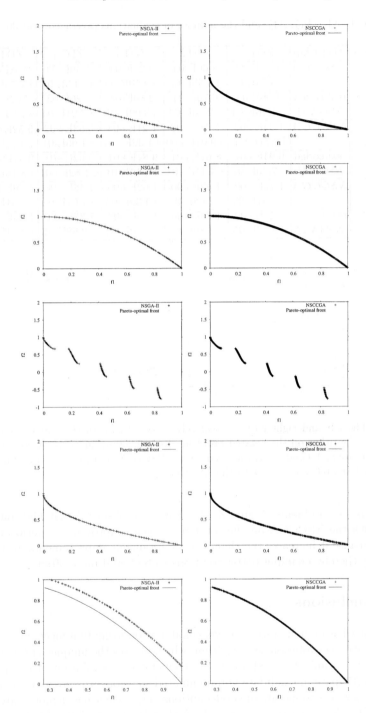

Fig. 2. From top to bottom are the typical non-dominated fronts of ZDT1, ZDT2, ZDT3, ZDT4, and ZDT6 for the NSGA-II and NSCCGA.

11. Potter, M. A., De Jong K.: A Cooperative Coevolutionary Approach to Function Optimization. In: Proc. of Parallel Problem Solving From Nature III (PPSN III), Springer-Verlag, Berlin Germany (1995) 249–257
12. Mao, J., Hirasawa, K., Hu, J., Murata, J.: Genetic Symbiosis Algorithm for Multiobjective Optimization Problems. In: Beyer, H., Cantu-Paz, E., Goldberg, D., Parmee, I., Spector, L., Whitley, D. (eds.): Proc. 2001 Genetic and Evolutionary Computation Congress (GECCO'01). Morgan Kaufmann, (2001) 771
13. Barbosa, H. J. C., Barreto, A. M. S.: An Interactive Genetic Algorithm with Co-evolution of Weights for Multiobjective Problems. In: Beyer, H., Cantu-Paz, E., Goldberg, D., Parmee, I., Spector, L., Whitley, D. (eds.): Proc. 2001 Genetic and Evolutionary Computation Congress (GECCO'01). Morgan Kaufmann, (2001) 203–210
14. Parmee, I. C., Watson, A. H.: Preliminary Airframe Design Using Co-evolutionary Multiobjective Genetic Algorithms. In: Banzhaf, W., Daida, J., Eiben, A. E., Garzon, M. H., Honavar, V., Jakiela, M., Smith, R.E. (eds.): Proc. 1999 Genetic and Evolutionary Computation Congress (GECCO'99). Morgan Kaufmann, (1999) 1657–1665
15. Lohn, J., Kraus, W. F., Haith, G. L.: Comparing a Coevolutionary Genetic Algorithm for Multiobjective Optimization. In: Fogel, D., et. al. (eds.): Proc. 2002 Congress on Evolutionary Computation (CEC'02). IEEE Press, Piscataway NJ (2002) 1157–1162
16. Lohn, J., Haith, G., Colombano, S., Stassinopoulos, D.: A Comparison of Dynamic Fitness Schedules for Evolutionary Design of Amplifiers. In: Stoica, A., Keymeulen, D., Lohn, J. (eds.): The First NASA/DoD Workshop on Evolvable Hardware. IEEE Press, (1999) 87–92
17. Deb, K., Agrawal, R. B.: Simulated Binary Crossover for Continuous Search Space. In: Complex Systems, Vol. 9, No. 2. (1995) 115–148
18. Deb, K., Kumar, A.: Real-coded Genetic Algorithms with Simulated Binary Crossover: Studies on Multi-modal and Multi-objective Problems. In: Complex Systems, Vol. 9, No. 6. (1995) 431–454
19. Deb, K., Goyal, M.: A Combined Genetic Adaptive Search (GENEAS) for Engineering Design. In: Computer Science and Informatics, Vol. 26, No. 4. (1995) 30–45
20. Zitzler, E., Deb, K. and Thiele, L.: Comparison of multiobjective evolutionary algorithms: Empirical results. *Evolutionary Computation*, 8(2):173-195, April (2000).
21. Salomon, R.: Re-evaluating Genetic Algorithm Performance Under Coordinate Rotation of Benchmark Functions: A Survey of Some Theoretical and Practical Aspects of Genetic Algorithms. In: Bio Systems, Vol. 39, No. 3. (1996) 263–278
22. Weicker, K., Weicker, N.: On the Improvement of Coevolutionary Optimizers by Learning Variable Interdependencies. In: Proc. 1999 Congress on Evolutionary Computation (CEC'99). IEEE Press, (1999) 1627–1632
23. Iorio, A. W., Li, X.: Parameter Control Within a Cooperative Coevolutionary Genetic Algorithm. In: Merelo Guervós, J.J., Adamidis, P., Beyer, H.-G., Fernández-Villacañas, J.-L., Schwefel, H.-P.(eds.): Proc. of Parallel Problem Solving From Nature VII (PPSN'02). Lecture Notes in Computer Science, Vol 2439. Springer-Verlag, Berlin Germany (2002) 247–256

Predicting Genetic Drift in 2 × 2 Games

Anthony M.L. Liekens, Huub M.M. ten Eikelder, and Peter A.J. Hilbers

Department of Biomedical Engineering
Technische Universiteit Eindhoven
P.O. Box 513, 5600MB Eindhoven, The Netherlands
{a.m.l.liekens,h.m.m.t.eikelder,p.a.j.hilbers}@tue.nl

Abstract. For the analysis of the dynamics of game playing populations, it is common practice to assume infinitely large populations. Infinite models yield predictions of fixed points and their stability properties.
However, these models cannot demonstrate the influence of genetic drift,
caused by stochastic sampling in small populations. Instead, we propose
Markov models of finite populations for the analysis of genetic drift in
games. With these exact models, we can study the stability of evolutionary stable strategies, and measure the influence of genetic drift in the
long run. We show that genetic drift can introduce significant differences
in the expectations of long term behavior.

1 Introduction

1.1 Evolutionary Game Theory

Evolutionary Game Theory (EGT, overviews can be found in [1,2]) studies the
dynamics and equilibriums of games played by populations of players. The strategies players employ in the games determine their interdependent payoff or fitness.
In contrast with the traditional applications of game theory, the players do not
act rationally when choosing their strategies, but act instead according to a pre-
programmed behavior pattern. In this paper, a pure strategy is encoded in an
individual's genome, which can evolve over time while repeatedly playing a game
against other players in a population.

A common model to study the dynamics of frequencies of strategies adopted
by these populations is based upon replicator dynamics. Replicator dynamics
assumes infinite populations, asexual reproduction, complete mixing, i.e., all
players are equally likely to interact in the game, and strategies breed true, i.e.,
strategies are transmitted to offspring proportionally to the payoff achieved.

We study two models from population genetics – a stochastic model of finite
populations and a deterministic model of infinite populations – and compare
their predictions in order to study the importance of finite population size when
the populations are involved in playing well-known symmetric 2 × 2 games, such
as the Hawk-Dove game, and the Prisoners' Dilemma. We show the importance
and influence of finite population size on the predicted behavior, as compared to
infinitely large populations. Evolutionary games with small search spaces – such
as 2 × 2 games – are sufficient to investigate this question. Nowak and Sigmund

K. Deb et al. (Eds.): GECCO 2004, LNCS 3102, pp. 549–560, 2004.
© Springer-Verlag Berlin Heidelberg 2004

[3] recently indicated the importance of finite population effects in EGT, and expect that the observations of finite models might question the importance of evolutionary stability of infinite models.

Both models studied in this paper assume discrete time steps, as compared to the continuous progression through time of the differential equations used in replicator dynamics. The discrete time steps denote the consecutive generations of the evolving populations. Both models are based on the simple or generational Genetic Algorithm (GA) where all individuals in the population of the current generation are replaced by a newly produced population of individuals in the next generation. As this creates a stochastic chain of events over time, where no memory is required, both models can be seen as Markov models. The infinite population model studied in this paper is based on the model studied by Vose in [4], where the finite population model is based on the Fischer-Wright model, originally described in [5,6], and its interpretation for the GA as outlined in [7, 8].

1.2   Finite Populations and Genetic Drift

When modeling evolving populations, the assumption of infinitely large populations vastly simplifies the computation and analysis of the predictions of the models under consideration. A population can be represented as a frequency distribution, a stochastic vector, over the set of strategies or available genotypes. The proportions of strategies at the next generation can easily be computed from a population, using deterministic methods. Each resulting frequency distribution then corresponds to an infinite population. In contrast, when studying finite population models, we have to consider every possible population and compute the limit or fixed point probability distribution over these possible states of the system using stochastic finite Markov chain techniques. The number of possible populations grows exponentially with population size and exceeds the number of possible strategies. This complicates the computation and study of the finite systems' behavior.

As most populations in nature are very large, and infinite populations resemble very large populations, the assumption of infinite populations seems to make sense. Note that predictions of the finite population model approximate predictions of infinite population models if the finite population size is sufficiently large. In this paper, we question the belief that predictions of infinite populations can indeed easily be translated to predictions of models that assume finite populations. When large, but finite populations are considered, the populations are expected to closely follow the dynamics as predicted by the infinite population model. Small perturbations in the frequencies of large populations, caused by elements of chance, are then also assumed to fade away easily in the following generations. A finite population model can test whether these assumptions are indeed acceptable for a given problem and finite population size.

Differences in the behavior of the models are expected due to stochastic sampling effects of finite systems. The element of chance when selecting and generating individuals has to be considered at the construction of each generation. The combinatorial effects of limited population size or variation are more pronounced in models that assume smaller population sizes. This element of chance can result in cumulative changes in the frequencies of evolutionary adopted genotypes or strategies. This effect is known as *genetic drift* and is part of the neutral theory of evolution [9]. With infinitely large populations, this element of chance does not exist. Instead of sampling a distribution over strategies in order to obtain the composition of a finite population at a generation, the distribution itself represents the next expected infinitely large population.

1.3   Games

In this section, we briefly introduce a selection of well-known symmetric 2 × 2 games: a Neutral game, the Hawk-Dove game and the Prisoners' Dilemma. These games are used to discuss the influence of finite population size and genetic drift on the predictions in finite models. All models assume a limited number of strategies (2) that can be employed. As such, the strategies can be represented as an atomic genotype with 2 alleles representing either strategy, which simplifies our analysis. Extended overviews of these games can be found in [1,2].

Fogel et al [10,11,12] and Ficici et al [13] have studied finite population effects of evolutionary dynamics on the stability of evolutionary stable strategies of the Hawk-Dove game empirically. Using simulations of the evolutionary systems, behaviors have been observed that are unrelated to an evolutionary stable strategy (ESS). They have suggested that ESSs may not provide a good expectation of a finite population's behavior. This paper presents a theoretical, Markov model approach to answer the same questions for a larger set of games, adapting Ficici's initial work [13]. We adopt genetic drift, and the causes of genetic drift, as an explanation of our theoretical observations.

All games are represented by a set of strategies Ω and a square payoff matrix $\mathbf{A}$. Each entry $A_{i,j}$ in this payoff matrix gives the payoff value for an individual adopting strategy i when confronted with an individual playing strategy j. We assume that all payoffs in matrix $\mathbf{A}$ are strictly positive.

Hawk-Dove game. In this game, a bird has a choice of 2 behaviors when a resource needs to be shared with another bird. It can either choose to act as an aggressive hawk or a pacific dove. If both players choose the hawk strategy, they fight and injure each other. If only one of both players chooses hawk, then this player defeats the pacific strategy of the dove. If both players play dove, there is a tie in profit, but the profit is lower than the profit of a hawk defeating a dove.

The Hawk-Dove game is also known as the snowdrift or chicken game.

The game can be modeled as a game with two strategies $\Omega = \{H, D\}$ (Hawk and Dove), with a payoff matrix $\mathbf{A}$ where $A_{H,H} < A_{D,H} < A_{D,D} < A_{H,D}$. Both pure strategies are unstable fixed points of the game if an infinite population without variation is assumed. There also exists a mixed strategy that is an ESS of the system if the proportion of Hawks in the population equals $\frac{A_{D,D} - A_{H,D}}{A_{H,H} + A_{D,D} - A_{H,D} - A_{D,H}}$.

Prisoners' Dilemma. Imagine two criminals who are arrested under the suspicion of a crime they have committed. The police doesn't have enough proof to convict them. The criminals are separately questioned. Both criminals must choose to either cooperate with each other or to defect. If either one of the criminals gives the police more evidence to convict the other, the defector is freed. If both players cooperate, they receive only a short time in jail. If both players tell out on each other, then the police has enough evidence to convict both. If one player defects her cooperating opponent, the defector receives a high payoff, and the cooperator spends a long time in jail.

Biological examples of the Prisoners' Dilemma can be found in the behavior of bacteriophage $\Phi 6$ and the evolution of ATP producing pathways [3].

Consequently, the game can be modeled with two strategies $\Omega = \{C, D\}$ (Cooperate and Defect) with a payoff matrix $\mathbf{A}$ where $A_{C,D} < A_{D,D} < A_{C,C} < A_{D,C}$. Both pure strategies are equilibrium strategies if no variation is assumed. The fixed point where the whole population adopts defection is stable, and the equilibrium where all players cooperate is unstable. There is no mixed strategy equilibrium for this game.

Neutral game. The last game we introduce is used for control measurements, and to show how the population behaves in the absence of selection. These predictions give us an idea of how strong genetic drift can become for certain parameters, as variation and sampling of the population are the only processes at work in systems with neutral selection.

We can model a Neutral game with two strategies $\Omega = \{0, 1\}$ with a payoff matrix $\mathbf{A}$ where $A_{0,0} = A_{0,1} = A_{1,0} = A_{1,1}$. If no variation is assumed, all pure and mixed strategies are fixed points of the game. If a variation operator – which is symmetric for both strategies – is assumed, only the mixed strategy at $1/2$ is a stable fixed point of the game.

2   Models and Methods

In this section, we give an overview of the reproduction schemes used in our evolutionary models. We also define the construction of new populations at a new generation, using sampling techniques if finite populations are considered.

2.1   Populations and Fitness of Individuals

Let P denote a population of individuals of type 0 and 1. These types correspond to the strategies in the games. Let $p(i|P)$ denote the proportion of genome $i \in \Omega$ in P. For now, we do not have to assume a size for the populations, and make the distinction between modeling of finite and infinite populations in a later section.

Let $f(i|P)$ with

$$f(i|P) = \sum_{j \in \Omega} A_{i,j} p(j|P)$$

denote the fitness of individual $i \in \Omega$. The fitness denotes the mean payoff received when the individual is matched against all individuals in the population, including itself. As such, the fitness of an individual denotes the expected payoff received when players are randomly chosen as opponents.

2.2   Selection

According to this fitness function, we can select an individual $i \in \Omega$ from population P with selection probability $s(i|P)$ with

$$s(i|P) = \frac{f(i|P)p(i|P)}{\sum_{j \in \Omega} f(j|P)p(j|P)}.$$

This selection method renders selected genotypes proportional to their fitness and abundance in population P. The denominator of the fraction is the expected fitness or payoff received by any individual in the population.

2.3   Reproduction

Commonly, the next step in the reproduction process would be to recombine these selected individuals in order to get recombined child individuals. However, since we are dealing with individuals with only one locus, there is no need for discussing recombination here. It suffices to consider mutation. Let $m(i|P)$ represent the probability that an individual $i \in \Omega$ is generated by selecting an individual from P, and then mutating it to i. More formally, if we assume a bit flip mutation probability μ with $0 \leq \mu \leq 1$, we can write $m(i|P)$ as $m(i|P) = \mu s(1 - i|P) + (1 - \mu)s(i|P)$. Note that $m(1 - i|P) = 1 - m(i|P)$ since we only have two possible individuals that can be generated. The resulting probability $m(i|P)$ now denotes the probability that genome i ends up in the population at the next generation.

2.4   Creating New Populations

Infinite populations. In the case of populations with an infinite number of individuals, the above reproduction scheme directly yields the proportion of the individuals in the population at the next generation. As such, population P' is generated from P in one generation with $p(i|P') = m(i|P)$. The reproduction of infinitely large populations is deterministic. The fixed points $\hat{P}$ of this system, with $p(i|\hat{P}) = m(i|\hat{P})$ can be derived, and their stability properties studied in order to investigate the long term behavior of the game under evolutionary selection and variation. The fixed point of 2 × 2 games can easily be found through iteration of the infinite population model. Note that the population with either $p(0|P) = 1$ or $p(1|P) = 1$ is a fixed point if $\mu = 0$, and not a fixed point otherwise.

Finite populations. In the case of populations with a fixed and finite number of individuals, we have to sample the results of reproduction r times in order to construct a population of size r at the next generation. With the introduction of a finite population size, the process is no longer deterministic and becomes stochastic. The probability that population P' with population size r is generated through sampling in one generation from population P is equal to

$$\Pr\left[\tau(P) = P'\right] = \binom{r}{rp(0|P')} \prod_{i \in \Omega} m(i|P)^{rp(i|P')}.$$

The binomial coefficient computes the number of possible arrangements for a population of size r whose proportion of 0 genomes equals $p(0|P')$. The other factors denote the probability that such an arranged population is sampled from the reproduction process. Note that $\forall i \in \Omega : rp(i|P') \in \mathbb{N}$, since $rp(i|P')$ represents the number of individuals with genotype i in P'. Note that the finite model thus samples the infinite model r times.

As the number of possible populations, i.e. $r+1$, is finite, we can study the resulting system as a finite Markov chain, over the state space of populations, with transition probability matrix T with entries $T_{P',P} = \Pr\left[\tau(P) = P'\right]$. As this system is irreducible and aperiodic, or ergodic, as $0 < \mu < 1$, we can obtain the system's limit behavior by computing the unique stochastic eigenvector of T with associated eigenvalue 1. This stochastic vector denotes the limit or steady state distribution over the states or possible populations of the system, and can be used to study the expected behavior of the system as a whole. The limit behavior of the system is undefined if $\mu = 0$ or $\mu = 1$. In the case of $\mu = 0$, the system becomes reducible, and the system ends in one of the populations that consist of a unique genome (i.e., either the population with all 0 or all 1 individuals). In the case of $\mu = 1$, the system becomes reducible and periodic, as the population consisting of all 0's can only become the population of all 1's in the next generation since all selected individuals are mutated from genome 0 to 1. Vice versa, this also holds when the system is started with a population of all 1's, which results in a periodic system. In practice, we only assume $0 < \mu \leq 1/2$, since mutation probabilities above $1/2$ work counterproductive for the evolutionary process. Otherwise, all selected individuals would have a too high probability of being mutated to less optimal genotypes.

2.5   Transitions

Figure 1 depicts the stochastic transition probability matrices of the finite model and the deterministic transition functions of the infinite model, for $r = 20$ and $\mu = 0.1$. The payoffs of the Hawk-Dove game are chosen with $A_{H,H} = 1 < A_{D,H} = 2 < A_{D,D} = 3 < A_{H,D} = 4$, and for the Prisoner's dilemma the utility function used in this paper is given by $A_{C,D} = 1 < A_{D,D} = 2 < A_{C,C} = 3 < A_{D,C} = 4$. Each of the columns of the transition matrices of the finite population models sums up to 1, as each column represents the probability distribution over the states at the next generation. As the population size increases, the sampling of the population at the next generation becomes more stable, and the stochastic model better resembles the deterministic infinite population model. On the other hand, if the rate of mutation is increased, the finite population model resembles the infinite model less. This is due to higher probabilities to end up in other population configurations then the most probable ones.

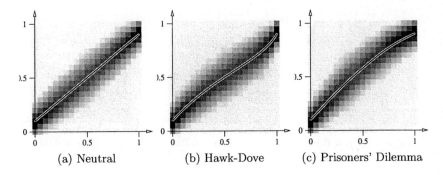

Fig. 1. Transition matrices of the finite population model ($r = 20$), overlayed by the deterministic map diagram (graph with white background) of the infinite population model, for $\mu = 0.1$. The horizontal axis represents the current proportion of (a) 0, (b) Hawk or (c) Defect genomes in the population, the vertical axis represents the proportion at the next generation. Each gray scaled box represents the transition probability between states in one generation for the finite model. Darker grays represent higher probabilities.

3   Results

3.1   Neutral Game

Given the Neutral game from section 1.3, figure 2 depicts the limit or fixed point distributions of the finite population model, for a small set of population sizes and mutation rates[1]. For all possible parameters, the weighted mean of the distribution is equal to the predicted fixed point of the infinite model, namely at $1/2$. Figure 2(a) shows a typical distribution for (relatively) large population sizes and large mutation rates. In this type, the system is most likely to end up with highly diverse populations. Figure 2(c) shows a typical distribution for systems with a small population size and a small mutation rate. In these cases, a run of the system will most likely end up in either one of the populations filled exclusively with either genome 0 or 1. Note that with these parameter settings, the system prefers extremes of the state space, and avoids the predicted "stable" fixed point of the infinite population model. Figure 2(b) shows a snapshot of the transition from the first type to the second. Note that the behavior depicted in these distributions is structurally very different, although the infinite population model predicts the same stable fixed point for all of these evolutionary systems. These differences in predicted behavior are due to genetic drift around this stable fixed point. Drift is stronger as populations become smaller (more sampling effects) or the mutation rate decreases (convergence due to low genetic diversity).

The weighted standard deviation σ of the steady state distributions' means can be employed to discuss the importance of genetic drift in our games with

---

[1] As the predicted stable fixed point of the infinite model equals the weighted mean of the distribution, the lines cannot be distinguished in these figures, but become important as other games than the Neutral game are introduced.

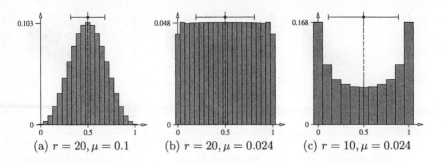

(a) $r = 20, \mu = 0.1$ (b) $r = 20, \mu = 0.024$ (c) $r = 10, \mu = 0.024$

Fig. 2. Limit or steady state distributions for the Neutral game for 3 different parameter settings. The bars denote the probability of ending up in the population with the given proportion of 0 genomes. The dashed vertical line denotes the weighted mean of the distribution, the error bars the weighted standard deviation for this mean. The dash-dotted vertical line gives the fixed point of the infinite population model. Note that there is no standard deviation in the infinite model as the system is deterministic.

finite populations, and we are able to predict how the parameters are influencing the system's behavior. Later on, we use similar techniques to discuss genetic drift in games with selective pressure, and can use the Neutral game as a control for our predictions and expectations.

Influence of population size. Assume a fixed mutation rate μ with $0 < \mu < 1$ throughout this section. As population size r increases, the standard deviation σ decreases. This is due to the genetic drift introduced by stochastic sampling of the finite population, which becomes more deterministic for larger populations. Consequently, as the population size becomes larger, the finite population model behaves more similar to the prediction of the unique stable fixed point of the deterministic infinite population model, where $\sigma = 0$. Indeed, if we keep on increasing the population size, then σ comes closer to 0, and the mean of the distribution converges to the fixed point of the infinite population model[2]. Note that this holds unless μ equals 0 or 1. If $\mu = 0$, the limit behavior of the finite model gives 2 attracting states, where the infinite model predicts a mixed strategy fixed point.

Influence of mutation rate. In this section, we assume a fixed and finite population size r. As the mutation rate in the system of evolving populations for the Neutral game is sufficiently decreased, a run of the system most probably ends up in either one of the populations with only one genome. Indeed, if a mutation rate of 0 is assumed, the transition matrix of the evolutionary system

---

[2] In the Neutral game, the fixed point of the infinite model and the mean of the finite population are always the same. This statement is thus trivial for this game, but becomes more important as there will be a difference between these two predictions as other games are considered.

becomes reducible, and the steady state distribution of the Markov chain is no longer unique. If $\mu = 0$, two linearly independent stochastic eigenvectors of the transition probability matrix with corresponding eigenvalue 1 exist, and either one of these eigenvectors represents a distribution where the evolutionary system ends up with a population containing only one genotype. As the mutation rate is sufficiently decreased toward 0, the predictions of the finite population model better resemble the extreme situation where $\mu = 0$, independent of the game considered, as the selective pressure becomes negligible.

As the variational pressure for the Neutral game is increased, more random individuals are generated by the reproduction process. In the extreme case, where $\mu = 1/2$, each generation renders a new random population with the probability of either individual in this population being $1/2$. At each step, a distribution over the state space is constructed that is a binomial distribution. The probability of encountering a population with n out of r individuals being of type 0 is then given by $\binom{r}{n} \frac{1}{2}^r$. As this is the case for each of the generations during a run of the system, it is also the limit or steady state distribution. Consequently, if a mutation rate of $1/2$ is assumed, the expected proportion of either genome is $1/2$, and σ becomes $\frac{1}{2\sqrt{r}}$. As the mutation rate is increased toward $1/2$, the finite population model better resembles the extreme situation where $\mu = 1/2$, independent of the game considered.

3.2   Hawk-Dove

The payoffs of the Hawk-Dove game have been chosen with $A_{H,H} = 1 < A_{D,H} = 2 < A_{D,D} = 3 < A_{H,D} = 4$. The stable fixed point of the infinite model with no variation for these parameters lies at $1/2$. Even more, if variation is assumed, the stable fixed point remains at $1/2$ and no other fixed points exist. Consequently, the evolutionary system with this game is similar to the Neutral game, in only having a stable mixed strategy fixed point at $1/2$. When choosing extremely small or large parameters for population size and mutation rate, so are the predictions. Under those parameters, the forces of genetic drift are much stronger, or much weaker, than those of selection according to the payoffs in the game. In the Hawk-Dove game however, selection is asymmetric to either genome. This allows a finite population to wander away from the infinite model's projected "stable" fixed point, which on its turn may result in genetic drift of the population. We can study this effect, and the balance between selective and variational pressure, by examining the differences in expected behavior.

Figure 3 represents the steady state distribution of the finite population model for the Hawk-Dove game, for three parameter settings of the system. Figures 3(a) and (b) show how the system balances between the selection around the fixed point on one hand, and the influence of genetic drift which forces the population to either extreme of its state space on the other hand. Figure 3(c) shows how genetic drift can force the expected behavior of the finite population model relatively far away from the infinite population's predicted stable fixed point, toward higher proportions of the Dove strategy. As the fixed points, means and standard deviations of the systems have been determined by exact techniques,

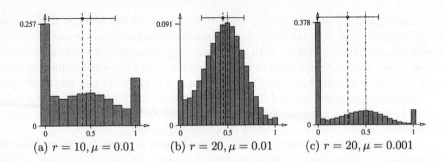

Fig. 3. Limit or steady state distributions for the Hawk-Dove game for 3 different parameter settings. The horizontal axis represents the proportion of Hawk genomes in the population. The vertical dashed line represents the mean of the distribution, where the dash-dotted line represents the fixed point of the infinite model, as in figure 2.

it is clear that genetic drift can introduce significantly different behavior when finite population sizes are considered. These predictions lose significance as population size r or mutation rate μ is increased.

We need an explanation why small populations drift to higher proportions of Dove. Consider two finite systems, with the same population size r and mutation rate μ. The first system is initialized with a population with $r/2 - k$ Hawks, the other is initialized with $r/2 + k$ Hawks, with k strictly positive. The probability of moving from these initial states to the state with a proportion of $1/2$ Hawks in n steps can be computed. The probability of reaching this state is higher when starting with $r/2 + k$ Hawks, where the system started with more Doves remains longer stuck. On average, the overall system thus remains longer in states that have a higher proportion of Doves. Genetic drift pushes the system to higher proportions of Doves as compared to the infinite model. The observation that populations drift toward higher proportions of Dove (and not the other way) is similar to Ficici's [13] observation of this effect in simulation runs and an infinite model of the Hawk-Dove game.

We can thus summarize that predictions of long term behavior differ significantly when infinite and finite models are compared, and that genetic drift gives a viable explanation of these deviations.

3.3   Prisoners' Dilemma

Genetic drift can also be observed in other games, such as the Prisoners' Dilemma. However, the influence of drift differs from the previous games. If no variation is assumed, there is one stable pure strategy (Defect) and there is no mixed strategy ESS. As a result, we expect the populations to contain a lot of the Defect genomes on the long run, even if variation is assumed. The effects of genetic drift observed in the previous games are different in the Prisoners' Dilemma. In the previous games, selective pressure pulls the populations toward diverse populations, as the stable ESS of those games is a mixed strategy. At the other end, genetic drift pushes instantiations of the system to less diverse

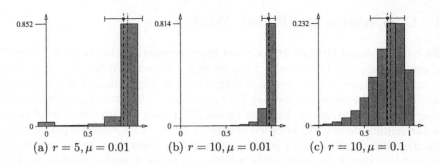

Fig. 4. Limit or steady state distributions for the Prisoners' Dilemma game for 3 different parameter settings. The horizontal axis represents the proportion of Defect genomes in the population.

populations, such that one strategy becomes prominently abundant in the population. In the Prisoners' Dilemma, the behavior in the finite model is expected to concentrate on populations with one strategy (Defect), where genetic drift moves populations to more diverse configurations.

Figure 4 depicts steady state distributions of the finite population model when the individuals are involved in the Prisoners' Dilemma, for a number of parameter settings. As predicted by the infinite model, the distributions are expected to have a large proportion of Defect genomes. As the population size is increased, the expected behavior of the system better resembles the expected infinite population behavior, and the standard deviation σ around the weighted mean of the expected distribution over the states decreases. As we increase the rate of mutation, σ increases, as the generation of random individuals tends to push the populations to more diverse configurations. Note that this observation contrasts with the expected behavior in the previous games, where higher variation resulted in predictions that better resembled the infinite population model. Of course, in these other games, selective pressure and a high mutation rate both guide the system to more diverse populations. For small population sizes and small mutation rates, the predictions of the infinite population model are thus more stable for the Prisoners' Dilemma as compared to the influence of genetic drift in the Neutral and Hawk-Dove game.

In the Prisoners' Dilemma, the Cooperate strategy is rationally the optimal strategy, if all other players in the game also opt for this strategy. In an evolutionary system, however, this pure strategy is an unstable ESS. Only for small population sizes and extremely small mutation rates, the finite population model predicts a noticeable proportion of Cooperate genomes in the populations. Figure 4(a) gives an example of a small probability of ending up in a population filled with the Cooperate genome.

4   Conclusions and Future Work

We have proposed stochastic models of finite populations to study the stability of evolutionary stable strategies for a set of 2×2 games. When the assumption of infinitely large populations in evolutionary models is discarded, statistically significant differences in expected behavior can be observed. In particular, the long term expectations of the finite model differ from the predicted fixed points of the infinite model. We adopted genetic drift to give a viable explanation for the deviation from the fixed points predicted by infinite population models. We have shown that finite population models can be used, and extended, to study the stability of evolutionary "stable" strategies in finitely sized populations.

In future work, we intend to investigate the relationship between mutation rate and population size in terms of genetic drift. We are currently studying larger games and larger populations to check the scalability of our observations. Similarly, we are interested in the influence of differing selective pressure on the amount of genetic drift, where a fixed selective pressure was chosen in this paper.

References

1. Weibull, J.W.: Evolutionary Game Theory. MIT Press (1995)
2. Hofbauer, J., Sigmund, K.: Evolutionary Games and Population Dynamics. Cambridge University Press (1998)
3. Nowak, M.A., Sigmund, K.: Evolutionary dynamics of biological games. Science **303** (2004) 793–799
4. Vose, M.D.: The Simple Genetic Algorithm. MIT Press (1999)
5. Fischer, R.A.: The Genetical Theory of Natural Selection. Clarendon (1930)
6. Wright, S.: Evolution in mendelian populations. Genetics **16** (1931) 97–159
7. Nix, A.E., Vose, M.D.: Modelling genetic algorithms with markov chains. Annals of Mathematics and Artificial Intelligence (1992) 79–88
8. Liekens, A.M.L., ten Eikelder, H.M.M., Hilbers, P.A.J.: Modeling and simulating diploid simple genetic algorithms. In: FOGA VII. (2003)
9. Kimura, M.: The Neutral Theory of Molecular Evolution. Cambridge University Press (1986)
10. Fogel, D.B., Fogel, G.B.: Evolutionary stable strategies are not always stable under evolutionary dynamics. In: Evolutionary Programming IV. (1995) 565–577
11. Fogel, D.B., Fogel, G.B., Andrews, P.C.: On the instability of evolutionary stable strategies. BioSystems **44** (1997) 135–152
12. Fogel, G.B., Andrews, P.C., Fogel, D.B.: On the instability of evolutionary stable strategies in small populations. Ecological Modelling **109** (1998) 283–294
13. Ficici, S.G., Pollack, J.B.: Effects of finite populations on evolutionary stable strategies. In: Proceedings of the 2000 Genetic and Evolutionary Computation Conference. (2000)

Similarities Between Co-evolution and Learning Classifier Systems and Their Applications

Ramón Alfonso Palacios-Durazo[1] and Manuel Valenzuela-Rendón[2]

[1] Lumina Software,
apd@luminasoftware.com,
http://www.luminasoftware.com/apd
Washington 2825 Pte C.P. 64040,
Monterrey N.L., Mexico
[2] Centro de Sistemas Inteligentes,
Instituto Tecnológico y de Estudios Superiores de Monterrey,
valenzuela@itesm.mx,
http://www-csi.mty.itesm.mx/~mvalenzu
Sucursal de Correos J, C.P. 64849
Monterrey, N.L., Mexico

Abstract. This article describes the similarities between learning classifier systems (LCSs) and coevolutionary algorithm, and exploits these similarities by taking ideas used by LCSs to design a non-generational coevolutionary algorithm that incrementally estimates fitness of individuals. The algorithm solves some of the problems known to exist in coevolutionary algorithms: it does not loose gradient and is successful in generating an arms race. It is tested on MAX 3-SAT problems, and compared to a generational coevolutionary algorithm and a simple genetic algorithm.

1 Introduction

Coevolution refers to the simultaneous evolution of two or more genetically distinct species. This evolution may be such that the species cooperate or compete. The application of competitive coevolution to problem solving has been of interest in the genetic algorithm research community because *competition*, in its most general sense, encourages the generation of better *competitors*. The implementation of this idea, in the form of a competition between possible solutions and instances of a problem, has been reported with varying degrees of success: Hillis [5] used a coevolutionary approach to find sorting nets for sorting arrays of 16 elements. He implemented a coevolutionary algorithm where sorting nets competed against permutations (each permutation is an instance of a sorting problem). Pollack and Rosin [8,10] generated playing strategies for games, Ficici [4] used a coevolutionary algorithm for generating predictors and Cliff [3] for evolving persecution and evasion strategies. Many other applications have been reported.

In the competitive coevolutionary approach to problem solving, the individuals in two populations are made to compete to determine their fitness. The fitness

K. Deb et al. (Eds.): GECCO 2004, LNCS 3102, pp. 561–572, 2004.

of an individual depends on its performance against opponents in the current generation of the competing population [9]. Since the competition is similar to the relationship between predator and prey, or parasites and their hosts, the populations in a coevolutionary algorithm are usually named as such. For the purpose of this article, *hosts* will refer to a population of possible solutions, and *parasites* will refer to a population of instances of or parts of a problem.

The central idea of coevolution lies in the fact that the fitness of an individual depends on its performance against the current individuals of the opponent population. This is the competition that we hope will generate better solutions. However, this simple idea gives rise to not-so-simple implications: the most simple form of coevolution is inherently unstable, costly in computer effort, and holds no guarantee it will work [12,9,4].

Unnoticed by most researchers are the similarities between coevolutionary algorithms (CEAs) and learning classifier systems (LCSs). The purpose of this article is to highlight these similarities, to address the issue of *loss of gradient*, and to introduce a new paradigm in CEAs based on the way LCSs work: incrementally adjusting fitness of individuals in a non-generational genetic algorithm. We devised a CEA based on these premises and test and compare it with a simple generational CEA and a simple genetic algorithm (SGA).

The remainder of the article is organized as follows: Section 2 defines a simple CEA and describes goals and obstacles CEAs face. Section 3 derives the incremental and non-generational algorithm. Section 4 describes the experimentation and results and Section 5 concludes this article.

2   Goals and Challenges of Coevolutionary Algorithms

In its most simple form, coevolution is implemented in the manner shown in Figure 1 in an algorithm we will call the *simple coevolutionary algorithm* (SCA).

Variations of the SCA are used by researchers [8,4], mostly by changing the form of the competition. The manner in which a competition is performed depends on the application as well as how the "winner" is determined. Some applications may not have an absolute winner, more likely a degree of winning, or a score. The way the fitness of an individual is calculated is also application dependant. The number of wins may be substituted by a sum of scores or other appropriate measures of performance.

Likewise, the opponents chosen for competition are not necessarily the complete population. Many researches use a subset of the opponent population for purposes of fitness calculation, either randomly or specifically selected. The SCA, and the variations found to be used by researchers encourage the selection and reproduction of individuals that perform well against the fitness landscape represented by the opponent population.

Watson and Pollak[12] clearly stated three interesting and useful goals and potential pitfalls for a coevolutionary algorithm:

- Providing a "hittable" target (gradient). The individuals in both populations should be relatively of the same quality and should improve roughly at the

```
(*Initialize populations*)
Generate random host population
Generate random parasite population

(* Main cycle*)
repeat

    (* Competition cycle*)
    for-each p ∈ parasites
        for-each h ∈ hosts
            h and p compete
        end-for
    end-for
    Fitness of parasites and hosts calculated
        based on competitions won

    One generation of a GA is applied to hosts
        selection, crossover, and mutation

    One generation of a GA is applied to parasites
        selection, crossover, and mutation

until termination criteria met
```

Fig. 1. The simple coevolutionary algorithm

same rate. If one population outperforms the other drastically, the latter will have no opportunity to learn and improve.

- Providing a relevant target (focusing). The opponents must represent something worth beating. Different strengths and weaknesses should emerge. These strengths are also known as *specialization niches* and have been described by Rosin and Belew [10].
- Providing a progressive moving target (open-endedness). The gradual improvement of both populations should be such that individuals *progress*. It is possible for individuals to "forget" strengths and fall into what is known as a mediocre stable state [1].

Methods and techniques have been proposed to compensate the challenges faced by CEAs [9], mostly trying to solve each known problem one at a time.

3   An Incremental and Non-generational Coevolutionary Algorithm

In this article we take lessons learned from the field of learning classifier systems to propose a different approach to implementing coevolution that we call the *incremental coevolutionary algorithm* (ICA).

In ICA, the importance of the coexistence of individuals in the same population is as great as the individuals in the opponent population. This is similar to the problem faced by learning classifier systems (LCSs) [6] and multiobjective optimization as done by Valenzuela-Rendón and Uresti-Charre [11]. We take ideas from these algorithms and put them into the ICA. A non-generational genetic algorithm is used and an incremental approach is taken to estimate the fitness of an individual. With this new design, we are able to avoid some of the pitfalls faced by other CEAs.

The motivation to use ideas from LCSs comes from the similarity between the challenges LCSs and CEAs face:

- The performance of the algorithm depends on the coexistence of individuals in the population.
- A generational algorithm in which each new generation is composed of new individuals is very disruptive.
- There is a need both to explore and to remember.

In a simple genetic algorithm, the objective is to find the individual that has the best possible fitness as defined by the objective function. In a CEA, the fitness landscape depends on the opponent population, therefore it changes over time (every generation, in fact). The evolution of hosts depends on the existence of parasites. The fitness landscape presented by the parasites determines how the population of hosts is formed. Likewise, the population of parasites depends on the fitness landscape of the population of hosts. The individuals selected for reproduction are those more promising to perform better against the fitness landscape represented by the opponent population. However, if the complete population of parasites and hosts are recreated in every generation, the offspring of each new generation face a fitness landscape unlike the one they where bred to defeat. Clearly, a purely generational approach to coevolution can be too disruptive.

As the fitness landscape presented by the opponent population gradually changes, so does the value of an individual. Instead of calculating or estimating the fitness based on competitions against the current opponents, it makes more sense to incrementally adjust the fitness of an individual as it faces each new generation of opponents.

These two ideas define the main approach of the ICA: the use of a non-generational genetic algorithm and the incremental adjustment of the fitness estimation of an individual. The general design of the ICA is presented in Figure 2.

The manner in which the fitness of host and parasite are adjusted must be designed to insure that the fitness of parasites and hosts behave as desired, and can be done in a manner very similar to the adjustment of strengths of classifiers in an LCS.

If we call the result of a competition a *score*, and hosts are set to minimize the score, and parasites are set to maximize it, then the general equation for adjusting the fitness of a parasite should include an increment proportional to the score, but should also include some form of fitness reduction so the value does not grow indefinitely. Therefore, the fitness of a parasite (S_p) at a time

```
(*Initialize populations*)
Generate random host population
Generate random parasite population

(* Main cycle*)
repeat
    (* Competition cycle*)
    for c ← 1 to determined number of cycles
        p ← parasite selected proportionally to fitness
        h ← host selected proportionally to fitness
        h and p compete, their fitness is adjusted
            incrementally, depending on the score
    end-for c

    (* 1 step of a GA in the parasite population*)
    Select parasite parents proportionally to fitness
    Create parasite by doing crossover
        and mutation
    Delete parasite with worst fitness and
        substitute with new parasite

    (* 1 step of a GA in the host population*)
    Select host parents proportionally to fitness
    Create host by doing crossover
        and mutation
    Delete host with worst fitness and
        substitute with new host

until termination criteria met
```

Fig. 2. General form of the incremental coevolutionary algorithm

$t + 1$ depends on the fitness at time t, plus some form of reward for obtaining the *score*, minus a cost or taxation for competing:

$$S_p(t+1) = S_p(t) + \text{Reward} - \text{Taxation} \tag{1}$$

The reward should be in some way proportional to the score and the term for taxation is defined as a portion of an individuals fitness, similar to the manner done by LCSs:

$$\text{Reward} = C_1(\text{score}) \tag{2}$$

$$\text{Taxation} = C_2 S_p(t) \tag{3}$$

The value of the score should be limited in magnitude to avoid unstable behavior. We define a function $A(\text{score})$ that will limit the values of score that can be used in equation 2. We use $\tanh(x)$, because $\tanh(0) = 0$ and it grows

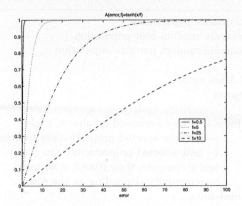

Fig. 3. Effect of the scaling factor

asymptotically to 1. Furthermore, it is possible to change the scale of $\tanh(x)$ adding a scaling factor h:

$$A(\text{score}, h) = \tanh\left(\frac{\text{score}}{h}\right) \tag{4}$$

The effect of the scaling factor can be seen in Figure 3. Equation 1 can now be fully written as:

$$S_p(t+1) = S_p(t) + C_1 A(\text{score}, H) - C_2 S_p(t) \tag{5}$$

which can be rewritten as:

$$S_p(t+1) = (1 - C_2)S_p(t) + C_1 A(\text{score}, h) \tag{6}$$

The equation for adjusting the fitness of a host is very similar.

$$S_h(t+1) = S_h(t) + \text{Reward} - \text{Taxation} \tag{7}$$

Taxation is also defined in a similar manner to parasites and the host's reward should be inversely related to the score:

$$\text{Reward} = C_3(1 - A(\text{score}, h)) \tag{8}$$

$$\text{Taxation} = C_4 S_h(t) \tag{9}$$

The complete equation for adjusting the fitness of a host is:

$$S_h(t+1) = S_h(t) + C_3(1 - A(\text{score}, h)) - C_4 S_h(t) \tag{10}$$

also expressed as:

$$S_h(t+1) = (1 - C_4)S_h(t) + C_3(1 - A(\text{score}, h)) \tag{11}$$

Do these equations accomplish what is wanted? A simple steady state analysis can show they do. When the algorithm stabilizes, the fitness of an individual should not change, therefore making $S(t+1) = S(t)$ for both hosts and parasites in equations 6 and 11. Doing trivial algebraic manipulation, we have that

$$S_p = \frac{C_1}{C_2} A(\text{score}, h) \tag{12}$$

and

$$S_h = \frac{C_3}{C_4} (1 - A(\text{score}, h)) \tag{13}$$

The fitness of parasites that cause low values of score will tend to 0, but parasites that cause high values of score will tend to C_1/C_2. Likewise, hosts that have high values of score will tend to 0, but hosts that have low score will tend to C_3/C_4. We can make $C_1 = C_2 = \alpha$ and $C_3 = C_4 = \beta$ for formula simplification:

$$S_p(t+1) = (1 - \alpha)S_p(t) + \alpha A(\text{score}, h) \tag{14}$$

$$S_h(t+1) = (1 - \beta)S_h(t) + \beta(1 - A(\text{score}, h)) \tag{15}$$

The equations will make the fitness of hosts and parasites behave as desired.

Two final modifications make the ICA work better: each parasite and host is initialized with a fitness equal to 1. This makes each individual start with a perfect fitness (which is obviously not true). This has the effect distributing the initial competitions evenly among the first generation. Second, each new individual is given a fitness equal to the average of the fitness of its parents. This is considered an initial estimate of the fitness of the new individual.

The ICA has some interesting properties. First of all, it is not generational. Each new individual faces a similar fitness landscape than its parents. The fitness landscape changes gradually, allowing an arms race to occur, avoiding loss of gradient. Second, each population works as a memory, deterring a fall into a mediocre stable state. Third, the fitness landscape each population sees is somewhat distorted in an interesting way: the actual score is modified by the function $A(\text{score}, h)$, but also, since opponents are chosen proportional to their fitness, an individual has a greater chance of facing good opponents. If a particular strength is found in a population, individuals that have it will propagate and will have a greater probability of coming into competition (both because more individuals carry the strength, and because a greater fitness produces a higher probability of being selected for competition). If the population overspecializes, another strength will propagate to maintain balance. Thus, a natural sharing occurs. The formal definition of the ICA can be seen in Figure 4 with the list of parameters defined in Table 1.

The ICA was tested on three different MAX 3-SAT problems, and its performance compared to that of a generational coevolutionary algorithm and a simple genetic algorithm.

```
(*Define A(x,h)*)
A(x, h) ← tanh(x/h)

(*Initialize populations*)
Generate random host population
Generate random parasite population

(* Initialize fitness*)
for-each  p ∈ parasites
  S_p ← 1
for-each  h ∈ hosts
  S_h ← 1

(* Main cycle*)
repeat
  (* Competition cycle*)
  for c ← 1 to N_c
    p ← parasite selected proportionally to S_p
    h ← host selected proportionally to S_h
    score ← competition between h and p
    S_p ← (1 − α)S_p + αA(score, h)
    S_h ← (1 − β)S_h + β(1 − A(score, h))
  end-for c

  (* 1 step of a GA in the parasite population*)
  Select parasite parents (p_1 and p_2) proportionally to S_p
  Create parasite p_0 crossover and mutation
  S_{p0} ← (S_{p1} + S_{p2})/2
  Delete parasite with worst fitness and substitute with p_0

  (* 1 step of a GA in the host population*)
  Select host parents (h_1 and h_2) proportionally to S_h
  Create host h_0 by crossover and mutation
  S_{h0} ← (S_{h1} + S_{h2})/2
  Delete host with worst fitness and substitute with h_0

until termination criteria met
```

Fig. 4. Incremental coevolutionary algorithm

4   Experimentation and Results

The ICA was run five times on each of three 3-SAT benchmark problems from SATLIB[7] (a total of fifteen times). All three problems have 960 clauses and 225 variables and have solutions. The problems are considered hard due to phase transition as described by Cheeseman, Kanefsky, and Taylor [2].

The hosts were coded as a direct variable instantiation (1 bit per variable). The parasites consisted of 5 segment chromosome, each segment representing one

Table 1. Incremental coevolutionary algorithm parameters

Parameter	Description
h	Score scaling factor.
α	Maximum increment ratio in host fitness.
β	Maximum increment ratio in parasite fitness.
N_c	Number of competitions between each step of the GA.
P_{cp}	Crossover probability of the parasite population.
P_{mp}	Mutation probability of the parasite population.
T_p	Parasite population size.
P_{ca}	Crossover probability of the host population.
P_{ma}	Mutation probability of the host population.
T_a	Host population size.

of the 225 variables. The score of the competition between hosts and parasites was the percentage of clauses that contained any one of the variables represented by the parasite that were solved by the variable instantiation represented by the host.

The same representation and competition was used in a generational approach to coevolution. En each step of the generational coevolutionary algorithm, each host was made to compete with each parasite. Fitness was calculated as the percentage of clauses solved by each host (and the reciprocal for the parasite).

Finally, the same number of tests were applied to a simple genetic algorithm. In this case, the fitness of each individual was directly the number of clauses solved. A few pilot runs were done with each algorithm to help determine the best configuration. The parameters used in all three algorithms can be seen in Tables 2, 3, and 4.

Figure 5 shows the results of these experiments. The graph shows the average and standard deviation of the best solution found for all problems. For both coevolutionary algorithms, at each generation step, the individual with best fitness was evaluated to determine how many clauses of the complete 3-SAT problem it solved. It is possible that a better solution existed in the population, but only the best host was evaluated.

It can be seen than the ICA was able to find better solutions and maintain progress. The gradual change provided by the non-generational nature of the algorithm, as well as the incremental adjustment of fitness allows a gradient to be maintained. In contrast, the generational CEA was unable to provide such an environment, so no progress occurs. It can be seen that very few competitions are required between steps of the genetic algorithm in the ICA. Computational effort is saved by using these competitions with good individuals.

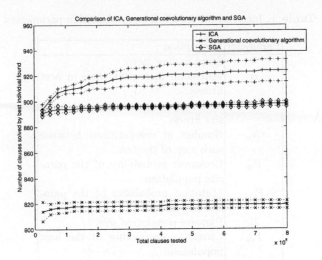

Fig. 5. Results of experimentation

Table 2. The ICA parameters

Parameter	Value
α	0.01
β	0.01
N_c	25
T_a	500
P_{ca}	0.95
P_{ma}	0.05
T_p	100
P_{cp}	0.65
P_{mp}	0.1
h	1

5   Conclusions

CEAs and LCSs have important similarities that have been unnoticed. A successful CEA can be designed following the guidelines of LCSs. The gradual change provided by a non-generational algorithm with incremental adjustment to fitness estimation provide an environment where a true competition can occur, with the desired benefits.

The incremental coevolutionary algorithm is quite robust, it does not fall into a stable mediocre state, it generates a successful arms race and niche specialization. It is simple and elegant and solves some of the problems known to happen in coevolutionary algorithms.

Table 3. Generational CEA parameters

Parameter	Value
T_a	500
P_{ca}	0.95
P_{ma}	0.05
T_p	100
P_{cp}	0.95
P_{mp}	0.5
Selection mechanism	Tournament (size 2)

Table 4. SGA parameters

Parameter	Value
Population size	500
P_c	0.9
P_m	0.05
Selection mechanism	Tournament (size 2)

References

1. Peter J. Angeline and Jordan B. Pollack. Competitive environments evolve better solutions for complex tasks. In *Proceedings of the 5th International Conference on Genetic Algorithms*, pages 264–270, 1994.
2. Peter Cheeseman, Bob Kanefsky, and William M. Taylor. Where the really hard problems are. In *Proceedings of the Twelfth International Joint Conference on Artificial Intelligence, IJCAI-91, Sidney, Australia*, pages 331–337, 1991.
3. Dave Cliff and Geoffrey F. Miller. Tracking the red queen: Measurements of adaptive progress in co-evolutionary simulations. In *European Conference on Artificial Life*, pages 200–218, 1995.
4. Sevan G. Ficici and Jordan B. Pollack. Challenges in coevolutionary learning: Arms-race dynamics, open-endedness, and mediocre stable states. In Christoph Adami, Richard K. Belew, Hiroaki Kitano, and Charles Taylor, editors, *Artificial Life VI: Proceedings of the Sixth International Conference on Artificial Life*, pages 238–247, Cambridge, MA, 1998. The MIT Press.
5. W. Daniel Hillis. Co-evolving parasites improve simulated evolution as an optimization procedure. In Christopher G. Langton, Charles Taylor, J. Doyne Farmer, and Steen Rasmussen, editors, *Artificial Life II*, volume X, pages 313–324. Addison-Wesley, Santa Fe Institute, NM, 1992.
6. John Holland. Escaping brittleness: The possibilities of general-purpose learning algorithms applied to parallel rule-based systems. *Machine Learning: An Artificial Intelligence Approach*, 2, 1986.
7. Holgar H. Hoos and Thomas Stützle. Satisfiability library. http://www.satlib.org, May 2001. Version 1.4.4.
8. Jordan B. Pollack, Alan D. Blair, and Mark Land. Coevolution of a backgammon player. In C. G. Langton, editor, *Proceedings of Artificial Life V*, Cambridge, MA, 1996. MIT Press.
9. Christopher D. Rosin. *Coevolutionary search among adversaries*. PhD thesis, University of California, San Diego, San Diego, CA, 1997.

10. Christopher D. Rosin and Richard K. Belew. Methods for competitive co-evolution: Finding opponents worth beating. In Larry Eshelman, editor, *Proceedings of the Sixth International Conference on Genetic Algorithms*, pages 373–380, San Francisco, CA, 1995. Morgan Kaufmann.

11. Manuel Valenzuela-Rendón and Eduardo Uresti-Charre. A non generational genetic algorithm for multiobjective optimization. In *Proceedings of the Seventh International Conference on Genetic Algorithms*, pages 658–665. Morgan Kaufmann, 1997.

12. Richard A. Watson and Jordan B. Pollack. Coevolutionary dynamics in a minimal substrate. In *Proceedings of the Genetic and Evolutionary Computation Conference (GECCO-2001)*, pages 702–709, San Francisco, California, USA, 7-11 2001. Morgan Kaufmann.

A Sensitivity Analysis of a Cooperative Coevolutionary Algorithm Biased for Optimization

Liviu Panait, R. Paul Wiegand, and Sean Luke

George Mason University, Fairfax, VA 22030
lpanait@cs.gmu.edu, paul@tesseract.org, and sean@cs.gmu.edu

Abstract. Recent theoretical work helped explain certain optimization-related pathologies in cooperative coevolutionary algorithms (CCEAs). Such explanations have led to adopting specific and constructive strategies for improving CCEA optimization performance by biasing the algorithm toward ideal collaboration. This paper investigates how sensitivity to the degree of bias (set in advance) is affected by certain algorithmic and problem properties. We discover that the previous static biasing approach is quite sensitive to a number of problem properties, and we propose a stochastic alternative which alleviates this problem. We believe that finding appropriate biasing rates is more feasible with this new biasing technique.

1   Introduction

Coevolutionary algorithms (CEAs) are popular augmentations of traditional evolutionary algorithms (EAs). The basic elements of these augmentations lay in the adaptive nature of fitness evaluation in coevolutionary systems: individuals are assigned fitness values based on direct interactions with other individuals. Of particular interest to us are cooperative coevolutionary algorithms (CCEAs), in which interacting individuals are rewarded when they perform well together as a team, and punished when they perform poorly.

At first blush, it would seem that CCEAs may do well on large domain spaces with certain structural properties among interacting components. The intuition behind this advantage is that the algorithm searches only projections of the space at any given time, thus presenting a narrower search domain in a particular generation. Unfortunately, though CCEAs search only a projection of the problem at a time, that projection is constantly changing. The result is that it is easy for the algorithm to get tricked by misleading information provided by poor samples of the projected space. This leads to algorithms that tend to prefer strategies in one population that will do well against many strategies in the other population(s) (so-called *robust resting balance*), whether or not the strategy is globally optimal.

Understanding this particular weakness of CCEAs has led to a very obvious mechanism for correcting it: bias the algorithm to seek ideal collaborations rather

K. Deb et al. (Eds.): GECCO 2004, LNCS 3102, pp. 573–584, 2004.

than robust resting balance. Early investigation into this idea has proved fruitful [1], but this study left at least two things unexplored. First, applying any kind of bias toward ideal collaboration would require learning that bias as the run proceeds, and it is unclear how best to do this. Second and specifically pertinent to this paper, it is not clear how sensitive the degree of bias is to various settings. Large amounts of bias may be unwise if ideal-partnership estimates are poor. But depending on problem sensitivity, small amounts of bias may have almost no effect. A smooth, relatively insensitive biasing procedure would then be a useful part of a successful cooperative coevolution technique.

The paper continues with a brief introduction to cooperative coevolution and to pathologies that may prevent it from achieving optimal results, followed by a description of our previous approach to improving coevolutionary search. We then perform a sensitivity study on a simplified version of the method, and show that results are greatly affected by problem domain properties. Next, we propose an alternative approach to biasing the coevolutionary search, and show that it is less sensitive to algorithm and problem features.

2   Background

In the Potter model of cooperative coevolution [2], which we use in this paper, each population contains individuals that represent a particular component of the problem, so that one member from each population is needed in order to assemble a complete solution. Evaluation of an individual from a particular population is performed by assembling the individual with collaborating partners from other populations. To combat noise in the evaluation process due to choice of partners, multiple evaluations are usually performed. An individual's fitness could be the maximum over such evaluations, or the mean, among other approaches [3]. Aside from evaluation, the populations are evolved independently. Applications of this method include optimization of inventory control systems [4], learning constructive neural networks [5], and rule learning [6,7].

Suppose we are optimizing a three argument function $f(x, y, z)$. One might assign individuals in the first population to represent the x argument, the second to represent y, and the third to represent z. Each population is evolved separately, except that when evaluating an individual in some population (e.g., x), collaborating representatives must be chosen from the other populations (e.g., y and z) in order to obtain an objective function value with a complete solution, $f(x, y, z)$. A simple example collaboration method is to choose representing members by using the most fit individual from those populations as determined by the previous round of evaluations. Another approach is to pick partners at random from the other population. Once a complete solution is formed, it can be evaluated and the resulting score can be assigned to the individual.

Unfortunately, as is often the case for coevolution, complications in the dynamical behaviors of these algorithms often arise in application. The algorithms frequently perform poorly on what seem to be relatively simple problems, falling into wells of mediocrity or collapsing prematurely due to sudden asymmetric losses in population diversities [8,9]. These problems have led researchers to focus on more sophisticated collaboration methods to attempt to address these pathologies [10,11,12].

Recent formal and empirical analysis of the CCEA has suggested a pathology special to CCEAs: *relative overgeneralization* [8]. This research states that cooperative coevolutionary systems are more attracted to points of robust resting balance than to ideal collaboration. From a game-theoretic viewpoint, this tendency can be described as an attraction to the Nash equilibria with the highest joint reward distributions, and not necessarily to those equilibria associated with the global optimum. In short, existing cooperative coevolutionary algorithms are not necessarily well-suited for static, single-objective optimization. Though no real analytical work exists for CCEAs on dynamic optimization problems, the same pathology will very likely confound CCEAs on these tasks, as well.

One straightforward approach to dealing with some pathologies is to bias the CCEA toward searching for ideal collaborations. This idea has been posited and explored at cursory level, with very encouraging results [1]. This existing study identifies bias rate sensitivity as a concern, but does not explore it.

In this paper we will perform this biasing by using the *actual* ideal collaborator for a given individual: in the problem space being tested we can compute this collaborator trivially. Of course, for most real-world problems this is hardly the case, and in the worst case the actual ideal collaborator can only be found via a full search over the individual's collaboration space. Thus in most cases one would need some approximation function for collaborators. In future work we hope to examine a variety of methods for approximating the ideal collaborator: for example, keeping histories of previous collaboration success, or performing small hill-climbing searches for a collaborator. But for purposes of this paper we concern ourselves with the theoretical bound: even *if* you knew the ideal collaborator, what problems in bias sensitivity still arise, and how might one go about solving them?

3   Biasing the CCEA

The biasing method employed here is relatively simple: assess the objective function value is a weighted sum of two terms. The first term is the immediate reward received when the individual component is teamed with partners from other populations chosen using some collaboration scheme. The second term is the reward that the individual would have received had it interacted with its ideal collaborators[1].

Our focus in this paper is on the fraction of reward due to each term and the factors influencing how this fraction is determined. Panait *et al* [1] propose

---

[1] Fitness for the individual involves teaming the individual with one or more partners, computing the objective function value in each case, and assigning a single fitness score in the manner stipulated by the collaboration scheme. The combination of the biased and interactive pieces of the objective function is orthogonal to the issue of combining objective function values when using multiple collaborators

a fitness assessment method that is a linear, weighted sum of these two components. Equation 1 below describes this idea mathematically. Here the fitness of argument x_a is being assessed by combining the result of the objective function, g, with an estimation of an ideal projection of that function, g'_a. The parameter δ controls the degree of emphasis the fitness places on the estimate of the ideal.

$$f(x_1, \ldots, x_a, \ldots, x_k) = (1 - \delta) \cdot g(x_1, \ldots, x_a, \ldots, x_k) + \delta \cdot g'_a(x_a) \qquad (1)$$

The intuitive high-level picture for this tradeoff is clear. At one extreme, when $\delta = 1$, the algorithm will trust only the estimate, and the states of the other populations are entirely irrelevant. It is no longer a coevolutionary system at all; it is k EAs searching the k-projected component spaces independently in parallel. This would be ideal if the estimate were completely trustworthy, though this will not typically be the case. At the other extreme, $\delta = 0$, the algorithm trusts only the collaborative assessment of the objective function. This is the traditional CCEA, with all the challenges that system faces.

Unfortunately, finding a good setting of δ is not an easy task. First of all, the coevolutionary algorithm may have very poor estimates for the ideal partnership. The result of high values of δ and such poor estimates could result in significantly reduced performance. Second, even supposing that such estimates are appropriately learned, it seems intuitive that different levels of δ may be appropriate at different stages in the coevolutionary search. This may be accomplished by either predefined or adaptive methods for setting it. Initially, the biased coevolutionary system does not have much information about the search space. As δ reflects the confidence in the accuracy of the estimated ideal partners, its initial values may be set relatively low. However, as the search progresses, these estimations get more and more accurate, and δ might be adaptively adjusted to higher values to help guide the search process to promising areas.

In this paper we use a simple, static value for δ throughout a run and focus our attention on how different static values affect final run-time performance. Ideally, we would like changes in δ to result in a smooth change in performance. Sadly, this is not guaranteed. The problem is that the effect of changing δ can be highly sensitive to domain or other parameters in a nonlinear fashion. When the transition in performance is very smooth and gradual, it should be easier to find appropriate settings; however, when the transition is sharp and sudden, setting δ appropriately may be quite difficult.

4   Problem and Algorithm Properties

We construct a class of problem domains designed to illustrate certain salient properties of the EC algorithm: the *Maximum of Two Quadratics* (or MTQ) can offer a range from simple to very difficult problem instances. Such problems instances can resolve or exacerbate the relative overgeneralization pathology of cooperative coevolution by adjusting the relative contributions of joint rewards implicitly [8]. It does this by providing a class of maximization problems with two peaks defined as

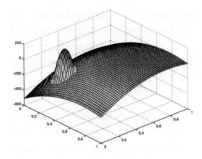 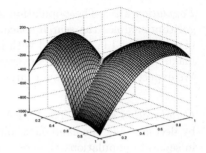

Fig. 1. Example *Maximum of Two Quadratics* problems, illustrating $(S_1 = 1.6, S_2 = 0.5)$, and $(S_1 = 0.5, S_2 = 0.5)$.

$$MTQ(x,y) \leftarrow \max \begin{cases} H_1 * (1 - \frac{16*(x-X_1)^2}{S_1} - \frac{16*(y-Y_1)^2}{S_1}) \\ H_2 * (1 - \frac{16*(x-X_2)^2}{S_2} - \frac{16*(y-Y_2)^2}{S_2}) \end{cases} \qquad (2)$$

where x and y take values ranging between 0 and 1. Figure 1 illustrates some example MTQ problem instances. Different settings for H_1, H_2, X_1, Y_1, X_2, Y_2, S_1 and S_2 affect the difficulty of the problem domain in one of the following aspects:

Peak height: H_1 and H_2 affect the heights of the two peaks. Higher peaks may increase the chances that the algorithm converges there.

Peak coverage: S_1 and S_2 affect the area that the two peaks cover: higher values for one of them result in a wider coverage of the specific peak. This makes it more probable that the coevolutionary search algorithm will converge to this peak, even though it may be suboptimal.

Peak relatedness: Different values for X_1, Y_1, X_2 and Y_2 result in changes in the locations of the centers of the two quadratics, which also affect the relatedness of the two peaks: similar values of the x or y coordinates for the two centers imply higher overlaps of the projections along one or both axes.

Aside from the impact of the problem domain properties, our sensitivity study targets three algorithmic settings:

Biasing rate: Altering the biasing rate allows us to study the result of CCEA runs at various fixed rates of $\delta \in [0,1]$. We are interested in how performance degrades during this transition.

Population size: The population size affects how much the coevolutionary algorithm samples the search space. It is easy to show that coevolution using a bias rate δ of 1.0, combined with infinite populations and perfect knowledge of maximal projections, will converge to the unique optimum with probability 1. We expect similar results for large populations as well.

Collaboration scheme: The CCEA algorithm tries to simplify the search process by decomposing the candidate solutions into components and coevolving them in separate populations. The only information such a population can get about the overall progress of the search process is through collaborators – samples usually representative of the status of the other populations. For some spaces, an increased number of collaborators may better capture the intricacies of the search space [10,12].

5   Bias Sensitivity

All experiments used the MTQ class of problems. The coevolutionary search process used two populations, one for each of the variables. Each such population used a real-valued representation, with individuals constrained to values between 0 and 1 inclusive. Non-adaptive Gaussian mutation (mean 0 and standard deviation 0.05) was the only variational operator. Each population used tournament selection of size 2, and the best individual survived automatically to the next generation. The search lasted for 50 generations, after which time the best individuals in each population were at, or very near, one of the two peaks. Each point in Figures 2-5 is computed over 250 independent runs. All experiments were performed with the ECJ system [13].

Unless stated otherwise, each population consisted of 32 individuals. The default collaboration scheme used two collaborators from each population: the best individual in the previous generation was always selected, and the other individual was chosen at random. The objective component of an individual's fitness was assessed as the better of its results when teamed with each of the collaborators. The default values of the parameters for the first (suboptimal) peak were $H_1=50$, $X_1=\frac{1}{4}$, $Y_1=\frac{1}{4}$, and $S_1=1.6$. The second (optimal) peak was characterized by $H_2=150$, $X_2=\frac{3}{4}$, $Y_2=\frac{3}{4}$, and $S_2=\frac{1}{32}$. With these settings, the two peaks are near at opposite corners of the search space.

5.1   Biasing and Domain Features

The first set of experiments investigated the relationship between the biasing rate and the three problem domain features previously described: the relative heights, coverages and locations of the peaks. There are 11 experimental groups for each property, one for each value of $\delta \in [0, 1]$ in increments of 0.1. Figure 2 shows the mean final results of these 33 groups.

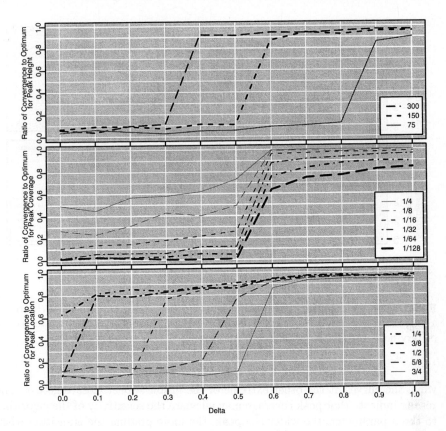

Fig. 2. Convergence ratios for peak height (top), peak coverage (center) and peak relatedness (bottom). x axis shows biasing rate δ, and y axis shows ratio of the 250 trials that converged to, or very near, the global optimum.

Peak height: We kept H_1 constant at 50, and set H_2 to 75, 150 and 300. The results indicate that less than 10% of runs converge optimally when the rate of biasing is low, while the ratio increases to more than 90% when using high biasing rates. Unfortunately, there is no smooth transition between these two extremes: rather, small modifications to the biasing rate may change the rate of convergence to optimum by as much as 70–80%. Moreover, the relative difference in peak height directly affects where these sudden jumps in performance appear. This suggests that δ may not only be quite sensitive to relative differences in peak height in the sense that small changes may have radical effects, but it may even be quite difficult to find the transition range of δ values at all.

Peak coverage: S_2 was set to $\frac{1}{128}$, $\frac{1}{64}$, $\frac{1}{32}$, $\frac{1}{16}$, $\frac{1}{8}$, and $\frac{1}{4}$, while S_1 was constantly 1.6. Here, the location of transition is more consistent among the various values, but the transitions themselves are still relatively abrupt. It also appears that the relative peak coverages cause more variation in results when the bias rate is small,

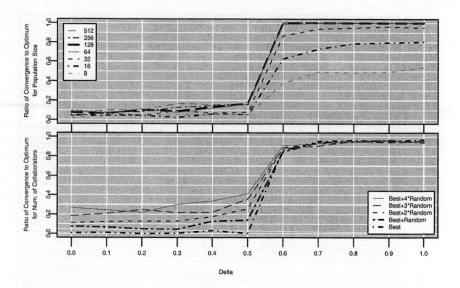

Fig. 3. Convergence ratios for population size (top) and collaboration scheme (bottom). x axis shows biasing rate δ, and y axis shows ratio of the 250 trials that converged to, or very near, the global optimum.

while the curves at the other extreme of the graph appear close together. The results indicate that peak coverage may alleviate the sensitivity of the algorithm to the δ parameter: the wider the peak, the more gradual the transition when varying the bias rate.

Peak relatedness: The bottom graph in Figure 2 shows the ratio of runs converged to optimum when varying the relative location of the two peaks: the Y_2 parameter was set to $\frac{1}{4}$, $\frac{3}{8}$, $\frac{1}{2}$, $\frac{5}{8}$ and $\frac{3}{4}$. These settings gradually transitioned the relative peak positions from completely opposite locations to ones aligned along one axis. Similar to peak height, the peak relatedness had a significant effect on the ratio of runs converged to optimum: the more related the peaks, the less biasing is required to assure good performance. However, the curves have an abrupt transition between lower and higher rates of convergence to optimum. Moreover, the location of this transition depends on the actual degree of peak relatedness, which suggests that δ may be highly sensitive to this parameter.

5.2   Biasing and Algorithm Features

A second set of experiments investigated the relationship between the biasing rate and the two other algorithm features previously described: the population size and the collaboration scheme. Again, there are 11 groups for each of these two parameters corresponding to each of the δ settings. The results are presented in Figure 3.

Population size: We set the size of each of the two populations to 8, 16, 32, 64, 128, 256 and 512. As expected, extremely small populations have a smaller chance of reaching the optimum even for high biasing rates. We observe the same abrupt shift in performance as we saw in the previous experiments. The results suggest that increasing the population size does not necessarily alleviate δ sensitivity.

Collaboration scheme: The default setting in all previous experiments used two collaborators to evaluate the fitness of each individual: the best performing individual from the other population in the previous generation, and also a random individual. To test sensitivity to this collaboration scheme, we varied the number of random individuals from 0 to 4; the best individual from each population in the previous generation was always used. The bottom graph in Figure 3 shows that the collaboration scheme has some influence over the performance of the algorithm at low biasing rates, but it has no effect when higher biasing rates are used. Again, the abrupt change in performance indicates that the δ parameter can be highly sensitive, regardless of the collaboration methodology.

6 A Probabilistic Biasing Alternative

In the previous section, we learned that the degree of bias rate, δ, can be very sensitive to certain problem properties, such as differences in relative peak height, while being less sensitive for others, such as peak coverage. Unfortunately, we also learned that certain coevolutionary algorithmic parameters, such as the size of the populations and the number of collaborators used during evaluation, may not be very helpful in making the task of setting δ more feasible. What can be done at the algorithm level to make δ more flexible?

To uncover a possible alternative, let's look again at the results of the peak height experiments. Recall that for larger differences in peak heights, a wider range of biasing rates results in a high ratio of convergence; however, when one peak is only slightly higher than the other, the range of high convergence ratios is much smaller. The transition is abrupt, but the location of the transition shifts depending on the peak height differences. Why might this be so?

Our hypothesis is that this extreme sensitivity of the biasing method to the relative peak heights is caused by the linear combination of the two fitness components: the fitness when teamed with collaborators, and the one when in combination with the ideal teammates. The higher the optimal peak, the lower the bias rate it needs to dominate the other term. However, if one peak is slightly higher than the other, the algorithm requires more biasing to locate the optimum.

We propose a new alternative biased coevolutionary algorithm to deal with this issue. The new algorithm does not combine the two components into a unique fitness. Rather, each individual is assigned two fitnesses: the objective one when combined with the collaborators from other populations, and another one indicating the performance of the individual when in combination with its ideal teammates. When comparing two individuals, with probability δ we will compare based on the first "fitness"; else we will compare based on the second.

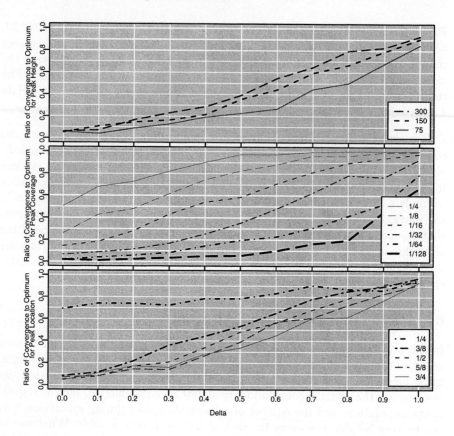

Fig. 4. Convergence ratios for probabilistic biasing when varying peak height (top), peak coverage (center) and peak relatedness (bottom). x axis shows biasing rate δ, and y axis shows ratio of the 250 trials that converged to, or very near, the global optimum.

We performed the same sensitivity analysis for the new algorithm. The results are presented in Figures 4 and 5. In all cases, the new algorithm does not exhibit the sudden jumps in performance that the original one did. This suggests that it is an improved algorithm that is significantly less sensitive to the settings we have investigated.

7 Conclusions and Future Work

Cooperative coevolution remains a powerful tool for a wide variety of problems. Unfortunately, it is fraught with difficulties, and understanding the reasons for these difficulties may help to improve CCEA algorithms for optimization. Biasing the CCEA toward optimal collaboration appears to be just such an improvement: it directly addresses one of the fundamental dynamical problems with cooperative coevolution. Still, it is important to understand how this augmenta-

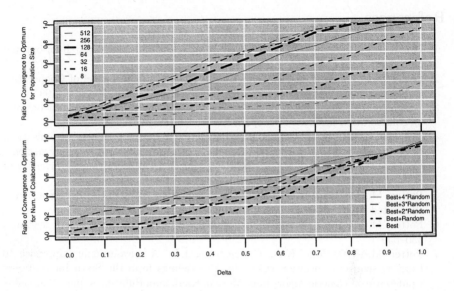

Fig. 5. Convergence ratios for probabilistic biasing when varying population size (top) and collaboration scheme (bottom). x axis shows biasing rate δ, and y axis shows ratio of the 250 trials that converged to, or very near, the global optimum.

tion works—what its benefits and drawbacks are. In particular, since the biasing method relies on a parameter that specifies the degree to which the algorithm trusts the bias over the collaborative interactions, understanding the nature and impact of this parameter is of paramount concern.

This paper takes a sizable step towards understanding how this δ parameter influences the biased CCEA by investigating its sensitivity. Starting with a model that combines traditional collaborative interaction and an optimization bias, we assume that the maximal projection of the problem is known and use this for the bias. The MTQ problem class helps elicit information about the relationship between certain problem properties and algorithmic parameters. Using this problem, we discovered that applying δ as a means to control the linear combination of the two measures results in high sensitivity to a variety of problem parameters. For two of those parameters, the location of δ sensitivity varied widely. This makes setting or adjusting δ a real challenge.

We offer a probabilistic alternative to the linear combination of bias and collaborative fitness assessment that seems to allow for more flexibility in the δ parameter. The result is an improved method that is much less sensitive to the minor differences in the degree of bias.

Our next step is to investigate tractable methods for learning estimations of ideal partners. In addition, we would like to study how δ might be dynamically or adaptively altered during the run to help CCEAs search more efficiently. Our hope is that a biased CCEA will help focus the power of cooperative coevolution more appropriately for optimization tasks.

References

1. Panait, L., Wiegand, R.P., Luke, S.: (Improving coevolutionary search for optimal multiagent behaviors) 653–658
2. Potter, M.: The Design and Analysis of a Computational Model of Cooperative CoEvolution. PhD thesis, George Mason University, Fairfax, Virginia (1997)
3. Wiegand, R.P., De Jong, K.A., Liles, W.C.: (The effects of representational bias on collaboration methods in cooperative coevolution) 257–268
4. Eriksson, R., Olsson, B.: Cooperative coevolution in inventory control optimisation. In Smith, G., Steele, N., Albrecht, R., eds.: Proceedings of the Third International Conference on Artificial Neural Networks and Genetic Algorithms, University of East Anglia, Norwich, UK, Springer (1997)
5. Potter, M., De Jong, K.: Cooperative coevolution: An architecture for evolving coadapted subcomponents. Evolutionary Computation **8** (2000) 1–29
6. Potter, M., De Jong, K.: (The coevolution of antibodies for concept learning) 530–539
7. Potter, M.A., De Jong, K.A., Grefenstette, J.J.: A coevolutionary approach to learning sequential decision rules. In: Proceedings from the Sixth International Conference on Genetic Algorithms, Morgan Kaufmann Publishers, Inc. (1995) 366–372
8. Wiegand, R.P.: An Analysis of Cooperative Coevolutionary Algorithms. PhD thesis, George Mason University, Fairfax, Virginia (2004)
9. Ficici, S., Pollack, J.: Challenges in coevolutionary learning: Arms–race dynamics, open–endedness, and mediocre stable states. In et al, A., ed.: Proceedings of the Sixth International Conference on Artificial Life, Cambridge, MA, MIT Press (1998) 238–247
10. Wiegand, R.P., Liles, W., De Jong, K.: (An empirical analysis of collaboration methods in cooperative coevolutionary algorithms) 1235–1242
11. Bull, L.: On coevolutionary genetic algorithms. Soft Computing **5** (2001) 201–207
12. Bull, L.: (Evolutionary computing in multi-agent environments: Partners) 370–377
13. Luke, S. ECJ 9: A Java EC research system. http://www.cs.umd.edu/projects/plus/ec/ecj/ (2002)

A Population-Differential Method of Monitoring Success and Failure in Coevolution

Ari Bader-Natal and Jordan B. Pollack

DEMO Lab, Brandeis University, Waltham, MA 02454 USA
{ari,pollack}@cs.brandeis.edu

Abstract. Coevolutionary algorithms require no domain-specific measure of objective fitness, enabling these algorithms to be applied to domains for which no objective metric is known or for which known metrics are too expensive. But this flexibility comes at the expense of accountability. Past work on monitoring has focused on measuring *success*, but has ignored *failure*. This limitation is due to a common reliance on "best-of-generation" (BOG) based analysis [1], and we propose a population-differential analysis based on an alternate "all-of-generation" (AOG) framework that is not similarly limited.

Coevolutionary analysis based on *generation tables* was introduced by Cliff and Miller as *CIAO data* [2]. In dual-population coevolution, the table's rows are assigned to the first population's generations, and columns to the second population. Internal entries contain a best-vs-best evaluation of the intersecting generations. This BOG approach appears particularly problematic for two reasons. First, analysis varies depending on the definition of "best" (within a population), but this definition has become arbitrarily fixed on the *Last Elite Opponent* criterion [3], while alternate definitions are equally viable. The coevolutionary algorithm under examination may itself define "best" differently (e.g. Pareto coevolution as "on the Pareto front") in which case LEO is inappropriate. Second, while BOG-based analysis may give useful insight into algorithmic dynamics of successful individuals (i.e. the "best",) it provides little about the population as a whole (i.e. the "rest",) and is therefore blind to many failures.

For an "all-of-generation" alternative, rather than identifying the "best" member of both populations and recording the outcome of their interaction, AOG records the outcome of all interactions between every pairs of individuals from the two populations, respectively. In the data provided below, we implement this population-grained evaluation $PEval$ as an averaging of all individual evaluations (each of which is either *win, tie,* or *lose*, which is denoted numerically as 1, 0, and -1, respectively. Next we construct the *population-differential analysis* measure, based on the insight that the progression of candidate generations ought to perform better over time with respect to a fixed test generation (and vice versa) if successful. First we define a single distinction with the population comparators (between current generation i and oldest generation in memory, j). We then collect all available such comparisons at each (where o is the oldest known generation) with the candidate and test performance metrics.

K. Deb et al. (Eds.): GECCO 2004, LNCS 3102, pp. 585–586, 2004.

Definition 1.

$$PC_{T_k}(C_i, C_j) = \begin{cases} 1, & \text{if } a > b \\ 0, & \text{if } a = b \\ -1, & \text{if } a < b \end{cases} \text{ and } PC_{C_k}(T_i, T_j) = \begin{cases} 1, & \text{if } c < d \\ 0, & \text{if } c = d \\ -1, & \text{if } c > d \end{cases}$$

where $i > j$, C are candidate generations, T are test generations, $a = PEval(C_i, T_k)$, $b = PEval(C_j, T_k)$, $c = PEval(C_k, T_i)$ and $d = PEval(C_k, T_j)$. The *PC-Performance* graphs displayed are simply the average of this $CPerf$ and $TPerf$.

Definition 2. $CPerf_i = \dfrac{\sum\limits_{T_k \in T} PC_{T_k}(C_i, C_o)}{|T|}$ and $TPerf_i = \dfrac{\sum\limits_{C_k \in C} PC_{C_k}(T_i, T_o)}{|C|}$

As evident in the graphs below, the population-differential analysis is able to the closely mirror behavior of an exteral evaluation of performance.

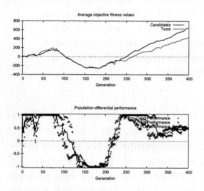

Fig. 1. Fitness-proportional coevolution on *intransitive numbers game* domain [4]

Fig. 2. AOG data from same simulation.

References

1. Ficici, S.G., Pollack, J.B.: A game-theoretic memory mechanism for coevolution. In E. Cantú-Paz et. al., ed.: Genetic and Evolutionary Computation – GECCO-2003. Volume 2723 of LNCS., Chicago (2003) 286–297
2. Cliff, D., Miller, G.F.: Tracking the red queen : Measurements of adaptive progress in co-evolutionary simulations. LNCS **929** (1995) 200–218
3. Sims, K.: Evolving 3d morphology and behavior by competition. In Brooks, R.A., Maes, P., eds.: Proceedings of the 4th International Workshop on the Synthesis and Simulation of Living Systems *ArtificialLifeIV*, Cambridge, MA, USA, MIT Press (1994) 28–39
4. Watson, R.A., Pollack, J.B.: Coevolutionary dynamics in a minimal substrate. In Lee Spector et. al., ed.: Proceedings of the 2001 Genetic and Evolutionary Computation Conference, Morgan Kaufmann (2001)

Cooperative Coevolution Fusion for Moving Object Detection

Sohail Nadimi and Bir Bhanu

Center for Research in Intelligent Systems
University of California, Riverside, 92521
{sohail, bhanu}@cris.ucr.edu

Abstract. In this paper we introduce a novel sensor fusion algorithm based on the cooperative coevolutionary paradigm. We develop a multisensor robust moving object detection system that can operate under a variety of illumination and environmental conditions. Our experiments indicate that this evolutionary paradigm is well suited as a sensor fusion model for different sensing modalities.

1 Introduction

A moving object detection system attempts to detect and distinguish objects such as moving pedestrians, vehicles, animals, motorcycles, bicyclists, and generally animated objects in a scene. The scene is assumed to be static and is generally referred to as background. Moving object detection systems generally use a single sensing modality such as a video camera, a near infrared camera, or a thermal infrared (IR) camera. The assumption of the static background is generally invalid for outdoor scenes. Unfortunately no single modality can operate under all environmental and illumination conditions. For example, during the nighttime, without active illumination, video cameras are useless, and during high noon in warm places, temperatures of certain background objects such as asphalts and concrete can reach as high as moving objects in the scene which makes the task of distinguishing objects difficult.

We overcome the above problems by combining multiple sensors in a seamless fashion. Each sensor operates at a different part of the electromagnetic spectrum and provides advantages that can complement other sensors.

2 Cooperative Coevolution Fusion

In order to detect a moving object in an image, first a background model for each pixel in an image must be constructed. The system uses both collected statistics and physics-based predictions of the signals from a video and thermal IR sensor. For each sensor, we represent each background pixel by a mixture of Gaussian distributions. The statistics are collected by simply observing a scene for a short period of time. During this time, various reflectance and thermal models, are applied to estimate the observed values. The physical models integrate contextual information such as time,

K. Deb et al. (Eds.): GECCO 2004, LNCS 3102, pp. 587–589, 2004.

location, sun's zenith angle, wind velocity, surface absorptivity and emissivity, into the system and provide predicted values for the observation from each sensor.

The degree of agreement between the physics-based predictions and the actual observations is the driving force in the coevolutionary adaptive module to generate the background model that can best describe the scene. The background models are input into a fusion module, which makes the decision (foreground pixel, background pixel) about the pixels in the current frame. This decision, along with a limited past history in turn provides training data for the co-evolutionary module. We have shown that a good estimate of the background can be achieved by observing a scene for a period of time where great deal of statistics can be collected [1].

In our approach, the cooperative coevolutionary algorithm [2], evolves populations of mixture of Gaussians. This evolution is affected by both observations and the physics-based predictions, which embed the environmental variables thorough a fitness function. The algorithm for building the background model starts by collecting some initial sample. Given initial samples, and the environmental conditions, physics-based modules predict both reflectance (for video sensor) and temperature (for IR sensor) of the observed values. The cooperative coevolutionary algorithm is then applied to search for the best representation (mixture of Gaussians) for both sensing modalities.

3 Experimental Results

Figure 1 shows the receiver operating characteristic (ROC) curves. These curves were obtained for two different periods of the day where IR performed better than video in early morning hours and vice-versa in the afternoon. As indicated, the fusion method clearly performed better consistently over the non-fused (single) sensors.

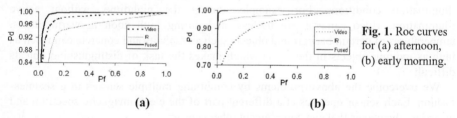

(a)   (b)

Fig. 1. Roc curves for (a) afternoon, (b) early morning.

4 Conclusion

A new sensor fusion method based on cooperative coevolutionary paradigm was introduced for integrating sensors of different modality. Experiments for moving object detection indicate that our fusion approach is robust to many environmental changes.

References

1. S. Nadimi and B. Bhanu, "Multistrategy fusion using mixture model for moving object detection," *Intl. Conf. Multisensor Fusion & Integration for Intelligent Systems*, pp. 317-322, 2001.
2. M.A Potter and K.A. De Jong, "Cooperative coevolution: an architecture for evolving coadapted subcomponents," *Evolutionray Computation*, Vol. 8, No. 1, pp. 1-29, 2000.

Learning to Acquire Autonomous Behavior
— Cooperation by Humanoid Robots —

Yutaka Inoue, Takahiro Tohge, and Hitoshi Iba

Department of Frontier Informatics, Graduate School of Frontier Sciences,
The University of Tokyo, 7-3-1 Hongo, Bunkyo-ku, Tokyo, 113-8656, Japan
{inoue,tohge,iba}@iba.k.u-tokyo.ac.jp

Abstract. In this paper, we describe a cooperative transportation to a
target position with two humanoid robots and introduce a machine learning approach to solving the problem. The difficulty of the task lies on
the fact that each position shifts with the other's while they are moving.
Therefore, it is necessary to correct the position in a real-time manner.
However, it is difficult to generate such an action in consideration of
the physical formula. We empirically show how successful the humanoid
robot HOAP-1's cooperate with each other for the sake of the transportation as a result of Q-learning. Furthermore, we show a result of the
experiment that transports an object cooperatively to a target position
using those robots.

1   Introduction

In this paper, we first clarify the practical difficulties we face from the cooperative
transportation task with two bodies of humanoid robots. Afterwards, we propose
a solution to these difficulties and empirically show the effectiveness both by
simulation and by real robots.

In recent years, many researches have been conducted upon various aspects
of humanoid robots [1][2]. Since humanoid robots have physical features similar to us, it is very important to let them behave intelligently like humans. In
addition, from the viewpoint of AI or DAI (Distributed AI), it is rewarding to
study how cooperatively humanoid robots perform a task just as we humans
can. However, there have been very few studies on the cooperative behaviors of
multiple humanoid robots. Thus, in this paper, we describe the emergence of
the cooperation between humanoid robots so as to achieve the same goal. The
target task we have chosen is a cooperative transportation, in which two bodies
of humanoids have to cooperate with each other to carry and transport an object
to a certain goal position.

As for the transportation task, several researches have been reported on the
cooperation between a human and a wheel robot [3][4] and the cooperation
among multiple wheel robots [5][6]. However, in most of these studies, the goal
was to let a robot perform a task instead of a human.

Research to realize collaboration with a legged robot includes lifting operations of an object with two robots [7] and box-pushing with two robots [8]. How-

K. Deb et al. (Eds.): GECCO 2004, LNCS 3102, pp. 590–602, 2004.
© Springer-Verlag Berlin Heidelberg 2004

ever, few studies have addressed cooperative work using similar legged robots. It is presumed that body swinging during walking renders cooperative work by a legged robot difficult [9]. Therefore, it is more difficult for a humanoid robot to carry out a transportation task, because it is capable of motions that are more complicated and less stable than a usual legged robot.

One hurdle in the case where multiple humanoid robots move carrying an object cooperatively is the disorder of cooperative motion by body swinging during walking. Therefore in this paper, learning is carried out to acquire behavior to correct a mutual position shift generated by this disorder of motion. For this purpose, we use two kinds of methods: (i) Classifier System [10] and (ii) Q-learning [11]. We will show that behavior to correct a position shift can be acquired based on the simulation results of this study. Moreover, according to this result, the applicability to a real robot is investigated. Furthermore, cooperative transportation to a target position is conducted.

This paper is organized as follows. The next section explains the clarified problem difficulties with the cooperative transportation. After that, Section 3 proposes our method to solve the problem. Section 4 presents an experimental result in the simulation and real robots environment. Then section 5 shows an experimental result of cooperative transportation with real robots. Section 6 discusses these results and future researches. Finally, a conclusion is given in Section 7.

2   Problem in Cooperative Transportation

Cooperative transportation by humanoid robots involves solving many difficult problems. It is different from the transportation by a single robot, in which another robot motion is negligible. On the other hand, in case of the cooperative transportation, one robot's motion has an influence on another robot to some extent. Thus, it is necessary to synchronize both robots' motions. However, the synchronization is not easily achieved because precise motions are not expected by humanoids due to the load weight or the floor friction.

We conducted an experiment assuming tasks to transport a lightweight object all around, aiming to extract specific problems from using two humanoid robots: HOAP-1 (manufactured by Fujitsu Automation Limited). Dimensions of a HOAP-1 are 223 x 139 x 483 mm (width, depth, and height) with a weight of 5.9 kg. It has two arm joints with 4 degrees of freedom each, and two leg joints with 6 degrees of freedom each: 20 degrees of freedom in all for right and left.

Actually, when a package is transferred, it seems to be more practical for two robots to have a single object. However, unless both robots move synchronously in the desirable direction, too much load will be given to the arms of robots, which may often cause the mechanical trouble in the arm and the shoulder. It is assumed in experiment that the arm movement can cancel the position shift, and that the distance and angle that can be cancelled would be in the space between two objects.

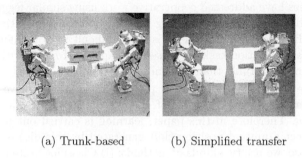

(a) Trunk-based   (b) Simplified transfer

Fig. 1. The target of cooperative transportation.

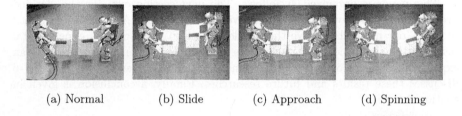

(a) Normal   (b) Slide   (c) Approach   (d) Spinning

Fig. 2. Normal positions and different kinds of positional shifts.

We assume the following task situation (see Fig. 1a): Each robot raises its platform, on which a brick, i.e., a transportation target, is to be placed. However, as a first step, we have removed the target for the sake of simplicity (Fig. 1b). The platform each robot raises is made of foam polystyrene and about 80 gram weigh. The size is about 150 mm wide, 150 mm deep and 200 mm high. This platform is larger than a conventional one because it has to bear the weight of the transportation target. A sponge grip is attached on each robot arm, so that an object would not slip off the arm during the experiment.

The two robots operate in Master-Slave mode. That is, the Master robot transmits data corresponding to each operation created in advance to the Slave robot; the two robots start a motion in the same direction simultaneously.

The experiment of several times was conducted using each motion. The results indicated that unintentional motions such as lateral movement (Fig. 2b) and back-and-forth movement (Fig. 2c) by sliding from the normal position (Fig. 2a), and rotation (Fig. 2d) occur frequently in basic transportation motions such as forward, backward, rightward, and leftward. This is considered mainly to result from swinging during walking and the weight of the object.

The following three factors can be considered the causes of these shifts in motion.

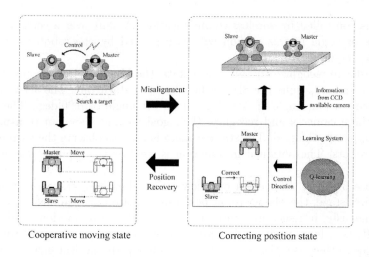

Fig. 3. Steps of the cooperative transportation.

- Swing when the robot moves
- Shift of the center of gravity by having an object
- Initialization error of robot's joint motors

Especially, in case of humanoid robots, we can think of motor vibration due to the body motion as its cause. This may affect the robot's translation or direction. In addition, the gravity change resultant from carrying an object may possibly cause some errors in the movement. When activating a robot, it is necessary to set the initial positions of each joint's motors manually. Thus, setting those initial values wrongly may result in fatal errors.

Such a position shift can be cancelled, if only slight, by installing a force sensor on a wrist and moving arms in the load direction. However, a robot's position must be corrected in case of a shift beyond the limitation of an arm. Improper correction may cause failure of an arm or a shoulder joint and breakage of an object.

3 Approach of Transportation Control

The practical problem of transporting an object is the possibility that a robot falls during movement, due to loss of body balance in connection with a load on the arm by a mutual position shift after moving. Therefore, it is important to acquire behavior for correcting the position shift generated from movement by learning algorithms.

A situation is assumed in which two robots move face to face while maintaining the distance within a range to transport an object stably. This motion can be divided into two stages: one in which the two robots move simultaneously, and one in which one robot corrects its position. Simultaneous movement of two robots is controlled by wireless communication. A shift over a certain limit of distance or angle in this motion will be corrected by one robot according to behavior acquired by learning.

In order to recognize an object or a state, the Master robot is equipped with an active camera, while the Slave robot carries a static one. The active camera works with a pan angle of ± 90[deg] and a tilt angle of ± 90[deg]. The robots rotate these cameras and recognize their goal so that they can transport the target object to the goal. The static camera is used to observe the current state of two robots. The obtained information is used as the input to the learning system.

Figure 3 shows the motion overview for conducting a transportation task. In the first stage, the Master robot performs a motion programmed in advance; simultaneously, it issues directions to perform the same motion to the Slave robot. If there is no position shift after movement, the process forwards to the next stage; otherwise, the position is corrected with the learning system. We have tried to realize a cooperative transportation task by repeating the series of this flow.

4   Learning to Correct Positioning

4.1   Learning Model

The learning for position correction is carried out with Q-learning and Classifier System.

Q-learning guarantees that the state transition in the environment of a Markov decision process converges into the optimal direction [12]. However, it requires much time until the optimal behavior obtains a reward in the early stage of learning. Thus, it takes time for the convergence of learning. Furthermore, because all combinations of a state and behavior are evaluated for a predetermined Q value, it is difficult to follow environmental change. Therefore, leaning by a real robot is extremely difficult because of the processing time.

On the other hand, Classifier System can learn a novel classification and to maintain the diversity by means of GA, which evolves a rule including # (don't care symbol). Thus, it enables the learning with relatively few trials so that the evolved robot may adapt the dynamic environment more effectively. However, too much generalization might result in the poor performance due to the overfitting.

We use these above two methods for the sake of simulation-based learning of the position correction and compare the obtained results.

The effective division of states and the selection of actions are very essential for the sake of efficient Q-learning and Classifier System. A static camera is attached to one robot to obtain information required for learning from the external environment. The external situation is evaluated with images from this static camera. Based on the partner robot's position projected on the image acquired by the static camera, a state space is arranged as shown in Fig. 4. It

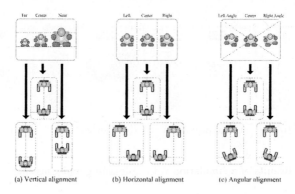

(a) Vertical alignment   (b) Horizontal alignment   (c) Angular alignment

Fig. 4. Different states (27-states).

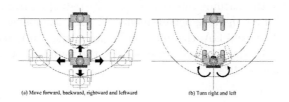

(a) Move forward, backward, rightward and leftward   (b) Turn right and left

Fig. 5. Different actions (6-actions).

is divided into three states: vertical, horizontal, and angular. Hence, the total number of states of the environment is 27. If the vertical, horizontal, and angular positions are all centered, the goal will be attained.

We assumed six behaviors that are especially important: forward, backward, rightward and leftward movement, and right and left turns. Figure 5 depicts all these motions.

4.2   Learning in Simulator

The learning model stated in the preceding subsection has been realized in a simulation environment. This simulator sets a target position at a place of constant distance from the front of the partner robot, which is present in a plane. A task will be completed if the learning robot reaches the position and faces the partner robot.

The target position here ranges in distance movable in one motion. In this experiment, back-and-forth and lateral distances and the rotational angle movable in one motion are assumed to be constant. That is, if the movable distance in one step is about 10 cm back-and-forth and 5 cm laterally, the range of the target point will be 50 cm². In this range, the goal will be attained if the learning robot is in place where it can face the partner robot with one rotation motion.

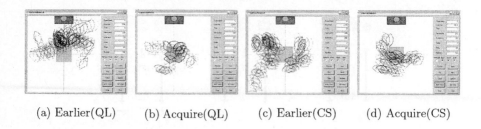

(a) Earlier(QL)   (b) Acquire(QL)   (c) Earlier(CS)   (d) Acquire(CS)

Fig. 6. Results of a simulation with Q-learning and Classifier System.

The Q-learning parameters for the simulation were as follows: the initial Q value, Q_0, was 0.0, the learning rate α was 0.01, the reduction ratio γ was 0.8 and the reward was 1.0 for the task achievement. We used the following parameters for Classifier Systems and GA: the initial value for a rule is 0.1, the tax is 0.001, the bid value is 0.01, the crossover rate is 0.95, the mutation ratio is 0.05, and the population size is 1,024.

4.3   Result of Simulator Learning

Behavior patterns obtained by simulation with the Q-learning approach in the early stage and acquired by learning are shown in Figs. 6a and 6b, respectively. In the early stage, motions are observed such as walking to the same place repeatedly and going to a direction different from the target position. Behavior approaching the target position is gradually observed as learning progresses; finally, behavior is acquired to move to the target position and turn to the front with relatively few motions.

As can be seen Classifier System simulation by in Figs. 6c and 6d, the trajectory divergence occurred at the earlier stage of learning. However, at the later generations, the effective actions were acquired so as to face the goal correctly.

Figure 7a plots the success rate of learning for 1,000 steps. Figure 7b gives the number of successful motions with generations. Both data were averaged over 10 runs. As can be seen, Q-learning is superior. This may be because it enables hill-climbing local search. Classifier System's performance goes up and down irregularly. However, this is considered to show the superiority in terms of the robust learning. As a result of this, numbers of motions are almost the same for both methods as the later stage of learning.

4.4   Experiments with Real Robots

Following the simulation results described in the previous subsection, we conducted an experiment with real robots to confirm their applicability. In this experiment, we have used the learning data obtained from Q-learning because the target task did not necessarily require the real-world experience.

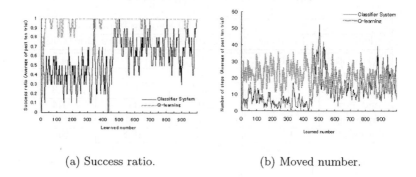

(a) Success ratio.   (b) Moved number.

Fig. 7. Q-learning vs. Classifier System.

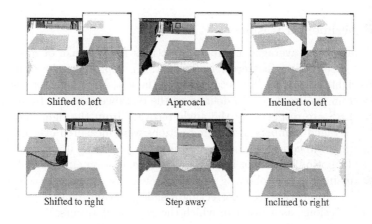

Shifted to left   Approach   Inclined to left

Shifted to right   Step away   Inclined to right

Fig. 8. Type of the experiments.

For the recovery from the horizontal left (right) slide, a humanoid robot was initially shifted leftward (rightward) against the opponent robot by 5.2 cm. On the other hand, it was initially moved forward (backward) from the correct position by 3.2 cm for the recovery from front (back) position. In case of the rotation failure, the robot was shifted either leftward or rightward by 5.2 cm and rotated toward the opponent by 20 degrees. The images of the static camera in each pattern are shown in Fig. 8. The actions used for the recovery were of six kinds, i.e., half forward, half backward, half rightward, half leftward, right turn and left turn.

For this experiment, robots started from one of the three patterns shown in Figs. 2b, 2c and 2d, which were classified as the failure of actions (see section 2). We employed two HOAP-1's, one of which used the learning results, i.e., the acquired Q-table, so as to generate actions for the sake of recovery from

Table 1. Numbers of average movement.

Failure	Recovery	Q-learning Iterations		
		1,000times	10,000times	100,000times
Horizontal	RL	4.6	4.4	4.4
slide	LR	5.4	5.2	5.2
Approach	NF	6.6	1.8	1.8
and away	FN	2.0	2.0	1.4
Spinning	RLS	9.4	9.4	8.6
around	LRS	11.4	16.8	10.2

the failure. Q-learning was conducted by simulation with different numbers of iterations, i.e., 1,000, 10,000, and 100,000 iterations. The learning parameters were the same as in the previous subsection.

4.5   Experimental Results

Table 1 shows the averaged numbers of actions for the sake of recovery from the above three failure patterns. In Table 1: RL represents the slide recovery from the right, LR is the slide recovery from the left, NF stands for the distance recovery from the front, FN is defined as the distance recovery from the back, RLS and LRS are respectively the angle recovery from the right and from the left. The averaged numbers of required actions were measured over five runs for each experimental condition, i.e., with different Q-learning iterations.

For slide motion, the robot learned an effective motion after 1,000 time steps. This is explained in the following way. A gap usually occurs even when a robot corrects a position. However, correcting a slide position requires only a simple sequence of actions, as a result of which the gap rarely occurs.

With 1,000 iterations, more actions were needed to recover from the front position to the back. This is because the robot had acquired the wrong habit of moving leftward when the opponent robot was approaching (see Fig. 9). This habit has been corrected with 10,000 iterations, so that much fewer actions were required for the purpose of repositioning.

The recovery from "spinning around" seems to be the most difficult among the three patterns. For this task, the movement from the slant to the front (see Fig. 10) was observed with 10,000 iterations, which resulted in the increase of required actions. This action sequence was not observed with 1,000 iterations. This is considered that the phenomenon is caused by the difference between simulation and a real-world environment.

5   Cooperative Transportation to Target Position

5.1   Experiments with Real Robots

The cooperative transportation task, i.e., two humanoid robots cooperate with each other to transport an object to a certain goal, is carried out by using

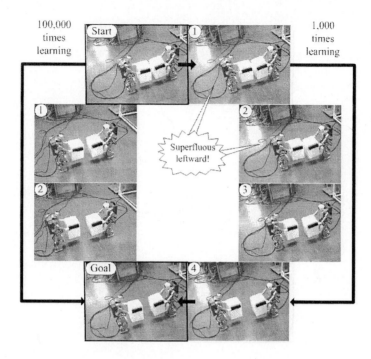

Fig. 9. Behavior of NF with short-time learning and full learning.

the obtained Q-learning data shown in the previous section. The transportation target is a sphere made of foam polystyrene. Its diameter is about 25 cm and 63 gram weigh. The goal is positioned in a place about 1m distant from each humanoid robot and is marked for the purpose of recognition.

The Master robot finds its mark using the active camera, and decides the transportation path to the destination. The path is derived as follows: first, move the Master robot forward or backward so that it is next to the goal; next, move the Master robot left or right to a position adjacent to the mark. In the meantime, if a positional shift occurs, the Slave robot recognizes its type and tries to recover from it. Afterward, the Master robot searches for a new path again and the transportation is restarted according to the new path.

5.2   Experimental Results

Figure 11 shows the transportation process with some recovery actions. As can be seen, two recovery actions were performed in case of side motions. As a result, the robots achieved the task successfully. In case of a position shift, the path to the goal was slightly changed. This was caused by each other's shift and its recovery. In order to reduce this anomaly and re-calculation of the path, two robots need to revise their positions simultaneously.

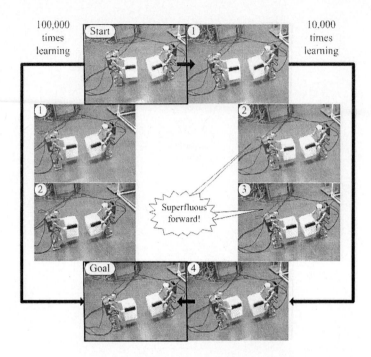

Fig. 10. Behavior of LRS with short-time learning and full learning.

Moreover, when the goal is seen overlapped with the opponent robot, the mark is difficult to recognize. In order to solve this difficulty, two robots should rotate cooperatively with the object on the platform or both robots should be equipped with active cameras for the recognition.

6   Discussion

In a real environment, at the earlier stage of learning, we have often observed the unexpected movement to a wrong direction by real humanoid robots; which was also the case with the simulation. In the middle of learning, the forward movement was more often observed from the slant direction. These types of movements, in fact, had resulted in the better learning performance by simulation, whereas in a real environment they prevented the robot from moving effectively. This is considered to be the distinction between simulation and a real-world environment.

In this paper, the position recovery was carried out by one robot. It is more desirable and efficient if both robots can do so. For this purpose, the learning of two robots in a real environment is essential. This is also important to nullify the difference between simulation and real-world environment. However, it is not easy using Q-learning because of the frequent loss of a goal or an opponent in

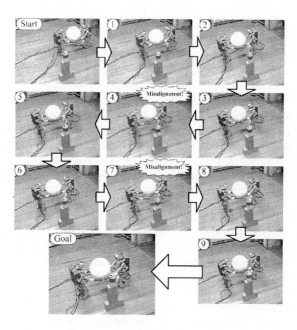

Fig. 11. Result of an experiment with real robots.

the early state of the learning in a real environment. Thus, we can conclude Classifier System is superior to Q-learning for the purpose of the cooperative learning in a real-world environment.

Moreover, we are now developing a methodology of filtering learning result by means of camera information from difference devices, for the purpose of applying the obtained result in a simulator to a real environment. This method is based on the evolutionary computation and probabilistic estimation. In order to solve the difficulty with the distinction, learning in the real world is essential. For this purpose, we are currently working on the integration of GP and Q-learning in a real robot environment [13].

7   Conclusion

Specific problems were extracted in an experiment using a practical system in an attempt to transport an object cooperatively with two humanoid robots. The result proved that both body swinging during movement and the shift in the center of gravity, by transporting an object, caused a shift in the position after movement.

Therefore, we have proposed a learning method to revise a position shift while the cooperative transportation, and established a learning framework in a simulation. In addition, the obtained results were verified by using real robots.

References

1. K. Yokoi, et al., "Humanoid Robot's Application in HRP", In *Proc. of IARP International Workshop on Humanoid and Human Friendly Robotics*, pp.134-141, 2002.
2. H. Inoue, et al., "Humanoid Robotics Project of MITI", *The 1st IEEE-RAS International Conference on Humanoid Robots*, Boston, 2000.
3. O. M. AI-Jarrah and Y. F. Zheng, "Armmanipulator Coordination for Load Sharing using Variable Compliance Control", In *Proc. of the 1997 IEEE International Conference on Robotics and Automation*, pp.895-900, 1997.
4. M. M. Rahman, R. Ikeura and K. Mizutani, "Investigating the Impedance Characteristics of Human Arm for Development of Robots to Cooperate with Human Operators", In *CD-ROM of the 1999 IEEE International Conference on Systems, Man and Cybernetics*, pp.676-681, 1999.
5. N. Miyata, J. Ota, Y. Aiyama, J.Sasaki and T. Arai, "Cooperative Transport System with Regrasping Car-like Mobile Robots", In *Proc. of the 1997 IEEE/RSJ International Conference on Intelligent Robots and Systems*, pp.1754-1761, 1997.
6. H. Osumi, H.Nojiri, Y.Kuribayashi and T.Okazaki, "Cooperative Control of Three Mobile Robots for Transporting A Large Object", In *Proc. of International Conference on Machine Automation (ICMA2000)*, pp.421-426, 2000.
7. M. J. Matarić, M. Nillson and K. T. Simsarian, "Cooperative Multi-robot Box-pushing", In *Proc. of the 1995 IEEE/RSJ International Conference on Intelligent Robots and Systems*, pp.556-561, 1995.
8. H.Kimura and G.Kajiura, "Motion Recognition Based Cooperation between Human Operating Robot and Autonomous Assistant Robot", In *Proc. of the 1997 IEEE International Conference on Robotics and Automation*, pp.297-302, 1997.
9. Y. Inoue, T. Tohge and H. Iba, "Cooperative Transportation by Humanoid Robots - Learning to Correct Positioning -", In *Proc. of the Hybrid Intelligent Systems (HIS2003)*, pp.1124-1133, 2003.
10. L. B. Booker, D. E. Goldberg, and J. H. Holland, "Classifier Systems and Genetic Algorithms", In *Machine Learning: Paradigms and Methods, MIT Press*, 1990.
11. R. S. Sutton and A. G. Barto, "Reinforcement Learning", *MIT Press*, Boston, 1998.
12. C. J. C. H. Watkins and P. Dayan, "Q-learning", *Machine Learning*, Vol. 8, pp.279-292, 1992.
13. S. Kamio, H. Mitsuhashi and H. Iba, "Integration of Genetic Programming and Reinforcement Learning for Real Robots", In *Proc. of the Genetic Computation Conference (GECCO2003)*, pp.470-477, 2003.

Evolved Motor Primitives and Sequences in a Hierarchical Recurrent Neural Network

Rainer W. Paine and Jun Tani

RIKEN Brain Science Institute, Laboratory for Behavior and Dynamic Cognition,
2-1 Hirosawa, Wako-shi, Saitama, 351-0198, Japan
{rpaine, tani}@brain.riken.jp

Abstract. This study describes how complex goal-directed behavior can evolve in a hierarchically organized recurrent neural network controlling a simulated Khepera robot. Different types of dynamic structures self-organize in the lower and higher levels of a network for the purpose of achieving complex navigation tasks. The parametric bifurcation structures that appear in the lower level explain the mechanism of how behavior primitives are switched in a top-down way. In the higher level, a topologically ordered mapping of initial cell activation states to motor-primitive sequences self-organizes by utilizing the initial sensitivity characteristics of nonlinear dynamical systems. A further experiment tests the evolved controller's adaptability to changes in its environment. The biological plausibility of the model's essential principles is discussed.

1 Introduction

It is widely believed that behavior systems develop certain hierarchical or level structures for achieving goal-directed complex behaviors, and that these structures should self-organize through interactions with the environment. It is reasonable to assume that an abstract event sequence is represented in a higher level, while its detailed motor program is generated in a lower level. Arbib [1] proposed the idea of movement primitives (also referred to as perceptual-motor primitives, motor schemas, or motor programs), which are a compact representation of action sequences for generalized movements that accomplish a goal. Evidence of such primitives in animals has been found ([2], [3]), and human studies also indicate their role in complex movement generation [4]. Once such motor primitives develop early in the life of an organism, diverse behaviors can emerge by learning to combine them in a multitude of complex sequences.

In this paper, we study the dynamic adaptation process of multiple levels of continuous time recurrent neural networks (CTRNNs) applied to a navigation task using a simulated robot. Through the experiments, we will demonstrate that a hierarchically organized network can perform well in adapting to complex tasks through combination with a genetic algorithm (GA). We will focus on how motor primitives are self-organized in the lower level, and how they are manipulated in the higher level.

2 Methods

The neural network model utilized in the current paper consists of two levels of fully connected CTRNNs ([5], [6]). The lower level network, as shown in Figure 1, receives sensory inputs and generates motor commands as outputs. This network is supposed to encode multiple sensory-motor primitives, such as moving straight down a corridor, and turning left or right at intersections or to avoid obstacles in the navigation task adopted in this study.

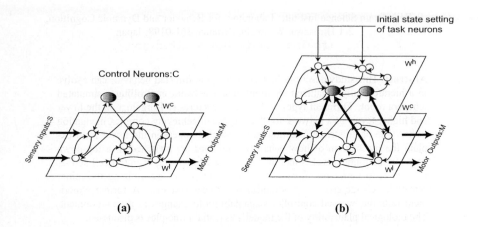

Fig. 1. Conceptual diagram of network architecture

A set of external neural units, called the "control neurons", are bidirectionally connected to all neurons in the lower level network. The control neurons influence lower level network functions and favor the generation of particular motor primitives. Through evolution of both the lower level internal synaptic weights (W_l) and the interface weights (W_c) between the control neurons and the lower level neurons, a mapping between the control neurons' activities and the sensory-motor primitives stored in the lower level network is self-organized. Modulation of the control neurons' activities causes shifts between generating one primitive and another. The scheme is analogous to the idea of the parametric bias [7] and the command neuron concept ([8], [9], [10]). How might more complex tasks, such as navigation in an environment, be generated? Such tasks require generating sequences of motor primitives. We propose that a higher level network may modulate the activities of the control neurons through time to generate sequences of lower level movement primitives (Figure 1b).

The higher level network evolves to encode abstract behavior sequences utilizing the control neurons. It is assumed that the desired sequences will be generated if adequate nonlinear dynamics can be self-organized in the higher level network. As will be described in detail later, the robot becomes able to navigate to multiple goal positions when starting from the same initial position in the maze environment. Therefore, the higher level network must encode multiple sequence patterns which have to be retrieved for the specified goal.

We utilize the initial sensitivity characteristics of nonlinear dynamic systems in order to initiate different sequences. When the robot is placed at the initial position in the environment, the internal values of all the higher level neurons are set to 0.0, except for two neurons called the task neurons (Figure 1b). The initial activity values of the task neurons determine the subsequent turn sequence, and the goal which is found. These goal-specific initial task neuron activities were evolved through the same genetic algorithm that yielded the network's synaptic weights.

3 Experiments

All experiments reported here were executed using a simulated Khepera II robot in the Webots 3 robot simulator (www.cyberbotics.com). The same CTRNN equations, parameters, and genome encoding were used as in [6], with the addition of the initial task neuron activities, $\gamma_{task}(0) \in [-10,10]$, to the genome for experiment 2. A standard GA was employed, with a 2% bit-mutation rate, an 80 robot population with the best 20 reproducing, and no mutation of one offspring of the best parent ([6], [14], [15]).

3.1 Experiment 1: T-Maze Task

Experiment 1 is designed to evolve a bottom level network which contains movement primitives of left and right turning behavior at intersections as well as collision-free straight movement in corridors. The same lower level and control neuron weights are used for both right and left turns. The only difference between the left and right turn controllers is in the bias values (θ) of the two control neurons. Intuitively, this might correspond to different sets of cortical "control" neurons becoming associated with each of the lower level movement primitives. Parallel connections from the bottom level to the control neurons might develop, yielding the same weights to both sets of control neurons. Intrinsic differences in the control neurons' responses (as through the different θ bias values used here) to the lower level signals would determine with which motor primitives each set of control neurons became associated.

The evolutionary runs consisted of up to 200 generations with 2 epochs and 3 trials per robot. Each trial was run for 500 time steps, starting at the same position at the bottom of the T maze. Different bias values (θ) evolved in the control nodes for the left and right turning tasks in epochs 1 and 2, respectively. All other parameters were identical in the left and right turning tasks. In epoch 1, fitness was awarded to robots that turned to the left at the intersection based on the following fitness rule. In epoch 2, fitness was awarded to robots that turned to the right. Each robot ran 3 trials per epoch.

Experiment 1 uses a two-component fitness rule. The first component consists of a reward for straight and fast movements with obstacle avoidance [16]. The second component of the fitness rule rewards the robot for finding a goal. The goal is located to the left of the intersection for epoch 1, and to the right for epoch 2. The robot is linearly rewarded, based on its position, for approaching and reaching the goal. Greater reward per time step is received linearly as the robot approaches the goal, starting at the middle of the top of the T maze.

At the start of each trial, the robot was placed at the same starting position at the bottom of the T maze. Three different starting orientations (facing 135°, 90°, and 45°; that is, left, straight, and right, respectively) were employed, one for each of the three trials

rons' outputs then passed the turn threshold, leading to a premature left turn when the robot finally arrived at the intersection (Figure 6). This result is not surprising, since no direct sensory input reaches the higher level network. It would be interesting to explore in future research the capability of the control neurons, which are connected more closely to sensory signals via their connections with the bottom level network, to modify the higher level activity based on environmental changes.

5 Discussion/Conclusion

The work presented here describes a novel hierarchical model of behavioral sequence memory and generation. It recalls in general terms the hierarchical organization of movements in the primate spinal cord, brainstem, and cortical regions. Different types of dynamic structures self-organize in the lower and higher levels of the network. A parametric bifurcation in the control neurons' interaction with the lower level allows top-down behavioral switching of the primitives embedded in the lower level. Utilizing the initial sensitivity characteristics of nonlinear dynamic systems [17], a topologically ordered mapping of initial task neuron activity to particular behavior sequences self-organizes throughout the development of the network. The interplay of task-specific top-down and bottom-up processes allows the execution of complex navigation tasks.

One unique feature of the current model is the hierarchical organization of the network and its training. The bottom level network represents movement primitives, such as collision avoidance and turning at intersections. Since it must directly deal with quickly changing environmental stimuli, its time constants (τ) have become small through adaptation so that the neuronal activity of the output neurons ($\tau_0 = 1$, $\tau_1 = 1$) can change rapidly to drive the robot's movement in real time. In contrast, the higher level represents sequences of the lower level primitives over longer time spans. Accordingly, the task neuron time constants have adapted to be large ($\tau_{task0} = 70$, $\tau_{task1} = 52$) so that neuronal activity changes much more gradually and is less affected by short-term sensory changes.

The neurons of the higher level receive no direct sensory inputs, but are gradually influenced by them through the control neurons, which are fully connected to the input-receiving bottom level. This system is reminiscent of the organization of sequence generation in primates ([13], [18]). In [13], cellular activity in monkeys' supplementary motor area (SMA) was found to be selective for the sequential order of forthcoming movements, much as the task neurons' initial activities determine future movement order in the current model. In [18], distinct groups of cells in the lateral prefrontal cortices (LPFC) of monkeys were found to integrate the physical and temporal properties of sequentially reached objects, in a manner analogous to integration of higher level sequential information and lower level sensory input by the control neurons in the present model.

Although other models of sequence generation have been trained in a modular fashion because it was felt necessary to achieve the task [5], the current work begins by explicitly evolving simple movement primitives, such as straight movements, collision avoidance, and turning at corners. The next level of the hierarchy subsequently develops to utilize the lower level primitives in complex movement sequences. One can envision further levels of complexity, with higher levels representing sequences of sequences for different sets of tasks, in a manner analogous to the "chunking" phenomenon observed in human memory of data sequences [19]. The beauty of this system is that the synaptic connections need not grow without bound as the number and complexity of sequences increases. As shown here, a *single* network can represent *multiple* complex movements through modulation of the activities of a small number of "task" neurons.

Although the initial sensitivity of the movement sequences generated to task neuron activations was an emergent feature of the system found by self-organization of network parameters through a genetic algorithm, the model architecture was predetermined, and the details of the network training influenced the specific functions that were assumed by different components of the architecture. Given that the current network architecture is loosely based upon the primate motor system's hierarchical design, one might expect it to perform better than a less biologically plausible giant first-order network that encompasses both simple movement primitives as well as their combination into complex sequences. This assumption will be tested in future work.

As seen in the results of experiment 3, one limitation of the current model is its inability to respond effectively to environmental changes without further weight modification. When the length of the corridors was doubled, the robot usually failed to reproduce the turn sequence which had been learned in the smaller environment. The dynamics of the higher level network were essentially independent of the external environment, as if the robot were executing a learned sequence by rote at a fixed speed. When environmental changes prevented the robot from turning at the usual time, the top network activity continued to progress toward the next turn in the sequence, skipping a turn instead of merely delaying it until the next intersection.

The higher level network's influence on the outputs of the lower level is disproportionately large compared to the lower level's influence on it. The higher level's relative isolation from the "real world's" sensory input is in stark contrast to the rich flow of both physical and temporal sensory information which is integrated in the primate lateral prefrontal cortex during the learning of movement sequences [18]. Although both the model's control neurons and primate LPFC neurons integrate both temporal sequence and physical sensory information, the monkey can modulate the speed of its sequence generation, whereas the current model cannot. Future work will therefore explore the possibility of better modulating the activity of the higher level through bottom-up connections in a way which reflects environmental changes.

References

1. Arbib, M. A.: Perceptual structures and distributed motor control. In: Brooks, V. B. (ed.): Handbook of Physiology, Section 2: The Nervous System, Vol. II, Motor Control, Part 1. American Physiological Society (1981) 1449-1480
2. Giszter, S. F., Mussa-Ivaldi, F. A., Bizzi, E.: Convergent force fields organized in the frog's spinal cord. Journal of Neuroscience, 13(2) (1993) 467-491

3. Mussa-Ivaldi, F. A., Giszter, S. F., Bizzi, E.: Linear combination of primitives in verte-brate motor control. Proceedings of the National Academy of Sciences, USA, 91 (1994) 7535-7538

4. Thoroughman, K. A., Shadmehr, R.: Learning of action through combination of motor primitives. Nature, 407 (2000) 742-747

5. Yamauchi, B., Beer, R.D.: Sequential behavior and learning in evolved dynamical neural networks. Adaptive Behavior, 2(3) (1994) 219-246

6. Blynel, J., Floreano, D.: Levels of dynamics and adaptive behavior in evolutionary neural controllers. In: Hallam, B., Floreano, D., Hallam, J., Hayes, G., Meyer, J.A. (eds.): From Animals to Animats 7: Proceedings of the Seventh International Conference on Simulation of Adaptive Behavior. MIT Press, Bradford Books, Cambridge, MA (2002)

7. Tani, J.: Learning to generate articulated behavior through the bottom-up and the top-down interaction processes. Neural Networks, 16(1) (2003) 11-23

8. Aharonov-Barki, R., Beker, T., Ruppin, E.: Spontaneous Evolution of Command Neurons, Place Cells and Memory Mechanisms in Autonomous Agents. In: Advances in Artificial Life, ECAL '99, Vol. 1674. Lecture Notes in Artificial Intelligence, Springer Verlag (1999)

9. Edwards, D. H., Heitler, W. J., Krasne, F. B.: Fifty years of a command neuron: the neu-robiology of escape behavior in the crayfish. Trends in Neurosciences, 22(4) (1999) 153-161

10. Teyke, T., Weiss, K. R., Kupfermann, I.: An identified neuron (CPR) evokes neuronal re-sponses reflecting food arousal in Aplysia. Science, 247, (1990) 85-87

11. Nishimoto, R., Tani, J.: Learning to Generate Combinatorial Action Sequences Utilizing the Initial Sensitivity of Deterministic Dynamical Systems. Proc. of the 7th International Work-Conference on Artificial and Natural Neural Networks (IWANN'03). Springer, (2003) 422-429

12. Blynel, J.: Evolving Reinforcement Learning-Like Abilities for Robots. In: Tyrrell, A, Haddow, P.C., and Torresen, J. (eds.): Evolvable Systems: From Biology to Hardware: 5th International Conference, ICES (2003)

13. Tanji, J., Shima, K.: Role for supplementary motor area cells in planning several move-ments ahead. Nature, 371 (1994) 413-416

14. Mitchell, M.: An Introduction to Genetic Algorithms. MIT Press, Cambridge, MA (1998)

15. Goldberg, D.E.: The Design of Innovation: Lessons from and for Competent Genetic Al-gorithms. Kluwer Academic Publishers, Boston, MA (2002)

16. Floreano, D., Mondada, F.: Automatic creation of an autonomous agent: genetic evolution of a neural-network driven robot. In: Cliff, D., Husbands, P., Meyer, J., Wilson, S.W. (eds.): From Animals to Animats 3: Proceedings of the Third Conference on Simulation of Adaptive Behavior. MIT Press, Bradford Books, Cambridge, MA (1994)

17. Fan, J., Yao, Q., Tong, H.: Estimation of Densities and Sensitivity Measures in Nonlinear Dynamical Systems. Biometrika, 83(1) (1996) 189-206

18. Ninokura, Y., Mushiake, H., Tanji, J.: Representation of the Temporal Order of Visual Objects in the Primate Lateral Prefrontal Cortex. Journal of Neurophysiology, 89 (2003) 2868-2873

19. Sakai, K., Kitaguchi, K., Hikosaka, O.: Chunking during human visuomotor sequence learning. Experimental Brain Research, 152(2) (2003) 229-242

Robot Trajectory Planning Using Multi-objective Genetic Algorithm Optimization

E.J. Solteiro Pires[1], J.A. Tenreiro Machado[2], and P.B. de Moura Oliveira[1]

[1] Universidade de Trás-os-Montes e Alto Douro, Dep. de Engenharia Electrotécnica,
Quinta de Prados, 5000–911 Vila Real, Portugal,
{epires,oliveira}@utad.pt, http://www.utad.pt/~epires
http://www.utad.pt/~oliveira
[2] Instituto Superior de Engenharia do Porto, Dep. de Engenharia Electrotécnica,
Rua Dr. António Bernadino de Almeida, 4200-072 Porto, Portugal
jtm@dee.isep.ipp.pt, http://www.dee.isep.ipp.pt/~jtm

Abstract. Generating manipulator trajectories considering multiple objectives and obstacle avoidance is a non trivial optimization problem. In this paper a multi-objective genetic algorithm is proposed to address this problem. Multiple criteria are optimized up to five simultaneous objectives. Simulations results are presented for robots with two and three degrees of freedom, considering two and five objectives optimization. A subsequent analysis of the solutions distribution along the converged non-dominated Pareto front is carried out, in terms of the achieved diversity.

1 Introduction

In the last twenty years genetic algorithms (GAs) have been applied in a plethora of fields such as: control, system identification, robotics, planning and scheduling, image processing, pattern recognition and speech recognition [1]. This paper addresses the planning of trajectories, meaning the development of an algorithm to find a continuous motion that takes the manipulator from a given starting configuration to a desired end position in the workspace without colliding with any obstacle.

Several single-objective methods for trajectory planning, collision avoidance and manipulator structure definition have been proposed. A possible approach consists in adopting the differential inverse kinematics, using the Jacobian matrix, for generating the manipulator trajectories [2,3]. However, this algorithm must take into account the kinematic singularities that may be hard to tackle. To avoid this problem, other algorithms for the trajectory generation are based on the direct kinematics [4,5,6,7,8].

Chen and Zalzala [2] propose a GA method to generate the position and the configuration of a mobile manipulator. In this report the inverse kinematics scheme is applied to optimize the least torque norm, the manipulability, the torque distribution and the obstacle avoidance. Davidor [3] also applies GAs to the trajectory generation by searching the inverse kinematics solutions to pre-defined end effector robot paths. Kubota et al. [4] study a hierarchical trajectory planning method for a redundant manipulator with a virus-evolutionary GA, running simultaneously two processes. One process calculates some manipulator collision-free positions and the other generates a collision free trajec-

K. Deb et al. (Eds.): GECCO 2004, LNCS 3102, pp. 615–626, 2004.
© Springer-Verlag Berlin Heidelberg 2004

tory by combining these intermediate positions. Rana and Zalzala [5] develope a method to plan a near time-optimal, collision-free, motion in the case of multi-arm manipulators. The planning is carried out in the joint space and the path is represented as a string of via-points connected through cubic splines. Chocron and Bidaud [9] propose an evolutionary algorithm to perform a task-based design of modular robotic systems. The system consists in a mobile base and an arm that may be built with serially assembled links and joints modules. The optimization design is evaluated with geometric and kinematic performance measures. Kim and Khosha [10] present the design of a manipulator that is best suited for a given task. The design consists of determining the trajectory and the length of a three degrees of freedom (*dof*) manipulator. Han et al. [11] describe a design method of a modular manipulator that uses the kinematic equations to determine the robot configuration and, in a second phase, adopts a GA to find the optimal length.

Gacôgne [12] presents a problem involving obstacle avoidance. He looks for an emergence of rules system for a mobile robot to have a good road-holding behavior in different playgrounds. A multi-objective genetic algorithm is used to find a short and readable solutions for every concrete problem.

Multi-objective techniques using GAs have been increasing in relevance as a research area. In 1989, Goldberg [13] suggested the use of a GA to solve multi-objective problems and since then other investigators have been developing new methods, such as multi-objective genetic algorithm (*MOGA*) [14], non-dominated sorted genetic algorithm (*NSGA*) [15] and niched Pareto genetic algorithm (*NPGA*) [16], among many other variants [17].

In this line of thought, this paper proposes the use of a multi-objective method to optimize a manipulator trajectory. This method is based on a GA adopting direct kinematics. The optimal manipulator front is the one that minimizes the objectives without any collision with the obstacles in the workspace. Following this introduction, the paper is organized as follows: section 2 formulates the problem and the GA-based method for its resolution. Section 3 presents several simulations results involving different robots, objectives and workspace settings. Finally, section 4 outlines the main conclusions.

2   Problem and Algorithm Formulation

This study considers robotic manipulators that are required to move from an initial point up to a given final configuration. Two and three *dof* planar manipulators (*i.e.* 2R and 3R robots) are used in the experiments with link lengths of one meter and rotational joints which are free to rotate 2π *rad*. To test a possible manipulator/obstacle collision, the arm structure is analyzed in order to verify if it is inside of any obstacle. The trajectory consists in a set of strings representing the joint positions between the initial and final robot configurations.

2.1   Representation

The path for a *iR* manipulator ($i = 2, 3$), at generation T, is directly encoded as vectors in the joint space to be used by the GA as:

$$[\{q_1^{(\Delta t,T)}, .., q_i^{(\Delta t,T)}\}, \{q_1^{(2\Delta t,T)}, .., q_i^{(2\Delta t,T)}\}, .., \{q_1^{((n-2)\Delta t,T)}, .., q_i^{((n-2)\Delta t,T)}\}] \quad (1)$$

where i is the number of *dof* and Δt the sampling time between two consecutive configurations.

The joints values $q_l^{(j\Delta t,0)}$ ($j = 1, \dots, n-2$; $l = 1, \dots, i$) are randomly initialized in the range $]-\pi, +\pi]$ *rad*. It should be noted that the initial and final configurations have not been encoded into the string because they remain unchanged throughout the trajectory search. Without losing generality, for simplicity, it is adopted a normalized time of $\Delta t = 0.1$ *sec*, because it is always possible to perform a time re-scaling.

2.2   Operators in the Multi-objective Genetic Algorithm

The initial population of strings is randomly generated. The search is then carried out among this population. Three different operators are used in the genetic planning: selection, crossover and mutation, as described in the sequel.

In what concerns the selection operator, the successive generations of new strings are reproduced on the basis of a Pareto ranking [13] with $\sigma_{\text{share}} = 0.01$ and $\alpha = 2$. To promote population diversity a metric count is used. This metric uses all solutions in the population independently of their rank to evaluate every fitness function. For the crossover operator it is used the simulated binary crossover (*SBX*)[15]. After crossover, the best solutions (among both parents and children) are chosen to form the next population. The mutation operator replaces one gene value with a given probability using the equation:

$$q_i^{(j\Delta t,T+1)} = q_i^{(j\Delta t,T)} + N(0, 1/\sqrt{2\pi}) \quad (2)$$

at generation T, where $N(\mu, \sigma)$ is the normal distribution function with average μ and standard deviation σ.

2.3   Evolution Criteria

Five indices $\{q, \dot{q}, p, \dot{p}, E_a\}$ (3) are used to qualify the evolving trajectory robotic manipulators. These criteria are minimized by the planner to find the optimal Pareto front. Before evaluating any solution all the values such that $|q_i^{((j+1)\Delta t,T)} - q_i^{(j\Delta t,T)}| > \pi$ are readjusted, adding or removing a multiple value of 2π, in the strings.

$$q = \sum_{j=1}^{n} \sum_{l=1}^{i} \left(\dot{q}_l^{(j\Delta t,T)}\right)^2 \quad (3a)$$

$$\dot{q} = \sum_{j=1}^{n} \sum_{l=1}^{i} \left(\ddot{q}_l^{(j\Delta t,T)}\right)^2 \quad (3b)$$

Table 1. Fronts parameters statistics

	Pareto front			Local front		
	κ	α	β	κ	α	β
Median	13.46	−8.32	−10.77	19.23	49.28	−13.02
Average	13.45	−7.40	−9.95	19.18	49.48	−13.19
Standard deviation	0.37	2.71	1.82	0.30	3.65	0.76

$$p = \sum_{j=2}^{n} d(p_j, p_{j-1})^2 \tag{3c}$$

$$\dot{p} = \sum_{j=3}^{n} \{d(p_j, p_{j-1}) - d(p_{j-1}, p_{j-2})\}^2 \tag{3d}$$

$$E_a = (n-1)\Delta t \, P_a = \sum_{j=1}^{n} \sum_{l=1}^{i} |\tau_l . \Delta q_l^{(j\Delta t, T)}| \tag{3e}$$

The joint distance q (3a) is used to minimize the manipulator joints travelling distance. For a function $y = g(x)$ the curve length is defined by:

$$\int [1 + (dg/dx)^2] dx \tag{4}$$

and, consequently, to minimize the curve length distance is adopted the simplified expression:

$$\int (dg/dx)^2 dx = \int \dot{g}^2 dx. \tag{5}$$

The joint velocity $\dot{q}$ is used to minimize the ripple in time evolution. The cartesian distance p (3c) minimizes the total arm trajectory length, from the initial point up to the final point, where p_j is the robot j intermediate arm cartesian position and $d(\cdot, \cdot)$ is a function that gives the distance between two arguments. The cartesian velocity is responsible for reducing the ripple in arm time evolution. Finally, the energy E_a in expression (3e), where τ_l are the robot joint torques, is computed assuming that power regeneration is not available by motors doing negative work, that is, by taking the absolute value of the power [18].

3   Simulation Results

In this section results of various experiments are presented. In this line of thought, subsections 3.1 and 3.2 present the trajectory optimization for the 2R and 3R robots respectively, for two objectives (2D). Finally, subsection 3.3 shows the results of a five dimensional (5D) optimization for a 2R robot.

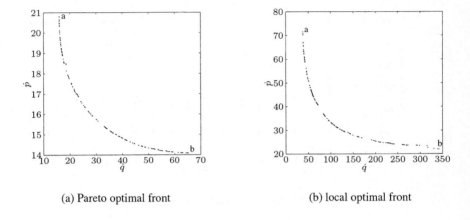

(a) Pareto optimal front   (b) local optimal front

Fig. 1. Optimal fronts for the 2R robot

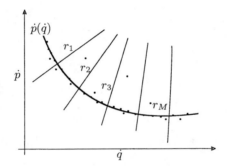

Fig. 2. Normal straight lines to the front obtained with $p(q)$ function

3.1 2R Robot Trajectory with 2D Optimization

The experiments consist on moving a 2R robotic arm from the starting configuration, defined by the joint coordinates $A \equiv \{-1.149, 1.808\}$ *rad*, up to the final configuration, defined by $B \equiv \{1.181, 1.466\}$ *rad*, in a workspace without obstacles. The objectives used in this section to optimize are the joint velocity $\dot{q}$ (3b) and the cartesian velocity $\dot{p}$ (3d).

The simulations results achieved by the algorithm, with $n = 9$ configurations, $T_t = 15000$ generations and $pop_{size=300}$, converge to two optimal fronts. One of the fronts (fig. 1(a)) corresponds to the movement of the manipulator around its base in the counterclockwise direction. The other front (fig. 1(b)) is obtained when the manipulator moves in the clockwise direction. The solutions a and b, shown in fig. 1 represent the best solution found for the $\dot{q}$ and $\dot{p}$ objectives, respectively.

In 66.6% of the 21 total number of runs, the Pareto optimal front was found. In all simulations for both cases the solutions converged to a front type which can be modelled by the following equation ($\kappa, \alpha, \beta \in \mathbb{R}$):

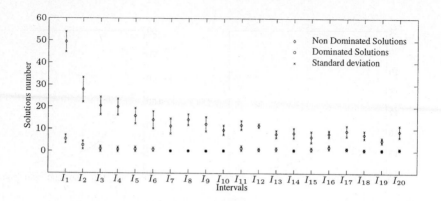

Fig. 3. Solution distribution statistics for the 2R robot

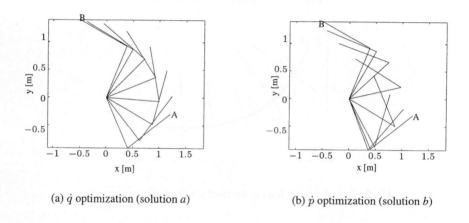

(a) $\dot{q}$ optimization (solution a)   (b) $\dot{p}$ optimization (solution b)

Fig. 4. Successive 2R robot configurations

Table 2. Range objectives in the 5D optimization

	q	$\dot{q}$	E_a	p	$\dot{p}$
min	79.8	18.2	1056.7	83.5	15.0
max	182.3	101.7	4602.7	121.8	56.4

$$p(\dot{q}) = \kappa \frac{\dot{q} + \alpha}{\dot{q} + \beta} \qquad (6)$$

The achieved median, average and standard deviation for the parameters κ, α and β of (6) are shown in table 1, both for the Pareto optimal and local fronts.

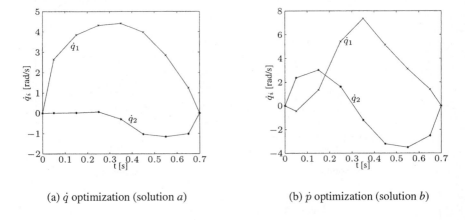

(a) $\dot{q}$ optimization (solution a)   (b) $\dot{p}$ optimization (solution b)

Fig. 5. Joint time evolution versus time for the 2R robot

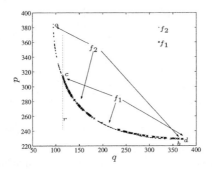

Fig. 6. Pareto optimal fronts, angular distance *vs.* cartesian distance optimization: $f_1 = \widehat{ab}$ – workspace without obstacles; $f_2 = \widehat{cd}$ – workspace with one obstacle

To study the solution front diversity, the approximated front was split into several intervals, limited by normal straight lines r_m (fig. 2), such that the front curve length is equal for all intervals. For any two consecutive normal straight lines an interval I_m ($m = 1, \ldots , 19$) is associated, and the solutions located between these lines are counted. Figure 3 shows the solution distribution statistics achieved by all simulation runs. In this chart non-dominated and dominated solutions are represented, namely its average and its standard deviation. From this chart, it can be seen that the solutions are distributed by all intervals. However, the distribution is not uniform. This is due to the use of a sharing function in the attribute domain in spite of the objective domain. Moreover, the algorithm does not incorporate any mechanism to promote the development of well distributed solutions in the objective domain.

The results obtained for solutions a and b, of the Pareto optimal front in figure 1(a), are presented in figures 4 and 5. Comparing figures 4(a) and 5(a) with figures 4(b) and

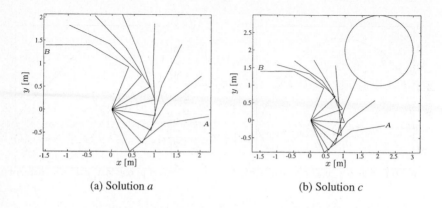

(a) Solution a (b) Solution c

Fig. 7. Successive $3R$ robot configurations

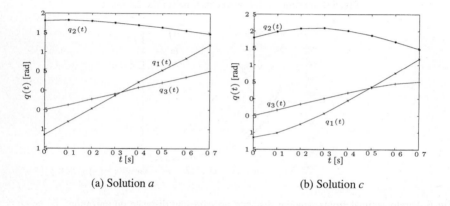

(a) Solution a (b) Solution c

Fig. 8. Joint position of trajectory *vs.* time for the $3R$ robot

5(b) it is clear that the joint/cartesian time evolution for the optimal solutions a and b, respectively, is significantly different due to the objective considered. Between these extreme optimal solutions several others were found, that have a intermediate behavior, and which can be selected according with the importance of each objective.

3.2 3R Robot Trajectory with 2D Optimization

In this subsection a $3R$ robot trajectory is optimized using the objectives q (3a) and p (3c) in a workspace which may include a circle obstacle with center at $(x, y) = (2, 2)$ and radius $\rho = 1$. The initial and final configurations are $A \equiv \{-1.15, 1.81, -0.50\}$ *rad* and $B \equiv \{1.18, 1.47, 0.50\}$ *rad*, respectively. The T_t and pop_{size} parameters used are identical to those adopted in the previous subsection. The trajectories which collide with the obstacle are assigned a very high fitness value.

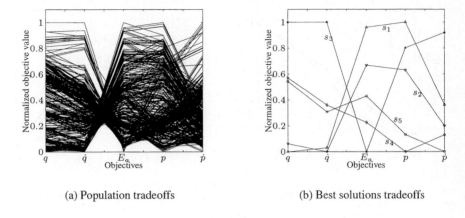

(a) Population tradeoffs   (b) Best solutions tradeoffs

Fig. 9. Normalized tradeoffs between q, $\dot{q}$, E_a, p and $\dot{p}$

For an optimization without any obstacle in the workspace the $f_2 = \widehat{ab}$ front (fig. 6) is obtained. However, when the obstacle is introduced the front is reduced to the $f_1 = \widehat{cd}$. Thus, only the objective q if affected by the introduction of the obstacle (figures 7 and 8). The solutions $\{a, b\}$ and $\{c, d\}$ represent the best solution found for the objectives $\{q, p\}$ for the 3R manipulator without and with obstacles, respectively.

3.3   2R Robot Trajectory with 5D Optimization

Here, the 2R manipulator trajectory is optimized considering the five objectives described by equations (3). Figure 9 and 10 show the optimization results achieved with $T_t = 50000$ generations and $pop_{size} = 1000$.

Table 2 contains the objectives range values archived in one simulation. Although, the 5D algorithms can't obtain so goods solutions as the 2D algorithm due to the significantly increase in the search complexity, the solutions have a good distribution (figure 9(a)) with values near to the ones for the 2D respective simulation. Figure 9 shows the normalized tradeoffs for the entire population and best solutions $s_i = \{s_1, s_2, s_3, s_4, s_5\}$ for each objective $O_i = \{q, \dot{q}, E_a, p, \dot{p}\}$. From figure 9(b) it can be concluded that q and $\dot{q}$ or p and $\dot{p}$ are conflicting objectives with a relative low tradeoff between them. On the other hand, the E_a objective presents the highest tradeoff among the others objectives. Figures 10 and 11 show the best solutions obtained for the objectives s_i. For the studied trajectory, the results indicate that as the manipulator moves near to its basis the energy consumed is lower (figures 11(a), 11(c) and 11(d)).

4   Summary and Conclusions

A multi-objective genetic algorithm robot trajectory planner, based on the kinematics approach, was proposed. The multi-objective genetic algorithm is able to reach optimal

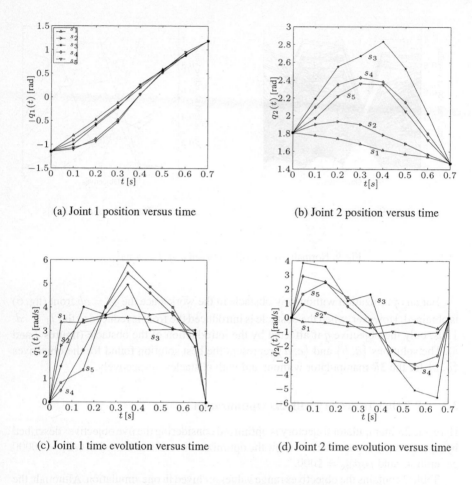

(a) Joint 1 position versus time

(b) Joint 2 position versus time

(c) Joint 1 time evolution versus time

(d) Joint 2 time evolution versus time

Fig. 10. Behavior of the best solutions

solutions regarding the optimization of multiple objectives. Simulation results were presented considering the optimization of two and five simultaneous objectives. The results obtained indicate that obstacles in the workspace may reduce the Pareto front length and for the case studied the single obstacle considered do not represent a difficulty for the algorithm to reach optimal solutions. Furthermore, the algorithm determines the non-dominated front maintaining a good distribution of solutions along the Pareto front.

Acknowledgment. This paper is partially supported by the grant Prodep III (2/5.3/2001) from FSE.

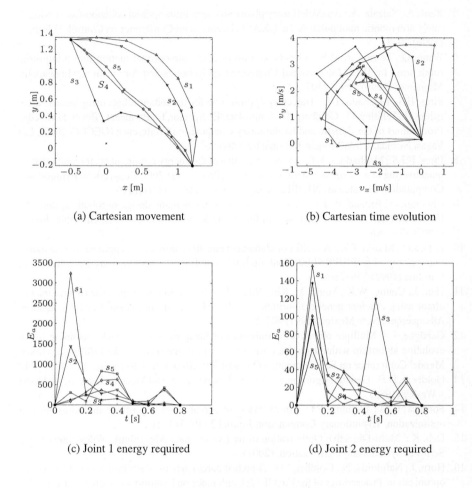

(a) Cartesian movement

(b) Cartesian time evolution

(c) Joint 1 energy required

(d) Joint 2 energy required

Fig. 11. Behavior of the best solutions (cont.)

References

1. Bäck, T., Hammel, U., Schwefel, H.P.: Evolutionary computation: Comments on the history and current state. IEEE Trans. on Evolutionary Computation **1** (1997) 3–17
2. Chen, M., Zalzala, A.M.S.: A genetic approach to motion planning of redundant mobile manipulator systems considering safety and configuration. Journal Robotic Systems **14** (1997) 529–544
3. Davidor, Y.: Genetic Algorithms and Robotics, a Heuristic Strategy for Optimization. Number 1 in Series in Robotics and Automated Systems. World Scientific Publishing Co. Pte Ltd (1991)
4. Kubota, N., Arakawa, T., Fukuda, T.: Trajectory generation for redundant manipulator using virus evolutionary genetic algorithm. In: IEEE International Conference on Robotics and Automation, Albuquerque, New Mexico (1997) 205–210

5. Rana, A., Zalzala, A.: An evolutionary planner for near time-optimal collision-free motion of multi-arm robotic manipulators. In: UKACC International Conference on Control. Volume 1. (1996) 29–35
6. Wang, Q., Zalzala, A.M.S.: Genetic control of near time-optimal motion for an industrial robot arm. In: IEEE International Conference on Robotics and Automation, Minneapolis, Minnesota (1996) 2592–2597
7. Pires, E.S., Machado, J.T.: Trajectory optimization for redundant robots using genetic algorithms. In Whitley, D., Goldberg, D., Cantu-Paz, E., Spector, L., Parmee, I., Beyer, H.G., eds.: Proceedings of the Genetic and Evolutionary Computation Conference (GECCO-2000), Las Vegas, Nevada, USA, Morgan Kaufmann (2000) 967
8. Pires, E.J.S., Machado, J.A.T.: A GA perspective of the energy requirements for manipulators maneuvering in a workspace with obstacles. In: Proc. of the 2000 Congress on Evolutionary Computation, Piscataway, NJ, IEEE Service Center (2000) 1110–1116
9. Chocron, O., Bidaud, P.: Evolutionary algorithms in kinematic design of robotic system. In: IEEE/RSJ International Conference on Intelligent Robotics and Systems, Grenoble, France (1997) 279–286
10. Kim, J.O., Khosla, P.K.: A multi-population genetic algorithm and its application to design of manipulators. In: IEEE/RSJ Int. Conf. on Intelligent Robotics and Systems, Raleight, North Caroline (1992) 279–286
11. Han, J., Chung, W.K., Youm, Y., Kim, S.H.: Task based design of modular robotic manipulator using efficient genetic algorithm. In: IEEE Int. Conf. on Robotics and Automation, Albuquerque, New Mexico (1997) 507–512
12. Gacôgne, L.: Multiple objective optimization of fuzzy rules for obstacles avoiding by an evolution algorithm with adaptative operators. In: In Proceedings of the Fifth International Mendel Conference on Soft Computing (Mendel'99), Brno, Czech Republic (1999) 236–242
13. Goldberg, D.E.: Genetic Algorithms in Search, Optimization, and Machine Learning. Addison – Wesley (1989)
14. Fonseca, C.M., Fleming, P.J.: An overview of evolutionary algorithms in multi-objective optimization. Evolutionary Computation Journal 3 (1995) 1–16
15. Deb, K.: Multi-Objective Optimization using Evolutionary Algorithms. Wiley-Interscience Series in Systems and Optimization. (2001)
16. Horn, J., Nafploitis, N., Goldberg, D.: A niched pareto genetic algorithm for multi-objective optimization, Proceedings of the First IEEE Conference on Evolutionary Computation (1994) 82–87
17. Coello, C., Carlos, A.: A comprehensive survey of evolutionary-based multiobjective optimization techniques. Knowledge and Information Systems 1 (1999) 269–308
18. Silva, F., Tenreiro Machado, J.: Energy analysis during biped walking. In: Proc. IEEE Int. Conf. Robotics and Automation, Detroit, Michigan (1999) 59–64

Evolution, Robustness, and Adaptation of Sidewinding Locomotion of Simulated Snake-Like Robot

Ivan Tanev [1], Thomas Ray[2,3], and Andrzej Buller[1]

[1] ATR Network Informatics Laboratories, 2-2-2 Hikaridai,
"Keihanna Science City", Kyoto 619-0288, Japan
[2] ATR Human Information Science Laboratories, 2-2-2 Hikaridai,
"Keihanna Science City", Kyoto 619-0288, Japan
[3] Department of Zoology, 730 Van Vleet Oval, Room 314,
University of Oklahoma Norman, Oklahoma 73019, USA
{i_tanev,ray,buller}@atr.jp

Abstract. Inspired by the efficient method of locomotion of the rattlesnake *Crotalus cerastes*, the objective of this work is automatic design through genetic programming, of the fastest possible (sidewinding) locomotion of simulated limbless, wheelless snake-like robot (Snakebot). The realism of simulation is ensured by employing the Open Dynamics Engine (ODE), which facilitates implementation of all physical forces, resulting from the actuators, joints constrains, frictions, gravity, and collisions. Empirically obtained results demonstrate the emergence of sidewinding locomotion from relatively simple motion patterns of morphological segments. Robustness of the sidewinding Snakebot, considered as ability to retain its velocity when situated in unanticipated environment, is illustrated by the ease with which Snakebot overcomes various types of obstacles such as a pile of or burial under boxes, rugged terrain and small walls. The ability of Snakebot to adapt to partial damage by gradually improving its velocity characteristics is discussed. Discovering compensatory locomotion traits, Snakebot recovers completely from single damage and recovers a major extent of its original velocity when more significant damage is inflicted. Contributing to the better understanding of sidewinding locomotion, this work could be considered as a step towards building real Snakebots, which are able to perform robustly in difficult environments.

Keywords: genetic programming, locomotion, snake-like robot

1   Introduction

Wheelless, limbless snake-like robots (Snakebots) feature potential robustness characteristics beyond the capabilities of most wheeled and legged vehicles – ability to traverse terrain that would pose problems for traditional wheeled or legged robots, and insignificant performance degradation when partial damage is inflicted. Some useful features of Snakebots include smaller size of the cross-sectional areas, stability, ability to operate in difficult terrain, good traction, high redundancy, and complete sealing of the internal mechanisms [2,3,12]. Robots with these properties open up several critical applications in exploration, reconnaissance, medicine and inspection. However, com-

K. Deb et al. (Eds.): GECCO 2004, LNCS 3102, pp. 627–639, 2004.
© Springer-Verlag Berlin Heidelberg 2004

pared to the wheeled and legged vehicles, Snakebots feature (i) smaller payload, (ii) more difficult thermal control, (iii) more difficult control of locomotion gaits and (iv) inferior speed characteristics. Considering the first two drawbacks as beyond the scope of our work, and focusing on the drawbacks of control and speed, we intend to address the following challenge: how to develop control sequences of Snakebot's actuators, which allow for achieving the fastest possible speed of locomotion.

Although for many tasks, handcrafting the robot locomotion control code can be seen as a natural approach, it might not be feasible for developing the control code of Snakebot due to its morphological complexity. While the overall locomotion gait of Snakebot might emerge from relatively simply defined motion patterns of morphological segments of Snakebot, neither the degree of optimality of the developed code nor the way to incrementally improve the code is evident to the human designer [7]. Thus, an automated mechanism for solution evaluation and corresponding rules for incremental optimization of the intermediate solution(s) are needed [6]. The proposed approach of employing genetic programming (GP) implies that the code, which governs the locomotion of Snakebot is automatically designed by a computer system via simulated evolution through selection and survival of the fittest in a way similar to the evolution of species in the nature. The use of an automated process to design the control code opens the possibility of creating a solution that would be better than one designed by a human.

Evolving a Snakebot's locomotion (and in general, behavior of any robot) could be performed as a first step in the sequence of simulated off-line evolution (phylogenetic learning) on the software model, followed by on-line adaptation (ontogenetic learning) of evolved code on a physical robot situated in a real environment [8]. Off-line software simulation facilitates the process of Snakebot's controller design because the verification of behavior on physical Snakebot is extremely time consuming, costly and often dangerous for Snakebot and surrounding environment. Moreover, in some cases it is appropriate to initially model not only the locomotion, but also to co-evolve the most appropriate morphology of the artifact (i.e. number of phenotypic segments; types and parameters of joints which link segments; actuators' power; type, amount and location of sensors; etc.) [1,9,10] and only then (if appropriate) to physically implement it as hardware. The software model, used to simulate Snakebot should fulfill the basic requirements of being quickly developed, adequate, and fast running [4]. Typically slow development time of GP stems from the highly specific semantics of the main attributes of GP (e.g. representation, genetic operations, fitness evaluation) and can be significantly reduced through incorporating off-the-shelf software components and open standards in software engineering. To address this issue, we developed a GP framework based on open XML standard and ensure adequacy and runtime efficiency of Snakebot simulation, we applied the Open Dynamic Engine (ODE) freeware software library for simulation of rigid body dynamics.

The *objectives* of our work are (i) to explore the feasibility of applying GP for automatic design of the fastest possible locomotion of realistically simulated Snakebot and (ii) to investigate the robustness and adaptation of such locomotion to unanticipated environmental conditions and degraded abilities of Snakebot. Inspired by the fast sidewinding locomotion of the rattlesnake *Crotalus cerastes*, this work is motivated by our desires (i) to better understand the mechanisms underlying sidewinding

locomotion of natural snakes, (ii) to explore the phenomenon of emergence of locomotion of complex bodies from simply defined motion patterns of the morphological
segments comprising these bodies, (iii) to verify the feasibility of employing ODE for
realistic software simulation of a Snakebot, and (iv) to investigate the practicality of
building real Snakebots.

The remainder of this document is organized as follows. Section 2 emphasizes
the main features of the GP proposed for evolution of locomotion of simulated Snakebot. Section 3 presents empirical results of evolving locomotion gaits of Snakebots
and discusses the emergence of sidewinding. The same section elaborates on robustness and adaptation of sidewinding to unanticipated environmental conditions and
partial damage of Snakebot. Finally, Section 4 draws a conclusion.

2 Approach

2.1 Representation of Snakebot

Snakebot is simulated as a set of identical spherical morphological segments ("vertebrae"), linked together via universal joints. All joints feature identical (finite) angle
limits and each joint has two attached actuators ("muscles"). In the initial, standstill
position of Snakebot the rotation axes of the actuators are oriented vertically (vertical
actuator) and horizontally (horizontal actuator) and perform rotation of the joint in the
horizontal and vertical planes respectively (Figure 1). Considering the representation
of Snakebot, the task of designing the fastest locomotion can be rephrased as developing temporal patterns of desired turning angles of horizontal and vertical actuators
of each segment, that result in fastest overall locomotion of Snakebot.

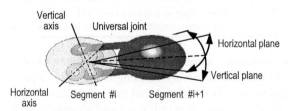

Fig. 1. Morphological segments of Snakebot linked via universal joint. Horizontal and vertical
actuators attached to the joint perform rotation of the segment #i+1 in vertical and horizontal
planes respectively.

2.2 Algorithmic Paradigm

GP. GP [5] is a domain-independent problem-solving approach in which a population
of computer programs (individuals' genotypes) is evolved to solve problems. The
simulated evolution in GP is based on the Darwinian principle of reproduction and
survival of the fittest. The fitness of each individual is based on the quality with which

the phenotype of the simulated individual is performing in a given environment. The major attributes of GP - function set, terminal set, fitness evaluation, genetic representation, and genetic operations are elaborated in the remaining of this Section.

Function Set and Terminal Set. In applying GP to evolution of Snakebot, the genotype is associated with two algebraic expressions, which represent the temporal patterns of desired turning angles of both the horizontal and vertical actuators of each morphological segment. Since locomotion gaits are periodical, we include the trigonometric functions sin and cos in the GP function set in addition to the basic algebraic functions. The choice of these trigonometric functions reflects our intention to verify the hypothesis (first expressed by Petr Miturich in 1920's) that undulative motion mechanisms could yield efficient gaits of snake-like artifacts operating in air, land, or water. Terminal symbols include the variables time, index of morphological segment of Snakebot, and two constants: Pi, and random constant within the range [0, 2]. The main parameters of the GP are summarised in Table 1.

Table 1. Main parameters of GP

Category	Value
Function set	{sin, cos, +, -, *, /}
Terminal set	{time, segment_ID, Pi, random constant, ADF}
Population size	200 individuals
Selection	Binary tournament, ratio 0.1
Elitism	Best 4 individuals
Mutation	Random subtree mutation, ratio 0.01
Fitness	Velocity of simulated Snakebot during the trial
Trial interval	180 time steps, each time step account for 50ms of "real" time
Termination criterion	(Fitness >100) *or* (Generations>30) *or* (no improvement of fitness for 16 generations)

The rationale of employing automatically defined function (ADF) is based on empirical observation that the evolvability of straightforward, independent encoding of desired turning angles of both horizontal and vertical actuators is poor, although it allows GP to adequately explore the search space and ultimately, to discover the areas which correspond to fast locomotion gaits in solution space. We discovered that (i) the motion patterns of horizontal and vertical actuators of each segment in fast locomotion gaits are highly correlated (e.g. by frequency, direction, etc.) and that (ii) discovering and preserving such correlation by GP is associated with enormous computational effort. ADF, as a way of introducing modularity and reuse of code in GP [5] is employed in our approach to allow GP to explicitly evolve the correlation between motion patterns of horizontal and vertical actuators as shared fragments in algebraic expressions of desired turning angles of actuators. Moreover, the best result was obtained by (i) allowing the use of ADF as a terminal symbol in algebraic expression of desired turning angle of vertical actuator only, and (ii) by evaluating the value of ADF by equalizing it to the value of currently evaluated algebraic expression of desired turning angle of horizontal actuator.

Fitness Evaluation. The fitness function is based on the velocity of Snakebot, estimated from the distance which the center of the mass of Snakebot travels during the trial. The real values of the raw fitness, which are usually within the range (0, 2) are multiplied by a normalizing coefficient in order to deal with integer fitness values within the range (0, 200). A normalized fitness of 100 (one of the termination criteria shown in Table 1) is equivalent to a velocity which displaced Snakebot a distance equal to twice its length. The fitness evaluation routine is shown in Figure 2.

Representation of Genotype. Inspired by its flexibility, and the recently emerged widespread adoption of document object model (DOM) and extensible markup language (XML), we represent evolved genotypes of simulated Snakebot as DOM-parse trees featuring equivalent flat XML-text. Our approach implies that both (i) the calculation of the desired turning angles during fitness evaluation (functions EvalHorizontalAngle and EvalVerticalAngle, shown in Figure 2, lines 18 and 20 respectively) and (ii) the genetic operations are performed on DOM-parse trees using off-the shelf, platform- and language neutral DOM-parsers. The corresponding XML-text representation (rather than S-expression) is used as a flat file format, feasible for migration of genetic programs among the computational nodes in an eventual distributed implementation of the GP. The benefits of using DOM/XML-based representations of genetic programs are (i) fast prototyping of GP by using standard built-in API of DOM-parsers for traversing and manipulating genetic programs, (ii) generic support for the representation of grammar of strongly-typed GP using W3C-standardized XML-schema; and (iii) inherent Web-compliance of eventual parallel distributed implementation of GP.

The slight performance degradation in computing the desired turning angles of actuators by traversing the DOM/XML-based representation of genetic programs during fitness evaluation is not relevant for the overall performance of GP. The performance profiling results indicate that fitness evaluation routine consumes more than 99% of GP runtime, however, even for relatively complex genetic programs featuring a few hundred tree nodes, most of the fitness evaluation runtime at each time step is associated with the relatively enormous computational cost of the physics simulation (actuators, joint limits, friction, gravity, collisions, etc.) of phenotypic segments of the simulated Snakebot (routine dWorldStep in Figure 2, line 32), rather than computing the desired turning angles of actuators.

Genetic Operations. Binary tournament selection is employed – a robust, commonly used selection mechanism, which has proved to be efficient and simple to code. Crossover operation is defined in a strongly typed way in that only the DOM-nodes (and corresponding DOM-subtrees) of the same data type (i.e. labeled with the same tag) from parents can be swapped. The sub-tree mutation is allowed in strongly typed way in that a random node in genetic program is replaced by syntactically correct subtree. The mutation routine refers to the data type of currently altered node and applies randomly chosen rule from the set of applicable rewriting rules as defined in the grammar of strongly typed GP.

ODE. We have chosen Open Dynamics Engine (ODE) [11] to provide a realistic simulation of physics in applying forces to phenotypic segments of Snakebot, for simulation of Snakebot locomotion. ODE is a free, industrial quality software library for simulating articulated rigid body dynamics. It is fast, flexible and robust, and it has built-in collision detection. The ODE-related parameters of simulated Snakebot are summarized in Table 2.

```
 1. function Evaluate(GenH, GenV: TGenotype): real;
 2. // GenH and GenV is a pair of algebraic expressions, which define the
 3. // turning angle of the horizontal and vertical actuators at the joints
 4. // of simulated Snakebot. GenH and GenV represent the evolved genotype.
 5. Const
 6.   TimeSteps          =180; // duration of the trial
 7.   SegmentsInSnakebot=15;   // # of phenotypic segments in simulated Snakebot
 8. var
 9.   t, s               : integer;
10.   AngleH, AngleV     : real;    // desired turning angles of actuators
11.   CurrAngleH, CurrAngleV: real; // current turning angles of actuators
12.   InitialPos, FinalPos : 3DVector;// (X,Y,Z)
13. begin
14.   InitialPos:=GetPosOfCenterOfMassOfSnakebot;
15.   for t:=0 to TimeSteps-1 do begin
16.    for s:=0 to SegmentsInSnakebot-1  do begin
17.      // traversing XML/DOM-based GenH using DOM-parser:
18.      AngleH    := EvalHorizontalAngle(GenH,s,t);
19.      // traversing XML/DOM-based GenV using DOM-parser:
20.      AngleV    := EvalVerticalAngle(GenV,s,t);
21.      CurrAngleH := GetCurrentAngleH(s);
22.      CurrAngleV := GetCurrentAngleV(s);
23.      SetDesiredVelocityH(CurrAngleH-AngleH,s);
24.      SetDesiredVelocityV(CurrAngleV-AngleV,s);
25.     end;
26.    // detect collisions between the objects (phenotypic segments,
27.    // ground plane, etc.):
28.    dSpaceCollide;
29.    // Obtain new properties (position, orientation, velocity
30.    // vectors, etc.) of morphological segments of Snakebot as a result
31.    // of applying all forces:
32.    dWorldStep;
33. end;
34. FinalPos := GetPosOfCenterOfMassOfSnakebot;
35. return GetDistance(InitialPos, FinalPos)/(TimeSteps);
36. end;
```

Fig. 2. Fitness evaluation routine

3   Results

This section discusses empirical results verifying the feasibility of applying GP for evolution of the fastest possible locomotion gaits of Snakebot for various fitness and environmental conditions. In addition, it investigates the properties of the fastest lo-comotion gait, evolved in an unconstrained environment from two perspectives: (i) robustness to various unanticipated environmental conditions and (ii) gradual adapta-tion to degraded mechanical abilities of Snakebot. These challenges are considered as relevant for successful accomplishment of various practical tasks during anticipated exploration, reconnaissance, medicine and inspection missions.

Table 2. ODE-related parameters of simulated Snakebot

Parameter	Value
Number of phenotypic segments in snake	15
Model of segment	Sphere, R=0.2
Type of joint between segments	Universal
Initial alignment of segments in Snakebot	Along Y-axis of the world
Number of actuators per joint	2
Orientation of axes of actuators	Horizontal – along X-axis and Vertical – along Z-axis of the world
Operational mode of actuators	dAMotorEuler
Max force of actuators	12
Actuators stops (angular limits)	±50°
Friction between segments and surface (μ)	5
Sampling frequency of simulation	20 Hz

3.1 Evolution of Fastest Locomotion Gaits

Figure 3 shows the fitness convergence characteristics of 10 independent runs of GP (Figure 3a) and sample snapshots of evolved best-of-run locomotion gaits (Figure 3b and Figure 3c) when fitness is measured in *any* direction in an unconstrained environment. Despite the fact that fitness is unconstrained and measured as velocity in any direction, *sidewinding* locomotion (defined as locomotion predominantly perpendicular to the long axis of Snakebot) emerged in all 10 independent runs of GP, suggesting that it provides superior speed characteristics for Snakebot morphology. The dynamics of evolved turning angles of actuators in sidewinding locomotion result in characteristic circular motion pattern of segments around the center of the mass as shown in Figure 4a. The circular motion pattern of segments and the characteristic track on the ground as a series of diagonal lines (Figure 4b) suggest that during sidewinding the shape of Snakebot takes the form of a rolling helix. Figure 4 demonstrates that the simulated evolution of locomotion via GP is able to invent the improvised "wheel" of the sidewinding Snakebot to achieve fast locomotion.

In order to verify the superiority of velocity characteristics of sidewinding locomotion for Snakebot morphology we compared the fitness convergence characteristics of evolution in unconstrained environment for the following two cases: (i) unconstrained fitness measured as velocity in any direction (as discussed above and illustrated in Figure 3a), and (ii) fitness, measured as velocity in forward (non-sidewinding) direction only. The results of evolution of forward locomotion, shown in Figure 5 indicate that non-sidewinding motion, compared to sidewinding, features much inferior velocity characteristics.

Fig. 7. Snapshots of sample evolved best-of-run standstill postures featuring elevated head of Snakebot: front view (left) and view from above (right).

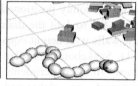

Fig. 8. Snapshots illustrating the robustness of sidewinding in clearing a pile of boxes: initial (left), intermediate (middle) and final (right) stages of the trial

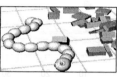

Fig. 9. Snapshots illustrating the robustness of sidewinding in emerging from burial under a stack of boxes: initial (left), intermediate (middle) and final (right) stages of the trial

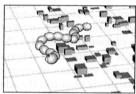
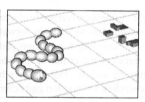

Fig. 10. Snapshots illustrating the robustness of sidewinding in rugged terrain area: initial (left), intermediate (middle) and final (right) stages of the trial

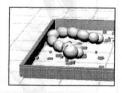
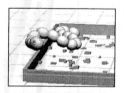
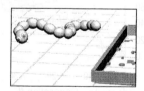

Fig. 11. Snapshots illustrating the ability of simulated sidewinding Snakebot in clearing walls forming a "pen": initial (left), intermediate (middle) and final (right) stages of the trial. Height of the walls is equal to the diameter of cross-section of simulated Snakebot.

3.3 Adaptation

The ability of sidewinding Snakebot to adapt to partial damage to 1, 2, 4 and 8 (out of 15) segments by gradually improving its velocity by simulated evolution via GP is shown in Figure 12. Demonstrated results are averaged over 4 independent runs for each case, where GP is initialized with a population comprising 190 randomly created individuals, plus 10 best-of-run genetic programs obtained from experiments with evolving sidewinding in an unconstrained environment as elaborated in Section 3.1. The damaged segments are evenly distributed along the body of Snakebot. Damage inflicted to a particular segment implies a complete loss of functionality of both horizontal and vertical actuators of the corresponding joint. As Figure 12a illustrates, Snakebot completely recovers from damage to single segment in 25 generations, attaining its previous velocity, and recovers to average of 94% of its previous velocity in the case where 2 (13% of total amount of 15) segments are damaged. With 4 (27%) and 8 (53%) damaged segments the degree of recovery is 77% (23% degradation) and 64% (36% degradation) respectively. Figure 12b shows a snapshot of frontal view of sidewinding Snakebot adapted to damage of a single segment. Compared to the side-winding locomotion of Snakebot before the adaptation (Figure 12c), the adapted locomotion gait features much higher elevation of the middle part of the body. This elevation compensates the complete lack of functionality of actuators in the damaged segment.

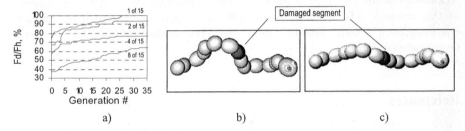

a)   b)   c)

Fig. 12. Adaptation of sidewinding Snakebot to damage of 1, 2, 4 and 8 segments (a), snapshots of frontal view of sidewinding, adapted to damage of single segment (b) and sidewinding before the adaptation (c). Fd is the best fitness in evolved population of damaged snakebots, and Fh is the best fitness of 10 best-of-run healthy sidewinding Snakebots.

4   Conclusion

We presented an approach to automatic design through genetic programming, of sidewinding locomotion of simulated limbless, wheelless artifacts. The software model used to simulate Snakebot should fulfill the basic requirements of being quickly developed, adequate, and fast running. To address the first of these issues, we employed an XML-based GP framework. To address the issues of adequacy and runtime efficiency of Snakebot simulation we applied the Open Dynamic Engine (ODE) – a freeware software library for simulation of rigid body dynamics. The empirically obtained results demonstrate that the complex locomotion of sidewinding emerges from relatively simple motion patterns of phenotypic segments (vertebrae). The evolved locomotion pattern of each segment is such that the segment is rotating in a circle-like trajectory around the center of the mass of the simulated Snakebot. This suggests that evolved sidewinding locomotion can be viewed as a process of rolling of the body of the simulated Snakebot in a helix shape, effectively inventing a kind of improvised wheel. The efficiency of sidewinding locomotion is much superior to locomotion in the forward direction, suggesting that sidewinding is the fastest possible locomotion for the simulated limbless wheelless robots with the characteristics used in this study (morphology, limits of actuator forces, joint type, joint movement limits, etc.). Robustness of the sidewinding Snakebot, initially evolved in unconstrained environment (considered as ability to retain its velocity when situated in unanticipated environment) was illustrated by the ease with which Snakebot overcomes various types of obstacles such as piles of and burial under boxes, rugged terrain and walls. The ability of Snakebot to adapt to partial damage by gradually improving its velocity characteristics was discussed. Discovering compensatory locomotion traits, Snakebot recovers completely from single damage and recovers a major extent of its original velocity when more significant damage is inflicted. Contributing to the better understanding of sidewinding locomotion, this work could be considered as a step towards building real limbless, wheelless robots, which featuring unique engineering characteristics are able to perform robustly in difficult environments.

Acknowledments. The authors thank Katsunori Shimohara for his immense support of this research. The research was conducted as part of "Research on Human Communication" with funding from the Telecommunications Advancement Organization of Japan.

References

1.   Bongard, J. C., R. Pfeifer: Evolving Complete Agents Using Artificial Ontogeny, in Hara, F. & R. Pfeifer, (eds.), Morpho-functional Machines: The New Species, Springer-Verlag, pp. 237-258 (2003)
2.   Dowling, K.: Limbless Locomotion: Learning to Crawl with a Snake Robot, doctoral dissertation, tech. report CMU-RI-TR-97-48, Robotics Institute, Carnegie Mellon University (1997)
3.   Hirose, S.: Biologically Inspired Robots: Snake-like Locomotors and Manipulators, Oxford University Press (1993)
4.   Jacobi, N.: Minimal Simulations for Evolutionary Robotics. Ph.D. thesis, School of Cognitive and Computing Sciences, Sussex University (1998)
5.   Koza, J.R.: Genetic Programming 2: Automatic Discovery of Reusable Programs, The MIT Press, Cambridge, MA (1994)
6.   Mahdavi, S., Bentley, P.J.: Evolving Motion of Robots with Muscles. In Proc. of Evo-ROB2003, the 2nd European Workshop on Evolutionary Robotics, EuroGP 2003 (2003) 655-664
7.   Morowitz, H.J.: The Emergence of Everything: How the World Became Complex, Oxford University Press, New York (2002)
8.   Meeden, L., Kumar, D.: Trends in Evolutionary Robotics, Soft Computing for Intelligent Robotic Systems, edited by L.C. Jain and T. Fukuda, Physica-Verlag, New York, NY (1998) 215-233
9.   Sims, K.: Evolving 3D Morphology and Behavior by Competition, Artificial Life IV Proceedings, MIT Press (1994) 28-39

10. Ray, T.: Aesthetically Evolved Virtual Pets, Leaonardo, Vol.34, No.4 (2001) 313 – 316
11. Smith, R.: Open Dynamics Engine (2001-2003) http://q12.org/ode/
12. Zhang, Y., Yim, M. H., Eldershaw, C., Duff, D. G., Roufas, K. D.: Phase automata: a programming model of locomotion gaits for scalable chain-type modular robots. IEEE/RSJ International Conference on Intelligent Robots and Systems (IROS 2003); October 27 - 31; Las Vegas, NV (2003)

Evolution Tunes Coevolution: Modelling Robot Cognition Mechanisms

Michail Maniadakis and Panos Trahanias

Inst. of Computer Science, FORTH, Crete, Greece, and
Dep. of Computer Science, University of Crete, Greece
{mmaniada,trahania}@ics.forth.gr

Abstract. We introduce a framework for brain modelling tasks, following a collaborative coevolutionary approach. A new coevolutionary scheme is also proposed which emphasizes collaborator selection issue. The proposed approach is employed to construct a computational model of brain motor areas, which is tested in driving a simulated robot.

1   Introduction

The problem of brain modelling fits very well to collaborative coevolutionary approaches, since the mammalian central nervous system consists of distinct interconnected modules [1]. Thus, separate coevolved species can be used to perform design decisions for each partial brain model, enforcing both a performance similar to reality and the cooperation within brain modules. However, there are open issues in the area of coevolutionary processes. One major problem concerns how collaborators are chosen among species [3]. We propose a two level collaborative coevolutionary strategy, aiming at a systematic method to approach collaborator selection. An additional evolutionary algorithm which performs in a higher level is employed to evolve collaborator schemes. In conjunction with the enhancement of single individuals performed in coevolved species, higher level evolution also performs the enhancement of successful assemblies of partial solutions.

2   Two Level Collaborative Coevolution

We implemented a general purpose genotype for both the evolution of species, and the higher-level collaborator selection process. Each individual is assigned an identification number, and encodes two kinds of variables. The first kind is SetVariables which are allowed to get a value from a unordered set. The second kind is termed RangeVariables and it is allowed to get a value within a range. Higher level evolutionary process performs on a population of individuals consisting only of SetVariables. Each SetVariable is joined with one lower level species. SetVariable's value can be any identification number of the individuals from the species it is joined with. In order to test the performance of individuals, the population at the higher level is sequentially accessed, and SetVariable's values are used as guides to select collaborators among species.

K. Deb et al. (Eds.): GECCO 2004, LNCS 3102, pp. 640–641, 2004.

Some individuals of the lower level species may be multiply selected to participate in various combinations. Unused individuals are utilised to decrease the heavy multiplicity of collaborations, by a novel genetic operator termed "Replication". For each non-collaborative individual x of a species, replication identifies the fittest individual y with more than max_c collaborations. The genome of y is then copied to x, and x is assigned $max_c - 1$ collaborations of y, by updating the appropriate individuals of the population at the higher level. After replication, individuals x and y are allowed to evolve separately.

3   Results

We employ the computational model presented in [2] to supply a computational structure for brain modelling. The latter consists of a neural cortical module to represent brain areas and a link module to support information flow within them. The computational model learns in two modes. The first mode represents phylogenesis (simulated by the coevolutionary process) and the second represents epigenesis (simulated by synaptic adjustment during environmental interaction).

The connectivity of neural network structures is illustrated in Fig 1(a). The whole model consists of 5 subcomponents (2 Modules and 3 Links) which have to cooperate to accomplish the desired performance. A higher-level evolutionary process with genomes of 5 SetVariables tunes the coevolution of all 5 species following the method presented in section 3. A population of 150 individuals evolved subcomponent species, while a population of 300 individuals evolved higher-level collaborator selection process. A sample result of phylogenetic process regarding the learning of a wall avoidance behaviour is illustrated in Fig 1(b).

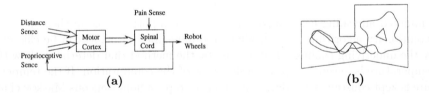

(a)   (b)

Fig. 1.   (a) Schematic overview of the model. (b) Sample result of robot navigation.

References

1. E.R. Kandel, J.H. Schwartz, T.M. Jessell: Principles of Neural Science, 3rd edition, Appleton & Lange, 2000.
2. M.Maniadakis, P. Trahanias: A computational model of neocortical-hippocampal cooperation and its application to self-localization. Proc. ECAL 03.
3. R.P. Wiegand, W.C. Liles, K. A. De Jong: An empirical analysis of collaboration methods in cooperative coevolutionary algorithms. Proc. GECCO 01.

Now, the relations expressed in (13) and (14) can be summarized to

Lemma 2 *A single step of the expansion of* $\sum_{f \notin M_0} \mathbf{a}_f(k)$ *results in*

$$\sum_{f \notin M_0} \mathbf{a}_f(k) = \sum_{f \notin M_0} \mathbf{a}_f(k-1) - \sum_{f \in M_1} \frac{r(f)}{|\mathcal{N}_f|} \cdot \mathbf{a}_f(k-1) +$$

$$+ \sum_{f' \in M_1} \sum_{i=1}^{s(f')} \varphi(f_i, f', 1) \cdot \mathbf{a}_{f_j}(k-1).$$

The diminishing factor $\left(1 - r(f)/|\mathcal{N}_f|\right)$ appears by definition for all elements of M_1. At subsequent reduction steps, the factor is "transmitted" successively to all probabilities from higher distance levels M_i because any element of M_i has at least one neighbour from M_{i-1}. The main task is now to analyze how this diminishing factor changes when it is transmitted to higher distance levels. We denote

$$(17) \quad \sum_{f \notin M_0} \mathbf{a}_f(k) = \sum_{f \notin M_0} \mu(f, v) \cdot \mathbf{a}_f(k-v) + \sum_{f' \in M_0} \mu(f', v) \cdot \mathbf{a}_{f'}(k-v),$$

i.e., the coefficients $\mu(\tilde{f}, v)$ are the factors at probabilities after v steps of an expansion of $\sum_{f \notin M_0} \mathbf{a}_f(k)$. Starting from step $(k-1)$, the probabilities $\mathbf{a}_{f'}(k-v)$, $f' \in M_0$, from (17) are expanded in the same way as the probabilities for all other $f \notin M_0$. We establish a recursive relation for the coefficients $\mu(\tilde{f}, v)$ defined in (17), where we apply the same expansion that resulted in equation (13) to the products $\mu(\tilde{f}, v) \cdot \mathbf{a}_{\tilde{f}}(k-v)$. For neighbouring elements we use $f' < \tilde{f}$ if $\mathcal{Z}(f') < \mathcal{Z}(\tilde{f})$, and $f' > \tilde{f}$ for the reverse relation of the objective function. Thus, taking into account (15) and (16), we obtain the following parameterized representation:

Lemma 3 *The following recurrent relation is valid for the coefficients* $\mu(\tilde{f}, v)$:

$$(18) \quad \mu(\tilde{f}, v) = \mu(\tilde{f}, v-1) \cdot D_{\tilde{f}}(k-v) + \sum_{f' < \tilde{f}} \frac{\mu(f', v-1)}{|\mathcal{N}_{\tilde{f}}|} +$$

$$+ \sum_{f'' > \tilde{f}} \mu(f'', v-1) \cdot \varphi(f'', \tilde{f}, v).$$

For $f \notin M_0$, we consider $\nu(f, v) = 1 - \mu(f, v)$ instead of $\mu(f, v)$ itself; for elements from M_0 we take the original value. When $\mu(f, v)$ is substituted in (18) by $1 - \nu(f, v)$, we obtain the same relation for $\nu(f, v)$ because the sum of transition probabilities equals 1 within the neighbourhood $\mathcal{N}_f$. We consider in more details the terms associated with elements of M_0 and M_1. We assume a representation $\mu(f', v-1) = \sum_{u'} T'_{u'}$ and $\nu(f, v-1) = \sum_u T_u$, where $T'_{u'}$ and T_u are arithmetic terms that have been generated at previous steps from the elementary terms listed in Lemma 3 for $v = 1$. Since there are no $f'' < f'$ for $f' \in M_0$, we obtain:

$$(19) \quad \mu(f', v) = D_{f'}(k - v) \cdot \sum_{u'} T'_{u'}(f') + \sum_{f > f'} \left(1 - \sum_u T_u(f) \right) \cdot \varphi(f, f', v)$$

$$(20) \qquad = \sum_{f > f'} \varphi(f, f', v) + D_{f'}(k - v) \cdot \sum_{u'} T'_{u'}(f') -$$

$$- \sum_{f > f'} \sum_u T_u(f) \cdot \varphi(f, f', v).$$

As can be seen, the term $\sum_{f > f'} \varphi(f, f', v)$ is generated at each time step v with the corresponding $\varphi(f, f', v)$. When (18) is written for $\nu(f, v)$, we obtain in the same way for elements of M_1:

$$(21) \qquad \nu(f, v) = \frac{r(f)}{|\mathcal{N}_f|} + D_f(k - v) \cdot \nu(f, v - 1) +$$

$$+ \sum_{f'' > f} \nu(f'', v - 1) \cdot \varphi(f'', f, v) - \sum_{f' < f} \frac{\sum_{u'} T'_{u'}(f')}{|\mathcal{N}_f|},$$

where $r(f)/|\mathcal{N}_f|$ is from $\sum_{f' < \tilde{f}} 1/|\mathcal{N}_{\tilde{f}}|$. The term $r(f)/|\mathcal{N}_f|$ appears in all recursive equations of $\nu(f, v)$, $f \in M_1$ and $v \geq 1$, and the same is valid for the value $\sum_{f > f'} \varphi(f, f', v)$ in all $\mu(f', v)$, $f' \in M_0$. Therefore, all arithmetic terms T are derived from terms of the type $r(f)/|\mathcal{N}_f|$ and $\sum_{f > f'} \varphi(f, f', v)$. We try to keep track for each individual term that is generated by a recursive step as given in (18). For this purpose, the coefficients $\nu(f, v)$ are represented by a sum $\sum_i T_i$ of arithmetic terms (as in the derivation of (19), ..., (21)), and we are now going to define the terms T_i in more details by an inductive procedure.

Definition 3 *The terms $r(f)/|\mathcal{N}_f|$ (the first in (21)), $f \in M_1$, and $\sum_{f > f'} \varphi(f, f', v)$ (the first sum in (20)), $f' \in M_0$, are called source terms of $\nu(f, v)$ and $\mu(f', v)$, respectively, where $v \geq 1$.*

During an expansion of $\sum_{f \notin M_0} \mathbf{a}_f(k)$ backwards according to (17), the source terms are distributed permanently to higher distance levels M_j as well as to elements from M_0. That means, in the same way as for M_1, the calculation of $\nu(f, v)$ ($\mu(f', v)$ for M_0) is repeated almost identically at any step, only the "history" of generations becomes longer. We introduce a counter $\mathbf{r}(f)$ to terms T that indicates the step at which the term has been generated from source terms. The value $\mathbf{r}(f)$ is called the rank of a term and we set $\mathbf{r}(f) = 1$ for source terms T from Definition 3. Basically, the rank $\mathbf{r}(f) \geq 1$ indicates the number of factors when T is represented by the subsequent multiplications according to the recurrent generation rules (20) and (21).

Let $\mathcal{T}_j(\tilde{f}, v)$ be the set of jth rate arithmetic terms from $\nu(\tilde{f}, v)$ with the same rank $\mathbf{r}(f)$, where $\tilde{f} \in \mathcal{M}_{d\_m} \backslash M_0$. We set

$$(22) \qquad \mathbf{S}_j(\tilde{f}, v) := \sum_{T \in \mathcal{T}_j(\tilde{f}, v)} T.$$

The same notation is used in case of $f' = \tilde{f} \in M_0$ with respect to $\mu(f', v)$. Now, the coefficients $\nu(\tilde{f}, v)$, $\mu(f', v)$, can be represented by

$$(23) \qquad \nu(\tilde{f}, v) = \sum_{j=1}^{v} \mathbf{S}_j(\tilde{f}, v) \quad \text{and} \quad \mu(f', v) = \sum_{j=1}^{v} \mathbf{S}_j(f', v).$$

We compare the computation of $\nu(f, v)$ and $\mu(f', v)$ for two different values $v = k_1$ and $v = k_2$, i.e., $\nu(f, v)$ is calculated backwards from k_1 and k_2, respectively. Let $\mathbf{S}_j^1$ and $\mathbf{S}_j^2$ denote the corresponding sums of terms related to two different starting steps k_1 and k_2. From Definition 3 we see that the source term $r(f)/|\mathcal{N}_f|$ does not depend on k. For the second type of source terms, we employ the simple equation $k_2 - (k_2 - k_1 + v) = k_1 - v$, which leads to

Lemma 4 *Given $k_2 \geq k_1 \geq K_0$ and $1 \leq j \leq k_1$, then for each $f \in \mathcal{M}$:*

$$\mathbf{S}_j^1(f, v) = \mathbf{S}_j^2(f, k_2 - k_1 + v).$$

We use (17) and obtain:

$$(24) \quad \sum_{f \notin M_0} \mathbf{a}_f(k_1) = \sum_{f \notin M_0} \big(\mathbf{a}_f(k_1) - \mathbf{a}_f(k_2) \big) + \sum_{f \notin M_0} \mathbf{a}_f(k_2)$$

$$(25) \qquad = \sum_{f \notin M_0} \big(\nu(f, k_2 - k_1) - \nu(f, 0) \big) \cdot \mathbf{a}_f(k_1) +$$

$$+ \sum_{f' \in M_0} \big(\mu(f', 0) - \mu(f', k_2 - k_1) \big) \cdot \mathbf{a}_{f'}(k_1) + \sum_{f \notin M_0} \mathbf{a}_f(k_2).$$

For the first part of the sum we obtain:

$$(26) \qquad \sum_{f \notin M_0} \big(\nu(f, k_2 - 1) - \nu(f, 0) \big) \cdot \mathbf{a}_f(k_1)$$

$$= \sum_{f \notin M_0} \Big(\sum_{j=1}^{k_2 - k_1} \mathbf{S}_j^2(f, k_2 - k_1) - \mathbf{S}_0^1(f, 0) \Big) \cdot \mathbf{a}_f(k_1),$$

and Lemma 4 leads to:

$$(27) \quad \sum_{f \notin M_0} \big(\nu(f, k_2 - k_1) - \nu(f, 0) \big) \cdot \mathbf{a}_f(k_1) = \sum_{f \notin M_0} \sum_{j=1}^{k_2 - k_1} \mathbf{S}_j^2(f, k_2 - k_1) \cdot \mathbf{a}_f(k_1).$$

The same applies to configurations $f' \in M_0$.

To find upper bounds for (27), we estimate $\mathbf{a}_f(k_1)$ for configurations different from global and local minima, and the $\mathbf{S}_j^2(f, k_2 - k_1)$ are then estimated for global and local minima separately. To distinguish between the two cases is necessary since for small j and f different from global and local minima, the values $\mathbf{S}_j^2(f, k_2 - k_1)$ are relatively large (cf. Definition 3). We note that the recursive application of (18) generates negative summands in the representation of values $\mathbf{S}_j(f, v)$, as can be seen from Definition 3 (cf. also (20) and (21)). We set $\mathbf{S}_j(f, v) = \mathbf{S}_j^+(f, v) - \mathbf{S}_j^-(f, v)$ and $\mathbf{S}_j(f', v) = \mathbf{S}_j^+(f', v) - \mathbf{S}_j^-(f', v)$ for $f \in M_1$ and $f' \in M_0$, where the partial sums consist of positive products only.

When $\mathbf{S}_j(f, v)$, $f \in M_1$, and $\mathbf{S}_j(f', v)$, $f' \in M_0$, are calculated, the negative products of $\mathbf{S}_{j-1}(f, v-1)$ become positive for $\mathbf{S}_j(f', v)$, and the negative products of $\mathbf{S}_{j-1}(f', v-1)$ become positive for $\mathbf{S}_j(f, v)$; see (20) and (21). The negative products of $\mathbf{S}_{j-1}(f, v-1)$ remain negative in the calculation of $\mathbf{S}_j(\tilde{f}, v)$, $\tilde{f} \in M_2$, and the same applies to higher distance levels. Hence, the negative and positive products can be considered separately at all distance levels. Thus, we concentrate on upper bounds of $\mathbf{S}_j^+(f, v)$ only. To simplify notations, we use $\mathbf{S}_j(f, v)$ instead of $\mathbf{S}_j^+(f, v)$. Furthermore, we use instead of $n+1$ from $N_{\tilde{f}} \leq n+1$ (see (1)) the value $n' = n+1$, and for convenience n again for n'. We set $\widehat{\mathcal{M}} := \{f : r(f) \geq 1\}$, and for a constant $a > 0$ we can prove

$$(28) \qquad \sum_{f \in L_h \cap \widehat{\mathcal{M}}} \mathbf{a}_f(k) < \frac{2 \cdot (n+1-h) \cdot n^a}{(k+2-n^a)^\gamma}.$$

Now, we estimate $\mathbf{S}_j(t, v)$ specifically for local and global minima. Here, we use the property that backwards expansions "entering" a local or global minimum are multiplied by $1/(k+2-v)^\gamma$, i.e., the upper bound is of the the type $\Pi/(k+2-v)^\gamma$, where Π represents the sum of products leading from M_1 (or M_0) to the local or global minimum. From Lemma 3 we conclude

$$(29) \quad \mathbf{S}_j(f, v) = \sum \Phi_1 \cdot \Phi_2 \cdots \Phi_j \leq \sum_{\substack{[d,g,h_1,h_2] \text{ Possible Positions of} \\ D, G, H_1, H_2}} D^d \cdot G^g \cdot H_1^{h_1} \cdot H_2^{h_2},$$

$d + g + h_1 + h_2 = j$, where D is the probability to stay in a local minimum, G corresponds to steps decreasing the objective function (we recall, that we are going backwards in the expansion of $\mathbf{a}_f(k - v)$), H_1 is associated with steps increasing the objective function, and H_2 is from the probability to stay in the same configuration which is not a local minimum.

For $f \in L_h$ we set $h(f) = h$ and we consider $f \in M \backslash \widehat{\mathcal{M}}$. We set $k_1 := k + 2 - v + j$ and $k_2 := k + 2 - v$, and by induction on j we show

Lemma 5 *For $f \in M \backslash \widehat{\mathcal{M}}$, $\Gamma > 3$, and $k \geq n^{2 \cdot \Gamma}$, the following inequality holds:*

$$(30) \qquad \mathbf{S}_j(f, v) < e^{-\frac{j}{k_1^{3 \cdot \gamma}}} \cdot \left(1 + \frac{1}{k_2^\gamma}\right)^{h(f)}.$$

Based on (27) and Lemma 5, we derive an upper bound for $\sum_{f \in M \backslash \widehat{\mathcal{M}}} \mathbf{a}_f(k)$, which leads to

$$(31) \qquad \sum_{f \in M \backslash \widehat{\mathcal{M}}} \mathbf{a}_f(k) < \frac{n^b}{(k+2-n^b)^\gamma}, \quad b = \text{const.} > 0.$$

Now, (28) and (31) are used to prove for $c \geq \max\{a+1, b\}$:

$$(32) \qquad |\sum_{f \notin M_0} (\nu(f, k_2 - k_1) - \nu(f, 0)) \cdot \mathbf{a}_f(k_1)| < O(\frac{n^c}{(k+2-n^c)^\gamma}).$$

Here, we consider $\mathbf{S}_j^+$, and (27) has been applied to these values only. But the same holds for $\mathbf{S}_j^-$, with even a smaller first factor of the expansion, see Lemma 3. Thus, in the same way we obtain the corresponding upper bound for $(\mu(f', 0) - \mu(f', k_2 - k_1))$, and we finally complete the

Proof of Theorem 2: We utilise (24) until (26) and employ Theorem 1, i.e., if the constant Γ from (6) is sufficiently large, the inhomogeneous simulated annealing procedure defined by (2), (3), and (4) tends to the global minimum of $\mathcal{Z}$ on $\mathcal{F}$. The value k_2 from (32) is larger but independent of $k_1 = k$, i.e., we can take a $k_2 > k$ such that

$$\sum_{\tilde{f} \notin M_0} \mathbf{a}_{\tilde{f}}(k_2) < \frac{\delta}{3}.$$

Additionally, we require that both differences $\sum_{\tilde{f} \notin M_0} (\nu(\tilde{f}, k_2 - k) - \nu(\tilde{f}, 0))$ and $\sum_{f' \in M_0} (\mu(f', 0) - \mu(f', k_2 - k))$ are smaller than $\delta/3$. From (32) we obtain the condition

$$O(\frac{n^c}{(k+2-n^c)^\gamma}) < \frac{\delta}{3}.$$

We finally arrive at

$$k > (\frac{n}{\delta})^{O(\Gamma)} \geq n^c - 2 + O(\frac{3 \cdot n^c}{\delta})^\Gamma.$$

q.e.d.

4   Computational Experiments

We are given a set $\mathcal{S} \subseteq \{0, 1\}^n$ of uniformly distributed binary n-tuples $\tilde{\eta} = \eta_1 \cdots \eta_n$ that represent negative examples for an unknown target conjunction $C_\ell = x_{i_1}^{\sigma_{i_1}} \& x_{i_2}^{\sigma_{i_2}} \& \cdots \& x_{i_\ell}^{\sigma_{i_\ell}}$ (here, we use $x^1 \equiv x$ and $x^0 \equiv \bar{x}$, i.e., $x^0 = 1$ for $x = 0$, and $x^0 = 0$ for $x = 1$), and a single positive example $\tilde{\sigma} = \sigma_1 \cdots \sigma_n$: $C_\ell(\tilde{\sigma}) = 1$ and $\forall \tilde{\eta}(\tilde{\eta} \in \mathcal{S} \to C_\ell(\tilde{\eta}) = 0)$. The task is to find a conjunction C_l of length $l \leq \ell$ that matches all of the samples, i.e., from C_ℓ generating the samples we do know only the length ℓ; cf. [8] and the Example in Section 2.

As explained in Section 2, we have $\Gamma \leq \lceil \log n \rceil$ for the problem to find a conjunction of length $\ell = \lceil \log n \rceil$. We implemented the search procedure for $m = 32$ negative examples, and for each element of $\mathcal{F}$ we counted the number of occurrences during the search procedure, in particular, for $\mathcal{F}_{\min}$. The calculations were repeated three times, and we present the average values (we obsereved only small deviations). The constant c in $O(\Gamma) = c \cdot \Gamma$ was set to $c = 1$.

		$n = 8$ and $\Gamma = 3$		$n = 16$ and $\Gamma = 4$	
δ	$1 - \delta$	k according to Theorem 2 $(c = 1)$	Frequency of $f \in \mathcal{F}_{\min}$	k according to Theorem 2 $(c = 1)$	Frequency of $f \in \mathcal{F}_{\min}$
0.50	0.50	4096	0.739	1048576	0.786
0.25	0.75	32768	0.812	16777216	0.895
0.10	0.90	512000	0.945	655360000	0.953
0.01	0.99	512000000	0.996	———	——

Frequencies of $f \in \mathcal{F}_{\min}$.

As we can see, the experimental results are in compliance with Theorem 2 for the small constant $c = 1$.

References

1. E.H.L. Aarts and J.H.M. Korst. *Simulated Annealing and Boltzmann Machines: A Stochastic Approach*, Wiley & Sons, New York, 1989.
2. S. Azencott (editor). *Simulated Annealing: Parallelization Techniques*. Wiley & Sons, New York, 1992.
3. O. Catoni. Rough Large Deviation Estimates for Simulated Annealing: Applications to Exponential Schedules. *Annals of Probability*, 20(3):1109 – 1146, 1992.
4. O. Catoni. Metropolis, Simulated Annealing, and Iterated Energy Transformation Algorithms: Theory and Experiments. *J. of Complexity*, 12(4):595 – 623, 1996.
5. V. Černy. A Thermodynamical Approach to the Travelling Salesman Problem: An Efficient Simulation Algorithm. Preprint, Inst. of Physics and Biophysics, Comenius Univ., Bratislava, 1982 (see also: *J. Optim. Theory Appl.*, 45:41 – 51, 1985).
6. B. Hajek. Cooling Schedules for Optimal Annealing. *Mathem. Oper. Res.*, 13:311 – 329, 1988.
7. W.E. Hart. A Theoretical Comparison of Evolutionary Algorithms and Simulated Annealing. In *Proc. of the 5th Annual Conf. on Evolutionary Programming*, pp. 147-154, 1996.
8. M. Kearns, M. Li, L. Pitt, and L.G. Valiant. Recent Results on Boolean Concept Learning. In *Proc. 4th Int. Workshop on Machine Learning*, pp. 337 – 352, 1987.
9. S. Kirkpatrick, C.D. Gelatt, Jr., and M.P. Vecchi. Optimization by Simulated Annealing. *Science*, 220:671 – 680, 1983.
10. F. Romeo and A. Sangiovanni-Vincentelli. A Theoretical Framework for Simulated Annealing. *Algorithmica*, vol. 6, no. 3, pp. 302 – 345, 1991.
11. E. Seneta. *Non-negative Matrices and Markov Chains*. Springer-Verlag, New York, 1981.
12. A. Sinclair and M. Jerrum. Approximate Counting, Uniform Generation, and Rapidly Mixing Markov Chains. *Information and Computation*, 82:93 – 133, 1989.
13. A. Sinclair and M. Jerrum. Polynomial-Time Approximation Algorithms for the Ising Model. *SIAM J. Comput.*, 22(5):1087 – 1116, 1993.

Actuator Noise in Recombinant Evolution Strategies on General Quadratic Fitness Models*

Hans-Georg Beyer

Department of Computer Science XI,
University of Dortmund, D-44221 Dortmund, Germany
hans-georg.beyer@cs.uni-dortmund.de

Abstract. This paper addresses the influence of actuator noise on the steady state behavior of multirecombinant evolution strategies (ES) on general quadratic fitness functions. Actuator noise degrades the ES's ability to locate the global optimizer. After a certain transient time the ES approaches a steady state behavior characterized by an expected fitness deviation from the global optimum. This expected value is calculated and the predictions are compared with ES runs on quadratic test functions.

1 Introduction

Actuator noise is a phenomenon widely observed in practice when trying to control the behavior of a device or machine by a set of control parameters which cannot be tuned exactly. While the control parameters can be prescribed exactly, its actual realization on the machine is disturbed by random perturbations such as vibrations (ground motion, turbulence effects, etc.) or other sources of noise (e.g., resistor noise, recombination noise, burst noise in electronic devices like resistors or transistors). If one wants to optimize the performance of such devices or machines, taking the actuator noise into account, one has to deal with goal functions which are intrinsically random functions. That is, optimizing such functions by neglecting its randomness can lead to a false optimal object parameter set.

Another problem domain with similar implications concerns *robust design* and optimization. Here one seeks to find optimal solutions which are robust with respect to random perturbation of design parameters [4,13,14]. The main application area is in the field of coping with production tolerances. There is only a limited degree of accuracy by which devices can be produced. Optimizing the design of a device has to serve two goals: On the one hand the device's performance, production costs, etc. should be maximal/minimal, on the other hand it must be "producable", i.e., the production process must allow for production tolerances. This can be achieved at the level of product design by superimposing random perturbations on the design variable modeling the impact of the production tolerances. Therefore, the aim of the design optimization process is

---

* This work was supported by the Deutsche Forschungsgemeinschaft (DFG) as part of the Collaborative Research Center (SFB) 531.

K. Deb et al. (Eds.): GECCO 2004, LNCS 3102, pp. 654–665, 2004.

basically not finding an optimum solution represented by a sharp peak in the fitness landscape, but rather optimal solutions which are less sensitive to small changes of the design parameters.

It is widely believed that evolutionary algorithms (EA) are good at such design tasks. These algorithms seem rather suited for finding large optimum attractors than finding the "needle in the haystack" peak. Up to now, there are only a few references addressing the question whether this "folklore" belief can be substantiated by hard and provable facts. Most investigations done on this topic are mainly of empirical nature [4,13] or consider special one-dimensional cases [14] without analyzing the EA's behavior on the test functions proposed. Only recently an attempt has been made to understand the behavior of evolution strategies (ES) on simple N-dimensional test functions disturbed by actuator noise [3,12]. These investigations revealed interesting behaviors such as (actuator) noise-induced bistabilities on a unimodal fitness landscape [12] and the appearance of an optimum localization error on a sphere model with actuator noise [3]. In this paper we apply a technique proposed in [3] to investigate the behavior of $(\mu/\mu_I, \lambda)$-ES[1] on *general* quadratic fitness functions disturbed by actuator noise. The results to be presented here extend the findings obtained for the simple (i.e. symmetrical) sphere model to a more realistic situation of a general quadratic fitness model. Such models can be regarded as local attractor models of real-world objective functions.

The rest of the paper is organized as follows. First, we will introduce the actuator noise model. Second, the steady state condition for the $(\mu/\mu_I, \lambda)$-ES on this fitness model will be derived. In Section 4 we compare the theoretical predictions with real ES runs. Finally, in the concluding section a short summary will be given including an outlook to future research.

2   The Actuator Noise Model

The actuator noise model was introduced in [3] to account for object parameter fluctuations like actuator jittering which are beyond the control of the user and the optimization algorithm, respectively. The model considered was the quadratic sphere. This paper investigates an *arbitrary* N-dimensional quadratic function $Q(\mathbf{y})$ (to be maximized)

$$Q(\mathbf{y}) := \mathbf{b}^{\mathrm{T}}\mathbf{y} - \mathbf{y}^{\mathrm{T}}\mathbf{Q}\mathbf{y} \tag{1}$$

with the N-dimensional real-valued vectors $\mathbf{b}$ and $\mathbf{y}$ and the symmetric (positive definite) matrix $\mathbf{Q}$. Given an object (or actuator) vector $\mathbf{y}$, the actually observed objective value, i.e. the fitness, is defined by the *actuator noise model*

$$F_{\mathrm{a}}(\mathbf{y}) := Q(\mathbf{y} + \mathbf{z}), \qquad \text{where} \qquad \mathbf{z} \sim \mathcal{N}(\mathbf{0}, \varepsilon^2 \mathbf{1}). \tag{2}$$

That is, each object parameter component is disturbed by independent normally distributed random events z_i with the same standard deviation ε.

---

[1] For the definitions of the evolution strategies used, see Appendix A.

3   Determination of the Steady State of the ES

The analysis of the steady state behavior follows the decomposition technique proposed in [3]. The basic idea is to transform the random function $F_a(\mathbf{y})$ in such a way that it appears as a sum of two parts, one carrying the stochastics and the other being deterministic. That is, the transformed problem appears as a fitness function with *additive* fitness noise. This transformation is admissible because the ES acts as a black-box algorithm which only uses the fitness information but not structural information from the fitness function. If we were able to make the transformation in such a way that we obtain a fitness noise model already analyzed then we are done. Therefore, the aim of the next section is to derive such a (approximative) model. As a result we will obtain the general quadratic model with (approximately) normally distributed fitness noise. Treating this model in Section 3.2 using techniques from [2] will yield the desired expected steady state fitness deviation from the global optimum.

3.1   Reducing the Actuator Noise Model to the General Quadratic Noisy Fitness Model

In order to obtain an approximative fitness noise model one has to decompose (2) into a deterministic part in terms of (1) and a normally distributed additive noise term. Since the $(\mu/\mu_I, \lambda)$-ES with isotropic mutations is considered, it is reasonable to express the fitness model and its decomposition in the eigensystem of the matrix $\mathbf{Q}$. Let q_i be the eigenvalues of $\mathbf{Q}$ and $\mathbf{e}_i$ the corresponding eigenvectors of length 1, i.e. $q_i \mathbf{e}_i = \mathbf{Q}\mathbf{e}_i$, the entire actuator noise model (1), (2) can be rewritten by a principal axes transformation as

$$F_a(\mathbf{y}) = \sum_{i=1}^{N} [b_i(y_i + z_i) - q_i(y_i + z_i)^2]$$

$$F_a(\mathbf{y}) = \sum_{i=1}^{N} [b_i y_i - q_i y_i^2] + \sum_{i=1}^{N} [(b_i - 2q_i y_i)z_i - q_i z_i^2], \qquad (3)$$

where $z_i \sim \mathcal{N}(0, \varepsilon^2)$ and $b_i = \mathbf{e}_i^{\mathrm{T}}\mathbf{b}$, $y_i = \mathbf{e}_i^{\mathrm{T}}\mathbf{y}$. Since it is the aim to decompose $F_a(\mathbf{y})$ in such a manner that

$$F_a(\mathbf{y}) \simeq E[F_a|\mathbf{y}] + \mathcal{N}(0, \mathrm{Var}[F_a|\mathbf{y}]) + \dots, \qquad (4)$$

one has to calculate $E[F_a|\mathbf{y}]$ and $\mathrm{Var}[F_a|\mathbf{y}]$. For the first conditional moment we easily obtain (recall $E[z_i] = 0$, $E[z_i^2] = \varepsilon^2$, $\sum q_i = \mathrm{Tr}[\mathbf{Q}]$)

$$E[F_a|\mathbf{y}] = \sum_{i=1}^{N} [b_i y_i - q_i y_i^2] - \sum_{i=1}^{N} q_i \overline{z_i^2} = Q(\mathbf{y}) - \varepsilon^2 \mathrm{Tr}[\mathbf{Q}]. \qquad (5)$$

For $\mathrm{Var}[F_a|\mathbf{y}]$ one obtains (recall that $E[z_i^3] = 0$, $E[z_i^4] = 3\varepsilon^4$, $\sum q_i^2 = \mathrm{Tr}[\mathbf{Q}^2]$)

$$\mathrm{Var}[F_a|\mathbf{y}] = \sum_{i=1}^{N} \mathrm{Var}[(b_i - 2q_i y_i)z_i - q_i z_i^2]$$

$$\text{Var}[F_a|\mathbf{y}] = \sum_{i=1}^{N} \left[\text{E}[((b_i - 2q_iy_i)z_i - q_iz_i^2)^2] - (\text{E}[(b_i - 2q_iy_i)z_i - q_iz_i^2])^2 \right]$$

$$= \sum_{i=1}^{N} \left[(b_i - 2q_iy_i)^2 \varepsilon^2 + 2q_i^2 \varepsilon^4 \right]$$

$$= \varepsilon^2 \sum_{i=1}^{N} (b_i - 2q_iy_i)^2 + 2\varepsilon^4 \text{Tr}[\mathbf{Q}^2].$$

$$= 4\varepsilon^2 \sum_{i=1}^{N} q_i^2 \left(y_i - \frac{b_i}{2q_i} \right)^2 + 2\varepsilon^4 \text{Tr}[\mathbf{Q}^2]. \tag{6}$$

Taking into account that the optimal state $\hat{\mathbf{y}}$ of $Q(\mathbf{y})$ is easily obtained from (5)

$$\hat{y}_i = \frac{b_i}{2q_i} \quad \text{and} \quad \hat{Q} := \max[Q] = \sum_{i=1}^{N} \frac{b_i^2}{4q_i}, \tag{7}$$

one gets $\text{Var}[F_a|\mathbf{y}] = 4\varepsilon^2 \sum_{i=1}^{N} q_i^2 (y_i - \hat{y}_i)^2 + 2\varepsilon^4 \text{Tr}[\mathbf{Q}^2]$. This can be written in vector notation

$$\text{Var}[F_a|\mathbf{y}] = 4\varepsilon^2 \|\mathbf{Q}(\hat{\mathbf{y}} - \mathbf{y})\|^2 + 2\varepsilon^4 \text{Tr}[\mathbf{Q}^2]. \tag{8}$$

Inserting (5) and (8) into (4) yields finally

$$F_a(\mathbf{y}) \simeq Q(\mathbf{y}) - \varepsilon^2 \text{Tr}[\mathbf{Q}] + \underbrace{\varepsilon \sqrt{4\|\mathbf{Q}(\hat{\mathbf{y}} - \mathbf{y})\|^2 + 2\varepsilon^2 \text{Tr}[\mathbf{Q}^2]}}_{=\sigma_\delta} \mathcal{N}(0,1). \tag{9}$$

Note, the constant term $-\varepsilon^2 \text{Tr}[\mathbf{Q}]$ (w.r.t. $\mathbf{y}$) in (9) is without relevance for the derivation of the evolution criterion because this term does not depend on the location in the object parameter space. While this is true for the derivations to be presented below, the effect of this term with respect to the attainable objective function values is of considerable importance because it degrades the maximal fitness independent of the ES used. Even if one were able to determine the optimal object parameter vector $\hat{\mathbf{y}}$, the expected value of the maximal fitness (7) $\hat{Q}$ will still be reduced by the term $\varepsilon^2 \text{Tr}[\mathbf{Q}]$.

3.2   Deriving the Evolution Criterion

Due to the (approximate) decomposition (9) we have reduced our problem to a case already known: Equation (14) in [2] characterizes the steady state behavior of the $(\mu/\mu_I, \lambda)$-ES on an arbitrary ellipsoidal function $Q(\mathbf{y})$ with Gaussian *fitness* noise of strength σ_δ

$$\|\mathbf{Q}(\hat{\mathbf{y}} - \mathbf{y})\|^2 \geq \frac{\sigma_\delta \text{Tr}[\mathbf{Q}]}{4\mu c_{\mu/\mu,\lambda}} \tag{10}$$

As one can see, the average deviation from the optimum without actuator noise comprises two terms: the constant term $\varepsilon^2 \text{Tr}[\mathbf{Q}]$ independent of the ES used and a strategy specific part $\text{E}[\Delta F]$ given by the equal sign in (21). $\text{E}[\tilde{\Delta} F]$ can be easily tested in ES runs: Since $Q(\hat{\mathbf{y}})$ is known for the models considered, it suffices to calculate the mean fitness over *all* offspring generated after reaching the vicinity of the steady state.[2]

4   Comparision with Experiments

The behavior of the $(\mu/\mu_I, \lambda)$-ES on the actuator noise function class (1), (2) have been tested on three ellipsoidal test functions given in Table 1 for dimensionality $N = 30$ and $N = 100$. Q_1 and Q_2 are axes-parallel ellipsoids. Q_3 has a

Table 1. Definitions and properties of the actuator noise test functions.

	Q_1	Q_2	Q_3
$Q(\mathbf{y}) :=$	$-\sum_{i=1}^{N} i y_i^2$	$-\sum_{i=1}^{N} i^2 y_i^2$	$-\sum_{j=1}^{N} \left(\sum_{i=1}^{j} y_i\right)^2$
$(\mathbf{Q})_{i,k} =$	$i\delta_{ij}$	$i^2 \delta_{ij}$	$\min[N - i + 1, N - j + 1]$
$\text{Tr}[\mathbf{Q}]_{N=30} =$	465	9455	465
$\text{Tr}[\mathbf{Q}^2]_{N=30} =$	9455	5273999	144305
$\text{Tr}[\mathbf{Q}]_{N=100} =$	5050	338350	5050
$\text{Tr}[\mathbf{Q}^2]_{N=100} =$	338350	2050333330	17003350

certain non-parallel orientation. Since we are using $(\mu/\mu_I, \lambda)$-ES (see Appendix A for its definition) with isotropic mutations, the orientation of the ellipsoid does not influence the performance of the strategy. However, Q_3 possesses a dominating eigenvalue, such that the shape of this ellipsoid resembles a distorted discus.

Similar to observations made on the behavior of ES on ellipsoidal test functions with fitness noise in [2], the σ control rule based on cumulative step-length adaptation (CSA) [6,7,8,9] does not work well on the test functions when the non-sphericity gets too large. This is shown in Fig. 1. The CSA-ES is not able to get close to the steady state but exhibits premature convergence: The mutation strength σ quickly reaches values too small for further object parameter evolution. There is a remedy to prevent this behavior by keeping σ above a certain limit σ_0. However, choosing σ_0 is a nontrivial task. Clearly, one should consider the covariance matrix adaptation (CMA-ES, [8]) instead, however, this is beyond the scope of this paper.

Figure 2 compares the predictive quality of (27) using (21) with ES runs. The data points (displayed as dots) have been obtained by recording the fitness values

---

[2] This assumes that the mutation strenght σ is sufficiently small. If this is not fulfilled, the fitness of the parental centroid must be evaluated at each generation in order to obtain $\tilde{\Delta} F$.

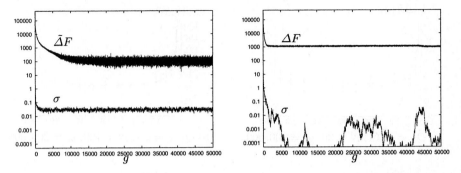

Fig. 1. Evolution dynamics of the $(20/20_I, 60)$-ES on the test function Q_2, $N = 30$, with actuator noise strength $\varepsilon = 0.1$. Adaptation of mutation strength σ is by σSA-ES (left-hand side) and by CSA (right-hand side). One observes the typical behavior of EAs on noisy problems: The fitness values reach a certain steady state distribution the expected value of which deviates from the optimum.

of the parental centroid states over a number of 200,000 generations starting after a number of generations g_0 (transient time for reaching the vicinity of the steady state). Since CSA-ES can exhibit premature convergence, the σSA-ES has been used. As one can see, the theory predicts the steady state behavior of the ES on Q_1 and Q_2 well (leaving aside the cases $\mu = 1$ and $\mu/\lambda \approx 1$). Unfortunately this does not hold for Q_3. In [2] the same test functions have been investigated, however, disturbed by fitness noise. There the authors found a good predictive quality on Q_3. Therefore, the reason for the deviations observed must be in the approximative decomposition (4): It has been assumed that the stochastics can be well approximated by a normal distribution. While this is indeed correct for Q_1 and Q_2 (actually, both functions reach normality exactly for $N \to \infty$) this is not the case for Q_3. As have already been mentioned, Q_3 has an eigenvalue spectrum where the ratio of the largest eigenvalue q_1 to the second largest eigenvalue q_2 approaches 9 (from below) as $N \to \infty$. This is in contrast to Q_1 and Q_2 where this ratio goes to 1. Even worse, considering the ratio $q_1 / \sum_{i=2}^{N} q_i$ one finds (numerically) that it approaches ≈ 4.279. In other words, the isolated large eigenvalue q_1 prevents Q_3 from reaching normality for $N \to \infty$ by violating the Lindeberg condition (see, e.g., [5]) and the central limit theorem of statistics does not apply. That is why, we do not observe an improved prediction quality for the $N = 100$ case compared to $N = 30$. The fitness noise produced by Q_3 has a high degree of skewness. The corresponding theory for non-Gaussian noise remains still to be developed.

5   Conclusions and Outlook

In this paper the impact of actuator noise on the steady state behavior of $(\mu/\mu_I, \lambda)$-ES optimizing general quadratic fitness functions has been analyzed. It has been shown that the decomposition method of [3] together with the equipar-

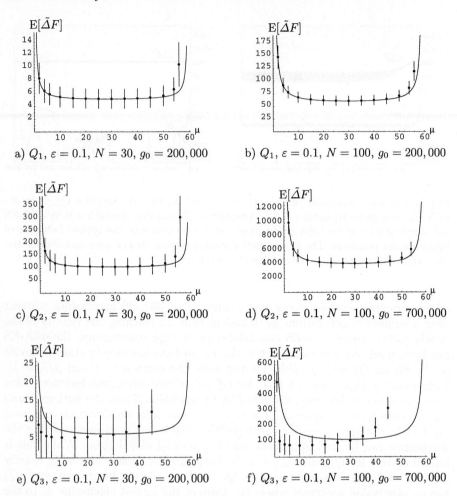

a) Q_1, $\varepsilon = 0.1$, $N = 30$, $g_0 = 200,000$

b) Q_1, $\varepsilon = 0.1$, $N = 100$, $g_0 = 200,000$

c) Q_2, $\varepsilon = 0.1$, $N = 30$, $g_0 = 200,000$

d) Q_2, $\varepsilon = 0.1$, $N = 100$, $g_0 = 700,000$

e) Q_3, $\varepsilon = 0.1$, $N = 30$, $g_0 = 200,000$

f) Q_3, $\varepsilon = 0.1$, $N = 100$, $g_0 = 700,000$

Fig. 2. Dependence of the expected fitness error $E[\tilde{\Delta}F]$ on the parent numbers $\mu = 1, 2, 4, 6, 10, 15, 20, 25, 30, 35, 40, 45, 50, 54, 56, 58, 59$ given fixed offspring number $\lambda = 60$. The vertical bars indicate the measured $\pm$ standard deviation of $\tilde{\Delta}F$. Missing data points are due to divergence (for μ/λ near 1) and premature convergence (for $\mu = 1$), respectively. The curves are the predictions made by (27) using the equal sign in (21).

tition assumption of [2] can be used to predict accurately the final fitness error (provided that the fitness noise induced is approximately normally distributed).

From the results obtained one can derive recommendations concerning the population sizing in order to get a minimal steady state fitness error $E[\tilde{\Delta}F]$: Looking at Fig. 2 one sees that – assuming normality of the actuator induced fitness noise (i.e., skipping Q_3) – $\mu/\lambda = 1/2$ yields minimal $E[\tilde{\Delta}F]$. On the other hand, $\mu/\lambda = 1/2$ is not the optimal population ratio for maximal progress toward the steady state. From sphere model theory we know that for $N \to \infty$ the ratio

$\mu/\lambda \approx 0.27$ should be preferred. Taking the behavior on Q_3 into account, a good population ratio compromise seems to be in the interval $0.2 \ldots 0.3$.

The results obtained might have more far-reaching implications. Consider the general quadratic fitness model as a local attractor model of real-world objective functions under actuator noise. The long term behavior of an ES at the end of an evolution process might be well described by such a fitness model and the steady state predictions of the theory might be valid for more complicated objective functions. Therefore, additional investigations are needed to determine the limitations of the model analysis presented.

As has been mentioned, the CSA-ES using isotropic mutations is not good at these ellipsoidal test functions with noise. It is reasonable to use non-isotropic mutations instead. This leads to the problem of adapting a full covariance matrix describing the distribution of the mutations. This is usually done by the CMA method [8]. Investigating the behavior of the CMA-ES should be one of the next steps in future research.

References

1. H.-G. Beyer. *The Theory of Evolution Strategies*. Natural Computing Series. Springer, Heidelberg, 2001.
2. H.-G. Beyer and D. V. Arnold. The Steady State Behavior of $(\mu/\mu_I, \lambda)$-ES on Ellipsoidal Fitness Models Disturbed by Noise. In E. Cantú-Paz et al., editors, *GECCO-2003: Proceedings of the Genetic and Evolutionary Computation Conference*, pages 525–536, Berlin, Germany, 2003. Springer.
3. H.-G. Beyer, M. Olhofer, and B. Sendhoff. On the Behavior of $(\mu/\mu_I, \lambda)$-ES Optimizing Functions Disturbed by Generalized Noise. In K. De Jong, R. Poli, and J. Rowe, editors, *Foundations of Genetic Algorithms, 7*, pages 307–328, San Francisco, CA, 2003. Morgan Kaufmann.
4. J. Branke. *Evolutionary Optimization in Dynamic Environments*. Kluwer Academic Publishers, Dordrecht, 2001.
5. M. Fisz. *Wahrscheinlichkeitsrechnung und mathematische Statistik*. VEB Deutscher Verlag der Wissenschaften, Berlin, 1971.
6. N. Hansen and A. Ostermeier. Adapting Arbitrary Normal Mutation Distributions in Evolution Strategies: The Covariance Matrix Adaptation. In *Proceedings of 1996 IEEE Int'l Conf. on Evolutionary Computation (ICEC '96)*, pages 312–317. IEEE Press, NY, 1996.
7. N. Hansen and A. Ostermeier. Convergence Properties of Evolution Strategies with the Derandomized Covariance Matrix Adaptation: The $(\mu/\mu_I, \lambda)$-CMA-ES. In H.-J. Zimmermann, editor, *5th European Congress on Intelligent Techniques and Soft Computing (EUFIT'97)*, pages 650–654, Aachen, Germany, 1997. Verlag Mainz.
8. N. Hansen and A. Ostermeier. Completely Derandomized Self-Adaptation in Evolution Strategies. *Evolutionary Computation*, 9(2):159–195, 2001.
9. A. Ostermeier, A. Gawelczyk, and N. Hansen. A Derandomized Approach to Self-Adaptation of Evolution Strategies. *Evolutionary Computation*, 2(4):369–380, 1995.
10. I. Rechenberg. *Evolutionsstrategie '94*. Frommann-Holzboog Verlag, Stuttgart, 1994.

11. H.-P. Schwefel. *Numerical Optimization of Computer Models*. Wiley, Chichester, 1981.
12. B. Sendhoff, H.-G. Beyer, and M. Olhofer. On Noise Induced Multi-Modality in Evolutionary Algorithms. In L. Wang, K.C. Tan, T. Furuhashi, J.-H. Kim, and F. Sattar, editors, *Proceedings of the 4th Asia-Pacific Conference on Simulated Evolution and Learning – SEAL*, volume 1, pages 219–224, 2002.
13. S. Tsutsui and A. Ghosh. Genetic Algorithms with a Robust Solution Searching Scheme. *IEEE Transactions on Evolutionary Computation*, 1(3):201–208, 1997.
14. D. Wiesmann, U. Hammel, and T. Bäck. Robust Design of Multilayer Optical Coatings by Means of Evolutionary Algorithms. *IEEE Transactions on Evolutionary Computation*, 2(4):162–167, 1998.

A   Description of the ESs Used

For the simulation of the dynamic behavior of $(\mu/\mu_I, \lambda)$-ES the ES must control the endogenous strategy parameter σ. We used the two standard approaches to this control problem: the σ self-adaptation [11,10] and alternatively the cumulative step size adaptation (CSA) [6,8].

The σ self-adaptation technique is based on the coupled inheritance of object and strategy parameters. Using the notation

$$\langle \mathbf{a} \rangle^{(g)} := \frac{1}{\mu} \sum_{m=1}^{\mu} \mathbf{a}_{m;\lambda}^{(g)} \tag{28}$$

for intermediate recombination (centroid calculation, i.e., averaging over the $\mathbf{a}$ parameters of the μ best offspring individuals), the $(\mu/\mu_I, \lambda)$-σSA-ES can be expressed in "offspring notation"

$$\forall l = 1, \ldots, \lambda : \begin{cases} \sigma_l^{(g+1)} := \langle \sigma \rangle^{(g)} e^{\tau \mathcal{N}_l(0,1)} \\ \mathbf{y}_l^{(g+1)} := \langle \mathbf{y} \rangle^{(g)} + \sigma_l^{(g+1)} \mathcal{N}_l(\mathbf{0}, \mathbf{1}). \end{cases} \tag{29}$$

That is, each offspring individual (indexed by l) gets its own mutation strength σ. And this mutation strength is used as mutation parameter for producing the offspring's object parameter. In (29) the log-normal update rule for mutating the mutation strength has been used. As learning parameter $\tau = 1/\sqrt{N}$ has been chosen in the simulations.

While in evolutionary self-adaptive ES each individual get its own set of endogenous strategy parameters, cumulative step-size adaptation (CSA) uses a single mutation strength parameter σ per generation to produce all the offspring. This σ is updated by a deterministic rule which is controlled by certain statistics gathered over the course of generations. The statistics used is the so-called (normalized) cumulative path-length $\mathbf{s}$. If $\|\mathbf{s}\|$ is greater than the expected length of a random path, σ is increased. In the opposite situation, σ is decreased. The update rule reads

$$\forall l = 1, \ldots, \lambda : \mathbf{y}_l^{(g+1)} := \langle \mathbf{y} \rangle^{(g)} + \sigma^{(g)} \mathcal{N}_l(\mathbf{0}, \mathbf{1})$$

$$\left. \begin{aligned} \mathbf{s}^{(g+1)} &:= (1 - c)\mathbf{s}^{(g)} + \sqrt{(2 - c)c} \frac{\sqrt{\mu}}{\sigma^{(g)}} \left(\langle \mathbf{y} \rangle^{(g+1)} - \langle \mathbf{y} \rangle^{(g)} \right) \\ \sigma^{(g+1)} &:= \sigma^{(g)} \exp \left(\frac{\|\mathbf{s}^{(g+1)}\| - \overline{\chi}_N}{D\overline{\chi}_N} \right) \end{aligned} \right\}, \tag{30}$$

where $\mathbf{s}^{(0)} = \mathbf{0}$ is chosen initially. The recommended standard settings for the cumulation parameter c and the damping constant D are used, i.e., $c = 1/\sqrt{N}$ and $D = \sqrt{N}$. For the expected length of a random vector comprising N standard normal components, the approximation $\overline{\chi}_N = \sqrt{N}(1 - 1/4N + 1/21N^2)$ was used.

B   The Progress Coefficient $c_{\mu/\mu,\lambda}$

The progress coefficient $c_{\mu/\mu,\lambda}$ is defined as the expectation of the average over the μ largest samples out of a population of λ random samples from the standard normal distribution. According to [1, p. 247], $c_{\mu/\mu,\lambda}$ can be expressed by a single integral

$$c_{\mu/\mu,\lambda} = \frac{\lambda - \mu}{2\pi} \binom{\lambda}{\mu} \int_{-\infty}^{\infty} e^{-t^2} \left(\Phi(t) \right)^{\lambda - \mu - 1} \left(1 - \Phi(t) \right)^{\mu - 1} dt, \tag{31}$$

where $\Phi(t)$ is the cumulative distribution function of the standard normal variate. The special $c_{\mu/\mu,\lambda}$ values used in this paper are given in the table below.

μ	1	2	4	6	10	15	20
$c_{\mu/\mu,60}$	2.31928	2.12722	1.88199	1.71349	1.47183	1.25171	1.07569
μ	25	30	35	40	45	50	56
$c_{\mu/\mu,60}$	0.924168	0.787546	0.66012	0.537847	0.417235	0.294366	0.134428
μ	58	59	60				
$c_{\mu/\mu,60}$	0.0733524	0.0393098	0				

Convergence Examples of a Filter-Based Evolutionary Algorithm

Lauren M. Clevenger[1] and William E. Hart[2]

[1] University of New Mexico, lmcleve@aol.com
[2] Sandia National Laboratories, Discrete Mathematics and Algorithms Dept.,
P. O. Box 5800, MS 1110, Albuquerque, NM 87185-1110; Phone: 505-844-2217
Fax: 505-845-7442; wehart@cs.sandia.gov; http://www.cs.sandia.gov/~wehart/

Abstract. We describe and critique the convergence properties of filter-based evolutionary pattern search algorithms (F-EPSAs). F-EPSAs implicitly use a filter to perform a multi-objective search for constrained problems such that convergence can be guaranteed. We provide two examples that illustrate how F-EPSAs may generate limit points other than constrained stationary points. F-EPSAs are evolutionary pattern search methods that employ a finite set of search directions, and our examples illustrate how the choice of search directions impacts an F-EPSA's search dynamics.

1 Introduction

Although evolutionary algorithms (EAs) have been successfully applied to many unconstrained optimization applications, the investigation of constrained EAs has received far less attention [6]. Despite this, handling constraints in EAs is necessary for their application to many problem domains. Thus the development of provably robust EAs is crucial to ensure that these methods can be effectively applied to a wide range of problems, including linear, non-linear, equality and inequality constraints [4].

Of the many different constraint-handling techniques used with EAs, the most common are penalty functions. Although penalty functions can have good convergence properties for specific problems, some penalty functions require an initial feasible solution that must be provided by the user and penalty approaches may also require extra parameters that can be hard to choose, especially when they are problem-dependent. Alternatives to penalty functions tend to be developed for very specific problems and problems in which estimating good penalty functions and generating even a single feasible solution are difficult. Some of the techniques surveyed by Coello [6] include approaches that use problem-specific representations and operators, algorithms that repair infeasible points to make them feasible, and approaches that separate objectives and constraints (e.g. multi-objective optimization techniques). Unfortunately, these methods sometimes have difficulty preserving diversity and avoiding stagnation. Additionally, some of these approaches require the generation of an initial feasible point (or population), which is often NP-hard [13].

K. Deb et al. (Eds.): GECCO 2004, LNCS 3102, pp. 666–677, 2004.

Considering all of these challenges for handling constraints in EAs, an approach that minimizes these difficulties and still maintains good convergence results is very desirable. We propose filter-based evolutionary algorithms (FEAs), which use a constraint-handling technique that is similar to multi-objective EAs. The optimization problem that we consider is

$$\min_{x \in \mathbf{R}^n} \ f(x)$$

$$\text{s.t.} \ \ C(x) \leq 0$$

$$l \leq x \leq u$$

where $f : \mathbf{R}^n \to \mathbf{R} \cup \{\infty\}$ and $C : \mathbf{R}^n \to (\mathbf{R} \cup \{\infty\})^m$ are the constraint functions with $C = (C_1, \ldots, C_m)^T$; $u, l \in \mathbf{Q}^n$ define upper and lower bounds on each dimension.

Clevenger, Ferguson and Hart [8,4] have recently developed a filter-based evolutionary pattern search algorithm (F-EPSA) for constrained optimization. This F-EPSA is closely related to pattern search methods that do not attempt to estimate a derivative in its search process. The main qualitative difference between pattern search methods and common real-coded EAs is that pattern search methods restrict the search in each iteration to a finite pattern of trial points, while most real-coded EAs employ continuous random variables to generate mutation steps. However, this restriction provides mathematical leverage for demonstrating convergence of pattern search methods, which has been effectively translated to EAs [8,11].

We evaluate the convergence behavior of the F-EPSA defined by Clevenger, Ferguson, and Hart [4,8]. This F-EPSA uses a pattern of search steps so the search behavior is dependent on this pattern. We provide two examples that illustrate this dependence. In both cases the F-EPSA converges to a limit point, but the limit points are not constrained stationary points. The following section describes F-EPSAs. Section 3 discusses these two examples.

2   Algorithmic Formulation

A filter-based optimizer uses a nonnegative continuous function to aggregate the constraint violations and then treats the resulting bi-objective problem (e.g., see Fletcher et al. [9,10]). In other words, a filter-based optimizer tries to minimize both the objective function and the aggregate constraint violation function simultaneously. Since a feasible solution is desired, priority is usually given to the aggregate function until a feasible solution is found. We give two definitions that are very similar to those stated in Audet and Dennis [1], which will be used throughout this paper.

Definition 1 *Given objective functions $f_i()$, $i = 0, \ldots, k$, if $f_i(x_1) \leq f_i(x_2)$ for every $i \in 0, \ldots, k$ and there is at least one j such that $f_j(x_1) < f_j(x_2)$, then x_1 is said to* dominate x_2. *This is denoted by $x_1 \prec x_2$. Also, $x_1 \preceq x_2$ denotes that either $x_1 \prec x_2$ or $x_1 = x_2$.*

Definition 2 *Given a set of points S, a point $s \in S$ is said to be a non-dominated point if there does not exist $x \in S$ such that $s \prec x$.*

A *filter* F is a (finite) set of points in $\mathbf{R}^n$ such that no pair x, x' in the filter are in the relation $x \prec x'$. That is, no point in F dominates or is dominated by any other point in F. Filter-based optimizers employ a filter that is used to eliminate trial points from consideration if they are dominated by points in the filter either by having a worse function value or worse aggregate constraint violation.

2.1   A Filter-Based EA

Figure 1 presents the basic steps of Algorithm A, the F-EPSA introduced by Clevenger and Hart [8]. This EA evolves a set of points $W_t = Y_t \bigcup X_t$, where Y_t are infeasible and X_t are feasible. We say $x \prec x'$ if and only if $(f(x), h(x)) \prec (f(x'), h(x'))$, where $f(x)$ is the objective function and $h(x) = \sum_{i=1}^m \max[0, C_i(x)]^2$. Note that $h(x) = \infty$ if any of the constraint function values at x are infinite.

This F-EPSA implicitly uses a filter, which is the subset of W_t containing the best non-dominated infeasible solutions Y_t and the best feasible solution x_t^*. Let x_t^* be the point in X_t with the best function value, and y_t^* be the point in Y_t with the minimal constraint violation (as defined by h). If two points have the same minimal constraint violation, y_t^* is the point with the minimal function value.

Intuitively, a point is locally optimal if it cannot be improved. In the context of an F-EPSA, a point cannot be improved if the mutation steps about it are dominated by the points in the filter contained in W_t.

Suppose that all points in $\mathcal{N}(x, \Delta_t, D_t) = \{x + \Delta_t d \mid d \in D_t\}$ have been generated. If either x_t^* or y_t^* is in this set, then x is locally optimal if $(f(x), h(x))$ equals $(f(x_t^*), h(x_t^*))$ or $(f(y_t^*), h(y_t^*))$, so generating this mutation step does not give a simple improvement in either x_t^* or y_t^*. Suppose that neither x_t^* nor y_t^* is in $\mathcal{N}(x, \Delta_t, D_t)$. Then all of the points in $\mathcal{N}(x, \Delta_t, D_t)$ are dominated by either x_t^* or y_t^*. Further, in subsequent iterations these values will only improve so for all $t' > t$, the points in $\mathcal{N}(x, \Delta_t, D_t)$ are dominated by either $x_{t'}^*$ or $y_{t'}^*$

The following provides further details about the definition of Algorithm A:

- X_1 and Y_1 could be simply initialized by randomly generating P points within the bound constraints, and then applying the standard update rule. However, in practice this initialization could exploit domain knowledge of the structure of the constraints.
- D is a finite set of mutation offsets that can be applied. All subsets $D_t \subseteq D$ must be selected to ensure that D_t is a positive spanning set (i.e. non-negative linear combinations of points in D_t generate $\mathbf{R}^n$).
- The determination of whether x_{t+1}^* or y_{t+1}^* is locally optimal is not made with respect to the current population W_t. Instead, this requires the explicit cataloging of the history of mutation steps about these points.

Given Δ_0, $\tau > 1$ $(\tau \in Q)$ and mutation directions D
Randomly initialize X_0 and Y_0; $W_0 = X_0 \bigcup Y_0$
Select $D_0 \subseteq D$
For $t = 0, \ldots, \infty$
    For $j = 1, \ldots, P$
        Randomly select $d \in D_t$ and $w \in W_t$
        $\hat{w}_j = \Delta_t d + w$
        Evaluate $\hat{w}_j$
    End For
    Update X_{t+1}, Y_{t+1}; $W_{t+1} = X_{t+1} \bigcup Y_{t+1}$
    Update x^*_{t+1} and y^*_{t+1}
    If $(f(x^*_{t+1}) < f(x^*_t))$ or
        $(h(y^*_{t+1}) < h(y^*_t))$ or
        $((h(y^*_{t+1}) = h(y^*_t))$ and $(f(y^*_{t+1}) < f(y^*_t)))$ Then
        $\Delta_{t+1} = \Delta_t \tau^\nu$, where $0 \leq \nu \leq \nu_{max}$
        Select $D_{t+1} \subseteq D$
    Else If x^*_{t+1} or y^*_{t+1} is locally optimal Then
        $\Delta_{t+1} = \Delta_t \tau^\nu$, where $\nu_{min} \leq \nu < 0$
        Select $D_{t+1} \subseteq D$
    Else
        $\Delta_{t+1} = \Delta_t$ and $D_{t+1} = D_t$
    Terminate if $\Delta_{t+1} < \Delta_{min}$
End For

Fig. 1. Pseudo-code for Algorithm A. For simplicity, we have not included the checks to see if either x^*_{t+1} or y^*_{t+1} are not defined because a feasible or infeasible point has not been encountered by iteration $t+1$. These checks would be used in all of the conditional statements after x^*_{t+1} and y^*_{t+1} are updated.

- Algorithm A updates the step length Δ_t by (a) possibly increasing it if some new point dominates either x^*_t or y^*_t, or (b) decreasing it if x^*_{t+1} or y^*_{t+1} are locally optimal (and thus no progress can be made about these points using D_t).
- Algorithm A terminates if the step length shrinks below some predetermined threshold, which is the termination rule commonly used with pattern search methods.

2.2 Related Constrained EAs

The concept of a filter is directly analogous to the notion of an *archive* of pareto optimal solutions, which has been used in a wide range of evolutionary algorithms (e.g., see Knowles and Corne [12]). The method of constraint handling proposed here shares some commonalities with a few of the techniques surveyed by Coello [6]. Since Algorithm A separates constraints from objectives, it is most similar to approaches that also use this separation. Consider the Similarity of Feasible Points technique proposed by Deb [7]. Deb gives three rules for comparing points:

1. A feasible point is always preferred over an infeasible one.
2. Between two feasible points, the one having a better objective function value is preferred.
3. Between two infeasible points, the one having a smaller constraint violation is preferred.

Deb's method also includes a selection procedure that only performs pairwise comparisons so that no penalty factor is required [6]. Similarly, Algorithm A performs pairwise comparison for selection and follows rules 2 and 3 of Deb's method. It does not necessarily follow the first rule because we want to keep infeasible points to ensure a robust search.

Algorithm A is also similar to some of the multi-objective optimization techniques surveyed by Coello [5]. The most closely related technique is the one proposed by Camponogara and Talukdar [2]. Their procedure restates a single optimization problem to consider two objectives: the optimization of the original objective function and the optimization of

$$\Phi(x) = \sum_{i=1}^{n} \max[0, C_i(x)].$$

Thus Φ is the analogue of h using the L_1 norm instead of the squared L_2 norm. Camponogara and Talukdar use pareto sets (implicitly using a filter) to impose dominance-based selection, which is used to estimate new search directions. The technique we propose implicitly uses a filter to impose dominance-based selection, but it is not used to generate new search directions. Instead, the filter is used to determine when step lengths are expanded and contracted (by imposing conditions for local optimality).

3   Convergence Analysis

Although Algorithm A is quite similar to several existing EAs, the structure of this F-EPSA ensures that with probability one, some subsequence of the points $\{x_t^*, y_t^*\}$ generated by Algorithm A provably converges. Let X_t and Y_t be the stochastic processes, defined on some probability space (Ω, F, P), that describe the behavior of Algorithm A for some problem and for some set of algorithmic parameters. We make the standard assumption that the processes X_t and Y_t generate points that lie in a compact set.

Ferguson and Hart [8] summarize a convergence theory for which Algorithm A generates a convergence subsequence with probability one. They assume that f is strictly differentiable at the limit point, which implies that $\nabla f(x)$ exists and

$$\nabla f(x)^T \omega = lim_{y \to x, t \downarrow 0} \frac{f(y + t\omega) - f(y)}{t}$$

for all $\omega \in \mathbf{R}^n$ [3]. If the limit point is strictly feasible, then the limit point is a first-order stationary point. Otherwise, the algorithm may converge to a constrained local optimimum for a problem that is implicitly defined by the set of search directions in D. Let $\hat{x}$ be a limit point of a convergent subsequence generated by Algorithm A. A convergent subsequence (for some set of indices K) is said to be refining if $lim_{k \in K} \Delta_t = 0$. Ferguson and Hart [8] describe the following result, which illustrates how an L-EPSA converges near a constraint boundary.

Theorem 1. *Let $\hat{x}$ be a limit point of a refining subsequence generated by Algorithm A that is not strictly feasible. Let $D' \subseteq D$ be the set of all the associated directions of all the refining subsequences that converge to the limit point $\hat{x}$ in such a manner that the constraint violation is constant. If f is strictly differentiable at $\hat{x}$, then $\nabla f(\hat{x})$ belongs to the polar of the cone generated by D'. If h is strictly differentiable at $\hat{x}$ then $\nabla h(\hat{x}) = 0$.*

Theorem 1 is not quite as strong as would be desired, since it does not guarantee convergence to a first-order constrained stationary point. In particular, this result depends on the set of search directions D that are defined, since this ultimately limits the cone that contains $\nabla f(\hat{x})$. Thus Algorithm A will perform a more robust search for constrained local minima as the number of search directions in D is increased.

None of these results ensures convergence to a globally optimal feasible point. It is not clear that such a convergence theory exists for methods like Algorithm A that dynamically adapt their search step lengths without imposing fundamental limitations on their adaptive dynamics (e.g. lower bounds on the step lengths). Our analysis provides insight into mechanisms that facilitate robust local convergence without concern of the global search dynamics. However, the efficacy of the global search is clearly influenced by the algorithmic choice, and we expect that methods like Algorithm A will perform a more global search than the pattern search methods discussed by Audet and Dennis [1].

The following examples illustrate the implications of Theorem 1 on two test problems. Specifically, these examples illustrate how the choice of search directions can limit the ability of Algorithm A to converge to constrained stationary points. In all of our examples we consider the case where Algorithm A is used with a single search pattern throughout the search, which is consistent with the manner in which most pattern search methods are employed. We discuss this point further in the next section.

3.1   Example I

Consider the problem

$$\min -ab$$
$$\text{s.t.} \ \ a^2 + b^2 \le 16$$
$$-4 \le a, b \le 4$$

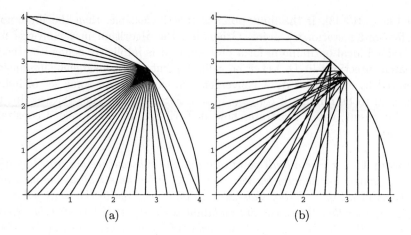

Fig. 2. Illustration of the convergence of Algorithm A for Example I when (a) $D_t = D = \{\pm(1,1), \pm(1,-1)\}$ and (b) $D_t = D = \{\pm(1,0), \pm(0,1)\}$. The two axes represent the coordinates of the solutions in this two-dimensional space. The lines in these figures connect the initial point and final feasible point. The lack of symmetry in (b) is due to the fact that ties are broken arbitrarily.

for which the optimal solution is $x^* = (a^*, b^*) = (2\sqrt{2}, 2\sqrt{2})$. Consider the behavior of Algorithm A using the search pattern $D_t = D = \{\pm(1,1), \pm(1,-1)\}$. We simplify our presentation by assuming that $\mu_F = \mu_I = 1$ and $P = 8$, so all mutation steps are generated in each iteration. We consider feasible starting points, so in initial iterations P is effectively equal to 4.

Figure 2a illustrates the convergence behavior of Algorithm A when started from a set of points along the x- and y-axes. The lines in this figure connect an initial point and final feasible point, and it is clear that in every case the F-EPSA converges to the optimal solution. Note that $-\nabla f(x) = (b, a)$. Now suppose that Algorithm A generates a limit point $\hat{z} = (a, b)$ on the constraint boundary for which $b > a$. It follows that the directions $(-1, -1)$ and $(1, -1)$ are the associated directions at this limit point (because the constraint violation function is constant in these directions), and they define the cone C_s. The polar cone C_s^o is defined by the directions $(-1, 1)$ and $(1, 1)$. However, at $\hat{z}$ we have $-\nabla f(\hat{z}) = (b, a)$ which is not in this cone if $b > a$ (note that $(-1, 1)^T (b, a) = a - b < 0$). Consequently, the only limit point that Algorithm A could generate that satisfies the conditions of Theorem 1 is the point (a^*, b^*).

The contrast between these two search patterns highlights the degree to which the choice of search pattern can impact how closely Algorithm A converges to constrained stationary points. If the search pattern is selected well, you may be able to ensure that a constrained stationary point is generated, but if the search pattern is selected poorly then any point on the nonlinear constraint boundary may be a limit point. Furthermore, it is clear that if the search pattern $D_t = D = \{\pm(1,1), \pm(1,-1)\}$ were perturbed slightly then this F-EPSA could converge to points other than the constrained stationary point. Consequently, this method may be sensitive to numerical instabilities such as round-off errors.

3.2   Example II

Consider the problem

$$\begin{aligned} \min \quad & f(a,b) \\ \text{s.t.} \quad & \tfrac{1}{5}a + \tfrac{4}{5} \geq b \\ & 5a - 4 \leq b \\ & 0 \leq a, b \end{aligned}$$

where $f : \mathbf{R}^2 \to \mathbf{R}$ is an arbitrary function. The solution to this problem lies within the feasible region, but we consider the convergence of Algorithm A starting from an initial point (λ, λ) for some $\lambda > 1$. The following analysis shows that Algorithm A converges to the point $(1,1)$ on the constraint boundary, regardless of whether this is a constrained stationary point. In fact, all iterates remain infeasible on this problem, so f could even be minimized at a strictly feasible point. Again, we assume that $\mu_F = \mu_I = 1$, and that all mutation steps are generated in each iteration (so we are taking the best of all neighboring points).

Figure 3 illustrates the initial point and the three search directions in the search pattern used in this example. From a point (a, b) the solution set steps during the search are 120 degrees apart from one another, given by

$$\left(a + \Delta \cos\left(\frac{\pi}{4}\right), b + \Delta \sin\left(\frac{\pi}{4}\right) \right) = \left(a + \Delta\sqrt{2}/2, b + \Delta\sqrt{2}/2 \right)$$

$$\left(a + \Delta \cos\left(\frac{-5\pi}{12}\right), b + \Delta \sin\left(\frac{-5\pi}{12}\right) \right) = \left(a + \Delta\sqrt{2}\,\omega_2, b - \Delta\sqrt{2}\,\omega_1 \right)$$

$$\left(a + \Delta \cos\left(\frac{11\pi}{12}\right), b + \Delta \sin\left(\frac{11\pi}{12}\right) \right) = \left(a - \Delta\sqrt{2}\,\omega_1, b + \Delta\sqrt{2}\,\omega_2 \right)$$

where $\omega_1 = (\sqrt{3}+1)/4$ and $\omega_2 = (\sqrt{3}-1)/4$. We label these points $\bar{a}$, $\bar{b}$ and $\bar{c}$ respectively, and we label the initial point $\bar{x}$.

We denote constraint (1) to be $b \geq \tfrac{1}{5}a + \tfrac{4}{5}$ and constraint (2) to be $b \leq 5a - 4$. Let $D_1^{\bar{x}}$ be the shortest squared distance from $\bar{x}$ to constraint (1), and let $D_2^{\bar{x}}$ be the shortest squared distance from $\bar{x}$ to constraint (2). We define similar values for $\bar{a}$, $\bar{b}$, and $\bar{c}$. To compute these values, we need to be able to compute the shortest squared distance from a point to the constraints that point is violating. The following lemma defines the point on a line that is closest to a given point.

Lemma 1. *The shortest squared distance between a point* (r, s) *and a line* $y = mx + b$ *is at* $x = \frac{r + (s-b)m}{m^2 + 1}$.

Proof. Let $f(x) = (x-r)^2 + (mx+b-s)^2$, which is the squared distance between (r, s) and the point on the line $(x, mx + b)$. To find the minimal distance, we minimize $f(x)$, which occurs when $f'(x) = 0$. $f'(x) = 2(x-r) + 2m(mx+b-s)$, which is zero at $x = \frac{r + (s-b)m}{m^2 + 1}$.

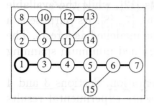

1 3 4 5 14	1 2 8
1 2 8	1 3 9 10
1 3 4 11 12 13	1 3 4 11 12 13
1 3 4 5 6 7	1 3 4 5 14
1 3 9 10	1 3 4 5 6 7
1 3 4 5 15	1 3 4 5 15

(a) A graph with a spanning tree indicated by thick edges

(b) The main chains of the spanning tree

(c) The main chains properly grouped

Fig. 1. A graph with a spanning tree and its main chains

The pairs must be disposed in the list in the same order they are in the intermediate representation, considering the chains from top to bottom and, in each chain, the nodes from left to right (see Figure 2(c)).

(a) Properly grouped main chains of a spanning tree

1 2 8
1 3 9 10
1 3 4 11 12 13
1 3 4 5 14
1 3 4 5 6 7
1 3 4 5 15

(b) Representation by main chains without repeated nodes

1 2 8
3 9 10
4 11 12 13
5 14
6 7
15

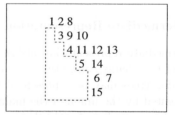

$$\begin{bmatrix} \text{depth} \\ \text{node} \end{bmatrix} \quad \begin{bmatrix} 0 & 1 & 2 & 1 & 2 & 3 & 2 & 3 & 4 & 5 & 3 & 4 & 4 & 5 & 4 \\ 1 & 2 & 8 & 3 & 9 & 10 & 4 & 11 & 12 & 13 & 5 & 14 & 6 & 7 & 15 \end{bmatrix}$$

(c) Node-depth Representation

Fig. 2. Node-depth encoding from the intermediate representation

2.3   Forest Encoding

The proposed forest encoding is composed by the union of the encodings of all trees of a forest. In this way, the forest data structure can be easily implemented using an array of pointers, where each pointer indicates the node-depth encoding of a tree of the forest.

3   Operators

This Section presents two operators (called **operator 1** and **operator 2**) to generate new spanning forests using the node-depth encoding. Both operators generate a spanning forest F' of a graph G when they are applied to another spanning forest F of G.

The results produced by the application of both operators are similar. The application of the operator 1 (or 2) to a forest is equivalent to transfer a subtree from a tree T_{from} to another tree T_{to} of the forest. Applying operator 1, the root of the pruned subtree will be also the root of this subtree in its new tree (T_{to}). On the other hand, the transferred subtree will have a new root (any node of the subtree different from the original root) when applying operator 2.

In this way, the operator 1 can produce simple and small changes in the forest. The operator 2 can generate larger and more complex alterations.

The operator 1 requires a set with two nodes previously determined: the prune node p, which indicates the root of the subtree to be transferred; and the adjacent node a, which is a node of a tree different from T_{from} and that is also adjacent to p in G.

The operator 2 requires a set with three nodes: the prune node p, the adjacent node a, and the new root node r of the subtree.

In the following, we explain both operators considering that the required set of nodes were previously determined. We show how to efficiently obtain these sets of nodes in Section 4.

3.1   Operator 1

In the description of the operator 1, we consider that the nodes p and a were previously chosen and that the node-depth representation were implemented using arrays. Besides, we assume the indices of p (i_p) and a (i_a) in the arrays T_{from} and T_{to}, respectively, are also known.

The operator 1 can be described by the following steps (see Figures 3(a), 3(b) and 3(c)):

1. Determine the range (i_p-i_l) of indices in T_{from} corresponding to the subtree rooted at the node p. Since we know i_p, we only need to find i_l. The range (i_p-i_l) corresponds to the consecutive nodes x in the array T_{from} such that $i_x \geq i_p$ and $d_x \geq d_p$ (the dashed lines in Figure 3(a)), where d_x is the depth of the node x;
2. Copy the data in the range i_p-i_l from T_{from} into a temporary array T_{tmp} (containing the data of the subtree being transferred), see Figure 3(b). The depth of each node x from the range i_p-i_l is updated as follows: $d_x = d_x - d_p + d_a + 1$;

3. Create an array T'_{to} containing the nodes of T_{to} and T_{tmp} (i.e., generate a new tree connecting the pruned subtree to T_{to}), see Figure 3(c);
4. Construct an array T'_{from} comprising the nodes of T_{from} without the nodes of T_{tmp};
5. Copy the forest data structure F to F' exchanging the pointers to the arrays T_{from} and T_{to} for pointers to the arrays T'_{from} and T'_{to}, respectively.

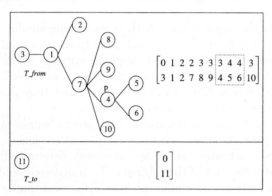

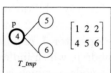

(a) T_{to}, T_{from} and their node-depth representations

(b) T_{tmp} and its node-depth representation

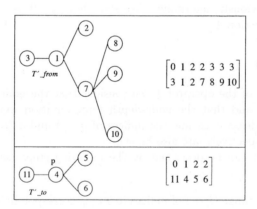

(c) T'_{to}, T'_{from} and their node-depth representations

Fig. 3. Example of application of the operator 1

3.2   Operator 2

The operator 2 requires a set of three nodes: the prune node p, the new root node r and the adjacent node a. The nodes p, r are in the tree T_{from} and a is in T_{to}.

The differences between operator 1 and operator 2 are in the steps 2 and 3 (see the operator-1 procedure, Section 3.1), i.e. only the formation of pruned subtrees and their storing in temporary arrays are different.

In the sequel, we described the steps 2 and 3 for the operator 2. Figures 4(a), 4(b) and 4(c) provide an illustrative example of these steps.

The procedure of copy of the pruned subtree for the operator 2 can be divided in two steps: The first step is similar to the step 2 for the operator 1 and differs from it in the exchanging of i_p by i_r. The array returned by this procedure is called T_{tmp1}.

The second step considers the nodes in the chain from r to p (i.e. r_0, r_1, $r_2, \ldots, r_n$, where $r_0 = r$ and $r_n = p$) as roots of subtrees (see the highlighted nodes in Figure 4(a)). The subtree rooted at r_1 contains the subtree rooted at r_0. The subtree rooted at r_2 contains the subtree rooted at r_1, and so on (see Figure 4(a)). The algorithm for the second step should copy the subtrees rooted at r_i $(i = 1, \ldots, n)$ without the subtree rooted at r_{i-1} (see Figure 4(b)) and store the resultant subtrees in a temporary array T_{tmp2} (see Figure 4(c)).

The step 3 of the operator 1 creates an array T'_{to} from T_{to} and T_{tmp}. On the other hand, the operator 2 utilizes both temporary arrays T_{tmp1} and T_{tmp2} to construct T'_{to}.

3.3   Operators for One-Tree Forests

The proposed operators require a forest with at least two trees. However, it is possible to utilize the same operators for forests with one tree. First, we add to the original forest with one tree (denoted T_{uniq}) an auxiliar tree T_{aux} containing only one node.

Second, the application of the operator 1 (2), given p and a (p, r, and a), is divided into two steps. Initially, the operator 1 is utilized to transfer the pruned subtree to the auxiliar tree ($T_{from} = T_{aux}$) using the node a equal to the unique node in T_{aux}. Afterward, we apply the operator 1 (or 2) to transfer the subtree from T_{aux} (T_{from}) to the tree T_{uniq} (T_{to}) using the original value of a.

4   Determination of the Nodes p, r, and a

As described in Section 3, the operators 1 and 2 require a set of predefined nodes and their positions in F. Next, we present a strategy to locate a given node in a forest F. Subsection 4.2 describes an efficient procedure to find adequate nodes p, r, and a.

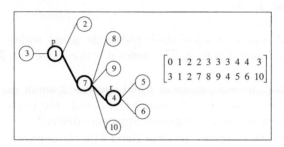

(a) T_{to} and its node-depth representation

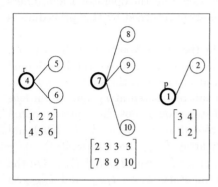

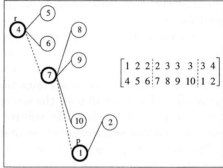

(b) Subtrees rooted at the nodes in the chain from r to p

(c) Node-depth Representation of the pruned subtree

Fig. 4. Example of determination of T_{tmp2}. The thick lines highlight the nodes in the chain from r to p. The depth values shown in this Figure consider the depth of the node a equal to zero

4.1   The Node Position in F

The determination of the position of a node in F can be efficiently achieved using auxiliar matrices, here named Π_x's, and a vector, here named π. Each node x of G possesses its correspondent matrix Π_x. For the original spanning forest F_0 of G, Π_x is a column matrix: $\Pi_x = \begin{bmatrix} 0 \\ i_0 \\ j_0 \\ k_0 \end{bmatrix}$, where i_0 is the index of the tree that contains x (T_{i_0}), j_0 is the index corresponding to x in the array T_{i_0} and k_0 is the depth of x in its tree.

Suppose a forest F_h is being generated from F_g ($g < h$) and x is in the subtree that will be transferred to a new tree generating F_h. Then, x will have a new position in F_h different from its position in F_g. So, we insert a new column in Π_x with the indices correspondent to this new position. The altered matrix results in $\Pi_x = \begin{bmatrix} 0 & h \\ i_0 & i_h \\ j_0 & j_h \\ k_0 & k_h \end{bmatrix}$. The position update is carried out for all nodes of the transferred subtree (see Section 3).

The vector π stores the forest g, from which the forest F_h was generated, at the rank h of π, i.e. $\pi(h) = g$. The parent of g is $\pi(g)$, the parent of $\pi(g)$ is $\pi(\pi(g))$, and so on. This constitutes a linked list with all precedessors of F_h. Obviously, the last position change of x occured in one of predecessors of h. In this way, we can look for the predecessors of h in the columns of Π_x. We start searching for $\pi(h)$. If this column is not found in Π_x, we try the column $\pi(\pi(h))$, and so on. The process of looking for such columns in Π_x can be achieved efficiently by running a binary search [13] on the list given by $\Pi_x(0, \cdot)$ (the first row of Π_x).

Once identified a column with a predecessor of h, we only need to read the position indices of x stored in the same column.

4.2   Choice of the Nodes p, r, and a

The proposed operators require a special set of nodes in order to generate a spanning forest F' of G based on another spanning forest F of G.

For the operator 1, this set can be efficiently obtained by the following strategy:

1. Pick up, randomly, an index of a tree in F and, for this tree, pick up, randomly, a node index that is not the tree root. Call p the correspondent node.
2. Pick up, randomly, a node adjacent to p (using the node adjacent list of G). Call this node a. If $a \notin T$, determine its position in F using the vector π and matrix Π_a; otherwise pick up, randomly, another a or return to step 1.

The strategy for the determination of p and a for the operator 2 works as follows:

1. Pick up randomly an index of a tree in F and, for this tree, pick up randomly a node index that is not a root. Call p the correspondent node.
2. Determine the range of nodes in the subtree rooted at p as in the step 1 of operator 1.Choose randomly an index of the selected range. Call r the correspondent node;
3. Pick up randomly a node adjacent to r (using the node adjacent list of G). Call this node a. If $a \notin T$, determine its position in F using the vector π and matrix Π_a; else pick up randomly another a or return to step 1.

5   Tests

This Section presents an evaluation of the proposed procedure for the degree-constrained minimum spanning tree problem [7]. The tests consider 11 complete graphs with number of vertices from 15 to 1000. For each graph, the constraint degree varies from 3 to 5. The edge weights were randomly obtained from the interval ranging from 1 to the number of vertices. The proposed approached was also compared with a Genetic Algorithm using the Prufer Encoding (GAPE) [6], [4].

Table 5 shows the results, where best cost is the mean of the best individuals in 20 trails and time is the mean running time corresponding to these individuals. Tournament was used for selection in both methodologies to reduce the running time.

The results suggest that the proposed procedure can deal with the degree constrained minimum spanning tree. Besides this methodology seems adequate to work with large graphs.

Table 1. Results from the proposed algorithm and GAPE applied to the degree constrained minimum spanning tree problem

Graph	Vertices	MST degree	Proposed Algorithm		GAPE	
			BestCost	Time(s)	Best Cost	Time(s)
1	15	3	23.0	0.14	37.5	0.63
		4	23.0	0.14	34.4	0.63
		5	23.0	0.13	35.0	0.64
2	20	3	36.0	0.15	70.4	0.71
		4	36.0	0.14	64.5	0.71
		5	35.5	0.12	67.4	0.70
3	25	3	41.5	0.16	103.4	0.79
		4	41.6	0.16	102.4	0.82
		5	41.3	0.16	95.7	0.80
4	30	3	51.7	0.18	136.5	0.89
		4	53.0	0.20	131.0	0.90
		5	53.7	0.18	127.5	0.92
5	50	3	107.6	0.21	419.1	1.42
		4	112.2	0.21	398.3	1.43
		5	112.3	0.21	404.7	1.43
6	100	3	477.1	0.28	1937.1	3.74
		4	495.5	0.28	1822.9	4.06
		5	509.0	0.28	1816.8	4.08
7	200	3	3006.3	0.51	9868.9	15.72
		4	2838.0	0.49	9628.3	17.24
		5	2776.4	0.49	9702.2	17.55
8	300	3	9216.0	0.94	26712.2	36.96
		4	9394.0	0.93	25838.5	41.07
		5	9407.1	0.93	25650.5	41.79
9	400	3	21074.0	1.39	53089.5	69.10
		4	20802.4	1.38	52114.0	73.80
		5	20820.0	1.38	51834.2	77.72
10	500	3	37445.4	1.88	89886.2	113.16
		4	37518.9	1.88	88421.2	123.67
		5	37445.4	1.88	87649.3	127.12
11	1000	3	247474.0	6.49	424783.2	525.71
		4	245404.0	6.49	420479.1	571.83
		5	247605.0	6.49	417868.4	583.01

6   Conclusions

EAs for network layout problems require special chromossome encoding. This paper presented a forest representation, named node-depth encoding. Based on this representation, we have developed two new operators capable to manipulate a forest generating a new one.

The proposed approach was evaluated for the degree-constrained minimum spanning tree problem. The results suggest that the proposed technique can deal with this problem and work with large graphs using relatively small running time. In this way, this paper may encourage the development of new EA approaches using the node-depth encoding for other NDPs.

References

1. Kershenbaum, A.: Telecommunications Network Design Algorithms. McGraw-Hill, New York (1993)
2. Chou, H.H., Premkumar, G., Chu, C.H.: Genetic algorithms for communications network design - an empirical study of the factors that influence performance. IEEE Transactions on Evolutionary Computation **5** (2001) 236–249(3)
3. Harary, F., Gupta, G.: Dynamic graph models. Mathl. Comput. Modelling **25** (1997) 79–87
4. Gen, M., Li, Y.Z., Ida, K.: Solving multiobjective transportatin problem by spanning tree-based genetic algorithm. IEICE Transactions on Fundamental of Electronics Communications and Computer Sciences **E82A** (1999) 2802–2810
5. Reijmers, T.H., Wehrens, R., Daeyaert, F.D., Lewi, P.J., Buydens, L.M.C.: Using genetic algorithm for the construction of phylogenetic trees: Application to g-protein coupled receptor sequences. Biosystems **49** (1999) 31–43
6. Gen, M., Cheng, R.: Genetic Algorithms and Engineering Design. Ashikaga Institute of Technology, Ashikaga, Japan (1997)
7. Knowles, J., Corne, D.: A new evolutionary approach to the degree-constrained minimum spanning tree problem. IEEE Transaction on Evolutionary Computation **4** (2000) 125–134(2)
8. Carvalho, P.M.S., Ferreira, L.A.F.M., Barruncho, L.M.F.: On spanning tree recombination in evolutionary large-scale network problems - application to electrical distribution planning. IEEE Transactions on Evolutionary Computation **5** (2001) 623–630(6)
9. Delbem, A.C.B., de Carvalho, A., Bretas, N.G.: Optimal energy restoration in radial distribution systems using a genetic approach and graph chain representation. Electric Power Systems Researchs **67/3** (2003) 197–205
10. Palmer, C., Kershenbaum, A.: An approach to a problem in network design using genetic algorithms. Networks **26** (1995) 101–107
11. Droste, S., Wiesmann, D.: On representation and genetic operators in evolutionary algorithms. Technical Report CI–41/98, Fachbereich Informatik, Universität Dortmund, 44221 Dortmund (1998)
12. Delbem, A.C.B., de Carvalho, A.: A forest encoding for evolutionary algorithms applied to design problems. Genetic Algorithm and Evolutionary Computation Conference 20003, Lecture Notes in Computer Science **2723** (2003) 634–635
13. Goodaire, E.G., Parmenter, M.M.: Discrete Mathematics with Graph Theory. Prentice Hall, Upper Saddle River, USA (1998)

Reducing Fitness Evaluations Using Clustering Techniques and Neural Network Ensembles

Yaochu Jin and Bernhard Sendhoff

Honda Research Institute Europe
Carl-Legien-Str. 30
63073 Offenbach/Main, Germany
yaochu.jin@honda-ri.de

Abstract. In many real-world applications of evolutionary computation, it is essential to reduce the number of fitness evaluations. To this end, computationally efficient models can be constructed for fitness evaluations to assist the evolutionary algorithms. When approximate models are involved in evolution, it is very important to determine which individuals should be re-evaluated using the original fitness function to guarantee a faster and correct convergence of the evolutionary algorithm. In this paper, the k-means method is applied to group the individuals of a population into a number of clusters. For each cluster, only the individual that is closest to the cluster center will be evaluated using the expensive original fitness function. The fitness of other individuals are estimated using a neural network ensemble, which is also used to detect possible serious prediction errors. Simulation results from three test functions show that the proposed method exhibits better performance than the strategy where only the best individuals according to the approximate model are re-evaluated.

1 Introduction

Many difficulties may arise in applying evolutionary algorithms to solving complex real-world optimization problems. One of the main concerns is that evolutionary algorithms usually need a large number of fitness evaluations to obtain a good solution. Unfortunately, fitness evaluations are often very expensive or highly time-consuming. Take aerodynamic design optimization as an example, one evaluation of a given design based on the 3-Dimensional computational fluid dynamics (CFD) simulation will take hours on a high-performance computer.

To alleviate this problem, computationally efficient models can be constructed to approximate the fitness function. Such models are often known as approximate models, meta-models or surrogates, refer to [9] for an overview of this topic. It would be ideal if an approximate model can fully replace the original fitness function, however, researchers have come to realize that it is in general necessary to combine the approximate model with the original fitness function to ensure the evolutionary algorithm to converge correctly. To this end, re-evaluation of some individuals using the original fitness function, also termed as *evolution control* in [7], is essential.

K. Deb et al. (Eds.): GECCO 2004, LNCS 3102, pp. 688–699, 2004.
© Springer-Verlag Berlin Heidelberg 2004

Generation-based or individual-based evolution control can be implemented. In the generation-based approach [15,2,7,8], some generations are evaluated using the approximate model and the rest using the original fitness function. In individual-based evolution control, part of the individuals of each generation are evaluated using the approximation model and the rest using the original fitness function [7,3,18,1]. Generally speaking, the generation-based approach is more suitable when the individuals are evaluated in parallel, where the duration of the optimization process depends to a large degree on the number of generations needed. By contrast, the individual-based approach is more desirable when the number of evaluations is limited, for example, when an experiment needs to be done for a fitness evaluation.

On the other hand, individual-based evolution control provides more flexibility in choosing which individuals need to be re-evaluated. In [7], it is suggested that one should choose the best individuals according to the approximate model rather than choosing the individuals randomly. In [3], not only the estimated function value but also the estimation error are taken into account. The basic idea is that individuals having a larger estimation error are more likely to be chosen for re-evaluation. Other uncertainty measures have also been proposed in [1].

In [10], the population of a genetic algorithm is grouped into a number of clusters and only one representative individual of each cluster is evaluated using the fitness function. Other individuals in the same cluster are estimated according to their Euclidean distance to the representative individuals. Obviously, this kind of estimation is very rough and the local feature of the fitness landscape is completely ignored. In this paper, we also group the population into a number of clusters, and only the individual that is closest to the cluster center is evaluated using the original fitness function. In contrast to the distance-based estimation method [10], we use the evaluated individuals (centers of the clusters) to create a neural network ensemble, which is used for estimating the fitness values of the remaining individuals. Both the structure and the parameters of the neural networks are optimized using an evolutionary algorithm with Lamarckian inheritance.

The remainder of the paper is organized as follows. Section 2 presents population clustering using the k-means algorithm. The construction of neural network ensembles using an evolutionary algorithm is described in Section 3. The proposed algorithm is applied to the optimization of three test functions in Section 4. A summary of the paper is provided in Section 5.

2   Population Clustering

A variety of clustering techniques have been proposed for grouping similar patterns (data items) [4]. Generally, they can be divided into hierarchical clustering algorithms and partitional clustering algorithms. A hierarchical algorithm yields a tree structure representing a nested grouping of patterns, whereas a partitional clustering algorithm generates a single partition of the patterns. Among the partitional clustering methods, the k-means is the simplest and the most commonly used clustering algorithm. It employs the squared error criterion and its computational complexity is $O(n)$, where n is the number of patterns. A standard k-means algorithm is given in Fig. 1. A typical stopping criterion is that the decrease in the squared error is minimized.

A major problem of the k-means clustering algorithm is that it may converge to a local minimum if the initial partition is not properly chosen. Besides, the number of clusters needs to be specified beforehand, which is a general problem for partitional clustering algorithms [4].

---

1. Choose k patterns randomly as the cluster centers

2. Assign each pattern to its closest cluster center

3. Recompute the cluster center using the current cluster members

4. If the convergence criterion is not met, go to step 2; otherwise stop

---

Fig. 1. The k-means algorithm.

To assess the validity of a given cluster, the silhouette method [17] can be used. For a given cluster, $X_j, j = 1, ..., k$, the silhouette technique assigns the i-th member $(x_{ij}, i = 1, ..., n_j)$ of cluster X_j a quality measure (*silhouette width*):

$$s_{ij} = \frac{b_i - a_i}{\max\{a_i, b_i\}},\tag{1}$$

where a_i is the average distance between x_{ij} and all other members in X_j and b_i denotes the minimum of $a_i, i = 1, 2, ..., n_j$, where n_j is the number of patterns in cluster X_j and naturally, $n_1 + ... + n_k$ equals n if each pattern belongs to one and only one cluster, n is the number of patterns to be clustered. It can be seen that s_{ij} has a value between -1 and 1. If s_{ij} equals 1, it means that s_{ij} is in the proper cluster. If s_{ij} is 0, it indicates that x_{ij} may also be grouped in the nearest neighboring cluster and if x_{ij} is -1, it suggests that x_{ij} is very likely in the wrong cluster. Thus, a global silhouette width can be obtained by summing up the silhouette width over all patterns:

$$S = \frac{1}{k} \sum_{j=1}^{k} \sum_{i=1}^{n_j} s_{ij}.\tag{2}$$

Consequently, this value can be used to determine the proper number of clusters.

3   Construction of Local Neural Network Ensemble

After the population is grouped into a number of clusters, only the individual that is closest to each cluster center will be evaluated using the original fitness function. In [10], the fitness value of all other individuals are estimated based on their Euclidean distance to the cluster center. Obviously, this simplified estimation ignores the local feature of the fitness landscape which can be extracted from the evaluated cluster centers.

In our previous work [7,8], a fully connected multi-layer perceptron (MLP) neural network has been constructed using the data generated during optimization. The neural network model is trained off-line and further updated when new data are available. One problem that may occur is that as the number of samples increases, the learning efficiency may decrease. To improve the learning efficiency, weighted learning [8] and off-line structure optimization of the neural networks have been shown to be promising. In this work, we attempt to further improve the approximation quality in two aspects. First, structure optimization of the neural network is carried out on-line and only the data generated in the most recent two generations are used. This makes it possible to have an approximate model that reflects the local feature of the landscape. Second, an ensemble instead of a single neural network will be used to improve the generalization property of the neural networks.

The benefit of using a neural network ensemble originates from the diversity of the behavior of the ensemble members on unseen data. Generally, diverse behavior on unseen data can be obtained by using various initial random weights, varying the network' architecture, employing different training algorithm, supplying different training data by manipulating the given training data, generating data from different sources, or encouraging diversity [13], decorrelation [16] or negative correlation [11,12] between the ensemble members.

In this work, a genetic algorithm with local learning [6] has been used to generate the neural network ensemble, which can provide two sources of diversity: both the architecture and the final weights of the neural networks are different. Since the goal of the neural networks is to learn the local fitness landscape, we only use the data generated in the two most recent generations instead of using all data.

Assume that the λ individuals in the population are grouped into ξ clusters, thus ξ new data will be generated in each generation. Accordingly, the fitness function for evolutionary neural network generation can be expressed as follows:

$$F = \frac{1}{\xi} \left\{ \alpha \cdot \sum_{i=1}^{\xi} (y_i - y_i^d(t))^2 + (1 - \alpha) \cdot \sum_{i=1}^{\xi} (y_i - y_i^d(t-1))^2 \right\}, \qquad (3)$$

where $0.5 < \alpha \le 1$ (set to 0.7 in this work) is a coefficient giving more importance to the newest data, $y_i^d(t), i = 1, ..., \xi$ are the data generated in the current generation and $y_i^d(t-1)$, $i = 1, ..., \xi$ are those generated in the last generation and y_i is the network output for the i-th data set.

Given N neural networks, the final output of the ensemble can be obtained by averaging the weighted outputs of the ensemble members:

$$y^{EN} = \sum_{k=1}^{N} w^{(k)} y^{(k)}, \tag{4}$$

where $y^{(k)}$ and $w^{(k)}$ are the output and its weight of the k-th neural network in the ensemble. If all the weights are equally set to $1/N$, it is termed basic ensemble method (BEM). Otherwise, it is termed generalized ensemble method (GEM). In this case, the expected error of the ensemble is given by:

$$E^{EN} = \sum_{i=1}^{N} \sum_{j=1}^{N} w^{(i)} w^{(j)} C_{ij}, \tag{5}$$

where C_{ij} is the error correlation matrix between network i and network j in the ensemble:

$$C_{ij} = E[(y_i - y_i^d)(y_j - y_j^d)], \tag{6}$$

where $E(\cdot)$ denotes the mathematical expectation.

It has been shown [14] that there exists an optimal set of weights that minimizes the expected prediction error of the ensemble:

$$w^{(k)} = \frac{\sum_{j=1}^{N} (C_{kj})^{-1}}{\sum_{i=1}^{N} \sum_{j=1}^{N} (C_{ij})^{-1}}, \tag{7}$$

where $1 \leq i, j, k \leq N$.

However, a reliable estimation of the error correlation matrix is not straightforward because the prediction errors of different networks in an ensemble are often strongly correlated. A few methods have been proposed to solve this problem [5,19,20]. Genetic programming is applied to the search for an optimal ensemble size in [20] whereas the recursive least-square method is adopted to optimize the weights in [19]. In [19], a GA is also used to search for an optimal subset of the neural networks in the final population as ensemble members.

To reduce the computational complexity, only a small number of networks (three to five) has been tried in this work. A canonical evolution strategy is employed to find the optimal weights to minimize the expected error in Eq. (5).

The algorithm for constructing the neural network ensemble and the entire evolutionary optimization algorithm are sketched in Fig. 2 and Fig. 3, respectively.

4 Empirical Results

4.1 Experimental Setup

In the simulations, optimization runs are carried out on three well known test functions, the Ackley function, the Rosenbrock function and the Sphere function.

1. Prepare the training and test data

2. Generate N (ensemble size) neural networks using GA

3. Calculate the error correlation between the ensemble members

4. Determine the optimal weight for each network by using ES

Fig. 2. Algorithm for constructing neural network ensemble.

1. Initialize λ individuals, evaluate all individuals using the original fitness function

2. For each generation

    a) select the best μ individuals
    b) generate λ offspring individuals by recombination and mutation
    c) evaluate
            – clustering the λ invividuals using the k–means algorithm
            – evaluate the ξ individuals closest to the cluster centers using the original fitness function
            – construct the neural network ensemble
            – calculate the fitness of the rest $\lambda - \xi$ individuals using the neural network ensemble

3. Go to step 2 if the termination condition is not met

4. Stop

Fig. 3. The proposed evolutionary optimization algorithm.

The dimension of the test functions are set to 30. A standard $(5, 30)$ evolution strategy (ES) is used in all simulations. The maximal number of fitness evaluations is set to 2000 in the optimization for all three test functions.

Before we implement the evolutionary optimization with approximate fitness models, we need to determine a few important parameters, such as the number of clusters and the number of neural networks in the ensemble.

The first issue is the number of clusters. This number is relevant to performance of the clustering algorithm, the quality of the approximate model, and eventually the convergence property of the evolutionary algorithm.

A few preliminary optimization runs are carried out with only a single neural network being used for fitness approximation on the 30-dimensional Ackley function. It is found that with the clustering algorithm, the evolutionary algorithm is able to converge correctly when about one third of the population is re-evaluated using the original fitness function. When the number of the re-evaluated individuals is much fewer than one third of the population, the performance of the evolutionary algorithm becomes unpredictable, that is, the evolutionary algorithm may converge to a false minimum.

We then evaluate the clustering performance when the number of clusters is set to be one third of the population. Fig. 4 shows the global silhouette width when the cluster number is 10 and the population size is 30 on the 30-dimensional Ackley function. It can be seen that the clustering performance is acceptable according to the discussions in Section 2.

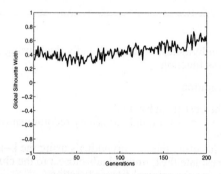

Fig. 4. Global silhouette width when the number of cluster is set to 10 and the population size is 30 on the 30-dimensional Ackley function.

Next, simulations are conducted to investigate the ensemble size. So far, the ensemble size has been determined heuristically in most applications. In [20], the optimal size turns out to be between 5 and 7. Considering the fact that a large ensemble size will increase computational cost, we compare two cases where the ensemble size is 3 and 5 on 200 samples collected in the first 20 generations of an optimization run on the 30-dimensional Ackley function. The ensemble output versus that of a single network is plotted in Fig. 5, where in Fig. 5(a) the ensemble size is 3 and in Fig. 5(b) the ensemble size is 5. Note that the more points locate in the right lower part of the figure the more effective the ensemble. It can be seen from the figure that no significant performance improvement has been achieved when the ensemble size is changed from 3 to 5. Thus, we fix the ensemble size to 3.

It seems that the use of an ensemble has not improved the prediction accuracy significantly. Thus, the motivation to employ an ensemble becomes questionable. In the following, we will show that an ensemble is important not only in that it is able to improve prediction. In this work, the equally important reason for introducing the ensemble is to estimate the prediction accuracy based on the different behaviors of the ensemble members, i.e., the variance of the members in the ensemble. To demonstrate this, Fig. 6(a) shows the relationship between the standard deviation of the predictions of the ensemble members and the estimation error of the ensemble. These data are also collected in the first 20 generations of an evolutionary run of the Ackley function. Additional function

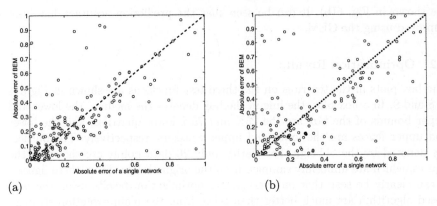

(a)   (b)

Fig. 5. (a) Ensemble size equals 3. (b) Ensemble size equals 5.

evaluations are carried out to get the prediction error. Of course, they are neither used in neural network training nor in optimization. It can be seen that a large standard deviation most probably indicates a large prediction error, although a small standard deviation does not guarantee a small prediction error. Encouraged by this close correlation between a large deviation and a large prediction error, we try to predict the model error. When the standard deviation is larger than a threshold (1 in this example), we replace the model prediction with the fitness of the individual closest to the cluster center, which is a very rough but feasible approximation.

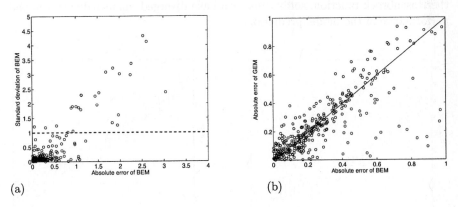

(a)   (b)

Fig. 6. (a) Prediction error versus the standard deviation. (b) Prediction error of the BEM versus that of the GEM.

Finally, we have run a standard evolution strategy with a population size of (3,15) for 100 generations to optimize the weight of the ensemble members. The predictions of the GEM, where the weights are optimized, and that of a BEM are shown in Fig. 6(b). It can be seen that the prediction accuracy has been improved using the GEM.

4.2   Optimization Results

The box plots of the ten runs on the three test functions are shown in Figures 7, 8 and 9. In a box plot, the line in the box denotes the median, the lower and upper bounds of the box are the 25% and 75% lower quartiles, and the lower and upper fences are the lower and upper whiskers, respectively. The outliers are denoted by the '+' sign. For clarity, only 20 data points are shown in the figures, which are uniformly sampled from the original data. From these figures, it can clearly be seen that on average, the optimization results using the proposed algorithm are much better than those from the plain evolution strategy on all test function. Meanwhile, they are also much better than the results reported in [7], where no clustering of the population has been implemented. As we mentioned, without clustering, the evolutionary algorithm does not converge correctly if only one third of the population is re-evaluated using the original fitness function. Nevertheless, we also notice that for the Ackley function, the result from one of the 10 runs using the proposed method is much worse than the average performance, even a little worse than the average result when the plain ES is used, refer to Fig. 7(a).

To show the benefit of using the neural network ensemble, the box plots of results using only a single neural network (where no remedy of large prediction errors is included) on the three test functions are provided in Figures 10, 11 and 12. Similarly, only 20 data points are presented for the clarity of the figures. Compared with the results shown in Figures 7, 8 and 9, they are much worse. In the Rosenbrock function, some runs even have diverged, mainly due to the bad performance of the model prediction.

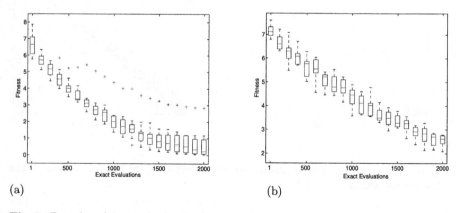

(a)   (b)

Fig. 7. Box plot of the results for the 30-dimensional Ackley function. (a) The proposed algorithm. (b) Plain ES. Notice that the scales in (a) and (b) are not the same.

Table 1. Statistical results obtained by our SES for the 13 test functions with 30 independent runs. A result in **boldface** means global optimum solution found.

Problem	Optimal	Best	Mean	Median	Worst	St. Dev.
			Statistical Results of the New SES with the Improved Diversity Mechanism			
g01	−15.00	**−15.00**	**−15.00**	**−15.00**	**−15.00**	0
g02	0.803619	0.803601	0.785238	0.792549	0.751322	1.67E-2
g03	1.00	**1.00**	1.00	**1.00**	**1.00**	2.09E-4
g04	−30665.539	**−30665.539**	**−30665.539**	**−30665.539**	**−30665.539**	0
g05	5126.498	5126.599	5174.492	5160.198	5304.167	50.05E+0
g06	−6961.814	**−6961.814**	−6961.284	**−6961.814**	−6952.482	1.85E+0
g07	24.306	24.327	24.475	24.426	24.843	1.32E-1
g08	0.095825	**0.095825**	**0.095825**	**0.095825**	**0.095825**	0
g09	680.630	680.632	680.643	680.642	680.719	1.55E-2
g10	7049.25	7051.90	7253.05	7253.60	7638.37	136.0E+0
g11	0.75	**0.75**	0.75	0.75	0.75	1.52E-4
g12	1.00	**1.00**	1.00	1.00	1.00	0
g13	0.053950	0.053986	0.166385	0.061873	0.468294	1.76E-1

Table 2. Comparison of results between the new SES and the old one proposed in [6]. "-" means no feasible solutions were found. A result in **boldface** means a better value obtained by our new approach.

Problem	Optimal	Best Result		Mean Result		Worst Result	
		NEW SES	OLD	NEW SES	OLD	NEW SES	OLD
g01	−15.00	−15.00	−15.00	−15.00	−15.00	−15.00	−15.00
g02	0.803619	**0.803601**	0.803569	**0.785238**	0.769612	**0.751322**	0.702322
g03	1.00	1.00	1.00	1.00	1.00	1.00	1.00
g04	−30665.539	−30665.539	−30665.539	−30665.539	−30665.539	−30665.539	−30665.539
g05	5126.498	**5126.599**	−	**5174.492**	−	**5304.167**	−
g06	−6961.814	**−6961.814**	−6961.814	−6961.284	−6961.814	−6952.482	−6961.814
g07	24.306	24.327	24.314	24.475	24.419	24.843	24.561
g08	0.095825	0.095825	0.095825	**0.095825**	0.095784	**0.095825**	0.095473
g09	680.630	**680.632**	680.669	**680.643**	680.810	**680.719**	681.199
g10	7049.25	**7051.90**	7057.04	**7253.05**	10771.42	**7638.37**	16375.27
g11	0.75	0.75	0.75	0.75	0.75	0.75	**0.76**
g12	1.00	1.00	1.00	1.00	1.00	1.00	1.00
g13	0.053950	0.053986	0.053964	**0.166385**	0.264135	**0.468294**	0.544346

Table 3. Comparison of the new version of the SES with respect to the Homomorphous Maps (HM) [11]. "-" means no feasible solutions were found. A result in **boldface** means a better value obtained by our new approach.

Problem	Optimal	Best Result		Mean Result		Worst Result	
		New SES	HM	New SES	HM	New SES	HM
g01	−15.00	**−15.00**	−14.7886	**−15.00**	−14.7082	**−15.00**	−14.6154
g02	0.803619	**0.803601**	0.79953	0.785238	0.79671	0.751322	0.79119
g03	1.00	**1.00**	0.9997	**1.00**	0.9989	**1.00**	0.9978
g04	−30665.539	**−30665.539**	−30664.5	**−30665.539**	−30655.3	**−30665.539**	−30645.9
g05	5126.498	**5126.599**	−	**5174.492**	−	**5304.167**	−
g06	−6961.814	**−6961.814**	−6952.1	**−6961.284**	−6342.6	**−6952.482**	−5473.9
g07	24.306	**24.327**	24.620	**24.475**	24.826	**24.843**	25.069
g08	0.095825	0.095825	0.0958250	**0.095825**	0.0891568	**0.095825**	0.0291438
g09	680.63	**680.632**	680.91	**680.643**	681.16	**680.719**	683.18
g10	7049.25	**7051.90**	7147.9	**7253.05**	8163.6	**7638.37**	9659.3
g11	0.75	0.75	0.75	0.75	0.75	0.75	0.75
g12	1.00	**1.00**	0.999999	**1.00**	0.999134	**1.00**	0.991950
g13	0.053950	0.053986	*NA*	0.166385	*NA*	0.468294	*NA*

Table 4. Values of ρ for the 13 test problems chosen.

Problem	n	Function	ρ	LI	NI	LE	NE
g01	13	quadratic	0.0003%	9	0	0	0
g02	20	nonlinear	99.9973%	1	1	0	0
g03	10	nonlinear	0.0026%	0	0	0	1
g04	5	quadratic	27.0079%	0	6	0	0
g05	4	nonlinear	0.0000%	2	0	0	3
g06	2	nonlinear	0.0057%	0	2	0	0
g07	10	quadratic	0.0000%	3	5	0	0
g08	2	nonlinear	0.8581%	0	2	0	0
g09	7	nonlinear	0.5199%	0	4	0	0
g10	8	linear	0.0020%	3	3	0	0
g11	2	quadratic	0.0973%	0	0	0	1
g12	3	quadratic	4.7697%	0	9^3	0	0
g13	5	nonlinear	0.0000%	0	0	1	2

population with only a 40% of the value obtained by the following formula (where n is the number of decision variables): $\sigma_i(0) = 0.4 \times (\Delta x_i / \sqrt{n})$ where Δx_i is approximated with the expression (suggested in [9]), $\Delta x_i \approx x_i^u - x_i^l$, where $x_i^u - x_i^l$ are the upper and lower limits of the decision variable i. For the experiments we used the following parameters: $(100+300)$-ES, number of generations $= 800$, number of objective function evaluations $= 240,000$.

To increase the exploitation feature of the global crossover operator we combine discrete and intermediate crossover. Each gene in the chromosome can be processed with any of these two crossover operators with a 50% of probability. This operator is applied to both, strategy parameters (sigma values) and decision variables of the problem. Note that we do not use correlated mutation. To deal with equality constraints, a parameterless dynamic mechanism originally proposed in ASCHEA [10] and used in [5] and in [6] is adopted. The tolerance value ϵ is decreased with respect to the current generation using the following expression: $\epsilon_j(t+1) = \epsilon_j(t)/1.00195$. The initial ϵ_0 was set to 0.001. For problem g13, ϵ_0 was set to 3.0 and, in consequence, the factor to decrease the tolerance value was modified to $\epsilon_j(t+1) = \epsilon_j(t)/1.0145$. Also, for problems $g03$ and $g13$ the initial stepsize required a more dramatic decrease of the stepsize. They were defined as 0.01 (just a 5% instead of the 40%) for $g03$ and 0.05 (a 2.5% instead of the 40%) for g13. These two test functions seem to provide better results with very smooth movements. It is important to note that these two problems share the following features: moderately high dimensionality (five or more decision variables), nonlinear objective function, one or more equality constraints, and moderate size of the search space (based on the range of the decision variables). These common features suggest that for this type of problem, finer movements provide a better sampling of the search space using an evolution strategy.

The statistical results of this new version of the SES with the improved diversity mechanism are summarized in Table 1. The comparison of the improved version against the previous one [6] is presented in Table 2. We compared our approach against the previous version of the SES [6] in Table 2 and against three state-of-the-art approaches: the Homomorphous Maps (HM) [11] in Table 3, Stochastic Ranking (SR) [9] in Table 5 and the Adaptive Segregational Constraint Handling Evolutionary Algorithm (ASCHEA) [10] in Table 6.

The Homomorphous Maps performs a homomorphous mapping between an n-dimensional cube and the feasible search region (either convex or non-convex). The main idea of this approach is to transform the original problem into another (topologically equivalent) function that is easier to optimize by the EA. Both, the Stochastic Ranking and ASCHEA are based on a penalty function approach. SR sorts the individuals in the population in order to assign them a rank value. However, based on the value of a user-defined parameter, the comparison between two adjacent solutions will be performed using only the objective function. The remaining comparisons will be performed using only the penalty value (the sum of constraint violation). ASCHEA uses three combined mechanisms: (1) an adaptive penalty function, (2) a constraint-driven recombination that forces to select a feasible individual to recombine with an infeasible one and (3) a segregational selection based on feasibility which maintains a balance between feasible and infeasible solutions in the population. ASCHEA also requires a niching mechanism to improve the diversity in the population. Each mechanism requires the definition by the user of extra parameters.

Table 5. Comparison of our new version of the SES with respect to Stochastic Ranking (SR) [9]. A result in **boldface** means a better value obtained by our new approach.

Problem	Optimal	Best Result		Mean Result		Worst Result	
		New SES	SR	New SES	SR	New SES	SR
g01	−15.00	−15.00	−15.000	−15.00	−15.000	−15.00	−15.000
g02	0.803619	**0.803601**	0.803515	**0.785238**	0.781975	**0.751322**	0.726288
g03	1.00	1.00	1.000	1.00	1.000	1.00	1.000
g04	−30665.539	−30665.539	−30665.539	−30665.539	−30665.539	−30665.539	−30665.539
g05	5126.498	5126.599	5126.497	5174.492	5128.881	5304.165	5142.472
g06	−6961.814	−6961.814	−6961.814	**−6961.284**	−6875.940	**−6952.482**	−6350.262
g07	24.306	24.327	24.307	24.475	24.374	24.843	24.642
g08	0.095825	0.095825	0.095825	0.095825	0.095825	0.095825	0.095825
g09	680.63	680.632	680.630	**680.643**	680.656	**680.719**	680.763
g10	7049.25	**7051.90**	7054.316	**7253.05**	7559.192	**7638.37**	8835.655
g11	0.75	0.75	0.750	0.75	0.750	0.75	0.750
g12	1.00	1.00	1.00	1.00	1.00	1.00	1.00
g13	0.053950	0.053986	0.053957	0.166385	0.057006	0.468294	0.216915

5   Discussion of Results

As described in Table 1, our approach was able to find the global optimum in seven test functions (g01, g03, g04, g06, g08, g11 and g12) and it found solutions very close to the global optimum in the remaining six (g02, g05, g07, g09, g10, g13). Compared with its previous version [6] (Table 2) this new diversity mechanism improved the quality of the results in problems g02, g05, g09 and g10. Also, the robustness of the results was better in problems g02, g05, g08, g09, g10 and g13.

Table 6. Comparison of our new version of the SES with respect to ASCHEA [10]. NA = Not Available. A result in **boldface** means a better value obtained by our new approach.

Problem	Optimal	Best Result		Mean Result		Worst Result	
		New SES	ASCHEA	New SES	ASCHEA	New SES	ASCHEA
g01	−15.0	**−15.00**	−15.0	**−15.00**	−14.84	−15.00	NA
g02	0.803619	**0.803601**	0.785	**0.785238**	0.59	0.751322	NA
g03	1.00	1.00	1.0	**1.00**	0.99989	1.00	NA
g04	−30665.539	−30665.539	30665.5	−30665.539	30665.5	−30665.539	NA
g05	5126.498	5126.599	5126.5	5174.492	5141.65	5304.167	NA
g06	−6961.814	−6961.814	−6961.81	−6961.284	−6961.81	−6952.482	NA
g07	24.306	**24.327**	24.3323	**24.475**	24.66	24.843	NA
g08	0.095825	0.095825	0.095825	0.095825	0.095825	0.095825	NA
g09	680.630	680.632	680.630	680.643	680.641	680.719	NA
g10	7049.25	**7051.90**	7061.13	7253.05	7193.11	7638.37	NA
g11	0.75	0.75	0.75	0.75	0.75	0.75	NA
g12	1.00	1.00	NA	1.00	NA	1.00	NA
g13	0.053950	0.053986	NA	0.166385	NA	0.468294	NA

When compared with respect to the three state-of-the-art techniques previously indicated, we found the following: Compared with the Homomorphous Maps (Table 3) the new SES found a better "best" solution in ten problems (g01, g02, g03, g04, g05, g06, g07, g09, g10 and g12) and a similar "best" result in other two (g08 and g11). Also, our technique reached better "mean" and "worst" results in ten problems (g01, g03, g04, g05, g06, g07, g08, g09, g10 and g12). A "similar" mean and worst result was found in problem g11. The Homomorphous maps found a "better" mean and worst result in function g02. No comparisons were made with respect to function g13 because such results were not available for HM.

With respect to Stochastic Ranking (Table 5), our approach was able to find a better "best" result in functions g02 and g10. In addition, it found a "similar" best solution in seven problems (g01, g03, g04, g06, g08, g11 and g12). Slightly better "best" results were found by SR in the remaining functions (g05, g07, g09 and g13). The new SES found better "mean" and "worst" results in four test functions (g02, g06, g09 and g10). It also provided similar "mean" and "worst" results in six functions (g01, g03, g04, g08, g11 and g12). Finally, SR found again just slightly better "mean" and "worst" results in functions g05, g07 and g13.

Compared against the Adaptive Segregational Constraint Handling Evolutionary Algorithm (Table 6), our algorithm found "better" best solutions in three problems (g02, g07 and g10) and it found "similar" best results in six functions (g01, g03, g04, g06, g08, g11). ASCHEA found slightly "better" best results in function g05 and g09. Additionally, the new SES found "better" mean results in four problems (g01, g02, g03 and g07) and it found "similar" mean results in three functions (g04, g08 and g11). ASCHEA surpassed our mean results in four functions (g05, g06, g09 and g10). We did not compare the worst results because they were not available for ASCHEA. We did not perform comparisons with respect to ASCHEA using functions g12 and g13 for the same reason. As we can see, our approach showed a very competitive performance with respect to these three state-of-the-art approaches.

Our approach can deal with moderately constrained problems (g04), highly constrained problems, problems with low (g06, g08), moderated (g09) and high (g01, g02, g03, g07) dimensionality, with different types of combined constraints (linear, nonlinear, equality and inequality) and with very large (g02), very small (g05 and g13) or even disjoint (g12) feasible regions. Also, the algorithm is able to deal with large search spaces (based on the intervals of the decision variables) with a very small feasible region (g10). Furthermore, the approach can find the global optimum in problems where such optimum lies on the boundaries of the feasible region (g01, g02, g04, g06, g07, g09). This behavior suggests that the mechanism of maintaining the best infeasible solution helps the search to sample the boundaries of the feasible region.

Besides still being a very simple approach, it is worth reminding that our algorithm does not require the fine-tuning of any extra parameters (other than those used with an evolution strategy) since the only parameters required by the approach have remained fixed in all cases. In contrast, the Homomorphous maps require an additional parameter (called v) which has to be found empirically [11]. Stochastic ranking requires the definition of a parameter called P_f, whose value has an important impact on the performance of the approach [9]. ASCHEA also requires the definition of several extra parameters, and in its latest version, it uses niching, which is a process that also has at least one additional parameter [10].

The computational cost measured in terms of the number of fitness function evaluations (FFE) performed by any approach is lower for our algorithm with respect to the others to respect to which it was compared. This is an additional (and important) advantage, mainly if we wish to use this approach for solving real-world problems. Our new approach performed $240,000$ FFE, the previous version required $330,000$ FFE, the Stochastic Ranking performed $350,000$ FFE, the Homomorphous Maps performed $1,400,000$ FFE, and ASCHEA required $1,500,000$ FFE.

6   Conclusions and Future Work

An improved diversity mechanism added to a multimembered Evolution Strategy combined with some selection criteria based on feasibility were proposed to solve (rather efficiently) constrained optimization problems. The proposed approach does not require the use of a penalty function and it does not require the fine-tuning of any extra parameters (other than those required by an evolution strategy), since they assume fixed values. The proposed approach uses the self-adaptation mechanism of a multimembered ES to sample the search space in order to reach the feasible region and it uses three simple selection criteria based on feasibility to guide the search towards the global optimum. Moreover, the proposed technique adopts a diversity mechanism which consists of allowing infeasible solutions close to the boundaries of the feasible region to remain in the next population. This approach is very easy to implement and its computational cost (measured in terms of the number of fitness function evaluations) is considerably lower than the cost reported by other three constraint-handling techniques which are representative of the state-of-the-art in evolutionary optimization. Despite its lower computational cost, the proposed approach was able to match (and even improve) on the results obtained by the other algorithms with respect to which it was compared.

As part of our future work, we plan to evaluate the rate at which our algorithm reaches the feasible region. This is an important issue when dealing with real-world applications, since in highly constrained search spaces, reaching the feasible region may be a rather costly task. Additionally, we have to perform more experiments in order to establish which of the three mechanisms of the approach (diversity mechanism, combined crossover or the reduced stepsize) is mandatory or if only their combined effect makes the algorithm work.

Acknowledgments. The first author acknowledges support from the Mexican Consejo Nacional de Ciencia y Tecnología (CONACyT) through a scholarship to pursue graduate studies at CINVESTAV-IPN's. The second author acknowledges support from CONACyT through project number 34201-A.

References

1. Bäck, T.: Evolutionary Algorithms in Theory and Practice. Oxford University Press, New York (1996)
2. Michalewicz, Z., Schoenauer, M.: Evolutionary Algorithms for Constrained Parameter Optimization Problems. Evolutionary Computation **4** (1996) 1–32
3. Coello Coello, C.A.: Theoretical and Numerical Constraint Handling Techniques used with Evolutionary Algorithms: A Survey of the State of the Art. Computer Methods in Applied Mechanics and Engineering **191** (2002) 1245–1287
4. Smith, A.E., Coit, D.W.: Constraint Handling Techniques—Penalty Functions. In Bäck, T., Fogel, D.B., Michalewicz, Z., eds.: Handbook of Evolutionary Computation. Oxford University Press and Institute of Physics Publishing (1997)
5. Mezura-Montes, E., Coello Coello, C.A.: A Simple Evolution Strategy to Solve Constrained Optimization Problems. In Cantú-Paz, E., Foster, J.A., Deb, K., Davis, L.D., Roy, R., Reilly, U.M.O., Beyer, H.G., Standish, R., Kendall, G., Wilson, S., Harman, M., Wegener, J., Dasgupta, D., Potter, M.A., Schultz, A.C., Dowsland, K.A., Jonoska, N., Miller, J., eds.: Proceedings of the Genetic and Evolutionary Computation Conference (GECCO'2003), Heidelberg, Germany, Chicago, Illinois, Springer Verlag (2003) 640–641 Lecture Notes in Computer Science Vol. 2723.
6. Mezura-Montes, E., Coello Coello, C.A.: Adding a Diversity Mechanism to a Simple Evolution Strategy to Solve Constrained Optimization Problems. In: Proceedings of the Congress on Evolutionary Computation 2003 (CEC'2003). Volume 1., Piscataway, New Jersey, Canberra, Australia, IEEE Service Center (2003) 6–13
7. Jiménez, F., Verdegay, J.L.: Evolutionary techniques for constrained optimization problems. In Zimmermann, H.J., ed.: 7th European Congress on Intelligent Techniques and Soft Computing (EUFIT'99), Aachen, Germany, Verlag Mainz (1999) ISBN 3-89653-808-X.
8. Deb, K.: An Efficient Constraint Handling Method for Genetic Algorithms. Computer Methods in Applied Mechanics and Engineering **186** (2000) 311–338
9. Runarsson, T.P., Yao, X.: Stochastic Ranking for Constrained Evolutionary Optimization. IEEE Transactions on Evolutionary Computation **4** (2000) 284–294
10. Hamida, S.B., Schoenauer, M.: ASCHEA: New Results Using Adaptive Segregational Constraint Handling. In: Proceedings of the Congress on Evolutionary Computation 2002 (CEC'2002). Volume 1., Piscataway, New Jersey, IEEE Service Center (2002) 884–889

11. Koziel, S., Michalewicz, Z.: Evolutionary Algorithms, Homomorphous Mappings, and Constrained Parameter Optimization. Evolutionary Computation **7** (1999) 19–44
12. Schwefel, H.P.: Evolution and Optimal Seeking. John Wiley & Sons Inc., New York (1995)

Appendix: Test Functions

1. **g01**: Minimize: $f(x) = 5 \sum_{i=1}^{4} x_i - 5 \sum_{i=1}^{4} x_i^2 - \sum_{i=5}^{13} x_i$ subject to:

$$g_1(x) = 2x_1 + 2x_2 + x_{10} + x_{11} - 10 \leq 0, \quad g_2(x) = 2x_1 + 2x_3 + x_{10} + x_{12} - 10 \leq 0$$
$$g_3(x) = 2x_2 + 2x_3 + x_{11} + x_{12} - 10 \leq 0, \quad g_4(x) = -8x_1 + x_{10} \leq 0,$$
$$g_5(x) = -8x_2 + x_{11} \leq 0, \quad g_6(x) = -8x_3 + x_{12} \leq 0, \quad g_7(x) = -2x_4 - x_5 + x_{10} \leq 0,$$
$$g_8(x) = -2x_6 - x_7 + x_{11} \leq 0, \quad g_9(x) = -2x_8 - x_9 + x_{12} \leq 0$$

   where the bounds are $0 \leq x_i \leq 1$ $(i = 1, \ldots, 9)$, $0 \leq x_i \leq 100$ $(i = 10, 11, 12)$ and $0 \leq x_{13} \leq 1$. The global optimum is at $x^* = (1,1,1,1,1,1,1,1,1,3,3,3,1)$ where $f(x^*) = -15$. Constraints g_1, g_2, g_3, g_4, g_5 and g_6 are active.

2. **g02**: Maximize: $f(x) = \left| \frac{\sum_{i=1}^{n} \cos^4(x_i) - 2 \prod_{i=1}^{n} \cos^2(x_i)}{\sqrt{\sum_{i=1}^{n} i x_i^2}} \right|$ subject to:

$$g_1(x) = 0.75 - \prod_{i=1}^{n} x_i \leq 0, \quad g_2(x) = \sum_{i=1}^{n} x_i - 7.5n \leq 0$$

   where $n = 20$ and $0 \leq x_i \leq 10$ $(i = 1, \ldots, n)$. The global maximum is unknown; the best reported solution is [9] $f(x^*) = 0.803619$. Constraint g_1 is close to being active $(g_1 = -10^{-8})$.

3. **g03**: Maximize: $f(x) = (\sqrt{n})^n \prod_{i=1}^{n} x_i$ subject to: $h(x) = \sum_{i=1}^{n} x_i^2 - 1 = 0$ where $n = 10$ and $0 \leq x_i \leq 1$ $(i = 1, \ldots, n)$. The global maximum is at $x_i^* = 1/\sqrt{n}$ $(i = 1, \ldots, n)$ where $f(x^*) = 1$.

4. **g04**: Minimize: $f(x) = 5.3578547x_3^2 + 0.8356891x_1x_5 + 37.293239x_1 - 40792.141$ subject to:
$$g_1(x) = 85.334407 + 0.0056858x_2x_5 + 0.0006262x_1x_4 - 0.0022053x_3x_5 - 92 \leq 0$$
$$g_2(x) = -85.334407 - 0.0056858x_2x_5 - 0.0006262x_1x_4 + 0.0022053x_3x_5 \leq 0$$
$$g_3(x) = 80.51249 + 0.0071317x_2x_5 + 0.0029955x_1x_2 + 0.0021813x_3^2 - 110 \leq 0$$
$$g_4(x) = -80.51249 - 0.0071317x_2x_5 - 0.0029955x_1x_2 - 0.0021813x_3^2 + 90 \leq 0$$
$$g_5(x) = 9.300961 + 0.0047026x_3x_5 + 0.0012547x_1x_3 + 0.0019085x_3x_4 - 25 \leq 0$$
$$g_6(x) = -9.300961 - 0.0047026x_3x_5 - 0.0012547x_1x_3 - 0.0019085x_3x_4 + 20 \leq 0$$
   where: $78 \leq x_1 \leq 102$, $33 \leq x_2 \leq 45$, $27 \leq x_i \leq 45$ $(i = 3, 4, 5)$. The optimum solution is $x^* = (78, 33, 29.995256025682, 45, 36.775812905788)$ where $f(x^*) = -30665.539$. Constraints g_1 y g_6 are active.

5. **g05**: Minimize: $f(x) = 3x_1 + 0.000001x_1^3 + 2x_2 + (0.000002/3)x_2^3$ subject to: $g_1(x) = -x_4 + x_3 - 0.55 \leq 0$, $g_2(x) = -x_3 + x_4 - 0.55 \leq 0$
$$h_3(x) = 1000 \sin(-x_3 - 0.25) + 1000 \sin(-x_4 - 0.25) + 894.8 - x_1 = 0$$
$$h_4(x) = 1000 \sin(x_3 - 0.25) + 1000 \sin(x_3 - x_4 - 0.25) + 894.8 - x_2 = 0$$
$$h_5(x) = 1000 \sin(x_4 - 0.25) + 1000 \sin(x_4 - x_3 - 0.25) + 1294.8 = 0$$ where $0 \leq x_1 \leq 1200$, $0 \leq x_2 \leq 1200$, $-0.55 \leq x_3 \leq 0.55$, and $-0.55 \leq x_4 \leq 0.55$. The best known solution is $x^* = (679.9453, 1026.067, 0.1188764, -0.3962336)$ where $f(x^*) = 5126.4981$.

6. **g06**: Minimize: $f(x) = (x_1 - 10)^3 + (x_2 - 20)^3$ subject to: $g_1(x) = -(x_1 - 5)^2 - (x_2 - 5)^2 + 100 \leq 0$, $g_2(x) = (x_1 - 6)^2 + (x_2 - 5)^2 - 82.81 \leq 0$ where $13 \leq x_1 \leq 100$ and $0 \leq x_2 \leq 100$. The optimum solution is $x^* = (14.095, 0.84296)$ where $f(x^*) = -6961.81388$. Both constraints are active.

7. **g07**: Minimize: $f(x) = x_1^2 + x_2^2 + x_1 x_2 - 14x_1 - 16x_2 + (x_3 - 10)^2 + 4(x_4 - 5)^2 + (x_5 - 3)^2 + 2(x_6 - 1)^2 + 5x_7^2 + 7(x_8 - 11)^2 + 2(x_9 - 10)^2 + (x_{10} - 7)^2 + 45$
   subject to: $g_1(x) = -105 + 4x_1 + 5x_2 - 3x_7 + 9x_8 \leq 0$
   $g_2(x) = 10x_1 - 8x_2 - 17x_7 + 2x_8 \leq 0$, $g_3(x) = -8x_1 + 2x_2 + 5x_9 - 2x_{10} - 12 \leq 0$
   $g_4(x) = 3(x_1 - 2)^2 + 4(x_2 - 3)^2 + 2x_3^2 - 7x_4 - 120 \leq 0$
   $g_5(x) = 5x_1^2 + 8x_2 + (x_3 - 6)^2 - 2x_4 - 40 \leq 0$
   $g_6(x) = x_1^2 + 2(x_2 - 2)^2 - 2x_1 x_2 + 14x_5 - 6x_6 \leq 0$
   $g_7(x) = 0.5(x_1 - 8)^2 + 2(x_2 - 4)^2 + 3x_5^2 - x_6 - 30 \leq 0$
   $g_8(x) = -3x_1 + 6x_2 + 12(x_9 - 8)^2 - 7x_{10} \leq 0$ where $-10 \leq x_i \leq 10$ $(i = 1, \ldots, 10)$. The global optimum is $x^* = (2.171996, 2.363683, 8.773926, 5.095984, 0.9906548, 1.430574,$ $1.321644, 9.828726, 8.280092, 8.375927)$ where $f(x^*) = 24.3062091$. Constraints $g_1, g_2,$ g_3, g_4, g_5 and g_6 are active.

8. **g08**: Maximize: $f(x) = \dfrac{\sin^3(2\pi x_1) \sin(2\pi x_2)}{x_1^3 (x_1 + x_2)}$
   subject to: $g_1(x) = x_1^2 - x_2 + 1 \leq 0$, $g_2(x) = 1 - x_1 + (x_2 - 4)^2 \leq 0$ where $0 \leq x_1 \leq 10$ and $0 \leq x_2 \leq 10$. The optimum solution is located at $x^* = (1.2279713, 4.2453733)$ where $f(x^*) = 0.095825$.

9. **g09**: Minimize: $f(x) = (x_1 - 10)^2 + 5(x_2 - 12)^2 + x_3^4 + 3(x_4 - 11)^2 + 10x_5^6 + 7x_6^2 + x_7^4 - 4x_6 x_7 - 10x_6 - 8x_7$
   subject to: $g_1(x) = -127 + 2x_1^2 + 3x_2^4 + x_3 + 4x_4^2 + 5x_5 \leq 0$, $g_2(x) = -282 + 7x_1 + 3x_2 + 10x_3^2 + x_4 - x_5 \leq 0$
   $g_3(x) = -196 + 23x_1 + x_2^2 + 6x_6^2 - 8x_7 \leq 0$, $g_4(x) = 4x_1^2 + x_2^2 - 3x_1 x_2 + 2x_3^2 + 5x_6 - 11x_7 \leq 0$ where $-10 \leq x_i \leq 10$ $(i = 1, \ldots, 7)$. The global optimum is $x^* = (2.330499,$ $1.951372, -0.4775414, 4.365726, -0.6244870, 1.038131, 1.594227)$ where $f(x^*) = 680.6300573$. Two constraints are active (g_1 and g_4).

10. **g10**: Minimize: $f(x) = x_1 + x_2 + x_3$ subject to: $g_1(x) = -1 + 0.0025(x_4 + x_6) \leq 0$
    $g_2(x) = -1 + 0.0025(x_5 + x_7 - x_4) \leq 0$, $g_3(x) = -1 + 0.01(x_8 - x_5) \leq 0$
    $g_4(x) = -x_1 x_6 + 833.33252x_4 + 100x_1 - 83333.333 \leq 0$
    $g_5(x) = -x_2 x_7 + 1250x_5 + x_2 x_4 - 1250x_4 \leq 0$
    $g_6(x) = -x_3 x_8 + 1250000 + x_3 x_5 - 2500x_5 \leq 0$ where $100 \leq x_1 \leq 10000$, $1000 \leq x_i \leq 10000$, $(i = 2, 3)$, $10 \leq x_i \leq 1000$, $(i = 4, \ldots, 8)$. The global optimum is: $x^* = (579.19, 1360.13, 5109.92, 182.0174, 295.5985, 217.9799, 286.40, 395.5979)$, where $f(x^*) = 7049.25$. g_1, g_2 and g_3 are active.

11. **g11**: Minimize: $f(x) = x_1^2 + (x_2 - 1)^2$ subject to: $h(x) = x_2 - x_1^2 = 0$ where: $-1 \leq x_1 \leq 1$, $-1 \leq x_2 \leq 1$. The optimum solution is $x^* = (\pm 1/\sqrt{2}, 1/2)$ where $f(x^*) = 0.75$.

12. **g12**: Maximize: $f(x) = \dfrac{100 - (x_1 - 5)^2 - (x_2 - 5)^2 - (x_3 - 5)^2}{100}$ subject to: $g_1(x) = (x_1 - p)^2 + (x_2 - q)^2 + (x_3 - r)^2 - 0.0625 \leq 0$ where $0 \leq x_i \leq 10$ $(i = 1, 2, 3)$ and $p, q, r = 1, 2, \ldots, 9$. The feasible region of the search space consists of 9^3 disjointed spheres. A point (x_1, x_2, x_3) is feasible if and only if there exist p, q, r such the above inequality (12) holds. The global optimum is located at $x^* = (5, 5, 5)$ where $f(x^*) = 1$.

13. **g13**: Minimize: $f(x) = e^{x_1 x_2 x_3 x_4 x_5}$ subject to: $h_1(x) = x_1^2 + x_2^2 + x_3^2 + x_4^2 + x_5^2 - 10 = 0$ $h_2(x) = x_2 x_3 - 5x_4 x_5 = 0$, $h_3(x) = x_1^3 + x_2^3 + 1 = 0$ where $-2.3 \leq x_i \leq 2.3$ $(i = 1, 2)$ and $-3.2 \leq x_i \leq 3.2$ $(i = 3, 4, 5)$. The optimum solution is $x^* = (-1.717143, 1.595709,$ $1.827247, -0.7636413, -0.763645)$ where $f(x^*) = 0.0539498$.

Randomized Local Search, Evolutionary Algorithms, and the Minimum Spanning Tree Problem

Frank Neumann[1] and Ingo Wegener[2]*

[1] Inst. für Informatik und Prakt. Mathematik,
Christian-Albrechts-Univ. zu Kiel, 24098 Kiel, Germany
`fne@informatik.uni-kiel.de`
[2] FB Informatik, LS 2, Univ. Dortmund, 44221 Dortmund, Germany
`ingo.wegener@uni-dortmund.de`

Abstract. Randomized search heuristics, among them randomized local search and evolutionary algorithms, are applied to problems whose structure is not well understood, as well as to problems in combinatorial optimization. The analysis of these randomized search heuristics has been started for some well-known problems, and this approach is followed here for the minimum spanning tree problem. After motivating this line of research, it is shown that randomized search heuristics find minimum spanning trees in expected polynomial time without employing the global technique of greedy algorithms.

1 Introduction

The purpose of this paper is to contribute to the growing research area where randomized search heuristics are analyzed with respect to the expected time until they consider an optimal search point. Such an approach should support the understanding how these heuristics work, should guide the choice of the free parameters of the algorithms, and should support the teaching of heuristics. This is a growing research area, some general results can be found in Papadimitriou, Schäffer, and Yannakakis (1990) for randomized local search and Beyer, Schwefel, and Wegener (2002) and Droste, Jansen, and Wegener (2002) for evolutionary algorithms.

Search heuristics are mainly applied to problems whose structure is not well understood but the analysis has to start with problems whose structure is well understood. One cannot hope to beat the best problem-specific algorithms on these problems. Hence, the main purpose is to study the behavior of randomized search heuristics which find many applications in real-world optimization problems. For combinatorial optimization, this approach has been started only

---

* This work was supported by the Deutsche Forschungsgemeinschaft (DFG) as part of the Collaborative Research Center "Computational Intelligence" (SFB 531) and by the German-Israeli Foundation (GIF) in the project "Robustness Aspects of Algorithms".

K. Deb et al. (Eds.): GECCO 2004, LNCS 3102, pp. 713–724, 2004.

recently. There are results on sorting as the minimization of unsortedness and on shortest paths problems (Scharnow, Tinnefeld, and Wegener (2002)), on maximum matchings (Sasaki and Hajek (1988) for simulated annealing and Giel and Wegener (2003) for randomized local search and evolutionary algorithms), and on minimum graph bisections (Jerrum and Sorkin (1998) for the Metropolis algorithm).

Here we study the well-known problem of computing minimum spanning trees in graphs with n vertices and m edges. The problem can be solved by greedy algorithms. The famous algorithms due to Kruskal and Prim have worst-case run times of $O((n + m)\log n)$ and $O(n^2)$, respectively, see any textbook on efficient algorithms, e.g., Cormen, Leiserson, and Rivest (1990). Greedy algorithms use global ideas. Considering only the neighborhoods of two vertices u and v, it is not possible to decide whether the edge $\{u, v\}$ belongs to some minimum spanning tree. Therefore, it is interesting to analyze the run times obtainable by more or less local search heuristics like randomized local search and evolutionary algorithms. One goal is to estimate the expected time until a better spanning tree has been found. For large weights, there may be exponentially many spanning trees with different weights. Therefore, we also have to analyze how much better the better spanning tree is. This is indeed the first paper where the expected fitness increase is estimated for problems of combinatorial optimization.

As already argued, we do not and cannot hope to beat the best algorithms for the minimum spanning tree problem. This can be different for two generalizations of the problem. First, one is interested in minimizing the weight of restricted spanning trees, e.g., trees with bounded degree or trees with bounded diameter. These problems are NP-hard, and evolutionary algorithms are competitive, see Raidl and Julstrom (2003). Second, one is interested in the multi-objective variant of the problem. Each edge has k weights, and one looks for the Pareto optimal spanning trees with respect to the weight functions, see Hamacher and Ruhe (1994) for the general problem and Zhou and Gen (1999) for the design of evolutionary algorithms. Many polynomially solvable problems have NP-hard multi-objective counterparts, see Ehrgott (2000). None of these papers contains a run time analysis of the considered search heuristics. We think that it is essential to understand how the heuristics work on the unrestricted single-objective problem before one tries to analyze their behavior on the more difficult variants.

After having motivated the problem to analyze randomized search heuristics on the minimum spanning tree problem, we give a survey on the rest of this paper. In Section 2, we describe our model of the minimum spanning tree problem and, in Section 3, we introduce the randomized search heuristics which will be considered in this paper. The theory on minimum spanning trees is well established. In Section 4, we deduce some properties of local changes in non-optimal spanning trees which are applied in the run time analysis presented in Section 5. After the discussion of some generalizations in Section 6, we finish with concluding remarks.

2   Minimum Spanning Trees

This classical optimization problem has the following description. Given an undirected connected graph $G = (V, E)$ on n vertices and m weighted edges, find an edge set $E' \subseteq E$ of minimal weight, which connects all vertices. The weight of an edge set is the sum of the weights of the considered edges. Weights are positive integers. Therefore, the solution is a tree on V, a so-called spanning tree. One can also consider graphs which are not necessarily connected. Then the aim is to find a minimum spanning forest, i.e., a collection of spanning trees on the connected components. All our results hold also in this case. To simplify the notation we assume that G is connected.

There are many possibilities how to choose the search space for randomized search heuristics. This problem has been investigated intensively by Raidl and Julstrom (2003). Their experiments point out that one should work with so-called "edge sets". The search space equals $S = \{0, 1\}^m$, where each position corresponds to one edge. A search point $s \in S$ corresponds to the choice of all edges e_i, $1 \le i \le m$, where $s_i = 1$. In many cases, many search points correspond to non-connected graphs and others correspond to connected graphs with cycles, i.e., graphs which are not trees. If all graphs which are not spanning trees get the same "bad" fitness, it will take exponential time to find a spanning tree when we apply a general search heuristic. We will investigate two fitness functions w and w'. The weight of e_i is denoted by w_i. Let $w_{\max}$ be the maximum weight. Then $w_{ub} := n^2 \cdot w_{\max}$ is an upper bound on the weight of each edge set. Let

$$w(s) := (c(s) - 1) \cdot w_{ub}^2 + (e(s) - (n-1)) \cdot w_{ub} + \sum_{i \mid s_i = 1} w_i$$

be the first fitness function where $c(s)$ is the number of connected components of the graph described by s and $e(s)$ is the number of edges in this graph. The fitness function has to be minimized. The most important issue is to decrease $c(s)$ until we have graphs connecting all vertices. Then we have at least $n-1$ edges, and the next issue is to decrease $e(s)$ under the condition that s describes a connected graph. Hence, we look for spanning trees. Finally, we look for minimum spanning trees.

It is necessary to penalize non-connected graphs since the empty graph has the smallest weight. However, it is not necessary to penalize extra connections since breaking a cycle decreases the weight. Therefore, it is also interesting to investigate the fitness function

$$w'(s) := (c(s) - 1)w_{ub} + \sum_{i \mid s_i = 1} w_i.$$

The fitness function w' is appropriate in the black-box scenario where the scenario contains as little problem-specific knowledge as possible. The fitness function w contains the knowledge that optimal solutions are *trees*. This simplifies the analysis of search heuristics. Therefore, we always start with results on the fitness function w and discuss afterwards how to obtain similar results for w'.

3 Randomized Local Search and the (1+1) EA

Randomized local search (RLS) uses the following mutation operator:
– Choose $i \in \{1, \ldots, m\}$ randomly and flip the ith bit.
Here we use the notion "choose randomly" for a choice according to the uniform distribution. This operator is not useful for most graph problems. Often the number of ones (or edges) is the same for all good search points, e. g., for TSP or minimum spanning trees. Then all Hamming neighbors of good search points are bad implying that we have many local optima. Therefore, we work with the larger neighborhood of Hamming distance 2. This mutation operator has already been discussed for maximum matchings by Giel and Wegener (2003). Finally, RLS can be described as follows.

Algorithm 1 (Randomized Local Search (RLS))
1.) Choose $s \in \{0, 1\}^m$ randomly.
2.) Choose $b \in \{0, 1\}$ randomly. If $b = 0$, choose $i \in \{1, \ldots, m\}$ randomly and define s' by flipping the ith bit of s. If $b = 1$, choose $(i, j) \in \{(k, l) \mid 1 \le k < l \le m\}$ randomly and define s' by flipping the ith and the jth bit of s.
3.) Replace s by s' if $w(s') \le w(s)$.
4.) Repeat Steps 2 and 3 forever.

In applications, we need a stopping criterion. Here we are interested in the expected value of T_G, which measures the number of fitness evaluations until s is a minimum spanning tree. This is the expected optimization time (sometimes called expected first hitting or passage time) of RLS. Indeed, we will estimate $E(T_G)$ with respect to the parameters n, m, and $w_{\max}$.

The simple evolutionary algorithm called (1+1) EA differs from RLS in the chosen mutation operator.

Algorithm 2 (Mutation operator of (1+1) EA)
Define s' in the following way. Each bit of s is flipped independently of the other bits with probability $1/m$.

This is the perhaps most simple algorithm which can be called evolutionary algorithm. It is adopted from the well-known (1+1) ES (evolution strategy) for the optimization in continuous search spaces. In Section 6, we will argue why we believe that larger populations will be harmful. There it will also be discussed whether genetic algorithms based on crossover can be useful.

4 Properties of Local Changes of Spanning Trees

Our aim is to show the following. In the rest of this paper we denote by w_{opt} the weight of minimum spanning trees. For a non-optimal tree s, there are either many weight-decreasing local changes which, on the average, decrease $w(s)$ by an amount which is not too small with respect to $w(s) - w_{\mathrm{opt}}$, or there are few of these local changes which, on the average, cause a larger decrease of the weight. This statement will be made precise in the following lemma.

Lemma 1. *Let s be a search point describing a non-minimum spanning tree T. Then there exist some $k \in \{1, \ldots, n-1\}$ and k different accepted 2-bit flips such that the average weight decrease of these flips is at least $(w(s) - w_{\mathrm{opt}})/k$.*

Proof. This result follows directly from results in the literature on spanning trees. Kano (1987) has proved the following result by an existence proof and Mayr and Plaxton (1992) have proved the same result by an explicit construction procedure.

Let s^* be a search point describing a minimum spanning tree T^*. Let $E(T)$ and $E(T^*)$ be the edge sets of T and T^*, respectively. Let $k := |E(T^*) - E(T)|$. Then there exists a bijection $\alpha : E(T^*) - E(T) \to E(T) - E(T^*)$ such that $\alpha(e)$ lies on the cycle created in T by including e into T and the weight of $\alpha(e)$ is not smaller than the weight of e.

We consider the k 2-bit flips flipping e and $\alpha(e)$ for $e \in E(T^*) - E(T)$. They are accepted since e creates a cycle which is destroyed by the elimination of $\alpha(e)$. Performing all the k 2-bit flips simultaneously changes T into T^* and leads to a weight decrease of $w(s) - w_{\mathrm{opt}}$. Hence, the average weight decrease of these steps is $(w(s) - w_{\mathrm{opt}})/k$. □

The analysis performed in Section 5 will be simplified if we can ensure that we always have the same parameter k in Lemma 1. This is easy if we allow also non-accepted 2-bit flips whose weight decrease is defined as 0. We add $n - k$ non-accepted 2-bit flips to the set of the k accepted 2-bit flips whose existence is proven in Lemma 1. Then we obtain a set of exactly n 2-bit flips. The total weight decrease is at least $w(s) - w_{\mathrm{opt}}$ since this holds for the k accepted 2-bit flips. Therefore, the average weight decrease is bounded below by $(w(s) - w_{\mathrm{opt}})/n$. We state this result as Lemma 2.

Lemma 2. *Let s be a search point describing a spanning tree T. Then there exists a set of n 2-bit flips such that the average weight decrease of these flips is at least $(w(s) - w_{\mathrm{opt}})/n$.*

When analyzing the fitness function w' instead of w, we may accept non-spanning trees as improvements of spanning trees. Non-spanning trees can be improved by 1-bit flips eliminating edges of cycles. A 1-bit flip leading to a non-connected graph is not accepted and its weight decrease is defined as 0.

Lemma 3. *Let s be a search point describing a connected graph. Then there exist a set of n 2-bit flips and a set of $m - (n - 1)$ 1-bit flips such that the average weight decrease of these flips is at least $(w(s) - w_{\mathrm{opt}})/(m + 1)$.*

Proof. We consider all 1-bit flips concerning the non-T^* edges. If we try them in some arbitrary order we obtain a spanning tree T. If we consider their weight decrease with respect to the graph G' described by s, this weight decrease can be only larger. The reason is that a 1-bit flip, which is accepted in the considered sequence of 1-bit flips, is also accepted when applied to s. Then we apply Lemma 2 to T. At least the same weight decrease is possible by adding e_i and deleting a non-T^* edge with respect to G'. Altogether, we obtain at least a weight decrease of $w(s) - w_{\mathrm{opt}}$. This proves the lemma, since we have chosen $m + 1$ flips. □

5   The Analysis of RLS and (1+1) EA for the Minimization of Spanning Trees

First, it is rather easy to prove that RLS and (1+1) EA construct spanning trees efficiently.

Lemma 4. *The expected time until RLS or (1+1) EA working on one of the fitness function w or w' has constructed a connected graph is bounded by $O(m \log n)$.*

Proof. The fitness functions are defined in such a way that the number of connected components will never be increased in accepted steps. For each edge set leading to a graph with k connected components, there are at least $k - 1$ edges whose inclusion decreases the number of connected components by 1. Otherwise, the graph would not be connected. The probability of a step decreasing the number of connected components is at least $\frac{1}{2} \cdot \frac{k-1}{m}$ for RLS and $\frac{1}{e} \cdot \frac{k-1}{m}$ for (1+1) EA. Hence, the expected time until s describes a connected graph is bounded above by

$$em \left(1 + \cdots + \frac{1}{n-1}\right) = O(m \log n).$$

$\square$

Lemma 5. *If s describes a connected graph, the expected time until RLS or (1+1) EA has constructed a spanning tree for the fitness function w is bounded by $O(m \log n)$.*

Proof. The fitness function is defined in such a way that, starting with s, only connected graphs are accepted and that the number of edges does not increase. If s describes a graph with N edges, it contains a spanning tree with $n-1$ edges, and there are at least $N - (n - 1)$ edges whose exclusion decreases the number of edges. If $N = n - 1$, s describes a spanning tree. Otherwise, by the same arguments as in the proof of Lemma 4, we obtain an upper bound of

$$em \left(1 + \cdots + \frac{1}{m - (n-1)}\right) = O(m \log(m - n + 1)) = O(m \log n).$$

$\square$

This lemma holds also for RLS and the fitness function w'. RLS does not accept steps only including an edge or only including two edges if s describes a connected graph. Since RLS does not affect more than two edges in a step, it does not accept steps in which the number of edges of a connected graph is increased. This does not hold for (1+1) EA. It is possible that the exclusion of one edge and the inclusion of two or more edges creates a connected graph whose weight is not larger than the weight of the given graph.

Before we analyze the expected time to turn a spanning tree into a minimum spanning tree, we investigate an example (see Figure 1).

The example graph consists of a connected sequence of p triangles and the last triangle is connected to a complete graph on q vertices. The number of

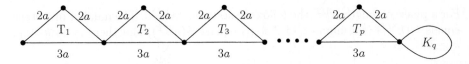

Fig. 1. An example graph with p connected triangles and a complete graph on q vertices with edges of weight 1.

vertices equals $n := 2p+q$ and the number of edges equals $m := 3p+q(q-1)/2$. We consider the case of $p = n/4$ and $q = n/2$ implying that $m = \Theta(n^2)$. The edges in the complete graph have the weight 1 and we set $a := n^2$. Each triangle edge has a weight which is larger than the weight of all edges of the complete graph altogether. Theorem 1 and Theorem 2 prove that this graph is a worst-case instance with polynomial weights.

Theorem 1. *The expected optimization time until RLS and (1+1) EA find a minimum spanning tree for the example graph equals $\Theta(m^2 \log n) = \Theta(n^4 \log n)$ with respect to the fitness functions w and w'.*

Proof. The upper bound is contained in Theorem 2. Here we prove the lower bound by investigating typical runs of the algorithm. We use the following notation. We partition the graph G into its triangle part T and its clique part C. Each search point x describes an edge set. We denote by $d(x)$ the number of triangles that are disconnected with respect to the edges chosen by x, by $b(x)$ the number of bad triangles (exactly one $2a$-edge and the $3a$-edge are chosen), by $g(x)$ the number of good triangles (exactly the two $2a$-edges are chosen), by $c(x)$ the number of complete triangles (all three edges are chosen), and by $\mathrm{con}_G(x)$, $\mathrm{con}_T(x)$, and $\mathrm{con}_C(x)$ the number of connected components in the different parts of the graph. We investigate four phases of the search. The first phase of length 1 is the initialization step producing the random edge set x. In the following, all statements hold with probability $1 - o(1)$.

Claim. After initialization, $b(x) = \Theta(n)$ and $\mathrm{con}_C(x) = 1$.

Proof. The statements can be proved independently since the corresponding parts of x are created independently. The probability that a given triangle is bad equals $1/4$. There are $n/4$ triangles and $b(x) = \Theta(n)$ by Chernoff bounds. We consider one vertex of C. It has $n/2 - 1$ possible neighbors. By Chernoff bounds, it is connected to at least $n/6$ of these vertices. For each other vertex, the probability to be not connected to at least one of these $n/6$ vertices is $(1/2)^{n/6}$. This is unlikely even for one of the remaining vertices. Hence, $\mathrm{con}_C(x) = 1$. $\square$

For the following phases, we distinguish the steps by the number k of flipping triangle edges and call them k-steps. Let p_k be the probability of a k-step. For RLS, $p_1 = \Theta(n^{-1})$, $p_2 = \Theta(n^{-2})$ and $p_k = 0$, if $k \geq 3$. For (1+1) EA and constant k

$$p_k = \binom{3n/4}{k} \left(\frac{1}{m}\right)^k \left(1 - \frac{1}{m}\right)^{3n/4-k} = \Theta(n^k m^{-k}) = \Theta(n^{-k}).$$

For a phase of length $n^{5/2}$, the following statements hold. The number of 1-steps equals $\Theta(n^{3/2})$, the number of 2-steps equals $\Theta(n^{1/2})$, and there is no k-step, $k \geq 3$.

Claim. Let $b(x) = \Theta(n)$ and $\text{con}_C(x) = 1$. In a phase of length $n^{5/2}$, a search point y where $b(y) = \Theta(n)$ and $\text{con}_G(y) = 1$ is produced.

Proof. By Lemma 4, the probability of creating a connected graph is large enough. Let y be the first search point where $\text{con}_G(y) = 1$. We prove that $b(y) = \Theta(n)$. All the 2-steps can decrease the b-value by at most $O(n^{1/2})$. A 1-step has two possibilities to destroy a bad triangle.

- It may destroy an edge of a bad triangle. This increases the con_G-value. In order to accept the step, it is necessary to decrease the con_C-value.
- It may add the missing edge to a bad triangle. This increases the weight by at least $2a$. No triangle edge is eliminated in this step. In order to accept the step, it is necessary to decrease the con_C-value.

However, $\text{con}_C(x) = 1$. In order to decrease this value, it has to be increased before. A step increasing the con_C-value can be accepted only if the con_T-value is decreased in the same step at least by the same amount. This implies that triangle edges have to be added. For a 1-step, the total weight is increased without decreasing the con_G-value and the step is not accepted. Hence, only the $O(n^{1/2})$ 2-steps can increase the con_C-value. By Chernoff bounds, the number of clique edges flipping in these steps is $O(n^{1/2})$. This implies that the number of bad triangles is decreased by only $O(n^{1/2})$. ☐

Claim. Let $b(y) = \Theta(n)$ and $\text{con}_G(y) = 1$. In a phase of length $n^{5/2}$, a search point z where $b(z) = \Theta(n)$, $\text{con}_G(z) = 1$, and $T(z)$ is a tree is produced.

Proof. Only search points x describing connected graphs are accepted, in particular, $d(x) = 0$. Let z be the first search point where $T(z)$ is a tree. Then $\text{con}_G(z) = 1$ and we have to prove that $b(z) = \Theta(n)$ and that z is produced within $n^{5/2}$ steps. A 1-step can be accepted only if it turns a complete triangle into a good or bad triangle. Such a step is accepted if no other edge flips. Moreover, $c(x)$ cannot be increased. In order to increase $c(x)$ it is necessary to add the missing edge to a good or bad triangle. To compensate this weight increase, we have to eliminate an edge of a complete triangle. Remember that we have no k-steps for $k \geq 3$. If $c(x) = l$, the probability of decreasing the c-value is at least $3l/(em)$ and the expected time to eliminate all complete triangles is $O(m \log n) = O(n^2 \log n)$. Hence, $n^{5/2}$ steps are sufficient to create z. The number of bad triangles can be decreased only in the $O(n^{1/2})$ 2-steps implying that $b(z) = \Theta(n)$. ☐

Claim. Let $b(z) = \Theta(n)$, $\text{con}_G(z) = 1$, and $T(z)$ be a tree. The expected time to find a minimum spanning tree is $\Omega(n^4 \log n)$.

Proof. First, we assume that only 2-steps change the number of bad triangles. Later, we complete the arguments. The expected waiting time for a 2-step flipping those two edges of a bad triangle that turn it into a good one equals $\Theta(n^4)$.

The expected time to decrease the number of bad triangles from b to $b-1$ equals $\Theta(n^4/b)$. Since b has to be decreased from $\Theta(n)$ to 0, we obtain an expected waiting time of

$$\Theta(n^4 \sum_{1 \le b \le \Theta(n)} (1/b)) = \Theta(n^4 \log n). \tag{$*$}$$

Similarly to the proof of the coupon collector's theorem we obtain that the optimization step if only 2-steps can be accepted equals $\Theta(n^4 \log n)$ with probability $1-o(1)$. Hence, it is sufficient to limit the influence of all k-steps, $k \ne 2$, within a time period of $\alpha n^4 \log n$ for some constant $\alpha > 0$. Again with probability $1-o(1)$, the number of 4-steps is $O(\log n)$ and there are no k-steps for $k \ge 5$. The 4-steps can decrease the number of bad triangles by at most $O(\log n)$. Because of the weight increase, a k-step, $k \le 4$, can be accepted only if it eliminates at least $\lceil k/2 \rceil$ triangle edges. Moreover, it is not possible to disconnect a good or a bad triangle. Hence, a 4-step cannot create a complete triangle. As long as there is no complete triangle, a 3-step or a 1-step has to disconnect a triangle and is not accepted. A 2-step can only be accepted if it changes a bad triangle into a good one. Hence, no complete triangles are created. The 4-steps eliminate $O(\log n)$ terms of the sum in $(*)$. The largest terms are those for the smallest values of b. We only have to substract a term of $O(n^4 \log \log n) = o(n^4 \log n)$ from the bound $\Theta(n^4 \log n)$ and this proves the claim. □

We have proved Theorem 1 since the sum of all failure probabilities is $o(1)$. □

In the following , we prove an upper bound of size $O(m^2(\log n + \log w_{\max}))$ on the expected optimization time for arbitrary graphs. This bound is $O(m^2 \log n)$ as long as $w_{\max}$ is polynomially bounded and it is always polynomially bounded with respect to the bit length of the input. Theorem 1 shows that the bound is optimal.

Theorem 2. *The expected time until RLS or (1+1) EA working on the fitness function w constructs a minimum spanning tree is bounded by $O(m^2(\log n + \log w_{\max}))$.*

Proof. By Lemmas 4 and 5, it is sufficient to investigate the search process after having found a search point s describing a spanning T. Then, by Lemma 2, there always exists a set of n 2-bit flips whose average weight decrease is at least $(w(s) - w_{\text{opt}})/n$. The choice of such a 2-bit flip is called a "good step". The probability of performing a good step equals $\Theta(n/m^2)$ and each of the good steps is chosen with the same probability. A good step decreases the difference between the weight of the current spanning tree and w_{opt} on average by a factor not larger than $1 - 1/n$. This holds independently from previous good steps. Hence, after N good steps, the expected difference of the weight of T and w_{opt} is bounded above by $(1 - 1/n)^N \cdot (w(s) - w_{\text{opt}})$. Since $w(s) \le (n-1) \cdot w_{\max}$ and $w_{\text{opt}} \ge 0$, we obtain the upper bound $(1 - 1/n)^N \cdot D$, where $D := n \cdot w_{\max}$.

If $N := \lceil (\ln 2) \cdot n \cdot (\log D + 1) \rceil$, this bound is at most $\frac{1}{2}$. Since the difference is not negative, by Markov's inequality, the probability that the bound is less than 1 is at least $1/2$. The difference is an integer implying that the probability of having found a minimum spanning tree is at least $1/2$. Repeating the same arguments, the expected number of good steps until a minimum spanning tree is found is bounded by $2N = O(n \log D) = O(n(\log n + \log w_{\max}))$.

By our construction, there are always exactly n good 2-bit flips. Therefore, the probability of a good step does not depend on the current search point. Hence, the expected time until r steps are good equals $\Theta(rm^2/n)$. Altogether, the expected optimization time is bounded by

$$O(Nm^2/n) = O(m^2(\log n + \log w_{\max})).$$

$\square$

Applying Lemma 3 instead of Lemma 2, it is not too difficult to obtain the same upper bound for the fitness function w'. The main difference is that a good 1-bit flip has a larger probability than a good 2-bit flip.

Theorem 3. *The expected time until RLS or (1+1) EA working on the fitness function w' constructs a minimum spanning tree is bounded by $O(m^2(\log n + \log w_{\max}))$.*

Proof. By Lemma 4, it is sufficient to analyze the phase after having constructed a connected graph. We apply Lemma 3. The total weight decrease of the chosen 1-bit flips and 2-bit flips is at least $w(s) - w_{\mathrm{opt}}$ if s is the current search point. If the total weight decrease of the 1-bit flips is larger than the total weight decrease of the chosen 2-bit flips, the step is called a 1-step. Otherwise, it is called a 2-step.

If more than half of the steps are 2-steps, we adapt the proof of Theorem 2 with $N' := 2N$ since we guarantee only an expected weight decrease by a factor of $1 - 1/(2n)$. Otherwise, we consider the good 1-steps which have an expected weight decrease by a factor of $1 - 1/(2m')$ for $m' = m - (n - 1)$. Choosing $M := \lceil 2 \cdot (\ln 2) \cdot m' \cdot (\log D + 1) \rceil$, we can apply the proof technique of Theorem 2 where M takes the role of N. The probability of performing a good 1-bit flip equals $\Theta(m'/m)$. In this case, we obtain the bound

$$O(Mm/m') = O(m(\log n + \log w_{\max}))$$

for the expected number of steps which is even smaller than the proposed bound.

$\square$

6   Generalizations

Theorems 1, 2, and 3 contain matching upper and lower bounds for RLS and (1+1) EA with respect to the fitness functions w and w'. The bounds are worst-case bounds and one can hope that the algorithms are more efficient for many graphs. Here we discuss what can be gained by other randomized search heuristics.

First, we introduce more problem-specific mutation operators. It is easy to construct spanning trees. Afterwards, it is good to create children with the same number of edges. The new mutation operators are:

- If RLS flips two bits, it chooses randomly a 0-bit and randomly a 1-bit.
- If s contains k 1-bits, (1+1) EA flips each 1-bit with probability $1/k$ and each 0-bit with probability $1/(m - k)$.

For spanning trees, the probability of a specific edge exchange is increased from $\Theta(1/m^2)$ to $\Theta(1/(n(m - n + 1)))$. It is easy to obtain the following result.

Theorem 4. *For the modified mutation operator, the bounds of Theorems 1, 2, and 3 can be replaced by bounds of size $\Theta(mn \log n)$ and $O(mn(\log n + \log w_{\max}))$ respectively.*

Using larger populations, we have to pay for improving all members of the population. This holds at least if we guarantee a large diversity in the population. The lower bound of Theorem 1 holds with overwhelming probability. Hence, we do not expect that large populations help. The analysis in the proof of Theorems 2 and 3 is quite precise in most aspects. There is only one essential exception. We know that the weight distance to w_{opt} is decreased on average by a factor of at most $1 - 1/n$ and we work under the pessimistic assumption that this factor equals $1 - 1/n$. For large populations or multi-starts the probability of having sometimes much larger improvements may increase for many graphs.

It is more interesting to "parallelize" the algorithms by producing more children in parallel. The well-known algorithm (1+λ) EA produces independently λ children from the single individual from the current population. The selection procedure selects an individual with the smallest w-value (or w'-value) among the parent and its children. In a similar way, we obtain λ-PRLS (parallel RLS) from RLS. In the proofs of Theorem 2 and Theorem 3 we have seen that the probability of a good step is $\Theta(n/m^2)$. Choosing $\lambda = \lceil m^2/n \rceil$, this probability is increased to a positive constant. We have seen that the expected number of good steps is bounded by $O(n(\log n + \log w_{\max}))$. This leads to the following result.

Theorem 5. *The expected number of generations until λ-PRLS or the (1+λ) EA with $\lambda := \lceil m^2/n \rceil$ children constructs a minimum spanning tree is bounded by $O(n(\log n + \log w_{\max}))$. This holds for the fitness functions w and w'.*

If we use the modified mutation operator defined above, the probability of a good step is $O(1/m)$ and we obtain the same bound on the expected number of generations as in Theorem 5 already for $\lambda := m$.

One-point crossover or two-point crossover are not appropriate for edge set representations. It is not possible to build blocks of all edges adjacent to a vertex. For uniform crossover, it is very likely to create graphs which are not spanning trees. Hence, only problem-specific crossover operators seem to be useful. Such operators are described by Raidl and Julstrom (2003). It is difficult to analyze heuristics with these crossover operators.

7   Conclusions

The minimum spanning tree problem is one of the fundamental problems which are efficiently solvable. Several important variants of this problem are difficult, and evolutionary algorithms have a good chance to be competitive on these problems. As a first step toward the analysis of evolutionary algorithms on these problems, randomized local search and simple evolutionary algorithms have been analyzed on the basic minimum spanning tree problem. The asymptotic worst-case (with respect to the problem instance) expected optimization time has been obtained exactly. The analysis is based on the investigation of the expected multiplicative weight decrease (with respect to the difference of the weight of the current graph and the weight of a minimum spanning tree).

References

1. Beyer, H.-G., Schwefel, H.-P., and Wegener, I. (2002). How to analyse evolutionary algorithms. Theoretical Computer Science 287, 101–130.
2. Cormen, T.H., Leiserson, C.E., and Rivest, R.L. (1990). *Introduction to Algorithms.* MIT Press.
3. Droste, S., Jansen, T., and Wegener, I. (2002). On the analysis of the (1+1) evolutionary algorithm. Theoretical Computer Science 276, 51–81.
4. Ehrgott, M. (2000). Approximation algorithms for combinatorial multicriteria optimization problems. Int. Transactions in Operational Research 7, 5–31.
5. Giel, O. and Wegener, I. (2003). Evolutionary algorithms and the maximum matching problem. Proc. of 20th STACS. LNCS 2607, 415–426.
6. Hamacher, H.W. and Ruhe, G. (1994). On spanning tree problems with multiple objectives. Annals of Operations Research 52, 209–230.
7. Jerrum, M. and Sorkin, G.B. (1998). The Metropolis algorithm for graph bisection. Discrete Applied Mathematics 82, 155–175.
8. Kano, M. (1987). Maximum and kth maximal spanning trees of a weighted graph. Combinatorica 7, 205–214.
9. Mayr, E.W. and Plaxton, C.G. (1992). On the spanning trees of weighted graphs. Combinatorica 12, 433–447.
10. Papadimitriou, C.H., Schäffer, A.A., and Yannakakis, M. (1990). On the complexity of local search. Proc. of 22nd ACM Symp. on Theory of Computing (STOC), 438–445.
11. Raidl, G.R. and Julstrom, B.A. (2003). Edge sets: an effective evolutionary coding of spanning trees. IEEE Trans. on Evolutionary Computation 7, 225–239.
12. Sasaki, G. and Hajek, B. (1988). The time complexity of maximum matching by simulated annealing. Journal of the ACM 35, 387–403.
13. Scharnow, J., Tinnefeld, K., and Wegener, I. (2002). Fitness landscapes based on sorting and shortest paths problems. Proc. of Parallel Problem Solving from Nature – PPSN VII. LNCS 2939, 54–63.
14. Zhou, G. and Gen, M. (1999). Genetic algorithm approach on multi-criteria minimum spanning tree problem. European Journal of Operational Research 114, 141–152.

An Evolution Strategy Using a Continuous Version of the Gray-Code Neighbourhood Distribution

Jonathan E. Rowe and Džena Hidović

School of Computer Science, University of Birmingham, Birmingham B15 2TT, Great Britain
{J.E.Rowe,D.Hidovic}@cs.bham.ac.uk

Abstract. We derive a continuous probability distribution which generates neighbours of a point in an interval in a similar way to the bitwise mutation of a Gray code binary string. This distribution has some interesting scale-free properties which are analogues of properties of the Gray code neighbourhood structure. A simple (1+1)-ES using the new distribution is proposed and evaluated on a set of benchmark problems, on which it performs remarkably well. The critical parameter is the *precision* of the distribution, which corresponds to the string length in the discrete case. The algorithm is also tested on a difficult real-world problem from medical imaging, on which it also performs well. Some observations concerning the scale-free properties of the distribution are made, although further analysis is required to understand why this simple algorithm works so well.

1   Introduction

There are two different approaches to solving continuous-value optimisation problems using Evolutionary Computation. The first is to represent points in the search space using real numbers and to generate new points using some continuous probability distribution (typically Gaussian or Cauchy). The second approach is to discretise the space and represent real numbers as binary strings. One then mutates the strings by flipping one or more bits. It is known that there can be problems with this second approach if the standard binary encoding is used: there exist so-called Hamming cliffs — points that are neighbours according to the topology of the space, but are not neighbours when considered as binary strings. An alternative representation is to use a Gray code, in which all neighbours in the original space are also neighbours as strings. The trade-offs between using the standard binary and Gray representations have been studied in some detail by Whitley [1]. It can be shown that on some classes of optimisation problem, the Gray code representation has definite advantages. For example, it can be shown that a local search algorithm using this representation can solve a one-dimensional unimodal problem in quadratic time, and a clever variant can do it in linear time [2].

As part of the theoretical investigation of the use of Gray codes, one can ask about the distribution of neighbours that a point has, under this encoding. Suppose we use ℓ bits to represent the numbers $0, 1, \ldots, 2^\ell - 1$. Given a point x in this range, we want to know something about the set of its neighbours, generated by flipping exactly one bit of the Gray code representation of x. For example, if $\ell = 4$ and $x = 13 = 1011$ in Gray code, then the neighbours of x are $2 = 0011, 10 = 1111, 12 = 1010, 14 = 1001$. One way to charactise this question in general terms is to ask: given a point x, how many neighbours of x are within a given distance t? It can be shown that, on average, there will be $\lfloor \lg t + 1 \rfloor$ such neighbours [3].

It would be nice, from a theoretical point of view, if we could relate this method of local search to the standard Evolution Strategies (ES) which make use of real-valued representations and generate neighbours using continuous probability distributions. In other words, we ask the question: is there a continuous probability distribution such that the probability of generating a neighbour within a given interval is the same as if we used a Gray code representation and bitwise (point) mutation? This is the question we address, and answer, in section 2. We would also like to know what properties an Evolution Strategy using this new distribution has. In particular, are there theorems analogous to those already proved for the discrete Gray code local search algorithm? We investigate some of these properties in section 3.

Having developed this distribution and analysed the corresponding ES from a theoretical point of view, an obvious question arises: is it any good for optimisation? We study its performance on a collection of standard benchmark problems in section 4, comparing its performance (under various settings of the main contol parameter) with a recently published Evolutionary Programming algorithm (the "Improved Fast Evolutionary Programming" algorithm [4]) which makes use of a population and self-adaptive mutation rates and mixed Gaussian-Cauchy mutation distributions (see also [5]). The conclusion is that, remarkably, the simple (1+1)-ES with the new distribution is exceptionally good.

Of course, benchmark problems are one thing, and real-world applications another. So we conclude by presenting some results from a difficult problem in medical tissue optics: finding the values of structural parameters describing colon tissue that could give rise to observed colours in colonoscopy images. This is an important application area in medicine: the ability to distinguish normal from cancerous colon tissue optically would reduce the need for biopsies and assist clinicians in making diagnoses [6]. Again, the new algorithm is compared to the IFEP algorithm, and performs remarkably well (section 5).

2   The Continuous Version of the Gray Neighbourhood Distribution

In this section we will derive a continuous probability distribution which has properties directly analogous to the discrete Gray code representation under the usual definition of Hamming neighbourhoods. The key properties which we emulate are:

- mutations of a bit string generate moves with a *minimum* distance, specified by the precision of the code (or equivalently, the string length).
- the probability of producing a neighbour within a distance of t discrete points of the current point is, on average, $\lfloor \lg t + 1 \rfloor / \ell$.
- the Gray code naturally represents a bounded interval.
- the Gray code "wraps around": the strings corresponding to $2^\ell - 1$ and 0 are Hamming neighbours. The *maximum* distance of a move is thus half the search space (in either direction).

To keep things simple, we will assume that we have a one-dimensional search space which is the interval $[-1, +1]$. Any other bounded interval can be mapped by an affine transformation into this standard interval.

$$\varphi : [a, b] \longrightarrow [-1, +1]$$

$$\varphi(x) = 2\left(\frac{x-a}{b-a}\right) - 1$$

We need to specify a *minimum step size* which we denote ε. Equivalently, we will define the *precision* to be $p = -\log \varepsilon$. The precision is an analogue of the string length of the Gray code. The maximum step size will be half the size of the interval (in either direction): that is, 1. We define a probability density function

$$f(x) = \begin{cases} \frac{1}{px} & \text{if } \varepsilon < x < 1 \\ 0 & \text{otherwise} \end{cases}$$

We will choose the distance between the current point and the new point (the neighbour) according to this density function, moving to the left or right with equal probability. Thus the probability of picking a neighbour within a distance τ of the current point is

$$\int_\varepsilon^\tau \frac{dx}{px} = \frac{\log \tau}{p} + 1$$

By analogy to the discrete Gray encoding, let t be the number of minimal steps needed to move a distance of τ. That is $\tau = \varepsilon t$. Then the probability of jumping within a distance τ becomes

$$\frac{\log \varepsilon t}{p} + 1 = \frac{\log t}{p}$$

This defines, therefore, a continuous probability distribution which distributes neighbours in a way exactly analagous to the Gray code representation. But what should we do if the distance to be moved takes us outside the range $[-1, +1]$? We will simply wrap around in the same way that the Gray code does.

All that remains to be able to write an Evolution Strategy based on this distribution is a method for generating random numbers according to the given distribution. To do this, we note that the cumulative distribution function is:

$$F(x) = \int_{-\infty}^x f(t)dt = 1 + \frac{\log x}{p}$$

We can generate a random number according to this distribution by first generating a random number u uniformly from $[0, 1]$ and then setting $\tau = F^{-1}(u) = \exp(-p(1-u))$. Equivalently, we can set $\tau = \exp(-pu)$ since $1 - u$ is also distributed uniformly in $[0, 1]$. See [7] for more details of this method.

Suppose $g : [-1, 1] \to \mathbb{R}$ is the objective function, and, without loss of generality, that we are minimising. We define our (1+1)-ES as follows:

1. Pick an initial point $x \in [-1, +1]$.
2. Generate a random number u uniformly in $[0, 1]$.
3. Set $\tau = \exp(-pu)$.
4. With probability half, set $y = x + \tau$, else $y = x - \tau$.
5. If $y < -1$ set $y = y + 2$. If $y > 1$ set $y = y - 2$.
6. If $g(y) < g(x)$, set $x = y$.
7. Go to 2.

3   Properties of the Distribution

We have shown that our new continuous probability distribution generates neighbours of a point that are distributed analagously to the discrete Gray code representation. We now look at some other properties that the Gray code has, and derive corresponding results for our new distribution.

The Gray code neighbourhood structure has some remarkable *scale invariant* properties. Firstly, it is clear from the recursive construction of the Gray code that any point has neighbours at all scales. That is, if the point is in one half of the search interval, it has a neighbour in the other half. Zooming in, if we consider the quarter of the search interval containing the point, then there is a neighbour in an adjacent quarter. One can continue to zoom in, throwing away half the interval at each step, and one will always find neighbours. When it comes to the continuous distribution, there is, of course, a non-zero probability of generating a neighbour right across the search interval. However, by a "scale free" distribution is meant one in which the probability of finding points at any distance is not vanishingly small. So the Gaussian distribution, for example, while assigning a non-zero probability across the range, has tails that shrink exponentially: it is therefore not scale-free. With our new distribution, however, one never has to wait too long for jumps of arbitrarily large size (up to the maximum). The maximum jump size is 1. The probability of making a jump bigger than $1 - \delta$ is

$$\int_{1-\delta}^{1} \frac{dx}{px} = -\frac{\log(1 - \delta)}{p}$$

so the expected waiting time is $O(p)$.

The second scale-invariant property shows up in the analysis of the steepest descent Gray code algorithm applied to a one-dimensional unimodal function, in which it takes a constant number of trials in order to disregard half of the remaining search interval under consideration. We have a similar result here. Suppose the current point is a distance z away from the optimum. The probability of making one jump that would take us within $z/2$ of the optimum is

$$\frac{1}{2} \int_{z/2}^{z} \frac{dx}{px} = \frac{\log 2}{2p}$$

That is, it is independent of the current position! The expected waiting time (and this is clearly an upper bound) is thus $2p/\log 2$. The number of steps required to get within δ of the optimum is therefore $O(p \log(1/\delta))$. This result might make one think that it is best to choose the precision p to be as small as possible, but of course, one needs a sufficiently small minimum step size to be able to approach the optimum as closely as desired.

4   Experiments with the ES

Having developed a simple search algorithm for theoretical purposes, it seemed worth trying it out on a range of test problems. Partly this was to investigate the effects of varying the precision parameter on the performance of the algorithm, but we also wished to see if its performance were comparable with other evolutionary optimisation algorithms. Consequently, we took eight benchmark problems from the paper [4] which introduced a new Evolutionary Programming algorithm called Improved Fast Evolutionary Programming (IFEP). We used a variant of the IFEP algorithm to provide a baseline performance against which we compared our algorithm. Specifically, we used a (15,45)-ES in which each population member produces three offspring. A single offspring is produced by mutating according to both Gaussian and Cauchy distributions and taking the best (thus each offspring requires two fitness evaluations). The mutations are self-adaptive, as described in the paper. The best 15 individuals are chosen from the offspring to form the next population.[1]

The test functions are taken from the same paper and are defined as follows (note that the minimum is zero in each case):

Sphere function
$$f_1(x) = \sum_{i=1}^{30} (x_i)^2 \qquad\qquad x_i \in [-100, 100]$$

Schwefel's problem 2.22
$$f_2(x) = \sum_{i=1}^{30} |x_i| + \prod_{i}^{30} |x_i| \qquad\qquad x_i \in [-10, 10]$$

Schwefel's problem 1.2
$$f_3(x) = \sum_{i=1}^{30} \left(\sum_{j=1}^{j=i} x_j \right)^2 \qquad\qquad x_i \in [-100, 100]$$

Schwefel's function 2.21
$$f_4(x) = \max\{|x_i|, 1 \le x_i \le 30\} \qquad\qquad x_i \in [-100, 100]$$

Generalised Rosenbrock's function
$$f_5(x) = \sum_{i=1}^{29} (100(x_{i+1} - x_i^2)^2 + (x_i - 1)^2) \qquad\qquad x_i \in [-30, 30]$$

Generalised Rastrigin's function
$$f_9(x) = \sum_{i=1}^{30} (x_i^2 - 10\cos(2\pi x_i) + 10) \qquad\qquad x_i \in [-5.12, 5.12]$$

Ackley's function
$$f_{10}(x) = -20\exp\left(-0.2\sqrt{\tfrac{1}{30}\sum_{i=1}^{30} x_i^2}\right) - \exp\left(\sum_{i=1}^{30}\cos 2\pi x_i\right)$$
$$+20 + e \qquad\qquad x_i \in [-32, 32]$$

Generalised Griewangk function
$$f_{11}(x) = \tfrac{1}{4000}\sum_{i=1}^{30} x_i^2 - \prod_{i=1}^{30}\cos\left(\tfrac{x_i}{\sqrt{i}}\right) + 1 \qquad\qquad x_i \in [-600, 600]$$

For multi-dimensional problems such as these, we have to adapt our Evolution Strategy slightly. At each iteration, we mutate *all* of the parameters of the current point simultaneously, using the method described in the previous section.

We conducted a number of experiments with these test functions, varying the precision parameter p through the values 25, 50, 100 and 200. We also considered the effect of the number of iterations allowed (1000, 10000, 100000 and 1000000). We adjusted

---

[1] Although this is a variant on the IFEP algorithm, the results we obtained are largely similar to those reported for IFEP.

the number of generations allowed to the IFEP algorithm accordingly, to get the same total number of function evaluations. Each experiment was run 30 times. The average results (function values) are shown in figure 1. Standard deviations are not shown, but nearly all differences are significant at the 99.9% level according to a two-tailed t-test. Note that each graph is shown on a log-log scale.

It is clear that our new (1+1)-ES has performed very well, especially over large iterations with a high precision value. What is also clear from the data is that it is often significantly better than the IFEP algorithm over a small number of iterations, when a lower precision value is used. It also gives results that are comparable to the best results reported for various evolutionary algorithms in [5][2]. We also ran some experiments with a (1+1)-ES with Gaussian mutation using the $1/5$ success rule. As might be expected, this algorithm performs extremely well on the sphere function. It performs moderately well on other unimodal functions. However it is terrible on multimodal functions — by construction, it is designed to converge rapidly to the nearest local optimum. The new algorithm (and indeed IFEP) is superior on such functions.

5   A Real-World Application

Having tested our algorithm on some standard benchmark problems, we then applied it to a difficult real-world problem, from medical image interpretation. An increasingly important application of image interpretation is the development of non-invasive techniques for studying tissue structure. Clinicians want to be able to deduce as much as possible about the structure of an organ from its visual images, to reduce the necessity of performing biopsies. One approach to this problem focuses on analysing the physics of image formation. The basic idea is to create a physics-based model of the tissue structure and to simulate the effect of shining white light onto the surface. As a result of that simulation, the amount of light that reemerges at the tissue surface (spectral reflectance), after interacting with the tissue structure, is calculated. By adjusting the parameters of the model, one tries to reproduce the optical spectra measured on real tissue, in order to analise the corectness of the model. One can then, in principle, match the spectra with the appropriate physical parameters and extract diagnostically valuable information about the tissue structure. However, it is rather difficult to establish all the relevant parameters and the corresponding value ranges for them. The initial stages of this research depend, therefore, on using optimisation algorithms to try to establish suitable parameter settings. For a more detailed description of the problem, see [6].

We have obtained a set of spectra from normal colon tissue taken during colonoscopy procedures, where an optical fibre bundle (which delivers and collects light) is passed through a working channal of a colonoscope, and placed against the colon wall of a patient [8]. In each case an observed spectrum is collected. We then use our optimisation algorithm to try to find parameter settings in the physics-based model which will account for the observed spectrum.

The physics-based model of colon tissue is developed so that it simulates the interaction of incident light with the structure and morphology of the real colon tissue, which

---

[2] Exact data are not presented in that paper. However, from the graphs shown it is clear that our new algorithm is comparable or better on all test functions than the algorithms presented in that paper.

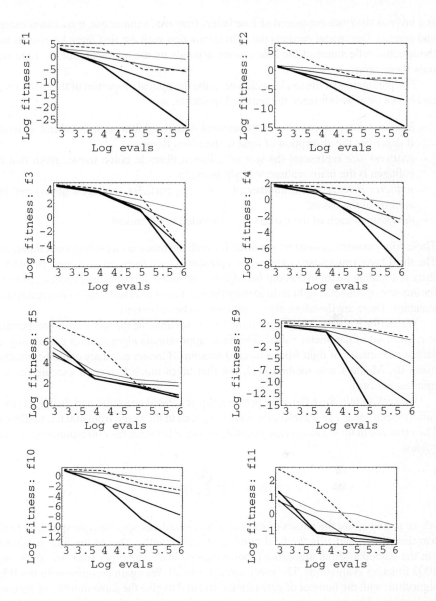

Fig. 1. Comparison of performance of the new (1+1)-ES (solid lines) and IFEP (dashed line). There are four different settings of the precision parameter ($p = 25, 50, 100, 200$), indicated by increasing thickness of the line used. Performance tends to improve with increasing precision.

is a layered structure composed of four layers (mucosa, submucosa, muscularis externa and serosa). Our model predicts the light interaction with the first three layers, because the spectral reflectance of the colon tissue depends on the interaction of the light with only those layers.

The parameters of the model which describe the optical properties of the colon layers and hence directly influence the remitted spectrum, are:

- *haemoglobin concentration* is the amount of hemoglobin per unit volume of tissue. It describes the absorption of light in the colon tissue.
- *scatterer size* represents the size of collagen fibres in colon tissue, given that the collagen is the main scatterer of light in colon.
- *scatterer density* is the number of scattering particles (collagen fibres) per unit volume of the tissue.
- *thickness* of each of the tissue layers included in the model

These four parameters must be specified for both the mucosa and submucosa separately. The third layer (muscularis externa) is represented by a fixed set of values. In addition there is a scaling factor to account for adjustments to the normalisation process, in which the amount of collected light is divided by the amount of light reflected from a reflectance standard. There are therefore nine parameters to be optimised.

We use the Kubelka-Munk algorithm, [9,10], to calculate the spectrum corresponding to a given set of parameter values. This is an approximate algorithm for calculating the diffuse reflectance of light from a layered structure. Greater accuracy could be obtained using the Monte Carlo method [11], but that takes much longer to execute (several minutes per run).

We seek to minimise the error between the generated spectrum and the target (measured) spectrum at 113 wavelengths equally spaced in the range from 400 nm to 624 nm. The error is calculated as average absolute distance between the corresponding spectral values:

$$d(y, z) = \frac{1}{n} \sum_{i=0}^{n} |y_i - z_i|$$

where y_i and z_i are the values of measured and simulated spectra corresponding to the wavelength w_i, and n is the total number of wavelengths. Due to the time it takes to run the Kubelka-Munk algorithm (approximately one second per run), we allow only 1000 function evaluations. The precision is set to 20. We again compare with the IFEP algorithm, with the number of generations adjusted to give the same number of function evaluations. The results are shown in table 1. The new Evolution Strategy is clearly superior (significance $> 99.99\%$ on a paired t-test). Some typical results are shown in figure 2.

6   Discussion and Conclusions

We have introduced a new Evolution Strategy, with a mutation probability distribution based on a continuous version of the Gray code neighbourhood structure. The distribution we have defined has certain *scale-free* properties which may be assisting its performance

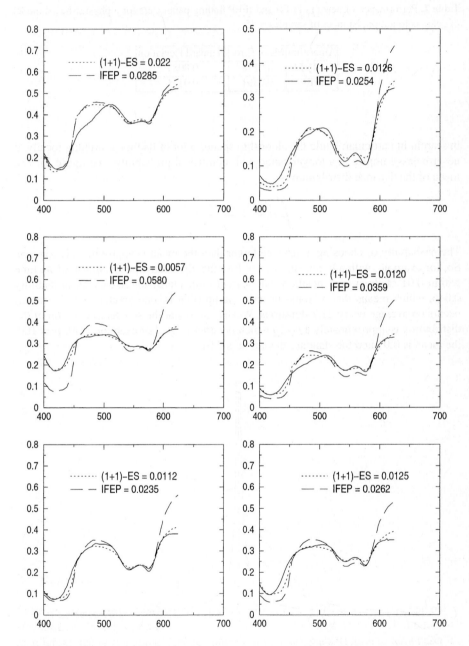

Fig. 2. Six example spectra from normal tissue(solid lines) obtained during colonoscopy. The new (1+1)-ES (dotted lines) and IFEP (dashed lines) algorithms were both used to find parameter settings to approximate these spectra, using the Kubelka-Munk method. Errors are measured as the average absolute distance from the generated curve and the target spectrum.

Table 1. Performance of new (1+1)-ES and IFEP finding parameters for a physics-based model of colon colouration. Number of samples = 45.

Algorithm	Mean Error	Standard Deviation
(1+1)-ES	0.0178	0.0081
IFEP	0.0391	0.0121

in search. In particular, while the algorithm spends a lot of its time searching locally, it nevertheless samples at a longer range with non-trivial probability (see figure 3). The mean of the distance distribution is

$$\int_\varepsilon^1 \frac{dx}{p} = \frac{1-\varepsilon}{p} \approx \frac{1}{p}$$

The probability of choosing a distance larger than the mean is approximately $\log p/p$. So, for example, with a precision of $p = 100$, the algorithm spends 95.4% of its time within 0.01 of its current position, but searches outside of this area with probability 0.046, which means that a "long-distance" jump (that is, one greater than the mean) occurs on average every 21.7 iterations. We also note that the standard deviation of the distribution is approximately $1/\sqrt{2p}$ which is rather large. For example, with $p = 100$, the mean is 0.01 and the standard deviation is 0.07.

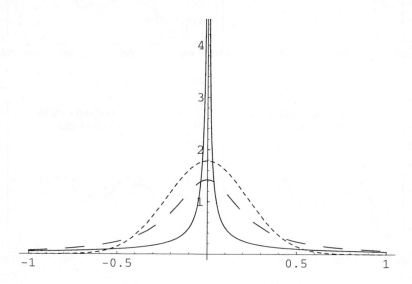

Fig. 3. The probability density function of the "Continuous-Gray" distribution for $p = 20$ (solid line). A Gaussian distribution with the same standard deviation (small dashes) and a Cauchy distribution with scale factor set to the same value (large dashes) are also shown. It can be seen that the new distribution strongly prefers small local moves, but with non-trivial probability of making larger jumps at all scales.

However, we are still some way from understanding the effects of changing the precision on the performance of the algorithm, apart from the simple one-dimensional unimodal case. Further experiment and analysis are required. From a theoretical point of view, it is interesting to refer this question back to the case of a local search algorthim using binary strings under Gray code: what is the effect of changing the precision (that is, the string length) in this case?

The search algorithm seems to be rather useful, especially when a relatively small number of function evaluations are allowed. This is often the case in real-world applications, such as the one presented above, when fitness can be very expensive to calculate. It seems that having a relatively low precision works well over a short number of iterations, although this is yet to be demonstrated theoretically. Of course it is possible that an adaptive scheme, with increasing precision over time, may be worth investigating.

One nice feature of the new algorithm is that it applies naturally to bounded optimisation problems, which are a very common class. The distance distribution, together with the wrapping around of the search interval (inherited from the discrete Gray code) means that new points are always generated within the required bounds. Algorithms that use Gaussian or Cauchy distibutions have to be artificially adjusted when invalid parameter values are generated (either by correcting the value to be the nearest bound, or by simply resampling until a legal value is obtained).

However, it is a well-known fact that if an algorithm is good for some problems, it must be bad for others. It is therefore worth considering situations in which the new algorithm would fail to perform well. If the problem has a number of narrow, well-separated optima, then any search algorithm maintaining a single point at each iteration is likely to be trapped in one of those optima — and it will be a matter of chance whether or not the right one is chosen. It is hard to see how this could be avoided without making use of a population. A population is also helpful if one wants to introduce some crossover. This can be a good idea if there exists some correlation between the variables of the problem. We have done some preliminary investigations into using the new distribution as a mutation operator in a steady-state GA, with crossover, and we have also looked at using it in a hybrid "memetic" style algorithm, with some success. One difficult landscape feature that is harder to address is the situation where there are "ridges" running at an angle to the axes specified by the parameters of the problem (e.g. if we rotated the axes for Rastrigin's function f_9). The most promising approach here would be to incorporate some sampling of the landscape so as to realign the search parameters with the ridges (e.g. by using a covariance mutation matrix). However, this kind of local modelling is itself quite expensive, and so we have a trade-off which may or may not be worthwhile.

Acknowledgements. Some of this work was done while Jon Rowe was visiting Prof. Darrell Whitley at Colorado State University, funded by National Science Foundation grant number IIS-0117209. The colonoscopy spectra were kindly given to us by Kevin Schomacker of MediSpectra, Lexington MA.

References

1. Whitley, L.D.: A free lunch proof for Gray versus binary encodings. In Banzhaf, W., Daida, J., Eiben, A.E., Garzon, M.H., Honavar, V., Jakiela, M., Smith, R.E., eds.: Proceedings of the Genetic and Evolutionary Computation Conference. Volume 1., Orlando, Florida, USA, Morgan Kaufmann (1999) 726–733
2. Whitley, L.D., Barbulescu, L., Watson, J.P.: Convergence results for high precision Gray codes. In Martin, W.N., Spears, W., eds.: Foundations of Genetic Algorithms VI, Morgan Kaufmann (2001) 295–311
3. Whitley, L.D., Bush, K., Rowe, J.E.: Subtheshold-seeking behaviour and robust local search. In: Proceedings of the Genetic and Evolutionary Computation Conference. (2004) To appear.
4. Yao, X., Liu, Y., Lin, G.: Evolutionary programming made faster. IEEE Transactions on Evolutionary Computation 3 (1999) 82–102
5. Bäck, T., Schwefel, H.P.: An overview of evolutionary algorithms for parameter optimization. Evolutionary Computation 1 (1993) 1–23
6. Hidović, D., Rowe, J.E.: Validating a model of colon colouration using an evolution strategy with adaptive approximations. In: Proceedings of the Genetic and Evolutionary Computation Conference. (2004) To appear.
7. Saucier, R.: Computer generation of statistical distributions. Technical Report ARL-TR-2168, Army Research Laboratory (2000) http://ftp.arl.mil/random/.
8. Ge, Z., Schomacker, K.T., Nishioka, N.S.: Identification of colonic dysplasia and neoplasia by diffuse reflectance spectroscopy and pattern recognition techniques. Applied Spectroscopy 52 (1998) 833–839
9. Egan, W.G., Hilgeman, T.W.: Optical Properties of Inhomogeneous Materials. Academic Press (1979)
10. Kubelka, P., Munk, F.: Ein beitrag zur optik der farbanstriche. Zeitschrift fur Technishen Physik 12 (1931) 593–601
11. Prahl, S., Keijzer, M., Jacques, S., Welch, A.: A Monte Carlo model of light propagation in tissue. In Mueller, G., Sliney, D., eds.: SPIE Proceedings of Dosimetry of Laser Radiation in Medicine and Biology. Volume IS 5. (1989) 102–111

A Novel Multi-objective Orthogonal Simulated Annealing Algorithm for Solving Multi-objective Optimization Problems with a Large Number of Parameters

Li-Sun Shu[1], Shinn-Jang Ho[1], Shinn-Ying Ho[2], Jian-Hung Chen[1], and Ming-Hao Hung[1]

[1] Department of Information Engineering and Computer Science
Feng China University, Taichung, Taiwan 407, ROC
{p860048@knight, syho@, p8800146@knight,
p8800043@knight}.fcu.edu.tw
[2] National Huwei Institute of Technology, Huwei, Yunlin, Taiwan 632, ROC
Department of Automation Engineering
sjho@nhit.edu.tw

Abstract. In this paper, a novel multi-objective orthogonal simulated annealing algorithm MOOSA using a generalized Pareto-based scale-independent fitness function and multi-objective intelligent generation mechanism (MOIGM) is proposed to efficiently solve multi-objective optimization problems with large parameters. Instead of generate-and-test methods, MOIGM makes use of a systematic reasoning ability of orthogonal experimental design to efficiently search for a set of Pareto solutions. It is shown empirically that MOOSA is comparable to some existing population-based algorithms in solving some multi-objective test functions with a large number of parameters.

1   Introduction

Many real-word applications usually involve simultaneous consideration of multiple performance criteria that are often incommensurable and conflict in nature. It is very rare for these applications to have a single solution, but rather a set of alternative solutions. These Pareto-optimal solutions are those for which no other solution can be found which improves along a particular objective without detriment to one or more other objectives. Multi-objective evolutionary algorithms (MOEAs) for solving mult-objective optimization problems gain significant attention from many researchers in recent years [1]-[8]. These optimizers not only emphasize the convergence speed to the Pareto-optimal solutions, but also the diversity of solutions. Niching techniques, such as fitness sharing and mating restriction, are employed for finding uniformly distributed Pareto-optimal solutions [2]-[3], and elitism is incorporated for improving the convergence speed to the Pareto front [4].

In recent years, many MOEAs employing local search strategies for further improving convergence speed have been successively proposed [4]-[7]. Population-

K. Deb et al. (Eds.): GECCO 2004, LNCS 3102, pp. 737–747, 2004.

based MOEAs have a powerful ability to extensively explore candidate solutions in a whole search space and painstakingly exploit candidate solutions in a local region, in parallel. In the neighborhood of each individual, it is beneficial for MOEAs to use local search strategies to exploit better solutions. However, local search strategies increase computation time in each generation. In order to avoid wasting time in un-necessary local searches, MOEAs must choose good individuals from the population for further exploiting non-dominated solutions [8]. However, it is difficult to deter-mine which individual is good for exploit.

Knowles and Corne [5] proposed a non-population based method, Pareto archived evolution strategy (PAES), to find a Pareto font. It employs a local search strategy for the generation of new candidate solutions, and utilizes elite set information to aid in the calculation of the solution quality. However, the local search strategy is based on generate-and-test methods that cannot efficiently solve large multi-objective optimi-zation problems (MOOPs) with a large and complex search space.

Recently, an efficient sigle-objective orthogonal simulated annealing algorithm OSA is proposed [9]. High performance of OSA mainly arises from an intelligent generation mechanism (IGM) which applies orthogonal experimental design to speed up the search. IGM can efficiently generate a good candidate solution for next move of OSA by using a systematic reasoning method. In this paper, a novel multi-objective orthogonal simulated annealing algorithm MOOSA using a generalized Pareto-based scale-independent fitness function and multi-objective IGM (MOIGM) is proposed to efficiently solve multi-objective optimization problems with large parameters. Instead of generate-and-test methods, MOIGM makes use of a systematic reasoning ability of orthogonal experimental design to efficiently search for a set of Pareto solutions. It is shown empirically that MOOSA is comparable to some existing population-based algorithms in solving some multi-objective test functions [1] with a large number of parameters.

2   Orthogonal Experimental Design [9]

MOOSA with a multi-objective intelligently generation mechanism (MOIGM) is based on orthogonal experimental design (OED). The basic concepts of OED are briefly introduced in Section 2.1. The orthogonal array and factor analysis of OED used in MOIGM are described in Section 2.2.

2.1   Concepts of OED

An efficient way to study the effects of several factors simultaneously is to use OED based on orthogonal array and factor analysis [10], [11]. Many design experiments use OED for determining which combinations of factor levels to use for each experiment and for analyzing the experimental results. The factors are the variables (parameters), which affect the chosen response variables (objective functions), and a setting (or a discriminative value) of a factor is regarded as a level of the factor. The term 'main effect' designates the effect on the response variable that one can trace to a design parameter [10].

Orthogonal array is a factional factorial matrix, which assures a balanced comparison of levels of any factor or interaction of factors. In the context of experimental matrices, orthogonal means statistically independent. The array is called orthogonal because all columns can be evaluated independently of one another, and the main effect of one factor dose not bother the estimation of the main effect of another factor [11]. Factor analysis using the orthogonal array's tabulation of experimental results can allow the main effects to be rapidly estimated, without the fear of distortion of results by the effects of other factors. Factor analysis can evaluate the effects of solution factors on the evaluation function, rank the most effective factors, and determine the best level for each factor such that the evaluation function is optimized.

Orthogonal experimental design can provide near-optimal quality characteristics for a specific objective. Furthermore, there is a large saving in the experimental effort. OED uses well-planned and controlled experiments in which certain factors are systematically set and modified, and then main effect of factors on the response can be observed. OED specifies the procedure of drawing a representative sample of experiments with the intention of reaching a sound decision [10]. Therefore, OED using orthogonal array and factor analysis is regarded as a systematic reasoning method.

2.2 Orthogonal Array and Factor Analysis

The three-level orthogonal array (OA) used in intelligent generation mechanism is described as follows. Let there be N factors with three levels for each factor. The number of total experiments is 3^N for the popular "one-factor-at-once" study. All the optimization parameters are generally partitioned into N groups.

Table 1. Orthogonal array $L_9(3^4)$

Experiment no. j	Factor i				Fitness value f_j
	1	2	3	4	
1	1	1	1	1	f_1
2	1	2	2	2	f_2
3	1	3	3	3	f_3
4	2	1	2	3	f_4
5	2	2	3	1	f_5
6	2	3	1	2	f_6
7	3	1	3	2	f_7
8	3	2	1	3	f_8
9	3	3	2	1	f_9

One group is regarded as a factor. To use an OA of N factors with three levels, we obtain an integer $M = 3^{\lceil \log_3(2N+1) \rceil}$, build a three-level OA $L_M(3^{(M-1)/2})$ with M rows and $(M-1)/2$ columns, use the first N columns, and ignore the other $(M-1)/2-N$ columns. Table 1 illustrates an example of OA $L_9(3^4)$. OA can reduce the number of experiments for factor analysis. The number of OA experiments required to analyze all solution factors is only M, where $2N+1 \le M \le 6N-3$. An algorithm of constructing OA can be found in [12]. After proper tabulation of experimental results, the summarized data are analyzed using factor analysis to determine the relative effects of levels of various factors as follows.

Let f_j denote a fitness value of the combination corresponding to the experiment j, where $j = 1, ..., M$. Define the main effect of factor i with level k as S_{ik} where $i = 1, ..., N$ and $k = 1, 2, 3$:

$$S_{ik} = \sum_{j=1}^{M} f_j \cdot AF_j, \qquad (1)$$

where $AF_j = 1$ if the level of factor i of experiment j is k; otherwise, $AF_j = 0$. Considering the case that the a fitness value is to be minimized, the level k is the best one when $S_{ik} = \min\{S_{i1}, S_{i2}, S_{i3}\}$. The main effect reveals the individual effect of a factor.

After the best one of three levels of each factor is determined, an intelligent combination consisting of all factors with the best levels can be easily derived. OED is effective for development design of efficient search for the intelligent combination of factor levels, which can yield a high-quality a fitness value compared with all values of 3^N combinations, and has a large probability that the reasoned value is superior to those of M representative combinations.

3   Multi-objective Orthogonal Simulated Annealing Algorithm MOOSA

MOOSA with MOIGM based on orthogonal experimental design (OED) can effectively solve intractable engineering problems comprising lots of parameters. A MOIGM uses a generalized Pareto-based scale-independent fitness function (GPSIFF) to efficiently evaluate the performance of solutions. GPSIFF evaluation procedure is described in Section 3.1. An MOIGM operation is briefly introduced in Section 3.2. A MOOSA using MOIGM is described in Section 3.3.

3.1   Use a Proposed GPSIFF

The fitness values for a set P of participant solutions to be evaluated are derived using a GPSIFF evaluation procedure at the same time in an objective space. GPSIFF makes direct use of general Pareto dominance relationship to obtain a single measurement of solutions. Simply, one solution has a higher score if it dominates more solutions. On the contrary, one solution has a lower score if more solutions dominate it.

Let a fitness value of a candidate solution be a tournament-like score obtained from all participant solutions in P. The fitness value of X can be given by the following score function:

$$\text{score}(X) = \left\{ p - q + c \mid p = |A|, \ q = |B|, c = |P| \ s.t. \ X \prec A, B \prec X, A \subseteq P \text{ and } B \subseteq P \right\}, \qquad (2)$$

where $\prec$ stands for *domination*, c is the size of P, p is the number of solutions of a set A which can be dominated by X, and q is the number of solutions of a set B which can dominate X in the objective space. It is noted that the GPSIFF scores for the non-dominated solutions as well as dominated solutions are not always identical.

GPSIFF uses a pure Pareto-ranking fitness assignment strategy, which differs from the traditional Pareto-ranking methods, such as non-dominated sorting [16] and Zitzler and Thiele's method [1]. GPSIFF can assign discriminative fitness values to not only non-dominated individuals but also dominated ones.

3.2 Multi-objective Intelligently Generation Mechanism MOIGM

Consider a parametric optimization function of m parameters. According to a current solution $X=[x_1, ..., x_m]^T$ where x_i is a parameter value, an MOIGM generates two temporary solutions $X_1=[x_1^1, ..., x_m^1]^T$ and $X_2=[x_1^2, ..., x_m^2]^T$ from perturbing X, where x_i^1 and x_i^2 are generated by perturbing x_i as follows:

$$x_i^1 = x_i + \overline{x}_i \text{ and } x_i^2 = x_i - \overline{x}_i, \ i=1, ..., m. \tag{3}$$

The values of $\overline{x}=[\overline{x}_1, ..., \overline{x}_m]^T$ are generated by Cauchy-Lorentz probability distribution [21].

Using the same division scheme for X, X_1, and X_2, partition all the m parameters into N non-overlapping groups with sizes l_i, $i=1, ..., N$, such that

$$\sum_{i=1}^{N} l_i = m. \tag{4}$$

The proper value of N is problem-dependent. The larger the value of N, the more efficient the MOIGM is if the interaction effects among groups are weak. If the existing interaction effect is not weak, the larger the value of l_i, the more accurate the estimated main effect is. Considering the trade-off, an efficient bi-objective division criterion is to minimize the interaction effects between groups and maximize the value of N. To efficiently use all columns of OA, N is generally specified as $N=(3^{\lfloor \log_3(2m+1) \rfloor} - 1)/2$ and the used OA is $L_{2N+1}(3^N)$ excluding the study of intractable interaction effects. The $N-1$ cut points are randomly specified from the $m-1$ candidate cut points which separate solution parameters.

MOIGM employs an elite set E to hold a limited number of non-dominated solutions and aims at efficiently combining good parameters from solutions X, X_1, and X_2 to generate a good candidate solution $\overline{Q}$ for the next move. Let H be the number of objectives for the problem. How to perform an MOIGM operation on X with m parameters for a GPSIFF fitness value F and objective function values $f^1, ..., f^H$ is described as follows:

Step 1: Generate two temporary solutions X_1 and X_2 using X from Equ. (3).

Step 2: Adaptively divide each of X, X_1, and X_2 into N groups of parameters where each group is treated as a factor.

Step 3: Use the first N columns of an OA $L_M(3^{(M-1)/2})$, where $M = 3^{\lceil \log_3(2N+1) \rceil}$.

Step 4: Let levels 1, 2 and 3 of factor i represent the ith groups of X, X_1, and X_2, respectively.

Step 5: Add M combination experiments of the OA into E. Compute F_j and f^h of the generated combinations corresponding to the experiment j, where $h=1, ...,H$ and $j = 2, ..., M$. Note that F_j and f^h are the fitness value of $F(X)$ and the hth objectives function value of $f(X)$, respectively.

Step 6: Compute the main effect S_{ik}^G using GPSIFF. Determine the best one of three levels of each factor based on the main effect S_{ik}^G, where $i = 1, ..., N$ and $k = 1, 2, 3$.

Step 7: The solution Q is formed using the combination of the best groups from the derived corresponding solutions.

Step 8: Compute the main effect S_{ik}^h using the one of objective fitness values. Determine the best one of three levels of each factor based on the main effect S_{ik}^h, where $h = 1, ..., H$, $i = 1, ..., N$ and $k = 1, 2, 3$. The solutions $Q^1, ..., Q^H$ are formed.

Step 9: Add Q and $Q^1, ..., Q^H$ solutions into E. Recompute the value of F for all non-dominated solution in E.

Step 10: $\overline{Q}$ is selected from the best one of M-1 combination experiments except X, Q and $Q^1, ..., Q^H$ according the GPSIFF fitness value, except that $\overline{Q}$ is not equal X.

For an MOIGM operation, the number of objective function evaluations is $M+H$ which includes M-1 evaluations for combinations of OA experiments, one for the evaluation of Q, and H evaluations for $Q^1, ..., Q^H$.

3.3 Procedure of MOOSA

MOOSA is based on a simulated annealing algorithm (SA) for solving multi-objective optimization problems. There are four choices must be made in implementing a SA algorithm for solving an optimization problem: 1) solution representation, 2) objective function definition, 3) design of the generation mechanism, and 4) design of a cooling schedule. The choices 1 and 2 are problem-dependent. Designing an efficient generation mechanism plays an important role in developing SA algorithms. Generally, there are four parameters to be specified in designing the cooling schedule: 1) an initial temperature T_0, 2) a temperature update rule and 3) a stopping criterion of the SA algorithm.

MOOSA employs an elite set E which maintains the non-dominated solutions and MOIGM to efficiently search for a good candidate solution for the next move. Let a variable value N_s be the number of trials with the same solution X, a constant $\overline{N_s}$ be the max number of trials with the same solution. Without lose of generality, consider the case that the fitness value $F(X)$ and H objective function values $f^1, ..., f^H$ are to be minimized. The proposed MOOSA is described as follows:

Step 1: (Initialization) Randomly generate an initial solution X and compute $F(X)$ and $f^1, \ldots, f^H$. Initialize the temperature $T = T_0$, $N_T = N_0$, and cooling rate CR. $Count = 0$, $N_s = 0$.

Step 2: (Update Elitism) Remove the dominated solutions in E.

Step 3: (Selection) If the solution X is not improved during $\overline{N_s}$ iterations (i.e. $N_s = \overline{N_s}$), randomly select a solution X from E and reset $N_s = 0$.

Step 4: (Generation) Perform an MOIGM operation using X to generate a candidate solution $\overline{Q}$. Set $\overline{X} = X$.

Step 5: (Acceptance criterion) Accept $\overline{Q}$ to be the new solution X with probability $P(\overline{Q})$:

$$P(\overline{Q}) = \begin{cases} 1 & ,if\,(F(\overline{Q}) > F(X)) \\ \min\left(\exp(\dfrac{f^1(X) - f^1(\overline{Q})}{T}), \cdots, \exp(\dfrac{f^H(X) - f^H(\overline{Q})}{T})\right) & ,if\,(F(\overline{Q}) \le F(X)) \end{cases} \quad (5)$$

If a new solution X is equal to an old solution $\overline{X}$, increase the value of N_s by one.

Step 6: (Decreasing temperature) Let the new values of T be $CR \times T$.

Step 7: (Termination test) If a pre-specified stopping condition is satisfied, stop the algorithm. Otherwise, go to Step 2.

Let G be the number of iterations. The complexity of MOOSA is $G \times (M+H)$ function evaluations.

4   Simulation Results

The coverage ratio of two non-dominated solution sets, A and B, obtained by two algorithms is used for performance comparison of the two algorithms, which is defined as follows [1]:

$$C(A, B) = \frac{\left|\{a \in A; b \in B; b \succeq a\}\right|}{|B|}, \quad (6)$$

where $b \succeq a$ means that b is weakly dominated by a. The value $C(A,B)=1$ means that all solutions in B are weakly dominated by A. On the contrary, $C(A,B)=0$ denotes that none of solutions in B is weakly dominated by A. Because the C measure considers the weakly dominance relationship between two sets A and B, $C(A, B)$ is not necessarily equal to $1 - C(B,A)$.

Recently, Deb [18] has identified several problem features that may cause difficulties for multi-objective algorithms in converging to the Pareto-optimal front and maintaining population diversity in the current Pareto front. These features are multi-modality, deception, isolated optima and collateral noise, which also cause difficulties in single-objective GAs. Following the guidelines, Zitzler et al. [1] constructed six test problems ZDT_1-ZDT_6 involving these features, and investigated the performance of various popular MOEAs. The empirical results demonstrated that SPEA outperforms

NSGA [7], VEGA [13], NPGA [2], HLGA [14] and FFGA [22] in small-scale problems. Each of the test functions is structured in the same manner and consists of three functions f_1, g, h [18]:

$$\text{Minimize } F(X) = (f_1(X), f_2(X)), \qquad (7)$$
$$\text{subject to } f_2(X) = g(x_2, \ldots, x_m) \cdot h(f_1(x_1), g(x_2, \ldots, x_m)),$$
$$\text{where } X = [x_1, x_2, \ldots, x_m]^T.$$

where f_1 is a function consisted of the first decision variable x_1 only, g is a function of the remaining m-1 variables, and the two variables of the function h are the function values of f_1 and g. These test problems are listed in Table 2. ZDT_5 is excluded because MOOSA uses real numbers for encoding.

Table 2. Test problems.

Test Problems	Objective functions	Domain x_i	Optimal solutions
ZDT_1	$f_1(X) = x_1$ $f_2(X) = g(X) \times h(f_1(X), g(X))$ $g(X) = 1 + 9 \cdot \sum_{i=2}^{m} x_i / (m-1)$ $h(f_1(X), g(X)) = 1 - \sqrt{f_1/g}$	$x_i \in [0, 1],$ $i = 1, \cdots, m.$	$x_1 \in [0, 1],$ $x_i = 0,$ $i = 2, \cdots, m.$
ZDT_2	$f_1(X) = x_1$ $f_2(X) = g(X) \times h(f_1(X), g(X))$ $g(X) = 1 + 9 \cdot \sum_{i=2}^{m} x_i / (m-1)$ $h(f_1(X), g(X)) = 1 - (f_1(X)/g(X))^2$	$x_i \in [0, 1],$ $i = 1, \cdots, m.$	$x_1 \in [0, 1],$ $x_i = 0,$ $i = 2, \cdots, m.$
ZDT_3	$f_1 = x_1$ $f_2(X) = g(X) \times h(f_1(X), g(X))$ $g(X) = 1 + 9 \cdot \sum_{i=2}^{m} x_i / (m-1)$ $h(f_1(X), g(X)) = 1 - \sqrt{f_1(X)/g(X)} - (\frac{f_1(X)}{g(X)}) \sin(10\pi x_1)$	$x_i \in [0, 1],$ $i = 1, \cdots, m.$	$x_1 \in [0, 1],$ $x_i = 0,$ $i = 2, \cdots, m.$
ZDT_4	$f_1 = x_1$ $f_2(X) = g(X) \times h(f_1(X), g(X))$ $g(X) = 1 + 10(m-1) + \sum_{i=2}^{m} (x_i^2 - 10\cos(4\pi x_i))$ $h(f_1(X), g(X)) = 1 - \sqrt{f_1(X)/g(X)}$	$x_1 \in [0, 1],$ $x_i \in [-5, 5],$ $i = 2, \cdots, m.$	$x_1 \in [0, 1],$ $x_i = 0,$ $i = 2, \cdots, m.$
ZDT_6	$f_1 = 1 - \exp(-4x_1)\sin^6(6\pi x_1)$ $f_2(X) = g(X) \times h(f_1(X), g(X))$ $g(X) = 1 + 9 \cdot ((\sum_{i=2}^{m} x_i)/(n-1))^{0.25}$ $h(f_1(X), g(X)) = 1 - (f_1(X)/g(x))^2$	$x_i \in [0, 1],$ $i = 1, \cdots, m.$	$x_1 \in [0, 1],$ $x_i = 0,$ $i = 2, \cdots, m.$

There are m parameters in each test problem. Each parameter in chromosomes is represented by 30 bits. The experiments in Zitzler's study indicate that the test problems ZDT_4 and ZDT_6 cause difficulties to evolve a well-distributed Pareto-optimal front. In their experiments, the reports are absent about the test problems with a large number of parameters. As a result, the extended test problems with a large number of parameters (m=63) are further tested in order to compare the performance of various algorithms in solving large MOOPs. Thirty independent runs were performed using the same fitness evaluations for various algorithms, N_{eval} = 25000. The parameter settings of VEGA, NPGA, NSGAII [24] and SPEA2 [23] are the same in [1], summarized as follows: the generations is 250, the crossover rate is 0.8, the mutation rate is 0.1, t_{dom}=10, the sharing factor σ_{share} is 0.4886, and the population size is 100. The population size and the external population size of SPEA2 are 80 and 20. Let the parameters of MOOSA be $\overline{N_s}$=10, CR=0.99, T_0=150.

The direct comparisons of each independent run between MOOSA and all compared MOEAs based on the C metric for 30 runs are depicted in Fig. 1. The average numbers of non-dominated solutions for various algorithms is shown in Table 3.

Table 3. The average number of non-dominated solutions for 30 runs of various algorithms.

	MOOSA	SPEA2	NSGAII	NPGA	VEGA
ZDT_1	174.53	68.23	61.73	16.33	13.90
ZDT_2	194.93	40.53	35.33	9.16	5.53
ZDT_3	100.30	78.87	65.20	17.10	12.60
ZDT_4	4.90	4.83	3.83	6.10	4.80
ZDT_6	21.17	9.90	9.57	6.90	5.27

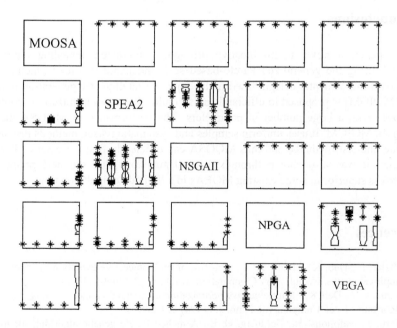

Fig. 1. Box plots based on the cover metric for multi-objective parametric problems. The leftmost box plot relates to ZDT_1, the rightmost to ZDT_6. Each rectangle refers to algorithm A associated with the corresponding row and algorithm B associated with the corresponding column and gives six box plots representing the distribution of the cover metric $C(A, B)$. The scale is 0 at the bottom and 1 at the top per rectangle.

For test problems ZDT_1, ZDT_2 and ZDT_3, MOOSA, SPEA2 and NSGAII evolved well-distributed Pareto fronts, and MOOSA is very close to the Pareto-optimal fronts. For the multimodal test problem ZDT_4, only MOOSA obtained a better Pareto front which is much closer to the Pareto-optimal front than those of the other algorithms. The well-distributed non-dominated solutions resulted from that OGM has well-distributed by-products which are candidate non-dominated solutions at that time. For ZDT_6, MOOSA also obtained a widely distributed front and MOOSA's solutions dominate all the solutions obtained by the other algorithms. Finally, it can be observed from [1] and our experiments that when the number of parameters increases, difficulties may arise in evolving a well-distributed non-dominated front. Moreover, it is observed that VEGA obtained some excel solutions in the objective f_1 in some runs of ZDT_2 and ZDT_6. This phenomenon agrees with [19], [20] that VEGA may converge to solution champion solutions only.

As shown in Table 3, the average number of non-dominated solutions obtained by MOOSA are more than the one obtained by others algorithms. As shown in Fig. 1, the quality of solutions obtained by MOOSA is superior to those of SPEA2, NSGAII, NPGA, and VEGA in terms of the number of non-dominated solutions, the distance between the obtained Pareto front and Pareto-optimal front, and the distribution of solutions.

5   Conclusions

In this paper, a novel multi-objective orthogonal simulated annealing algorithm MOOSA using the generalized Pareto-based scale-independent fitness function and orthogonal experimental design-based multi-objective intelligently generation mechanism (MOIGM) is proposed to efficiently solve multi-objective optimization problems (MOOPs) with a large number of parameters. The performance of MOOSA mainly rises from MOIGM. It uses uniform samples and systematic reason methods instead of generate-and-test methods, and thus MOOSA can efficiently find out a set of Pareto-solutions. It was also shown through the test functions that the performance of MOOSA is superior to some existing MOEAs in a limited computation time.

References

1.   Zitzler, E., Deb, K., Thiele, L.: Comparison of multiobjecctive evolutionary algorithms: empirical results. Evolutionary Computation, vol. 8, no. 2, (2000) 173-195
2.   Srinivas, N., Deb, K.: Multiobjective optimization using nondominated sorting in genetic algorithms. Evol. Comput., vol. 2, no. 3, (1994) 221-248
3.   Horn, J., Nafpliotis, N., Goldberg, D. E.: A niched Pareto genetic algorithm for multi-objective optimization. Proc. 1st IEEE Conf. Evol. Comput., Orlando, FL, June 27-19, (1994) 82-87
4.   Zitzler, E., Deb, K., Thiele, L.: Comparsion of multiobjective evolutionary algorithms: Empirical results. Evol. Comput., vol. 8, no. 2, (2000) 173-195

5. Knowles, J. D., Corne, D. W.: The Pareto archived evolution strategy: A new basedline algorithm for Pareto multiobjecitve optimization. Proc. 1999 Congress on Evol. Comput., Washington, DC, July 6-9, (1999) 98-105
6. Zitzler, E., Thiele, L.: Multiobjective evolutionary algorithms: A comparative case study and strength Pareto approach. IEEE trans. Evol. Comput., vol. 3, (1999) 257-271
7. Deb, K., Pratap, A., Agarwal, S., Meyarivan, T.: A fast and elitist multiobjective algorithms: NSGA-II. IEEE trans. Evol. Comput., vol. 6, (2002) 182-197
8. Ishibuchi, H., Yoshida, T., Murata, T.: Balance between genetic search and local search in memetic algorithms for multiobjective permutation flowshop scheduling. IEEE trans. Evol. Comput., vol. 7, no. 2, (2003) 204-223
9. Shu, L.-S., Ho, S.-J., and Ho, S.-Y.: OSA: Orthogonal Simulated Annealing Algorithm and Its Application to Designing Mixed $H2/H_\infty$ Optimal Controllers. IEEE Trans. Systems, Man, and Cybernetics—Part A to appear
10. Bagchi, T.-P.: Taguchi Methods T.-P. Bagchi, Taguchi Methods Explained: Practical Steps to Robust Design. Prentice-Hall, (1993)
11. Phadke, M.-S.: Quality Engineering Using Robust Design, Englewood Cliffs. NJ: Prentice-Hall
12. Leung, Y.-W., Wang, Y.: An orthogonal genetic algorithm with quantization for global numerical optimization. IEEE Trans. Evol. Comput., vol. 5, (2001) 41-53
13. J Schaffer, D.: Multi-objective optimization with vector evaluated genetic algorithms. Proc. 1st Int. Conference Genetic Algorithms, J. J. Grefenstette, Ed. Hillsdale, NJ:Lawrence Erlbaum, (1985) 93-100
14. Hajela, P., Lin, C.-Y.: Genetic search strategies in multicriterion optimal design. Structural Optimization, no. 4, (1992) 99-107
15. Ishibuchi, H., Murata, T.: A multi-objective genetic local search algorithm and its application to flowshop scheduling. IEEE Trans. SMC-Part C: Applications and Reviews, vol. 28, no.3, (1998) 392-403
16. Osyczka, A., Kundu, S.: A modified distance method for multicriteria optimization, using genetic algorithms. Computers and Industrial Engineering, vol. 30, no. 4, (1996) 871-882
17. Goldberg, D. E.: Genetic Algorithms in Search, Optimization and Machine Learning. Addison – Wesley Publishing Company, (1989)
18. Deb, K.: Multi-objective genetic algorithms: problem difficulties and construction of test problems. Evol. Comput., vol. 7, no. 3, (1999) 205-230
19. Coello, C. A. C.: A comprehensive survey of evolutionary-based multiobjective optimization techniqures. International Journal of Knowledge and Information System, vol. 1, no. 3, (1999) 269-308
20. Deb, K.: Multi-Objective Optimization Using Evolutionary Algorithms. John Wiley & Sons, (2001)
21. Szu, H., Hartley, R.: Fast simulated annealing. Physics Letters, vol. 122, (1987) 157-162
22. Fonseca, C. M., Fleming, P. J.: Genetic algorithms for multiobjective optimization: formulation, discussion and generalization. Proc. fifth Int. Conference Genetic Algorithms, S. Forrest, Ed. San Mateo, CA: Morgan-Kaufmann, (1993) 416-423
23. Zitzler, E., Laumanns, M., and Thiele, L.: SPEA2: Improving the strength Pareto evolutionary algorithm. Technical Report 103, Computer Engineering and Communication Networks Lab (TIK), Swiss Federal Institute of Technology (ETH) Zurich, Gloriastrasse 35, CH-8092 Zurich (2001)
24. Deb, K., Pratap, A., Agarwal, S., and Meyarivan, T.: A fast and elitist multiobjective genetic algorithm: NSGA-II. IEEE Trans. Evol. Comput., vol. 6, no. 2, (2002) 182-197

On the Choice of the Population Size

Tobias Storch

Dept. Computer Science II, Univ. Dortmund
44221 Dortmund, Germany
tobias.storch@uni-dortmund.de

Abstract. Evolutionary Algorithms (EAs) are population-based randomized optimizers often solving problems quite successfully. Here, the focus is on the possible effects of changing the parent population size. Therefore, new functions are presented where for a simple mutation-based EA even a decrease of the population size by one leads from an efficient optimization to an enormous running time with an overwhelming probability. This is proven rigorously for all feasible population sizes. In order to obtain these results, new methods for the analysis of the EA are developed.

1   Introduction

Evolutionary Algorithms (EAs) are a broad class of general randomized search heuristics. The probably best-known types of EAs are Genetic Algorithms and Evolution Strategies. Their area of application is as huge as their variety and they have been applied successfully in numerous situations. Here, we consider the problem to maximize pseudo-Boolean functions $f_n : \{0,1\}^n \to \mathbb{R}_0^+$. We remark that analysis in discrete search spaces differs substantially from that in continuous ones.

With regard to populations, the problems how to choose its size and how to find a method to preserve the diversity are well known. If the size of the population or its diversity are too small, the EA is likely to stagnate in local optima. On the other hand, the EA is likely to waste much time on the evaluation of unnecessary elements, if the population or diversity are too large. Many ideas have been presented to cope with the difficulty of the correct choice of these parameters and they all have shown their usefulness in experiments, e.g., niching methods, multistarts, and many more. In order to understand the success of EAs, theory often investigates the behavior of simple EAs on typical or constructed problems. These artificial problems are often developed to illustrate particular effects of EAs or one of their components at best. Our aim is to illustrate conveniently that the choice of the parent population size may be critical. Therefore, we develop functions where even a decrease of the parent population size by one leads from an efficient optimization with an overwhelming probability to an enormous running time.

We estimate the efficiency of a randomized algorithm. Therefore, let T_{A,f_n} be the random number of function evaluations until algorithm A first evaluates an

K. Deb et al. (Eds.): GECCO 2004, LNCS 3102, pp. 748–760, 2004.

optimal search point of f_n. If the expected value of T_{A,f_n} is polynomially bounded in the dimension of the search space n, we call A *efficient* on f_n and *inefficient*, if the expected value of T_{A,f_n} is at least exponentially bounded. Finally, we call A *totally inefficient* on f_n, if after exponential many steps the probability that an optimal search point has been evaluated, remains exponentially small. In this particular situation a polynomially bounded number of (parallel) (independent) multistarts of A is inefficient. Moreover, we are interested in asymptotical results with respect to n.

We investigate one of the best-known EAs. This is the so-called $(\mu+\lambda)$ EA working with a parent population of size $\mu \geq 1$ and an offspring population of size $\lambda \geq 1$. Surprisingly, on many typical functions even the $(1+1)$ EA is quite efficient. Indeed, Jansen and De Jong (2002) considered the role of the offspring population size. They presented functions where a decrease of this parameter leads to enormous differences in the optimization time. Jansen and Wegener (2001b) have shown something less strong for the role of the parent population size. Witt (2003) improved this result.

We develop functions $f_{n,d}$ where the considered mutation-based $(\mu+1)$ EA is totally inefficient, if the parent population has size $\mu \leq d$. However, if the population has size $\mu > d$, the EA is efficient. We introduce such functions for all $d \in \{1, \ldots, n^c\}$ and every constant $c > 0$. And we call d the *threshold value of the population size*. In order to prove these results rigorously, we present simple but powerful methods to analyze this EA. They extend the so-called *method of f-based partitions* and help to upper bound the expected optimization time of the $(\mu+1)$ EA on a particular function (see Wegener (2002)).

The paper begins in Section 2 with an introduction of the investigated steady-state $(\mu+1)$ EA. Section 3 presents the desired extensions of the method of f-based partitions and Section 4 exhibit our first results. These results handle only the threshold value of the population size one and do not satisfy all the desired properties but they illustrate some of the main effects which occur. We divide the possible threshold values of the population size d into three domains. For convenience, we consider them in an unnatural order later.

- The first domain encloses $d \in \{1, \ldots, \lceil n/(c_1 \log n) \rceil - 1\}$ for some constant $c_1 > 0$. These are investigated in Section 7.
- The second domain encloses $d \in \{\lceil n/(c_1 \log n) \rceil, \ldots, \lceil n/c_2 \rceil\}$ for some constant $c_2 > 0$. These are investigated in Section 5.
- And the third domain encloses $d \in \{\lceil n/c_2 \rceil + 1, \ldots, n^c\}$ for every constant $c > 0$. These are investigated in Section 6.

We finish with some conclusions.

2   The Steady-State $(\mu+1)$ EA

The considered mutation-based steady-state $(\mu+1)$ EA works with a natural and weak method to preserve diversity. It just avoids duplicates of elements in the population. This technique can be understood as a special niching method. Moreover, in this case the population structure is not only a multiset but a set.

(μ+1) EA

1. Choose μ different individuals $x_i \in \{0,1\}^n$, $1 \le i \le \mu$, uniformly at random. These individuals constitute the population $\mathcal{P}$, i.e., $\mathcal{P} = \{x_1, \ldots, x_\mu\}$.

2. Choose an individual x from the population $\mathcal{P}$ uniformly at random. Create y by flipping each bit in x independently with probability $1/n$.

3. If $y \notin \mathcal{P}$, i.e., $y \ne x_i$ for all i, $1 \le i \le \mu$,
   then let $z \in \mathcal{P} \cup \{y\}$ be randomly chosen among those individuals with the worst f-value and let the population be $\mathcal{P} \cup \{y\} - \{z\}$, goto 2.,
   else let the population be $\mathcal{P}$, goto 2.

Obviously, only populations of size $\mu \le 2^n$ are possible. In Step 2, the parameter $1/n$ is the standard choice for mutations.

We remark that the theorems of Sections 5, 6 and 7 hold, if fitness-proportional selection (x_i, $1 \le i \le \mu$, is chosen with probability $f(x_i)/\sum_{k=1}^{\mu} f(x_k)$) instead of uniform selection (x_i, $1 \le i \le \mu$, is chosen with probability $1/\mu$) is used in Step 2. Furthermore, it is irrelevant which of the elements with smallest f-value is deleted in Step 3.

3   Methods for Upper Bounds on the Expected Optimization Time for the (μ+1) EA

We present two extensions for the (μ+1) EA of the method of f-based partitions (see Wegener (2002)). These extensions can easily be combined. At first, we recall the original method of f-based partitions that is a simple proof technique which helps to upper bound the expected running time of the (1+1) EA to optimize a particular function.

Given $A, B \subseteq \{0,1\}^n$, $A, B \ne \emptyset$, the *relation* $A <_f B$ holds, iff $f(a) < f(b)$ for all $a \in A$, $b \in B$ and a pseudo-Boolean function f. Moreover, we call $(A_1, \ldots, A_m; f)$ an *f-based partition*, iff $A_1, \ldots, A_m$ is a partition of $\{0,1\}^n$ and $A_1 <_f \cdots <_f A_m$ holds. Furthermore, A_m merely consists of optimal elements a, i.e., $f(a) = \max\{f(b) \mid b \in \{0,1\}^n\}$. Finally, let $p(a)$, $a \in A_i$, $i < m$, be the probability that a mutation of a creates some $b \in A_{i+1} \cup \cdots \cup A_m$ and let $p(A_i) := \min\{p(a) \mid a \in A_i\}$, $i < m$, i.e., $p(A_i)$ constitutes a lower bound on the probability to leave A_i. Given an f-based partition $(A_1, \ldots, A_m; f)$, the expected optimization time of the (1+1) EA is bounded above by

$$1 + p(A_1)^{-1} + \cdots + p(A_{m-1})^{-1} \quad .$$

Our first extension of this method for the (μ+1) EA allows to disregard areas of the search space at the expense of the population size. It is even not necessary to know the areas, only their sizes. Therefore, let $(A_1, \ldots, A_m; f; A)$ be the variant (1) of f-based partitions where $A_1, \ldots, A_m$ is only a partition of $A \subseteq \{0,1\}^n$ but all other constraints still hold.

Theorem 1. *Let $A_0 \subset \{0,1\}^n$ and $(A_1, \ldots, A_m; f; \{0,1\}^n - A_0)$ be given. The expected optimization time of the (μ+1) EA, where $\mu \ge |A_0| + 1$, is bounded above by*

$$\mu\left(1 + p(A_1)^{-1} + \cdots + p(A_{m-1})^{-1}\right) \quad .$$

Proof. The initialization evaluates μ different elements. There always exist at least $\mu - |A_0| \geq 1$ individuals in the population which do not belong to A_0. These are at worst all elements of A_1 after the initialization. Once the $(\mu+1)$ EA has left A_i, $i < m$, this area will never be reached again, i.e., at least one element of the population belongs to $A_{i+1} \cup \cdots \cup A_m$. Since the probability to select an individual of the population that belongs to A_i, $i < m$, is lower bounded by $1/\mu$, the expected number of fitness evaluations is bounded by $\mu p(A_i)^{-1}$ until an element of $A_{i+1} \cup \cdots \cup A_m$ is created. At worst, this is an element of A_{i+1}. □

If $A_0 = \emptyset$ and $\mu = 1$, we have the original method for the $(1+1)$ EA.

For the original method it is essential that $A_i <_f A_{i+1}$, $i < m$, holds. We weaken this condition at the expense of the population size. Therefore, let $(A_1, \ldots, A_m; f; b_1, \ldots, b_{m-1})$, $b_i \geq i+1$, be the variant (2) of f-based partitions where $A_1, \ldots, A_m$ is a partition of $\{0,1\}^n$ but it just holds $A_i <_f A_{b_i}$, $i < m$. Furthermore, A_m still merely consists of optimal elements.

Theorem 2. $(A_1, \ldots, A_m; f; b_1, \ldots, b_{m-1})$ *is given and let*

$$v_i := \sum_{1 \leq j < i,\, b_j > i} |A_j| \quad \text{for all } i < m \quad .$$

The expected optimization time of the $(\mu+1)$ EA, where $\mu \geq \max\{v_j \mid j < m\}+1$, is bounded above by

$$\mu\left(1 + p(A_1)^{-1} + \cdots + p(A_{m-1})^{-1}\right) \quad .$$

Proof. After initialization, the population contains at worst only elements that belong to A_1. At most the elements of A_j where $j < i$ and $b_j > i$ have an f-value at least as large as the worst element of A_i, $i < m$, but belong to $A_1 \cup \cdots \cup A_{i-1}$. By definition, these are at most v_i elements. Therefore and since $\mu \geq \max\{v_j \mid j < m\} + 1 \geq v_i + 1$, once the $(\mu+1)$ EA has reached A_i, $i < m$, this area will never be given up again, i.e., at least one element of the population belongs to $A_i \cup \cdots \cup A_m$. The expected number of fitness evaluations is bounded by $\mu p(A_i)^{-1}$ until an element $a \in A_j$, $i+1 \leq j \leq m$, is created. Since a is either optimal or it holds $\mu \geq v_j + 1$, the element a will be inserted in the successive population. □

If $b_i = i + 1$, $i < m$, and $\mu = 1$, we have the original method for the $(1+1)$ EA again.

4   First Results

Here, we only consider the threshold value of the population size one. The presented functions in this section do not satisfy all the desired properties. But we make observations that help to identify functions in the following sections where all these properties hold. To obtain these results for all threshold values of the population size, the functions have to be modified in different manners.

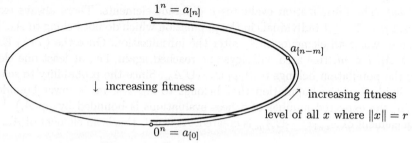

Fig. 1. An illustration of PONEP$_n$.

Our example functions for the threshold value of the population size one consist of one *global optimum* a_{global} and one *local optimum* a_{local} which is the second-best search point. We call a_{local} a *peak*, too. The *Hamming distance* $\mathbf{H}(x,y)$ of two search points x and y equals the number of indices i where $x_i \neq y_i$. If $\mathbf{H}(x,y) = 1$, we call x and y *Hamming neighbors*. A *path (of length r)* is a sequence of search points $a_{[0]}, \ldots, a_{[r-1]}$ where $a_{[i]}$ and $a_{[i+1]}$ are Hamming neighbors and the elements are pairwise distinct. The two search points a_{global} and a_{local} are lying on a path $a_{[0]}, \ldots, a_{[n]}$ (of length $n+1$) that leads from the zero string to the one string. It is $a_{\text{global}} := a_{[n]}$ and $a_{\text{local}} := a_{[n-\lceil n/3 \rceil]}$ and they have (inevitably) Hamming distance $\lceil n/3 \rceil$. The functions have the additional property that with an overwhelming probability the path is first entered in front of a_{local}. Therefore, during a typical run of the investigated EA a_{local} is created before a_{global} is reached. If the population consists only of the second-best search point a_{local}, the probability is extremely small to produce a_{global}, since their Hamming distance is large. But if the population consists of at least one more individual, these elements search forward on the path and find the global optimum efficiently. We remark that the functions are influenced by the short path functions of Jansen and Wegener (2001a).

To define the functions, let 0^k denote the string of k zeros and $|x|$ the number of ones in x. To simplify the notation, let $m := \lceil n/3 \rceil$. Now, we can give a complete definition of PONEP$_n$ (Path with ONE Peak) illustrated in Fig. 1.

$$\text{PONEP}_n(x) := \begin{cases} n+i & \text{if } x = 0^{n-i}1^i =: a_{[i]}, \, 0 \le i \le n \text{ and } i \neq n-m \\ 2n-1/2 & \text{if } x = 0^m 1^{n-m} =: a_{[n-m]} \\ n-|x| & \text{otherwise} \end{cases}$$

We begin our considerations with populations of size $\mu \ge 2$. The result is proven in two different ways to demonstrate different interpretations of the local optimum. The first proof uses Theorem 1 while the second one uses Theorem 2.

Theorem 3. *The expected time until the $(\mu+1)$ EA, where $\mu \ge 2$, has optimized* PONEP$_n$ *is bounded above by* $\mathcal{O}(\mu n^2)$.

Proof. (using Theorem 1) Let $A_0 := \{a_{[n-m]}\}$ and $(A_1, \ldots, A_{2n-1}; \text{PONEP}_n; \{0,1\}^n - A_0)$ be the variant (1) of f-based partitions where

$$A_i := \begin{cases} \{a \mid \text{PONEP}_n(a) = i\} & \text{if } 1 \le i < n \\ \{a_{[i-n]}\} & \text{if } n \le i < 2n - m \\ \{a_{[i-n+1]}\} & \text{if } 2n - m \le i < 2n \end{cases}.$$

Hence, for the element of A_{2n-m-1} a special 2-bit mutation creates the element of A_{2n-m} and for the elements of A_i, $i < 2n - 1$ and $i \ne 2n - m - 1$, at least one special 1-bit mutation creates an element of A_{i+1}. Therefore, it holds

$$p(A_i) \ge \begin{cases} (1/n^2)(1 - 1/n)^{n-2} \ge 1/(en^2) & \text{if } i = 2n - m - 1 \\ (1/n)(1 - 1/n)^{n-1} \ge 1/(en) & \text{otherwise} \end{cases}.$$

It holds $\mu \ge |A_0| + 1 = 2$. Hence, an application of Theorem 1 leads to an expected optimization time of at most $\mu\big(1 + en^2 + (2n - 3)en\big) = \mathcal{O}(\mu n^2)$. □

Proof. (using Theorem 2) Let $(A_1, \dots, A_{2n}; \text{PONEP}_n; b_1, \dots, b_{2n-1})$ be the variant (2) of f-based partitions where

$$A_i := \begin{cases} \{a \mid \text{PONEP}_n(a) = i\} & \text{if } 1 \le i < n \\ \{a_{[i-n]}\} & \text{if } n \le i \le 2n \end{cases}, \quad b_i := \begin{cases} 2n & \text{if } i = 2n - m \\ i + 1 & \text{otherwise} \end{cases}.$$

Hence, it holds

$$v_i = \begin{cases} |\emptyset| = 0 & \text{if } 1 \le i \le 2n - m \\ |A_{2n-m}| = 1 & \text{otherwise} \end{cases}$$

and $p(A_i) \ge 1/(en)$, $1 \le i < 2n$, since for every element of A_i, $i < 2n$, at least one special 1-bit mutation creates an element of A_{i+1}. It holds $\mu \ge \max\{v_j \mid j < 2n\} + 1 = 2$. Hence, by Theorem 2 the expected optimization time is bounded above by $\mu\big(1 + (2n - 1)en\big) = \mathcal{O}(\mu n^2)$. □

Theorem 4. *With a probability of $1 - \mathcal{O}(1/n)$ the (1+1) EA needs an exponential time $2^{\Omega(n)}$ to optimize PONEP_n. The expected time until the (1+1) EA has optimized PONEP_n is bounded below by $2^{\Omega(n)}$.*

Proof. By Chernoff bounds (see, e.g., Motwani and Raghavan (1995)), the probability is exponentially small that the initial element consists of more than $n - m - \lceil n/7 \rceil$ ones. From then on, at most an element of $\{a \mid |a| \le n - m - \lceil n/7 \rceil\} \cup \{a_{[0]}, \dots, a_{[n]}\}$ is accepted as population. Each $a_{[i]}$, $i \ge n - m$, consists of at least $n - m$ ones. Hence, a mutation of an element of $\{a \mid |a| \le n - m - \lceil n/7 \rceil\}$ has to change at least $\lceil n/7 \rceil$ bits to create $a_{[i]}$, $i \ge n - m$. The probability for such a mutation is bounded by $(1/n)^{\lceil n/7 \rceil} = 2^{-\Omega(n)}$. Therefore, when first an element of $\{a_{[n-m]}, \dots, a_{[n]}\}$ is produced this happens with an exponentially small failure probability by a mutation of $a_{[n-m-k]}$, $k \ge 1$. Exactly one k-bit mutation of $a_{[n-m-k]}$ generates $a_{[n-m]}$ and one $(k + l)$-bit mutation produces $a_{[n-m+l]}$, $1 \le l \le m$. The probability of the k-bit mutation equals $(1/n^k)(1 - 1/n)^{n-k} =: q_1$ and of all the $(k+l)$-bit mutations together $\sum_{l=1}^{m}(1/n^{k+l})(1 - 1/n)^{n-k-l} =: q_2$. Since $q_2/q_1 = \mathcal{O}(1/n)$, the probability to create $a_{[n-m]}$ before $a_{[i]}$, $i > n - m$, is altogether bounded by $1 - \mathcal{O}(1/n)$. But if $a_{[n-m]}$ is the individual of the population, just a special m-bit mutation generates the global optimum $a_{[n]}$. This is the only element that has an at least as large f-value as the local optimum. The probability for such a mutation is $(1/n)^m (1 - 1/n)^{n-m} = 2^{-\Omega(n)}$. □

5   Medium Threshold Values of the Population Size

Now, we consider larger threshold values of the population size. Therefore, we enlarge the peak. Of course, as a result, it is not a real peak at all but rather a plateau. We play with the definition of PONEP_n and identify threshold values of the population size that satisfy the desired properties.

Let $\text{PONEP}_{n,d}$, $1 \leq d \leq m := \lceil n/3 \rceil$, be the variant of PONEP_n where the d search points $a_{[n-m]}, \ldots, a_{[n-m+d-1]}$ have f-value $2n - 1/2$. Thereby, these elements form the new peak.

Similar to Theorem 3, the expected optimization for the $(\mu+1)$ EA on $\text{PONEP}_{n,d}$, if $\mu \geq d + 1$, is bounded by $\mathcal{O}(\mu n^2)$. In the proof using Theorem 2 it holds $\max\{v_j \mid j < 2n\} + 1 = d + 1$, since an appropriate choice of b_i is now

$$b_i := \begin{cases} 2n & \text{if } 2n - m \leq i \leq 2n - m + d - 1 \\ i+1 & \text{otherwise} \end{cases}.$$

We investigate the situation $\mu \leq d$ and assume $d \leq \lceil n/c_2 \rceil$ for an appropriate constant $c_2 > 0$. Due to the proof of Theorem 4 the probability is exponentially small that an element of the initial population consists of more than $n - m - \lceil n/7 \rceil$ ones. Furthermore, $q_2/q_1 = \mathcal{O}(1/n^d)$, since we have to compare $(k+l)$-bit mutations, $0 \leq l < d$, with $(k + d + l)$-bit mutations, $0 \leq l \leq m - d$. After this, one element of $\{a_{[n-m]}, \ldots, a_{[n-m+d-1]}\}$ and some elements of $\{a \mid |a| \leq n - m - \lceil n/7 \rceil\} \cup \{a_{[0]}, \ldots, a_{[n-m-1]}\}$ constitute the population. We claim that it typically takes longer to reach the end of the path than to fill up the peak. More precisely, we lower bound the failure probability that within $\lceil 2e\mu n^2/c_2 \rceil$ steps the population consists of some elements of $\{a_{[n-m]}, \ldots, a_{[n-m+d-1]}\}$ only. Since we suppress the chance to create $a_{[n]}$ during these steps, we also have to lower bound the probability for this event. Both probabilities are exponentially small. So, with a failure probability of $\max\{\mathcal{O}(1/n^d), 2^{-\Omega(n)}\}$ the $(\mu+1)$ EA, where $\mu \leq d$, needs an exponential time to optimize $\text{PONEP}_{n,d}$.

Since it holds $\mu \leq d$, there exists an element $a_{[n-m+l]} \notin P$, $0 \leq l < d$, at least until the population consists of elements of $\{a_{[n-m]}, \ldots, a_{[n-m+d-1]}\}$ only. A special 1-bit mutation of a correct individual of the population creates an element $a_{[n-m+l]} \notin P$, $0 \leq l < d$. This element is also inserted into the population. Therefore, we are in a similar situation as in an experiment where the success probability is bounded by $1/(e\mu n)$ in each of $\lceil 2e\mu n^2/c_2 \rceil$ trials. By Chernoff bounds the probability is exponentially small that less than μ, $\mu \leq d \leq \lceil n/c_2 \rceil$, successes occur.

The probability is bounded above by $1/n^j$ to produce $a_{[l+j]}$, $j \geq 0$ and $0 \leq l + j \leq n$, by a mutation of a where $|a| \leq l$. At least j special bits have to change for this event. Hence, let $a_{[n-m+k]} \in P$, $k \geq 0$, and $a_{[n-m+k+j]} \notin P$, for all $j \geq 1$. Since the population consists of elements of $\{a \mid |a| \leq n - m - \lceil n/7 \rceil\} \cup \{a_{[0]}, \ldots, a_{[n-m+k]}\}$ and $n - m - \lceil n/7 \rceil \leq n - m + k - (\mu - 1)$, there is at most one individual in the population that consists of at most $n - m + k - l$, $0 \leq l \leq \mu - 1$, ones. So, the probability to create $a_{[n-m+k+j]}$ is bounded by $\sum_{l=0}^{\mu-1} 1/(\mu n^{j+l}) \leq 2/(\mu n^j)$. At the beginning $k = d - 1$ holds at best. By Chernoff bounds the probability is exponentially small to create $a_{[n]}$ within $\lceil 2e\mu n^2/c_2 \rceil$ steps, if c_2 is large enough.

Theorem 5. *With a probability of* $1 - 2^{-\Omega(n)}$ *the* $(\mu+1)$ *EA, where* $\mu \leq d$, *needs* $2^{\Omega(n)}$ *steps to optimize* $\text{PONEP}_{n,d}$, $\lceil n/(c_1 \log n) \rceil \leq d \leq \lceil n/c_2 \rceil$ *for every constant* $c_1 > 0$ *and an appropriate constant* $c_2 > 0$. *The expected optimization time of the* $(\mu+1)$ *EA, where* $\mu \geq d + 1$, *is bounded above by* $\mathcal{O}(\mu n^2)$.

6   Large Threshold Values of the Population Size

We consider even larger threshold values of the population size. Therefore, we enlarge the peak more and play more extensively with the definition of PONEP_n. When we investigated the $(\mu+1)$ EA, $\mu \leq d$, on $\text{PONEP}_{n,d}$, we claimed that with an overwhelming probability the peak is filled up before the end of the path is reached. In order to retain this property now, we slow down the arrival at the end of the path, since the peak is larger and thereby, it takes longer to fill it up.

To simplify the notation let $m := \lceil n/3 \rceil$ and $s := \lceil n/c_2 \rceil$ where c_2 is the positive constant of the previous section. Let $\text{PONEP}_{n,d,c}$, $s < d \leq n^c$, for a constant integer $c \geq 1$, be the variant of PONEP_n where $a_{[n-m+s+l]}$, $0 \leq l \leq m - s - 1$ and $l \mod (c+1) \not\equiv -1$, have f-value $n - |a_{[n-m+s+l]}|$. Thus, the path behind the peak consists of $\lfloor (m-s)/(c+1) \rfloor$ gaps of size c and possibly one further gap of smaller size. These gaps slow down the arrival at the global optimum. But the path is not a real one at all. Furthermore, beside $a_{[n-m]}, \ldots, a_{[n-m+s-1]}$ the elements $a_{(l)}$, $0 < l \leq d - s$, have f-value $2n - 1/2$. Thereby, all these elements form the new peak. But before we describe the appearance of $a_{(l)}$, $1 \leq l \leq d - s$, we remember the Gray Code. The $(\ell$-digit) *Gray Code* $\mathbf{G}_\ell$, $\ell \in \mathbb{R}$, maps the integer x, $0 \leq x \leq 2^\ell - 1$, (bijective) to the binary space $\{0,1\}^\ell$. But in contrast to *Binary Code* the values x and $x + 1$ always have Hamming distance one, $\mathbf{H}(\mathbf{G}_\ell(x), \mathbf{G}_\ell(x+1)) = 1$ for all $0 \leq x < 2^\ell - 1$. Similar to Binary Code $\mathbf{G}_\ell(0) = 0^\ell$ holds. We define $a_{(l)}$. The element $a_{(l)}$, $1 \leq l \leq d-s$, equals $g_{s-1} \cdots g_0 0^{m-s} 1^{n-m}$ if $\mathbf{G}_s^{-1}(g_{s-1} \cdots g_0) = l$. It holds $l \leq n^c \leq 2^s - 1$, if n is large enough. The mentioned properties of the Gray Code claim that $\mathbf{H}(a_{(l)}, a_{(l+1)}) = 1$, $1 \leq l < d - s$, and $\mathbf{H}(a_{[n-m]}, a_{(1)}) = 1$.

We investigate the situation $\mu \geq d+1$. An application of Theorem 2 leads to an expected optimization time of $\mathcal{O}(\mu n^{c+2})$. To show this, we define a sequence of the elements of the path and the peak

$$S := (a_{[0]}, \ldots, a_{[n-m-1]}, a_{(d-s)}, \ldots, a_{(1)}, a_{[n-m]}, a_{[n-m+1]}, \ldots, a_{[n-m+s-1]},$$
$$a_{[n-m+s-1+(c+1)]}, \ldots, a_{[n-m+s-1+\lfloor (m-s)/(c+1) \rfloor (c+1)]}, a_{[n]})$$
$$=: (s_0, \ldots, s_{n-m+d+\lfloor (m-s)/(c+1) \rfloor}) \quad .$$

We choose the partition induced by the areas

$$A_i := \begin{cases} \{a \mid \text{PONEP}_{n,d,c}(a) = i\} & \text{if } 1 \leq i < n \\ \{s_{i-n}\} & \text{if } n \leq i \leq 2n - m + d + \lfloor (m-s)/(c+1) \rfloor \end{cases}$$

and analogously to the previous section for an appropriate choice of the b_i it holds $\max\{v_j \mid j < 2n - m + d + \lfloor (m-s)/(c+1) \rfloor\} + 1 = d + 1$. Hence, for $i = 2n - m - 1$ we consider the Hamming distance of $s_{i-n} = a_{[n-m-1]}$ and $s_{i+d-s-n} = a_{[n-m]}$ and otherwise of the element of A_i and an appropriate element of A_{i+1}. So,

$$p(A_i) \geq \begin{cases} 1/(en) & \text{if } 1 \leq i < 2n - m + d - 1 \\ 1/(en^{c+1}) & \text{otherwise} \end{cases} \quad .$$

Now, we investigate the situation $\mu \leq d$. Similar to the arguments that led to Theorem 5, the probability is exponentially small that an element of $\{a_{[n-m+s-1+k(c+1)]} \mid 1 \leq k \leq \lfloor (m-s)/(c+1) \rfloor\} \cup \{a_{[n]}\}$ is created before an element of the peak $\{a_{[n-m]}, \ldots, a_{[n-m+s-1]}, a_{(1)}, \ldots, a_{(d-s)}\}$. Furthermore, by Chernoff bounds the failure probability is exponentially small that after $\lceil 2e\mu n^{c+1} \rceil$ steps the population consists of elements with f-value $2n - 1/2$ only. Hence, let $a_{[n-m+s-1+k(c+1)]} \in \mathcal{P}$, $k \geq 1$, and $a_{[n-m+s-1+(k+j)(c+1)]} \notin \mathcal{P}$, for all $j \geq 1$. Since the Hamming distance of $a_{(l)}$, $1 \leq l \leq d-s$, and $a_{[n-m+s-1+(k+j)(c+1)]}$ is at least s, the probability is bounded by $2/(\mu n^{j(c+1)})$ to create $a_{[n-m+s-1+(k+j)(c+1)]}$. By Chernoff bounds the probability is exponentially small that $a_{[n]}$ is created within $\lceil 2e\mu n^{c+1} \rceil$ steps.

Theorem 6. *With a probability of $1 - 2^{-\Omega(n)}$ the $(\mu+1)$ EA, where $\mu \leq d$, needs $2^{\Omega(n)}$ steps to optimize $\mathrm{PONEP}_{n,d,c}$, $\lceil n/c_2 \rceil < d \leq n^c$ for an appropriate constant $c_2 > 0$ and every constant integer $c \geq 1$. The expected optimization time of the $(\mu+1)$ EA, where $\mu \geq d+1$, is bounded above by $\mathcal{O}(\mu n^{c+2})$.*

7   Small Threshold Values of the Population Size

At first, we consider again the threshold value of the population size one only. But here, the presented functions satisfy the desired properties. After these are proven, we extend our observations up to threshold values of the population size of $\lceil n/(c_1 \log n) \rceil - 1$ for an appropriate constant $c_1 > 0$. Our results for PONEP_n do not satisfy the desired properties, since the probability to jump over the peak is just $\mathcal{O}(1/n)$. We modify PONEP_n that such a situation occurs numerous times. More precisely, our example functions consist of many peaks and paths between them that are also called *bridges*. Together with elements leading to the first peak these form the new path. The global optimum is again located at the end of this path. Typically, the path is first entered in front of the first peak and no shortcuts are taken. This means that never a peak and the bridge located behind it are jumped over. Thus, with an overwhelming probability, at least once a peak is produced before the global optimum is found. Similar to the behavior on PONEP_n, the probability is exponentially small to leave a peak, if the population consists of this peak only. But if there is at least one more individual in the population, these elements search forward on the path and find the next peak efficiently. This goes on until the global optimum is found.

At first, we define the peaks. Therefore, we divide an element x of length n into $\lceil \log n \rceil + 1$ disjoint blocks. Block j, $0 \leq j \leq \lceil \log n \rceil - 1$, encloses the $\lceil n/(4 \log n) \rceil =: s$ bits $x_{js+1}, \ldots, x_{(j+1)s}$ and the last block $\lceil \log n \rceil$ the remaining bits $x_{\lceil \log n \rceil s+1}, \ldots, x_n$. With block j, $0 \leq j \leq \lceil \log n \rceil - 1$, we associate a bit

$$x_{(j)} := \begin{cases} x_{js+1} & \text{if } x_{js+1} = \cdots = x_{(j+1)s} \\ \texttt{undefined} & \text{otherwise} \end{cases}.$$

Let $a_{[i]}$, $0 \leq i \leq 2^{\lceil \log n \rceil} - 1$, be the element where each $a_{[i](j)}$, $0 \leq j \leq \lceil \log n \rceil - 1$, is not $\texttt{undefined}$, furthermore $\mathbf{G}^{-1}_{\lceil \log n \rceil}(a_{[i](\lceil \log n \rceil - 1)} \cdots a_{[i](0)}) = i$ and block

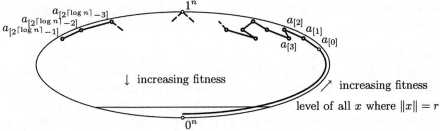

Fig. 2. An illustration of PLINPs_n.

$\lceil \log n \rceil$ consists of ones only. Therefore, in exactly all bits of one block j, $0 \leq j \leq \lceil \log n \rceil - 1$, the elements $a_{[i]}$ and $a_{[i+1]}$, $0 \leq i < 2^{\lceil \log n \rceil} - 1$, differ. The bridge between $a_{[i]}$ and $a_{[i+1]}$ consists of $a_{[i,k]} := a_{[i],1} \cdots a_{[i],js+k} a_{[i+1],js+k+1} \cdots a_{[i+1],n}$, $0 < k < s$. Finally, the elements $a_{[-1,k]} := 0^{n-k}1^k$, $0 \leq k < n - \lceil \log n \rceil s$, lead from 0^n to the first peak $a_{[0]}$. We remark that the functions are influenced by the long path functions of Rudolph (1997), but ours are short, of course.

$$S := \big(a_{[-1,0]}, \ldots, a_{[-1,n-\lceil \log n \rceil s-1]}, a_{[0]}, a_{[0,1]}, \ldots, a_{[0,s-1]}, a_{[1]}, a_{[1,1]}, \ldots,$$
$$a_{[2^{\lceil \log n \rceil}-3,s-1]}, a_{[2^{\lceil \log n \rceil}-2]}, a_{[2^{\lceil \log n \rceil}-2,1]}, \ldots, a_{[2^{\lceil \log n \rceil}-2,s-1]},$$
$$a_{[2^{\lceil \log n \rceil}-1]}\big) =: \big(s_0, \ldots, s_{n+s(2^{\lceil \log n \rceil}-\lceil \log n \rceil-1)}\big)$$

describes the whole path. Now, we can give a complete definition of PLINPs_n (Path with LInear in n many Peaks) illustrated in Fig. 2.

$$\text{PLINPs}_n(x) := \begin{cases} n+s+i & \text{if } x = s_i \text{ and } x = a_{[j]} \text{ for some } j \\ n+i & \text{if } x = s_i \text{ and } x \neq a_{[j]} \text{ for every } j \\ n-|x| & \text{otherwise} \end{cases}$$

Theorem 7. *The expected time until the $(\mu+1)$ EA, where $\mu \geq 2$, has optimized PLINPs_n is bounded above by $\mathcal{O}(\mu n^3/\log n)$.*

Proof. The proof is similar to that of Theorem 3 using Theorem 2. We choose the partition induced by the areas

$$A_i := \begin{cases} \{a \mid \text{PLINPs}_n(a) = i\} & \text{if } 1 \leq i < n \\ \{s_{i-n}\} & \text{if } n \leq i \leq 2n + s(2^{\lceil \log n \rceil} - \lceil \log n \rceil - 1) \end{cases}.$$

Hence, it holds $\mu \geq \max\{v_j \mid j < 2n + s(2^{\lceil \log n \rceil} - \lceil \log n \rceil - 1)\} + 1 = 2$ if

$$b_i := \begin{cases} i+1 & \text{if } i \neq 2n - (\lceil \log n \rceil + j)s \text{ for every } j \\ i+s & \text{if } i = 2n - (\lceil \log n \rceil + j)s \text{ for some } j \end{cases}$$

and $1 \leq i < 2n + s(2^{\lceil \log n \rceil} - \lceil \log n \rceil - 1)$. If we consider the areas A_i and A_{i+1}, it holds $p(A_i) \geq 1/(en)$. So, by Theorem 2 the expected optimization time is bounded above by $\mu\big(1 + (2n + s(2^{\lceil \log n \rceil} - \lceil \log n \rceil - 1) - 1)en\big) = \mathcal{O}(\mu n^3/\log n)$. $\square$

We consider a technical lemma that summarizes one main property of PLINPs_n.

Lemma 8. *For* $0 \le i \le 2^{\lceil \log n \rceil} - 1$ *and all* c *where* $\mathrm{PLINPs}_n(c) > \mathrm{PLINPs}_n(a_{[i]})$ *it holds* a) $\mathbf{H}(a_{[i]}, c) \ge s$ *and* b) $\mathbf{H}(a_{[i-1,k]}, c) \ge s$ *for arbitrary* k.

Proof. The element c can only be $a_{[i+l]}$, $l \ge 1$, or $a_{[i+l,j]}$ for arbitrary j.
a) The elements $a_{[i]}$ and $a_{[i+l]}$ differ in all bits of at least one block. Therefore, it is $\mathbf{H}(a_{[i]}, a_{[i+l]}) \ge s$ and since by construction of a_i all bits in each block have the same value, it holds $\mathbf{H}(a_{[i]}, a_{[i+l,j]}) \ge \min\{\mathbf{H}(a_{[i]}, a_{[i+l]}), \mathbf{H}(a_{[i]}, a_{[i+l+1]})\} \ge s$.
b) Due to the situation described in a) it is $\mathbf{H}(a_{[i-1,k]}, a_{[i+l]}) \ge \min\{\mathbf{H}(a_{[i-1]}, a_{[i+l]}), \mathbf{H}(a_{[i]}, a_{[i+l]})\} \ge s$ and it is $\mathbf{H}(a_{[i-1,k]}, a_{[i+l,j]}) \ge \min\{\mathbf{H}(a_{[i-1,k]}, a_{[i+l]}), \mathbf{H}(a_{[i-1,k]}, a_{[i+l+1]}) \ge \min\{\mathbf{H}(a_{[i-1]}, a_{[i+l]}), \mathbf{H}(a_{[i-1]}, a_{[i+l+1]}), \mathbf{H}(a_{[i]}, a_{[i+l]}), \mathbf{H}(a_{[i]}, a_{[i+l+1]})\} \ge s$. □

Theorem 9. *With a probability of* $1 - 2^{-\Omega(n)}$ *the* $(1+1)$ *EA needs* $2^{\Omega(n)}$ *steps to optimize* PLINPs_n.

Proof. When first an element of $\{s_{n-\lceil \log n \rceil s} = a_{[0]}, \dots, s_{n+s(2^{\lceil \log n \rceil} - \lceil \log n \rceil - 1)}\}$ is produced this happens similar to the proof of Theorem 4 with an exponentially small failure probability by a mutation of $a_{[-1,n-\lceil \log n \rceil s - k]}$, $k \ge 1$. We analyze the situation that the population is $a_{[-1,n-\lceil \log n \rceil s - k]}$, $k \ge 1$, or $a_{[i,s-k]}$, $i \ge 0$. By Lemma 8 the probability is bounded by $|S|(1/n)^s = 2^{-\Omega(n)}$ to create an arbitrary element c where $\mathrm{PLINPs}_n(c) > \mathrm{PLINPs}_n(a_{[i+1]})$. Furthermore, again similar to the proof of Theorem 4, the probability to create $a_{[i+1,l]}$ for an arbitrary l before $a_{[i+1]}$ is bounded by $\mathcal{O}(1/n)$. If the population is $a_{[i]}$, $0 \le i < 2^{\lceil \log n \rceil} - 1$, by Lemma 8 the probability is exponentially small to create an element $c \ne a_{[i]}$ where $\mathrm{PLINPs}_n(c) \ge \mathrm{PLINPs}_n(a_{[i]})$. Hence, the probability to produce the global optimum before an element $a_{[i]}$, $0 \le i < 2^{\lceil \log n \rceil} - 1$, is bounded by $2^{-\Omega(n)} + \mathcal{O}(1/n)^{2^{\lceil \log n \rceil} - 1} = 2^{-\Omega(n)}$. □

We consider threshold values of the population size of up to $\lceil n/(c_1 \log n) \rceil - 1$ for an appropriate constant $c_1 > 0$. Therefore, we play with the definition of PLINPs_n. This is done similar to the changings of PONEP_n that led to $\mathrm{PONEP}_{n,d}$. We enlarge the peak. Let $\mathrm{PLINPs}_{n,d}$, $1 \le d < \lceil n/(c_1 \log n) \rceil$, be the variant of PLINPs_n where beside $a_{[i]}$, $0 \le i < 2^{\lceil \log n \rceil} - 1$, the elements $a_{[i,k]}$, $1 \le k < d$, have f-value $\mathrm{PLINPs}_n(a_{[i]})$. So, these elements form the new peaks.

The arguments that led to Theorems 5 and 7 bound the expected optimization time for the $(\mu+1)$ EA on $\mathrm{PLINPs}_{n,d}$, if $\mu \ge d + 1$, by $\mathcal{O}(\mu n^3 / \log n)$.

The result of Theorem 9 also holds for the $(\mu+1)$ EA on $\mathrm{PLINPs}_{n,d}$, if $\mu \le d$ and $d < \lceil n/(c_1 \log n) \rceil$ for an appropriate constant $c_1 > 0$. The path is reached at its beginning. If the population consists only of elements of the peak $\{a_{[i]}, a_{[i,1]}, \dots, a_{[i,d-1]}\}$, $0 \le i < 2^{\lceil \log n \rceil} - 1$, by Lemma 8 the probability is exponentially small to create an element $c \notin \{a_{[i]}, a_{[i,1]}, \dots, a_{[i,d-1]}\}$ where $\mathrm{PLINPs}_{n,d}(c) \ge \mathrm{PLINPs}_{n,d}(a_{[i]})$, if c_1 is large enough. Otherwise, let $s_k \in \mathcal{P}$ but $s_j \notin \mathcal{P}$ for all $j > k$. If $s_k = a_{[i,l]}$, $l \ge d$, similar to the arguments that led to Theorems 5 and 9, the probability to produce an element of $\{a_{[i+1,d]}, \dots, a_{[i+1,s-1]}\}$ before an element of $\{a_{[i+1]}, a_{[i+1,1]}, \dots, a_{[i+1,d-1]}\}$ is bounded by $\mathcal{O}(1/n^d) = $

$\mathcal{O}(1/n)$. If $s_k = a_{[i]}$ or $s_k = a_{[i,l]}$, $l < d$, the failure probability is bounded by $2^{-\Omega(n/\log n)} = \mathcal{O}(1/n)$ that after $\lceil 2e\mu n^2/(c_1\log n)\rceil$ steps the population consists of elements of $\{a_{[i]}, a_{[i,1]}, \ldots, a_{[i,d-1]}\}$ only. And the probability is also bounded by $\mathcal{O}(1/n)$ that within these steps an element of $\{a_{[i+1]}, a_{[i+1,1]}, \ldots, a_{[i+1,s-1]}\}$ is created. Therefore and since for both situations of s_k the probability is exponentially small to produce an arbitrary element c where $\mathrm{PLINPs}_{n,d}(c) > \mathrm{PLINPs}_{n,d}(a_{[i+1]})$, the probability to produce the global optimum before the population consists only of some elements of some peak $\{a_{[i]}, a_{[i,1]}, \ldots, a_{[i,d-1]}\}$, $0 \le i < 2^{\lceil\log n\rceil} - 1$, is again bounded by $2^{-\Omega(n)} + \mathcal{O}(1/n)^{2^{\lceil\log n\rceil}-1} = 2^{-\Omega(n)}$.

Theorem 10. *With a probability of* $1-2^{-\Omega(n)}$ *the* $(\mu+1)$ *EA, where* $\mu \le d$, *needs* $2^{\Omega(n)}$ *steps to optimize* $\mathrm{PLINPs}_{n,d}$, $1 \le d < \lceil n/(c_1\log n)\rceil$ *for an appropriate constant* $c_1 > 0$. *The expected optimization time of the* $(\mu+1)$ *EA, where* $\mu \ge d+1$, *is bounded above by* $\mathcal{O}(\mu n^3/\log n)$.

Conclusions

We have proved that functions exist where a simple mutation-based EA is efficient if the population size $\mu > d$ and is totally inefficient if $\mu \le d$. This has been proven rigorously by specifying some functions for all values of d polynomially bounded in the dimension of the search space. These results form a typical so-called hierarchy result. We have developed methods to analyze the investigated EA. These help to upper bound the expected optimization time. The question if the smallest possible increase of the population size may be advantageous has been answered positively. However, in most cases of application such a sensitive decrease of the population size does not have such enormous effects. But these results support the importance of a correct choice of the population size.

Acknowledgements. This research was supported by a Grant from the G.I.F., the German-Israeli Foundation for Scientific Research and Development. The author thanks Ingo Wegener for his help while preparing this paper.

References

Jansen, T. and De Jong, K. (2002). *An Analysis of the Role of the Offspring Population Size in Evolutionary Algorithms*. Proceedings of the Genetic and Evolutionary Computation Conference (GECCO) 2002, 238–246.

Jansen, T. and Wegener, I. (2001a). *Evolutionary Algorithms – How to Cope with Plateaus of Constant Fitness and when to Reject Strings with the Same Fitness*. IEEE Transactions on Evolutionary Computation 5, 589–599.

Jansen, T. and Wegener, I. (2001b). *On the Utility of Populations in Evolutionary Algorithms*. Proceedings of the Genetic and Evolutionary Computation Conference (GECCO) 2001, 1034-1041.

Motwani, R. and Raghavan, P. (1995). *Randomized Algorithms*. Cambridge University Press, Cambridge.

(Jansen and De Jong (2002)) and some variants of $(\mu+\mu)$ EAs (He and Yao (2002)).

The aim of this paper is to contribute to a theory of standard $(\mu+\lambda)$ EAs, where $\mu > 1$. We start with the simple case $\lambda = 1$, considering a $(\mu+1)$ EA that is a generalization of the (1+1) EA for the search space $\{0,1\}^n$, and follow the research line started for this (1+1) EA. We study the behavior of the $(\mu+1)$ EA on example functions and compare the obtained results with those for the (1+1) EA. To this end, a new and general proof technique for bounding the expected runtime of the $(\mu+1)$ EA from below is developed. An advantage of the new method is that it has not been designed for a special mutation operator. In particular, we are able to analyze the $(\mu+1)$ EA with a global search operator that may flip many bits. Often, the analysis of EAs is much more difficult with a global than with a local search operator (see, e. g., Wegener and Witt (2003)).

The paper is structured as follows. In Sect. 2, we define the $(\mu+1)$ EA and the considered example functions. Moreover, we introduce the tool of family trees, which is essential throughout the paper. In Sect. 3, simple upper bounds on the expected runtime of the $(\mu+1)$ EA on the example functions are presented. In Sect. 4, we describe the new lower bound technique completely but omit the proofs of technical lemmas due to space limitations; a full version of the paper is available as a technical report. In Sect. 5, we apply the technique to prove lower bounds on the expected runtime and bounds on the success probability. These bounds are tight for two of the examples. Moreover, they show that here the $(\mu+1)$ EA is never more efficient than the (1+1) EA. However, it is a common belief that a population helps to better explore the search space, and it is important to find an example where the $(\mu+1)$ EA with $\mu > 1$ outperforms the (1+1) EA. Therefore, a function where an increase of μ by a sublinear factor decreases the expected runtime drastically, namely from exponential to polynomial, is identified in Sect. 6. We finish with some conclusions.

2   Definitions

We obtain the $(\mu+1)$ EA for the maximization of functions $f\colon \{0,1\}^n \to \mathbb{R}$ as a generalization of the well-known (1+1) EA (see Droste, Jansen and Wegener (2002)). As for continuous search spaces, a pure $(\mu+1)$ evolution strategy should do without recombination and should employ a uniform selection for reproduction. As usual, a truncation selection is applied for replacement. The mutation operator should be able to search globally, i. e., to flip many bits in a step. Therefore, a standard mutation flipping each bit with probability $1/n$ seems the most sensible. These arguments lead to the following definition of the $(\mu+1)$ EA.

1. Choose μ individuals $x^{(i)} \in \{0,1\}^n$, $i \in \{1,\ldots,\mu\}$, uniformly at random. Let the multiset $X^{(0)} = \{x^{(1)},\ldots,x^{(\mu)}\}$ be the initial population at time 0.
2. Repeat infinitely
   a) Choose an x from the population $X^{(t)}$ at time t uniformly at random.
   b) Create x' by flipping each bit of x independently with probability $1/n$. Let X' be the population obtained by adding x' to $X^{(t)}$.

c) Create $X^{(t+1)}$, the current population at time $t + 1$, by deleting an individual from X' with lowest f-value uniformly at random. Set $t := t + 1$.

We have kept the (μ+1) EA as simple as possible and refrain from employing diversity-maintaining mechanisms. The (μ+1) EA with $\mu = 1$ is very similar to the (1+1) EA, but differs in one respect. If an individual created by mutation has the same f-value as its father, either of both is retained with equal probability.

As usual in theoretical investigations, we leave the stopping criterion of the (μ+1) EA unspecified. We analyze the number of iterations (also called *steps*) of the infinite loop until the current population for the first time contains an optimal individual, i. e., one that maximizes f. The sum of this number and the population size μ is denoted as the *runtime* of the (μ+1) EA and corresponds to the number of function evaluations (a common approach in black-box optimization, cf. Droste, Jansen, Tinnefeld and Wegener (2002)). Throughout the paper, we consider only $\mu = poly(n)$, i. e., values of μ bounded by a polynomial of n.

We study the (μ+1) EA on the following example functions. The well-known function $\text{ONEMAX}(x) = x_1 + \cdots + x_n$ counts the number of ones of a string $x \in \{0, 1\}^n$ and $\text{LEADINGONES}(x) = \sum_{i=1}^n \prod_{j=1}^i x_j$ counts the number of leading ones. The function $\text{SPC}(x)$ *(short path with constant fitness)* introduced by Jansen and Wegener (2001a) equals $n - \text{ONEMAX}(x)$ if x cannot be written as $1^i 0^{n-i}$ for any i. It equals $2n$ if $x = 1^n$ and $n + 1$ otherwise. SPC is of special interest since EAs have to cross a plateau of constant fitness to find the optimum.

To elucidate the utility of the (μ+1) EA's population, throughout the paper, we compare the (μ+1) EA with μ parallel runs of the (1+1) EA. The total cost (neglecting initialization cost) of t steps of the (μ+1) EA corresponds to the cost raised by μ parallel runs of the (1+1) EA up to time t/μ. Thus, if we consider a (μ+1) EA at time t, we denote μ parallel runs of the (1+1) EA considered at time t/μ as the *corresponding parallel run*. In order to derive runtime bounds for the (μ+1) EA, it is helpful to consider the so-called *family trees* of the individuals from the initial population (this concept has been introduced in a different context by Rabani, Rabinovich and Sinclair (1998)). Fix an arbitrary such individual x. If x is mutated, a descendant of x is produced. More generally, we can visualize the descendants of x and their descendants by the family tree $T_t(x)$ at time t as follows. $T_0(x)$ contains only the root x. $T_t(x)$ contains $T_{t-1}(x)$ and the additional edge $\{v, w\}$ if w is the result of a mutation of the individual v at time $t - 1$ and v is contained in $T_{t-1}(x)$. Note that the tree $T_t(x)$ may contain individuals that have already been deleted from the corresponding population.

3   Upper Bounds

The following upper bounds on the runtime are not too difficult to obtain.

Theorem 1. *Let $\mu = poly(n)$. Then the expected runtime of the (μ+1) EA on* LEADINGONES *is bounded above by* $\mu + 3en \cdot \max\{\mu \ln(en), n\} = O(\mu n \log n + n^2)$.

Proof. We measure the progress to the optimum by the potential L, defined as the maximum LEADINGONES value of the current population's individuals. To increase L, it is sufficient to select an individual with maximum value and to flip the leftmost zero. The selection and the mutation operator of the $(\mu{+}1)$ EA are independent. Hence, if there are i individuals with maximum value, the probability of the considered event is at least $\frac{i}{\mu} \cdot \frac{1}{n} \cdot \left(1 - \frac{1}{n}\right)^{n-1} \geq \frac{i}{e\mu n}$, and the waiting time is at most $e\mu n/i$. The potential has to increase at most n times. Estimating $i \geq 1$ would lead to an upper bound $\mu{+}e\mu n^2$ on the expected runtime.

However, the $(\mu{+}1)$ EA can produce replicas of individuals with maximum function value. If their number is i, the probability of creating a further replica is at least $(i/\mu)(1 - 1/n)^n \geq i/(2e\mu)$. Furthermore, if $i < \mu$, this replica replaces a worse individual and increases the number of best ones. Assume pessimistically that L stays fixed until we have at least $\min\{n/\ln(en), \mu\}$ replicas. The expected time for this is, by elementary calculations, at most $2e\mu \ln(en)$. Now the expected time to increase L is at most $e\mu n/(\min\{n/\ln(en), \mu\})$. Altogether, the expected runtime is at most $\mu{+}n(2e\mu \ln(en)+\frac{e\mu n}{\min\{n/\ln(en),\mu\}}) \leq \mu{+}3en \cdot \max\{\mu \ln(en), n\}$. $\qquad\square$

Theorem 2. *Let $\mu = poly(n)$. Then the expected runtime of the $(\mu{+}1)$ EA on* ONEMAX *is bounded above by $\mu + 5e\mu n + en \ln(en) = O(\mu n + n \log n)$.*

Proof. The proof idea is similar as in Theorem 1. Let L be the maximum ONE-MAX value of the current population. In contrast to LEADINGONES, the probability of increasing L depends on L itself. Since each individual has at least $n{-}L$ zeros, the considered probability is bounded below by $\frac{i}{\mu} \cdot \frac{n-L}{n} \cdot \left(1 - \frac{1}{n}\right)^{n-1} \geq \frac{i(n-L)}{e\mu n}$ if the population contains at least i individuals with maximum value.

The expected time until the population contains at least $\min\{n/(n-L), \mu\}$ replicas of an individual with value L is bounded by $2e\mu \ln(en/(n-L))$ if L does not increase before. If we sum up these expected waiting times for all values of L, we obtain (using Stirling's formula) a total expected waiting time of at most

$$2e\mu \sum_{L=0}^{n-1} \ln\left(\frac{en}{n-L}\right) = 2e\mu \ln\left(\frac{e^n n^n}{n!}\right) \leq 2e\mu \ln(e^{2n}) = 4e\mu n.$$

After the desired number of replicas has been obtained, the expected time for increasing L is at most $\frac{e\mu n}{\min\{\mu, n/(n-L)\} \cdot (n-L)} = \frac{e\mu n}{\min\{\mu(n-L), n)\}}$. By elementary calculations, the expected waiting time for all L-increases is at most $en \ln(en) + e\mu n$, and the total expected runtime, therefore, at most $\mu{+}en \ln(en){+}5e\mu n$. $\qquad\square$

For SPC, we can only prove a (seemingly) trivial upper bound.

Theorem 3. *Let $\mu = poly(n)$. Then the expected runtime of the $(\mu{+}1)$ EA on* SPC *is bounded by $O(\mu n^3)$.*

Sketch of proof. For each individual x from the initial population, we consider paths in its family tree directed from the root to a node v. If the individual corresponding to v has been deleted, we call the path dead, and alive otherwise. There is always at least one alive path in some family tree.

We want to show that the following property P holds for every initial individual x. The expected time until at least one of x's paths reaches length k or until all of its paths are dead is bounded by $4e\mu k$ for all k. This will imply the theorem for the following reasons. By similar arguments as in the proof of Theorem 1, one can show that the (μ+1) EA reaches a situation where the entire population contains individuals of shape 1^i0^{n-i} with $i \neq n$, i.e., from the plateau of constant fitness, after $O(\mu n \log n)$ expected steps (or is done before). Afterwards, we can ignore steps of the (μ+1) EA creating individuals outside the plateau since these individuals are deleted immediately after creation. Since the (μ+1) EA chooses for deletion uniformly from the worst individuals, the event that a path dies is independent of the individual at the path's end provided it is from the plateau. Hence, any path of plateau points has the same properties as a path of plateau points drawn by a run of the (1+1) EA on SPC. By the results of Jansen and Wegener (2001a), such a path contains an optimal individual after an expected length of $O(n^3)$, i.e., after $O(\mu n^3)$ expected steps according to P.

To prove P, we assume w.l.o.g. that there is always at least one alive path for x. Consider the potential L, denoting the length of the currently longest alive path leading to an x' that will always have an alive descendant. There must be such an x' according to our assumptions. Moreover, L cannot shrink in the run, and there is the following sufficient condition for increasing L. An individual x' defining the current L-value is mutated, a child from the plateau is created, and x' is deleted before its child is deleted. The probability is $1/\mu$ for the first event, at least $(1 - 1/n)^n \geq 1/(2e)$ for the second event since producing a replica is sufficient, and $1/2$ for the third one since the considered individuals have equal fitness. Hence, the expected time to increase L is at most $4e\mu$, implying P. $\square$

4   A General Lower Bound Technique

For lower bounds on the runtime, we consider the growth of the family tree for any initial individual of the (μ+1) EA. Upper bounds on the depth of family trees always follow from the selection mechanism of the (μ+1) EA, which selects the individual to be mutated uniformly from the current population. Therefore, it is possible to model the stochastic process growing a family tree as follows.

Definition 1 (1/μ-tree). *Let $p := p_{t,u}$, $t, u \geq 0$, be a sequence of probability distributions s.t. the support of $p_{t,u}$ is $\{0, 1, \ldots, u\}$. A p-tree at time 0 consists only of the root. A p-tree T_t at time $t \geq 1$ is obtained from a p-tree T_{t-1} as follows. Let u be the number of nodes of T_{t-1}. Sample v by $p_{t-1,u}$. If $v > 0$, append a new leaf to the v-th inserted node of T_{t-1}; otherwise, let $T_t := T_{t-1}$.*

A p-tree is called a 1/μ-tree if $p_{t,u}(v) \leq 1/\mu$ for all $v > 0$.

A 1/μ-tree at time t can have less than $t + 1$ nodes since p can put some probability on 0. If we model family trees by 1/μ-trees, we do not specify the distributions $p_{t,u}$ exactly since it is too difficult to predict whether and, if so, which individuals corresponding to nodes are deleted. In the $(\mu+1)$ EA, deleted nodes have probability 0 of being chosen and alive nodes have probability $1/\mu$.

The following lemma contains an interesting result for the depth of $1/\mu$-trees.

Lemma 1. *Let $D(t)$ denote the depth of a $1/\mu$-tree at time t. For all $t \geq 0$ and $d \geq 0$, $\mathrm{Prob}(D(t) \geq d) \leq (t/\mu)^d/d!$. Moreover, $\mathrm{Prob}(D(t) \geq 3t/\mu) = 2^{-\Omega(t/\mu)}$.*

Lemma 1 states that, with overwhelming probability, a family tree of the $(\mu+1)$ EA becomes asymptotically no deeper than the total number of mutations performed in a single run of the corresponding parallel run. The tree can become wide, but a flat tree means that few mutations lie on any path from the root to a node in the tree. Hence, if the depth is small, this means that a leaf is an individual that is likely to be similar to the root. This makes the optimization of even simple functions very unlikely if the tree is not deep enough. The following result is tight for some simple functions such as ONEMAX (if μ is not too small).

Theorem 4. *Let $\mu = poly(n)$ and let f be a function with a unique global optimum. Then the expected runtime of the $(\mu+1)$ EA on f is $\Omega(\mu n + n \log n)$. Moreover, the success probability within some $c\mu n$ steps, $c > 0$, is $2^{-\Omega(n)}$.*

Sketch of proof. The lower bound of $\Omega(n \log n)$ follows for $\mu \leq \log n/2$ by a generalization of the coupon collector's theorem described by Droste, Jansen and Wegener (2002) for the considered class of functions and the $(1+1)$ EA. For the lower bound $\Omega(\mu n)$, we set up a phase of length $s := \lfloor c\mu n \rfloor$ for some constant $c > 0$ and show that the $(\mu+1)$ EA requires at least s steps with probability $1 - 2^{-\Omega(n)}$ if c is small enough. The proof idea is as follows. In s steps, a family tree created by the $(\mu+1)$ EA with high probability has to reach a certain depth to optimize f; however, the probability of reaching this depth is very small.

Let x be an arbitrary initial individual x. We consider the infinite random process of building its family tree. Let $T_t(x)$ denote the tree at time t. According to Lemma 1, the probability of $T_s(x)'s$ depth reaching at least $3cn$ is $2^{-\Omega(n)}$. Now the aim is to prove that with probability $1 - 2^{-\Omega(n)}$, a depth of at least $3cn$ is necessary for optimization (if c is small enough).

During the process building the trees $T_t(x)$, we consider the event that a node v with optimal f-value is inserted. Consider the path p_v from x to v. We claim that with probability $1 - 2^{-\Omega(n)}$, its length is at least $n/4$. By Chernoff bounds (see Motwani and Raghavan (1995)), the root x has Hamming distance at least $n/3$ to the unique optimal string (represented by v) with probability $1 - 2^{-\Omega(n)}$. Moreover, consider a sequence of $n/4$ strings where each string is the result of a mutation of its predecessor by means of the $(\mu+1)$ EA's mutation operator. The expected Hamming distance of any two strings in this sequence is at most $n/4$, and, by Chernoff bounds, it is less than $n/3$ with probability $1 - 2^{-\Omega(n)}$. Since the nodes on each path in the trees $T_t(x)$ form such a random sequence of strings, the claim follows. Moreover, $T_s(x)$ contains at most $s = poly(n)$ paths, and there are at most polynomially many choices for x since $\mu = poly(n)$. Therefore, the probability that there is a node with optimal f-value at depth less than $n/4$ in a family tree at time s is still $2^{-\Omega(n)}$. If c is small enough, $n/4$ is at least $3cn$. Since the sum of all failure probabilities is $2^{-\Omega(n)}$, the proof of the theorem is complete.   □

Theorem 4 covers the wide range of unimodal functions. For some unimodal functions (e. g., linear functions), the (1+1) EA's expected runtime is $O(n \log n)$. For such functions, Theorem 4 states that the (μ+1) EA is (for large μ) at most by a factor of $O(\log n)$ more efficient than the corresponding parallel run.

For more difficult functions (meaning that the (1+1) EA's expected runtime is $\omega(n \log n)$), the proof concept of Theorem 4 can be carried over to show larger lower bounds also for the (μ+1) EA. However, we have to derive better lower bounds on the depth of family trees. Therefore, more structure of the function f and the encountered individuals comes into play. Although all nodes of a family tree are different individuals, many individuals may represent the same string $x \in \{0,1\}^n$. For an individual x^*, we call the $x \in \{0,1\}^n$ associated with x^* the *string of x^** or say that *x^* is the string x*. We also call the string of an individual its *color*. This leads to the following definition.

Definition 2 (Monochromatic Subtree (MST)). *A connected subgraph of a family tree is called a* monochromatic subtree *if all its nodes are the same string.*

Obviously, all nodes in an MST have equal f-value. It is interesting that the stochastic process creating an MST sometimes equals the process for a so-called random recursive tree (RRT), a model of random trees well known from the literature (e. g., Smythe and Mahmoud (1995)). This will allow us to apply the known theory on RRTs. We obtain an RRT by the following stochastic process.

Definition 3 (Random Recursive Tree (RRT)). *An RRT at time 0 consists only of the root. An RRT T_t at time $t \geq 1$ is obtained from an RRT T_{t-1} by choosing uniformly at random one of its nodes and appending a new leaf to it.*

Note that the RRT at time $t \geq 0$ consists of exactly $t+1$ nodes. The processes generating MSTs and RRTs coincide only if the (μ+1) EA can choose uniformly from the set of nodes of the MST. Since deleted individuals are nevertheless kept in the family tree, this property can only be guaranteed if the individuals of the considered MST are still present in the population. To prove the following lemma, one exploits that considering MSTs, the event of appending a node whose color is different from that of the father is independent of the choice of the father.

Lemma 2. *Let T^* be a monochromatic subtree of a family tree and let V be the set of nodes of T^*. If the (μ+1) EA does not delete any individual from V until the creation of the last node of T^* then T^* is an RRT.*

If the (μ+1) EA deletes individuals of an MST from the population, it chooses these, by the definition of an MST and the (μ+1) EA, uniformly from the alive nodes of the MST. Hence, the earliest inserted nodes have the highest chances of having been deleted by any fixed time t. Early inserted nodes are close to the root. This implies that an MST that is affected by deletion steps is typically deeper than an RRT of the same size. We can make this precise by considering generalized RRTs, namely so-called p-marked random trees (p-marked RTs).

Definition 4 (*p*-marked RT). *Let $p_{t,u}$, $t, u \geq 0$, be a sequence of probability distributions s.t. the support of $p_{t,u}$ is $\{0, \ldots, u\}$. A p-marked RT at time 0 consists only of the unmarked root. A p-marked RT T_t at time $t \geq 1$ is obtained from a p-marked RT T_{t-1} in two steps. First, an unmarked node is chosen uniformly at random and a new, unmarked leaf is appended. Let U denote the set of unmarked nodes after this step. Then u^* is sampled according to $p_{t-1,|U|-1}$, a subset $S^* \subseteq U$ of size u^* is chosen uniformly, and all nodes in S^* are marked.*

Again, a tree at time t has exactly $t + 1$ nodes, only the unmarked ones of which can become fathers of new nodes. It is crucial that for all $p_{t,u}$, the set of newly marked nodes is, by definition, uniform over the yet unmarked ones and that always at least one node remains unmarked.

Lemma 3. *A monochromatic subtree of a family tree is a p-marked RT.*

By technical analyses, one can show that the probability of a *p*-marked RT with t nodes reaching depth d is, for any p, at least as large as the respective probability of an RRT. Let for a *p*-marked RT and an RRT at time t the measures $D^*(t, i)$ resp. $D(t, i)$ denote the depth of the node that was inserted at time i.

Lemma 4. *For all $t, i, d \geq 0$ and $i \leq t$, $\mathrm{Prob}(D^*(t, i) \geq d) \geq \mathrm{Prob}(D(t, i) \geq d)$.*

Since lower bounds on the depth of ordinary RRTs are well known (Pittel (1994)), we have developed new tools for lower bounding the depth of MSTs and, therefore, of family trees. Upper bounds are still provided by Lemma 1.

5   More Special Lower Bounds

We apply the proof method developed in the last section to a well-studied function. Here, the method can also be considered as a generalization of the proof method of artificial fitness layers (e. g., Droste, Jansen and Wegener (2002)).

Theorem 5. *Let $\mu = poly(n)$. Then the expected runtime of the $(\mu+1)$ EA on* LEADINGONES *is $\Omega(\mu n \log n + n^2)$. Moreover, the success probability within some $c\mu n \log n$ steps, $c > 0$, is $2^{-\Omega(n)}$.*

Sketch of proof. The bound $\Omega(n^2)$ follows by applying the analysis of LEADING-ONES and of the $(1+1)$ EA by Droste, Jansen and Wegener (2002) to the potential L from the proof of Theorem 1. The basic idea for the bound $\Omega(\mu n \log n)$ is the same as in Theorem 4. We show that for some small enough constant $c > 0$, the $(\mu+1)$ EA requires at least $s := \lfloor c\mu n \log n \rfloor$ steps with probability $1 - 2^{-\Omega(n)}$. Now we consider the family tree $T_s(x)$ obtained after s steps for an arbitrary initial individual x. By Lemma 1, it suffices to show that a depth of at least $3cn \log n$ is necessary for optimization with probability $1 - 2^{-\Omega(n)}$.

For notational convenience, let $f :=$ LEADINGONES. During the process of building the trees $T_t(x)$, we consider the event that a node v with optimal f-value is inserted. Since initial individuals are uniform over $\{0,1\}^n$, the root x has an

f-value of at most $n/2$ with probability $1 - 2^{-\Omega(n)}$. Consider the path p_v from x to v. By standard arguments from the analysis of the (1+1) EA on f (Droste, Jansen and Wegener (2002)), the bits after the leftmost zero are, in each string on p_v, uniformly distributed. W.l.o.g., the f-value is non-decreasing along p_v. Since the f-value has to increase by at least $n/2$ along p_v with probability $1 - 2^{-\Omega(n)}$, the mentioned arguments imply that at least $n/6$ different strings lie on p_v with probability $1 - 2^{-\Omega(n)}$. We call the nodes that are different strings than their fathers *subtree roots*. For a subtree root r, by $T^*(r)$ we denote the maximal MST rooted at r. Now we work under the assumption that p_v contains at least $n/6$ subtree roots.

Fix an arbitrary subtree root $r \neq v$ and the next subtree root r' on p_v. By Lemma 3, the MST $T^*(r)$ is a p-marked RT, and r' is some node inserted into (but not not attributed to) a p-marked RT. Considering the construction of $T^*(r)$, we prove that r' was likely to be created late during this process. The probability of mutating a string with value $f(r)$ to a better string is bounded above by $1/n$. Hence, with probability at least $1/2$, the first $n/2$ steps that choose a father in the already existing MST create nodes with at most the same value as the root. Since producing a replica of a string has probability $(1-1/n)^n \geq 1/(2e)$, the expected number of replicas within $n/2$ steps is at least $n/(4e)$. By Chernoff bounds, with probability at least $1/2 - 2^{-\Omega(n)}$, $T^*(r)$ receives at least $n/(8e)$ nodes before an individual with larger value than $f(r)$ is appended. Hence, with probability at least $1/2 - 2^{-\Omega(n)}$, the node r' has a distance to r that is bounded below by the depth of the at least $n/(8e)$-th node of a p-marked RT.

How deep is the k-th node such that $k \geq n/(8e)$ within a p-marked RT? We know it if the tree is an ordinary RRT. Then, by the theory on RRTs (Smythe and Mahmoud (1995)), the depth is at least $(\log n)/2$ with probability at least $1/2$ (for n large enough). By Lemma 4, the same statement holds also for a p-marked RT. Altogether, the distance of r and r' on p_v is at least $(\log n)/2$ with probability at least $1/4 - o(1)$. Since the process creating $T^*(r')$ is independent of the process creating $T^*(r)$, we can apply Chernoff bounds. Since at least $n/6$ choices for r are available on p_v, at least $n/25$ subtree roots have their successive subtree roots at distance at least $(\log n)/2$ with probability $1 - 2^{-\Omega(n)}$. Altogether, the length of p_v is at least $n(\log n)/50$ with probability $1 - 2^{-\Omega(n)}$.

Since at time s, the number of all nodes in all trees is bounded by a polynomial, the probability that there is a node with f-value n at depth less than $n(\log n)/50$ in a family tree is $2^{-\Omega(n)}$. If c is small enough, the bound $n(\log n)/50$ is at least $3cn \log n$. Finally, the sum of all failure probabilities is $2^{-\Omega(n)}$.   □

The method of lower bounding the depth of MSTs can also be used to lower bound the expected runtime of the $(\mu+1)$ EA on the function SPC. It is easy to see that, due to the plateau of constant fitness, there are with high probability even $\Omega(n^2)$ subtree roots on any path leading to an optimal node in a family tree. Hence, a straightforward application of the proof of Theorem 5 would lead to a lower bound of $\Omega(\mu n^2 \log n)$ on the expected runtime. However, one can improve on this by considering the number of alive nodes in MSTs (which a p-marked RTs and at least as deep as ordinary RRTs) more carefully. One can

$2^{(n^{1/2}+o(1))/(700 \log n)}$. The constants have been chosen such that the product of the waiting time and the probability of reaching the local optimum is $o(1)$.   □

Conclusions

We have presented a first analysis of the $(\mu+1)$ EA for pseudo-Boolean functions by studying the expected runtime on three well-known example functions. For two of these, we have derived asymptotically tight bounds, showing that $\mu = 1$ leads asymptotically to the lowest runtime. In contrast to this, we have identified a function where the $(\mu+1)$ EA outperforms the $(1+1)$ EA and its multistart variants drastically provided that $\mu \geq n/\ln(en)$.

To prove lower bounds, we have developed a new technique. This technique is not only limited to the $(\mu+1)$ EA. The upper bounds on the depth of family trees are independent of the mutation operator and even of the search space, and the lower bounds derived in the proofs of Theorem 4 and Theorem 5 hold for every selection operator choosing uniformly from individuals of the same fitness. For different selection-for-reproduction mechanisms, the concept of $1/\mu$-trees can be adapted. Nevertheless, the most interesting direction seems to be an extension to $(\mu+\lambda)$ strategies by a combination with the existing theory on the $(1+\lambda)$ EA.

Acknowledgements. Thanks to Thomas Jansen and Ingo Wegener for discussions on the proof techniques and to the anonymous referees for helpful comments.

References

1. Droste, S., Jansen, T., Wegener, I.: On the analysis of the $(1+1)$ evolutionary algorithm. Theoretical Computer Science **276** (2002) 51–81
2. Droste, S., Jansen, T., Tinnefeld, K., Wegener, I.: A new framework for the valuation of algorithms for black-box optimization. In: Proc. of Foundations of Genetic Algorithms 7 (FOGA 2002), Morgan Kaufmann (2003) 253–270
3. Garnier, J., Kallel, L., Schoenauer, M.: Rigorous hitting times for binary mutations. Evolutionary Computation **7** (1999) 173–203
4. Giel, O.: Expected runtimes of a simple multi-objective evolutionary algorithm. In: Proc. of the 2003 Congress on Evol. Computation, IEEE Press (2003) 1918–1925
5. He, J., Yao, X.: From an individual to a population: An analysis of the first hitting time of population-based evolutionary algorithms. IEEE Transactions on Evolutionary Computation **6** (2002) 495–511
6. Jansen, T., De Jong, K.: An analysis of the role of offspring population size in EAs. In: Proc. of GECCO 2002. (2002) 238–246
7. Jansen, T., Wegener, I.: Evolutionary algorithms – how to cope with plateaus of constant fitness and when to reject strings of the same fitness. IEEE Transactions on Evolutionary Computation **5** (2001a) 589–599
8. Jansen, T., Wegener, I.: On the utility of populations. In: Proc. of GECCO 2001. (2001b) 1034–1041

f-value of at most $n/2$ with probability $1 - 2^{-\Omega(n)}$. Consider the path p_v from x to v. By standard arguments from the analysis of the (1+1) EA on f (Droste, Jansen and Wegener (2002)), the bits after the leftmost zero are, in each string on p_v, uniformly distributed. W. l. o. g., the f-value is non-decreasing along p_v. Since the f-value has to increase by at least $n/2$ along p_v with probability $1 - 2^{-\Omega(n)}$, the mentioned arguments imply that at least $n/6$ different strings lie on p_v with probability $1 - 2^{-\Omega(n)}$. We call the nodes that are different strings than their fathers *subtree roots*. For a subtree root r, by $T^*(r)$ we denote the maximal MST rooted at r. Now we work under the assumption that p_v contains at least $n/6$ subtree roots.

Fix an arbitrary subtree root $r \neq v$ and the next subtree root r' on p_v. By Lemma 3, the MST $T^*(r)$ is a p-marked RT, and r' is some node inserted into (but not not attributed to) a p-marked RT. Considering the construction of $T^*(r)$, we prove that r' was likely to be created late during this process. The probability of mutating a string with value $f(r)$ to a better string is bounded above by $1/n$. Hence, with probability at least $1/2$, the first $n/2$ steps that choose a father in the already existing MST create nodes with at most the same value as the root. Since producing a replica of a string has probability $(1-1/n)^n \geq 1/(2e)$, the expected number of replicas within $n/2$ steps is at least $n/(4e)$. By Chernoff bounds, with probability at least $1/2 - 2^{-\Omega(n)}$, $T^*(r)$ receives at least $n/(8e)$ nodes before an individual with larger value than $f(r)$ is appended. Hence, with probability at least $1/2 - 2^{-\Omega(n)}$, the node r' has a distance to r that is bounded below by the depth of the at least $n/(8e)$-th node of a p-marked RT.

How deep is the k-th node such that $k \geq n/(8e)$ within a p-marked RT? We know it if the tree is an ordinary RRT. Then, by the theory on RRTs (Smythe and Mahmoud (1995)), the depth is at least $(\log n)/2$ with probability at least $1/2$ (for n large enough). By Lemma 4, the same statement holds also for a p-marked RT. Altogether, the distance of r and r' on p_v is at least $(\log n)/2$ with probability at least $1/4 - o(1)$. Since the process creating $T^*(r')$ is independent of the process creating $T^*(r)$, we can apply Chernoff bounds. Since at least $n/6$ choices for r are available on p_v, at least $n/25$ subtree roots have their successive subtree roots at distance at least $(\log n)/2$ with probability $1 - 2^{-\Omega(n)}$. Altogether, the length of p_v is at least $n(\log n)/50$ with probability $1 - 2^{-\Omega(n)}$.

Since at time s, the number of all nodes in all trees is bounded by a polynomial, the probability that there is a node with f-value n at depth less than $n(\log n)/50$ in a family tree is $2^{-\Omega(n)}$. If c is small enough, the bound $n(\log n)/50$ is at least $3cn \log n$. Finally, the sum of all failure probabilities is $2^{-\Omega(n)}$.   □

The method of lower bounding the depth of MSTs can also be used to lower bound the expected runtime of the $(\mu+1)$ EA on the function SPC. It is easy to see that, due to the plateau of constant fitness, there are with high probability even $\Omega(n^2)$ subtree roots on any path leading to an optimal node in a family tree. Hence, a straightforward application of the proof of Theorem 5 would lead to a lower bound of $\Omega(\mu n^2 \log n)$ on the expected runtime. However, one can improve on this by considering the number of alive nodes in MSTs (which a p-marked RTs and at least as deep as ordinary RRTs) more carefully. One can show that p-marked RTs become the deeper the less alive nodes they contain. Considering SPC, one can analyze the random walk describing the number of alive individuals in MSTs. As with LEADINGONES, $\Theta(n)$ expected nodes are added to an MST before the first relevant node with different color is created. Since the probability of deleting a node from and of adding a node to an MST are almost equal, we can bound the number of alive nodes before this creation by $O(n^{1/2+\varepsilon})$ with high probability. This leads to a depth of $\Omega(n^{1/2-\varepsilon})$ with probability $\Omega(1)$. One can even refine this analysis to show an $\Omega(n^{1-\varepsilon})$ bound.

Theorem 6. *Let $\mu = poly(n)$. Then the expected runtime of the $(\mu+1)$ EA on SPC is $\Omega(\mu n^{3-\varepsilon})$ for any constant $\varepsilon > 0$. Moreover, the success probability within some $c\mu n^{3-\varepsilon}$ steps, $c > 0$, is $2^{-\Omega(n^{\varepsilon/4})}$.*

6 An Example Where $\mu > 1$ Is Essential

In the previous sections, we have shown for example functions that the $(\mu+1)$ EA can only be slightly more efficient than its corresponding parallel run. Moreover, it is never more efficient than a single run of the $(1+1)$ EA on two of these functions, and it becomes less and less efficient for increasing values of μ.

However, it is believed that populations help to better explore search spaces. We can make this precise in some respect for an example function similar to that considered by Witt (2003) for a GA with fitness-proportional selection. Suppose that in a subspace $\{0,1\}^{\ell}$ of the search space, an optimal setting for LEADING-ONES is sought, while in the subspace $\{0,1\}^{n-\ell}$, the optimum for ONEMAX is sought. If ℓ is not too small, the $(1+1)$ EA normally finds the optimal setting for ONEMAX faster than for LEADINGONES. On the other hand, by the results from Sections 3–5, the expected runtime of the $(\mu+1)$ EA is $O(\mu\ell\log n + \ell n)$ for the LEADINGONES part and $\Omega(\mu(n-\ell))$ for the ONEMAX part. For $\ell = \sqrt{n}$ and $\mu = \Omega(n)$, e. g., this means that the $(\mu+1)$ EA is faster on the LEADINGONES part. This can be explained since now the subspace of the ONEMAX part is better explored but less exploited than the other subspace. If the function leads to an isolated local optimum if the ONEMAX part is optimized first, the $(1+1)$ EA is expected to behave inefficiently. Moreover, if a global optimum is reached if the LEADINGONES part is optimized first, we expect the $(\mu+1)$ EA to be efficient.

The following function has been defined according to this idea. Let strings $x \in \{0,1\}^n$ be divided into a prefix $(x_1, \ldots, x_m)$ of length m and a suffix $(x_{m+1}, \ldots, x_n)$ of length ℓ. Let $\ell := \lceil n^{1/2} \rceil$, i. e., $m = n - o(n)$. For $x \in \{0,1\}^n$, we define $PO(x) := x_1 + \cdots + x_m$ as the number of so-called **prefix** ones. Let $LSO(x) := \sum_{i=0}^{\ell-1} \prod_{j=0}^{i} x_{m+1+j}$ be the number of leading suffix ones. Finally, let $b := 2m/3 + \lceil n^{1/2}/(700\log^2 n) \rceil$. Then let

$$f(x) := \begin{cases} PO(x) + n^2 \cdot LSO(x) & \text{if } PO(x) \leq 2m/3, \\ n^2\ell - n \cdot |PO(x) - b| + LSO(x) & \text{otherwise.} \end{cases}$$

We discuss the structure of f. The first case occurs if x has few POs. Then the f-value is strongly influenced by the number of LSOs. The optimum f-value of

$n^2\ell + 2m/3$ holds if $\mathrm{LSO}(x) = \ell$ and $\mathrm{PO}(x) = 2m/3$. However, if $\mathrm{PO}(x) \leq 2m/3$ and $\mathrm{LSO}(x) < \ell$, the f-value is at most $n^2(\ell - 1) + 2m/3$, which is less than $n^2\ell - nb$, a lower bound on the value in the second case ($\mathrm{PO}(x) > 2m/3$). If $\mathrm{PO}(x) = b$ and $\mathrm{LSO}(x) = \ell$, we have a locally optimal string with f-value $n^2\ell + \ell$. The Hamming distance to any better string is $b - 2m/3 = \Omega(n^{1/2}/\log^2 n)$. In fact, the (1+1) EA is likely to get stuck here, and even multistarts do not help.

Theorem 7. *With probability* $1 - 2^{-\Omega(n^{1/2}/\log n)}$*, the runtime of the (1+1) EA on f is $2^{\Omega(n^{1/2}/\log n)}$.*

Sketch of proof. We show that the (1+1) EA is likely to create b POs before ℓ LSOs. Then it has to overcome a Hamming distance at least $b - 2m/3$ in one step to reach the optimum. This takes $2^{\Omega(n^{1/2}/\log n)}$ steps with high probability.

We estimate the probability p^* of creating ℓ LSOs before reaching b POs as follows. With high probability, $O(n)$ steps suffice to create $b = m - \Omega(n)$ POs whereas increasing the LSO-value takes $\Omega(n)$ steps with probability $\Omega(1)$. Since $\Omega(n^{3/2})$ steps are necessary for ℓ LSOs with high probability, p^* is very small. $\square$

Theorem 8. *Let $n/\ln(en) \leq \mu = poly(n)$. With probability* $1 - 2^{-\Omega(n^{1/2}/\log n)}$*, the runtime of the ($\mu$+1) EA on f is $O(\mu n^{3/2}/\log n)$. Its expectation is $O(\mu n)$.*

Sketch of proof. For the first claim, we use the idea of Theorem 1. Assume all individuals to have always at most $2m/3$ POs. Then we use the potential L, denoting the maximum number of LSOs in the population. By the definition of f, L cannot decrease, and no individual with L LSOs can be deleted if there are individuals with less LSOs. Hence, as a corollary of Theorem 1 for our choice of μ, the expected time until creating an individual with ℓ LSOs is at most $3e\mu\lceil n^{1/2}\rceil \ln(en)$. Moreover, is is easy to see that the time is $O(\mu n)$ with probability $1 - 2^{-\Omega(n^{1/2}/\log n)}$. Afterwards, there is always at least one individual with ℓ LSOs in the population. It is sufficient to reach the optimum by increasing the number of POs of such an individual to m. The expected time for this is bounded by $O(\mu n)$, and the time is $O(\mu n^{3/2}/\log n)$ with probability $1 - 2^{-\Omega(n^{1/2}/\log n)}$.

We estimate the probability that no individual ever has more than $2m/3$ POs within $s := \lfloor c\mu n\rfloor$ steps, $c > 0$ a constant, using the approach from Sect. 4. By Lemma 1, no family tree reaches a depth of at least $3cn$ with probability $1 - 2^{-\Omega(n)}$. No initial individual has at least $7m/12$ POs with probability $1 - 2^{-\Omega(n)}$. If c is chosen small enough, the probability of $\lceil 3cn\rceil$ mutations flipping a total number of at least $m/12$ bits is $2^{-\Omega(n)}$. Altogether, the probability of more than $2m/3$ POs within s steps is $2^{-\Omega(n)}$. Since the sum of all considered failure probabilities is $2^{-\Omega(n^{1/2}/\log n)}$, this proves the theorem's first statement.

For the statement on the expected runtime, we have to consider the case that an individual has more than $2m/3$ POs at some time. It is easy to see that then a locally optimal individual is created after $O(\mu n)$ expected steps. Since the Hamming distance to a locally optimal individual is bounded by $b - 2m/3 \leq n^{1/2}/(700\log^2 n) + 1$, the expected time until overcoming this distance is at most

$2^{(n^{1/2}+o(1))/(700\log n)}$. The constants have been chosen such that the product of the waiting time and the probability of reaching the local optimum is $o(1)$.   □

Conclusions

We have presented a first analysis of the $(\mu+1)$ EA for pseudo-Boolean functions by studying the expected runtime on three well-known example functions. For two of these, we have derived asymptotically tight bounds, showing that $\mu = 1$ leads asymptotically to the lowest runtime. In contrast to this, we have identified a function where the $(\mu+1)$ EA outperforms the $(1+1)$ EA and its multistart variants drastically provided that $\mu \geq n/\ln(en)$.

To prove lower bounds, we have developed a new technique. This technique is not only limited to the $(\mu+1)$ EA. The upper bounds on the depth of family trees are independent of the mutation operator and even of the search space, and the lower bounds derived in the proofs of Theorem 4 and Theorem 5 hold for every selection operator choosing uniformly from individuals of the same fitness. For different selection-for-reproduction mechanisms, the concept of $1/\mu$-trees can be adapted. Nevertheless, the most interesting direction seems to be an extension to $(\mu+\lambda)$ strategies by a combination with the existing theory on the $(1+\lambda)$ EA.

Acknowledgements. Thanks to Thomas Jansen and Ingo Wegener for discussions on the proof techniques and to the anonymous referees for helpful comments.

References

1. Droste, S., Jansen, T., Wegener, I.: On the analysis of the (1+1) evolutionary algorithm. Theoretical Computer Science **276** (2002) 51–81
2. Droste, S., Jansen, T., Tinnefeld, K., Wegener, I.: A new framework for the valuation of algorithms for black-box optimization. In: Proc. of Foundations of Genetic Algorithms 7 (FOGA 2002), Morgan Kaufmann (2003) 253–270
3. Garnier, J., Kallel, L., Schoenauer, M.: Rigorous hitting times for binary mutations. Evolutionary Computation **7** (1999) 173–203
4. Giel, O.: Expected runtimes of a simple multi-objective evolutionary algorithm. In: Proc. of the 2003 Congress on Evol. Computation, IEEE Press (2003) 1918–1925
5. He, J., Yao, X.: From an individual to a population: An analysis of the first hitting time of population-based evolutionary algorithms. IEEE Transactions on Evolutionary Computation **6** (2002) 495–511
6. Jansen, T., De Jong, K.: An analysis of the role of offspring population size in EAs. In: Proc. of GECCO 2002. (2002) 238–246
7. Jansen, T., Wegener, I.: Evolutionary algorithms – how to cope with plateaus of constant fitness and when to reject strings of the same fitness. IEEE Transactions on Evolutionary Computation **5** (2001a) 589–599
8. Jansen, T., Wegener, I.: On the utility of populations. In: Proc. of GECCO 2001. (2001b) 1034–1041

9. Jansen, T., Wegener, I.: Real royal road functions – where crossover provably is essential. In: Proc. of GECCO 2001. (2001c) 375–382
10. Motwani, R., Raghavan, P.: Randomized Algorithms. Cambr. Univ. Press (1995)
11. Pittel, B.: Note on the heights of random recursive trees and random m-ary search trees. Random Structures and Algorithms 5 (1994) 337–348
12. Rabani, Y., Rabinovich, Y., Sinclair, A.: A computational view of population genetics. Random Structures and Algorithms 12 (1998) 313–334
13. Smythe, R.T., Mahmoud, H.M.: A survey of recursive trees. Theory of Probability and Mathematical Statistics 51 (1995) 1–27
14. Storch, T., Wegener, I.: Real royal road functions for constant population size. In: Proc. of GECCO 2003. (2003) 1406–1417
15. Wegener, I., Witt, C.: On the optimization of monotone polynomials by the (1+1) EA and randomized local search. In: Proc. of GECCO 2003. (2003) 622–633
16. Witt, C.: Population size vs. runtime of a simple EA. In: Proc. of the 2003 Congress on Evol. Computation. Volume 3., IEEE Press (2003) 1996–2003

Program Evolution by Integrating EDP and GP

Kohsuke Yanai and Hitoshi Iba

Dept. of Frontier Informatics, Graduate School of Frontier Science,
The University of Tokyo.
7-3-1 Hongo, Bunkyo-ku, Tokyo 113-8654, Japan
{yanai,iba}@iba.k.u-tokyo.ac.jp

Abstract. This paper discusses the performance of a hybrid system which consists of EDP and GP. EDP, Estimation of Distribution Programming, is the program evolution method based on the probabilistic model, where the probability distribution of a program is estimated by using a Bayesian network, and a population evolves repeating estimation of distribution and program generation without crossover and mutation. Applying the hybrid system of EDP and GP to various problems, we discovered some important tendencies in the behavior of this hybrid system. The hybrid system was not only superior to pure GP in a search performance but also had interesting features in program evolution. More tests revealed how and when EDP and GP compensate for each other. We show some experimental results of program evolution by the hybrid system and discuss the characteristics of both EDP and GP.

1 Introduction

1.1 Program Evolution Using Probability Distribution

Recently, attention has been focused on evolutionary algorithms based on a probabilistic model. These are called Estimation of Distribution Algorithms (EDA) [Larranage and Lozano02] or Probabilistic Model Building Genetic Algorithms. EDA is a search method that eliminates crossover and mutation from the Genetic Algorithm (GA) and places more emphasis on the relationship between gene loci. Much research has been performed on this. However, there have been almost no researches on its application to program evolution problems (see Section 4.3).

We have proposed EDP, Estimation of Distribution Programming, based on a probability distribution expression using a Bayesian network [Yanai03a] . EDP is a population based search method and evolves a population by repeating estimation of distribution and program generation. In program evolution experiments, EDP showed different characteristics from GP and could solve GP's weak problems. On the other hand, in GP standard problems, for example, a function regression problem or a boolean problem, GP was far superior to EDP. Therefore, we built the hybrid system of GP and EDP and tried to test it in a function regression problem. If the performance of this hybrid system is worse than pure GP, we can conclude that EDP is useless in this GP standard problem. However, contrary to our expectation, experimental results indicated interesting

K. Deb et al. (Eds.): GECCO 2004, LNCS 3102, pp. 774–785, 2004.
© Springer-Verlag Berlin Heidelberg 2004

tendencies. Although pure GP was superior in younger generations, the performance of the hybrid system overtook GP on the evolution and was better in later generations.

We were interested in the robustness of this hybrid system's make-up and what causes the "overtaking." This paper discusses the performance and the characteristics of the hybrid system according to various experiments and considers GP's and EDP's defects and how GP and EDP compensate for each other.

This paper is organized as follows: Section 2 describes the algorithm of the hybrid system and the details of estimation of distribution and program generation. Section 3 indicates the performance difference due to the hybrid ratio of GP to EDP and discusses whether the "overtaking" is significant. In Section 4, we show experiments of 2 systems: a system which changes the hybrid ratio for each generation and a system which estimates distribution independent of a past state, and thoroughly analyze the systems. On the basis of these three experiments, an important conclusion about EDP's function is reached. Section 5 summarizes this paper and considers future work.

1.2   Features of EDP

From comparative experiments with GP in a max problem [Langdon02] and a boolean 6-multiplexer problem, the following characteristics of EDP are obtained [Yanai03b].

1. In a max problem, EDP is superior to GP.
2. When adding a harmful node, which is the source of introns in a max problem, EDP is far superior to GP.
3. In a boolean 6-multiplexer problem, EDP cannot search as well as GP.
4. In both, a max problem and a boolean 6-multiplexer problem, EDP can find a better solution than a random search.
5. It is expected that EDP can control introns effectively because it keeps the occurrence probability of harmful nodes low.
6. EDP has positional restriction and useful part trees cannot shift their position, while GP's crossover can move part trees to another position in the tree.

The 6th point is EDP's critical defect and makes its performance low in a boolean 6-multiplexer problem. A radical improvement is under consideration in order to eliminate this defect. The hybrid system introduced in the next Section, which is an easy extension, can overcome this difficulty. In brief, it leaves the shifting of part trees to GP's crossover.

2   Hybrid System of EDP and GP

2.1   Algorithm of Hybrid System

In this Section, we explain our hybrid system which consists of EDP and GP. This hybrid system carries out a search using the following procedure:

Step 1 Initialize a population.
Step 2 Evaluate individuals and assign fitness values.
Step 3 If a termination criterion is satisfied, then go to Step 9.
Step 4 Estimate the probability distribution.
Step 5 Use the elitist strategy.
Step 6 Generate new $rM - E_S$ individuals with GP operator.
Step 7 Generate new $(1 - r)M$ individuals with EDP operator.
Step 8 Replace the population and go to Step 2.
Step 9 Report the best individual.

In Step 1, according to function node generation probability P_F and terminal node generation probability P_T $(= 1 - P_F)$, initial M individuals are generated randomly, where M is the population size. However, if tree size limitation is reached, terminal nodes are generated. For example, the probabilities of function node "+" and terminal node "x" are given:

$$\text{if tree size limitation is not reached,} \tag{1}$$

$$\begin{cases} P(X = " + ") & = P_F \times \frac{1}{N_F} \\ P(X = "x") & = P_T \times \frac{1}{N_T} \end{cases} \tag{2}$$

$$\text{if tree size limitation is reached,} \tag{3}$$

$$\begin{cases} P(X = " + ") & = 0 \\ P(X = "x") & = \frac{1}{N_T} \end{cases} \tag{4}$$

where N_F is the number of function nodes and N_T is the number of terminal nodes.

Next, each individual in the current population is evaluated by a fitness function and assigned its fitness value (Step 2). If a termination criterion is met, then go to Step 9. Usually a termination criterion is a previously specified maximum number of generations (Step 3).

In Step 4, superior individuals with high fitness values are selected within sampling size S_S, and a new distribution is estimated based on those selected individuals (see Section 2.3). We use the elitist strategy in Step 5, i.e., elite E_S individuals are selected from the population in the order of fitness superiority and copied to the new population, where E_S is the elite size.

In Step 6, nearly $100r\%$ $(0 \leq r \leq 1)$ of the population, precisely $rM - E_S$ individuals, is generated by standard GP operators: crossover and mutation. It selects superior individuals of GP operator's target by tournament selection with tournament size T_{gp} and performs mutation with the mutation probability P_M or crossover with the probability $1 - P_M$. Note that mutation and crossover which violate tree size limitation are not performed, and generated individuals are under tree size limitation.

Then in Step 7, the remaining $100(1-r)\%$ of the population, that is $(1-r)M$ individuals, is generated by using a newly acquired distribution (see Section 2.4). This new distribution is considered better than the previous one because it samples superior individuals in the population.

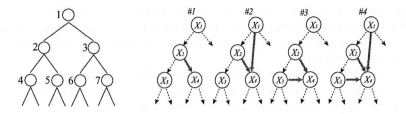

Fig. 1. Program tree.   **Fig. 2.** Efficient network topology.

This process is repeated until a termination criterion is met. Finally in Step 9, the best individual is reported as the solution to the problem.

r is the most important parameter, it decides the system behavior and the ratio of GP to EDP in an individual generation, called the hybrid ratio. Through the combination of EDP and GP, the difficulty indicated in Section 1.2 might be overcome. However, it is not obvious whether GP gains anything from hybridization. In Section 3, we test the system performance in a function regression problem changing the hybrid ratio r from 0 to 1.

2.2 Distribution Model

We use a Bayesian network as the distribution model of programs. Values of probabilistic variables are symbols for each node in the program tree. Assign the index numbers to each node of evolving programs as in Fig. 1, the range of probabilistic variable X_i is the symbols of node i, that is, $X_i \in T \cup F$, where F is the function node set, T is the terminal node set.

For instance, assume $F = \{+, -, *, /\}$ and $T = \{x_1, x_2\}$,

$$P(X_5 = "+" | X_2 = "/") = \frac{2}{7} \qquad (5)$$

means that the conditional probability that node 5 becomes " $+$ " is $\frac{2}{7}$ if node 2 is " $/$ ". C_i is the set of probabilistic variables which X_i is dependent on. In the former example, $C_5 = \{X_2\}$.

Although there are several efficient topologies of a Bayesian network as indicated in Fig. 2, the simplest one, that is, #1 in Fig. 2, is used for our experiments. The topology of a Bayesian network is tree-like and it is the same as the program's topology.

2.3 Estimation of Distribution

The probability distribution is updated incrementally [Baluja94] as follows:

$$P_{t+1}(X_i = x | C_i = c) = (1 - \eta)\hat{P}(X_i = x | C_i = c) + \eta P_t(X_i = x | C_i = c) \qquad (6)$$

where $P_t(X_i = x|C_i = c)$ is the distribution of the tth generation and $\hat{P}(X_i = x|C_i = c)$ is the distribution estimated based on superior individuals in the $(t+1)^{th}$ population, η is the learning rate which means dependence degree on the previous generation.

$\hat{P}(X_i = x|C_i = c)$ is estimated as follows. At first, S_S individuals are sampled by tournament selection with tournament size T_{edp}, and maximum likelihood estimation is performed based on these selected individuals. Therefore,

$$\hat{P}(X_i = x|C_i = c) = \frac{\sum_{j=1}^{S_S} \delta(j, X_i = x, C_i = c)}{\sum_{j=1}^{S_S} \sum_{x \in FUT} \delta(j, X_i = x, C_i = c)}, \tag{7}$$

where

$$\delta(j, X_i = x, C_i = c) = \begin{cases} 1 & \text{if } X_i = x \text{ and } C_i = c \\ & \text{at the individual } j \ . \\ 0 & \text{else} \end{cases} \tag{8}$$

2.4   Program Generation

At first, the acquired distribution $P_t(X_i = x|C_i = c)$ is modified like Laplace correction [Cestnik90] by

$$P_t'(X_i = x|C_i = c) = (1 - \alpha)P_t(X_i = x|C_i = c) + \alpha P_{bias}(X_i = x|C_i = c), \tag{9}$$

where α is a constant that expresses the Laplace correction rate, $P_{bias}(X_i = x|C_i = c)$ is the probability to bias distribution. This modification makes all occurrence probabilities of node symbols positive. Next, according to $P_t'(X_i = x|C_i = c)$, node symbols are decided in sequence from root to terminals.

2.5   Parameter Control

Table 1 indicates the parameters used for experiments.

3   Performance Difference Due to the Hybrid Ratio

3.1   Function Regression Problem

Consider a function regression problem. $prog_i$ is a function expressed by a program tree and f_{obj} is the function to be approximated. The fitness value is given with the following formula:

$$fitness = 1000 - 50 \sum_{j=1}^{30} |prog(X_j) - f_{obj}(X_j)|, \tag{10}$$

where

$$X_j = 0.2(j - 1). \tag{11}$$

Table 1. Parameters for a function regression problem.

Common parameters for EDP and GP	
M: population size	1000
E_S: elite size	5
F: function node set	$\{+, -, *, /, \cos, \sin\}$
T: terminal node set	$\{x, 0.05, 0.10, 0.15, \cdots, 1.00\}$
N_F: the number of function nodes	6
N_T: the number of terminal nodes	21
P_F: generation probability of function node	0.8
P_T: generation probability of terminal node	0.2
Tree size limitation in initializing population	max depth = 6
EDP parameters	
α: Laplace correction rate	0.3
$P_{bias}(X_i = x \mid C_i = c)$: the probability to bias distribution	$\frac{1}{N_F + N_T}$
η: learning rate	0.2
S_S: sampling size	200
T_{edp}: tournament size for sampling	20
Tree size limitation	max depth = 6
GP parameters	
P_M: mutation probability	0.1
T_{gp}: tournament size for GP operator	5
Tree size limitation	max depth = 6

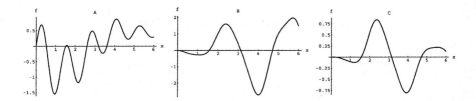

Fig. 3. Objective functions.

Objective functions are

$$A : f_{obj}(x) = (2 - 0.3x)\sin(2x)\cos(3x) + 0.01x^2 \tag{12}$$

$$B : f_{obj}(x) = x\cos(x)\sin(x)(\sin^2(x)\cos(x) - 1) \tag{13}$$

$$C : f_{obj}(x) = x^3\cos(x)\sin(x)e^{-x}(\sin^2(x)\cos(x) - 1) \tag{14}$$

which are plotted in Fig. 3. Objective function C is cited from [Salustowicz97].
Although B is obtained from simplification of C, B is more difficult to search (see
fitness values in Fig. 5 and 6). A is our original function and the most difficult
of the three objective functions.

Fig. 4, 5, and 6 show the mean of max fitness values for 100 runs, that is,

$$\bar{f}_{max}m = \frac{1}{100}\sum_{k=1}^{100} f_{max}k,m \tag{15}$$

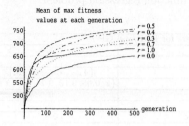
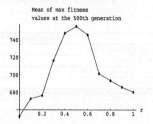

Fig. 4. Results for objective function A.

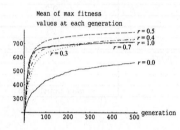
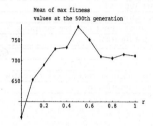

Fig. 5. Results for objective function B.

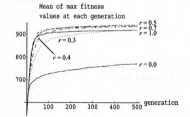
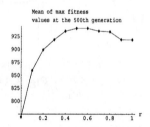

Fig. 6. Results for objective function C.

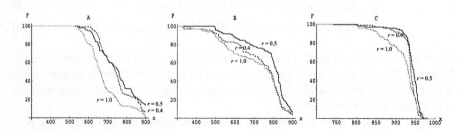

Fig. 7. $F(x)$: frequency of max fitness at the 500 th generation greater than x, with objective functions A and B.

where $f_{max k,m}$ is the maximum fitness value in a population of the mth generation at the kth run. Note that $\bar{f}_{max m}$ is not a mean fitness value of a population, but a mean value of the maximum fitness value $f_{max k,m}$. The solution in an evolutionary computing is given by an individual who has the maximum fitness value in a population. Therefore, system performances should be compared in maximum fitness values.

Fig. 7 shows the frequency of runs in which the maximum fitness value at the 500th generation is over x, that is,

$$F(x) = \sum_{k=1}^{100} \delta(x \leq f_{max k,500}) \tag{16}$$

where

$$\delta(x \leq a) = \begin{cases} 1 & : x \leq a \\ 0 & : x > a \end{cases}. \tag{17}$$

Fig. 4, 5, 6, and 7 indicate the similar tendency in each case. Although the $r = 1.0$ system which is pure GP, demonstrated the best performance in younger generations, gradually hybrid systems overtook pure GP one after another. The "overtaking" was conspicuous when $r = 0.3$ or $r = 0.4$. At the 500th generation, the performance of the $r = 0.5$ system was the best in all cases. The system performances at the 500th generation reached a peak at $r = 0.5$, and got worse as the hybrid ratio was biased.

Mean cannot give adequate information for system performances, hence we showed Fig. 7. Fig. 7 demonstrates that the hybrid system is also superior to pure GP in the success rate of a search. For instance, in the case of A, the probabilities that the maximum fitness value at the 500th generation is over 700 are $\frac{63}{100}$ with $r = 0.5$ and $\frac{30}{100}$ with pure GP respectively.

3.2 Analysis of the Results

The system performances are estimated by $\bar{f}_{max m}$. However, in order to conclude that the differences of these values are statistically significant and reliable, not only mean but also standard deviation and sample size ($= 100$) should be taken into consideration. We used Welch's test for the obtained experimental results. By means of Welch's test, it can be judged whether 2 data sets are samples from the same statistical population or not. As a result of Welch's test with 10% significance level, the differences between the $r = 0.5$ system and pure GP at the 500th generation were significant in all cases. Statistically speaking, the null hypothesis that data in the $r = 0.5$ system and in pure GP were sampled from the same statistical population was rejected (the probability that the null hypothesis is correct is less than 10%). In the case of objective function C, although the difference in values was slight, standard deviation was negligible (see Fig. 7); Welch's test concluded that the differences were significant.

In the $r = 0.5$ hybrid system, the updating times of the maximum fitness values at each generation of the EDP operator and the GP operator are counted respectively. Surprisingly, the EDP operator hardly contributes to construction of the best individual directly, and only the GP operator does.

The summary of results is as follows:

1. The success probability of the hybrid system in a search is higher.
2. Statistical testing proved that the $r = 0.5$ system was superior to pure GP $(r = 1.0)$ at the 500th generation.
3. In any case, the same tendencies, the "overtaking", pure GP's superiority in younger generations and so on, were found.
4. Pure EDP was worse.
5. The obtained graphs were consistent and well formed.
6. The EDP operator could not produce better individuals, but played some invisible roles.

We consider these features of the hybrid system to be universal. In other words, the parameter r characterizes the system behavior and the performance. Besides, hybridization helps GP and EDP compensate for their defects, and build a better evolutionary system.

Here are some follow-up questions:

1. Why is the hybrid system superior to pure GP? What are EDP's roles?
2. Is the low performance in younger generations important?
3. How should r be controlled? What method is the best?

The next section will answer some of these.

4   Discussion

4.1   Change of Hybrid Ratio at Each Generation

This section investigates the hybrid system's performance, changing the hybrid ratio r at each generation. In Fig. 4, until the 50th generation, the higher the GP ratio of the system is, the better its performance is. Therefore, the system that has a high GP ratio in younger generations and decreases the ratio later is expected to have higher performance.

Comparative experiments were carried out in 8 systems, shown in Fig. 8. Objective function is A given in the formula (12). In system D, the GP ratio is linearly increased from 0, at the 0th generation, to 1.0, at the 500th generation. On the other hand, the system E decreases the ratio linearly. System G switches the ratio from 1.0 to 0.3 at the 205th generation because the $r = 0.3$ system overtook pure GP at the 205th generation, as shown in Fig. 4. System H was prepared in the same manner as G. Therefore, H and G are the top favorites in these systems.

Fig. 9 and 10 show the results of comparative experiments. Surprisingly, system A overtook G (B also overtook H). As a result of Welch's test with

System	r
A: classical hybrid	$r = 0.3$
B: classical hybrid	$r = 0.5$
C: pure GP	$r = 1.0$
D: linear increasing	$r = \dfrac{i}{500}$
E: linear decreasing	$r = 1 - \dfrac{i}{500}$
F: random	r is a random value from 0 to 1 and different for each generation.
G: switching	$r = \begin{cases} 1.0 & : i < 205 \\ 0.3 & : i \geq 205 \end{cases}$
H: switching	$r = \begin{cases} 1.0 & : i < 40 \\ 0.5 & : i \geq 40 \end{cases}$

Fig. 8. Systems with changing r, where i is the generation number.

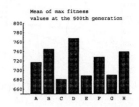

Fig. 9. Mean of max fitness values at the 500th generation.

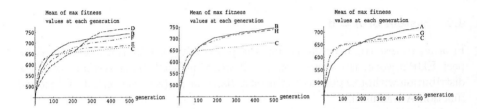

Fig. 10. Mean of max fitness values at each generation.

10% significance level, the differences were significant. This result means that population states of A and G are far different in spite of close performance at the 205th generation. In other words, EDP's behavior before the 205th generation likely has a good influence later.

Another interesting result is that system D was superior to all other systems, especially E. As a result of Welch's test with 10% significance level, the differences were significant. Although it was expected that D would be worse than E, judging from Fig. 4, the result was quite the opposite. This point is evidence that EDP functions well in early generations.

How does the hybrid system transmit EDP's work in an early stage of evolution to posterity?

1. The probability distribution (Bayesian network) learned incrementally memorizes the past population state.
2. With EDP, the diversity of the population is maintained at each generation and useful part structures can survive.

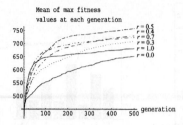
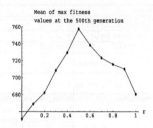

Fig. 11. System of $\eta = 0$

3. There is diversity inside individuals. Individuals constructed by EDP have more multifarious part structures. These various structures are put together in later generations of evolution.

The next section considers these possibilities.

4.2   System of $\eta = 0$

In order to test the hypothesis that the probability distribution memorizes the past EDP's work, the system of $\eta = 0$ was simulated. This system estimates distribution without referring to the past distribution (see Section 2.3). Objective function A was used.

As indicated in Fig. 11, the characteristic of the hybrid system was kept. The "overtaking" still took place and the $r = 0.5$ system was the best. Therefore, the past information accumulated in the probability distribution does not cause the high performance of the hybrid system.

The result shown in Fig. 9 suggests the third possibility mentioned in Section 4.1. This is because system D, which has the best performance of all, cannot benefit from EDP in later generations. However, in order to obtain more reliable evidence, we are currently working on testing the second possibility.

4.3   Related Work

Probabilistic Incremental Program Evolution (PIPE) [Salustowicz97] was used to perform a program search based on a probabilistic model. However, PIPE assumes the independence of program nodes and differs from our approach using a Bayesian network in this assumption. The merits of having probabilistic dependency relationship are as follows:

1. Because an occurrence probability of a node symbol is dependent on its parent node, estimation and generation are serial from a parent node to a child. Therefore, it can derive and generate building blocks.
2. The past dominant structure can survive after switching the probability distribution based on a parent node symbol.

On the other hand, optimization using a Bayesian network is much researched. [Larranaga et al.00a] [Larranaga et al.00b]. However, their application is limited to fixed length array search problems.

5   Conclusion

In this paper, we proposed a hybrid system of EDP and GP, and demonstrated that the hybrid system was superior to both pure GP and pure EDP. The experimental results indicated that EDP worked effectively in early generations and contributed to later high performance. It turned out that pure GP could not generate enough various part trees in early generations to build excellent individuals. On the other hand, EDP cannot shift useful part trees to another position in the tree. Hybridization helps EDP and GP compensate for each other.

However, it is not clear how EDP works in the hybrid system. In future work, the detail of EDP's function in early generations will be researched. We are also interested in the greatest control of the hybrid ratio r and the robustness of the behavior that the hybrid system exposed in our experiments.

References

[Salustowicz97] Rafal Salustowicz and Jurgen Schmidhuber (1997) "Probabilistic Incremental Program Evolution," Evolutionary Computation 5(2):123-141.

[Baluja94] Baluja S. (1994) "Population Based Incremental Learning: A Method for Integrating Genetic Search Based Function Optimization and Competitive Learning," Technical Report No. CMU-CS-94-163, Carnegie Mellon University, Pittsburgh, Pennsylvania.

[Larranage and Lozano02] Pedro Larranage and Jose A. Lozano (2002) "Estimation of Distribution Algorithms," Kluwer Academic Publishers

[Larranaga et al.00a] Larranaga, P., Etxeberria, R. Lozano, J. A. and Pena, J.M. (2000) "Combinatorial Optimization by Learning and Simulation of Bayesian Networks," Proceedings of the Sixteenth Conference on Uncertainty in Artificial Intelligence, Stanford, pp343-352.

[Larranaga et al.00b] Larranaga, P., Etxeberria, R. Lozano, J. A. and Pena, J.M. (2000) "Optimization in continuous domains by Learning and simulation of Gaussian networks," Proceedings of the 2000 Genetic and Evolutionary Computation Conference Workshop Program, pp201-204.

[Cestnik90] Cestnik, B. (1990) "Estimating probabilities: A Crucial Task in Machine Learning," Proceedings of the European Conference in Artificial Intelligence, pp.147-149.

[Langdon02] William B. Langdon, Riccardo Poli (2002) "Foundations of Genetic Programming," Springer-Verlag Berlin Heidelberg, pp175-176.

[Yanai03a] Kohsuke Yanai and Hitoshi Iba (2003) "Estimation of Distribution Programming based on Bayesian Network," In Proc. of Congress on Evolutionary Computation (CEC) 2003, pp1618-1625.

[Yanai03b] Kohsuke Yanai and Hitoshi Iba (2003) "Program Evolution using Bayesian Network," In Proc. of The First Asian-Pacific Workshop on Genetic Programming (ASPGP03), pp16-23.

A Step Size Preserving Directed Mutation Operator

Stefan Berlik

Universität Dortmund, Computer Science, Chair I,
44221 Dortmund, Germany
Stefan.Berlik@Uni-Dortmund.de
http://ls1-www.cs.uni-dortmund.de/

Abstract. Using a directed mutation can improve the efficiency of processing many optimization problems. The first mutation operators of this kind proposed by Hildebrand [1], however, suffer from the asymmetry parameter influencing the mutation step size. Extreme asymmetry can lead to infinite step size. The operator presented here overcomes this drawback and preserves the step size.

The main idea of the directed mutation is to focus on mutating into the most beneficial direction by using a customizable asymmetrical distribution. In this way the optimization strategy can adopt the most promising mutation direction over the generations. It thus becomes nearly as flexible as with Schwefel's correlated mutation [2] but causes only linear growth of the strategy parameters instead of quadratic growth.

A normalization function is introduced to decouple asymmetry from the variance, i.e. the step size. By incorporating the normalization function the variance becomes independent of the asymmetry parameter. Given below are the definitions of the density function for the normalized directed mutation and its normalization function:

$$
f_{\sigma,a}(x) = \begin{cases}
\sqrt{\dfrac{2}{\pi}} \dfrac{\sqrt{1-a}}{\left(1+\sqrt{1-a}\right)\sigma_{\text{norm}}(a)\sigma} e^{-\frac{x^2}{2\left(\sigma_{\text{norm}}(a)\sigma\right)^2}} & \text{for } a \leq 0,\, x \leq 0 \\[4mm]
\sqrt{\dfrac{2}{\pi}} \dfrac{\sqrt{1-a}}{\left(1+\sqrt{1-a}\right)\sigma_{\text{norm}}(a)\sigma} e^{-\frac{(1-a)x^2}{2\left(\sigma_{\text{norm}}(a)\sigma\right)^2}} & \text{for } a \leq 0,\, x > 0 \\[4mm]
\sqrt{\dfrac{2}{\pi}} \dfrac{\sqrt{1+a}}{\left(1+\sqrt{1+a}\right)\sigma_{\text{norm}}(a)\sigma} e^{-\frac{(1+a)x^2}{2\left(\sigma_{\text{norm}}(a)\sigma\right)^2}} & \text{for } a > 0,\, x \leq 0 \\[4mm]
\sqrt{\dfrac{2}{\pi}} \dfrac{\sqrt{1+a}}{\left(1+\sqrt{1+a}\right)\sigma_{\text{norm}}(a)\sigma} e^{-\frac{x^2}{2\left(\sigma_{\text{norm}}(a)\sigma\right)^2}} & \text{for } a > 0,\, x > 0
\end{cases}
\tag{1}
$$

$$
\sigma_{\text{norm}}(a) = \sqrt{\frac{\pi\left(1+|a|\right)}{4\left(\sqrt{1+|a|}-1\right)+|a|(\pi-2)+\pi\left(2-\sqrt{1+|a|}\right)}}.
\tag{2}
$$

Formulas for the expected value and variance of a random variable X distributed according to the normalized asymmetrical distribution take the following form:

K. Deb et al. (Eds.): GECCO 2004, LNCS 3102, pp. 786–787, 2004.
© Springer-Verlag Berlin Heidelberg 2004

$$E(X) = \sqrt{\frac{2}{\pi}} \frac{a \sigma_{\text{norm}}(a) \sigma}{1 + |a| + \sqrt{1 + |a|}}, \qquad V(X) = \sigma^2.$$

(3)

To get a notion of this distribution the following figure shows some graphs of the density function and distribution for different asymmetry settings.

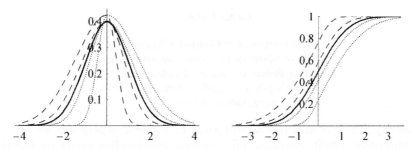

Fig. 1. Density function and distribution of the *normalized asymmetrical mutation* for $\sigma = 1$. Asymmetry parameters: $a = -10$, $a = -1$ *(dashed)*; $a = 0$ *(solid)*; $a = 1$, $a = 10$ *(dotted)*.

Random numbers distributed according to the normalized asymmetrical distribution can be generated by multiplying its inverse function with uniformly distributed random numbers. The inverse function is defined by

$$\overline{F}_{\sigma,a}(y) = \begin{cases} \sqrt{2}\,\sigma_{\text{norm}}(a)\,\sigma\,\text{inverf}\!\left(y\left(1+\dfrac{1}{\sqrt{1-a}}\right)-1\right) & \text{for } a \le 0,\ y \le \dfrac{\sqrt{1-a}}{1+\sqrt{1-a}} \\[2ex] \dfrac{\sqrt{2}\,\sigma_{\text{norm}}(a)\,\sigma}{\sqrt{1-a}}\,\text{inverf}\!\left(y\left(1+\sqrt{1-a}\right)-\sqrt{1-a}\right) & \text{for } a \le 0,\ y > \dfrac{\sqrt{1-a}}{1+\sqrt{1-a}} \\[2ex] \dfrac{\sqrt{2}\,\sigma_{\text{norm}}(a)\,\sigma}{\sqrt{1+a}}\,\text{inverf}\!\left(y\left(1+\sqrt{1+a}\right)-1\right) & \text{for } a > 0,\ y \le \dfrac{1}{1+\sqrt{1+a}} \\[2ex] \sqrt{2}\,\sigma_{\text{norm}}(a)\,\sigma\,\text{inverf}\!\left(y\left(1+\dfrac{1}{\sqrt{1+a}}\right)-\dfrac{1}{\sqrt{1+a}}\right) & \text{for } a > 0,\ y > \dfrac{1}{1+\sqrt{1+a}}. \end{cases}$$

(4)

Using the normalized directed mutation has shown to be very effective in optimizing test functions, as well as real world problems [3]. Taking into account that the application of the operator itself is quite fast, e.g. compared to the correlated mutation, the use of the directed mutation might be quite beneficial for many problems.

References

1. Hildebrand, L.: Asymmetrische Evolutionsstrategien. PhD thesis, Department of Computer Science, Universität Dortmund (2002)
2. Schwefel, H.-P.: Evolution and Optimum Seeking. John Wiley & Sons, New York (1994)
3. Berlik, S.: A Polymorphic Mutation Operator for Evolution Strategies. In: Proc. of the 3rd Int. Conf. in Fuzzy Logic and Technology, EUSFLAT'03, Zittau, Germany (2003)

A Comparison of Several Algorithms and Representations for Single Objective Optimization

Crina Grosan

Department of Computer Science
Faculty of Mathematics and Computer Science
Babeş-Bolyai University, Kogălniceanu 1
Cluj-Napoca, 3400, Romania.
cgrosan@cs.ubbcluj.ro

In this paper we perform two experiments. In the first experiment we analyze the convergence ability to using different base for encoding solutions. For this purpose we use the bases 2 to 16. We apply the same algorithm (with the same basic parameters) for all considered bases of representation and for all considered test functions. The algorithm is an (1+1) ES. In the second experiment we will perform a comparison between three algorithms which use different bases for solution representation. Each of these algorithms uses a dynamic representation of the solutions in the sense that the representation is not fixed and is changed during the search process. The difference between these algorithms consists in the technique adopted for changing the base over which the solution is represented. These algorithms are: Adaptive Representation Evolutionary Algorithms (AREA) [1], Dynamic Representation Evolution Strategy (DRES) and Seasonal Model Evolution Strategy (SMES) [2].

AREA change the alphabet if the number of successive harmful mutations for an individual exceeds a prescribed threshold. In DRES algorithm the base is changed at the end of each generation with a fixed probability. In SMES algorithm the base in which solution is encoded is changed after a fixed (specified) number of generations.

Test functions used in these experiments are are well known benchmarking problems ([3]): Ackley's function (f_1), Griewangk's function (f_2), Michalewicz function (f_3), Rosenbrock's function (f_4), Rastrigin's function(f_5) and Schwefel's function (f_6).

The essential role of these experiments is to show that using only one base for solution encoding (without change it during the search process) there are cases when the optimum cannot be found. Changing the representation base provides a new way of searching through the solution space. The second experiment show us which technique used for changing the base is suitable.

The number of space dimension was set to 30 for each test function. Each algorithm is run 100 times for each test function in each experiment and with any considered parameters.

In first experiment for test functions f_1, f_2 and f_4 the best results are obtained using binary encoding.

For test functions f_3, f_5 and f_6 the best result is obtained by encoding solutions in the base 4.

K. Deb et al. (Eds.): GECCO 2004, LNCS 3102, pp. 788–789, 2004.
© Springer-Verlag Berlin Heidelberg 2004

In second experiment a comparison between AREA DRES and SMES is performed. The difference between these algorithms consists in the strategy used for changing the current base in which the solutions is encoded with another one. All considered algorithms used a population with a single individual. Parameters used for AREA are: *Number of alphabets* = 31; MAX_HARMFUL_MUTATION = 50 and *Number of mutation / chromosome* = 2. DRES and SMES use the same parameters as AREA uses. The probability of changing an alphabet in DRES is 0.02. The number of generations after SMES changes the alphabet is 50. The results obtained by these three algorithms are presented in Table 1.

Table 1. Results obtained by AREA, DRES and SMES for test functions f_1-f_6

Function	Mean best		
	AREA	**DRES**	**SMES**
f_1	1.6510	2.3978	2.1838
f_2	0.6328	0.7085	0.8307
f_3	-26.803	-26.6668	-26.949
f_4	146.756	161.53	156.776
f_5	8.1164	9.3483	10.2461
f_6	-11894.6	-11986.1	-11956.9

However, AREA significantly outperforms the standard evolutionary algorithms on the well-known difficult (multimodal) test functions. This advantage of AREA makes it very suitable for real-world applications where we have to deal with highly multi-modal functions. Had only one base been used for solution encoding the gain of AREA over standard ES would have been minimal. Thus, the AREA individuals use a dynamic system of alphabets that may be changed during (and without halting) the search process. If an individual gets stuck in a local optimum - from where it is not able to "jump"- , the individual representation is changed, hoping that this new representation will help the individual to escape from the current position and to explore farther and more efficiently the search space.

References

1. Grosan C., Oltean M.: Adaptive Representation Evolutionary Algorithm – a new technique for single objective optimization. In Proceedings of First Balcanic Conference in Informatics (BCI), Thessaloniki, Greece (2003) 345-355
2. Kingdon J, Dekker L. "The Shape of Space", Proceedings of the First IEE/IEEE International Conference on Genetic Algorithms in Engineering Systems: Innovations and Applications (GALESIA '95) IEE, London, (1995) 543-548
3. Yao, X., Liu, Y., lin, G.: Evolutionary programming made faster. IEEE Transaction on Evolutionary Computation, Vol. 3(2) (1999) 82-102

Towards a Generally Applicable Self-Adapting Hybridization of Evolutionary Algorithms

Wilfried Jakob[1], Christian Blume[2], and Georg Bretthauer[1]

[1] Forschungszentrum Karlsruhe, Institute for Applied Computer Science, Postfach 3640,
76021 Karlsruhe, Germany
{wilfried.jakob, georg.bretthauer}@iai.fzk.de

[2] University of Applied Sciences, Cologne, Campus Gummersbach, Am Sandberg 1,
51643 Gummersbach, Germany
blume@gm.fh-koeln.de

Abstract. Practical applications of Evolutionary Algorithms (EA) frequently use some sort of hybridization by incorporating domain-specific knowledge, which turns the generally applicable EA into a problem-specific tool. To overcome this limitation, the new method of HyGLEAM was developed and tested extensively using eight test functions and three real-world applications. One basic kind of hybridization turned out to be superior and the number of evaluations was reduced by a factor of up to 100.

1   Introduction

When applied to real-world problems, the powerful optimization tool of Evolutionary Algorithms frequently turns out to be too time-consuming due to elaborate fitness calculations that are often based on run-time-intensive simulations. Incorporating domain-specific knowledge by problem-tailored heuristics or local searchers is a commonly used solution, but turns the generally applicable EA into a problem-specific tool. The new method of hybridization implemented in HyGLEAM (Hybrid GeneraL purpose Evolutionary Algorithm and Method) [1, 2] is aimed at overcoming this limitation and getting the best of both algorithm classes: a *fast*, *global searching* and *robust* procedure with the *convergence reliability* of evolutionary search being maintained. The basic idea of the concept can be summarized in two points:

1. Usage of generally applicable local search algorithms instead of the commonly used problem-specific ones for hybridization.
2. Usage of a convergence-dependent control mechanism for distributing the computational power between the basic algorithms for suitable kinds of hybridization.

The first point may appear simple, but it is a matter of fact that nearly all real-world applications and investigations are based on problem-specific local searchers. Appropriate local search algorithms for parameter optimization must be derivative-free and able to handle restrictions in order to be generally applicable. The Rosenbrock procedure and the Complex algorithm, two well-known powerful local searchers [3], were chosen, as they fulfill these requirements. GLEAM (General Learning Evolutionary Algorithm and Method) [4] was used as an EA, but it must be noted that the method can be applied easily to every other population-based EA.

K. Deb et al. (Eds.): GECCO 2004, LNCS 3102, pp. 790–791, 2004.
© Springer-Verlag Berlin Heidelberg 2004

2   Experiments and Conclusions

The test cases comprised real, integer, and mixed parameter optimization, combinatorial and multi-objective optimization as well as parameter strings of dynamic length. They are described in more detail together with references in [2, 5]. In most cases, the results were based on an average of 100 runs per algorithm and parameterization. Four basic kinds of hybridization were investigated:

1. Pre-optimization of the start population: The idea is that the evolution can start with solutions of more or less good quality. It works pretty well (up to 24 times less evaluations) in some cases, but not always and more evaluations may be required.

2. Post-optimization of the EA results: As EAs are known to converge slowly, an improvement may result from stopping the evolution after approaching the area of attraction of the (global) optimum and leaving the rest to the local search. The appropriate switching point is determined by the convergence-dependent control procedure mentioned above. This approach improves the EA results, but does not fulfill the expectation of reliably finding the solution.

3. Direct integration: Optimizing every or the best offspring of one mating only causes the EA to operate over the peaks of the fitness landscape exclusively rather than to treat the valleys and slopes, too. The offspring's genotype can be updated (Lamarckian evolution) or left unchanged (Baldwinian evolution). This works with the Rosenbrock procedure in all cases, yielding up to 77 times less evaluations. Using the Complex procedure instead does not always work, but if it does, better results may be obtained (up to 104 times less evaluations). Lamarckian evolution and the improvement of the best offspring of one mating only proved to be the best choice in almost all cases.

4. Delayed direct integration: This variant of direct integration, where the evolution works on its own until a certain convergence of the population is reached, produced better results in some cases (e.g. up to 90 times less evaluations instead of 77).

As no common settings for important strategy parameters like population size, termination threshold of the Rosenbrock procedure or the choice of the local searcher for the (delayed) direct integration could be extracted from the experiments, a new concept of an *adaptive direct integration* has been developed. It is described in [5] and will be subject of future work.

References

1. Jakob, W.: HyGLEAM – An Approach to Generally Applicable Hybridization of Evolutionary Algorithms. In: Merelo, J.J., et al. (eds): Conf. Proc. PPSN VII. LNCS 2439, Springer Verlag, Berlin (2002) 527-536
2. Jakob, W.: Eine neue Methodik zur Erhöhung der Leistungsfähigkeit Evolutionärer Algorithmen durch die Integration lokaler Suchverfahren. Doctoral thesis, FZKA 6965, University of Karlsruhe (in German) (2004), see also: www.iai.fzk.de/~jakob/HyGLEAM/
3. Schwefel, H.-P.: Evolution and Optimum Seeking. John Wiley & Sons, New York (1995)
4. Blume, C.: GLEAM - A System for Intuitive Learning. In: Schwefel, H.P., Männer, R. (eds): Conf. Proc. of PPSN I. LNCS 496, Springer Verlag, Berlin (1990) 48-54
5. Jakob, W., Blume, C., Bretthauer, G.: Towards a Generally Applicable Self-Adapting Hybridization of Evolutionary Algorithms. In: Deb. K. (ed): GECCO -2004, Vol. Late Breaking Papers (2004)

High Temperature Experiments for Circuit Self-Recovery

Didier Keymeulen, Ricardo Zebulum, Vu Duong, Xin Guo[*], Ian Ferguson, and
Adrian Stoica

Jet Propulsion Laboratory   4800 Oak Grove Drive,
Pasadena, CA 91109, USA
didier.keymeulen@jpl.nasa.gov

Abstract. Temperature and radiation tolerant electronics, as well as long life
survivability are key capabilities required for future NASA missions. Current
approaches to electronics for extreme environments focus on component level
robustness and hardening. Compensation techniques such as bias cancellation
circuitry have also been employed. However, current technology can only
ensure very limited lifetime in extreme environments. This paper presents a
novel approach, based on evolvable hardware technology, which allows
adaptive in-situ circuit redesign/reconfiguration during operation in extreme
environments. This technology would complement material/device
advancements and increase the mission capability to survive harsh
environments. The approach is demonstrated on a mixed-signal programmable
chip, which recovers functionality until 280°C. We show in this paper the
functionality recovery at high temperatures for a variety of circuits, including
rectifiers, amplifiers and filters.

1   Introduction

In-situ planetary exploration requires extreme-temperature electronics able to operate
in low temperatures, such as below −220°C on Neptune (-235°C for Triton and Pluto)
or high temperatures, such as above 470°C as needed for operation on the surface of
Venus. Extrapolations of current developments indicate that hot electronics
technology for >400°C environments may not be ready in time for the 2006-2007
missions, except possibly for "grab-and-go" or "limited life" operations [1]. For
extended missions, innovative approaches are needed. Terrestrial applications include
combustion systems, well logging, nuclear reactors and dense electronic packages.

The maximum working temperature for semiconductors can be estimated from
their intrinsic carrier density, which depends on the band-gap of the material. When
the intrinsic density reaches the doping level of the devices, electrical parameters are
expected to change drastically [2]. For the high-voltage regime (1000V), the
theoretical limit for silicon is 150°C; for discrete devices below 100V, it is expected
about 250°C [2]. Materials used up to 300°C include bulk silicon and silicon-on-
insulator (SOI) technologies; for higher temperatures, gallium arsenide (GaAs),
silicon carbide (SiC), and diamond show promise, and devices have been

---

[*] Chromatech Alameda CA 94501

K. Deb et al. (Eds.): GECCO 2004, LNCS 3102, pp. 792–803, 2004.
© Springer-Verlag Berlin Heidelberg 2004

demonstrated at 500°C [3]. A survey of high-temperature effects and design considerations is found in [4]. A review of the physical limits and lifetime limitations of semiconductor devices at high-temperatures is found in [2].

In addition to material/device solutions, circuit solutions for the compensation of the effects of temperature have also been employed. Circuit solutions that compensate offset voltage and current leakage problems are described for example in [3], where several circuit topologies for high-temperature design, including a continuous-time auto-zeroed OpAmp and an A/D circuit that uses error suppression to overcome high-temperature leakages, are given. Another circuit for high-temperature operation with current leakage compensation is presented in [5]. Bias cancellation techniques for high-temperature analog application are presented in [6].

All the above solutions are fixed circuit design solutions, and satisfy the operational requirements only over a given temperature range. Once the limits of the range are exceeded, the performance deteriorates and cannot be recovered. In this paper, we propose the use of reconfigurable chips, which allow for a large number of topologies to be programmed, some more suitable for high-temperature. The interconnections between components can be changed, and new circuits can be configured, in an arrangement that uses the on-chip components/devices *at the new operational point on their characteristics*. In essence, a new design process takes place automatically, in-situ, under the control of a search algorithm. The configurations could be determined either before launch - part of the original design (which would identify good configurations and store them in a memory) - or in-situ. At the higher temperatures, once the performance of the current topology starts to deteriorate, the system would switch to a more suitable topology.

Reconfiguration can be controlled by evolutionary algorithms, a research area called Evolvable Hardware (EHW). Evolvable hardware technology is particularly significant to future NASA autonomous systems, providing on-board resources for reconfiguration to self-adapt to situations, environments and mission changes. It would enable future space missions using concepts of spacecraft surviving in excess of 100 years as well as electronics for temperatures over 460°C as encountered on the surface of Venus which pose challenges to current electronics. In addition, this technology can be used to reduce massive amounts of sensor data to lean data sent to centralized digital systems.

As part of an effort to develop evolution-oriented devices for Evolvable Hardware experiments, we designed and fabricated a series of Field Programmable Transistor Array (FPTA) chips in 0.5 micron and 0.18 micron bulk CMOS. These chips are reconfigurable at transistor level and were used to demonstrate on-chip evolution/synthesis of a variety of conventional building blocks for electronic circuits such as logical gates, transconductance amplifiers, filters, Gaussian neurons, data converters, etc [7], [8].

We present results on using evolution to recover the functionality of FPTA-mapped circuits affected by changes in temperature In this paper we present a more detailed account of the evolutionary recovery, and explain how temperature degradation can fundamentally impact the intended function of the IC. The examples chosen include analog circuits whose behavior deteriorates as the temperature increase, thus totally altering the intended analog function. Evolution is able to find alternate circuits that perform correctly at the higher temperature.

The paper is organized as follows: Section 2 presents the details on a FPTA-2 chip developed as an evolution-oriented architecture for reconfigurable hardware, and introduces the experimental high temperature testbed. Section 3 presents experiments that illustrate that evolution-guided reconfiguration can recover functionality deteriorated/altered by increased temperature.   Section 4 concludes the work.

2   Structure of Evolvable Systems

An evolvable hardware system is constituted of two main components: reconfigurable hardware (RH) and an evolutionary processor (EP) that acts as a reconfiguration mechanism. In the evolvable systems we built for this effort, the EP was implemented and ran on a stand-alone DSP board. The RH was embodied in the form of a Field Programmable Transistor Array (FPTA-2) architecture,   a custom made chip fabricated in silicon. This section will refer to the general characteristics of the two components and will also describe the Evolvable System testbed for high temperature experiments.

2.1   The FPTA

The FPTA is an evolution-oriented reconfigurable architecture (EORA) [8] with configurable granularity at the transistor level. It can map analog, digital and mixed signal circuits. The architecture is cellular, with each cell having a set of transistors, which can be interconnected by other "configuration transistors". For brevity, the "configuration transistors" are called switches. However, unlike conventional switches, these can be controlled for partial opening, with appropriate voltage control on the gates, thus allowing for transistor-resistor type topologies.

Cells are interconnected to local neighbors with switches. A variety of simple circuits can be mapped onto this device or on multiple devices by cascading them. Its design was inspired by observing a variety of analog designs in which transistors often come in rows of pairs of transistors (for various current mirrors, differential pairs etc.), and have an average of four rows between VDD and ground. More rows can be ensured cascading cells, while fewer rows can be mapped by closing some switches to bypass rows.

The FPTA-2 is a third generation of reconfigurable chips designed at JPL, consisting of an 8x8 array of re-configurable cells. It was fabricated using TSMC 0.18u/1.8V technology. Each cell has a transistor array as well as a set of programmable resources, including programmable resistors and static capacitors. Figure 1 provides a broad view of the chip architecture together with a detailed view of the reconfigurable transistor array cell. The re-configurable circuitry consists of 14 transistors connected through 44 switches. The re-configurable circuitry is able to implement different building blocks for analog processing, such as two and three stages OpAmps, logarithmic photo detectors, or Gaussian computational circuits. It includes three capacitors, Cm1, Cm2 and Cc, of 100fF, 100fF and 5pF respectively. Control signals come on the 9-bit address bus and 16-bit data bus, and access each individual cell providing the addressing mechanism for downloading the bit-string configuration of each cell. A total of ~5000 bits is used to program the whole chip. The pattern of interconnection between cells is similar to the one used in commercial FPGAs: each cell interconnects with its north, south, east and west neighbors.

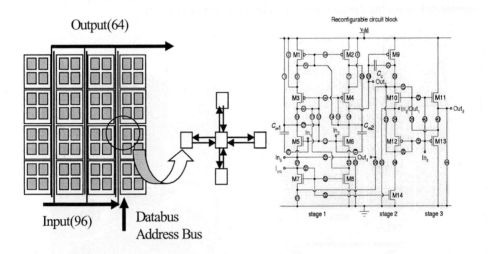

Fig. 1. FPTA 2 architecture (left) and schematic of cell transistor array (right). The cell contains additional capacitors and programmable resistors (not shown).

2.2 A Stand-Alone Board-Level Evolvable System

A complete stand-alone board-level evolvable system (SABLES) was built by integrating the FPTA-2 and a DSP implementing the EP. The system is connected to the PC only for the purpose of receiving specifications and communicating back the result of evolution for analysis. The system fits in a box 8" x 8" x 3". Communication between DSP and FPTA is very fast with a 32-bit bus operating at 7.5MHz. The FPTA can be attached to a Zif socket attached to a metal electronics board to perform extreme temperature experiments. The evaluation time depends on the tests performed on the circuit. Many of the tests attempted here require less than two milliseconds per individual, and runs of populations of 100 to 200 generations require only 20 seconds.

2.3 Extreme Temperature Testbed

The purpose of this testbed is to achieve temperatures exceeding 350°C on the die of the FPTA-2 while staying below 280°C on the package. It is necessary to stay below 280°C on the package in order not to destroy the interconnects and package integrity. Die temperatures should stay below 400°C to make sure die attach epoxy does not soften and that the crystal structure of the aluminum core does not soften. To achieve these high temperatures the testbed includes an Air Torch system. The Air Torch is firing hot compressed air through a small hole of a high temperature resistance ceramic protecting the chip. To measure temperature Thermocouples were used.

Figure 2 shows the Air Torch apparatus electronically controlled by PID controller, which maintains a precision of ±10° C up to 1000° C. Figure 2 shows also the ceramic protecting the die connections and the package. The Temperature was measured above the die and under the die using thermocouples.

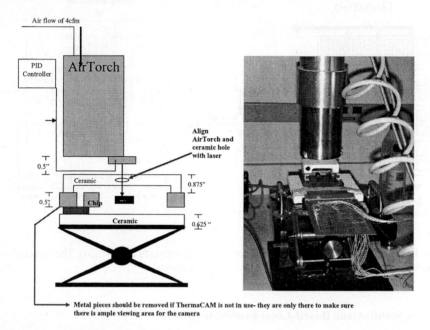

Fig. 2. Experimental Setup for Extreme Temperature Experiments for the FPTA.

3   Extreme Temperature Experiments

We describe here experiments for evolutionary recovery of the functionality of the following circuits:

-   Halfwave Rectifier at 280°C
-   Closed Loop OpAmp at 245°C
-   Low Pass Filters at 230°C

The rationale of these experiments was of first evolving the proposed circuits at room temperature. After the functionality is achieved the temperature is increased using the apparatus shown in Figure 2, until the functionality is degraded. When the device characteristics change with temperature, one can preserve the function by finding a different circuit (topology) solution, which exploits the altered/modified characteristics. Therefore, in order to recover the functionality, the evolutionary process is started again at high temperature. Evolution can obtain a circuit that works at high temperature if the search process is carried on at the temperature in which the circuit is supposed to work.

One limitation of these experiments is the fact that we assume the Evolutionary Processor as fault-free, i.e. the DSP implementing the Evolutionary Algorithm is always at room temperature. Further studies should be performed to investigate the effect of high temperature in the device implementing the Evolutionary Algorithm.

3.1 Half Wave Rectifier on FPTA-2 at 280°C

The objective of this experiment is to recover functionality of a half wave rectifier for a 2kHz sine wave of amplitude 2V using only two cells of the FPTA-2 at 280°C. The fitness function given below does a simple sum of error between the target function and the output from the FPTA.

The input was a 2kHz excitation sine wave of 2V amplitude, while the target waveform was the rectified sine wave. The fitness function rewarded those individuals exhibiting behavior closer to target (by using a sum of differences between the response of a circuit and the target) and penalized those farther from it. The fitness function was:

$$F = \sum_{t_s=0}^{n-1} \begin{cases} R(t_s) - S(t_s) & \text{for } (t_s < n/2) \\ R(t_s) - V_{max}/2 & \text{otherwise} \end{cases} \qquad (1)$$

where $R(t_s)$ is the circuit output, $S(t_s)$ is the circuit stimulus, n is the number of sampled outputs, and V_{max} is 2V (the supply voltage). The output must follow the input during half-cycle, staying constant at a level of half way between the rails (1V) in the other half.

After the evaluation of 100 individuals, they were sorted according to fitness and a 9% (elite percentage) portion was set aside, while the remaining individuals underwent crossover (70% rate), either among themselves or with an individual from the elite, and then mutation (4% rate). The entire population was then reevaluated. Only two cells of the FPTA were allocated and used in this experiment.

The fitness function in equation (1) should be minimized. At room temperature the initial population has an average fitness of 100,000 and the final solution achieves a fitness around 4,500 [10]. At increased temperatures the fitness of the circuit solution is slightly worse, being around 6,000.

Figure 3 depicts the response of the evolved circuit at room temperature and the degraded response at high temperature. Figure 4 shows the response of circuit obtained by running evolution at 280°C, where we can see that the functionality is recovered.

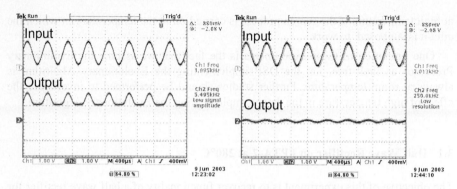

Fig. 3. Input and output waves of the half-wave rectifier. At the left we show the response of the circuit evolved at 27°C. At the right we show the degraded response of the same circuit when the temperature is increased to 280°C.

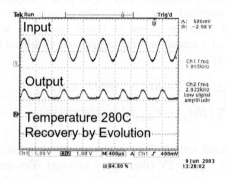

Fig. 4. The solution for the Half wave rectifier at 280°C.

3.2   Amplifier Circuit Using Closed Loop OpAmp at 245°C

The objective of this experiment is to recover by evolution the functionality of a circuit that can provide a gain using compensation circuit introduced in the feedback loop of a conventional OpAmp implemented on the FPTA-2. Amplifiers are a very important building block in sensor circuits and it has been verified in this experiment that three FPTA cells can accomplish this task. One sine wave of 50mV amplitude and 1kHz frequency was applied as stimulus and the target output was a sine wave of twice the amplitude. The fitness encompassed the absolute sum of errors between the FPTA output, R(t), and the target, T(t) as shown below.

$$F = \sum_{t=0}^{n-1} \mid (R(t) - T(t)) \mid \qquad (2)$$

The other Evolutionary Algorithm parameters were similar to the ones described in the previous section.

Figure 5 illustrates the block diagram of the circuit in the FPTA-2. One cell of the chip implements a conventional OpAmp, while 3 re-configurable cells in a feedback loop have their configurations changed by evolution to achieve a compensating structure providing a voltage gain of 2.

Figure 6 shows the response of the circuit evolved at room temperature and the degraded functionality. Figure 7 depicts the recovered response of the circuit evolved at 245°C.

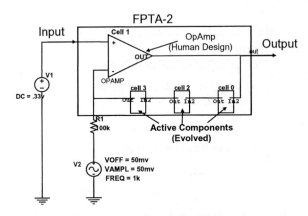

Fig. 5. Block diagram of the closed-loop amplifier implemented in the FPTA-2. Cell 1 realizes a conventional OpAmp; cells 0, 2 and 3 are evolved to provide an amplification gain of 2.

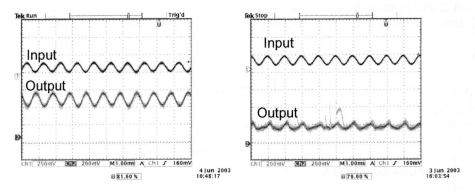

Fig. 6. Degradation of the amplifier circuit. At the left response at room temperature, at the right response at 245°C.

3.3 Low Pass Filter at 230°C

The objective of this experiment is to recover the functionality of a low-pass filter given ten cells of the FPTA-2. The fitness function given below performs a sum of error between the target function and the output from the FPTA in the frequency domain.

$$F = \sum_{f_s=0}^{n-1}(R(f)-T(f)) \qquad (3)$$

Given two tones at 1kHz and 10kHz, the objective is to have at the output only the lowest frequency tone (1kHz). This hardware evolved circuit demonstrated that the FPTA-2 is able to recover active filters with some gain at 230°C. Ten FPTA cells were used in this experiment.

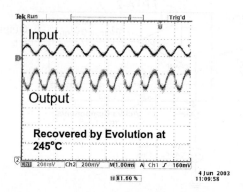

Fig. 7. The solution for the recovered amplifier circuit

Figure 8 shows the response of the evolved filter at room temperature and degradation at 230°C. Figure 9 displays the same information in the frequency domain. Figure 10 shows the time and frequency response of the recovered circuit evolved at 230°C.

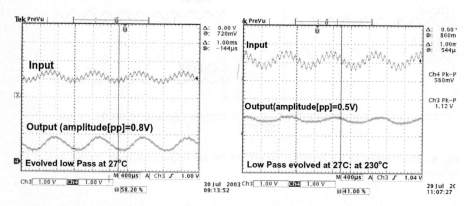

Fig. 8. Low-Pass Filter. The graph in the left displays the input and output signals in the time domain. The graph in the right shows the input and output in the time domain when the FPTA-2 was submitted to temperature of 230°C (Circuit stimulated by two sine waves: 1kHz and 10kHz).

LPF evolved at 27C: Bode Plot using FFT

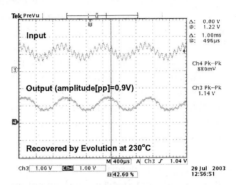

Fig. 9. Low-Pass Filter. The graph in the left displays the frequency response of the output signal at room temperature. The graph in the right shows the frequency response of the output when the FPTA-2 was submitted to temperature of 230°C. (Circuit was stimulated by a sine wave with a frequency sweeping from 1kHz and 10kHz).

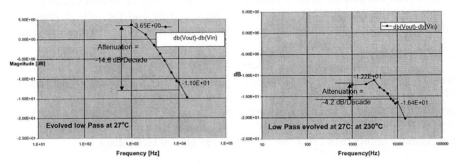

LPF evolved at 230C: Bode Plot using FFT measurement at 230C

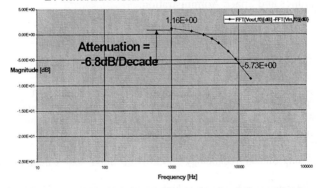

Fig. 10. Low-Pass Filter. The graph at the top shows the circuit input stimulus and response in the time domain (Time response was obtained using a stimulation signal made of two sine waves: 1kHz and 10kHz – Frequency response was obtained by sweeping a frequency from 1kHz to 10kHz). The graph at the bottom displays the Bode diagram of the output signals.

At room temperature, the originally evolved circuit presents a gain of 3dB at 1kHz and a roll-off of -14dB/dec. When the temperature is increased to 230°C, the roll-off goes to -4dB/dec and the gain at 1kHz falls to -12dB. In the recovered circuit at high temperature the gain at 1kHz is increased back to 1dB and the roll-off goes to -7dB/dec. Therefore the evolved solution at high temperature is able to restore the gain and to partially restore the roll-off.

All the recovery experiments described above were performed for a maximum temperature of 280°C. For temperatures higher than 280°C, it was observed that the FPTA switch elements work as a fixed resistance (partly closed switch) regardless of the control voltage. This incorrect behavior, due to increase of parasitic currents at high temperature, turns unfeasible evolutionary recovery experiments for temperatures higher than 280°C for this particular type of technology.

4 Conclusions

The experiments demonstrate the possibility of using evolutionary self-configuration to recover functionality lost at extreme temperatures (up to 280°C). In addition, evolutionary design can be used to create designs targeted to the extreme temperatures. One should mention here that while a device may work at a certain temperature, the real limiting factors for applications will be failure rates and lifetimes. The experiments were performed on bulk CMOS because of the convenience and low cost of fabricating in this technology. For maximum performance, evolvable hardware should make use of an enhancing technique combined with materials/devices more appropriate for extreme temperatures, such as SiC, etc.

Acknowledgements. The work described in this paper was performed at the Jet Propulsion Laboratory, California Institute of Technology and was sponsored by the National Aeronautics and Space Administration.

References

1. Proceedings of the NASA/JPL Conference on Electronics for Extreme Environments, Feb. 9-11, 1999; Pasadena, CA. In http://extremeelectronics.jpl.nasa.gov/conference".
2. Wondrak, W. Physical Limits and Lifetime, "Limitations of Semicondctor Devices at High Temperatures", Microelectronics Reliability 39 (1999) 1113-1120.
3. J. Haslett, F. Trofimenkoff, I. Finvers, F. Sabouri, and R. Smallwood, "High Temperature Electronics Using Silicon Technology", 1996 IEEE Solid State Circuits Conf., pp. 402-403.
4. Shoucair, F., "Design considerations in high temperature analog MOS integrated circuits". IEEE Transactions on Components, Hybrids, and Manufacturing Technology, 9(3):242, 1986.
5. Mizuno, K., N. Ohta, F. Kitagawa , H., Nagase, E., "Analog CMOS Integrated Circuits for High-Temperature Operation with Leakage Current Compensation", 4th International High Temperature Electronics Conf., Albuquerque, p. 41, 1998.

6.   F. Shi, "Analyzing Bias Cancellation Techniques for High temperature Analog Applications", 4th International High Temperature Electronics Conf., Albuquerque, pp. 172-175, 1998.
7.   A. Stoica,   "Toward evolvable hardware chips: experiments with a programmable transistor array", Proceedings of 7th International Conference on Microelectronics for Neural, Fuzzy and Bio-Inspired Systems, Granada, Spain, April 7-9, IEEE Comp Sci. Press, 1999, 156-162.
8.   A. Stoica, R. Zebulum, D. Keymeulen, R. Tawel, T. Daud, and A. Thakoor, "Reconfigurable VLSI Architectures for Evolvable Hardware: from Experimental Field Programmable Transistor Arrays to Evolution-Oriented Chips", IEEE Transactions on VLSI Systems, February 2001,227-232.
9.   A. Stoica, D. Keymeulen, and R. Zebulum, "Evolvable Hardware Solutions for Extreme Temperature Electronics", Third NASA/DoD Workshop on Evolvable Hardware, Long Beach, July, 12-14, 2001, (pp.93-97), IEEE Computer Society.
10. A. Stoica, R. Zebulum, M.I. Ferguson, D. Keymeulen, V. Duong, "Evolving Circuits in Seconds: Experiments with a Stand-Alone Board Level Evolvable System", pp. 67- 74. 2002 NASA/DoD Conf. on Evolvable Hardware, Virginia, USA, July, 2002, IEEE Computer Society.

The Emergence of Ontogenic Scaffolding in a Stochastic Development Environment

John Rieffel and Jordan Pollack

DEMO Lab, Brandeis University, Waltham MA, 02454, USA
{jrieffel,pollack}@cs.brandeis.edu
http://demo.cs.brandeis.edu

Abstract. Evolutionary designs based upon Artificial Ontogenies are beginning to cross from virtual to real environments. In such systems the evolved genotype is an indirect, procedural representation of the final structure. To date, most Artificial Ontogenies have relied upon an error-free development process to generate their phenotypic structure. In this paper we explore the effects and consequences of developmental error on Artificial Ontogenies. In a simple evolutionary design task, and using an indirect procedural representation that lacks the ability to test intermediate results of development, we demonstrate the emergence of ontogenic mechanisms which are able to cope with developmental error.

1 Introduction

Recently, evolved designs have begun to cross the boundary from the virtual to the real [1,2]. Many of these designs are based upon Artificial Ontogenies [3, 4], which use an *indirect encoding* of the evolved object. Between genotype and phenotype lies some developmental process responsible for assembling the phenotypic structure by interpreting instructions contained in the genotype.

While many such systems take noisy physics into account when evaluating the fully developed phenotype [5,6,7], the problem of noise during development is yet to be addressed, and to date, Artificial Ontogenies have not been shown to be adaptive to errors caused by noisy development environments. With the real-world assembly of evolved designs in mind, our interest here is on the ability of Artificial Ontogenies to adapt to error during development. This is a line of inquiry intimated by Stanley and Miikkulainen in their recent survey [4].

As we show, developmental error can complicate an otherwise trivial design task. Error during development results in a stochastic process wherein each genotype, instead of reliably developing into a single phenotype, develops into an entire *distribution* of heterogeneous phenotypic structures, with a corresponding range of fitness values. As such, a credit-assignment problem arises: when a genotype develops into a variety of heterogeneous phenotypes, how should the entire range of related fitnesses be attributed to that genotype?

In this paper we begin to explore whether, without incorporating tests into the developmental system, there is enough information available to the evolutionary process to allow for mechanisms to emerge which can cope with stochastic

K. Deb et al. (Eds.): GECCO 2004, LNCS 3102, pp. 804–815, 2004.

development. We first evolve an indirect encoding in an error-free development environment and demonstrate its failure when assembled in a stochastic environment. We then incorporate noise into the development environment used *within* the evolutionary process. In this setup we are able to observe the emergence of ontogenic mechanisms capable of overcoming developmental error.

2   Theory and Background

Artificial Embryogenies [4] distinguish themselves from other forms of evolutionary computation by treating the genotype as an *indirect*, or *procedural* encoding of the phenotype. The genotype is decoded and transformed into a phenotype by means of some developmental process. As a result, a single-point change to the genotype can have multiple (or zero) effects upon the phenotype. This abstraction layer between genotype and phenotype allows for quite a bit of flexibility during evolution, and has several demonstrated advantages [8,9,4,3,10]. An advantage of indirect encodings that we are particularly interested in is their ability to specify intermediate morphological elements that are useful for ontogenesis, but that do not exist in the final phenotype.

2.1   Genotypes as Assembly Plans

In distinguishing the direct encodings used in traditional GAs from the *indirect* encodings used by Artificial Ontogenies, it is informative to consider the distinction between a blueprint and an assembly plan. A direct encoding is a *descriptive* representation. It is like a blueprint in the sense that it conveys what the phenotype should look like, but carries no information about how to build it (or whether in fact it can be built at all.) Examples of evolved direct encodings include Lipson's Golems [2] and Fune's LEGO structures [11]. Indirect encodings, on the other hand, provide no information about what the final structure should look like. Rather they are a *procedural* representation, and like an assembly plan, give specific instructions on how to build the structure step by step.

When their genomes are described as assembly plans, Artificial Ontogenies can be considered a form of Genetic Programming (GP) [12]. The genome, either linear or in the form of the tree, consists of loci which are instructions to some ontogenic agent. This agent (which is not necessarily external to the developing structure), interprets each instruction and builds the emerging structure from raw materials accordingly. In the case of Hornby [8,7,2], the instructions are commands to a LOGO-like turtle which builds three dimensional structures out of voxels. In the case of Toussaint [9], the instructions are for a system which draws three-dimensional plants from component stems and leaves. Assembly plans can be categorized as either *ballistic* or *adaptive*. Ballistic assembly plans have no internal feedback mechanisms - they proceed uninterrupted until done, regardless of the results of each action. Adaptive assembly plans, on the other hand, are able to measure the results of their executed instructions, and change their behavior accordingly.

2.2   The Effects of Noise During Development

Most Artificial Embryogenies (Fig. 1) rely upon a deterministic development process. As such, there is a one-to-one relation between genotype and phenotype: a given genotype will always develop into the same phenotype.

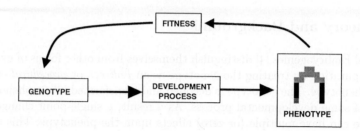

Fig. 1. In a simple Artificial Ontogeny with a deterministic development, each genotype consistently develops into a single phenotype and associated fitness

Introducing error into development causes a one-to-many genotype/phenotype relationship. Since the result of each stage of the ontogeny is predicated upon the result of the previous stage, an early error can drastically affect the outcome of the final phenotype. Under these conditions a genotype may produce any number of phenotypically heterogeneous results, as illustrated by Fig. 2. [13] provides a more nuanced treatment of this phenomenon.

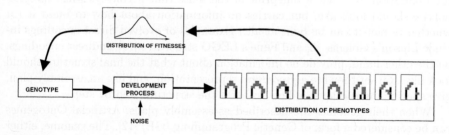

Fig. 2. In a Artificial Ontogeny with a noisy development environment, each genotype can develop into an entire range of phenotype, with a corresponding range of fitnesses

One possibility for overcoming developmental error is to include some form of test into the genotype's set of primitive instructions. However, incorporating tests into each step of an ontogeny can be time consuming, particularly in the context of an evolutionary search spanning thousands or millions of generations. Another way to handle stochastic ontogenies might be to use systems capable of modularity and parallelism such as generative grammars [8] or genetic regulatory networks [10]. Like tests, however, such methods come at the expense of simplicity of the ontogenic process.

Before exploring more complex, albeit powerful, genotypes and ontogenies it is worthwhile to first explore the capabilities and limits of a simple linear, ballistic assembly plan, whose only feedback exists at the evolutionary scale.

2.3   Measuring Fitness Distributions

Rather than give each genotype only one chance to stochastically develop into a phenotype, it may be more informative to allow each genotype multiple stochastic developments. A genotype will then produce an entire *distribution* of phenotypes, with a corresponding range of fitness values, per Fig. 3. Statistical measurements of the resulting distribution can then be used to measure the fitness of the genotype.

Fig. 3. A noisy development environment leads to a distribution of phenotypic fitnesses. Yield is the frequency with which the distribution reaches the maximum fitness

In the case where there is an achievable maximum fitness, gradient can be further induced by considering *yield*: the frequency with which the maximum fitness is attained (see Fig. 3). To illustrate this, consider the case where there is a particular evolutionary goal in mind, such as the pre-defined letter shapes on a grid in Kumar and Bentley's recent work [3]. In this context, yield can be described as the percentage of times that a given assembly plan is able to successfully generate the goal phenotype.

With such a range of different statistical measurements available to compare genotypes, choosing a specific scalar fitness function which somehow weighs and combines the measurements into a single informative value can be difficult. In this situation, Evolutionary Multi-Objective Optimization (EMOO) [14,15] can prove useful.

EMOO allows each measurement to exist as an independent objective. Instead of a scalar fitness value, each genotype is given a set of fitness values, one for each objective. When comparing two sets of objective values, one is said to *Pareto dominate* the other when it is at least as good in all objective dimensions, and better in at least one dimension. Given a population of individuals, the *Pareto front* consists of the set of individuals that are not dominated by any other individuals. A more detailed mathematical explanation of EMOO can be found in [14] and [15].

3   Experiments

The goals of our experiments are twofold: first to demonstrate that "naive" indirect encodings evolved in an error-free development environment are brittle in the face of error during ontogeny; and secondly to show how indirect encodings evolved *within* a stochastic environment are able to adapt to error, and reliably produce fit phenotypes.

We phrase our problem as a type of Genetic Programming [12] in which we are evolving a linear assembly plan to build a predefined "goal" structure. In this case, we chose an arch (Fig. 4), in part for the expected level of difficulty, and in part for historical reasons - its presence in Winston's seminal work on Machine Learning [16].

Fig. 4. The goal structure. Note: vertical bricks are black, horizontal bricks are grey

The genotype consists of a linear set of parameterized instructions for a LOGO-turtle like builder, the ontogenic agent. The turtle is capable of moving along a vertical 2-D plane, and placing and removing $2x1$ bricks within the plane. Table 1 lists the instructions used. Note that assembly plans are completely ballistic: there are no instructions that can test the state of the world or the results of the most recent instruction.

Table 1. Parameterized Assembly Instructions

Instruction	Parameters
(M)ove	+2, +1, -1, -2
(R)otate	+90, -90, +180
(P)ut Brick	(a)head, to (r)ight, to (l)eft, (b)ehind
(T)ake Brick	*(none)*

Because genotypes are linear sequences of instructions, they are amenable to both crossover and mutation. In order to allow for a broader syntactic range of acceptable genotypes, the builder is tolerant of redundant instructions (such as putting a brick where a brick already exists), as well as instructions which would force it off of the grid.

3.1   Physics

Bricks placed by the turtle are subject to a simple physics model. They must either be supported by the floor of the plane or by another brick. Bricks unsupported from below will fall until they hit a supporting surface.

By adding noise to the physics of the development environment, we can induce developmental errors. Bricks placed vertically on a surface have a 50% chance of staying in place, and a 50% chance of falling to either side. Similarly, bricks placed horizontally such that they are cantilevered have a 50% chance of remaining in place and a 50% chance of falling. Naturally, surrounding bricks may act as supports, and reduce the chance that a brick will fall. Bricks that fall will drop until they find a resting place. Once a brick has settled it is considered "glued" in place until it is removed or one of its supporting bricks is removed. Table 2 summarizes the rules of the stochastic physics. Note that the turtle itself is imperturbable. Its position on the plane remains constant regardless of whether the brick it has placed falls or not.

Table 2. Basic Rules for Stochastic Physics

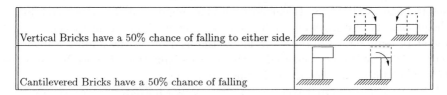

| Vertical Bricks have a 50% chance of falling to either side. | |
| Cantilevered Bricks have a 50% chance of falling | |

The developmental error of our assembly is therefore of a very specific nature: each instruction in the assembly plan is always reliably executed by the builder, but the *result* of that instruction may vary.

3.2   Algorithm

As mentioned above, we chose to phrase the problem as one of Evolutionary Multi-Objective Optimization (EMOO) [14,15]. The specific objectives vary between experimental setups, and are discussed in detail for each.

Evaluation. Individuals are evaluated by interpreting their assembly plans within the specified environment and measuring the properties of the resulting structure. For non-stochastic environments, each assembly plan only needs to be build once. For stochastic environments, assembly plans are built several times in order to gather statistical properties of their phenotypic distribution.

Generation and Selection. Population size is variable - new children are added and evaluated until the population is doubled. New individuals are generated by a combination of two-point crossover (70%) and single-point mutation (30%). Once the new population has been generated and evaluated, the population is culled by and keeping only non-dominated individuals, i.e. the Pareto front.

3.3   Evolving Without Developmental Noise

As a first demonstration, consider a "naive" assembly plan evolved in an error-free development environment. The objectives used for this run are as follows:

- length of genome (shorter is better)
- genotypic diversity
- number of squares missing from goal structure (fewer is better, 0 is best)
- sum of number of missing squares and extra squares(fewer is better, 0 is best)

The length objective exists in order to find minimal solutions, as well as a deterrent to bloat [17,18]. Because of the small number of objectives, and due to the propensity of the system to find a large number of genotypically similar, and therefore redundant solutions, we follow the lead of [17] by adding a diversity metric. This metric is calculated as the average hamming distance between the genome and all other genotypes in the population.

Treating the goal and result structures as 2-D bitmaps, the third objective can be calculated as the sum of the bitwise AND of the goal and the inverse of the result, $\sum_{i,j}(goal(i,j) \otimes \neg result(i,j))$, and the fourth objective as the sum of the bitwise XOR of the goal structure and the result: $\sum_{i,j}(goal(i,j) \oplus result(i,j))$. As an example, consider the leftmost structure in Fig. 6: three squares are absent from the goal structure, and there are nine extra squares. The third objective would therefore be 3, and the fourth objective would be 12.

The last metric, which adds the number of missing square and extra squares, may seem cumbersome, but earlier attempts which simply tried to minimize the number extraneous bricks ended up rewarding long, diverse assembly plans which simply moved about but did not place any bricks. By combining missing squares and extra squares, this behavior is avoided.

Results. With this set-up, the system is able to find a minimal assembly plan capable of building the arch in Fig. 4, as shown in the sequence of frames in Fig. 5. The corresponding genotype is: [R(+90) M(-2) P(r) P(a) P(l) M(-1) R(+90) M(-2) P(r) M(+2) P(l) P(b) P(r) P(b) M(-2) P(b)]

Not surprisingly, when that same minimal assembly plan is then built with a noisy development environment it completely fails to build the goal structure - even given repeated attempts. Figure 6 shows a sample of the resulting phenotypes.

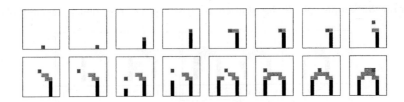

Fig. 5. "Naive" Assembly Plan for Arch. Frames are read left-to-right, top to bottom. The dark grey square is location of the builder

Fig. 6. A sample of the resulting phenotypes when built with noise

3.4   Evolving with Developmental Noise

Our second approach is to integrate a noisy development environment into the evolutionary process itself - such that every candidate genotype is evaluated in the noisy physics. Instead of being built once, each assembly plan is evaluated 50 times, and statistical measures used as evolutionary objectives. The set of measurements that most consistently yielded the best results are:

- length (shorter is better)
- number of missing squares:
  - best, average and yield percentage (no missing bricks)
- sum of extra squares and missing squares:
  - best, average and yield percentage (perfect structure)

Note the absence of the diversity metric used in the first experiment. Such a large number of objectives here results in a relatively large Pareto front with a sufficient amount of diversity.

Results. The evolutionary system described above is typically able to generate assembly plans with yields above 70%. The result we present below is 82 instructions long, and reached a 70% yield during its 50 evolutionary evaluations. When evaluated a further 500 times, its yield drops to 65%. This discrepancy can be attributed to the relatively small sample size used in evolution. Table 3 below shows some samples of the range of phenotypes produced by this assembly plan over the course of multiple developments. It is able to perfectly build the goal structure (far left) 65% of the time. It was able to produce a structure without any squares missing from the structure (middle figures) an additional 8% of the time. The remainder of results (right hand figures) contained some, but not all, of the goal structure.

Table 3. Samples of the distribution of phenotypes of the robust assembly plan

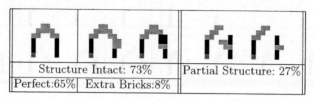

Structure Intact: 73%		Partial Structure: 27%
Perfect:65%	Extra Bricks:8%	

In a typical run, by the time a genotype with 64% yield is achieved, the evolution has run through 26300 generations, and more than 100,000,000 genotype evaluations (where each genotype is evaluated 50 times!), and the population consists of more than 3000 individuals. Beyond this point we therefore suspect that the limitation on further maximizing yield lies largely in the computational effort involved in evaluating such large populations.

Emergence of Ontogenic Scaffolding. When a genome's fitness is based upon the statistical properties of its phenotypic distribution we can think of the role of evolution as learning to shift phenotypic fitness distributions, rather than individual values, towards the optimal. For instance, given two genotypes, the one that on average produces more fit individuals can be considered the better one. In this context, the value of the indirect encoding as assembly plan comes into play. Because assembly plans have the ability to describe *how* a structure is to be built, they can include instructions which place intermediate elements into the structure whose role is to ensure that later elements of the structure stay in place. We call these elements *ontogenic scaffolding*. Once all of the elements of the final structure have been placed, the ontogenic scaffolding can be removed, leaving behind a stable final structure. This ontogenic scaffolding is evident in the results above.

Consider the frames in Figs. 7 through 9 below, which show a typical development from the robust assembly plan discussed above. (Animated versions of these images can be found at
http://www.cs.brandeis.edu/~jrieffel/arches.html)

The assembly begins with Fig. 7. The assembly plan first places horizontal bricks to the left and right of what will become the first leg of the structure. Their presence guarantees that the leg will stay in place. The plan then places the first and second vertical bricks - both parts of the goal structure. Note the "redundant" instruction in the sixth frame for Fig. 7. Although it appears extraneous in this particular sequence, it proves useful in situations where the first attempt at laying the second brick fails: in which case the fallen brick ends up acting as scaffolding for the subsequent attempt.

In the following frames of Fig. 7 the assembly plan proceeds to lay scaffolding for what will be the leftmost leg and leftmost cantilever of the arch.

The assembly continues in Fig. 8 as the plan continues to lay bricks that are simultaneously scaffolding for the leftmost cantilever and for the left leg of the arch. Once scaffolding is laid on both sides, both vertical bricks of the left leg

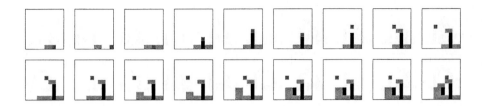

Fig. 7. Robust Assembly Plan Steps 1-18: In the first steps, the builder lays scaffolding

are placed. By the final frames of Fig. 8 all the bricks of the final structure are in place.

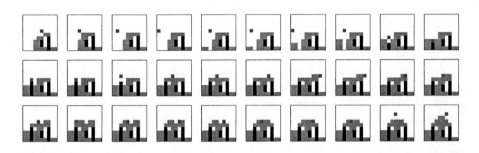

Fig. 8. Robust Assembly Plan Frames 19-49: more scaffolding is lain and the arch is completed

All that remains is for the builder to remove the scaffolding, as it does in Fig. 9, leaving, finally, the complete goal structure.

4   Conclusion

We have demonstrated that using only evolutionary-scale feedback, ballistic assembly plans can be evolved to overcome a noisy development environment. They are able to do this largely my means of ontogenic scaffolding - intermediate and temporary structural elements necessary for reliable assembly of the goal structure. Our result of an assembly plan capable of 70% yield is typical of our system. Running the evolution for longer can likely result in higher yields, but search grows harder over time as the size of the Pareto-front population increases - a consequence of using multi-objective optimization.

It is worth noting that the assembly of the structure shown in Figs. 7- 9 falls into two distinct ontogenic phases - in the first phase the structure is built with the aid of scaffolding, and in the second, the scaffolding is removed. The presence of two distinct phases, as opposed to a process in which scaffolding is created and removed for each element of the final structure, is likely due to the specific search

Fig. 9. Robust Assembly Plan Frames 50-80: scaffolding is removed

gradient created by the two objectives which compare the assembled structure to the goal structure. Evolved assembly plans can first improve along the dimension of missing bricks until they begin to reliably generate all of the parts of the goal structure. Once this is achieved, then can then focus on minimizing the number extraneous bricks in the structure.

Ontogenic scaffolding, while demonstrably useful, provides a challenge for evolutionary design. To begin with, assembly plans which place and then remove scaffolding will by necessity be longer than those that don't. Secondly, any intermediate assembly plan which places scaffolding but doesn't remove it may incur a penalty for the extraneous structure - the cost of exploration, therefore, tends to be high. Finally, for sufficiently complex structures, the scaffolding itself may require meta-scaffolding. These conditions, among others, combine to make the evolution of ontogenic scaffolding, even in simple environments, a non-trivial task.

Our next step will be to explore methods of evaluating assembly plans in noisy environments without a goal structure provided *a priori*. Without the ability to measure yield, the task is complicated quite a bit. Ultimately, we suspect that more powerful and versatile encodings - such as generative representations [8, 9], or gene-regulatory networks [10], equipped with ontogenic-level feedback, will be better able to adapt to stochastic assembly.

References

1. Lohn, J., Crawford, J., Globus, A., Hornby, G., Kraus, W., Larchev, G., Pryor, A., Sriviastava, D.: Evolvable systems for space applications. In: International Conference on Space Mission Challenges for Information Technology (SMC-IT). (2003)
2. Pollack, J.B., Lipson, H., Hornby, G., Funes, P.: Three generations of automatically designed robots. Artifial Life **7** (2001) 215–223
3. Kumar, S., Bentley, P.J.: Computational embryology: past, present and future. In: Advances in evolutionary computing: theory and applications. Springer-Verlag New York, Inc. (2003) 461–477
4. Stanley, K.O., Miikkulainen, R.: A taxonomy for articial embryogeny. Artificial Life **9** (2002) 93–130

5. Jakobi, N., Husbands, P., Harvey, I.: Noise and the reality gap: The use of simulation in evolutionary robotics. In: Proc. of the Third European Conference on Artificial Life (ECAL'95), Granada, Spain (1995) 704–720

6. Sims, K.: Evolving virtual creatures. In: Proceedings of the 21st annual conference on Computer graphics and interactive techniques, ACM Press (1994) 15–22

7. Hornby, G.S.: Generative Representations for Evolutionary Design Automation. PhD thesis, Brandeis University, Dept. of Computer Science, Boston, MA, USA (2003)

8. Hornby, G.S., Pollack, J.B.: The advantages of generative grammatical encodings for physical design. In: Proceedings of the 2001 Congress on Evolutionary Computation CEC2001, COEX, World Trade Center, 159 Samseong-dong, Gangnam-gu, Seoul, Korea, IEEE Press (2001) 600–607

9. Toussaint, M.: Demonstrating the evolution of complex genetic representations: An evolution of artificial plants. In: 2003 Genetic and Evolutionary Computation Conference (GECCO 2003). (2003)

10. Bongard, J.C., Pfeifer, R.: Repeated structure and dissociation of genotypic and phenotypic complexity in artificial ontogeny. In Spector, L., Goodman, E.D., Wu, A., Langdon, W.B., Voigt, H.M., Gen, M., Sen, S., Dorigo, M., Pezeshk, S., Garzon, M.H., Burke, E., eds.: Proceedings of the Genetic and Evolutionary Computation Conference (GECCO-2001), San Francisco, California, USA, Morgan Kaufmann (2001) 829–836

11. Funes, P.: Evolution of Complexity in Real-World Domains. PhD thesis, Brandeis University, Dept. of Computer Science, Boston, MA, USA (2001)

12. Koza, J.R.: Genetic Programming: on the Programming of Computers by Means of Natural Selection. MIT Press: Cambridge, MA (1992)

13. Viswanathan, S., Pollack, J.: On the evolvability of replication fidelity in stochastic construction. Technical Report CS-04-248, Brandeis University (2003)

14. Coello, C.A.C.: An updated survey of evolutionary multiobjective optimization techniques: State of the art and future trends. In Angeline, P.J., Michalewicz, Z., Schoenauer, M., Yao, X., Zalzala, A., eds.: Proceedings of the Congress on Evolutionary Computation. Volume 1., Mayflower Hotel, Washington D.C., USA, IEEE Press (1999) 3–13

15. Fonseca, C.M., Fleming, P.J.: Genetic algorithms for multiobjective optimization: Formulation, discussion and generalization. In: Genetic Algorithms: Proceedings of the Fifth International Conference, Morgan Kaufmann (1993) 416–423

16. Winston, H.P.: Learning By Analyzing Differences. In: Artificial Intelligence: Third Edition. Addison-Wesley, Reading MA (1993) 349–364

17. De Jong, E.D., Watson, R.A., Pollack, J.B.: Reducing bloat and promoting diversity using multi-objective methods. In Spector, L., Goodman, E., Wu, A., Langdon, W., Voigt, H.M., Gen, M., Sen, S., Dorigo, M., Pezeshk, S., Garzon, M., Burke, E., eds.: Proceedings of the Genetic and Evolutionary Computation Conference, GECCO-2001, San Francisco, CA, Morgan Kaufmann Publishers (2001) 11–18

18. Langdon, W.B.: The evolution of size in variable length representations. In: 1998 IEEE International Conference on Evolutionary Computation, Anchorage, Alaska, USA, IEEE Press (1998) 633–638

A Reconfigurable Chip for Evolvable Hardware

Yann Thoma* and Eduardo Sanchez

Swiss Federal Institute of Technology at Lausanne (EPFL), Lausanne, Switzerland
yann.thoma@epfl.ch

Abstract. In the recent years, Xilinx devices, like the XC6200, were the preferred solutions for evolving digital systems. In this paper, we present a new System-On-Chip, the POEtic chip, an alternative for evolvable hardware. This chip has been specifically designed to ease the implementation of bio-inspired systems. It is composed of a microprocessor, and a programmable part, containing basic elements, like every standard Field Programmable Gate Array, on top of which sits a special layer implementing a dynamic routing algorithm. Online on-chip evolution can then be processed, as every configuration bit of the programmable array can be accessed by the microprocessor. This new platform can therefore replace the Xilinx XC6200, with the advantage of having a processor inside.

1 Introduction

Engineers and scientists have much to learn from nature, in term of design capabilities. Living beings are capable of evolution, learning, growth, and self-repair, among others. Each of these fields can serve as inspiration to build systems that are more robust and adaptable. Three life axis define what makes nature a good candidate from which we can draw inspiration: Phylogenesis (P), Ontogenesis (O), and Epigenesis (E).

Phylogenesis is the way species are evolving, by transmitting genes from parents to children, after a selection process. Based on the principles of the neo-darwinian theory, scientists have designed evolutionary algorithms, and more particularly genetic algorithms [1], that are used to solve complex problems for which a deterministic algorithm can not find a solution in an acceptable period of time.

Ontogenesis corresponds to the growth of an organism. In living beings, after fertilization, a single cell, the zygote, contains the genome that describes the entire organism and starts dividing, until the organism is totally created. Ontogenesis takes also care of self-healing, a very important feature of living beings, that prevents them from dying after a light injury. In electronics, self-repair based on ontogenetic principles has been applied to building more robust systems [2,3,4,5].

Finally, epigenesis deals with learning capabilities. A brain, or more generally a neural network, is the way life solved the learning problem. Taking inspiration

---

* Corresponding author

K. Deb et al. (Eds.): GECCO 2004, LNCS 3102, pp. 816–827, 2004.

of real neurons, scientists have designed a huge variety of neural networks, to solve different tasks, like pattern recognition [6] and robot learning [7].

These three life axis have often been considered separately for designing systems, or as a conjunction of learning and evolution. Until now, no real POE system has been constructed. The POEtic project is therefore the logical continuity of bio-inspired systems. A new chip has been specially designed to ease the development of such systems. It contains a microprocessor, and a reconfigurable array offering capabilities of dynamically creating paths at runtime.

This paper focuses on the way POEtic, a promising alternative to the XC6200, can be used as a platform for evolvable hardware [8,9]. Next section presents briefly the principles of evolvable hardware and why field programmable gate arrays are good candidates for such systems. Section 3 describes the POEtic chip, with an emphasis on its usefulness for evolvable hardware. Section 4 presents the way POEtic will be used for this purpose, and finally section 5 concludes.

2   Evolvable Hardware

Evolvable hardware (EHW), on the phylogenetic axis, deals with the design of analog or digital circuits using genetic algorithms. This technic replaces an engineer in the design task, and can act in many different areas. For instance, basic systems like adders or multipliers can be built, while robot control can also be generated. EHW processes can be evolved in simulation in many cases, but software implementations are very slow, and cannot always fit real conditions. Therefore, hardware platforms are needed, to generate operating circuits, in case of analog design, and to speed up the entire process, in case of digital design.

2.1   FPGAs and the Xilinx XC6200 Family

Field Programmable Gate Arrays (FPGAs) [10] are digital circuits that can be reconfigured, and thus make them excellent candidates for implementing EHW. Every commercial FPGA is based on a 2-dimensional array of cells, in which it is possible to define the cells' functionalities and the routing. The most widely used for EHW, the Xilinx Virtex XC6200 family, has been utilized in many experiments [11,12,13,14,15], due to its routing implementation based on multiplexers rather than on anti-fuse or memory bits (short circuits can be generated in almost every other types of FPGAs). The architecture of the XC6200 is very simple, with cells based on some multiplexers and a flip-flop. Moreover, the configuration bits arrangement is public, giving a programmer total control over the configuration. Unfortunately, these devices are not available any more, and no equivalent FPGA is available as of today.

The inherent parallelism of FPGAs allows to rapidly test individuals to evaluate their fitness, but a problem remains: the configuration is very slow. One of the last family of Xilinx devices, the Virtex II Pro, embeds a microprocessor that can access a reconfigurable array, but without the capability of reconfiguring it. The POEtic chip, as explained in the next section, will be a new hardware platform that solves this last drawback.

3   The POEtic Chip

The POEtic chip has been specifically designed to ease the development of bio-inspired applications. It is composed of two main parts: a microprocessor, in the environmental subsystem, and a 2-dimensional reconfigurable array, called the organic subsystem (figure 1). This array is made of small elements, called molecules, that are mainly a 4-input look-up table, and a flip-flop. In the organic subsystem, a second layer implements a dynamic routing algorithm that will allow multi-chip designs, letting the user work with a bigger reconfigurable virtual array.

The next section presents some features of the on-chip microprocessor. The subsequent section describes the reconfigurable array, with a special emphasis on how the different parts of the basic elements can be used to build an EHW system similar to the XC6200.

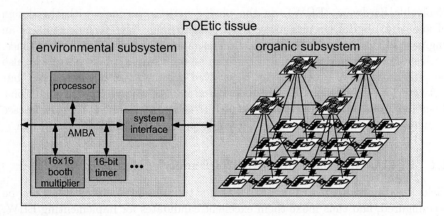

Fig. 1. The POEtic chip, showing the microprocessor and the reconfigurable array. Many elements connected to the AMBA bus, like a second timer, serial and parallel ports, are omitted in order to simplify the schematics. On the right, the organic subsystem shows molecules on the bottom, and routing units on the top.

3.1   The Microprocessor

The microprocessor is a 32-bit RISC processor, specially designed for the POEtic chip. It exposes 57 instructions, two of which give access to a hardware pseudo-random number generator, that can be very useful for evolutionary processes. This small number of instructions limits the size of the processor, leaving more room for the reconfigurable array.

An AMBA bus [16] allows communication with all internal elements, as shown in figure 1, as well as with the external world. It also permits to connect many POEtic chips together, in order to have a bigger reconfigurable virtual array.

The microprocessor can configure the array, and also retrieve its state. The access is made in a parallel manner, the array being mapped on the microprocessor address space. As a result, so that it is very fast to configure, or to partially reconfigure the array, since the configuration of one molecule requires only three write instructions. For instance, when dealing with evolutionary processes, the retrieved state can be used to calculate the fitness of an individual and evolution can be performed by the microprocessor, avoiding fastidious data transmission with a computer.

A C compiler, as well as an assembler, has been developed, letting a user easily write programs for this microprocessor. Furthermore, an API will be supplied, in order to rapidly build a genetic algorithm by choosing the type of crossing-over, the selection process, and so on. Special functions will also simplify the reconfigurable array configuration.

3.2   The Reconfigurable Array

The reconfigurable array is composed of two planes. The first one is a grid of basic elements, called molecules, based on a 4-input look-up table, and a flip-flop. The second one is a grid of routing units, that can dynamically create paths at runtime between different points of the circuit. They implement a distributed dynamic routing algorithm, based on addresses. It can be used to create connections between cells in a cellular system (e.g. a neural network), to connect chips together, or simply to create long-distance connections at runtime (interested readers can see a description of this algorithm in [17]).

The so-called molecules (see figure 2) execute a function, according to an operational mode, defined for each molecule by three configuration bits (for more details, see [17]). The eight operational modes are:

- **4-LUT**: The molecule is a 4-input LUT.
- **3-LUT**: The molecule is divided into two 3-input LUTs.
- **Shift memory**: The molecule is considered like a 16-bit shift register.
- **Comm**: The molecule is divided into a 3-input LUT and a 8-bit shift register.
- **Configure**: The molecule has the possibility of partially reconfigure its neighborhood.
- **Input**: The molecule is an input from the routing plane.
- **Output**: The molecule is an output to the routing plane.
- **Trigger**: This mode is used to synchronize the dynamic routing algorithm.

3.3   Molecular Communication

In addition to its functional part, a molecule contains a switch box for inter-molecular communication. Like in the Xilinx XC6200 family, inter-molecular communication is implemented with multiplexers. This feature, although being

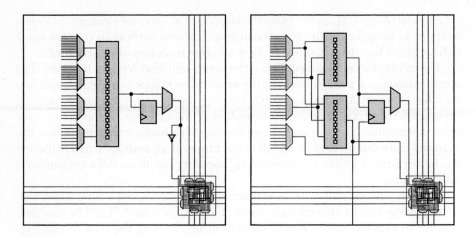

Fig. 2. A "molecule" can act in eight different operational modes, the mode being defined by three configuration bits. The left drawing shows a molecule in 4-LUT mode, while the right depicts a molecule in 3-LUT mode.

more expensive in term of space and delays, avoids short circuits that could happen when partially reconfiguring a molecule, or during an unconstrained evolution process.

Every molecule is directly connected to its four neighbors, sending them its output, while long-distance connections are implemented by the way of switch boxes (figure 3). There are two input lines from each cardinal direction, and two corresponding outputs. Each output can be selected from the six input lines from the other cardinal directions, or from the output of the molecule (or the inverse).

As there are eight possible configurations for an output multiplexer, three configuration bits are necessary for each output. The total lets the switch box being defined by (2 outputs by 4 directions by 3 bits =) 24 bits. These 24 bits could be part of the evolutionary process, or fixed, depending on the kind of system we want to evolve.

For instance, in order to use the POEtic chip like a Xilinx XC6200, every switch box should be configured as in figure 3. By fixing some configuration bits to '0', we can choose to only deal with one line to each direction, as shown in the right of the figure.

In every of its operational modes, a molecule needs up to four inputs. Multiplexers are taking care of the selection of these inputs, like the two first inputs shown in figure 4. An input can basically come from any long-distance line, but each multiplexer has special features. Some can retrieve the flip-flop value, some the direct neighbors output, and so on. By fixing some configuration bits, we can for instance force the selection of a signal coming from N0_in, E0_in, S0_in, W0_in. Therefore, only two bits are necessary to completely define every input.

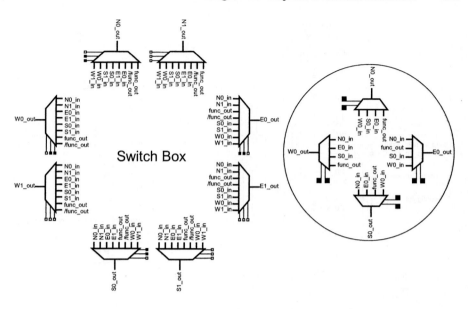

Fig. 3. On the left, the switch box contained in every molecule, as shown in figure 2, and on the right a subset of the possible configurations to reduce the genome size. Black boxes represent configuration bits that can be used for evolution, while white boxes are "don't care bits" for such applications.

This way, every input has the same potential as the others, which would not be the case if every configuration bit could be modified.

3.4   Configuration Bits

One of the advantages of the POEtic chip is the possibility to define any of the 76 configuration bits. These bits are split into five blocks, as shown in table 1. The first bit of each block indicates whether the block has to be reconfigured or not, in case of a partial reconfiguration coming from a neighbor molecule. As mentioned before, the microprocessor can access (read/write) the configuration bits with a 32-bit bus. For EHW, this feature is very important in terms of execution time. Since only two clock cycles are needed for a write and three words of 32 bits define a molecule, the configuration of the entire array or of only a part of it is very fast. In comparison with standard FPGAs, like a Xilinx with JBits [18,19], in which the entire configuration bitstream must be sent each time in serial, the reconfiguration, like the first configuration, is made in parallel, allowing a huge gain in term of time. Moreover, compared to a total reconfiguration, if we only evolve the switch box or the LUT, loading time can be divided by three, as only part of the molecule configuration needs to be reloaded.

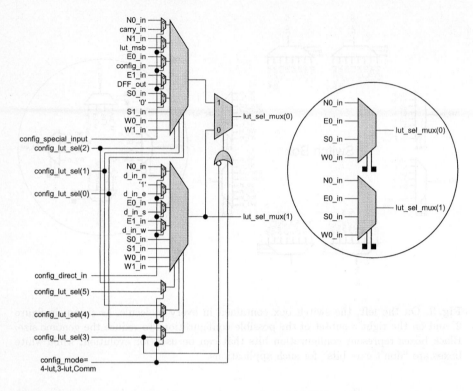

Fig. 4. The two first inputs of the molecule. The signals `config_*` are configuration bits, the two right outputs (`lut_sel_mux(X)`) are the two first inputs of the LUT, and all other signals are inputs that can be selected. The right figure shows a subset of the possible inputs, obtained by fixing some configuration bits (black boxes represent configuration bits that can be used for evolution).

4   Evolvable Hardware on POEtic

In last section, we showed different parts of the reconfigurable array that can be used in an evolvable process. The final chip being not yet available, we will not present experimental results, but concepts that will be used later to demonstrate the full potential of the POEtic chip. First we will have a look at what kind of EHW is supported by POEtic, and secondly, we will describe how we can directly evolve the bitstream as the genome.

4.1   POEtic Evolvable Characteristics

Following the classification developed by Torresen in [20,21], we can now precisely identify the capabilities of POEtic:

- The microprocessor can run a **Genetic Algorithm**.
- The target technology is **Digital**.

Table 1. The five blocks of configuration bits (the first three bits cannot be partially configured by a neighbor molecule).

Number of bits	Description
1	global partial configuration enable
2	configuration input origin
1	lut partial configuration enable
16	lut(15 downto 0) (cf. figure 2)
1	lut inputs configuration enable
14	selection of the lut inputs (cf. figure 4)
1	switchbox partial configuration enable
8x3	3 bits for each of the 8 multiplexers (cf. figure 3)
1	mode partial configuration enable
3	operational mode (cf. section 3.2)
1	other bits partial configuration enable
1	sequential or combinational output
1	flip-flop reset value
1	dff enable used or not
1	clock edge
3	local reset origin
1	local reset enable
1	asynchronous/synchronous reset
1	molecule enable
1	value of the flip-flop

- The architecture applied in evolution can be **Complete Circuit Design,** where building blocks and routing are evolved, or **Circuit Parameter Tuning**, where only configurable parameters are evolved.
- The possible building blocks can be **Gates** (the LUTs), or **Functions** (neurons, ...).
- The evolution is made **Online**, because every individual will be tested using the reconfigurable array.
- The evolution is **On-chip**, as the microprocessor is incorporated into the reconfigurable chip.
- The scope of evolution can be **Static**, or **Dynamic**, depending on the type of application.

POEtic, with its dynamic routing capability, could show function level evolution that involves sine generators, adders, multipliers, artificial neurons, or others. However, in this paper we only present gate level evolution, that involves OR/AND gates, or in our example, look-up tables.

Basically, an unconstrained evolution could be executed with the entire configuration bitstream, since it is impossible to create a short-circuit. However, 76 bits for each molecule signify a huge genome, if, for instance, we deal with a 10 by 10 array. Therefore, in many cases, only part of the bitstream will be evolved, in order to reduce the search space.

The experiments made by Thompson using the Xilinx XC6200 are based on the same principle of avoiding to evolve the entire bitstream. They only deal with 18 bits per element, in order to evolve oscillators, for instance. The same types of applications could be resolved with 22 bits using the POEtic chip.

4.2   Genome Representation

In the approach chosen in this paper, we evolve a system at the gate level, by evolving the routing or the function of molecules. Therefore, it is natural to directly evolve the configuration stream of the chip. Since there are 76 configuration bits, and the bus has a width of 32 bits, only three words define a molecule. In order to evolve routing and functionality, we do not want to evolve the entire bitstream, but only part of it. By using very simple logical operations we can modify the entire genome, without modifying fixed parts, as shown in figure 6.

In our example, the routing uses half of its capabilities, with the subset shown in figure 3. The molecule inputs are the same as shown in figure 4, and the operational mode is fixed to the 3-LUT mode. Therefore, the functionality can be any 3 inputs function. This case corresponds to an evolution of the basic cells of figure 5.

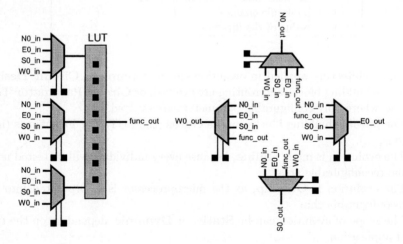

Fig. 5. The basic element, subset of the molecule, that can be evolved, defined by only 22 bits. Black boxes are configuration bits that are evolved.

The full genome is composed, for each molecule, of 96 bits (3x32), 76 defining configuration bits. However, in our example, only 22 bits really represent information used to define the phenotype, the 74 other bits being fixed. Compared to the 18 bits used by Thompson with a XC6200, we deal with 4 more bits, because we totally evolve the look-up table content, rather than just some multiplexers. This way the genome is bigger, but each element has more flexibility.

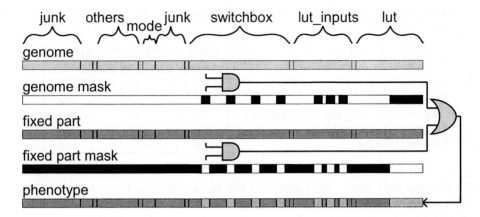

Fig. 6. This figure depicts the way a phenotype can be generated, from a variable genome and a fixed part. A line represents the 3x32=96 bits where 20 are unused bits and 76 are configuration bits of a molecule. These 76 bits are divided into the five blocks described in table 1. The first line is the genome evolved using crossing-over and mutation. The genome mask is used in a logical "and" operation with the genome. It contains '1' at every place the genome is defined by the evolutionary algorithm. Only 22 bits are relevant to define the phenotype: 8 bits for the switch box, 6 bits for the molecular inputs, and 8 bits for the 3-input LUT. The fixed part, combined with its mask (the inverse of the genome mask) corresponds to every configuration bits not defined by the evolution. By simply using an "or" operation on the two results of "and" operations, we obtain the phenotype that is the real configuration of the molecule.

In the evolution process, crossing-over and mutation will be applied to the entire configuration stream, and the very simple logical operations will erase parts of it with the fixed bits. This way, there is no need to use complex transformation, from a smaller virtual bitstream to a real one, saving execution time. Moreover, the fixed parts can be viewed like junk DNA in living beings, in which a large part of the genome is simply unused.

5 Conclusion

In this paper we presented how the POEtic chip can be useful as an EHW platform. The conjunction of a custom microprocessor and a reconfigurable array is perfect to implement an on-chip evolution process. Moreover, compared to a Xilinx Virtex II Pro where there is also a microprocessor, the advantage of POEtic is the fact it is aware of the entire memory map of the configuration bits, and that the microprocessor can configure the chip. Finally, compared to a Xilinx XC6200, POEtic has the advantage of having a microprocessor inside, allowing fast configuration of the reconfigurable array. Table 2 summarizes the features of the XC6200, the Virtex II Pro, and POEtic.

At present, a test chip is being fabricated. After functional tests on this small chip (it only contains the microprocessor and 12 molecules), the final POEtic

Table 2. Comparison of features useful for EHW between a XC6200, a Virtex II Pro and the POEtic chip.

Feature	Xilinx XC6200	Xilinx Virtex II Pro	POEtic
Impossible to short-circuit	Yes	No	Yes
Processor inside	No	Yes	Yes
Processor accessing the configuration bits	No	No	Yes
Bitstream detail available	Yes	No	Yes
Dynamic routing	No	No	Yes

chip, containing about 200 molecules, will be designed and sent to fabric. As soon as it is available, the concepts described in this paper will be tested with the real hardware, to show the promising usefulness of the POEtic chip as a powerful replacement of the Xilinx XC6200 for EHW.

Acknowledgements. This project is funded by the Future and Emerging Technologies programme (IST-FET) of the European Community, under grant IST-2000-28027 (POETIC). The information provided is the sole responsibility of the authors and does not reflect the Community's opinion. The Community is not responsible for any use that might be made of data appearing in this publication. The Swiss participants to this project are supported under grant 00.0529-1 by the Swiss government.

References

1. Holland, J.: Genetic algoritms and the optimal allocation of trails. In: SIAM Journal of Computing. Volume 2:2. (1973) 88–105
2. Kitano, H.: Building complex systems using developmental process: An engineering approach. In: Proc. 2nd Int. Conf. on Evolvable Systems (ICES'98). Volume 1478 of LNCS, Berlin, Springer Verlag (1998) 218–229
3. Mange, D., Sipper, M., Stauffer, A., Tempesti, G.: Towards robust integrated circuits: The embryonics approach. In: Proceedings of the IEEE. Volume 88:4. (2000) 516–541
4. Ortega, C., Tyrell, A.: MUXTREE revisited: Embryonics as a reconfiguration strategy in fault-tolerant processor arrays. In: Proc. 2nd Int. Conf. on Evolvable Systems (ICES'98). Volume 1478 of LNCS, Berlin, Springer Verlag (1998) 206–217
5. Pearson, H.: The regeneration gap. Nature **414** (2001) 388–390
6. Dayhoff, J.: Pattern recognition with a pulsed neural network. In: Proceedings of the conference on Analysis of neural network applications, New York, NY, USA, ACM Press (1991) 146–159
7. Grossmann, A., Poli, R.: Continual robot learning with constructive neural networks. In Birk, A., Demiris, J., eds.: Proceedings of the Sixth European Workshop on Learning Robots. Volume 1545 of LNAI, Brighton, England, Springer-Verlag (1997) 95–108
8. Gordon, T.G.W., Bentley, P.J.: On evolvable hardware. In Ovaska, S., Sztandera, L., eds.: Soft Computing in Industrial Electronics, Heidelberg, Physica-Verlag (2002) 279–323

9. Yao, X., Higuchi, T.: Promises and challenges of evolvable hardware. IEEE Trans. on Systems, Man, and Cybernetics – Part C: Applications and Reviews **29** (1999) 87–97

10. Brown, S., Francis, R., Rose, J., Vranesic, Z.: Field Programmable Gate Arrays. Kluwer Academic Publishers (1992)

11. Fogarty, T., Miller, J., Thomson, P.: Evolving digital logic circuits on xilinx 6000 family fpgas. In Chawdrhy, P., Roy, R., Pant, R., eds.: Soft Computing in Engineering Design and Manufacturing, London, Springer Verlag (1998) 299–305

12. Huelsbergen, L., Rietman, E., Slous, R.: Evolution of astable multivibrators *in Silico*. In Sipper, M., Mange, D., Pérez-Uribe, A., eds.: ICES'98. Number 1478 in Lecture Notes in Computer Science, Berlin Heidelberg, Springer-Verlag (1998) 66–77

13. Tangen, U., McCaskill, J.: Hardware evolution with a massively parallel dynamically reconfigurable computer: Polyp. In Sipper, M., Mange, D., Pérez-Uribe, A., eds.: ICES'98. Volume 1478 of LNCS, Berlin Heidelberg, Springer-Verlag (1998) 364–371

14. Thompson, A.: Silicon evolution. In Koza, J.R., Goldberg, D.E., Fogel, D.B., Riolo, R.L., eds.: Genetic Programming 1996: Proceedings of the First Annual Conference, Stanford University, CA, USA, MIT Press (1996) 444–452

15. Thompson, A.: On the automatic design of robust electronics through artificial evolution. In Sipper, M., Mange, D., Pérez-Uribe, A., eds.: ICES'98. Volume 1478 of LNCS, Berlin Heidelberg, Springer-Verlag (1998) 13–24

16. ARM: AMBA specification, rev 2.0. advanced RISC machines Ltd (ARM). http://www.arm.com/armtech/AMBA_Spec (1999)

17. Thoma, Y., Sanchez, E., Arostegui, J.M.M., Tempesti, G.: A dynamic routing algorithm for a bio-inspired reconfigurable circuit. In Cheung, P.Y.K., Constantinides, G.A., de Sousa, J.T., eds.: Proc. of the 13th International Conference on Field Programmable Logic and Applications (FPL'03). Volume 2778 of LNCS, Berlin, Heidelberg, Springer Verlag (2003) 681–690

18. Guccione, S.A., Levi, D., Sundararajan, P.: Jbits: A java-based interface for reconfigurable computing. In: 2nd Annual Military and Aerospace Applications of Programmable Devices and Technologies Conference (MAPLD). (1999)

19. Hollingworth, G., Smith, S., Tyrell, A.: Safe intrinsic evolution of virtex devices. In: proceedings of 2nd NASA/DoD Workshop on Evolvable Hardware. (2000) 195–204

20. Torresen, J.: Possibilities and limitations of applying evolvable hardware to real-world applications. In Hartenstein, R., Grünbacher, H., eds.: FPL 2000. Volume 1896 of LNCS, Berlin Heidelberg, Springer-Verlag (2000) 230–239

21. Torresen, J.: Evolvable hardware as a new computer architecture. In: Proc. of the International Conference on Advances in Infrastructure for e-Business, e-Education, e-Science, and e-Medicine on the Internet. (2002)

Experimental Evaluation of Discretization Schemes for Rule Induction

Jesus Aguilar–Ruiz[1], Jaume Bacardit[2], and Federico Divina[3]

[1] Dept. of Computer Science, University of Seville, Seville, Spain
aguilar@lsi.us.es
[2] Intelligent Systems Research Group, Universitat Ramon Llull, Barcelona, Spain
jbacardit@salleURL.edu
[3] Dept. of Computer Science, Vrije Universiteit, Amsterdam, The Netherlands
F.Divina@few.vu.nl

Abstract. This paper proposes an experimental evaluation of various discretization schemes in three different evolutionary systems for inductive concept learning. The various discretization methods are used in order to obtain a number of discretization intervals, which represent the basis for the methods adopted by the systems for dealing with numerical values. Basically, for each rule and attribute, one or many intervals are evolved, by means of ad–hoc operators. These operators, depending on the system, can add/subtract intervals found by a discretization method to/from the intervals described by the rule, or split/merge these intervals. In this way the discretization intervals are evolved along with the rules. The aim of this experimental evaluation is to determine for an evolutionary–based system the discretization method that allows the system to obtain the best results. Moreover we want to verify if there is a discretization scheme that can be considered as generally good for evolutionary–based systems. If such a discretization method exists, it could be adopted by all the systems for inductive concept learning using a similar strategy for dealing with numerical values. Otherwise, it would be interesting to extract relationships between the performance of a system and the discretizer used.

1 Introduction

The task of learning a target concept in a given representation language, from a set of positive and negative realizations of that concept (examples) and some background knowledge, is called inductive concept learning (ICL). Real life learning tasks are often described by nominal as well as continuous, real-valued, attributes. However, most inductive learning systems treat all attributes as nominal, hence cannot exploit the linear order of real values. This limitation may have a negative effect not only on the execution speed but also on the learning capabilities of such systems.

In order to overcome these drawbacks, continuous-valued attributes are transformed into nominal ones by splitting the range of the attribute values in a finite

K. Deb et al. (Eds.): GECCO 2004, LNCS 3102, pp. 828–839, 2004.

number of intervals. The so found intervals are then used for treating continuous-valued attributes as nominal. Alternatively, the intervals can be determined during the learning process. This process, called discretization, is supervised when it uses the class labels of examples, and unsupervised otherwise. Discretization can be applied prior or during the learning process (global and local discretization, respectively), and can either discretize one attribute at a time (univariate discretization) or take into account attribute interdependencies (multivariate discretization) [1].

Researchers in the Machine Learning community have introduced many discretization algorithms. An overview of various types of discretization algorithms can be found, e.g., in [2]. Most of these algorithms perform an iterative greedy heuristic search in the space of candidate discretizations, using different types of scoring functions for evaluating a discretization.

In [3,4,5,6] various multivariate local discretization methods are introduced and embedded into systems for rules induction. The idea behind the methods is similar. A number of basic discretization intervals are used in order to evolve the best discretization for each rule. A discretization interval for an attribute in a rule is formed by the union of a number of basic discretization intervals.

In this paper we want to experimentally evaluate the effect of using different basic discretization intervals. In order to do this we use a multivariate discretization method inside three evolutionary rule induction systems: HIDER* [5], ECL [7] and GAssist [4]. All these systems take as input a set of discretization intervals, and adapt them during the learning process, by means of ad–hoc genetic operators.

The paper is structured in the following way. In Section 2 we give a brief description of the discretization methods used for finding the basic discretization interavals. Section 3 contains the experimental evaluation of the various discretization methods. First an overview of the rules induction systems is given, then the experiment settings are described and the results of the experiments are presented and discussed. Section 4 summarizes important conclusions and the future work. Finally, in Section 5 some related work is presented.

2   Discretization Methods

All the systems used in this paper treat numerical values locally. Starting from a set of basic discretization intervals the systems evolve the discretization intervals for each rule. At this end some operators are used, which can merge the basic discretization intervals. Thus, at the end of the evolution the discretization intervals present in the evolved rule are the union of n basic discretization intervals, where $n \geq 1$.

The basic discretization intervals are the results of the application of a discretization scheme. In this paper we used the following discretization method for finding the basic discretization intervals:

1. The method used by ID3 [8], as no pruned version of the Fayyad & Irani's algorithm (which is described below). The values of each continuous attribute

A are sorted in increasing order. The midpoints of two successive values of *A* occuring in examples with different classes are considered as potential cut points. The cut points are then found by recursively choosing the potential cut points that minimizes the entropy, until all the intervals determined in this way contains values relative to examples of the same class;

2. USD [9] divides the continuous attributes in a finite number of intervals with maximum goodness, so that the average-goodness of the final set of intervals will be the highest. The main process is divided in two different parts: first, it calculates the initial intervals by means of projections, which will be refined later, depending on the goodnesses obtained after carrying out two possible actions: to join or not adjacent intervals. The main features of the algorithm are: it is deterministic, does not need any user–parameter and its complexity is subquadratic;

3. Fayyad & Irani's algorithm [10]. This supervised recursive algorithm uses the class information entropy of candidate intervals to select the boundaries of the bins for discretization. Given a set S of instances, an attribute p, and a partition bound t, the class information entropy of the partition induced by t is given by: $E(p, t, S) = Entropy(S_1)\frac{|S_1|}{|S|} + Entropy(S_2)\frac{|S_2|}{|S|}$ where S_1 is the set of instances whose values of p are in the first half of the partition and S_2 the set of instances whose values of p are in the second half of the partition. Moreover $|S|$ denotes the number of elements of S and *Entropy* is defined as: $Entropy(S) = -p_+ \cdot log_2(p_+) - p_- \cdot log_2(p_-)$ with p_+ and p_- the proportion of positive and negative examples in S respectively.
   For a given attribute p the boundary t which minimizes $E(p, t, S)$ is selected as a binary discretization boundary. The method is then applied recursively to both the partitions induced by the selected boundary t^* until a stopping criterion is satisfied. The MDL principle [11] is used to define the stopping criterion;

4. Random discretizer. In this paper we have considered using a random discretizer as a baseline for the tests. This discretizer selects, for each test, a random subset of all the midpoints between the values in the attribute domain;

5. Equal interval width method. In this method the continuous values are simply divided into n equal sized bins, where n is a parameter. In this paper we consider values of n equal to $5, 10, 15, 20$;

6. Equal frequency method. In this method the continuous values are divided into n bins, each bin containing the same number of values. Thus, the regions of the attribute domain with more density of values have more intervals. Again, n is a parameter, considering for this paper the values $5, 10, 15, 20$;

3   Experimental Evaluation

In this section we first give a brief description of the three rule induction systems used in the experiments. The results of the experiments are then presented and discussed. We have omitted the results from equal–width5, equal–width15, equal–freq5 and equal–freq15 because they are very similar to those described in Table 2 with equal–width10, equal–width20, equal–freq10 and equal–freq20, respectively.

3.1 Rule Induction Systems

ECL [7] is a hybrid evolutionary algorithm for ICL. The systems evolves rules by means of the repeated application of selection, mutation and optimization. The mutation operators applied do not act randomly, but consider a number of mutation possibilities, and apply the one yielding the best improvement in the fitness of the individual. The optimization phase consists in a repeated application of mutation operators until the fitness of the individual does not worsen, or until a maximum number of optimization steps has been reached. In the former case the last mutation applied is retracted.

Numerical values are handled by means of inequalities, which describes discretization intervals. Inequalities can be initialized to a given discretization interval, e.g., found with the application of the Fayyad & Irani's algorithm. ECL can modify inequalities using class information, however for allowing a fair comparison with the other systems, this feature is not used here. Instead, inequalities are modified during the learning process, by mutation operators that can add or subtract a basic discretization interval to the interval described by an inequality.

HIDER* [12] is a tool that produces a hierarchical set of rules. When a new example is going to be classified, the set of rules is sequentially evaluated according to the hierarchy, so if the example does not fulfil a rule, the next one in the hierarchy order is evaluated. This process is repeated until the example matches every condition of a rule and then it is classified with the class that such rule establishes. An important feature of HIDER* is its encoding method [5]: each attribute is encoded with only one gene, reducing considerably the length of the individuals, and therefore the search space size, making the algorithm faster while maintaining its prediction accuracy.

GAssist [4] is a Pittsburgh Genetic–Based Machine Learning system descendant of *GABIL* [13]. It evolves individuals that are ordered variable–length rule sets. The control of the bloat effect is performed by a combination of a rule deletion operator and hierarchical selection [14]. The knowledge representation for real–valued attributes is called Adaptive Discretization Intervals rule representation (*ADI*) [4]. This representation uses the semantics of the *GABIL* rules (Conjuntive Normal Form predicates), but using non–static intervals formed by joining several neighbour discretization intervals. These intervals can evolve through the learning process splitting or merging among them. The representation can also combine several discretizations at the same time, allowing the system to choose the correct discretizer for each attribute. This feature will not be used in this paper, to allow a fair comparison with the other two rule induction systems tested.

The three evolutionary approaches used in this paper are different in the way they look for solutions within the search space. GAssist encodes variable–length individuals, which will represent a whole set of decision rules. ECL and HIDER encode single rules, i.e. each individual is one decision rule. However, ECL finds the entire decision rule set at the final generation, whereas HIDER finds a single rule at the end of the evolutionary process. Therefore, HIDER needs to be run several times, until all the examples are covered by any decision rule, following a sequential covering methodology.

An example of these diffences on encoding is shown in Figure 1. We have selected a simple rule set composed by only two rules from Wisconsin dataset. The genetic representation of this rule set is illustrated for each system. The cutpoints have been obtained with ID3.

Attributes in GAssist rules codify the full attribute domain as one or many intervals. Each interval is formed by a subset of consecutive basic discretization intervals. The semantical definition of the rule is formed by the intervals with value 1. HIDER encodes every attribute with only one natural number, as it is described in [5]. Every possible interval defined by two cutpoints is associated to a natural number, so genetic operators are designed to handle this method efficiently. ECL uses a high level representation, where a rule is represented as a list of predicates, variables, constants and inequalities.

3.2   Experiments Settings

Table 1 shows the features of the datasets used in the experiments. These datasets were taken from the UCI Machine Learning repository [15]. We have chosen these datasets because they contain only numerical attributes, and no nominal attributes. For this reason they represent a good testing for the discretization schemes.

In the experiments a 10–fold cross–validation is used. Each dataset is divided in ten disjoint sets of approximately similar size; one of these sets is used as test set, and the union of the remaining nine forms the training set.

For the random discretization method, we have run the 10–fold cross–validation 15 times with different random seeds. Therefore, 7 datasets, with 8

Table 1. Features of the datasets used in the experiments. For each dataset the number of examples and the number of continuous attributes is given.

Code	Name	Examples (+,-)	Continuous
ION	ionosphere	351 (225,126)	34
LIV	liver	345 (145,200)	6
PIM	pima-indians	768 (500,268)	8
SON	sonar	208 (97,111)	59
WD	wdbc	569 (212,357)	30
WIS	wisconsin	699 (458,241)	10
WP	wpbc	198 (47,151)	33

Rule 1: If AT1<6.5 and (AT2<4.5 or (7.5<AT2<9.5)) and ... and AT9 is irrelevant --> class is 0
Rule 2: Else --> class is 1 DEFAULT RULE

Cut points for AT1: {1.5, 2.5, 3.5, 4.5, 5.5, 6.5, 7.5, 8.5}
Cut points for AT2: {1.5, 2.5, 3.5, 4.5, 5.5, 6.5, 7.5, 8.5, 9.5}
Cut points for AT9: {1.5, 2.5, 3.5, 4.5, 5.5, 6.5, 7.5, 9.5}

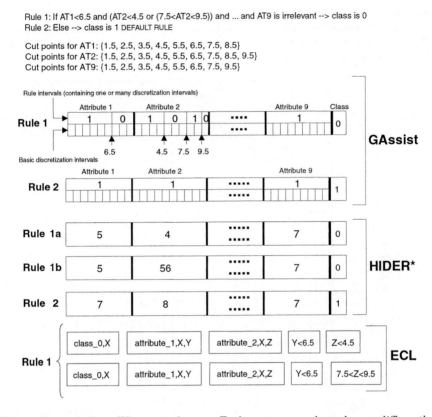

Fig. 1. Example from Wisconsin dataset. Each system encodes rule sets differently.

discretization methods (one of them 15 times) and using 10–fold cross–validation means 1540 runs for each system.

Each system uses its usual configuration settings, defined in previous work
[4,5,6]. Common values of the population size, generations, etc. are not suitable
in this case because we are testing systems with diverse structure (One *GA* run
generating a complete rule set vs. sequential *GA* runs learning one rule at a
time). As a consequence, the search space size for each system can vary, and this
leads to each system needing a specific set of parameters.

3.3   Results

We here report the results obtained by the three systems on each dataset for all
the discretization schemes. We also report the average performance achieved by
each discretization schemes, as a way of summarizing the results.

Table 2 reports the results obtained by ECL, *GAssist* and HIDER*, respectively. For each dataset, the average accuracy and the average number of rules
are reported for each discretization method used for obtaining the basic discretization intervals, including standard deviations.

Table 2. Average accuracies (Acc.) and average number of rules contained in the solution (# rules) obained by each sistems on the datasets with the different discretization methods. Best and worst results are highlighted. EW stands for equal width, EF stands for equal frequency.

ID	System		ID3	USD	Fayyad	Rand	EW10	EW20	EF10	EF20
ION	ECL	Acc.	**88.1**±4.4	74.4±4.9	87.4±3.1	**69.8**±4.0	71.6±4.5	73.3±5.0	72.6±4.3	71.2±5.4
		# rules	12.5±1.5	15.5±0.7	9.8±1.3	14.7±1.7	18.2±1.8	15.5±5.7	12.6±1.7	12.5±1.8
	GAssist	Acc.	90.4±5.0	90.7±4.9	**93.3**±4.3	90.2±5.1	92.3±4.1	92.5±3.7	90.0±4.8	**89.5**±4.5
		# rules	3.4±1.2	3.5±1.3	2.3±0.7	4.2±1.5	2.9±1.0	2.7±0.9	4.0±1.4	4.0±1.4
	HIDER*	Acc.	74.3±7.7	74.9±7.2	**89.5**±5.6	70.1±3.6	**58.4**±11.5	60.0±12.9	86.7±5.2	83.8±3.1
		# rules	33.2±1.9	32.3±2.5	2.0±0.0	5.6±2.8	1.0±0.0	1.1±0.3	10.1±3.4	22.0±2.6
LIV	ECL	Acc.	61.5±5.0	**66.6**±4.5	63.2±4.5	57.9±4.9	58.3±4.5	59.4±4.8	**57.1**±4.4	59.2±3.4
		# rules	13.5±1.7	11.3±1.3	1.3±0.7	12.0±1.5	7.9±1.8	8.6±1.7	9.3±1.8	12.7±1.3
	GAssist	Acc.	**65.5**±7.8	65.0±7.6	**59.5**±6.2	64.5±8.6	63.7±7.7	63.7±8.3	64.3±7.9	65.3±7.9
		# rules	9.1±2.0	8.7±1.9	2.9±1.4	8.2±1.9	7.7±1.8	8.0±1.7	8.8±2.0	9.3±2.2
	HIDER*	Acc.	58.3±6.5	**65.2**±3.9	61.2±5.1	**51.9**±4.5	59.4±6.9	58.6±7.5	60.6±7.1	62.4±5.8
		# rules	3.9±0.5	4.3±0.6	1.8±0.8	4.1±0.7	3.0±0.0	3.0±0.0	5.0±0.6	4.4±0.8
PIM	ECL	Acc.	**71.4**±1.9	69.7±4.2	68.3±2.3	**60.9**±4.5	63.2±2.6	62.4±1.9	63.9±1.5	61.7±3.9
		# rules	75.5±5.2	20.9±2.1	2.2±0.6	72.5±7.5	25.0±1.6	24.1±1.0	21.5±3.4	37.7±7.3
	GAssist	Acc.	73.9±3.9	73.7±4.0	**72.5**±4.6	73.6±4.4	73.6±4.3	73.8±4.7	73.6±4.0	**74.3**±3.8
		# rules	5.4±0.8	5.3±0.7	5.1±0.5	5.2±0.6	5.3±0.8	5.2±0.6	5.5±1.1	5.3±0.8
	HIDER*	Acc.	**74.2**±1.8	**74.2**±4.1	72.8±2.8	**71.7**±2.6	73.3±3.6	73.7±3.0	74.1±3.1	72.4±4.0
		# rules	5.8±0.6	4.9±0.5	2.7±1.2	4.2±0.8	4.1±0.8	3.8±0.7	4.2±0.6	4.4±1.1
SON	ECL	Acc.	64.4±5.1	72.0±6.9	**76.4**±2.1	**63.1**±4.8	69.3±5.9	66.4±5.1	72.4±4.0	68.3±4.2
		# rules	30.0±3.6	17.9±5.8	3.2±0.6	29.1±8.8	9.9±3.6	14.8±4.1	13.6±4.7	21.4±7.4
	GAssist	Acc.	73.3±9.8	73.4±9.1	74.5±9.4	73.1±10.2	74.3±10.1	**74.8**±8.8	72.8±9.6	**72.4**±9.8
		# rules	8.8±2.1	8.9±2.1	6.4±0.7	8.7±2.1	8.3±1.7	8.6±2.0	9.3±2.0	9.7±2.1
	HIDER*	Acc.	68.8±4.7	68.7±6.6	**72.6**±6.5	66.4±4.4	**62.4**±11.4	64.8±8.6	66.9±12.6	64.7±10.2
		# rules	30.9±3.0	24.4±3.1	4.4±1.3	16.5±3.9	11.6±1.9	14.6±0.9	60.3±5.3	63.2±3.4
WD	ECL	Acc.	91.4±4.3	93.3±4.1	**94.2**±3.1	**88.2**±4.8	91.2±3.6	89.3±5.3	89.9±5.7	90.0±4.3
		# rules	15.5±5.7	22.7±2.8	5.6±2.0	24.5±13.6	6.2±2.1	7.4±2.8	7.0±2.4	9.3±4.1
	GAssist	Acc.	**94.2**±3.1	93.9±3.1	94.0±3.1	93.7±3.5	**93.6**±3.2	94.0±3.2	**94.2**±3.1	94.0±3.1
		# rules	4.0±1.1	3.9±1.1	3.7±0.8	4.2±1.3	3.5±0.8	3.6±0.8	4.0±1.0	4.1±1.2
	HIDER*	Acc.	**85.6**±5.9	92.8±6.1	**93.5**±1.5	88.8±2.5	91.3±3.5	89.6±4.9	90.7±3.2	88.3±5.0
		# rules	22.3±2.1	16.8±1.0	3.9±0.7	14.3±1.5	4.9±0.5	5.5±0.5	7.9±1.3	11.2±1.2
WIS	ECL	Acc.	94.7±2.2	**95.6**±2.4	93.4±2.2	94.7±2.4	94.6±2.6	95.0±2.6	93.8±2.7	**93.3**±3.0
		# rules	13.7±2.1	3.3±0.8	6.1±1.3	5.6±1.6	11.4±2.3	14.5±2.1	11.3±2.8	11.2±2.7
	GAssist	Acc.	95.4±2.4	**96.0**±2.2	95.1±2.5	**95.1**±2.5	95.8±2.3	95.9±2.2	95.7±2.2	95.8±2.0
		# rules	2.5±0.7	2.2±0.5	3.3±0.5	2.6±0.6	2.3±0.6	2.4±0.6	2.7±0.7	2.5±0.6
	HIDER*	Acc.	96.4±2.3	**96.6**±1.8	95.8±2.4	**93.4**±2.0	96.3±2.1	96.4±2.0	96.4±1.6	95.4±2.1
		# rules	3.7±0.6	2.0±0.0	4.0±0.0	2.5±0.8	3.5±0.8	3.7±0.6	3.7±0.6	3.9±1.1
WP	ECL	Acc.	**76.9**±3.5	74.5±3.6	76.4±3.7	74.2±5.1	72.9±4.0	72.8±4.2	**70.2**±5.4	76.5±5.8
		# rules	21.6±2.1	20.6±1.7	2.2±0.6	21.3±0.9	15.5±2.3	20.2±2.3	19.6±1.1	20.3±2.1
	GAssist	Acc.	**72.7**±7.4	72.8±8.6	74.0±3.5	74.2±7.4	**75.7**±6.7	75.3±7.3	75.2±7.2	74.5±7.6
		# rules	5.5±1.6	5.4±1.4	2.0±0.2	5.2±1.3	3.8±1.4	3.6±1.3	4.5±2.3	5.0±2.0
	HIDER*	Acc.	69.2±6.5	**74.9**±2.8	72.9±3.1	**62.2**±7.1	74.3±3.6	73.8±9.8	68.0±8.3	67.7±7.8
		# rules	18.0±1.4	14.8±2.1	14.6±1.8	13.5±1.5	3.6±1.1	5.2±1.0	11.2±1.2	16.2±2.1

3.4   Analysis of Results

The aim of this analysis is to show if there is one or a group of discretization methods that present better performance than the others for evolutionary–based learning systems. It is important to note that the output of each discretization method is handled as a set of boundary points to define conditions over attributes in the decision rules provided by the evolutionary approaches. Those outputs will define the search space. The quality of solutions generated by the evolutionary systems will depend on how many and how good are those boundary points, and how well the evolutionary systems are able to join discretization intervals by means of genetic operators.

From Table 2, it can be noticed that ECL is sensitive to the discretization method used for obtaining the basic discretization intervals, thus the difference in the quality of the solutions depends strongly on the discretization method used.

ECL obtained the worst results when the random discretizer is used for determining the basic discretization intervals. This is due to the mutation operators used for modifying discretization intervals inside rules. In fact, these operators can add or subtract only one basic discretization interval at a time to the intervals the operators are applied to. In this way, if the number of basic discretization intervals is high, like in the case of the random discretizator, it is likely that the individuals are evolved very slowly. This is because only a little change in the discretization intervals can be applied by each mutation. Moreover, if this mutation was applied during the optimization phase, and had negative effects on the fitness of the individual, then the mutation is retracted. This cause the production of individuals that are too specific. However, when there are few discretization intervals, it is more likely that ECL can establish a good discretization interval for a rule.

In general ECL encountered problems with unsupervised discretizers that produces many intervals, e.g., the random discretizer, while it produces good results when the basic discretization intervals were produced with supervised discretizers. And the worst results are always obtained when the basic discretization intervals are determined by an unsupervised discretization method. On average ECL obtained the best results when the Fayyad & Irani's algorithm was used for producing the basic discretization intervals.

As far as the number of rules is concerned, in general ECL obtained the simpliest results when the Fayyad & Irani's algorithm was used. This is due to the fact the this algorithm produces a limited number of basic discretization intervals, and ECL can not generate the same diversity of rules that it can generate when other discretization methods that produces more basic discretization intervals, e.g., the ID3 method, are used.

The results for GAssist show a different behavior. Looking at the results we can see that the sensitivity to the discretizer is much smaller than in ECL or HIDER* . Moreover, there is not a clear correlation between the number of cut points of each discretizer and the performance of the system, showing that GAssist can explore successfully the search space.

Another important observation is that the discretizer performing worst in GAssist is the Fayyad & Irani's one, which was the best method for ECL. The reason of the poor performance of this discretizer in GAssist is due to the *hierarchical selection* operator used to control the *bloat effect* [16]. This operator introduces a bias towards compact individuals, in both rules and intervals (as this system uses rules in Conjunctive Normal Form, the number of intervals per rule can vary). While this feature is usually good, the difference between a well generalized solution and a too simple one is very small, specially if we use a discretizer that generates few cut points (like the Fayyad & Irani's). Therefore, it is easier to get the system stuck in a local optimum.

Table 3. Average results for each discretization method. From left to right, average accuracy, average average number of rules in the solutions, number of times the discretizator obtained the best and worst performance, rank of the discretizator and average number of cut points per attribute.

Method	Avg. Accuracy	Avg. # rules	# best	# worst	Rank	Cut points
ID3	78.1±11.9	16.1±16.3	6	2	3	83.2±48.1
USD	79.0±11.1	11.9±8.6	7	0	2	72.1±42.4
Fayyad	80.0±12.1	4.3±3.0	6	2	1	1.3±1.5
Random	75.1±13.2	13.3±15.1	0	9	7	106.1±97.5
EW10	76.5±13.2	7.6±5.9	1	3	6	9
EW20	76.5±13.0	8.4±6.3	1	0	6	19
EF10	77.6±12.4	11.2±12.0	1	2	4	9
EF20	77.2±12.1	13.8±13.8	1	3	5	19

The group of supervised discretization methods have better performance with HIDER* than the others. HIDER* expands the limits of intervals within rules slowly, similar to ECL, so it needs the boundary points to be accurate, i.e. they might be possible interval limits for conditions in decision rules. As ID3 and USD generates more intervals, boundary points can be more precise but the evolutionary system needs more time to find them. HIDER* has a specific mutation operator to remove attributes from rules, so when there are a lot of attributes it is more difficult to remove any if the number of intervals is high (ionosphere and sonar datasets are good examples). In contrast, when the number of discretization intervals is small, as provided by Fayyad & Irani's method, results can be better.

Finally, it is at least interesting to remark the performance of the random discretizer in GAssist, as it does perform quite well. Probably the reason is that the average number of cut points used across all domains was 106.1. This number is large enough to minimize the loss of information intrinsical in any discretization algorithm.

For summarizing the results, we propose two tables. In Table 3, we present the average results obtained by the discretization methods on all datasets. We also report the number of time a discretizator obtained the best and the worst performance on a dataset for a system. For instance, ID3 obtained six times the best results and three times the worst results. It can be seen that the supervised methods performed better than the unsupervised methods. In particular the random discretizer obtained the worst performance.

The second table, Table 4, proposes a ranking of the discretization methods. For each row of results from Table 2, the results are ranked. At the end of the table the average ranks are computed and the final rank is assigned to the discretization methods. From the Table 4 it emerges that USD results as the best discretizer. This is because USD has a stable behavior. USD never obtained the worst results, while in seven cases (out of 21, three methods by seven datasets) it obtained the best results. The worst rank is assigned to the random discretizer.

In fact it never obtained the best results, and in nine cases it obtained the worst ones.

Table 4. Ranking of the discretization methods. For each dataset the single results obtained by the sistems with each discretization methods are ranked. At the end the average of the ranks is computed and the final rank is assigned. E, G and H mean ECL, GAssist and HIDER*, respectively.

Discretizer	ION			LIV			PIM			SON			WD			WIS			WP			Average	Rank
	E	G	H	E	G	H	E	G	H	E	G	H	E	G	H	E	G	H	E	G	H		
ID3	1	5	5	3	1	7	1	2	1	7	5	2	3	1	8	3	6	2	1	8	5	3.67	3
USD	3	4	4	1	3	1	2	4	1	3	4	3	2	6	2	1	1	1	4	7	1	2.76	1
Fayyad	2	1	1	2	8	3	3	8	6	1	2	1	1	3	1	7	7	6	3	6	4	3.62	2
Random	8	6	6	7	4	8	8	5	8	8	6	5	8	7	6	3	8	8	5	5	8	6.52	8
EW10	6	3	8	6	6	5	5	5	5	4	3	8	4	8	3	5	3	5	6	1	2	4.81	6
EW20	4	2	7	4	6	6	6	3	4	6	1	6	7	3	5	2	2	2	7	2	3	4.19	4
EF10	5	7	2	8	5	4	4	5	3	2	7	4	6	2	4	6	5	2	8	3	6	4.67	5
EF20	7	8	3	5	2	2	7	1	7	5	8	7	5	3	7	8	3	7	2	4	7	5.14	7

4   Conclusions and Future Work

In general, all of the systems provided good performance for all of the discretization methods. However, GAssist differs from ECL and HIDER* with respect to the representational methodology. GAssist generates one solution with a set of decision rules, while HIDER* uses a sequential covering technique to find one rule at a time, removing examples covered by previous rules, and ECL evolves a population of rules from which a subset of rules representing the final solution is extracted.

It seems that the performance of Pittsburgh–based methodology is more stable, especially when the number of attributes and discretization intervals is high, which leads to a better exploration of the search space. This means that GAssist is less sensitive to the effect of discretization choices. On the other hand, supervised discretization methods, e.g., ID3, USD and Fayyad & Irani's algorithm, seem more appropriate for evolutionary algorithms where an individual encodes a single rule, which simplifies the search space. This reason justifies the fact that neither ECL or HIDER* find better results with non–supervised discretization methods.

There is no doubt that intrinsic properties of datasets might have influence on the results, so our future research directions will include to analyze the relationship between the discretizer outputs and the dataset features, and also, how this can give us a clue to choose the evolutionary approach (Pittsburgh–based approach or sequential covering methodology).

5   Related Work

Discretization is not the only way to handle real–valued attributes in Evolutionary Computation–based Machine Learning systems. Some examples are induction of decision trees (either axis–parallel or oblique), by either generating a full tree by means of genetic programming operators [17] or using a heuristic method to generate the tree and later a Genetic Algorithm or an Evolutionary Strategy to optimize the test performed at each node [18]. Other examples are inducing rules with real–valued intervals [19] or generating an instance set used as the core of a *k-NN* classifier [17]. Other approaches to concept learning, e.g., Neural Network, do not need any discretization for handling numerical values.

Also, several discretization algorithms are reported in the literature. Some examples not tested in this paper are the Mántaras discretizer [20] which is similar to that of Fayyad & Irani's, but using a different formulation of the entropy minimization. Another example is ChiMerge [21]. This discretizer creates an initial pool of cut points containing the real values in the domain to discretize, and iteratively merges neighbour intervals that make true a certain criterion based on the χ^2 statistical test.

Acknowldegments. This work was partially supported by the Spanish Research Agency Comisión Interministerial de Ciencia y Tecnología (CICYT) under Grants TIC2001-1143-C03-02, TIC2002-04160-C02-02 and TIC2002-04036-C05-03. The second author acknowledges the support provided by the "Departament d'Universitats, Recerca i Societat de la Informació de la Generalitat de Catalunya" under grants 2001FI 00514 and 2002SGR 00155.

References

1. Dougherty, J., Kohavi, R., Sahami, M.: Supervised and unsupervised discretization of continuous features. In: International Conference on Machine Learning. (1995) 194–202
2. Liu, H., Hussain, F., Tan, C., Dash, M.: Discretization: An enabling technique. Journal of Data Mining and Knowledge Discovery **6** (2002) 393–423
3. Bacardit, J., Garrel, J.M.: Evolution of adaptive discretization intervals for a rule-based genetic learning system. In: GECCO 2002: Proceedings of the Genetic and Evolutionary Computation Conference, Morgan Kaufmann Publishers (2002) 677
4. Bacardit, J., Garrel, J.M.: Evolving multiple discretizations with adaptive intervals for a pittsburgh rule-based genetic learning classifier system. In: GECCO 2003: Proceedings of the Genetic and Evolutionary Computation Conference, Springer (2003) 1818–1831
5. Giráldez, R., Aguilar-Ruiz, J., Riquelme, J.: Natural coding: A more efficient representation for evolutionary learning. In: GECCO 2003: Proceedings of the Genetic and Evolutionary Computation Conference, Chicago, Springer-Verlag Berlin Heidelberg (2003) 979–990

6. Divina, F., Keijzer, M., Marchiori, E.: A method for handling numerical attributes in GA-based inductive concept learners. In: GECCO 2003: Proceedings of the Genetic and Evolutionary Computation Conference, Chigaco, Springer (2003) 898–908

7. Divina, F., Marchiori, E.: Evolutionary concept learning. In: GECCO 2002: Proceedings of the Genetic and Evolutionary Computation Conference, New York, Morgan Kaufmann Publishers (2002) 343–350

8. Quinlan, J.R.: Induction of decision trees. In Shavlik, J.W., Dietterich, T.G., eds.: Readings in Machine Learning. Morgan Kaufmann (1986) Originally published in *Machine Learning* 1:81–106, 1986.

9. Giraldez, R., Aguilar-Ruiz, J., Riquelme, J., Ferrer-Troyano, F., Rodriguez, D.: Discretization oriented to decision rules generation. Frontiers in Artificial Intelligence and Applications **82** (2002) 275–279

10. Fayyad, U., Irani, K.: Multi-interval discretization of continuos attributes as preprocessing for classification learning. In: Proceedings of the 13th International Join Conference on Artificial Intelligence, Morgan Kaufmann Publishers (1993) 1022–1027

11. Rissanen, J.: Stochastic Complexity in Statistical Inquiry. World Scientific, River Edge, NJ. (1989)

12. Aguilar-Ruiz, J., Riquelme, J., Toro, M.: Evolutionary learning of hierarchical decision rules. IEEE Transactions on Systems, Man, and Cybernetics, Part B: Cybernetics **33(2)** (2003) 324–331

13. DeJong, K.A., Spears, W.M.: Learning concept classification rules using genetic algorithms. Proceedings of the International Joint Conference on Artificial Intelligence (1991) 651–656

14. Bacardit, J., Garrell, J.M.: Bloat control and generalization pressure using the minimum description length principle for a pittsburgh approach learning classifier system. In: Proceedings of the 6th International Workshop on Learning Classifier Systems, (in press), LNAI, Springer (2003)

15. Blake, C., Merz, C.: UCI repository of machine learning databases (1998)

16. Langdon, W.B.: Fitness causes bloat in variable size representations. Technical Report CSRP-97-14, University of Birmingham, School of Computer Science (1997) Position paper at the Workshop on Evolutionary Computation with Variable Size Representation at ICGA-97.

17. Llorà, X., Garrell, J.M.: Knowledge-independent data mining with fine-grained parallel evolutionary algorithms. In: Proceedings of the Genetic and Evolutionary Computation Conference (GECCO-2001), Morgan Kaufmann (2001) 461–468

18. Cantu-Paz, E., Kamath, C.: Inducing oblique decision trees with evolutionary algorithms. IEEE Transactions on Evolutionary Computation **7** (2003) 54–68

19. Stone, C., Bull, L.: For real! xcs with continuous-valued inputs. Evolutionary Computation Journal **11** (2003) 298–336

20. De Mántaras, R.L.: A distance-based attribute selection measure for decision tree induction. Machine Learning **6** (1991) 81–92

21. Kerber, R.: Chimerge: Discretization of numeric attributes. In: Proc. of AAAI-92, San Jose, CA (1992) 123–128

Real-Coded Bayesian Optimization Algorithm: Bringing the Strength of BOA into the Continuous World

Chang Wook Ahn[1], R.S. Ramakrishna[1], and David E. Goldberg[2]

[1] Department of Information and Communications
Kwang-Ju Institute of Science and Technology, Gwangju 500-712, Korea
{cwan,rsr}@kjist.ac.kr
http://parallel.kjist.ac.kr/~cwan/
[2] Department of General Engieering
University of Illinois, Urbana, IL 61801, USA
deg@illigal.ge.uiuc.edu
http://www-illigal.ge.uiuc.edu/goldberg/d-goldberg.html

Abstract. This paper describes a continuous estimation of distribution algorithm (EDA) to solve decomposable, real-valued optimization problems quickly, accurately, and reliably. This is the *real-coded Bayesian optimization algorithm* (rBOA). The objective is to bring the strength of (discrete) BOA to bear upon the area of real-valued optimization. That is, the rBOA must properly decompose a problem, efficiently fit each subproblem, and effectively exploit the results so that correct linkage learning even on nonlinearity and probabilistic building-block crossover (PBBC) are performed for real-valued multivariate variables. The idea is to perform a Bayesian factorization of a mixture of probability distributions, find maximal connected subgraphs (i.e. substructures) of the Bayesian factorization graph (i.e., the structure of a probabilistic model), independently fit each substructure by a mixture distribution estimated from clustering results in the corresponding partial-string space (i.e., subspace, subproblem), and draw the offspring by an independent subspace-based sampling. Experimental results show that the rBOA finds, with a sublinear scale-up behavior for decomposable problems, a solution that is superior in quality to that found by a mixed iterative density-estimation evolutionary algorithm (mIDEA) as the problem size grows. Moreover, the rBOA generally outperforms the mIDEA on well-known benchmarks for real-valued optimization.

1   Introduction

In the community of evolutionary computation, *estimation of distribution algorithms* (EDAs), also known as *probabilistic model building genetic algorithms* (PMBGAs), have attracted due attention of late [1], [2]. Incorporating (automated) linkage learning techniques into a graphical probabilistic model, EDAs exploit a feasible probabilistic model of selected (promising) solutions found so

K. Deb et al. (Eds.): GECCO 2004, LNCS 3102, pp. 840–851, 2004.
© Springer-Verlag Berlin Heidelberg 2004

far while efficiently traversing the search space [2]. EDAs iterate the three steps listed below, until some termination criterion is satisfied:

1. Select good candidates (i.e., solutions) from a (initially randomly generated) population (of solutions).
2. Estimate the probability distribution from the selected individuals.
3. Generate new candidates (i.e., offspring) from the estimated distribution.

It must be noted that the third step uniquely characterizes EDAs because it replaces traditional recombination and mutation operators employed by simple genetic algorithms (sGAs). Although the sGAs (with well-designed mixing operator) and EDAs deal with solutions (i.e., individuals) in quite different ways, it has been theoretically shown (and empirically observed) that their performances are quite close to each other [1], [2]. Moreover, EDAs ensure an effective mixing and reproduction of building blocks (BBs) due to their ability to accurately capture the BB structure of a given problem, thereby solving GA-hard problems with a linear or sub-quadratic performance in terms of (fitness) function evaluations (i.e., sublinear scale-up behavior) [2]-[5]. However, there is a trade-off between the accuracy of the estimated distribution and the efficiency of computation [4], [5]. For instance, a complicated, accurate model is recommended if the fitness function to be evaluated is computationally expensive.

A large number of EDAs have been proposed for discrete and real-valued (i.e., continuous) variables [1]-[6]. Depending on how intricate and involved the probabilistic models are, they are divided into three categories: *no dependencies*, *pairwise dependencies*, and *multivariate dependencies*. Among them, the category of multivariate dependencies endeavors to use general probabilistic models, thereby solving many difficult problems quickly, accurately, and reliably [2]. The more complex the probabilistic model the harder as well is the task of finding the best structure. At the expense of some computational efficiency (with regard to learning the model), they can significantly improve the overall time complexity for large decomposable problems due to their ability to largely reduce the number of (computationally expensive) fitness function evaluations [2]. *Extended compact genetic algorithm* (ecGA), *factorized distribution algorithm* (FAD), and *Bayesian optimization algorithm* (BOA) for discrete variables and *estimation of multivariate normal algorithm* (EMNA) and (*mixed*) *iterative density-estimation evolution algorithms* ((m)IDEAs) for real-valued variables belong to this category [1]-[6].

Note that the BOA is perceived to be an important effort that employs general probabilistic models for discrete variables [3], [4]. It employs techniques for modeling multivariate data by Bayesian networks so as to estimate the joint probability distribution of promising solutions. The BOA is very effective even on large decomposable (discrete) problems with tight BBs. It is only natural that the principles of BOA be tried on continuous (i.e., real-valued) variables. This attempt led to (m)IDEAs [5], [6] which exploit Bayesian Information Criterion (BIC) (that is a penalized maximum likelihood metric) for selecting a probabilistic model and employ a mixture of normal distributions for fitting the (chosen) model. Like the BOA, they do not require any problem dependent information. There is a general, but simple factorization mixture selection among the (m)IDEAs with regard to model accuracy and computational efficiency. This is called 'mIDEA' in this paper. The mIDEA clusters the selected individuals and subsequently estimates a factorized probability distribution in each cluster separately [5], [6]. It allows the mIDEA to efficiently model nonlinear dependencies by breaking up the nonlinearity between the variables and recognizing only linear relations in each cluster. This results in a better performance on epistatic and nonlinear (real-valued) problems.

It is noted that the power of BOA arises from modeling any type of dependency and realizing *probabilistic building-block crossover* (PBBC) that approximates *population-wise building-block crossover* by a probability distribution estimated from the results of proper decomposition [4]. Analogously to (one-bit) uniform crossover, the PBBC may shuffle as many superior partial solutions (i.e., BBs) as possible in order to bring about an efficient and reliable search for the optimum. However, the mIDEA cannot realize the PBBC although learning various types of dependency is possible. This is explained below.

BBs can be defined by groups of real-valued variables, each having values in some neighborhood (i.e., small interval), that break up the problem into smaller chunks which can be intermixed to reach the optimum. As the mIDEA clusters the selected individuals on the problem dimension itself, preserving and breeding BBs is quite difficult unless clusters contain many BBs at the same time. However, the probability of coming up with clusters is very small and decreases exponentially as the problem size grows. In other words, the mIDEA can hardly find an optimal solution without maintaining at least one cluster that contains most of the superior BBs. It follows that the mIDEA may not be very effective on large decomposable problems. It is noteworthy that many real-world optimization problems are bounded difficult: the problems can be (additively) decomposed into subproblems of a certain/bounded order [4], [5].

In this paper, we propose a real-coded BOA (rBOA) along the lines of BOA. The rBOA can solve various types of decomposable problems in an efficient and scalable manner, and also find a high quality solution to traditional benchmark cases.

The rest of the paper is organized as follows. Section 2 explains the rBOA in detail and Section 3 presents the experimental results obtained with the algorithms. The paper concludes with a summary in Section 4.

2   Proposed Real-Coded BOA

This section describes rBOA as a tool to efficiently solve problems of bounded difficulty with a sublinear scale-up behavior. Fig. 1 presents the pseudocode of rBOA.

Parameters. $\mathcal{P}$ ($\mathcal{S}$, $\mathcal{O}$): population (selected individuals, offspring), $\mathbf{Z}^i$: i-th subproblem $\mathcal{S}(\mathbf{Z}^i)$ ($\mathcal{O}(\mathbf{Z}^i)$): selected individuals (offspring) that contain the genes corresponding to $\mathbf{Z}^i$, K: number of mixtures, n: population size, q: offspring size, c_i: number of clusters on $\mathbf{Z}^i$, $\mathbf{C}_i^j$: partial-individuals in j-th cluster over $\mathbf{Z}^i$.

The rest of parameters are described in the body of the paper.

Step 1. Randomly generate initial population $\mathcal{P}$

  for $i \leftarrow 0$ **to** $n-1$ **do**

   $\mathcal{P} \leftarrow \textbf{\textit{RandomVector}}()$;

Step 2. Select τ portion individuals $\mathcal{S}$ from the population $\mathcal{P}$

   $\mathcal{S} \leftarrow \textbf{\textit{Selection}}(\tau, \mathcal{P})$;

Step 3. Search a probabilistic model structure and decompose the problem from the structure

   $\zeta \leftarrow \textbf{\textit{Search}}(K, \mathcal{S})$;

   $(\mathbf{Z}^0, \cdots, \mathbf{Z}^{m-1}) \leftarrow \textbf{\textit{Decomposition}}(\mathbf{Y}, \zeta)$

Step 4. Fit each submodel by mixing the clustered normal distributions of subspace

  for $i \leftarrow 0$ **to** $m-1$ **do**

   $(\mathbf{C}_i, c_i) \leftarrow \textbf{\textit{Clustering}}(\mathcal{S}(\mathbf{Z}^i))$;

   for $j \leftarrow 0$ **to** $c_i - 1$ **do**

    $\beta_{ij} \leftarrow |\mathbf{C}_i^j| / \sum_{k=0}^{c_i-1} |\mathbf{C}_i^k|$;

    $\theta_j^{\mathbf{Z}^i} \leftarrow \textbf{\textit{Estimation}}(\mathbf{C}_i^j)$;

   $f_{(\zeta, \theta)}(\mathbf{Y}) \leftarrow \prod_{i=0}^{m-1} \sum_{j=0}^{c_i-1} \beta_{ij} f_{(\zeta^{\mathbf{Z}^i}, \theta_j^{\mathbf{Z}^i})}(\mathbf{Z}^i)$;

Step 5. Generate offspring $\mathcal{O}$ from the joint pdf $f_{(\zeta, \theta)}(\mathbf{Y})$ on the basis of subproblems

  for $i \leftarrow 0$ **to** $m-1$ **do**

   for $k \leftarrow 0$ **to** $q-1$ **do**

    $I_c \leftarrow \textbf{\textit{ChooseCluster}}(\beta_{i,0}, \cdots, \beta_{i,c_i-1})$;

    $\mathcal{O}^{\mathbf{Z}^i} \leftarrow \textbf{\textit{Sampling}}(\zeta^{\mathbf{Z}^i}, f_{(\zeta^{\mathbf{Z}^i}, \theta_{I_c}^{\mathbf{Z}^i})}(\mathbf{Z}^i))$;

Step 6. Create a new population $\mathcal{P}$ by replacing some individuals with $\mathcal{O}$

   $\mathcal{P} \leftarrow \textbf{\textit{PartiallyReplace}}(\mathcal{O})$;

Step 7. If the termination criteria are not met, go to **Step 2**.

Fig. 1. Pseudocode of rBOA.

2.1   Model Selection

A *factorization* (or a *factorized probability distribution*) is a probability distribution that can be described as a product of generalized probability density functions (gpdfs) [5]. *Bayesian factorizations*, also known as *Bayesian factorized probability distributions* come under a general class of factorizations [5], [7]. A Bayesian factorization estimates a joint gpdf for multivariate (dependent) variables by a product of univariate conditional gpdfs of each random variable. The Bayesian factorization is represented by a directed acyclic graph, called a Bayesian factorization graph, in which nodes (vertices) and edges (arcs) identify the corresponding variables (in the data set) and the conditional dependencies between variables, respectively [5].

An l-dimensional real-valued optimization problem is considered. In general, a pdf is represented by a probabilistic model $\mathcal{M}$ that consists of a structure ζ and an associated vector of parameters $\boldsymbol{\theta}$ (i.e., $\mathcal{M} = (\zeta, \boldsymbol{\theta})$) [5], [6]. As the rBOA employs the Bayesian factorization, the joint pdf of a problem can be encoded as

$$f_{(\zeta,\boldsymbol{\theta})}(\mathbf{Y}) = \prod_{i=0}^{l-1} f_{\dot{\boldsymbol{\theta}}^i}(Y_i | \Pi_i) \tag{1}$$

where $\mathbf{Y} = (Y_0, \cdots, Y_{l-1})$ presents a vector of real-valued random variables, Π_i is the set of parents of Y_i (i.e., the set of nodes from which there exists an edge to Y_i), and $f_{\dot{\boldsymbol{\theta}}^i}(Y_i | \Pi_i)$ is the conditional pdf of Y_i conditioned on Π_i with its parameters $\dot{\boldsymbol{\theta}}^i$.

There are two basic factors behind any scheme for learning the structure of a probabilistic model (i.e., model selection): a scoring metric and a search procedure [3]-[6]. The scoring metric measures the quality of the structure of Bayesian factorization graph and the search procedure efficiently traverses the space of all feasible structures for finding the best one with regard to a given scoring metric. It may be noted that BOA and mIDEA employ a *Bayesian information criterion* (BIC) as the scoring metric and an incremental greedy algorithm as the search procedure.

Let $\boldsymbol{S}$ be the set of selected individuals, viz., $\boldsymbol{S} = (\mathbf{y}^0, \mathbf{y}^1, \cdots, \mathbf{y}^{|\boldsymbol{S}|-1})$. The BIC metric that should be minimized is formulated as follows:

$$BIC\left(f_{(\zeta,\boldsymbol{\theta})}(\mathbf{Y}), \boldsymbol{S}\right) = -\ln \left(\prod_{j=0}^{|\boldsymbol{S}|-1} f_{(\zeta,\boldsymbol{\theta})}(\mathbf{y}^j) \right) + \lambda \ln\left(|\boldsymbol{S}|\right) |\boldsymbol{\theta}|$$

$$= -\sum_{j=0}^{|\boldsymbol{S}|-1} \ln\left(f_{(\zeta,\boldsymbol{\theta})}\left(\mathbf{y}^j\right)\right) + \lambda \ln\left(|\boldsymbol{S}|\right) |\boldsymbol{\theta}| \tag{2}$$

[5], [6] where λ regularizes the extent of penalty. In (2), the first and second terms represent the model fitting error and the model complexity, respectively. Since minimal negative log-likelihood is equivalent to minimal entropy, (2) is rewritten as

$$BIC\left(f_{(\zeta,\boldsymbol{\theta})}(\mathbf{Y}), \boldsymbol{S}\right) = |\boldsymbol{S}| h\left(f_{(\zeta,\boldsymbol{\theta})}(\mathbf{Y})\right) + \lambda \ln\left(|\boldsymbol{S}|\right) |\boldsymbol{\theta}| \tag{3}$$

[5], [6] where $h\left(f_{(\zeta,\boldsymbol{\theta})}(\mathbf{Y})\right)$ represents the differential entropy of $f_{(\zeta,\boldsymbol{\theta})}(\mathbf{Y})$.

Although the BIC fails to exactly capture the types of interaction between variables, the important point is to have a knowledge of the variables which are dependent regardless of linearity or nonlinearity. The reason for this assertion is that the dependent type itself is learned in the model fitting phase (in Section

2.2). However, the BIC might lead to incorrect factorization if there is some kind of symmetry in the selected individuals. In other words, there is a high possibility that the dependent variables are learned as independent ones. In order to avoid this problem as well as to enhance the reliability of learning dependency, a (joint) mixture distribution is employed for modeling the selected individuals. With this in view, the BIC in (3) can be modified as (4)

$$BIC\left(K, f_{(\zeta,\theta)}(\mathbf{Y}), \mathcal{S}\right) = \sum_{i=1}^{K} \left\{|\mathcal{S}_i| \, h\left(f_{(\zeta,\theta_i)}(\mathbf{Y})\right)\right\} + K\lambda \ln\left(|\mathcal{S}|\right)|\boldsymbol{\theta}_i| \quad (4)$$

where K is the number of mixture components, $|\mathcal{S}_i|$ is the expected number of selected individuals drawn from a probability distribution $f_{(\zeta,\theta_i)}(\mathbf{Y})$, and $\boldsymbol{\theta}_i$ is parameters of ith mixture component.

The incremental greedy algorithm starts with an empty graph with no edges, and proceeds by (incrementally) adding an edge that maximally improves the metric until no more improvement is possible [3]-[6]. The greedy algorithm does not find an optimal structure in general because searching for the structure is an NP-complete problem. However, the computed structure is good enough for encoding most important interactions between variables of the problem [3], [4].

A Bayesian factorization graph that represents a probabilistic model structure is obtained after factorization and application of the incremental greedy algorithm.

2.2   Model Fitting and Sampling

It may be noted that maximal connected subgraphs of a Bayesian factorization graph are the component subproblems. A hard optimization problem can thus be reduced to the problem of solving several easy subproblems (if the resulting graph consists of several maximally connected subgraphs). Any graph search algorithm can be applied to extract the maximally connected subgraphs. As the BBs in real space are defined as a set of variables with some neighborhood that can be intermixed for finding an optimum, the variables of each subproblem can eventually build up BBs in view of their close interactions.

The BOA models any type of dependency because it maintains all the conditional probabilities, without losing any information due to the finite cardinality (of the set of variables). Moreover, the BOA naturally performs the PBBC because it strictly separates and independently treats the maximally connected subgraphs all through the (probabilistic) model selection, model fitting, and (offspring) sampling phases. Hence, the BOA can solve difficult problems quickly, accurately, and reliably.

On the other hand, the mIDEA clusters the selected individuals for breaking up the nonlinear dependencies between variables and subsequently estimates a factorized probability distribution in each cluster [5], [6]. However, it cannot realize the PBBC even though any type of mixture distribution, depending on the pdf used in each cluster, can be constructed. The reason is discussed below.

The clustering is performed on the problem space itself (instead of the sub-problem space) and the mixture distribution is constructed from a linear combination of factorized pdfs (estimated in clusters). In the sampling phase, an entire individual is drawn from a proportionally chosen pdf. Hence, at least one cluster must contain almost all the (superior) BBs of the problem if the aim is to find an optimal solution. In order to get such clusters, however, a huge population and a very large number of clusters are required. It may result in an exponential scale-up behavior, even if the problem is decomposable into subproblems of bounded order. In other words, the mIDEA may easily be misled by many (deceptive) suboptima as it cannot efficiently capture and boost (superior) BBs.

After finding a feasible probabilistic model through the Bayesian factorization (in Section 2.1), the promising subproblems whose variables have linear and/or nonlinear interactions are obtained by extracting maximal connected subgraphs from the resulting factorization graph. Next to the problem decomposition, a clustering is performed on each subproblem (i.e., subspace). Although any clustering algorithm can be used, a computationally inexpensive algorithm is desirable. The purpose of clustering is twofold: comprehending the nonlinearity and searching the space effectively. That is, the clustering can model the nonlinear dependencies using a combination of piecewise linear interaction models resulting from breaking up the nonlinearity. In addition, the clustering has an effect of partitioning each subspace for effective search. Thereby, the rBOA can treat each subproblem independently and then fit it efficiently by mixing pdfs even in the presence of nonlinearly dependent variables. Although Bosman [5] suggested a framework from which to carve the model as a matter of course, the focus is somewhat different from that of rBOA.

Let $\mathbf{Z}^i = \left\{ Z_0^i, \cdots, Z_{|\mathbf{Z}^i|-1}^i \right\}$ be a vector of random variables of the ith subproblem (viz., $\bigcup_i \mathbf{Z}^i = \mathbf{Y}$ and $\bigcap_i \mathbf{Z}^i = \phi$), in which the variables are already topologically sorted for drawing new partial-individuals corresponding to the subproblem. Let $\zeta^{\mathbf{Z}^i}$ and $\theta^{\mathbf{Z}^i}$ indicate a (probabilistic model) structure of the variables $\mathbf{Z}^i$ (i.e., substructure) and its associated parameters, respectively (viz., $\mathcal{M}^{\mathbf{Z}^i} = (\zeta^{\mathbf{Z}^i}, \theta^{\mathbf{Z}^i}))$, and $f_{\left(\zeta^{\mathbf{Z}^i}, \theta_j^{\mathbf{Z}^i}\right)}(\mathbf{Z}^i)$ represent a joint pdf (i.e. probability distribution) with parameters $\theta_j^{\mathbf{Z}^i}$ that are estimated from jth cluster over $\mathbf{Z}^i$ (i.e., the ith subspace). Therefore, a joint pdf of $\mathbf{Y}$ can be constructed by a product of linear combinations of subproblem pdfs as given by

$$f_{(\zeta,\theta)}(\mathbf{Y}) = \prod_{i=0}^{m-1} \sum_{j=0}^{c_i-1} \beta_{ij} f_{\left(\zeta^{\mathbf{Z}^i}, \theta_j^{\mathbf{Z}^i}\right)}(\mathbf{Z}^i) \tag{5}$$

where m is the number of subproblems, c_i is the number of clusters of $\mathbf{Z}^i$, β_{ij} is the mixture coefficients, $\beta_{ij} \geq 0$, and $\sum_{j=0}^{c_i-1} \beta_{ij} = 1$ for all i. In general, the mixture coefficient β_{ij} is proportional to the number of individuals of the jth cluster of $\mathbf{Z}^i$ [5], [6].

If the problem is non-decomposable, the operational mechanism of rBOA is not much different from that of mIDEA except that the mIDEA can construct a different probabilistic model for each cluster. However, the encouraging fact is that many (real-world) problems are indeed decomposable into several subproblems of bounded order [4].

As preparation for drawing new individuals (i.e., offspring), univariate conditional pdfs of the proposed mixture distribution must be derived on the basis of subproblems. Therefore, (5) is rewritten as (6) from a sampling point of view:

$$
f_{(\zeta,\boldsymbol{\theta})}(\mathbf{Y}) = \prod_{i=0}^{m-1} \sum_{j=0}^{c_i-1} \beta_{ij} f_{\left(\zeta^{\mathbf{z}^i},\boldsymbol{\theta}_j^{\mathbf{z}^i}\right)}(\mathbf{Z}^i) = \prod_{i=0}^{m-1} \sum_{j=0}^{c_i-1} \beta_{ij} \prod_{k=0}^{|\mathbf{Z}^i|-1} f_{\ddot{\boldsymbol{\theta}}_j^k}\left(Z_k^i | \Pi_{Z_k^i}\right) \quad (6)
$$

Let $\mathbf{X} = \left\{ Z_k^i, \Pi_{Z_k^i} \right\}$, and $\boldsymbol{\mu}$, Σ be the mean vector and the symmetric covariance matrix of $\mathbf{X}$, respectively. Employing the normal pdf due to its inherent advantages - close approximation and simple analytic properties - the univariate conditional pdfs in (6) can be obtained from

$$
f_{\ddot{\boldsymbol{\theta}}_j^k}\left(Z_k^i | \Pi_{Z_k^i}\right) = f_{\mathcal{N}}\left(X_0 | X_1, \cdots, X_{|\mathbf{X}|-1}\right) = \frac{1}{\sqrt{2\pi}\widetilde{\sigma}} e^{-\frac{(X_0-\widetilde{\mu})^2}{2\widetilde{\sigma}^2}} \quad (7)
$$

[5], [6] where $\widetilde{\sigma} = \frac{1}{\sqrt{(\Sigma^{-1})_{0,0}}}$, $\widetilde{\mu} = \mu_0 - \frac{\sum_{i=1}^{|\mathbf{X}|-1}(X_i-\mu_i)\left(\Sigma^{-1}\right)_{i,0}}{(\Sigma^{-1})_{0,0}}$.

Sampling the new individuals from the resulting factorization of (6) is straightforward [5]. At first, the normal pdf over the jth cluster of the normal mixture estimate for the ith subproblem is selected with probability β_{ij}. Subsequently, a multivariate string (i.e., partial-individual) corresponding to $\mathbf{Z}^i$ can be drawn by simulating the univariate conditional pdfs (i.e., eqn. (7)) of the chosen normal pdf which models one of the promising partitions (i.e., a superior BB) of a subspace (i.e., subproblem). By repeating this for all the subproblems, superior BBs can be mixed and bred for subsequent search.

3   Experiments and Discussion

This section investigates the performance of rBOA by comparing it with that of the mIDEA through computer experiments. Solution quality returned by the fixed number of function evaluations is taken to be a performance measure. For simplicity, one normal distribution is used for Bayesian factorization (i.e., $K = 1$). The BEND leader algorithm (with a threshold value of 0.3) is used as the clustering algorithm due to its speed and flexibility [5], [6]. Truncation selection with $\tau = 0.5$ and BIC with $\lambda = 0.5$ have been invoked. Since no prior information about the problem structure is available in practice, we set $|\mathbf{Y}| - 1$ for the number of allowable parents. Each experiment is terminated when the number of function evaluations reaches $100n$, where n is the population size. All the results were averaged over 100 runs.

3.1   Test Problems

Two types of problem are considered for investigating the performance of rBOA on decomposable problems. The first test problem is a real-valued deceptive problem (RDP) composed of real-valued trap functions. The problem is given by

$$F_{RDP}(\mathbf{y}) = \sum_{i=0}^{m-1} f_{trap}\left(y_{i \cdot k}, \cdots, y_{i \cdot k + (k-1)}\right) \tag{8}$$

where $y_j \in [0,1]$, $\forall j$, k and m are the subproblem size and the number of subproblems, respectively, and $f_{trap}\left(y_{i \cdot k}, \cdots, y_{i \cdot k + (k-1)}\right)$ is a k-dimensional trap function defined by

$$f_{trap}\left(y_{i \cdot k}, \cdots, y_{i \cdot k + (k-1)}\right) = \begin{cases} 1.0, & \text{if } 0.8 \le y_{i \cdot k + j} \le 1.0, \forall j, \\ 0.8 - \sqrt{\frac{\sum_{j=0}^{k-1} y_{i \cdot k + j}^2}{k}}, & \text{otherwise.} \end{cases} \tag{9}$$

The second test problem is a real-valued nonlinear problem (RNP) that is constructed by concatenating Rosenbrock functions. The problem is formulated as

$$F_{RNP}(\mathbf{y}) = \sum_{i=0}^{m-1} f_R\left(y_{i \cdot k}, \cdots, y_{i \cdot k + (k-1)}\right) \tag{10}$$

where $y_j \in [-5.12, 5.12]$, $\forall j$, and $f_R\left(y_{i \cdot k}, \cdots, y_{i \cdot k + (k-1)}\right)$ is a k-dimensional Rosenbrock function defined as Table 1.

Note that the variables of the subproblem (i.e., real-valued trap function, Rosenbrock function) strongly interact with each other.

Moreover, four traditional benchmark problems that do not have any obvious 'decomposibility' features are also investigated. They are shown in Table 1 and should be minimized.

3.2   Experimental Results and Discussion

Fig. 2(a) compares the proportion of correct BBs for the algorithms as applied to the RDPs with $k = 2$ and varying m. Since a RDP consists of m subproblems, the effective problem difficulty is proportional to m in general. Hence, the population size is supplied by a linear model $\alpha \cdot m$, viz., $n = 100m$. The results show that the solution found by the rBOA is much better than that returned by the mIDEA. It is also seen that the rBOA achieves stable quality of solutions while the performance of mIDEA rapidly deteriorates as the problem size increases. That is, the rBOA exhibits a linear scale-up behavior for (additively)

Table 1. Traditional benchmark problems for numerical optimization

Problem	Function	Range
Sphere	$\sum_{j=0}^{l-1} y_j^2$	$y_j \in [-5, 5]$
Griewank	$\frac{1}{4000} \sum_{j=1}^{l-1} (y_j - 100)^2 - \prod_{j=0}^{l-1} \cos\left(\frac{y_j - 100}{\sqrt{j+1}}\right) + 1$	$y_j \in [-600, 600]$
Michalewicz	$\sum_{j=0}^{l-1} sin(y_j) sin^2 0 \left(\frac{(j+1) \cdot y_j^2}{\pi}\right)$	$y_j \in [0, \pi]$
Rosenbrock	$\sum_{j=1}^{l-1} \left\{ 100 \cdot (y_j - y_{j-1}^2)^2 + (1 - y_{j-1})^2 \right\}$	$y_j \in [-5.12, 5.12]$

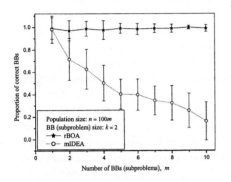 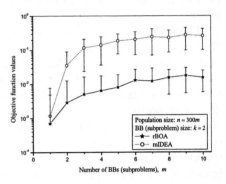

(a) Performance on F_{RDP} with $k = 2$ and varying m.

(b) Performance on F_{RNP} with $k = 2$ and various m.

Fig. 2. Comparison of the rBOA and mIDEA on decomposable problems.

decomposable deceptive problems; while the mIDEA has an exponential scalability. Fig. 2(b) depicts the objective function values returned by the algorithms when applied to the RNP with $k = 2$ and varying m. A linear model is also used for supplying population, i.e., $n = 300m$. It is seen that the performance of both algorithms gracefully deteriorate as the number of subproblems grows. It implies that the scale-up behavior of both algorithms becomes sublinear for decomposable nonlinear problems. However, the results show that the rBOA outperforms the mIDEA rather substantially with regard to the quality of solution. From Figs. 2(a) and (b), we may conclude that the rBOA finds a better solution with a sublinear scale-up behavior for decomposable problems than does the mIDEA. Note that the good solution and the sublinear scale-up behavior of the rBOA bring about the problem decomposition (in the model selection), and subspace-wise model fitting and (offspring) sampling operations.

Table 2 compares the solutions found by the algorithms as applied to the test functions of Table 1. The results show that the rBOA is also superior to

Table 2. Performance of the algorithms on the benchmarks ($l = 5$)

Problem Type	Population Size	mIDEA		rBOA	
		Mean	STD	Mean	STD
Sphere	$n = 500$	0.000310	0.001758	$< 10^{-7}$	-
Griewank	$n = 2000$	0.067267	0.018433	0.063001	0.016415
Michalewicz	$n = 500$	-4.606095	0.066925	-4.813710	0.019322
Rosenbrock	$n = 5000$	0.003899	0.010477	0.017825	0.091988

the mIDEA except when working on the Rosenbrock function. This is explained below.

The variables of the Rosenbrock function are highly nonlinear. In other words, they strongly interact around a curved valley. Also, it is symmetric. It is clear that incorrect factorizations (i.e., no dependencies between variables) are encountered at an early stage of rBOA and mIDEA. Due to the incorrect structure, they try to solve the problems by treating the variables in isolation. Of course, finding an optimum in this way is difficult because any given algorithm does not cross the intrinsic barrier. After a few generations, individuals start to collect round the curved valley. In fact, the rBOA can easily capture such a nonlinear symmetric dependency. However, the factorization employed one normal distribution in this experiment. It may bring about incorrect linkage learning (i.e., independent interaction) due to symmetry. This is no cause for concern when more than one normal distribution is used for factorization (the situation arising of more than one distribution is not shown in this paper, though). On the other hand, the mIDEA can cope with the cancellation effect to some extent by the use of clustering in the overall problem space.

As a result, the proposed rBOA finds a high quality solution with a sublinear scale-up behavior for decomposable problems while finding acceptable solutions to popular test problems.

4   Conclusion

This paper has presented a real-coded BOA as a continuous EDA. Decomposable problems were the prime targets. Sublinear scale-up behavior (of rBOA) was a major objective. This was achieved by linkage learning and probabilistic building-block crossover (PBBC) on real-valued variables. As a step in this direction, Bayesian factorization was performed by means of a mixture of pdfs, the substructures were extracted from the resulting Bayesian factorization graph (i.e., problem decomposition), and each substructure was fitted by mixing normal pdfs whose parameters were estimated from the subspace-based (e.g., subproblem-based) clusters. In the sampling phase, offspring were generated by a subproblem-wise sampling procedure.

Experimental studies demonstrated that the rBOA finds a better solution and exhibits a superior scale-up behavior (i.e., sublinear) (vis-a-vis the mIDEA) while encountering decomposable problems regardless of inherent problem characteristics such as deception and/or nonlinearity. Moreover, the solution of rBOA is generally better than that of mIDEA for traditional real-valued benchmarks.

Although more work needs to be done, rBOA's strategy of decomposing problems, modeling the resulting building blocks, and then searching for better solutions appears to have certain advantages over clustered model building that has been suggested and used elsewhere. Certainly, there is much work to be done in exploring the method of decomposition, the types of models utilized, as well as their computational implementation and speed, but this path appears to lead to a class of practical procedures that should find widespread use in many engineering and scientific applications.

References

1. P. Larrañaga and J. A. Lozano, *Estimation of Distribution Algorithms: A New Tool for Evolutionary Computation*, Kluwer Academic Publishers, 2002.
2. M. Pelikan, D. E. Goldberg, and F. G. Lobo, "A Survey of Optimization by Building and Using Probabilistic Models," *Computational Optimization and Applications*, vol. 21, pp. 5–20, 2002.
3. M. Pelikan, D. E. Goldberg, and E. Cantú-Paz, "BOA: The Bayesian optimization algorithm," *Proceedings of GECCO'99*, pp. 525–532, 1999.
4. M. Pelikan, *Bayesian Optimization Algorithm: From Single Level to Hierarchy*, Ph. D. Thesis, University of Illinois at Urbana-Champaign, Urbana, IL, 2002.
5. P. A. N. Bosman, *Design and Application of Iterated Density-Estimation Evolutionary Algorithms*, Ph. D. Thesis, Utrecht University, TB Utrecht, The Netherlands, 2003.
6. P. A. N. Bosman and D. Thierens, "Advancing Continuous IDEAs with Mixture Distributions and Factorization Selection Metrics," *Proceedings of OBUPM workshop at GECCO'01*, pp. 208–212, 2001.
7. S. L. Lauritzen, *Graphical Models*, Clarendon Press, Oxford, 1996.

Training Neural Networks with GA Hybrid Algorithms

Enrique Alba and J. Francisco Chicano

Departamento de Lenguajes y Ciencias de la Computación
University of Málaga, SPAIN
{eat,chicano}@lcc.uma.es

Abstract. Training neural networks is a complex task of great importance in the supervised learning field of research. In this work we tackle this problem with five algorithms, and try to offer a set of results that could hopefully foster future comparisons by following a kind of standard evaluation of the results (the Prechelt approach). To achieve our goal of studying in the same paper population based, local search, and hybrid algorithms, we have selected two gradient descent algorithms: Backpropagation and Levenberg-Marquardt, one population based heuristic such as a Genetic Algorithm, and two hybrid algorithms combining this last with the former local search ones. Our benchmark is composed of problems arising in Medicine, and our conclusions clearly establish the advantages of the proposed hybrids over the pure algorithms.

1 Introduction

The interest of the research in Artificial Neural Networks (ANNs) resides in the appealing properties they exhibit: adaptability, learning capability, and ability to generalize. Nowadays, ANNs are receiving a lot of attention from the international research community with a large number of studies concerning training, structure design, and real world applications, ranging from classification to robot control or vision [1].

The neural network training task is a capital process in supervised learning, in which a pattern set made up of pairs of inputs plus expected outputs is known beforehand, and used to compute the set of weights that makes the ANN to learn it. One of the most popular training algorithms in the domain of neural networks is the Backpropagation (or generalized delta rule) technique [2], a gradient-descent method. Other techniques such as evolutionary algorithms (EAs) have been also applied to the training problem in the past [3,4], trying to avoid the local minima that so often appear in complex problems. Although training is a main issue in ANN's design, many other works are devoted to evolve the layered structure of the ANN or even the elementary behavior of the neurons composing the ANN. For example, in [5] a definition of neurons, layers, and the associated training problem is analyzed by using parallel genetic algorithms; also, in [6] the architecture of the network and the weights are evolved by using the EPNet evolutionary system. It is really difficult to perform a revision of this

K. Deb et al. (Eds.): GECCO 2004, LNCS 3102, pp. 852–863, 2004.

topic; however, the work of Yao [7] represents an excellent starting point to get acquired of the research in training ANNs.

The motivation of the present work is manyfold. First, we want to perform a standard presentation of results that promotes and facilitates future comparisons. This sounds common sense, but it is not frequent that authors follow standard rules for comparisons such as the structured Prechelt's set of recommendations [8], a "de facto" standard for many ANN researchers. A second contribution is to include in our study, not only the well known Genetic Algorithm (GA) and Backpropagation algorithm, but also the Levenberg-Marquardt (LM) approach [9], and two additional hybrids. The potential advantages coming from an LM utilization merit a detailed study. We have selected a benchmark from the field of Medicine, composed of three classification problems: diagnosis of breast cancer, diagnosis of diabetes in Pima Indians, and diagnosis of heart disease.

The remainder of the article is organized as follows. Section 2 introduces the Artificial Neural Network computation model. Next, we give a brief description of the algorithms under analysis (Section 3). The details of the experiments and their results are shown in Section 4. Finally, we summarize our conclusions and future work in Section 5.

2   Artificial Neural Networks

Artificial Neural Networks are computational models naturally performing a parallel processing of information [10]. Essentially, an ANN can be defined as a pool of simple processing units (*neurons*) which communicate among themselves by means of sending analog signals. These signals travel through weighted connections between neurons. Each of these neurons accumulates the inputs it receives, producing an output according to an internal activation function. This output can serve as an input for other neurons, or can be a part of the network output. In Fig. 1 left we can see a neuron in detail.

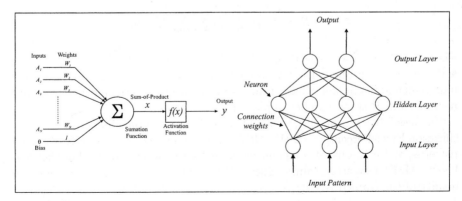

Fig. 1. An artificial neuron (left) and a multilayer perceptron (right)

There is a set of important issues involved in the ANN design process. As a first step, the architecture of the network has to be decided. Initially, two major options are usually considered: *feedforward* networks and *recurrent* networks (additional considerations regarding the *order* of the ANN exist, but are out of our scope). The feedforward model comprises networks in which the connections are strictly feedforward, i.e., no neuron receives input from a neuron to which the former sends its output, even indirectly. The recurrent model defines networks in which feedback connections are allowed, thus making the dynamical properties of a capital importance. In this work we will concentrate on the first and simpler model: the feedforward networks. To be precise, we will consider the so-called *multilayer perceptron* (MLP) [11], in which units are structured into ordered layers, and connections are allowed only between adjacent layers in an input-to-output sense (see Fig. 1 right).

For any MLP, several parameters such as the number of layers and the number of units per layer must be defined. After having done this, the last step in the design is to adjust the weights of the network, so that it produces the desired output when the corresponding input is presented. This process is known as *training* the ANN or *learning* the network weights. Network weights comprise both the previously mentioned connection weights, as well as a *bias* term for each unit. The latter can be viewed as the weight of a constant saturated input that the corresponding unit always receives. As initially stated, we will focus on the learning situation known as *supervised* training, in which a set of input/desired-output patterns is available. Thus, the ANN has to be trained to produce the desired output according to these examples. The input and output of the network are both real vectors in our case.

In order to perform a supervised training we need a way of evaluating the ANN output error between the actual and the expected output. A popular measure is the Squared Error Percentage (SEP). We can compute this error term just for one single pattern or for a set of patterns. In this last case, the SEP is the average value of the patterns individual SEP. The expression for this global SEP is:

$$SEP = 100 \cdot \frac{o_{max} - o_{min}}{P \cdot S} \sum_{p=1}^{P} \sum_{i=1}^{S} (t_i^p - o_i^p)^2 \ . \tag{1}$$

where t_i^p and o_i^p are, respectively, the i-th components of the expected vector and the actual current output vector for the pattern p; o_{min} and o_{max} are the minimum and maximum values of the output neurons, S is the number of output neurons, and P is the number of patterns.

In classification problems, we could use still an additional measure: the Classification Error Percentage (CEP). CEP is the percentage of incorrectly classified patterns, and it is a usual complement to any of the other two (SEP or the well-known MSE) raw error values, since CEP reports in a high-level manner the quality of the trained ANN.

3   The Algorithms

We use for our study several algorithms to train ANNs: the Backpropagation algorithm, the Levenberg-Marquardt algorithm, a Genetic Algorithm, a hybrid between Genetic Algorithm and Backpropagation, and a hybrid between Genetic Algorithm and Levenberg-Marquardt. We briefly describe them in the following paragraphs.

3.1   Backpropagation

The Backpropagation algorithm (BP) [2] is a classical domain-dependent technique for supervised training. It works by measuring the output error, calculating the gradient of this error, and adjusting the ANN weights (and biases) in the descending gradient direction. Hence, BP is a gradient-descent local search procedure (expected to stagnate in local optima in complex landscapes).

First, we define the squared error of the ANN for a set of patterns:

$$E = \sum_{p=1}^{P}\sum_{i=1}^{S}(t_i^p - o_i^p)^2 \ . \tag{2}$$

The actual value of the previous expression depends on the weights of the network. The basic BP algorithm (without momentum in our case) calculates the gradient of E (for all the patterns in our case) and updates the weights by moving them along the gradient-descendent direction. This can be summarized with the expression $\Delta \mathbf{w} = -\eta \nabla E$, where the parameter $\eta > 0$ is the learning rate that controls the learning speed. The pseudo-code of the BP algorithm is shown in Fig. 2.

```
InitializeWeights;
while not StopCriterion do
    for all i,j do
        w_{ij} := w_{ij} - η ∂E/∂w_{ij} ;
    endfor;
endwhile;
```

Fig. 2. Pseudo-code of the BP algorithm

3.2   Levenberg-Marquardt

The Levenberg-Marquardt algorithm (LM) [9] is an approximation to the Newton method used also for training ANNs. The Newton method approximates the error of the network with a second order expression, which contrasts to the Backpropagation algorithm that does it with a first order expression. LM is popular in the ANN domain (even it is considered the first approach for an unseen MLP training task), although it is not that popular in the metaheuristics field. LM updates the ANN weights as follows:

$$\Delta \mathbf{w} = - \left[\mu I + \sum_{p=1}^{P} J^p(\mathbf{w})^T J^p(\mathbf{w}) \right]^{-1} \nabla E(\mathbf{w}) \ . \tag{3}$$

where $J^p(\mathbf{w})$ is the Jacobian matrix of the error vector $\mathbf{e}^p(\mathbf{w})$ evaluated in $\mathbf{w}$, and I is the identity matrix. The vector error $\mathbf{e}^p(\mathbf{w})$ is the error of the network for pattern p, that is, $\mathbf{e}^p(\mathbf{w}) = \mathbf{t}^p - \mathbf{o}^p(\mathbf{w})$. The parameter μ is increased or decreased at each step. If the error is reduced, then μ is divided by a factor β, and it is multiplied by β in other case. Levenberg-Marquardt performs the steps detailed in Fig. 3. It calculates the network output, the error vectors, and the Jacobian matrix for each pattern. Then, it computes $\Delta \mathbf{w}$ using (3) and recalculates the error with $\mathbf{w} + \Delta \mathbf{w}$ as network weights. If the error has decreased, μ is divided by β, the new weights are maintained, and the process starts again; otherwise, μ is multiplied by β, $\Delta \mathbf{w}$ is calculated with a new value, and it iterates again.

```
InitializeWeights;
while not StopCriterion do
    Calculates e^p(w) for each pattern;
    e1 := ∑_{p=1}^{P} e^p(w)^T e^p(w);
    Calculates J^p(w) for each pattern;
    repeat
        Calculates Δw;
        e2 := ∑_{p=1}^{P} e^p(w + Δw)^T e^p(w + Δw);
        if (e1 <= e2) then
            μ := μ * β;
        endif;
    until (e2 < e1);
    μ := μ/β;
    w := w + Δw;
endwhile;
```

Fig. 3. Pseudo-code of the LM algorithm

3.3   Genetic Algorithm

A GA [12] is a stochastic general search method. It proceeds in an iterative manner by generating new populations of individuals from the old ones. Every individual is the encoded (binary, real, etc.) version of a tentative solution. The canonical algorithm applies stochastic operators such as selection, crossover, and mutation on an initially random population in order to compute a new population. In *generational* GAs all the population is replaced with new individuals. In *steady-state* GAs (used in this work) only one new individual is created and it replaces the worst one in the population if it is better. The pseudo-code of the GA we are using here can be seen in Fig. 4. The search features of the GA contrast with those of the BP and LM in that it is not trajectory-driven, but population-driven. The GA is expected to avoid local optima frequently by promoting exploration of the search space, in opposition to the exploitative trend usually allocated to local search algorithms like BP or LM.

```
t := 0;
Initialize:        P(0) := {a₁(0),...,aμ(0)} ∈ Iμ;
Evaluate:          P(0) : {Φ(a₁(0)),...,Φ(aμ(0))};
while ι(P(t)) ≠ true do          //Reproductive loop
       Select:     P'(t) := s_Θ_s (P(t));
       Recombine:  P''(t) := ⊗_Θ_c (P'(t));
       Mutate:     P'''(t) := m_Θ_m (P''(t));
       Evaluate:   P'''(t) : {Φ(a₁'''(t)),...,Φ(aλ'''(t))};
       Replace:    P(t+1) := r_Θ_r (P'''(t) ∪ Q);
       t := t + 1;
endwhile;
```

Fig. 4. Pseudo-code of a Genetic Algorithm

3.4   Hybrid Algorithms

Here, the hybridization refers to the inclusion of problem-dependent knowledge in a general search template [13,14]. We can distinguish two kinds of hybridization: *strong* and *weak* hybridization. In the first one, the knowledge is included using specific operators or representations. In the latter, several algorithms are combined somehow. In this last case, an algorithm can be used to improve the results of another one separately or it can be used as an operator of the other.

The hybrid algorithms that we use in this work are combinations of two algorithms (weak hybridization), where one of them acts as an operator in the other. We combine a GA with the BP algorithm (GABP), and a GA with LM (GALM). In both cases the problem-specific algorithm (BP and LM) is used as a mutation-like operation of the general search template (GA). Therefore, GAxx is a GA (Fig. 4) in which the mutation has been replaced by the "xx" algorithm that is applied with probability p_t.

4   Empirical Study

After discussing the algorithms, we present in this section the experiments performed and their results. The benchmark for training and the parameters of the algorithms are presented in the next subsection. The analysis of the results is shown in Subsection 4.2.

4.1   Computational Experiments

We tackle three classification problems. These problems consist in determining the class that a certain input vector belongs to. Each pattern from the training pattern set contains an input vector and its desired output vector. These vectors are formed by real numbers. However, in classification problems, the output of the network must be interpreted as a class. Such interpretation can be performed in different ways [8]. One of them consists in assigning an output neuron to each class. When an input vector is presented to the network, the network response is the class associated with the output neuron with the larger value. This method is known as *winner-takes-all* and it is employed in this work.

The instances solved here belong to the PROBEN1[1] benchmark [8]: Cancer, Diabetes, and Heart. We now briefly detail them:

- **Cancer**: Diagnosis of breast cancer. Classify a tumor as either benign or malignant based on cell descriptions gathered by microscopic examination. There are 699 examples that were obtained by Dr. William H. Wolberg at the University of Wisconsin Hospitals, Madison [15,16,17,18].
- **Diabetes**: Diagnose diabetes of Pima Indians. Based on personal data and the results of medical examinations, decide whether a Pima Indian individual is diabetes positive or not. There are 768 examples from the *National Institute of Diabetes and Digestive and Kidney Diseases* by Vincent Sigillito [19].
- **Heart**: Predict heart disease. Decide whether at least one of four major vessels is reduced in diameter by more than 50%. This decision is made based on personal data and results of medical examinations. There are 920 examples from four different sources: Hungarian Institute of Cardiology in Budapest (Andras Janosi, M.D.), University Hospital of Zurich in Switzerland (William Steinbrunn, M.D.), University Hospital of Basel in Switzerland (Mathhias Pfisterer, M.D.), V.A. Medical Center of Long Beach and Cleveland Clinic Foundation (Robert Detrano, M.D., Ph.D.) [20,21].

The structure of the MLP used for any problem accounts for three layers (input-hidden-output) having six neurons in the hidden layer. The number of neurons in the input and output layers depends on the concrete instance. The activating function of the neurons is the sigmoid function. Table 1 summarizes the network architecture for each instance.

To evaluate an ANN, we split the pattern set into two subsets: the training one and the test one. The ANN is trained with all the algorithms by using the training pattern set, and then it is evaluated on the unseen test pattern set. The training set for each instance is approximately made of the first 75% of the examples, while the last 25% constitutes the test set. The exact number of patterns for each instance is presented in Table 1 to ease future comparisons.

After presenting the problems, we now turn to describe the parameters for the algorithms (Table 2). To get the parameters of the pure algorithms we performed

---

[1] Available from ftp://ftp.ira.uka.de/pub/neuron/proben1.tar.gz.

Table 1. MLP architecture and patterns distribution for all instances

Instance	Architecture	Patterns	
		Training	Test
Cancer	9 - 6 - 2	525	174
Diabetes	8 - 6 - 2	576	192
Heart	35 - 6 - 2	690	230

some preliminary experiments and defined those with the best results. The hybrid algorithms GABP and GALM use the same parameters as their elementary components. However, the mutation operator of the GA is not applied; instead, it is replaced by BP or LM, respectively. The BP and LM are applied with an associated probability p_t only to one individual generated after recombination at each iteration. When applied, BP/LM only performs one single *epoch*.

Table 2. Parameters for the algorithms

		BC	DI	HE
BP	Epochs	1000	1000	500
	η	0.01	0.01	0.001
LM	Epochs	1000	1000	500
	μ	0.001	0.001	0.001
	β	10	10	10
GA	Population size	64		
	Selection	Roulette (2 inds.)		
	Recombination	SPX ($p_c = 1.0$)		
	Mutation	Bit-Flip ($p_m = 1/lenght$)		
	Replacement	Elitist		
	Stop criterion	1064 evals.		
GAxx	p_t	1.0	1.0	0.5
	Epochs of xx	1	1	1

As to the representation of the individuals, the weights are encoded as binary vectors. These vectors allocate 16 bit substrings to represent a real value in the interval $[-1, +1]$. The weights associated to any link arriving to a neuron (and the neuron bias) are placed together in the chromosome.

Finally, we need to define a fitness function to guide the search of the GA (either pure or hybrid). The fitness function (to be maximized) is the inverse of the SEP for the training set.

4.2   Analysis of the Results

In this section we present the results obtained after the application on the three instances of the five algorithms. We report the mean and the standard deviation of the CEP for the test pattern set after performing 50 independent runs. Table 3 and Fig. 5 show the results.

Table 3. Results of the ANN training

CEP(%)		BP	LM	GA	GABP	GALM
Cancer	$\overline{x}$	0.91	3.17	16.76	1.43	0.02
	σ_n	0.28	1.29	6.15	4.87	0.11
Diabetes	$\overline{x}$	21.76	25.77	36.46	36.46	28.29
	σ_n	0.38	3.26	0.00	0.00	1.15
Heart	$\overline{x}$	27.41	34.73	41.50	54.30	22.66
	σ_n	1.48	3.68	14.68	20.03	0.82

A first conclusion is that the GA obtains always a higher CEP than BP, LM and the hybrids (except for Heart and GABP). This is not a surprising fact, since the GA performs a rather explorative search in this kind of problems. BP is slightly more accurate than LM for all the instances, what we did not expect after the accurate behavior of LM in other studies.

With respect to the hybrid algorithms, the results do confirm our hypothesis of work: GALM is more accurate than GABP. In fact, this is noticeable since BP performed better than LM. Of course, we are not saying that this holds for any ANN training problem. However, we do state a clear claim after these results, i.e., GABP has received "too much" attention from the community, while maybe GALM could have worked out lower error percentages. To help the reader we also display these results in a graph in Fig. 5.

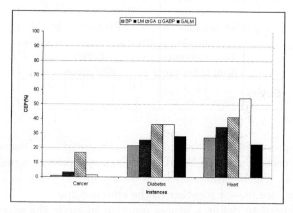

Fig. 5. Comparison among the algorithms (CEP)

We have traced the evolution of each algorithm for the Cancer instance to better explain how the different algorithms work (Fig. 6). We measure the SEP of the network in each epoch of the algorithm. For population-based algorithms (GA, GABP and GALM) we trace the SEP of the best fitness network. Each trace line represents the average SEP over 50 independent runs. We can observe that LM is the faster algorithm, followed by BP, what confirms intuition on the velocity of local search compared to GAs and hybrids. BP an LM clearly stagnate before 200 epochs in a solution. The GA is the slowest algorithm, and its hybridization with BP, and especially with LM, shows an acceleration of the evolution. An interesting observation is that the algorithms with the lowest SEP (BP and LM) do not always get the lowest CEP (best classification) for the test patterns. For example, GALM, which exhibits the lowest CEP, has only a modest value of SEP in the training process. This is due to the overtraining of the network in the BP and the LM algorithms, and confirms the necessity of reporting both, ANN errors and classification percentages in this field of research.

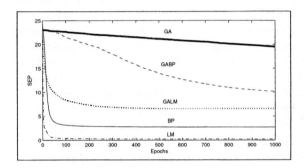

Fig. 6. Average evolution of SEP for the algorithms on the Cancer instance

There are many interesting works related to neural network training that also solve the instances tackled here. But unfortunately, some of the results are not comparable with ours, because they use a different definition of the training and test sets; this is why we consider a capital issue to adhere to any standard way of evaluation like the one proposed by Prechelt [8]. However, we did find some works for meaningful comparisons.

For the Cancer instance we find that the best mean CEP [22] is 1.1%, which represents a lower accuracy compared to our 0.02% obtained with the GALM hybrid. In [23], a CEP close to 2% for this instance is achieved, while our GALM is one hundred times more accurate. The mentioned work uses 524 patterns for the training set and the rest for the test set, that is, almost exactly our configuration with only one pattern changed (a minor detail), and therefore the results can be compared. The same occurs for the work of Yao and Liu [6], where their EPNet algorithm works out neural networks of a lower quality (1.4% of CEP).

For the Diabetes instance, a CEP of 30.11% is reached in [24] (outperformed by our BP, LM, and GALM) with the same network architecture as in our work. In [6] we found for this instance a 22.4% of CEP (outperformed by our BP with a 21.76%).

Finally, in [24] we found a 45.71% of CEP for the Heart instance using the same architecture. In this case, all our algorithms outperform their CEP measure (except GABP).

In summary, while we have found some of the more accurate results for the three instances, it is still needed to get ahead on other instances, always keeping in mind the importance of reporting results in a standardized manner.

5   Conclusions

In this work we have tackled the neural network training problem with five algorithms: two well-known problem-specific algorithms such as Backpropagation and Levenberg-Marquardt, a general metaheuristic such as a Genetic Algorithm, and two hybrid algorithms combining the Genetic Algorithm with the problem-specific techniques. To compare the algorithms we solve three classification problems from the domain of Medicine: the diagnosis of breast cancer, the diagnosis of diabetes in the Pima Indians, and the diagnosis of heart disease.

Our results show that the problem-specific algorithms (BP and LM) get lower classification error than the genetic algorithm, and thus confirm numerically what intuition can only suggest. The hybrid algorithm GALM outperforms in two of the three instances the classification error of the problem-specific algorithms. This makes GALM look as a promising algorithm for neural network training. On the other hand, many of the classification errors obtained in this work are below those found in the literature, what represents a cutting-edge result. As a future work we plan to add new algorithms to the analysis, and to apply them to more instances, especially in the domain of Bioinformatics.

Acknowledgments. This work has been partially funded by the Ministry of Science and Technology and FEDER under contract TIC2002-04498-C05-02 (the TRACER project, http://tracer.lcc.uma.es).

References

1. Alander, J.T.: Indexed Bibliography of Genetic Algorithms and Neural Networks. Technical Report 94-1-NN, University of Vaasa, Department of Information Technology and Production Economics (1994)
2. Rumelhart, D., Hinton, G., Williams, R.: Learning Representations by Backpropagation Errors. Nature **323** (1986) 533–536
3. Cotta, C., Alba, E., Sagarna, R., Larrañaga, P.: Adjusting Weights in Artificial Neural Networks using Evolutionary Algorithms. In Larrañaga, P., Lozano, J., eds.: Estimation of Distribution Algorithms. A New Tool for Evolutionary Computation, Kluwer Academic Publishers (2001) 357–373
4. Cantú-Paz, E.: Pruning Neural Networks with Distribution Estimation Algorithms. In Erick Cantú-Paz et al., ed.: GECCO 2003, LNCS 2723, Springer-Verlag (2003) 790–800
5. Alba, E., Aldana, J.F., Troya, J.M.: Full Automatic ANN Design: A Genetic Approach. In Mira, J., Cabestany, J., Prieto, A., eds.: New Trends in Neural Computation, Springer-Verlag (1993) 399–404
6. Yao, X., Liu, Y.: A New Evolutionary System for Evolving Artificial Neural Networks. IEEE Transactions on Neural Networks **8** (1997) 694–713

7. Yao, X.: Evolving Artificial Neural Networks. Proceedings of the IEEE **87** (1999) 1423–1447
8. Prechelt, L.: PROBEN1 — A Set of Neural Network Benchmark Problems and Benchmarking Rules. Technical Report 21, Fakultät für Informatik Universität Karlsruhe, 76128 Karlsruhe, Germany (1994)
9. Hagan, M.T., Menhaj, M.B.: Training Feedforward Networks with the Marquardt Algorithm. IEEE Transactions on Neural Networks **5** (1994)
10. McClelland, J.L., Rumelhart, D.E.: Parallel Distributed Processing: Explorations in the Microstructure of Cognition. The MIT Press (1986)
11. Rosenblatt, F.: Principles of Neurodynamics. Spartan Books, New York (1962)
12. Holland, J.H.: Adaptation in Natural and Artificial Systems. The University of Michigan Press, Ann Arbor, Michigan (1975)
13. Davis, L., ed.: Handbook of Genetic Algorithms. Van Nostrand Reinhold, New York (1991)
14. Cotta, C., Troya, J.M.: On Decision-Making in Strong Hybrid Evolutionary Algorithms. Tasks and Methods in Applied Artificial Intelligence, Lecture Notes in Artificial Intelligence **1415** (1998) 418–427
15. Bennett, K.P., Mangasarian, O.L.: Robust Linear Programming Discrimination of Two Linearly Inseparable Sets. Optimization Methods and Software **1** (1992) 23–34
16. Mangasarian, O.L., Setiono, R., Wolberg, W.H.: Pattern Recognition via Linear Programming: Theory and Application to Medical Diagnosis. In Coleman, T.F., Li, Y., eds.: Large-Scale Numerical Optimization. SIAM Publications, Philadelphia (1990) 22–31
17. Wolberg, W.H.: Cancer Diagnosis via Linear Programming. SIAM News **23** (1990) 1–18
18. Wolberg, W.H., Mangasarian, O.L.: Multisurface Method of Pattern Separation for Medical Diagnosis Applied to Breast Cytology. In: Proceedings of the National Academy of Sciences. Volume 87., U.S.A (1990) 9193–9196
19. Smith, J.W., Everhart, J.E., Dickson, W.C., Knowler, W.C., Johannes, R.S.: Using the ADAP Learning Algorithm to Forecast the Onset of Diabetes Mellitus. In: Proceedings of the Twelfth Symposium on Computer Applications in Medical Care, IEEE Computer Society Press (1988) 261–265
20. Detrano, R., Janosi, A., Steinbrunn, W., Pfisterer, M., Schmid, J., Sandhu, S., Guppy, K., Lee, S., Froelicher, V.: International Application of a New Probability Algorithm for the Diagnosis of Coronary Artery Disease. American Journal of Cardiology (1989) 304–310
21. Gennari, J.H., Langley, P., Fisher, D.: Models of Incremental Concept Formation. Artificial Intelligence **40** (1989) 11–61
22. Ragg, T., Gutjahr, S., Sa, H.: Automatic Determination of Optimal Network Topologies Based on Information Theory and Evolution. In: Proceedings of the 23rd EUROMICRO Conference, Budapest, Hungary (1997)
23. Land, W.H., Albertelli, L.E.: Breast Cancer Screening Using Evolved Neural Networks. In: IEEE International Conference on Systems, Man, and Cybernetics, 1998. Volume 2., IEEE Computer Society Press (1998) 1619–1624
24. Erhard, W., Fink, T., Gutzmann, M.M., Rahn, C., Doering, A., Galicki, M.: The Improvement and Comparison of Different Algorithms for Optimizing Neural Networks on the MasPar MP-2. In Heiss, M., ed.: Neural Computation – NC'98, ICSC Academic Press (1998) 617–623

Growth Curves and Takeover Time in Distributed Evolutionary Algorithms

Enrique Alba and Gabriel Luque

Departamento de Lenguajes y Ciencias de la Computación
E.T.S.I. Informática, Campus Teatinos, 29071 Málaga (España)
{eat,gabriel}@lcc.uma.es

Abstract. This paper presents a study of different models for the growth curves and takeover time in a distributed EA (dEA). The calculation of the takeover time and the dynamical growth curves is a common analytical approach to measure the selection pressure of an EA. This work is a first step to mathematically unify and describe the roles of the migration rate and the migration frequency in the selection pressure induced by the dynamics of dEAs. In order to achieve these goals we evaluate the appropriateness of the well-known logistic model and of a hypergraph model for dEAs. After that, we propose a corrected hypergraph model and two new models based in an extension of the logistic one. Our results show that accurate models for growth curves can be defined for dEAs, and explain analytically the migration rate and frequency effects.

1 Introduction

The increasing availability of clusters of machines has endorsed the fast development of parallel EAs (PEAs) [1]. Most popular PEAs split the whole population in separate subpopulations that are dealt with independently (islands). A sparse exchange of information among the component subalgorithms leads to a whole new class of algorithms that do not only perform faster (more steps by unit time), but that often lead to superior numerical performance [2,3].

In the core of these parallel EAs we can find a spatially structured distributed EA (dEA) that has been implemented (usually) in parallel on a cluster of machines interconnected by a communication network. Many interesting parallel issues can be defined and studied in PEAs, but in this work we are interested in the distributed algorithm model using multiple populations, that is really the responsible of the search features. In this article we concentrate on the dynamics of the distributed EA, in particular in developing a mathematical description for the takeover time, i.e., the time for the best solution to completely fill up all the subpopulations of the dEA. We first will propose and analyze several models for the induced growth curves, and then address the calculation of the takeover time. In this work, only tournament selection is considered. Also, since we only focus on selection, we expect an easy extension of the results to many other EAs.

In order to design a dEA we must take several decisions. Among them, a chief decision is to determine the migration policy: topology, migration rate (number

K. Deb et al. (Eds.): GECCO 2004, LNCS 3102, pp. 864–876, 2004.

of individuals that undergo migration in every exchange), migration frequency (number of steps in every subpopulation between two successive exchanges), and the selection/replacement of the migrants. In general, decisions on these choices are made by experimental studies. Therefore, it would be interesting if we could provide an analytical basis for such decisions.

Several works have studied the takeover time and growth curves for other classes of structured EAs [4,5,6,7,8,9,10]. In general, these works are oriented to study cellular EAs (with the important exception of the Sprave's one [10]) and it really exists a gap in the studies about dEAs from which something could be gained for other researchers or applications.

In the present work we focus on the influence of migration rate and migration frequency in the takeover time and in the growth curves. To achieve this goal we fix the topology to a simple and, at the same time, widely used one: a static directional ring. We also preset the policies of selection/replacement of the migrants. The emigrants are selected by binary tournament while the immigrants are included in the target population only if they are fitter than the worst-existing solution. In our analysis we will use the binary tournament mechanism with an elitist replacement (concretely, we use a $(\mu + \mu)$-dEA). We defer for a future work the theoretical analysis on other topologies and selection methods. Our contribution is to put to work the logistic model [5] and the hypergraph models [10], since they have never been tested and compared in practice (to the best of our knowledge). Then, we will propose three new mathematical models for the dynamics of selection: a corrected hypergraph plus two extended logistic models. Our aim is to improve on the accuracy (low error) of the initially tested models and consequently compute takeover times.

This paper is organized as follows. Section 2 is an introduction containing some preliminary background about previous works. Section 3 studies the effects of the migration frequency in the resulting growth curves; just after that, we extend the analysis by considering also the migration rate (Section 4). In Section 5, we analyze the predicted takeover times provided by the models. In the last section we summarize the conclusions and give some hints on the future work.

2   Performance of the Existing Theoretical Models

A common analytical approach to study the selection pressure of an EA is to characterize its takeover time [11], i.e., the number of generations it takes for the best individual in the initial population to fill the entire population under selection only. The growth curves are another important issue to analyze the dynamics of the dEAs. These growth curves are functions that associate the number of generations of the algorithm with the proportion of the best individual in the whole population. In this section we describe briefly the main models found in the literature defining the behavior of structured population EAs.

2.1   The Logistic Model

Let us begin by discussing the work of Sarma and De Jong (1997) for cellular EAs. In that work, they performed a detailed empirical analysis of the effects of the neighborhood size and shape for several local selection algorithms. They proposed a simple quantitative model for cellular EAs based in the logistic family of curves already known to work for panmictic EAs [11]. In summary, the proposed equation is (1):

$$P(t) = \frac{1}{1 + \left(\frac{1}{P(0)} - 1\right) e^{-at}} \, . \tag{1}$$

where a is a growth coefficient and $P(t)$ is the proportion of the best individual in the population at time step t. This model threw accurate results for synchronous updates of square shaped cellular EAs. Recently, for the asynchronous case, improved models has been proposed in [9] not following a logistic growth. Anyway, using a logistic curve represents an interesting precedent that however should be validated for dEAs. In brief, we will do so in this article.

2.2 The Hypergraph Model

Sprave (1999) has proposed a unified description for any non-panmictic population structured EA, that could even end in an accurate model for panmictic populations (since they can be considered as fully connected structured populations). He modelled the population structure by means of *hypergraphs*. A hypergraph is an extension of a canonical graph. The basic idea of a hypergraph is the generalization of edges from pairs of vertices to arbitrary subsets of vertices.

He developed a method to estimate growth curves and takeover times. This method is based on the calculation of the diameter of the actual population structure and on the probability distribution induced by the selection operator. In fact, Chakraborty et al. (1997) previously calculated the success probabilities for the most common selection operators (p_{select}), what represents an interesting complement for putting hypergraphs to work in practice. A complete description of the hypergraph model can be found in [10].

2.3 Other Models

Although the logistic model is relatively well known, and hypergraphs could play an important role in the field, they are not the only existing models that can inspire or influence the present study. Gorges-Schleuter (1999) also accomplished a theoretical study about takeover times for a cellular ES algorithm. In her analysis, she studied the propagation of information over time through the entire population. She finally obtained a linear model for a ring population structure and a quadratic model for a torus population structure.

In a different work, Rudolph (2000) carried out a theoretical study on the takeover time in populations with array and ring topologies. He derived lower bounds for arbitrary connected neighborhood structures, lower and upper bounds for array-like structures, and an exact closed form expression for a ring topology.

Later, Cantú-Paz (2000) studied the takeover time in dEAs where the migration occurs in every iteration, which is the lower bound of the migration frequency value. He generalized the panmictic model presented in [11] by adding a policy-dependent term. That term represents the effects of the policy used to select migrants and the individuals that they replace at the receiving island.

Giacobini et al. (2003) studied the takeover time in cellular EAs that use asynchronous cell update policies. The authors presented quantitative models for the takeover time in asynchronous cellular EAs with a ring topology.

In the present work, we focus on the described models: logistic and hypergraphs. The first one (logistic) is based on biological processes and it is well known in the case of cellular EAs, the other type of structured EAs. The second model (hypergraphs) posses a unique unified-like feature for all non-panmictic algorithms. We do not use the results of the other works directly since either they are linked to specialized algorithms or have a different focus (selection policy).

3   Effects of the Migration Frequency

In this section we analyze the effects of the migration frequency over the growth of the best individual in dEAs. In this aim we begin by performing an experimental set of tests for several migration frequencies. First, we describe the parameters used in these experiments, and later we analyze the obtained results.

3.1   Parameters

We have performed several experiments with different values of migration frequencies: 1, 2, 4, 8, 16, 32, and 64 generations. In general, researchers use frequencies in this range. Notice that a low frequency value (e.g., 1) means high coupling, while a high value (e.g., 64) means loose coupling (large gap). The rest of parameters are kept constant. In the experiments, we use a $(\mu + \mu)$-dEA with 8 islands (512 individuals per island), binary tournament selection, a static ring topology, and a preset moderate migration rate (8 individuals chosen by binary tournament are exchanged at each migration step). For all tests, we use randomly generated populations with individual fitness between 0 and 1023. Then we introduce a single best individual (fitness = 1024) in a randomly selected island. In hypergraphs we have used an expected level of accuracy of $\varepsilon = 2.5 \cdot 10^{-4}$. For the actual curves we have performed 100 independent runs.

In order to compare the accuracy of the models we proceeded to calculate the mean square error (2) between the actual values and the theoretically predicted ones (where k is the number of points of the predicted curve). The MSE gives the error for an experiment. But we also define a metric that summarizes the error for all experiments, thus allowing to perform a quantitative comparison between the different models easily. We studied several statistical values (mean, median, standard deviation, etc.) but finally we decided to use the $\| \cdot \|_1$ (3) that represents the area below the MSE curve (E is the number of experiments).

$$MSE(model) = \frac{1}{k} \sum_{i=1}^{k} (model_i - experimental_i)^2 \ . \tag{2}$$

$$\|model\|_1 = \sum_{i=1}^{E} |MSE(model)| \ . \tag{3}$$

3.2   Analysis of the Results

Now, we analyze the curves that have been obtained in the experiments. Figure 1 contains the lines of the actual takeover time for different migration frequencies. This figure shows that, for low frequency values, the dEA resembles the panmictic case [11]. This is common sense, since there exists high interaction among the subalgorithms. However, for higher frequency values (uncoupled search), the observed behavior is different: the subpopulations in the islands having the best solution converge quickly, and then the global convergence of the algorithm stops progressing (flat lines) until a migration of the best individual happens to take place. The observed effect is that of a *stairs*-like curve. The time span of each step in such a curve is governed by the migration frequency. The higher the migration frequency value, the largest the span of the step.

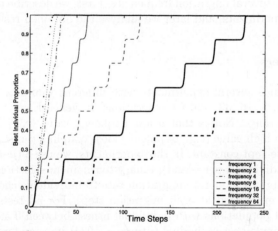

Fig. 1. Actual growth curves for several migration frequencies (100 independent runs)

Once we have understood the basic regularities behind, our goal is to find a mathematical model that allows an accurate fitting to all these curves. We begin this task by trying to use the mentioned logistic and hypergraph models. Let us first address the logistic case:

$$P(t) = \frac{1}{1 + a \cdot e^{-b \cdot t}} \ . \tag{4}$$

To strictly adhere to the original work of Sarma and De Jong for cellular EAs, the a parameter should be defined as a constant value ($a = \frac{1}{P(0)} - 1$). We call this model LOG1, and we plot its accuracy in Fig. 2a. We can quickly arrive to the conclusion that for low values of migration frequency (panmictic-like scenario), the error is small, what means good news for a logistic fitting.

However, as the interaction among the islands decreases, it turns to be very inaccurate. One could think that trying a fitting with the a parameter fixed is provoking such an inaccuracy, and this is why we propose a new variant of the logistic model called LOG2 (see Fig. 2b). In this case, we consider a and b as free variables (in the previous model LOG1 a was a constant parameter). LOG2 allows a fitting with a smaller error than LOG1, but it still seems harmful since the actual steps are ignored both in LOG1 and LOG2.

Then, a clear conclusion is that the basic logistic model, even when enhanced, cannot be used for distributed EAs, as the existing literature also claim for most non canonical cellular EAs [6,7,8,9].

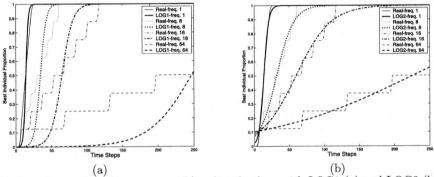

(a)   (b)

Fig. 2. Comparison between actual/predicted values with LOG1 (a) and LOG2 (b)

Therefore, we now turn to consider the hypergraph approach. In fact we present two variants of hypergraphs: the one in which p_{select} (5) accounts only for the probability of selection (HYP1), and the one where this probability (6) accounts both for selection and for replacement (HYP2). We introduce such distinction since in the seminal work [12] this second choice (combining selection and replacement within a probability) is said to be more exact.

$$p_{select1}(i, N) = 2 \cdot \frac{i}{N} - \left(\frac{i}{N}\right)^2 . \tag{5}$$

$$p_{select2}(i, N) = \frac{i}{N} + \left(1 - \frac{i}{N}\right) p_{select1}(i, N) . \tag{6}$$

where N is the population size and i denotes the total number of best individuals in the population.

When the hypergraph model is put to work, we can notice a clear improvement over the logistic models, obtaining an almost perfect curve fitting. As expected, HYP1 (Fig. 3a) generates a slightly worse fit than the HYP2 (Fig. 3b), because HYP1 is not accounting for the replacement effects.

To end this section we introduce a more accurate extension of the logistic model (N is the number of islands in the dEA):

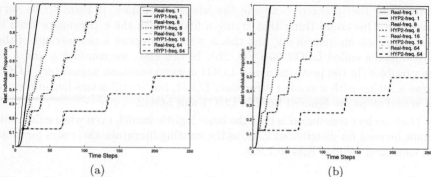

(a)   (b)

Fig. 3. Comparison between actual/predicted values with HYP1 (a) and HYP2 (b)

$$P(t) = \sum_{i=1}^{i=N} \frac{1/N}{1 + a \cdot e^{-b \cdot (t - freq \cdot i)}} \cdot \qquad (7)$$

This expression is an extension of the logistic model based in the idea that each island converges according to a logistic model, and that the entire population grows up as a sum of the growth of each component island. If the subpopulations are in turn structured in some way this assumption could not hold, but much must be said on this special subject (e.g., on distributed cellular EAs [3]), and thus it is left for a future work. To find the takeover time, we simply iterate the model until it reaches 1. We should notice that since it is an extension of the logistic approach, two variants could be also defined as we did before with LOG1 and LOG2. The first (SUM1) in which a is constant ($a = \frac{1}{P(0)/N} - 1$), and the second (SUM2) where a and b are adjustable parameters.

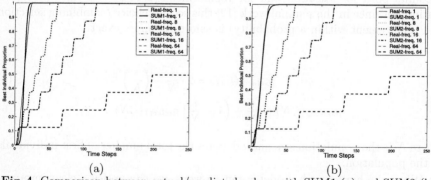

(a)   (b)

Fig. 4. Comparison between actual/predicted values with SUM1 (a) and SUM2 (b)

In figure 4 we plot the behavior of such a model. For low values of the migration frequency, SUM1 shows a less accurate behavior with respect to SUM2, but they two are equally or more precise than any other existing model, in particular with respect to the basic logistic and hypergraph variants. For the rest of migration frequencies, the SUMx models outperform the rest.

We conclude this section with a summary of the results. We have fitted the actual growth curves of a dEA with six different theoretical models. In Fig. 5 we graph together the mean square error and the $\| \cdot \|_1$ of that error for all the models under different migration frequencies. We can notice that in the case of low values of the migration frequency, most of models obtain a large error (with the exceptions of LOG2 and SUM2 models). In general, the second variant of each model is always better than the first one. The LOGx models show a stable behavior for all frequencies, but, while LOG1 is always very inaccurate (it is the worst model), the LOG2 performs well, and it is only worse than SUM2 model. The SUMx and HYPx models reduce their errors as the migration frequency enlarges, although the SUMx models are always more accurate than the HYPx ones.

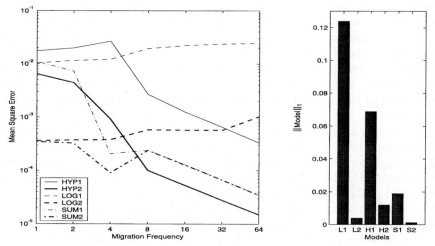

Fig. 5. Error (MSE) and $\| \cdot \|_1$ between actual and predicted values for all the models

4   Effects of the Migration Rate

In the previous section, we analyzed the effects of the migration frequency over the growth curves of a dEA. However, the number of individuals undergoing migration was fixed at a given value (rate = 8). Now, in this section we answer a common sense question: are the results somehow biased by the utilization of a concrete migration rate? To extend the previous study we now also analyze the effects of the migration rate. As we made for the migration policy, we begin by inspecting the induced behavior by different values of the migration rate.

As done before, we first compute the proportion of the best individual in a dEA when utilizing these values of the migration rate: 1, 2, 4, 8, 16, 32, and 64.

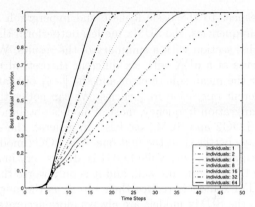

Fig. 6. Actual growth curves for several migration rates (100 independent runs)

The rest of the parameters are similar to the presented in Subsection 3.1. We proceed to perform a comparative study with the models discussed before.

In Fig. 6 we plot the way in which the migration rate influences the growth curve (frequency = 1). From this figure we can infer that the value of the migration rate determines the slope of the curve. The reason is that, when the migration rate value is high, the probability of migrating the best individual increases, and then the target island converges faster than if the migration rate were smaller.

Let us now proceed with the fitting of these curves with all the considered mathematical models. In Fig. 7 (left) we show the error for any combined value of the migration frequency and migration rate. To interpret the graph you must notice that the first seven points of a line correspond to the error incurred by the associated predictive model for the seven different values of the migration rate at the same frequency, and that there exists seven groups of such points, one for each migration frequency from 1 to 64 (from left to right in the horizontal axis).

We can see in Fig. 7 some behavioral patterns of the models with respect to the final MSE error they exhibit. First, we observe that the logistic behavior is very similar to the showed previously (Fig. 5), i.e., both variants are very stable; LOG1 obtains always larger errors while LOG2 is very accurate for all configurations. Second, the hypergraph model obtains low error for larger frequency values, while their inaccuracy is more evident for smaller values of migration frequency (high coupling). The proposed SUMx model also is somewhat sensitive to low frequency values, but it is quite stable and accurate for larger values of migration frequency. Both, the HYPx and the SUMx models seem to perform a cycle: reduction/enlargement (respectively) of error as the migration rate enlarges (for any given frequency). The $\| \cdot \|_1$ summarizes quantitatively the MSE results in a single value per model. We observe that the LOG1 model is the most inaccurate one. Also, LOG2 obtains very accurate results and only is worse than the SUM2 model. Although the HYPx models are very accurate for lager frequency values, they show high errors for smaller values, thus making its $\| \cdot \|_1$ value larger than the rest. Clearly, SUM2 obtains the lowest overall error.

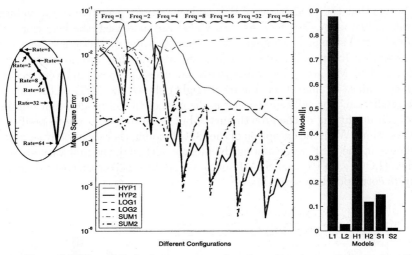

Fig. 7. Error (MSE) and $\| \cdot \|_1$ between actual and predicted growth curve for all the values of migration rate and frequency

5   Takeover Time Analysis

In the previous sections we have studied the effects of the migration frequency and rate over the takeover growth curves. Now, we analyze the effect of these parameters over the takeover times themselves. Figure 8 contains the value of the actual takeover time for different migration frequencies and rates. We can notice that the takeover value increases for higher frequencies and smaller rates. However, the rate effect over the takeover time is smoother than the frequency one.

<table>
<tr><td colspan="8" align="center">takeover time</td></tr>
<tr><td></td><td colspan="7" align="center">frequency</td></tr>
<tr><td></td><td>1</td><td>2</td><td>4</td><td>8</td><td>16</td><td>32</td><td>64</td></tr>
<tr><td>1</td><td>51</td><td>57</td><td>72</td><td>72</td><td>123</td><td>234</td><td>457</td></tr>
<tr><td>2</td><td>46</td><td>49</td><td>59</td><td>64</td><td>120</td><td>232</td><td>456</td></tr>
<tr><td>4</td><td>38</td><td>44</td><td>52</td><td>63</td><td>119</td><td>231</td><td>455</td></tr>
<tr><td>rate 8</td><td>34</td><td>36</td><td>42</td><td>62</td><td>118</td><td>230</td><td>454</td></tr>
<tr><td>16</td><td>28</td><td>32</td><td>37</td><td>61</td><td>117</td><td>229</td><td>453</td></tr>
<tr><td>32</td><td>23</td><td>27</td><td>36</td><td>60</td><td>116</td><td>228</td><td>452</td></tr>
<tr><td>64</td><td>20</td><td>23</td><td>36</td><td>60</td><td>116</td><td>228</td><td>452</td></tr>
</table>

Fig. 8. Actual takeover time values for all configurations (100 independent runs)

Once we have observed the effect of the migration rate and frequency over the takeover time, we analyze the predicted values provided by the models. Figure 9 shows the error of the predicted takeover time with all the models. We can notice that the predictions of the models are very sensitive to low values of migration frequency. However, as such values are increased, the behavior of the models is more stable. The logistic models obtain worse predictions as we enlarge the migration frequency value, and for lager values, they are no longer useful. HYPx and SUMx are quite stable (with small oscillations) and very accurate for lager values of migration. Figure 9 (right) shows the MSE of the takeover time prediction. Specially interesting is the case of LOG2 model, that obtains a very accurate fitting of the growth curves but is quite inaccurate to predict the takeover time. With the exception of the LOG2 model, there is not any significant difference in the MSE value among the models, although the first variants of the models (LOG1, HYP1, and SUM1) obtain a slightly better takeover time value than its corresponding second one.

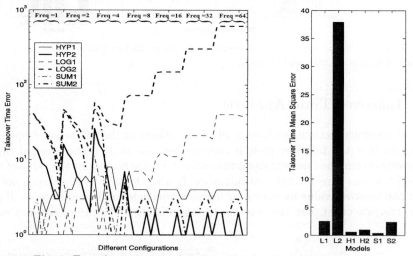

Fig. 9. Error between actual and predicted values of takeover time

We conclude this section by showing a closed equation for the takeover time calculation for the new models presented in this paper: SUMx models. This formula (8) is derived from the growth curve equation (7) of these models:

$$Takeovertime = freq \cdot (N - 1) - \frac{1}{b} \cdot Ln \left(\frac{\varepsilon \cdot N}{a \cdot (1 - \varepsilon \cdot (N - 1))} \right) . \quad (8)$$

where $freq$ is the migration frequency, N is the number of islands and ε is the the expected level of accuracy (a small value near zero).

6   Conclusions

In this paper we have performed an analysis of the growth curves and takeover regimes of distributed evolutionary algorithms. We compared the well-known logistic model, a hypergraph model and a newly proposed model consisting in a sum of logistic definition of the component takeover regimes. A second variant of each model has been also proposed for the shake of accuracy. In this article we have shown how the models appropriately captured the effects of both migration frequency and migration rate.

Although every model has its own advantages, either simplicity (LOGx), extensibility (HYPx), or accuracy (SUMx), SUM2 is the model that better fitting obtained, while its predicted takeover time was of a similar accuracy with respect to the rest. However, much needs still to be said on them, since subtle factors could provoke deviations from the predicted behavior, like it may occur for the takeover regime near to the moment in which one subpopulation is completely filled with the optimum solution.

As a future work we plan to check the results presented in this paper on additional topologies and selection methods. Dealing with bounding cases of *widely spread* algorithms is hopefully the right way to improve on our knowledge of the algorithms that most researchers are using in practice.

Acknowledgments. This work has been partially funded by the Spanish MCyT and FEDER under contract TIC2002-04498-C05-02 (the TRACER project).

References

1. Alba, E., Tomassini, M.: Parallelism and Evolutionary Algorithms. IEEE Transactions on Evolutionary Computation **6** (2002) 443–462
2. Gordon, V.S., Whitley, D.: Serial and Parallel Genetic Algorithms as Function Optimizers. In Forrest, S., ed.: Proceedings of the Fifth International Conference on Genetic Algorithms, Morgan Kaufmann (1993) 177–183
3. Alba, E., Troya, J.M.: Influence of the Migration Policy in Parallel dGAs with Structured and Panmictic Populations. Applied Intelligence **12** (2000) 163–181
4. Sarma, J., De Jong, K.: An Analysis of the Effects of Neighborhood Size and Shape on Local Selection Algorithms. In Voigt, H.M., Ebeling, W., Rechenberg, I., Schwefel, H.P., eds.: PPSN IV. Volume 1141 of LNCS., Springer (1996) 236–244
5. Sarma, J., De Jong, K.: An Analysis of Local Selection Algorithms in a Spatially Structured Evolutionary Algorithm. In Bäck, T., ed.: Proceedings of the 7th International Conference on Genetic Algorithms, Morgan Kaufmann (1997) 181–186
6. Gorges-Schleuter, M.: An Analysis of Local Selection in Evolution Strategies. In Banzhaf, W., Daida, J., Eiben, A.E., Garzon, M.H., Honavar, V., Jakiela, M., Smith, R.E., eds.: Proceedings of the Genetic and Evolutionary Computation Conference. Volume 1., Orlando, Florida, USA, Morgan Kaufmann (1999) 847–854
7. Rudolph, G.: Takeover Times in Spatially Structured Populations: Array and Ring. In Lai, K.K., Katai, O., Gen, M., Lin, B., eds.: 2nd Asia-Pacific Conference on Genetic Algorithms and Applications, Global-Link Publishing (2000) 144–151
8. Giacobini, M., Tettamanzi, A., Tomassini, M.: Modelling Selection Intensity for Linear Cellular Evolutionary Algorithms. In Liardet, P., et al., eds.: Artificial Evolution, Sixth International Conference, Springer Verlag (2003)
9. Giacobini, M., Alba, E., Tomassini, M.: Selection Intensity in Asynchronous Cellular Evolutionary Algorithms. In Cantú-Paz, E., ed.: Proceedings of the Genetic and Evolutionary Computation Conference, Chicago, USA (2003) 955–966

10. Sprave, J.: A Unified Model of Non-Panmictic Population Structures in Evolutionary Algorithms. In Angeline, P.J., Michalewicz, Z., Schoenauer, M., Yao, X., Zalzala, A., eds.: Proceedings of the Congress of Evolutionary Computation. Volume 2., Mayflower Hotel, Washington D.C., USA, IEEE Press (1999) 1384–1391

11. Goldberg, D.E., Deb, K.: A Comparative Analysis of Selection Schemes Used in Genetic Algorithms. In Rawlins, G.J., ed.: Foundations of Genetic Algorithms. Morgan Kaufmann, San Mateo, CA (1991) 69–93

12. Chakraborty, U.K., Deb, K., Chakraborty, M.: Analysis of Selection Algorithms: A Markov Chain Approach. Evolutionary Computation 4 (1997) 133–167

13. Cantú-Paz, E.: 7. Migration, Selection Pressure, and Superlinear Speedups. In: Efficient and Accurate Parallel Genetic Algorithms. Kluwer (2000) 97–120

Simultaneity Matrix for Solving Hierarchically Decomposable Functions

Chatchawit Aporntewan and Prabhas Chongstitvatana

Chulalongkorn University, Bangkok 10330, Thailand
Chatchawit.A@student.chula.ac.th Prabhas.C@chula.ac.th

Abstract. The simultaneity matrix is an $\ell \times \ell$ matrix of numbers. It is constructed according to a set of ℓ-bit solutions. The matrix element m_{ij} is the degree of linkage between bit positions i and j. To exploit the matrix, we partition $\{0, \ldots, \ell - 1\}$ by putting i and j in the same partition subset if m_{ij} is significantly high. The partition represents the bit positions of building blocks (BBs). The partition is used in solution recombination so that the bits governed by the same partition subset are passed together. It can be shown that by exploiting the simultaneity matrix the hierarchically decomposable functions can be solved in a polynomial relationship between the number of function evaluations required to reach the optimum and the problem size. A comparison to the hierarchical Bayesian optimization algorithm (hBOA) is made. The hBOA uses less number of function evaluations than that of our algorithm. However, computing the matrix is 10 times faster and uses 10 times less memory than constructing Bayesian network.

1 Introduction

For some conditions [6, Chapter 7–11], the success of genetic algorithms (GAs) can be explained by the schema theorem and the building-block hypothesis [4]. The schema theorem states that the number of solutions that match the above average, short defining-length, and low-order schemata grows exponentially. The optimal solution is hypothesized to be composed of the above average schemata or the building blocks (BBs). However, in simple GAs only short defining-length and low-order schemata are permitted to the exponential growth. The other schemata are more disrupted due to the single-point crossover. When the good BBs are more disrupted, it is said to be a GA-hard problem. Trap function [1] is an adversary function for studying BBs and linkage problems in GAs [7]. The general k-bit trap functions are defined as:

$$F_k(b_0 \ldots b_{k-1}) = \begin{cases} f_{\text{high}} & ; \text{ if } u = k \\ f_{\text{low}} - u\frac{f_{\text{low}}}{k-1} & ; \text{ otherwise,} \end{cases} \tag{1}$$

where $b_i \in \{0, 1\}$, $u = \sum_{i=0}^{k-1} b_i$, and $f_{\text{high}} > f_{\text{low}}$. Usually, f_{high} is set at k and f_{low} is set at $k - 1$. The additively decomposable functions (ADFs), denoted by $F_{m \times k}$, are defined as:

K. Deb et al. (Eds.): GECCO 2004, LNCS 3102, pp. 877–888, 2004.
© Springer-Verlag Berlin Heidelberg 2004

$$F_{m \times k}(B_0 \ldots B_{m-1}) = \sum_{i=0}^{m-1} F_k(B_i), \ B_i \in \{0,1\}^k. \tag{2}$$

The m and k are varied to produce a number of test functions. The ADFs fool gradient-based optimizers to favor zeroes, but the optimal solution is composed of all ones. Trap function is a fundamental unit for designing test functions that resist hill-climbing algorithms. The test functions can be effectively solved by composing BBs. Several discussions of the test functions can be found in [9,21, 22].

The BBs are inferred from a population of highly-fit individuals [6, pp. 60–61]. A population of highly-fit individuals (5×3-trap function) is shown in Table 1. The dependency between variables b_i, b_{i+1}, b_{i+2} ($i = 0, 3, 6, 9, 12$) can be detected by means of a statistical method. An inference might be that the highly-fit individuals are composed of triple zeroes and triple ones. It is said that the triple zeroes and triple ones are common traits or BBs. We aim to identify these BBs.

Table 1. A population of highly-fit individuals (5×3-trap function)

Individual no.	$b_0 b_1 b_2$	$b_3 b_4 b_5$	$b_6 b_7 b_8$	$b_9 b_{10} b_{11}$	$b_{12} b_{13} b_{14}$	Fitness
1	111	111	000	111	000	13.0
2	000	000	111	000	111	12.0
3	111	000	000	111	000	12.0
4	000	000	000	000	111	11.0
5	000	000	000	000	000	10.0

Thierens raised the scalability issue of simple GAs [20]. He used the uniform crossover so that the solutions are randomly mixed. The objective function is the $m \times 5$-trap functions. The analysis shows that either the computational time grows exponentially with the number of 5-bit trap functions or the population size must be exponentially increased. It is clear that scaling up the problem size requires information about the BBs so that the solutions are efficiently mixed. In addition, the performance of simple GAs relies on the ordering of solution bits. The ordering may not pack the dependent bits close together. Such an ordering results in poor mixing. Therefore the BBs need to be identified to improve the scalability issue.

Many strategies in the literature use the bit-reordering approach to pack the dependent bits close together, for example, inversion operator [4], messy GAs [5], and linkage learning [7]. The bit-reordering approach does not explicitly identify BBs, but it successfully delivers the optimal solution. Several works explicitly identify BBs. An approach is to find a partition of bit positions. For instance, Table 1 infers the partition:

$$\{\{0, 1, 2\}, \{3, 4, 5\}, \{6, 7, 8\}, \{9, 10, 11\}, \{12, 13, 14\}\}. \tag{3}$$

In the case of nonoverlapped BBs, partition is a clear representation [8,11,12,13]. Note that Kargupta [12] computes Walsh's coefficients which imply the partition. The bits governed by the same partition subset are passed together to prevent BB disruption.

Identifying BBs is somewhat related to building a distribution of solutions [8,14,15]. The basic concept of optimization by building a distribution is to start with a uniform distribution of solutions. Next, a number of solutions is drawn according to the distribution. Some good solutions (winners) are selected, and the distribution is adjusted toward the winners (the winners-like solutions will be drawn with higher probability in the next iteration). These steps are repeated until the optimal solution is found or reaching a termination condition. The works in this category are referred to as *probabilistic model-building genetic algorithms (PMBGAs)*. For a particular form of distribution used in the extended compact genetic algorithm (ECGA), building the distribution is identical to searching for a partition [8]. The Bayesian optimization algorithm (BOA) uses Bayesian network to represent a distribution [14]. Pelikan showed that if the problem is composed of k-bit trap functions, the network will be fully connected sets of k nodes [17, pp. 54]. In addition, the Bayesian network is able to represent joint distributions in the case of overlapping BBs. The hierarchical BOA (hBOA) is the BOA enhanced with decision tree/graph and a niching method called restricted tournament replacement [17]. The hBOA can solve the hierarchically decomposable functions (HDFs) in a scalable manner [17]. Successful applications for BB identification are financial applications [10], cluster optimization [19], maximum satisfiability of logic formulas (MAXSAT) and Ising spin glass systems [18].

The Bayesian network is able to identify common structures in a population. Nevertheless, building the network is time-consuming. This paper presents a BB identification algorithm that is simpler and faster than that of the hBOA. In addition, our algorithm uses less memory. The algorithm is named building-block identification by simultaneity matrix (BISM) [2]. The BISM input is a set of ℓ-bit solutions. The BISM output is a partition of $\{0, \ldots, \ell - 1\}$. Algorithm BISM consists of two parts: simultaneity matrix construction (SMC) and partitioning (PAR) algorithms. The SMC constructs the matrix according to a set of solutions. Next, PAR searches for a partition for the matrix. The remainder of the paper is organized as follows. Section 2 defines the hierarchically decomposable functions. Section 3 describes the SMC algorithm. Section 4 describes the PAR algorithm. Section 5 presents the experimental results and discussions. Section 6 concludes the paper.

2   Hierarchically Decomposable Functions

To solve ADFs, the BBs need to be identified so that the solutions are efficiently mixed. The hierarchically decomposable functions (HDFs) are far more difficult than the ADFs. First, BBs in the lowest level need to be identified. The solution quality is improved by exploiting the identified BBs in solution recombination. Next, the improved population reveals larger BBs. Again the BBs in higher levels need to be identified. Identifying and exploiting BBs are repeated many times until reaching the optimal solution. Commonly used HDFs are hierarchically if-and-only-if (HIFF), hierarchical trap 1 (HTrap1), and hierarchical trap 2 (HTrap2) functions. Due to page limitations, the original definitions of HDFs can be found in [21,17].

3   Simultaneity Matrix Construction (SMC) Algorithm

The SMC input is a set of ℓ-bit binary string denoted by:

$$S = \{s_0, \ldots, s_{n-1}\}, \tag{4}$$

where s_i is the ith string, $0 \le i \le n - 1$. The $s_i[j]$ denotes the jth bit of s_i, $0 \le j \le \ell - 1$. Algorithm SMC outputs an $\ell \times \ell$ symmetric matrix of numbers, denoted by $M = (m_{ij})$, $0 \le i, j \le \ell - 1$. A closed form of m_{ij} is shown in Equation 5.

$$m_{ij} = \begin{cases} 0 & ; \text{ if } i = j \\ \text{Count}_S^{00}(i,j)\text{Count}_S^{11}(i,j) + \text{Count}_S^{01}(i,j)\text{Count}_S^{10}(i,j) & ; \text{ otherwise,} \end{cases} \tag{5}$$

where $\text{Count}_S^{ab}(i,j) = |\{x \in \{0, \ldots, n-1\} : s_x[i] = a \text{ and } s_x[j] = b\}|$ for all $0 \le i, j \le \ell - 1$, $(a, b) \in \{0,1\}^2$.

Algorithm SMC is shown in Figure 1. Step 1 constructs only the upper triangle of the matrix by using Equation 5. Step 2 perturbs the matrix so that there are no identical elements. This matrix, in which all the elements are distinct, is greatly helpful in partitioning. The perturbation does not totally change the matrix because each element is incremented by a small real random number ranging between 0 and 1. The perturbation by adding an integer with a real number is practical for a random number generator with a sufficiently large period because it is hardly possible to produce identical random numbers. Step 3 copies the upper triangle $\{m_{ij} \mid i < j\}$ to the lower triangle $\{m_{ij} \mid i > j\}$. Step 4 returns the simultaneity matrix $M = (m_{ij})$. The time complexity of SMC is $O(\ell^2 n)$.

The matrix element m_{ij} is proportional to the probability that 2-bit BBs at bit positions i and j will be disrupted by the uniform crossover. All cases for mixing 2-bit BBs are enumerated. Mixing "00" with "11" results in "01" and "10." Mixing "01" with "10" results in "00" and "11." Only mixing in the two cases must be done carefully because the processing BBs will be lost. Mixing 2-bit BBs in the other cases gives the same BBs. Therefore Algorithm SMC counts a pair of 2-bit BBs that are complement to each other. To exploit the matrix, the bits at positions i and j are passed together every time performing crossover if the matrix element m_{ij} is significantly high. The 3-bit BBs are identified by inserting k to $\{i, j\}$. If the matrix elements m_{ij}, m_{jk}, and m_{ik} are significantly high, i, j, k should be in the same partition subset. Larger BBs can be identified in a similar fashion.

The trap functions embedded in the HDFs bias the population to two aligned chunks of zeroes and ones, that are complementary to each other. Certainly, the dependency between every pair of bits in a chunk is stored in the matrix. The matrix is not limited to the cases where the two aligned chunks are complementary

```
Algorithm SMC(S)
1.  for i = 0 to ℓ − 1 do
        m_ii ← 0;
        for j = i + 1 to ℓ − 1 do
            m_ij ← Count_S^{00}(i, j) × Count_S^{11}(i, j) + Count_S^{01}(i, j) × Count_S^{10}(i, j);
2.  for i = 0 to ℓ − 1 do
        for j = i + 1 to ℓ − 1 do
            m_ij ← m_ij + Random(0, 1);
3.  for i = 0 to ℓ − 1 do
        for j = i + 1 to ℓ − 1 do
            m_ji ← m_ij;
4. return M = (m_ij);
```

Fig. 1. SMC algorithm

to each other. In the other cases, the matrix does not detect unnecessary dependency. For instance, the bits at positions of $\{0, 1, 2, 3, 4\}$ are mostly "$b_0 b_1 000$" and "$b_0 b_1 111$" where $b_i \in \{0, 1\}$. The dependency among five bits is obvious, but passing the bits governed by $\{2, 3, 4\}$ together is sufficient to guarantee that "$b_0 b_1 000$" and "$b_0 b_1 111$" will exist in the next generation with a high probability. In summary, the matrix records only dependency that is actually necessary for preserving BBs.

4   Partitioning (PAR) Algorithm

The PAR input is an $\ell \times \ell$ simultaneity matrix. The PAR outputs the partition:

$$P = \{B_0, \dots, B_{|P|-1}\}, \quad \bigcup_{i=0}^{|P|-1} B_i = \{0, \dots, \ell-1\}, \; B_i \cap B_j = \emptyset \text{ for all } i \neq j. \quad (6)$$

The B_i is called partition subset. There are several definitions of the desired partition, for example, the definitions in the senses of nonmonotonicity [13], GEMGA [11], Walsh coefficients [12], and entropy measurement [8]. We develop a definition in the sense of simultaneity matrix. Algorithm PAR searches for a partition P such that

1. $P \neq \{\{0, \dots, \ell-1\}\}$.
2. For all $B \in P$ such that $1 < |B| < \ell$, for all $b \in B$, the largest $|B| - 1$ matrix elements in row b are founded in columns of $B \setminus \{b\}$.
3. For all $B \in P$ such that $1 < |B| < \ell$, $H_{max} - H_{min} < \alpha(H_{max} - L_{min})$ where $\alpha \in [0, 1]$,
    $H_{max} = max(m_{ij} \mid (i, j) \in B^2, \; i \neq j)$,
    $H_{min} = min(m_{ij} \mid (i, j) \in B^2, \; i \neq j)$,
    $L_{min} = min(m_{ij} \mid i \in B, \; j \in \{0, \dots, \ell-1\} \setminus B)$.

4. There are no partition P_x such that for some $B \in P$, for some $B_x \in P_x$, P and P_x satisfy the first, the second, and the third conditions, $B \subset B_x$.

An example of the simultaneity matrix is shown in Figure 2. The perturbation is omitted because the values of $\{m_{ij} \mid i < j\}$ are distinct. The first condition does not allow the coarsest partition because it is not useful in solution recombination. The second condition makes i and j, in which m_{ij} is significantly high, in the same partition subset. For instance, $P_1 = \{\{0, 1, 2\}, \{3, 4, 5\}, \{6, 7, 8\}, \{9, 10, 11\}, \{12, 13, 14\}\}$ satisfies the second condition because the largest two elements in row 0 are found in columns of $\{1, 2\}$, the largest two elements in row 1 are found in columns of $\{0, 2\}$, the largest two elements in row 2 are found in columns of $\{0, 1\}$, and so on. However, there are many partitions that satisfy the second condition, for example, $P_2 = \{\{0, 1, 2\}, \{3, 4, 5, 6, 7, 8\}, \{9, 10, 11\}, \{12, 13, 14\}\}$. There is a dilemma between choosing the fine partition (P_1) and the coarse partition (P_2). Choosing the fine partition prevents the emergence of large BBs, while the coarse partition results in poor mixing. To overcome the dilemma, the coarse partition will be acceptable if it satisfies the third condition. The fourth condition says choosing the coarsest partition that is consistent with the first, the second, and the third conditions.

By the third condition, the partition subset $\{3, 4, 5\}$ is acceptable because the values of matrix elements governed by $\{3, 4, 5\}$ are close together (see Figure 2). Being close together is defined by $H_{max} - H_{min}$ where H_{max} and H_{min} is the maximum and the minimum of the nondiagonal matrix elements governed by a partition subset. The $H_{max} - H_{min}$ is a degree of irregularities of the matrix. The main idea is to limit $H_{max} - H_{min}$ to a threshold. The threshold, $\alpha(H_{max} - L_{min})$, is defined relatively to the matrix elements because the threshold cannot be fixed for a problem instance. The partition subset $\{3, 4, 5\}$ gives $H_{max} = 71543$, $H_{min} = 70172$, and $L_{min} = 61115$. L_{min} is the minimum of the nondiagonal matrix elements in rows of $\{3, 4, 5\}$. The third condition limits $H_{max} - H_{min}$ to $100 \times \alpha$ percent of the difference between H_{max} and L_{min}. An empirical study showed that α should be set at 0.75 for both ADFs and HDFs. Choosing $\{3, 4, 5, 6, 7, 8\}$ yields ($H_{max} = 73739, H_{min} = 68064, L_{min} = 61115$) which does not violate the third condition. The fourth condition prefers a coarse partition $\{\{3, 4, 5, 6, 7, 8\}, \ldots\}$ to a fine partition $\{\{3, 4, 5\}, \ldots\}$ so that the partition subsets can be grown to compose larger BBs in higher levels.

Algorithm PAR is shown in Figure 3. A trace of the algorithm is shown in Table 2. The outer loop processes row 0 to $\ell - 1$. In the first step, the columns of the sorted values in row i are stored in array R. For $i = 0$, array $R[\ ] = \{2, 1, 8, 6, 12, 5, 4, 7, 3, 10, 13, 11, 9, 14, 0\}$. Next, the inner loop tries a number of partition subsets by enlarging B_1 ($B_1 \leftarrow B_1 \cup \{R[j]\}$). If B_1 satisfies the second and the third conditions, B_1 will be saved to B_2. Finally, P is the partition that satisfies the four conditions. Checking the second and the third conditions is the most time-consuming section. It can be done in $O(\ell^2)$. The checking is done at most ℓ^2 times. Therefore the time complexity of PAR is $O(\ell^4)$.

elements governed by {3, 4, 5}

elements governed by {3, 4, 5, 6, 7, 8}

	Col 0	Col 1	Col 2	Col 3	Col 4	Col 5	Col 6	Col 7	Col 8	Col 9	Col 10	Col 11	Col 12	Col 13	Col 14
Row 0	0	70220	70451	61129	61841	62405	63493	61560	63968	60455	61065	60472	62699	60534	60272
Row 1	70220	0	70130	61115	62569	61972	63075	62080	61943	61290	60002	61259	63515	60205	61223
Row 2	70451	70130	0	62233	63643	62571	64586	64432	64146	61489	61774	61260	63214	61133	62010
Row 3	61129	61115	62233	0	70999	70172	68228	68722	68782	61817	62222	62241	63219	62016	61715
Row 4	61841	62569	63643	70999	0	71543	68738	68474	68064	63443	63244	62739	65128	62765	62995
Row 5	62405	61972	62571	70172	71543	0	68715	68567	68727	62289	62683	62613	63685	62914	62791
Row 6	63493	63075	64586	68228	68738	68715	0	72764	73739	63571	63877	63976	65485	63230	62969
Row 7	61560	62080	64432	68722	68474	68567	72764	0	73045	63215	62996	63359	64957	62862	62538
Row 8	63968	61943	64146	68782	68064	68727	73739	73045	0	63289	63623	63590	66003	63272	63170
Row 9	60455	61290	61489	61817	63443	62289	63571	63215	63289	0	70259	70527	62390	62794	62619
Row 10	61065	60002	61774	62222	63244	62683	63877	62996	63623	70259	0	70457	61318	63258	61094
Row 11	60472	61259	61260	62241	62739	62613	63976	63359	63590	70527	70457	0	63025	61219	63465
Row 12	62699	63515	63214	63219	65128	63685	65485	64957	66003	62390	61318	63025	0	70316	71092
Row 13	60534	60205	61133	62016	62765	62914	63230	62862	63272	62794	63258	61219	70316	0	70832
Row 14	60272	61223	62010	61715	62995	62791	62969	62538	63170	62619	61094	63465	71092	70832	0

Fig. 2. Simultaneity matrix

Table 2. A trace of the PAR algorithm

i	j	B_1	2nd cond.	3rd cond.	B_2
0	0	{0, 2}	True	True	{0, 2}
0	1	{0, 2, 1}	True	True	{0, 1, 2}
0	2	{0, 2, 1, 8}	False	False	{0, 1, 2}
0	3	{0, 2, 1, 8, 6}	False	False	{0, 1, 2}
0	4	{0, 2, 1, 8, 6, 12}	False	False	{0, 1, 2}
0	5	{0, 2, 1, 8, 6, 12, 5}	False	False	{0, 1, 2}
0	6	{0, 2, 1, 8, 6, 12, 5, 4}	False	False	{0, 1, 2}
0	7	{0, 2, 1, 8, 6, 12, 5, 4, 7}	False	False	{0, 1, 2}
0	8	{0, 2, 1, 8, 6, 12, 5, 4, 7, 3}	False	False	{0, 1, 2}
0	9	{0, 2, 1, 8, 6, 12, 5, 4, 7, 3, 10}	False	False	{0, 1, 2}
0	10	{0, 2, 1, 8, 6, 12, 5, 4, 7, 3, 10, 13}	False	False	{0, 1, 2}
0	11	{0, 2, 1, 8, 6, 12, 5, 4, 7, 3, 10, 13, 11}	False	False	{0, 1, 2}
0	12	{0, 2, 1, 8, 6, 12, 5, 4, 7, 3, 10, 13, 11, 9}	False	False	{0, 1, 2}

5   Experimental Results

5.1   Methodology

Most papers report the performance in terms of function evaluations required to reach the optimum. Such a performance measurement is affected by selection method, solution recombination, and the other factors. At present, research community does not provide a formal framework for measuring the effectiveness of a BB identification algorithm regardless of the other factors we have mentioned. Inevitably, we have to make a comparison in terms of function evaluations. We have presented the building-block identification by simultaneity matrix (BISM). An optimization algorithm that exploits the BISM is needed. We customize simple GAs as follows. Every generation, the simultaneity matrix is constructed.

```
Algorithm PAR(M)
P ← ∅;
for i = 0 to ℓ − 1 do
    if i ∉ B for all B ∈ P then
        array T = {matrix elements in row i sorted in descending order};
        for j = 0 to ℓ − 1 do
            R[j] = x where m_ix = T[j];
        endfor
        B₁ ← {i};
        B₂ ← {i};
        for j = 0 to ℓ − 3 do
            B₁ ← B₁ ∪ {R[j]};
            if {B₁} satisfies the second and the third conditions then
                B₂ ← B₁;
            endif
        endfor
        P ← P ∪ {B₂};
    endif
endfor
return P;
```

Fig. 3. PAR algorithm

The PAR algorithm is executed to find the partition. Two parents are chosen by the roulette-wheel method. The solutions are reproduced by a restricted uniform crossover – bits governed by the same partition subset must be passed together. The mutation is turned off. The diversity is maintained by the rank-space method [23, pp. 520–523]. The population size is determined empirically by the bisection method [17, pp. 64]. The bisection method performs binary search for the minimal population size. There might be 10% different between the population size used in the experiments and the minimal population size that ensures the optimal solution in all independent 10 runs.

5.2   A Visualization of the Simultaneity Matrix

To illustrate how the matrix changes over time, a matrix element is represented by a square. The square intensity is proportional to the value of matrix element (see Figure 4). In the early generation (A), the matrix elements are nearly identical because the initial population is generated at random. After that (B), the matrix elements become more distinct. The BBs in the lowest level are detected. The solution recombination is more speculative. Multiple bits are passed together, and therefore forming larger BBs. A few generations later (C), higher-level BBs are revealed. Finally (D), the population begins to lose diversity. The matrix elements are going to be identical. Note that the bits governed by the same BB do not need to be packed close together. It is done for the ease of presentation.

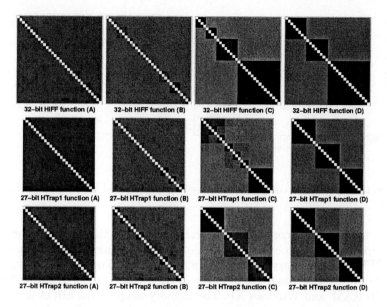

Fig. 4. Simultaneity matrices (HIFF, HTrap1, and HTrap2 functions)

5.3   A Comparison to the hBOA

Our algorithm is compared to the hBOA [17, pp. 164–165]. Figure 5 shows the number of function evaluations required to reach the optimum. The HTrap2 result is not shown because it is identical to that of the HTrap1. The linear regression in log scale indicates a polynomial relationship between the number of function evaluations and the problem size. The degree of polynomial can be approximated by the slope of linear regression. It can be seen that the hBOA and BISM can solve the HDFs in a polynomial time. The hBOA performs better than the BISM. However, the performance gap narrows as the problem becomes harder (HIFF, HTrap1, and HTrap2 functions respectively).

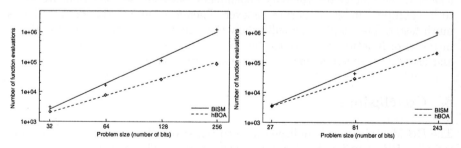

Fig. 5. Performance comparison: HIFF (left) and HTrap1 (right)

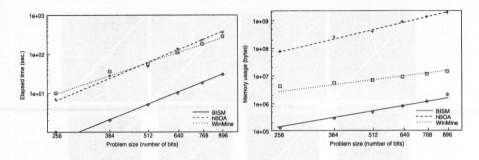

Fig. 6. Performance comparison: elapsed time (left) and memory usage (right)

We make another comparison in terms of elapsed time and memory usage. The elapsed time is an execution time of a call on `constructTheNetwork` subroutine [16]. The memory usage is the number of bytes dynamically allocated in the subroutine. The hardware platform is HP NetServer E800, 1GHz Pentium-III, 2GB RAM, and Windows XP. The memory usage in the hBOA is very large because of inefficient memory management in constructing Bayesian network. A fair implementation of the Bayesian network is the WinMine Toolkit [3]. The WinMine is a set of tools that allow you to build statistical models from data. It constructs Bayesian network with decision tree that is similar to that of the hBOA. The WinMine's elapsed time and memory usage are measured by an execution of `dnet.exe` – a part of the WinMine that constructs the network. All experiments are done with the same biased population that is composed of aligned chunks of zeroes and ones. The parameters of the hBOA and WinMine Toolkit are set at default. The population size is set at three times greater than the problem size.

The elapsed time and memory usage averaged from 10 independent runs are shown in Figure 6. The Bayesian network is a powerful tool that builds statistical models from data. However, constructing the network is time-consuming. This is because the network gathers all dependency between bit variables. In contrast, the matrix records only dependency between two bits that are likely to be disrupted in the uniform crossover. Therefore the matrix computation is much faster. The empirical results show that the hBOA outperforms the BISM in terms of function evaluations, but computing the matrix is 10 times faster and uses 10 times less memory than constructing Bayesian network.

6   Conclusions

The BB identification is indispensable to the scalability of GAs. We have presented a BB identification by simultaneity matrix. The matrix element m_{ij} is proportional to the probability that 2-bit BBs at positions i and j will be disrupted by the uniform crossover. The matrix does not detect all dependency between bit variables. We have shown that there might be dependency between bits at positions i and j that cannot be detected by the matrix. Such dependency is not necessary because the 2-bit BBs at positions i and j are very likely to survive in the next generation regardless of the solution recombination methods. Exploiting the matrix is simply passing the bits at positions i and j together if m_{ij} is significantly high. More formally, we search for a partition of bit positions. The bits governed by the same partition subset are passed together every time performing crossover. It can be shown that the BISM can solve the hierarchical problem in a polynomial relationship between the number of function evaluations and the problem size. More importantly, the matrix computation is simple, fast, and memory efficient. The partition may not fully take advantages of the matrix. The matrix could be exploited in another way rather than partitioning. Future work is to combine the strengths of Bayesian network and the simultaneity matrix.

References

1. Ackley, D. H. (1987). A Connectionist Machine for Genetic Hillclimbing. Kluwer Academic Publishers, Boston, MA.
2. Aporntewan, C., and Chongstitvatana, P. (2003). Building-block identification by simultaneity matrix. In CantúPaz, E. et al., editors, Proceedings of the Genetic and Evolutionary Computation, page 1566–1567, Springer-Verlag, Heidelberg, Berlin.
3. Chickering, D. M. (2002). The WinMine Toolkit. Technical Report MSR-TR-2002-103, Microsoft Research, Redmond, WA.
4. Goldberg, D. E. (1989). Genetic Algorithms in Search Optimization and Machine Learning. Addison Wesley, Reading, MA.
5. Goldberg, D. E., Korb, B., and Deb, K. (1989). Messy genetic algorithms: Motivation, analysis and first results. Complex Systems, Vol. 3, No. 5, page 493–530, Complex Systems Publications, Inc., Champaign, IL.
6. Goldberg, D. E. (2002). The Design of Innovation: Lessons from and for Competent Genetic Algorithms. Kluwer Academic Publishers, Boston, MA.
7. Harik, G. R. (1997). Learning linkage. In Belew, R. K., and Vose, M. D., editors, Foundation of Genetic Algorithms 4, page 247–262, Morgan Kaufmann, San Francisco, CA.
8. Harik, G. R. (1999). Linkage learning via probabilistic modeling in the ECGA. Technical Report 99010, Illinois Genetic Algorithms Laboratory, University of Illinois at Urbana-Champaign, Champaign, IL.
9. Holland, J. H. (2000). Building blocks, cohort genetic algorithms, and hyperplane-defined functions. Evolutionary Computation, Vol. 8, No. 4, page 373–391, MIT Press, Cambridge, MA.
10. Kargupta, H., and Buescher, K. (1995). The gene expression messy genetic algorithm for financial applications. In Proceedings of the IEEE/IAFE Conference on Computational Intelligence for Financial Engineering, page 155–161, IEEE Press, Piscataway, NJ.
11. Kargupta, H. (1996). The gene expression messy genetic algorithm. In Proceedings of the IEEE International Conference on Evolutionary Computation, page 814–819, IEEE Press, Piscataway, NJ.
12. Kargupta, H., and Park, B. (2001). Gene expression and fast construction of distributed evolutionary representation. Evolutionary Computation, Vol. 9, No. 1, page 43–69, MIT Press, Cambridge, MA.

13. Munetomo, M., and Goldberg, D. E. (1999). Linkage identification by non-monotonicity detection for overlapping functions. Evolutionary Computation, Vol. 7, No. 4, page 377–398, MIT Press, Cambridge, MA.
14. Pelikan, M., Goldberg, D. E., and Cantú-Paz, E. (1999). BOA: The Bayesian optimization algorithm. In Banzhaf, W. et al., editors, Proceedings of Genetic and Evolutionary Computation Conference, Vol. 1, page 525–532, Morgan Kaufmann, San Francisco, CA.
15. Pelikan, M., Goldberg, D. E., and Lobo, F. (1999). A survey of optimization by building and using probabilistic models. Computational Optimization and Applications, Vol. 21, No. 1, page 5–20, Kluwer Academic Publishers.
16. Pelikan, M. (2000). A C++ implementation of the Bayesian optimization algorithm (BOA) with decision graph. Technical Report 2000025, Illinois Genetic Algorithms Laboratory, University of Illinois at Urbana-Champaign, Champaign, IL.
17. Pelikan, M. (2002). Bayesian optimization algorithm: From single level to hierarchy. Doctoral dissertation, University of Illinois at Urbana-Champaign, Champaign, IL.
18. Pelikan, M., and Goldberg, D. E. (2003). Hierarchical BOA solves Ising Spin Glasses and MAXSAT. In Cant-úPaz, E. et al., editors, Proceedings of Genetic and Evolutionary Computation Conference, page 1271–1282, Springer-Verlag, Heidelberg, Berlin.
19. Sastry, K., and Xiao, G. (2001). Cluster optimization using extended compact genetic algorithm. Technical Report 2001016, Illinois Genetic Algorithms Laboratory, University of Illinois at Urbana-Champaign, Champaign, IL.
20. Thierens, D. (1999). Scalability problems of simple genetic algorithms. Evolutionary Computation, Vol. 7, No. 4, page 331–352, MIT Press, Cambridge, MA.
21. Watson, R. A., and Pollack, J. B. (1999). Hierarchically consistent test problems for genetic algorithms. In Angeline, P. J., Michalewicz, Z., Schoenauer, M., Yao, X., and Zalzala, A., editors, Proceedings of Congress on Evolutionary Computation, page 1406–1413, IEEE Press, Piscataway, NJ.
22. Whitley, D., Rana, S., Dzubera, J., and Mathias, K. E. (1996). Evaluating evolutionary algorithms. Artificial Intelligence, Vol. 85, No. 1–2, page 245–276, Elsevier.
23. Winston, P. H. (1992). Artificial Intelligence, third edition. Addison-Wesley, Reading, MA.

Metaheuristics for Natural Language Tagging

Lourdes Araujo[1], Gabriel Luque[2], and Enrique Alba[2]

[1] Dpto. Sistemas Informáticos y Programación,
Facultad de Informática, Univ. Complutense,
28040 Madrid, SPAIN
lurdes@sip.ucm.es
[2] Dpto. de Lenguajes y Ciencias de la Computación,
E.T.S. Ingeniería Informática,
Campus Teatinos, 29071, Málaga, SPAIN
{eat,gabriel}@lcc.uma.es

Abstract. This work compares different metaheuristics techniques applied to an important problem in natural language: tagging. Tagging amounts to assigning to each word in a text one of its possible lexical categories (tags) according to the context in which the word is used (thus it is a disambiguation task). Specifically, we have applied a classic genetic algorithm (GA), a CHC algorithm, and a Simulated Annealing (SA). The aim of the work is to determine which one is the most accurate algorithm (GA, CHC or SA), which one is the most appropriate encoding for the problem (integer or binary) and also to study the impact of parallelism on each considered method. The work has been highly simplified by the use of MALLBA, a library of search techniques which provides generic optimization software skeletons able to run in sequential, LAN and WAN environments. Experiments show that the GA with the integer encoding provides the more accurate results. For the CHC algorithm, the best results are obtained with binary coding and a parallel implementation. SA provides less accurate results than any of the evolutionary algorithms.

1 Introduction

Part of speech tagging is one of the basic tasks in natural language processing. Tagging amounts to assigning to each word of a sentence one of its possible lexical categories according to the context in which the word is used. For instance, the word *can* can be a noun, an auxiliary verb or a transitive verb. The category assigned to the word will determine the structure of the sentence in which it appears and thus its meaning. In fact, tagging is a necessary step for parsing, for information retrieval systems, for speech recognition, etc. Moreover, tagging is a difficult problem since, many words belong to more than one lexical class. To give an idea, according to [7], over 40% of the words appearing in the hand-tagged Brown corpus [11] are ambiguous.

Because of the importance and difficulty of this task, a lot of work has been carried out to produce automatic taggers. Automatic taggers [5,10,12], usually based on Hidden Markov Models, rely on statistical information to establish the

K. Deb et al. (Eds.): GECCO 2004, LNCS 3102, pp. 889–900, 2004.

probabilities of each scenario. The statistical data are extracted from previously tagged texts, called *corpus*. These stochastic taggers neither require knowledge of the rules of the language nor try to deduce them, and thus they can be applied to texts in any language, provided they can be trained on a corpus for that language previously.

The context in which the word appears helps to decide which is its more appropriate tag, and this idea is the basis for most taggers. For instance, consider the sentence in Figure 1, extracted from the Brown corpus. The word *questioning* can be disambiguated as a common name if the preceding tag is disambiguated as an adjective. But it might happen that the preceding word were ambiguous, so there may be many dependencies which must be resolved simultaneously.

This	the	therapist	may	pursue	in	later	questioning	.
DT	AT	NN	NNP	VB	RP	RP	VB	.
QL			MD	VBP	NNP	RB	NN	
					RB	JJ	JJ	
					NN	JJR		
					FW			
					IN			

Fig. 1. Tags for the words in a sentence extracted from the Brown corpus. Underlined tags are the correct ones, according to the Brown corpus. Tags correspond to the tag set defined in the Brown corpus: DT stands for determiner/pronoun, AT for article, NN for common noun, MD for modal auxiliary, VB for uninflected verb, IN for preposition, JJR for comparative adjective, etc.

The statistical model considered in this work amounts to maximize a global measure of the probability of the set of contexts (a tag and its neighboring tags) corresponding to a given tagging of the sentence. Then, we need a method to perform the search of the tagging which optimizes this measure of probability.

The aim of this work is to check and compare two different variants of evolutionary algorithms to perform such a search: a classic genetic algorithm (GA) and a CHC algorithm. Genetic algorithms have been previously applied to the problem [3,4], obtaining accuracies as good of those of typical algorithms used for stochastic tagging (such as the widely used of Viterbi [5]) or even better [4]. CHC is a non-traditional genetic algorithm, which presents some particular features. CHC guarantees the survival of the best individuals found by putting the children and parents together and applying selection among them. Similar individuals are not allowed to mate in order to improve diversity. Crossover also differs in such a way that two parents exchange exactly half of the differing parental genes and instead of traditional mutation, CHC re-initializes the population when stagnation is detected.

One of the aims of this work is to investigate if the particular mechanism of CHC for diversity can improve the selection of different sets of tags. From previous work, it has been observed that words incorrectly tagged are usually those which require one of their more rare tags, or which appear in an infrequent context. The fitness function is based on the probability of the contexts of a sequence of tags assigned to a sentence. Therefore, it is difficult for the GA to change tags within high probability contexts. CHC allows changing simultaneously several tags of the sequence, which can lead to explore combinations of tags very different from those of the ancestors. Thus, it is interesting to study what is more advantageous, the quiet exploration of the GA or the more disruptive one of CHC. We have also compared the results of the GAs with those obtained from Simulated Annealing (SA), in order to ascertain the suitability of the evolutionary approach compared with other optimization methods.

For most tagging applications, the whole process of search is time consuming, what made us to include in the study a parallel version of the algorithms.

The work has been highly simplified by the use of MALLBA [1], a library of search techniques, which provides generic optimization software skeletons that run in sequential, LAN and WAN environments.

The rest of the paper proceeds as follows: Section 2 is devoted to present the MALLBA system, under which the algorithms have been implemented. Sections 3, 4 and 5 describe the GA, CHC, and SA algorithms, and Section 6 discusses the parallel version of these algorithms. Section 7 presents the details of the algorithms as applied to tagging, including the genetic operators. Section 8 describes and discusses the experimental results, and Section 9 draws the main conclusions of this work.

2   MALLBA System

The MALLBA project [1] provides, in an integrated way, a library of skeletons for combinatorial optimization (including exact, heuristic and hybrid methods) which can deal with parallelism in a user-friendly and, at the same time, efficient manner. Its three target environments are sequential computers, LANs of workstations and WANs. Skeletons are generic templates which must be instantiated with the features of the problem to solve. The features related to the selected generic resolution method and its interaction with the particular problem are implemented by the skeleton.

Skeletons are implemented by a set of *required* and *provided* C++ classes that represent an abstraction of the entities participating in the solver method. The *provided* classes implement internal aspects of the skeleton in a problem-independent way. The *required* classes specify information and behavior related to the problem. This conceptual separation allows us to define required classes with a fixed interface but without any implementation, so that provided classes can use required classes in a generic way. MALLBA is publicly available at http://neo.lcc.uma.es/mallba/easy-mallba/index.html.

3   Genetic Algorithm

Genetic Algorithms (GAs) [6] are stochastic search methods that have been successfully applied in many real applications of high complexity. A GA is an iterative technique that applies stochastic operators on a pool of individuals (tentative solutions). An evaluation function associates a value to every individual indicating its suitability to the problem. Traditionally, GAs are associated to the use of a binary representation, but nowadays you can find GAs that use other types of representations. A GA usually applies a recombination operator on two solutions, plus a mutation operator that randomly modifies the individual contents to promote diversity.

```
1   t = 0
2   initialize P(t)
3   evaluate structures in P(t)
4   while not end do
5       t = t + 1
6       select: C(t) = P(t-1)
7       for each pair (p1,p2) in C(t)
8           if 'incest prevention condition'
9               add to C'(t) HUX(p1,p2)
10      evaluate structures in C'(t)
11      replace P(t) from C"(t) and P(t-1)
12      if convergence(P(t))
13          re-start P(t)
```

Fig. 2. Scheme of the CHC algorithm

4   CHC Algorithm

CHC [8] is a variant of genetic algorithm with a particular way of promoting diversity. It uses a highly disruptive crossover operator to produce new individuals maximally different from their parents. It is combined with a conservative selection strategy which introduces a kind of inherent elitism. Figure 2 shows a scheme of the CHC algorithm, whose main features are:

- The mating is not restricted to the best individuals, but parents are randomly paired in a mating pool $C(t)$ (line 6 of Figure 2). However, recombination is only applied if the Hamming distance between the parents is above a certain threshold, a mechanism of *incest prevention* (line 8 of Figure 2).
- CHC uses a *half-uniform crossover* (HUX), which exchanges exactly half of the differing parental genes (line 9 of Figure 2).
- Traditional selection methods do not guarantee the survival of best individuals, though they have a higher probability to survive. On the contrary, CHC guarantees survival of the best individuals selected from the set of parents $(P(t-1))$ and offsprings $(C'(t))$ put together (line 11 of Figure 2).
- Mutation is not applied directly as an operator.
- CHC applies a re-start mechanism if the population remains unchanged for some number of generations (lines 12-13 of Figure 2). The new population includes one copy of the best individual, while the rest of the population is generated by mutating some percentage of bits of such best individual.

5   Simulated Annealing

Simulated Annealing (SA) [9] is a stochastic optimization technique, which has its origin in statistical mechanics. It is based upon a cooling procedure used in industry. This procedure heats the material to a high temperature so that it becomes a liquid and the atoms can move relatively freely. The temperature is then slowly lowered so that at each temperature the atoms can move enough to begin adopting the most stable configuration. In principle, if the material is cooled slowly enough, the atoms are able to reach the most stable (optimum) configuration. This smooth cooling process is known as *annealing*. Figure 3 shows a scheme of SA. First at all, the parameter T, called the temperature, and the solution, are initialized (lines 2-4). The solution $s1$ is accepted as the new current solution if $\delta = f(s1) - f(s0) < 0$. Stagnations in local optimum are prevented by accepting also solutions which increase the objective function value with a probability $exp(-\delta/T)$ if $\delta > 0$. This process is repeated several times to obtain good sampling statistics for the current temperature. The number of such iterations is given by the parameter $Markov\_Chain\_length$, whose name alludes the fact that the sequence of accepted solutions is a Markov chain (a sequence of states in which each state only depends on the previous one). Then the temperature is decremented (line 14) and the entire process repeated until a frozen state is achieved at T_{min} (line 15). The value of T usually varies from a relatively large value to a small value close to zero.

```
1    t = 0
2    initialize(T)
3    s0 = Initial_Solution()
4    v0 = Evaluate(s0)
5    repeat
6        repeat
7            t = t + 1
8            s1 = Generate(s0,T)
9            v1 = Evaluate(s0,T)
10           if Accept(v0,v1,T)
11               s0 = s1
12               v0 = v1
13       until t mod Markov_Chain_length == 0
14       T = Update(T)
15   until 'loop stop criterion' satisfied
```

Fig. 3. Scheme of the Simulated Annealing (SA) algorithm

6   Parallel Heuristics

A parallel EA (PEA) is an algorithm having multiple component EAs, regardless of their population structure. Each component (usually a traditional EA)

subalgorithm includes an additional phase of *communication* with a set of sub-algorithms [2]. In this work, we have chosen a *distributed EA* (dEA) because of its popularity and because it can be easily implemented in clusters of machines. In distributed EAs (also known as Island Model) there exists a small number of islands performing separate EAs, and periodically exchanging individuals after a number of isolated steps (*migration frequency*). Concretely, we use a static ring topology in which the best individual is migrated, and asynchronously included in the target populations only if it is better than the local worst-existing solutions.

For the parallel SA (PSA) there also exist multiple asynchronous component SAs. Each component SA, start off from a different random solution, exchanges the best solution found (*cooperation* phase) with its neighbor SA in the ring.

7   The Model for Tagging

The generated tagger must be able to learn from a training corpus so as to produce a table of rules (contexts) called *training table*. Evaluation of tentative solutions is done according to the training table. This table records the different contexts of each tag. The table can be computed by going through the training text and recording the different contexts and the number of occurrences of each of them for every tag in the training text. For example, if we consider contexts with two tags on the left and two tags on the right, the entry in the table for tag *JJ* could have the form:

```
JJ 4557 9519

VBD AT JJ NN IN 37
IN PP$ JJ NNS NULL 20
PPS BEZ JJ TO VB 18
NN IN JJ NN WDT 3
        . . .
```

denoting that *JJ* has 4557 different contexts and appears 9519 times in the text, and that in one of those contexts, which appears 37 times, *JJ* is preceded by tags *VBD* and *AT* and followed by *NN* and *IN*, and so on until all the 4557 different contexts have been listed.

The search process is run for each sentence in the text to be tagged. Improvement steps aim to maximize the total probability of the tagging of the sentences in the test corpus. The process finishes either if the fitness deviation lies below a threshold value (convergence) or if the evolutionary process has been running for a maximum number of generations.

7.1   Individuals

Tentative solutions are sequences of genes which correspond to each word in the sentence to be tagged. Figure 4 shows some possible individuals for the sentence in Figure 1. Each gene represents a tag and additional information useful in the

Sentence	This	the	therapist	may	pursue	in	later	questioning	.
Ind. 1:	DT AT	NN	NNP VBP	IN	JJ	VB	.
Ind. 2:	DT AT	NN	MD	VB	RB	RB	NN	.
Ind. 3:	QL AT	NN	NNP	VB	FW	JJ	JJ	.

Fig. 4. Potential individuals for the sentence in Figure 1

evaluation of the chromosome, such as counts of contexts for this tag according to the training table. Each gene's tag is represented by an index to a vector which contains the possible tags of the corresponding word. The composition of the genes depends on the chosen coding, as Figure 5 shows. In the integer coding the gene is just the integer value of the index. In the binary coding the gene is the binary representation of the index. As in the texts we have used for experiments the maximum number of tags per word is 6, we have used both a binary code of 7 and another one of 4 bits.

word	tag index						integer	binary(7)	binary(4)
	0	1	2	3	4	5 ···			
This	DT	QL					0	0000000	0000
the	AT						0	0000000	0000
therapist	NN						0	0000000	0000
may	NNP	MD					1	0000001	0001
pursue	VB	VBP					0	0000000	0000
in	RP	NNP	RB	NN	FW	IN	5	0000101	0101
later	RP	RB	JJ	JJR			3	0000011	0011
questioning	VB	NN	JJ				1	0000001	0001

Fig. 5. Integer and binary codings (7 and 4 bits) of a possible selection of tags chosen for the words of a sentence extracted from the Brown corpus. The selected tags appear underlined.

Initial Population. For a given sentence of the test corpus, the chromosomes forming the initial population are created by randomly selecting from a dictionary one of the valid tags for each word, with a bias to the most probable tag. Words not appearing in the dictionary are assigned the most probable for its corresponding context, according to the training text.

7.2   Fitness Evaluation

The fitness of an individual is a measure of the total probability of its sequence of tags, according to the data from the training table. It is computed as the sum of the fitness of its genes, $\sum_i f(g_i)$. The fitness of a gene is defined as

$$f(g) = \log P(T|LC, RC)$$

where $P(T|LC, RC)$ is the probability that the tag of gene g is T, given that its context is formed by the sequence of tags LC to the left and the sequence

RC to the right (the logarithm is taken in order to make fitness additive). This probability is estimated from the training table as

$$P(T|LC, RC) \approx \frac{occ(LC, T, RC)}{\sum_{T' \in \mathcal{T}} occ(LC, T', RC)}$$

where $occ(LC, T, RC)$ is the number of occurrences of the list of tags LC, T, RC in the training table, and $\mathcal{T}$ is the set of all possible tags of g_i. For example, if we are evaluating the individual 1 of Figure 4 and we are considering contexts composed of one tag on the left and one tag on the right of the position evaluated, the fourth gene (word *may*), for which there are two possible tags, NNP (the one chosen in this individual) and MD, will be evaluated as

$$\frac{\#(\text{NN NNP VBP})}{[\#(\text{NN NNP VBP}) + \#(\text{NN MD VBP})]}$$

where # represents the number of occurrences of the context.

A particular sequence LC, T, RC may not be listed in the training table, either because its probability is strictly zero (if the sequence of tags is forbidden for some reason) or, most likely, because there is insufficient statistics. In these cases we proceed by successively reducing the size of the context, alternatively ignoring the rightmost and then the leftmost tag of the remaining sequence (skipping the corresponding step whenever either RC or LC are empty) until one of these shorter sequences matches at least one of the training table entries or until we are left simply with T. In this latter case we take as fitness the logarithm of the frequency with which T appears in the corpus (also contained in the training table).

7.3   Genetic Operators

For the GA, we use a one point crossover, i.e. a crossover point is randomly selected and the first part of each parent is combined with the second part of the other parent thus producing two offsprings. Then, a mutation point is randomly selected and the tag of this point is replaced by another of the valid tags of the corresponding word. The new tag is randomly chosen according to its probability (the frequency it appears in the corpus).

The CHC algorithm applies HUX crossover, randomly taking from each parent half of the tags in which they differ and exchanging them.

Individuals resulting from the application of genetic operators along with the old population are used to create the new one.

8   Experiments

We have used as the set of training texts for our taggers the Brown corpus [11], one of the most widespread in linguistics. The tag set of this corpus is not too large, what favours the accuracy of the system. Moreover, this tag set has been reduced by grouping some related tags under a unique name tag, what improves statistics. For instance, different kinds of adjectives (JJ, $JJ + JJ$, JJR, $JJR + CS$, JJS, JJT) distinguished in the corpus have been grouped under the common tag JJ.

The CHC algorithm has been run with a crossover rate of 50%, without mutation. Whenever convergence is achieved, 90% of population is renewed. The GA applies the recombination operator with a rate of 50%, and the mutation operator with a rate of 5%. In the parallel version, the migration occurs every 10 generations. We made several tests with different parameter settings for determining the best values for each algorithms. The analysis of other specific operators is defered for a future work.

Table 1. Tagging accuracy obtained with the CHC algorithm for a test text of 2500 words. PS stands for Population Size.

Context	CHC-Int				CHC-Bin(7)				CHC-Bin(4)			
	PS = 20		PS = 56		PS = 20		PS = 56		PS = 20		PS = 56	
	Seq.	Par.	Seq.	Par.	Seq.	Par.	Seq.	Par.	Seq.	Par.	Seq.	Par.
1-0	89.96	90.15	89.34	89.34	92.08	92.17	91.04	90.95	91.94	**92.53**	91.18	91.35
2-0	91.41	91.68	90.91	91.32	93.34	93.43	92.35	91.90	93.38	**93.52**	92.04	92.40
3-0	92.58	92.89	91.68	91.72	93.74	93.97	92.98	93.07	93.92	**93.97**	93.16	93.25
1-1	93.12	93.48	92.39	92.58	94.78	94.97	93.88	94.06	**95.14**	94.82	94.06	94.10
2-1	93.56	93.70	93.07	93.21	94.51	94.51	93.83	94.06	94.47	**94.65**	93.83	94.15
2-2	94.51	94.11	94.29	93.98	94.61	94.73	94.78	94.78	95.01	**95.23**	94.06	94.87

Tables 1 and 2 show the results obtained with the CHC and GA algorithms, using both, integer and binary codings. In order to study the impact of the length of the code for the binary representation, we have used 7 bits and a 4 bits codes (which are enough, because the maximum number of possible tags of a word is 6). Each row in tables corresponds to a kind of context: 1-0 is a context which considers only the tag of the preceding word, 1-1 considers the tag of the preceding and succeeding words, etc. Figures represent the best result out of twenty independent runs. The globally best result for each row appears in boldface. **Int** stands for the integer representation, **Bin(7)** for the binary representation with a code of 7 bits, and **Bin(4)** for binary with a code of 4 bits. Measures have been taken for different population sizes. Furthermore, sequential and parallel versions with 4 islands are analyzed. The population size of each island is the global population size divided by the number of islands.

Looking at Table 1, the first conclusion is that the binary coding always achieves a higher accuracy than the integer one. Moreover, the shorter the binary code to represent the tag, the better. This suggests that the integer representation is not appropriate for CHC, probably because the low number of genes of the latter limits the CHC mechanism to avoid crossover between similar individuals. Regarding the parallel executions, we can observe that the parallel version usually provides more accurate results, particularly for the population of 20 individuals, because for such a small population the higher diversity introduced by parallelism is beneficial.

Table 2. Accuracy obtained with the GA for a test text of 2500 words. PS stands for Population Size.

Context	GA-Int				GA-Bin(7)				GA-Bin(4)			
	PS = 20		PS = 56		PS = 20		PS = 56		PS = 20		PS = 56	
	Seq.	Par.	Seq.	Par.	Seq.	Par.	Seq.	Par.	Seq.	Par.	Seq.	Par.
1-0	**93.30**	93.12	92.94	92.67	92.84	92.48	92.53	92.39	93.07	92.44	92.35	92.21
2-0	**93.97**	93.70	93.43	93.88	93.83	93.56	93.61	93.16	93.43	92.89	93.61	93.07
3-0	**94.47**	94.38	93.88	93.79	94.19	94.01	94.28	93.65	93.97	93.52	93.83	93.70
1-1	94.78	94.87	94.92	94.42	**95.14**	94.37	94.64	94.42	94.64	94.01	94.64	94.06
2-1	94.69	**95.19**	94.69	94.87	94.87	94.69	94.55	94.78	94.64	94.37	94.69	94.28
2-2	95.19	95.23	95.14	94.83	**95.54**	94.91	94.96	95.14	95.41	94.64	94.73	95.00

Table 2 shows the results obtained with the GA. In this case, the integer representation provides the best results. The parallel version for a population size of 20 individuals is not able to improve the sequential results, because of the small size of the islands. However, for the population size of 56 individuals the parallel results improve the sequential ones for some contexts.

Comparing both tables, 1 and 2, we can observe that the GA has reached the globally best results for most kinds of contexts (1-0, 2-0, 3-0, 2-1 and 2-2), though the differences are small. This shows that the exploration of the search space given by the classical crossover and mutation operators are enough for this specific problem.

Table 3. Accuracy obtained with the SA algorithm for a test text of 2500 words (best result out of twenty independent runs)

	Context type					
	1-0	2-0	3-0	1-1	2-1	2-2
Seq.	91.32	92.40	92.98	94.24	94.06	94.60
Par.	91.00	92.53	92.67	93.47	94.15	94.33

Table 3 presents the data obtained with the SA algorithm. The SA algorithm performs 5656 iterations using a Markov chain of length 800 and with a decreasing factor of 0.99. In the parallel version, each SA component exchanges the best solution found with its neighbor SA in the ring, every 100 iterations. We can observe that SA provides worse results than any of the evolutionary algorithms, thus proving the advantages of the evolutionary approach.

Table 4 presents the average and standard deviation for the codings which provide the best results of each algorithm (integer for GA and binary of 4 bits for CHC), in both the sequential and the parallel implementations. We can observe that fluctuations in the accuracy of different runs are within a 1% interval, so we can claim that the algorithm is very robust.

Another feature of the results that is worth mentioning is that the accuracy obtained, around 95%, is a very good result [5] according to the statistical model used. To our knowledge, the results reported here outperforms in accuracy and efficiency to any existing work on tagging English texts with heuristics methods.

Table 4. Average and standard deviation of the accuracy for a population of 20 individuals and a test text of 2500 words

Context	GA-Int		CHC-Bin(4)	
	Seq.	Par.	Seq.	Par.
1-0	93.07 ± 0.24	92.81 ± 0.23	91.79 ± 0.16	91.96 ± 0.30
2-0	93.68 ± 0.23	93.37 ± 0.18	92.91 ± 0.28	93.31 ± 0.16
3-0	94.15 ± 0.32	94.07 ± 0.20	93.66 ± 0.16	93.76 ± 0.18
1-1	94.54 ± 0.13	94.41 ± 0.23	94.61 ± 0.31	94.49 ± 0.24
2-1	94.31 ± 0.26	94.27 ± 0.25	94.23 ± 0.21	94.44 ± 0.17
2-2	94.91 ± 0.26	94.72 ± 0.35	94.76 ± 0.15	94.82 ± 0.32

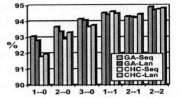

We must take into account that the accuracy is limited by the statistical data provided to the search algorithm. Moreover, the goal of the model is to maximize the probability of the context composed by the tags assigned to a sentence, but it is only an approximate model. The correct tag for a word is not always (though most times) the most probable one, and the algorithm captures this fact, but sometimes it is not the one which provides the most probable context either, and it is just in these cases when the tagger fails.

Table 5. Ratios of the execution times of the different versions of the algorithms with respect to the execution time of the sequential GA with integer representation and 1-0 context (17.2805 s.)

Context	GA-Int		GA-Bin(4)		CHC-Int		CHC-Bin(4)	
	Seq	Par.	Seq	Par.	Seq	Par.	Seq	Par.
1-0	1	0.899	1.999	1.349	1.002	0.603	1.849	1.308
2-0	4.085	2.160	11.043	6.258	3.749	1.832	10.027	5.436
3-0	18.695	6.138	55.394	18.998	17.003	5.384	47.328	16.490
1-1	4.493	1.866	13.152	5.502	4.450	1.857	11.850	5.480
2-1	21.250	7.206	67.692	23.467	21.014	5.566	60.878	20.655
2-2	96.434	25.344	268.559	52.860	95.742	26.349	247.645	68.255

Table 5 shows the average execution time for the integer and binary (4 bits) codings of the GA and CHC algorithms, which respectively provided the best results for each of them. We can observe that the execution time increases with the size of the context. We can also observe that CHC is slightly faster than GA when using the same codification. Probably, the lack of mutation in CHC compensates its additional computations. Binary codings are slower than integer ones, because they require a decodification step to apply the fitness function. The table also shows that the parallel implementation reduces the execution time, and this reduction is increasingly beneficial with the size of the context.

9   Conclusions

This work compares different optimization methods to solve an important natural language task: the selection for each word in a text of one of its possible lexical categories. The optimization methods considered here have been a classic genetic algorithm (GA), a CHC algorithm and a simulated annealing (SA). The implementation of each of them has been carried out with MALLBA.

Results obtained allow extracting a number of conclusions, such as that the integer coding performs better than the binary one for the GA, while the binary one is the best for the CHC algorithm. Furthermore, the shorter the binary code for the tag of each word, the better the performance. Parallelism has also proven useful, allowing to obtain the best results even with small populations in the case of the CHC, and to reduce the execution time in any algorithm. The GA has been found to be slightly better than CHC, indicating that the exploration of the search space achieved by the classical genetic operators is enough for this problem. The two evolutionary algorithms have outperformed SA.

For the future, we plan to extend this study to additional corpus such as the Susanne and Penn Treebank.

Acknowledgments. The first author has been supported by the project TIC2003-09481-C04. The two last authors have been partially funded by the Spanish MCyT and FEDER under contracts TIC2002-04498-C05-02 (the TRACER project).

References

1. E. Alba, F. Almeida, M. J. Blesa, J. Cabeza, C. Cotta, M. Díaz, I. Dorta, J. Gabarró, C. León, J. Luna, L. M. Moreno, C. Pablos, J. Petit, A. Rojas, and F. Xhafa. MALLBA: A library of skeletons for combinatorial optimisation. In *Euro-Par*, pages 927–932. Springer, 2002.
2. E. Alba and M. Tomassini. Parallelism and evolutionary algorithms. *IEEE Transactions on Evolutionary Computation*, 6(5):443–462, 2002.
3. L. Araujo. Part-of-speech tagging with evolutionary algorithms. In *Proc. of the Int. Conf. on Intelligent Text Processing and Computational Linguistics (CICLing-2002), LNCS 2276*, pages 230–239. Springer-Verlag, 2002.
4. L. Araujo. Studying the advantages of a messy evolutionary algorithm for natural language tagging. In *Proc. of the Int. Genetic and Evolutionary Computation Conference (GECCO), LNCS 2724*, pages 1951–1962. Springer-Verlag, 2003.
5. E. Charniak. *Statistical Language Learning*. MIT press, 1993.
6. L. Davis. *Handbook of Genetic Algorithms*. Van Nostrand Reinhold,1991.
7. S. J. DeRose. Grammatical category disambiguation by statistical optimization. *Computational Linguistics*, 14:31–39, 1988.
8. L. J. Eshelman. The CHC adaptive search algorithm: How to have safe search when engaging in nontraditional genetic recombination. In *Proceedings of the First Workshop on FOGA*, pages 265–283, San Mateo, CA, 1991. Morgan Kauffman.
9. S. Kirkpatrick, C. D. Gelatt, and M. P. Vecchi. Optimization by simulated annealing. *Science, Number 4598, 13 May 1983*, 220, 4598:671–680, 1983.
10. B. Merialdo. Tagging English text with a probabilistic model. *Computational Linguistics*, 20(2):155–172, 1994.
11. F. W. Nelson and H. Kucera. Manual of information to accompany a standard corpus of present-day edited American English, for use with digital computers. Technical report, Department of Linguistics, Brown University., 1979.
12. H. Schutze and Y. Singer. Part of speech tagging using a variable memory Markov model. In *Proc. of the 1994 of the Association for Computational Linguistics*. Association for Computational Linguistics, 1994.

An Effective Real-Parameter Genetic Algorithm with Parent Centric Normal Crossover for Multimodal Optimisation

Pedro J. Ballester and Jonathan N. Carter

Imperial College London, Department of Earth Science and Engineering,
RSM Building, Exhibition Road, London SW7 2AZ, UK
{p.ballester,j.n.carter}@imperial.ac.uk

Abstract. Evolutionary Algorithms (EAs) are a useful tool to tackle real-world optimisation problems. Two important features that make these problems hard are multimodality and high dimensionality of the search landscape.

In this paper, we present a real-parameter Genetic Algorithm (GA) which is effective in optimising high dimensional, multimodal functions. We compare our algorithm with two previously published GAs which the authors claim gives good results for high dimensional, multimodal functions. For problems with only few local optima, our algorithm does not perform as well as one of the other algorithm. However, for problems with very many local optima, our algorithm performed significantly better. A wider comparison is made with previously published algorithms showing that our algorithm has the best performance for the hardest function tested.

1   Introduction

Many real-world optimisation problems, particularly in engineering design, have a number of key features in common: the parameters are real numbers; there are many of these parameters; and they interact in highly non-linear ways, which leads to many local optima in the objective function. Clearly it is useful to have optimisers that are effective at solving problems with these characteristics. It has been shown elsewhere [1] that Genetic Algorithms (GAs) are good at solving multimodal functions. In this work, we describe and demonstrate a GA that appears to be good at solving problems where the objective function is characterised as being: high dimensional; real variable; continuous and smooth; many local optima.

Deb et al. [2] have recently produced a comprehensive review of optimisation methods. They included in their study: real parameter GAs, Self-Adaptive ESs (Evolution Strategies), DE (Differential Evolution) and GMs (Gradient Methods). All of these methods were tested on a set of high dimensional analytical test functions. They concluded that a real-parameter GA known as G3-PCX had the best overall performance. It is worth noting that G3-PCX was shown

K. Deb et al. (Eds.): GECCO 2004, LNCS 3102, pp. 901–913, 2004.

to have better convergence than GMs on some unimodal functions. G3-PCX also obtained the best results, in the published literature, for the Schwefel and Rosenbrock functions. An excellent result on the Rastrigin function was also reported.

Ballester and Carter [1] investigated the performance of a variety of GA formulations over a set of 2-variable multimodal functions. It was found that GAs which used random selection with crowding replacement strategies were robust optimisers. The same authors showed [3] that one of those GAs (named SPC-vSBX) was also effective in optimising high dimensional real-variable functions. Of particular importance were the results obtained with the Rastrigin and rotated Rastrigin functions, the hardest of all tested in terms of number of local minima. On these functions, SPC-vSBX achieved the best performance, starting with a skewed initialisation, reported in the literature. A version of SPC-vSBX has been also successful in a real-world optimisation problem [4]. The algorithm was applied to the direct inversion of a synthetic oil reservoir model with three free parameters. In experiments where the optimal model was a priori known, it was observed that the algorithm was able to find the global minimum

In this paper, a new family of crossovers is presented which, unlike vSBX, are not biased with respect to the coordinate directions. One of this crossovers will be studied in combination with the SPC model. A benchmark will be set up to test the behaviour of the new algorithm together with G3-PCX and SPC-vSBX. This benchmark consists of a set of analytical test functions that are known to be difficult to many optimisation algorithms. These functions have been widely used, which will allow a wider comparison with previously published studies.

We arrange the rest of the paper as follows. Section 2 discusses the approach to testing algorithms. Section 3 describes the structure of the proposed GA. In Sect. 4, the experimental setup is explained. Results are presented and a comparison with G3-PCX and SPC-vSBX made in Sect. 5. Section 6 reviews the presented results with respect to past studies. Lastly, we present our conclusions and discuss future work in Sect. 7.

2   Testing Algorithms on Analytical Functions

When tackling a real-world problem, the conventional approach is to test first the algorithm on a set of analytical functions. Most real-world applications involve objective functions considerably more expensive to evaluate than an analytical one. Consequently, it is usually unviable to test the effectiveness of an algorithm directly on the real problem. There is the assumption that the chosen set of functions share some characteristics with the target problem. Based on this assumption, one applies the algorithm that performed well on the benchmark in the expectation that it will do also well on the real-world problem.

This approach to evaluating algorithms is not without drawbacks. As pointed out by Whitley et al. [5], there is the potential danger that algorithms 'become overfitted to work well on benchmarks and therefore that good performance on benchmarks does not generalize to real world problems'. An example of this is algorithms that exploit benchmark symmetries unlikely to be present on the target problem. For instance, many algorithms are tested by initialising the population symmetrically around the global optimum. The algorithm might have an inherent tendency to create children near the centroid of the parents (eg. mean-centric recombination in GAs). Deb et al. [2] argued that this is unfair since: 'a mean-centric recombination of two solutions at either side of $x_j = 0$ is likely to result in a children near $x_j = 0$. Moreover, in most real-world problems, the knowledge of the exact optimum is usually not available, and the performance of an Evolutionary Algorithm (EA) on a symmetric initialisation may not represent the EA's true performance in solving the same problem with a different initialisation or other problems'. Consequently it is important that algorithms are tested with skewed initialisations, so as to give a better indication of their performance on real-world problems. There are a number of additional studies that have also pointed out the need of using a skewed initialisation [6, 7,8,9,10,3]. In this work, we have chosen a skewed initialisation that does not bracket the global minimum. This has been done to test the algorithms under the hardest situation. However, we feel that an initialisation bracketing the global minimum, but in a sufficiently asymmetrical way, is also a valid approach.

A different example of a bias that results in improved performance is given by Ballester and Carter [3]. In that study, the vSBX crossover was used. vSBX has a preference for searching along the coordinate directions. It was pointed out that this may give the GA an advantage on test functions with minima aligned with the axis. The latter is a property of the Rastrigin function. The algorithm success in solving a 50-variable Rastrigin could have benefitted from this characteristic. Consequently, the authors introduced a rotation in the function to neutralise the algorithm's advantage. A significantly inferior performance on the rotated Rastrigin was reported.

3   GA Description

Our real parameter GA uses a steady state population model. In each generation, two parents are selected from the current population to produce λ children through crossover. Offspring and current populations are then combined so that the population remains at a constant size.

This GA combines the following features: parental selection is not fitness biased, a self-adaptative unbiased crossover operator, implicit elitism and locally scaled probabilistic replacement. We will refer to this GA as SPC-PNX (Scaled Probabilistic Crowding Genetic Algorithm with Parent Centric Normal crossover). Below we describe the details of SPC-PNX selection, replacement and crossover schemes:

3.1   Selection

We use uniform random selection, without replacement, to select two parents from the current population. Unusually for a GA, fitness is not taken into account during the selection process.

3.2   Scaled Probabilistic Crowding Replacement

We use a scaled probabilistic crowding scheme for our replacement policy. First, NREP individuals from the current population are selected at random. These individuals then compete with the offspring for a place in the population.

In the probabilistic crowding scheme [11], the closest preselected individual (x^{cst}) enters a probabilistic tournament with the offspring (x^{ofp}), with culling likelihoods (survival, if we were in a maximisation problem) given by

$$p(x^{ofp}) = \frac{f(x^{ofp})}{f(x^{ofp}) + f(x^{cst})}, \quad p(x^{cst}) = \frac{f(x^{cst})}{f(x^{ofp}) + f(x^{cst})} \quad . \tag{1}$$

where $f(x)$ is the objective function value for an individual x.

If the differences in function values across the population are small with respect to their absolute values, these likelihoods would be very similar in all cases. The scaled probabilistic crowding replacement is introduced to avoid this situation. It operates with culling likelihoods

$$p(x^{ofp}) = \frac{f(x^{ofp}) - f_{best}}{f(x^{ofp}) + f(x^{cst}) - 2f_{best}}, \quad p(x^{cst}) = \frac{f(x^{cst}) - f_{best}}{f(x^{ofp}) + f(x^{cst}) - 2f_{best}} \quad . \tag{2}$$

where f_{best} is the function value of the best individual in the offspring and selected group of NREP individuals.

This replacement scheme has several beneficial features. The fittest individual does not always win, which helps to prevent premature convergence. Crowding schemes such as this promote the creation of subpopulations that explore different regions of the search space. This has been shown [1] [3] to be beneficial for creating multiple optimal solutions and to increase the effectiveness in finding the global minimum. It implements elitism in an implicit way. If the best individual in either offspring or current parent population enters this replacement competition will have probability zero of being culled.

3.3   Crossovers

In this work, we test two crossovers in combination with the SPC model: a version of the Simulated Binary Crossover (SBX) [12] [13] called vSBX [1] [3] and a new crossover called PNX. These crossovers are self-adaptative in the sense that the spread of the possible offspring solutions depends on the distance between the parents, which decreases as the population converge.

In SBX, children have zero probability of appearing in some regions of the parameter space, as shown in Fig 1. vSBX does not exclude any regions, while preserving the good SBX properties. This may allow a better exploration of the search space. It should be noted that SBX and vSBX preferentially search along the coordinate directions. This may give an advantage on test functions where minima are aligned along coordinate directions.

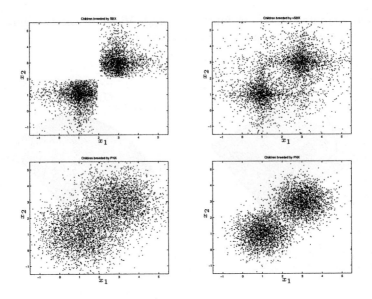

Fig. 1. Children bred from parents $x^{(1)} = (1,1)$ and $x^{(2)} = (3,3)$ for (clockwise starting from upper left plot) a) SBX ($\eta = 1$), b) vSBX ($\eta = 1$), c) PNX ($\eta = 3$) and d) PNX ($\eta = 2$)

Like vSBX, PNX does not exclude any regions, while creating offspring close to the parents. However, unlike SBX and vSBX, PNX does not preferentially search along the axis and hence it is not biased towards coordinate directions. In PNX, for each of the λ children, we proceed as follows to determine its jth gene (y_j). First, we draw a single random number $w \in [0,1]$, we use the form $y_j^{(1)}$ if $w < 0.5$ and $y_j^{(2)}$ if $w \geq 0.5$. Once this choice is made, the same selected form is used for every component j. The forms are

$$y_j^{(1)} = N(x_j^{(1)}, |x_j^{(2)} - x_j^{(1)}|/\eta), \quad y_j^{(2)} = N(x_j^{(2)}, |x_j^{(2)} - x_j^{(1)}|/\eta) \ . \tag{3}$$

where $N(\mu, \sigma)$ is a random number drawn from a gaussian distribution with mean μ and standard deviation σ, $x_j^{(i)}$ is the jth component of the ith parent and η is a tunable parameter. The larger is the value of η the more concentrated is the search around the parents.

4   Experimental Setup

We use the same experimental setup as in Deb et al. [2], allowing a direct comparison with their results. The stopping criteria are: either a maximum of 10^6 function evaluations or an objective value of 10^{-20} is obtained.

Our benchmark consists in six analytical 20-variable functions: ellipsoidal (f_{elp}), Schwefel (f_{sch}), Generalized Rosenbrock (f_{ros}), Ackley (f_{ackl}), Rastrigin

(f_{rtg}) and a rotated Rastrigin function (f_{rrtg}). Views of the two-dimensional versions of these functions are given in Figs. 2 to 7.

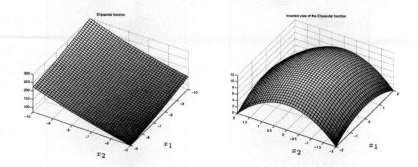

Fig. 2. Initialisation (left) and global minimum (right, inverted view) for the Ellipsoidal function. $f_{elp}(x) = \sum_{j=1}^{M} j x_j^2$

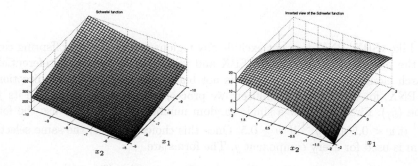

Fig. 3. Initialisation (left) and global minimum (right, inverted view) for the Schwefel function: $f_{sch}(x) = \sum_{j=1}^{M} \left(\sum_{k=1}^{j} x_k \right)^2$

These functions were selected for several reasons. First, they have been widely used, which will allow an extensive comparison with previously published algorithms. Also, these functions have a number of features that are known to be hard for optimisation algorithms and believed to be present in many real-world problems. The Ellipsoidal is a unimodal function with different weights for each variable. This will serve to test the algorithms with a badly scaled objective function. The Schwefel function is also unimodal, but its variables are correlated. The Generalized Rosenbrock has been regarded as a unimodal function, but there is

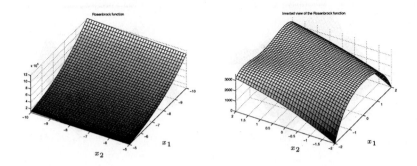

Fig. 4. Initialisation (left) and global minimum (right, inverted view) for the Rosenbrock function: $f_{ros}(\boldsymbol{x}) = \sum_{j=1}^{M-1}(100(x_j^2 - x_{j+1})^2 + (x_j - 1)^2)$

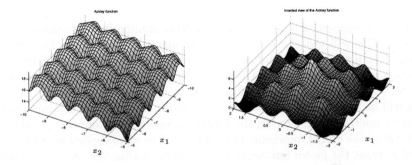

Fig. 5. Initialisation (left) and global minimum (right, inverted view) for the Ackley function: $f_{ackl}(\boldsymbol{x}) = 20 + e - 20\exp(-0.2\sqrt{\frac{1}{M}\sum_{j=1}^{M}x_j^2}) - \exp(\frac{1}{M}\sum_{j=1}^{M}\cos(2\pi x_j)))$

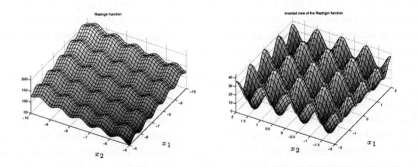

Fig. 6. Initialisation (left) and global minimum (right, inverted view) for the Rastrigin function: $f_{rtg}(\boldsymbol{x}) = 10M + \sum_{j=1}^{M}(x_j^2 - 10\cos(2\pi x_j))$

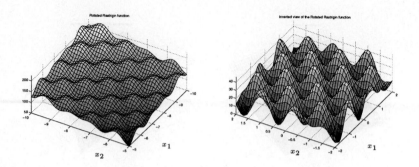

Fig. 7. Initialisation (left) and global minimum (right, inverted view) for the Rotated Rastrigin function: $f_{rrtg}(\boldsymbol{y}) = 10M + \sum_{j=1}^{M}(y_j^2 - 10\cos(2\pi y_j))$, $\boldsymbol{y} = \boldsymbol{Ax}$ with $A_{j,j} = 4/5$, $A_{j,j+1} = 3/5$ (j odd), $A_{j,j-1} = -3/5$ (j even), $A_{j,k} = 0$ (the rest)

evidence [2] suggesting that it contains several minima in high dimensional instances. This will test the behaviour of the algorithm with objective functions having a couple of minima and an almost flat region near the global. The Ackley function is highly multimodal. The basin of these local minima increase in size as one moves away from the global minimum, as discussed by [8]. Thus, this function will be useful to study the behaviour of the algorithms when initialised in a highly multimodal region. In the Rastrigin function, the opposite is observed. Away from the global minimum, the landscape has a parabolic structure. As we move towards the global minimum, the size of the basins increase [8]. Therefore, an algorithm has to discard many local minima of similar quality before reaching the global minimum. This is known to be difficult for many optimisation algorithms, specially in high dimensional Rastrigin instances. Lastly, a rotation is carried out on the Rastrigin to make it non-separable, while still being highly multimodal. The resulting rotated function has no longer local minima arranged along the axis. The rotated Rastrigin function is expected to help to avoid overestimating the performance of algorithms using separable objective functions. All functions have a single global minimum with value zero. The global minimum is located at $x_j = 1$ (Rosenbrock) or $x_j = 0$ (the rest).

As an experiment has some dependence on the initial conditions, we repeat each experiment, each time with a different initial population. We do not initialise the population symmetrically around the global minimum, all variables are initialised at random within $[-10, -5]$. The purpose of this skewed initialisation is two fold. First, it ensures that the algorithms generally have to overcome a number of local minima before reaching the global minimum. Second, it neutralises the advantage enjoyed by algorithms that have an inherent tendency to create solutions near the centroid of the parents.

5   Discussion of the Results

SPC-vSBX and SPC-PNX contain four tunable parameters: N, λ, NREP and η. In this study, we fix $\eta = 0.01$ (for vSBX), $\eta = 2.0$ (for PNX) and NREP=2. G3-PCX's results for the Ellipsoidal, Schwefel, Generalized Rosenbrock and Rastrigin are extracted from the original study [2]. For the rest of functions, we use the G3-PCX code downloaded from the KanGAL website [14]. The procedure consists in doing some preliminary runs to determine the best N and λ for each function. Due to the limited computing precision, the accuracy for the Ackley function was set as 10^{-10}.

In Table 1, we compare G3-PCX, SPC-vSBX and SPC-PNX. G3-PCX reports the best results in the literature for the Schwefel and Generalised Rosenbrock. It also obtained an excellent result for the Rastrigin starting with a skewed initialisation. For the Ellipsoidal, only a Gradient Method (the BFGS quasi-Newton algorithm with a mixed quadratic-cubic polynomial line search approach achieved a solution in the order of 10^{-24} in 6,000 function evaluations [2]) was shown to outperform it. Over the unimodal functions SPC-vSBX and SPC-PNX are not competitive in terms of number of function evaluations, although they reached the required accuracy in all runs.

In the 20-variable Rosenbrock function, there are two known local minima [2] with function values of 3.986624 and 65.025362. G3-PCX found solutions better than 10^{-20} in 36 out of 50 runs, but in the other 14 got stuck in the best local minimum. SPC-vSBX only found a best solution of 10^{-4} in 50 runs. It found solutions below the best local minimum (ie. within the global basin) in 48 out of 50 runs. SPC-PNX found a best solution of 10^{-10} in 50 runs. It found solutions below the best local minimum in 38 out of 50 runs. By incrementing N, SPC-PNX reached the global basin in 47 out of 50 runs. Since the SPC model is not strongly fitness biased, we conjecture that the slow convergence observed is due to the function's flat regions.

In the 20-variable Ackley function, G3-PCX was not able to find the global basin in any of the ten runs, finding a best value of 3.959. In most of the runs, the algorithm could not escape the highly multimodal initialisation region. Whereas SPC-vSBX and SPC-PNX found the global minimum in all runs, with SPC-PNX outperforming SPC-vSBX in terms of required number of evaluations.

In the 20-variable Rastrigin function, SPC-vSBX could find a solution better than 10^{-20} in 6 out of 10 runs, whereas G3-PCX was reporting an overall best solution of 15.936 within the prescribed limits. In the other 4 runs, SPC-vSBX always found one of the best local minima with value 0.9949591. However, the authors of this algorithm warned that vSBX may benefit of an advantage when applied to the Rastrigin function because of its preferential search along the axis. To neutralise this advantage, SPC-vSBX was tested on the rotated Rastrigin function and a best value of 8.955 was found, whereas a best value of 309.429 was reported with G3-PCX. By contrast, SPC-PNX found best values of 4.975 and 3.980 for the Rastrigin and rotated Rastrigin, respectively.

It has been generally observed that by incrementing N, in both SPC-vSBX and SPC-PNX, better results are found at a cost of taking longer to converge.

Table 1. Performance comparison between G3-PCX, SPC-vSBX and SPC-PNX over the test functions. The best, median and worst columns refer to the number of function evaluations required to obtain a value of 10^{-20}. If the target is not reached then the best found function value within 10^6 evaluations is given. 'Success' refers to how many runs reach the target accuracy (unimodal) or end up within the global basin (multimodal). The latter is determined by checking if the best found solution is below the function's best local minimum. '?' accounts for information not specified in the original study [2].

Model	Crossover	(N,λ)	Function	Best	Median	Worst	Best Found	Success
G3	PCX-(0.1,0.1)	(100,2)	Elp	5,826	6,800	7,728	10^{-20}	10/10
SPC	vSBX-0.01	(6,1)	Elp	49,084	50,952	57,479	10^{-20}	10/10
SPC	PNX-2.0	(35,1)	Elp	36,360	39,360	40,905	10^{-20}	10/10
G3	PCX-(0.1,0.1)	(150,2)	Sch	13,988	15,602	17,188	10^{-20}	10/10
SPC	vSBX-0.01	(6,1)	Sch	260,442	294,231	334,743	10^{-20}	10/10
SPC	PNX-2.0	(35,1)	Sch	236,342	283,321	299,301	10^{-20}	10/10
G3	PCX-(0.1,0.1)	(150,4)	Ros	16,508	21,452	25,520	10^{-20}	36/50
SPC	vSBX-0.01	(12,1)	Ros	10^6	-	-	10^{-4}	48/50
SPC	PNX-2.0	(35,1)	Ros	10^6	-	-	10^{-10}	38/50
SPC	PNX-2.0	(80,1)	Ros	10^6	-	-	10^{-6}	47/50
G3	PCX-(0.1,0.1)	(150,2)	Ackl	10^6	-	-	3.959	0
SPC	vSBX-0.01	(8,1)	Ackl	57,463	63,899	65,902	10^{-10}	10/10
SPC	PNX-2.0	(50,1)	Ackl	45,736	48,095	49,392	10^{-10}	10/10
G3	PCX-(?,?)	(?,?)	Rtg	10^6	-	-	15.936	0
SPC	vSBX-0.01	(20,3)	Rtg	260,658	306,819	418,482	10^{-20}	6/10
SPC	vSBX-0.01	(40,3)	Rtg	639,102	721,401	800,754	10^{-20}	10/10
SPC	PNX-2.0	(400,4)	Rtg	10^6	-	-	4.975	0
G3	PCX-(0.1,0.1)	(300,3)	Rot. Rtg	10^6	-	-	309.429	0
SPC	vSBX-0.01	(75,3)	Rot. Rtg	10^6	-	-	8.955	0
SPC	PNX-2.0	(400,4)	Rot. Rtg	10^6	-	-	3.980	0

Also, a restricted search (through a higher value of η) seems to be beneficial in the highly multimodal functions. Based on these observations, we investigate the performance of SPC-PNX with different combinations of N, λ and η, allowing a higher number of function evaluations and using the same initialisation. As a result, we solved the 20-variable Rastrigin (N=2,000, $\lambda = 4$, PNX-3.0 and $2 \cdot 10^6$ evaluations, obtaining a function value of $3.634 \cdot 10^{-12}$) and the 20-variable rotated Rastrigin (N=2,500, $\lambda = 3$, PNX-3.0 and $2.25 \cdot 10^6$ evaluations, obtaining a function value of $2.438 \cdot 10^{-2}$).

6   Review of Results with Respect to Other Studies

In this section, other previous studies reporting results on the used test functions are reported. This will allow a wider comparison with previously published studies.

Eiben and Bäck [7] used an (μ, λ)-ES to optimise 30-variable Schwefel, Ackley and Rastrigin functions. On the Schwefel, the ES was initialised within [60,65] and a best solution greater than 1.0 was reported. The initialisation for the Ackley function was [15,30] and the best found values was greater than 10^{-13}. The Rastrigin was initialised within [4,5] and a solution better than 10.0 was reported. Storn and Price [15] used DE on a testbed including Ackley and Rastrigin functions with symmetric initialisations. The 30-variable Ackley, 100-variable Ackley, 20-variable Rastrigin and 100-variable Rastrigin functions were solved. However, as the authors admit for these multimodal functions: 'As many symmetries are present, the main difficulty of these test functions lies in their dimensionality'. Chellapilla and Fogel [8] solved the 10-variable Rastrigin and Ackley functions starting from [8,12]. Compared to a symmetric initialisation, this study showed negative improvement in best function values with the skewed initialisation. Patton et al. [9] used also a skewed initialisation (but bracketting the global minimum) and solved the 10-variable instances of the Schwefel, Rosenbrock, Ackley and Rastrigin. Wakunda and Zell [16] apply a number of CMA-ESs (Covariance Matrix Adaptation Evolutionary Strategies) and solved the 20-variable Ellipsoidal, Schwefel, Rosenbrock and Ackley functions. As the initialisation was not stated, it is not possible to compare with these results. Kita [17] using a real-parameter GA (known as MGG-UNDX) and a ES solves the 20-variable Rosenbrock function. Also, the symmetric initialisation [-5.12,5.12] was used to solve a 5-variable rotated Rastrigin. Hansen and Ostermeier [18] applied a CMA-ES to the Ellipsoidal, Schwefel, Rosenbrock and Rastrigin. Starting from a unit away from the global minimum, the Ellipsoidal, Schwefel and Rosenbrock functions were solved with up to 320 variables. The algorithm was applied on the 20-variable Rastrigin. Starting with a solution initialised in [-5.12,5.12], function values within 30.0 and 100.0 were found.

7   Conclusions and Future Work

We have presented a GA (SPC-PNX) which has been shown to be effective in optimising high dimensional real-variable functions. This algorithm incorporates the new parent-centric crossover PNX (Parent-centric Normal Crossover).

SPC-PNX's performance has been tested on a set of high dimensional real-variable functions. These functions have a number of features that are known to be hard for optimisation algorithms and believed to be present in many real-world problems. By using PNX instead of vSBX, a better convergence was obtained while maintaining practically the same average performance. This was observed for all test functions but the Rastrigin, where SPC-vSBX is known to enjoy an advantage. In comparison with G3-PCX, SPC-PNX does not perform as well as this algorithm for problems with few local minima. However, for problems with very many local optima, our algorithm performed significantly better. In the hardest test function (rotated Rastrigin) in terms of separability and multimodality, SPC-PNX widely overcomes the performance previously reported.

In future work, we will investigate the effect of varying the parameter NREP, which seems to affect the ability of maintaining several subpopulations during the GA run. Also, we plan to apply it to carry out the calibration of the model parameters corresponding to a real petroleum reservoir.

References

1. Ballester, P.J., Carter, J.N.: Real-parameter genetic algorithms for finding multiple optimal solutions in multi-modal optimization. In: Genetic and Evolutionary Computation Conference, Lecture Notes in Computer Science 2723. (2003)
2. Deb, K., Anand, A., Joshi, D.: A computationally efficient evolutionary algorithm for real-parameter optimization. Evolutionary Computation 10 (2002) 371–395
3. Ballester, P.J., Carter, J.N.: An effective real-parameter genetic algorithms for multimodal optimization. In Parmee, I.C., ed.: Proceedings of the Adaptive Computing in Design and Manufacture VI. (2004) In Press.
4. Carter, J.N., Ballester, P.J., Tavassoli, Z., King, P.R.: Our calibrated model has no predictive value: An example from the petroleum industry. In: Proceedings of the Sensitivity Analysis and Model Output Conference (SAMO-2004), Santa Fe, New Mexico, U.S.A. (2004) In Press.
5. Whitley, D., Watson, J., Howe, A., Barbulescu, L.: Testing, evaluation and performance of optimization and learning systems. In Parmee, I.C., ed.: Proceedings of the Adaptive Computing in Design and Manufacture V, Springer Verlag (2002) 27–39
6. Fogel, D.B., Beyer, H.G.: A note on the empirical evaluation of intermediate recombination. Evolutionary Computation 3 (1996) 491–495
7. Eiben, A.E., Bäck, T.: Empirical investigation of multiparent recombination operators in evolution strategies. Evolutionary Computation 5 (1998) 347–365
8. Chellapilla, K., Fogel, D.B.: Fitness distributions in evolutionary computation: Analysis of local extrema in the continuous domain. In Angeline, P.J., Michalewicz, Z., Schoenauer, M., Yao, X., Zalzala, A., eds.: Proceedings of the Congress on Evolutionary Computation. Volume 3., Mayflower Hotel, Washington D.C., USA, IEEE Press (1999) 1885–1892
9. Patton, A.L., Goodman, E.D., III, W.F.P.: Scheduling variance loss using population level annealing for evolutionary computation. In Angeline, P.J., Michalewicz, Z., Schoenauer, M., Yao, X., Zalzala, A., eds.: Proceedings of the Congress of Evolutionary Computation. Volume 1., Mayflower Hotel, Washington D.C., USA, IEEE Press (1999) 760–767
10. Deb, K., Beyer, H.G.: Self-adaptive genetic algorithms with simulated binary crossover. Evolutionary Computation 9 (2001) 197–221
11. Mengshoel, O., Goldberg, D.: Probabilistic crowding: Deterministic crowding with probabilistic replacement. In Banzhaf, W., Daida, J., Eiben, A., Garzon, M., V.Honavar, Jakiela, M., Smith, R., eds.: Proceedings of the Genetic and Evolutionary Computation Conference, Morgan Kaufmann (1999) 409–416
12. Deb, K., Agrawal, S.: Simulated binary crossover for continous search space. Complex Systems 9 (1995) 115–148
13. Deb, K., Kumar, A.: Real-coded genetic algorithms with simulated binary crossover: Studies on multi-modal and multi-objective problems. Complex Systems 9 (1995) 431–454
14. KanGAL: (January 2004) http://www.iitk.ac.in/kangal/soft.htm.

15. Storn, R., Price, K.: Differential evolution a simple and efficient heuristic for global optimisation over continuous spaces. Journal of Global Optimization **11** (1997) 341–359
16. Wakunda, J., Zell, A.: Median-selection for parallel steady-state evolution strategies. In Schoenauer, M., Deb, K., Rudolph, G., Yao, X., Lutton, E., Merelo, J.J., Schwefel, H.P., eds.: Proceedings of the 6th Internation Conference "Parallel Problem Solving from Nature - PPSN VI. Volume 1917 of Lecture Notes in Computer Science., Paris, Fance, Springer Verlag (2000) 405–414
17. Kita, H.: A comparison study of self-adaptation in evolution strategies and real-coded genetic algorithms. Evolutionary Computation **9** (2001) 223–241
18. Hansen, N., Ostermeier, A.: Completely derandomized self-adaptation in evolution strategies. Evolutionary Computation **9** (2001) 159–195

Looking Under the EA Hood with Price's Equation

Jeffrey K. Bassett[1], Mitchell A. Potter[2], and Kenneth A. De Jong[1]

[1] George Mason University, Fairfax, VA 22030
{jbassett, kdejong}@cs.gmu.edu

[2] Naval Research Laboratory, Washington, DC 20375
mpotter@aic.nrl.navy.mil

Abstract. In this paper we show how tools based on extensions of Price's equation allow us to look inside production-level EAs to see how selection, representation, and reproductive operators interact with each other, and how these interactions affect EA performance. With such tools it is possible to understand at a deeper level how existing EAs work as well as provide support for making better design decisions involving new EC applications.

1 Introduction

Evolutionary algorithm design is difficult for a number of reasons, not the least of which is that the choices of selection, representation, and reproductive operators interact in non-linear ways to affect EA performance. As a consequence, EA designers are often faced with their own "black box" optimization problem in terms of finding combinations of design choices that improve EA performance on their particular application domain.

Results from the EC theory community continue to provide new insights into this difficult design process, but often are obtained by make simplifying assumptions in order to make the mathematics tractable. This often leaves open the question as to whether particular theoretical results apply in practice to actual EAs being used and/or to newly designed EAs.

The results presented here are part of an ongoing effort that tries to bridge this gap by using theoretical results to help us build useful tools that can be used to look inside actual EAs in order to better understand what's happening "under the hood". In particular, we have been exploring the use of some theoretical work done by Price (1970) to obtain deeper insights into the interactions of selection, representation, and the reproductive operators of crossover and mutation. In this respect we are indebted to Lee Altenberg who for some time now has been encouraging the EC community to pay more attention to Price's Theorem (Altenberg 1994; Altenberg 1995).

In section 2 we provide a brief overview of Price's Theorem and how it can be extended in a way to provide a useful analysis tool. Section 3 describes how Price's Theorem can be further extended to provide additional insights. We illustrate these ideas in section 4 by using the developed tools to instrument actual production-level EAs and showing how two EAs with similar "black box" behavior look quite different "under the hood". Finally, in section 5 we conclude with some observations and future work.

K. Deb et al. (Eds.): GECCO 2004, LNCS 3102, pp. 914–922, 2004.

2   Background

In 1970, George Price published the article *Covariance and Selection* (Price 1970) in which he presented an equation that has proved to be a major contribution to the field of evolutionary genetics (Frank 1995). The equation describes the existence of a covariance relationship between the number of successful offspring that an individual produces and the frequency of any given gene in that individual. If this covariance value is high, then the existence of that gene is a good predictor of selection.

2.1   Price's Equation

Although Price focused on gene frequency, his equation is more general and can be used to estimate the change in any measurable attribute from the parent population to the child population, and separate the change attributable to selection from the change attributable to the genetic operators. Specifically,

$$\Delta Q = \frac{Cov(z,q)}{\overline{z}} + \frac{\sum z_i \Delta q_i}{N\overline{z}}, \tag{1}$$

where q_i is the measurement of some attribute of parent i such as the number of occurrences of a particular gene or combination thereof, z_i is the number of children to which parent i contributed genetic material, $\overline{z}$ is the average number of children produced by each parent, N is the number of parents, and Δq_i is the difference between the average q value of the children of i and the q value of parent i.

Price's Equation combines the effect of all the genetic operators into a single term. To further separate the effects of the individual reproductive operators, Potter et al. (2003) extended the equation as follows:

$$\Delta Q = \frac{Cov(z,q)}{\overline{z}} + \sum_{j=1}^{P} \frac{\sum z_i \Delta q_{ij}}{N\overline{z}}, \tag{2}$$

where P is the number of genetic operators and Δq_{ij} is the difference between the average q value of the children of i measured before and after the application of operator j.

2.2   Previous Applications in Evolutionary Computation

Altenberg (1995) was one of the first in the EC community to call attention to Price's Equation. He demonstrated Price's assertion that gene frequency is not the only attribute of the individuals which can be predicted by the equation, and identified several different measurement functions which could be useful, including mean fitness from both the biological and evolutionary computation perspectives, frequency of schemata, and evolvability.

More recently, Langdon and Poli (2002) showed how measuring gene frequencies is equivalent to determining the frequency of use of the available primitives in the evolving solution trees, and used Price's Equation to diagnose the probable causes of poorer performing runs.

Finally, Potter, Bassett, and De Jong (2003) concentrated on using fitness as a measurement function and applied Price's Equation to the visualization of the dynamics of evolvability, that is, the ability of an EA to continue to make improvements in fitness over time.

3   Variance of Operator Effects

Each term in Price's equation calculates the contribution of a different operator as a mean of the attribute being measured. Although visualizations based on the decomposition of delta mean fitness into operator-based components certainly provides more information than simple best-so-far curves (Potter et al. 2003), focusing on the mean can sometimes be misleading. In particular, the mean may be close to zero, leading one to believe that an operator is making a minimal contribution, when in fact it is a critical component of the evolutionary process.

In fact, the average individuals are not the ones driving evolution forward. It is the occasional exceptional individuals created by crossover and mutation that enable the population to continue to improve over time. The best and worst individuals are at the upper and lower tails of the population fitness distributions. Therefore, if we want to get some sense of how often an operator creates above or below average individuals, we need to look at the variance of the Δq values for mutation and crossover, not just the mean.

We should emphasize that we are interested in the variance of the effect of an operator on the individuals of a population, not the variance of the mean effect given multiple runs. Specifically, let $E[X]$ be the expected change in the measurement of q due to a particular operator such as crossover. For simplicity we will assume for the moment that only a single operator is used. Expanding the second term of equation 1 we have

$$E[X] = \frac{\sum_{i=1}^{N} z_i \Delta q_i}{N\overline{z}}$$

$$= \frac{\sum_{i=1}^{N} z_i \frac{\sum_{k=1}^{z_i}(q_{ik}-q_i)}{z_i}}{N\overline{z}}$$

$$= \frac{\sum_{i=1}^{N} \sum_{k=1}^{z_i}(q_{ik}-q_i)}{N\overline{z}},$$

where q_{ik} is the measured q of the kth child of parent i, N is the number of parents, z_i is the number of children produced by parent i, and $\overline{z}$ is the average number of children produced by each parent. From this expansion we see that the random variable of interest is $X = q_{ik} - q_i$, and

$$Var[X] = \frac{\sum_{i=1}^{N} \sum_{k=1}^{z_i}(q_{ik}-q_i)^2}{N\overline{z}} - \left(\frac{\sum_{i=1}^{N} \sum_{k=1}^{z_i}(q_{ik}-q_i)}{N\overline{z}}\right)^2. \tag{3}$$

This can be extended to multiple operators by expanding equation 2, resulting in $X = q_{ijk} - q_{i(j-1)k}$, where q_{ijk} is the measured q of the kth child of parent i after the application of operator j.

4   Looking Under the Hood

To illustrate how to use these ideas to look under the hood, we will compare the performance of our evolutionary algorithm using two different mutation operators. In particular we will focus on the importance of the fitness variance in relation to Price's Equation.

4.1   Example Problem

To help in our illustration, we have chosen a standard problem from the function optimization literature introduced by Schwefel (1981). The objective function

$$f(x) = 418.9829n + \sum_{i=1}^{n} x_i \sin\left(\sqrt{|x_i|}\right)$$

defines a landscape covered with a lattice of large peaks and basins. The predominant characteristic of the Schwefel function is the presence of a second-best maximum far away from the global maximum, intended to trap optimization algorithms on a suboptimal peak. The best maximums are near the corners of the space. In this formulation of the problem, the global minimum is zero and the global maximum is $837.9658n$. In our experiments the problem has thirty independent variables constrained to the range $(-500.0, 500.0)$.

4.2   Algorithm Design

Typically as one makes design decisions for an evolutionary algorithm they go through a process of trial and error. Most of us simply exchange components, perform some runs, and then compare the best-so-far curves. We find a component that seems to work well for our problem, and then move on to other design decisions or parameter tuning, hoping that all of our choices will complement each other.

Let us assume that we have already made a series of design decisions, including using a real-valued representation and (μ, λ) selection of (500, 1000). We've chosen these unusually large population sizes in order to increase the sample sizes and reduce the amount of noise when calculating the terms in Price's equation. We've also decided to use two-point crossover at a rate of 0.6 and we've implemented it all with the ECKit Java class library developed by Potter (1998).

Now we are left with one final decision, the mutation operator. We are going to use a Gaussian mutation which is applied to all genes in the genome, but we want to decide between using a fixed standard deviation or one where the standard deviations are adapted, as described in (Bäck and Schwefel 1993).

4.3   Comparing Mutation Operators

In order to compare operators, we perform 100 runs using the fixed Gaussian mutation (with a standard deviation of 1.0), and 100 runs with the adaptive gaussian mutation. The best-so-far curves for each are plotted Figure 1. Based on what we see here, both

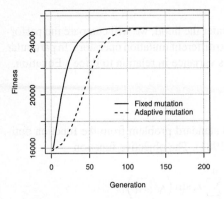
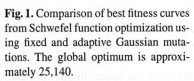
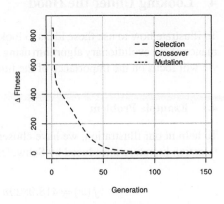

Fig. 1. Comparison of best fitness curves from Schwefel function optimization using fixed and adaptive Gaussian mutations. The global optimum is approximately 25,140.

Fig. 2. The mean contributions from selection, crossover and mutation during optimization of the Schwefel function using fixed Gaussian mutation. These were calculated using the extended Price's equation.

appear to be able to reach the optimum consistently, but the fixed Gaussian mutation operator seems to be able to do so more quickly. Without any other information, this is the operator we would choose.

But do we have any idea why this operator is better than the other? The EA is essentially a black box, and we have very little idea of what is happening inside. If we could get more information about how the operators are interacting and what effects they are having, we could make more intelligent decisions about how to improve the performance of our EA.

By applying the extended version of Price's Equation (equation 2) to the EA while it is running, we can calculate the average effect that each operator has on the population. Figure 2 shows the mean effects of selection, crossover and mutation when using the fixed Gaussian mutation operator. The plot shows that the crossover and mutation operators have an average effect near zero (which may be difficult to read in the plot), while selection seems to be doing all of the work. What exactly does this mean? These average values are clearly not telling us the whole story. Recall that in section 3 we claimed that the average individuals created by the genetic operators are not as important as the exceptional ones because it is the exceptional individuals which are chosen by selection.

We want to look at the standard deviations of the Δq values, but instead of plotting them in a separate plot, we have combined it with the existing plots of the delta mean fitnesses. Figure 3 gives an example of this type of plot for the crossover operator. For each generation a gray band is drawn starting from the mean effect of crossover (which is very close to zero) up to plus one standard deviation and down to minus one standard deviation. This of course assumes that the fitness distribution of the Δq values is normal, which is not always a good assumption.

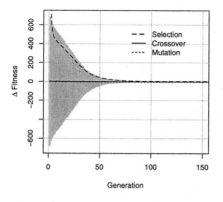

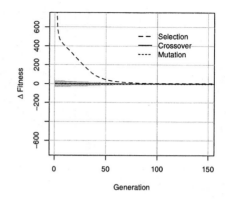

Fig. 3. Standard deviation of delta fitness effects caused by the crossover operator. The experiment was run using a fixed Gaussian mutation with standard deviation of 1.0 on the Schwefel function. Results are averaged over 100 runs.

Fig. 4. Standard deviation of delta fitness effects caused by the mutation operator. The experiment was run using a fixed Gaussian mutation with standard deviation of 1.0 on the Schwefel function. Results are averaged over 100 runs.

The advantage of this type of plot is that one can compare the relative effects of the delta mean fitness to the effects of the variance. In other words, an operator may on average be quite disruptive (low mean fitness), but still have a high variance. This would mean that the operator could still be effective. One should keep in mind though that we are plotting only one standard deviation. Genetic operators can often create individuals with fitnesses that are two to three standard deviations from the mean when run on this problem.

Getting back to our example, Figures 3 and 4 show the variance effects of crossover and fixed Gaussian mutation. Whereas before the two operators were indistinguishable, now we can see a large difference between them. Notice that the upper edge of the crossover standard deviation curve follows the curve representing mean contribution from selection very closely. It seems clear that crossover is contributing more to the search process than mutation, at least in the early generations. The variance of the mutation operator is much lower, as can be seen by the very narrow gray band along the x-axis. It is unlikely mutation does much more than local hill climbing.

Moving on to the adaptive Gaussian mutation operator, Figure 5 shows the mean effects of selection, crossover and mutation when using this operator. Here we see that once again the average effect of crossover is close to zero, but this time the average effects of mutation are much more negative. The adaptive mutation operator appears to be much more disruptive than the fixed Gaussian mutation.

But we should not draw too many conclusions before we see the variance effects of the operators, which are plotted in Figures 6 and 7. One of the first things to note in Figure 7 is that although mutation is very disruptive on average, the upper edge of the standard deviation curve still comes above the x-axis by a fair margin, which indicates

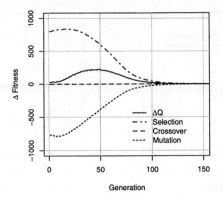

Fig. 5. The mean contributions from selection, crossover and mutation during optimization of the Schwefel function using adaptive Gaussian mutation. These were calculated using the extended Price's equation.

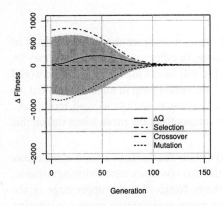

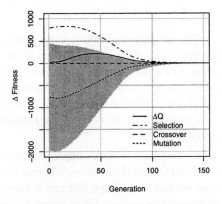

Fig. 6. Standard deviation of delta fitness effects caused by the crossover operator. The experiment was run using an adaptive Gaussian mutation on the Schwefel function. Results are averaged over 100 runs.

Fig. 7. Standard deviation of delta fitness effects caused by the mutation operator. The experiment was run using an adaptive Gaussian mutation on the Schwefel function. Results are averaged over 100 runs.

that it is contributing to the search process more than the fixed gaussian mutation operator did.

There is something more interesting though. Now we can get some insight into why it takes longer to reach the optimum using the adaptive Gaussian mutation than it does using the fixed Gaussian mutation. Compare the crossover standard deviations in Figure 6 with the ones in Figure 3. In conjunction with the adaptive mutation, crossover continues to have high variances out to generation 50, as opposed to the rapidly decreasing variances we see when fixed Gaussian mutation is used. The disruptive effects of mutation are so high that crossover is having to spend more time repairing individuals. It cannnot make headway in the search process until the standard deviations for mutation have been reduced.

With this knowledge we can now make a more informed decision about choosing our mutation operator. It is clear that crossover can take care of exploration on its own (in this domain), so we do not need a mutation operator which performs search also, especially when it comes with the undesirable side effects of disuption.

5   Conclusions and Future Work

In this paper we have shown how tools based on extensions of Price's equation allow us to look inside production-level EAs to see how selection, representation, and reproductive operators interact with each other, and how these interactions affect EA performance. In particular, we have shown how these extensions can provide insight into the way in which reproductive operator variance provides the exploratory power needed for good performance.

The reported results focused on ES-like EAs. We are in the process of completing a similar study for GA-like EAs. The interesting preliminary results suggest that in this case crossover and mutation interact internally in quite different ways. With results such as these we believe that it is possible to understand at a deeper level how existing EAs work as well as provide support for making better design decisions involving new EC applications.

Acknowledgments. We thank Donald Sofge, Magdalena Bugajska, Myriam Abramson, and Paul Wiegand for helpful discussions on the topic of Price's Equation. The work reported in this paper was supported by the Office of Naval Research under work request N0001403WX20212.

References

[Altenberg 1994] Altenberg, L. (1994). The evolution of evolvability in genetic pro-gramming. In K. E. K. Jr. (Ed.), *Advances in Genetic Programming*, Chapter 3, pp. 47–74. MIT Press.

[Altenberg 1995] Altenberg, L. (1995). The schema theorem and Price's theorem. In L. D. Whitley and M. D. Vose (Eds.), *Foundations of Genetic Algorithms III*, pp. 23–49. Morgan Kaufmann.

[Bäck and Schwefel 1993] Bäck, T. and H.-P. Schwefel (1993). An overview of evolu-tionary algorithms for parameter optimization. *Evolutionary Computation 1(1)*, 1–23.

[Frank 1995] Frank, S. A. (1995). George price's contributions to evolutionary genetics. *Journal of Theoretical Biology 175*, 373–388.

[Langdon and Poli 2002] Langdon, W. B. and R. Poli (2002). *Foundations of Genetic Programming*. Berlin Heidelberg: Springer-Verlag.

[Potter 1998] Potter, M. A. (1998). Overview of the evolutionary computation toolkit. http://cs.gmu.edu/ mpotter/.

[Potter, Bassett, and De Jong 2003] Potter, M. A., J. K. Bassett, and K. A. De Jong (2003). Visualizing evolvability with price's equation. In *Proceedings of the 2003 Congress on Evolutionary Computation*, pp. 2785–2790. IEEE.

[Price 1970] Price, G. (1970). Selection and covariance. *Nature 227*, 520–521.

[Schwefel 1981] Schwefel, H.-P. (1981). *Numerical optimization of Computer models*. Chichester: John Wiley & Sons, Ltd.

Distribution of Evolutionary Algorithms in Heterogeneous Networks

Jürgen Branke, Andreas Kamper, and Hartmut Schmeck

Institute AIFB, University of Karlsruhe, 76128 Karlsruhe,Germany
{branke,kamper,schmeck}@aifb.uni-karlsruhe.de

Abstract. While evolutionary algorithms (EAs) have many advantages, they have to evaluate a relatively large number of candidate solutions before producing good results, which directly translates into a substantial demand for computing power. This disadvantage is somewhat compensated by the ease of parallelizing EAs. While only few people have access to a dedicated parallel computer, recently, it also became possible to distribute an algorithm over any bunch of networked computers, using a paradigm called "grid computing". However, unlike dedicated parallel computers with a number of identical processors, the computers forming a grid are usually quite heterogeneous. In this paper, we look at the effect of this heterogeneity, and show that standard parallel variants of evolutionary algorithms are significantly less efficient when run on a heterogeneous rather than on a homogeneous set of computers. Based on that observation, we propose and compare a number of new migration schemes specifically for heterogeneous computer clusters. The best found migration schemes for heterogeneous computer clusters are shown to be at least competitive with the usual migration scheme on homogeneous clusters. Furthermore, one of the proposed migration schemes also significantly improves performance on homogeneous clusters.

Keywords. Evolutionary Algorithm, Heterogeneous Networks, Parallelization, Island Model, Grid Computing

1 Introduction

Evolutionary algorithms (EA) are randomized search techniques inspired by natural evolution. They have proven to work successfully on a wide range of optimization problems. While they have many advantages, they are computationally expensive since they have to evaluate a relatively large number of candidate solutions before producing good results. This drawback is partially compensated by the apparent ease of parallelizing EAs.

Consequently, there is a wide range of publications on how to best parallelize EAs. However, basically all these publications assume identical (homogeneous) processors, as it is usually the case if one has access to a dedicated parallel machine. More recently, however, it became possible to also harness the combined computing power of any heterogeneous set of networked computers, also known as "computer grids" (see e.g. [10]). These computer grids make the power of parallelization available to a much larger

K. Deb et al. (Eds.): GECCO 2004, LNCS 3102, pp. 923–934, 2004.

group of people, as for example most companies have a network of (heterogeneous) PCs, and also the Internet is basically a huge computer network. The power of computer grids has been demonstrated e.g. by the project seti@home [1], which connected thousands of computers to search for extraterrestrial life, or by the many applications in drug design [7]. Companies like Parabon [2] or United Devices [3] commercialize the idea of networked computing. Clearly, computer grids have the potential to resolve the problem of high computer resources required by EAs, and would help pave their way to an even more widespread application.

However, as we show in this paper, approaches to parallelize EAs for homogeneous parallel computers can not readily be applied in heterogeneous environments, or at least their efficiency drops significantly. Our goal therefore is to develop parallelization schemes particularly suitable for heterogeneous computer clusters. More precisely, in this paper we consider an island model and examine different aspects of migration, like the connectivity pattern or the time for migration.

The paper is structured as follows: In the following section, we will provide a brief overview on related work. Then, we will explain the experimental setup and show how heterogeneity affects standard parallelization approaches. Section 5 looks at the influence of the connectivity pattern, Section 6 compares sender- and receiver-initiated migration, and Section 7 suggests migration based on a population's level of convergence. The paper concludes with a summary and some ideas for future work.

2   Related Work

As has already been mentioned in the introduction, parallelization of EAs is relatively straightforward. The genetic operators crossover and mutation as well as the evaluation can be performed independently on different individuals, and thus on different processors. The main problem is the selection operator, where global information is required to determine the relative performance of an individual with respect to all others in the current population. There is a vast amount of literature on how to parallelize EAs. The approaches can be grouped into three categories:

1. Master-slave: Here, a single processor maintains control over selection, and uses the other processors only for crossover, mutation and evaluation of individuals. It is useful only for few processors or very large evaluation times, as otherwise the strong communication overhead outweighs the benefit from parallelization.
2. Island model: In this model, every processor runs an independent EA, using a separate sub-population. In regular intervals, *migration* takes place: The processors cooperate by exchanging good individuals. The island model is particularly suitable for computer clusters, as communication is limited.
3. Diffusion model: Here, the individuals are spatially arranged, and mate with other individuals from the local neighborhood. When parallelized, there is a lot of inter-processor communication (every individual has to communicate with its neighbors in every iteration), but the communication is only local. Thus this paradigm is particularly suitable for massively parallel computers with a fast local intercommunication network.

A detailed discussion of parallelization approaches is out of the scope of this paper. The interested reader is referred to e.g. [13,8,5].

In loosely coupled networks such as computer grids, due to slow communication, the island model seems to be the most promising approach. Therefore, in the remainder of this paper, we will focus exclusively on the island model.

Almost all papers on parallelizing EAs assume equally powerful (homogeneous) processors, only very few deal with heterogeneity. While in a homogeneous network, populations usually exchange individuals in a synchronized way, this doesn't make sense in a heterogeneous network, as the faster processors would always have to wait for the slowest one. Chong [9] was among the first to look at this aspect and naturally concluded that communication should be non-blocking (i.e. asynchronous) and buffered, a result that has been confirmed in [11]. In [4], the performance of the island model is compared on a number of small homogeneous and heterogeneous network clusters, again with asynchronous migration. The DREAM project [6] distributes EAs over the Internet and thus naturally has to deal with heterogeneous computers. Again, communication is asynchronous, but otherwise the aspect of heterogeneity is not addressed explicitly.

3   Experimental Setup

We base our experiments on the standard island model, with a more or less standard EA running on each island. Each island has a population size of 25, and a $(25 + 12)$ reproduction scheme is used, i.e. in every generation, 12 offspring are generated and compete for survival with the individuals from the old population. Crossover type is two-point, mutation is Gaussian with step size $\sigma = 1.0$. Islands are connected in a ring. In a migration step, each processor sends a copy of its best individual to its two neighboring populations, where it replaces the worst individual. Unless stated otherwise, migration is executed always at the end of a generation.

In order to have full control over all aspects of the environment, we simulate the underlying computer network. The network's computing power is assumed to be such that a total of 1920 generations (equivalent to $1920 \cdot 12 = 23040$ evaluations) can be computed per virtual time unit. In this paper, we ignore communication time and bandwidth restrictions and focus primarily on the aspect of different processing speeds.

The performance measure is the best solution quality obtained after 300 time units, which corresponds to almost 7 Mio. evaluations.

In the case of *homogeneous* processors, we assume to have 64 processors of equal speed, i.e. each processor computes 30 generations per virtual time unit. Because all processors have the same speed, all populations exchange migrants synchronously.

For the *heterogeneous* case, we assume 63 processors with a scale-free distribution of processing power, i.e. there is 1 processor capable of computing 320 generations per virtual time unit, 2 processors capable of computing 160 generations, 4 processors capable of computing 80 generations, down to 32 processors capable of computing 10 generations. The total computing power is the same as in the homogeneous case. In heterogeneous networks, islands communicate asynchronously using a buffer: Each island has a buffer, and a population which wants to send migrants to their neighbors simply puts them in the respective buffer. After each generation, a population checks its buffer for possible immigrants and, if there are any, integrates. This scheme largely corresponds to the asynchronous migration scheme proposed in [11], except that our buffers are unlimited. Note that in the case of neighboring islands with very different speeds, the buffer may contain more individuals than the island's population size.

For our comparisons, we use the following 30-dimensional test function which has once been suggested by Michalewicz for a conference competition:

$$f(x) = \sum_{i=0}^{29} \sin(y_i) * \left(\sin\left(\frac{i+1}{\pi} y_i^2 \right) \right)^{20}$$

$$y_i = \begin{cases} x_i * \cos(\frac{\pi}{6}) - x_{i+1} * \sin(\frac{\pi}{6}) & : \quad i = 0, 2, 4, \ldots, 28 \\ x_i * \cos(\frac{\pi}{6}) + x_{i-1} * \sin(\frac{\pi}{6}) & : \quad i = 1, 3, 5, \ldots, 29 \end{cases}$$

$$-2 \leq x_i \leq 2 \quad i = 0 \ldots 29$$

When comparing different migration schemes, whether one approach works better than another may strongly depend on the parameter settings. We assume that the most critical parameter in our setting is the migration interval, i.e. the number of generations between migrations. Therefore, for each migration scheme, we tested migration intervals of 10, 50, 100, 150, 200, 250, 300, 350, 400, and 450 generations, and usually report on the results of the best parameter setting for the respective migration scheme. All results reported are averaged over 120 runs.

4   EAs in Homogeneous and Heterogeneous Networks

Figure 1 compares the performance of the standard island model on the homogeneous and the heterogeneous network. Since we expected that the sorting of the processors in the ring could influence the result, for each of the 120 test runs, the sorting was generated at random. As can be seen, the heterogeneous network converges much faster, but the final solution quality is significantly worse. The faster convergence is due to the fact that the faster processors can perform a larger number of generations per time unit, thereby faster advancing search. However, this is at the expense of diversity: The good individuals from the fast processors will soon migrate to slower populations and dominate them, while the individuals from the slower processors are usually not good enough to compete with the individuals from other populations, and their genetic material is lost.

The steps in the curve of the homogeneous network are a result of the synchronous communication at migration. All populations export and import at the same time, and whenever new genetic material is introduced by migration, search seems to benefit and it is likely to quickly find better solutions. So even in the average of 120 runs we can see these bursts of improvement after every migration step. In the heterogeneous network, the communication is asynchronous, and the performance boosts due to migration average out. The steps that are visible are at points in time when all islands happen to migrate at the same time.

The final results of all test runs are also reported at the end of this paper in Table 1. Table 2 contains a pairwise comparison of the different migration schemes and results of a t-Test to determine significance. As can be seen, the difference between runs on homogeneous and heterogeneous networks is highly significant.

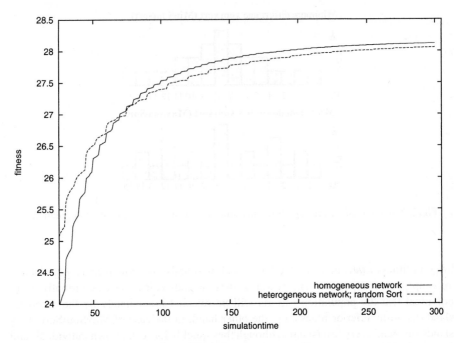

Fig. 1. Convergence plot, comparison of EA in homogeneous and heterogeneous networks, export-oriented migration.

5   Sorting

The first attempt to improve performance in a heterogeneous environment was to look at the sorting of the processors on the ring topology. In the tests reported before, we used a random sorting. Now we introduce two extreme ways of sorting the islands.

The first way is called "minimal difference sum sort" (MinSumSort) in which we minimize the sum of all speed differences between adjacent islands. The other way is to maximize those differences (MaxSumSort). These sortings are not unique. There are different possibilities for a given sum to arrange the islands. Figure 2 shows, exemplary for 15 populations, the sortings that we used for the following tests.

Figure 3 compares the convergence curves for the three chosen sortings MinSumSort, MaxSumSort, and random sorting, together with the homogeneous case for reference. Again, there seems to be a trade-off between fast convergence to an inferior solution, and slower convergence to a better solution. The MinSumSort converged quickest, but to the worst solution, while MaxSumSort is the slowest in the beginning, but yielding the best solution in the end. The random sort, as expected, is somewhere in between. Actually, the good performance of MaxSumSort came as a surprise to us, as we assumed that in the case of neighboring processors having quite different speeds, diversity would be lost much faster. However, a closer examination showed that the slow processors are, by themselves, not competitive. For MinSumSort, it takes a very long time for a good individual to travel from the fastest processor to the slowest islands. This results in very

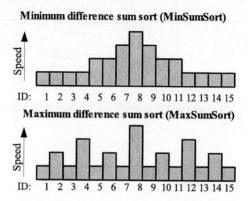

Fig. 2. Minimum and maximum difference sum sorting at the example of 15 populations

different fitness levels between the fast and slow islands, the slow islands' populations are simply not competitive. The power of these islands is lost because virtually every individual that the population exports will be rejected immediately by the receiving island due to its inferior fitness. On the other hand, in the case of MaxSumSort, slow islands surround every fast island. Although they quickly loose their own individuals and can not really follow their own search direction, they serve as a buffer between faster populations, and since they always work on relatively good individuals (migrated from its faster neighbors), even with their low computing power, they sometimes generate good individuals. They may even be regarded as melting pots, where good individuals from the two faster neighbors can meet on "neutral ground".

Another interesting result when comparing the test runs for the different sortings is that the optimal migration interval is different: it increases from 100 generations for MinSumSort over 150 generations for random sorting to 200 generations with MaxSumSort. If communication time is significant, that is another advantage for MaxSumSort, requiring only half the number of communications as MinSumSort. The different optimal migration intervals also confirm the above observations that for MinSumSort, the distance between fast and slow processors is too large, which is partially compensated by a short migration interval. On the other hand, MaxSumSort benefits from larger migration intervals, which help to maintain diversity and give the slower populations a chance to improve on the imported individuals.

Looking at Table 2 confirms that the differences between the three considered sortings are significant. The best sorting, MaxSumSort, is almost as good as the homogeneous network; the difference is not significant.

Overall, if the sorting can be controlled, MaxSumSort is clearly the sorting to be preferred. However, in many practical grid environments, the use of the computers is not exclusive, and the load and thus the preferable sorting can change dynamically.

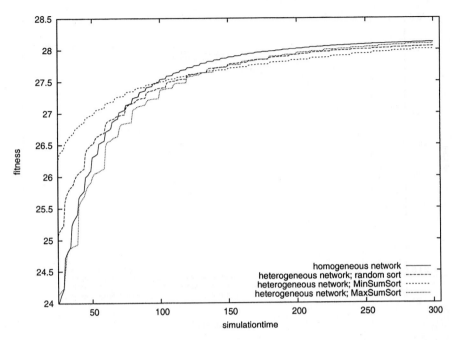

Fig. 3. Comparison of different sorting strategies (with homogeneous network for comparison)

6   Import, Export, Time

In a homogeneous network, it doesn't matter whether importing or exporting islands initiate the migration. For heterogeneous networks however, faster populations would like to migrate much more often than slower populations. In the previous experiments, as in all other publications we are aware of, we assumed that migration is initiated by the exporting island. That is, after an island has reached its migration interval, it exports a copy of its best individual to the buffers of its neighbors. If the speed difference between two islands is high, the buffer must be very large or old individuals must be removed to make space for new ones. In such a situation it is possible that all individuals in a population are replaced by individuals from the buffer at migration. Good search directions could be destroyed because of a continuous flow of new individuals form a faster island. An alternative way of communication would be to have migration initiated by the importing population. In that case, a population that has reached its migration interval would ask its neighbors to send over their current best individual. Thus, the number of individuals imported depends on an island's own speed and not on the speed of its neighbors. Good individuals or genes from fast processors will still spread, but presumably slower than by an export-oriented strategy. Figure 4 compares the import and export oriented migration strategies. There seems to be relatively little difference in performance, with the import-oriented yielding a slightly higher final solution quality, and export-oriented converging slightly faster. According to Table 2, the difference of the final fitness is not significant. Note that the import-oriented strategy has a slightly

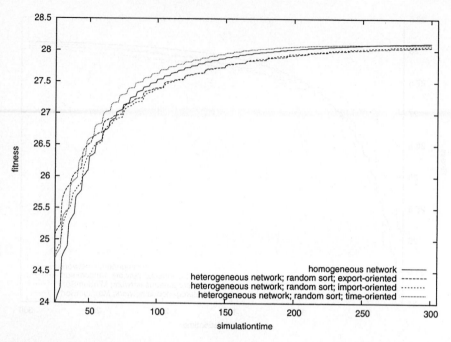

Fig. 4. Comparison of import- and export- and time-based migration strategies (with homogeneous network for comparison)

higher communication overhead because a population has to request immigrants. On the other hand, memory usage is reduced because there is no need for a buffer anymore.

Besides the just discussed import-initiated and export-initiated migration, migration could be triggered independently by clock time. In that case, communication would again be synchronous, as in the homogeneous network, where all population communicate at the same time. We tested this strategy again for different migration intervals. Best results for this testfunction were achieved for migration every 6 time steps, which corresponds to an average of 183 generations per population, i.e. the migration interval is in the same range as was optimal for export- or import-oriented migration schemes. The result is also included in Figure 4. As can be seen, the runs with time-based migration are slightly better than either export-based or import-based migration. Its main advantage however, is a much faster convergence. Overall, synchronous time-based migration seems to be the best choice, performing just as good as synchronous migration on a homogeneous network, but with a somewhat faster convergence.

7   Convergence-Based Migration

The experiments have shown that migration is an important feature, and the infusion of new genetic material into a population at the right time boosts performance. If good individuals are imported too early, the new individuals take over the population and

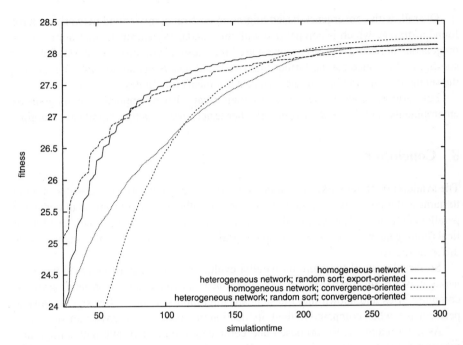

Fig. 5. Comparison of export-oriented and convergence based migration on homogeneous and heterogeneous environments.

are likely to re-direct the current (possibly good) search focus. On the other hand, if migration is done too late, search stagnates due to loss of diversity.

Therefore, it seems reasonable to have each island decide independently on the time it would like to import new individuals from its neighbors, based on its degree of convergence. Munetomo et al. [12] have previously suggested to initiate migration (import and export) whenever the standard deviation of a population's fitness distribution falls below a certain threshold. In this section, we propose to only initiate import, and use the number of generations without improvement[1] as an indicator for convergence.

That way, even slow populations get a chance to continue undisturbed and to explore their current search focus until they get stuck. Then, they can ask for new genetic material from their neighbors.

Note that this method may be applied in heterogeneous as well as homogeneous networks and leads to asynchronous communication in any case.

Figure 5 compares the results of the different approaches and the reference curves. As can be seen, the convergence-based migration improves over the standard generation-based strategies in the heterogeneous as well as the homogeneous network. Differences in final solution quality are significant (cf. Table 2). While overall convergence of these strategies is slower than by all other variations, they yield the best final results.

---

[1] We tested the same numbers of generations without improvement as the migration intervals for the generation-based strategies. Again, we only report on the best results

Overall, with convergence-based migration on a randomly sorted network we have found a strategy which is competitive with the standard synchronous migration on homogeneous networks. If convergence-based migration is combined with MaxSumSort sorting, performance can be further improved, and results get significantly better than the standard migration in homogeneous environments (cf. Tables 1 and 2).

Furthermore, convergence-based migration can also be applied in homogeneous environments, and then outperforms all other tested strategies by a significant margin.

8 Conclusion

The availability of networked computer clusters, so-called "grids", is a great possibility to harness the power of evolutionary algorithms without having to invest in dedicated parallel computers. However, while parallel computers usually have a number of identical (homogeneous) processors, computer grids usually consist of computers with very different speeds.

So far, almost all literature on parallel evolutionary algorithms assumed homogeneous clusters. In this paper, we have examined the island model on heterogeneous computer grids and shown that the standard migration schemes suffer significantly in performance when compared to their application in homogeneous environments.

As a consequence, we proposed and compared a number of alternative migration schemes for the island model, specifically targeted at heterogeneous computer grids. We demonstrated that the solution quality and convergence speed are influenced by the sorting of computers in a ring structure, and that specific sortings perform better than others. We have compared strategies which base the time of migration on the importing population, the exporting population, or an external clock. Furthermore, we proposed to make migration adaptive by allowing each population to request new immigrants from its neighbors whenever it has converged. This strategy, which is also applicable in homogeneous networks, turned out to perform best.

Overall, the paper provided new insights into the role of migration and the behavior of the island model on heterogeneous computer clusters. We have shown that by sorting the computers appropriately in the ring structure, results competitive to homogeneous networks can be obtained. Furthermore, the suggested convergence-based migration, which is also applicable in homogeneous networks, leads to a further significant improvement in either case.

There are several avenues for future work. First of all, the presented results should be examined for their sensitivity to changes in the problem instance, the heterogeneity of the computers, the number of computers in the network, the island model topology, etc. Furthermore, strategies should be explored to cope with dynamic computer speeds, as in a computer grid, the computers can usually not be used exclusively, and the effective speed of a computer may vary with its workload. Finally, communication costs like latencies should be considered.

References

1. http://setiathome.ssl.berkeley.edu/.
2. http://www.parabon.com.
3. http://www.ud.com/home.htm.
4. E. Alba, A. Nebro, and J. Troya. Heterogeneous computing and parallel genetic algorithms. *Journal of Parallel and Distributed Computing*, pages 1362–1385, 2002.
5. E. Alba and M. Tomassini. Parallelism and evolutionary algorithms. *IEEE Transactions on Evolutionary Computation*, 6(5):443–461, 2002.
6. M. Arenas, P. Collet, A. Eiben, M. Jelasity, J. Merelo, B. Paechter, M. Preuß and M. Schoenauer. A framework for distributed evolutionary algorithms. In *Parallel Problem Solving from Nature*, pages 665–675. Springer, 2002.
7. R. Buyya, K. Branson, J. Gidy, and D. Abramson. The virtual laboratory: a toolset to enable distributed molecular modelling for drug design on the world-wide grid. *Concurrency and Computation: Practice and Experience*, 15:1–25, 2003.
8. E. Cantu-Paz. *Efficient and Accurate Parallel Genetic Algorithms*. Kluwer, 2000.
9. F. S. Chong. Java based distributed genetic programming on the internet. Technical report, School of Computer Science, University of Birmingham, B15 2TT, UK, 1999.
10. I. Foster and C. Kesselman, editors. *The Grid: Blueprint for a New Computing Infrastructure*. Morgan-Kaufmann, 1999.
11. P. Liu, F. Lau, and J. Lewisand C. Wang. Asynchronous parallel evolutionary algorithm for function optimization. In *Parallel Problem Solving from Nature*, pages 405–409. Springer, 2002.
12. M. Munetomo, Y. Takai, and Y. Sato. An efficient migration scheme for subpopulation-based asynchronously parallel genetic algorithms. In S. Forrest, editor, *International Conference on Genetic Algorithms*, page 649. Morgan Kaufmann, 1993.
13. H. Schmeck, U. Kohlmorgen, and J. Branke. Parallel implementations of evolutionary algorithms. In A. Zomaya, F. Ercal, and S. Olariu, editors, *Solutions to Parallel and Distributed Computing Problems*, pages 47–66. Wiley, 2001.

Appendix: Numerical Results

Table 1. Fitness of different migration strategies after 100, 200, and 300 time steps. Numbers are averages over 120 runs.

Strategy	Fitness after 100 time units	Fitness after 200 time units	Fitness after 300 time units
Heterogeneous network			
random sort export-oriented	27.39	27.92	28.05
minSumSort export-oriented	27.49	27.83	28.02
maxSumSort export-oriented	27.35	27.95	28.10
random sort import-oriented	27.37	27.94	28.07
random sort time-based	27.85	28.04	28.10
random sort convergence-based	26.54	27.91	28.13
maxSumSort convergence-based	26.11	27.96	28.15
Homogeneous network			
Standard EA	27.53	28.02	28.12
convergence-based	26.43	28.03	28.21

Table 2. Pairwise comparison of the different migration strategies and cluster characteristics based on the fitness after 300 time steps (values in brackets). Upper triangular matrix depicts T-value of a two-sided t-Test, lower triangular matrix contains a "+" if the results differ significantly with significance level 97.5%, and "-" otherwise.

| | | | Heterogeneous | | | | | | | Homogeneous | |
| | | | random | | | | MinSum | MaxSum | | | |
			export (28.05)	import (28.07)	time (28.10)	conv. (28.13)	export (28.02)	export (28.10)	conv.(28.15)	standard (28.12)	conv. (28.21)
heterogeneous	random	export (28.05)		1.25	3.10	4.92	2.50	3.31	7.98	4.68	11.98
		import (28.07)	-		1.71	3.77	3.91	1.85	6.59	3.24	11.52
		time (28.10)	+	-		2.10	6.17	0.0	4.74	1.44	9.27
		conv. (28.13)	+	+	-		8.19	2.3	2.24	0.90	7.64
	minSum	export(28.02)	+	+	+	+		6.61	12.17	8.19	16.74
	maxSum	export(28.10)	+	-	-	+	+		5.16	1.56	10.28
		conv.(28.15)	+	+	+	+	+	+		3.52	6.10
hom.	standard(28.12)		+	+	-	-	+	-	+		8.9
	convergence (28.21)		+	+	+	+	+	+	+	+	

A Statistical Model of GA Dynamics for the OneMax Problem

Bulent Buyukbozkirli[1,3] and Erik D. Goodman[2,3]

[1] Department of Mathematics, Michigan State University
[2] Department of Electrical and Computer Engineering, Michigan State University
[3] Genetic Algorithms Research and Applications Group (GARAGe)
Michigan State University
East Lansing, MI 48824
buyukboz@msu.edu, goodman@egr.msu.edu

Abstract. A model of the dynamics of solving the counting-ones (OneMax) problem using a simple genetic algorithm (GA) is developed. It uses statistics of the early generations of GA runs to describe the dynamics of the problem for all time, using a variety of crossover and mutation rates. The model is very practical and can be generalized to cover other cases of the OneMax, such as weighted OneMax, as well as the deceptive function problem, for *high enough* crossover rates. Proportional selection with and without Boltzmann scaling have been modeled; however the Boltzmann extensions are not described here. In the development of the model, we introduce a new quantity that measures the effect of the crossover operation in the counting-ones problem and is independent of generation, for practical purposes.

1 Introduction

Theoretical models of Genetic Algorithms (GAs) fall into three main categories. The Markov chain model, as developed by Nix, Vose and Liepins [1][2], completely describes the probabilistic behavior of the GA. However, this model is too costly to implement computationally for problems with realistic population size and chromosome length. The statistical mechanics approach, developed by Prügel-Bennett, Shapiro and Rattray [3][4], gives fairly good results in modeling the OneMax problem with Boltzmann scaling, for a crossover rate of 100%, however it is not developed for lower crossover rates or to handle other benchmark problems of GA such as deceptive functions. The approach of modeling GAs by considering building blocks (Goldberg [7] and Goldberg, Deb, Thierens [6]), on the other hand, gives us a good idea about the appropriate population size or the convergence time of the OneMax and help us determine the failure boundaries in the "control maps". But the question of finding the most appropriate crossover or mutation rate is answered, so far, only by experimental results. We still lack a model that describes the behavior of the OneMax problem for different crossover and mutation rates together and allows us to choose the best parameters.

K. Deb et al. (Eds.): GECCO 2004, LNCS 3102, pp. 935–946, 2004.
© Springer-Verlag Berlin Heidelberg 2004

Studying the OneMax problem is important not for solution of that problem, per se, but because many real-world problems solved via genetic algorithms consist of a set of separable sub-problems for which the optimum is to optimize each individually, which is reminiscent of OneMax.

In this paper, we develop a model that describes the mean allele dynamics of the OneMax problem for very high crossover rates. Then, we modify the model by using statistics of very early generations from GA runs, to describe the complete dynamics for different (lower) crossover rates. The model is developed to estimate the average GA dynamics, but it can be used for an individual run of the GA and has the potential to apply to other cases of GA-based solution of the OneMax problem, such as using Boltzmann scaling, and the weighted OneMax, or to benchmark problems involving deceptive functions. The authors hope to extend the approach to model solution of more representative real-world problems with various degrees of OneMax similarity and various amounts of deception.

2   Problem Description and Visual Representation of GA Dynamics

We consider the simple genetic algorithm in which two-point crossover, fitness-proportional selection and mutation are applied in the order given. We develop a model on the OneMax problem with a population consisting of P chromosomes of length L. Let $S(t)$ be the set of all chromosomes at time t, *chrom* an element of this set, and *chrom(i)* the allele at the ith locus of this chromosome. The fitness of a chromosome, *chrom*, will be denoted as *f(chrom)*, which is equal to $\sum_i chrom(i)$ for the simple OneMax problem. The variables of the population that we are interested in are the mean fitness $\kappa_1(t)$, the variance of the fitness $\kappa_2(t)$, and the set of the mean of the alleles at each locus i $\{\alpha_i(t)\}_{i=1,...,L}$, at time t. Define $A_h(t)$ to be the number of $\alpha_i(t)$'s whose value is less than or equal to h,

$$A_h(t) = \#\left\{ \alpha_i(t) \,\middle|\, \alpha_i(t) \le h, i = 1,...,L \right\}. \tag{1}$$

By this definition, for example, $A_0(t)$ gives the number of loci where all of the chromosomes have value 0, while $A_{0.6}(t)$ gives the number of loci where at most 60% of the chromosomes have value 1.

The values of the variables $(\kappa_1(t), \kappa_2(t), \{\alpha_i(t)\}_{i=1,...,L})$ and $A_h(t)$ change from one GA experiment to another even if we have the same initial population. In terms of experimental results, we run a GA, with fixed parameters of selection, mutation and crossover, many times. For each run of the GA, we measure these quantities at each generation and take the average over all of the runs. The goal of our model is to estimate average values of these variables, hence the average behavior of the GA. In order to simplify the notation, we will use the same symbols $(\kappa_1(t), \kappa_2(t), \{\alpha_i(t)\}_{i=1,...,L})$ and $A_h(t)$ for values of a specific run of a GA, or of an experimental average of these values, or of the estimated theoretical average in our model. Which one is denoted will be clear from the context. We will use the superscripts c, cs or csm in order to distinguish these variables after crossover, selection or mutation is applied, respectively. So, $\alpha_i^{cs}(t)$ represents the mean of alleles at the ith locus at the tth generation after the crossover and selection are applied, and $\alpha_i^{csm}(t)$ equals $\alpha_i(t+1)$. Note that the crossover operation does not change the mean allele values. Thus, $\alpha_i(t)$ equals $\alpha_i^c(t)$.

The study of the time evolution of $A_h(t)$'s for several values of h gives a very practical insight into the behavior of the GA. Figure (1) shows the graphs of $A_h(t)$ for $h = 0, 0.1, \ldots 0.9$, the crossover rate is 25% and the mutation rate is 0.1%. The population size is taken as 50 chromosomes and the chromosome length is 100 genes. Each time slice of such a graph can be seen as a "bar graph" of the mean allele distribution at the given instant. In other words, the vertical distance between two curves gives the number of gene locations at which the mean allele is between the corresponding values, averaged across runs. For example, at t=100, at about 18 gene locations, none of the chromosomes (i.e. h=0) have value 1; at about 5 locations from 1 to 5 chromosomes (i.e. 1% to 10% of the population, i.e. 0<h 0.1) have a 1 and the rest have a 0; and at about 50 locations from 45 to 50 chromosomes (i.e. 91% to 100% of the population, i.e., 0.9<h 1) have a 1 and the rest have a 0, etc. The closer the curves are to each other, the smaller the variation in the population. We observe that although the population converges to a more or less stable configuration after 100 generations, there is still some variation within the population due to the existence of mutation, which has the potential of creating new chromosomes.

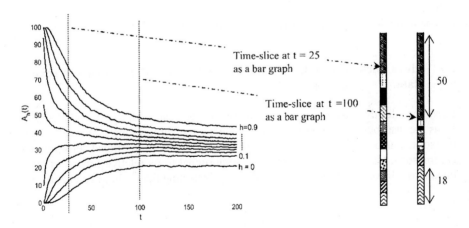

Fig. 1. The experimental average values of $A_h(t)$ as a function of time for $h =0, 0.1, 0.2, \ldots 0.9$, and the bar graph interpretation. The population size is 50 and the chromosome length is 100 genes. The GA parameters are $p_c = 0.25$, $p_m = 0.001$. The average is taken over 100 experiments

When h is quantized with a gap of 0.1 between two consecutive values as above, we get 10 regions formed between the graphs, including the region above the top graph. We will use the index h' to count these regions, $h' = 1, 2, \ldots, 10$, given by

$$R_{h'} = \left\{ (y,t) \,\middle|\, A_{(h'-1)/10}(t) \le y \le A_{h'/10}(t) \right\}, \tag{2}$$

where $A_1(t)$ is defined as the constant function 1.

3   The Model

The model is developed first for the case with a "high enough" crossover rate. Then it is modified to model the cases with lower crossover rates. The first case involves three main steps. First, mean alleles after the crossover and the selection are estimated assuming that the crossover rate is "high enough". Then, the effect of mutation on the mean allele is determined. The last step involves the estimation of fitness variance given the mean allele values.

The second case, in which the crossover rate takes more realistic values, is modeled by observing some statistical properties of the GA at early generations.

At any GA stage, the mean fitness, κ_1, is always the sum of the mean alleles across loci at that moment, i.e.

$$\kappa_1(t) = \sum_i \alpha_i(t) . \tag{3}$$

The following sections describe each of these steps in detail.

3.1   Mean Allele After Selection and Crossover with "High Enough" Crossover Rate

First, consider the case in which the crossover rate is so high that the alleles at any locus are distributed essentially randomly among the chromosomes. We will call this crossover rate a "high enough" crossover rate. In fitness-proportional selection, each chromosome has a selection probability proportional to its relative fitness within the population. If we denote as q_j the probability of selecting the jth chromosome, then

$$q_j = \frac{f_j}{\sum_{k=1}^{P} f_k} , \tag{4}$$

where f_k is the fitness of the kth chromosome.

Let p_i be the probability that a chromosome that is selected randomly with the above probability scheme after the application of crossover, has 1 at its ith locus. As with the other symbols, we will use the notation $p_i(t)$ for values of a specific run of a GA at time t, or of an experimental average of these values at time t, or of the estimated theoretical average in our model, depending on the context. It is easy to estimate $p_i(t)$ theoretically in terms of $\kappa_1(t)$ and $\alpha_i(t)$ when the crossover rate is "high enough". For this purpose, define the subsets $S_0^i(t)$ and $S_1^i(t)$ of $S(t)$ as

$$S_0^i(t) = \{chrom \in S(t) \mid chrom(i) = 0\} \quad \text{and} \qquad (5)$$

$$S_1^i(t) = \{chrom \in S(t) \mid chrom(i) = 1\} .$$

Then, we have

$$\overline{\sum_{chrom \in S_0^i(t)} f(chrom)} = P(1 - \alpha_i(t))(\kappa_1(t) - \alpha_i(t)) \quad \text{and} \qquad (6)$$

$$\overline{\sum_{chrom \in S_1^i(t)} f(chrom)} = P\alpha_i(t)(1 + \kappa_1(t) - \alpha_i(t)) ,$$

where the bar over the summation means the average over all possible configurations of gene distributions, in which we assume that the genes are distributed randomly satisfying the given mean allele values, since the crossover rate is "high enough". Thus, the estimated average value of $p_i(t)$ is

$$p_i(t) = \frac{P\alpha_i(t)(1 + \kappa_1(t) - \alpha_i(t))}{P\alpha_i(t)(1 + \kappa_1(t) - \alpha_i(t)) + P(1 - \alpha_i(t))(\kappa_1(t) - \alpha_i(t))} , \qquad (7)$$

which simplifies to

$$p_i(t) = \alpha_i(t) + \frac{(1 - \alpha_i(t))\alpha_i(t)}{\kappa_1(t)} . \qquad (8)$$

In the process of fitness proportional selection, we apply selection of chromosomes P times with replacement. Each time, the probability that the selected chromosome has 1 as its ith allele, is $p_i(t)$. So, the expected number of 1's at the ith locus, after the selection is over, can be obtained by using a binomial distribution. Let $B(n,P,p_i)$ denote the probability of having n successes after P trials, when the success probability is p_i for each trial. Then, the expected theoretical value of $\alpha_i^{cs}(t)$ is

$$\alpha_i^{cs}(t) = \frac{1}{P} \sum_{n=1}^{P} n \cdot B(n, P, p_i(t)) , \qquad (9)$$

when the crossover rate is "high enough".

3.2   Mean Allele After Mutation

In this section, we want to estimate $\alpha_i^{csm}(t)$ given the values of $\alpha_i^{cs}(t)$. Each gene of a chromosome has the probability p_m of changing its value from 1 to 0 or from 0 to 1 by mutation. When we consider the possible changes at the ith locus only, the expected number, N, of total allele changes due to mutation can be found by using a binomial distribution as

$$N = \sum_{n=1}^{P} n \cdot B(n, P, p_m) \ . \tag{10}$$

Since the percentage of 1's at the ith locus is $\alpha_i^{cs}(t)$, $\alpha_i^{cs}(t)N$ of these changes are going to be from 1 to 0, and $(1-\alpha_i^{cs}(t))N$ of the changes are from 0 to 1, on the average. This means that the number of 1's at the ith locus, which is $P\alpha_i^{cs}(t)$, will become $P\alpha_i^{cs}(t)- \alpha_i^{cs}(t)N+(1-\alpha_i^{cs}(t))N$ after the mutation. Simplifying this quantity and dividing by P gives the mean allele for the next generation as

$$\alpha_i(t+1) = \alpha_i^{csm}(t) = \alpha_i^{cs}(t) + \frac{1-2\alpha_i^{cs}(t)}{P}N \ . \tag{11}$$

3.3   Estimation of Fitness Variance for "High Enough" Crossover Rates

The fitness variance by definition is

$$\kappa_2(t) = \frac{1}{P}\left(\sum_{k=1}^{P} f(chrom_k)^2 \right) - \kappa_1(t)^2 \ . \tag{12}$$

If we write the fitness of $chrom_k$ as the sum of its gene values a_k^i and change the order of summation after expanding the square sign above, we obtain

$$\kappa_2(t) = \kappa_1(t) + \frac{1}{P}\sum_{i \neq j}^{L} \sum_{k=1}^{P} a_k^i a_k^j - \kappa_1(t)^2 \ . \tag{13}$$

The term $\sum_{k=1}^{P} a_k^i a_k^j$ in Equation (13), counts the number of chromosomes in which loci i and j both contain 1's. In the case of "high enough" crossover rates, this count is estimated by using α_i^{cs} and α_j^{cs} as follows. The probability, $p(i,j,n)$, that locations i and j have n common 1's is found by $\begin{pmatrix} P\alpha_i^{cs} \\ n \end{pmatrix} \times \begin{pmatrix} P - P\alpha_i^{cs} \\ P\alpha_j^{cs} - n \end{pmatrix} \div \begin{pmatrix} P \\ P\alpha_j^{cs} \end{pmatrix}$, where n could take any value between $max(0, P\alpha_i^{cs} + P\alpha_j^{cs} -P)$ and $min(P\alpha_i^{cs}, P\alpha_j^{cs})$ and the product of P with α's is rounded to the nearest integer in order to calculate the combinations. Thus, the estimation of the fitness variance in the case of "high enough" crossover rates is found by using

$$\kappa_2^{cs}(t) = \kappa_1^{cs}(t) + \frac{1}{P}\sum_{i \neq j}^{L} \sum_{n} n \cdot p(i,j,n) - \kappa_1^{cs}(t)^2 \ . \tag{14}$$

The estimation of fitness variance after mutation is done by Prügel-Bennett and Shapiro, [5]. Their formula gives us

$$\kappa_2^{csm} = \left(1-2p_m\right)^2 \kappa_2^{cs} + \left(1-\frac{1}{P}\right) p_m \left(1-p_m\right) \sum_{i=1}^{L} w_i^2 \; , \tag{15}$$

where w_i is the weight of the ith locus. In other words, the fitness of a chromosome $(a_1, a_2, \ldots, a_L)$ is calculated by the weighted summation $\sum w_i a_i$. In our special case, the values of w_i's are all 1. So, we use the formula

$$\kappa_2^{csm}(t) = \left(1-2p_m\right)^2 \kappa_2^{cs}(t) + \left(1-\frac{1}{P}\right)\left(p_m - p_m^2\right)\sum_{i=1}^{L} 1 \; , \tag{16}$$

$$= \left(1-2p_m\right)^2 \kappa_2^{cs} + L\left(1-\frac{1}{P}\right)\left(p_m - p_m^2\right) = \kappa_2(t+1) \; ,$$

to estimate the fitness variance after mutation.

3.4   Lower Crossover Rates

Equation (8) gives the probability p_i when the crossover rate is very high. In such a case, as in Section (3.1), we are able to treat the 1's at a fixed locus of different chromosomes as identical to each other in terms of their roles in selection because of the high mixing rate of the crossover operator, which makes chromosomes look similar to each other, on the average. However, for lower and more realistic crossover rates, there will be some correlation between alleles within a chromosome and Equation (8) will no longer hold. Let's keep the usage of notation $p_i(t)$ for the probability of selecting a 1 at the ith locus in the case of the "high enough" crossover rate and denote the corresponding probability in the case of a lower crossover rate by $\tilde{p}_i(t)$. To remedy this situation and estimate $\tilde{p}_i(t)$ correctly, we consider imaginary weights, $c_i(t)$, for each locus in order to reflect the average change in the role of 1's played in the selection process due to correlation between alleles. The correction weights, $c_i(t)$, are defined implicitly by

$$\tilde{p}_i(t) = \alpha_i(t) + \frac{\left(1-\alpha_i(t)\right)\alpha_i(t)c_i(t)}{\kappa_1(t)} \; . \tag{17}$$

The reason why we defined the correction weights as in Equation (17) is because if we write Equation (8) for a fitness function of the form $f(chrom) = \sum_i w_i \cdot chrom(i)$, with weights w_i, we would get an equation exactly like Equation (17) with c_i replaced by w_i. Our correction weights play a similar role at each locus as w_i's would, except that c_i's change over time.

The next step will be to estimate the c_i's statistically by means of some data gathered from experiments. In order to do this, the GA is run with fixed rates of p_m and p_c up to a pre-selected generation, say t_0 . Let us call this generation G_0. The crossover operation with the current rate, p_c, is applied to G_0 many times. Each time,

$\tilde{p}_i(t)$ values are calculated from the experimental data for each locus i, and the corresponding c_i values are found using Equation (17). This process is repeated for many runs of the GA to obtain statistical measures. It is observed that the value of c_i strongly depends on the values of α_i, as expected. Because of this dependence, it makes more sense to group the c_i according to their corresponding α_i values before finding the statistics of the data gathered from experiments. So, define $C_k^h(t_0)$ as the set $\left\{ c_i \text{ values of the } k^{th} \text{ experiment of GA such that } h \le \alpha_i(t_0) < h+0.1 \right\}$, for $h = 0$, $0.1, \ldots 0.9$. The mean of the correction weights is obtained by finding $\mu(h',t_0) = \underset{k}{mean}\left(mean\left(C_k^h(t_0)\right)\right)$, $h' = 1, 2, \ldots, 10$, where the relationship between the index h and h' is given by $h = (h'-1)/10$, to be consistent with definition (2). In order to measure how much the correction weights vary from one experiment to another, we also calculate the standard deviation $\sigma(h',t_0) = \underset{k}{std}\left(mean\left(C_k^h(t_0)\right)\right)$.

The experimental results show that, when p_c is not too low (below about 4%), μ and σ remain more or less at the same value regardless of the time, t_0. Moreover, μ shows a linear-like behavior while σ shows a quadratic-like behavior as a function of h'. This behavior of the crossover operator allows us to use the linear approximation of $\mu(h',5)$ to predict $\tilde{p}_i(t)$ for the following generations. Figure (2) shows the graphs of μ for two different rates of crossover with t_0 at generations 5, 15 and 30. We have observed that the inclusion of μ in our model is good enough for describing the effects of c_i distributions and the information coming from σ does not play a significant role in the counting-ones problem. However, for other problems, such as OneMax with Boltzmann scaling, σ might be needed in the model. The deviations from the linear behavior, in Figure (2), at $h'=1$ or 10 are due to the statistical averaging in which there were not enough data points available for these border values.

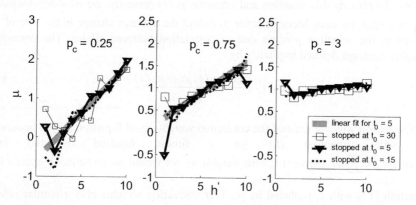

Fig. 2. The mean value of the correction weights as a function of mean allele levels, h', for crossover rates $p_c = 0.25$, 0.75 and 3. The statistical average is found over 100 experiments of GA. Population size is 50 and the chromosome length is 100 genes

4   Simulation of the Model and Comparison of the Results with the Experiments

The simulation of the model for high enough crossover rates starts with selecting a set of $a_i(0)$ values chosen by considering a binomial distribution for each locus in which we have P selections with a 50% chance of selecting a 1 each time. Equations (9) and (11) are applied to estimate the mean alleles after the crossover, selection and mutation operations. This process is iterated for each generation to obtain a dynamic simulation of the mean allele. At any moment, the mean fitness is estimated by Equation (3), and the fitness variance in the case of "high enough" crossover rates is estimated using Equations (14) or (16), depending on whether we are considering the variance right after the selection process or after the mutation, respectively. The following simulation results are obtained by taking an average over 10 runs of this model.

In the case of normal crossover rates Equation (9) is replaced by Equation (17), in which the c_i-values are pre-determined by the linear approximation of the data gathered at the 5th generation of a set of GA runs as described in Section (3.4), Figure (2).

In Figures (3)-(6), we see the comparison of the experimental results with the simulation of the model for population size 50 and the chromosome length 100 genes. In all these graphs, the black lines represent the experimental results, which are obtained by averaging 100 runs of the GA, and the thick gray lines represent the results obtained by the simulation of our model. The "high enough" crossover rate in our case is $p_c = 300\%$, which means that 100% crossover is applied 3 times in a row before the selection. This rate of crossover is verified experimentally as "high enough" by observing that there is no significant change in the graphs of $A_h(t)$, $\kappa_1(t)$ and $\kappa_2(t)$ if a higher value of p_c is used. It can also be verified from Figure (2), since the correction weights, when $p_c = 300\%$, are all very close to 1. In Figure (3), the time variation of A_h is shown for crossover rates of $p_c = 4\%$, 25%, 75% and 300%. We observed in our experiments with the model that when the crossover rate is to low, such as $p_c = 4\%$, the statistics of correction weights taken only from the 5th generation is not enough and we needed adjustment by using the statistics at the 15th generation. Figure (3.a) shows the graph with this adjustment. For $p_c = 25\%$ and 75%, the statistics from only the 5th generation are used. The simulation for $p_c = 300\%$ is obtained by taking all c_i's as 1. In all four cases, no mutation is applied — i.e., $p_m = 0$. The graphs look quite similar to each other, except that there is a slight variation in the value to which the lines converge as time goes to 200 generations. We see that the limit value decreases from around 40 to 30 as p_c increases from 4% to 300%. This slight decrease observed in the experimental graphs is well captured by the model simulations.

In Figure (4), the time evolution of A_h is shown for two different mutation rates, in both of which p_c is kept constant at 50%. In the first case the mutation rate is very low at $p_m = 0.1\%$, while in the second case it is $p_m = 2\%$. In both cases, the model predicts the mean allele behavior very well.

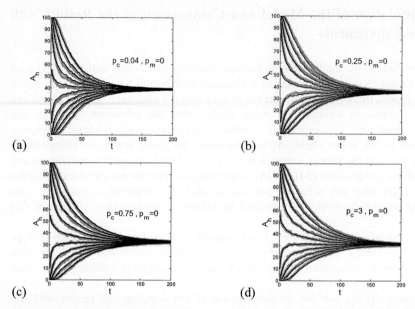

Fig. 3. A_h as a function of time for four different cases where the crossover rate is 4, 25, 75 and 300 percent, respectively. There is no mutation in all four cases. Black lines are the experimental averages over 100 GA runs and thick gray lines are the results obtained by model simulations. Population size is 50 and the chromosome length is 100 genes

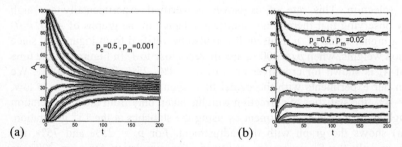

Fig. 4. A_h as a function of time for two different cases where the mutation rate is 0.1 and 2 percent, respectively. The crossover rate is 50% in both cases. Black lines are the experimental averages over 100 GA runs and thick gray lines are the results obtained by model simulations

The experimental results and the model estimations of mean fitness for several values of p_m with crossover rates of $p_c = 25\%$, 75% and 300% are shown in Figure (5). The impact on the mean fitness of changing p_c from 300% to 25% is more visible when p_m is low, around 0 or 0.1%. The effect of higher mutation rates dominates the dynamics of mean fitness evolution, and decreases the amount by which different crossover rates affect the mean fitness. The estimation of the fitness variance when $p_c = 300\%$ is shown in Figure (6) for various mutation rates, together with experimental averages. In all these graphs, we see that these dynamics of mean fitness and fitness variance are well captured by the simulation of the model.

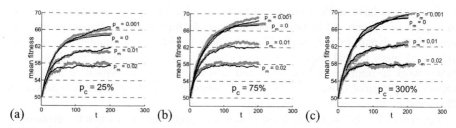

Fig. 5. The mean fitness for three different crossover cases, namely $p_c = 25\%$, 75% and 300%. Each figure shows the graphs for four different mutation rates, $p_m = 0\%$, 0.1%, 1% and 2%. Black lines are the experimental averages obtained by averaging over 100 GA runs and thick gray lines are the results obtained by model simulations. Population size is 50 and the chromosome length is 100 genes

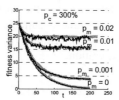

Fig. 6. The fitness variance for four different rates of mutation, $p_m = 0\%$, 0.1%, 1% and 2%, with crossover rate at 300%. Black lines are the experimental averages obtained by averaging over 100 GA runs and thick gray lines are the results obtained by model simulations

5   Conclusions and Future Work

In this paper, we have developed a new and very practical model for the GA dynamics of the OneMax problem, which, for modeling the case of typical crossover rates, uses some statistics of early generations of the GA in order to predict the rest of the evolution. The simulation results in Section (4) show that the model describes the GA dynamics for the OneMax problem very well for different crossover and mutation rates with fitness proportional selection. The correction weights introduced by Equation (17) are a new way of analyzing the crossover operator, and they work very well for two-point crossover, in our GA problem.

Note that our model for "high enough" crossover rates is covering a different case than the statistical mechanics model of Prügel-Bennett and Shapiro [3]. The maximum entropy assumption of Prügel-Bennett and Shapiro essentially models a situation in which the crossover operator is assumed to be effective enough to allow a relocation of the alleles which is probabilistically most likely to occur, under the constraints of the given mean fitness and fitness variance, when the alleles move freely. On the other hand, our model of "high enough" crossover rates does not assume any constraint in relation to how much the allele can be mixed. The lower crossover rates are modeled relative to this extreme case using correction measures.

The authors have applied the model to the weighted OneMax[1] problem as well as to a class of deceptive functions; for the case of "high enough" crossover rates, it predicts the dynamics very well. They are now modifying the model to cover the effects of more typical, lower crossover rates for these benchmark problems. The future work to improve the model would also include the estimation of the fitness variance for normal crossover rates, and an investigation of the predictive power of the model in the presence of external noise.

The model can be applied to these benchmark problems even when the crossover, mutation and the selection rates (in the case of Boltzmann scaling) are changed at predetermined generations during a GA run. Because of this capability of the model, it is unique, to the best of the author's knowledge, among the current models of the GA.

The method of building blocks for modeling parallel genetic algorithms is applied by Cantú-Paz [8] in the case where the migration occurs only when all the populations are converged. Since our model estimates the mean value at each locus at any generation, it can be used to determine a suitable migration time as well as the migration rate for parallel genetic algorithms (in the island model case) when migrations are allowed at any generation.

Acknowledgment. The authors would like to thank Prof. Charles R. MacCluer for helpful and inspiring discussions.

References

1. Nix, A.E., Vose, M.D.: Modeling Genetic Algorithms with Markov Chains. Ann. Math. Art. Intell., (1991), 5:79-88
2. Vose, M.D., Liepins, G.E.: Punctuated Equilibria in Genetic Search. Complex Systems, (1991), 5:31-44
3. Prügel-Bennett, A., Shapiro, J.L.: An Analysis of Genetic Algorithms Using Statistical Mechanics, Phys. Rev. Lett., (1994), 72(9):1305-1309
4. Rattray, L.M.: Modeling the Dynamics of Genetic Algorithms Using Statistical Mechanics. PhD thesis, University of Manchester, Manchester, U.K., (1996)
5. Prügel-Bennett, A., Shapiro, J.L.: The Dynamics of a Genetic Algorithm for Simple Random Ising Systems. Physica D, (1997), 104:75-114
6. Goldberg, D.E.: The Design of Innovation. Kluwer Academic Publishers, Boston, Dordrecht, London, (2002)
7. Goldberg, D.E., Deb, K., Thierens, D.: Toward a Better Understanding of Mixing in Genetic Algorithms. Journal of the Society of Instrument and Control Engineers, (1993), 32(1), 10-16
8. Cantú-Paz, E.: Efficient and Accurate Parallel Genetic Algorithms. Kluwer Academic Publishers, (2001)

---

[1]   The fitness function is given as a weighted sum of alleles.

Adaptive Sampling for Noisy Problems

Erick Cantú-Paz

Center for Applied Scientific Computing
Lawrence Livermore National Laboratory
Livermore, CA 94551
cantupaz@llnl.gov

Abstract. The usual approach to deal with noise present in many real-world optimization problems is to take an arbitrary number of samples of the objective function and use the sample average as an estimate of the true objective value. The number of samples is typically chosen arbitrarily and remains constant for the entire optimization process. This paper studies an adaptive sampling technique that varies the number of samples based on the uncertainty of deciding between two individuals. Experiments demonstrate the effect of adaptive sampling on the final solution quality reached by a genetic algorithm and the computational cost required to find the solution. The results suggest that the adaptive technique can effectively eliminate the need to set the sample size a priori, but in many cases it requires high computational costs.

1 Introduction

Evolutionary algorithms (EAs) are considered relatively robust to the noise present in the evaluation of the objective function of many real-world optimization problems [1]. The usual approach to deal with noise is to take an arbitrary number of samples of the objective function and use the average as an estimate of the true objective value. If many samples are taken, the estimates will be very accurate, but it may be a waste of computing resources. On the other hand, if too few samples are taken, the algorithm may incorrectly select inferior solutions and this might lead to failure. This paper presents results with an adaptive sampling technique that adjusts the number of samples to the uncertainty in deciding between two specific individuals. This adaptive approach eliminates the need to selecting the number of samples a priori.

The objective of this paper is to study the effect of adaptive sampling on the final solution quality and execution time. This paper presents a systematic study of the impact of the sampling using a simple test problem (a 100-bit onemax function). The experiments demonstrate that the adaptive sampling can find better solutions than an arbitrary fixed sample and it can save time over an excessively large fixed sample.

An alternative way to deal with noise in the fitness evaluations is to increase the population size [2,3]. The paper shows that increasing the population size seems a very effective and efficient way to deal with noise. However, in practice

K. Deb et al. (Eds.): GECCO 2004, LNCS 3102, pp. 947–958, 2004.

it may be difficult to determine the correct population size because current population sizing models require knowledge of problem-specific parameters and the noise intensity [2,3]. If the parameters are unknown or if the noise levels are not uniform, it will be difficult to use these models accurately.

The next section provides some background on previous work in solving noisy problems with EAs. Section 3 presents in detail the adaptive sampling method. Section 4 shows the results of experiments. Finally, section 5 concludes the paper and discusses future work.

2   Background

The robustness of evolutionary algorithms to noise in the evaluation of solutions has been recognized for a long time. Recent research suggests that the use of populations is the cause of the robustness of EAs in noisy environments [1]. Harik et al. [2] presented models to determine the size of the populations required to solve certain types of problems and considered the case where the fitness evaluations are noisy. Miller [4] extended Harik et al.'s model to account for sampling the objective function, and later presented models to optimize the sample size [3].

There is still controversy about the tradeoff between increasing the sample size or increasing the size of the populations. Sampling the objective function n times increases the computation time by a factor of n, but reduces the standard deviation of the estimate by a factor of only $\sqrt{n}$. Fitzpatrick and Grefenstette [5] argue in favor of increasing the population size rather than the sample size. Arnold and Beyer's [6] calculations for the sphere model agree that increasing the population is beneficial in evolution strategies with intermediate recombination. On the other hand, Beyer [7] argues that in a $(1, \lambda)$-ES, the sample size should be increased, rather than λ. Hammel and Bäck [8] verify Beyer's result and show that there is no benefit of increasing the parent population size.

The previous works assume that the sample size is fixed beforehand and remains constant during the execution of the EA. Aizawa and Wah [9] were probably the first to introduce a method that allocates different number of samples to different individuals. Their objective is to find an allocation that minimizes a pre-defined loss function. They minimize the expected estimation error as the loss function, which has the effect of drawing more samples from better individuals and spending less time in the inferior ones. Branke and Schmidt [10] also proposed an adaptive sampling method that takes additional samples of both individuals participating in a tournament until the normalized fitness difference between the two individuals falls below some threshold. The normalized fitness difference is obtained dividing the difference of the observed fitnesses by the standard deviation of the difference: $(\bar{f}_x - \bar{f}_y)/\sigma_d$.

The approach presented in the present paper differs from Aizawa and Wah's in that our objective is to take the smallest number of samples necessary to make a decision between competing individuals during the selection process. Our approach is very similar to Branke and Schmidt's, but differs in that we take samples one at a time from the individual with the highest observed variance, and we use standard statistical tests to select the winner of the tournament with certain probability. In addition, Branke did not examine the impact of the technique on the final solution quality, only on the probability of selecting the correct individual.

Somewhat similar to our approach, Teller and Andre [11] proposed a method that allocates varying numbers of fitness cases to evaluate individuals in genetic programming. With their algorithm, individuals are initially evaluated on a small number of fitness cases, and are further evaluated only if there is some chance that the outcome of the tournaments they participate in can change. There is no point on refining evaluations of individuals that are so much better (worse) than their competitors that they are not likely to lose (win) their tournaments. A similar algorithm was developed independently by Giacobini et al. [12].

3   Adaptive Sampling

We consider pairwise tournament selection, where the best of two randomly chosen individuals is selected to continue in the algorithm. Without loss of generality we consider maximization problems. The noisy fitness F' of an individual can be described as

$$F' = F + N, \tag{1}$$

where F is the true fitness and N is the added noise. In this paper, we use normally distributed noise: $N \sim N(0, \sigma_N^2)$, but the same approach can be used with other noise distributions.

Assume that we want to compare two individuals x and y. In a noisy environment, their fitnesses are the random variables $F_x \sim N(\mu_x, \sigma_x^2)$ and $F_y \sim N(\mu_y, \sigma_y^2)$. If $\mu_x > \mu_y$, we would like to select individual x. However, the true means are unknown and we estimate them using the averages $\bar{f}_x = \frac{1}{n_x} \sum_i^{n_x} f_i$. and $\bar{f}_y$ of multiple samples of the fitness function. The true variances are also unknown, so we approximate them with the observed variances s_x^2 and s_y^2. When dealing with noise, it is common to use an arbitrary number of samples and choose the individual corresponding to the highest mean. However, this simple approach may lead to choosing the wrong individual, because it does not take into consideration the uncertainty in the estimations. If too few samples are taken, the estimates will be inaccurate and may lead to failures. If too many samples are taken, computational resources will be wasted.

Let $d = \bar{f}_x - \bar{f}_y$ be the difference between the observed means of individuals x and y. By the central limit theorem, as the number of samples increases the distributions of $\bar{f}_x$ and $\bar{f}_y$ approach normal distributions, regardless of the noise distribution. The distribution of d also approaches a normal $d \sim N(\bar{f}_x - \bar{f}_y, s_x^2 + s_y^2)$. The probability P_c of choosing the individual with the highest quality is

$$P_c = \Phi \left(\frac{\bar{f}_x - \bar{f}_y}{\sqrt{s_x^2 + s_y^2}} \right), \tag{2}$$

where $\Phi(\cdot)$ is the cumulative distribution function of a standard normal distribution. Note that deciding correctly between the two individuals becomes more difficult as d becomes smaller and the observed variances become larger.

The proposed adaptive sampling method is simply to estimate the mean fitnesses and variances using a small number of samples initially and then sample the fitness of the individual with the highest variance until a statistical test can decide between the individuals with some certainty or a limit in the number of samples is reached. There are many possible variations on this idea depending on the initial number of samples, and the way to determine the certainty of the decision. Branke and Schmidt [10] used 10 initial samples of each individual and sample both individuals until the normalized fitness difference falls below some user-specified threshold. We test the method taking only two initial samples from each individual. Instead of using an arbitrary threshold, we use a statistical test to determine when to stop sampling. Branke and Schmidt studied the effect of sampling in the probability of selecting between two individuals, but did not present results of the impact of the adaptive sampling on the final solutions or the cost associated with finding those solutions. We extend their study in those directions.

For the experiments in this paper, we take only two initial samples from each individual. Then the means and variances of the fitness of each individual are estimated. As additional samples are taken, the means will approach a normal distribution, but since we have a small number of initial samples, we decide between the two individuals using a one-sided t test. We conservatively use the minimum of the number of samples of x and y as the degrees of freedom in the test. If the p value of the test is greater or equal to (an arbitrarily chosen) 0.9, then we consider that the test discriminates between the individuals with sufficient certainty. If the p value is below our threshold, then we take an additional sample from the individual with the highest variance and repeat the test.[1] Resampling continues until the p value meets the threshold.

Figure 1 shows the probability that the adaptive sampling procedure chooses the best individual as a function of the difference of their fitnesses. The experiments consisted of using the adaptive sampling to compare an individual with real fitness of 100 to an individual with fitness $100 - i$ for $i = 1, ..., 20$ adding unbiased normal noise with standard deviation of 5, 10, and 20. For each fitness difference we execute 10000 trials and report the fraction of trials where the individual with true higher fitness was selected along with 95% confidence intervals. If the fitnesses of the two individuals are equal, the decision is a random choice. As the fitness difference increases, the probability of deciding correctly quickly approaches 1.0. This suggests that it is easy to decide between individuals with large fitness differences and/or low variance. Therefore we should spend additional computational resources in dealing with individuals with small differences

---

[1] Strictly, re-testing using the same samples might lead to elevated type-I errors and sequential testing methods might be required. However, as the experiments demonstrate (see figure 1), the proposed procedure chooses the correct individual very consistently, so we opted for the simpler re-testing.

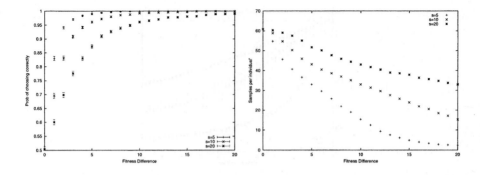

Fig. 1. The probability of correctly choosing the best individual using the adaptive sampling method (left) depends on the difference of the fitnesses of the two individuals. The right graph displays the number of samples required by the adaptive sampling.

and/or large variances. The figure also shows the number of samples necessary to decide correctly with the proposed technique.

The next section examines the effect of using this method on the final solution quality as well as the computational cost. We examine the effect of varying different parameters of interest, such as the noise levels, the population size, and the thresholds for the p values.

4   Experiments

The experiments use a simple generational GA with uniform crossover applied with probability 1.0. No mutation was used to try to limit the source of randomness to the exogenous noise added to the fitness function. The GA used pairwise tournament selection without replacement. The population sizes and noise levels are indicated in each experiment. The GA was terminated when all the members of the population were identical. The random number generator was a Mersenne Twister [13] initialized with 32 bit unsigned integers obtained from www.random.org. All results presented are over 100 repetitions of each parameter setting, and the graphs include 95% confidence intervals.

The results of the adaptive sampling method will be compared against a GA that always uses 10 samples to estimate the objective value and against a GA that ignores the noise and selects tournament winners based on a direct comparison of a single function evaluation.

4.1   Noise and Selection Intensity

A way to measure progress in solving a problem is to examine the average fitness of the population as a run progresses. The increase in average fitness depends on the intensity of the selection method used. Miller suggested that noise reduces the selection intensity [14]. Selecting incorrectly the individuals that have

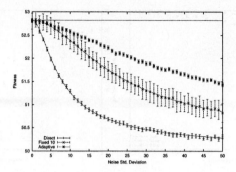

Fig. 2. Effect of noise on the average (real) fitness of individuals selected by pairwise tournament selection. The adaptive method remains closer to the expected fitness than the fixed-size sampling.

lower fitness values has the effect of reducing the average fitness of the selected individuals. The selection intensity is defined as

$$I = \frac{\mu_{\text{sel}} - \mu_F}{\sigma_F},$$

where μ_{sel} is the mean fitness of the selected individuals and μ_F and σ_F are the mean fitness and standard deviation of the original population. Each selection method has a different intensity, and for pairwise tournament selection $I = 0.5642$. Miller [3] established that the expected fitness of the selected individuals in a noisy environment is:

$$\mu_{\text{sel}} = \mu_F + I \frac{\sigma_F^2}{\sqrt{\sigma_F^2 + \sigma_N^2}}. \tag{3}$$

This result can be easy extended to account for the reduction in the uncertainty that comes from taking a number n of multiple samples:

$$\mu_{\text{sel}} = \mu_F + I \frac{\sigma_F^2}{\sqrt{\sigma_F^2 + \sigma_N^2/n}}. \tag{4}$$

Figure 2 shows the effect of noise on the mean fitness of the individuals selected by pairwise tournament selection. To obtain these results a population of 1000 individuals was initialized randomly and we measured the average fitness of the individuals selected using tournament selection. The continuous lines in the figure were obtained with equation 4. The graph shows that ignoring the noise (using one sample) can reduce the selection intensity quite strongly. With 10 fixed samples, the reduction is much less severe as expected. The best results are obtained with the adaptive sampling.

It is not clear, however, what will be the effect of the adaptive sampling on the overall performance of the algorithm. The next two subsections examine this.

4.2   Noise and Solution Quality

Figure 3 shows the means of the real fitnesses at the end of the runs of a GA that uses direct comparisons and a GA with the adaptive sampling method. The experiments considered different noise levels and the population size was varied between 2 and 100 individuals. As expected, the figure shows that the final solution quality improves with larger populations, but with direct comparisons the quality degrades as the noise level increases. The adaptive sampling method shows much smaller quality degradations that are significant at the 0.95 significance level only with high noise strengths of $\sigma = 10$ and 20.

The effect of noise in the final quality of solutions can be observed more directly in figures 4 and 5. In these figures, the population size is kept constant at $N = 20$ and $N = 40$ individuals while the noise level is varied. The left panels show the *observed* fitness values at the end of the runs and the left panels show the *real* fitness. As expected, ignoring the noise and making direct comparisons based on one sample has the worst results: The algorithm is misled with very large apparent fitness values that belong to individuals with very low real fitnesses. The adaptive method always finds the solutions with the best real fitnesses and appears much more robust to increased noise levels than the other two methods (i.e., the real fitness decreases much less with increasing noise).

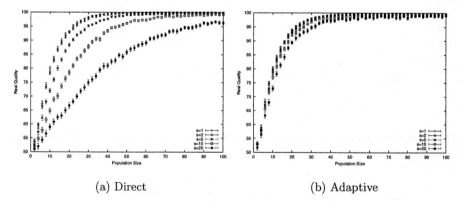

(a) Direct   (b) Adaptive

Fig. 3. Final real fitnesses for different noise levels varying the population size. The quality degradation is much smaller in the adaptive algorithm. Error bars denote 95% confidence intervals.

In the adaptive method, the user does not have to specify the number of samples per individual, but has to specify the p value of the test. The results in figure 6 show that the final solution quality is not affected greatly for a range of commonly used values of p, and the assignment $p = 0.9$ used in the experiments seems an appropriate choice.

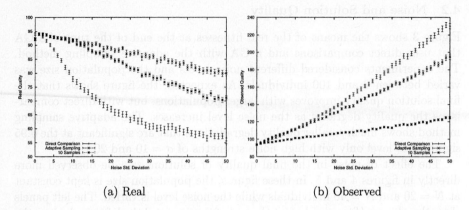

(a) Real

(b) Observed

Fig. 4. Real and observed fitnesses for a 100-bit onemax problem for different noise levels and a population of 20 individuals.

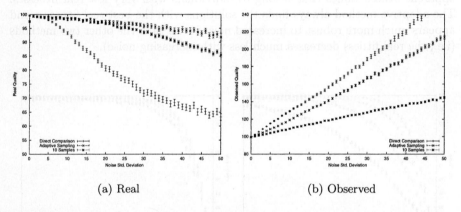

(a) Real

(b) Observed

Fig. 5. Real and observed fitnesses for a 100-bit onemax problem for different noise levels and a population of 40 individuals.

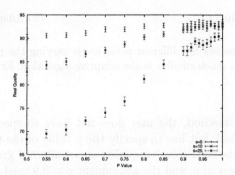

Fig. 6. Mean real quality for different noise levels varying the parameter p.

4.3   Computational Cost

So far we have seen that adaptive sampling can eliminate the need to fix the number of samples without sacrificing solution quality. However, eliminating the guess of the right sample size comes at a significant computational cost.

Figure 7 shows the means of the number of samples per individual and the number of generations until convergence of the experiments corresponding to figure 3. The number of samples appear to be dependent of the noise level and independent of the population size, except for very small populations. Similar observations can be made for the number of generations.

While the results of the previous subsection present a favorable outcome of the sampling methods, a fairer comparison of these algorithms should take into account the total computational cost. The total number of function evaluations taken by each experiment is the product of the population size, the number of generations, and the number of samples. Figures 8 and 9 present the number of samples and generations corresponding to the experiments in figures 4 and 5. Figure 10 compares the total cost incurred by each algorithm to reach solutions of a particular quality. The results show that the sampling methods need an order of magnitude more computations to reach the same solutions than simply ignoring the noise and using larger populations.

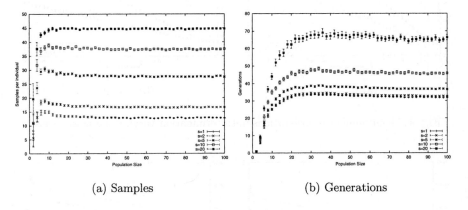

(a) Samples   (b) Generations

Fig. 7. Mean number of samples and generations for different noise levels and varying the population size. The results shown are of the adaptive sampling algorithm.

As the run progresses we can expect the individuals to become more alike each other, reducing their fitness differences and making the decisions more difficult. Albert and Goldberg studying a different but related problem conclude that the sample size should increase over the run [15]. The results in figure 11 confirm this recommendation. We tracked the average number of samples used per generation in a problem with $N = 20$. In all cases, there is a clear upward tendency in the number of samples over time.

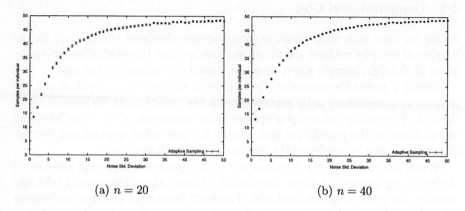

(a) $n = 20$ (b) $n = 40$

Fig. 8. Mean number of samples per individual varying the noise levels.

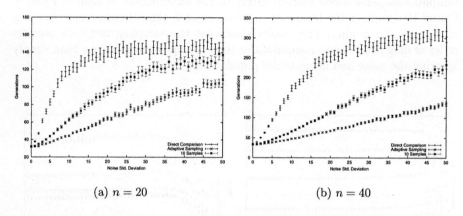

(a) $n = 20$ (b) $n = 40$

Fig. 9. Mean number of generations until termination varying the noise levels.

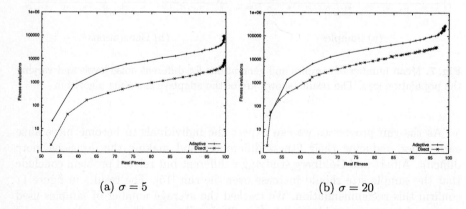

(a) $\sigma = 5$ (b) $\sigma = 20$

Fig. 10. Number of fitness evaluations required to reach different solution qualities.

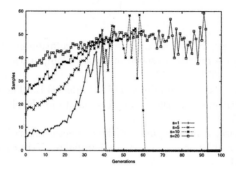

Fig. 11. Number of samples per generation under different noise levels. $N = 20$.

5   Conclusions

Adapting the number of samples to the uncertainty of the decision between two specific individuals eliminates the need to set the sample size a priori. We demonstrated that the adaptive method used in this paper can adapt to a large range of noise levels. For a fixed population size, the experiments showed that the quality reached with adaptive sampling is much higher than ignoring the noise or taking too few samples. The method was also demonstrated to be fairly robust to its only parameter, the p value of the test. For commonly used p values (0.85–1.0), the final quality does not differ much.

The main drawback of the method is that the convenience of adapting the sample size requires a substantial amount of computations. The experiments suggest that increasing the population size slightly and using direct comparisons (no resampling) might be the best strategy. However, the adaptive method is useful in practice, where there might be no knowledge of the noise level or the domain-dependent parameters necessary to use existing population sizing models. Also, in some applications the noise varies over time or the noise may not be uniform in the entire search space, and adapting the sample size will be especially beneficial in those situations.

Future work should include a more detailed investigation of the adaptive sampling with other test functions and adding highly biased noise. Although the assumption of normality is warranted for sufficiently large numbers of samples, it may be possible to mislead the algorithm if the observed fitness difference is small because very few samples will be taken. It is not clear what will be the effect of these mistakes on the overall performance of the algorithm.

Acknowledgments. UCRL-CONF-203216. This work was performed under the auspices of the U.S. Department of Energy by University of California Lawrence Livermore National Laboratory under contract No. W-7405-Eng-48.

References

1. Arnold, D.V., Beyer, H.G.: A comparison of evolution strategies with other direct search methods on the presence of noise. Computational Optimization and Applications **24** (2003) 135–159

2. Harik, G., Cantuú-Paz, E., Goldberg, D.E., Miller, B.L.: The gambler's ruin problem, genetic algorithms, and the sizing of populations. Evolutionary Computation **7** (1999) 231–253

3. Miller, B.L.: Noise, sampling, and efficient genetic algorithms. doctoral dissertation, University of Illinois at Urbana-Champaign, Urbana (1997) Also IlliGAL Report No. 97001.

4. Miller, B.L., Goldberg, D.E.: Optimal sampling for genetic algorithms. Proceedings of the Artificial Neural Networks in Engineering (ANNIE '96) conference **6** (1996) 291–297

5. Fitzpatrick, J.M., Grefenstette, J.J.: Genetic algorithms in noisy environments. Machine Learning **3** (1988) 101–120

6. Arnold, D.V., Beyer, H.G.: Local performance of the $(\mu/\mu_i, \lambda)$-ES in a noisy environment. In Martin, W., Spears, W., eds.: Foundations of Genetic Algorithms, Morgan Kaufmann (2000) 127–142

7. Beyer, H.G.: Toward a theory of evolution strategies: Some asymptotical results from the $(1,^+ \lambda)$-Theory. Evolutionary computation **1** (1993) 165–188

8. Hammel, U., Bäck, T.: Evolution strategies on noisy functions: How to improve convergence properties. In Davidor, Y., Schwefel, H.P., Männer, R., eds.: Parallel Problem Solving fron Nature, PPSN III, Berlin, Springer-Verlag (1994) 159–168

9. Aizawa, A.N., Wah, B.W.: Scheduling of genetic algorithms in a noisy environment. Evolutionary Computation **2** (1994) 97–122

10. Branke, J., Schmidt, C.: Selection in the presence of noise. In Cantú-Paz, E., Foster, J.A., Deb, K., Davis, D., Roy, R., O'Reilly, U.M., Beyer, H.G., Standish, R., Kendall, G., Wilson, S., Harman, M., Wegener, J., Dasgupta, D., Potter, M.A., Schultz, A.C., Dowsland, K., Jonoska, N., Miller, J., eds.: Genetic and Evolutionary Computation – GECCO-2003, Berlin, Springer-Verlag (2003) 766–777

11. Teller, A., Andre, D.: Automatically choosing the number of fitness cases: The rational allocation of trials. In Koza, J.R., Kalyanmoy, D., Dorigo, M., Fogel, D.B., Garzon, M., Iba, H., Riolo, R.L., eds.: Genetic Programming 97, San Francisco, CA, Morgan Kaufmann Publishers (1997) 321–328

12. Giacobini, M., Tomassini, M., Vanneschi, L.: Limiting the number of fitness cases in genetic programming using statistics. In et al., J.J.M., ed.: Parallel Problem Solving from Nature (PPSN VII), Berlin, Springer Verlag (2002)

13. Matsumoto, M., Nishimura, T.: Mersenne twister: A 623-dimensionally equidistributed uniform pseudorandom number generator. ACM Transactions on Modeling and Computer Simulation **8** (1998) 3–30

14. Miller, B.L., Goldberg, D.E.: Genetic algorithms, selection schemes, and the varying effects of noise. Evolutionary Computation **4** (1996) 113–131

15. Albert, L.A., Goldberg, D.E.: Efficient discretization scheduling in multiple dimensions. In Langdon, W.B., Cantú-Paz, E., Mathias, K., Roy, R., Davis, D., Poli, R., Balakrishnan, K., Honavar, V., Rudolph, G., Wegener, J., Bull, L., Potter, M.A., Schultz, A.C., Miller, J.F., Burke, E., Jonoska, N., eds.: GECCO 2002: Proceedings of the Genetic and Evolutionary Computation Conference, New York, Morgan Kaufmann Publishers (2002) 271–278

Feature Subset Selection, Class Separability, and Genetic Algorithms

Erick Cantú-Paz

Center for Applied Scientific Computing
Lawrence Livermore National Laboratory
Livermore, CA 94551
cantupaz@llnl.gov

Abstract. The performance of classification algorithms in machine learning is affected by the features used to describe the labeled examples presented to the inducers. Therefore, the problem of feature subset selection has received considerable attention. Genetic approaches to this problem usually follow the wrapper approach: treat the inducer as a black box that is used to evaluate candidate feature subsets. The evaluations might take a considerable time and the traditional approach might be impractical for large data sets. This paper describes a hybrid of a simple genetic algorithm and a method based on class separability applied to the selection of feature subsets for classification problems. The proposed hybrid was compared against each of its components and two other feature selection wrappers that are used widely. The objective of this paper is to determine if the proposed hybrid presents advantages over the other methods in terms of accuracy or speed in this problem. The experiments used a Naive Bayes classifier and public-domain and artificial data sets. The experiments suggest that the hybrid usually finds compact feature subsets that give the most accurate results, while beating the execution time of the other wrappers.

1 Introduction

The problem of classification in machine learning consists of using labeled examples to induce a model that classifies objects into a set of known classes. The objects are described by a vector of features, some of which may be irrelevant or redundant and may have a negative effect on the accuracy of the classifier. There are two basic approaches to feature subset selection: wrapper and filter methods [1]. Wrappers treat the induction algorithm as a black box that is used by the search algorithm to evaluate each candidate feature subset. While giving good results in terms of the accuracy of the final classifier, wrapper approaches are computationally expensive and may be impractical for large data sets. Filter methods are independent of the classifier and select features based on properties that good feature sets are presumed to have, such as class separability or high correlation with the target. Although filter methods are much faster than wrappers, filters may produce disappointing results, because they completely ignore the induction algorithm.

K. Deb et al. (Eds.): GECCO 2004, LNCS 3102, pp. 959–970, 2004.
© Springer-Verlag Berlin Heidelberg 2004

This paper presents a hybrid algorithm that combines the strengths of filters and wrappers and attempts to avoids their weaknesses. The hybrid consists of a simple genetic algorithm (sGA) used in its traditional role as a wrapper, but initialized with the output of a filter method based on a class separability metric. The objective of this study is to determine if the hybrid method presents advantages over simple GAs and conventional feature selection algorithms in terms of accuracy or speed when applied to feature selection problems. The experiments described in this paper use public-domain and artificial data sets. The classifier was a Naive Bayes, a simple classifier that can be induced quickly, and that has been shown to have good accuracy in many problems [2].

Our target was to maximize the accuracy of classification. The experiments demonstrate that, in most cases, the proposed hybrid algorithm finds subsets that result in the best accuracy (or in an accuracy not significantly different from the best), while finding compact feature subsets, and performing faster than the wrapper methods.

The next section briefly reviews previous applications of EAs to feature subset selection. Section 3 describes the class separability filter and its hybridization with a GA. Section 4 describes the algorithms, data sets, and the fitness evaluation method used in the experiments reported in section 5. Section 6 concludes this paper with a summary and a discussion of future research directions.

2   Feature Selection

Reducing the dimensionality of the vectors of features that describe each object presents several advantages. As mentioned above, irrelevant or redundant features may affect negatively the accuracy of classification algorithms. In addition, reducing the number of features may help decrease the cost of acquiring data and might make the classification models easier to understand.

There are numerous techniques for dimensionality reduction. Some common methods seek transformations of the original variables to lower dimensional spaces. For example, principal components analysis reduces the dimensions of the data by finding orthogonal linear combinations with the largest variance. In the mean square error sense, principal components analysis yields the optimal linear reduction of dimensionality. However, it is not necessarily true that the principal components that capture most of the variance are useful to discriminate among objects of different classes. Moreover, the linear combinations of variables make it difficult to interpret the effect of the original variables on class discrimination. For these reasons, in the remainder of this paper we ignore methods that transform the features and we focus on techniques that select subsets of the original variables.

Among the feature subset algorithms, wrapper methods have received considerable attention. Wrappers are attractive because they seek to optimize the accuracy of a classifier, tailoring their solutions to a specific inducer and a domain. They search for a good feature subset using the induction algorithm to evaluate the merit of candidate subsets. Numerous search algorithms have been used to search for feature subsets [3]. Genetic algorithms are usually reported to deliver good results, but exceptions have been reported where simpler (and faster) algorithms result in higher accuracies on particular data sets [3].

Applying GAs to the feature selection problem is straightforward: the chromosomes of the individuals contain one bit for each feature, and the value of the bit determines whether the feature will be used in the classification. Using the wrapper approach, the individuals are evaluated by training the classifiers using the feature subset indicated by the chromosome and using the resulting accuracy to calculate the fitness. Siedlecki and Sklansky [4] were the first to describe the application of GAs in this way. GAs have been used to search for feature subsets in conjunction with several classification methods such as neural networks [5,6], decision trees [7], k-nearest neighbors [8,9,10,11], rules [12], and Naive Bayes [13, 14].

Besides selecting feature subsets, GAs can extract new features by searching for a vector of numeric coefficients that is used to transform linearly the original features [8,9]. In this case, a value of zero in the transformation vector is equivalent to avoiding the feature. Raymer et al. [10,15] combined the linear transformation with explicit feature selection flags in the chromosomes, and reported an advantage over the pure transformation method.

More sophisticated Distribution Estimation Algorithms (DEAs) have also been used to search for optimal feature subsets. DEAs explicitly identify the relationships among the variables of a problem by building a model of selected individuals and using the model to generate new solutions. In this way, DEAs avoid the disruption of groups of related variables that might prevent a simple GA from reaching the global optimum. However, in terms of accuracy, the DEAs do not seem to outperform simple GAs when searching for feature subsets [13, 14,16,17]. For this reason, we limit this study to simple GAs.

The wrappers' evaluation of candidate feature subsets can be computationally expensive on large data sets. Filter methods are computationally efficient and offer an alternative to wrappers. Genetic algorithms have been used as filters in regression problems to optimize a cost function derived from the correlation matrix between the features and the target value [18]. GAs have also been used as a filter in classification problems minimizing the inconsistencies present in subsets of the features [19]. An inconsistency between two examples occurs if the examples match with respect to the feature subset considered, but their class labels disagree. Lanzi demonstrated that this filter method efficiently identifies feature subsets that were at least as predictive as the original set of features (the results were never significantly worse). However, the accuracy on the reduced subset is not much different (better or worse) than with all the features. In this study we show that the proposed method can reduce the dimensionality of the data and increase the predictive accuracy considerably.

3   Class Separability

The idea of using a measure of class separability to select features has been used in machine learning and computer vision [20,21]. The class separability filter that we propose calculates the class separability of each feature using the Kullback-Leibler (KL) distance between histograms of feature values. For each feature, there is one histogram for each class. Numeric features are discretized using $\sqrt{|D|}/2$ equally-spaced bins, where $|D|$ is the size of the training data. The histograms are normalized dividing each bin count by the total number of elements to estimate the probability that the j-th feature takes a value in the i-th bin of the histogram given a class n, $p_j(d = i | c = n)$. For each feature j, we calculate the class separability as

$$\Delta_j = \sum_{m=1}^{c} \sum_{n=1}^{c} \delta_j(m, n), \tag{1}$$

where c is the number of classes and $\delta_j(m, n)$ is the KL distance between histograms corresponding to classes m and n:

$$\delta_j(m, n) = \sum_{i=1}^{b} p_j(d = i | c = m) \log \left(\frac{p_j(d = i | c = m)}{p_j(d = i | c = n)} \right), \tag{2}$$

where b is the number of bins in the histograms. Of course, other distribution distance metrics could be used instead of KL distance.

The features are then sorted in descending order of the distances Δ_j (larger distances mean better separability). Heuristically, we consider that two features are redundant if their distances differ by less than 0.0001, and we eliminate the feature with the smallest distance. We eliminate irrelevant non-discriminative features with Δ_j distances less than 0.001.

The heuristics used to eliminate redundant and irrelevant features were calibrated using artificial data sets that are described later. We recognize that these heuristics may fail in some cases if the thresholds chosen are not adequate to a particular classification problem. However, perhaps the major disadvantage of the method is that it ignores pairwise (or higher) interactions among variables. It is possible that features that appear irrelevant (not discriminative) when considered alone are relevant when considered in conjunction with other variables. For example, consider the two-class data displayed in figure 3. Each of the features alone does not have discriminative power, but taken together the two features perfectly discriminate the two classes.

To explore combinations of features we decided to use a genetic algorithm. After running the filter algorithm, we have some knowledge about the relative importance of each feature considered individually. This knowledge is incorporated into the GA by using the relative distances to initialize the GA. The distances Δ_j are linearly normalized between 0.1 and 0.9 to obtain the probability p_j that the j-th bit in the chromosomes is initialized to 1 (and thus that the corresponding feature is selected). By making the lower and upper limits of p_j different from 0 and 1, we are able to explore combinations that include features that the filter had eliminated as redundant or irrelevant. It also allows a chance to delete features that the filter identified as important.

After the GA is initialized with the output of the filter, the GA runs as a wrapper feature selection algorithm. The GA manipulates a population of candidate feature subsets using conventional GA operators. Each candidate solution is evaluated using an estimate of the accuracy of a classifier on the feature subset indicated in the chromosome and the best solution is reported to the user.

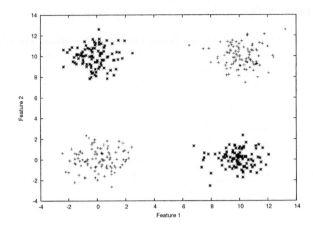

Fig. 1. Example of a data set where each feature considered alone does not discriminate between the two classes, but the two features taken together discriminate the data perfectly.

4   Methods

This section describes the algorithms and the data used in this study as well as the method used to evaluate the fitness.

4.1   Algorithms and Data Sets

The GA used uniform crossover with probability 1.0, and mutation with probability $1/l$, where l was the length of the chromosomes and corresponds to the total number of features in each problem. The population size was set to $\lfloor 3\sqrt{l} \rfloor$, following the Gambler's ruin model for population sizing that asserts that the population size required to reach a solution of a particular quality is $O(\sqrt{l})$ [22]. Promising solutions were selected with pairwise binary tournaments without replacement. The algorithms were terminated after observing no improvement of the best individual over consecutive generations. Inza et al. [13] and Cantú-Paz [14] used similar algorithms and termination criterion.

We compare the results of the class separability filter and the GAs with two traditional greedy feature selection algorithms. Greedy feature selection algorithms that add or delete a single feature from the candidate feature subset are common. There are two basic variants: sequential forward selection (SFS) and sequential backward elimination (SBE). Forward selection starts with an empty set of features. In the first iteration, the algorithm considers all feature subsets with only one feature. The feature subset with the highest accuracy is used as the basis for the next iteration. In each iteration, the algorithm tentatively adds to the basis each feature not previously selected and retains the feature subset that results in the highest estimated performance. The search terminates after

Table 1. Description of the data used in the experiments.

Domain	Instances	Classes	Numeric Feat.	Nominal Feat.	Missing
Anneal	898	6	9	29	Y
Arrhythmia	452	16	206	73	Y
Euthyroid	3163	2	7	18	Y
Ionosphere	351	2	34	–	N
Pima	768	2	8	–	N
Segmentation	2310	7	19	–	N
Soybean Large	683	19	–	35	Y
Random21	2500	2	21	–	N
Redundant21	2500	2	21	–	N

the accuracy of the current subset cannot be improved by adding any other feature. Backward elimination works in an analogous way, starting from the full set of features and tentatively deleting each feature not deleted previously.

The classifier used in the experiments was a Naive Bayes (NB). This classifier was chosen for its speed and simplicity, but the proposed hybrid method can be used with any other supervised classifiers. In the NB, the probabilities for nominal features were estimated from the data using maximum likelihood estimation (their observed frequencies in the data) and applying the Laplace correction. Numeric features were assumed to have a normal distribution. Missing values in the data were skipped.

The algorithms were developed in C++ and compiled with g++ version 2.96 using -O2 optimizations. The experiments were executed on a single processor of a Linux (Red Had 7.3) workstation with dual 2.4 GHz Intel Xeon processors and 512 Mb of memory. A Mersenne Twister random number generator [23] was used in the GA and the data partitioning.

The data sets used in the experiments are described in table 1. With the exception of Random21 and Redundant21, the data sets are available in the UCI repository [24]. Random21 and Redundant21 are two artificial data sets with 21 features each and were proposed originally by Inza [13]. The target concept of these two data sets is whether the first nine features are closer to $(0,0,...,0)$ or $(9,9,...,9)$ in Euclidean distance. The features were generated uniformly at random in the range [3,6]. All the features in Random21 are random, and the first, fifth, and ninth features are repeated four times each in Redundant21.

4.2   Measuring Fitness

Since we are interested in classifiers that generalize well, the fitness calculations must include some estimate of the generalization of the Naive Bayes using the candidate subsets. We estimate the generalization of the network using crossvalidation. In k-fold crossvalidation, the data D is partitioned randomly into k non-overlapping sets, $D_1, ..., D_k$. At each iteration i (from 1 to k), the classifier is trained with $D \backslash D_i$ and tested on D_i. Since the data are partitioned randomly, it is likely that repeated crossvalidation experiments return different results. Although there are well-known methods to deal with "noisy" fitness evaluations in EAs [25], we chose to limit the uncertainty in the accuracy estimate by repeating 10-fold crossvalidation experiments until the standard deviation of the accuracy estimate drops below 1% (or a maximum of five repetitions). This heuristic was proposed by Kohavi and John [2] in their study of wrapper methods for feature selection, and was adopted by Inza et al. [13]. We use the accuracy estimate as our fitness function.

Even though crossvalidation is expensive computationally, the cost was not prohibitive in our case, since the data sets were relatively small and the NB classifier is very efficient. If larger data sets or other inducers were used, we would have to deal with the uncertainty in the evaluation by other means, such as increasing slightly the population size (to compensate for the noise in the evaluation) or by sampling the training data. We defer a discussion of possible performance improvements until the final section.

Our fitness measure does not include any term to bias the search toward small feature subsets. However, the algorithms found small subsets, and with some data the algorithms consistently found the smallest subsets that describe the target concepts. This suggests that the data sets contained irrelevant or redundant features that decreased the accuracy of the Naive Bayes.

5   Experiments

To evaluate the generalization accuracy of the feature selection methods, we used 5 iterations of 2-fold crossvalidation (5x2cv). In each iteration, the data were randomly divided in halves. One half was input to the feature selection algorithms. The final feature subset found in each experiment was used to train a final NB classifier (using the entire training data), which was then tested on the other half of the data. The accuracy results presented in table 2 are the mean and standard deviations of the ten tests.

To determine if the differences among the algorithms were statistically significant, we used a combined F test proposed by Alpaydin [26]. Let $p_{i,j}$ denote the difference in the accuracy rates of two classifiers in fold j of the i-th iteration of 5x2 cv, $\bar{p} = (p_{i,1} + p_{i,2})/2$ denote the mean, and $s_i^2 = (p_{i,1} - \bar{p})^2 + (p_{i,2} - \bar{p})^2$ the variance, then

$$f = \frac{\sum_{i=1}^{5} \sum_{j=1}^{2} (p_{i,j})^2}{2 \sum_{i=1}^{5} s_i^2}$$

is approximately F distributed with 10 and 5 degrees of freedom. We rejected the null hypothesis that the two algorithms have the same error rate at a 0.95 significance level if $f > 4.74$ [26]. Care was taken to ensure that all the algorithms used the same training and testing data in the two folds of the five crossvalidation experiments.

Table 2 has the mean accuracies obtained with each method. The best observed result in the table is highlighted in **bold** type as well as those results that according to the combined F test are not significantly different from the best at a 0.95 significance level. There are two immediate observations that we can make from the results. First, the feature selection algorithms result in an improvement

Table 2. Means and standard deviations of the accuracies found in the 5x2cv experiments. The best result and those not significantly different from the best are displayed in **bold**.

Domain	Naive	Filter	FilterGA	sGA	SFS	SBE
Anneal	89.93 2.72	**93.43** 1.44	**93.07** 2.89	92.47 1.69	90.36 2.37	**93.47** 2.71
Arrhythmia	56.95 3.18	**62.08** 2.52	**64.16** 2.13	59.78 3.51	58.67 3.25	59.73 2.33
Euthyroid	87.33 3.23	89.06 0.41	**94.20** 2.02	**94.92** 0.74	**94.57** 0.54	**94.48** 0.42
Ionosphere	83.02 2.04	**89.57** 1.29	**90.54** 0.83	88.95 2.14	85.23 2.76	**89.17** 1.73
Random21	93.89 0.81	82.24 2.32	**95.41** 1.06	92.45 3.96	82.12 1.70	80.61 2.13
Pima	**74.87** 2.55	**74.45** 2.23	**75.49** 2.49	**75.29** 2.57	73.46 1.77	**74.45** 1.71
Redundant	77.12 0.33	80.29 1.09	**83.68** 2.94	**86.70** 2.73	79.74 2.54	80.32 1.03
Segment	79.92 0.73	85.40 1.11	**87.97** 1.12	84.73 2.37	**90.85** 1.02	**91.28** 0.93
Soybean	84.28 4.72	**86.01** 4.89	81.23 5.73	81.79 6.12	78.63 3.23	**86.27** 5.00

Table 3. Means and standard deviations of the sizes of final feature subsets. The best result and those not significantly different from the best are in **bold**.

Domain	Original	Filter	FilterGA	sGA	SFS	SBE
Anneal	38	23.8 3.97	12.8 2.04	22.1 3.81	**5.4** 0.92	16.4 9.54
Arrhythmia	279	212.5 16.30	86.2 6.42	138.9 4.99	**3.9** 1.76	261.1 28.2
Euthyroid	25	**1.0** 0.00	6.3 1.68	13.7 1.55	**1.3** 0.64	1.2 0.40
Ionosphere	34	33.0 0.00	11.2 2.04	16.0 1.95	**4.4** 1.56	30.9 1.76
Pima	8	4.3 2.87	**2.9** 0.83	4.9 0.70	**1.6** 0.66	5.3 1.00
Random21	21	**10.2** 3.60	**10.3** 1.10	13.6 2.06	**9.3** 0.90	12.6 4.48
Redundant	21	**8.8** 0.40	**8.1** 1.70	10.6 1.43	**8.6** 0.92	9.1 0.70
Segmentation	19	11.0 0.00	9.9 1.51	9.6 1.69	**4.0** 0.63	7.7 2.79
Soybean Large	35	32.9 1.51	19.50 2.11	21.7 2.15	**10.6** 2.01	30.7 2.28

of accuracy over using a NB with all the features. However, this difference is not always significant (Soybean Large, Pima). Second, the proposed hybrid always reaches the highest accuracy or accuracies that are not significantly different from the highest. The simple GA with random initialization also performs very well, reaching results that are not significantly different from the best for all but two data sets.

In terms of the size of the final feature subsets (table 3), forward sequential selection consistently found the smallest subsets. This was expected, since this algorithm is heavily biased toward small subsets (because it starts from an empty set and adds features only when they show improvements in accuracy). However, in many cases SFS resulted in significantly worse accuracies than the proposed GA hybrid. The proposed hybrid found significantly—and substantially—smaller feature subsets than the filter alone or the sGA.

Table 4 shows the mean number of feature subsets examined by each algorithm. In most cases, the GAs examine fewer subsets than SFS and SBE, and the FilterGA examined fewer subsets than the GA initialized at random. This suggests that the search of the FilterGA was highly biased toward good solutions.

Table 4. Means and standard deviations of the number of feature subsets examined by each algorithm. The best result and those not significantly different from the best are in **bold**.

Domain	FilterGA	sGA	SFS	SBE
Anneal	**38.84** 19.31	48.08 32.24	225.50 29.46	569.20 185.49
Arrhythmia	**105.23** 26.98	120.26 40.09	1356.0 480.76	4706.9 6395.16
Euthyroid	**36.00** 28.62	37.50 18.06	55.8 14.46	324.8 0.40
Ionosphere	**38.48** 21.85	**41.98** 23.73	170.5 45.73	131.5 53.18
Pima	**12.73** 6.84	20.36 6.79	**18.5** 3.77	24.1 4.83
Random21	**35.74** 20.58	64.61 34.81	168.0 10.68	147.9 59.35
Redundant21	**32.99** 23.17	42.62 46.19	159.9 11.85	193.9 6.43
Segmentation	**37.92** 32.27	**30.08** 23.43	84.8 9.17	160.3 21.35
Soybean Large	**42.60** 25.35	**42.60** 22.73	342.5 47.39	171.5 67.46

Table 5. Execution time (in CPU seconds) of the 5x2cv experiments with each algorithm. The Filter method is always the fastest algorithm (denoted with **bold** type). The results in *italics* type correspond to the second fastest algorithm.

Domain	Filter	FilterGA	sGA	SFS	SBE
Anneal	**0.28**	44.2	66.4	*26.1*	190
Arrhythmia	**4.37**	926.0	1322.9	*775*	32497
Euthyroid	**0.31**	62.4	91.9	*21.2*	290.3
Ionosphere	**0.12**	*9.9*	12.8	10.4	22.1
Pima	**0.03**	2.1	2.8	*0.9*	2.3
Random21	**0.46**	*44.8*	80.6	71.9	119.6
Redundant21	**0.45**	*44.0*	54.6	67.1	148.6
Segmentation	**0.64**	77.3	65.5	*31.6*	138.6
Soybean Large	**1.81**	*94.5*	99.7	137.2	293.4

The number of examined subsets can be used as a coarse surrogate for the execution time, but the actual times depend on the number of features present in each candidate subset and may vary considerably from what we might expect. The execution times (user time in CPU seconds) for the entire 5x2cv experiments are reported in table 5. For the filter method, the time reported includes the time to compute and sort class separabilities and the time to evaluate the naive Bayes on the feature subset found by the filter method. The proposed filter method is by far the fastest algorithm, beating its closest competitor by two orders of magnitude. However, the filter found significantly less accurate results for four of the nine datasets. Among the wrapper methods, SFS and the hybrid of the filter and the GA are the fastest.

Figure 5 summarizes the tradeoff between accuracy (table 2) and execution time (table 5) for six of the data sets. The other data sets were omitted to reduce clutter. The graph clearly shows that the filter is two orders of magnitude faster than the other methods, but the wrappers usually result in accuracy improvements.

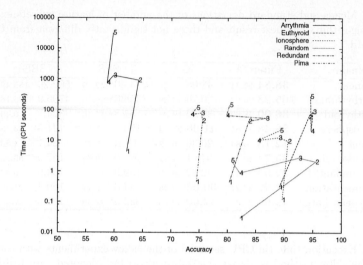

Fig. 2. Plot of accuracies vs. execution time for five data sets. The algorithms are identified by labels as follows: (1) Filter, (2) FilterGA, (3) sGA, (4) SFS, (5) SBE.

6   Conclusions

This paper presented experiments with a proposed GA-Filter hybrid for feature selection in classification problems. The results were compared against a simple GA, two traditional sequential methods, and a filter method based on a simple class separability metric. The experiments considered a Naive Bayes classifier and public-domain and artificial data sets. In the data sets we tried, the proposed method always found the most accurate solutions or solutions that were not significantly different from the best. The proposed method usually found the second smallest feature subsets (behind SFS) and performed faster than simple GAs, SFS, and SBE methods.

This work can be extended with experiments with other evolutionary algorithms, classification methods, additional data sets, and alternative class distance metrics. In particular, it would be interesting to explore methods that consider more than one feature at a time to calculate class separabilities.

There are numerous opportunities to improve the computational efficiency of the algorithms to deal with much larger data sets. In particular, subsampling the training sets and parallelizing the fitness evaluations seem like promising alternatives. Note that SFS and SBE are inherently serial methods and cannot benefit from parallelism as much as GAs. In addition, future work should explore efficient methods to deal with the noisy accuracy estimates, instead of using the relatively expensive multiple crossvalidations that we employed.

Acknowledgments. I would like to thank Martin Pelikan and Chandrika Kamath for useful discussions on this topic.

UCRL-CONF-202041. This work was performed under the auspices of the U.S. Department of Energy by University of California Lawrence Livermore National Laboratory under contract No. W-7405-Eng-48.

References

1. John, G., Kohavi, R., Phleger, K.: Irrelevant features and the feature subset problem. In: Proceedings of the 11th International Conference on Machine Learning, Morgan Kaufmann (1994) 121–129
2. Kohavi, R., John, G.: Wrappers for feature subset selection. Artificial Intelligence **97** (1997) 273–324
3. Jain, A., Zongker, D.: Feature selection: evaluation, application and small sample performance. IEEE Transactions on Pattern Analysis and Machine Intelligence **19** (1997) 153–158
4. Siedlecki, W., Sklansky, J.: A note on genetic algorithms for large-scale feature selection. Pattern Recognition Letters **10** (1989) 335–347
5. Brill, F.Z., Brown, D.E., Martin, W.N.: Genetic algorithms for feature selection for counterpropagation networks. Tech. Rep. No. IPC-TR-90-004, University of Virginia, Institute of Parallel Computation, Charlottesville (1990)
6. Brotherton, T.W., Simpson, P.K.: Dynamic feature set training of neural nets for classification. In McDonnell, J.R., Reynolds, R.G., Fogel, D.B., eds.: Evolutionary Programming IV, Cambridge, MA, MIT Press (1995) 83–94
7. Bala, J., De Jong, K., Huang, J., Vafaie, H., Wechsler, H.: Using learning to facilitate the evolution of features for recognizing visual concepts. Evolutionary Computation **4** (1996) 297–311
8. Kelly, J.D., Davis, L.: Hybridizing the genetic algorithm and the K nearest neighbors classification algorithm. In Belew, R.K., Booker, L.B., eds.: Proceedings of the Fourth International Conference on Genetic Algorithms, San Mateo, CA, Morgan Kaufmann (1991) 377–383
9. Punch, W.F., Goodman, E.D., Pei, M., Chia-Shun, L., Hovland, P., Enbody, R.: Further research on feature selection and classification using genetic algorithms. In Forrest, S., ed.: Proceedings of the Fifth International Conference on Genetic Algorithms, San Mateo, CA, Morgan Kaufmann (1993) 557–564
10. Raymer, M.L., Punch, W.F., Goodman, E.D., Sanschagrin, P.C., Kuhn, L.A.: Simultaneous feature scaling and selection using a genetic algorithm. In Bäck, T., ed.: Proceedings of the Seventh International Conference on Genetic Algorithms, San Francisco, Morgan Kaufmann (1997) 561–567
11. Kudo, M., Sklansky, K.: Comparison of algorithms that select features for pattern classifiers. Pattern Recognition **33** (2000) 25–41
12. Vafaie, H., De Jong, K.A.: Robust feature selection algorithms. In: Proceedings of the International Conference on Tools with Artificial Intelligence, IEEE Computer Society Press (1993) 356–364
13. Inza, I., Larrañaga, P., Etxeberria, R., Sierra, B.: Feature subset selection by Bayesian networks based optimization. Artificial Intelligence **123** (1999) 157–184
14. Cantú-Paz, E.: Feature subset selection by estimation of distribution algorithms. In Langdon, W.B., Cantú-Paz, E., Mathias, K., Roy, R., Davis, D., Poli, R., Balakrishnan, K., Honavar, V., Rudolph, G., Wegener, J., Bull, L., Potter, M.A., Schultz, A.C., Miller, J.F., Burke, E., Jonoska, N., eds.: GECCO 2002: Proceedings of the Genetic and Evolutionary Computation Conference, San Francisco, CA, Morgan Kaufmann Publishers (2002) 303–310

15. Raymer, M.L., Punch, W.F., Goodman, E.D., Kuhn, L.A., Jain, A.K.: Dimensionality reduction using genetic algorithms. IEEE Transactions on Evolutionary Computation **4** (2000) 164–171

16. Inza, I., Larrañaga, P., Sierra, B.: Feature subset selection by Bayesian networks: a comparison with genetic and sequential algorithms. International Journal of Approximate Reasoning **27** (2001) 143–164

17. Inza, I., Larrañaga, P., Sierra, B.: Feature subset selection by estimation of distribution algorithms. In Larrañaga, P., Lozano, J.A., eds.: Estimation of Distribution Algorithms: A new tool for Evolutionary Computation. Kluwer Academic Publishers (2001)

18. Ozdemir, M., Embrechts, M.J., Arciniegas, F., Breneman, C.M., Lockwood, L., Bennett, K.P.: Feature selection for in-silico drug design using genetic algorithms and neural networks. In: IEEE Mountain Workshop on Soft Computing in Industrial Applications, IEEE Press (2001) 53–57

19. Lanzi, P.: Fast feature selection with genetic algorithms: a wrapper approach. In: IEEE International Conference on Evolutionary Computation, IEEE Press (1997) 537–540

20. Guyon, I., Elisseeff, A.: An introduction to variable and feature selection. Journal of Machine Learning Research **3** (2003) 1157–1182

21. Oh, I.S., Lee, J.S., Suen, C.: Analysis of class separation and combination of class-dependent features for handwritting recognition. IEEE Transactions on Pattern Analysis and Machine Intelligence **21** (1999) 1089–1094

22. Harik, G., Cantú-Paz, E., Goldberg, D.E., Miller, B.L.: The gambler's ruin problem, genetic algorithms, and the sizing of populations. Evolutionary Computation **7** (1999) 231–253

23. Matsumoto, M., Nishimura, T.: Mersenne twister: A 623-dimensionally equidistributed uniform pseudorandom number generator. ACM Transactions on Modeling and Computer Simulation **8** (1998) 3–30

24. Blake, C., Merz, C.: UCI repository of machine learning databases (1998)

25. Miller, B.L., Goldberg, D.E.: Genetic algorithms, selection schemes, and the varying effects of noise. Evolutionary Computation **4** (1996) 113–131

26. Alpaydin, E.: Combined 5×2cv F test for comparing supervised classification algorithms. Neural Computation **11** (1999) 1885–1892

Introducing Subchromosome Representations to the Linkage Learning Genetic Algorithm

Ying-ping Chen[1] and David E. Goldberg[2]

[1] Department of Computer Science and Department of General Engineering
University of Illinois, Urbana, IL 61801, USA
ypchen@illigal.ge.uiuc.edu
[2] Department of General Engineering
University of Illinois, Urbana, IL 61801, USA
deg@uiuc.edu

Abstract. This paper introduces subchromosome representations to the linkage learning genetic algorithm (LLGA). The subchromosome representation is utilized for effectively lowering the number of building blocks in order to escape from the performance limit implied by the convergence time model for the linkage learning genetic algorithm. A preliminary implementation to realize subchromosome representations is developed and tested. The experimental results indicate that the proposed representation can improve the performance of the linkage learning genetic algorithm on uniformly scaled problems, and the initial implementation provides a potential way for the linkage learning genetic algorithm to incorporate prior linkage information when such knowledge exists.

1 Introduction

Linkage learning, which makes genetic algorithms (GAs) capable of detecting associations among genes and properly arranging these closely related genes to form building blocks, is one of the key challenges of the genetic algorithm design. In order to ensure a genetic algorithm works well, the building blocks represented on the chromosome have to be tightly linked. Otherwise, studies [1, 2] have shown that a genetic algorithm may fail to solve problems without such prior knowledge. One way to alleviate the burden of choosing an appropriate chromosome representation for genetic algorithm users is to employ the genetic linkage learning technique. Among the existing linkage learning methods, such as perturbation-based techniques [3,4], model builders [5,6,7], and linkage learners [8,9], is the linkage learning genetic algorithm (LLGA), which uses an evolvable genotypic structure capable of learning genetic linkage during the evolutionary process through its special expression mechanism.

While LLGA achieved successful linkage learning on problems with badly scaled building blocks, it was less successful on problems consisting of uniformly scaled building blocks. The convergence time model for LLGA [10] explains the difficulty faced by LLGA and indicates the performance limit of LLGA on uniformly scaled problems. This paper seeks to enhance the design of LLGA

K. Deb et al. (Eds.): GECCO 2004, LNCS 3102, pp. 971–982, 2004.

based on the time models in order to improve the performance of LLGA on uniformly scaled problems.

In particular, this paper introduces subchromosome representations to LLGA. The subchromosome representation is developed to avoid the performance limit implied by the convergence time model for LLGA. This paper presents a preliminary implementation of the proposed representation and verifies the performance improvement with empirical results. The objective of this study is to initiate a better design of LLGA that can lead to scalable genetic linkage learning.

This paper is organized as follows. The next section gives a brief review of the linkage learning genetic algorithm. Section 3 describes the subchromosome representation proposed in this paper in detail. The experiments for observing the effect of using subchromosomes and the experimental results are presented in Sect. 4. Finally, we outlined future research directions followed by conclusions.

2   Review of the Linkage Learning Genetic Algorithm

This section reviews key elements of the linkage learning genetic algorithm (LLGA) [11]. LLGA is capable of learning genetic linkage in the evolutionary process without the help of extra measurements and techniques. A modified version of LLGA working with *promoters* [12] is used in this study and described in this section. Readers may consult other materials [11,12] for detailed information.

2.1   Chromosome Representation

The LLGA's chromosome representation is mainly composed of moveable genes, non-coding segments, probabilistic expression, and promoters. Moveable genes are encoded as (*gene number, allele*) pairs on the LLGA chromosome, and an LLGA chromosome is considered as a circle. These genes are allowed to move around and reside anywhere in any order on the chromosome. Non-coding segments are inserted into the chromosome to create an evolvable genotype capable of learning linkage. Non-coding segments act as non-functional genes residing between functional genes to form gaps for precisely expressing genetic linkage.

Probability expression (PE) was proposed to preserve building-block level diversity. For each gene, all possible alleles coexist in a PE chromosome at the same time. For the purpose of evaluation, a chromosome is *interpreted* with a *point of interpretation* (POI). The allele for each gene is determined by the order according to which the chromosome is traversed clock-wisely from the point of interpretation. A complete string is then expressed and evaluated.

Consequently, each PE chromosome represents not just a single solution but a probability distribution over the range of possible solutions. If different points of interpretation are selected, a PE chromosome might be interpreted as different solutions. Furthermore, the probability of a PE chromosome to be expressed as a particular solution depends on the length of the non-coding segment between genes critical to that solution. It is the essential technique of LLGA to capture the knowledge about linkage and to prompt the evolution of linkage.

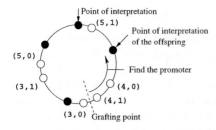

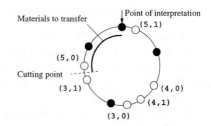

Fig. 1. After selecting the grafting point on the recipient, the nearest promoter *before* the grafting point is then the point of interpretation of the offspring.

Fig. 2. After selecting the cutting point on the donor, the genetic material *after* the cutting point and before the current point of interpretation is transferred.

The use of *promoters*, which were called *start expression genes*, was proposed [12] in LLGA to handle separation inadequacy and to improve nucleation potential. Promoters are special non-functional elements on the chromosome. While in LLGA without promoters, all genes and non-coding segments can be the points of interpretation of the child created by crossover, only promoters can be the points of interpretation in LLGA with promoters.

2.2   Exchange Crossover

The exchange crossover operator is another key mechanism to make LLGA capable of learning genetic linkage. It is defined on a pair of chromosomes. One of the two chromosomes is the *donor*, and the other is the *recipient*. Exchange crossover cuts a *random* segment of the donor, selects a grafting point on the recipient, and grafts the segment onto the recipient. The grafting point is the point of interpretation of the offspring. Starting from the point of interpretation, redundant genetic materials caused by injection are removed right after crossover to ensure the validity of the offspring.

In LLGA with promoters, although the grafting point can still be any genes or non-coding segments, the point of interpretation of the offspring is no longer the grafting point. Instead, the new point of interpretation is the nearest promoter *before* the grafting point on the chromosome. After the grafting point is randomly chosen, the first promoter in front of the grafting point is the point of interpretation of the offspring. The genetic material is then transferred in the following order: (1) the segment between the promoter and the grafting point, (2) the segment chosen from the donor, and (3) the rest of the recipient. Figure 1 shows how promoters work, the black filled circles are promoters of the chromosome. Exchange crossover in LLGA with promoters selects only one cutting point at random. The other cutting point is always the element (either functional or non-functional) just before the point of interpretation of the donor. Figure 2 shows the genetic materials to be transferred during a crossover event.

2.3   Linkage Learning Mechanisms

With the integration of PE and exchange crossover, LLGA is capable of solving difficult problems without prior knowledge of good linkage. Traditional GAs have been shown to perform poorly on difficult problems [1,2] without such knowledge. To better decompose and understand the behavior of LLGA, two key mechanisms of linkage learning, *linkage skew* and *linkage shift*, have been identified and analyzed [11]. Both mechanisms make the building block's linkage tighter. With these two mechanisms, the linkage of building blocks can evolve, and tightly linked building blocks are formed during the process.

Quantifying Linkage. For studying the linkage learning process, a proposed definition for quantifying linkage [11] is adopted. The linkage is the sum of the square of the inter-gene distances of a building block, considering the chromosome to be a circle of circumference 1. The definition is appropriate in that the linkage specifies a measure directly proportional to the probability for a building block to be preserved under exchange crossover.

Linkage skew. Linkage skew, the first linkage learning mechanism [11], occurs when an optimal building block is successfully transferred from the donor onto the recipient. The conditions for an optimal building block to be transferred are (1) the optimal building block resides in the cut segment, and (2) the optimal building block gets expressed before an inferior one does. The effect of linkage skew was found to make linkage distributions move toward higher linkages by eliminating less fit individuals. Linkage skew does not make the linkage of a building block of any particular individual tighter. Instead, it drives the whole linkage distribution to a higher state.

Linkage shift. Linkage shift is the second linkage learning mechanism [11]. It occurs when an optimal building block resides in the recipient and survives a crossover event. For the optimal building block to survive, there cannot be any gene contributing to a deceptive building block transferred. Linkage shift gets the linkage of a building block in an individual higher with deletion of duplicate genetic material caused by injection of exchange crossover. Compared to linkage skew, linkage shift gets linkage of building blocks in each individual higher.

2.4   Time Models

LLGA has been studied on problems containing multiple building blocks in two forms—the uniformly scaled problem and the exponentially scaled problem—not only because of their prevalence in the literature but also because they are abstract versions of many decomposable problems [13]. Uniformly scaled problems resemble those with subproblems of equal importance, and exponentially scaled problems represent those with subproblems of distinguishable importance. As reported previously [11], when the building blocks of a problem are exponentially scaled, LLGA can solve the problem in a linear time function of the number of building blocks. However, when the building blocks are uniformly scaled, LLGA either needs a population size that grows exponentially with the problem size or takes exponential time to converge. In order to explain LLGA's seemingly inconsistent behavior, the following time models were previously proposed to understand how LLGA works.

Tightness Time. Tightness time was proposed [14] to model the time for learning genetic linkage of a single building block of order-k. By extending linkage skew and linkage shift, tightness time was expressed as

$$t'_\ell(\epsilon) = \frac{k^2}{2c_s} \log \frac{\epsilon}{\epsilon_0} , \tag{1}$$

where $\epsilon = 1 - \lambda$, λ is the given linkage, k is the order of the building block, c and c_s are constants. The model shows that tightness time is proportional to the square of the order, k, and to the logarithm of the desired linkage.

Convergence Time. Based on the tightness time model for a single building block, a convergence time model for LLGA was developed [10] to model the LLGA convergence on multiple uniformly scaled building blocks. First, the sequential behavior of LLGA, which indicates that LLGA works on both uniformly scaled and exponentially scaled building blocks one by one, was identified. Then, the *first-building-block model*, which assumes that the convergence time is an accumulation of the time to tighten the first building block, was proposed to model the sequential behavior. Considering the effect and interaction of coexisting uniformly scaled building blocks by using the probability of linkage learning events, the connection between tightness time and the sequential behavior were established. By integrating these models, the LLGA convergence time model for certain desired linkage can be presented as

$$t_C(m, \epsilon) = \left(\frac{k^2(k+1)}{2c_s\sqrt{2\pi}} \log \frac{\epsilon}{\epsilon_0} \right) \sum_{i=1}^{m} \frac{2^i}{i\sqrt{i}} + t_{C0} , \tag{2}$$

where c_s and t_{C0} are constants, m is the number of uniformly scaled building blocks, k is the order of a building block, $\epsilon = 1 - \lambda$, and λ is the desired linkage.

2.5   Limit to LLGA's Competence

Although the proposed LLGA convergence time model [10] describes the way LLGA works on uniformly scaled problems and explains LLGA's inconsistent behavior on problems composed of building blocks of different scalings, it also reveals a critical limit to the competence in LLGA that the time for LLGA to solve uniformly scaled problems grows exponentially with the number of building blocks. Because the parameters involved in the convergence time model are the properties of the problem to solve, little guidance can be obtained from the model for setting the existing algorithmic parameters of LLGA. Therefore, in-

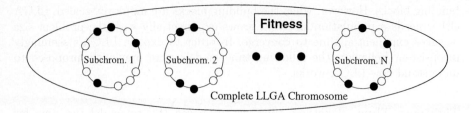

Fig. 3. The structure of a subchromosome is identical to that of an LLGA chromosome. Each subchromosome contains moveable genes, non-coding segments, as well as promoters and is interpreted with probabilistic expression. The union of all subchromosomes belonging to one individual forms a complete LLGA chromosome. The fitness corresponding to the solution obtained from interpreting the complete chromosome is considered the fitness of each subchromosome.

stead of adjusting those algorithmic parameters, another way to improve LLGA's performance on uniformly scaled problems has to be taken. This paper seeks a new design to enhance LLGA based on the insight provided by the convergence time model and takes an initial step to realize the design.

3   Subchromosome Representations

According to the LLGA convergence time model described by Equation (2), one possible way to enhance LLGA's performance on uniformly scaled problems is to modify the chromosome representation used by LLGA such that the number of building blocks, m, is effectively lowered at run time. Thus, the exponential growth of convergence time can be reduced. This section introduces the subchromosome representation to the linkage learning genetic algorithm. The subchromosome representation is first described in detail, and then, the exchange crossover operator for handling subchromosome representations is discussed.

3.1   Chromosome Representation

The subchromosome representation in LLGA separates a LLGA chromosome into several parts, called *subchromosomes*. The structure of a subchromosome is identical to that of an LLGA chromosome. Like an LLGA chromosome described in Sect. 2, a subchromosome contains moveable genes, non-coding segments, as well as promoters and is interpreted with probabilistic expression. The union of all subchromosomes belonging to one individual forms a complete LLGA chromosome. In subchromosome representations, there is no separate fitness measurement for each subchromosome. The fitness that corresponds to the solution obtained from interpreting the complete chromosome is used by all subchromosomes. Figure 3 shows an LLGA chromosome consisting of subchromosomes.

The goal of subchromosome representations is to create a flexible encoding mechanism that makes LLGA chromosomes capable of grouping closely related

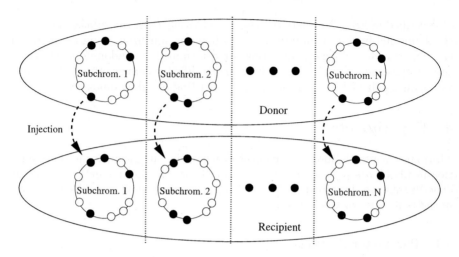

Fig. 4. For a pair of subchromosomes, the exchange crossover operator works as it does on conventional LLGA chromosomes. The operator cuts the genetic materials from the subchromosome of the donor and injects them into the corresponding subchromosome of the recipient. The transferred genetic materials are determined at random for each pair of subchromosomes.

building blocks to form higher-level building blocks in addition to moving genes together on the chromosome to form the first-level building blocks. Similar to linkage learning, the process of forming higher-level building blocks should be integrated with the evolutionary and problem-solving process. The subchromosomes of an LLGA chromosome shown in Fig. 3 are building blocks of the second level. The subchromosome representation can be designed to hierarchically express building blocks of even higher levels, such as the third level, and so on.

However, as an initial step to realize this representation scheme and as a pilot study of the effect of using subchromosomes in LLGA, only subchromosomes of the second level are implemented and examined in this paper. Moreover, the groups of building blocks are pre-defined, and genes on each subchromosome do not migrate to other subchromosomes. Within a subchromosome, genes and non-coding segments are still randomly distributed in initialization as they are on an LLGA chromosome without subchromosomes.

3.2   Exchange Crossover

Due to the adoption of the new representation, the exchange crossover operator is modified to handle subchromosomes. Since in this paper, the subchromosomes are pre-defined and do not exchange genetic materials with one another as discussed in the previous section, for simplicity, after determining the donor and the recipient, the exchange crossover operator works on subchromosomes one by one. For a pair of subchromosomes, one from the donor and the other from the recipient, exchange crossover works as it does on conventional LLGA chromosomes as described in Sect. 2. It cuts the genetic materials from the subchromosome of the donor and injects them into the corresponding subchromosome of the recipient. The transferred genetic materials are determined at random for each pair of subchromosomes. Figure 4 shows how the modified exchange crossover operator works on a pair of LLGA chromosomes consisting of subchromosomes.

4   Experiments

The experiments to observe the effect of using the subchromosome representation in LLGA are presented in this section. First, the parameter settings of the experiments are described in detail. Then, the experimental results are shown in the remainder of this section.

4.1   Parameter Settings

In this paper, trap functions [15] are used for examining the effect of adopting subchromosome representations in LLGA because trap functions provide decent linkage structures among variables, and good linkage is necessary for solving problems consisting of traps. The experiments in this study were done for order-4 traps. An order-k trap function can be described by

$$\text{trap}_k(u) = \begin{cases} u & u = k \\ k - 1 - u & \text{otherwise} \end{cases},$$

where u is the number of ones in the bitstring. In order to simulate the infinite-length chromosome, we let one order-4 building block embedded in 250 genes, including functional and non-functional genes. For example, for five order-4 building blocks, the 20 genes are embedded in a 1250-gene chromosome with 1230 non-functional elements. Table 1 lists all experiments conducted in this paper. The total number of building blocks in one experiment is n_{bb} (the number of building blocks per subchromosome) $\times$ n_s (the number of of subchromosomes). From 2 to 8 building blocks per subchromosome, all conditions for the total number of building blocks less than or equal to 60 are included in the experiments.

The gambler's ruin model [16] is utilized in the present work for population sizing, which can be approximated with the following formula:

$$\text{population size } n = -2^{k-1} \ln(\alpha) \frac{\sigma_{bb} \sqrt{\pi(m-1)}}{d}, \tag{3}$$

where k is the order of building blocks, α is the failure probability, σ_{bb} is the standard deviation of the fitness of a building block, m is the total number of building blocks, and d is the signal, which is adjusted for tournament size $s = 3$ with the equation [16]

$$d' = d + \Phi^{-1}\left(\frac{1}{s}\right) \sigma_{bb},$$

where $\Phi^{-1}(1/s)$ is the ordinate of a unit normal distribution.

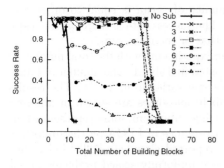

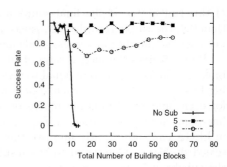

Fig. 5. Results of the experiments with less than or equal to 60 building blocks. The number of building blocks distributed on each subchromosome varies from 2 to 8. "No Sub" indicates LLGA without the subchromosome representation.

Fig. 6. Instead of every pair, only one pair of subchromosomes is randomly chosen for applying exchange crossover to reduce building block disruption. The results indicate that the representation works in the range of these experiments.

Other parameters are set as follows. The crossover rate is 1.0 such that the crossover event always happens. The maximum number of generation is 100,000. The number of promoters on each subchromosome is set to $2m$, where m is the number of building blocks on the subchromosome. Finally, each experiment was repeated with 50 independent runs.

4.2 Experimental Results

For each experiment listed in Table 1, the success rate is calculated according to the results obtained in the 50 independent runs. In this paper, a success is determined by the solution quality. The solution quality is the ratio between the number of correctly solved building blocks in the end of the run and that of the total building blocks in the trial. For example, if in a particular run for solving 20 building blocks, 12 building blocks are correctly solved, the solution quality

Table 1. All experiments conducted in this paper. From 2 to 8 building blocks per subchromosome, all conditions for the total number of building blocks less than or equal to 60 are included in the experiments of this study.

BBs per Subchromosome (n_{bb})	Number of Subchromosomes (n_S)
2	2, 3, 4, 5, 6, ... , 26, 27, 28, 29, 30
3	2, 3, 4, 5, 6, ... , 16, 17, 18, 19, 20
4	2, 3, 4, 5, 6, ... , 11, 12, 13, 14, 15
5	2, 3, 4, 5, 6, 7, 8, 9, 10, 11, 12
6	2, 3, 4, 5, 6, 7, 8, 9, 10
7	2, 3, 4, 5, 6, 7, 8
8	2, 3, 4, 5, 6, 7

of this run is 0.6. If the final solution quality of a run is equal to or greater than 0.9, the run is recorded as a success. The success rate is therefore the ratio between the number of success trials and that of the total runs.

Figure 5 gives the success rates of all the experiments with the total number of building blocks less than or equal to 60 as listed in Table 1. The number of building blocks distributed on each subchromosome varies from 2 to 8. The results for each number of building blocks on subchromosomes are shown in different line-point styles. As shown in Fig. 5, utilizing subchromosome representations in LLGA can significantly improve the performance of LLGA on the uniformly scaled problems. Compared to the results reported elsewhere [12], LLGA with the subchromosome representation can solve uniformly scaled problems about five times larger in terms of the number of building blocks than that can be solved by LLGA without subchromosomes.

Figure 5 also shows that the limit for LLGA with the subchromosome representation to solve uniform scaled problems in terms of the total number of building blocks seems to be around 50. Even with different numbers of building blocks distributed on subchromosomes, no success trial was found among all the experiments with totally 60 building blocks. However, the procedure to apply exchange crossover on subchromosomes in these experiments causes serious building block disruption, as it does in LLGA without subchromosomes. While the original design of LLGA prevents us from lowering the disruption rate and maintaining the mixing rate at the same time, LLGA with the subchromosome representation provides us a viable way to appropriately adjust the probability for applying the operator. Therefore, the crossover operator is slightly modified as follows. Instead of every pair, only one pair of subchromosomes is now randomly chosen for applying exchange crossover to reduce building block disruption. The previous experiments were repeated for $n_{bb} = 5$ and 6 to check the effect of adjusting the probability. The results are shown in Fig. 6 and indicate that the representation works in the range of these experiments.

5   Future Work

These results are tantalizing in that parallel evolution of linkage of large number of gene groupings has been demonstrated provided the appropriate genes are associated in the same linkage group. The key challenge left is to develop mechanisms to permit or encourage the evolution of these proper associations. Different mechanisms can be imagined for this purpose, and the potential of each is briefly outlined in the following paragraphs.

Gene migration: Gene migration moves genes among subchromosomes within one chromosome. Proper associations can be achieved through gene migration and favored by the the evolutionary process.

Gene duplication or redundant genes: Redundant genes can provide higher probabilities to form correct gene groups or clusters within subchromosomes.

Adaptive expression: Adaptive expression can resolve the conflicts caused by gene duplication and promote those identified building blocks.

Before running off to do more mass quantities of computation, however, we should think carefully about the key lessons of this paper. In going from the limited results of Fig. 5 to the much better results of Fig. 6, we recall that the primary difference was the limited amount of mechanical disruption that was permitted in the modified subchromosome crossover. In looking back over all studies of LLGA to this point, it is clear that these procedures can only tolerate a certain amount of *rearrangement disruption*. Large amounts of fitness variance are not problematic because they can be overcome through larger populations; however, attempts to move too much material around have always caused a combinatorial overload that cannot be sorted out. In designing mechanisms to move genes around either physically or virtually, we must recognize that the overall structure can assimilate only so much movement at any one time. Attention to this should guide the design of mechanisms to realize the potential of LLGA.

6   Summary and Conclusions

This paper started with a brief review of the linkage learning genetic algorithm, including the chromosome representation, exchange crossover, linkage learning mechanisms, and time models. The subchromosome representation was developed and employed in the linkage learning genetic algorithm for effectively lowering the number of building blocks to escape from the limit implied by the convergence time model. An initial step to realize subchromosome representations in the linkage learning genetic algorithm was taken in this work. The preliminary experimental results of using subchromosomes in the linkage learning genetic algorithm indicated that the proposed scheme can improve the performance of the linkage learning genetic algorithm on uniformly scaled problems.

In addition to showing that the subchromosome representation helps the linkage learning genetic algorithm to solve larger uniformly scaled problems, the initial step for implementing the proposed representation in the current work also leads a possible way in making the linkage learning genetic algorithm capable of incorporating prior linkage information. With the use of subchromosomes, the distribution of genes, non-coding segments, and building blocks can be determined according to the available linkage information of the problem. In the linkage learning genetic algorithm without subchromosomes, utilizing prior linkage information is extremely difficult if not impossible. Overall, the results reveal a promising path for achieving scalable genetic linkage learning techniques.

Acknowledgments. The work was sponsored by the Air Force Office of Scientific Research, Air Force Materiel Command, USAF, under grant F49620-03-1-0129. The US Government is authorized to reproduce and distribute reprints for Government purposes notwithstanding any copyright notation thereon.

The views and conclusions contained herein are those of the authors and should not be interpreted as necessarily representing the official policies or endorsements, either expressed or implied, of the Air Force Office of Scientific Research or the U.S. Government.

References

1. Thierens, D., Goldberg, D.E.: Mixing in genetic algorithms. *Proceedings of the Fifth International Conference on Genetic Algorithms (ICGA-93)* (1993) 38–45

2. Goldberg, D.E., Deb, K., Thierens, D.: Toward a better understanding of mixing in genetic algorithms. *Journal of the Society of Instrument and Control Engineers* **32** (1993) 10–16

3. Kargupta, H.: The gene expression messy genetic algorithm. *Proceedings of the 1996 IEEE International Conference on Evolutionary Computation* (1996) 814–819

4. Munetomo, M., Goldberg, D.E.: Linkage identification by non-monotonicity detectio for overlapping functions. *Evolutionary Computation* **7** (1999) 377–398

5. Baluja, S.: *Population-based incremental learning: A method for integrating genetic search based function optimization and competitive learning.* Tech. Rep. No. CMU-CS-94-163, Carnegie Mellon University, Pittsburgh, PA (1994)

6. Mühlenbein, H., Mahnig, T.: FDA - a scalable evolutionary algorithm for the optimization for the optimization of additively decomposed functions. *Evolutionary Computation* **7** (1999) 353–376

7. Pelikan, M., Goldberg, D.E., Cantú-Paz, E.: BOA: The bayesian optimization algorithm. *Proceedings of Genetic and Evolutionary Computation Conference 1999 (GECCO-99)* (1999) 525–532

8. Levenick, J.R.: Metabits: Generic endogenous crossover control. *Proceedings of the Sixth International Conference on Genetic Algorithms (ICGA-95)* (1995) 88–95

9. Smith, J., Fogarty, T.C.: Recombination strategy adaptation via evolution of gene linkage. *Proceedings of the 1996 IEEE International Conference on Evolutionary Computation* (1996) 826–831

10. Chen, Y.-p., Goldberg, D.E.: Convergence time for the linkage learning genetic algorithm. *Proceedings of the 2004 Congress on Evolutionary Computation (CEC2004)* (2004) N/A (To appear).

11. Harik, G.R.: *Learning gene linkage to efficiently solve problems of bounded difficulty using genetic algorithms.* PhD thesis, University of Michigan, Ann Arbor, MI (1997)

12. Chen, Y.-p., Goldberg, D.E.: Introducing start expression genes to the linkage learning genetic algorithm. *Proceedings of the Seventh International Conference on Parallel Problem Solving from Nature (PPSN VII)* (2002) 351–360

13. Goldberg, D.E.: *The Design of Innovation: Lessons from and for Competent Genetic Algorithms.* Kluwer Academic Publishers (2002)

14. Chen, Y.-p., Goldberg, D.E.: Tightness time for the linkage learning genetic algorithm. *Proceedings of Genetic and Evolutionary Computation Conference 2003 (GECCO-2003)* (2003) 837–849

15. Deb, K., Goldberg, D.E.: Analyzing deception in trap functions. *Foundations of Genetic Algorithms 2* (1993) 93–108

16. Harik, G., Cantú-Paz, E., Goldberg, D.E., Miller, B.L.: The gambler's ruin problem, genetic algorithms, and the sizing of populations. *Proceedings of the 1997 IEEE International Conference on Evolutionary Computation* (1997) 7–12

Interactive One-Max Problem Allows to Compare the Performance of Interactive and Human-Based Genetic Algorithms

Chihyung Derrick Cheng and Alexander Kosorukoff

University of Illinois at Urbana-Champaign, Urbana, Illinois 61801
{cdcheng,kosoruko}@uiuc.edu

Abstract. Human-based genetic algorithms (HBGA) use both human evaluation and innovation to optimize a population of solutions (Kosorukoff, 2001). The novel contribution of HBGAs is an introduction of human-based innovation operators. However, there was no attempt to measure the effect of human-based innovation operators on the overall performance of GAs quantitatively, in particular, by comparing the performance of HBGAs and interactive genetic algorithms (IGA) that do not use human innovation. This paper shows that the mentioned effect is measurable and further focuses on quantitative comparison of the efficiency of these two classes of algorithms. In order to achieve this purpose, this paper proposes an interactive analog of the one-max problem, suggests human-based innovation operators appropriate for this problem, and compares convergence results of HBGA and IGA for the same problem.

1 Introduction

Interactive genetic algorithms (IGA) had extended evolutionary computation (EC) to the areas where computational evaluation is not possible. This extension was made through the use of selection that is based on human evaluation (Herdy, 1996). Nowadays there is a growing field of IGA applications from music composition to architectural design (Takagi, 2001). A human-based genetic algorithm (HBGA) did a similar thing by extending EC even further to the area where it is hard or impossible to find a good representation and usable computational innovation operators (Kosorukoff, 2000). The task of searching solutions expressed in natural language is one of such problems. It could not be approached by IGAs simply because we do not know how to do computational recombination in a natural language. HBGAs have solved this problem by outsourcing innovation operators to humans in the same way as IGAs did earlier with outsourcing evaluation. However, so far it has remained unclear if HBGAs and human-based innovation operators are strictly limited to such areas where human recombination is the only option or they can play a role in a field where IGAs are applied. In this paper, we suggest an example of a problem for which both HBGAs and IGAs are applicable. Although it is not very useful practically,

K. Deb et al. (Eds.): GECCO 2004, LNCS 3102, pp. 983–993, 2004.

this problem makes a good simple benchmark for both IGAs and HBGAs. Hence it can play a role of the one-max problem for genetic algorithms using human interaction. In this paper, we describe this problem in detail and perform a set of experiments to compare the speed of convergence of those algorithms.

The rest of this paper is organized as follows: section 2 reviews recent research on HBGAs, section 3 describes the application, section 4 describes the problem we use for experiments, section 5 describes experimental settings, and section 6 contains results of experiments. Finally, we conclude with discussion of the results in section 7.

2   Background

Free Knowledge Exchange (FKE) project (3form.com, 1998) was the first application of a technique now known as *human-based genetic algorithm* (HBGA) for evolutionary knowledge management and collaborative problem solving. The main goal of the project was efficient knowledge discovery, sharing, and innovation. The technology got its current name and was systematically analyzed after it had received several successful evaluations from its users (Kosorukoff, 2000; Kosorukoff, 2001). The idea of HBGA was mainly borrowed from a business practice of outsourcing where operations that are not a part of the core competence of an organization are delegated to some external agents which are competent enough to choose their own method of executing those operations such that the purpose of the operations can be accomplished.

Kosorukoff and Goldberg (2002) discuss HBGAs among two approaches to evolutionary organization and innovation. Defaweux et al. (2003) discuss an application of HBGA for storytelling and developing of marketing slogans, examines similarities between HBGA and a traditional GA. Based on experiments with small groups of people, the authors show that HBGA stimulates creativity and consensus, concluding that it has a great potential in fields like marketing, industrial design, and creative writing. Goldberg et al. (2003) discuss HBGA as a part of distributed innovation support system.

3   HBGA for Color Fitting

It is appropriate to say that the first application of HBGA has unintentionally limited the range of its further applications to the tasks connected with evolution of text messages expressed in some natural language. However, in this section we will try to show that actually there is no such limitation. Our simple example will be from a different domain which is classical to an IGA, but new to HBGA: finding a color according to a user preference.

We do not have a problem with finding a genetic representation for color. The most common way to represent color is a triple of R, G, B values, where 8 bits are allocated for each component, making 24 bits in total (Foley 1990; Hunt, 1992). We will use this standard representation for our genetic coding. Actually, RGB colormap is not the best for the purpose of designing a practical

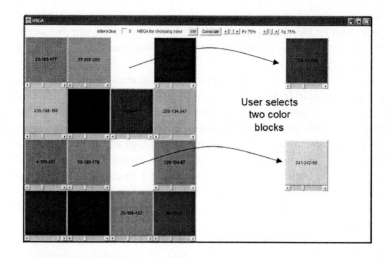

Fig. 1. An interface of HBGA for the interactive one-max problem

application, we would use HSV or LUV for this purpose (Douglas, 1996; Schwarz, 1987). However, since colormap conversions is far from the topic of this paper the performance was sacrificed for simplicity and clarity. Here is a straightforward example how purely white color $(R, G, B) = (255, 255, 255)$ is coded:

$$11111111\ 11111111\ 11111111$$

It is clear that we can use a classic set of genetic operators to work with this representation.

User interface is an important part of any interactive program. Repetitive use of the same interface in interactive evolutionary computation makes its design especially important, so we describe it in detail. A user is presented with 16 (population size) of colors (Figure 1). There are two panels for this purpose, the left one is for unselected colors, the right one is for selected colors. Each panel is divided into 4 by 4 grid and each cell color corresponds to an individual in the population. The user selects color by clicking on the left panel with a mouse that makes the selected cell move into the right panel. The cell in the right panel can be deselected by another click of a mouse and moved back into the left panel. This is all controls that IGA needs. HBGA, however, should allow user to create new colors in addition to those created by usual mutation/crossover operators (Goldberg, 1998) over the binary representation described earlier.

There are many ways to create an innovation interface. FKE application (Kosorukoff, 2000) uses cut and paste controls to allow for human-based recombination of textual information. This way would work here by allowing human to modify genotype directly, though this is not a natural way for people to modify color. In a practical HBGA application, different controls to adjust such parameters as brightness, hue, and saturation would be present, as well as a control element providing the opportunity to mix two or more colors as artists usually do. However, some of these operators when applied to our task will allow convergence in one or two iterations and this is not good for our research purpose of algorithm comparison. Therefore, we had chosen operators from classical GAs for research purpose only.

Before we will proceed with description of human-based innovation operators, we have to clarify how their computational peers are performed in our algorithm. There is essentially no difference in crossover, except that parents are stored together with individuals because they will be used later. The computational mutation operation is the following. The program will first select a parent from the population, then another 24-bit binary string (mutation mask) will be generated randomly such that the probability of 1 in each position equals to the probability of mutation p_m. A mutant is generated by XOR operation over two strings. Both parents and a mutation mask are stored with an individual for further use.

For human-based operators, each color cell in our HBGA has a slider associated with it. If a cell represents a result of one-point crossover, the slider can be used to move the locus of crossover among 23 possible positions. If a cell represents a result of mutation, the slider can rotate the mutation mask (a binary string in which each locus of mutation is marked as 1, other loci are 0) to the right or to the left depending where slider is moved. These operations allow human user to modify the result of genetic operators, so innovation in this algorithm is a collaborative result of human-computer interaction.

There are two modes of operation of the user interface. When we are in the IGA mode sliders are not shown and we are only allowed to select. When the HBGA mode is enabled we are allowed both to select and to modify individual colors in the population. According to our experience, it does not take many generations to play with this toy and converge to the colors of our preference. This shows that both algorithms are working but we want something more than just user satisfaction, something that allows for a more precise comparison.

4   Interactive One-Max Problem

In a usual IGA or HBGA application each user follows her own goal determined by her preferences. However, in order to get some measurable results we need to fix the goal as we cannot compare different preferences which can require different amount of effort to achieve them. We had borrowed our goal from the one-max problem, so we require to converge to the white color $(255, 255, 255)$ or:

$$11111111\ 11111111\ 11111111$$

Despite of its apparent similarity with the one-max problem, the interactive one-max problem has also some important differences:

1. People rarely can distinguish colors that are close to each other in the color space, especially if they are not placed close to each other during comparison. This is unlike the usual notion of fitness which distinguish very fine grades of solution quality.

2. The actual form of fitness function is not known to us. One user can follow the brightness of colors, another can follow hue and saturation. Many strategies of achieving the goal are possible. The efficiency of these strategies will not be the same, but for the purpose of this paper we are interested only in the averaged performance and how it depends on the type of the algorithm that was used.

We have to address these two issues and we do it in the following two subsections.

4.1   Defining White

We address the first difference by allowing a certain range of colors to be "white" as determined by a margin. The RGB value of pure white color is (255, 255, 255). We relax our requirement considering any color with all components greater than 245 to be white. The ability of people to distinguish colors is not the same and we wanted that most of the users could achieve the goal with reasonable amount of effort. The range of colors that fall within a margin is generally dependent on the choice of progress measure. Criteria will be described shortly, but as a general rule we treat color as white if it is not inferior to (245,245,245) according to our selected criteria.

4.2   Progress Criteria

In our case, we do not have a direct access to a fitness function that humans use when doing their selections. However, it is convenient to have some value similar to fitness that we can track to know how close we are to achieve the goal. We have three candidates for this purpose. They are brightness (M1), Euclidean metric (M2), and minimum component metric (MS):

$$M1(R, G, B) = L1((255, 255, 255), (0, 0, 0)) - L1((255, 255, 255), (R, G, B))$$
$$= 255 * 3 - ((255 - R) + (255 - G) + (255 - B)) = R + G + B$$

$$M2(R, G, B) = L2((255, 255, 255), (0, 0, 0)) - L2((255, 255, 255), (R, G, B))$$
$$= 255 * \sqrt{3} - \sqrt{(255 - R)^2 + (255 - G)^2 + (255 - B)^2}$$

$$MS(R, G, B) = LS((255, 255, 255), (0, 0, 0)) - LS((255, 255, 255), (R, G, B))$$
$$= 255 - \max(255 - R, 255 - G, 255 - B) = min(R, G, B)$$

We have no preference among them at this point, so we will use all three in our experiments. The identification of the fitness function people actually use in our experimental task could be a topic of another research.

5   Experiment Settings

We compare IGA and HBGA in two categories: generational and steady state. Each user is asked to achieve the white color through a sequence of selections only (IGA), and through a sequence of selections and modifications (HBGA).

The program will generate 16 color cells at each generation and present them to a user. Probability of computational crossover and mutation were as usual $p_x = 0.75$, $p_m = 1/24$. Another parameter F_p determines the probability of selection of preferred individuals (the right panel) for reproduction, and thus determines the selective pressure. In our experiment $F_p = 0.9$, i.e. colors picked by a user have 90% chance of being selected for reproduction, while other colors have only 10% chance.

We invited 10 people to compare the algorithms and use intra-subject experimental design that suggests that each person tries to achieve white color using four algorithms in the following sequence:

- IGA-Generational
- HBGA-Generational
- IGA-SteadyState
- HBGA-SteadyState

None of the users had any background in evolutionary computation and supposedly did not understand the mechanism behind evolutionary programs they are using. We had recorded every color in each generation, so that the best fit in brightness, Euclidean and MS metric can be calculated. Then we had averaged experimental results over ten replications.

6   Results

Figures 2-7 present the graphs of progress metrics: brightness (M1), Euclidean metric (M2), and minimum component metric (MS). For each metric first IGA-Generational and HBGA-Generational, then IGA-Steady State and HBGA-Steady State are compared. In each plot, there are two horizontal lines. The upper one represents the optimum values. The pure-white color $(255, 255, 255)$ should behave like this line. Similarly, the bottom one represents the margin cut-off. It is based on our minimum requirement of white color $(245, 245, 245)$. The experiment is considered successful if the experiment sample passes through the cut-off margin. The same information in summarized form is presented in Table 1.

Overall, the curves of four algorithms increase over the generations and show convergence. The curves for HBGA usually show an advantage starting from the first generation. Both HBGA and IGA are able to accomplish the task successfully. However, the number of generations spent differs in about 3 times favoring HBGA.

Measuring convergence in numbers of iterations does not give a complete picture. Time comparison in Table 2 adds the missing part. A generation of our

Table 1. Performance comparison for different progress metrics

Progress metric	Generational			Steady State		
	HBGA	IGA	HBGA Speedup	HBGA	IGA	HBGA Speedup
M1	4	13	3.25	4	13	3.25
M2	6	14	2.33	5	14	2.8
MS	11	27	2.45	6	19	3.17

Table 2. Time performance comparison. Generations are shown according to MS criterion

	Generational			Steady State		
	HBGA	IGA	HBGA Speedup	HBGA	IGA	HBGA Speedup
Generation time (s)	9.2	5.5	-	14.2	6.5	-
Generations	11	27	-	6	19	-
Time to converge (s)	103.8	147.6	1.42	92.3	125.7	1.36

HBGA is approximate 2 times slower than a generation of an IGA. The higher speed of convergence in terms of generations, however, compensates the slower time of each HBGA generation, so our experiments show that HBGA has an advantage in time to converge as well as the number of generations to converge.

The time of each generation in HBGA is dependent on a strategy that a particular user chooses when using HBGA. It is clear that one can use HBGA exactly as IGA just by ignoring all color modification sliders, in this case the time of each generation and the number of generations to converge should be equal to those of IGA. If one will use color modification heavily, then the time of each generation will likely to grow, but the number of generations needed to converge decreases. We didn't investigate the influence of different users' strategies in this research.

There is a discrepancy in the average time of the same algorithm in generational and steady state mode, which is quite high and unexpected. There are two possible reasons for those discrepancies. One possible reason connected with the sequence of our experiments and human factors (Berk, 1982), in particular, the fact that user fatigue increases over time, so we see slower response on successive experiments. We cannot reliably establish this fact from our set of 40 experiments which were not designed to take user fatigue into account and have no meaning to measure it.

7   Discussion

This paper has suggested a way to measure the performance of human-based genetic algorithms quantitatively. We compared the results of these two algorithms in their generational and steady-state implementations. The experimental results

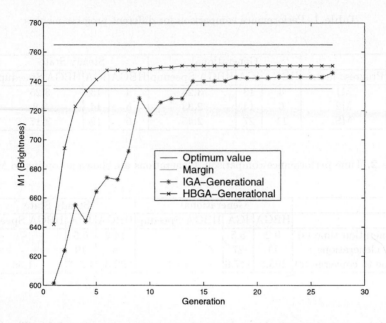

Fig. 2. Brightness (M1) of the brightest individual in the population for generational type of IGA and HBGA. The speedup factor of HBGA is 3.25

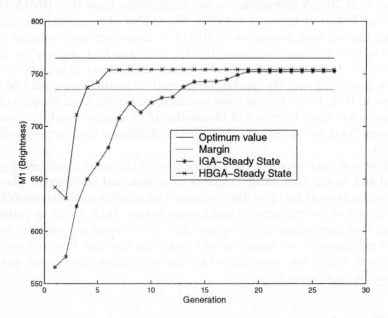

Fig. 3. Brightness (M1) of the brightest individual in the population for generational type of IGA and HBGA. The speedup factor of HBGA is 3.25

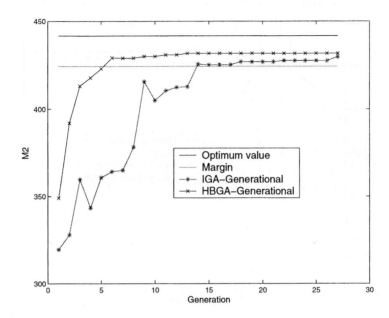

Fig. 4. Measure M2 for generational type of IGA and SGA. The speedup factor of HBGA is 2.33

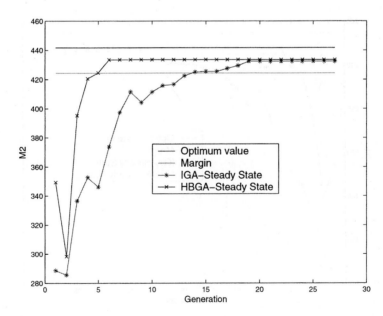

Fig. 5. Measure M2 for steady state type of IGA and HBGA. The speedup factor of HBGA is 2.8

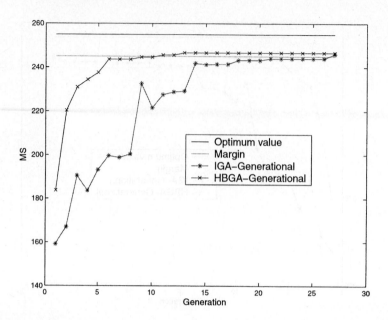

Fig. 6. Measure MS for generational type of IGA and HBGA. The speedup factor of HBGA is 2.45

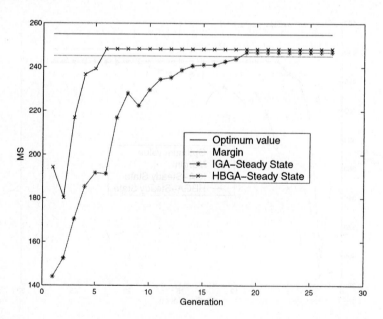

Fig. 7. Measure MS for steady state type of IGA and HBGA. The speedup factor of HBGA is 3.17

strongly support the idea that using human-based innovation operators can be advantageous even in the area where we actually can perform all innovations computationally. Algorithm using human-based innovation operators requires 2-3 times less generations to converge. Supplemental materials for this research can be found online at
http://www.derrickcheng.com/Project/HBGA/index.html.

Acknowledgements. The authors would like to thank Alexander Kirlik and Sarah Miller for sharing their expertise in human factors and experimental design. We also thankful to all people who helped us to evaluate the performance of the algorithms examined in our experiments. Finally, we want to express our thankfulness to the reviewers of GECCO for their valuable suggestions.

References

1. Berk, T., Brownston L. and Kaufman, A. (1982). A human factors study of color notation systems for computer graphics. *Communications of the ACM 25, 8*, pp. 547-550.
2. Defaweux, A., Grosche, T., Karapatsiou, M., Moraglio, A., Shenfield, A. (2003). Automated Concept Evolution. Tech Report
3. Douglas S., Kirkpatrick T. (1996). Do color models really make a difference?. *Proceedings of CHI-1996.*
4. Foley, J., van Dam, A., Feiner, S. and Hughes, J. (1990). Computer Graphics: Principles and Practice, 2nd. ed. Addison Wesley.
5. Goldberg, D., Welge, M., Llora', X. (2003). Distributed Innovation and Scalable Collaboration In Uncertain Settings. IlliGAL Report No. 2003017.
6. Herdy, M. (1996). Evolution strategies with subjective selection. In *Parallel Problem Solving from Nature, PPSN IV*, Volume 1141 of *LNCS*, pp. 22-31.
7. Hunt, R. (1992). Measuring Color, 2nd. ed. Ellis Horwood.
8. Kosorukoff, A. (2000). Human based genetic algorithm. Online at
   http://www.hbga.org/hbga.html.
9. Kosorukoff, A. (2001). Human based genetic algorithm. *IEEE Transactions on Systems, Man, and Cybernetics, SMC-2001*, 3464-3469.
10. Kosorukoff, A., & Goldberg, D. (2002). *Genetic algorithm as a form of organization, Genetic and Evolutionary Computation Conference, GECCO-2002.*
11. Takagi, H. (2001). Interactive Evolutionary Computation: Fusion of the Capacities of EC Optimization and Human Evaluation. *Proceedings of the IEEE 89, 9*, pp. 1275-1296.

Polynomial Approximation of Survival Probabilities Under Multi-point Crossover

Sung-Soon Choi and Byung-Ro Moon

School of Computer Science and Engineering,
Seoul National University,
Seoul, 151-742 Korea
{sschoi,moon}@soar.snu.ac.kr

Abstract. We propose an analytic approach to approximate the survival probabilities of schemata under multi-point crossover and obtain its closed form. It gives a convenient way to mathematically analyze the disruptiveness of multi-point crossover. Based on the approximation, we describe a geometric property of the survival probability under multi-point crossover and show the relationship between the survival probability and the distribution of the specific symbols in schemata.

1 Introduction

Crossover is a major operator that plays an important role in the design of genetic algorithms. Discussion on the effect of crossover dates back to the Schema Theorem [9] which is one of the milestone theorems in the area of genetic algorithms. The theorem says that the probability that a schema survives through generations depends on not only its fitness but also the disruptiveness of crossover. So considerable attention has been given to estimating the disruptiveness of crossover.

A variety of crossover operators have been proposed and the disruptiveness of crossovers has also been a hot research topic. For one-dimensional chromosomes, multi-point crossover and uniform crossover are two representative crossovers that have been much studied in the area of genetic algorithms [15] [11]. Recently, several crossovers for non-linear chromosomes were proposed and studies on the disruptiveness of those crossovers were conducted [7] [1] [4] [13].

In uniform crossover, the allele at any position in an offspring is determined by the allele of the first parent with probability p or by the allele of the second parent with probability $1-p$. One property of this crossover is that it is simple to estimate the disruptiveness of the crossover; the survival probability for any order schema is easily obtained in a closed form. Based on this, Syswerda provided an initial analysis of the disruptive effects of uniform crossover for the case of $p = 0.5$ [15].

There have been more studies for the disruptiveness of multi-point crossover. De Jong [10] first observed that the defining length is not necessarily a dominating factor of a schema's survival probability when multi-point crossovers are used. He provided an exact expression for the survival probability for the

K. Deb et al. (Eds.): GECCO 2004, LNCS 3102, pp. 994–1005, 2004.
© Springer-Verlag Berlin Heidelberg 2004

2nd-order schemata. Later, De Jong and Spears [11] extended this to provide a detailed analysis of the survival probabilities of higher-order schemata. Bui and Moon [3] [5] proposed a new schema model which is convenient for dealing with the distribution of specific symbols within schemata. In the model, Bui and Moon investigated the characteristics of schemata that are prone to have good survival probabilities under multi-point crossovers. They also provided an exact expression for the survival probability for multi-point crossover.

Despite these results, the closed form for the survival probability of a schema under multi-point crossover has not been discovered. Due to this, the researches for multi-point crossover depended primarily on empirical analyses. The objective of this paper is to introduce an analytic approach to approximate the survival probabilities of schemata under multi-point crossover and obtain a closed form for the survival probability. As stated below, it gives a convenient way to mathematically analyze the disruptiveness of multi-point crossover and, consequently, leads to a deeper understanding of multi-point crossover.

The rest of this paper is organized as follows. In Section 2, we present some previous works on the disruptiveness of crossover, mainly, multi-point crossover. In Section 3, we propose an approach to approximate the survival probability of a schema under multi-point crossover to a polynomial related to the distribution of the specific symbols in the schema. Based on such an approximation, in Section 4, we show the convex property of the survival probability and prove the relationship between the survival probability of a schema and the distribution of the specific symbols in the schemata. Finally, we make our conclusions in Section 5.

2   Survival Probabilities

At first, we summarize basic terminologies of genetic algorithms for the completeness of the paper. A *chromosome* is a sequence of gene values. Each gene has a value from an alphabet $\mathcal{A}$. A *schema* of a chromosome of length n can be represented as an n-tuple $< s_1, s_2, \ldots, s_n >$ where $s_i \in \mathcal{A} \cup \{*\}$ for $i = 1, 2, \ldots, n$. In a schema, the symbol "$*$" represents don't-care positions and non-$*$ symbols (called *specific symbols*) specify defining positions of the schema and their corresponding gene values. The *distance* between two genes is the distance between their positions. The *defining length* of a schema is defined to be the distance from the leftmost specific symbol to the rightmost specific symbol in that schema. The *order* of a schema is the number of specific symbols in the schema.

The analysis of schemas' survival probabilities based on defining lengths works well on 1-point crossover models, but does not generally work well on multi-point crossover models. De Jong [10] found an exact expression for the survival probabilities of 2nd-order schemata under multi-point crossover and observed that the survival probabilities of 2nd-order schemata are not necessarily affected by their defining lengths when multi-point crossovers are used. Extending this, De Jong and Spears [11] investigated the relationship between the survival probabilities and the defining lengths of higher-order schemata. (See [14] for more detailed discussion.) Their key observation is that a schema is not disrupted when an even number (including 0) of crossover points fall between every pair of adjacent specific symbols.

Let $s_0, s_1, \ldots, s_{r-1}$ be the specific symbols of an rth-order schema from left to right. Let d_i be the distance between s_0 and s_i for $i = 1, 2, \ldots, r-1$ and let n be the length of the chromosome. They provided a recursive equation to calculate $P_{k,even}(r)$, the probability that an even number of crossover points fall between each of the defining positions of the rth-order schema (consisting of s_0 through s_{r-1}) by a k-point crossover as follows:

$$P_{k,even}(r) = \sum_{i=0}^{\lfloor k/2 \rfloor} \binom{k}{2i} \left(\frac{d_{r-1}}{n}\right)^{2i} \left(\frac{n - d_{r-1}}{n}\right)^{k-2i} P_{2i,even}(r-1). \tag{1}$$

The above equation assumes that the crossover points are independent of one another. They are actually dependent because no two crossover points can fall onto the same position. But this approximation causes little harm as long as $k \ll n$, which is true in most cases.

Bui and Moon [3] [5] investigated the relationship between the inner structures of schemata and their survival probabilities. They concentrated on the specific-symbol clusters and introduced a new type of schema, called *clustered schema* or *c-schema*. A *c*-schema is defined as $D_0 C_1 D_1 \cdots C_q D_q$, where $C_i \in \mathcal{A}^+, i = 1, \ldots, q$, $D_i \in \{*\}^+, i = 1, \ldots, q-1$, and $D_0 D_q \in \{*\}^*$. Let the length of the string D_i be $|D_i|$. If we define $P_k(D_0 C_1 D_1 \cdots C_q D_q)$ to be the probability that the *c*-schema $D_0 C_1 D_1 \cdots C_q D_q$ is not disrupted by a k-point crossover, then the following holds [3] [5]:

$$P_k(D_0 C_1 D_1 \cdots C_q D_q) = \sum_{i_1 + \cdots + i_q \leq \lfloor k/2 \rfloor} \frac{\binom{|D_1|+1}{2i_1} \cdots \binom{|D_{q-1}|+1}{2i_{q-1}} \binom{|D_0 D_q|}{k-2(i_1+\cdots+i_{q-1})}}{\binom{n-1}{k}}. \tag{2}$$

Based on this equation, Bui and Moon suspected that the survival probability of a schema considerably depends on the distribution of the specific symbols. They empirically supported this to show that, the more clustered the specific symbols of a schema are, the higher survivial probability it has.

Although the equations (1) and (2) measure the survival probability of a given schema, they involve complicated summations of binomial coefficients. It seems to be impossible to mathematically handle the equations to derive some other results. On the other hand, as stated above, it is easy to obtain the closed form for the survival probability of any order schema under uniform crossover. If we denote the probability parameter by p, the survival probability of an rth-order schema under uniform crossover, $P_u(r)$, is of a simple form as follows:

$$P_u(r) = p^r + (1 - p)^r.$$

In the next section, we propose a way to approximate the survival probability of a schema under multi-point crossover and obtain its closed form.

Note that a schema can survive even when the above conditions for the equations (1) and (2) are not satisfied if all lost specific symbols in one parent are accidently recovered by the other parent. Our results can be generalized to handle the cases in the same manner as in [11]. In this discussion, we ignore such cases for simplicity.

3   Polynomial Approximations

3.1   Uniform Convergence and Linkage Distribution

Let $\langle f_n \rangle$ be a sequence of functions defined on a set X and with range in $\mathbb{R}$. We say the sequence $\langle f_n \rangle$ *uniformly converges* to the function f on the set X if and only if for each $\epsilon > 0$ there is a number N independent of x such that

$$|f_n(x) - f(x)| < \epsilon \text{ for all } x \in X \text{ and all } n > N.$$

Uniform convergence differs from ordinary pointwise convergence in that the integer N does not depend on x, although naturally it depends on ϵ. Roughly speaking, uniform convergence means that all the values of f_n on X converge to the value of f on X in the same rate, as n increases. More strictly, the uniform convergence of $\langle f_n \rangle$ to f on X implies that $\sup_X |f_n - f| \to 0$ as $n \to \infty$, and vice versa. We refer to [12] for more details.

Let $s_0, s_1, \dots, s_q$ be the specific symbols of a $(q+1)^{th}$-order schema from left to right and n be the chromosome length. We denote the distance between two symbols s_i and s_j by $d(s_i, s_j)$. Let $x_i = \frac{d(s_{i-1}, s_i)}{n-1}, i = 1, \dots, q$, which is the fraction of the distance between the adjacent specific symbols s_{i-1} and s_i over the chromosome length minus one. Then, $(x_1, \dots, x_q)$ represents the relative distances of the specific symbols of the $(q+1)^{th}$-order schema. Note that, under multi-point crossover, the survival probability of a schema depend little on the absolute positions of the specific symbols; it depends highly on their relative positions. So the notation $(x_1, \dots, x_q)$ is sufficient to represent the $(q+1)^{th}$-order schema in analyzing the survival probability. We call $(x_1, \dots, x_q)$ the *linkage distribution* of the $(q+1)^{th}$-order schema. We denote $x = \sum_{i=1}^{q} x_i$ and call it the *linkage sum* of the schema. The linkage sum of a schema is the defining length of the schema normalized by the chromosome length. In representing a 2nd-order schema, we simply use the notation x instead of x_1.

We denote by $P_{k,n}(x_1, \dots, x_q)$ the survival probability of a $(q+1)^{th}$-order schema with linkage distribution $(x_1, \dots, x_q)$ in a chromosome of length n under k-point crossover. For a 2nd-order schema, we simply use the notation $P_{k,n}(x)$ as above. Note that $P_{k,n}(x_1, \dots, x_q)$ is a function of the chromosome length n. In this section, we show that the sequence $\langle P_{k,n}(x_1, \dots, x_q) \rangle$ with respect to n uniformly converges to a polynomial of $x_1, \dots, x_q$ on the set $\{(x_1, \dots, x_q) | 0 < x = \sum_{i=1}^{q} x_i \leq 1\}$ that contains the linkage distributions of the $(q+1)^{th}$-order schemata. This supports that the survival probability may be approximated to the polynomial for sufficiently large n.

3.2 2nd-Order Schemata

Let $\delta(i) = 1$ if $i = 0$, and $\delta(i) = 0$ otherwise. Then the following holds.

Theorem 1. *The sequence $\langle P_{k,n}(x) \rangle$ uniformly converges on the set $\{x|0 < x \le 1\}$ to the kth-order polynomial of x, $P_k(x)$, with the form*

$$P_k(x) = \sum_{i=0}^{k} c_i x^i,$$

where $c_i = (-1)^i \binom{k}{i} 2^{i-1+\delta(i)}$ for $i = 0, 1, 2, \dots$.

Proof: Let d be the defining length of a 2nd-order schema. Then,

$$P_{k,n}(x) = P_{k,n}(\frac{d}{n}) = \sum_{i=0}^{\lfloor k/2 \rfloor} \binom{d}{2i} \binom{n-1-d}{k-2i} / \binom{n-1}{k}.$$

The numerator in the right hand side can be rewritten as follows:

$$\sum_{i=0}^{\lfloor k/2 \rfloor} \binom{d}{2i} \binom{n-1-d}{k-2i} = \sum_{i=0}^{\lfloor k/2 \rfloor} \frac{d^{2i}}{2i!} \frac{(n-d)^{k-2i}}{(k-2i)!} + O(n^{k-1}).$$

Dividing both sides by $\binom{n-1}{k}$,

$$\sum_{i=0}^{\lfloor k/2 \rfloor} \binom{d}{2i} \binom{n-1-d}{k-2i} / \binom{n-1}{k} = \sum_{i=0}^{\lfloor k/2 \rfloor} \frac{k!}{2i!(k-2i)!} \frac{d^{2i}(n-d)^{k-2i}}{(n-1)^k} + O(\frac{1}{n}).$$

So, we get

$$P_{k,n}(x) = \sum_{i=0}^{\lfloor k/2 \rfloor} \binom{k}{2i} x^{2i}(1-x)^{k-2i} + O(\frac{1}{n}), \tag{3}$$

which means that $P_{k,n}(x)$ uniformly converges to a kth-order polynomial of x. From the equation, the coefficient of x^i is

$$(-1)^i \sum_{j=0}^{\lfloor i/2 \rfloor} \binom{k}{2j} \binom{k-2j}{i-2j} = (-1)^i \sum_{j=0}^{\lfloor i/2 \rfloor} \binom{k}{i} \binom{i}{2j} = (-1)^i \binom{k}{i} 2^{i-1+\delta(i)}$$

using the fact that $\binom{r}{m}\binom{m}{k} = \binom{r}{k}\binom{r-k}{m-k}$ and the binomial theorem [8].   □

Here are some examples.

$$P_1(x) = -x + 1,$$
$$P_2(x) = 2x^2 - 2x + 1,$$
$$P_3(x) = -4x^3 + 6x^2 - 3x + 1,$$
$$P_4(x) = 8x^4 - 16x^3 + 12x^2 - 4x + 1.$$

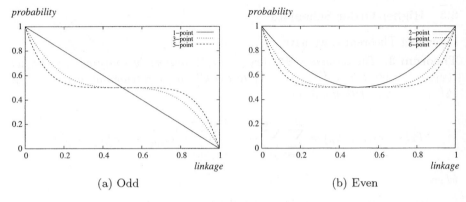

Fig. 1. Survival probabilities of 2nd-order schemata under k-point crossover

On the other hand, from the equation (3),

$$P_k(x) = \sum_{i=0}^{\lfloor k/2 \rfloor} \binom{k}{2i} x^{2i}(1-x)^{k-2i}$$

$$= \binom{k}{0} x^0(1-x)^k + \binom{k}{2} x^2(1-x)^{k-2} + \cdots + \binom{k}{2\lfloor \frac{k}{2} \rfloor} x^{2\lfloor \frac{k}{2} \rfloor}(1-x)^{k-2\lfloor \frac{k}{2} \rfloor}.$$

Then, by the binomial theorem [8], the following holds.

$$2P_k(x) = (x + (1-x))^k + (-x + (1-x))^k = 1 + (1-2x)^k.$$

Therefore, we have the following simple form.

Theorem 2.

$$P_k(x) = \frac{1 + (1-2x)^k}{2}.$$

Note that Theorem 1 can be derived directly from Theorem 2. Figure 1 shows the survival probabilities of 2nd-order schemata under k-point crossover with respect to linkage x. For odd k, the graph $y = P_k(x)$ is symmetric around the point $(x, y) = (0.5, 0.5)$, and, for even k, the graph $y = P_k(x)$ is symmetric about $x = 0.5$. These facts are easily checked from Theorem 2. We also have that, for $x \in (0, 1)$, the survival probability of a 2nd-order schema with linkage x pointwise converges to $y = \frac{1}{2}$, the same as that of uniform crossover with $p = 0.5$, as k increases.

3.3 Higher-Order Schemata

Extending Theorem 1, we have

Theorem 3. *The sequence* $\langle P_{k,n}(x_1,\ldots,x_q)\rangle$ *uniformly converges on the set* $\{(x_1,\ldots,x_q)|0 < x = \sum_{i=1}^{q} x_i \le 1\}$ *to the* k^{th}-*order polynomial of* $x_1,\ldots,x_q$, $P_k(x_1,\ldots,x_q)$, *with the form*

$$P_k(x_1, x_2, \ldots, x_q) = \sum_{i_1=0}^{k} \sum_{i_2=0}^{k-i_1} \cdots \sum_{i_q=0}^{k-(i_1+\cdots+i_q)} c_{i_1,i_2,\ldots,i_q} x_1^{i_1} x_2^{i_2} \cdots x_q^{i_q},$$

where

$$c_{i_1,i_2,\ldots,i_q} = (-1)^{i_1+\cdots+i_q} \frac{(i_1 + \cdots + i_q)!}{i_1! \cdots i_q!} \binom{k}{i_1 + \cdots + i_q} 2^{i_1+\cdots+i_q-q+\sum_{j=1}^{q} \delta(i_j)}$$

for $i_1,\ldots,i_q = 0, 1, 2, \ldots$.

Proof: The proof is analogous to that of Theorem 1. We omit the proof by space limitation. See [6] for details.   □

Again, in an analogous way to Theorem 2, we have the following.

Theorem 4.

$$P_k(x_1,\ldots,x_q) = \frac{1}{2^q}(1 + (1 - 2x_1)^k + \cdots + (1 - 2x_q)^k$$
$$+ (1 - 2x_1 - 2x_2)^k + (1 - 2x_1 - 2x_3)^k + \cdots + (1 - 2x_{q-1} - 2x_q)^k$$
$$+ \cdots$$
$$+ (1 - 2x_1 - 2x_2 - \cdots - 2x_q)^k).$$

Proof: We omit the proof by space limitation. See [6] for details.   □

It is clear that $P_k(x_1,\ldots,x_q)$ is symmetric in $x_1,\ldots,x_q$ in that $P_k(x_1,\ldots,x_q)$ is left fixed by all permutations of $x_1,\ldots,x_q$. Here are some examples.

$$\begin{aligned}
P_1(x_1,\ldots,x_q) &= (1 - x), \\
P_2(x_1,\ldots,x_q) &= (1 - x)^2 + (x_1^2 + \cdots + x_q^2), \\
P_3(x_1,\ldots,x_q) &= (1 - x)^3 + 3(1 - x)(x_1^2 + \cdots + x_q^2), \\
P_4(x_1,\ldots,x_q) &= (1 - x)^4 + 6(1 - x)^2(x_1^2 + \cdots + x_q^2) \\
&\quad + 6(x_1^2 x_2^2 + x_1^2 x_3^2 + \cdots + x_{q-1}^2 x_q^2) \\
&\quad + (x_1^4 + \cdots + x_q^4).
\end{aligned} \tag{4}$$

As shown in the proofs of the above theorems, the difference between $P_{k,n}(x_1,\ldots,x_q)$ and the limiting polynomial $P_k(x_1,\ldots,x_q)$ is $O(\frac{1}{n})$, which means that the approximation is acceptable even for not very large n. Consider two 8th-order schemata H_1 and H_2 in a chromosome of length n. Figure 2 shows the appearances of the two schemata, where the symbol $*$ represents

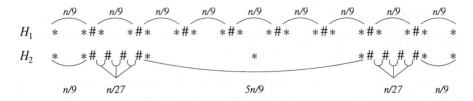

Fig. 2. Example schemata H_1 and H_2

don't-care positions and the symbol # represents specific positions. H_1 is the schema with linkage distribution $(\frac{1}{9}, \frac{1}{9}, \frac{1}{9}, \frac{1}{9}, \frac{1}{9}, \frac{1}{9}, \frac{1}{9})$ and H_2 is the schema with linkage distribution $(\frac{1}{27}, \frac{1}{27}, \frac{1}{27}, \frac{5}{9}, \frac{1}{27}, \frac{1}{27}, \frac{1}{27})$. As shown in the figure, the specific symbols are evenly distributed in H_1 and are more clustered in H_2. For more concrete experiments, we set the number of the don't-care symbols between adjacent specific symbols to be $\lfloor \frac{n-8}{9} \rfloor$ in H_1. And, in H_2, we set the number of the don't-care symbols between adjacent specific symbols in a cluster to be $\lfloor \frac{n-8}{27} \rfloor$ and make the defining length of H_2 be the same as that of H_1.

For $n = 100$, the exact values of the survival probabilities of H_1 and H_2 under 2-point crossover are 0.1270 and 0.3391, respectively.[1] Using the equation (4), we get the corresponding approximate values as follows:

$$P_2(H_1) = P_2(\frac{11}{99}, \frac{11}{99}, \frac{11}{99}, \frac{11}{99}, \frac{11}{99}, \frac{11}{99}, \frac{11}{99}) = 0.1358,$$

$$P_2(H_2) = P_2(\frac{4}{99}, \frac{4}{99}, \frac{4}{99}, \frac{53}{99}, \frac{4}{99}, \frac{4}{99}, \frac{4}{99}) = 0.3458.$$

The approximation errors of $P_2(H_1)$ and $P_2(H_2)$ are 0.0088 and 0.0067, respectively, which are fairly small. Table 1 shows how well such an approximation

Table 1. Survival probabilities of H_1 and H_2 under 2-point crossover

n	H_1			H_2		
	Exact	Approx.	Δ	Exact	Approx.	Δ
200	0.1323	0.1367	0.0044	0.3413	0.3446	0.0033
400	0.1350	0.1371	0.0021	0.3574	0.3590	0.0016
600	0.1359	0.1373	0.0014	0.3629	0.3639	0.0010
800	0.1343	0.1354	0.0011	0.3636	0.3644	0.0008
1,000	0.1349	0.1358	0.0009	0.3656	0.3663	0.0006

scales with the chromosome length n. In this table, "Exact" and "Approx." indicate the exact values and the approximate values of the survival probabilities of H_1 and H_2 under 2-point crossover, respectively. Δ indicates the difference

---

[1] We get the values, for example, using the equation (2).

between these values, i.e., the approximation error. As expected, the approximation errors seem to decrease in the rate of $O(\frac{1}{n})$. In particular, the errors for H_1 and H_2 decrease almost at the same rate as n increases, which relates to the uniform convergence of the approximation.

Despite the same defining length, on the other hand, the survival probability of H_2 is much higher than that of H_1 under 2-point crossover. As claimed in [3] and [5], this is because the distribution of the specific symbols are different in H_1 and H_2. In the next section, we show how the survival probabilities of the schemata with the same defining length relate to the distributions of the specific symbols in the schemata.

4   Convex Property of Survival Probability

In this section, we assume that the chromosome length n is sufficiently large, and that, based on the assumption, $P_k(x_1, \ldots, x_q)$ is the survival probability of a schema with linkage distribution $(x_1, \ldots, x_q)$ under k-point crossover and $\{(x_1, \ldots, x_q)|0 < x = \sum_{i=1}^{q} x_i \leq 1\}$ is the set of the linkage distributions of the $(q+1)^{th}$-order schemata.

A set S is *convex* if the line segment between any two points in S lies in S, i.e., if for any $\mathbf{x}, \mathbf{y} \in S$ and any θ with $0 \leq \theta \leq 1$ we have $\theta\mathbf{x} + (1-\theta)\mathbf{y} \in S$. Roughly speaking, a set is convex if every point in the set can be seen by every other point, along an "unobstructed" straight path between them. We call a point of the form $\theta_1\mathbf{x}_1 + \cdots + \theta_k\mathbf{x}_k$, where $\theta_1 + \cdots + \theta_k = 1$ and $\theta_i \geq 0, i = 1, \ldots, k$, a *convex combination* of the points $\mathbf{x}_1, \ldots, \mathbf{x}_k$. The *convex hull* of a set C, denoted $\mathbf{conv}C$, is the set of all convex combinations of points in C:

$$\mathbf{conv}C = \{\theta_1\mathbf{x}_1 + \cdots + \theta_k\mathbf{x}_k | \mathbf{x}_i \in C, \theta_i \geq 0, i = 1, \ldots, k, \theta_1 + \cdots + \theta_k = 1\}.$$

As the name suggests, the convex hull of a set is always convex. We let

$$S_{x,q} = \{(x_1, \ldots, x_q)| \sum_{i=1}^{q} x_i = x, x_i \geq 0, i = 1, \ldots, q\}$$

for a fixed $x \in (0, 1]$, which is the set of the linkage distributions of the $(q+1)^{th}$-order schemata with linkage sum x. Let $\mathbf{e}_i$ be the vector of dimension q all whose elements are zero except that the ith element is one. Then, it holds that $S_{x,q} = \mathbf{conv}\{x\mathbf{e}_i | i = 1, \ldots, q\}$ and, consequently, $S_{x,q}$ is a convex set.

A function $f : \mathbb{R}^n \to \mathbb{R}$ is *convex* if the domain of f is a convex set and, for all $\mathbf{x}, \mathbf{y}$ in the domain and θ with $0 \leq \theta \leq 1$, we have

$$f(\theta\mathbf{x} + (1-\theta)\mathbf{y}) \leq \theta f(\mathbf{x}) + (1-\theta)f(\mathbf{y}). \tag{5}$$

Geometrically, this inequality means that the line segment between $(\mathbf{x}, f(\mathbf{x}))$ and $(\mathbf{y}, f(\mathbf{y}))$ lies above the graph of f. A function f is *strictly convex* if strict

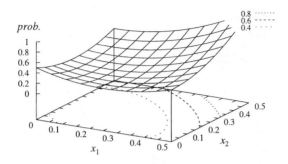

Fig. 3. The strict convexity of $P_3(x_1, x_2, x_3)$ with $x = 0.5$ under 3-point crossover

inequality holds in the inequality (5) whenever $\mathbf{x} \neq \mathbf{y}$ and $0 < \theta < 1$. A nonnegative weighted sum of convex functions is convex. Similarly, a nonnegative, nonzero weighted sum of strictly convex functions is strictly convex. We refer to [2] for more details.

From Theorem 2, $P_k(x) = \frac{1+(1-2x)^k}{2}$. Since $P_k''(x) = 2k(k-1)(1-2x)^{k-2}$, $P_k(x)$ is strictly convex for even k. We define $P_0(x_1, \ldots, x_q) = 1$. Then, in the same manner as in [11], we get the following recurrence which we omit the proof in this paper:

$$P_k(x_1, \ldots, x_q) = \sum_{i=0}^{\lfloor k/2 \rfloor} x^{2i}(1-x)^{k-2i} P_{2i}\left(\frac{x_1}{x}, \ldots, \frac{x_{q-1}}{x}\right).$$

Hence, by induction, $P_k(x_1, \ldots, x_q)$ is strictly convex for even k and any $q \geq 1$, since a nonnegative weighted sum of strictly convex functions is also strictly convex. For odd k and $q \geq 2$, for the same reason, $P_k(x_1, \ldots, x_q)$ is strictly convex in the set $S_{x,q}$ for a fixed $x \in (0, 1]$. Therefore, we get the following.

Theorem 5. *For even k, $P_k(x_1, \ldots, x_q)$ is strictly convex in $\mathbb{R}^q$. And, for any odd k and $q \geq 2$, $P_k(x_1, \ldots, x_q)$ is strictly convex in the set $S_{x,q}$ for a fixed $x \in (0, 1]$.*

Figure 3 shows the survival probabilities of the 4th-order schemata (x_1, x_2, x_3) with linkage sum $x = 0.5$ under 3-point crossover. In the figure, we plot the graph of $y = P_3(x_1, x_2, x_3)$ with respect to x_1 and x_2 since x_3 is determined by x_1 and x_2. The strict convexity of the survival probability is clear from the graph. Note that $y = P_3(x_1, x_2, x_3)$ is not convex any longer if x is not fixed.

An important result for convexity is that a strictly convex function has one global minimum and has at least one global maximum at the boundary of the domain. In particular, if its domain is the convex hull of a set, the function can have global maxima only at the points in the set. The following theorem shows how the survival probability depends on the distribution of the specific symbols of the schema.

Theorem 6. *Over the $(q+1)^{th}$-order schemata with linkage distribution $(x_1, \ldots, x_q)$ and linkage sum $x \in (0, 1]$, the survival probability is minimized*

difference

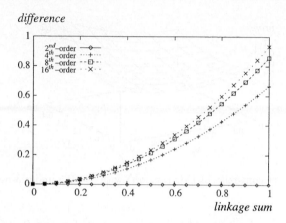

linkage sum

Fig. 4. The difference between the maximum and minimum values of the survival probabilities under 2-point crossover

when the variance of x_i's is minimized, and maximized when the variance is maximized.

Proof: We use the convexity of the survival probability and Lagrange multipliers. We omit the proof by space limitation. See [6] for details. $\qquad \square$

From Theorem 4, the maximum value of the survival probabilities of $(q+1)^{th}$-order schemata under k-point crossover is $P_k(0,\ldots,0,x) = \frac{2^{q-1}+(1-2x)^k 2^{q-1}}{2^q} = \frac{1+(1-2x)^k}{2} = P_k(x)$. The minimum value of the survival probability is $P_k(\frac{x}{q},\ldots,\frac{x}{q}) = \frac{1}{2^q q^k} \sum_{i=0}^{q} \binom{q}{k}(q-2xi)^k$. Figure 4 shows the difference between the maximum and minimum values of the survival probabilities under 2-point crossover with respect to linkage sum x. It is seen that the difference is fairly large for the higher-order schemata with relatively large defining lengths.

5   Conclusion

We proposed an analytic approach to approximate the survival probabilities of schemata under multi-point crossover. By adapting the approach, we obtained a closed form for the survival probabilities of schemata, which can be calculated in polynomial time with the distribution of the specific symbols. Such an approximation provides with a convenient way to analyze the disruptiveness of multi-point crossover mathematically. This also enables us to understand multi-point crossover more deeply. It becomes also possible to investigate more complex schemata in genetic studies.

We showed the convex property of the survival probabilities under multi-point crossover. We then proved how the survival probability of a schema relates to the distribution of the specific symbols in the schema. This confirms the previous results that were empirically shown: the more clustered specific symbols a schema has, the higher survival probability it has.

We are currently working on investigating another properties of multi-point crossover based on the approximation. We are going to applying such an approximation approach to other crossovers to analyze the survival probabilities under the crossovers.

Acknowledgment. This work was supported by Brain Korea 21 Project. The ICT at Seoul National University provided research facilities for this study.

References

1. C. Anderson, K. Jones, and J. Ryan. A two-dimensional genetic algorithm for the ising problem. *Complex Systems*, 5:327–333, 1991.
2. S. Boyd and L. Vandenberghe. *Convex Optimization.* Cambridge University Press, 2003.
3. T. N. Bui and B. R. Moon. Analyzing hyperplane synthesis in genetic algorithms using clustered schemata. In *Parallel Problem Solving from Nature*, pages 108–118, 1994.
4. T. N. Bui and B. R. Moon. On multi-dimensional encoding/crossover. In *Sixth International Conference on Genetic Algorithms*, pages 49–56, 1995.
5. T. N. Bui and B. R. Moon. GRCA: A hybrid genetic algorithm for circuit ratio-cut partitioning. *IEEE Trans. on CAD*, 17(3):193–204, 1998.
6. S. S. Choi and B. R. Moon. An analytic approach for approximation of survival probability under multi-point crossover. In preparation, 2004.
7. J. Cohoon and D. Paris. Genetic placement. In *IEEE International Conference on Computer-Aided Design*, pages 422–425, 1986.
8. R. L. Graham, D. E. Knuth, and O. Patashnik. *Concrete Mathematics, Second Edition.* Addison Wesley, 1994.
9. J. Holland. *Adaptation in Natural and Artificial Systems.* University of Michigan Press, 1975.
10. K. A. De Jong. *An Analysis of the Behavior of a Class of Genetic Adaptive Systems.* PhD thesis, University of Michigan, Ann Arbor, MI, 1975.
11. K. A. De Jong and W. M. Spears. A formal analysis of the role of multi-point crossover in genetic algorithms. *Annals of Mathematics and Artificial Intelligence Journal*, 5:1–26, 1992.
12. W. Rudin. *Principles of Mathematical Analysis, Third Edition.* McGraw-Hill, 1976.
13. D. I. Seo and B. R. Moon. Voronoi quantized crossover for traveling salesman problem. In *Genetic and Evolutionary Computation Conference*, pages 544–552, 2002.
14. W. M. Spears. *The Role of Mutation and Recombination in Evolutionary Algorithms.* PhD thesis, George Mason University, Fairfax, Virginia, 1998.
15. G. Syswerda. Uniform crossover in genetic algorithms. In *Third International Conference on Genetic Algorithms*, pages 2–9, 1989.

Evolving Genotype to Phenotype Mappings with a Multiple-Chromosome Genetic Algorithm

Rick Chow

Division of Mathematics and Computer Science
University of South Carolina Spartanburg
800 University Way, Spartanburg, SC, 29303, U.S.A.
rchow@uscs.edu
http://faculty.uscs.edu/rchow

Abstract. This paper presents an evolutionary coding method that maps genotype to phenotype in a genetic algorithm. Unlike traditional genetic algorithms, the proposed algorithm involves mating and reproduction of cells that have multiple chromosomes instead of single chromosomes. The algorithm also evolves the mapping from genotype to phenotype rather than using a fixed mapping that is associated with one particular encoding method. The genotype-to-phenotype mapping is conjectured to explicitly capture important schema information. Some empirical results are presented to demonstrate the efficacy of the algorithm with some GA-Hard problems.

1 Background

Genetic Algorithms (GAs) are a class of powerful algorithms for solving optimization problems using evolution as a search strategy. Although GAs can successfully solve many difficult optimization problems, some problems are still hard for GAs to solve and such problems are called *GA-Hard problems*. Particularly, there is a class of problems called deceptive problems that exploit the weaknesses in the encodings of chromosomes [1], [2]. This paper purposes a GA based method to solve GA-Hard problems by evolving a genotype-to-phenotype mapping that also capture schema information.

Let's understand how an optimization problem can be deceptive by understanding basic schema theory, which attempts to explain the ability of GAs to perform global search in a large and usually high dimensional problem space [3]. The theory hypothesizes the existence of building blocks that are called *schemata* in the chromosomes. A schema is a chromosome pattern that consists of three symbols, 0, 1, and *, where 0 and 1 are the fixed *defining bits* and the symbol * is a wildcard symbol that matches a 0 or 1. While an 8-bit binary chromosome 10010011 presents a single point in the solution space, the schema 10*1***1 represents a *hyper-plane* in the solution space that consists of 16 data points because each * symbol could be a 0 or 1. A schema is said to *sample* its hyper-plane because as it survives from generation to generation, data points in its hyper-plane are tested repeatedly to be as good solutions in the solu-

K. Deb et al. (Eds.): GECCO 2004, LNCS 3102, pp. 1006–1017, 2004.

tion space. Through genetic recombination, mutation, and selection, promising hyper-planes in the solution space are given higher sampling rates by an exponential increase in quantities of the schemata in the chromosome population.

A binary chromosome may consist of chunks of binary sequences each of which is called a *gene* that corresponds to a parameter of the optimization function. Usually, a particular binary number encoding scheme is chosen to translate such a binary se-quence into a numeric value for the parameter. For example, 00001111 can be trans-lated to 15 (decimal) using two's complement coding method. The binary chromo-some represents the *genotype* whereas the decoded parameter list represents the *phe-notype*. Note that a parameter may not be storing numeric information; it could also represent structural information of a phenotype, e.g., architectural information of a neural network.

1.1 Binary Encoding Methods

This paper focuses on binary encodings where a parameter in the phenotype is en-coded as a binary number. The simplest encoding method is the traditional two's complement binary encoding method. Unfortunately, such an encoding method may cause discontinuity in the search process because a single step in the phenotypic space from, say, 15 to 16, would require 5 steps in the genotypic space from 00001111 to 00010000. In this example, the Hamming distance between 00001111 and 00010000 is 5. There were many attempts to use alternative encoding methods such as Gray coding to alleviate this problem [4], [5]. The encoding and decoding steps of Gray code are detailed in [6]. The significance of Gray code is that consecutive binary numbers differ by only one digit, that is, the Hamming distance between two consecu-tive numbers is one. For example, numbers 0 to 8 in Gray code are as follows: 0000, 0001, 0011, 0110, 0111, 0101, 0100. However, even with Gray coding, a problem space can still be difficult for GAs to search in.

1.2 Deceptive Problems

Some of the problems that are hard for GAs are called deceptive problems. Before defining what a deceptive problem is, we need to understand what kind of schemata could lead to deception. Detailed discussions of deception problems can be found in [1], [2], [7]. A schema's fitness depends on the fitness values of its sampling points. For instance, the schema 10*1***1 would probably have a high fitness value if most (if not all) of the 16 data points carry high fitness values. Schemata with high fitness values have a higher chance to survive into the next generation. A schema with N defining bits is called an order-N schema. For example, the schemata *0**0, *0**1, *1**0, and *1**1 are all order-2 schemata. Moreover, schemata that are of the same order and have defining bits at the same locations are essentially *primary competitors* in the selection process because such schemata compete as orthogonal planes in the hyperspace. A lower order hyper-plane also contains a collection of higher order hy-per-planes that share common defining bits with the lower order hyper-plane. For example, the order-2 hyper-plane 0***0 contains some order-3 hyper-planes such as 00**0 and 01**0 whereas 1***0 contains another set of order-3 hyper-planes such as 10**0 and 11**0. An order-N schema and the order-K schemata (N<K) that it contains are said to be *relevant* to one another. When the lower order schema competes in a selection process, all of its own relevant higher order schemata also participate in the competition. For example, the competition between the order-2 schemata 0***0, 0***1, 1***0, and 1***1 also involves the competitions between the order-3 schemata 00**0, 01**0, 00**1, 01**1, 10**0, 11**0, 10**1, and 11**1. On the other hand, the competitions between higher hyper-planes also involve competition of their relevant lower order hyper-planes.

A deception occurs when a hyper-plane competition of higher order K leads to a global winner that is radically different, in terms of Hamming distance, from the global winner of the competitions among the relevant order-N hyper-planes, where N<K. For example, consider the fitness values for the following deceptive function f1:

Table 1. A Deceptive Function f1

Function f1 Values	
f1(000) = 28	f1(001) = 26
f1(010) = 22	f1(100) = 14
f1(110) = 0	f1(011) = 0
f1(101) = 0	f1(111) = 30

With function f1, the global winner of an order-3 hyper-plane competition is 111. Consider 111's relevant hyper-plane *11, when it competes with *00, *01, and *10 in order-2 competitions, the average fitness of *11 is 15 ((0+30)/2) versus the average fitness values of *00, *01, and *10 are 21, 13, and 11, respectively. The schemata *00 will most likely win the order-2 hyper-plane competition. The radical difference, or long Hamming distance, between the two global winners 111 and *00 creates deception. We are now ready to define a deception problem:

A *deceptive problem* is any problem of a specific order K that involves deception in one or more relevant lower order N hyper-plane competitions where N<K.

A deceptive problem misleads a GA to converge incorrectly to a region in the problem space called the *deceptive attractor*.

2   Previous Work on Solving Deceptive Problems

Different techniques have been proposed to tackle deceptive problems. Some techniques focus on the encoding methods. According to the work in [1],[8], a tagged bit that specifies the phenotypic bit location is attached to each binary bit in the chromosome. A binary string 11011001 is represented as a list of order pairs of bit value-bit location: ((1 1) (2 1) (3 0) (4 1) (5 1) (6 0) (7 0) (8 1)). The order of the ordered pairs may be rearranged or randomized but bits on two chromosomes are first lined up in the same order before a crossover operation. However, these approaches involve an additional search space for finding the right permutation of the bit ordering that is as large as the original function optimization space. In addition, one of the bit orderings of the parents is randomly chosen as the ordering of the siblings. As suggested in a related work [12], gene positions, instead of bit positions, could also be rearranged to preserve building blocks.

In [9], a 2-level GA is proposed. Each low level gene in the chromosome is tagged with a high level control gene that turns the gene on or off. This 2-level GA approach is very similar to a redundant gene approach in [10] where multiple genotypic bits would vote to turn on or off a phenotypic bit. In [1], [11], multiple populations were maintained and migration was allowed among different sub-populations. The primary goal was to maintain diversity. So long as at least one subpopulation does not converge to the deceptive attractor, there is still a chance that the global optimum can be found.

3   An Evolutionary Mapping Method

The proposed method in this paper is based on a single population GA model although it could be extended easily to use multiple populations. A population of cells instead of chromosomes is maintained. Each cell has two chromosomes, a binary *data chromosome* that stores the genotype for the optimization function and a value-coded *mapping chromosome* that stores bit locations as integers. For example, an initial cell may have two chromosomes as follow:

| 1 | 1 | 0 | 1 | 1 | 1 | 0 | 0 |

| 0 | 1 | 2 | 3 | 4 | 5 | 6 | 7 |

The chromosome to the left is the data chromosome that stores genotypic bits while the mapping chromosome on the right hand side defines the mapping from a genotype to a phenotype. For example, the above mapping chromosome maps the genotype to a phenotype that is exactly the same as the genotype itself. However, a different mapping chromosome (7,6,0,1,2,3,4,5) will map the above data chromosome to a phenotype (0,0,1,1,0,1,1,1), which may be a parameter of an optimization function. In the evolution process, the data chromosomes and mapping chromosomes undergo separate genetic operations with their counter parts in the other cells. The mapping chromosomes also go through additional permutation operations to alter the bit ordering. While the genetic operations for the data chromosomes are traditional 2-point crossover and mutation, different implementations of the genetic operations for the mapping chromosomes may result in significantly different outcomes. If the bit ordering on the mapping chromosome is constantly maintained as a permutation of the bit locations, then the approach will be very similar to the tagged bit approach as in Genitor [1] and Goldberg's work [8]. However, this permutation method by itself is shown not to be very effective against deceptive problems.

This paper suggests a decoupling of the genotype from the mapping process by allowing the mapping chromosomes to cross and mutate freely using integer genetic operators without maintaining a one-to-one mapping between a genotypic bit and a phenotypic bit.

3.1 Genetic Operators

As mentioned earlier, the data chromosomes undergo the traditional 2-point crossover and mutation operations. On the other hand, the permutation operator rearranges bits on the mapping chromosomes by swapping and shifting bits. The crossover operator for the mapping chromosomes is a traditional 2-point integer crossover operator. The mutation operator is an integer mutation operator that randomly switches a gene to one of the bit locations. In addition, a gene replacement operation is also included in the mutation process. Through this gene replacement operation, a gene on the mapping chromosome can be copied to another gene. The whole genetic operation process will break the one-to-one mapping between a genotypic bit and a phenotypic bit. A genotypic bit on the data chromosome may be mapped to more then one bit on the phenotype. For example, given a data chromosome $(1,0,0,1,1,1,0,0)$ and a mapping chromosome $(1,4,0,1,2,7,4,5)$, a phenotype $(0,1,1,0,0,0,1,1)$ can be constructed. Bit locations 3 and 6 on the data chromosomes are actually not used.

The rationales of this approach are: First, this approach will force the data chromosomes to concentrate on exploring and surviving in the genotypic space where genomic materials constantly undergo construction and destruction of schemata. The mapping chromosomes on the other hand, will concentrate on the effort of obtaining an optimal mapping between a genotype and a phenotype. Second, with cells of multiple chromosomes, different mappings can evolve at the same time because each mapping is associated with a particular genotype. On the contrary, some previous approaches evolved only one mapping for the entire population. Third, perhaps the most important reason is that the introduction of non-one-to-one mapping creates the possibility of capturing useful schemata as described below.

3.2 Capturing Schemata

Suppose an order N schema consists of r 0 bits and s 1 bits. The 0-bits together form a subschema of order r called 0-subschema whereas the 1-bits form a subschema of order s called 1-subschema. For example, the schema 1*0*1*0*1 consists of an order-3 1-subschema 1***1***1 and an order-2 0-subschema **0***0**. As the defining bits of a subschema emerge during evolution, the system will try to maintain the same bit values at those locations.

Because the mapping chromosomes are allowed to evolve freely to form non-one-to-one mapping between a genotype and a phenotype, the selection pressure should map one defining bit location to another defining bit location in the same subschema. The result is an explicit formation of a schema. For example, the phenotypic subschema 1***1***1 can still be reconstructed from the genotypic pattern 1***0***0 if the mapping chromosome is $(0,1,2,3,0,5,6,0)$.

Once a subschema is explicit formed, the change of one genotypic bit may trigger the changes of multiple defining bits on the phenotype. The ability to alter multiple bit values greatly shortens the Hamming distance between two distant hyper-planes in the solution space. For example, the distance between 1***1***1 and 0***0***0 can be reduced to one. Hence the search may be able to get away from a deceptive attractor or go toward a global optimum much faster. In deceptive problems, the global optimum is frequently the complement of the deceptive attractor. Therefore, the explicit formation of 1-subschema or 0-schschema may help the search to jump from a deceptive attractor straight to the global optimum.

The schema theory proposed that schemata of higher order and with longer length are less likely to survive because they are more likely to be destroyed by genetic operators [3]. On the contrary, as shown in the above section, the higher the order of subschema, the greater in reduction of Hamming distance between two distant hyperplanes.

4 Experiments

Two sets of experiments were setup to test the dynamic mapping approach. The first set of experiments is primarily based on the deceptive functions constructed in [1] while the second set is based on functions used in [14], [15].

4.1 First Set of Experiments

The first step in these experiments was to construct some deceptive functions. First, the following algorithm from [1] was applied to construct an order-4 deceptive function:

Step 1: Select a global optimum pattern.
Step 2: Sort all binary patterns by their Hamming distance to the optimum pattern. The group of patterns that have the same distances can be placed in any order within the same group
Step 3: The first pattern is pattern 1, the optimum pattern and the last pattern N will be the deceptive attractor. Assign fitness values according to:
fitness(pattern X) = fitness(pattern X−1) + C
where C is a constant and fitness(pattern 2) is another constant, say, 0.

Using the above algorithms, two deceptive functions f2 and f3 were defined as shown in Table 2. The function f2 was directly obtained from [1] while the function f3 was newly constructed using the above algorithm. Note that the optimum of f3 1001 has equal numbers of 1 and 0.

Next, an "ugly" 40-bit function could be constructed based on the discussion in [1] by spreading the four bits of the parameter of function f2 to the chromosome bit locations 0, i+10, i+20, i+40, respectively, for 10 times where $0 \leq i \leq 9$. For instance, based on the parameter 0101, the function value of
ugly(0000000000 1111111111 0000000000 1111111111)

is 16. Similarly, another "bad" 40-bit function could be constructed using f3. In the experiments, the 40-bit ugly and bad functions were also extended to 60-bit by repeating the spreading the four parameter bits 15 times.

Table 2. Deceptive Functions f2 and f3

f2				f3			
f2(1111)	30	f2(0110)	14	f3(1001)	30	f3(0101)	14
f2(0111)	0	f2(0101)	16	f3(1011)	0	f3(0011)	16
f2(1011)	2	f2(0011)	18	f3(1101)	2	f3(0000)	18
f2(1101)	4	f2(1000)	20	f3(1000)	4	f3(0111)	20
f2(1110)	6	f2(0100)	22	f3(0001)	6	f3(1110)	22
f2(1100)	8	f2(0010)	24	f3(1111)	8	f3(0100)	24
f2(1010)	10	f2(0001)	26	f3(1100)	10	f3(0010)	26
f2(1001)	12	f2(0000)	28	f3(1010)	12	f3(0110)	28

Two additional algorithms were also included for comparison. The first one was simply a regular GA without any special operators. The second algorithm was similar to the tagged bit approach in which the one-to-one genotype-phenotype mapping was maintained. In all experiments, for the data chromosomes, the crossover rate was set to 0.6 and mutation rate was set to 1/(length of chromosome). For the mapping chromosome, the crossover rate was 0.6 and the mutation was set to 0.005, which was about 1/5 of the mutation rate for the data chromosome. The permutation rate was 0.025. Several tests were set up to test the algorithms using the 40-bit ugly function, 60-bit ugly function, 40-bit bad function, and 60-bit bad function. The results are obtained summarized in Table 3.

Table 3. Experimental Results using the ugly and bad Functions

	Traditional GA		Tagged Bits		Dynamic Mapping	
	Percentage of Optimal Runs	Avg. Gen.	Percentage of Optimal Runs	Avg. Gen.	Percentage of Optimal Runs	Avg. Gen.
ugly 40 bits	0%	n/a	0%	n/a	100%	1387
ugly 60 bits	0%	n/a	0%	n/a	100%	2471
bad 40 bits	0%	n/a	0%	n/a	100%	4240
bad 60 bits	0%	n/a	0%	n/a	80%	8931

Each result is based on 5 independent runs. The dynamic mapping approach proposed in this paper successfully found the global optimum in all of the ugly function tests and the 40-bit bad function tests and it found the global optimum of the 60-bit bad function 80% of the time. The average numbers of generations for the dynamic mapping approach to reach the optimum were listed in the last column. The averages were calculated for successful runs only. The search missed the optimum two times out of ten runs. In addition, the time to find the optimum almost doubled when the size of the problem was increased by 50% from 40 bits to 60 bits. The other two approaches got stuck in the deceptive attractors and never found the optima. The failures of the other approaches simply verify how deceiving the functions were for traditional GAs.

4.2 Second Set of Experiments

The second set of experiment is based on the some of the most difficult GA-Hard problems listed in [14], [15] although other difficult functions were first proposed in [13]. The chosen functions are listed in Table 4 and they are considered GA-Hard because the search spaces are multimodal with large number of local optima. The Rastrigin's function is particularly difficult to optimize because it contains millions of local optima in the interval of consideration. Many search method failed to converge to the global optimum at $x_i = 0$. The Schwefel's function has a global optimum at $x_i = 420.9687$. In implementation, the Schwefel's function was modified as $418.9828872721624 - f_7$ to avoid negative fitness values. The optima of these optimization functions are indeed needles in haystacks.

Table 4. Additional Optimization Functions

Name	Function	Interval
Rastrigin's	$f_6(\vec{x}) = \sum_1^{20} [x_i^2 - 10\cos(2\pi x_i) + 10]$	$-5.12 \le x_i \le 5.12$
Schwefel's	$f_7(\vec{x}) = \sum_1^{10} -x_i \, \sin(\sqrt{\lvert x_i \rvert})$	$-500 \le x_i \le 500$

The experiments were set up to compare the dynamic mapping approach and the traditional GA approach. Because both approaches had problems to converge to an exact optimal point in the solution space, additional local hill climbing was incorporated to help the algorithms to perform local search to zoom into the optimum. Note that the optimization functions are defined in the real number domain with infinite precision. Therefore, if $\mathbf{x}^*$ is the optimum point in the solution space and $\mathbf{x}$ is the current search point, the search will stop as soon as $\lvert f(\mathbf{x}) - f(\mathbf{x}^*)\rvert < \varepsilon$, where ε is a small comparison threshold. In the experiments, the threshold ε was set to 10^{-16}. However, using this small threshold, the Rastrigin's problem was not 100% solved. The traditional GA kept wondering in vicinity of the exact optimal point while the dynamic mapping algorithm keeps converging to the optimal point at an ever decreasing rate.

The results of the experiments are summarized below in Table 5. Each of the results is based on data averaged over 5 separate runs. The maximum number of evolutions for the experiments was set to 10,000 and 20,000 generations for Schwefel's function and Rastrigin's function, respectively. For the Schwefel's function, both approaches found the optimum. The fitness values were in the order of 10^{-17} when the evolution was stopped. Again, the dynamic mapping approach performed better than the traditional version. The Rastrigin's function is the hardest among all the functions presented here. If the comparison threshold ε is set to 10^{-16}, no trial runs were considered successful in finding an optimum. However, if the threshold ε is lowered to 10^{-6}, 40% of the dynamic mapping algorithm found solutions that are close enough to the optimum. The traditional GA did not find any optimum even with the lowered threshold.

Table 5. Experimental Results using some GA-Hard Functions

	Traditional GA		Dynamic Mapping	
	Percentage of Optimal Runs	Avg. Gen.	Percentage of Optimal Runs	Avg. Gen.
Schwefel's	100%	2231	100%	1198
Rastrigin's	0% / 0%	n/a	0% / 40%	n/a / 15826

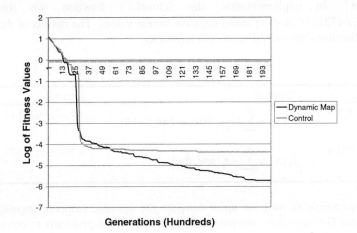

Fig. 1. Fitness Values of the All-Time-Best Cells

The fitness values from the all-time-best cell were averaged and plotted in Figure 1. The traditional GA converged very quickly in the beginning and at times produced better solutions earlier than the dynamic mapping approach; however, it got stuck in some local optima. The search by the dynamic mapping approach continued until the cut-off point at 20,000 generations. Since we did not use any elitist approach in any experiment, the all-time-best cells were not reintroduced into the populations. However, the current best cells in each generation also behaved similarly as shown in Figure 2.

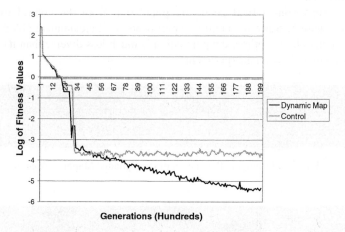

Fig. 2. Fitness Values of the Current Best Chromosomes

4.3 Population Diversity

The final populations of the above experiments were also examined. The following figures, Figures 3 to 5, are population dumps for the 40-bit ugly experiments. The populations are shown as matrices with 256 columns and 40 rows. A column in a matrix is a 40-bit chromosome and there are 256 chromosomes in a population. A 0-bit is shown as a black dot whereas a 1-bit is shown as a white dot.

Fig. 3. A Phenotypic Population of the Dynamic Mapping Approach

Figure 3 shows the final phenotypic population for an experiment of the dynamic mapping approach. A perfect solution with all 1's resides near the far right. The rest of the population consists of many local optimal solutions with all 0's since they were attracted to the deceptive local optima.

The corresponding genotypic population as shown in Figure 4 is quite different from the phenotypic population. There are many different individuals and the diversity in the genotypic population prevents the population from converging prematurely. However, some bit locations do consist of predominantly the same bit values across almost the whole population and such locations may be locations of defining bits of some schemata.

As a comparison, a final genotypic population from the control experiment is shown in Figure 5. Since the control experiments used a canonical GA that got stuck in the deceptive local optima, the population is much less diverse than its counter part from the dynamic mapping experiments.

Fig. 4. A Genotypic Population of the Dynamic Mapping Approach

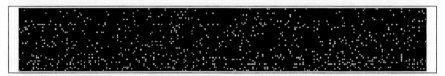

Fig. 5. A Genotypic Population of the Control Experiment

5   Conclusions

The proposed algorithm is every effective in solving some of the deception problems. It is also rather effective in solving some of the GA-Hard problems listed in [14], [15]. The dynamic mapping genetic algorithm is a novel approach that utilizes multiple chromosomes in a single cell for mating with another cell within a single population. The mapping from genotype-to-phenotype is explicitly evolved and maintained. Defining bits of 0-subschema and 1-subschema are supposed to be explicitly captured and stored in the mapping chromosomes. The direct accessibility to such subschema allows fast navigation to complementary hyper-planes and the Hamming distance between two such hyper-planes is greatly shortened to one. As a comparison, the Dual Genetic Algorithm approach in [16], [17], a meta-gene is used to flip gene values to their complements and hence reducing the number of steps needed to transform a chromosome to its genetic opposite.

Although this initial study has shown good potential in solving some GA-Hard problems, more study is needed to better understand and analyze the inner working of the algorithm. The 0/1-subschemata are also limited because they consist of defining bits of the same values. More research is needed to define a structure to capture the whole schema effectively.

References

1. Whitley, L. D.: Fundamental principles of deception in genetic search. In: Rawlins, G., (ed.): Foundations of Genetic Algorithms. Morgan Kaufmann, San Mateo, CA, (1991) 221-241
2. Goldberg, D.: Genetic Algorithms and Walsh Functions: Part II, Deception and its Analysis. Complex Systems 3 (1989) 153-171
3. Holland, J.H.: Adaptation in Natural and Artificial Systems. Ann Arbor, MI: University of Michigan Press (1975).
4. Mathias K., Whitley, L.D.: Transforming the search space with gray coding. In Proc. IEEE Int'l Conference on Evolutionary Computation (1994) 513-518
5. Yokose, Y., Cingoski, V., Kaneda K., and Yamashita, H.: Performance Comparison Between Gray Coded and Binary Coded Genetic Algorithms for Inverse Shape Optimization of Magnetic Devices, Applied Electromagnetics, (2000) 115-120
6. Wolfram, Gray Code, http://mathworld.wolfram.com/GrayCode.html
7. Goldberg, D.: Genetic Algorithms and Walsh Functions: Part I, A Gentle Introduction. Complex Systems 3 (1989) 129-152
8. Goldberg, D., Bridges, C. An Analysis of a Reordering Operator on a GA-Hard Problem. Biological Cybernetics 62 (1990) 397-405
9. Dasgupta, D.: Handling deceptive problems using a different genetic search. In Proceedings of the First IEEE Conference on Evolutionary Computation (1994) 807-811
10. Shackleton M., Shipman, R., and Ebner, M.: An investigation of redundant genotype-phenotype mappings and their role in evolutionary search. In Proceedings of Congress on Evolutionary Computation (2000) 493-500
11. Whitley, D., Rana, S., Heckendorn, R.B.: Exploiting Separability in Search: The Island Model Genetic Algorithm. Journal of Computing and Information Technology, v. 7, n. 1, (1999) 33-47 (Special Issue on Evolutionary Computing)
12. Sehitoglu, O.T., Ucoluk, G.: A building block favoring reordering method for gene positions in genetic algorithms. In Spector, L., Goodman, E.D., Wu, A. et.al., (eds.), Proceedings of the Genetic and Evolutionary Computation Conference (2001) 571-575
13. De Jong, K.A.: An Analysis of the Behavior of a Class of Genetic Adaptive Systems, Ph.D. Thesis, University of Michigan (1975)
14. Digalakis J., Konstantinos, M.: An Experimental study of Benchmarking Functions for Evolutionary Algorithms. International Journal of Computer Mathemathics, Vol. 79, (2002) 403-416
15. Salomon. R.: Reevaluating Genetic Algorithm Performance under Coordinate Rotation of Benchmark Functions. BioSystems vol. 39, Elsevier Science (1995) 263-278
16. Collard P., Aurand J.-P.: DGA: an efficient Genetic Algorithm. In Proceedings of the 11th European Conference on Artificial Intelligence (ECAI'94) (1994) 487-492
17. Collard P., Clergue M., Defoin P.M.: Artificial Evolution: In Proceedings of the Fourth European Conference (AE'99), Lecture Notes in Computer Sciences 1829, Springer-Verlag Ed. (2000) 254-265

Despite the covariance of (20), the facility of its analysis as well as its physical interpretation are basis-dependent. The dynamics is governed by the mutation matrix $M_I{}^J$, the tensor $\lambda_I{}^{JK}(M)$, the mask probability distribution $p(M)$ and the fitness values f_I, hidden inside P'_I. In this sense the evolutionary algorithm is a "black box" whose output depends on a large set of parameters. It therefore behooves us to look for symmetries and regularities that may be exploited to effect a natural coarse graining, making manifest the effective degrees of freedom of the dynamics.

5.1   Recombination

We will now consider recombination in the δ- and BB bases. For a discussion of recombination in the Walsh basis, in a much different context, see [11].

Recombination in the δ ("string") basis. In this basis, P_I is the probability (relative population) of the string I. For each mask M, there are generally several pairs of parent strings $\{J, K\}$ that produce I as their child. The tensor $\lambda(M)$ in Eq. (19) is given by

$$\lambda_I{}^{JK}(M) = \prod_{r=1}^{N} [1 + i_r + j_r + m_r(j_r + k_r)] \quad \text{mod } 2 \qquad (21)$$

which is 1 if the first child of the recombination of J, K, with mask M, is I, and zero otherwise (we use the convention that a 0 in the mask denotes that the first child obtains the corresponding bit from the first parent). Then $\lambda_I{}^{KJ}(M) = \lambda_I{}^{JK}(\bar{M})$ checks whether I is being produced as a second child. One may define a mask-independent average λ_I by $\lambda_I{}^{JK} = \sum_M p(M)\lambda_I{}^{JK}(M)$ whereupon (19) becomes, in matrix notation,

$$G_I(t) = \mathbf{P}^T R_I \mathbf{P}, \qquad R_I \equiv \frac{1}{2}\left(\lambda_I + \lambda_I^T\right). \qquad (22)$$

Notice that the second of (22) is valid in all bases, since both matrix indices of λ_I are contravariant (upper). For reasons explained in Sect. 5.2, λ_I is a more convenient object to work with than R_I. Again, the covariance of (22) guarantees its validity in all bases, with $R_I{}^{JK}$ transforming, along with $\lambda_I{}^{JK}$, as a rank-three tensor (see Ex. 5 below). Ignoring selection and mutation, Eq. (20) then becomes

$$P_I(t + 1) = (1 - p_c)P_I + p_c \mathbf{P}^T R_I \mathbf{P}. \qquad (23)$$

Example 4 $N = 2$ *recombination in the δ-basis*

We fix $I = (11)$ and take $p(M) = 1/4$ (independent of M). From (21) we compute

$$\lambda_{(11)}\big((00)\big) = \begin{pmatrix} 0\,0\,0\,0 \\ 0\,0\,0\,0 \\ 0\,0\,0\,0 \\ 1\,1\,1\,1 \end{pmatrix}, \qquad \lambda_{(11)}\big((01)\big) = \begin{pmatrix} 0\,0\,0\,0 \\ 0\,0\,0\,0 \\ 0\,1\,0\,1 \\ 0\,1\,0\,1 \end{pmatrix}, \qquad (24)$$

while $\lambda_{(11)}\big((10)\big) = \lambda_{(11)}\big((01)\big)^T$ and $\lambda_{(11)}\big((11)\big) = \lambda_{(11)}\big((00)\big)^T$. Then

$$\lambda^\delta_{(11)} = \frac{1}{4} \begin{pmatrix} 0\,0\,0\,1 \\ 0\,0\,1\,2 \\ 0\,1\,0\,2 \\ 1\,2\,2\,4 \end{pmatrix} , \tag{25}$$

where we have reinstated a (so far suppressed) superscript δ to remind us of the basis, and $R^\delta_{(11)} = \lambda^\delta_{(11)}$. Eq. (23) then gives

$$P_{(11)}(t+1) = P_{(11)} + \frac{p_c}{2} \left(P_{(10)} P_{(01)} - P_{(11)} P_{(00)} \right) . \tag{26}$$

The equations for the other strings are obtained by renaming the indices.   □

Recombination in the monomial ("building block") basis. As the above example shows, recombination is rather complicated in the δ-basis. A more efficient organization of the various terms that contribute to $G_I(t)$ can be achieved if one thinks in terms of Building Block *schemata*. For example, (11) can be obtained by recombining the schemata (1∗) and (∗1), where a ∗ denotes *any* bit. Each string gives rise to 2^N schemata associated with it, by all possible substitutions of its bits by ∗'s — the corresponding set of schemata constitutes the BBB for that string. For example, (11) generates the basis $\{(\ast\ast), (\ast 1), (1\ast), (11)\}$. Recombination involves the interaction of conjugate schemata only[4], so one expects some sort of "skew diagonalization" of the process in this basis. To connect with the discussion in Sect. 3.2, notice that substitution of a particular bit by a ∗ corresponds, at the level of CF's, to substitution of a coordinate x_i (or $\bar{x}_i$) by the unit function e. It is then clear that the CF's of the Building Blocks are exactly the elements of the monomial basis of Sect. 3.2. We conclude that *the Taylor basis is dual to the BBB.*

The CF corresponding to a schema is the sum of the CF's of all vertices (strings) that the schema matches. On the other hand, it is clear that the probability of a certain schema is likewise the sum of the probabilities of all strings that the schema matches. This implies that, in going from one basis to another, probabilities transform like CF's — in particular

$$\mathbf{P}^{\mathrm{m}} = \Lambda_N \mathbf{P}^\delta . \tag{27}$$

Example 5 $N = 2$ *recombination in the monomial basis*

One can calculate the mask-averaged interaction term in the BBB, $(\lambda^{\mathrm{m}})_I{}^{JK}$, by transforming λ^δ as a rank-three tensor,

$$(\lambda^{\mathrm{m}})_I{}^{JK} = (\lambda^\delta)_{I'}{}^{J'K'} (\Lambda_2)_I{}^{I'} (\Lambda_2^{-1})_{J'}{}^J (\Lambda_2^{-1})_{K'}{}^K , \tag{28}$$

to find, for example, for $\lambda^{\mathrm{m}}_{(11)}$

---

[4] We define the conjugation ⊼ of schemata: the string $R = (r_1 r_2 \dots r_N)$ generates the basis $\{(\ast\ast\dots\ast), (\ast\ast\dots r_N), \dots, (r_1 r_2 \dots r_N)\}$ and $\bar{r}_i = \ast$ while $\bar{\ast} = r_i$, if the ∗ is in position i — the conjugate of a schema is the schema with conjugate bits.

$$\lambda^{m}_{(11)} = \frac{1}{4} \begin{pmatrix} 0\,0\,0\,1 \\ 0\,0\,1\,0 \\ 0\,1\,0\,0 \\ 1\,0\,0\,0 \end{pmatrix}. \tag{29}$$

As expected, it is skew diagonal. The dynamical equation for $P_{(11)}(t)$ is

$$P_{(11)}(t+1) = (1 - \frac{p_c}{2})P_{(11)} + \frac{p_c}{2}P_{(*1)}P_{(1*)}, \tag{30}$$

which by substituting $P_{(*1)} = P_{(11)} + P_{(01)}$, and analogously for $P_{(1*)}$, can be seen to coincide with (26). $\qquad\qquad\qquad\qquad\qquad\qquad\qquad\qquad\qquad\qquad\qquad\square$

The above result generalizes to arbitrary N (see Sect. 5.2 below)

$$(\lambda^{m})_{(11...1)}{}^{JK} = 2^{-N}\delta^{J,\,2^{N}+1-K}. \tag{31}$$

In the δ-basis, the equations for the other elements of the basis can be obtained from the one for $(11\ldots1)$ by renaming the indices. In the monomial basis, the situation is even simpler: one obtains, for example, the equation for $(11*)$ from the one for (11) simply by attaching an extra $*$ to all indices - this generalizes in the obvious way to any number of $*$'s in any position, so that (31), inserted in (23), gives essentially the equations for all basis elements, for all N.

5.2   The Tensor Product Structure of Recombination

As we have seen above, the dynamics of recombination is controlled by the tensor $\lambda(M)$, which contains the information about which parents may give rise to a particular child. In deciding this, one needs to perform a bit-by-bit test, the outcome for the entire string being the logical AND of the individual bit tests (see Eq. 21, where AND corresponds to multiplication). The fact that the value of $\lambda(M)$ factorizes in this manner reflects itself in that $\lambda_I{}^{JK}(M)$, for a length-N string, is the tensor product of the λ's of its bits,

$$\lambda_I(M) = \prod_{r=1}^{N} \lambda_{(i_r)}{}^{(j_r)(k_r)}((m_r)), \tag{32}$$

or, in matrix notation,

$$\lambda_I{}^{JK}(M) = \lambda_{(i_1)}((m_1)) \otimes \lambda_{(i_2)}((m_2)) \otimes \ldots \otimes \lambda_{(i_N)}((m_N)). \tag{33}$$

A simple calculation then shows that the same is true for the mask-independent λ, i.e., in matrix notation, $\lambda_I = \lambda_{(i_1)} \otimes \ldots \otimes \lambda_{(i_N)}$. Finally, given that Λ_N is itself the N-th tensor power of the 1-bit Λ_1, we conclude that the above statements about λ are valid in all bases. Notice that $R_I{}^{JK}$ does not factorize in this manner — this is because checking for the first or the second child, for $N > 1$, is not a bit-wise operation.

Example 6 *The tensor product structure in the string and Building Block bases*

Consider $N = 1$ recombination in the string basis. We find

$$\lambda_{(0)}^{\delta} = \frac{1}{2}\begin{pmatrix} 2 & 1 \\ 1 & 0 \end{pmatrix}, \qquad \lambda_{(1)}^{\delta} = \frac{1}{2}\begin{pmatrix} 0 & 1 \\ 1 & 2 \end{pmatrix}. \tag{34}$$

Transforming to the BBB we find

$$\lambda_{(*)}^{m} = \begin{pmatrix} 1 & 0 \\ 0 & 0 \end{pmatrix}, \qquad \lambda_{(1)}^{m} = \frac{1}{2}\begin{pmatrix} 0 & 1 \\ 1 & 0 \end{pmatrix}. \tag{35}$$

The second equation in (35) above clearly shows that $\lambda_{(11...1)}^{m} = (\lambda_{(1)}^{m})^{\otimes N}$ is skew-diagonal for all N. Notice also that the $\lambda_{(11)}^{\delta}$ given in Eq. (25) is just the tensor square of $\lambda_{(1)}^{\delta}$ given in Eq. (34) above. □

Much of our discussion so far referred to the case of equally probable masks. The above results however are also valid for the case of uniform crossover, where the first child gets the i-th bit from the first parent with probability p_i, resulting in the mask probability distribution $p(M) = \prod_{i \in I_0} p_i \prod_{j \in I_1}(1 - p_j)$, where I_α is the subset of indices in I with value α. For example, for $N = 2$, we get

$$p_{(00)} = p_1 p_2, \quad p_{(01)} = p_1(1-p_2), \quad p_{(10)} = (1-p_1)p_2, \quad p_{(11)} = (1-p_1)(1-p_2).$$

Then the average λ still factorizes, with the string basis 1-bit factors

$$\lambda_{(0)}^{\delta} = \begin{pmatrix} 1 & p_i \\ 1 - p_i & 0 \end{pmatrix}, \qquad \lambda_{(1)}^{\delta} = \begin{pmatrix} 0 & 1 - p_i \\ p_i & 1 \end{pmatrix}, \tag{36}$$

and their BBB counterparts

$$\lambda_{(*)}^{m} = \begin{pmatrix} 1 & 0 \\ 0 & 0 \end{pmatrix}, \qquad \lambda_{(1)}^{m} = \begin{pmatrix} 0 & 1 - p_i \\ p_i & 0 \end{pmatrix}, \tag{37}$$

where i denotes the position of the bit. We see again that $\lambda_{(11...1)}^{m}$ is skew-diagonal. It is easy to see that this generalizes to *any* probability distribution $p(M)$ since $\lambda_{(11...1)}^{m}(M)$ itself is skew-diagonal (a fact that does not depend on $p(M)$) and λ_I^{JK} is a sum over such matrices. It is for this reason that BBB makes manifest the effective degrees of freedom for recombination — the Building Block schemata themselves.

5.3 Selection and Mutation

In Eq. (20), selection is "hidden" inside $P_I'(t)$. In the absence of any further information, all that can be said is that $\mathbf{P}'$ transforms like a vector. When $\mathbf{P}'$ is given in terms of a fitness matrix, $P_I' = \sum_J F_I{}^J P_J$, we may infer that, under a change of basis, $F \rightarrow \Lambda F \Lambda^{-1}$. In the δ-basis, F is taken to be diagonal. In the Walsh basis, F is complicated, the number of non-zero elements depending on the degree of epistasis in the landscape. In the BBB, $F^m = \Lambda F^\delta \Lambda^{-1}$ is not diagonal, however, it can be shown that $F^m = F^{m\prime} + A$, where $F^{m\prime}$ is diagonal and $AP^m = 0$, hence the dynamics is given essentially by a diagonal matrix, as in the δ-basis.

For proportional selection, $(F^{m\prime})_I{}^J = (f_I(t)/\bar{f}(t))\delta_I{}^J$, where $f_I(t)$ is the fitness of the Building Block I and is population- (and hence time-) dependent. Interestingly enough, $AP^m = 0$, hence the dynamics is given essentially by a diagonal matrix, as in the δ-basis. However, the algebraic relation between the two sets of diagonal elements is non-trivial. Note also that only in the very restrictive case of a multiplicative fitness landscape can one generate the N-bit problem from the tensor product of N 1-bit problems.

The mutation matrix transforms like $M \to \Lambda M \Lambda^{-1}$. When the mutation probability p_i of the i-th bit is independent of the other bits, the N-bit mutation matrix factorizes in 1-bit factors,

$$M_N = M(p_1) \otimes M(p_2) \otimes \ldots \otimes M(p_N), \tag{38}$$

where $M(p_i) \equiv \begin{pmatrix} (1-p_i) & p_i \\ p_i & (1-p_i) \end{pmatrix}$. The factorizability of M_N is then preserved in all bases,

$$M_N \to \Lambda_N M_N \Lambda_N{}^{-1} = \Lambda_1 M(p_1)\Lambda_1{}^{-1} \otimes \ldots \otimes \Lambda_1 M(p_N)\Lambda_1{}^{-1}. \tag{39}$$

In the Walsh basis, $M_1^W = \begin{pmatrix} 1 & 0 \\ 0 & 1-2p_i \end{pmatrix}$, while in the BBB, $M_1^m = \begin{pmatrix} 1 & 0 \\ p_i & 1-2p_i \end{pmatrix}$. Thus, as is well known, in the Walsh basis the N-bit mutation matrix is diagonal, while in the BBB it is triangular. In both cases the eigenvalues can simply be read off from the diagonal.

6   Conclusions

We presented GA dynamics in a covariant form, showing how different existing formulations — string, Walsh mode, Building Block schemata — can be related by linear coordinate transformations. It was shown that the N-bit transformation matrices are the N-th tensor power of the corresponding 1-bit matrix. The manifest covariance of the dynamical equations guarantees their validity in all bases — nevertheless, the analysis and its interpretation can be greatly simplified by choosing the basis best adapted to the genetic operator under study. The string basis is convenient for selection-dominated dynamics, while that of Walsh is natural for dynamics dominated by mutation. In this paper we concentrated on the most complicated operator — recombination — showing how the BBB offered the most natural description, the effective degrees of freedom of recombinative dynamics being Building Block schemata. Introducing a description in terms of characteristic functions in configuration space, we showed that the BBB is dual to the standard Taylor basis — the presented mathematical framework, we believe, clarifies several conceptually obscure points. A thorough analysis of the factorizability of the various operators was given, resulting in an enormous simplification of their calculation in the different bases. With the unification program for EC in mind, straightforward generalizations to the case of higher cardinality alphabets and variable-length strings have been alluded to. Given the great similarity between the coarse-grained formulations of GA's and GP, it is reasonable to expect that the above coordinate transformations have analogues in the GP case.

Acknowledgements. The authors acknowledge support from DGAPA-PAPIIT projects IN114302 (CC), ES100201 (CRS) and CONACyT projects 41208-F (CC), 30422-E (CRS).

References

1. W. B. Langdon and R. Poli. *Foundations of Genetic Programming*. Springer Verlag, Berlin, New York, 2002.
2. C. R. Stephens and R. Poli. E C theory - in theory: Towards a unification of evolutionary computation theory. In A. Menon, editor, *Frontiers of Evolutionary Computation*, pages 129–156. Kluwer Academic Publishers, 2004.
3. C. R. Stephens and A. Zamora. EC theory: A unified viewpoint. In Erick Cantú Paz, editor, *Proceedings of GECCO 2003*, pages 1394–1402, Berlin, Germany, 2003. Springer Verlag.
4. Michael D. Vose. *The simple genetic algorithm: Foundations and theory*. MIT Press, Cambridge, MA, 1999.
5. C. R. Stephens and H. Waelbroeck. Schemata evolution and building blocks. *Evol. Comp.*, 7:109–124, 1999.
6. Riccardo Poli. Exact schema theory for genetic programming and variable-length genetic algorithms with one-point crossover. *Genetic Programming and Evolvable Machines*, 2(2):123–163, 2001.
7. C. R. Stephens and H. Waelbroeck. Effective degrees of freedom of genetic algorithms and the block hypothesis. In T. Bäck, editor, *Proceedings of ICGA97*, pages 31–41, San Francisco, CA, 1997. Morgan Kaufmann.
8. Christopher R. Stephens. The renormalization group and the dynamics of genetic systems. *Acta Phys. Slov.*, 52:515–524, 2003.
9. M.A. Akivis and V.V. Goldberg. *An Introduction to Linear Algebra and Tensors*. Dover Publications, Mineola, NY, 1977.
10. Edward D. Weinberger. Fourier and Taylor series on fitness landscapes. *Biological Cybernetics*, 65:321–330, 1991.
11. Alden H. Wright. The exact schema theorem. http://www.cs.umt.edu/CS/FAC-/WRIGHT/papers/schema.pdf, January 2000.

Exploiting Modularity, Hierarchy, and Repetition in Variable-Length Problems

Edwin D. de Jong and Dirk Thierens

Decision Support Systems Group, Universiteit Utrecht, PO Box 80.089
3508 TB Utrecht, The Netherlands
{dejong,dirk.thierens}@cs.uu.nl

Abstract. Current methods for evolutionary computation can reliably address problems for which the dependencies between variables are limited to a small order k. Furthermore, several recent methods can address certain hierarchical problems which feature dependencies between all variables. In addition to modularity and hierarchy, a third problem feature that can be exploited when present is repetition. To enable the study of these problem features in isolation, two test problems for modularity and hierarchy detection by variable length problems are introduced. To explore how a variable length method can exploit these three problem features, a module formation algorithm is investigated. It is found that the algorithm identifies all three forms of problem structure to a substantial degree, leading to significant performance improvements for both the hierarchical and repetitive test problems. The experimental results indicate that the simultaneous exploitation of hierarchy and repetition will require both position-specific module testing and position-independent module use.

Modularity, hierarchy, repetition, SEQ problem, HSEQ problem

1   Introduction

Currently, evolutionary computation can reliably address problems for which the order of the dependencies between variables is limited to a small number k, where two variables are called *dependent* if the fitness contribution of one variable depends on the setting of the other variable and the order of the dependencies is the largest number of interdependent variables. Methods that can address this class of order-k limited problems have been called *competent GA's* [3] and include the fast messy GA [4], the extended compact GA [5], the Bayesian Optimization Algorithm (BOA) [11], LFDA [9], and EBNA [2].

Apart from order-k limited problems, there are certain specific problems with higher-order dependencies that can also be addressed. Specifically, problems with hierarchical structure can feature dependencies up to order $k = n$. These dependencies are limited to specific relations, and by virtue of this hierarchical problems can still be solvable in a scalable manner. Examples of hierarchical problems that have been described so far include H-IFF [15], H-TRAP [10], and

K. Deb et al. (Eds.): GECCO 2004, LNCS 3102, pp. 1030–1041, 2004.

H-XOR [15]. Methods such as SEAM [16] and H-BOA [10] can solve difficult hierarchical problems by exploiting their hierarchical structure.

The class of hierarchical fixed-length problems is of interest because it is the most complex problem class, measured by the order of dependencies between variables, that may still be efficiently addressed by currently known evolutionary algorithms. The class of feasible problems may be further extended however if *variable-length* problems can be addressed.

One reason for employing *variable length* methods is that the length of optimal solutions in a problem may not be known in advance. Furthermore, variable length methods facilitate the use of *translocation* [7], i.e. applying optimized settings from one set of variables to other variables. Translocation exploits the problem feature of *repetition*, and can be used to address increasingly large problem spaces, performing directed exploration of very large search spaces without considering an exponentially increasing number of states; see also [6].

The above suggests that variable length methods exploiting modularity, hierarchy, and repetition would provide a valuable extension of the arsenal of methods currently available. A potential in this direction is demonstrated by the DevRep algorithm [1]; this method was reported to address a 1024-bit version of the HXOR problem. While HXOR [15] features modularity, hierarchy, and repetition, the presence of these multiple features leaves open the question of how these problem features can be exploited *in isolation*. This question will be explored here.

To enable the study of modularity, hierarchy and repetition in isolation, two test problems are introduced: the Sequence problem (SEQ), and the Hierarchical Sequence problem (HSEQ). SEQ features modularity, but no hierarchy or repetition. HSEQ features hierarchy and thereby modularity, but no repetition. To study repetition, we employ the OneMax problem [12]. We investigate how modularity, hierarchy, and repetition can be exploited, and develop a variable length algorithm for module formation. The operation of the algorithm on the SEQ, HSEQ, and OneMax problems is studied in experiments. Control experiments are performed to analyze the necessity of different features of the algorithm.

The structure of the article is as follows. First, a dependency-based classification of problems is provided, discussing modularity and hierarchy. Next, the SEQ and HSEQ problems are introduced. In section 3.1, measures for evaluating the detection of modularity, hierarchy, and repetition are discussed. The module formation that will be employed is described in section 4. Results are provided in section 5, followed by conclusions.

2   A Dependency-Based Classification of Problems

Discrete optimization problems can be classified based on the dependencies between the variables in the problem. Two variables are interdependent if and only if the fitness contribution of one variable depends on the setting of the other variable. Below, we discuss a classification of problems based on the dependencies they feature. The notions of modularity and hierarchy employed here are discussed, and correspond with the different problem classes.

The class of problems that is easiest to address is that for which no dependencies are present between variables. Problems in this class can be solved in linear time by optimizing each variable in turn. If dependencies up to a limited order k are present, a variety of modern genetic algorithms can be used to address the problem in a reliable way. This criterion is closely related to our notion of modularity. Our modularity concept is based on the criterion that the number of settings of a module can be reduced, called *decomposability* [14]. A subset $\mathcal{M}$ of the variables in a problem will be called a *module* if the number of settings of $\mathcal{M}$ that maximize fitness for at least one setting of the remaining variables is less than the number of possible settings for $\mathcal{M}$. There is a clear relation between modularity and order k schemata; in both cases, the dependency of the selected variables on the remaining variables is reduced.

Hierarchical problems may have dependencies up to order $k = n$ while still being solvable in an efficient manner. Thus, exploiting hierarchical structure can permit solving difficult problems that cannot be addressed otherwise.

We will now discuss the notion of hierarchy in more detail. Recall that a module is defined as a set of variables whose number of settings can be safely reduced because only some settings occur in optimal solutions. We view hierarchy as the recursive application of this same principle. Thus, a combination of two or more modules can be viewed as a single composite module if and only if this permits a further reduction of the number of possible settings of the variables involved. A clear demonstration of this principle is given by the Hierarchical IF-and-only-iF (H-IFF) function [15], which can be addressed efficiently by the fixed length methods of SEAM [16] and H-BOA [10]. Other examples of hierarchical problems that can be addressed efficiently include H-XOR [15] and H-TRAP [10].

The notions of modularity and hierarchy that have been described are characteristics of a *problem*. In the literature, the concepts of modularity and hierarchy are often used to describe the operation of a *method*. The problem-based notions of modularity and hierarchy provided here are intended to capture the modules that modular and hierarchical methods should ideally find. Thus, an algorithm is expected to have good performance if the modules identified by the method correspond to the modules of the problem.

The modularity and hierarchy concepts provide a strict criterion for determining which combinations of variables may be called modules; since any combination of variables or modules that is called a module must further reduce the number of possible variable settings that must be considered, the identification of modularity and hierarchy effectively reduces the size of the search space that must be visited, and therefore permits a performance gain.

Repetition is viewed as the presence of optimal substrings that occur multiple times within an individual. If a problem features repetition, translocation can in principle confer an advantage. Translocation is normally avoided in genetic algorithms, as the theoretical foundation aimed that explains the operation of the genetic algorithm requires the propagation of schemata at given locations. If the order of the variables on a genome is non-random however, then patterns of adjacent bits may carry information that can be usefully applied in other parts of the string by means of translocation. For example, if a bitstring encodes natural text using four bits for each letter, then the identification of a string in which certain letters are much more frequent than others may benefit from translocation.

3   The SEQ and HSEQ Problems

In this section, two new test problems for the study of modularity and hierarchy in variable length methods are introduced. Several test problems exist that permit testing whether a method not using translocation can identify modularity and hierarchy; examples include HIFF and HXOR [15]. Since these problems feature some degree of repetition however, translocation is expected to be beneficial in addressing these problems. While the combined exploitation of hierarchy and repetition may be very successful on such problems, as found e.g. in [1], our aim is to study how modularity and hierarchy may be identified in isolation by methods permitted to use translocation.

First, we consider the definition of a non-repetitive test problem featuring modularity. The requirement that the problem should not feature repetition can be guaranteed by ensuring that within across all optimal individuals, any combination of two consecutive values may occur in at most one position. This requirement can be only be satisfied for problems of non-trivial size by using an arity that is greater than two.

The SEQ problem or Sequence problem is defined as follows. For a string of length n, there are two global optima: the ascending string $A = 0, 1, \ldots, n-1$ and the descending string $D = n-1, n-2, \ldots, 0$. In this n-ary problem, the ith variable provides a fitness contribution of 1 if it equals either A_i or D_i. In addition, any two consecutive variables $2j, 2j+1$ for $0 < j < \frac{n}{2}$ provide an extra fitness contribution of 2 if they equal $A_{2j}A_{2j+1}$ or $D_{2j}D_{2j+1}$. Thus, there are two levels that contribute fitness: the level of single variables and the level of consecutive pairs of variables.

The HSEQ problem or Hierarchical Sequence problem is defined by extending the SEQ problem to a higher number of levels; consecutive pairs of ascending modules form ascending modules at the next level, and likewise for the descending modules. By continuing this principle, the highest level features two modules that equal the global optima A and D. The HSEQ problem is analogous to HXOR and HIFF; all three problems combine pairs of consecutive modules into higher-level modules, thereby reducing the number of optimal settings for the constituent modules from four to two. The HSEQ problem is different however in that it employs an arity of n for a length-n string. As a result, it features a much larger search space; the size of the search space is n^n, and thus grows super-exponentially as a function of the string length n.

It is important to note that to address HSEQ problem effectively, it is necessary to maintain *multiple* settings for each combination of variables (two, in this case) until a global optimum is found; if subsets of the variables are allowed to independently converge to one setting (ascending or descending numbers), it becomes increasingly likely with increasing problem size that different subsets will converge to different choices, thereby ruling out the possibility of finding an optimum. Part of the structure in the problem can already be exploited by identifying pairs of variables, and reducing the number of possible settings from the initial n^2 to two; this would amount to the use of modularity. The resulting number of possibilities that must be maintained in this case is still exponential in the number of pairs of variables. By using hierarchy however, the problem can be addressed efficiently. The full potential for efficiency improvement in this hierarchical problem is exploited by *recursively* applying the principle of identifying the optimal settings for each module, thereby identifying correct settings for modules of exponentially increasing size.

3.1   Measuring Modularity, Hierarchy, and Repetition

In measuring the degree two which the three problem features investigated are identified, two criteria are of interest. First, the number of correct modules identified should be high; the more modules are identified, the higher the computational benefit that can be gained. Second, the number of modules formed that are not modules of the problem should be as low as possible since the construction of modules influences the *exploration distribution* [13], i.e. the distribution of individuals that will be visited.

To measure repetition, a slightly different approach is necessary; if a method uses translocation, then maintaining a single instance of a repetitive element is sufficient. Thus, the number of modules identified does not reflect the degree to which repetition is exploited. Therefore, we measure the extent to which the pattern of interest is repeated *within* the modules formed. Since the length of the repeated pattern equals one in the case of OneMax, this reduces to measuring the frequency of the most frequent bit (zero or one) within modules.

4   A Variable-Length Method Exploiting Modularity, Hierarchy, and Repetition

In the following, we develop a variable-length method designed to exploit modularity, hierarchy, and repetition. The algorithm maintains a population of individuals. Individuals are sequences of modules. Initially, the set of available modules contains the primitives of the problem; $\{0, 1\}$ for a binary problem, or $\{0, 1, \ldots n - 1\}$ for the n-variable SEQ or HSEQ problems.

Periodically, a module formation step is performed. The basis for module formation is provided by the notion of hierarchical modularity that has been described. Thus, the aim is to identify combinations of variables for which the number of optimal settings is reduced compared to the number of possible settings. For reasons of computational efficiency, two main restrictions are placed on the modules that can be formed: modules can only consist of consecutive variables, and a module always contains precisely two elements. The main loop of the algorithm is as follows:

Module Formation Algorithm()
1.   pop:=generate_random_individuals(pop_size)
2.   **for** generation=1:no_generations{
3.   **if** generation mod frequency $==$ 0
4.   module_formation(pop,front)
5.   pop:=evolve(pop)
6.   pop:=local_search(pop)
7.   **if** average_length(pop) $<$ size_factor * initial_length
8.   update_lengths(pop)
9.   front:=front∪non_dominated(pop)
10. }

Fig. 1. Module Formation Algorithm

The operation of the algorithm will now be detailed. The module_formation procedure first ensures that for any composite modules C=AB present in the module set, occurrences of the modules combination AB in individuals are replaced by a reference to module C, thereby shortening the length of individuals by one element for each occurrence. Modules only replace occurrences of their elements however if the absolute position of the occurrence, measured in terms of the number of preceding primitives (typically bits, but integers here), is identical to the absolute position at which the module itself was located when it was formed. The relative position of a module in the sequence of modules that specifies an individual is simply its index in the sequence.

Next, the module formation set randomly selects two consecutive elements $[AB]$ from a randomly chosen individual that is part of the *front* set, and forms a candidate module $C = AB$. Thus, the pairs of modules selected as candidate modules reflect the distribution of module pairs in the population. If multiple objectives are used, the front accumulates the non-dominated individuals identified over time. In the current experiments, only fitness is used as an objective, and *front* thus accumulates individuals that have improved the maximum fitness achieved at some point.

If the candidate module C does not already exist as a module, it is tested as follows. For each individual x, the primitives represented by C temporarily replace the corresponding primitives of the individual and the individual is evaluated. Next, the same primitives are replaced by alternative settings for the same absolute positions. Specifically, the module C=AB is compared to all alternative modules combinations A* and *B, where the asterisk (*) is filled in by every existing module of the same size. If any of these alternative settings result in an increased evaluation for any objective, the candidate module will not be formed. If this test is passed for all individuals, the module C=AB is formed and added to the module set. Furthermore, the module replacement procedure is invoked for the newly formed module, i.e. occurrences of AB in individuals at the same absolute position are replaced by C. The use of a position-specific module-formation test ensures that a particular combination of values is tried out for the same set of variables in all compared individuals.

After module formation, the algorithm performs a generation of evolution, using Mahfoud's deterministic crowding algorithm [8]. The operators of variation are as follows. The crossover operator used respects the absolute position of the modules that make up the individuals. It does so by considering all module boundaries shared by the two parents (i.e. determining which relative positions correspond in terms of their absolute positions), and selecting a crossover point randomly from these options.

The first mutation operator randomly selects an element in an individual and replaces it with a randomly selected module of the same length. The second mutation operator randomly selects a module and, if possible, replaces the corresponding elements in the individual at the module's original position. Next, optionally, local search is applied. Local search takes each individual, and optimizes each of its elements in a random order. Optimization consists of considering all alternative modules of the same length, and randomly selecting one that achieves the maximum attainable fitness given the remaining variables.

Finally, a mechanism is used to enable the growth of individuals over time. Initially, individuals contain initial_length elements. When elements are replaced by modules, the effective length of the individual remains the same, but the actual length diminishes as a result of the more compact representation facilitated by the use of modules. To allow the effective length of individuals to increase gradually over time, the update_length procedure restores the length of individuals back to the original initial_length by adding random elements. This operation is performed when the average actual length of the population drops below a threshold.

The algorithm that has been described bears a close relation to existing algorithms designed to exploit hierarchy, such as SEAM and DevRep. A main difference with SEAM is the use of a variable length representation, while a main difference with DevRep is in the use of a position-specific module-formation test.

5   Results

In this section, we investigate the ability of the module formation algorithm to identify modularity, hierarchy, and repetition in test problems that feature these characteristics in isolation as far as possible. The experimental settings are as follows. For methods employing module formation, the initial_length parameter is set to 16. The population size is 100. Crossover is used with $P = 0.8$, and both mutation operators with $P = 0.1$. Module formation is performed every 25 generations. The parameter size_factor $= 0.5$.

5.1   Modularity

The maximum baseline performance for a module formation method that is not able to detect modularity is 50% of correct modules; if module formation randomly selects pairs of consecutive modules, this maximum performance is obtained if the population solely contains global optima.

The results are shown in figure 2. The graph shows that the number of correct modules is substantially higher than the number of incorrect modules, and thus well exceeds the baseline performance. Thus, according to the measurement criterion employed, the method is able to correctly identify modularity in the problem. Preliminary experiments suggest that the effect of identifying correct modules in SEQ on algorithm performance is limited however; the computational benefit of identifying interdependent sets of variables is expected to be most clear in deceptive problems.

5.2   Hierarchy

To test the ability of the module formation algorithm to identify hierarchy, the method is applied to the HSEQ problem. Again, the evaluation criterion for module formation consists of counting the number of correct and incorrect modules formed. The results are given in Figure 2, and show that the number of correct modules formed grows at a steady pace and then quickly flattens. The number of incorrect modules formed remains low. While it might be expected that the correct modules in HSEQ are more difficult to detect, as the modules range from size 2 to modules representing complete individuals (size 128), the relation between correct and incorrect modules is in fact higher than for the SEQ problem. The total number of correct non-primitive modules for the 128-variable HSEQ problems equals 254.[1] The number of correct modules identified (228) is thus higher, both in absolute and relative terms, than for the SEQ problem; 90% of the modules present in the 128-variable HSEQ problem are identified.

The search space for 128-HSEQ is of size $128^{128} = 2^{896}$, and thus corresponds to a 896-bit problem. In addition to being large, there is very little gradient in this space; in a random string, only 1 in every 64 variables is expected to make a fitness contribution. Therefore, local search is used to speed up the search; this renders the problem more comparable to a 128-bit problem again while maintaining the property that no repetition is present in the problem.

Figure 3 show the performance of the module formation method on the 128-variable HSEQ problem. The method achieves near optimal performance on average. Inspection of the runs showed that 8 out of ten runs reached the global

---

[1] The 64 pairs for each direction (ascending and descending) form the leaves of a binary tree, which therefore has 63 internal nodes.

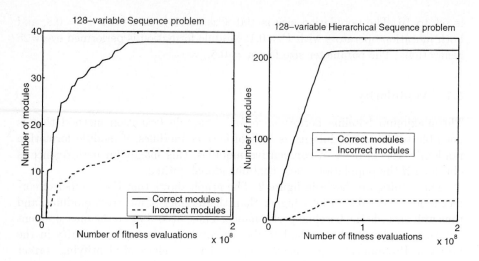

Fig. 2. Number of correct and incorrect modules formed on the SEQ (left) and HSEQ (right) problems.

optimum of 1024. Interestingly, the abrupt change in the module formation graph (Figure 2) corresponds precisely to the moment at which a global optimum is reached on average; this suggests that module formation ends when all correct modules present in the population have been formed.

As a comparison, we apply a genetic algorithm that does not perform module formation to the same task. To focus on the effect of module formation, the algorithm is identical to the module formation algorithm, except that no modules are formed after the module set has been initialized. As a result, the length of individuals cannot increase, and the initial length of individuals is therefore made equal to the length required for the problem, i.e. 128 in this case. Thus, the genetic algorithm receives some prior information about the problem, namely the required length for the problem, which the module formation algorithm does not receive. As figure 3 shows, the performance of the genetic method is significantly less however, and module formation can thus be concluded to have a positive effect on performance.

Two variants of the module formation algorithm are compared. The first variant uses position-independent operators by using variants of the crossover and mutation operators that do not respect the alignment of elements of individuals, and can thus translocation. In the second control experiment, the module formation does not test a candidate module at the position where it occurred, but at all positions within the individual.

Figure 3 shows the results of the control experiments. The use of position-independent operators does affect performance, but still permits the formation of useful modules; this algorithm is still very different from the genetic algorithm in figure 2, as individuals have an initial length of 16, and the substantial perfor-

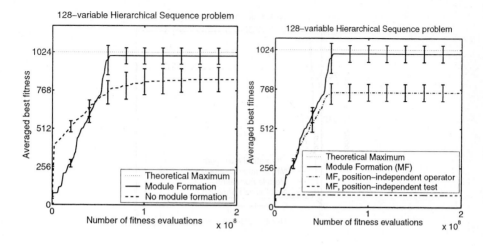

Fig. 3. Performance on the 128-variable HSEQ problem for various methods.

mance increase shows that many modules are still formed. This is different for the control experiment in which the module acceptance test does not respect the positions; for this method, no modules are formed, and performance is therefore limited to a maximum of 80, which is quickly attained but not exceeded.

5.3 Repetition

Finally, we study the ability of the module formation method to identify and exploit repetition. To exploit repetition, translocation is required. Thus, the position-independent operators used in the previous experiments are used again. Since the OneMax problem is binary, we use a larger, 1024-bit version, and local search is not required.

Figure 4 demonstrates the potential benefit of translocation. The standard module formation algorithm already improves substantially over the genetic algorithm not performing module formation. When translocation is employed, maximum performance is reached almost immediately. Inspection of the degree of repetition achieved, as measured by the average relative frequency of the most frequent primitive, showed that the maximum degree repetition, i.e. 1.0, was reached and maintained.

It appears that the recursive combination of strings of ones into larger and larger modules can produce correct solutions to one-max of sizes growing exponentially in the number of generations, but this hypothesis remains to be tested. Likewise, information about the scalability of the method on modular and hierarchical problems is of central interest in further evaluating the potential of the method that has been described.

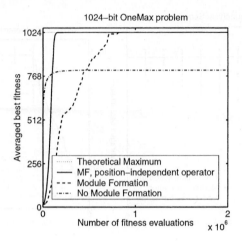

Fig. 4. Performance on the 1024-bit OneMax problem.

6   Conclusions

A wide range of problems can currently be reliably addressed by methods from evolutionary computation. Furthermore, particular subclasses of the remaining problems can also be addressed efficiently. One such subclass which has received recent attention is that of problems featuring hierarchy. Two other problem features that can improve the efficiency of a search method when exploited are modularity and repetition. We explore how these problem features may be detected and used to benefit by variable length algorithms.

To study the identification and exploitation of modularity, hierarchy, and repetition, two new test-problems have been introduced: the SEQ and HSEQ problems. Existing test problems contain a combination of these features. In contrast, the new test problems, and the existing OneMax problem, enabled the study of these problem features in isolation.

A variable length algorithm employing module formation has been described. In experiments, it was demonstrated that the modules formed by the method correspond to the modules present in the problems, and the method can thus be said to detect modularity, hierarchy, and repetition to a substantial degree. For the HSEQ and OneMax problems, a significant performance gain was achieved as a result of module formation. While translocation was seen to be useful in the presence of repetition and no insurmountable obstacle in the hierarchical HSEQ problem, a position-specific module-acceptance test was found crucial in the latter problem. These findings suggest that successful independent exploitation of hierarchy and repetition requires both position-specific module testing and position-independent module use.

References

1. Edwin D. De Jong. Representation development from Pareto-coevolution. In E. Cantú-Paz et al., editor, *Proceedings of the Genetic and Evolutionary Computation Conference, GECCO-03*, pages 262–273, Berlin, 2003. Springer.
2. R. Etxeberria and P. Larrañaga. Global optimization using bayesian networks. In A. Ochoa Rodriguez, M. Soto Ortiz, and R. Santana Hermida, editors, *Proceedings of the Second Symposium on Artificial Intelligence CIMAF*, 1999.
3. David E. Goldberg. *The design of innovation. Lessons from and for competent genetic algorithms.* Kluwer Academic Publishers, 2002.
4. David E. Goldberg, K. Deb, H. Kargupta, and G. Harik. Rapid, accurate optimization of difficult problems using fast messy genetic algorithms. In *Proceedings of the Fifth International Conference on Genetic Algorithms*, pages 56–64, 1993.
5. Georges Harik. Linkage learning via probabilistic modeling in the ECGA. Technical Report Illigal report no. 99010, University of Illinois at Urbana-Champain, Urbana, IL, 1999.
6. Inman Harvey. The SAGA cross: the mechanics of recombination for species with variable-length genotypes. In R. Männer and B. Manderick, editors, *Parallel Problem Solving from Nature, PPSN-II*, volume 2, pages 269–278, Amsterdam, 1992. North-Holland.
7. John H. Holland. *Adaptation in Natural and Artifical Systems.* University of Michigan Press, Ann Arbor, MI, 1975.
8. Samir W. Mahfoud. *Niching Methods for Genetic Algorithms.* PhD thesis, University of Illinois at Urbana-Champaign, May 1995. IlliGAL Report 95001.
9. Heinz Mühlenbein and Thilo Mahnig. FDA - A scalable evolutionary algorithm for the optimization of additively decomposed functions. *Evolutionary Computation*, 7(4):353–376, 1999.
10. Martin Pelikan and David E. Goldberg. Escaping hierarchical traps with competent genetic algorithms. In L. Spector et al., editor, *Proceedings of the Genetic and Evolutionary Computation Conference, GECCO-01*, pages 511–518. Morgan Kaufmann, 2001.
11. Martin Pelikan, David E. Goldberg, and Erick Cantu-Paz. BOA: The bayesian optimization algorithm. In Wolfgang Banzhaf, Jason Daida, Agoston E. Eiben, Max H. Garzon, Vasant Honavar, Mark Jakiela, and Robert E. Smith, editors, *Proceedings of the Genetic and Evolutionary Computation Conference*, volume 1, pages 525–532, San Francisco, CA, 13-17 July 1999. Morgan Kaufmann.
12. J.D. Schaffer and L.J. Eshelman. On crossover as an evolutionary viable strategy. In R.K. Belew and L.B. Booker, editors, *Proceedings of the 4th International Conference on Genetic Algorithms*, pages 61–68, 1991.
13. Marc Toussaint. The structure of evolutionary exploration: On crossover, buildings blocks, and Estimation-Of-Distribution algorithms. In *2003 Genetic and Evolutionary Computation Conference (GECCO 2003)*, Berlin, 2003. Springer.
14. Richard A. Watson. *Compositional Evolution: Interdisciplinary Investigations in Evolvability, Modularity, and Symbiosis.* PhD thesis, Brandeis University, 2002.
15. Richard A. Watson, Gregory S. Hornby, and Jordan B. Pollack. Modeling building-block interdependency. In A.E. Eiben, Th. Bäck, M. Schoenauer, and H.-P. Schwefel, editors, *Parallel Problem Solving from Nature, PPSN-V.*, volume 1498 of *LNCS*, pages 97–106, Berlin, 1998. Springer.
16. Richard A. Watson and Jordan B. Pollack. A computational model of symbiotic composition in evolutionary transitions. *Biosystems*, 69(2-3):187–209, May 2003. Special Issue on Evolvability, ed. Nehaniv.

Optimal Operating Conditions for Overhead Crane Maneuvering Using Multi-objective Evolutionary Algorithms

Kalyanmoy Deb and Naveen Kumar Gupta

Kanpur Genetic Algorithms Laboratory (KanGAL)
Department of Mechanical Engineering
Indian Institute of Technology Kanpur
Kanpur, PIN 208 016, INDIA
deb@iitk.ac.in
http://www.iitk.ac.in/kangal/deb.htm

Abstract. While operating a crane for maximum productivity, the time of operation and the required energy are two important conflicting factors faced by a crane operator. In such a case, trying to reach the destination too quickly demands a large energy supply, while a small powered motion requires longer time. In this paper, we consider such a problem for two different pairs of objectives and employ a multi-objective genetic algorithm for the task. Besides finding a set of trade-off optimized solutions (operating conditions), an analysis of these solutions reveals salient operating principles, which would be difficult to achieve by other means. The methodology demonstrated in this paper can be used for other similar engineering design and application problems.

Keywords: Crane maneuvering, Multi-objective GAs, Optimal trade-off, Dynamics of cranes.

1 Introduction

Overhead cranes are used in the industries, workshops, factories, docks, and other places to transport the heavy components from one place to another. In order to increase the productivity, individual such operation must be optimized to find what speed the overhead trolley must be moved so that the supplied energy to the crane and the overall operation time are minimum. The overall operation time has two components: (i) the trolley time which denotes the time needed for the trolley to move from the starting position to the destination point and (ii) the sway time which denotes the time needed by the hanging load to damped out its oscillation to a critical acceptable limit. Although not obvious, these two objectives have a conflicting effect. If an operator tries to reach the destination too fast (by spending too much energy) the trolley time will be saved, but the sudden stopping of the trolley will cause the remaining energy to be transferred to the hanging load, thereby starting a large-amplitude oscillation to the hanging

K. Deb et al. (Eds.): GECCO 2004, LNCS 3102, pp. 1042–1053, 2004.

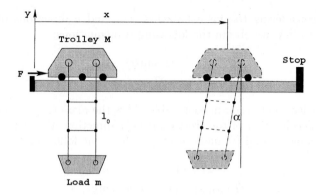

Fig. 1. A schematic of the overhead crane consisting of a trolley and a swaying load.

load. On the other hand, a careful and slow motion towards the destination point will, though not impart a large-amplitude oscillation to the hanging load, require a larger trolley time. Thus, it is important to know what and how to move the trolley right from the starting position so that certain goals are achieved.

Although there exist a number of classical multi-objective optimization techniques [8,1,6], multi-objective evolutionary algorithms (EMO) [3,9,2] have gained tremendous popularity in solving different kinds of engineering problems. In this paper, rather than finding one solution to the problem, we employ a multi-objective genetic algorithm — the elitist non-dominated sorting GA or NSGA-II [4] — to first find a set of trade-off optimal solutions for two conflicting objectives of operation. Thereafter, the obtained solutions are analyzed to reveal important operating principles for the task. By considering two different pairs of objectives, important information about the optimal crane operations are found. The methodology used in this study can be followed in handling similar other engineering problems.

2   Modeling the Dynamics of Crane Operation

Figure 1 shows a schematic model of the crane used in this study. In this simplified model, the cable connecting the trolley and the hanging load is considered of fixed length, however in practice such cables are varied in length while moving in order to lower or raise the load [5]. The fixed length assumption reduces one degree-of-freedom of the system, however, a similar study without this assumption can also be made. The system has two degrees-of-freedom: (i) x denoting the linear motion of the trolley along x-direction and (ii) α denoting the angular motion of the hanging mass.

In the model, we assume that there is a time-varying force $F(t)$ applied to the trolley in the direction of the destination (along positive x-direction). The trolley experiences a friction force $f = \mu N$ (μ being the coefficient of friction between the trolley and the guide and N is the normal force) opposite to its motion. By considering the force balances in x and y directions of all forces acted on the trolley, we obtain the following two equations:

$$M\ddot{x} = F + 2T\sin(\alpha) - \mu N,$$
$$N = Mg + 2T\cos(\alpha),$$

where, T is twice the tension in each cable, M is the mass (in kg) of the trolley, and $\ddot{x}$ is the acceleration of the trolley in the x-direction. Performing a similar task for the hanging load (of mass m kg), we have the following two equations:

$$-2T\sin(\alpha) - c\dot{x}_1 = m\ddot{x}_1,$$
$$2T\cos(\alpha) - c\dot{y}_1 - mg = m\ddot{y}_1,$$

where, c is the coefficient of damping arising due to several factors on to the cable and the hanging mass. The variables x_1 and y_1 are the displacement of the hanging load in the x and y directions, respectively.

In addition, the following relationships between trolley and the hanging load motions can be written with variables along x and y directions:

$$x_1 = x + l_o\sin(\alpha), \qquad\qquad y_1 = -l_o\cos(\alpha),$$
$$\dot{x}_1 = \dot{x} + l_o\dot{\alpha}\cos(\alpha), \qquad\qquad \dot{y}_1 = l_o\dot{\alpha}\sin(\alpha),$$
$$\ddot{x}_1 = \ddot{x} + l_o\ddot{\alpha}\cos(\alpha) - l_o\dot{\alpha}^2\sin(\alpha). \qquad \ddot{y}_1 = l_o\ddot{\alpha}\sin(\alpha) + l_o\dot{\alpha}^2\cos(\alpha).$$

Here, l_o is the length of the cable, $\dot{\alpha}$ $\ddot{\alpha}$ are the angular velocity and acceleration of the cable. By eliminating T, x_1 and y_1 from the above expressions, we get the following two equations of motion of the trolley and the hanging mass:

$$\ddot{x} = \left[F - c\dot{x}\sin^2(\alpha) + ml_o\sin(\alpha)\dot{\alpha}^2 + mg\sin(\alpha)\cos(\alpha) - f(ml_o\cos(\alpha)\dot{\alpha}^2\right.$$
$$\left. -c\dot{x}\sin(\alpha)\cos(\alpha) - mg\sin^2(\alpha))\right] / (M + m\sin^2(\alpha) - fm\sin(\alpha)\cos(\alpha))(1)$$

$$\ddot{\alpha} = -\left(\ddot{x} + r\dot{x} + g\tan(\alpha)\right)\frac{\cos(\alpha)}{l_o} - r\dot{\alpha}, \tag{2}$$

where, r is the ratio of c to m. These two equations can be solved using a numerical integration technique and the variation of x and α with time can be found. Here, we use an adaptive scheme for stable solutions to the above equations.

3   Energy and Time Minimizations

A little thought over the problem makes it clear that the two objectives (i) total energy supplied to the system and (ii) the total time for the block-load system to reach at the desired position and stabilize are the two conflicting objectives. The supplied energy will be minimum for the case of moving ever slowly towards the destination. But such a solution will require quite a long time to complete the task. On the other hand, reaching the destination with a large velocity and suddenly stopping at the destination would be a quick way to reach the destination, however some time needs to be spent for the sway of the load to diminish. Although such a solution may not be the quickest overall time solution, there would exist a solution with a reasonable velocity which would minimize the overall time.

In this paper, we use the following parameters for the crane system:

$$M = 20,000 \text{ kg}, \quad m = 30,000 \text{ kg}, \quad \mu = 0.1$$
$$l_o = 25 \text{ m}, \quad c = 50 \text{ N-s/m}.$$

Here, we use NSGA-II [4] for minimizing the above two objectives. Constraints are handled using the constraint-domination approach suggested elsewhere [3]. We use the following NSGA-II parameters: (i) population size = 150, (ii) maximum number of generations = 1,000, (ii) probability of crossover = 0.9, and (iv) probability of mutation = 0.01.

The decision variables in the crane operating problem are the magnitude and sequence of application of forces on the trolley till it reaches the destination. To keep matters simple, we have used a force F_0 and a sequence of Boolean variables to denote the application of the force. A typical NSGA-II solution is as follows:

$$(F_0 \quad (1110010100))$$

Each time step is assumed to be of $\Delta t = 4$ sec duration. Thus, in the above example, the trolley reaches the destination in (10×4) or in 40 sec. The force F_0 is always applied at the beginning of the first time step. Thus, in the binary string representing the sequence of operation, the first bit is always a 1. It is clear that with the above representation scheme, every solution may have a different size of the binary string. To avoid this problem of coding, we maintain a fixed length (of large size, $l_{max} = 150$) string and use the front part of the string. Thus, a NSGA-II solution will have a maximum trolley time of 150×4 or 600 sec. The motion of the trolley-load is simulated with the pattern of application of force as dictated by a string and the corresponding F_0 and as soon as the trolley reaches the destination, the string is not used further.

The force parameter is also treated as a binary string of length $l_F = 10$ initialized in the range $[100, 1670]$ N. Thus, the total string length of a NSGA-II solution is $l_F + l_{max} = 10 + 150$ or 160. The overall binary string is operated by a single-point crossover and a bit-wise mutation [7].

Three different implementations are adopted for the energy-time minimizations:

Approach 1: The magnitude of force is varied and the pattern of the application of force is kept periodic, such as on, off, on, off, etc (or (1010 $\cdots$)).

Approach 2: The pattern of application of force is varied as a series of Boolean variables (on or off) and the force is kept to a constant value (F_0).

Approach 3: Both the pattern of application of force and the magnitude of force is varied.

In each case, the maximum string length for representing the pattern of force is kept to be 150. We discuss each of the above implementations and obtained results in the following subsections.

3.1   Approach 1: Variation of Magnitude of Force

Here, a fixed pattern of application of force is applied and the magnitude of the force is dictated by a GA solution. Thus, for each solution we keep track of the total work being done to move the trolley to the final destination and call it the energy to be supplied. The second objective is the total time required to reach the destination and to have the load sway to reduce to a permissible value. Here we call the system is stable if the angular sway is reduced to $\alpha_c = 0.002$ rad.

Figure 2 shows the optimized non-dominated front, trading-off the two conflicting objectives. We observe that a small energy solution takes longer time

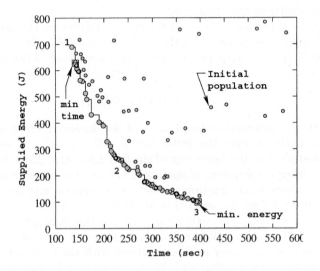

Fig. 2. Optimized trade-off solutions for Approach 1 are compared with the initial random population. Single-objective optimized solutions are also marked.

and a fast solution requires large energy. Although such a trade-off which was anticipated at the start of the study, the figure quantifies the terms and shows a number of such trade-off solutions. It then depends on the operator to choose a particular solution depending on the available time and energy at his/her disposal. The figure also shows the initial random population. This population is shown to get a clear idea of the extent of progress of NSGA-II in the objective space.

Figure 2 also shows two individual minima for the objectives. Since the individual minima comes on the the optimized front, it can concluded that the obtained NSGA-II solutions are the true non-dominated solutions or are very close to them.

To show the trade-off further, we choose three solutions from the optimized front (marked as 1, 2 and 3 in Figure 2) and show the time-variation of the

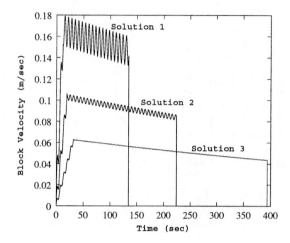

Fig. 3. The velocity variation of the trolley as it moves towards the destination for three different non-dominated solutions. The trade-off in their variations is clear.

trolley's velocity with time in Figure 3. Since a periodic pattern is used for applying the force, a periodic variation in the velocity is observed. Solution 1 is the minimum-time solution and hence require a large energy. On the other hand, solution 3 is the minimum energy solution. This solution suggests the smallest velocity of the trolley as it moves towards the destination, but requires the longest time to reach there.

3.2   Approach 2: Variation of Pattern of Application of Force

Here, we keep a constant force $F_0 = 500$ N and vary the pattern of application of force. Such an implementation is quite practical as with a fixed energy source it can be assumed that the force applied to the trolley would be identical at different time steps and the user only needs to know with what sequence the force is to be applied.

Figure 4 shows the obtained optimized front for the same two objectives. The initial population is also marked. The inset figure shows the trade-off between supplied energy and the time more clearly. Individual minima are also shown in the plot. NSGA-II solutions are found to be non-dominated with these solutions.

Some interesting observations can be made when we investigate the force patterns, as shown in Table 1. With the increase in supplied energy the strings get smaller, meaning that the system reaches the destination quickly. In almost all solutions the force is continuously applied in the initial few time steps. Once the trolley has acquired the required energy to overcome the friction and other

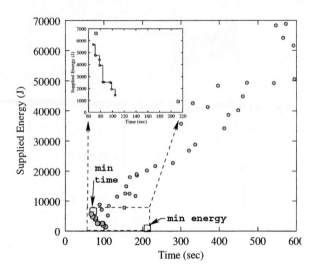

Fig. 4. Optimized trade-off solutions for Approach 2 are compared with the initial random population. Single-objective optimized solutions are also marked.

dynamics, only occasionally the force is required to be applied. A general pattern for an optimal maneuvering seems to be to apply the force early on and let the frictional force reduce the motion of the trolley later on and till the trolley reaches its destination. Such a pattern is not hard-coded in NSGA-II. Such a property of the optimized solutions emerge as a desired mode of operation in an optimal manner. Such informations are useful to the operators and can be quite useful in real-world applications.

3.3   Approach 3: Variation of Force and Application Pattern

Next, we keep both the force and the application of force pattern as variables. Figure 5 shows the obtained non-dominated front. Once again, the individual minima (obtained using a single-objective GA with identical representation scheme and operators as in NSGA-II) are also shown on the plot. It can be observed that the NSGA-II frontier is non-dominated with these individual minima. Also, since the representation involves both F_0 and force pattern, a wider non-dominated front is discovered compared to Figures 2 and 4.

In Table 2, we show the force and its application pattern for a few obtained trade-off solutions. It is clear that quicker solutions require more energy and a larger magnitude of force. Importantly, the force is required to be applied early on and then the trolley moves with its acquired energy till it reaches the destination. Smaller force solutions consume smaller energy and moves slowly to the destination. A similar pattern was also found in Approach 2.

To show the above pattern and force graphically, we have chosen five solutions from the entire range of the obtained solutions and the force variation is

Table 1. Trade-off solutions and corresponding patterns of application of force.

Time (sec)	Work (J)	Pattern
104.5	1440.446	111111000000000000000000
98.5	1938.642	111111000001000000000
97.2	2487.745	101111000100010000000100
95.5	2520.041	111111000001000100000
85.8	2531.828	111111000011000000000
83.7	2539.399	111111010100000000000
79.0	3933.450	111111000011100100000
78.5	3937.728	111111000111100000000
78.4	4399.622	111111000111000100000
71.9	4759.727	111111101110000100000
71.1	4779.473	111111111100001000
68.0	5670.620	111111111110011000

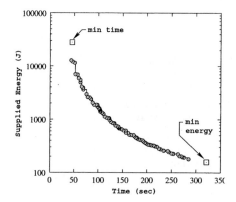

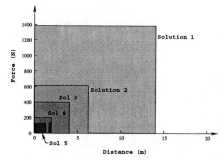

Fig. 6. Force versus distance moved by the trolley for five widely-distributed solutions on the obtained front.

Fig. 5. Optimized trade-off solutions for Approach 3. Single-objective optimized solutions are also marked.

plotted with distance moved by the trolley in Figure 6. Since the area covered by such variations is related to the supplied energy, it is clear that the minimum time solution (solution 1 marked in Figure 6) requires the largest energy to complete the task, while solution 5 requires a small fraction of the energy needed in solution 1 to complete the task, but the time taken to reach the destination is 6.5 times more.

4 Trolley Time and Sway Time Minimizations

Now, we consider two components of the overall time of completion of the task: (i) minimization of the trolley time — time needed by the trolley to reach the

Table 2. Objective values and corresponding solutions for a few non-dominated solutions.

Time (sec)	Force (N)	Pattern
285.0	176.9	(11101110000000000000000000000000000000000000000000000000000000000)
266.1	189.2	(1110111000000000000000000000000000000000000000000000000000000000)
213.7	237.0	(11101110000000000000000000000000000000000000000000000)
191.3	264.7	(111011100000000000000000000000000000000000000000)
167.5	297.0	(1110111000000000000000000000000000000000000)
116.6	432.5	(1110111000000000000000000000)
98.0	484.8	(111101110000000000000000)
75.6	717.3	(111011100000000000)
68.9	792.8	(1110111000000000)
63.0	926.7	(11101110000000)
51.2	1060.7	(11111111000)
43.8	1493.3	(1110111000)

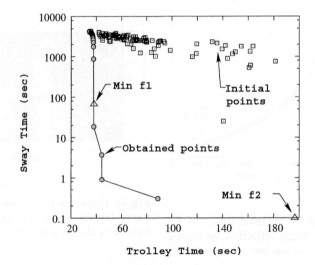

Fig. 7. Optimized trade-off solutions for the trolley time ($f1$) and sway time ($f2$) minimizations. Single-objective optimized solutions are also marked.

destination and (ii) sway time — time needed for the hanging load to stabilize (maximum angular displacement comes within a small limit $\alpha_c = 0.0002$ rad). In this case, we consider $\Delta t = 6.67$ sec and a maximum string length of $\ell_{\max} = 750$, so that a maximum trolley time of 750×6.67 or $5,000$ sec can be achieved. For this case, the a NSGA-II solution is 760 bit long.

In this case, we only show the approach in which both the force and the pattern of application of the force are varied. Figure 7 shows the initial population,

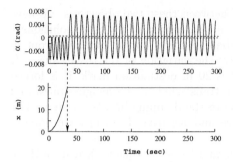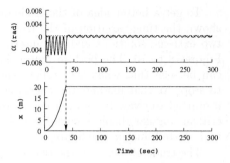

Fig. 8. Minimum trolley-time solution (sol. 1). Sway stabilizes after 4,248.1 sec.

Fig. 9. Intermediate solution (solution 2). Sway stabilizes after 870.6 sec.

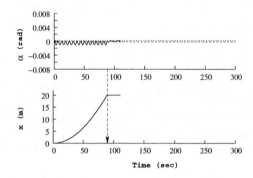

Fig. 10. Minimum sway time solution (solution 3). Sway stabilizes after 0.3 sec.

Table 3. Objective values and corresponding solutions for a few of the non-dominated solutions.

Trolley Time (sec)	Sway Time (sec)	Force (N)	Pattern
34.6	4248.1	1673.460	(11111)
36.5	3800.2	1618.035	(11110)
37.8	1758.1	1413.269	(11111)
37.9	870.6	1404.032	(11111)
38.0	18.4	1402.492	(11111)
44.3	3.7	1031.451	(111111)
88.9	0.3	260.117	(11100101011010011)

obtained trade-off front by NSGA-II, and two individual minima for the objectives obtained using a single-objective GA. A comparison with the individual minima indicates that the obtained NSGA-II solutions are non-dominated with the individual minima.

To get a better idea of the trade-off between these two objectives, we have shown the time-variation of the trolley and hanging load for three solutions – two extreme solutions and one intermediate solution on the obtained front in Figures 8 to 10. The bottom figure shows the variation of the trolley mass. It can be seen that when the trolley reaches 20 m destination mark, it is forced to stop. At that instant, the energy gets transferred to the hanging mass and it started to sway. Its motion decays due to the damping into the system. The critical angular displacements in positive and negative directions are marked with dotted lines in the figure.

The trade-off between the trolley time and the sway time is clear from these figures. Solution 1 requires smallest trolley time (34.6 sec), but gets into a larger amplitude oscillation which requires 4,248.1 sec to get damped to the critical α_c. On other hand, solution 2 (intermediate one) takes slightly more trolley time (37.9 sec), but gets damped out to the limit within 870.6 sec. Solution 3 reaches the destination slowly requiring as large as 88.9 sec. But the slow arrival at the destination causes the hanging mass to get damped out to the limit almost immediately (in only 0.3 sec).

Table 3 shows the objective values and corresponding solutions for a few of the obtained trade-off solutions.

It can be observed that the applied force is inversely proportional to the elapsed trolley time, that is, for a quick (small time) arrival at the destination, more force (hence more energy) must be applied. Although most solutions require an early application of the force, as dictated by the pattern in the table, the smallest sway time requires a careful on-off application of the force till it reaches the destination.

If the time of completing the task is important, the summation of trolley and sway times can be optimized. For the solutions mentioned in the above table, the second-last solution seems to be the optimal solution. Although the above consideration of two-objective minimization would usually include this optimized solution, it also provides other useful information about the problem which would be useful to the operators or users.

5   Conclusions

In this paper, we attempted to find optimal operating conditions of an overhead crane in carrying a load over a distance. First, the task is optimized for two conflicting goals of design: the supplied energy and the task completion time. Using a multi-objective GA (NSGA-II), we have obtained a number of trade-off solutions. It has been observed that an operation requiring minimal time of completion demands for a large energy; on the other hand, an operation requiring minimal energy demands for a longer time of completion. In each case, an optimization of individual objectives has been performed to build confidence on the obtained non-dominated front.

Moreover, an investigation of the obtained trade-off solutions reveal the following important operating principles:

1. The bang-bang force model used in the study requires the forces to be applied early on so that the system acquire enough energy to complete the task. Although not obvious, such a strategy would enable the trolley to reach its destination with a minimal energy so that when stopped suddenly at the destination the hanging load does not sway much.
2. The applied force is inversely proportional to the time to reach the destination.

In another case study, the trolley time and the sway time are minimized using NSGA-II and a trade-off relationship between them is observed.

Although the application study considered here is a specific one related to the overhead crane operating conditions, the methodology used here can be used in other engineering design and applications. A consideration of more than one objective (with a conflict in them) in the optimization process is expected to produce a set of trade-off solutions. An investigation of such trade-off solutions should reveal important information about the problem, which may be difficult to obtain by any other means.

References

1. V. Chankong and Y. Y. Haimes. *Multiobjective Decision Making Theory and Methodology*. New York: North-Holland, 1983.
2. D. Corne, J. Knowles, and M. Oates. The Pareto envelope-based selection algorithm for multiobjective optimization. In *Proceedings of the Sixth International Conference on Parallel Problem Solving from Nature VI (PPSN-VI)*, pages 839–848, 2000.
3. K. Deb. *Multi-objective optimization using evolutionary algorithms*. Chichester, UK: Wiley, 2001.
4. K. Deb, S. Agrawal, A. Pratap, and T. Meyarivan. A fast and elitist multi-objective genetic algorithm: NSGA-II. *IEEE Transactions on Evolutionary Computation*, 6(2):182–197, 2002.
5. G. Dissanayake and G. Fang. Minimum-time trajectory for reducing load sway in quay-cranes. *Engineering Optimization*, 33:643–662, 2001.
6. M. Ehrgott. *Multicriteria Optimization*. Berlin: Springer, 2000.
7. D. E. Goldberg. *Genetic Algorithms for Search, Optimization, and Machine Learning*. Reading, MA: Addison-Wesley, 1989.
8. K. Miettinen. *Nonlinear Multiobjective Optimization*. Kluwer, Boston, 1999.
9. E. Zitzler, M. Laumanns, and L. Thiele. SPEA2: Improving the strength pareto evolutionary algorithm for multiobjective optimization. In *Evolutionary Methods for Design Optimization and Control with Applications to Industrial Problems*, pages 95–100, 2001.

Efficiently Solving: A Large-Scale Integer Linear Program Using a Customized Genetic Algorithm

Kalyanmoy Deb[1] and Koushik Pal[2]

[1] Kanpur Genetic Algorithms Laboratory (KanGAL), Department of Mechanical Engineering, Indian Institute of Technology Kanpur, Kanpur, PIN 208016, India, deb@iitk.ac.in, http://www.iitk.ac.in/kangal/deb.htm
[2] Department of Mathematics and Scientific Computing, Indian Institute of Technology Kanpur, Kanpur, 208016, India, kapal@iitk.ac.in

Abstract. Many optimal scheduling and resource allocation problems involve large number of integer variables and the resulting optimization problems become integer linear programs (ILPs) having a linear objective function and linear inequality/equality constraints. The integer restrictions of variables in these problems cause tremendous difficulty for classical optimization methods to find the optimal or a near-optimal solution. The popular branch-and-bound method is an exponential algorithm and faces difficulties in handling ILP problems having thousands or tens of thousands of variables. In this paper, we extend a previously-suggested customized GA with four variations of a multi-parent concept and significantly better results are reported. We show variations in computational time and number of function evaluations for 100 to 100,000-variable ILP problems and in all problems a near-linear complexity is observed. The exploitation of linearity in objective function and constraints through genetic crossover and mutation operators is the main reason for success in solving such large-scale applications. This study should encourage further use of customized implementations of EAs in similar other applications.

Keywords: Integer linear programs, customized GAs, Large-scale optimization, computational time

1 Introduction

Optimal scheduling and resource allocation problems often arise in different real-world activities and are routinely solved using classical search and optimization algorithms including linear programming methods. The difficulties often faced in solving such problems are (i) dimensionality of the search space and (ii) integer restriction of the decision variables. If the resulting problem is linear (that is, the objective function and constraints are all linear functions of the decision variables), the linear programming (LP) approaches are ideal candidates to solve such problems (Taha, 1989). Although the first difficulty is not a matter for solving such problems using an LP, the second difficulty requires an LP approach

K. Deb et al. (Eds.): GECCO 2004, LNCS 3102, pp. 1054–1065, 2004.

to be used with an integer programming approach, such as the *branch-and-bound* method. Since the branch-and-bound approach requires branching every non-integer variable into two different LPs, the presence of a large number of integer decision variables demands an exponentially large number of function evaluations to solve the problem to optimality.

For past few decades, such problems have also been solved using various non-traditional methods, such as simulated annealing, genetic algorithms, tabu search etc. (Kirkpatrick, Gelatt and Vecchi, 1983, Goldberg, 1989; Glover, 1997). In some of these methods, although the second difficulty of handling integer variables is not a matter, the first difficulty of handling a large number of decision variables is not well researched. A recent paper (Deb, Reddy and Singh, 2003) clearly demonstrated that a direct use of a genetic algorithm (binary-coded or real-coded) with generic crossover and mutation is too expensive to handle integer linear programs. The study also suggested a customized GA for handling such problems and problems having as large as *one million* variables were solved with a less-than-quadratic computational time complexity. That study was probably the first attempt to solve such a large-scale application using an evolutionary algorithm.

This paper is motivated from this earlier study and a multi-parent customized GA is suggested and tested on very large-scale ILP problems. Four different variations of the proposed multi-parent recombinative GA are compared with each other and with the previous study in terms of computational time complexity and in terms of required function evaluations. For each case, efficient parent and population sizes are found experimentally by solving a casting scheduling problem, which is an ILP and a representative problem to many other real-world optimization problems. The advantage of developing a customized GA for large-scale application and the systematic parametric approach followed in this study should encourage readers to pay attention to population-based customized optimization techniques for solving real-world optimization problems.

2   Casting Scheduling as an Integer Linear Program

In a typical foundry, casting of various sizes are made from a *heat* by melting metal in a large crucible. For convenience, more than one crucible are usually used. In this study, we shall assume a foundry using two crucible of different sizes (W_I and W_{II}), so that each crucible is used on alternate days. Depending on the crucible size used on the j-th day, the number of heats (H_j) allowed per day vary. We assume that the total number of castings (R) to be made is so large that a multi-day schedule requiring a total of H heats is necessary. Let us also assume that r_k copies of order k having a weight w_k kg are to be made, such that the total number of castings is $R = \sum_{k=1}^{K} r_k$ (where K is the total number of orders) and the total amount of metal required to make all castings is $M = \sum_{k=1}^{K} r_k w_k$ kg. With these parameters, one can optimize a casting sequencing for multiple days by introducing decision variables x_{ki} denoting the number of copies of order k made from the i-th heat. The equality constraints ensure that all desired copies of each order k are made from the combination of H heats. The inequality constraints arise to ensure that total amount of metal used for pouring in each heat is less than or equal to the size of the crucible (can be either W_I or W_{II} depending on the day). We form the following ILP:

$$\left.\begin{array}{ll} \text{Maximize } \frac{1}{H}\sum_{i=1}^{H} 100 \sum_{k=1}^{K} w_k x_{ki}/W_i, & \\ \text{Subject to } \sum_{i=1}^{H} x_{ki} = r_k, & \text{for } k = 1, 2, \dots, K, \\ \quad\quad\quad \sum_{k=1}^{K} w_k x_{ki} \leq W_i, & \text{for } i = 1, 2, \dots, H, \\ \quad\quad\quad x_{ki} \geq 0, & \\ \quad\quad\quad x_{ki} \text{ is an integer.} & \end{array}\right\} \quad (1)$$

A sequence of operation can be evaluated based on the utilization of the molten metal in each heat. Since $\sum_{k=1}^{K} w_k x_{ki}$ kg of molten metal is utilized during the i-th heat, the percentage utilization of molten metal of the i-th heat is $100 \sum_{k=1}^{K} w_k x_{ki}/W_i$.

As can be seen from the above equation that the objective function and constraints are all linear in terms of the decision variable x_{ki}. There are two difficulties that an LP solver can face in order to solve the above problem:

1. The number of decision variables and constraints are usually too large to solve them using classical methods, and
2. All decision variables are integer-valued.

The total number of decision variables are $n = HK$ and total number of constraints are $(H+K)$. However, since an LP solver will first convert the inequality constraints into equality constraints using *slack* variables (in one approach), the total number of decision variables for an LP solver become $n = (K + 1)H$. For example, for $K = 50$ orders requiring $H = 200$ heats to complete all castings, the LP problem involves a total of 10,000 integer-valued decision variables, 200 slack variables and a total of 250 equality constraints. Besides, the discreteness of the decision variables causes the major hurdle. The commonly-used technique to handle such problems is the branch-and-bound (BB) method (Deb, 1995; Reklaitis, Ravindran and Ragsdell, 1983). As shown in the sketch in Figure 1, the BB method first relaxes the integer restrictions and finds the real-parameter optimal solution. Thereafter, it chooses one particular variable (for which an non-integer value is obtained) and divides (branches) the original problem into two new problems (nodes) with an additional constraint to each problem. These procedure of branching into subproblems is terminated (bounded) when feasible integer-valued optimized solutions are found on a node or when a violation of some other optimality criteria cannot justify the continuation of the node. It is clear that as the number of decision variables increases, exponentially more such branching into new LPs are required, thereby making the overall approach computationally expensive.

3   Previously-Suggested Customized GA

It was clearly shown in another study (Deb, Reddy and Singh, 2003) that canonical binary-coded or real-coded GAs with generic crossover and mutation oper-

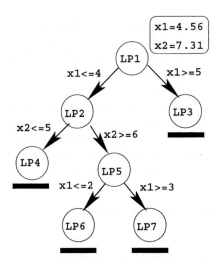

Fig. 1. A sketch of the working principle of the branch-and-bound method for a two-variable integer program. Every LP having a non-integer solution is branched into two LPs by adding an extra constraint on the variable.

ators are not at all adequate for handling more than 500 variable version of the above casting scheduling problem. To solve the above problem close to optimality, that study introduced a customized GA. We first briefly describe that algorithm here.

3.1   Customized Initial Population

The initial population, instead of being created at random, is created in a way so as to satisfy all the equality constraints in each solution. Although it may sound impossible, the linearity of the equality constraints can be exploited and a simple procedure can be adopted. First, every variable x_{ki} is initialized within $[0, a]$ (a being an upper limit to the number of copies of an order k that can be made from one heat). Thereafter, we normalize each entry x_{ki} such that the total number of copies allocated for an order k is the same as that desired (r_k):

$$x_{ki} \leftarrow \frac{x_{ki}}{\sum_{i=1}^{H} x_{ki}} r_k. \tag{2}$$

3.2   Customized Recombination Operator

The main purpose of a recombination operator is to recombine partial good information of two or more solutions and create an offspring solution. In the context of the ILP problem stated in equation 1, a solution can be called good, if the overall utilization of molten metal is close to 100%, so that the corresponding casting strategy causes minimal waste of energy. However, a good solution may

two approaches in Figures 8 and 9, respectively. The corresponding time complexities are $O(n^{1.025})$ and $O(n^{1.026})$. The corresponding complexities on the required number of function evaluations are $O(n^{0.894})$ and $O(n^{0.928})$, respectively.

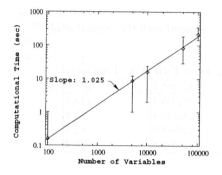

Fig. 8. Computational time complexity for the steady-state approach 2.

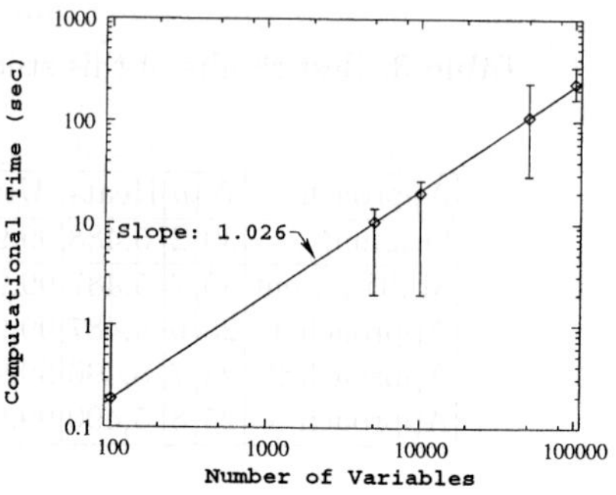

Fig. 9. Computational time complexity for the steady-state approach 3.

To put the above results into a better perspective, the best approach of this paper finds a near-optimum solution of the 100,000-variable ILP problem within only 130 sec on a Pentium IV processor. Solving ILP problems as large as 100,000 variables in an almost linear computational time complexity is rarely demonstrated in the EA literature. By performing extensive parametric studies for different algorithms proposed in this paper, we believe to have developed very efficient population-based steady-state optimization algorithms for solving ILP problems which would be difficult to match in terms of computational time or overall function evaluations by classical point-by-point search approaches.

5 Conclusions

Despite many applications of evolutionary algorithms (EAs) to various search and optimization problems, there exists very few studies where EAs have been tested on very large sized problems, such as problems having hundreds or thousands of variables. An earlier study demonstrated the use of a customized GA for the purpose, in which initialization and genetic operators are all customized using the problem knowledge. In this paper, we have extended the idea further to solve integer linear programs often arising in scheduling and resource allocation problems. A multi-parent version and three steady-state GAs have been suggested and extensive parametric studies have been performed to develop efficient optimization algorithms. Following conclusions can be made from this study:

1. The parametric study on population size indicates an efficient population size varying between 25 to 30, indicating that a population-based approach

(not a point-by-point approach) is a more efficient procedure for solving the ILP problems considered here.

2. The parametric study on parent size indicates that the usual consideration of two parent recombination is not the optimal approach. However, there lies an optimal parent size (about six to eight parents) with which the suggested algorithms work the best. This is due to the exploration and exploitation balance which must be honored in a population-based search algorithm in order to constitute a successful search process.

3. The multi-parent steady-state approaches have been found to be better than the multi-parent generational approach both in terms of computational time and amount of function evaluations.

4. The results of this study are better than those reported in the original study in which two parents were used. The best computational time complexity of this study is $O(n^{1.025})$, whereas the best reported in the original study was $O(n^{1.789})$.

This paper has demonstrated the power of a recombinative GA in solving large-sized problems to near-optimality, as the use of multiple good parents to construct a new solution may often provide *jumps* towards the minimum solution in the case of linear problems, a matter which is not possible to obtain with point-by-point classical approaches. Hopefully, this study will motivate the design and application of EAs to similar other large-scale real-world optimization problems. The approach used here may be directly applicable to other scheduling problems (with discrete variables) having a set of linear equality and inequality constraints, such as the commonly-used knap-sack problems, however, the idea of using a good initial population, a steady-state approach, a multi-parent recombination, and the use of knowledge-based genetic operators is important in solving very large-sized problems.

References

Deb, K. (1995). *Optimization for engineering design: Algorithms and examples.* New Delhi: Prentice-Hall.

Deb, K., Reddy, A. R., and Singh, G. (2003). Optimal scheduling of casting sequence using genetic algorithms. *Journal of Materials and Manufacturing Processes, 18*(3). 409–432.

Glover, F. and Laguna, M. (1997). *Tabu search.* Boston: Kluwer.

Goldberg, D. E. (1989). *Genetic Algorithms for Search, Optimization, and Machine Learning.* Reading, MA: Addison-Wesley.

Kirkpatrick, S., Gelatt, C. D., Vecchi, M. P. (1983). Optimization by simulated annealing. *Science, 220*, 671–680.

Reklaitis, G. V., Ravindran, A. and Ragsdell, K. M. (1983). *Engineering Optimization Methods and Applications.* New York: Wiley.

Taha, H. A. (1989): *Operations Research.* New York: Macmillan.

Using a Genetic Algorithm to Design and Improve Storage Area Network Architectures

Elizabeth Dicke, Andrew Byde*, Paul Layzell, and Dave Cliff

Hewlett-Packard Labs Europe
Filton Road, Bristol, BS34 8QZ, UK
{andrew.byde|dave.cliff|paul.layzell}@hp.com

Abstract. Designing storage area networks is an NP-hard problem. Previous work has focused on traditional algorithmic techniques to automatically determine fabric requirements, network topology, and flow routes. This paper presents work performed with a genetic algorithm to both improve designs developed with heuristic techniques and to create new designs. For some small networks (10 hosts, 10 devices, and single-layered) we find that we can create networks which result in savings of several thousand dollars over previously established methods. This paper is the first publication, to our knowledge, to describe the successful application of this technique to storage area network design.

1 Introduction

As IT systems and employees become more geographically distributed and it becomes more and more important to access shared data, *Storage Area Networks* (SANs) become the choice of companies looking for efficient, distributed storage solutions. A SAN is a set of *fabric elements* connecting a set of *hosts* – from which data is requested – to a set of storage *devices* – on which data is stored (see figures 2 & 3). The fabric elements are *fabric nodes*, which route data through the network, *ports* on the nodes and *links* physically connecting the ports. A link has a port at each end and a port is the terminal of at most one link. SANs allow for efficient use of storage related resources such as hardware and maintenance personnel, resulting in a storage solution that is more effective than local storage, in addition to being more scalable.

Once purchased, installed, and configured appropriately, a SAN can be a cost effective solution to the storage problem. Recent work has focused on automating this process, since solutions designed by hand to support specified data flow requirements tend to over provision resources by a considerable margin [6]. Efficiency is an important issue because the physical components of a storage area network can cost millions of dollars; An over-provisioned design can waste anywhere from thousands to millions of dollars, depending on the size of the network.

A SAN problem is specified by providing a list of hosts, a list of devices, a list of possible types of fabric nodes, and a description of the network's data flow

---

* to whom correspondence should be addressed.

K. Deb et al. (Eds.): GECCO 2004, LNCS 3102, pp. 1066–1077, 2004.

requirements. Each host, device, and fabric node has a cost, a maximum number of ports that are available to accept links, and a maximum amount of data that may pass through it, called its *bandwidth*. The network's data flow requirements are specified by a list of *flows*, each of which is defined by a source host, a destination device, and a bandwidth requirement. A flow may not be routed through a fabric element which does not have enough remaining bandwidth.

A SAN design specifies a list of each fabric element and its connectivity along with a path for each flow. The aim of an automated SAN designer is to find the cheapest SAN that supports the specified flows, while satisfying the port constraints, bandwidth constraints and non-splitting of flows.

The problem of SAN design can be compared to that of design of other types of networks, as well as the problem of routing data within those networks. However, SAN design is more difficult than other network design problems because there are the additional limitations of not being able to split a data flow from host through to device, the limited number of ports available on the nodes, the limited amount of bandwidth associated with each node and port, and the fact that the network topology is not pre-determined. It is NP-hard to find the minimal cost network[6], and best-known algorithms on state-of-the-art machinery take days to complete for moderate sized problems.

Hewlett-Packard's Appia project [6] has shown that traditional algorithmic optimisation techniques can quickly specify a topology that both satisfies the design requirements and competes with designs created by human SAN experts. While able quickly to determine a possible SAN topology, the Appia algorithms are not guaranteed to find the optimal solution. As a result of the need to find a solution within minutes, the algorithms presented by the Appia group build usable networks following heuristic procedures that have previously shown to yield good networks.

This paper seeks to explore SAN design using genetic algorithms (GAs) to produce well designed SANs. We will discuss work using a GA to evolve SAN topologies which will result in both original buildable designs and improvements to previous designs.

We will show that the use of a genetic algorithm can result in SAN architectures which cost thousands of dollars less than designs created either by traditional heuristic methods, or by Appia. This paper is, to the best of our knowledge, the first publication to describe the successful application of these biologically inspired techniques to SAN design.

In the next section we will discuss previous work relating to network design and routing of data through a network both in terms of other types of networks and in relation to SANs specifically. In Section 3, we will introduce the specific configuration and design of our genetic algorithm and discuss the results obtained both with creation of a storage area network from scratch and given an input of a previously designed network. We will then conclude with suggestions for further work.

2   Background

Automating SAN design is a relatively new research area and prior work specifically relating to Storage Area Networks is limited. In this section we will discuss research undertaken in network topology and network routing, both of which relate to SAN design. We will encompass work relating to both SANs and other types of networks which may be similar in structure and constraints.

2.1   Previous Automated SAN Design Work

Much work has been done by the Decision Technologies Department and the Storage Content and Distribution Department within HP Labs Palo Alto, in order to automate the process of SAN design [6]. They have concentrated on two different algorithms called *FlowMerge* and *QuickBuilder*, which each have different strengths and weaknesses in terms of finding efficient solutions to the SAN fabric design problem. Each will be described in brief, for more details, see [6]. The FlowMerge algorithm begins with a SAN connecting each host to its required device, given a set of flow specifications or requirements. This configuration typically results in a large number of port violations (i.e. a node has more links than available ports). These are gradually reduced by considering individual *flowsets*. Each flow is initially considered to be in its own flowset. With each iteration of the algorithm, two flowsets are merged together, choosing an appropriate fabric node, and links to connect hosts and devices appropriately. Each iteration results in a reduction of the number of port violations, or, if that is not possible, a reduction in the cost of the design. The algorithm continues until there are no possible improvements on the design, or there are no other flowsets that may be merged.

Ward, et al. [6] show that FlowMerge is one of the faster performing algorithms for the smaller 10 host, 10 device networks, especially those which have 20 to 30 flows spread fairly evenly throughout the network.

The QuickBuilder algorithm also begins with a SAN connecting each host to an associated device as given by a set of flow requirements. However in this case, the initial SAN configuration includes assignment to a particular port on each device. The configuration is then arranged into port groups, which consist of all connected ports. Each port group is then analysed separately in order to determine fabric node requirements.

While FlowMerge tends to find solutions with many small port groups, the QuickBuilder algorithm tends to find SAN configurations with larger port groups when necessary. This algorithm tends to result in cost effective designs for large networks and those that are more densely populated. QuickBuilder is also faster than the FlowMerge algorithm for large problems (10 times as fast for the largest problems consisting of 50 hosts and 100 devices).

2.2   Automated Design of Other Types of Networks

There appears to be limited published work relating to automated design of Storage Area Networks specifically, aside from that referenced above. Much of the automated design work has been done for other types of networks, and will be discussed in the following sections.

Automated network design is not a new research field. Several researchers have attacked this problem with traditional techniques, for networks with varying constraints. However, there is no other network problem which also contains all of the constraints placed on network design for SANs [6]. For example, Gavish ([3]) expresses the network design and routing problem as a combinatorial optimisation problem and uses Langrangean relaxation to obtain close to optimal networks. However, although Gavish's work includes restrictions on node cost and does not allow flows to be split, it does not take into account node capacity issues, as the SAN design problem must.

Network Design With a Genetic Algorithm. Intending to improve on the work done with traditional techniques, several researchers have attempted network design with genetic algorithms. Chu et al. [1] describe work done using Genetic Algorithms to design a degree-constrained minimal spanning tree (DCMST). A minimal spanning tree is a collection of edges that joins together all vertices in a set with a minimum sum of weighted edge values. The degree-constrained modifier implies that there is a maximum number of edges connected to a particular vertex. Like heuristic design of SANs, traditional programming approaches to DCMST design do not scale well. As network size increases, the number of constraints increases exponentially and realistic problems become difficult to solve with traditional mathematics. Encoding the connected components of the network within the genome, both valid and invalid solutions are evolved for a network of n nodes, where each node has varying degree constraints. Invalid solutions may be specified by the genome, in which case an attempt will be made to modify the network in order to make the solution viable. This process, which they call *chromosome repair*, may or may not be successful in producing a valid network. However, it acts as an effective local search mechanism for the genetic algorithm. The fitness of each specified network is measured as the cost of connecting the connected nodes together as specified. It was found that the GA could produce more optimal solutions than the traditional minimisation algorithms supplied, but at significantly higher computational cost. Knowles and Corne [4] also use a genetic algorithm to design DCMSTs, however in their case, they use a genome encoding that only permits generation of valid networks, effectively narrowing the search space to a much more manageable size. They find similar results in that the GA outperforms other compared design methods.

Raidl and Julstrom [5] also use a GA but for designing a bounded-diameter minimum spanning tree (BDMST). A bounded-diameter tree is one which has a maximum number of edges connecting any two vertices in the graph. They also restrict their generated genomes to only specify valid networks. With this type of network, this GA implementation can outperform the other compared heuristic techniques.

Design of DCMSTs and BDMSTs is similar to SAN design in that both require connecting nodes when each node has a limited amount of connections available. Both problems also require the minimisation of some cost value. Additionally Chu, et al. allow the production of invalid networks. However the SAN design problem has the added issue of the data flow through the network. Each component in the network has its own limit on the amount of bandwidth it has available. Moreover a SAN does not need to be fully connected. Valid, cost-effective solutions will not have all nodes connected to each other. A flow must not be split between separate fabric elements (i.e. the network is non-bifurcated). Furthermore, with SAN design the set of nodes is not known in advance.

3   Methods

We now describe the specific implementation of the genetic algorithm used. We will then present the results of experiments to both further optimise Appia designs and to create new networks. We will show that for small networks, optimisation of Appia designs is possible and we can save over 40% of the original FlowMerge cost. We will also show that it is possible to design networks with a direct-connection initialised GA, although the performance does not always equal that of the Appia improved designs.

3.1   Genome Encoding

We commence with a list of flow requirements for the desired SAN (see Table 1 for an example), and a pool of available fabric nodes, containing a number of different types[1], each node having its own unique identification number, ranging for convenience from 1 to n where n is the number of fabric nodes available.

The genome encoding used here is limited to expression of single-layered networks only. It contains one locus for each flow requirement as specified in the list. Each locus specifies as an integer the fabric node, if any, that the flow from host to device should be routed through. A direct connection is specified with a fabric node number of 0.

For example, if we have a fabric node pool of two switches, four flows (as defined in Table 1), three hosts, and two devices, we would represent a possible solution as in Figure 1(a). In this example, there is a direct connection between host0 and device0.

3.2   Solution Evaluation

A candidate network is created from the genome representation in several steps. First, we determine which fabric nodes from the pool are being used (that is those that have flows routed through them). For example, The network built

---

[1] In this paper we consider two types of fabric node: *switch* and *hub*. For our purposes, a hub differs from a switch in three ways: the cost of the fabric node itself, the cost of the ports on the fabric node, and the amount of incoming bandwidth that the fabric node can handle. The hub itself is cheaper than a switch node. Ports on a hub do not cost anything, but the ports on a switch do. However, a switch's bandwidth is only limited to the sum of bandwidth supported by its ports, a hub has additional incoming bandwidth restrictions that are less than the bandwidth supported by the ports.

Table 1. Flow requirements for an arbitrary example SAN design problem. There are three hosts, and two devices

Name	Source	Destination	Bandwidth Required (MB)
flow0	host0	device0	1.0e07
flow1	host1	device1	5.4e07
flow2	host0	device1	6.8e07
flow3	host2	device1	9.7e07

0	1	2	1		0	1	1	1		0	1	0	1

host0 host1 host0 host2   host0 host1 host0 host2   host0 host1 host0 host2
device0 device1 device1 device1   device0 device1 device1 device1   device0 device1 device1 device1

      (a)   (b)   (c)

Fig. 1. Possible genomes representing a solution for the SAN design problem whose flow requirements are specified in Table 1. The fabric node pool has two switches. Genomes (b) and (c) are mutations of (a) where the gene at locus 'host0-device1' is mutated

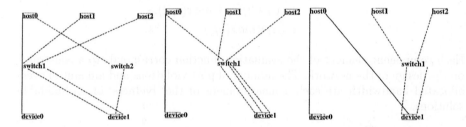

Fig. 2. SANs built from the specification given in Figure 1, from left to right (a), (b), and (c). The solid line represents a direct connection between a host and a device. The dashed lines are links that connect to a fabric node on one end or the other. Each link has a capacity of 10e07MB. Each host and device has two available ports. The switch has 16 available ports. In each network device1 therefore has a port violation of 1

from the genome in Figure 1(c) only uses one of the fabric nodes, while the network represented by the genome in Figure 1(a) uses two. Next, the number of links needed to support the flows is determined. For each flow, its path is determined from the genome specification. If the flow is routed directly between its host and device, a link is created between its source and destination. If the flow is routed through a fabric node, the algorithm first checks to see if there is already a link between the specified source and fabric node that can support the bandwidth needed by the flow. If there is, then that link is used, otherwise a new link is created between the source and fabric node. For link allocation, we will only be constrained by available bandwidth; we will permit port violations at this stage. Following this method, the genomes in Figure 1 would be built as illustrated in Figure 2.

Once a network has been built, the topology and routing of the flows can then be evaluated. A formula for the *overall cost* C associated with the production of a particular design is given by Equation 1.

$$C = w_1 c_m + w_2 p_{hd} + w_3 p_f + w_4 b. \tag{1}$$

The terms c_m, p_{hd}, p_f and b are normalisations of, respectively, the monetary cost of each of the components necessary, the number of host/device port violations, the number of fabric node port violations, and the amount of bandwidth which is required but not available. The constants w_n, which are set at the start of each run, allow the relative importance of each term to be configured.

The terms c_m, p_{hd}, p_f and b are normalised to lie between 0 and 1 by dividing by an over-approximation of their worst-case values. The worst-case monetary cost c_{m_w} is approximated with the formula expressed in Equation 2, in which n_h, n_d and n_f are the number of hosts, devices, or flows in the problem and c_h, c_d, c_l, c_f and c_p is the monetary cost of a host, device, fibre cable, fabric node, or port.

$$
\begin{aligned}
c_{m_w} = & (n_h * c_h) + (n_d * c_d) \\
& + (n_f * 2)(c_l + max(c_P)) \\
& + (n_f)(max(c_f))
\end{aligned}
\tag{2}
$$

Each component element of the evaluation function corresponds to a constraint on the design of the network. The number of port violations and amount of over-allocated bandwidth are each a measurement of the 'badness' of un-buildable solutions.

3.3   The Genetic Algorithm

A GA with rank selection, single-point crossover (probability 0.05), and elitism was used for all runs. When mutation occurs at a particular locus (probability 0.01), a random number is chosen between 0 and n, to represent a new route for a flow. A mutation always results in a new value for a particular gene.

4   Experimental Results

The GA was tested using two different initialisation methods. The first is to initialise each member of the population so that each flow requirement is met by directly connecting its source host to its destination device. This method is called "direct connection initialisation". This is similar to the initial step in the FlowMerge algorithm. The second method is to initialise each genome with a buildable, though potentially sub-optimal solution from one of the Appia algorithms. This method is called "Appia initialisation". In each case, over successive generations, the GA will evolve new networks, routing the flows through available fabric nodes.

4.1   Test Data

The Appia project [6] has generated a set of random test cases classified into nine distinct groups. Each test case has a possible solution, though the optimal solution is not necessarily known. Each group has two specific characteristics, one which represents the number of hosts and the number of devices, and the other which categorises the number of flows between host, device pairs. There were three possible categories of size: problems with 10 hosts and 10 devices, 20 hosts and 100 devices, and 50 hosts and 100 devices. The results presented in this paper are only for 10 by 10 problems. The flows were then characterised by three labels: sparse (a few number of flows generally uniformly distributed across possible host-device pairs), dense (a large number of flows generally uniformly distributed) or clustered (a small number of host-device pairs carry most of the flow requirements).

This same test set was applied to the GA described above, in order to measure its effectiveness against the more traditional algorithms developed and applied in the Appia project. Each grouping of sparse, clustered, and dense problems was numbered from 1 to 30. The first 10 represent problems whose hosts/devices have a higher maximum percentage of *port saturation* (i.e the proportion of the maximum bandwidth that may be used on a particular host or device). The last 10 have a higher number of maximum flows per individual host or device.

4.2   Results

Direct Connection Initialisation. The GA was initialised with genomes representing a network with all direct connections. It was then run for 1000 generations with a population size of 100. The weights corresponding to Equation 1 were set to $(w_1, \ldots, w_4) = (1, 10000, 1000, 100)$. Since the weighting is applied after the normalisation, this tiered weighting ensures that a solution with host/device port violations is always worse than one without host/device port violations, even if the solution with no host/device port violations has the maximum possible number of fabric node violations. The idea is that as solutions are evolved the number of host or device port violations will be decreased first until there are none. Only then will a focus on decreasing the number of fabric node port violations occur, and so with bandwidth and then monetary cost. This ensures that the GA will find buildable solutions first, and only then will monetary cost become a consideration.

We ran the GA for each of 30 sparse problems, 30 clustered problems, and 30 dense problems for a 10 host, 10 device problem. The GA generated a buildable solution in 62% of the problems. There were 27 buildable sparse solutions, 29 buildable clustered solutions but only 10 buildable dense solutions.

Table 2 compares the average solution monetary cost of each algorithm, for those cases where the GA was able to construct a solution.

Figure 4 shows the percent improvement of the GA's solutions over the Appia solutions for those cases in which the GA was able to find a solution within 1000 generations. In some cases, the direct connection initialised GA is able to create better networks than either of the Appia algorithms, especially for problems characterised as clustered or dense. But in general it is outperformed by them.

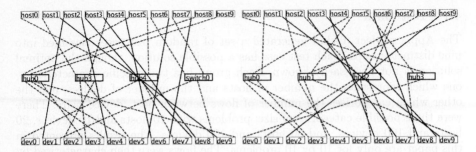

Fig. 3. On the left, an example of a network designed by the GA initialised with direct connections. After 1000 generations, monetary cost is $52,470. The network on the right is the corresponding Appia QuickBuilder solution, cost is $28,470.

Table 2. Comparison between average solution monetary costs for direct-connection initialised GA, FlowMerge, and QuickBuilder Algorithms. Negative values indicate the GA did not perform as well as the specified Appia algorithm

| Type | n | Average Cost | | | min(FM,QB)-GA | |
		GA	FM	QB	Average	Standard Deviation
Sparse	27	$65,088	$46,629	$51,113	$-21,023	$17,584
Clustered	29	$50,512	$50,309	$54,912	$-2,004	$20,844
Dense	10	$88,470	$100,404	$147,700	$11,934	$27,308

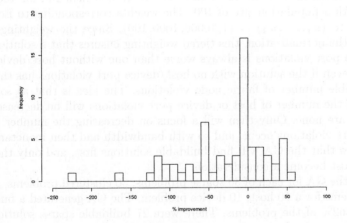

Fig. 4. Percent improvement of direct-connection initialised GA evolved networks over the lower of the Appia QuickBuilder or FlowMerge costs

Appia Initialisation. The GA approach to SAN design appears to be most effective when initialised with a design which has been provided by one of the Appia algorithms. To illustrate this, we initialised each member of the population with the cheaper solution produced by either the FlowMerge or QuickBuilder algorithm. The GA then evolved modified solutions, resulting in equal or lower cost designs. In many of these cases, the GA can find a solution that is less expensive than the cheapest Appia solution.

For these experiments, we used the same 30 sparse, 30 clustered, and 30 dense sample problems as in the direct connection initialised experiments described in Section 4.2. The weights in Equation 1 were, however, changed to $(w_1, \ldots, w_4) = (1, 10, 15, 1)$. This puts slightly more emphasis on the fabric node port violations, over the host/device port violations. This GA was able to find better solutions 71% of the time. That is, in 23 of the 30 sparse problems, 18 of the 30 clustered problems, and 16 of the available 20 dense problems[2]. Table 3 shows the quantified ability of the GA to redesign the SAN topology and routing so that the monetary cost of the new SAN is cheaper.

Table 3. Comparison between average solution monetary costs for Appia initialised GA, FlowMerge, and QuickBuilder Algorithms

| Type | n | Average Cost | | | min(FM,QB)-GA | |
		GA	FM	QB	Average	Standard Deviation
Sparse	30	$45,142	$48,735	$53,923	$1,286	$1,423
Clustered	30	$46,328	$51,722	$57,206	$3,713	$7,550
Dense	20	$90,280	$94,766	$126,146	$4,062	$3,864

These improvements over the Appia algorithms, summarised in Table 3 are shown graphically in Figure 5. The improvement in design over the Appia solutions results generally from a slight re-arrangement in flows in order to take advantage of already available components. For example, a GA solution to sparse problem 1 takes advantage of an available path through an existing switch, instead of creating an additional direct connection between a host and its corresponding device. The use of an already available route is cheaper than the use of an unnecessary host or device port and additional link and in this case, gives savings of $620.

Other improved solutions will result in the use of a smaller number of fabric nodes, which leads to a more significant cost savings. For example, Figure 6 shows an Appia solution and a resulting improved solution found by the GA for clustered problem 1. The improved solution uses one less switch resulting in a monetary cost difference between the two designs of $33, 220$.

The experiments described in this paper have concentrated solely on input designs which were single layered. More complex multi-layer designs were not considered. This has limited us to the exploration of relatively simple SANs.

---

[2] The remaining 10 dense problems had Appia solutions which were for multi-layer networks. Since our genome representation only encompasses single layer networks there was no way to initialise the population with these solutions. Therefore, there was no data collected on improvement of Appia generated designs for these particular dense networks.

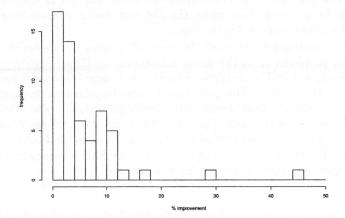

Fig. 5. Percent improvement of Appia initialised GA evolved networks over the lower of the Appia QuickBuilder or FlowMerge costs

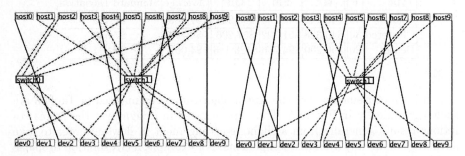

Fig. 6. On the left, SAN designed by the Appia FlowMerge algorithm at a cost of $80,990. On the right, a lower monetary cost SAN designed by a GA given the design on the left as input. The monetary cost savings is $33,320

However, the above-described results have shown that in many cases, we can improve the design of a SAN generated by Appia to decrease the monetary cost needed to support the required flows. For a detailed exploration of the nature of the fitness landscapes produced in this work, the reader is referred to [2].

5   Conclusion

We have found that a GA which encoded single layered networks could evolve buildable networks for small, 10 host by 10 device problems. These solutions, when evolved from a directly connected network, do not generally outperform the Appia FlowMerge solutions, but can, about half of the time, outperform the QuickBuilder solutions. However, when the GA is initialised with an Appia solution, in most cases it can quickly evolve a better solution, and is hence of immediate utility as a tool for SAN optimisation. The use of either technique can result in solutions which provide significant monetary cost savings over Appia designed networks. There are several different directions in which the work presented here could be taken in the future. First in terms of the problem representation and solution space, we made a specific decision that we limit possible solutions to include only single-layer networks. This may prevent the GA implementation from finding a more optimal solution. We now intend to extend the approach to multi-layered networks. We also demonstrated that the GA was an effective tool for optimising already designed networks. One of the problems in SAN design is that flow requirements have a tendency to change over the life-time of a SAN solution. The GA presented here should also be an effective tool for re-designing SANs when the requirements change, including an increase in the number of flows.

References

1. Chao-Hsien Chu, G. Premkumar, Carey Chou, and Jianzhong Sun. Dynamic degree constrained network design: A genetic algorithm approach. In *Proceedings of GECCO-99 (Genetic and Evolutionary Computation Conference 1999)*, pages 141–148, 1999.
2. Elizabeth Dicke, Andrew Byde, Dave Cliff, and Paul Layzell. Using a genetic algorithm to design improved storage area network architectures. Technical Report HPL-2003-221, HP Laboratories, Bristol, November 2003.
3. Bezalel Gavish. Topological design of computer communication networks - the overall design problem. *European Journal of Operational Research*, 58:149–172, 1992.
4. Joshua Knowles and David Corne. A new evolutionary approach to the degree-constrained minimum spanning tree problem. *IEEE Transactions on Evolutionary Computation*, 4(2):125–134, July 2000.
5. Gunther R. Raidl and Bryant A. Julstrom. Greedy heuristics and an evolutionary algorithm for the bounded-diameter minimum spanning tree problem. In G. Lamont et al., editor, *Proceedings of the 2003 ACM Symposium on Applied Computing*, pages 747–752, 2203.
6. Julie Ward, Michael O'Sullivan, Troy Shahoumian, and John Wilkes. Appia: automatic storage area network fabric design. In *Proceedings of the FAST 2002 Conference on File and Storage Technologies*, pages 203–217, January 2002.

Distributed Constraint Satisfaction, Restricted Recombination, and Hybrid Genetic Search

Gerry Dozier, Hurley Cunningham, Winard Britt, and Funing Zhang

Department of Computer Science and Software Engineering
Auburn University, AL 36849-5347
gvdozier@eng.auburn.edu

Abstract. In this paper, we present simple and genetic forms of an evolutionary paradigm known as a society of hill-climbers (SoHC). We compare these simple and genetic SoHCs on a test suite of 400 randomly generated distributed constraint satisfaction problems (DisCSPs) that are composed of asymmetric constraints (referred to as DisACSPs). Our results show that all of the genetic SoHCs dramatically outperform the simple SoHC even at the phase transition where the most difficult DisACSPs reside.

1 Introduction

Evolutionary Computation (Bäck 1997; Fogel 2000) is the field of study devoted towards the design, development, and analysis of problem solvers based on simulated genetic and/or social evolution. Evolutionary computations (ECs) have been successfully used to solve a wide variety of problems in the areas of robotics, engineering, scheduling, planning, and machine learning, just to name a few (Bäck 1997; Fogel 2000).

In the Evolutionary Constraint Satisfaction community, we have seen a migration away from pure evolutionary computations for constraint satisfaction towards the hybridization of ECs with traditional CSP techniques and/or the incorporation of heuristics and problem specific knowledge (Dozier, Bowen, and Homaifar 1998) in an effort to solve CSPs more efficiently. To date, much of this research has focused on centralized CSPs. Little research has been conducted by the evolutionary constraint satisfaction community on the development of ECs for solving distributed constraint satisfaction problems (DisCSPs) (Dozier 2002).

A DisCSP (Yokoo 2001) can be viewed as a 4-tuple (X, D, C, A), where X is a set of n variables, D is a set of n domains (one domain for each of the n variables), C is a set of constraints that constrain the values that can be assigned to the n variables, and A is a set of agents for which the variables and constraints are distributed. Constraints between variables belonging to the same agent are referred to as intra-agent constraints while constraints between the variables of more than one agent are referred to as inter-agent constraints. The objective in solving a DisCSP is to allow the agents in A to develop a consistent distributed

K. Deb et al. (Eds.): GECCO 2004, LNCS 3102, pp. 1078–1087, 2004.

solution by means of message passing. The constraints are considered private and are not allowed to be communicated to fellow agents due to privacy, security, or representational reasons (Yokoo 2001). When comparing the effectiveness of DisCSP-solvers the number of communication cycles (through the distributed algorithm) needed to solve the DisCSP at hand is more important than the number of constraint checks (Yokoo 2001) .

Many real world problems have been modeled and solved using DisCSPs (Bejar et al. 2001; Calisti and Faltings 2001; Freuder, Minca, and Wallace 2001; Silaghi et al. 2001; Zhang 2002); however, many of these approaches use mirrored (symmetric) inter-agent constraints. Since these inter-agent constraints are known by the agents involved in the constraint, they can not be regarded as private. If these constraints were truly private then the inter-agent constraints of one agent would be unknown to the other agents involved in those constraints. In this case the DisCSP would be composed of asymmetric constraints. To date, with the exception of (Freuder, Minca, and Wallace 2001) and (Silaghi et al. 2001), little research has been done on distributed asymmetric CSPs (DisAC-SPs).

In this paper, we demonstrate how distributed restricted forms of uniform crossover can be used to improve the effectiveness of a previously developed EC for solving DisACSPs known as a society of hill-climbers (SoHC) (Dozier 2002). We refer to these new algorithms as a genetic SoHCs (GSoHCs). Our results show that the GSoHCs dramatically outperform the simple SoHC even at the phase transition where the hardest DisACSPs reside.

2   Constraint Networks, Asymmetric Constraints, and the Phase Transition

Constraint satisfaction problems (CSPs) are based on constraint networks (Bowen and Dozier, 1996; Mackworth 1997). A constraint network is a triple $\langle X, D, C \rangle$ where X is set of variables, D is set of domains where each $x_i \in X$ takes its value from the corresponding domain $d_i \in D$, and where C is a set of r constraints. Consider a binary constraint network (one where each constraint constrains the value of exactly two variables)[1]

Constraint networks possess two additional attributes: tightness and density. The tightness of a constraint is the ratio of the number of tuples disallowed by the constraint to the total number of tuples in $d_i \times d_j$. The average constraint tightness of a binary constraint network is the sum of the tightness of each constraint divided by the number of constraints in the network. The density of a constraint network is the ratio of the number of constraints in the network to the total number of constraints possible.

---

[1] In this paper, we only consider binary constraint networks because any constraint involve more than one varible can be transformed into a set of binary constraints.

2.1   Asymmetric Constraints

Constraints in a binary constraint network may also be represented as two directional constraints referred to as arcs (Mackworth 1977; Tsang 1993). For example, the symmetric constraint c_{EF} can be represented as $c_{EF} = \{c_{\underset{EF}{\rightarrow}}, c_{\underset{EF}{\leftarrow}}\}$, where $c_{\underset{EF}{\rightarrow}} = c_{\underset{EF}{\leftarrow}} = \{\langle\, \text{e1},\text{f2}\,\rangle, \langle\, \text{e1},\text{f3}\,\rangle, \langle\, \text{e2},\text{f2}\,\rangle, \langle\, \text{e3},\text{f2}\,\rangle\,\}$, where $c_{\underset{EF}{\rightarrow}}$ represents the directional constraint imposed on variable F by variable E, and where $c_{\underset{EF}{\leftarrow}}$ represents the directional constraint imposed on variable E by variable F. This view of a symmetric binary constraint admits the possibility of an asymmetric binary constraint between variables E and F as one where $c_{\underset{EF}{\rightarrow}} \neq c_{\underset{EF}{\leftarrow}}$.

2.2   Predicting the Phase Transition

Classes of randomly generated CSPs can be represent as a 4-tuple $(n,m,p1,p2)$ (Smith 1994) where n is the number of variables in X, m is the number of values in each domain, $d_i \in D$, $p1$ represents the constraint density, the probability that a constraint exists between any two variables, and $p2$ represents the tightness of each constraint.

Smith (Smith 1994) developed a formula for determining where the most difficult symmetric randomly generated CSPs can be found. This equation is as follows, where $\hat{p2}_{crit_S}$ is the critical tightness at the phase transition for n, m, $p1$.

$$\hat{p2}_{crit_S} = 1 - m^{\frac{-2}{p1(n-1)}} \tag{1}$$

Randomly generated symmetric CSPs of the form $(n,m,p1,\hat{p2}_{crit_S})$ have been shown to be the most difficult because they have on average only one solution. Problems of this type are at the border (phase transition) between those classes of CSPs that have solutions and those that have no solution. Classes of randomly generated symmetric CSPs for which $p2$ is relatively small compared to $\hat{p2}_{crit_S}$ are easy to solve because they contain a large number of solutions. Similarly, classes of CSPs where $p2$ is relatively large compared to $\hat{p2}_{crit_S}$, are easy to solve because the constraints are so tight that simple backtrack-based CSP-solvers (Smith 1994) can quickly determine that no solution exists. Thus, for randomly generated CSPs, one will observe an easy-hard-easy transition as $p2$ is increased from 0 to 1.

Smith's equation can be modified (Dozier 2002) to predict the phase transition in randomly generated asymmetric CSPs as well. This equation is as follows where $p1_\alpha$ represents the probability that an arc exits between two variables and where $\hat{p2}_{crit_A}$ is the critical tightness at the phase transition for n, m, and $p1_\alpha$.

$$\hat{p2}_{crit_A} = 1 - m^{\frac{-1}{p1_\alpha(n-1)}}. \tag{2}$$

3   Society of Hill-Climbers

A society of hill-climbers (SoHC) (Dozier 2002; Sebag and Schoenauer 1997) is a collection of hill-climbers that communicate promising (or futile) directions of search to one another through some type of external collective structure. In the society of hill-climbers that we present in this paper, the external collective structure which records futile directions of search comes in the form of a distributed list of breakout elements, where each breakout element corresponds to a previously discovered nogood[2] of a local minimum (Morris 1993). Before presenting our society of hill-climbers, we must first discuss the hill-climber that makes up the algorithm. In this section, we first introduce a modified version of Yokoo's distributed breakout algorithm with broadcasting (DBA+BC) (Yokoo 2001) which is based on Morris' Breakout Algorithm (Morris 1993). After introducing the modified DBA+BC algorithm (mDBA) we will describe the framework of a SoHC.

For the mDBA, each agent $a_i \in A$ is responsible for the value assignment of exactly one variable. Therefore agent a_i is responsible for variable $x_i \in X$, can assign variable x_i one value from domain $d_i \in D$, and has as constraints $C_{\overrightarrow{x_i x_j}}$ where $i \neq j$. The objective of agent a_i is to satisfy all of its constraints $C_{\overrightarrow{x_i x_j}}$. Each agent also maintains a breakout management mechanism (BMM) that records and updates the weights of all of the breakout elements corresponding to the nogoods of discovered local minima. This distributed hill-climber seeks to minimize the number of conflicts plus the sum of all of the weights of the violated breakout elements.

3.1   The mDBA

The mDBA used in our SoHCs is very similar to Yokoo's DBA+BC with the major exception being that each agent broadcasts to every other agent the number of conflicts that its current value assignment is involved in. This allows the agents to calculate the total number of conflicts (fitness) of the current best distributed candidate solution (dCS) and to know when a solution has been found (when the fitness is equal to zero). The mDBA, as outlined in Figure 1, is as follows.

Initially, each agent, a_i, randomly generates a value $v_i \in d_i$ and assigns it to variable x_i. Next, each agent broadcasts its assignment, $x_i = v_i$, to its neighbors $a_k \in Neighbor_i$ where $Neighbor_i$[3] is the set of agents that a_i is connected with via some constraint. Each agent then receives the value assignments of every neighbor. This collection of value assignments is known as the **agent_view** of an agent a_i (Yokoo 2001). Given the agent_view, agent a_i computes the number of conflicts that the assignment $(x_i = v_i)$ is involved in. This value is denoted as γ_i.

---

[2] A nogood is a tuple that causes a conflict.
[3] In this paper, $Neighbor_i = A - \{a_i\}$.

Once the number of conflicts, γ_i, has been calculated, each agent a_i randomly searches through its domain, d_i, for a value $b_i \in d_i$ that resolves the greatest number of conflicts (ties broken randomly). The number of conflicts that an agent can resolve by assigning $x_i = b_i$ is denoted as r_i. Once γ_i and r_i have been computed, agent a_i broadcasts these values to each of its neighbors.

When an agent receives the γ_j and r_j values from each of its neighbors, it sums up all γ_j including γ_i and assigns this sum to f_i where f_i represents the fitness of the current dCS. If agent a_i has the highest r_i value of its neighborhood then agent a_i sets $v_i = b_i$, otherwise agent a_i leaves v_i unchanged. Ties are broken randomly using the commonly seeded tie-breaker[4] that works as follows: if t(i) > t(j) then a_i is allowed to change otherwise a_j is allowed to change where t(k) = (k+$rnd()$) mod $|A|$, and where $rnd()$ is a commonly seeded random number generator used exclusively for breaking ties.

If r_i for each agent is equal to zero, i.e. if none of the agents can resolve any of their conflicts, then the current best solution is a local minimum and all agents a_i send the nogoods that violate their constraints to their BMM_i. An agent's BMM will create a breakout element for all nogoods that are sent to it. If a nogood has been encountered before in a previous local minimum then the weight of its corresponding breakout element is incremented by one. All weights of newly created breakout elements are assigned an initial value of one. Therefore the task for mDBA is to reduce the total number of conflicts plus the sum of all breakout elements violated.

After the agents have decided who will be allowed to change their value and invoked their BMMs (if necessary), the agents check their f_i value. If $f_i > 0$ the agents begin a new cycle by broadcasting their value assignments to each other. If $f_i = 0$ the algorithm terminates with a distributed solution.

3.2   The Simple and Genetic SoHCs

The SoHCs reported in this paper are based on mDBA. Each simple SoHC runs ρ mDBA hill-climbers in parallel, where ρ represents the society size. Each of the ρ hill-climbers communicate with each other indirectly through a distributed BMM. In a SoHC, each agent, a_i assigns values variables $x_{i1}, x_{i2}, \cdots, x_{i\rho}$ where each variable x_{ij} represents the ith variable for the jth dCS. Each agent, a_i, has a local BMM (BMM_i) which manages the breakout elements that correspond to the nogoods of its constraints.

There are a total of 4 genetic SoHCs (GSoHCs) reported in this paper. They differ only in the type of recombination operator used. They are as follows. GSoHC$_{spx}$ is similar to the simple SoHC (SoHC) except that it uses a distributed restricted single-point crossover operator (dSPX-μ), where μ is the mutation rate. The dSPX-μ operator works as follows. On each cycle, each dCS$_j$ that has an above average number of conflicts is replaced with an offspring by recombining

---

[4] In case of a tie between two agents a_i and a_j, Yokoo's DBA+BC will allow the agent with the lower agent address number is allowed change its current value assignment. We refer to this as the deterministic tie-breaker (DTB) method

```
procedure mDBA(Agent a_i)
{

    Step 0: randomly assign v_i ∈ d_i to x_i;
    do
        {
            Step 1: broadcast (x_i = v_i) to other agents;
            Step 2: receive assignments from other agents, agent_view_i;
            Step 3: assign conflicts_i the number conflicts that (x_i = v_i)
                    is involved in;
            Step 4: randomly search d_i for a value b_i that minimizes the number
                    of conflicts of x_i (ties broken randomly),
            Step 5: let r_i equal the number of conflicts resolved by (x_i = b_i);
            Step 6: broadcast conflicts_i and r_i to other agents;
            Step 7: receive conflicts_j and r_j from other agents,
                    let f = ∑ conflicts_k;
            Step 8: if (max(r_k) == 0)
                        for each conflict, (x_i = v, x_j = w)
                            update_breakout_elements(BMM_i,(x_i = v, x_j = w));
            Step 9: if (r_i == max(r_k))†                          v_i = b_i;
        } while (f > 0)
}

    † Ties are broken with randomly with a synchronized tie-breaker.
```

Fig. 1. mDBA Agent Protocol

dCS_j with dCS_q, which is created as follows. With probability μ, agent a_i will randomly assign v_{ij} a value from d_i. With probability 1-μ, a cut point, cp, is selected using a commonly seeded random number generator from 1 to N-1, where N is the number of agents of an individual. All agents a_i where i ¡ cp will assign v_{ij} the value v_{iq}. This takes place with probability $\frac{1-\mu}{2}$. With probability $\frac{1-\mu}{2}$, All agents a_i where $i \geq cp$ will assign v_{ij} the value v_{iq}. In this fashion, the new dCS_j will have values up to cp from dCS_q and values from cp to N from dCS_j or it will have have values up to cp from dCS_j and values from cp to N from dCS_q which is the way single-point crossover works on centralized CSs.

GSoHC$_{tpx}$ uses a distributed restricted two-point crossover operator (dTPX-μ) that is based on two-point crossover, while GSoHC$_{mtpx}$ uses a modified dTPX-μ (referred to as dMTPX-μ) where the first and third segments of dCS_j are assigned values from dCS_q.

GSoHC$_{ux}$ works exactly like a the other GSoHCs except that on each cycle a distributed restricted uniform crossover operator is applied as follows. Each distributed candidate solution that has an above average number of conflicts, dCS_j, is replaced with an offspring that is a recombination of the best individual, dCS_q, and dCS_j as follows. An agent a_i will assign v_{ij} the value from v_{iq} with probability $\frac{1-\mu}{2}$ and will leave v_{ij} unchanged with probability $\frac{1-\mu}{2}$. With probability μ agent a_i will randomly assign v_{ij} a value from d_i the domain of values for variable x_i. We refer to this form of recombination as distributed restricted uniform crossover (dRUC-μ).

The simple and genetic SoHCs compared in this paper all use a society size of 32. The GSoHCs all use $\mu = 0.06^5$.

4   Results

4.1   Experiment I

In our first experiment, our test suite consisted of 400 instances of randomly generated DisACSPs of the form <30,6,1.0,p2>. In this experiment, $p2$ took on values from the set {0.03, 0.04, 0.05, 0.06} for the 400 instances (100 instances for each class of DisACSP) of <30,6,1.0,p2>, where $\hat{p2}_{crit_A} \approx 0.06$. Each of the 30 agents randomly generated 29 arcs where each arc contained approximately 1.08, 1.44, 1.88, and 2.16 nogoods respectively for $p2$ values of 0.03, 0.04, 0.05, and 0.06. The arcs were generated according to a hybrid between Models A & B in (Macintyre et al. 1998). This method of constraint generation is as follows. If each arc was to have 1.08 nogoods (which is the case when $p2 = 0.03$) then every arc received at least 1 nogood and was randomly assigned an additional nogood with probability 0.08. Similarly, if the average number of nogoods needed for each constraint was 2.16, (which is the case when $p2 = 0.06$) then every constraint received at least 2 nogoods and was randomly assigned and additional nogood with probability 0.16. The probability that an arc existed was determined with probability $p1_\alpha$.

In this section, we compare SoHC and the GSoHCs on the 400 randomly generated DisACSPs described earlier. Table 1 presents the performance results of these four SoHCs. In Tables 1a-1d, the first column represents the algorithm, the second column represents the success rate of an algorithm when given a maximum of 2000 cycles to solve each of the 100 problems within a class, and the third column represents the average number of cycles needed to solve the problems within a class.

In Tables 1a-1d, one can see that the GSoHCs outperform the SoHC on each of the four classes of DisACSPs. This suggests that using restricted distributed crossover results in improved performance. For each of the GSoHCs, we developed a 'headless' version (HSoHC) and the GSoHCs all had a statistically significant better performance (Zhang, F. 2003). In the 'headless' form of the distributed restricted recombination operators, the best dCS is crossed with a randomly generated individual. The purpose of these 'headless' operators is to validate the effectiveness of operator. If a HSoHC using 'headless' recombination outperforms a similar GSoHC that uses distributed restricted recombination then one can conclude that the recombination operator is not an effective for

---

[5] In (Zhang, F. 2003), GSoHCs using distributed restricted forms of single-point, two-point, and uniform crossover were compared. The society sizes, ρ, and mutation rates, μ, for the GSoHCs were taken from the sets {2,4,8,16,32} and {0.0,0.03,0.06,0.12,0.25} respectively. The best overall society size and mutation rate for the GSoHCs was $\rho = 32$ and $\mu = 0.06$. Therefore we only show the results of the GSoHCs where $\rho = 32$ and $\mu = 0.06$.

Table 1. Performances on the <30,6,1.0,0.03>, <30,6,1.0,0.04> <30,6,1.0,0.05> and <30,6,1.0,0.06> DisACSPs

Alg.	SR	Cycles
SoHC	1.00	17.93
GSoHC$_{spx}$	1.00	14.41
GSoHC$_{tpx}$	1.00	14.00
GSoHC$_{mtpx}$	1.00	13.15
GSoHC$_{ux}$	1.00	14.32

Alg.	SR	Cycles
SoHC	1.00	54.28
GSoHC$_{spx}$	1.00	31.01
GSoHC$_{tpx}$	1.00	30.35
GSoHC$_{mtpx}$	1.00	28.32
GSoHC$_{ux}$	1.00	31.29

(a) On the <30,6,1.0,0.03> DisACSPs (b) On the <30,6,1.0,0.04> DisACSPs

Alg.	SR	Cycles
SoHC	0.52	1323.80
GSoHC$_{spx}$	0.93	402.28
GSoHC$_{tpx}$	0.94	359.4
GSoHC$_{mtpx}$	0.98	273.62
GSoHC$_{ux}$	0.96	329.81

Alg.	SR	Cycles
SoHC	0.02	1981.06
GSoHC$_{spx}$	0.12	1861.59
GSoHC$_{tpx}$	0.10	1880.95
GSoHC$_{mtpx}$	0.11	1901.72
GSoHC$_{ux}$	0.09	1881.65

(c) On the <30,6,1.0,0.05> DisACSPs (d) On the <30,6,1.0,0.06> DisACSPs

the types of problems within the test suite and that actually macromutation is responsible for the performance improve over the simple SoHC (Jones 1995).

Notice also that GSoHC$_{mtpx}$ has the best performance in term of SR and Cycles for all classes of DisACSPs except for when $p2 = 0.06$. It seems that by taking 67% of the genes from the best individual in the population improves search performance on the easier classes of DisACSPs. However, at the phase transitions this performance improvement disappears. It would be interesting to see how a biased uniform crossover operator would perform on this test suite. Instead of taking values from the best individual for 50% of the genes, it would be interesting to see if a bias of 67% would improve performance the way that the modified two-point crossover did. It would also be interesting to see if a bias of less than 50% (but not 0%) would improve the performance on the DisACSPs located at the phase transition.

4.2 Discussion

The increased performance of the GSoHCs over SoHC is primarily due to the way in which their operators intensify search around the current best individual in the population. The basic assumption made by anyone applying an EC to a problem is that optimal (or near optimal) solutions are surrounded by good solutions. However, this assumption is not always true for constrained problems. Even for problems where this is the case ECs typically employ local search in an effort exploit promising regions. Thus, the EC will intensify search periodically in some region. Actually, the search behavior of the GSoHCs is no different. The hill-climbers associated with individuals that are involved in a below average number of conflicts are allowed to continue their search (via distributed hill-climbing) while hill-climbers associated with individuals that are involved in an above average number of conflicts have their associated individuals replaced by offspring that more closely resemble the current best individual in the population.

5 Conclusions

In this paper, we have introduced the concept of DisACSPs and have demonstrated how distributed restricted operators can be used to improve the search of a society of hill-climbers on easy and difficult DisACSPs. We also provided a brief discussion of some of the reasons why the performances of the GSoHCs were so dramatically superior to SoHC. We also showed that a modified version of two-point crossover outperforms two-point crossover (and all other recombination operators) on DisACSPs that are near the phase transition. We discussed how this result may lead to the development of a biased dRUC operator. Perhaps the bias may be adapted to the problem type. For easy problems the bias should be higher than 0.5 and for harder problems the bias should be lower than 0.5.

Acknowledgements. This research was supported by the National Science Foundation under grant #IIS-9907377.

References

Bäck, T., Hammel, U., and Schwefel, H.-P. (1997). "Evolutionary Computation: Comments on the History and Current State", *IEEE Transactions on Evolutionary Computation*, 1:1-23, IEEE Press.

Bejar, R., Krishnamachari, B., Gomes, C., and Selman, B. (2001). "Distributed Constraint Satisfaction in a Wireless Sensor Tracking System", *Proceedings of the IJCAI-2001 Workshop on Distributed Constraint Reasoning*, pp. 81-90.

Bowen, J. and Dozier, G. (1996). "Constraint Satisfaction Using A Hybrid Evolutionary Hill-Climbing Algorithm That Performs Opportunistic Arc and Path Revision," *Proceedings of AAAI-96*, pp. 326-331, AAAI Press / The MIT Press.

Calisti, M., and Faltings, B. (2001). "Agent-Based Negotiations for Multi-Provider Interactions", *Journal of Autonomous Agents and Multi-Agent Systems*.

Dozier, G. (2002). "Solving Distributed Asymmetric CSP Using via a Society of Hill-Climbers", *Proceedings of the 2002 International Conference on Artificial Intelligence*, pp. 949-953, June 24-27, Las Vegas, NV, CSREA.

Dozier, G., Bowen, J. and Homaifar, A. (1998). "Solving Constraint Satisfaction Problems Using Hybrid Evolutionary Search", *IEEE Transactions on Evolutionary Computation*, Vol. 2, No. 1, pp. 23-33, April 1998, Institute of Electrical & Electronics Engineers.

Fogel, D. B. (2000). *Evolutionary Computation: Toward a New Philosophy of Machine Intelligence*, 2nd Edition, IEEE Press.

Freuder, E. C., Minca, M., and Wallace, R. J. (2001). "Privacy/Efficiency Tradeoffs in Distributed Meeting Scheduling by Constraint-Based Agents", *Proceedings of the IJCAI-2001 Workshop on Distributed Constraint Reasoning*, pp. 63-71.

Jones, T. (1995). "Crossover, Macromutation and Population-based Search", *Proceedings of The Sixth International Conference (ICGA 1995)* pp. 73-80, Morgan Kaufmann.

MacIntyre, E., Prosser, P., Smith, B., and Walsh, T. (1998). "Random Constraint Satisfaction: Theory Meets Practice," *The Proceedings of the 4th International Conference on Principles and Practices of Constraint Programming (CP-98)*, pp. 325-339, Springer-Verlag.

Mackworth, A. K. (1977). "Consistency in networks of relations". *Artificial Intelligence*, 8 (1), pp. 99-118.

Modi, P. J., Jung, H., Tambe, M., Shen, W.-M., and Kulkarni, S. (2001). "Dynamic Distributed Resource Allocation: A Distributed Constraint Satisfaction Approach" *Proceedings of the IJCAI-2001 Workshop on Distributed Constraint Reasoning*, pp. 73-79.

Morris, P. (1993). "The Breakout Method for Escaping From Local Minima," *Proceedings of AAAI'93*, pp. 40-45.

Sebag, M. and Shoenauer, M. (1997). "A Society of Hill-Climbers," *The Proceedings of the 1997 International Conference on Evolutionary Computation*, pp. 319-324, IEEE Press.

Silaghi, M.-C., Sam-Haroud, D., Calisti, M., and Faltings, B. (2001). "Generalized English Auctions by Relaxation in Dynamic Distributed CSPs with Private Constraints", *Proceedings of the IJCAI-2001 Workshop on Distributed Constraint Reasoning*, pp. 45-54.

Smith, B. (1994). "Phase Transition and the Mushy Region in Constraint Satisfaction Problems," *Proceedings of the 11th European Conference on Artificial Intelligence*, A. Cohn Ed., pp. 100-104, John Wiley & Sons, Ltd.

Solnon, C. (2002). "Ants can solve constraint satisfaction problems", *IEEE Transactions on Evolutionary Computation*, Vol. 6, No. 4, pp. 347-357, August, IEEE Press.

Tsang, E. (1993). *Foundations of Constraint Satisfaction*, Academic Press, Ltd.

Yokoo, M. (2001). Distributed Constraint Satisfaction, Springer-Verlag Berlin Heidelberg.

Zhang, F. (2003). "A Comparison of Distributed Restricted Recombination Operators for Genetic Societies of Hill-Climbers: a DisACSP Perspective," *Auburn University Masters Thesis*, Department of Computer Science & Software Engineering.

Zhang, X. and Xing, Z. (2002). "Distributed Breakout vs. Distributed Stochastic: A Comparative Evaluation on Scan Scheduling," *AAMA-02 Third International Workshop on Distributed Constraint Reasoning*, July 16, 2002, Bologna, Italy, pp.192-201.

Analysis of the (1+1) EA for a Noisy OneMax

Stefan Droste*

LS Informatik 2, Universität Dortmund, 44221 Dortmund, Germany
stefan.droste@udo.edu

Abstract. In practical applications evaluating a fitness function is frequently subject to noise, i.e., the "true fitness" is disturbed by some random variations. Evolutionary algorithms (EAs) are often successfully applied to noisy problems, where they have turned out to be particularly robust. Theoretical results on the behavior of EAs for noisy functions are comparatively very rare, especially for discrete search spaces. Here we present an analysis of the (1+1) EA for a noisy variant of OneMax and compute the maximal noise strength allowing the (1+1) EA a polynomial runtime asymptotically exactly. The methods used in the proofs are presented in a general form with clearly stated conditions in order to simplify further applications.

1 Introduction

When trying to optimize problems in practice, it is rarely possible to determine the fitness value of a search point exactly. Most often noise changes the fitness value by some (typically small) amount because of interferences during the experiment resulting in the fitness value. Evolutionary algorithms (EAs) are often good at coping with noise due to the use of a population of search points. But there are only few theoretical results about the effects of noise on the performance of EAs for discrete search spaces, while continuous search spaces found more attention (see [AB03] for an overview).

In the last years the rigorous runtime analysis of EAs (e.g. see [DJW02], [JW99] or [WW03]) has proven to be a useful approach for extending our knowledge about EAs, besides other well-known approaches like schema theory or macroscopic, statistical, fitness landscape resp. local analysis. Here we extend this approach to the optimization of noisy functions by EAs. Since this is the first rigorous runtime analysis for noisy functions, we start with a simple EA and fitness function: the (1+1) EA for a noisy variant of OneMax. The (1+1) EA uses only one individual, changed by mutation only, and has been analyzed for a number of different fitness functions (e.g. see [Rud97] or [DJW02]), greatly extending the knowledge about the random processes underlying the (1+1) EA and appropriate proof techniques. Note that all these fitness functions are static and noise-free (just recently, dynamic variants of OneMax have been analyzed, see [Dro02] and [Dro03]), which is not very common in practical applications.

This research was partly supported by the Deutsche Forschungsgemeinschaft as part of the Collaborative Research Center "Computational Intelligence" (531).

K. Deb et al. (Eds.): GECCO 2004, LNCS 3102, pp. 1088–1099, 2004.

In this paper we analyze the influence of the noise strength on the runtime of the (1+1) EA until the optimum of ONEMAX is found. We show that noise, which changes one uniformly chosen bit before evaluation with probability p, makes an efficient optimization of the (1+1) EA impossible if and only if p grows asymptotically faster than $\log(n)/n$ (where n is the dimension of the search space). This is especially interesting since $\log(n)/n$ is also the critical value for the (1+1) EA on a dynamic ONEMAX where in each step with probability p one uniformly chosen bit of the target bit string changes. Although the two processes have no obvious interpretation in terms of each other, we can generalize the methods used to analyze the dynamic ONEMAX (see [Dro02]) to noisy ONEMAX.

In the next section, we formally define the (1+1) EA, our noise model, and the runtime. Section 3 presents the methods used in Section 4 to analyze the runtime of the (1+1) EA for noisy ONEMAX. This separation should increase comprehensibility, because the methods in Section 3 clearly state the conditions necessary for applying them, while Section 4 shows that the (1+1) EA on noisy ONEMAX fulfills these conditions. We finish with some conclusions.

2   The (1+1) EA and Noisy ONEMAX

In this section, we formally define the (1+1) EA and the noise model investigated. The (1+1) EA is the most basic EA since it uses only one individual changed by mutation only. This simplicity makes it an ideal starting point for theoretical analyzes (see [DJW02]). Furthermore, the (1+1) EA has proven to be surprisingly efficient for some functions compared to more involved EAs (see [FM92]), making its analysis worthwhile in itself.

The (1+1) EA for noisy functions differs from the (1+1) EA for noise-free evaluation only in one point: since the fitness-value of the parental search point evaluated in a generation before is not guaranteed to be the correct one, *both* the parent and the child are evaluated in each generation. Without this resampling a noisy evaluation can only be corrected if a copy of the search point is evaluated (e. g. if the mutation has no effect). Let $f_N : \{0,1\}^n \to \mathbb{R}$ denote the noisy function. Then the (1+1) EA looks as follows:

Definition 1 ((1+1) EA for maximization of noisy functions).

1. Set $t := 0$ and choose $x_t \in \{0,1\}^n$ randomly uniformly.
2. Set $x' := x_t$ and independently flip each bit of x' with prob. $1/n$.
3. If $f_N(x') \geq f_N(x_t)$, set $x_{t+1} := x'$, else $x_{t+1} := x_t$.
4. Set $t := t + 1$ and go to step 2.

In the following we will look at the case that f_N results from the *true fitness function* $f : \{0,1\}^n \to \{0,1\}$ by randomly changing the argument, i. e. $f_N(x) = f(N(x))$, where the random function $N : \{0,1\}^n \to \{0,1\}^n$ represents the noise. Hence, our noise model considers noise which takes effect before evaluation. We assume that the noisy evaluations are independent, i. e., the result of previous evaluations does not influence the actual evaluation.

We are interested in the number of generations the (1+1) EA needs to evaluate a maximum of the true fitness function f for the first time. Since there are two fitness evaluations per generation, the number of generations uniquely determines the number of fitness evaluations, which is often the most costly operation in practice. Considering an optimum of the true fitness function as an optimum of the noisy function seems to be a more natural choice than investigating the number of generations until a search point with maximal noisy fitness value is evaluated, since such a point can be "far away" from a true maximum. Hence, we analyze the number $T_{f,N}$ of generations the (1+1) EA needs to find a maximum of the underlying true fitness function f under noise N:

Definition 2 (Runtime of the (1+1) EA for noisy functions). *The runtime $T_{f,N}$ of the (1+1) EA for a noisy function f_N is the number of generations until a maximum of the true fitness function f is found:*

$$T_{f,N} := \min\{t \in \mathbb{N}_0 \mid f(x_t) = \max\{f(x) \mid x \in \{0,1\}^n\}\}.$$

We analyze the noise model N_p^1 changing with probability p exactly one uniformly chosen bit of x, while no bit is changed with probability $1 - p$. Such a model can be appropriate for applications where the genotype-phenotype mapping is error-prone, but the evaluation of the phenotype exact.

Definition 3 (One-bit noise N_p^1). *Let $p \in [0, 1]$. The random noise function $N_p^1 : \{0,1\}^n \to \{0,1\}^n$ is defined as follows:*

$$\forall x, y \in \{0,1\}^n : P(N_p^1(x) = y) = \begin{cases} 1 - p & \text{if } x = y, \\ p/n & \text{if } H(x,y) = 1, \\ 0 & \text{if } H(x,y) > 2. \end{cases}$$

We say that noisy evaluation increases resp. decreases the fitness of a point $x \in \{0,1\}^n$ if $f_N(x) > f(x)$ resp. $f_N(x) < f(x)$. For $f = \text{ONEMAX}$ (with $\text{ONEMAX}(x_1, \ldots, x_n) := x_1 + \cdots + x_n$) the noise function N_p^1 can only change the fitness of a point by one, and the probability of N_p^1 increasing resp. decreasing the fitness of x is directly proportional to $n - \text{ONEMAX}(x)$ resp. $\text{ONEMAX}(x)$.

Our goal is to determine the critical noise strength $p = p(n)$ such that the runtime of the (1+1) EA is polynomial in n with high probability if the noise is at most p, but super-polynomial with high probability if the noise is asymptotically larger than p. Although only one bit of the search point to be evaluated is changed by N_p^1, this can make the (1+1) EA accept a child x' whose true fitness $\text{ONEMAX}(x')$ is by two smaller than the true fitness $\text{ONEMAX}(x_t)$ of the parent x_t (see Section 4). This differs from the (1+1) EA for a dynamically changing ONEMAX, where the target bit string is changed by the operator N_p^1 and the Hamming distance to the target is to be minimized (see [Dro02] for an analysis that the expected runtime is polynomial if and only if $p = O(\log(n)/n)$). Although in the latter process the distance to the optimum can only increase by one in one step, our techniques can also cope with the first process and show that in this case $p = O(\log(n)/n)$ is the critical noise strength, too.

To distinguish more clearly between the key properties of the (1+1) EA on noisy ONEMAX resulting in polynomial resp. super-polynomial runtime and the necessary technical proofs, we show some general methods for proving polynomial upper and super-polynomial lower bounds in the next section.

3   Techniques for Bounding Markov Chains

In this section, we present some general results that can be applied to analyze the (1+1) EA for the noisy ONEMAX presented in the last section. They are presented in a general form with clearly stated conditions in order to make them easily applicable to other problems. Furthermore, this should make their proofs more comprehensible, since they need not to take care of the specific details of the EA analyzed as long as the EA fulfills the general conditions.

The (1+1) EA on noisy ONEMAX is a Markov process on the state space $\{0, \ldots, n\}$, where state i represents that the actual individual x_t has true fitness $\text{ONEMAX}(x_t) = i$. Hence, we identify the process by $(n+1)^2$ transition probabilities $p_{\cdot,\cdot} = (p_{i,j})_{i,j \in \{0,\ldots,n\}}$, where $p_{i,j} \in [0,1]$ is the probability of moving from state i to state j in one step. Let $T_{i,j}$ be the number of steps to come from state i to j for the first time, i.e. $T_{i,n}$ is the runtime when starting in state i.

3.1   Upper Bound Techniques

Our first result, implicitly already used in [Dro02], upper bounds the runtime of a process $p_{\cdot,\cdot}$ that can only move from i to $i-1$, i or $i+1$ in one step (a so-called $\{-1, 0, +1\}$-process) by a polynomial:

Lemma 1. *Let $p_{\cdot,\cdot}$ be a $\{-1, 0, +1\}$-process on $\{0, \ldots, n\}$. If there are two constants $c^+, c^- \in \mathbb{R}^+$ such that*

$$\forall i \in \{0, \ldots, n\}: \ p_{i,i+1} \geq c^+ \cdot \frac{n-i}{n} \ \text{ and } \ p_{i,i-1} \leq c^- \cdot \frac{\log(n)}{n} \cdot \frac{i}{n},$$

then for all $i \in \{0, \ldots, n-1\}$

$$E(T_{i,i+1}) \leq \frac{n^{1+c^-/(c^+ \cdot \ln(2))}}{c^+ \cdot (n-i)} \ \text{ and } \ E(T_{0,n}) = O\left(n^{1+c^-/(c^+ \cdot \ln(2))} \cdot \log(n)\right).$$

Proof. It is well known (e.g. see [DJW00]) that $E(T_{i,i+1})$ for $\{-1, 0, +1\}$-processes is $\sum_{k=0}^{i} \frac{1}{p_{k,k+1}} \cdot \prod_{l=k+1}^{i} \frac{p_{l,l-1}}{p_{l,l+1}}$. Utilizing the bounds for $p_{i,i-1}$ and $p_{i,i+1}$, we get:

$$E(T_{i,i+1}) \leq \sum_{k=0}^{i} \frac{n}{c^+(n-k)} \cdot \prod_{l=k+1}^{i} \frac{n}{c^+(n-l)} \cdot \frac{c^- \cdot \log(n) \cdot l}{n^2}$$

$$= \sum_{k=0}^{i} \frac{n}{c^+(n-k)} \cdot \left(\frac{c^-}{c^+} \frac{\log(n)}{n}\right)^{i-k} \cdot \frac{i!}{k!} \cdot \frac{(n-i-1)!}{(n-k-1)!}$$

$$= \frac{n}{c^+} \cdot \sum_{k=0}^{i} \left(\frac{c^-}{c^+} \frac{\log(n)}{n}\right)^{i-k} \frac{\binom{i}{k}}{\binom{n-k}{i}(n-i)} \leq \frac{n}{c^+(n-i)} \left(1 + \frac{c^-}{c^+} \frac{\log(n)}{n}\right)^i$$

$$\leq \frac{n}{c^+(n-i)} \cdot \exp\left(\frac{c^-}{c^+} \frac{\log(n)}{n} \cdot i\right) \leq \frac{n^{1+c^-/(c^+ \cdot \ln(2))}}{c^+(n-i)}.$$

Summing up these values for all $i \in \{0, \ldots, n-1\}$ gives the desired upper bound on $E(T_{0,n})$ because $\sum_{i=1}^{n} 1/i$ is $O(\log(n))$. □

To upper bound a Markov process $p_{.,.}$, which is no $\{-1, 0, +1\}$-process (i. e. replacing it by a process whose finite runtime to come from i to n stochastically dominates the runtime $T_{i,n}$ of the old process for all $i < n$) , we can proceed as follows: first, we "delete" all improvements by more than one, i. e., we set all probabilities $p_{i,i+d}$ for $d \geq 2$ to zero and increase $p_{i,i}$ by $\sum_{d=2}^{n-i} p_{i,i+d}$. This leads to a process whose runtime stochastically dominates the runtime of the old process (see [Dro03]), where a random variable X stochastically dominates a random variable Y if for all values d of X and Y: $P(X \geq d) \geq P(Y \geq d)$.

Afterwards, we have a process that can move from state i only to states $0, \ldots, i+1$, a so-called ≤ 1-process. To upper bound such a ≤ 1-process $p_{.,.}$ by a $\{-1, 0, 1\}$-process $\tilde{p}_{.,.}$, we can use the following lemma proven in [Dro03]:

Lemma 2. *Let $(p_{.,.})$ be a ≤ 1-process and $(\tilde{p}_{.,.})$ be a $\{-1, 0, 1\}$-process on the state space $\{0, \ldots, n\}$. If the following conditions hold*

1. *$\forall i \in \{1, \ldots, n-1\} : p_{i,0} \leq \prod_{k=1}^{i} \tilde{p}_{k,k-1}$,*
2. *$\forall i \in \{1, \ldots, n-1\} : \forall j \in \{1, \ldots, i-1\} : p_{i,j} \leq \tilde{p}_{j,j} \prod_{k=j+1}^{i} \tilde{p}_{k,k-1}$,*
3. *$\forall i \in \{0, \ldots, n-1\} : p_{i,i+1} \geq \tilde{p}_{i,i+1}$,*

then $E(T_{i,n}) \leq E(\tilde{T}_{i,n})$ for all $i \in \{0, \ldots, n\}$.

Now we can use Lemma 1 to upper bound the expected runtime of this slower $\{-1, 0, +1\}$-process $\tilde{p}_{.,.}$ by a polynomial if the transition probabilities fulfill the conditions. We will see in Section 4 that the process resulting from the (1+1) EA on noisy ONEMAX for small noise strength $p = O(\log(n)/n)$ has this form.

3.2   Lower Bound Techniques

If the probabilities $p_{i,i-1}$ of the process $p_{.,.}$ are only a little bit larger than necessary for Lemma 1, i. e. by a non-constant factor $\alpha(n)$, the expected runtime of the process is super-polynomially. To be more exact, the runtime is polynomial only with super-polynomially small probability $o(1/poly(n))$, i. e. smaller than $1/q(n)$ for any polynomial q (see also [Dro02] and [Dro03]):

Lemma 3. *Let $p_{.,.}$ be a Markov process on $\{0, \ldots, n\}$. If for a function $\alpha : \mathbb{N}^+ \to \mathbb{R}$ with $\alpha(n) \overset{n \to \infty}{\longrightarrow} \infty$ and $\alpha(n) \leq n/\log(n)$ the conditions*

1. *$\forall i \geq 0, \ d = \omega(\log(n)) : p_{i,i+d} = o(1/poly(n))$,*
2. *$\forall i \geq n - \alpha(n)\log(n), \ d = \omega(1) : p_{i,i+d} = o(1/poly(n))$, and*
3. *$\exists c^+, c^- \in \mathbb{R}^+ : \forall i \geq n - \alpha(n)\log(n) :*

$$\sum_{j=1}^{n-i} p_{i,i+j} \leq c^+ \cdot \frac{n-i}{n} \quad \text{and} \quad p_{i,i-1} \geq c^- \cdot \alpha(n) \cdot \frac{\log(n)}{n} \cdot \frac{i}{n}$$

hold then for all $i \leq n - \alpha(n)\log(n)$ the probability that $T_{i,n}$ is polynomial is super-polynomially small, i. e. $o(1/poly(n))$.

Proof. Let $I_t \in \{0, \ldots, n\}$ be the state of the process at time step t. Since for all i the probability of a direct step from i to $i + d$ with $d = \omega(\log(n))$ is super-polynomially small, we assume that only improving steps by $O(\log(n))$ happen. Hence, regardless of the initial state $i \leq n - \alpha(n) \log(n)$, the process reaches state n only via a state between $n - \alpha(n) \log(n)$ and $n - \alpha(n) \log(n)/2$. Since a constant factor does not matter for our definition of $\alpha(n)$, for an arbitrary constant $a \in {]}0, 1{[}$ there have to be time steps $t_1 < t_2$, such that $I_t \in \{n - \alpha(n)^a \log(n), \ldots, n-1\}$ for all $t \in \{t_1, \ldots, t_2-1\}$, and $I_{t_2} = n$ (a similar technique is used in [RRS95]). We show that it is super-polynomially unlikely to "bridge the gap" from state $n - \alpha(n)^a \log(n)$ to n in polynomially many steps. Hence, we can assume that $i \geq n - \alpha(n)^a \log(n)$ and that no improvements by more than $\alpha(n)^d$ for a constant $d \in {]}0, 1{[}$ do happen.

Since $(n - \alpha(n)^a \log(n))/n$ converges to 1, we bound $p_{i,i-1}$ and $\sum_{d=1}^{n-i} p_{i,i+d}$ for all n larger than a constant n_0, i.e. *n large enough*:

$$p_{i,i-1} \geq \frac{c^-}{2} \cdot \alpha(n) \frac{\log(n)}{n} \text{ and } \sum_{d=1}^{n-i} p_{i,i+d} \leq c^+ \cdot \alpha(n)^a \frac{\log(n)}{n}.$$

Therefore, the probability of an improving step under the condition that the step changes the state (i.e. the step is *effective*) is at most

$$\frac{c^+ \alpha(n)^a \log(n)/n}{c^+ \alpha(n)^a \log(n)/n + c^- \alpha(n) \log(n)/(2n)} \leq \frac{2c^+}{c^-} \cdot \frac{1}{\alpha(n)^{1-a}}.$$

Hence, the expected number of improving steps during t effective steps is at most $(2c^+/c^-) \cdot t/\alpha(n)^{1-a}$. However, to bridge the gap of size $\alpha(n)^a \log(n)$ in t effective steps consisting of t^+ improving and t^- decreasing steps, we must have

$$\alpha(n)^d \cdot t^+ - t^- \geq \alpha(n)^a \cdot \log(n) \iff t^+ \geq \frac{\alpha(n)^a \log(n) + t}{\alpha(n)^d + 1}.$$

The last term is at least $t/\alpha(n)^{c+d}$ for every constant $c > 0$ and n large enough. If we choose a, c, and d with $c + d < 1 - a$, the number of improving steps necessary to reach state n is by a non-constant factor larger than its expected number. For n large enough, this factor is at least 2.

By Chernoff bounds (see [MR95]) the probability that the sum of binary random variables is by a constant factor $1 + \delta$ larger than its expected value μ is upper bounded by (we use $\delta = 1$ and $\mu = (2c^+/c^-)\alpha(n)^{a-1}t$)

$$\left(\frac{\exp(\delta)}{(1+\delta)^{1+\delta}} \right)^\mu = \left(\frac{\exp(1)}{4} \right)^{(2c^+/c^-)\alpha(n)^{a-1}t}.$$

If t is small, this bound is not super-polynomially small. But since the process must bridge a gap of size $\alpha(n)^a \log(n)$ and any improvement by more than $\alpha(n)^d$ is excluded, the number t of effective steps must be at least $\alpha(n)^{a-d} \log(n)$. Hence, the above bound is at most $(\exp(1)/4)^{(2c^+/c^-)\alpha(n)^{2a-d-1} \log(n)}$, which is super-polynomially small if $2a - d - 1 > 0$. Since $a = 2/3$, $c = 1/7$, and $d = 1/7$ fulfill $2a - d - 1 > 0$ and $c + d < 1 - a$, the probability of every phase of polynomial length reaching the optimum is super-polynomially small. □

If the noise strength is too large, this result cannot be applied to lower bound the (1+1) EA on noisy ONEMAX. In this case a step decreasing the state is by a constant factor more likely than a step increasing it if the process is close to the optimal state n. If furthermore an improvement by d is by a non-constant factor more likely than an improvement by $d + 1$, we can apply the following result:

Lemma 4. *Let $p_{.,.}$ be a Markov process on $\{0, \ldots, n\}$ such that*

1. *$\forall i \geq 0, \ d = \omega(\log(n))$: $p_{i,i+d} = o(1/poly(n))$,*
2. *$\exists \delta, \varepsilon, \gamma > 0$: $\forall i \geq n - n^\delta, d \in \{1, \ldots, n - i\}$:*

$$\sum_{j=1}^{i} p_{i,i-j} \geq (1+\varepsilon) \cdot \sum_{j=1}^{n-i} p_{i,i+j} \quad and \quad \frac{p_{i,i+d}}{p_{i,i+d+1}} \geq n^\gamma.$$

Then for all $i \leq \lfloor n - n^\delta \rfloor$ the probability of $T_{i,n}$ being polynomially is super-polynomially small.

Proof. Analogously to the proof of Lemma 3, we can assume that to reach the target state n, there has to be a phase ranging from time steps t_1 to t_2, such that $I_{t_2} = n$ and for all $t \in \{t_1, \ldots, t_2 - 1\}$ we have $I_t \in \{\lfloor n - n^\delta \rfloor, \ldots, n - 1\}$. We show that every phase of polynomial length is super-polynomially unlikely.

Let $X_t \in \{-(\lfloor n - n^\delta \rfloor), \ldots, \lceil n^\delta \rceil\}$ be the change of state in time step t, i.e. $P(X_t = d) = p_{i,i+d}$ if the process is in state i at time step t. The phase can only be successful if $X_{t_1} + \cdots + X_{t_2} = n^\delta$, which will be shown to be exponentially unlikely. Since $X_t < 0$ is at least by a factor $1 + \varepsilon$ more likely than $X_t > 0$, we make the process only faster if we assume that $P(X_t < 0) = (1 + \varepsilon) \cdot P(X_t > 0)$ and that every step decreasing the number of ones decreases it by exactly one.

Since $p_{i,i+d}/p_{i,i+d+1}$ is at least n^γ for $d > 0$, we replace the step size d of an improving step by the value of a geometrically distributed random variable Y with success probability $1 - n^{-\gamma}$: this leads to a stochastically dominating process because $P(Y = d)/P(Y = d+1)$ is exactly n^γ. All in all, we have replaced every step of $p_{.,.}$ in the considered phase by the value of a $\{-1, \ldots, n^\gamma\}$-valued random variable X' defined by

$$P(X' = d) = \begin{cases} \frac{1+\varepsilon}{2+\varepsilon} & \text{if } d = -1 \\ \frac{1}{2+\varepsilon} \cdot (n^{-\gamma})^{d-1} \cdot (1 - n^{-\gamma}) & \text{if } d \geq 1 \end{cases}$$

(Note that $X' < 0$ is by a factor $1+\varepsilon$ more likely than $X' > 0$.) Hence, the random variable $S_{t(n)} = X'_1 + \cdots + X'_{t(n)}$ stochastically dominates the improvement of the process $p_{.,.}$ during a phase of length $t(n)$ starting from a state at least $\lfloor n - n^\delta \rfloor$. As the expected value of Y is $1/(1 - n^{-\gamma})$, for n large enough we have

$$E(X') = -\frac{1+\varepsilon}{2+\varepsilon} + \frac{1}{2+\varepsilon} \cdot \frac{1}{1 - n^{-\gamma}} = -\frac{1+\varepsilon}{2+\varepsilon} + \frac{1}{2+\varepsilon} \cdot \frac{n^\gamma}{n^\gamma - 1} \leq -\frac{\varepsilon/2}{2+\varepsilon}.$$

Hence, the expected value of $S_{t(n)}$ is for n large enough and a constant $\varepsilon' > 0$ at most $-t(n)\varepsilon'$. In order for a phase of length $t(n)$ to be successful $S_{t(n)}$ has to be at least n^γ. We show that even $P(S_{t(n)} > 0)$ is super-polynomially unlikely for all possible phase lengths $t(n)$. To estimate $P(S_{t(n)} > 0)$ we cannot use Chernoff bounds, since $S_{t(n)}$ is no sum of $\{0, 1\}$-valued, but $\{-1, \ldots, n^\gamma\}$-valued random variables. In this situation we use Hoeffding's inequality ([Hoe63]), a generalization of Chernoff bounds, stating in our notation

$$P\big(S_{t(n)} - E(S_{t(n)}) \geq t(n)\varepsilon\big) \leq \exp\left(\frac{-2t(n)\varepsilon^2}{z^2}\right)$$

where z is the number of different values of X_i'. Because of Condition 1 we can assume that no improvements by more than $\log(n)^2 - 2$ happen implying $z = \log(n)^2$ and $t(n) \geq n^\gamma / \log(n)^2$. Therefore, Hoeffding's inequality gives us an exponentially small upper bound on the probability, that a phase of polynomial length is successful.   □

4   Application of the Techniques for the (1+1) EA on Noisy ONEMAX

Now we apply the techniques presented in the previous section to determine the maximal noise strength p under which the (1+1) EA is still able to optimize ONEMAX with single-bit noise in expected polynomial time. Hence, let $Mut_i \in \{-i, \ldots, n-i\}$ be the random variable denoting the change of the number of ones after a mutation of a search point x with $\text{ONEMAX}(x) = i$. Depending on the value $d \neq 0$ of Mut_i we determine if and when this mutation leads to a change of the number of ones of the actual individual by d:

- If $Mut_i = d$ with $d \geq 2$, the mutation will be accepted regardless of the noisy evaluations. Hence, for all $i \in \{0, \ldots, n-2\}$ and $d \in \{2, \ldots, n-i\}$:

$$p_{i,i+d} = P(Mut_i = d).$$

- If $Mut_i = 1$, the mutation will be accepted except in case that the noisy evaluation of the parent increases its value while the noisy evaluation of the child decreases its value. Hence, for all $i \in \{0, \ldots, n-1\}$:

$$p_{i,i+1} = P(Mut_i = 1) \cdot \left(1 - p\frac{n-i}{n} \cdot p\frac{i+1}{n}\right). \tag{1}$$

- If $Mut_i = -1$, the mutation will be accepted if and only if one of the three following cases happens:
    - the noisy evaluation of the parent decreases its value and the evaluation of the child is not disturbed by noise,
    - the noisy evaluation of the parent decreases its value and the noisy evaluation of the child increases it value, or
    - the evaluation of the parent is not changed by noise and the noisy evaluation of the child increases its value.

  Hence, for all $i \in \{1, \ldots, n\}$:

$$p_{i,i-1} = P(Mut_i = -1) \cdot \left(p\frac{i}{n}\left(1 - p + p\frac{n-i+1}{n}\right) + (1-p)p\frac{n-i+1}{n}\right). \tag{2}$$

- If $Mut_i = -2$, the mutation will be accepted only if the noisy evaluation of the parent decreases its value and the noisy evaluation of the child increases its value. Hence, for all $i \in \{2, \dots, n\}$:

$$p_{i,i-2} = P(Mut_i = -2) \cdot \left(p\frac{i}{n} p \frac{n-i+2}{n} \right). \tag{3}$$

As every mutation decreasing the number of ones by at least two is not accepted, $p_{i,i-d} = 0$ for all $d \geq 2$.

It is obvious that the probability $P(Mut_i = d)$ is essential when analyzing the (1+1) EA on noisy ONEMAX. We can easily show, that this probability is $O(((n-i)/n)^d)$ for $d > 0$ and $O((i/n)^d)$ for $d < 0$:

$$d > 0 : P(Mut_i = d) = \sum_{k=d}^{n-i} \binom{n-i}{k} \cdot \binom{i}{k-d} \cdot \left(\frac{1}{n}\right)^{2k-d} \left(1 - \frac{1}{n}\right)^{n-2k+d}$$

$$\leq \left(\frac{n-i}{n}\right)^d \left(1 - \frac{1}{n}\right)^{n-d} \frac{1}{d!} + \sum_{k=d+1}^{n-i} \left(\frac{n-i}{n}\right)^k \frac{1}{k!}. \tag{4}$$

$$d < 0 : P(Mut_i = d) = \sum_{k=d}^{i} \binom{i}{k} \cdot \binom{n-i}{k-d} \cdot \left(\frac{1}{n}\right)^{2k-d} \left(1 - \frac{1}{n}\right)^{n-2k+d}$$

$$\leq \left(\frac{i}{n}\right)^d \cdot \left(1 - \frac{1}{n}\right)^{n-d} \cdot \frac{1}{d!} + \sum_{k=d+1}^{i} \left(\frac{i}{n}\right)^k \cdot \frac{1}{k!}. \tag{5}$$

These bounds are asymptotically tight as long as d is a constant:

$$d > 0 : P(Mut_i = d) = \sum_{k=d}^{n-i} \binom{n-i}{k} \cdot \binom{i}{k-d} \cdot \left(\frac{1}{n}\right)^{2k-d} \left(1 - \frac{1}{n}\right)^{n-2k+d}$$

$$\geq \binom{n-i}{d} \left(\frac{1}{n}\right)^d \cdot \exp(-1) \geq \left(\frac{n-i}{n}\right)^d \cdot \frac{\exp(-1)}{d^d}. \tag{6}$$

$$d < 0 : P(Mut_i = d) = \sum_{k=d}^{i} \binom{i}{k} \cdot \binom{n-i}{k-d} \cdot \left(\frac{1}{n}\right)^{2k-d} \left(1 - \frac{1}{n}\right)^{n-2k+d}$$

$$\geq \binom{i}{d} \left(\frac{1}{n}\right)^d \cdot \exp(-1) \geq \left(\frac{i}{n}\right)^d \cdot \frac{\exp(-1)}{d^d}. \tag{7}$$

4.1   A Polynomial Upper Bound for $p = O(\log(n)/n)$

First, we want to upper bound the expected runtime of the (1+1) EA on noisy ONEMAX for $p = O(\log(n)/n)$. Therefore, we "delete" all transitions going from i to $i+d$ with $d \geq 2$ (making the process only slower) and bound the remaining transition probabilities in the following way for n large enough:

$$p_{i,i+1} \overset{(1),(6)}{\geq} \frac{n-i}{n} \cdot \exp(-1) \cdot \left(1 - p^2 \frac{(n-i)(i+1)}{n^2}\right) \geq \frac{n-i}{n} \cdot \frac{3\exp(-1)}{4},$$

$$p_{i,i-1} \overset{(2)}{\leq} \frac{i}{n} \cdot 2p \ \text{ and } \ p_{i,i-2} \overset{(3)}{\leq} \binom{i}{2} \left(\frac{1}{n} \right)^2 p^2 = \frac{i(i-1)}{n^2} \cdot \frac{p^2}{2}.$$

To upper bound $p_{.,.}$, we replace it by the following $\{-1, 0, +1\}$-process $\tilde{p}_{.,.}$:

$$\tilde{p}_{i,i+1} = \frac{n-i}{n} \cdot \frac{3 \exp(-1)}{4} \ \text{ and } \ \tilde{p}_{i,i-1} := \frac{i}{n} \cdot 4p.$$

It is obvious that $\tilde{p}_{i,i+1} \geq p_{i,i+1}$. Since $3 \exp(-1)/4 < 1/2$ and $4p$ converges to 0, $\tilde{p}_{i,i}$ is at least $1/2$ for n large enough. Hence, $\tilde{p}_{i,i-1} \cdot \tilde{p}_{i-1,i-1} \geq p_{i,i-1}$ and $\tilde{p}_{i,i-1} \cdot \tilde{p}_{i-1,i-2} \cdot \tilde{p}_{i-2,i-2} \geq p_{i,i-2}$ hold, implying that Lemma 2 guarantees that $\tilde{p}_{.,.}$ upper bounds $p_{.,.}$. Finally, applying Lemma 1 to $\tilde{p}_{.,.}$ gives us:

Theorem 1. *The expected runtime T_{ONEMAX, N_p^1} of the (1+1) EA on ONEMAX with noise function N_p^1 is polynomial for all $p = O(\log(n)/n)$.*

4.2 A Super-polynomial Lower Bound for $p = \omega(\log(n)/n)$

Let us now look at the case that p grows asymptotically faster than $\log(n)/n$, i.e. $p \geq \gamma(n) \log(n)/n$ for some function $\gamma : \mathbb{N}^+ \to \mathbb{R}$ with $\gamma(n) \overset{n \to \infty}{\longrightarrow} \infty$. In the following, we want to show that the conditions of Lemma 3 hold for $p_{.,.}$ with $\alpha(n) := \min\{\log(n), \gamma(n)\}$, if $p \leq 1 - \alpha(n) \log(n)/n$.

It is obvious that $\alpha(n) \overset{n \to \infty}{\longrightarrow} \infty$. Condition 1 holds, since the probability $P(Mut_i = d)$ is at most $1/d!$, which is super-polynomially small for $d = \omega(\log(n))$. Furthermore, $p_{i,i+d} = o(1/poly(n))$ for all $i \geq n - \log(n)\alpha(n)$ and non-constant d because of the upper bound (4) for $P(Mut_i = d)$. Hence, condition 2 holds. We can upper bound $\sum_{d=1}^{n-i} p_{i,i+d}$ (using (4)) by

$$\sum_{d=1}^{n-i} \sum_{k=d}^{n-i} \left(\frac{n-i}{n} \right)^k \cdot \frac{1}{k!} = \sum_{k=1}^{n-i} \left(\frac{n-i}{n} \right)^k \cdot \frac{1}{(k-1)!} = O\left(\frac{n-i}{n} \right).$$

In turn, we lower bound $p_{i,i-1}$ for $i \geq n - \alpha(n) \log(n)$ and n large enough by:

$$p_{i,i-1} \overset{(2),(7)}{\geq} \frac{i}{n} \cdot \exp(-1) \cdot p \cdot \left(\frac{i}{n} \left(1 - p + p \frac{n-i+1}{n} \right) + (1-p) \frac{n-i+1}{n} \right)$$

$$\geq \frac{i}{n} \cdot \exp(-1) \cdot \frac{1}{4} \cdot \frac{\alpha(n) \log(n)}{n}.$$

The last inequality holds due to the following case inspection: if $p \leq 1/2$, we use $p \geq \alpha(n) \log(n)/n$, $i/n \geq 1/2$, $1-p \geq 1/2$ and replace all other terms by zero. If $1/2 < p \leq 1 - \alpha(n) \log(n)/n$, we use $p > 1/2$, $i/n \geq 1/2$, $1-p \geq \alpha(n) \log(n)/n$, and replace all other terms by zero. Hence, all conditions of Lemma 3 hold.

Let us now look at the case $p > 1 - \alpha(n) \log(n)/n$. We cannot lower bound $p_{i,i-1}$ by $\Theta(\alpha(n)(\log(n)/n)(i/n))$ anymore: e.g. for $p = 1$ decreasing the number of ones implies that the evaluation of the parent decreases its fitness, while the evaluation of the child increases its fitness, i.e. $p_{i,i-1} = P(Mut_i = -1) \cdot (i/n) \cdot (n - i + 1)/n$, which is smaller than $\alpha(n)(\log(n)/n)(i/n)$ for i close to n.

So, we use Lemma 4 to lower bound the runtime for $p > 1 - \alpha(n)\log(n)/n$. First, we show that decreasing the number of ones is by a constant factor more likely than increasing it, if $i \geq n - n^{1/3}$:

$$\sum_{d=1}^{\bar{n-i}} p_{i,i+d} \overset{(4)}{\leq} \sum_{d=1}^{\bar{n-i}} P(Mut = d) \leq \frac{n-i}{n} \cdot \left(1 - \frac{1}{n}\right)^{-1} + \sum_{d=2}^{\bar{n-i}} O\left(\left(\frac{n-i}{n}\right)\right)$$

$$= \frac{n-i}{n} \cdot \left(1 - \frac{1}{n}\right)^{-1} + O\left(\frac{1}{n^{4/3}}\right),$$

$$p_{i,i-1} \overset{(2)}{\geq} P(Mut = -1) \cdot p^2 \cdot \frac{i}{n}\frac{n-i}{n} \overset{(7)}{\geq} \left(\frac{i}{n} \cdot p\right)^2 \left(1 - \frac{1}{n}\right)^{-1} \cdot \frac{n-i}{n},$$

$$p_{i,i-2} \overset{(3)}{\geq} P(Mut = -2) \cdot p^2 \cdot \frac{i}{n}\frac{n-i}{n} \overset{(7)}{\geq} \frac{i^2(i-1)}{2n^3} \cdot p^2 \cdot \left(1 - \frac{1}{n}\right)^{-2} \cdot \frac{n-i}{n}.$$

Hence, we have

$$\frac{\sum_{d=1}^{n-i} p_{i,i+d}}{p_{i,i-1} + p_{i,i-2}} \leq \frac{\frac{n-i}{n} \cdot \left(1 - \frac{1}{n}\right)^{n-1} + O\left(\frac{1}{n^{4/3}}\right)}{\left(\frac{i}{n} \cdot p\right)^2 \left(1 - \frac{1}{n}\right)^{n-1} \cdot \frac{n-i}{n} + \frac{i^2(i-1)}{2n^3} \cdot p^2 \cdot \left(1 - \frac{1}{n}\right)^{n-2} \cdot \frac{n-i}{n}}$$

$$= \frac{1 + O\left(\frac{1}{n^{1/3}}\right)}{\left(\frac{i}{n} \cdot p\right)^2 + \frac{i^2(i-1)}{2n^3} \cdot p^2 \cdot \left(1 - \frac{1}{n}\right)^{-1}}.$$

Since the denominator converges to $3/2$ and the nominator to 1, the last term is at least $4/3$ for n large enough. Hence, an impairment is by a constant more likely than an improvement for all $i \geq n - n^{1/3}$ and n large enough.

To apply Lemma 4 we must show additionally that p_{i+d}/p_{i+d+1} is at least n^γ for all $i \geq n - n^{1/3}$, $d \geq 1$, and a constant $\gamma > 0$:

$$\frac{p_{i+d}}{p_{i+d+1}} \overset{(1),(4),(6)}{\geq} \frac{\Omega\left(\left(\frac{n-i}{n}\right)^d\right)}{O\left(\left(\frac{n-i}{n}\right)^{d+1}\right)} = \Omega\left(\frac{n}{n-i}\right) = \Omega(n^{2/3}).$$

Hence, Lemma 4 tells us that the runtime of the $(1+1)$ EA on noisy ONEMAX for $p \geq 1 - \alpha(n)\log(n)$ is polynomial only with super-polynomially small probability:

Theorem 2. *The runtime T_{ONEMAX, N_p^1} of the $(1+1)$ EA on ONEMAX with noise function N_p^1 is polynomial with super-polynomially small probability for all $p = \omega(\log(n)/n)$.*

5   Conclusions

We have analyzed the runtime of the $(1+1)$ EA on a variant of ONEMAX where every evaluation is subject to noise with probability p and noise changes one uniformly chosen bit. In this case, the $(1+1)$ EA optimizes ONEMAX with high probability in polynomial time if and only if p is $O(\log(n)/n)$. This is the first rigorous analysis of an EA for a noisy fitness function without any assumptions. Moreover, the paper presents the methods used to prove this result in a general and modular way in order to help understanding the ideas and to simplify application to other EAs and/or noise models. For instance, generalizing the result for a noise model where every bit is changed by noise with some probability p is an obvious goal for future research.

Acknowledgements. Hereby I thank Jens Jägersküpper, Tobias Storch, and Carsten Witt for valuable advice and discussions.

References

[AB03]   Dirk V. Arnold and H.-G. Beyer. A comparison of evolution strategies with other direct search methods in the presence of noise. *Computational Optimization and Applications*, 24:135 – 159, 2003.

[DJW00]   S. Droste, Th. Jansen, and I. Wegener. Dynamic parameter control in simple evolutionary algorithms. In *Proceedings of FOGA 2000*, pages 275–294, 2000.

[DJW02]   S. Droste, Th. Jansen, and I. Wegener. On the analysis of the $(1 + 1)$ EA. *Theoretical Computer Science*, (276):51–81, 2002.

[Dro02]   S. Droste. Analysis of the (1+1) EA for a dynamically changing OneMax-variant. In *Proceedings of CEC 2002*, pages 55–60, 2002.

[Dro03]   S. Droste. Analysis of the (1+1) EA for a dynamically bit-wise changing OneMax-variant. In *Proceedings of GECCO 2003*, pages 909–921, 2003.

[FM92]   S. Forrest and M. Mitchell. Relative building block fitness and the building block hypothesis. In *Proceedings of FOGA 1992*, pages 198–226, 1992.

[Hoe63]   W. Hoeffding. Probability inequalities for sums of bounded random variables. *Journal of the American Statistical Association*, 58:13–30, 1963.

[JW99]   Th. Jansen and I. Wegener. On the analysis of evolutionary algorithms – a proof that crossover really can help. In *Proceedings of ESA 1999*, pages 184–193, 1999.

[MR95]   R. Motwani and P. Raghavan. *Randomized Algorithms*. Cambridge University Press, 1995.

[RRS95]   Y. Rabani, Y. Rabinovich, and A. Sinclair. A computational view of population genetics. In *Proceedings of STOC 1995*, pages 83–92, 1995.

[Rud97]   G. Rudolph. *Convergence Properties of Evolutionary Algorithms*. Verlag Dr. Kovač, 1997.

[WW03]   I. Wegener and C. Witt. On the optimization of monotone polynomials by the (1+1) EA and randomized local search. In *Proceedings of GECCO 2003*, pages 622–633, 2003.

A Polynomial Upper Bound for a Mutation-Based Algorithm on the Two-Dimensional Ising Model

Simon Fischer

FB Informatik, LS2, Univ. Dortmund, 44221 Dortmund, Germany
simon.fischer@cs.uni-dortmund.de

Abstract. Fitness functions based on the Ising model are suited excellently for studying the adaption capabilities of randomised search heuristics. The one-dimensional Ising model was considered a hard problem for mutation-based algorithms, and the two-dimensional Ising model was even thought to amplify the difficulties. While in one dimension the Ising model does not have any local optima, in two dimensions it does. Here we prove that a simple search heuristic, the Metropolis algorithm, optimises on average the two-dimensional Ising model in polynomial time.

Keywords: Ising model, expected optimisation time, adaption

1 Introduction

In 1925, Ising [5] developed the Ising model in order to study the effects of ferromagnetism. Naudts and Naudts [7] introduced a class of interesting fitness functions based on this model to the genetic algorithm community. In its most general form the fitness function is defined on colourings of the vertices of weighted, undirected graphs. Given an undirected graph $G = (V, E)$ and a weight function $w : E \mapsto \mathbb{R}$, the fitness of a colouring is defined as the sum of the weights of all monochromatic edges, i. e. edges connecting vertices of equal colour. This function is to be maximised. In its general form, the Ising model is NP-hard, though there exist quite efficient algorithms for this problem (see for example [8,6,2]).

In this article we use the constant weight function 1 for all edges and two colours 0 and 1, which yields a trivial optimisation problem, but an interesting fitness function for the investigation of the adaption capabilities of randomised search heuristics. While nowadays GAs and search heuristics are mainly used as optimisers (see e. g. [3]), they were designed as adaption systems by Holland [4].

In general, most of the graphs are of no interest as regards physics. Here, we focus on a two-dimensional Ising model, or torus, which is – from a physicist's point of view – the most relevant one. Usually, a torus is defined as a set of vertices placed on a $X \times Y$-lattice with edges connecting horizontal and vertical neighbours plus the wrap-around edges connecting vertices in the last column

K. Deb et al. (Eds.): GECCO 2004, LNCS 3102, pp. 1100–1112, 2004.

(row) with the vertex in the first column (row) of the same row (column), respectively. Here we also use diagonal edges, which is natural since the Ising model was designed as a model for the behaviour of interacting particles, where the strength of the interaction decreases with distance. Throughout this paper, we will restrict ourselves to quadratic toruses, i.e. with n vertices and $X = Y = \sqrt{n}$.

GA experiments on the one-dimensional Ising model and similar fitness functions have been reported by Van Hoyweghen [9] stating that "the Ising model can be seen as an archetypical problem where symmetry prevents an unspecialized GA from solving the problem quickly." Similar statements can be found in [10] and [11].

We aim to analyse a randomised search heuristic on the torus. It is easy to see that there exist non-optimal colourings from which a hill climber can hardly escape. Imagine a torus coloured 1 in the upper half and 0 in the lower half. Flipping only a few bits, the fitness can only decrease, since the flipping bits must produce islands or dents, introducing new non-monochromatic edges. In this situation hill climbers like the (1+1)EA have exponential waiting time. The simplest non-hill climbing selection strategy is used by the Metropolis algorithm:

1. Among all colourings of V, choose x uniformly at random.
2. Repeat
   a) Create x' from x by flipping the colour of a single vertex selected uniformly at random.
   b) Replace x by x' with probability $e^{\min\{0,(f(x')-f(x))/T\}}$.

For $T \to 0$ the algorithm never accepts offspring with lesser fitness and becomes randomised local search (RLS). We apply a local search operator, which is common for the Metropolis algorithm. Using a global search operator that flips each bit with probability $1/n$ independently, as is usually done for evolutionary algorithms, would also be possible. While the analysis gives little further insight, using global mutation inflates a tedious case differentiation to several pages.

A step flipping exactly one bit can change the fitness by at most 8, since each vertex has exactly 8 neighbours. If we set the parameter T to $-2/\ln p_T$ for some $p_T \in [0,1]$, the probability of accepting a step decreasing the fitness by i is at most $p_T^{i/2}$. We will use this fact to limit the number of fitness-decreasing steps.

Section 2 introduces characteristic properties of colourings which decide whether or not it is easy to improve the fitness. Section 3 and 4 deal with two classes of colourings which do not require acceptance of individuals with lesser fitness. Section 5 will describe how the algorithm can escape from the local optima we already mentioned. We finish with some conclusions.

Note that in all images in this article vertices are represented by squares; the implicit edges are omitted. Gray indicates colour 1, white indicates colour 0.

2   Structure of the Torus and Sketch of Analysis

The optimisation process can be seen as a struggle between the two colours. A connected area can dominate another area, if it surrounds it completely. We will show that islands being surrounded by vertices of the other colour shrink and finally vanish. As this happens, the fitness increases from time to time. At their outermost borders, islands are delimited by blocks which can grow and shrink, where shrinking is at least as probable as growing. We can describe this process as a random walk of the block length. When there are none of these blocks or dents left, the colouring consists of rings winding around the torus once or more, and there is no dominating colour. Still the size of a ring may vary and perform a random walk, unless all of the borders of the ring are exactly parallel to edges. This is the only situation in which fitness decreasing steps are necessary. We will now give formal definitions of these notions. If we place the vertices of the torus in a lattice, we can talk about neighboured vertices using the terms "left", "right", "above", and "below". With respect to a directed path p, we can also talk of neighbours of path vertices as being situated left or right of the path.

Definition 1 (Border). *Let $G = (V, E)$ be a torus and $C \subseteq V$ a maximal connected set of vertices coloured c. A path p is called* border *of C if all vertices of p lie in C and all vertices left of p do not lie in C and p cannot be extended without violating the first two conditions.*

With respect to the lattice all edges in p can be assigned a vector from $\{-1, 0, 1\}^2$ specifying its direction in the lattice. The path p is called x-monotone (y-monotone) if for all edges in p the x-component (y-component) of the vector associated with the edge is either always non-negative or always non-positive. p is called monotone *if it is x-monotone and y-monotone. Otherwise p is* bent.

Since the torus is finite, all borders are loops. Note that monotone borders meet their starting point after making, e. g., $\sqrt{n}$ steps up and running into their starting point from below. This is why we call the areas delimited by monotone borders "rings."

We will now define two promising classes of colourings. The first definition is straightforward and covers the dents or blocks delimiting an island while the second definition covers special cases.

Definition 2 (Delimiting block). *Let $G = (V, E)$ be a torus and $B \subseteq V$ a connected set of vertices situated on a horizontal or vertical line in the lattice. Let N be the set of vertices that are neighbours of B-vertices situated in any fixed half-plane defined by this line. B is called a* delimiting c-block *of length $|B|$, iff all vertices in B are coloured c and all vertices in N are coloured $1 - c$.*

Since the outermost vertices in a delimiting block B have at least four neighbours that are coloured complimentarily, these vertices can always flip and hence decrease the size of B. Possibly the block can grow, but the probability of this event is at most as large as the probability for shrinking, if $|B| \geq 2$.

In the first part of the analysis we want to show that bent borders vanish leaving only rings behind. We do so by showing that the fitness increases as long as we have delimiting blocks. Unfortunately, not all colourings with bent borders do also contain delimiting blocks. In a checkerboard-like colouring, borders run counter-clockwisely around the individual fields and are therefore bent, though not containing any delimiting blocks. Despite of this, it is not hard to improve the fitness by flipping vertices near the corners, where borders intersect each other. We will show that for subgraphs of this type fitness improvement is easy.

Definition 3 (Intersection). *In a coloured torus the following subgraphs and their horizontal and vertical mirages as well as rotations are called* intersection:

$$\text{Type 1: } \frac{101}{011} \qquad \text{Type 2: } \frac{100}{011} \qquad \text{Type 3: } \frac{101}{010}$$

We will show that in a colouring with bent borders there must exist at least one delimiting block or an intersection. In both cases a fitness improvement is easy.

Since we will frequently encounter random walks, we will start by providing for an important tool heavily used throughout the analysis. Here we consider a Markov chain that is "at least fair."

Lemma 1. *Consider a Markov chain with transition probabilities $p(i, i+1) \geq 1/2$ and $p(i, i-1) = 1 - p(i, i+1) \leq 1/2$. Starting at state 0 the probability of reaching a state $m \geq n$ after at most cn^2 steps is bounded below by a positive constant, given that c is large enough.*

Proof. Among the $N = cn^2$ steps the probability that the number X of steps that decrease the position equals k is $P(X = k) \leq \binom{N}{k} \cdot 2^{-n}$. The maximum of this term is reached for $k = N/2$ and by Stirling's formula it is bounded from above by $aN^{-1/2}$ for an appropriate a. Since $P(X = k)$ is decreasing left and right of the maximum, the probability of deviating by more than $n/2$ from $N/2$ is bounded above by the product $n/2 \cdot aN^{-1/2} = (na)/(2cn^2) = a/(2\sqrt{c})$. Conversely, the probability that X is at least $(1/2)N + (1/2)n$, meaning that we reach a state $m \geq n$, is at least $1 - a/(2\sqrt{n}) \geq \varepsilon > 0$ if c is large enough.   □

As our second main technique, we show that it is sufficient to find a fitness-increasing mutation sequence of constant length to bound the success probability within $\mathcal{O}(n)$ steps by a positive constant from below.

Lemma 2. *Let $v_1, \ldots, v_j$ a sequence of vertices with $j = \mathcal{O}(1)$. If the sequence of mutations flipping $v_1, \ldots, v_j$ increases the fitness and is acceptable, then RLS increases the fitness within $\mathcal{O}(n)$ steps with positive constant probability.*

Proof. Denote by N all vertices connected to a vertex in $\{v_1, \ldots, v_j\}$. It holds that $|N| = \mathcal{O}(1)$. Since RLS does not accept fitness decreasing steps, we do not have to care about steps modifying vertices outside N. A mutation flipping v_i is accepted if $v_1, \ldots, v_{i-1}$ already flipped and none of the vertices in N flipped. For all i, v_i is the first vertex to flip in $N \cup \{v_i\}$ with probability at least $1/(|N| + 1)$. The probability for this to hold for all $i \in \{1, \ldots, j\}$ is at least $(|N| + 1)^{-j} \geq \varepsilon > 0$. The expected time for $\mathcal{O}(1)$ fixed bits to flip is $\mathcal{O}(n)$.   □

3   Delimiting Blocks and Intersections

In this section, we will prove that bent borders can ensure a fitness gain within $\mathcal{O}(n^2)$ steps. Therefore we show that, whenever there are bent borders, there are also delimiting blocks or intersections. In this as well as in the following section we analyse RLS instead of the Metropolis algorithm, because it is likely to behave like RLS if p_T is small.

Lemma 3. *If there exists a bent border within a colouring of a torus, there also exists a delimiting block or an intersection.*

Proof. W.l.o.g. the border is not y-monotone and coloured 1. Then there must exist a path section starting with a non-zero y-component y_s and ending with a non-zero y-component $y_t = -y_s$, where the edges in between all have zero y-components. W.l.o.g. $y_s = 1$ and $y_t = -1$. The path section has the form

$$(x_s, +1), (x_h, 0), \ldots, (x_h, 0), (-x_h, 0), \ldots, (-x_h, 0), (x_t, -1).$$

The $(-x_h, 0)$ edges are usually absent. There are two cases:

1. $x_h = +1$, i.e. the path is bent to the right. Since all vertices left of the path are coloured 0, the vertices in the horizontal section form a delimiting block.
2. $x_h = -1$, i.e. the path is bent to the left. The vertices below the horizontal part of the border are coloured 0 and are a delimiting block candidate. Their upper, left, and right neighbours are border vertices and coloured 1, as is required by the definition of the delimiting block. However, the diagonal neighbours of the outermost 0-vertices are not necessarily coloured 1. If they are not, we have found an intersection.

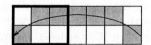

In the image, we see a border (arrow) and a type 2 intersection (heavy frame). Type 1 and 3 are possible, if the length of the horizontal path section is 1.

Hence, a bent border contains at least one delimiting block or intersection.   □

We will use Lemma 2 to show that fitness improvements are probable for all three types of intersections and for all situations, in which certain types of "failures" can occur during the random walk of the block length, which is analysed using Lemma 1.

Lemma 4. *If the current colouring contains an intersection, for RLS the probability of increasing the fitness within a phase of $\mathcal{O}(n)$ steps is bounded below by a positive constant.*

Proof. For all three types of intersections we will show that a sequence of not more than four mutations can increase the fitness by two, which, by Lemma 2 suffices to prove this lemma. In the following images, intersections are surrounded by heavy frames. The colouring of the neighbourhood results from considerations below.

Type 1   Type 2   Type 3

x	w	w'	w''
x'	v	v'	
x''	u		

x	w	w'	w''
x'	v	v'	v''
x''	u	u'	
x'''			

x	w	w'	w''
x'	v	v'	v''
x''	u	u'	

Type 1 intersections. v' has four fixed 1-neighbours. Either v' can increase the fitness or the remaining three vertices w, w', and w'' above v' are coloured 0. Similarly, either v can increase the fitness or all three vertices x, x', and x'' left of v are coloured 1. If u and v' flip subsequently, the fitness increases.

Type 2 intersections. By definition of the intersection, the 1-vertex v has at least one fixed 1-neighbour. Either v can increase the fitness or it has three additional 1-neighbours. Since the 0-vertex and v-neighbour v' already has three fixed 1-neighbours, it can either increase the fitness or it may only have one additional 1-neighbour. Hence, at most one of the vertices w and w' may be coloured 1. Then at least two of the left neighbours x, x', and x'' of v must have colour 1. Since the situation is symmetrical, the same holds for the vertices below and left of u: Either u or u' can increase the fitness or at least two of the three vertices left of u are coloured 0. Hence, exactly one of the vertices x' and x'' are coloured 1 and the other is coloured 0, while x is coloured 1 and x''' is coloured 0. Now v can flip and u' gets a fifth 0-neighbour. Flipping u' increases the fitness.

Type 3 intersections. By definition of the intersection five of the neighbours of v' have a fixed colour. There are three cases:

1. v' has five or six 1-neighbours. v' can flip and increase the fitness.
2. All of the remaining neighbours of v' are coloured 0 and hence, v' has three 1-neighbours. Then the three 1-vertices of the intersection v, u', and v'' can flip subsequently and increase the fitness.
3. v' has exactly four 1-neighbours. We can place the remaining 1-neighbour of v' above the intersection at positions w, w' and w''. Let $w^* \in \{w, w', w''\}$ be this 1-neighbour of v'. Because of symmetry we can neglect the case that $w^* = w''$. So far, three of the neighbours of v are fixed to colour 0. If v has five 0-neighbours, it can increase the fitness. Otherwise, at least two of the three remaining v-neighbours x, x' and x'' left of v are coloured 1. If both x' and x'' are coloured 1, the vertices u and v' can increase the fitness if they flip subsequently (because v' has the 1-neighbour w^*). If at most one of the vertices x' and x'' is coloured 1, then v and u' can flip now. If $w^* = w'$, it can also flip. Finally, v'' has five 0-neighbours and can increase the fitness.

For all three types of intersections and all possible colourings of neighbours we found sequences of not more than four single-bit-mutations. □

Lemma 5. *If the current colouring contains a delimiting block, for RLS the probability of increasing the fitness within a phase of $\mathcal{O}(n^2)$ steps is bounded below by a positive constant.*

Proof. We have already seen that the outermost vertices of the delimiting block can always flip and decrease the block length by 1. Pessimistically, we assume that the outer neighbours of these vertices can also flip and enlarge the block length by 1. Unfortunately, there are some mutations which flip the neighbours of our block such that it loses its property of being surrounded by complementary vertices. This event F is referred to as a "failure." We will show that the failure probability is not much larger than the probability of increasing the fitness ahead of time, which is referred to as event E. The phase terminates after $\mathcal{O}(n^2)$ steps or when the event $E \cup F$ occurs. We will show that $P(E|E \cup F) \geq \varepsilon > 0$.

For now, let us take this inequality for granted and ignore all events from F. The block length performs a random walk where shrinking is at least as probable as growing, if $|B| \geq 2$. Since it cannot grow beyond $\sqrt{n}$, by Lemma 1 $\mathcal{O}(n)$ essential steps, i.e. steps changing the block length, suffice to reduce it to 1 with positive constant probability. The probability of an essential step is at least $1/n$. Hence with positive constant probability there is a sufficiently large number of essential steps among a phase of cn^2 steps if c is large enough. If $|B| = 1$, the probability of increasing the fitness by flipping the single vertex within $\mathcal{O}(n)$ steps is at least a positive constant. Multiplying all success probabilities still yields at least a positive constant.

The only thing left to do is to find a lower bound for $P(E|E \cup F)$. In order to do so, we apply the techniques used in the treatment of intersections. We will show that, whenever a failure is acceptable, there exists a sequence of mutations increasing the fitness. W. l. o. g. our delimiting block has colour 1 and is horizontal. Let B denote the set of vertices belonging to the block and let N denote the neighbours of B which are required to have colour 0.

There are two types of failures: First, a B-vertex can flip. However, this subdivides our block into two which is actually a shortening of B. Second, a vertex in N can flip. Denote the leftmost vertex of B by v and denote its right neighbour by v'. There are five different locations for an N-vertex: above one of the inner vertices of B (this covers actually $|B| - 4$ locations), above v, above v', left above v, and left of v (plus the locations at the right counterparts of v and v', respectively).

Before analysing the five cases we make a preliminary remark: By definition of the delimiting block, v has at least four 0-neighbours. If any of the three vertices below v has colour 0, v can flip and increase the fitness. Knowing that, we only have to cover colourings in which all lower neighbours of v are coloured 1. Furthermore, we can assume that the block length is at least 5. Otherwise, the mutation sequence flipping these vertices subsequently can increase the fitness.

Denote the row of the block by row 0, the row above by row 1, etc. (Apart from v and v' the scope of all names assigned to vertices in the case differentiation below is limited to the case they are defined in.)

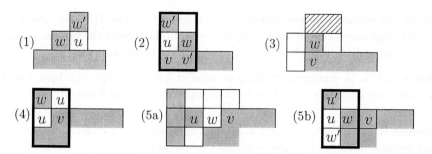

1. A vertex w above an inner vertex (i.e. not v or v') of B flips. This step is only acceptable if w has at least one 1-neighbour w' in row 2. Now there is a 0-vertex u in row 1 which is a neighbour of both w and w'. Since v is located above an inner vertex of B, u also has three 1-neighbours from B in row 0. Altogether, u has five 1-neighbours and can increase the fitness.

2. The vertex w above v' flips. This case only differs from the first case, if the 1-neighbour w' of w in row 2 is in the column of v. Then we have a type 1 intersection (rotated by $90°$) and we found a fitness-increasing mutation sequence sequence in the proof of Lemma 4.

3. The vertex w above v flips. By definition of the delimiting block, two neighbours of w are fixed to colour 1 and three are fixed to 0. In order for this step to be acceptable it is necessary that at least two of the three w-neighbours in row 2 (hatched) are coloured 1. In particular, one of the two vertices directly above and right above w must be coloured 1 which means that the 0-vertex u right of w has five 1-neighbours and can increase the fitness.

4. The vertex w left above v flips. Again, this is a rotated type 1 intersection.

5. The vertex w left of v flips. We have a failure iff the vertex u left of w or u' left above v is coloured 1. There are two cases.
   a) u is coloured 1. We analyse the situation *before* the flip of w. Including u, four of the neighbours of w are fixed to colour 1. Either the flip of w increases the fitness or its remaining neighbours are coloured 0. A similar argument holds for the remaining three neighbours of u. In this situation w and the vertex below u can flip subsequently and increase the fitness.
   b) u' is coloured 1. Furthermore, u is coloured 0 (otherwise we are in case 1). Independently of the colour of the the vertex w' below u, again, we have a type 1 intersection rotated by $90°$.

Altogether we found fitness increasing mutation sequences that have a probability of at least $e > 0$ in $\mathcal{O}(n)$ steps for all situations in which failures are acceptable. Since the probability of the failure happening in the same time is at most 1, we have $P(E|E \cup F) \geq \frac{e}{e+1} > 0$, finishing the proof.   □

4   Random Walk of Rings

We have seen that bent borders vanish and we are left with rings delimited by monotone borders or stairs. Usually, the size of rings can perform a random walk and finally vanish, which is stated by the following lemma. However, there are special rings which cannot grow or shrink without decreasing the fitness. We call these rings and their borders "stable" and analyse them in Section 5.

Definition 4. *A monotone border p is called* stable *if all edges in p have exactly the same direction, i. e. all edges are horizontal, vertical or diagonal.*

Lemma 6. *If the current search point does not contain any stable borders and is non-optimal, then for RLS the probability of increasing the fitness within $\mathcal{O}(n^3)$ steps is bounded below by a positive constant.*

Proof. If we ever reach a colouring with bent borders, Lemmas 4 and 5 together with Lemma 3 state that we reach our goal of increasing the fitness within $\mathcal{O}(n^2)$ steps with positive constant probability. In the following we assume that the current search point is non-optimal and that there exists a monotone, but not stable border. We show that the optimum is reached within cn^3 steps with positive constant probability.

Let C be a monochromatic area delimited by monotone borders that are not stable. Without decreasing the fitness we may only flip vertices with at least four neighbours of complementary colour. Along monotone borders, these vertices are the corner vertices of steps as depicted below. Here, v and w can flip.

If the step has length at least 2, both corner vertices can change their colour, one of them increasing $|C|$ and one decreasing $|C|$. For every step of the stair we find one such pair of vertices. The number of possible $|C|$-decreasing and $|C|$-increasing steps is therefore identical and $|C|$ performs a random walk. If the step is part of a stable subsection of the border, these vertices cannot flip. Since there is at least one step which is not stable, the probability of an essential, i. e. $|C|$-changing step is at least $1/n$. By Lemma 1, cn^2 essential steps suffice for $|C|$ to shrink to 0 (which increases the fitness) with positive constant probability if c is large enough, which it is (again with positive constant probability) if we wait $c'n^3$ steps with c' a constant large enough.   □

Taking into account that stairs typically have $\sqrt{n}$ steps we could save a factor of $\mathcal{O}(\sqrt{n})$. However, this would require us to show that not many of the steps are stable. The following theorem states that it is possible to transfer our previous results on RLS to the Metropolis algorithm.

Theorem 1. *Within $\mathcal{O}(n^4)$ steps, the Metropolis algorithm finds an optimum or a colouring without bent borders, if $p_T = \mathcal{O}(n^{-3})$ is sufficiently small.*

Proof. From Lemmas 4, 5, and 6 we know that RLS increases the fitness by at least 2 within $\mathcal{O}(n^3)$ steps with positive constant probability p_+, as long as the colouring does not entirely consist of stable rings. Also, by our choice of p_T

the probability that the Metropolis algorithm behaves exactly like RLS within $\mathcal{O}(n^3)$ steps is bounded below by a positive constant p_{RLS} which can be chosen arbitrarily close to 1. A phase ends after $\mathcal{O}(n^3)$ steps or when a fitness decreasing step is accepted. The expected fitness gain in each phase is bounded below by

$$p_{RLS} \cdot p_+ \cdot 2 + p_{RLS} \cdot (1 - p_+) \cdot 0 - (1 - p_{RLS}) \cdot 8,$$

since the fitness can decrease by at most 8 in one step. Choosing p_{RLS} large enough, both expectation value and the probability of observing a positive constant fitness gain are bounded below by a positive constant.

The fitness performs a random walk with a positive constant bias. Since negative steps are limited by 8 the expected number of phases until the fitness reaches its maximum of $4n$ is at most $\mathcal{O}(n)$. The probability of observing at most this number of phases is also a positive constant. Multiplying with the phase length $\mathcal{O}(n^3)$ yields a total number of $\mathcal{O}(n^4)$ steps.

These arguments hold as long as bent borders exist. If bent borders vanish, we accomplish the theorem ahead of time. □

5   Escape from Rings

In the preceding sections we did not make use of the property of the Metropolis algorithm of accepting search points with lesser fitness. Contrarily, we avoided it. Here we use this property for escaping from stable rings.

Theorem 2. *The expected optimisation time of the Metropolis algorithm with $p_T = \Theta(n^{-3})$ small enough is bounded above by $\mathcal{O}(n^{4.5})$.*

Proof. By Theorem 1 we know that the expected number of steps for reaching a colouring in which all borders are stable is bounded above by $\mathcal{O}(n^4)$. Then all mutations decrease the fitness. We start the phase by waiting for a step flipping a vertex near the border of a ring. The expected time until a step is accepted is hence p_T^{-1}. The probability that the vertex flipped in this step hits a border is at least $n^{-1/2}$ since the length of a monotone border is at least $\sqrt{n}$. If it does not hit a border it is likely to flip back before anything else happens. For now, let us assume that all bits flipping without hitting a border flip back before they cause any harm. Furthermore let us assume that no fitness decreasing steps are accepted within the phase described in the next paragraph.

Here we describe the case of a horizontal or vertical ring, but it should be clear that diagonal rings can be treated in exactly the same manner. The fitness decreasing mutation produced a delimiting block, or rather two: one of length 1 and one of length $\sqrt{n} - 1$. As we already know, the outer vertices of the blocks can flip lengthening or shortening it without decreasing the fitness. With probability $1/3$ the single-vertex block does not flip back to its original state before growing to length 2. When the length of the block is at least 2, growing is at least as probable as shrinking and we can apply Lemma 1 again to obtain a positive constant lower bound on the probability of reaching length $\sqrt{n}$, and thereby changing the thickness of the original ring by 1, within $N = cn$ steps, c sufficiently large. However, now we may not reach length 1 or 0 *in between*, which seems very probable. We can calculate the success probability by applying the Ballot Theorem [1]. Given that there are $N/2 + \sqrt{n}/2$ steps lengthening the block and $N/2 - \sqrt{n}/2$ steps shortening it, the probability that the block never is shorter than 2 is

$$\frac{(N/2 + \sqrt{n}/2) - (N/2 - \sqrt{n}/2)}{(N/2 + \sqrt{n}/2) + (N/2 - \sqrt{n}/2)} = \frac{\sqrt{n}}{cn} = \frac{1}{c\sqrt{n}},$$

which is small, but not too small. In order to provide for cn essential steps with positive constant probability we need $c'n^2$ steps, c' large enough. The total length of the phase is dominated by the $\mathcal{O}(n^3)$ steps for waiting for the initial creation of the dent. Multiplying with the inverse of the success probability, we obtain $\mathcal{O}(n^{7/2})$ as a lower bound on the expected number of steps for changing the thickness of the ring by 1.

As we see, the thickness of the ring does now perform a random walk. In a colouring with i rings there exists, by the pigeon hole principle, at least one ring with thickness not greater than $\lfloor \sqrt{n}/i \rfloor$. Again, we apply Lemma 1 to retrieve a positive constant lower bound on the probability for this ring to vanish within $\mathcal{O}(\sqrt{n}^2/i^2)$ phases. This we call a meta-phase. Altogether, the expected length of a meta-phase is bounded above by $\mathcal{O}(\sqrt{n}^2/i^2) \cdot \mathcal{O}(n^{7/2}) = \mathcal{O}(n^{9/2}/i^2)$.

We still have to treat situations in which fitness decreasing mutations other than those described above are accepted. As long as vertices that decreased the fitness are isolated, they can always flip back which happens with probability $1 - \mathcal{O}(n^{-2})$ before the next step decreases the fitness. If it does not, there are four possibilities:

1. The flipping vertex never interferes with the process described above.
2. The flipping vertex collides with a delimiting block which is performing its random walk. The probability that a fitness decreasing vertex flips during the $\mathcal{O}(n^{5/2})$ steps of the phase in which there exists a delimiting block ($\mathcal{O}(\sqrt{n})$ trials with $\mathcal{O}(n^2)$ steps each) and that it is near the border is at most $\mathcal{O}(n^{5/2} \cdot n^{-3} \cdot n^{-1/2}) = \mathcal{O}(n^{-1})$. The probability that this happens at least once in a meta-phase (of at most $\mathcal{O}(n/i^2) = \mathcal{O}(n)$ phases) is at most $\mathcal{O}(1)$ which is a constant smaller than 1 if p_T is small enough.
3. The flipping vertex connects two rings. This is not a failure, but it is precisely what we are waiting for. With a similar argument we can conclude that with error probability $o(1)$ the width of the bridge is at most a constant. Then, a sequence of a constant number of acceptable mutations can increase the thickness of this bridge and the fitness. The expected time for this to happen is bounded above by $\mathcal{O}(n)$. Steps decreasing the width of the bridge also decrease the fitness and hence are not accepted with probability $1 - \mathcal{O}(n^{-3/2})$ within $\mathcal{O}(n^{3/2})$ steps which is the expected time necessary for filling the entire area in between the former rings now connected by the bridge.
4. There are two or more competing bridges at a time, i.e. one 1-bridge connecting ring A and B and a 0-bridge bypassing one of the rings A and B. Given

that there is one bridge present, we have seen that the phase terminates within $\mathcal{O}(n^{3/2})$ steps is bounded below by a positive constant. The probability that within this time a second bridge (for which a fitness decreasing step is necessary) evolves, again, is bounded by $\mathcal{O}(n^{-3/2})$.

When a meta-phase is successful, it increases the fitness by at least $2\sqrt{n}$. Altogether, this happens with positive constant probability. Whenever a failure occurs in a meta-phase, we know from Theorem 1, that within $\mathcal{O}(n^4)$ steps, a situation with stable rings is restored, and that within this time the expected fitness gain is positive. Altogether the expected fitness gain of the meta-phase is bounded below by $\Omega(\sqrt{n})$. Summing up the lengths of the meta-phases for all values of i we retrieve an expected optimisation time of

$$\sum_{i=1}^{\sqrt{n}} \mathcal{O}(n^{9/2}/i^2) = \mathcal{O}(n^{4.5})$$

steps.   □

6   Conclusion

The two-dimensional Ising model is an interesting model for the investigation of adaption processes with spatial interaction. We have proved that even a simple mutation-based search heuristic is able to leave local optima and converge to a global optimum within polynomial time by the means of random walk arguments on several levels. This is a surprising result since mutation-based algorithms were thought to be very slow even on the one-dimensional Ising model which can be seen as a subproblem frequently solved during the optimisation of the two-dimensional Ising model.

References

1. W. Feller. *An Introduction to Probability Theory and Its Applications*, volume 1. Wiley, New York, 1971.
2. A. Galluccio, M. Loebl, and J. Vondrak. A new algorithm for the Ising problem: Partition function for finite lattice graphs. *Physical Review Letters*, 84:5924–5927, 2000.
3. D. E. Goldberg. *Genetic Algorithms in Search, Optimization, and Machine Learning*. Addision-Wesley, Reading, MA, 1989.
4. J. H. Holland. *Adaptation in Natural and Artificial Systems*. University of Michigan, MI, 1975.
5. E. Ising. Beitrag zur Theorie des Ferromagnetismus. *Z. Physik*, 31:235–258, 1925.
6. M. Jerrum and A. Sinclair. Polynomial-time approximation algorithms for Ising model. In *Automata, Languages and Programming*, pages 462–475, 1990.
7. B. Naudts and J. Naudts. The effect of spin-flip symmetry on the performance of the simple GA. In *Proceedings of the 5th Conference on Parallel Problem Solving from Nature*, volume 1498 of *LNCS*, pages 67–76, Berlin Heidelberg New York, 1998. Springer-Verlag.

8. M. Pelikan and D. E. Goldberg. Hierarchical BOA solves Ising spin glasses and MAXSAT. In *Proc. of the Genetic and Evolutionary Computation Conference (GECCO 2003)*, number 2724 in LNCS, pages 1271–1282, 2003.

9. C. Van Hoyweghen. *Symmetry in the Representation of an Optimization Problem.* PhD thesis, Univ. Antwerpen, 2002.

10. C. Van Hoyweghen, D. E. Goldberg, and B. Naudts. From twomax to the Ising model: Easy and hard symmetrical problems. In *Proc. of the Genetic and Evolutionary Computation Conference (GECCO 2002)*, pages 626–633, San Mateo, CA, 2002. Morgan Kaufmann.

11. Clarissa Van Hoyweghen, David E. Goldberg, and Bart Naudts. Building block superiority, multimodality and synchronization problems. In *Proceedings of the Genetic and Evolutionary Computation Conference (GECCO-2001)*, pages 694–701, San Francisco, California, USA, 7-11 2001. Morgan Kaufmann.

The Ising Model on the Ring: Mutation Versus Recombination

Simon Fischer and Ingo Wegener*

FB Informatik, LS2, Univ. Dortmund, 44221 Dortmund,
Germany
{simon.fischer,ingo.wegener}@uni-dortmund.de

Abstract. The investigation of genetic and evolutionary algorithms on Ising model problems gives much insight how these algorithms work as adaptation schemes. The Ising model on the ring has been considered as a typical example with a clear building block structure suited well for two-point crossover. It has been claimed that GAs based on recombination and appropriate diversity-preserving methods outperform by far EAs based only on mutation. Here, a rigorous analysis of the expected optimization time proves that mutation-based EAs are surprisingly effective. The $(1 + \lambda)$ EA with an appropriate λ-value is almost as efficient as usual GAs. Moreover, it is proved that specialized GAs do even better and this holds for two-point crossover as well as for one-point crossover.

Keywords: Ising model, mutation vs. recombination, expected optimization time, fitness sharing

1 Introduction

Nowadays, genetic algorithms (GAs) and evolutionary algorithms (EAs) are mainly applied as optimization algorithms. Holland [3] has designed GAs as adaptation systems. The building block hypothesis (see Goldberg [2]) claims that GAs work by combining different building blocks in different individuals by crossover (or recombination). There is a long debate on the role of mutations in this context.

Naudts and Naudts [7] have presented the Ising model as an interesting subject for the investigation of GAs and EAs. Ising [4] has described the model now called Ising model to study the theory of ferromagnetism. In its most general form, the model consists of an undirected graph $G = (V, E)$ and a weight function $w \colon E \to \mathbb{R}$. Each vertex $i \in V$ has a positive or negative spin $s_i \in \{-1, +1\}$. The contribution of the edge $e = \{i, j\}$ equals $f_s(e) := s_i \cdot s_j \cdot w(e)$. The fitness $f(s)$ of the state s equals the sum of all $f_s(e)$, $e \in E$, and has to be maximized.

---

* Supported in part by the Deutsche Forschungsgemeinschaft (DFG) as part of the Collaborative Research Center "Computational Intelligence" (SFB 531) and by the German Israeli Foundation (GIF) in the project "Robustness Aspects of Algorithms."

K. Deb et al. (Eds.): GECCO 2004, LNCS 3102, pp. 1113–1124, 2004.

The Ising problem in its general form is NP-hard. Nevertheless, there are quite efficient algorithms for this problem (Pelikan and Goldberg [8]). For the investigation of the adaptation capabilities of simple GAs and EAs, one is interested in the case where $w(e) = 1$ for all $e \in E$. By an affine transformation, we consider the state space $\{0,1\}^n$ instead of $\{-1,+1\}^n$. The fitness $f(s)$ equals the number of monochromatic edges, i.e. edges connecting vertices of equal spin or color. The states 0^n and 1^n are the only optimal states for connected graphs. Connected monochromatic subgraphs are schemata of high fitness and, therefore, building blocks. However, the fitness function has the property of spin-flip symmetry, i.e., $f(s) = f(\bar{s})$ for all states s and their bitwise complement $\bar{s}$. Therefore, 0-colored building blocks compete with 1-colored building blocks.

The Ising model on the ring is of particular interest. The ring is a graph on $V = \{1, \ldots, n\}$ with edges $\{i, i+1\}$, $1 \le i \le n-1$, between neighbored vertices and the turn-around edge $\{n, 1\}$. Building blocks are also blocks in the string (if the positions 1 and n are considered as neighbored) and two-point crossover can cut out a building block. Extensive experiments on GAs for this problem have been reported by van Hoyweghen [9], van Hoyweghen, Goldberg, and Naudts [12], and van Hoyweghen, Naudts, and Goldberg [11]. These papers contain also discussions how the algorithms work and some theoretical results but no run time analysis. In recent years, the rigorous run time analysis of EAs has led to interesting results. Here, this approach is applied to the Ising model on the ring.

Sections 2, 3, and 4 analyze mutation-based algorithms. Experiments have led to the conjecture that these algorithms are quite inefficient for the Ising model. The authors of the papers mentioned above do not explicitly state such a conjecture but they and many others have argued in discussions that mutation-based EAs will need exponential optimization time. In Section 2, we analyze randomized local search (RLS) flipping one bit per step and applying a plus-strategy for selection. This simple algorithm finds the optimum in an expected number of $O(n^3)$ steps and the constants in the O-term are surprisingly small. Based on this analysis, a similar bound is obtained in Section 3 for the $(1+1)$ EA.

In Section 4, we analyze parallel variants of the algorithms, parallel RLS (PRLS) and the $(1+\lambda)$ EA, respectively. They produce λ offspring per generation and select a best individual. For $\lambda = n/\log n$, the expected optimization time consists of $O(n^2 \log n)$ generations and $O(n^3)$ fitness evaluations. This analysis follows the line of research started by Jansen and De Jong [5] and Jansen, De Jong, and Wegener [6]. In Section 5, we compare our results with the experiments on GAs.

It would be even more interesting to obtain also bounds on the expected optimization time of GAs. We are not able to do this for the GAs used in experiments which apply an island model to preserve diversity. We analyze in Section 6 the GA introduced by Culberson [1] and known as GIGA (Gene Invariant GA) and in Section 7 an idealized GA with fitness sharing. Both algorithms are tailored to cope with the given problem and perform better than RLS and the $(1+1)$ EA. We finish with some conclusions.

2   Randomized Local Search

Randomized Local Search (RLS) chooses the first search point $x \in \{0,1\}^n$ uniformly at random. Afterwards, it chooses a position $i \in \{1, \ldots, n\}$ uniformly at random, computes x' by flipping bit i of x, and replaces x by x' iff $f(x') \geq f(x)$. We are interested in the expected number of f-evaluations until $x \in \{0^n, 1^n\}$.

Instead of maximizing f, we investigate the equivalent problem of minimizing the number i of monochromatic blocks on the ring. This number is even for non-optimal points and has to be decreased from at most n to 1. For $2 \leq i \leq n$ and i even, let $t_i(n)$ be the expected time until i is decreased if we start with a worst search point with i blocks. We estimate the expected run time by the sum of all $t_i(n)$ and the term 1 for the initialization step.

By the pigeon-hole principle, there is one block whose length is bounded above by $N := \lfloor n/i \rfloor$. We investigate a shortest block B of the first search point x. If i is not decreased, the length of B can change at most by 1 per step. We distinguish relevant steps (either decreasing i or changing the length of B) from the other steps called non-relevant. First, we only investigate the relevant steps. It is possible that some block $B' \neq B$ gets shorter than B and vanishes earlier. Pessimistically, we ignore this. Only if B grows to length $N + 1$ we switch our interest to another block whose length is at most N. Pessimistically, we assume that this length equals N. Then we obtain the following Markoff chain on $\{0, 1, \ldots, N\}$ where the state j describes the length of the considered block. If $j \in \{2, \ldots, N-1\}$, by symmetry, the transition probability $p(j, j-1) = p(j, j+1) = 1/2$. By the discussion above, state "$N + 1$" is replaced by N and $p(N, N-1) = p(N, N) = 1/2$. State 1 is untypical, since there are two bits whose flip increases the block length but only one decreasing it. Hence, $p(1, 0) = 1/3$ and $p(1, 2) = 2/3$. Let $T_N(j)$ be the expected time until reaching state 0 when starting in state j.

Lemma 1. *If $j \geq 1$, $T_N(j) = 4N - 1 + N(N-1) - (N - j + 1)(N - j) = 2Nj + 2N + j - j^2 - 1$.*

Proof. We fix N and omit the index N. We prove by induction that

$$T(j) = 2 \cdot (N - j + 1) + T(j - 1),$$

if $j \geq 2$. By the law of total probability,

$$T(N) = 1 + (1/2) \cdot T(N) + (1/2) \cdot T(N - 1)$$

implying that $T(N) = 2 + T(N - 1)$. If $j < N$, by induction hypothesis,

$$\begin{aligned} T(j) &= 1 + (1/2) \cdot T(j + 1) + (1/2) \cdot T(j - 1) \\ &= 1 + (N - j) + (1/2) \cdot T(j) + (1/2) \cdot T(j - 1). \end{aligned}$$

Solving for $T(j)$, this proves the claim. Finally,

$$\begin{aligned} T(1) &= 1 + (1/3) \cdot T(0) + (2/3) \cdot T(2) \\ &= 1 + (2/3) \cdot (2 \cdot (N - 1) + T(1)) \end{aligned}$$

implying that $T(1) = 4N - 1$. This proves the lemma for $j = 1$ and, if $j \geq 2$,

$$T(j) = 2 \cdot (N - j + 1) + 2 \cdot (N - j + 2) + \cdots + 2 \cdot (N - 1) + 4N - 1 \qquad (*)$$

which implies the lemma.   □

Equation $(*)$ implies that T_N is monotone increasing and concave, i.e.,

$$T_N(j + 1) - T_N(j) \le T_N(j) - T_N(j - 1).$$

In order to estimate the expected number of relevant steps, it is sufficient to sum up all $T_{\lfloor n/i \rfloor}(\lfloor n/i \rfloor), i \in I := \{j \mid 2 \le j \le n, j \text{ even}\}$. Since $T_N(N) = N^2 + 3N - 1$, we obtain

$$\sum_{i \in I} T_{\lfloor n/i \rfloor}(\lfloor n/i \rfloor) \le n^2 \sum_{i \in I}(1/i^2) + 3n \sum_{i \in I}(1/i) - \lfloor n/2 \rfloor$$

$$\le 0.411 \cdot n^2 + 1.5 \cdot n \ln n + n.$$

We are interested in the probability that a step is relevant. There are 4 positions such that the length of B changes if one of the corresponding bits flips and the length of B is at least 2. If B has length 1, there are only 3 such positions. The expected waiting time until one of k bits flips is exactly n/k. In order to get good bounds, we estimate the expected number of relevant steps where the block length equals 1. Since the probability of reaching state 0 and finishing a phase equals $1/3$, the expected number of steps in state 1 equals 3 independent of i. Hence, $(3/2)n$ of the relevant steps have to be multiplied by $n/3$ and the other ones by $n/4$ to obtain an upper bound on the expected run time.

Theorem 1. *The expected number of steps until RLS finds an optimum for the Ising model on the ring is bounded above by*

$$T_{\mathrm{RLS}}(n) = 0.103 \cdot n^3 + 0.375 \cdot n^2 \cdot (\ln n + 1).$$

This bound is pessimistic in the following aspects:

- the first search point can have less than the maximal number of blocks,
- the first search point with i blocks can contain a block which is shorter than $\lfloor n/i \rfloor$,
- larger blocks can get shorter than the considered block.

In any case, the bound of Theorem 1 is surprisingly small when considering the discussions about this problem. Experiments have shown that, in the case $i = 2$, the shorter block has an average block length of $0.28n$ when reaching this phase. It is easy to obtain the following result.

Theorem 2. *Starting with two blocks of length εn and $(1 - \varepsilon)n$, $0 < \varepsilon \le 1/2$ a constant, the expected number of steps until RLS finds an optimum for the Ising model on the ring is $\Theta(n^3)$.*

3   The (1+1) EA

The (1+1) EA can be considered as the simplest evolutionary algorithm. It works like RLS with the exception of the search operator. The mutant x' is obtained from x by flipping each bit of x independently of the others with probability $1/n$. Steps without flipping bits do not count since they do not lead to a fitness evaluation. Let $e = 2.718\ldots$ be the Eulerian constant.

Theorem 3. *The expected number of steps until the (1+1) EA finds an optimum for the Ising model on the ring is bounded above by $T_{(1+1)}(n) = (e - 1) \cdot (1 + o(1)) \cdot T_{\mathrm{RLS}}(n) \leq 0.177 \cdot n^3 + o(n^3)$.*

Proof. We use the same ideas as in the proof of Theorem 1. In particular, we concentrate our analysis on the length of one block and we consider first only relevant steps, i. e., steps changing the length of the chosen block. We investigate another block if the chosen block has a length larger than $\lfloor n/i \rfloor$. The main idea is that we do not estimate the number of steps directly but we compare the (1+1) EA with RLS. For this purpose, we investigate some stochastic processes "between" RLS and the (1+1) EA.

We start with RLS* which applies the search operator of the (1+1) EA but only mutants x' where exactly one bit has flipped are considered for selection. Then the expected run time increases by the expected waiting time for a step flipping exactly one bit. This waiting time equals $e - 1$ for a Poisson distribution with $\lambda = 1$. Here we obtain a factor of $(e - 1) \cdot (1 + o(1))$ since the number of flipping bits is asymptotically Poisson distributed. This indeed is the essential factor why the (1+1) EA is slower than RLS. If the number of blocks is not too large, the probability that a step flipping more than one bit is relevant is much less than the corresponding probability for steps flipping one bit. The reason is that the other flipping bits typically increase the number of blocks.

Nevertheless, there are relevant steps flipping more than one bit and there are relevant steps changing the length of the considered block by more than 1. For each search point x let $p_k^+(x)$ be the probability that the next step is accepted and produces a search point where the length of the considered block B has been increased by k and let $p_k^-(x)$ be the corresponding probability for decreasing the length of B. We know from Section 2 that $p_k^+(x)$ may be larger than $p_k^-(x)$. To simplify the analysis, we investigate two further stochastic processes called (1+1) $\mathrm{EA}_{\mathrm{sym}}$ and $\mathrm{RLS}_{\mathrm{sym}}$. They are based on the algorithms (1+1) EA and RLS, respectively, but, if $p_k^+(x) > p_k^-(x)$, the probability of increasing the length of B is decreased to $p_k^-(x)$. As before, we consider another block if the length of B is larger than $\lfloor n/i \rfloor$. First, $\mathrm{RLS}_{\mathrm{sym}}^*$ is obviously faster than RLS*. We show that the expected run time of the (1+1) $\mathrm{EA}_{\mathrm{sym}}$ is bounded by the upper bound proven for RLS* and, therefore, also for $\mathrm{RLS}_{\mathrm{sym}}^*$ and, later, we compare the (1+1) EA and the (1+1) $\mathrm{EA}_{\mathrm{sym}}$.

Let A_t be the algorithm working t steps like the (1+1) $\mathrm{EA}_{\mathrm{sym}}$ and afterwards like $\mathrm{RLS}_{\mathrm{sym}}^*$. We prove by induction on t that the expected run time of A_t is not larger than the upper bound obtained for RLS*. This is true for $t = 0$, since $A_0 = \mathrm{RLS}_{\mathrm{sym}}^*$. For the induction step, we compare A_t and A_{t+1}. They are identical for the first t steps and we consider the (random) search point x after t

steps. The probability of a relevant step is for the (1+1) EA$_{\text{sym}}$ not smaller than for RLS$^*_{\text{sym}}$. We compare the algorithms conditioned on some events and prove the claim for each of the cases. If the next step is neither relevant for A_t nor for A_{t+1}, the claim is obvious since the upper bound for RLS* only depends on the length of the considered block. The perhaps larger probability of a relevant step of A_{t+1} is only in favor of A_{t+1}. Finally, we have to compare the effect of relevant steps. Instead of having steps changing the length of B by $+1$ and -1 (with the same probability), we now may change the length of B by $+k$ and $-k$ (with the same probability). Afterwards, we apply RLS$^*_{\text{sym}}$ in both cases. The upper bound for RLS* (and also RLS$^*_{\text{sym}}$), namely the function T_N from Section 2 is increasing and concave. Therefore, a $\pm k$-step instead of a ± 1-step reduces the expected run time, i. e., $(T(j+k)+T(j-k))/2 < (T(j+1)+T(j-1))/2$, if $k \geq 2$. For $t \to \infty$, we obtain the claim.

Finally, we have to compare the (1+1) EA and the (1+1) EA$_{\text{sym}}$. We investigate a phase of length $n^{7/2}$. By Markoff's inequality, the probability that the (1+1) EA$_{\text{sym}}$ needs more than $n^{7/2}$ steps is $O(n^{-1/2}) = o(1)$. In this case, we repeat the arguments for the next phase leading to an additional $1 + o(1)$ factor. In the following, we investigate a phase of length $n^{7/2}$. Events which altogether have a probability of $o(1)$ can be ignored since then the phase can be considered as unsuccessful also leading to a $1 + o(1)$ factor.

Let k be the length of the considered shortest block B, w. l. o. g. a block of ones. If $k \geq 4$, the string contains $0^4111^{k-4}110^4$. We consider the substrings 0^411 and 110^4. The probability that a phase contains a step with at least four flipping bits at these positions is $o(1)$ and this event can be ignored. Steps with at most three flipping bits at these positions do not eliminate one of the blocks. The situation is symmetric with respect to lengthenings and shortenings of B.

We are left with the situation $k \leq 3$. Recalling the analysis of RLS in Section 2, it is easy to obtain the result that the (1+1)*EA$_{\text{sym}}$ has an expected number of $O(n)$ steps where $k \leq 3$. By Markoff's inequality, we can ignore runs where this number is larger than $n^{3/2}$. The probability that a phase contains a step with at least two flipping bits in the substring $0^k1^k0^k$ is $o(1)$.

Finally, decreasing the length of B from k to 0 does not imply that we decrease the number of blocks. A new block may be created somewhere else. The probability of no bit flipping elsewhere is at least e^{-1}. Hence, with a probability of $1-e^{-1}$ we are still in the same situation as before, i. e., we have the same values of $k \in \{1,2\}$. This happens on average $e/(e-1)$ times, each time increasing the expected run time by $O(n^2)$. Hence, we have proved the theorem. $\square$

It is worth noticing that we were not able to prove such a small bound by analyzing the (1+1) EA directly. It was helpful to analyze the simpler algorithm RLS and to compare RLS and the (1+1) EA.

Finally, we prove a lower bound similarly to the lower bound of Theorem 2.

Theorem 4. *Starting with two blocks of length εn and $(1 - \varepsilon)n$, $0 < \varepsilon \leq 1/2$ a constant, the expected number of steps until the $(1 + 1)$ EA finds an optimum for the Ising model on the ring is $\Theta(n^3)$.*

We omit the proof here. The essential argument is that the function T_N from Lemma 1 is concave but the curvature is not strong. In particular, $T_N(j - k) +$

$T_N(j+k) = T_N(j-1) + T_N(j+1) - c_k$ where c_k only depends on k. Since steps changing the block length by k have a probability of $\Theta(n^{-k})$, we do not save too much by steps changing the block length by at least 2.

4 Parallel RLS and the (1+λ) EA

A GA works with a population of $s(n)$ individuals and, in most cases, run time is defined as the number of generations. The number of fitness evaluations is larger by a factor of $s(n)$. For RLS and the (1+1) EA, the number of generations equals the number of fitness evaluations. In order to have a fair comparison with GAs, we consider population-based RLS and (1+1) EA. Parallel RLS (PRLS) or (1+λ) RLS creates λ children from the parent x using the search operator of RLS. The children are created independently. Selection chooses x if all children are worse and chooses one of the fittest children uniformly at random otherwise.

For $s(n) = n$ we get an expected number of $O(n^2 \log n)$ generations and $O(n^3 \log n)$ fitness evaluations. While reducing $s(n)$ to $n/\log n$ does not affect the number of generations, it reduces the number of fitness evaluations to $O(n^3)$.

Theorem 5. *The expected number of generations until* $(1+\lfloor n/\log n \rfloor)$ *RLS finds the optimum for the Ising model on the ring is bounded above by* $O(n^2 \log n)$ *and the expected number of fitness evaluations by* $O(n^3)$.

Proof. It is sufficient to investigate the number of generations since each generation consists of $\lfloor n/\log n \rfloor$ fitness evaluations. Again, let B be the considered block. The probability that no child shortens or lengthens B equals $(1 - c/n)^{\lfloor n/\log n \rfloor} = 1 - \Theta(1/\log n)$. The expected waiting time for a generation with a child changing B equals $\Theta(\log n)$. If x contains i blocks, the expected number of children with the same number of blocks as x is $\Theta(i/\log n)$ and the probability that this number is bounded by $O(i/\log n)$ is at least $1/2$ (Markoff's inequality).

If $i \geq \log n$, the probability of choosing a child where B is changed, if such a child is created, is $\Omega(\log n/i)$. The waiting time for such a step is $O(i/\log n)$. Hence, each step has a probability of $\Omega(1/i)$ of being relevant. By Lemma 1, the expected number of relevant steps to decrease i is $O(n^2/i^2)$ and this takes $O(n^2/i)$ generations on average. For all i, $\log n \leq i \leq n$ and i even, we obtain a bound of $O(n^2 \log n)$.

If $i < \log n$ and one child changes B, the probability that all other children have more blocks equals $(1 - \Theta(i/n))^{\lfloor n/\log n \rfloor - 1}$ which is bounded below by a positive constant. Then the generation is relevant. Hence, the expected waiting time for a relevant generation equals $\Theta(\log n)$ and the expected number of generations is bounded by $O((n^2 \log n)/i^2)$. Considering all $i < \log n$ and even, this gives an additional term of $O(n^2 \log n)$. $\square$

The (1+λ) EA applies the search operator of the (1+1) EA and produces independently λ children from the parent which is the only individual of the current population. We have to be careful with the selection operator. It is likely that many children are a replica of the parent. In order to guarantee exploration of the search space, we select the parent x only if all children $y \neq x$ have a worse fitness than x. Otherwise, we randomly select an individual among the fittest children $y \neq x$.

Combining the methods from Section 3 and Theorem 5 we obtain the following result.

Theorem 6. *The expected number of generations until the $(1 + \lfloor n/\log n \rfloor)$ EA finds the optimum for the Ising model on the ring is bounded above by $O(n^2 \log n)$ and the expected number of fitness evaluations by $O(n^3)$.*

5   A Comparison with GA Experiments

We have no doubt that crossover can play an essential role for the Ising model on the ring. A theoretical fundament for this argument will be presented in Sections 6 and 7. Here, we want to argue that mutation-based EAs are better than expected in many papers. Van Hoyweghen [9] claims that "the presence of spin-flip symmetry in the one-dimensional Ising model prevents an unspecialized GA to find an optimum in a reasonable amount of time." Van Hoyweghen, Goldberg, and Naudts [10] indicate in this context that "the Ising model shows that for a certain class of optimization problems niching becomes a necessity for a GA to solve these problems." Our results have shown that unspecialized EAs solve this problem in reasonable time. The upper bounds on the expected run times of RLS ($0.103n^3 + 0.375n^2 (\ln n + 1)$ and even $117,957$ for $n = 100$) and of the $(1+1)$ EA (by a factor of 1.72 slower than RLS) show this even for populations of size 1. The time bounds are much better, namely $O(n^2 \log n)$, if $n/\log n$ children are generated in parallel. Hence, the optimization is finished in a reasonable amount of time without any niching. Van Hoyweghen [9] has considered the case of GAs for $n = 100$ and a population size of 100. The best parameters for tournament selection and two-point crossover lead to an average number of $35,857$ generations. This can be decreased to $10,881$ using SAWing (Stepwise Adaptation of Weights). With an Island model and a distributed GA there is a good chance that $10,000$ generations suffice. In all these cases a population of size $s(n) \geq 100$ is used. In general, it is claimed that a population size of $10.9n^{0.57}$ suffices. These algorithms need less generations than the mutation-based algorithms examined in this paper but they do not beat RLS with respect to the expected number of fitness evaluations (at least for $n = 100$).

6   The Expected Run Time of GIGA

Although mutation-based algorithms are surprisingly efficient for the Ising model on a ring, it is believed that GAs can be faster. It is difficult to analyze the effect of crossover if one is interested in the expected optimization time. We are not able to analyze distributed GAs. Therefore, we analyze GAs which are specialized to work on the Ising model on the ring. We cannot expect to obtain the same good time bounds for unspecialized GAs.

In this section, we analyze a simple variant of GIGA (Gene Invariant Genetic Algorithm) introduced by Culberson [1]. The population has size 2 and consists of a search point $x \in \{0, 1\}^n$ and its bitwise complement $\overline{x}$. In the initialization step, x is chosen uniformly at random. Later, a new pair of search points $(y, \overline{y})$ is produced from $(x, \overline{x})$ by crossover. Since $f(x) = f(\overline{x})$, the new pair $(y, \overline{y})$ replaces $(x, \overline{x})$ if $f(y) \geq f(x)$. Since we want to cut out a block in x and to replace it by its bitwise complement, two-point crossover seems to be the appropriate recombination operator. Let us consider the effect of crossover at the positions j and k, $0 \leq j < k < n$. A position p is called border of x, if $x_p \neq x_{p+1}$ or $x_n \neq x_1$ if $p = 0$. Let i be the number of blocks of x.

Case 1: The positions j and k are not borders. Then y has $i + 2$ blocks and $(y, \overline{y})$ is not accepted.

Case 2: Exactly one of the positions j and k is a border. Then y also has i blocks and $(y, \overline{y})$ is accepted but the fitness is not changed.

Case 3: The positions j and k are borders. If $i > 2$, y has $i - 2$ blocks. If $i = 2$, y has one block. In any case, $(y, \overline{y})$ is accepted and the fitness is improved.

As long as x is not optimal, $i \geq 2$ and there are $\binom{i}{2}$ among $\binom{n}{2}$ pairs of positions which lead to an improved fitness. Hence, the expected optimization time can be bounded above by (remember that $I = \{i \mid 2 \leq i \leq n, i \text{ even}\}$)

$$\sum_{i \in I} \binom{n}{2} / \binom{i}{2} \leq 0.70 \cdot n \cdot (n - 1).$$

With probability $1 - o(1)$, the initial value of i is at least $n/3$. We obtain the following results.

Theorem 7. *The expected number of steps until GIGA with two-point crossover finds an optimum for the Ising model on the ring is bounded above by $0.70 \cdot n \cdot (n - 1)$ and bounded below by $0.69 \cdot n^2 - o(n^2)$.*

We can generalize GIGA to $(1+\lambda)$ GIGA where λ offspring pairs are produced independently and a best one is chosen if it is not worse than the parent. We analyze the $(1+n)$ GIGA. The probability of producing a better offspring is bounded below by a positive constant, if $i > n^{1/2}$, and by $\Omega(i^2/n)$, otherwise. Hence, the expected number of generations equals $\Theta(n)$.

Theorem 8. *The expected number of generations until the $(1 + n)$ GIGA with two-point crossover finds an optimum for the Ising model on the ring equals $\Theta(n)$, the expected number of fitness evaluations equals $\Theta(n^2)$.*

Surprisingly, one-point crossover is almost as good as two-point crossover. The probability that two consecutive steps with one-point crossover decrease i is $\Theta(i^2/n^2)$ as for one step of two-point crossover. This leads to the following result.

Theorem 9. *The expected number of fitness evaluations until GIGA or the $(1 + n)$ GIGA with one-point crossover finds an optimum for the Ising model on the ring equals $\Theta(n^2)$. For the $(1 + n)$ GIGA, the number of generations equals $\Theta(n)$.*

7 The Expected Run Time of a GA with Fitness Sharing

The variant of GIGA analyzed in Section 6 is highly specialized. Diversity in the population of size 2 is guaranteed by choosing always individuals with the maximal Hamming distance. Here, we consider a GA with the unusually small population size 2 where diversity is supported by fitness sharing. Populations are multisets. In fitness sharing, the closeness of x and y is measured by

$$S(x,y) := \max\{1 - d(x,y)/\sigma, 0\}$$

where d is an appropriate distance measure and σ is a critical value deciding when x and y are so far from each other that they do not share fitness. In our case, d is the Hamming distance and $\sigma := n$ since we like to produce individuals with large Hamming distance. Then, for population P, $S(x,P)$ is the sum of all $S(x,y)$, $y \in P$. The shared fitness of x in the population P is defined by

$$f(x,P) := f(x)/S(x,P)$$

if f is the real fitness. Finally, $f(P)$ is defined as the sum of all $f(x,P)$, $x \in P$.

The following GA applies two-point crossover to produce two children and mutations flipping each bit independently with probability $1/n$.

Algorithm 1 *(Steady-state GA with population size 2 and fitness sharing)*

1.) *The initial population P consists of two individuals chosen independently and uniformly at random.*
2.) *Selection for reproduction. Both individuals x and y are chosen.*
3.) *Offspring creation. One of the Steps 3a and 3b is chosen uniformly at random.*
3a.) $x' := \text{mutate}(x)$, $y' := \text{mutate}(y)$, $P' := P \cup \{x', y'\}$.
3b.) $(\tilde{x}, \tilde{y}) := \text{two-point-crossover}(x, y)$, $x' := \text{mutate}(\tilde{x})$,
      $y' := \text{mutate}(\tilde{y})$, $P' := P \cup \{x', y'\}$.
4.) *Selection of the next generation. Choose a population $P \subseteq P'$ of size 2 with the maximal $f(P)$-value.*

Since we work with populations of very small size, it is not too time-consuming to choose in Step 4 a population with the largest f-value. This reflects the real idea behind fitness sharing. The shared fitness of the population should be large.

Let the population P consist of the individuals x and y with a Hamming distance of $d = d(P)$. Let $i(z)$ be the number of borders within the individual z and let $i = i(P) := i(x) + i(y)$. Then $f(z) = n - i(z)$ and

$$f(x,P) = \frac{n - i(x)}{1 - H(x,x)/n + 1 - H(x,y)/n} = \frac{n - i(x)}{2 - d/n}$$

and

$$f(P) = \frac{2n - i}{2 - d/n}.$$

Hence, we can increase P by decreasing i and/or by increasing d. As long as we do not decrease i, we hope to increase d. If $d = n$, we have two complementary individuals and two-point crossover is a good operator to decrease i. Since $0 \leq f(P) \leq 2n$ and f cannot decrease because of the plus-strategy for selection, we try to analyze the expected time until f has been increased at least by a constant additive term c. For this purpose, we classify the possible populations P:

- type OPT contains all P where $i \leq 1$, i.e., at least one individual is optimal,
- type $A(i)$, $i \geq 2$, contains all P where $i = i(P)$ and $d = n$,
- type B contains all P where $2 \leq i \leq n$ and $d < n$, and
- type C contains all P where $i > n$ and $d < n$.

Theorem 10. *The expected number of fitness evaluations until the steady-state GA with population size 2 and fitness sharing finds an optimum for the Ising model on the ring is bounded above by $O(n^2)$.*

Proof. All populations of type $A(i)$ have the same fitness $2n - i$. After having increased the fitness, we will never accept a population of type $A(i)$. Moreover, if $P = \{x, y\}$ is of type $A(i)$, then $y = \bar{x}$. The expected waiting time until two-point crossover creates a population P' of type $A(i-4)$ is bounded by $O(n^2/i^2)$, see Section 6. Then $f(P') - f(P) = 4$. By standard arguments, the expected time with populations of type A is bounded by $O(n^2)$.

For populations of type B or C, we prove that the probability of increasing the fitness by at least $1/4$ is bounded below by $\Omega(1/n)$. We have to wait for at most $8n$ of such steps which proves the theorem.

Let $P = \{x, y\}$ be of type B. Since $d < n$, $x \neq \bar{y}$. Let j be the rightmost position where $x_j = y_j$. Then $x_{j+1} \neq y_{j+1}$ (where $n+1$ is identified with 1 since we are on a ring). W. l. o. g. $x_j = x_{j+1}$ and $y_j \neq y_{j+1}$. With a probability of $\Omega(1/n)$, we choose Step 3a and only bit j is flipped when producing y'. Then $f(y') \geq f(y)$ and $H(x, y') = H(x, y) + 1$. The population $P' = \{x, y'\}$ is a possible successor population and

$$f(P') - f(P) \geq \frac{2n - i}{2 - (d+1)/n} - \frac{2n - i}{2 - d/n}$$
$$= \frac{(2n - i) \cdot (2 - d/n) - (2n - i) \cdot (2 - d/n - 1/n)}{(2 - (d+1)/n) \cdot (2 - d/n)} > \frac{1}{4},$$

since the numerator equals $2 - i/n \geq 1$ and the denominator is at most 4.

Type-C populations can be handled in a similar way. Since $i \geq n$, one individual has a block of length 1 which can be eliminated by a 1-bit flip. $\square$

Finally, we can consider a GA with population size 2 and fitness sharing which produces $P'_1, \ldots, P'_n$ by performing Step 3 n times independently in parallel. Then it selects a population $P \subseteq P'_i$ for some i which has the largest $f(P)$-value.

Theorem 11. *The expected number of generations until the GA with population size 2, fitness sharing, and n pairs of offspring per generation finds an optimum for the Ising model on the ring is bounded above by $O(n)$.*

Conclusions

The Ising model is a good model to analyze the adaptation capabilities of EAs and GAs. In particular, the Ising model on the ring leads to surprising results. Mutation-based algorithms and even randomized local search are much more efficient than expected in the GA community. This is especially true if we consider the number of generations in the case of producing more than one offspring. Nevertheless, recombination can decrease the expected optimization time. This has been proved rigorously for two specialized GAs which work with very small populations. It is an open problem to analyze generic GAs with niching for the Ising model on the ring.

References

1. J. Culberson. Genetic invariance: A new paradigm for genetic algorithm design. Technical Report 92–02, University of Alberta, 1992.
2. D. E. Goldberg. *Genetic Algorithms in Search, Optimization, and Machine Learning.* Addision-Wesley, Reading, MA, 1989.
3. J. H. Holland. *Adaptation in Natural and Artificial Systems.* University of Michigan, MI, 1975.
4. E. Ising. Beiträge zur Theorie des Ferromagnetismus. *Z. Physik*, (31):235–258, 1925.
5. T. Jansen and K. De Jong. An analysis of the role of offspring population size in EAs. In *Proc. of the Genetic and Evolutionary Computation Conference (GECCO 2002)*, pages 238–246, San Mateo, CA., 2002. Morgan Kaufmann.
6. T. Jansen, K. De Jong, and I. Wegener. On the choice of the offspring population size in evolutionary algorithms. *Submitted to Evolutionary Computation*, 2003.
7. B. Naudts and J. Naudts. The effect of spin-flip symmetry on the performance of the simple GA. In *Proc. of 5th Conf. on Parallel Problem Solving from Nature (PPSN - V)*, number 1498 in LNCS, pages 67–76, 1998.
8. M. Pelikan and D. E. Goldberg. Hierarchical BOA solves Ising spin glasses and MAXSAT. In *Proc. of the Genetic and Evolutionary Computation Conference (GECCO 2003)*, number 2724 in LNCS, pages 1271–1282, 2003.
9. C. Van Hoyweghen. *Symmetry in the Representation of an Optimization Problem.* PhD thesis, Univ. Antwerpen, 2002.
10. C. Van Hoyweghen, D. E. Goldberg, and B. Naudts. Building block superiority, multimodality and synchronization problems. Technical report, Illinois Genetic Algorithms Laboratory, 2001. IlliGAL Rep. No. 2001020.
11. C. Van Hoyweghen, D. E. Goldberg, and B. Naudts. From twomax to the Ising model: Easy and hard symmetrical problems. In *Proc. of the Genetic and Evolutionary Computation Conference (GECCO 2002)*, pages 626–633, San Mateo, CA, 2002. Morgan Kaufmann.
12. C. Van Hoyweghen, B. Naudts, and D. E. Goldberg. Spin-flip symmetry and synchronization. *Evolutionary Computation*, (10):317–344, 2002.

Effects of Module Encapsulation in Repetitively Modular Genotypes on the Search Space

Ivan I. Garibay[1,2], Ozlem O. Garibay[1,2], and Annie S. Wu[1]

[1] University of Central Florida, School of Computer Science,
P.O. Box 162362, Orlando, FL 32816-2362, USA,
{igaribay,ozlem,aswu}@cs.ucf.edu,
http://ivan.research.ucf.edu
[2] University of Central Florida, Office of Research
Orlando Tech Center/ Research Park
12443 Research Parkway Orlando, FL 32826, USA.

Abstract. We introduce the concept of *modularity-preserving representations*. If a representation is modularity-preserving, the existence of modularity in the problem space is translated into a corresponding modularity in the search space. This kind of representation allows us to analyze the impact of modularity at the genomic level. We investigate the question of what constitutes a module at the genomic level of evolutionary search and provide a static analysis of how to identify good and bad modules based on their ability to reduce the search space, thus, biasing the search space towards a solution. We also prove, under a set of assumptions, that the systematic encapsulation of lower order modules into higher order modules does not change the size or bias of a search space and that this process produces a hierarchy of equivalent search spaces.

1 Introduction

The success of evolutionary algorithms for a given problem is heavily affected by the representation used [1,2,3]. The task of designing a good representation is, at this point, more an art than a science. In this paper, we analyze the effects that the choice of representation primitives and the encapsulation of primitives into modules can have on the size of a search space and bias towards a solution. We focus on two questions. Does the encapsulation and replacement of lower level modules or primitives with higher level counterparts, by itself, benefit evolutionary search? Under what circumstances does the encapsulation of lower level modules or primitives into higher level modules benefit evolutionary search and how? We offer a static mathematical analysis as well as experimental results to shed light on these two issues. We find that replacing lower level elements with encapsulated higher level counterparts, by itself, has no effect on the size or bias of the search space. This result is, in essence, a kind of No Free Lunch theorem [4] for genomic module encapsulation. We also find that the process of encapsulating primitives into modules has two static effects. It enlarges the search space, because it introduces a new element into the search space alphabet; and it biases this extended search space towards solutions that contain the

K. Deb et al. (Eds.): GECCO 2004, LNCS 3102, pp. 1125–1137, 2004.

encapsulated primitives. As a result, there is a trade-off between the gains and losses of introducing a new module in terms of search space size and bias. We provide a closed form expression, under certain assumptions, whether the creation of a module will be beneficial in terms of the size of the resulting search space size. We show that this bias is governed by the module size and by how many times the module appears in the solution string.

2   Modularity

The concept of modularity has been studied extensively in complex systems and also in evolutionary computation. Recently, issues such improving the "innovativeness" and the scalability of evolutionary search have attracted renewed interest to the study of modularity in the evolutionary computation community. For instance see [5,6,7]. From our perspective, modularity implies not only the hierarchical organization of components from one level of complexity to the next, but also the ability to freely reuse components. In Evolutionary Computation, various techniques have been used to incorporate modularity into the evolutionary search, for example, *compress* and *expand* operators [8], automatically defined functions [9], speciation [10], repetitive modularity [11,12], and coevolutionary methods [13], to name a few. All of these techniques seek to identify "good" modules for a given problem. In contrast with module definitions based on fitness [14,15,13], we study *location-independent genomic modularity*. For us, a *genomic module* is simply a pattern of consecutive genomic primitives or lower-level genomic modules occurring at any location in a genome.

We base our study of modular genomes on the following assumptions: first, that the class of problems with modular solutions is of interest; and second, that there exist representations that correlate modularity in solutions with corresponding modularity in genomes and that such representations can be found for any repetitively modular problem. We call this kind of representations *modularity-preserving representations(MPR)* and we call the second assumption the *Modula-rity-preserving representation hypothesis*. For the reminder of this paper, we assume that this hypothesis is true.

3   Mathematical Analysis

We introduce the following definitions using standard set theory and formal languages notation, i.e. see [16].

3.1   Modular Search Spaces

Primitives are the atomic components of problem representation that are used to encode a *candidate solution*. A candidate solution, also called an *individual*, may consist of both primitives and modules. *Modules* are substrings of interest. A module may contain two kinds of symbols: primitives and previously defined modules. The *search space* of a problem is the space of all possible candidate solutions. The structure of a search space is determined by its alphabet which can consist of both primitive and module symbols, and by the length of its individuals. The elements of the search space are all possible strings of a given length over the search space alphabet.

Definition 1 (Search space). *A search space S is a 3-tuple:*

$$S = \langle \Sigma, l, \mathcal{R} \rangle$$

Where, $\Sigma \subseteq \mathcal{P} \cup \mathcal{M}$ is the search space alphabet; $\mathcal{P}$ is a set of primitive symbols, and $\mathcal{M}$ is a set of module symbols; l is the length of the individuals; and $\mathcal{R}$ is the set of module defining rules.

Definition 2 (Module defining rules). *For a given search space $S = \langle \Sigma, l, \mathcal{R} \rangle$, if $r_i \in \mathcal{R}$ then r_i is of the form:*

$$r_i = M_i \rightarrow w_i$$

Where, $M_i \in \mathcal{M}$ is the module name; $w_i \in \{\mathcal{P} \cup \{M_1, M_2, ..., M_{i-1}\}\}^*$ is the module defining string; and $|w_i| \leq l$, since we consider modules to be substrings of candidate solutions.

There is one defining rule in $\mathcal{R}$ for each module symbol in $\mathcal{M}$, hence $|\mathcal{R}| = |\mathcal{M}|$. Note that rules are hierarchically defined in terms of primitives and previously defined modules; hence, circularity in definitions is not possible. Let us define a module *size* as the length of its defining string $|w_i|$. A module is of *order* zero if its defining substring consist solely of primitives, and it is of order n if its defining substring consist of primitives and symbols naming modules of at most order $n - 1$.

Definition 3 (Search space elements). *The elements of the search space S, denoted by $L(S)$, are all strings of length l over the search space alphabet Σ:*

$$L(S) = \{e \mid e \in \Sigma^* \wedge |e| = l\}$$

Since Σ can potentially contain primitives as well as modules, individuals can be *expanded* using the module definitions in $\mathcal{R}$. An individual is expanded by rewriting module names by module definitions until the individual consist of a string of only primitives.

Definition 4 (expanded form). *Let $e \in L(S)$ be an element of search space $S = \langle \Sigma, l, \mathcal{R} \rangle$. We define the expanded form of element e as:*

$$e_{exp} = Expand_{\mathcal{R}}(e)$$

Where $Expand_{\mathcal{R}}$ is the expanding function for module definitions $\mathcal{R}$. The expanding function applies the rewriting rules in $\mathcal{R}$ to its input e until a string solely over $\mathcal{P}$ is obtained, $Expand_{\mathcal{R}}$ outputs that string. In this case, we said that e_{exp} has been derived *from e using rewriting rules $\mathcal{R}$.*

Notice that $e \in \Sigma^*$ and $|e| = l$, while $e_{exp} \in \mathcal{P}^*$ and $|e_{exp}| \geq l$. Also, if e does not contain any module symbols, then $e = e_{exp}$.

Definition 5 (expanded search space). *Let* $\mathcal{S} = \langle \Sigma, l, \mathcal{R} \rangle$ *be a search space and* $L(\mathcal{S})$ *its elements. We define the expanded search space* $L_{exp}(\mathcal{S})$ *as follows:*

$$L_{exp}(\mathcal{S}) = \{g \mid g = Expand_{\mathcal{S}}(e) \wedge e \in L(\mathcal{S})\}$$

$L_{exp}(\mathcal{S})$ is the set of all individuals in the search space $\mathcal{S}$ in their expanded form. Elements of $L_{exp}(\mathcal{S})$ are variable length strings over $\mathcal{P}$. It is easy to show that the length of the expanded individuals, l_{exp}, is bounded by: $l \leq l_{exp} \leq l^{|M|+1}$.

3.2   Search Space Size and Bias

The size of a search space $\mathcal{S} = \langle \Sigma, l, \mathcal{R} \rangle$, denoted by $|L(\mathcal{S})|$, is $|L(\mathcal{S})| = |\Sigma|^l$. Notice that it is possible for two different individuals to expand to the same string of primitives, therefore, $|L_{exp}(\mathcal{S})| \leq |L(\mathcal{S})|$. As we are interested in the bias produced by module definitions, we measure the bias of a search space using the expanded form. If the expanded search space $L_{exp}(\mathcal{S})$ contains all possible strings of primitives of a given length t, then we said that the search space has no *structural bias* for length t because there are no unreachable strings of primitives of length t. The search space is structurally biased otherwise. Note that even in an structurally unbiased search space where all strings of primitives are reachable, some strings of primitives may have multiple derivations and therefore be preferred. We call this later case a *modularly biased* search space. [1]

Definition 6 (Search space bias). *For a given search space* $\mathcal{S} = \langle \Sigma, l, \mathcal{R} \rangle$, *if*

$$L_{exp}(\mathcal{S}) \supseteq \{e \mid e \in \mathcal{P}^* \wedge |e| = t\}$$

is true, we say that $\mathcal{S}$ *is not structurally biased for length* t, *or simply that it is not structurally biased in the case that* $t = l$. $\mathcal{S}$ *is structurally biased otherwise.*

A set of module definition rules $\mathcal{R}$ are *complete-$\mathcal{P}$* with respect to a set of primitives $\mathcal{P}$ it the rules are able to derive every string of a given length t over the alphabet of primitives $\mathcal{P}$, starting from strings containing only module names from $\mathcal{R}$.

Definition 7 (Complete-$\mathcal{P}$). *Lets* $\mathcal{R}$ *and* $\mathcal{P}$ *be a set of module definition rules and a set of primitive symbols respectively. We say that* $\mathcal{R}$ *is complete-$\mathcal{P}$ with respect to* $\mathcal{P}$ *for a given length* t *if the following expression holds true:*

$$\{Expand_{\mathcal{R}}(e) \mid e \in \mathcal{M}^*\} \supseteq \{e \mid e \in \mathcal{P}^* \wedge |e| = t\}$$

Where $\mathcal{M}$ *is the set of all module names from* $\mathcal{R}$

---

[1] For instance, the search space $\mathcal{S} = \langle \Sigma = \{1, 0, A\}, l = 8, \mathcal{R} = \{A \rightarrow 11\} \rangle$ is structurally unbiased since all strings of primitives of length 8 can be reached; however it is modularly biased since, for instance, the string "00000000" can only be derived from "00000000" and the string "111111111" can be derived from "A1111111", "1A111111", "11A11111", etc.

Definition 8 (Complete m-module set). $\mathcal{R}_c$ *is the complete m-module set of Q iff:*

$$\mathcal{R}_c = \{(M_i \to w) \mid w \in Q^* \wedge |w| = m \wedge i \text{ is a unique index for } w\}$$

Where, Q is an arbitrary set of symbols, $\mathcal{R}_c$ is a set of module defining rules, m is the size of all modules in $\mathcal{R}_c$, and i is an arbitrary index for strings w over Q.

Since there is one module defined on $\mathcal{R}_c$ per each string of size m over Q, we have $|\mathcal{R}_c| = |Q|^m$

Lemma 1. *Lets $\mathcal{R}$ and $\mathcal{P}$ be a set of module definition rules and a set of primitive symbols respectively. The following two expressions are true:*

$$(\mathcal{R} \supseteq \text{complete m-module set of } \mathcal{P}) \to (\mathcal{R} \text{ is complete-}\mathcal{P}) \quad (1)$$
$$(\mathcal{R} \supseteq \text{complete m-module set of a complete-}\mathcal{P} \text{ set}) \to (\mathcal{R} \text{ is complete-}\mathcal{P}) \quad (2)$$

Lemma 2. *Let $\mathcal{S} = \langle \Sigma, l, \mathcal{R} \rangle$ be a search space. If $\mathcal{S}$ is not structurally biased then one of the following is true:*

1. $\Sigma \supseteq \mathcal{P}$, or
2. $\Sigma \supseteq \mathcal{B}$, where $\mathcal{B}$ is the set of module names of a complete-$\mathcal{P}$ set of rules with respect to $\mathcal{P}$.

It is not difficult to prove Lemmas 1 and 2

3.3   Search Space Altering Operations

Module encapsulation or *module creation* is the process of naming a substring of interest with a new alphabet symbol. This process changes the structure of the search space by adding a new module symbol to the alphabet.

Definition 9 (Encapsulation). *Module encapsulation, $\mathcal{E} : \mathcal{S} \times r_k \mapsto \mathcal{S}$ is defined as follows:*

$$\mathcal{E}(\mathcal{S}, M_k \to w_k) = \langle \Sigma \cup \{M_k\}, l, \mathcal{R} \cup \{M_k \to w_k\}\rangle$$

Where, $\mathcal{S} = \langle \Sigma, l, \mathcal{R} \rangle$ is a search space, $M_k \to w_k$ is the rewriting rule defining the new module to be encapsulated, M_k is a new module symbol, w_k is a string over Σ, and $|w_k| \leq l$.

Strict-encapsulation of a search space $\mathcal{S}$ is the process of creating all possible modules of a given size m over the current search space alphabet Σ and then replacing the current alphabet with the newly created module names. This process evidently changes the structure of the search space by replacing completely the search space alphabet and the module defining rules. Notice that there are $|\Sigma|^m$ modules of size m that can be created over Σ. The new individual length is l/m, since individuals consist of only new modules of size m.

Definition 10 (Strict-encapsulation). Strict-encapsulation, $\mathcal{E}_s : S \mapsto S$ is defined as follows:

$$\mathcal{E}_s(S) = \langle \Sigma', l/m, \mathcal{R}' \rangle$$

Where, $S = \langle \Sigma, l, \mathcal{R} \rangle$ is a search space; $\mathcal{R}'$ is a complete m-module set for Σ ; and Σ' is the set of newly created module symbols from $\mathcal{R}'$.

Lemma 3. Let $S_1 = \langle \Sigma_1, l_1, \mathcal{R}_1 \rangle$ and $S_2 = \langle \Sigma_2, l_2, \mathcal{R}_2 \rangle$ be search spaces such that $S_2 = \mathcal{E}_s(S_1)$, then $|L(S_1)| = |L(S_2)|$.

Proof. We know that $|L(S_1)| = \Sigma_1{}^{l_1}$ and $|L(S_2)| = |\Sigma_2|^{l_2}$. Since $S_2 = \mathcal{E}_s(S_1)$ we have that $|\Sigma_2| = |\Sigma_1|^m$ and $l_2 = l_1/m$. Therefore $|\Sigma_2|^{l_2} = |\Sigma_1|^{m(l_1/m)} = |\Sigma_1|^{l_1}$.   □

Lemma 4. Let S_1 and S_2 be search spaces such that $S_2 = \mathcal{E}_s(S_1)$, then $(not\ structurally\ biased\ S_1) \to (not\ structurally\ biased\ S_2)$.

Proof. Let $S_1 = \langle \Sigma_1, l_1, \mathcal{R}_1 \rangle$ and $S_2 = \langle \Sigma_2, l_2, \mathcal{R}_2 \rangle$ be search spaces. We need to prove that:

$$(L_{exp}(S_1) \supseteq \{e \mid e \in \mathcal{P}_1{}^* \wedge |e| = l_1\}) \to (L_{exp}(S_2) \supseteq \{e \mid e \in \mathcal{P}_2{}^* \wedge |e| = l_2\})$$

Let us assume that S_1 is not structurally biased. We will prove that for that case S_2 is also not structurally biased. According to Lemma 2, we would need to consider two cases:
Case 1: $\Sigma_1 \supseteq \mathcal{P}$ hence $\Sigma_1 = \{\mathcal{P} \cup \mathcal{U}\}$ for some possibly empty set $\mathcal{U}$. Since $S_2 = \mathcal{E}_s(S_1)$, $\mathcal{R}_2$ is a complete m-module set of $\Sigma_1 = \{\mathcal{P} \cup \mathcal{U}\}$. It is easy to show that for this case:

$$\mathcal{R}_2 \supseteq complete\ \text{m-module set of}\ \mathcal{P}_2$$

by Lemma 1 then, $\mathcal{R}$ is complete-$\mathcal{P}$. Therefore, S_2 is also not structurally biased.
Case 2: $\Sigma_1 \supseteq \mathcal{B}$, where $\mathcal{B}$ is complete-$\mathcal{P}$. The prove is analogous to Case 1, but we need to use the second part of Lemma 1.   □

Lemma 5. Lets S_1 and S_2 be search spaces such that $S_2 = \mathcal{E}_s(S_1)$, then $(structurally\ biased\ S_1) \to (structurally\ biased\ S_2)$.

The proof for Lemma 5 is analogous to that of Lemma 4 and omitted for space constrains. Further details can be found at [17].

3.4   Hierarchy of Modular Search Spaces

Theorem 1. *Strictly-encapsulating lower-order modules into a complete set of higher-order modules does not change the search space size or structural bias.*

Proof. Lets S_1 and S_2 be two search spaces such that $S_2 = \mathcal{E}_s(S_1)$. We need to prove that

$$|L(S_1)| = |L(S_2)|$$
$$structurally\,bias\,S_1 \rightarrow structurally\,bias\,S_2$$
$$not\,structurally\,bias\,S_1 \rightarrow not\,structurally\,bias\,S_2$$

All three statements have been proved in Lemmas 3, 4, and 5.   □

Continuously strictly-encapsulating a search space produces a hierarchy of search spaces. At the bottom of this hierarchy we have a search space with individuals of size l and trivial modules of size one (l primitives). As we move up the hierarchy, we have search spaces with more modules and larger module sizes. At the top of this hierarchy, we have individuals of size one consisting of only one module of size l. We call this type of hierarchy a *modularity representation pyramid* for S.

Definition 11 (Modularity representation pyramid). *Let S be a search space with only primitives. Let us recursively define the modularity representation pyramid, $A = \{a_0, a_1, ...\}$, for S as:*

$$a_0 = S$$
$$a_{i+1} = \mathcal{E}_s(a_i)$$

The recursion terminates when the individual length for a given a_i is equal to one (the individuals can not be further encapsulated into modules).

Corollary 1. *Let S be a search space with only primitives and A be a modularity representation pyramid for S. Then the following is true for all levels of the pyramid:*

1. all search spaces are of equal size: Σ^l
2. all search spaces are structurally unbiased.

Proof. Base case: S is trivially structurally unbiased and of size Σ^l. Recursive step: by the previous theorem, a_{i+1} and a_i are of the same size and if a_i is structurally unbiased so does a_{i+1}.   □

4   What Makes Module Encapsulation Advantageous?

Replacing lower level modules with higher level modules does not produce any advantage unless there is some rationale for pruning some of the "undesired" higher level modules, hence reducing the search space. We define "good" modules to be modules that are present in a solution and "bad" modules to be modules that are not present in a solution. In this section, we analyze the effect on search space size of encapsulating good modules versus bad modules. Let us assume that the optimal solution is known. Consequently, the type and number of modules in the optimal solution is also known. Furthermore, we assume that the length of the individuals in each search space equals the shortest optimal individual length. Based on these assumptions, we can derive the relationship that must be satisfied in order for encapsulation to be beneficial to a search. Let $S_1 = \langle \Sigma_1, l_1, \mathcal{R}_1 \rangle$ and $S_2 = \langle \Sigma_2, l_2, \mathcal{R}_2 \rangle$ be search spaces such that S_2 is the product of encapsulating the module M into search space S_1; hence, $S_2 = \mathcal{E}(S_1, (M \to w))$. Let the optimal solution for search space S_1 be w_s. Let w_s contain x copies of the string w which defines module M. According to Definition 9, we have $\Sigma_2 = \Sigma_1 \cup \{M\}$, therefore:

$$|\Sigma_2| = |\Sigma_1| + 1 \tag{3}$$

Because the optimal solution can be expressed in terms of module $(M \to w)$ and because the optimal solution contains x copies of the module defining string w, then:

$$l_2 = l_1 - x(|w| - 1) \tag{4}$$

In order for the encapsulation of module $(M \to w)$ to produce an advantage in terms of search space size, we need:

(Search space size with M) $\leq$ (Search space size without M)

Which it is:

$$|L(S_2)| \leq |L(S_1)|$$
$$|\Sigma_2|^{l_2} \leq |\Sigma_1|^{l_1}$$

Using (3) and (4) on the previous equation we have:

$$(|\Sigma_1| + 1)^{l_1 - C} \leq |\Sigma_1|^{l_1} \tag{5}$$

where,

$$C = x(|w| - 1)$$

is a constant which depends only on the modular properties of the solution string w_s. Applying logarithms to both sides of (5) and rearranging the terms gives us:

$$C \geq l_1 \frac{\ln(1 + 1/|\Sigma_1|)}{\ln(|\Sigma_1| + 1)} \tag{6}$$

Therefore, there are three possible outcomes when encapsulating a good module. If (6) is satisfied as an inequality, we predict an advantageous reduction of search

Table 1. Details for the modularity representation pyramid A_1 used in experiment 1.

	Alphabet size	Module length	Solution length
Level 1	2	1	8
Level 2	4	2	4
Level 3	16	4	2
Level 4	256	8	1

space size when the module $(M \to w)$ is encapsulated. If it is satisfied as an equality, we predict no advantage or disadvantage because the search spaces will be of equal size. If (6) is not satisfied, then the encapsulation of module $(M \to w)$ will be disadvantageous since it will result in a larger search space. On the other hand, encapsulating a bad module is always disadvantageous. Since a bad module is not present in the optimal solution the value of the constant is $C = 0$, which renders (6) unsatisfiable.

5   Experimental Analysis

5.1   Objectives

In Sects. 3 and 4, we present a simple analysis of the effects of module encapsulation on the size and bias of a search space. We next present two experiments to empirically verify our conclusions. In both experiments, we evaluate the performance of a GA on a modular space in terms of the best fitness in the final generation. Performance evaluation is averaged over 40 runs. As our analyses focus only on the search space and do not take into account the characteristics or dynamics of any search algorithm—such as the GA that we use—, we expect only qualitative verification of our conclusions from the empirical tests.

The first experiment is an empirical validation of Theorem 1. We select a small modularity representation pyramid A_1 for $\mathcal{S}_1 = \langle \Sigma = \{0, 1\}, l = 8, \mathcal{R} = \{\} \rangle$. Each level of the pyramid is described in Table 1. Although GA dynamics at various levels of the search space may affect results to some degree. We expect to see a qualitative validation of Theorem 1 in the form of roughly comparable performance at all pyramid levels. In the second experiment, we attempt to validate (6) by comparing theoretically and experimentally obtained values of C. We follow the analysis in Sect. 4 using the following values: search space $\mathcal{S}_1 = \langle \Sigma_1 = \{0, 1\}, l_1 = 64, \mathcal{R}_1 = \{\} \rangle$, module to be encapsulated $M \to 00010111$, optimal solution w_s contain x, for $x \in \{0, 2, 3, 4, 5, 6\}$, copies of the string 00010111. We use (4) to calculate the value of the constant $C = x(|00010111| - 1) = 7x$. Equation 6 gives us the condition that must be satisfied for encapsulation to be beneficial: $C \geq 64 \frac{ln(1+1/2)}{ln(2+1)} = 23.62$. From these equations, we can calculate the theoretical threshold value $x \geq 3.37$. Therefore, according to the analysis in Sect. 4, a target solution must contain more than 3 copies of 00010111 for the encapsulation of module $M \to 00010111$ to be beneficial. If the target solution contains more than three copies of the module M,

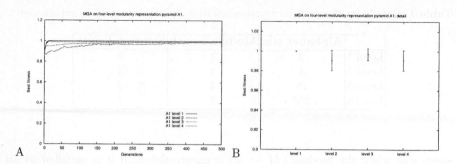

Fig. 1. MGA on random pattern matching for each level of the modularity representation pyramid A_1: (A) best fitness by generation averaged over 40 runs, (B) fitness of best solution found averaged over 40 runs and 95% confidence intervals. This empirical result validates Theorem 1 since the performance is comparable for all levels.

then we expect the search space size to decrease and performance to improve. If the target solution contains fewer than three copies of M then we expect the search space size to increase and the performance to degrade.

5.2   Settings

For all experiments we use a modular version of a genetic algorithm (GA) [18,19] called the *modular genetic algorithm* (MGA) [11]. MGA is a simple GA with the ability to encapsulate modules as described in Sect. 3. MGA genetic operators are analogous to GA operators, but work with strings over alphabets of any size. The following parameter settings are common for all experiments: the crossover type is two-point, the crossover rate is 0.9, the mutation rate is 0.005, selection type is fitness proportional, population sizes of 50 for the first and 500 for the second set of experiments, and the number of generations is 500. We perform 40 trials for all experiments and report average values with their 95% confidence intervals.

5.3   Results

Figure 1 shows the results from the first experiment. Figure 1(A) shows the best fitness for each generation averaged over 40 runs. Figure 1(B) shows the fitness of the best solution found averaged over 40 runs and 95% confidence intervals. For all four levels of the hierarchy of modular search spaces defined by the pyramid A_1 we obtain comparable performance as predicted by Theorem 1. Figure 2 shows the results from the second experiment. We compare the MGA with a traditional GA on a pattern matching problem. The goal is to generate a target solution w_s containing x copies of module M. Figure 2 shows the fitness of the best solution found averaged over 40 runs and 95% confidence intervals. These results indicate that GA performance is insensitive to the number of modules M contained in a solution. MGA performance with module M encapsulated is

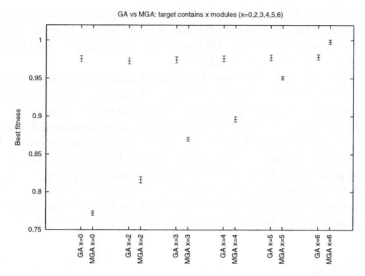

Fig. 2. MGA -vs- GA on pattern matching for target solution w_s containing $x \in \{0, 2, 3, 4, 5, 6\}$ copies of module M: Fitness of best solution found averaged over 40 runs and 95% confidence intervals.

linearly correlated with x. We observe that the experimental threshold for x is approximately $x \geq 6$. This experiment validates qualitatively our analysis of what makes the encapsulation of a module advantageous. For the case $x = 0$, no good module is defined, encapsulation is disadvantageous, and a GA outperforms the MGA. For the cases of $x = 2, 3, 4, 5$ there is a good module defined but the experimental threshold has not been reached; therefore, encapsulation remains disadvantageous. Finally, for $x = 6$, there is a good module defined, the threshold has been met, and encapsulation in this case is advantageous—the MGA outperforms a traditional GA.

6   Conclusions

In this paper, we investigate the effects of module encapsulation in repetitively modular genomes on the search space size and bias. We introduce the concept of *modularity-preserving representations*. If a representation is modularity-preserving, the existence of modularity in the problem space is translated into a corresponding modularity in the search space. We hypothesize that such representations can be found for any given modular problem, which allows us study the impact of modularity at the genomic level. In Sect. 1 we pose two questions to focus our work. We now discuss our conclusions with respect to those questions:

1. Does the encapsulation and replacement of lower level modules or primitives with higher level counterparts, by itself, benefit evolutionary search?

We prove, under a set of assumptions, that systematically encapsulating lower order modules into higher order ones does not change the size of a search space or its structural bias: Theorem 1. We also provide an experimental analysis in support of this analytical result. Therefore, the encapsulation of modules alone does not benefit evolutionary search.

2. Under what circumstances does the encapsulation of lower level modules or primitives into higher level modules benefit evolutionary search and how?

Adding a module to a search space increases the size of the alphabet and, consequently, the size of the search space. This increase can be countered if appropriate modules are selected which decrease the length of an optimal solution. We define good modules as modules that are present in the optimal solution and bad modules as modules that are not. We provide an expression in (6) to determine when module encapsulation is advantageous. Using this expression we can answer this question as follows: encapsulating bad modules is always detrimental; encapsulating good modules is advantageous only if (6) is satisfied.

References

1. Clark, A., Thornton, C.: Trading spaces: Computation, representation, and the limits of uninformed learning. Behavioral and Brain Sciences **20** (1997) 57–90
2. Jones, T., Forrest, S.: Fitness distance correlation as a measure of problem difficulty for gas. In Eshelman, L.J., ed.: Proc. 6th Int'l Conf. on GAs. (1995)
3. Mathias, K., Whitley, L.D.: Transforming the search space with gray coding. In: Proc. IEEE Int'l Conference on Evolutionary Computation. (1994) 513–518
4. Wolpert, D.H., Macready, W.G.: No free lunch theorems for search. Technical Report SFI-TR-95-02-010, The Santa Fe Institute, Santa Fe, NM (1995)
5. Watson, R.A.: Hierarchical module discovery. In: 2003 AAAI Spring Symposium Series. (2003) 262–267
6. Koza, J.R., Streeter, M., Keane, M.: Automated synthesis by means of genetic programming. In: 2003 AAAI Spring Symposium Series. (2003) 138–145
7. Garibay, I.I., Wu, A.S.: Cross-fertilization between proteomics and computational synthesis. In: 2003 AAAI Spring Symposium Series. (2003) 67–74
8. Angeline, P.J., Pollack, J.: Evolutionary module adquisition. In: Proceedings of the second annual conference on evolutionary programming. (1993) 154–163
9. Koza, J.R.: Genetic Programming II: Automatic Discovery of Reusable Programs. MIT Press, Cambridge, MA (1994)
10. Darwen, P.J., Yao, X.: Speciation as automatic categorical modularization. IEEE Transactions on Evolutionary Computation **1** (1997) 101–108
11. Garibay, O.O., Garibay, I.I., Wu, A.S.: The modular genetic algorithm: exploiting regularities in the problem space. In: Proc. of ISCIS 2003. (2003) 578–585
12. De Jong, E.D., Oates, T.: A coevolutionary approach to representation development. In: Proc. of the ICML-2002 WS on development of rep. (2002) 1
13. De Jong, E.D.: Representation development from pareto-coevolution. In: Proceedings of GECCO 2003. LNCS series, Springer-Verlag (2003) 265–276
14. Goldberg, D.E., Korb, B., Deb, K.: Messy genetic algorithms: Motivation, analysis, and first results. Complex Systems **3** (1989) 493–530
15. Watson, R.A., Pollack, J.: Symbiotic combination as an alternative to sexual recombination in genetic algorithms. In: Proc. of PPSN VI. (2003) 262–267

16. Hopcroft, J.E., Ullman, J.D.: Introduction to Automata Theory, Languages, and Computation. Addison Wesley (1979)
17. Garibay, O.O., Garibay, I.I., Wu, A.S.: No free luch theorem for modular genomes. Technical Report CS-TR-04-03, University of Central Florida (2004)
18. Holland, J.H.: Adaptation in Natural and Artificial Systems. University of Michigan Press, Ann Arbor, MI (1975)
19. Goldberg, D.E.: Genetic algorithms in search, optimization, and machine learning. Addison Wesley (1989)

Modeling Selection Intensity for Toroidal Cellular Evolutionary Algorithms

Mario Giacobini[1], Enrique Alba[2], Andrea Tettamanzi[3], and Marco Tomassini[1]

[1] Information Systems Department, University of Lausanne, Switzerland
{mario.giacobini, marco.tomassini}@unil.ch
[2] Department of Computer Science, University of Málaga, Málaga, Spain
eat@lcc.uma.es
[3] Information Technologies Department, University of Milano, Italy
andrea.tettamanzi@unimi.it

Abstract. We present quantitative models for the selection pressure of cellular evolutionary algorithms structured in two dimensional regular lattices. We derive models based on probabilistic difference equations for synchronous and several asynchronous cell update policies. Theoretical results are in agreement with experimental values and show that the selection intensity can be controlled by using different update methods.

1 Introduction

Cellular evolutionary algorithms (cEAs) use populations that are structured according to a lattice topology. The structure may be an arbitrary graph, but more commonly it is a one-dimensional or two-dimensional grid. This kind of evolutionary algorithm has become popular because it is easy to implement on parallel hardware. However, what really matters is the model, not its implementation. Thus, in this work we will focus on cEA models and on their properties without worrying about implementation issues.

Several results have appeared on selection pressure and convergence speed in cEAs. Sarma and De Jong performed empirical analyses of the dynamical behavior of cellular genetic algorithms (cGAs) [8,9], focusing on the effect that the local selection method, the neighborhood size, and neighborhood shape have on the global induced selection pressure. Rudolph and Sprave [7] have shown how cGAs can be modeled by a probabilistic automata network and have provided proofs of complete convergence to a global optimum based on Markov chain analysis for a model including a fitness threshold. Recently, Giacobini *et al.* [2] have successfully modeled the selection pressure curves in cEAs on one-dimensional ring structures, and a preliminary study of two-dimensional, torus-shaped grids has appeared in [1].

Our purpose here is to investigate in detail selection pressure in two-dimensional population structures for two kinds of dynamical systems: synchronous and asynchronous. For that purpose, we model the experimentally observed takeover-time curves with simple difference equations describing the propagation of the best individual under probabilistic conditions.

K. Deb et al. (Eds.): GECCO 2004, LNCS 3102, pp. 1138–1149, 2004.

The paper proceeds as follows. In section 2 we briefly describe synchronous and asynchronous cEAs. Section 3 introduces the concept of takeover time. In sections 4 and 5 we describe our mathematical models for synchronous and asynchronous updates. Theoretical predictions are compared with experimental results in section 6, and section 7 gives our conclusions.

2   Synchronous and Asynchronous cEAs

We consider cEAs defined on a square lattice of finite size $n \times n$. Let us call S the (finite) set of states that a cell (individual) can take up: this is the set of points in the (discrete) search space of the problem. The set N_i is the set of neighbors of a given cell i, and let $|N_i| = N$ be its size. The local transition function $\phi(\cdot)$ can then be defined as:

$$\phi : S^N \to S$$

which maps the state $s_i \in S$ of a given cell i into another state from S, as a function of the states of the N cells in the neighborhood N_i. The neighborhood we consider in this paper is the so-called von Neumann neighborhood, also called linear5, which is constituted by a central cell and the four first neighbor cells in the directions north, east, south, and west, and $|N_i| = 5$. Thus, the implicit form of the stochastic transition function $\phi(\cdot)$ is:

$$\phi(\cdot) = P\{x_i(t+1) \mid x_j(t) \in N_i\}$$

where P is the conditional probability that cell x_i will assume at the next time step $t + 1$ a certain value from the set S, given the current (time t) values of the states of all the cells in the neighborhood. We are thus dealing with probabilistic automata, and the set S should be seen as a set of values of a random variable. The probability P will be a function of the particular selection and variation methods; that is, it will depend on the genetic operators. In this paper we model cEAs using two particular selection methods: binary tournament and linear ranking, but the same framework could easily be extended to other selection strategies.

A cEA starts with the cells in a random state and proceeds by successively updating them using evolutionary operators, until a termination condition is met. Updating a cell in a cellular EA means selecting two parents in the individual's neighborhood, applying genetic operators to them, and finally replacing the individual if an offspring has a better fitness (different replacement policies can be used). Cells can be updated *synchronously* or *asynchronously*. In the synchronous case all the cells change their states simultaneously, while in the asynchronous case cells are updated one at a time in some order. There are many ways for sequentially updating the cells of a cEA. We consider four commonly used asynchronous update methods [10]:

- In *fixed line sweep* (LS), the n cells are updated sequentially from left to right and line after line starting from the upper left corner cell.

- In *fixed random sweep* (FRS), the next cell to be updated is chosen with uniform probability without replacement; this will produce a certain update sequence $(c_1^j, c_2^k, \ldots, c_n^m)$, where c_q^p means that cell number p is updated at time q and $(j, k, \ldots, m)$ is a permutation of the n cells. The same permutation is then used for all update cycles.
- The *new random sweep* method (NRS) works like FRS, except that a new random cell permutation is used for each sweep through the array.
- In *uniform choice* (UC), the next cell to be updated is chosen at random with uniform probability and with replacement. This corresponds to a binomial distribution for the updating probability.

A *time step* is defined as updating n times sequentially, which corresponds to updating *all* the n cells in the grid for LS, FRS and NRS, and possibly less than n different cells in the uniform choice method, since some cells might be updated more than once.

3   Takeover Time

The *takeover time* is defined as being the time it takes for a single best individual to take over the entire population. It can be estimated experimentally by measuring the propagation of the proportion of the best individual under the effect of selection only, without any variation operator. Shorter takeover times indicate a higher selection pressure, and thus a more exploitative algorithm. By lowering the selection intensity the algorithm becomes more explorative. Theoretical takeover times have been derived by Deb and Goldberg [3] for panmictic populations and for the standard selection methods. These times turn out to be logarithmic in the population size, except in the case of proportional selection, which is a factor of n slower, where n is the population size.

It has been empirically shown in [8] that as we move from a panmictic to a square grid population of the same size with synchronous updating of the cells, the selection pressure induced on the entire population is weaker.

A study on the selection pressure in the case of ring and array topologies in one dimensional cEAs has been done by Rudolph [6]. Abstracting from specific selection methods, he splits the selection procedure into two stages: in the first stage an individual is chosen in the neighborhood of each individual, and then, in the second stage, for each individual it is decided whether the previously chosen individual will replace it in the next time step. Using only replacement methods in which extinction of the best by chance cannot happen, i.e. non-extinctive selection, Rudolph derives the expected takeover times for the two topologies as a function of the population size and the probability that in the selection step the individual with the best fitness is selected in the neighborhood. This study has been followed by Giacobini *et al.* investigation of the asynchronous cases for the ring topology [2].

In the present paper we study in detail the two-dimensional case for both the synchronous and the asynchronous cell update mode. In the next section we introduce quantitative models for the growth of the best individual in the form of difference stochastic equations.

4   Models

Let us consider the random variables $V_i(t) \in \{0,1\}$ indicating the presence in cell $i \, (1 \leq i \leq n)$ of a copy of the best individual $(V_i(t) = 1)$ or of a worse one $(V_i(t) = 0)$ at time step t, where n is the the population size. The random variable

$$N(t) = \sum_{i=1}^{n} V_i(t)$$

denotes the number of copies of the best individual in the population at time step t. Initially $V_i(1) = 1$ for some individual i, and $V_j(1) = 0$ for all $j \neq i$.

Following Rudolph's definition [6], if the selection mechanism is non-extinctive, the expectation $E[T]$ with $T = \min\{t \geq 1 : N(t) = n\}$ is called the takeover time of the selection method. In the case of spatially structured populations the quantity $E_i[T]$, denoting the takeover time if cell i contains the best individual at time step 1, is termed the takeover time with initial cell i. Assuming a uniformly distributed emergence of the best individual among all cells, the takeover time is therefore given by

$$E[T] = \frac{1}{n} \sum_{i=1}^{n} E_i[T]$$

In the following sections we give the recurrences describing the growth of the random variable $N(t)$ in a cEA with torus topology for the synchronous and the four asynchronous update policies described in Section 2. We consider a non-extinctive selection mechanism that selects the best individual in a given neighborhood with probability $p \in (0,1)$.

5   Torus Structure

Sarma and De Jong [8] proposed a simple quantitative model for the study of the selection pressure curves for cEAs. They assumed that the diffusion of the best individual in the artificial evolution of a torus-structured population would follow a logistic curve. As suggested by Gorges-Schleuter in [4], in the artificial evolution of locally interacting, spatially structured populations, the assumption of a logistic growth doesn't hold anymore, if the local neighborhood is small enough. In fact, for a torus structure we have a quadratic growth. We complete here her analysis which holds for deterministic unrestricted growth, extending it to finite-size synchronously and asynchronously updated spatial populations using probabilistic selection.

As derived in [1], for a structured population let us consider the limiting case, which represents an upper bound on growth rate, in which the selection mechanism is deterministic, and a cell always chooses its best neighbor for updating.

In the case of a population of size n disposed on a torus grid of size $\sqrt{n} \times \sqrt{n}$ (assuming $\sqrt{n}$ odd) and the von Neumann neighborhood structure, the number of copies of the best individual can be described by the following recurrence:

$$\begin{cases} N(0) = 1 \\ N(t) = N(t-1) + 4t & , \quad for\ 0 \le t \le \frac{\sqrt{n}-1}{2} \\ N(t) = N(t-1) + 4(\sqrt{n} - t) , & for\ t > \frac{\sqrt{n}-1}{2} \end{cases}$$

This growth is described by a convex quadratic equation followed by a concave one, as the two closed forms of the recurrence clearly show:

$$\begin{cases} N(t) = 2t^2 + 2t + 1 & , \quad for\ 0 \le t \le \frac{\sqrt{n}-1}{2} \\ N(t) = -2t^2 + 2(2\sqrt{n} - 1)t + 2\sqrt{n} - n , & for\ t > \frac{\sqrt{n}-1}{2} \end{cases}$$

The described case of a deterministic growth of the number of copies of the best individual is shown in figure 1 in the case of a population of 81 individuals disposed on a 9×9 torus structure.

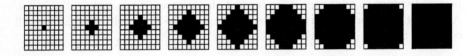

Fig. 1. Example of a deterministic growth of $N(t)$ for a population of 81 individuals on a 9×9 torus structure

Thus, a more accurate model should take into account a non-exponential quadratic growth followed by a quadratic saturation (crowding effect).

In the following sub-sections we will present models for the synchronous and the four asynchronous updates. To keep the models mathematically simple and understandable, some approximations have been made. We will see that the resulting recurrences still fit the experimental curves quite well.

In the limiting case the time t in the recurrences determines the measure of the half diagonal of the 45 degrees rotated square (see figure 1) containing the $N(t)$ copies of the best individual. Since we want to model probabilistic selection mechanisms, we can approximate the measures of the side s and the half diagonal d of the 45 degrees rotated square in the following way:

$$s = \sqrt{N(t)}, \quad d = \frac{\sqrt{N(t)}}{\sqrt{2}}$$

5.1   Synchronous Takeover Time

Let us consider the growth of such a region with a selection mechanism of probabilities p_1, p_2, p_3, p_4 and p_5 of selecting the best individual when there are respectively 1, 2, 3, 4 and 5 copies of it in the neighborhood.

Assuming that the region containing the copies of the best individual expands keeping the shape of a 45 degrees rotated square, we can model the growth of $N(t)$ with the following recurrence:

$$\begin{cases} N(0) = 1 \\ N(t) = N(t-1) + 4p_2 \dfrac{\sqrt{N(t-1)}}{\sqrt{2}} , & for\ N(t) \le \frac{n}{2} \\ N(t) = N(t-1) + 4p_2 \sqrt{n - N(t-1)} , & for\ N(t) > \frac{n}{2} \end{cases}$$

5.2   Asynchronous Fixed Line Sweep Takeover Time

This update method, that is meaningful in a ring topology, in the case of a toroidal topology can be criticized. In fact, there is no biological parallel for this update mechanism. A precise model for such update would be very complicated, since it is difficult to approximate the shape of the region containing the copies of the best individual. We have therefore decided, to keep the model simple and understandable, to roughly approximate the shape of the region with a square stretched to the south-east direction, growing with probability p_1 on the north-east side, p_2 on the south-east side, and p_1 in the south direction.

Let us suppose that in any line the cells containing a copy of the best individual at time step t have index r to s. In the next time step, the cell $r - 1$ will contain a copy of the best individual with probability p, while the cells $s + j$ (with $j = 1, \ldots, n - s$) will contain a copy of the best individual with probability p^j. The number of copies of the best individual in the considered line in the next time step is

$$p + \sum_{i=1}^{\sqrt{n} - j} p^i$$

For large n we can approximate this quantity by the limit $(2p - p^2)/(1 - p)$. Therefore, we can model the growth of $N(t)$ with the following recurrence:

$$\begin{cases} N(0) = 1 \\ N(t) = N(t-1) + \left(\dfrac{2p_2 - p_2^2}{1 - p_2} + 2\dfrac{2p_1 - p_1^2}{1 - p_1} \right) \sqrt{N(t-1)} , & for\ N(t) \le \frac{n}{2} \\ N(t) = N(t-1) + \left(\dfrac{2p_2 - p_2^2}{1 - p_2} + 2\dfrac{2p_1 - p_1^2}{1 - p_1} \right) \sqrt{n - N(t-1)} , & for\ N(t) > \frac{n}{2} \end{cases}$$

5.3   Asynchronous Fixed and New Random Sweep Takeover Time

The behaviors of fixed random sweep and new random sweep averaged over all possible permutations of grid individuals are equivalent. We therefore give only one model describing the growth of the random variable $N(t)$ for both policies.

In a time step the probability of one individual on the border of the region being taken over by the best is p_2, while an individual at distance 2 from the region can be replaced by the best if one or two of its neighbors have already been replaced during the sweep. One of its neighbors is replaced if

- only one neighbor comes before in the sweep (and it has been replaced)
- two neighbors come before in the sweep but just one has been replaced

Two of its neighbors are replaced if both come before in the sweep and both have been replaced. The average probability of an individual of being before another in a sweep is 1/2, therefore an individual at distance 2 from the region is replaced with probability

$$2\left(\frac{1}{2}\left(1-\frac{1}{2}\right)p_2p_1\right) + 2\left(\frac{1}{2}\frac{1}{2}p_2(1-p_2)p_2\right) + \frac{1}{2}\frac{1}{2}p_2^2p_2 = p_2p_1 + \frac{1}{4}(p_2 - 2p_1)p_2^2$$

At distance 3 or more the same reasoning can be done, but we have decided to model the growth up to distance 2 because, as it can been seen in figure 2, the probability at distances ≥ 3 become very small.

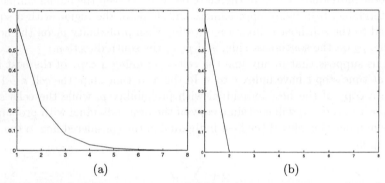

Fig. 2. Probability of an individual being replaced by a copy of the best individual (y axis) with respect to distance (x axis) from the region formed by copies of the best for asynchronous (a) Fixed Random Sweep and (b) Uniform Choice.

Thus, we can model the growth of $N(t)$ with the following recurrence:

$$\begin{cases} N(0) = 1 \\ N(t) = N(t-1) + 4\left(p_2p_1 + \frac{1}{4}(p_2 - 2p_1)p_2^2\right)\left(\sqrt{N(t-1)} - 1\right) + 4p_1, \\ \qquad\qquad\qquad\qquad\qquad\qquad\qquad\qquad\qquad for\ N(t) \leq \frac{n}{2} \\ N(t) = N(t-1) + 4\left(p_2p_1 + \frac{1}{4}(p_2 - 2p_1)p_2^2\right)\left(\sqrt{n - N(t-1)} - 1\right) + 8p_3, \\ \qquad\qquad\qquad\qquad\qquad\qquad\qquad\qquad\qquad for\ N(t) > \frac{n}{2} \end{cases}$$

5.4   Asynchronous Uniform Choice Takeover Time

The ways in which an individual can be replaced in a time step for this update case are the same as for fixed and new random sweep (see above). In the present case, the average probability of an individual coming before another in a time step is $1/n$, therefore an individual at distance 2 from the region is replaced with probability

$$\frac{1}{n}p_2p_1 + \frac{1}{n^2}(p_2 - 2p_1)p_2^2$$

The probability is already very small at distance 2 (see figure 2). Thus, in our model we only take into account individuals at distance 1 from the region.

In terms of time steps, the growth of $N(t)$ can be modeled with the following recurrence:

$$\begin{cases} N(0) = 1 \\ N(t) = N(t-1) + 4p_2\sqrt{N(t-1)} & , \quad for\ N(t) \leq \frac{n}{2} \\ N(t) = N(t-1) + 4p_2(\sqrt{n - N(t-1)} - 1) + 8p_3 , & for\ N(t) > \frac{n}{2} \end{cases}$$

6   Empirical Results

Since cEAs are good candidates for using selection methods that are easily extensible to small local pools, we use binary tournament and linear ranking in our experiments. Fitness-proportionate selection could also be used but it suffers from stochastic errors in small populations, and it is more difficult to model since it requires knowledge of the fitness distribution. The cEA structure has torus topology of size 32×32 with von Neumann neighborhood. Only the selection operator is active: for each cell it selects one individual in the cell neighborhood, and the selected individual replaces the old individual only if it has a better fitness.

6.1   Binary Tournament Selection

We have used the binary tournament selection mechanism described by Rudolph [6]: two individuals are randomly chosen with replacement in the neighborhood of a given cell, and the one with the better fitness is selected for the replacement phase.

Figure 3 shows the growth curves of the best individual for the panmictic, the synchronous and three asynchronous update methods. In all cases the same set of parameters has been used. The mean curves for the two asynchronous methods, fixed and new random sweep, show a very similar behavior, so we have decided to plot only the new random sweep results. The graph shows that the asynchronous update methods give an emergent selection pressure greater than that of the synchronous case, growing from the uniform choice to the line sweep, with the fixed random sweep in between.

The numerical values of the mean takeover times for the five update methods, together with their standard deviations are shown in Table 1, where it can be seen that the fixed random sweep and new random sweep methods give results that are statistically indistinguishable.

Since we use a von Neumann neighborhood, the probabilities p_1, p_2 and p_3 of selecting the best individual when there are respectively 1, 2 and 3 copies

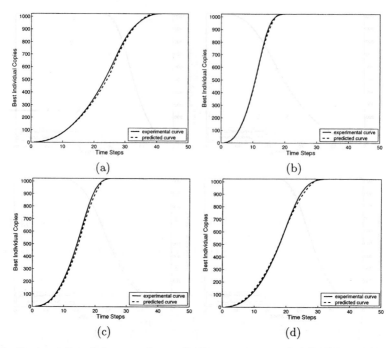

Fig. 5. Comparison of the experimental takeover time curves (full) with the model (dashed) in the case of linear ranking selection for four update methods: synchronous (a), asynchronous line sweep (b), asynchronous fixed random sweep (c), uniform choice (d).

7   Conclusions and Future Work

We have presented quantitative models describing the growth of a single best individual in cellular evolutionary algorithms structured as a torus with von Neumann neighborhood. New results have been obtained for synchronous and some asynchronous cell update policies. The models are given as probabilistic recurrence equations. We have studied two types of selection mechanisms that are commonly used in cEAs: binary tournament and linear ranking. With these selection methods, our results show that there is a good agreement between theory and experiment; in particular, we confirmed that asynchronous cell update methods give rise to different global selection intensity. This should allow the control of selection pressure in an easy and principled way, without using *ad hoc* parameters.

In the future, we intend to extend this type of analysis to larger neighborhoods, and to more complex topologies such as general graph structures, including random graphs. Moreover, we intend to investigate Markov chain modeling of our system and the relationships that may exist with probabilistic particle systems such as voter models [5].

Acknowledgements. Marco Tomassini and Mario Giacobini gratefully acknowledge financial support by the Fonds National Suisse pour la Recherche Scientifique under contract 200021-103732/1.

References

1. M. Giacobini, E. Alba, and M. Tomassini. Selection intensity in asynchronous cellular evolutionary algorithms. In E. Cantú-Paz et al., editor, *Proceedings of the genetic and evolutionary computation conference GECCO'03*, pages 955–966. Springer Verlag, Berlin, 2003.
2. M. Giacobini, M. Tomassini, and A. Tettamanzi. Modelling selection intensity for linear cellular evolutionary algorithms. In P. Liardet et al., editor, *Proceedings of the Sixth International Conference on Artificial Evolution, Evolution Artificielle 2003*. Springer Verlag, Berlin, 2003. To appear.
3. D. E. Goldberg and K. Deb. A comparative analysis of selection schemes used in genetic algorithms. In G. J. E. Rawlins, editor, *Foundations of Genetic Algorithms*, pages 69–93. Morgan Kaufmann, 1991.
4. M. Gorges-Schleuter. An analysis of local selection in evolution strategies. In W. Banzhaf, J. Daida, A. E. Eiben, M. Garzon, V. Honavar, M. Jakiela, and R. Smith, editors, *Genetic and evolutionary conference, GECCO99*, volume 1, pages 847–854. Morgan Kaufmann, San Francisco, CA, 1999.
5. H. Mühlenbein and R. Höns. Stochastic analysis of cellular automata with application to the voter model. *Advances in Complex Systems*, 5(2 & 3):301–337, 2002.
6. G. Rudolph. On takeover times in spatially structured populations: Array and ring. In K. K. Lai et al., editor, *Proceedings of the Second Asia-Pacific Conference on Genetic Algorithms and Applications*, pages 144–151. Global-Link Publishing Company, 2000.
7. G. Rudolph and J. Sprave. A cellular genetic algorithm with self-adjusting acceptance thereshold. In *First IEE/IEEE International Conference on Genetic Algorithms in Engineering Systems: Innovations and Applications*, pages 365–372, London, 1995. IEE.
8. J. Sarma and K. A. De Jong. An analysis of the effect of the neighborhood size and shape on local selection algorithms. In H. M. Voigt, W. Ebeling, I. Rechenberg, and H. P. Schwefel, editors, *Parallel Problem Solving from Nature (PPSN IV)*, volume 1141 of *Lecture Notes in Computer Science*, pages 236–244. Springer-Verlag, Heidelberg, 1996.
9. J. Sarma and K. A. De Jong. An analysis of local selection algorithms in a spatially structured evolutionary algorithm. In T. Bäck, editor, *Proceedings of the Seventh International Conference on Genetic Algorithms*, pages 181–186. Morgan Kaufmann, 1997.
10. B. Schönfisch and A. de Roos. Synchronous and asynchronous updating in cellular automata. *BioSystems*, 51:123–143, 1999.

Evolution of Fuzzy Rule Based Classifiers

Jonatan Gomez

Universidad Nacional de Colombia and The University of Memphis
jgomezpe@unal.edu.co, jgomez@memphis.edu

Abstract. The paper presents an evolutionary approach for generating fuzzy rule based classifier. First, a classification problem is divided into several two-class problems following a fuzzy unordered class binarization scheme; next, a fuzzy rule is evolved (not only the condition but the fuzzy sets are evolved (tuned) too) for each two-class problem using a Michigan iterative learning approach; finally, the evolved fuzzy rules are integrated using the fuzzy round robin class binarization scheme. In particular, heaps encoding scheme is used for evolving the fuzzy rules along with a set of special genetic operators (variable length crossover, gene addition and gene deletion). Experiments are conducted with different public available data sets.

Keywords: Fuzzy Rule Evolution, Fuzzy Set Tuning, Evolutionary Algorithm, Fuzzy Class Binarization

1   Introduction

Classification is a supervised learning technique that takes labeled data samples and generates a model (classifier) that classifies new data samples in different predefined groups or classes [1]. Classification has been extensively studied in machine learning and data mining [1,2,3], and has received particular attention of soft-computing techniques such as fuzzy Logic [4,5,6] and evolutionary algorithms [7,8,9]. Due to high interpretability of fuzzy rule based classifiers (**FRBC**) and the ability of evolutionary algorithms (**EA**) to find good solutions, some research work has focused on developing evolutionary techniques for generating FRBC [10,11,7,12,6,9]. These techniques receive the name of Genetic Fuzzy Rule Based Systems (**GFRBS**) [13,14]. Of several GFRBS approaches proposed, all of them differ from each other in at least one of the following aspects: number of fuzzy rules that each individual encodes, type of rule expression encoded by an individual, and scope of the evolutionary process [10,15,13].

1.1   Michigan and Pittsburgh

According to the number of crisp/fuzzy rules that each individual of the population encodes, GFRBS can be divided in two broad approaches: Pittsburgh and Michigan. Each one has its advantages and disadvantages [15]. In the Pittsburgh approach, each individual of the population encodes a set of rules that will compose the RBC [16,17,18]. It is possible to capture the rules interaction

K. Deb et al. (Eds.): GECCO 2004, LNCS 3102, pp. 1150–1161, 2004.

in the fitness function but the search space grows exponentially with respect to the number of rules encoded. In the Michigan approach, each individual of the population encodes one rule [19,20]. Although the search space is small compared with the Pittsburgh approach, only one rule is encoded, it is not possible to capture the rule interactions in the fitness function [15]. Michigan approaches are further divided in two groups: simple and iterative learning. In the simple approach, all the rules are evolved using a single EA run. It is done by introducing a niching strategy in the EA [21,22,23]. In the iterative learning approach, the set of rules is evolved in several runs of an EA - a rule is evolved in each run [12,7,6]. The number of EA runs, the type of rule evolved and the mechanism used for combining such rules depends on the particular approach [15]. Some approaches penalize rules evolved in previous iterations and stop the iterative process when the set of rules is adequate [12,6,13,24]. Other approaches run an EA as many times as the number of classes the problem has, each run with a different target class [25].

1.2 Rule Encoding

Several fuzzy rule encoding mechanisms have been proposed for GFRBS. Some of them are:

- **Conjunctions of simple terms**. The condition length is fixed. It is composed by atomic expressions connected with a fuzzy *and* logic operator. Such atomic expressions are the only elements evolved [11,7,12,6,9].
- **Fixed condition structure**. The condition is determined by a template where the logic operators and the tree structure are fixed. The atomic conditions are the only elements evolved [8].
- **Linear-tree representation with precedence of operators**. The tree structure of the condition is determined by priorities associated with each logic operator in the condition. Atomic expressions, logic operators and operator priorities are all evolved [25].
- **Heaps or complete binary tree structures**. The tree structure of the condition is always a heap (binary trees filled by levels from left to right). Atomic expressions and logic operators are evolved [10].

1.3 Evolution Scope

It is possible to use an EA for evolving (tuning) the fuzzy sets membership functions at the same time the fuzzy rule is evolved [7,26,14]. In [27], Murata proposes a binary encoding for evolving fuzzy sets in such a way that a bit on '1' indicates when a fuzzy set membership has the value 1.0. The closest bits in on, at the left and right of such bit, indicate that the fuzzy set membership has value 0.0. In this way, fuzzy sets are encoded along with the fuzzy rule being evolved. Karr proposed a mechanism for evolving triangular fuzzy sets in [28]. Karr encoded into the chromosome the two control points that define the base of the triangular fuzzy set. The highest point, the point that will take the maximum fitness value of 1.0, is defined as the middle point between the evolved control points. Each control point is encoded independently using a set of m bits.

This paper provides a fuzzy set tuning mechanism to the heap encoding scheme proposed by Gomez et al. in [10], simplifies the fitness function, and uses a fuzzy unordered class binarization scheme to divide a multi-class problem into several two class problems. This paper is divided in four sections. Section 2 describes the proposed approach: fuzzy rule encoding, fuzzy set tuning mechanism, fitness function and class binarization. Section 3 presents the experiments performed and the analysis of results. Section 4 draws some conclusions.

2   Proposed Approach

In order to evolve a FRBC, we developed a fuzzy round robin binarization technique and used an evolutionary algorithm (EA) for evolving a fuzzy rule for each two-class classification problem (Michigan approach). The EA takes as input the training data set (preprocessed to represent only two classes), applies the evolutionary strategies, and returns one fuzzy rule. Such a rule has the form : R: **IF** *condition* **THEN** data is *positive*. A data sample is classified as positive with the truth-value (**TV**) of the fuzzy rule R and classified as negative with TV equal to the fuzzy negation of the TV of R.

2.1   Fuzzy Unordered Class Binarization

An unordered class binarization transforms an m-class problem into m two-class problems, where the i-th classifier that is generated using the samples of class i as positive samples and samples of the other classes $(j=1..m, j\neq i)$ as negative samples [29]. Each classifier is generated using only samples of the two corresponding classes. Algorithm 1 presents a fuzzy version of unordered class binarization. This binarization scheme has been successfully applied with a maximum defuzzyfication technique in [10,30,25].

Algorithm 1 Fuzzy Unordered Classification

CLASSIFY($classifier[1..m]$, *sample*)
1. $winners = \emptyset$
2. **for** $i = 1$ **to** m **do**
3. $\quad winners = winners \cup \{ \mu_{positive} (classifier_i , sample) \}$
4. **return** DEFUZZY(*winners*)

2.2   Fuzzy Rule Encoding

Because an evolutionary algorithm is executed for each two class problem, and a single fuzzy rule is the classifier associated to it, it is not necessary to encode the class that the fuzzy rule is discriminating (it is always the positive class). Only the fuzzy expression that corresponds with the condition part of the fuzzy rule is encoded. In this paper, we extend the heap encoding scheme proposed by

Gomez et al. in [10] by allowing the EA to evolve the fuzzy sets associated with the atomic conditions. A heap tree is a binary tree that is filled completely on all the levels except possibly the last level that is filled from left to right [31], see figure 1.

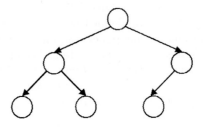

Fig. 1. Heap.

Gomez et al. [10] shown that it is possible to use a linear structure (individual chromosome) for representing such heap expression trees. This structure is defined as a list of genes, each gene encoding an atomic expression (defined as *fuzzy_variable [is/not] fuzzy_set*) and a logic operator (*and* or *or*). The logic operator encoded in the last gene is not taken into account because the number of logic operators is one less than the number of atomic expressions, see Figure 2.

$Gene_1$		...	$Gene_{n-1}$		$Gene_n$	
$Atom_1$	Op_1	...	$Atom_{n-1}$	Op_{n-1}	$Atom_n$	*
var $\in$ / $\notin$ set $\wedge/\vee$		...				*

Fig. 2. Linear Representation of heaps.

Given a chromosome with n genes, $A = a_1 a_2 .. a_n$, the encoded heap expression can be obtained using 1.

$$Tr(a_1 a_2 .. a_n) = \begin{cases} [\lambda,\, atomic\,(a_1),\, \lambda] & if\ n = 1 \\ rep\,(Tr\,(a_1 a_2 .. a_{n-1}),\, atomic\,(a_n),\, oper\,(a_n)) & other\ case \end{cases}$$

(1)

Here, $rep\,(T,\, A,\, O)$ replaces the first leaf node of T (using the level tree enumeration [31]), with the node $[first_{leaf}\,(T),\, O,\, A]$.

We use $\lceil \log_2(m) \rceil$ bits for encoding m possible attributes, one bit for the membership relation ($\in$ / $\notin$) and one bit for the logic operator ($\wedge/\vee$). The number of bits used for representing the fuzzy set depends on the scope of the evolutionary process; whether it is a fixed fuzzy space or fuzzy set tuning.

2.3   Fuzzy Set Tuning

Instead of encoding the index of a predefined fuzzy set into each gene as proposed by Gomez et al. in [10] where $\lceil \log_2 (m) \rceil$ bits are used for representing m predefined fuzzy sets, a set of parameters defining a fuzzy set can be encoded into each gene and allow the EA to tune it. Isosceles triangular fuzzy sets can be tuned by encoding two values, the points defining the base of the triangle [28]. Gaussian shaped fuzzy sets can be defined by encoding two values, the median and the standard deviation. In this paper, the fuzzy set tuning is restricted to trapezoidal fuzzy sets defined by two parameters. The attribute space is divided into m regions of the same length (m is a parameter given by the user). This division generates $m+1$ control points, see Figure 3.a. Given two control points, x and y ($x \le y$), it is possible to define the trapezoidal fuzzy set: ($max \left\{0, \frac{x-1}{m}\right\}$, $\frac{x}{m}$, $\frac{y}{m}$, $min \left\{1, \frac{y+1}{m}\right\}$), see Figure 3.b[1].

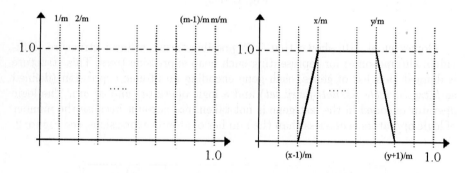

(a) Division of the space in m intervals   (b) Two control points trapezoid

Fig. 3. Tuning of Trapezoidal membership functions.

Therefore, $2 \lceil \log_2 (m + 1) \rceil$ bits are used for representing the two control points of the trapezoidal fuzzy set.

2.4   Genetic Operators

Variable Length Simple Point Crossover (VLSPX). Given two chromosomes $A = a_1 a_2 .. a_n$ and $B = b_1 b_2 .. b_m$, n and m are the size in bits of A and B respectively, the VLSPX selects a random point k in the interval $[2, min \{n, m\} - 1]$, and generates two offspring $C = a_1 a_2 .. a_k b_{k+1} .. b_m$ and $D = b_1 b_2 .. b_k a_{k+1} .. a_n$. When PFEs are encoded, it is possible that VLSPX does not only exchange genes but modifies one of them (if the crossover point is selected in the middle of such gene).

---

[1] It will define a triangular fuzzy set instead of a trapezoidal if $x = y$.

Single Bit Mutation (SBM). Given a chromosome $A = a_1 a_2 .. a_k .. a_n$, the SBM produces an offspring by flipping one random bit in it, $C = a_1 a_2 .. \overline{a_k} .. a_n$.

Gene Addition (ADD). Given a chromosome $A = a_1 a_2 .. a_n$ of n genes, the ADD operator produces an offspring by generating a random gene r and appending it to the end of the chromosome: $C = a_1 a_2 .. a_n r$.

Gene Deletion (DEL). Given a chromosome $A = a_1 a_2 .. a_n$ of $n \geq 2$ genes, the DEL operator produces an offspring by removing the last of the chromosome, $C = a_1 a_2 .. a_{n-1}$.

2.5   Fitness Function

The concept of fuzzy confusion matrix was introduced in [30] in order to determine the performance of an individual. This section proposes a simplification of such fitness function that has shown good performance in the classification problem.

Fuzzy Confusion Matrix. Values in a confusion matrix correspond with the cardinality of intersection sets. For example, PP is the number of positive samples that are classified (predicted) as positive, i.e., the cardinality of the set: Actual-Positive $\cap$ Predicted-Positive. These values can be calculated by using the membership function of the data samples to the actual and predicted data sets as:

$$PP = \sum_{i=1}^{n} \mu_A (d_i) \wedge \mu_B (d_i) \tag{2}$$

$$PN = \sum_{i=1}^{n} \mu_C (d_i) \wedge \mu_B (d_i) = \sum_{i=1}^{n} \overline{\mu_A (d_i)} \wedge \mu_B (d_i) \tag{3}$$

$$NP = \sum_{i=1}^{n} \mu_A (d_i) \wedge \mu_D (d_i) = \sum_{i=1}^{n} \mu_A (d_i) \wedge \overline{\mu_B (d_i)} \tag{4}$$

$$NN = \sum_{i=1}^{n} \mu_C (d_i) \wedge \mu_D (d_i) = \sum_{i=1}^{n} \overline{\mu_A (d_i)} \wedge \overline{\mu_B (d_i)} \tag{5}$$

Where, n is the number of samples used to test the classifier, A is the actual (real) positive set, B is the predicted positive set, C is the actual negative set, D is the predicted negative set, and d_i is the i-th data record sample in the data set.

Notice that, for a two-class classification problem, one only needs to know the membership of a data sample to the actual data set and to the predicted membership value of the positive set.

Equations 2, 3, 4 and 5 can be extended to fuzzy sets and fuzzy rules. The degree of membership of the data sample to the actual positive set is given by the data sample label and the predicted membership to the positive set is calculated as the truth-value of the condition part of the fuzzy rule. The confusion matrix generated by using these extensions is called **fuzzy confusion matrix**. Performance metrics like accuracy and true positives can be generated from the fuzzy confusion matrix. Such new performance metrics will be called **fuzzy performance metrics:** (fuzzy accuracy, fuzzy true positives, etc).

Definition. Since the goal of the evolutionary process is to generate a simple fuzzy rule that can discriminate the positive class from the negative, the fitness of an individual is defined by the fuzzy accuracy (**FAC**), and the fuzzy rule length (**FRL**), i.e. the number of atomic conditions defining the fuzzy rule. In this way, the optimization problem is a two-goal objective function: maximizing the FAC while minimizing the fuzzy rule length (FRL). Although there are several ways to deal with multi-goal objective functions, in this work, the weighted sum technique was used. Therefore, the fitness of an individual is calculated using equation 6.

$$
\begin{aligned}
fitness(R) &= w * FAC(R) + (1 - w) * \left(1 - \frac{FRL(R)}{M}\right) \\
&= w * \left(\frac{PP(R)+NN(R)}{PP(R)+PN(R)+NP(R)+NN(R)}\right) + (1 - w) * \left(1 - \frac{FRL(R)}{M}\right)
\end{aligned}
\tag{6}
$$

Here, w is the weight associated with the fuzzy accuracy reached by the individual and M is the maximum number of atomic expressions defining a fuzzy rule.

2.6   Rule Extraction

The best individual of the population, according to the fitness value, will determine the fuzzy rule that will be used for discriminating between the two classes under consideration.

3   Experimentation

3.1   Experimental Settings

Five benchmark data sets (publically available), were used as a test bed. See table 1. A 10-fold cross-validation was applied to each data set 5 different times. The reported results are the average over those 50 runs.

For each data set, we used the Hybrid Adaptive Evolutionary Algorithm (**HaEa**) proposed by Gomez [32] as the two-class evolutionary algorithm (which evolves the fuzzy rule associated with each two-class problem) . HAEA adapts the genetic operator rates while evolves the solution of the problem. HAEA was executed for 100 iterations using 100 individuals as population and VLSPX,

Table 1. Test bed

DATA SET	CLASSES	DIM	SAMPLES	
			TOTAL	PER CLASS
BREAST	2	9	699	{458, 241}
PIMA	2	8	768	{500, 268}
HEART	2	13	270	{150,120}
IRIS	3	4	150	{50, 50, 50}
WINE	3	13	178	{59, 71,4 8}

SBM, ADD and DEL as genetic operators. Individuals of the initial population were randomly generated with a length varying between 1 and the number of attributes defining the data set (9 for *Breast*, 8 for *Pima* and 13 for *Heart*). A 10 fold cross-validation technique was applied to each data set 5 different times. The reported results are the average over those 50 runs. We used the *average-and* $(TV\,(p \wedge q) = \frac{2*TV(p)*TV(q)}{TV(p)+TV(q)})$ as fuzzy *and* operator[2], *max* as fuzzy *or* operator, and $1.0 - x$ as fuzzy *not* operator. We compared the performance of the fuzzy rules evolved using the fuzzy set tuning mechanism (with 6 divisions) against the fuzzy rules evolved using a fixed collection of 5 well tuned fuzzy sets, as shown in figure 4. We set the number of division to 6 in order to match the division generated using the 5 predefined fuzzy sets.

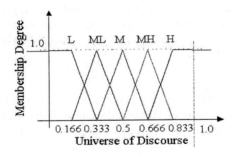

Fig. 4. Fixed Collection of Fuzzy Sets

3.2   Results and Analysis

Comparing fuzzy sets tuning against fixed fuzzy sets. Table 2 shows the performance of the fixed and tuning fuzzy sets approach.

---

[2] The *average-and* fuzzy logic operator produced better results than the *min-and* operator.

Table 2. Performance of fuzzy set tuning against fixed fuzzy sets.

	PRE-DEFINED	TUNING
BREAST	94.94±2.71	**94.85±2.41**
PIMA	73.79±5.75	**74.21±5.63**
HEART	78.44±7.79	**78.96±8.26**
IRIS	94.51±4.83	**95.20±5.97**
WINE	92.78±6.13	**93.18±6.72**

As shown, the performance reached by the proposed approach using tuning of fuzzy sets is better than the performance reached using a predefined collection of fuzzy sets in almost all the data sets (exception done with the *Breast* data set). These results indicate that the tuning mechanism can evolve fuzzy sets that approximate patterns hidden in the data set. Take for example the following fuzzy rule generated for the *Pima* data set in a sample run:

IF x_8 *is* $set_{2,3}$ **AND** x_1 *is* $set_{0,1}$ **OR** x_2 *is not* $set_{0,4}$ **THEN** *Diabetes-Disease*

Here, $set_{x,y}$ represents the trapezoidal fuzzy set:

$$\left(max\left\{ 0, \tfrac{x-1}{m} \right\}, \tfrac{x}{m}, \tfrac{y}{m}, min\left\{ 1, \tfrac{y+1}{m} \right\} \right)$$

with m being the number of divisions. In order to approximate the atomic expression x_8 *is* $set_{2,3}$ using a collection of predefined fuzzy sets, it will require other type of fuzzy or logic operator, like *Restricted-Sum Or*, and the condition x_8 *is* ML **OR** x_8 *is* M.

Fuzzy Rule Complexity. Figure 5 shows the evolution of the fuzzy rule length for both the best individual in the population and the average length of individuals in the population. Clearly, our approach using the tuning mechanism produces simple fuzzy rules as no more than 4 attributes are included in the condition part of the fuzzy rules.

Comparison with Results Reported in the Literature. We took the results produced by our approach, Tuning of fuzzy sets with Heaps encoding (**T-HEAP**), and compared them against results reported in the literature. See table 3^3. As shown, our results (first row) compare well.[4]

---

[3] Results reported for QDA, LDA, C4.5, kNN, SSV and FSM taken from [33]. Results for GAP taken from [27], where the number of fuzzy rules was close to the number of classes. Results for CTree are taken from [10].

[4] Although all of these results were obtained with different statistical validation methods (leave-one-out, or 10-cross-validation) or not statistical validation, the values reported here are an indicative of the performance of the proposed approach.

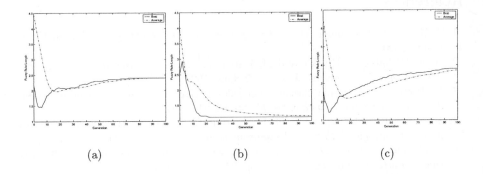

(a)   (b)   (c)

Fig. 5. Fuzzy Rule Length Evolution. (a) BREAST, (b) PIMA, (c) HEART.

Table 3. Comparative performance of the proposed approach

Method	BREAST	PIMA	HEART	WINE	Statistical Test
T-HEAPS	**94.85**	**74.21**	**78.96**	**93.18**	**10-cross-validation**
QDA	94.90	74.80	57.80	99.40	Leave-one-out
LDA	96.00	77.20	60.40	98.90	Leave-one-out
GAP-sel	-	-	-	97.20	None
GAP-par	-	-	-	93.30	None
C4.5	94.70	73.00	22.90	-	Leave-one-out
kNN	96.90	71.90	65.60	95.50	Leave-one-out
SSV	96.30	73.70	-	98.30	10-cross-validation
FSM	96.90	-	-	96.10	10-cross-validation
WM	87.10	71.30	-	-	-
GIL	90.10	73.10	-	-	-
ABD	96.00	75.90	-	-	-
ABA	95.10	74.80	-	-	random (50-50)%

4   Conclusions

We proposed a technique for evolving fuzzy rules that follows an iterative Michigan approach. The iterative process is performed using a fuzzy unordered class binarization scheme. A fuzzy set tuning mechanism was developed for the Heaps encoding strategy proposed by Gomez et al. in [10]. Also, a simplified fitness function was introduced. The results obtained indicate that the proposed approach is able to evolve good fuzzy rule based classifiers. In general, the quality of such evolved classifiers is higher when the fuzzy set tuning scheme was included in the evolutionary process.

References

1. J. Han and M. Kamber, *Data Mining: Concepts and Techniques.* Morgan Kaufmann, 2000.
2. R. S. Michalski, I. Bratko, and M. Kubat, *Machine Learning and Data Mining: Methods and Applications.* J. Wiley & Sons, 1998.
3. R. Holte, "Very simple classification rules perform well in most common used datasets," *Machine Learning*, no. 11, pp. 63–91, 1993.
4. Y.-C. Hu, R.-S. Chen, and T. G-H., "Finding fuzzy classification rules using data mining techniques," *Pattern Recognition Letters*, no. 24, pp. 509–519, 2003.
5. Q. Shen and A. Chouchoulas, "A rough-fuzzy approach for generating classification rules," *Pattern Recognition*, no. 35, pp. 2425–2438, 2002.
6. A. Gonzalez and R. Prez, "Completeness and consistency conditions for learning fuzzy rules," *Fuzzy Sets and Systems*, no. 96, pp. 37–51, 1998.
7. H. Ishibushi and T. Nakashima, "Liguistic rule extraction by genetics-based machine learning," in *Proceedings of the Genetic and Evolutionary Computation Conference GECCO'00*, pp. 195–202, 2000.
8. A. Giordana and L. Saitta, "Regal: An integrated system for learning relations using genetic algorithms," in *Proceedings of the Second International Workshop on Multi-strategy Learning*, pp. 234–249, 1993.
9. K. De Jong and W. Spears, "Learning concept classification rules using genetic algorithms," in *Proceedings of the Twelfth International Joint Conference on Artificial Intelligence*, pp. 651–656, 1991.
10. J. Gomez, D. Dasgupta, O. Nasraoui, and F. Gonzalez, "Complete expression trees for evolving fuzzy classifier systems with genetic algorithms," in *Proceedings of the North American Fuzzy Information Processing Society Conference NAFIPS-FLINTS 2002*, pp. 469–474, 2002.
11. M. V. Fidelis, H. S. Lopes, and A. A. Freitas, "Discovering comprehensible classification rules with a genetic algorithm," in *Proceedings of Congress on Evolutionary Computation (CEC)*, pp. 805–810, 2000.
12. J. Liu and J. Kwok, "An extended genetic rule genetic algorithm," in *Proceedings of Congress on Evolutionary Computation (CEC)*, pp. 458–263, 2000.
13. O. Cordon, A. Gonzalez, F. Herrera, and R. Perez, "Encouraging cooperation in the genetic iterative rule learning approach for quality modeling," in *Computing with Words in Intelligent/Information Systems 2. Applications, J. Kacprzyk, L. Zadeh (Eds.)*, Physica-Verlag, 1998.
14. O. Cordon and F. Herrera, "A general study on genetic fuzzy system," in *Genetic Algorithms in Engineering and Computer Sciences*, pp. 33–57, Jonh Wiley and Sons, 1995.
15. A. A. Freitas, "A survey of evolutionary algorithms for data mining and knowledge discovering," in *Advances in Evolutionary Computation. A. Ghosh and S. Tsutsui. (Eds.)*, Springer-Verlag, 2001.
16. K. De Jong, W. Spears, and D. F. Gordon, "Using genetic algorithms for concept learning," *Machine Learning Research*, no. 13, pp. 161–188, 1993.
17. C. Z. Janikow, "A knowledge-intensive genetic algorithm for supervised learning," *Machine Learning Research*, no. 13, pp. 189–228, 1993.
18. S. F. Smith, *A Learning System based on Genetic Adaptive Algorithms.* Ph. D. Thesis, University of Pittsburgh, 1980.
19. G. Giordana and F. Neri, "Search-intensive concept induction," *Evolutionary Computation*, no. 3(4), pp. 375–416, 1995.

20. L. Booker, *Intelligent Behaviour as an Adaption to the Task Environment*. Ph. D. Thesis, University of Michigan, 1982.
21. S. W. Mahfoud, "Crowding and preselection revisited," in *Proceedings Second Conference Parallel Problem Solving from Nature*, 1992.
22. D. Goldberg and J. J. Richardson, "Genetic algorithms with sharing for multimodal function optimization," in *Proceedings Second International Conference on Genetic Algorithm*, pp. 41–49, 1987.
23. J. H. Holland, *Adaptation in Natural and Artificial Systems*. The University of Michigan Press, 1975.
24. O. Cordon and F. Herrera, "A three-stage evolutionary process for learning descriptive and approximate fuzzy logic controller knowledge bases," *International Journal of Approximate Reasoning*, no. 17(4), pp. 369–407, 1997.
25. D. Dasgupta and F. Gonzalez, "Evolving complex fuzzy classifier rules using a linear tree genetic algorithm," in *Proceedings of the Genetic and Evolutionary Computation Conference GECCO'01*, pp. 299–305, 2001.
26. B. Carse, T. Fogarty, and A. Munro, "Evolving fuzzy rule based controllers using genetic algorithms," *Fuzzy Sets and Systems*, no. 80, pp. 273–294, 1996.
27. T. Murata, S. Kawakami, H. Nozawa, M. Gen, and H. Ishibushi, "Three-objective genetic algorithms for designing compact fuzzy rule-based systems for pattern classification problems," in *Proceedings of the Genetic and Evolutionary Computation Conference GECCO'01*, pp. 485–492, 2001.
28. C. Karr, "Genetic algorithms for fuzzy controllers," *AI Experts*, pp. 26–33, 1991.
29. J. Fürnkranz, "Round robin classification," *Machine Learning Research*, no. 2, pp. 721–747, 2002.
30. J. Gomez and D. Dasgupta, "Evolving fuzzy rules for intrusion detection," in *Proceedings of the Third Annual IEEE Information Assurance Workshop 2002 Conference*, pp. 68–75, 2002.
31. T. Cormer, C. Leiserson, and R. Rivest, *Introduction to Algorithms*. McGraw Hill, 1990.
32. J. Gomez, "Self adaptation of operator rates in evolutionary algorithms," in *Proceedings of the Genetic and Evolutionary Computation Conference (GECCO 2004)*, June 2004.
33. D. Wlodzislaw, "Data sets used for classification: Comparison of results," in *http://www.phys.uni.torun.pl/kmk/projects/datasets.html*.

Self Adaptation of Operator Rates in Evolutionary Algorithms

Jonatan Gomez

Universidad Nacional de Colombia and The University of Memphis
jgomezpe@unal.edu.co, jgomez@memphis.edu

Abstract. This work introduces a new evolutionary algorithm that adapts the operator probabilities (rates) while evolves the solution of the problem. Each individual encodes its genetic rates. In every generation, each individual is modified by only one operator that is selected according to the encoded rates. Such rates are updated according to the performance achieved by the offspring (compared to its parent) and a random learning rate. The proposed approach is augmented with a simple transposition operator and tested on a number of benchmark functions.

1 Introduction

Evolutionary algorithms (**EA**) are optimization techniques based on the principles of natural evolution [1]. Although EAs have been used successfully in solving optimization problems, the performance of this technique (measure in time consumed and solution quality) depends on the selection of the EA parameters. Moreover, the process of setting such parameters is considered a time-consuming task [2]. Several research works have tried to deal with this problem [3]. Some approaches tried to determine the appropriated parameter values by experimenting over a set of well-defined functions [4,5], or by theoretical analysis [6,7,8]. Another set of approaches, called Parameter Adaptation (**PA**), tried to eliminate the parameter setting process by adapting parameters through the algorithm's execution [9,10,11,8,12]. PA techniques can be roughly divided into centralized control techniques (central learning rule), decentralized control techniques, and hybrid control techniques.

In the centralized learning rule approach, genetic operator rates (such as mutation rate, crossover rate, etc) are adapted according to a global learning rule that takes into account the operator productivities through generations (iterations) [3]. Generally, only one operator is applied per generation, and it is selected based on its productivity. The productivity of an operator is measured in terms of good individuals produced by the operator. A good individual is one that improves the fitness measure of the current population. If an operator generates a higher number of good individuals than other operators then its probability is rewarded. Two well known centralized learning rule mechanism are the adaptive mechanism of Davis [9], and Julstrom [10]. In Davis's approach, the operators and parents are stored for each offspring. If an offspring is better than the current best individual in the population, the individual and the genetic op-

K. Deb et al. (Eds.): GECCO 2004, LNCS 3102, pp. 1162–1173, 2004.

erator used for generating it are rewarded. Some portion of the reward is given recursively to the parents, grand parents, and so on. After certain number of iterations the operator probabilities are updated according to the reward reached. In the approach of Julstrom, a similar mechanism to Davis is developed, but only information about recently generated chromosomes is maintained for each individual. Although applied with relative success, a centralized technique has two main disadvantages: First, it requires extra memory for storing information on the effect of each genetic operator applied to an individual, parents, grandparents, etc. The amount of memory required grows exponentially on the number of generations used. For example, if the operator productivity is defined using the latest n generations and a single byte is used per productivity information, the extra memory required will be approximated 2^n bytes per individual. Second, a centralized technique requires an algorithm that uses such information for calculating the operator productivity in a global sense; it cannot be defined at the individual level, only at the population level. Therefore, the time complexity of the operator productivity grows linearly on the number of generations used and is increased by the size of the population.

In decentralized control strategies, genetic operator rates are encoded in the individual and are subject to the evolutionary process [3,13]. Accordingly, genetic operator rates can be encoded as an array of real values in the semi-open interval [0.0,1.0), with the constraint that the sum of these values must be equal to one [11]. Since the operator rates are encoded as real numbers, special genetic operators, meta-operators, are applied to adapt or evolve them. Although decentralized approaches have been applied to many problems with relative success, these approaches have two main disadvantages: First, a set of meta genetic operators used for evolving the probabilities must be defined. It is not easy to determine which operators to use and how these operators will affect the evolution of the solution. Second, rates for such meta operators have to be given. Although these techniques are not very sensitive to the setting of the meta operator rates, not every set of values works well.

In this paper, a "hybrid" technique for parameter adaptation is proposed. Specifically, each individual encodes its own operator rates and uses a randomized version of a learning rule mechanism for updating them. Such a randomized learning rule mechanism is defined locally (per individual) and uses the productivity of the genetic operator applied and a "random" generated learning rate. If a non-unary operator is selected, the additional parents are chosen using a selection strategy. The operator rates are adapted according to the performance achieved by the offspring compared to its parent, and the random learning rate generated. Although the population size is an EA parameter that can be adapted, and each genetic operator has its own parameters that can be adapted, this work is only devoted to the adaptation of genetic operator rates.

This paper is divided in five sections. Section 2 presents the proposed hybrid adaptive evolutionary algorithm. Section 3 reports some experimental results on binary encoding functions. Section 4 reports some results on real encoding optimization problems. Section 5, draws some conclusions.

Algorithm 1 Hybrid Adaptive Evolutionary Algorithm (HAEA)

HAEA(λ, terminationCondition)
1. $t_0 = 0$
2. $P_0 = $ initPopulation(λ) ,
3. **while**(terminationCondition(t, P_t) is false) **do**
4. $P_{t+1} = \{\}$
5.   **for** each ind $\in P_t$ **do**
6.   rates = extract_rates(ind)
7. $\delta = $ random(0,1)   // *learning rate*
8.   oper = OP_SELECT(operators, rates)
9.   parents = PARENTSELECTION(P_t, ind)
10.   offspring = apply(oper, parents)
11.   child = BEST(offspring, ind)
12.   **if**(fitness(child) ¿ fitness(ind)) **then**
13.   rates[oper] = $(1.0 + \delta)$*rates[oper]   //*reward*
14.   **else**
15.   rates[oper] = $(1.0 - \delta)$*rates[oper]   //*punish*
16.   normalize_rates(rates)
17.   set_rates(child, rates)
18. $P_{t+1} = P_{t+1} \cup \{child\}$
19. $t = t + 1$

---

2 Hybrid Adaptive Control

Algorithm 1 presents the proposed Hybrid Adaptive Evolutionary Algorithm (**HaEa**). This algorithm is a mixture of ideas borrowed from Evolutionary Strategies (**ES**), decentralized control adaptation, and central control adaptation.

2.1 Selection Mechanism

In HAEA, each individual is "independently" evolved from the other individuals of the population, as in evolutionary strategies [14]. In each generation, every individual selects only one operator from the set of possible operators (line 8). Such operator is selected according to the operator rates encoded into the individual. When a non-unary operator is applied, additional parents (the individual being evolved is considered a parent) are chosen according to any selection strategy, see line 9. As can be noticed, HAEA does not generate a parent population from which the next generation is totally produced. Among the offspring produced by the genetic operator, only one individual is chosen as child (line 11), and will take the place of its parent in the next population (line 17). In order to be able to preserve good individuals through evolution, HAEA compares the parent individual against the offspring generated by the operator. The BEST selection mechanism will determine the individual (parent or offspring) that has the highest fitness (line 11). Therefore, an individual is preserved through evolution if it is better than all the possible individuals generated by applying the genetic operator.

2.2   Encoding of Genetic Operator Rates

The genetic operator rates are encoded into the individual in the same way as decentralized control adaptation techniques, see figure 1. These probabilities are initialized (into the *initPopulation* method) with values following a uniform distribution $U[0, 1]$. We used a roullete selection scheme for selecting the operator to be applied (line 8). To do this, we normalize the operator rates in such a way that their summation is equal to one (line 16).

SOLUTION	OPER$_1$	...	OPER$_n$
100101011..01	0.3	...	0.1

Fig. 1. Encoding of the operator probabilities in the chromosome

2.3   Adapting the Probabilities

The performance of the child is compared against its parent performance in order to determine the productivity of the operator (lines 12-15). The operator is rewarded if the child is better than the parent and punished if it is worst. The magnitude of reward/punishment is defined by a learning rate that is randomly generated (line 7). Finally, operator rates are recomputed, normalized, and assigned to the individual that will be copied to the next population (lines 16-17). We opted by generating the learning rate in a random fashion instead of setting it to a specific value for two main reasons. First, there is not a clear indication of the correct value that should be given for the learning rate; it can depend on the problem being solved. Second, several experiments encoding the learning rate into the chromosome showed that the behavior of the learning rate can be simulated with a random variable with uniform distribution.

2.4   Properties

Contrary to other adaptation techniques, HAEA does not try to determine and maintain an optimal rate for each operator. Instead, HAEA tries to determine the appropiate operator rate at each instance in time according to the concrete conditions of the individuals. If the optimal solution is reached by an individual in some generation, then the rates of the individual will converge to the same value in subsequent generations. This is true because no operator is able to improve the optimal solution. HAEA uses the same amount of extra information as a decentralized adaptive control; HAEA requires a matrix of $n * M$ doubles, where n is the number of different genetic operators and M is the population size. Thus, the space complexity of HAEA is linear with respect to the number of operators (the population size is considered a constant). Also, the time expended in calculating and normalizing the operator rates is linear with respect to the number of operators $n * M$ (lines 8 and 12-16). HAEA does not require special operators or additional parameter settings. Well known genetic operators can be used without modifications. Different schemes can be used in encoding the solution: binary, real, trees, programs, etc. The average fitness of the population grows monotonically each generation since an individual is replaced by other with equal or higher fitness.

3   Experiments Using Binary Encoding

We conducted experiments using binary encoding in order to determine the applicability of the proposed approach. In the binary encoding, the solution part of the problem is encoded as a string of binary symbols [1].

3.1   Test Functions

We used four well known binary functions: The 100 bits MaxOnes function where the fitness of an individual is defined as the number of its bits that are set to 1, the ten deceptive order-3 and ten bounded deceptive order-4 functions were developed by Golberg in 1989 [15] and the $(8 * 8)$ royal road function developed by Forrest and Mitchell [16].

3.2   Experimental Settings

For each function, a population size of 100 was used and the HAEA algorithm was executed a maximum of 10000 fitness function evaluations (100 generations). A tournament of size 4 was applied to determine the additional parent of crossover. The reported results are the average over 100 different runs. We compared the performance reached by HAEA using different combinations of three well known genetic operators: single point mutation, single point crossover, and a simple transposition. In the single bit mutation, one bit of the solution is randomly selected (with uniform distribution) and flipped. This operator always modifies the genome by changing only one single bit. In single point crossover, a cutting point in the solution is randomly selected. Parents are divided in two parts (left and right) using such cutting point. The left part of one parent is combined with the right part of the other. In the simple transposition operator, two points in the solution are randomly selected and the genes between such points are transposed [17]. Five different combinations of genetic operators were tested: Only mutation (M); mutation and crossover (MX); mutation and transposition (MT); crossover and transposition (XT); and mutation, crossover and transposition (MXT).

3.3   Results

Table 1 shows the performance reached by HAEA on the binary functions. A value $a \pm b\,[c]$ indicates that a maximum average performance of a, with standard deviation of b, was reached by the HAEA using c fitness evaluations in all the runs.

Table 1. Performance reached by HaEa on binary functions using different operators.

	MaxOnes	Royal Road	Deceptive-3	Deceptive-4
M	90.00±1.32 [10000]	19.12±4.37 [10000]	292.34±1.44 [9900]	34.14±0.81 [9900]
MX	100.00±0.00 [5000]	55.12±7.83 [10000]	297.92±1.98 [2400]	38.56±1.04 [9900]
MT	84.56±1.69 [10000]	19.60±4.44 [10000]	291.72±1.44 [9600]	33.64±0.79 [10000]
XT	100.00±0.00 [3800]	64.00±0.00 [4000]	300.00±0.00 [2600]	40.00±0.00 [3100]
MXT	100.00±0.00 [3900]	64.00±0.00 [4900]	300.00±0.00 [3000]	40.00±0.00 [3100]

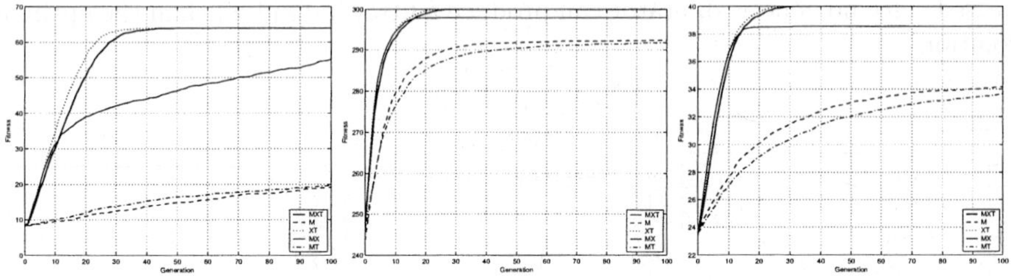

(a)   (b)   (c)

Fig. 2. Performance evolution using HaEa on binary functions. (a) Royal Road, (b) Deceptive-3 and (c) Deceptive-4.

Genetic Operators Analysis. Figure 2 shows the evolution of the fitness value for each variation of HaEa tested.

As expected, the performance of HaEa is highly affected by the set of operators used. While the performance reached by HaEa is low when using only mutation or mutation and transposition (M and MT), the performance is high and the optimal solution is found when all of the operators are used (MXT). These results indicate that HaEa is able to recognize the usefulness of each genetic operator. When crossover and transposition were used (MXT and XT), HaEa was able to find the optimal solution of each function. A possible explanation of this behavior is that transposition exploits the repetitive and symmetric structure definition of these test functions - any of these functions can be seen as the concatenation of small symmetric binary functions. Transposition can modify two of these small functions, the outer limits in the string being transposed, while maintaining other small functions, the inner part being transposed. It is not surprising that performance and convergence rate reached by HaEa when using crossover are higher than without using crossover. When crossover is not included (M and MT), HaEa is not able to find the optimal solution of any of the four functions. These results indicate that HaEa identifies the main role of crossover in the optimization process and allows it to exploit "good" regions of the search space. Figure 3 presents the HaEa(MXT) evolution of both solution and genetic operator rates for the tested functions. As shown, the crossover rate increases rapidly in early stages of the evolutionary process. This means that crossover produced better individuals in early stages than other operators. When, crossover was not enough for locating the optimal solution, its rate decreases allowing HAEA to try mutation and/or transposition. Notice that HAEA is not trying to determine and maintain an optimal probability for each genetic operator chromosome as other adaptation techniques. HAEA has temporal learning of operator rates. This adaptive property allows HAEA to have a higher chance to locate an optimal solution than another evolutionary algorithm. When the optimal solution was reached for all individuals in the population, the genetic operators probabilities converge to the same value. This behavior is expected since no genetic operator can produce a fitter individual than the optimal solution.

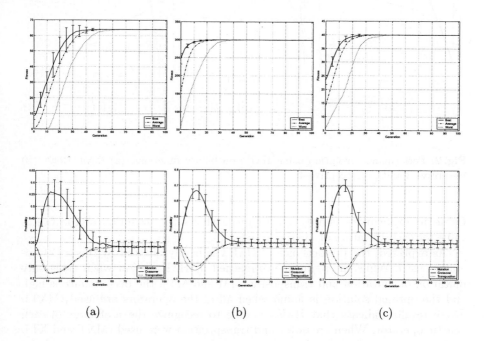

(a)   (b)   (c)

Fig. 3. Performance (first row) and rates evolution (second row) using HAEA(MXT) on binary functions. (a) Royal Road, (b) Deceptive-3 and (c) Deceptive-4.

Comparing HAEA against other Evolutionary Algorithms and Previous Reported Results. Two standard GA were implemented in order to compare the performance of the proposed approach (HAEA): a Generational Genetic Algorithm (GGA) and a Steady State Genetic Algorithm (SSGA). The GGA used a tournament selection of size 4 for selecting each individual of the parent population, a standard bit mutation and single point crossover as genetic operators. The mutation rate of each bit in the chromosome was fixed to $\frac{1}{l}$ where l is the length of the chromosome while the crossover rate was set to 0.7. These parameters have been considered good for solving binary encoding problems [14,2,11,8]. For the SSGA we used a tournament selection of size 4 for selecting the parent in the crossover operator, standard bit mutation and single point crossover as genetic operators, and *kill-worst* replacement policy. The crossover rate was fixed to 1.0 while the mutation rate of each bit was fixed to $\frac{1}{l}$ where l is the length of the chromosome. Table 2 summarizes the average performance reached by HaEa(MX), HaEa(MXT), our implementation of the GGA and SSGA, and compares them against the performance of some evolutionary algorithms reported in the literature.

Table 2. Comparison of HaEa performance on binary functions.

	MaxOnes	Royal Road	Deceptive-3	Deceptive-4
HaEa(MXT)	100.00±0.0[3900]	64.00±0.00[4900]	300.00±0.00[4800]	40.00±0.00[5000]
HaEa(MX)	100.00±0.0[5000]	55.12±7.83[10000]	297.92±1.98[2400]	38.56±1.04[9900]
GGA	100.00±0.0[4800]	49.52±9.24[10000]	293.52±3.01[10000]	37.08±1.29[2900]
SSGA	100.00±0.0[2800]	48.24±9.01[10000]	288.56±3.08[2000]	35.00±1.38[1100]
T-GGA [11]	99.96±0.20[7714]	35.52±6.02[7804]	289.68±2.41[5960]	-
T-SSGA[11]	100.00±0.0[2172]	40.64±7.65[3786]	289.12±3.08[3506]	-
T-D-GGA[11]	99.52±0.64[8438]	31.36±6.16[6086]	289.12±2.83[5306]	-
T-D-SSGA[11]	100.00±0.0[2791]	29.76±8.32[2428]	289.32±2.61[2555]	-
GA [8]	100.00±0.0[2500]	-	-	14.00 [10000]
PL-GA [8]	100.00±0.0[7400]	-	-	28.00 [10000]

Results in rows 5-8 are reported by Tuson and Ross in [11][1]. Results in rows 9 and 10 are reported by Lobo in [8][2]. HaEa(MXT) and HaEa(MX) outperformed other EA approaches in the three hard binary functions: Royal Road, Deceptive-3 and Deceptive-4[3]. In the simple MaxOnes problem, HaEa found the global optimal solution, but it took around twice the number of evaluations required by SSGA based approaches and the regular GA used by Lobo.

---

[1] These results are the average over 50 different runs. Row 5 presents the results obtained by Tuson and Ross using a Generational Genetic Algorithm (GGA) with fixed operator probabilities, row 6 using a Steady State Genetic Algorithm (SSGA), row 7 a GGA using distributed control parameter adaptation and row 8 a SSGA using distributed control. The population size used by Tuson and Ross was fixed to 100 individuals and the maximum number of iterations was set to 10000.

[2] These results are the average over 20 different runs. Row 9 presents the results obtained by Lobo using a regular GA with optimal parameters setting, population size of 100 individuals for the MaxOnes problem and 50 individuals for the deceptive-4 problem, , while row 10 presents the results using the Parameter-less GA proposed by Lobo. Results for the deceptive 4 function were approximated from [8], figure 4.4.

[3] The Parameter-less GA was able to find the solution after 4e+6 evaluations by increasing the population size

4   Experiments Using Real Encoding

We conducted experiments using real encoding in order to determine the applicability of the proposed approach. In the real encoding, the solution part of the problem is encoded as a vector of real values [14].

4.1   Experimental Settings

We used the real functions shown in table 3 as our testbed. In the last three functions, we fixed the dimension of the problem to $n = 10$. HAEA was executed a maximum of 20000 fitness evaluations (2000 iterations) with a population size of 100 individuals. The reported results are the average over 100 different runs. We implemented three different genetic operators: Gaussian mutation, Uniform mutation, Single real point crossover. The Gaussian mutation operator adds to one component of the individual a random number Δ that follows a Gaussian distribution $G(0, \sigma)$. We used $\sigma = \frac{max-min}{100}$, where max and min are the maximum and minumum values that can take each component. The uniform mutation operator replaces one of the component values of the individual with a random number following a uniform distribution $U(min, max)$. The single point real crossover generates two individuals by randomly selecting one component k and exchanging components $k, k + 1, .., n$ between the parent. Four different combinations of operators were used by HAEA: Gaussian and uniform mutations (GU); Crossover and Gaussian mutation (XG); Crossover and uniform mutation (XM); and Crossover, uniform and Gaussian mutations (XUG).

Table 3. Real functions tested

Name	Function	Feasible region
Rosenbrock	$f(x) = 100 * \left(x_1^2 - x_2\right)^2 + (1 - x_1)^2$	$-2.048 \leq x_i \leq 2.048$
Schwefel	$f(x) = 418.9829 * n + \sum_{i=1}^{n} \left[-x_i * sin\left(\sqrt{|x_i|}\right) \right]$	$-512 \leq x_i \leq 512$
Rastrigin	$f(x) = n * A + \sum_{i=1}^{n} \left[x_i^2 - A * cos\left(2\pi x_i\right) \right]$	$-5.12 \leq x_i \leq 5.12$
Griewangk	$f(x) = 1 + \sum_{i=1}^{n} \left[\frac{x_i^2}{4000} \right] - \prod_{i=1}^{n} \left[cos\left(\frac{x_i}{\sqrt{i}}\right) \right]$	$-600 \leq x_i \leq 600$

4.2   Analysis of Results

Table 4 summarizes the results obtained by the proposed approach (HAEA) with different set of genetic operators after 20000 fitness evaluations. As expected, XUG has the best performance among the variations of HAEA tested (GU performs better than XUG only for the Rosenbrock function while XG performs better than XUG only for the Griewangk function). Figure 4 shows the fitness and operator rates evolution using HaEa(XUG) variation. The behavior of the operator rates is similar to the behavior observed for binary functions; HaEa applies with high probability the best operator in early stages of the evolution while at the end operators are applied with similar probabilities. Clearly, crossover is very useful in locating the optimal solution of the Rastrigin, Schwefel, and Griewangk functions, but it is not so good for the Rosenbrock saddle function. This behavior can be due to the fact that crossover generates good offspring that are located in the narrow local solution valley of the Rosenbrock saddle function. Due to the adaptation mechanism, an individual can leave or move faster from such narrow optimal solution by trying other genetic operators.

We implemented two versions of a generational algorithm, GGA(XG) and GGA(XU), and two versions of steady state algorithm, SSGA(XG) and SSGA(XU), in order to compare the performance of HaEa. A tournament of size 4 was used as parent selection mechanims, with single point real crossover and Gaussian (Uniform) mutation as genetic operators. For the steady-state approaches, we replaced the worst individual of the population with the best offspring generated from the selected parent after crossover and possibly mutation. Many different mutation and crossover rates were analized and the best results, obtained with 0.5 mutation rate and 0.7 crossover rate, are reported. As can be noticed, both or none of the genetic operators may be applied to a single individual for both SSGA and GGA. Table 5 summarizes the results obtained by HaEa, our implementation of SSGA and GGA, and compares them against some results reported by Digalakis and Margaritis [18][4], and Patton et all [19][5]. As shown, HaEa compares well in solving real encoding optimization problems. Moreover, HaEa(XUG) outperforms the SSGA and GGA approaches in all the tested functions. Variations of HaEa outperform the SSGA and GGA approaches in the Rosenbrock saddle function and in all the tested functions when Gaussian mutation is used. In general, the behavior of HaEa(XU) is similar to the behavior of SSGA(XU) and GGA(XU) for the Schwefel, Rastrigin and Griewangk functions.

Table 4. Solutions found by the tested EAs with the real functions

	Rosenbrock	Schwefel	Rastrigin	Griewangk
XUG	0.000509±0.001013	0.005599±0.011702	0.053614±0.216808	0.054955±0.029924
XU	0.004167±0.004487	1.362088±0.932791	0.240079±0.155614	0.530857±0.227458
XG	0.001322±0.003630	140.5647±123.7203	7.731156±3.223595	0.050256±0.025888
GU	0.000160±0.000258	201.9162±81.28619	6.320374±1.462898	1.586373±0.383703

---

[4] Digalakis and Margaritis varied the population size, the mutation rate and the crossover rate for a generational and a steady state genetic algorithm. The results reported in row 7 are the best results reported by Digalakis and Margaritis.

[5] Last row (Patton) reports the best result obtained by Patton et all [19] after 100000 evaluations.

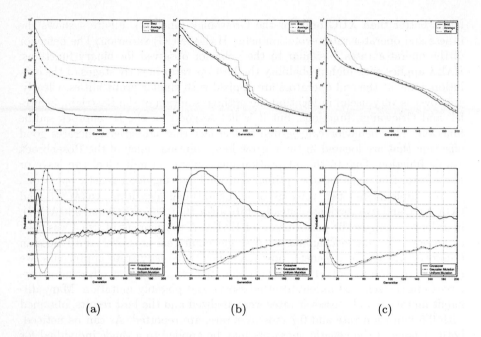

(a)   (b)   (c)

Fig. 4. Performance evolution (first row) and Rates evolution (second row) using HAEA(MXT) on real functions. (a) Rosenbrock, (b) Schwefel, (c) Rastrigin

Table 5. Solutions found by the tested EAs on real functions

EA	Rosenbrock	Schwefel	Rastrigin	Griewangk
HaEa(XUG)	0.00051±0.00101	0.00560±0.01170	0.05361±0.21681	0.05495±0.02992
GGA(XU)	0.17278±0.11797	2.00096±1.21704	0.26500±0.15951	0.63355±0.24899
GGA(XG)	0.03852±0.03672	378.479±222.453	12.1089±5.01845	0.05074±0.02577
SSGA(XU)	0.06676±0.08754	0.88843±0.57802	0.12973±0.07862	0.32097±0.13091
SSGA(XG)	0.04842±0.04624	659.564±277.334	19.7102±7.80438	0.04772±0.02991
Digalakis [18]	0.40000000	-	10.000	0.7000
Patton [19]	-	-	4.8970	0.0043

5   Conclusions

In this paper, a new evolutionary algorithm (HAEA) was introduced. HAEA evolves the operator rates at the same time it evolves the solution. HAEA was tested on a variety of functions (using binary and real encoding), and the results indicated that HAEA is able to find good solutions.

References

1. J. H. Holland, *Adaptation in Natural and Artificial Systems*. The University of Michigan Press, 1975.
2. M. Mitchell, *An intoduction to genetic algorithms*. MIT Press, 1996.
3. A. E. Eiben, R. Hinterding, and Z. Michalewicz, "Parameter control in evolutionary algorithms," *IEEE Transactions in Evolutionary Computation*, vol. 3(2), pp. 124–141, 1999.
4. K. De Jong, *An analysis of the Behavior of a class of genetic adaptive systems*. PhD thesis, University of Michigan, 1975.
5. J. Schaffer, R. Caruana, L. Eshelman, and R. Das, "A study of control parameters affecting online performance of genetic algorithms for function optimization," in *Third International Conference on Genetic Algorithms*, pp. 51–60, 1989.
6. D. Goldberg and K. Deb, "A comparative analysis of selection schemes used in genetic algorithms," *Foundations of Genetic Algorithms*, no. 1, pp. 69–93, 1991.
7. T. Back and H. Schwefel, "An overview of evolutionary algorithms for parallel optimization," *Evolutionary Computation*, vol. 1, no. 1, pp. 1–23, 1993.
8. F. Lobo, *The parameter-less genetic algorithm: rational and automated parameter selection for simplified genetic algorithm operation*. PhD thesis, Nova University of Lisboa, 2000.
9. L. Davis, "Adapting operator probabilities in genetic algorithms," in *Third International Conference on Genetic Algorithms and their Applications*, pp. 61–69, 1989.
10. B. Julstrom, "What have you done for me lately? adapting operator probabilities in a steady-state genetic algorithm," in *Sixth International Conference on Genetic Algorithms*, pp. 81–87, 1995.
11. A. Tuson and P. Ross, "Adapting operator settings in genetic algorithms," *Evolutionary Computation*, 1998.
12. W. Spears, "Adapting crossover in evolutionary algorithms," in *Evolutionary Programming Conference*, 1995.
13. M. Srinivas and L. M. Patnaik, "Adaptive probabilities of crossover and mutation in genetic algorithms," *Transactions on Systems, Man and Cybernetics*, vol. 24(4), pp. 656–667, 1994.
14. T. Back, *Evolutionary Algorithms in Theory and Practice: Evolution Strategies, Evolutionary Programming, Genetic Algorithms*. Oxford University Press, 1996.
15. D. Goldberg, B. Korb, and K. Deb, "Messy genetic algorithms: motivation, analysis, and first results," *Complex Systems*, vol. 3, pp. 493–530, 1989.
16. S. Forrest and M. Mitchell, "Relative building blocks fitness and the building block hypothesis," *Foundations of Genetic Algorithms*, no. 2, 1993.
17. A. Simoes and E. Costa, "Transposition: a biologically inspired mechanism to use with genetic algorithms," in *Fourth International Conference on Neural Networks and Genetic Algorithms*, pp. 612–619, 1999.
18. J. Digalakis and K. Margaritis, "An experimental study of benchmarking functions for genetic algorithms," in *IEEE Conferences Transactions, Systems, Man and Cybernetics*, vol. 5, pp. 3810–3815, 2000.
19. A. Patton, T. Dexter, E. Goodman, and W. Punch, "On the application of cohort-driven operators to continuous optimization problems using evolutionary computation," *Evolutionary Programming*, no. 98, 1998.

PolyEDA: Combining Estimation of Distribution Algorithms and Linear Inequality Constraints

Jörn Grahl and Franz Rothlauf

University of Mannheim
Dept. of Information Systems 1
68131 - Mannheim, Germany
joern.grahl@bwl.uni-mannheim.de
rothlauf@uni-mannheim.de

Abstract. Estimation of distribution algorithms (EDAs) are population-based heuristic search methods that use probabilistic models of good solutions to guide their search. When applied to constrained optimization problems, most evolutionary algorithms use special techniques for handling invalid solutions. This paper presents PolyEDA, a new EDA approach that is able to directly consider linear inequality constraints by using Gibbs sampling. Gibbs sampling allows us to sample new individuals inside the boundaries of the polyhedral search space described using a set of linear inequality constraints by iteratively constructing a density approximation that lies entirely inside the polyhedron. Gibbs sampling prevents the creation of infeasible solutions. Thus, no additional techniques for handling infeasible solutions are needed in PolyEDA. Due to its ability to consider linear inequality constraints, PolyEDA can be used for highly constrained optimization problems, where even the generation of valid solutions is a non-trivial task. Results for different variants of a constrained Rosenbrock problem show a higher performance of PolyEDA in comparison to a standard EDA using rejection sampling.

1 Introduction

Estimation of distribution algorithms (EDA) are population-based optimization methods that use probabilistic models to guide their search [1]. In contrast to genetic algorithms (GA), EDAs do not use classical search operators, but crossover and mutation are replaced by the following two steps:

1. A probabilistic model is built of selected solutions.
2. New random solutions are sampled from the probabilistic model.

EDAs have shown promising results [2,3] when used for combinatorial, discrete, and continuous problems. When using EDAs for continuous real-world planning and optimization problems, there are often a number of additional problem-specific constraints which significantly affect the performance of optimization algorithms [4]. For example, many variables in real-world problems have lower

K. Deb et al. (Eds.): GECCO 2004, LNCS 3102, pp. 1174–1185, 2004.

and upper bounds and the feasible regions of the search space are constrained using linear inequality constraints [5]. Linear inequality constraints are a powerful approach for describing problem-specific knowledge, are common and well understood in the field of mathematical programming, and form the basis of many traditional optimization techniques. During the last few years many methods have been proposed for handling constraints when using GAs or EDAs. The most common are: (1) methods that repair invalid solutions, (2) methods that use penalties, (3) methods which distinguish between feasible and infeasible solutions, and (4) methods that are based on decoders (compare [6]). The general idea when dealing with constraints is to penalize invalid solutions.

The purpose of this paper is to develop a new EDA approach, PolyEDA, that is able to consider linear inequality constraints without penalizing infeasible solutions. PolyEDA is designed in such a way that no infeasible solutions are created during the optimization process. Therefore, the different parts of an EDA like the construction of a probabilistic model, or the sampling of new solutions, must be modified such that the inequality constraints are satisfied. Consequently, PolyEDA uses factorizations of truncated multi-normal distributions that consider the inequality constraints for the building of probabilistic models. Furthermore, the sampling of new solutions according to the given constraints is done using Gibbs sampling [7,8]. Gibbs sampling allows us to sample inside the boundaries of the polyhedral search space described using a set of linear inequality constraints by iteratively constructing a density approximation which lies entirely inside the polyhedron. Therefore, the Gibbs sampler uses well known univariate conditional distributions instead of calculating the highly complicated multivariate constrained densities directly. In contrast to standard EDAs, PolyEDA is able to optimize highly constrained optimization problems, and example results on constrained Rosenbrock problems show a higher performance of PolyEDA in comparison to a standard EDA using rejection sampling.

The paper is structured as follows. The following section introduces some basic results from polyhedral theory that are relevant for linear inequality constraints. Section 2.2 outlines the Gibbs sampling approach for multivariate random number generation considering linear constraints. In section 3, the functionality of PolyEDA is outlined by discussing all of its elements, namely techniques for sampling the first generation (Sect. 3.2), the principles of model-selection (Sect. 3.3), the estimation of parameters (Sect. 3.4), and the generation of new solutions (Sect. 3.5). In section 4, PolyEDA is applied to the constrained Rosenbrock problem and its performance is compared to a standard EDA. The paper ends with some concluding remarks and a short outlook into future work.

The notation and symbols we use throughout this paper are based on the notation used in the IDEA-Framework (see [3]).

2 Polyhedrons and Gibbs Sampling

2.1 Polyhedral Theory

In most linear programming approaches, the feasible search space is described by using a set of linear inequality constraints. Using certain assumptions, the set of points in the search space that is feasible under a finite set of inequalities is a polyhedron. Many classical optimization methods like the simplex algorithm [9], cutting planes, or branch and cut techniques [10], are based on such a polyhedral description of the search space. The mathematical grounding of these approaches is the polyhedral theory. We introduce some basic definitions [10] from polyhedral theory that are relevant for PolyEDA and which are used in the later sections.

Definition 1. *A polyhedron $P \subseteq R^n$ is a set of points that satisfy a finite number of linear inequalities. The linear inequalities are described using $P = \{y \in R^n : Ax \leqslant c\}$, where (A, c) is an $m \times (n + 1)$ matrix.*

Definition 2. *A polyhedron is a convex set.*

Thus, the set of points that are feasible under a set of linear inequalities is a convex polyhedron. In the following we want to use linear inequalities with the structure

$$a \leqslant Dy \leqslant b, \tag{1}$$

where the $(n \times n)$ matrix D has rank n. The vectors a and b are $(n \times 1)$ each. Equation 1 allows the formulation of maximal n linearly independent inequalities. Because each system of linear inequalities can be transformed into a system with the structure $Ax \leqslant c$ the set of points that are feasible under (1) is a convex polyhedron.

2.2   Gibbs Sampling

Gibbs sampling is a statistical method that allows us to generate random variables from highly complicated distributions without calculating their density functions. Complex calculations like the recurrent evaluation of normal integrals can be replaced by a series of computational easier calculations. Gibbs sampling was introduced by [7]. Later, the approach was modified and improved by [11]. A good introduction into Gibbs sampling can be found in [12]. In this paper, the Gibbs sampler is used to generate random variables from multivariate normal distributions that are subject to linear constraints [13].

The Gibbs Algorithm. The following paragraphs describe how Gibbs sampling can be used for creating multivariate random numbers.

We assume that we want to draw an $n-$dimensional random vector $x = (x_1, x_2, \ldots, x_n)'$ from a multivariate density $f(x)$. In addition, we assume (e.g. due to the complexity of the density function) that there is no method available for performing this task directly. Gibbs sampling can be used if all of the following conditional distributions are known:

$$x_i | \{x_1, \ldots, x_{i-1}, x_{i+1}, \ldots, x_n\} \sim P_i(x_1, \ldots, x_{i-1}, x_{i+1}, \ldots, x_n), \tag{2}$$

where $i = 1 \ldots n$ and P_i denotes the conditional distribution of x_i given all other variables x_j $(j \neq i)$. A second condition for using Gibbs sampling is that a method is available that allows the efficient generation of random numbers from the conditional distributions P_i.

Based on these assumptions we can describe the functionality of Gibbs sampling: Let $x^{0'}$ be a $n-$dimensional (starting) point in the multivariate distribution $f(x)$. Then, Gibbs sampling iteratively creates the random variables:

$$x_i^1| \{x_1^1, \ldots, x_{i-1}^1, x_{i+1}^0, \ldots, x_n^0\} \sim P_i(x_1^1, \ldots, x_{i-1}^1, x_{i+1}^0, \ldots, x_n^0) \quad \forall (i = 1 \ldots n) \tag{3}$$

Having done this for the first time, exactly n random numbers have been generated. The creation of these n random number is the first iteration of the Gibbs sampling algorithm. The following iterations are carried out exactly the same. Therefore, after generating the i'th random number of the j'th iteration, we have:

$$x_i^j| \left\{x_1^j, \ldots, x_{i-1}^j, x_{i+1}^{j-1}, \ldots, x_n^{j-1}\right\} \sim P_i(x_i^j, \ldots, x_{i-1}^{j-1}, x_{i+1}^{j-1} \ldots, x_n^{j-1}) \tag{4}$$

After completing the jth iteration, the random vector has the following structure:

$$x^{j'} = (x_1^j, \ldots, x_n^j)' \tag{5}$$

The central element of Gibbs sampling is that with growing j the distribution of $x^{j'}$ converges against the correct multivariate distribution of x [14].

It should be noted, that it is not necessary to calculate this multivariate density directly. Instead, random numbers are drawn from conditional distributions. As this is often less complex, Gibbs sampling has become a popular method in many areas of statistics. The second aspect is that this vector does not constitute a complete sample, but merely an $n-$dimensional point of the multivariate density $f(x)$. In order to generate a sample of size k, the above steps can be repeated k times. Alternative methods for generating samples of a specific size can be found in [12].

Sampling from Truncated Multinormal Distributions. The method outlined in the previous paragraphs can be used to generate random vectors according to complex multivariate distributions. In the following paragraphs, we explain a Gibbs sampler for generating i.i.d. (identically, independently distributed) random vectors from multivariate normal distributions that are subject to linear constraints. The algorithm has been developed by [13]. For technical and mathematical details on truncated multinormal distributions compare [15, p. 204].

Supposing we want to generate $n-$dimensional random vectors x that follow a multivariate normal distribution, but at the same time consider linear inequality constraints:

$$x \sim \mathcal{N}(\mu, \Sigma), \quad \text{s.t.} \quad a \leqslant Dx \leqslant b, \tag{6}$$

where $x = (x_0, x_1, \ldots, x_{n-1})$ and $\mathcal{N}(\mu, \Sigma)$ is the $n-$variate multinormal distribution with mean vector μ and covariance matrix Σ. Furthermore, $-\infty$ and $+\infty$ may be elements of a and b and the matrix D is $(n \times n)$ and of rank n.

This allows the formulation of maximal n linearly independent inequality relationships. Generating random numbers from (6) is equal to the generation of random numbers from

$$z \sim \mathcal{N}(0, T), \qquad \alpha \leqslant z \leqslant \beta \tag{7}$$

with

$$T = D \Sigma D', \quad \alpha = a - D\mu, \quad \text{and} \quad \beta = b - D\mu. \tag{8}$$

The vector x can be calculated from z as

$$x = \mu + D^{-1} z$$

Many different methods have been developed for the sampling of random values from the distribution described by 7. An overview of such methods can be found in [16]. A problem all methods have to solve is that, in general, iterated calculations of the normal integral are necessary when sampling random numbers according to equation 7. Also, these methods are most often incapable of generating i.i.d. samples. By using a Gibbs sampler, i.i.d. samples from 7 can be generated without performing iterated calculations of the normal integral [13]. A necessary condition for using a Gibbs-sampler is previous knowledge regarding the conditional distribution of one random variable x_i on all other random variables x_j, $(i \neq j)$. These conditional distributions are truncated univariate normal. As a result, when using a Gibbs sampler the recurrent evaluation of normal integrals can be replaced by repeated sampling from a truncated univariate normal distribution. This is a less complex problem, and highly efficient techniques for this purpose exist [13].

3   PolyEDA: Combining EDAs with Linear Constraints

This section describes PolyEDA, an EDA that is able to consider a set of linear inequality constraints during optimization. It uses a continuous problem representation and a solution is represented by the vector $y = (y_1, y_2, \ldots, y_n)'$. PolyEDA is the result of combining EDA with a system of linear inequalities of the type

$$a \leqslant Dy \leqslant b. \tag{9}$$

The set of n−dimensional points that satisfy (9) is a polyhedron (compare section 2.1). PolyEDA is able to sample new solutions according to the boundaries of this search space and to consider problem-specific knowledge. Modeling problem-specific knowledge as a set of linear inequality constraints is a common technique in evolutionary computation [6] and mathematical programming [17].

To consider linear constraints in PolyEDA, the elements of the EDA (model selection, parameter estimation, and sampling) must be modified. The following sections outline the necessary adaptations.

3.1   Probabilistic Model

In PolyEDA we use a probabilistic model that consists of factorizations of truncated multinormal distributions. This probabilistic model is based on multivariate normal factorizations as used in the IDEA-Framework [3] and additionally incorporates a set of linear inequality constraints. The truncated distributions reflect the constrained nature of the search space. All n−dimensional points that are feasible under the linear inequality constraints have positive, non-negative probabilities. All infeasible points have a probability of 0. Then, the model can be formulated as

$$P_{(v,\theta)}(Y)(y) = \begin{cases} \dfrac{1}{Pr(y \in S)} \prod_{i=0}^{|v|-1} P_{(\mu_{v_i}, \Sigma_{v_i})}^{\mathcal{N}}(Y_{v_i})(y) & ; y \in S \\ 0 & ; y \notin S \end{cases} \tag{10}$$

$$S = \{ y \mid a \leqslant Dy \leqslant b \} \tag{11}$$

where a solution is represented by a vector $y = (y_0, y_1, \ldots, y_{(n-1)})'$ of random variables. y can be separated into subsets of random variables and all subsets follow multivariate truncated normal distributions. Random variables of different subsets are independent of each other; random variables in the same subset depend on each other. For indicating the different subsets we use node vectors v_i. The entries of each node vector are the indices of the variables of one subset (all variables that depend on each other) and $v_i \wedge v_j = ()$ for all $i \neq j$. This means, that each random variable occurs only in one node vector. The partition vector v consists of all v_i and describes the structure of the whole factorization. The set of parameters θ consists of the mean vectors μ_{v_i} and covariance matrices Σ_{v_i} of each node vector.

The matrix D is $(n \times n)$ and of rank n. Individual elements of a and b must be real values, using $\pm\infty$ is not allowed. This allows the formulation of n inequality constraints.

3.2   Sampling the Initial Population

In EDA, the first population is sampled uniformly over all possible solutions. Therefore, when using the model from equation 10, we have to uniformly sample i.i.d. solutions inside the polyhedron, which is described by the inequality constraints $a \leq Dy \leq b$. To the best of our knowledge, currently no efficient method is available for doing this. Thus, for sampling the initial population, we use rejection sampling, which means that we sample uniformly inside a cube that entirely covers the polyhedron. We calculate lower and upper bounds for each of the y_i and all randomly sampled solutions that lie outside the polyhedron are rejected and sampled again; all solutions that lie inside the polyhedron are accepted and make up the initial population.

As the performance of rejection-sampling depends mainly on its acceptance rate, we seek to maximize the number of accepted solutions by choosing close bounds for the y_i. The smallest rectangular region that entirely covers the polyhedral search space can by calculated by the Fourier-Motzkin-elimination (see [18]). This technique eliminates variables from the inequality system $a \leq Dy \leq b$. In order to generate the lower and upper bounds a_i and b_i for the random variable y_i, all variables $y_j, j \neq i$ are eliminated from the system, leaving a single inequality $a_i \leqslant y_i \leqslant b_i$. Doing this n times we get n inequalities which are used as lower and upper bounds for the sampling of the initial population.

3.3   Model Selection

In the model selection step, the structure v of the factorization that fits best to the current population is searched. If linear inequalities are considered, the factorization is subdivided into a variable part and a fixed part. The variable part depends on statistical properties of a population and describes the interactions between the variables due to the fitness function. The fixed part of the factorization depends on the linear inequalities and describes the interactions between different elements of v.

To generate the variable part of the factorization, we use the greedy factorization selection method outlined in [3]. This heuristic method uses local search steps, beginning with a univariate factorization. The decision between two candidate factorizations is based on a negative log-likelihood metric that penalizes complexity of the factorization. The variable part of the factorization has to be generated in every iteration of the EDA. It reflects the dependencies between the random variables in the current area of the search space.

The fixed part of the factorization can be generated from the inequality system $a \leq Dy \leq b$. To do this, the rows of the matrix D have to be examined. Let the sets S_j $(j = 1 \ldots n)$ denote the columns in which the matrix D has entries $\neq 0$ in the jth row. Then, all variables $y_{(S_j)}$ depend on each other following a truncated multivariate distribution.

The first n node vectors are determined by the n sets of variables $y_{(S_j)}$. Then, it is checked whether the same random variables occur in more than one node vector v_i. If this is the case, these vectors are merged and duplicate entries are deleted. This step is repeated, until every random variable appears in only one node vector. It should be noted, that the fixed part of the factorization needs only to be generated once. Since the matrix D does not change during the optimization, the fixed part can be generated before the optimization and the fixed factorization remains unchanged in later generations. It reflects the dependencies that are necessary in order to consider the linear constraints.

After generating the variable and the fixed part of the factorization, these two parts are combined to create the complete factorization. All dependencies between the variables that are a result of the fixed parts of the factorization are considered first. Then, the interactions that come from the variable part are considered by checking whether they do not already occur in the fixed part. Finally, it is examined whether some random variables occur in more than one node vector. If this is the case, these vectors are merged and duplicated entries are deleted. This step is repeated until every random variable appears in only one node vector.

3.4   Estimation of Parameters

The set of parameters that has to be estimated consists of the mean vector μ and the covariance matrix Σ. Well-known maximum likelihood estimators exist for the estimation of μ and Σ from a sample $S = (S_0, S_1, \ldots, S_{|S|-1})$ [15].

$$\hat{\boldsymbol{\mu}}_{\boldsymbol{v}_i} = \frac{1}{|\mathcal{S}|} \sum_{j=0}^{|\mathcal{S}|-1} (\mathcal{S}_j)_{\boldsymbol{v}_i},$$ (12)

$$\hat{\boldsymbol{\Sigma}}_{\boldsymbol{v}_i} = \frac{1}{|\mathcal{S}_j|} \sum_{j=0}^{|\mathcal{S}|-1} ((\mathcal{S}_j)_{\boldsymbol{v}_i} - \hat{\boldsymbol{\mu}}_{\boldsymbol{v}_i})((\mathcal{S}_j)_{\boldsymbol{v}_i} - \hat{\boldsymbol{\mu}}_{\boldsymbol{v}_i})'$$ (13)

Unfortunately, these estimators are designed for estimation from (multi)normal samples and do not consider truncated normal samples. Nonetheless, we use these estimators in PolyEDA being fully aware of the fact that modified estimators need to be developed that consider the truncation of the distributions. We believe that developing estimators that consider truncated distributions would result in a significant enhancement of PolyEDA.

3.5   Sampling New Solutions

The sampling of new solutions according to equation 10 is not trivial as i.i.d. random vectors must be generated from multinormal distributions that consider the linear inequality constraints $\boldsymbol{a} \leq \boldsymbol{Dy} \leq \boldsymbol{b}$. Therefore, PolyEDA uses the Gibbs sampling algorithm outlined in section 2.2 for sampling new solutions. This ensures that only feasible solutions are sampled and new populations lie entirely inside the given boundaries of the search space. As a result, PolyEDA does not need to use penalties to consider linear inequality constraints.

4   Experiments

In the following paragraphs PolyEDA is exemplarily applied to some versions of the Rosenbrock problem. We want to illustrate how PolyEDA considers linear inequality constraints during optimization in the probabilistic model and we show that the direct sampling of feasible solutions results in higher performance in comparison to standard EDAs using rejection sampling. We are aware of the fact that to fully evaluate the performance of PolyEDA more exhaustive tests on a large number of different test problems are necessary. However, as the emphasis of this paper is on introducing and explaining the functionality of PolyEDA we postpone exhaustive tests until future publications.

4.1   Problem Definition

The Rosenbrock's function is a highly non-linear function that is commonly used for the test of numerical optimization methods. Rosenbrock's function is defined as

$$\text{minimize:}\quad f(\boldsymbol{y}) = \sum_{i=0}^{l-2} \left[100 \left(y_{i+1} - y_i^2 \right)^2 + (y_i - 1)^2 \right],$$ (14)

where l is the dimensionality of the problem. The optimal solution $\boldsymbol{y}^*$ has fitness $f(\boldsymbol{y}^*) = 0$ and is located at $y_i^* = 1$ $(i = 1 \ldots l)$. In our test problem Rosenbrock's function is defined for $y_i \in [-5.12; 5.12]$. We used three test problems of dimension 10, 20, and 40. In each of these test problems, l linear inequalities of the following type have to be considered:

$$1.0 \leqslant y_i \leqslant 2.0 \quad \forall i = 1 \ldots l \tag{15}$$

These inequalities make up a rectangular feasible region. The optimal solution of Rosenbrock's function is at the edge of the search space.

4.2   Experimental Results

In our experiments, we compared a standard EDA to PolyEDA outlined in section 4.1. For both EDAs we used a population size of $N = 300$ of which the 100 best solutions are selected. A statistical model (compare section 3.1) is build from these 100 best solutions and in the next generation 300 offspring are generated according to this model using a sampling algorithm. In PolyEDA we used Gibbs sampling as described in section 2.2 for the creation of new solutions. The number of iterations j that has been used by the Gibbs sampler to approximate the truncated distributions has been set to $j = 100$.

Both, the standard EDA and PolyEDA use factorizations of multivariate normal distributions as outlined in [3]. The only difference lies in considering the linear inequalities. PolyEDA considers the linear inequality constraints when sampling new solutions by using Gibbs-sampling. In the standard EDA new solutions are sampled using the unconstrained multi-normal distributions and neglecting the linear constraints. To consider the additional constraints newly generated solutions that are infeasible are rejected and not considered for the creation of the statistical model (rejection-sampling). This means, that invalid solutions (solutions that were infeasible under the linear inequalities outlined in section 4.1), were rejected and sampled again (until the population is filled).

PolyEDA and the standard EDA with rejection-sampling have been compared on all three instances of the constrained Rosenbrock function. We performed 15 runs for every problem instance. Fig. 1 shows the mean of the average fitness of the best solution in a population (left) and the average fitness of the population (right) over the number of generations for different sizes of the Rosenbrock function.

The plots show that PolyEDA outperforms a standard EDA using rejection sampling for the constraint Rosenbrock problem. Although both approaches use the same population size ($N = 300$), PolyEDA results in better results in terms of average population fitness as well as average best fitness. Obviously, using a probabilistic model and a sampling technique that considers the given linear constraints is advantageous and lead to more efficient EDAs.

Standard EDAs using rejection sampling have the problem that many of the sampled individuals are infeasible. Table 1 shows the average ratio of infeasible solutions to all generated solutions. The results are averaged over all fifteen runs and all generations. When applying standard EDA with rejection sampling to the 40-dimensional constraint Rosenbrock function, more than 60 percent of all generated solutions are infeasible which results in a great overhead. This

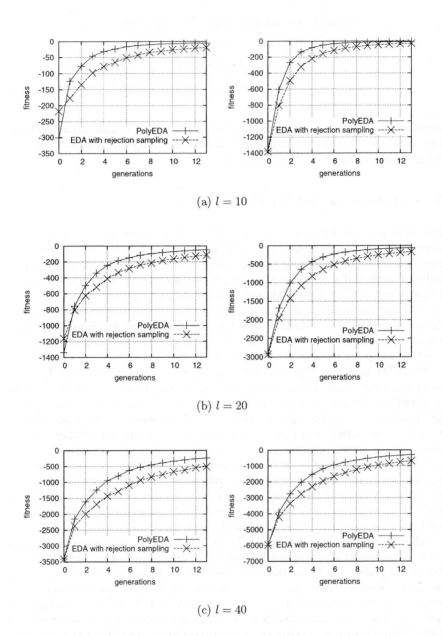

(a) $l = 10$

(b) $l = 20$

(c) $l = 40$

Fig. 1. The plots show the mean of the average fitness of the best solution (left) and the average fitness of a population (right) over the number of generations. Results are presented for the constraint Rosenbrock function of size $l = 10$ (Fig. 1(a)), $l = 20$ (Fig. 1(b)), and $l = 40$ (Fig. 1(c)). The constraints are chosen such that the optimal solution is at the edge of the feasible solution space. The results show that PolyEDA outperforms a standard EDA using rejection sampling independently of the size of the problem.

Table 1. Average percentage of infeasible solutions

dimension l	10	20	40
rejection sampling	8.58 %	23.1 %	60.8 %
PolyEDA		0 %	

situation can become even worse for highly constraint optimization problems. This problem of standard EDAs illustrates the advantage of PolyEDA which did not generate a single infeasible solution.

5   Conclusions and Further Work

This work presented the functionality of PolyEDA, an EDA that is able to consider linear inequality constraints during the optimization without penalizing infeasible solutions. In section 2.1 the paper reviewed some foundations of polyhedral theory and Gibbs sampling which are necessary for PolyEDA. Section 3 explained in detail PolyEDA and focused on the different aspects like sampling of the first generation, model selection, the estimation of parameters, and the generation of new solutions from the probabilistic model. Finally, section 4 compared exemplarily the performance of PolyEDA to a standard EDA using rejection sampling for some variants of the constrained Rosenbrock function.

PolyEDA is a new type of EDA that allows us to directly consider linear inequality constraints. It was designed in such a way that it avoids the creation of infeasible solutions. The used probabilistic model is based on factorizations of multivariate truncated normal distributions. In this model, all solutions that are feasible under the linear inequality constraints have positive probabilities, solutions that are infeasible have a probability of zero. For the sampling of feasible solutions from this model, a Gibbs sampler is used. Gibbs sampling allows the generation of random values from the multivariate truncated distribution without calculating their density. By using a Gibbs sampler, the complex calculation of the density can be avoided and replaced by the generation of random values from univariate truncated normal distributions.

This paper illustrated exemplarily the performance of PolyEDA for some small examples of the constraint Rosenbrock problem. In further work, PolyEDA should be applied to a larger variety of linearly constraint optimization problems and also compared to different other techniques that can be used for solving constrained optimization problems. Furthermore, PolyEDA should be continuously improved. The work presented here focused on the sampling of solutions from multivariate truncated distributions by using Gibbs sampling and neglected a proper estimation of truncated distributions. This aspect should be addressed in future work.

References

1. Pelikan, M., Goldberg, D.E., Lobo, F.: A survey of optimization by building and using probabilistic models. Technical Report IlliGAL Report 99018, University of Illinois at Urbana-Champaign (1999)
2. Pelikan, M.: Bayesian optimization algorithm: From single level to hierarchy. PhD thesis, University of Illinois at Urbana-Champaign, Dept. of Computer Science, Urbana, IL (2002) Also IlliGAL Report No. 2002023.
3. Bosman, P.A.N.: Design and Application of Iterated Density-Estimation Evolutionary Algorithms. PhD thesis, University of Utrecht, Institute of Information and Computer Science (2003)
4. Michalewicz, Z.: Heuristic methods for evolutionary computation techniques. Journal of Heuristics **1** (1995) 177–206
5. Michalewicz, Z., Deb, K., Schmidt, M., Stidsen, T.J.:   Towards understanding constraint-handling methods in evolutionary algorithms.   In Angeline, P.J., Michalewicz, Z., Schoenauer, M., Yao, X., Zalzala, A., eds.: Proceedings of the Congress on Evolutionary Computation. Volume 1., Mayflower Hotel, Washington D.C., USA, IEEE Press (1999) 581–588
6. Michalewicz, Z., Schoenauer, M.: Evolutionary computation for constrained parameter optimization problems. Evolutionary Computation **4** (1996) 1–32
7. Metropolis, N., Rosenbluth, A.W., Rosenbluth, M.N., Teller, A.H., Teller, E.: Equation of state calculations by fast computing machines. Journal of Chemical Physics **21** (1953) 1087–1091
8. Geman, S., Geman, D.: Stochastic relaxation, Gibbs distributions, and the bayesian restauration of images. IEEE Transactions on Pattern Analysis and Machine Intelligence **6** (1984) 721–741
9. Vanderbei, R.J.: Linear Programming, Foundations and Extensions. Volume 2 of International Series In Operations Research And Management Science.   Kluwer Academic Publishers, Boston (2001)
10. Nemhauser, G.L., Wolsey, L.A.: Integer and Combinatorial Optimization. Wiley-Interscience Series in Discrete Mathematics and Optimization (1999)
11. Hastings, W.K.:   Monte carlo sampling methods using markov chains and their applications. Biometrika **57** (1970) 97–109
12. Casella, G., George, E.I.: Explaining the gibbs sampler. The American Statistician **46** (1992) 167–174
13. Geweke, J.: Efficient simulation from the multivariate normal and student-t distributions subject to linear constraints and the evaluation of constraint probabilities. Technical report, University of Minnesote, Dept. of Economics (1991)
14. Gelfand, A.E., Smith, A.F.M.: Sampling-based approaches to calculating marginal densities. Journal of the American Statistical Association **87** (1990) 398–409
15. Kotz, S., Balakrishnan, N., Johnson, N.L.: Continuous Multivariate Distributions. Volume 2 of Wiley Series in Probability and Statistics. John Wiley and Sons (2000)
16. Hajivassiliou, V.A., McFadden, D. L.and Ruud, P.A.: Simulation of multivariate normal orthan probabilities: Methods and programs.   Technical report, M.I.T. (1990)
17. Williams, H.: Model Building in Mathematical Programming. Volume 4. Auflage. Wiley (1999)
18. Duffin, R.: On fourier's analysis of linear inequality systems. Mathematical Programming Study **1** (1974) 71–95

Improving the Locality Properties of Binary Representations

Adrian Grajdeanu and Kenneth De Jong

Department of Computer Science
George Mason University
Fairfax, VA 22030

Abstract. Choosing representations and operators that preserve locality between genotype and phenotype space is an important goal in EA design. In the GA literature there has been considerable discussion of this issue with respect to the choice between standard binary encoding and Gray codes. In this paper we argue that an important and unappreciated aspect of such discussions is the degree to which locality preservation is isotropic in phenotype space (i.e., independent of location in phenospace). We show that using a traditional bit-flip mutation operator with either of these two representations results in rather weak isotropic locality. These insights lead to the design of a new binary mutation operator that increases isotropic locality. The results from an initial set of experiments supports the hypothesis that this improvement in isotropic locality leads to improvements in GA performance as well.

1 Introduction

It is well-known in the EC community that the choice of representation and reproductive operators is critical to success of an application. One of the desirable features of such choices is that locality is preserved when mapping between the internal representation space (genospace) and the external application space (phenospace). The classical example of this in the GA literature involves the decision as to how best to map a problem into a binary representation. The most straightforward approach to representing ordered sets of phenotypic objects internally as binary strings is to assign the strings in order according to their binary value: 00...00, 00...01, 00...10, ..., 11...10, 11...11. So, for example, the interval [0.0,1.0] would be represented internally as binary strings whose length is dictated by the desired level of precision ϵ with the string 00...00 representing the real number 0.0 and the string 11...11 representing the real number 1.0. The most common mutation operator for binary representations is bit-flip mutation in which individual bits are stochastically selected to be flipped (switched from zero to one or vice versa). As a consequence, Hamming distance is the most natural distance metric in genospace while Euclidean distance is the most natural one for real-valued phenospaces.

The notion, then, of locality preservation is that small steps in one space correspond to small steps in the other space. Clearly, this is true for real numbers such as 0.0 and 0.0+ϵ and their corresponding strings 00...00 and 00...01.

K. Deb et al. (Eds.): GECCO 2004, LNCS 3102, pp. 1186–1196, 2004.
© Springer-Verlag Berlin Heidelberg 2004

However, it is clearly not true for the real numbers whose binary representations are 01...11 and 10...00. In this case, achieving a small step in phenospace requires a large step in genospace. In the GA literature the traditional name for this phenomenon is a "Hamming cliff".

A standard way of resolving this issue is to switch to a Gray code representation in which adjacent points in phenospace are assigned bit strings that differ in only a single bit position (see, for example, [1], [2] or [3]). Hence, every small step in phenospace corresponds to a small step in genospace, but the reverse is clearly not true since there are single bit flips that result in large steps in phenospace.

The effect this can have on EA performance is easily seen by using a family of artificially constructed Hamming cliff landscapes. This parameterized family of landscapes, $hc[a, b, \alpha]$, is defined as follows. Given a real-valued interval $[a, b]$ and a parameter α whose value lies somewhere in $[a, b]$, then let:

$$hc[a, b, \alpha](x) = \begin{cases} x - (\alpha + 1) + b - a, x <= \alpha \\ x - (\alpha + 1), x > \alpha \end{cases}$$

By varying α one can position the global maximum anywhere in $[a, b]$. In particular, setting $\alpha = (b - a)/2$ places it immediately to the right of a large Hamming cliff. Figure 1 illustrates this by plotting $hc[0, 31, 16](x)$. Higher dimension versions are easily constructed by summing n copies of a 1-dimensional version:

$$HC[a, b, \alpha, n](x) = \sum_{i=1}^{n} hc[a, b, \alpha](x)$$

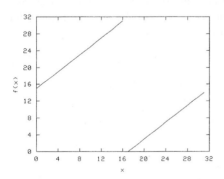

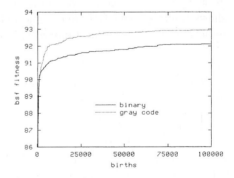

Fig. 1. Example of a 1-dimensional Hamming cliff: $hc[0, 31, 16](x)$.

Fig. 2. Best-so-far performance of a standard GA with binary and Gray coding on $HC[0, 31, 16, 3](x)$.

Figure 2 illustrates how the performance of a standard GA, in terms of average best-so-far (bsf) curves, can be improved by switching from a standard binary representation to a Gray code representation on a 3-dimensional version: $HC[0, 31, 16, 3](x)$. By contrast, when $\alpha = b$, no Hamming cliffs serve as barriers to the global optimum and both representations work equally well (Figures 3 and 4).

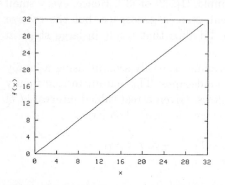

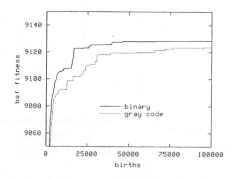

Fig. 3. The $hc[0, 31, 31](x)$ landscape.

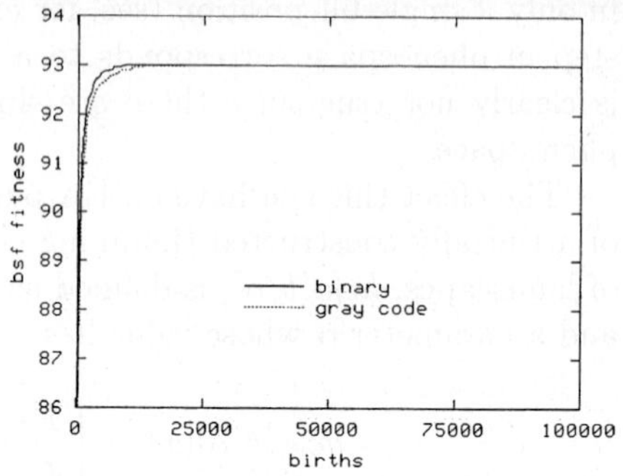

Fig. 4. Best-so-far performance of a standard GA with binary and Gray coding on $HC[0, 31, 31, 3]$.

Experiments such as these suggest that switching to Gray coding can improve performance when Hamming cliff barriers are encountered, and Gray coding doesn't seem to hurt performance much when no such barriers exist. Unfortunately, one doesn't need to look far to find counter examples. The Schwefel function [4] is a standard EC benchmark, and we shall consider it on a 2-dimensional $[0, 5000] \times [0, 5000]$ landscape, designated here as $Schwefel[0, 5000, 2]$. Figure 5 shows the negative effect that using a Gray code has on the average best-so-far performance of the same standard GA on this landscape.

Fig. 5. Best-so-far performance of a standard GA with binary and Gray coding on the Schwefel[0.5000,2] landscape.

A survey of the EC literature produces a similar collection of mixed results For example, [5] presents a statistical comparison of binary and Gray encodings that suggests that Gray encodings are generally superior while an analysis in [6] concludes just the opposite. Attempting to bring the theory more in line with empirical experience, Whitley argues ([7]) for the benefit of Gray codes by showing that Gray codes induce fewer local optima than the standard binary encoding on a certain class of problems, arguably those of interest to the application oriented researchers. By contrast, using a Markov model to study relative performance of binary and Gray coding in genetic algorithms, it was shown in [8] that Gray coding does not necessarily improve performance for functions which have fewer local optima in Gray representation than in binary. In a subsequent paper Whitley shows that there is a complementary bias in the Gray and binary neighborhood structures and explores the possibility of using a combination of the two [9].

The results presented in this paper approach this representation/operator choice issue from a somewhat different perspective. Our sense is that a critical feature of such choices is the extent to which they result in locality preservation *uniformly* throughout phenospace, i.e., the notion of isotropic locality. We show that both binary encodings when used in conjunction with standard bit-flip mutation exhibit weak isotropic locality. This leads to the design of a generalized bit-flip mutation operator which, when used with a standard binary encoding, has better isotropic locality than either the binary or Gray code with the standard bit-flip mutation operator. In addition, we present the results of a preliminary set of experiments that shows a corresponding improvement in GA performance as well.

2   Locality Properties of Encodings

Ideally, the choice of internal representation and reproductive operators would result in a distance-preserving mapping between genospace and phenospace. This is difficult to achieve in general. A less ambitious goal is to choose a mapping that is locality preserving (i.e., small steps in genospace produce small steps in phenospace and vice versa). We noted above that the primary motivation for choosing a Gray code representation over the standard binary encoding was to improve on this locality property. In this section we explore in more detail how this is achieved.

2.1   Phenotypic Effects of 1-Bit-Flip Mutation

The simplest way to obtain insight into locality issues is to focus on the effect that a one-bit-flip mutation in genospace has in phenospace. Suppose, for example, that the phenospace consists of the interval $[0.0, 1.0]$ and the desired precision results in a 9 bit encoding. Then, for each of the 2^9 phenotype values, one can plot its change in value as a result of each one of the 9 possible 1-bit-flip mutation.

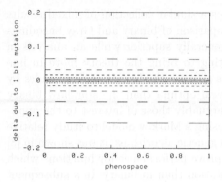
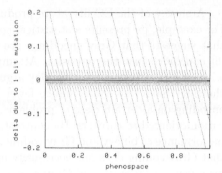

Fig. 6. A scatter plot for 1-bit-flip muta- **Fig. 7.** A scatter plot for 1-bit-flip mutation with a standard binary encoding.   tion using Gray code.

This analysis is presented in Figure 6 for the standard binary encoding and in
Figure 7 for the Gray code representation.

In both cases single bit flips can produce large changes in phenotype values.
However, since we are focusing on locality issues, we have zoomed in on the small
phenotypic changes. Hence the Y range depicted is the $(-0.2, 0.2)$ sub-range and
not the entire range of $(-1, 1)$. The degree of isotropic locality is represented by
the continuity of the horizontal lines. The lack of continuity indicates a dependency on location (i.e., non-isotropy), while a continuous horizontal line depicts
the ability of produce a particular change in phenotype value *regardless* of location in phenospace. Figures 6 and 7 show clearly that the Gray code produces
much more consistent locality (i.e., better isotropy) for *small* changes in phenotypic value, but at the expense of decreased isotropy as the size of the change in
phenotype values increases.

2.2   Quantifying Isotropic Locality

The scatter plots in the previous section are visually suggestive that a key difference between the two encodings is the degree to which the ability to take small
steps in phenospace varies as a function of where one is in phenospace. The
reverse of this, the notion of isotropic locality is - more formally - the ability
to take certain steps independent of location in phenospace. We can quantify
the degree to which an encoding is isotropic by calculating the probability of
taking a step of size δ using 1-bit-flip mutation for each point in phenospace,
and then looking at the mean and variance of these probabilities. This approach
is illustrated in Figures 8 and 9 using the same setup as in the previous section.

For clarity Figures 8 and 9 zoom in on the first 31 possible (positive) changes
in phenotypic value. They use box plots to display the mean, standard deviation,
and maximum and minimum values of the probability distributions for each
of the included positive changes δ. Hence, the degree of isotropic locality is
represented by the variances of the probability distributions (high variance means
low isotropy).

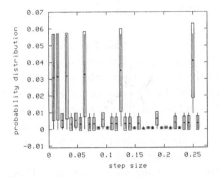
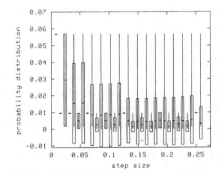

Fig. 8. Probability distribution for small step sizes δ using 1-bit-flip under the standard binary encoding.

Fig. 9. Probability distribution for small step sizes δ using 1-bit-flip under the Gray code representation.

By comparing these two figures one can see clearly the isotropic differences between the two encodings. For the standard binary encoding (Figure 8) notice the high variances associated with the first, second, fourth, etc. values for δ which correspond precisely to the existence of Hamming cliffs. By contrast the Gray code probability distributions for the same step sizes have zero variance (Figure 9).

Further comparison of these two figures shows that the improvements in locality obtained for Hamming cliffs using the Gray encoding are obtained at the expense of a reduction elsewhere. Notice how the variances oscillate as one increases the value of δ with the maximum probability values remaining *uniformly* high throughout the entire range.

These observations raise the question as to whether it is possible to improve isotropy in a more consistent and uniform manner that incorporates the good features of both representations while avoiding the bad ones. We answer that question in the affirmative in the next section.

3   A Bit Level Mutation Operator with Isotropic Locality

One possible approach to answering this question would be to invent a new type of binary representation. In this paper we explore an alternative approach, namely, by retaining the standard binary representation and modifying the bit-flip mutation operator.

Using the standard binary encoding of an interval $[a, b]$ at a particular level of precision results in an internal representation of points in $[a, b]$ as bit strings of length l. When a particular bit k is mutated, the classical bit-flip has the phenotypic effect in $[a, b]$ of adding or subtracting a quantity proportional to 2^k. The choice of addition or subtraction is dictated by the current value of bit k and this is precisely the reason for the lack of isotropy noted in the previous section. This fact suggests that isotropy could be improved by simply breaking this coupling.

We achieve this decoupling by modifying the standard bit-flip mutation operator as follows. When bit k is selected to be mutated, the binary value 2^k is genotypically added/subtracted independently of the current value of bit k. The choice of addition vs. subtraction is determined by flipping an unbiased coin. This results in exactly the same effects as standard bit-flip operation if bit k is a zero and addition is selected, or if bit k is a one and subtraction is selected. The remaining two cases are handled by performing standard binary addition or subtraction with overflow and underflow conditions ignored.

For example, suppose we are representing $[a, b]$ using 6-bit genotypes. Then, whenever bit 3 of a genotype is selected for mutation, a value of 001000 will be added to or subtracted from the binary value of the genotype undergoing the mutation, with the choice between addition or subtraction being determined for each such mutation by the flip of an unbiased coin. Hence, if the genotype undergoing a mutation at bit 3 is 010101, the result of applying the new (decoupled) mutation operator would be 011101 if addition were selected and 001101 in case of subtraction. Similarly, mutating the fifth bit of a genotype results in the addition or subtraction of 000010.

If multiple bits in the same genotype are selected for mutation, the application of this operator is sequential and cumulative (one can prove that the order in which bits undergo mutation is irrelevant). Extending this operator to multi-dimensional problems requires that each dimension be mapped into an independent binary gene allowing this new mutation operator to handle each gene (dimension) independently at the bit string level. In other words, overflow/underflow conditions do not propagate beyond gene boundaries.

3.1   Locality Properties of Decoupled Mutation

To see whether or not this decoupling idea truly improves isotropic locality, we performed the same analysis as we did earlier for bit-flip mutation using standard binary and Gray encodings. Figures 10 and 11 present the results.

By comparing these two figures with the earlier ones (Figures 6 - 9), one can see clearly the rather dramatic improvement in isotropic locality as reflected by the horizontal bars in Figure 10 and the lack of variance in the box plots in Figure 11. What remains to be seen is whether this improvement has a positive effect on performance.

4   Empirical Studies

To assess the effects that improved isotropic locality has on performance, we performed an initial set of empirical studies using a standard generational GA (popsize=100, 2-point crossover, fitness-proportional selection, and a mutation probability of $1/l$). We varied the representation and mutation operator as follows:

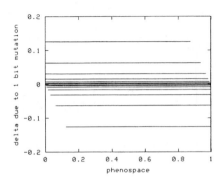

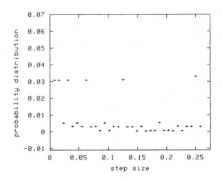

Fig. 10. A scatter plot for decoupled mutation with a standard binary encoding.

Fig. 11. Probability distribution for small step sizes δ using decoupled mutation with the standard binary encoding.

- Study 1 used the standard binary encoding and the standard bit-flip mutation operator.
- Study 2 used a Gray code representation and the standard bit-flip mutation operator.
- Study 3 used the standard binary representation with the decoupled mutation operator.

The set of landscapes used for these initial studies were HC[0,31,16,3], HC[0,31,31,3], and $Schwefel[0, 5000, 2]$. In all cases, 100 independent runs were performed and the results were averaged to obtain both mean and variance.

4.1   Hamming Cliff Landscape Results

Figures 12 - 15 present the results on the artificial Hamming cliff landscapes in terms of the effect on best-so-far performance curves. Figures 12 and 13 plot just the average best-so-far curves. What is striking is that the effects of the decoupled mutation operator operating on a standard binary representation are nearly identical to those of standard bit-flip mutation operating on a Gray code representation. Figures 14 and 15 include one standard deviation error bars to indicate the statistical advantage that both have over and EA using bit-flip mutation with a standard binary representation.

4.2   Schwefel Landscape Results

Unlike the artificial Hamming cliff landscapes, the Schwefel function provides a more realistic landscape on which to evaluate performance. Figure 16 shows the average best-so-far curves of the 3 studies on $Schwefel[0, 5000, 2]$. If we compare that with Figure 5, we see a rather striking result. On this landscape the decoupled mutation operator performed somewhat better than bit-flip mutation

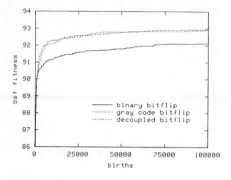

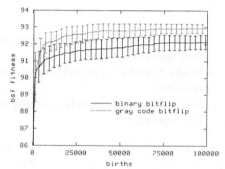

Fig. 12. Best-so-far performance on HC[0,31,16,3]

Fig. 13. Best-so-far performance on HC[0,31,31,3]

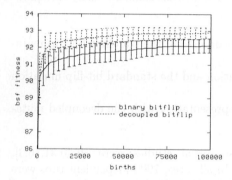

Fig. 14. Best-so-far performance on HC[0,31,16,3]

Fig. 15. Best-so-far performance on HC[0,31,16,3]

with a standard binary representation, and much better than bit-flip mutation on a Gray code representation.

Figures 17 and 18 include one standard deviation error bars and suggest why this is the case. The improved uniformity of the locality properties of decoupled mutation allow significantly more opportunity for local exploitation than bit-flip with a binary representation, resulting in faster and more consistent convergence to the global optimum. By contrast, the locality improvements of the Gray code representations are obtained at the cost of higher bias towards global exploration - with much less effectiveness.

5 Conclusions

We have illustrated some statistical properties of the effects of 1-bit-flip mutation under binary and Gray code encodings. These findings better highlight the trade-off introduced by adopting a Gray code mapping in order to improve operator locality, and result in a better understanding as to when Gray code mappings fail

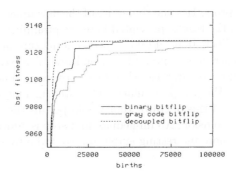

Fig. 16. Best-so-far performance on Schwefel[0,5000,2].

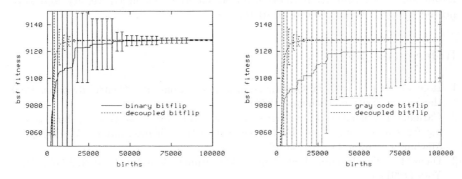

Fig. 17. Best-so-far performance on Schwefel[0,5000,2].

Fig. 18. Best-so-far performance on Schwefel[0,5000,2]

to outperform the standard binary mapping. This, in turn, inspired the design and implementation of the decoupled bit-flip mutation operator that has better isotropic locality properties. That is, the probability distribution of inducing a certain step size δ from a single application of the 1-bit-flip mutation operator, taken for all points in phenospace has zero variance. This property is necessary for the absence of any Hamming cliff anomalies. In addition the decoupled bit-flip mutation has a negative exponentially modulated propensity towards global exploration. This property is inherited from the classical bit-flip mutation and is unlike the Gray code mapping which has the same propensity uniformly modulated. Because of this, the decoupled bit-flip mutation is able to maintain a more effective balance between exploitation and effective exploration throughout the run.

6   Future Work

We are considering several continuations of this work, grouped in two major categories. The first area concerns itself with further study and better understanding of various properties and dynamics of the decoupled bit-flip mutation as introduced. Clearly, the results presented here are preliminary in nature and additional insights are likely to be obtained by extending this analysis and empirical study to various other landscapes. The second area revolves around extending the decoupled bit-flip operator to other phenospaces beyond the bounded numerical parameters. We are particularly interested in identifying the essential statistical properties of this operator that would permit the creation of a formal mechanism of such extension. An immediate first step is addressing rank ordered spaces, but a much more interesting extension would be to address partially ordered phenotypes.

Acknowledgments. This research was performed using the facilities of the Evolutionary Computation Laboratory at George Mason University and supported by ONR Grant N000140110193.

References

1. Hollstein, R.B.: Artificial Genetic Adaptation in Computer Control Systems. PhD thesis, University of Michigan (1971)
2. Caruana, R., Schaffer, J.D.: Representation and hidden bias: Gray vs. binary coding for genetic algorithms. In: Proceedings of the 5th International Workshop on Machine Learning. (1988) 153–161
3. Davis, L.: The Handbook of Genetic Algorithms. Van Nostrand Reinhold, New York (1991)
4. Schwefel, H.P.: Numerical Optimization of Computer Models. John Wiley & Sons, Ltd., Chichester (1981)
5. Hinterding, R., Gielewski, H., Peachey, T.C.: The role of mutation in the optimization of numerical functions by genetic algorithms. Technical report, Department of Computer and Mathematical Sciences, Victoria University of Technology, Victoria, Australia (1995)
6. Rothlauf, F.: The influence of binary representation of integers on the performance of selectorecombinative genetic algorithms. Technical Report 1, University of Bayreuth, Dept. of Information Systemsa (2002)
7. Whitley, D.: A free lunch proof for gray versus binary encodings. In: Proceedings of the Genetic and Evolutionary Computation Conference. (1999) 726–733
8. Chakraborty, U.K., Janikow, C.Z.: An analysis of gray versus binary encoding in genetic search. Information Sciences **156** (2003) 253–269
9. Whitley, D., Barbulescu, L., Watson, J.P.: Local search and high precision gray codes: Convergence results and neighborhoods. In: Foundations of Genetic Algorithms 6. (2000) 295–311

Schema Disruption in Chromosomes That Are Structured as Binary Trees

William A. Greene

Computer Science Department
University of New Orleans
New Orleans, LA 70148
bill@cs.uno.edu

Abstract. We are interested in schema disruption behavior when chromosomes are structured as binary trees. We give the definition of the disruption probability $dp(H)$ of a schema H, and also the relative diameter $rel\Delta(H)$ of H. We show that in the general case that $dp(H)$ can far exceed $rel\Delta(H)$, but when the chromosome is a complete binary tree then the inequality $dp(H) \leq rel\Delta(H)$ holds almost always. Thus the more compactly the tree chromosome is structured, the better is the behavior to be expected from geneticism.

1 Introduction

The field of Genetic Algorithms (GAs) is a heuristic problem-solving paradigm which is inspired by the machinations of evolution. There is some problem of interest at hand. There are candidate solutions to the problem, some of which are better (or fitter) than others. Usually the number of possible candidate solutions is enormous, too large to search exhaustively. In GAs, a population of candidate solutions is maintained; this population is a small sampling of the full solution space. The population is subjected to such evolutionary forces as survival of the fittest, mating with crossover, and mutation. The hope is that as the population evolves, fitter and fitter solutions will appear.

Candidate solutions differ one from another by having different property values for certain properties that are pertinent to the problem at hand. We represent (identify) the candidate solution with the property values that characterize it.

In classical Genetic Algorithms as invented by Holland [1] and popularized by Goldberg [2], candidate solutions (which we will begin to refer to as *individuals* and *chromosomes*) in the population are bits strings, all of the same length N. Thus, each is an element of the Cartesian product space $\{0, 1\}^N$. We here emphasize that the bits are arranged in a linear sequence. For a theoretical analysis of the convergence behavior of an evolving population, the notion of the *schema* is introduced. Let the symbol * be a don't-care symbol. Then a schema H (for hyperplane) is defined to be an element of $\{0, 1, *\}^N$. The positions in H which are the don't-care symbol are termed the *free positions*; the positions occupied by 0's or 1's are the *fixed positions*. A schema stands for an entire subspace of the possible bit strings, namely, those bit strings which agree with the schema at its fixed positions. An element of this subspace is termed a *repre-*

K. Deb et al. (Eds.): GECCO 2004, LNCS 3102, pp. 1197–1207, 2004.

sentative of schema H. The *defining length* $\delta(H)$ of a schema means the distance between its outermost fixed positions. For example, $\delta((*, *, 0, 1, *, *, *, 1, *))$ equals 5.

The term *building block* is used to signify a group of related bits, plus values (0's or 1's) for them, that enhance the fitness of an individual whose bits are so assigned. Clearly, a schema gives the characterizing properties of a building block.

Under *one-point crossover* for mating by two parental chromosomes, we uniformly randomly choose one of the $N - 1$ links between the N bits, cut both parents at that same point, and the parental fragments that result are interchanged to form the two children. A cutpoint is said to *disrupt* a schema if it falls between the two outermost fixed positions. The *disruption probability dp(H)* is the probability that this occurs, and clearly $dp(H)$ equals $\delta(H) / (N - 1)$. The term disruption is appropriate, since if one parent is a representative of H, and the cutpoint falls between the two outermost fixed positions, then it is possible that neither child is again a representative of H. Thus, building blocks can fall away under mating with crossover. Clearly the disruption probability of a schema (or building block) is diminished if the pertinent bits are located close together.

There is an obvious weakness of the standard linear arrangement of bits in a chromosome: a bit has two nearest bits, not more. What if it is in the nature of the problem at hand that a bit should be equally close to three, or four, or more, other bits?

We are interested in chromosomes whose bits are arranged in ways other than as a linear sequence. In particular we are interested in chromosomes structured as binary trees.

Alternative bit arrangements have frequently been used in applications written up in the literature. As for theoretical study, non-linear bit arrangements and a schema theory for such, also have been studied. For instance, Greene [3] has a non-linear schema theorem which may apply when the chromosome is structured as an arbitrary connected graph. As for chromosomes structured as trees, study of schema theory for them has principally come from those in the Genetic Programming (GP) community. In GP approaches, individuals are programs, specifically functions, realized as expression trees. Mating with crossover consists of clipping out and exchanging subtrees between the two parents. The individuals in a population can have quite different shapes, which fact complicates a number of issues, such as, what will be the definition of a schema, and what relation will hold between the locations of the cutpoints in the two parents? For Koza [4], O'Reilly [5], O'Reilly & Oppacher [6], and Whigham [7], schemas are expression fragments which incorporate don't-care symbols, and which are further characterized by not being anchored to some fixed position within the expression tree and moreover can be instantiated multiple times within the same individual. In Rosca [8], the innovation is that a schema is an expression fragment which is anchored at the root of the expression tree.

Our own interest in non-linear bit arrangements did not originate from a prior interest in genetic programming. Rather, our intuition has been that strictly linear bit arrangements are simply too confining and too inflexible for GAs. Furthermore, we envision a population of chromosomes that all have the same shape. From within the

GP community, the work that comes closest to our own efforts is that of Poli & Langdon [9]. Their definitions of schema, mutation, and crossover are the closest carryover to GP of the allied notions from standard GAs with linear bit arrangements. For Poli & Langdon, a schema is a rooted tree of symbols, where the root is to correspond to the root of an expression tree that is an individual in the population. Below, we will remark on similarities between our present research and the approach of Poli & Langdon.

In this paper we are interested specifically in the disruption probability $dp(H)$ of schemas H in chromosomes structured as binary trees. Knowing about the value of $dp(H)$ must figure in the statement and proof of any schema theorem akin to the classic one by Holland (confer [1] or [2]). A closed expression that exactly calculates $dp(H)$ for arbitrary H is likely hard to come up with. Hence we seek an expression which is more easily calculated and which may be an upper bound for $dp(H)$. An insight which should be carried over from the linear case is that we should explore how $dp(H)$ is related to how closely situated together are the fixed positions of schema H.

2   Binary Trees

We assume the reader is familiar with binary trees. Trees consist of *nodes*, connected by *edges*. All the binary trees we consider are finite. We use T to name a binary tree. The *level* of a node: the root of the tree is at level zero; the level of a non-root node is one greater than the level of its parent. A *level is full* if it contains the maximum possible number of nodes, which is 2^{level}. A binary *tree is full* if its every level is full. A binary *tree is complete* if its every level is full, except possibly the bottom level. (The usual definition of complete insists that the leaves on the bottom level are packed together without gaps off to the left, but we do not need this stipulation.)

There is a unique path between any two nodes in a binary tree. We define the distance between two tree nodes to be the length of that path. This notion of distance satisfies the triangle inequality, $dist(x, z) \leq dist(x, y) + dist(y, z)$. Given a set S of nodes in the tree, the diameter $\Delta(S)$ is the maximum distance between any two elements of S.

An individual whose bits are linked as the nodes in some binary tree is an easily comprehended concept. A *schema* will be the obvious analogue from linear chromosomes: imagine replacing some of the bits (0's or 1's) with the don't-care symbol. We define the *relative diameter rel*$\Delta(H)$ of a schema H to be the ratio $\Delta(fixed(H)) / \Delta(T)$, where *fixed(H)* is the set of fixed positions of H. We abbreviate $\Delta(fixed(H))$ to simply $\Delta(H)$. Moreover, we sometimes will use the same name H to designate just the fixed nodes of schema H. The relative diameter *rel*$\Delta(H)$ captures the notion of how closely together the fixed positions of H are situated in the host chromosome T.

We assume all our individuals have the same shape as binary trees. Cutting one edge in a tree divides the tree into two connected subtrees; these fragments are to be used to construct two children at crossover time.

Cutting a tree edge *separates two nodes* x and y if that edge is one lying in the unique path between x and y; in this case x and y end up in different fragmental subtrees. A cut *separates a set S* of tree nodes if there are two nodes in S which are separated by the cut. A cut *disrupts* a schema H if it separates the fixed positions of H. We intend to practice mating with crossover by uniformly randomly cutting one edge of chromosomal tree T, and interchanging parental fragments to form the children. Hence we will define the *disruption probability dp(H)* of schema H to be the fraction (number of edges that disrupt H) / (total number of edges in T).

Given a schema H of tree T, we let T_H denote the smallest subtree of T which contains all the fixed positions of H. T_H unequivocally exists, for it is the intersection of all the subtrees of T which contain the fixed positions of H. (Our research was done independently from Poli & Langdon [], but there is an overlap of ideas. The number of edges that disrupt H comes closest to what they term the defining length of a tree schema, and they give the name minimum tree fragment to T_H.)

We state without proof the following results.

Proposition 1: (a) Each leaf of T_H is an element of H;
  (b) $\Delta(H) = \Delta(T_H)$, and $rel\Delta(H) = rel\Delta(T_H)$;
  (c) The set of T-edges that separate H is the same as the set that separate T_H;
  (d) $dp(H) = dp(T_H)$.

We will be interested in if and when the relation $dp(H) \le rel\Delta(H)$ holds. First we observe that a linear sequence of bits (as in classical GAs) is in particular also a binary tree of bits, and in this case $dp(H)$ equals $rel\Delta(H)$ and both equal $\delta(H) / (N - 1)$ where, recall, $\delta(H)$ is the defining length of H and N is the number of bits in the sequence.

3   The General Case: Arbitrary Binary Trees

Proposition 2: There is no constant $k > 0$ which will make the inequality
$dp(H) \le k \cdot rel\Delta(H)$ hold for arbitrary schemas H in arbitrary binary trees T.

Proof: Consider the binary tree T illustrated in Figure 1. The left subtree T_l of root r is a full binary tree of height h. The (fixed nodes of) schema H consists of the nodes on the bottom level of subtree T_l. The rest of tree T besides T_l consists of the root r and the depicted nodes trailing off from it in a line to the right; we term all these (including r) as tail nodes and we let t be the number of them. The smallest subtree T_H of T which contains (the fixed nodes of) H is the left subtree T_l, and it has $2^{h+1} - 1$ nodes and therefore $2^{h+1} - 2$ edges. The tail nodes add another t edges to tree T, and it follows that $dp(H) = \dfrac{2^{h+1} - 2}{2^{h+1} - 2 + t}$. We will soon choose $t > h$, in which case $rel\Delta(H) = \dfrac{2h}{h + t}$. Now consider the ratio

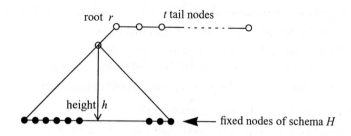

Fig. 1. Example of a wayward binary tree

$$R = \frac{dp(H)}{rel\Delta(H)} = \frac{2^{h+1}-2}{2^{h+1}-2+t} \cdot \frac{h+t}{2h} .$$

If we choose $t = 2^{h+1}$, then $\dfrac{2^{h+1}-2}{2^{h+1}-2+t} \approx \dfrac{1}{2}$ and hence ratio $R \approx \dfrac{h+t}{4h}$, which

can be made arbitrarily large. This proposition now follows.

4 Complete Binary Trees

Proposition 3: The inequality $dp(H) \le rel\Delta(H)$ holds for all schemas H and all complete binary trees T, with the exception of certain small trees T, and certain schemas H containing very large numbers of fixed positions.

    Proof: Recall the notation, T_H is the smallest subtree of T which contains (the fixed positions of) H. If 1 is the value of the fraction $dp(H) =$ (number of edges that separate T_H) / (number of edges in T), it follows that $T_H = T$, then $\Delta(H) = \Delta(T_H) = \Delta(T)$, then $rel\Delta(H) = 1$, so that $dp(H) = rel\Delta(H)$. Thus, the interesting case is when the fraction $dp(H)$ is strictly less than 1.

Given a certain diameter Δ, there are many schemas H which have that diameter. Some are large and some are small, and the same can be said for the enclosing subtree T_H of H. Now imagine the diameter value as a given. We will find an upper bound for fraction $dp(H)$, by calculating the most that its numerator can be, and then the least its denominator can be and still exceed the numerator.

The numerator of $dp(H)$ can be as large as the number of edges in the largest possible subtree T_H whose diameter is the given diameter value. Let h_d be a deepest node of H. We introduce some notation: let h be the height of tree T; let d be the depth of h_d; let $\delta = \Delta(H)$. Every node of T_H must be within distance $\delta = \Delta(H)$ of any given element of H, in particular this is so for h_d. We will count how many nodes can possibly be in our chromosomal tree T, be no deeper than h_d, and be at distance at most δ from h_d. Subtree T_H can be as big as that set of nodes. (For emphasis let us note that according to Proposition 1, also the set of fixed nodes of H can be that big.) After we have determined T_H, we will find the smallest complete binary tree T of height h such that $T \supseteq T_H$. These two, T and T_H, will determine the largest that fraction $dp(H)$ can be, for the given diameter value.

Either diameter δ is even or it is odd. And either $\delta \leq d$ or $\delta > d$. In our proof we will consider four cases. We will give full details for two of the cases and leave the details of the other two cases to the reader.

Case I: even $\delta \leq d$: (See Figure 3 for guidance.) Consider the path of length δ, consisting of ancestor nodes from h_d towards the root of T; denote the nodes on this path as $h_d = v_0, v_1, v_2, \dots, v_\delta$. Subtree T_H could contain all the nodes in a full subtree,

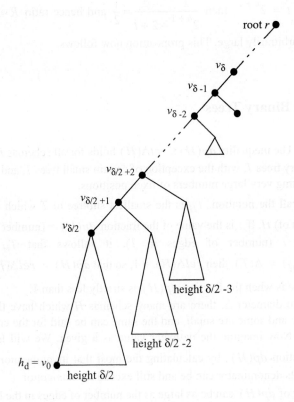

Fig. 2. δ is even, $\delta \leq d$

rooted at $v_{\delta/2}$ and having height $\delta/2$. Similarly, T_H could contain all the nodes in a full subtree, rooted at the second child of $v_{\delta/2+1}$, and having height $\delta/2-2$. Similarly for a full subtree, rooted at the second child of $v_{\delta/2+2}$, and having height $\delta/2-3$. And so on, back to a full subtree, rooted at the second child of $v_{\delta-2}$, and having height 1; then a full subtree of height 0 that consists exactly of the second child of $v_{\delta-1}$, and, finally, node v_δ by itself, of course. How many nodes have we to count? There are the $\delta/2$ full subtrees we have identified, and also the $\delta/2$ ancestor nodes $v_{\delta/2+1}, \ldots, v_\delta$ which are not inside the identified subtrees. The sizes of the full subtrees are the exceptional size $2^{\delta/2+1}-1$, then sizes $2^{\delta/2-1}-1$, $2^{\delta/2-2}-1$, down to 2^1-1. Altogether the number of nodes we are tallying is $2^{\delta/2+1}+(2^{\delta/2-1}+2^{\delta/2-2}+\ldots+2^1) = 2^{\delta/2+1}+(2^{\delta/2}-2) = 3 \cdot 2^{\delta/2}-2$. This latter value is the most nodes that T_H can have, hence the most edges that can be in T_H is one less, or $3 \cdot 2^{\delta/2}-3$.

Continuing with Case I, we now consider host chromosomal tree T. If the depth d of node h_d is less than the height h of T, then to be a complete tree of height h, T can be as small as having only one node on level h, in which case T has 2^h nodes and therefore 2^h-1 edges. But if the depth d of h_d equals the height h of T, then since we have allowed T_H to be as big as containing the full subtree of height $\delta/2$ rooted at node $v_{\delta/2}$, it follows that T will be required to have at least $2^{\delta/2}$ nodes on its bottom level h. Then T must have at least $2^h+2^{\delta/2}-1$ nodes and hence at least $2^h+2^{\delta/2}-2$ edges. Ergo, $dp(H)$ is bounded above by $\dfrac{3 \cdot 2^{\delta/2}-3}{2^h-1}$ if $d<h$, but

bounded above by $\dfrac{3 \cdot 2^{\delta/2}-3}{2^h+2^{\delta/2}-2}$, if $d=h$.

Since T is complete but not necessarily full, $\Delta(T)$ is either $2h$ or $2h-1$; in either event, $rel\Delta(H) \geq \dfrac{\delta}{2h}$.

Putting our two bounds together, and for now focusing on the possibility that $d<h$, our task is to determine if or when $\dfrac{3 \cdot 2^{\delta/2}-3}{2^h-1} \leq \dfrac{\delta}{2h}$, or equivalently, if or when

$\dfrac{3 \cdot 2^{\delta/2} - 3}{\delta/2} \le \dfrac{2^{h} - 1}{h}$. Define the function f by $f(n) = \dfrac{2^{n} - 1}{n}$. At issue is if or when

$3 \cdot f(\delta/2) \le f(h)$. Since function f is near-exponential, certainly this latter inequality will hold if δ is not too near its maximum value of $2h$, except the inequality may possibly fail for small values of h. On the other hand, in Case I, $\delta \le d \le h$ and so δ is not near $2h$.

We used a computer program to examine if or when the inequalities

$$\dfrac{3 \cdot 2^{\delta/2} - 3}{2^{h} - 1} \le \dfrac{\delta}{2h} \qquad \text{, for even } \delta \le d < h \text{, and}$$

$$\dfrac{3 \cdot 2^{\delta/2} - 3}{2^{h} + 2^{\delta/2} - 2} \le \dfrac{\delta}{2h} \qquad \text{, for even } \delta \le d = h$$

held, for T heights h in the range from 2 to 100 (T of height 1 amounts to trivialities), and found the following results. There were only three failures, involving trees of heights 2 and 3.

Case II: odd $\delta > d$. (See Figure 4 for guidance.) Note that since h_d is a deepest node in H, $d \ge \lceil \delta/2 \rceil$. This time we consider the path of length d, consisting of ancestor

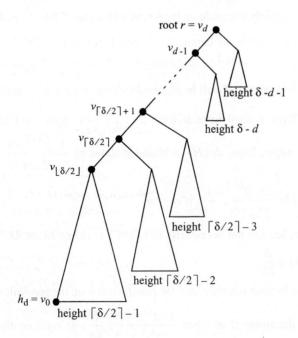

Fig. 3. δ is odd, $\delta > d$

nodes from h_d back to the root r of T; denote the nodes on this path as $h_d = v_0, v_1, v_2,$ ..., $v_d = r$. (See Figure 4.) Subtree T_H could contain all the nodes in a full subtree, rooted at $v_{\lfloor \delta/2 \rfloor}$ and having height $\lfloor \delta/2 \rfloor = \lceil \delta/2 \rceil - 1$. Similarly, T_H could contain all the nodes in a full subtree, rooted at the second child of $v_{\lfloor \delta/2 \rfloor + 1} = v_{\lceil \delta/2 \rceil}$, and having height $\delta - (\lfloor \delta/2 \rfloor + 2) = \lceil \delta/2 \rceil - 2$. Similarly for a full subtree, rooted at the second child of $v_{\lceil \delta/2 \rceil + 1}$, and having height $\lceil \delta/2 \rceil - 3$. And so on, back to a full subtree, rooted at the second child of v_{d-1}, and having height $\delta - d$; and, finally, a full subtree, rooted at the second child of T's root $r = v_d$, and having height $\delta - d - 1$. Thus, T_H could contain the $d - \lfloor \delta/2 \rfloor$ nodes $v_{\lceil \delta/2 \rceil}$, ..., v_d, plus the nodes in $d - \lfloor \delta/2 \rfloor + 1$ full subtrees of heights $\lceil \delta/2 \rceil - 1$, ..., $\delta - d - 1$. The number of nodes in T_H could be as great as $d - \lfloor \delta/2 \rfloor + ((2^{\lceil \delta/2 \rceil} - 1) + ... + (2^{\delta - d} - 1)) = 2^{\lceil \delta/2 \rceil + 1} - 2^{\delta - d} - 1$. The number of edges in T_H would be one less, or $2^{\lceil \delta/2 \rceil + 1} - 2^{\delta - d} - 2$. Reasoning as in Case I, we see that $dp(H)$ will be no greater than $rel\Delta(H)$ provided

$$\frac{2^{\lceil \delta/2 \rceil + 1} - 2^{\delta - d} - 2}{2^h - 1} \le \frac{\delta}{2h} \quad , \text{ for odd } \delta > d \text{ when } d < h, \text{ and}$$

$$\frac{2^{\lceil \delta/2 \rceil + 1} - 2^{\delta - d} - 2}{2^h + 2^{\lfloor \delta/2 \rfloor} - 2} \le \frac{\delta}{2h} \quad , \text{ for odd } \delta > d \text{ when } d = h.$$

At this point we can observe that in the event that the depth d of h_d equals the height h of T, and the diameter δ of H equals $2h - 1$, then the second inequality fails, since fraction $\dfrac{2^{\lceil \delta/2 \rceil + 1} - 2^{\delta - d} - 2}{2^h + 2^{\lfloor \delta/2 \rfloor} - 2}$ simplifies to 1, whereas $\dfrac{\delta}{2h} = \dfrac{2h - 1}{2h}$ is strictly less than 1.

A computer program which examined tree heights in the range from 2 to 100 revealed all the failure instances just commented upon. Beyond those, the run revealed only four other particular failures of the relation $dp(H) \le rel\Delta(H)$ and they involved trees of heights 3, 4, and 5.

Let us contemplate the failures that are arising when $d = h$ and $\delta = 2h - 1$. We calculated the largest that T_H can be. T_H can achieve our bound, but only if H contains as fixed nodes all the nodes on the bottom levels of the full subtrees we cited, in which case H contains $2^{\lceil \delta/2 \rceil - 1} + 2^{\lceil \delta/2 \rceil - 2} + ... + 2^{\delta - d - 1} = 2^{\lceil \delta/2 \rceil} - 2^{\delta - d - 1}$ fixed nodes. When $d = h$ and $\delta = 2h - 1$, this means H contains $3 \cdot 2^{h-2}$ fixed nodes, which is slightly more than half of the $2^h + 2^{\lfloor \delta/2 \rfloor} - 2 = 3 \cdot 2^{h-1} - 2$ nodes in chromosome T. We generally expect building blocks to be smaller portions of the chromosome than that. Put another way, the failures arising when $d = h$ and $\delta = 2h - 1$ are ones for quite atypically large schemas.

Case III: odd $\delta \le d$: Then $dp(H)$ will be no greater than $rel\Delta(H)$ provided

$$\frac{2^{\lceil \delta/2 \rceil + 1} - 3}{2^h - 1} \le \frac{\delta}{2h} \quad , \text{ for odd } \delta \le d < h \text{, and}$$

$$\frac{2^{\lceil \delta/2 \rceil + 1} - 3}{2^h + 2^{\lfloor \delta/2 \rfloor} - 2} \le \frac{\delta}{2h} \quad , \text{ for odd } \delta \le d = h$$

A computer program which examined if and when these inequalities held, for T heights h in the range from 2 to 100, revealed only three failures, involving trees of heights 2 and 3.

Case IV: even $\delta > d$: Then $dp(H)$ will be no greater than $rel\Delta(H)$ provided

$$\frac{3 \cdot 2^{\delta/2} - 2^{\delta - d} - 2}{2^h - 1} \le \frac{\delta}{2h} \quad , \text{ for even } \delta > d \text{ when } d < h \text{, and, and}$$

$$\frac{3 \cdot 2^{\delta/2} - 2^{\delta - d} - 2}{2^h + 2^{\delta/2} - 2} \le \frac{\delta}{2h} \quad , \text{ for even } \delta > d \text{ when } d = h.$$

We ran a computer program which examined if and when these inequalities held, for T heights h in the range from 2 to 100, with the following results. Invariably the first inequality failed when $h = d + 1$ and $\delta = 2d$. This class of failure, like that of Case II, arises when H contains a very large number of fixed nodes, approximately half the nodes of host chromosomal tree T. Beyond this class of failure, there were altogether only four other particular failures, and they involved trees of heights 3, 4, and 5.

Clearly there is a lesson to be learned from Propositions 2 and 3. The example tree given in the proof of Proposition 2 is irregular, a clump connected to a long string of bits, and for it $dp(H)$ can be made much greater than $rel\Delta(H)$. Proposition 3 shows us that $dp(H)$ is much better behaved when the binary tree is compact. If one is going to structure one's chromosome as a binary tree, then it is better to use as compact a tree as possible.

5   Conclusion

It is reasonable to say that schema disruption probability is better behaved the more closely it is related to how closely together are situated the fixed positions of the schema. That is, $dp(H)$ is better behaved the more closely it is approximated or dominated by $rel\Delta(H)$.

This paper contains two informative results, in Propositions 2 and 3. They show that if one is going to link the bits of a chromosome in the structure of a binary tree, then it is best to make the tree as compact as possible (in the context of the problem at hand).

The results of this paper are ones from a larger paper we are writing, in which we explore more types of trees and other issues as well. We will submit the larger paper elsewhere.

More generally we are interested in schema disruption behavior in chromosomes structured in other than the classical way as a linear sequence. After trees we plan to explore chromosomes that are 2- or 3- (or n-) dimensional grids. An example of the latter is to use as nodes the set $\{ (x, y, z) \in \Re^3 \mid x, y, z$ are integers between 0 and $K \}$, then connect each node to each of its up to six neighbors in the axial directions. To cut such a chromosome in two, we could use random planes in 3-space, or random planes which are parallel to an axis, or some other way. These considerations await further exploration.

References

1. Holland, John (1975). *Adaptation in Natural and Artificial Systems*. University of Michigan Press, Ann Arbor, MI.

2. Goldberg, David E. (1989). *Genetic Algorithms in Search, Optimization, and Machine Learning*. Addison-Wesley Publishing, Reading, MA.

3. Greene, William A. (2000). "A Non-Linear Schema Theorem for Genetic Algorithms," in Whitley, D. (Eds.) Proceedings of the Genetic and Evolutionary Computation Conference (GECCO-2000), pp. 189-194. Morgan Kaufmann Publishers, San Francisco, CA.

4. Koza, John R. (1992). *Genetic Programming: On the Programming of Computers by Natural Selection*. MIT Press, Cambridge, MA

5. O'Reilly, Una-May (1995). *An Analysis of Genetic Programming*. PhD thesis, Carleton University, Ottawa-Carleton Institute for Computer Science, Ottawa, Ontario, Canada, 22 September 1995.

6. O'Reilly, Una-May, and Franz Oppacher (1995). "The Troubling Aspects of a Building Block Hypothesis for Genetic Programming", in Whitley, D. and Vose, M. D. (eds.) *Foundations of Genetic Algorithms 3*. Morgan Kaufmann Publishers, San Francisco.

7. Whigham, Peter A. (1995) "A Schema Theorem for Context-Free Grammars", in *1995 IEEE Conference on Evolutionary Computation*, Vol 1, pp. 178-181. IEEE Press

8. Rosca, Justinian P. (1997). "Analysis of Complexity Drift in Genetic Programming", in Koza, John R. *et al*. (eds) *Genetic Programming 1997: Proceedings of the Second Annual Conference* (pp. 286-294). Morgan Kaufmann Publishers, San Francisco, CA.

9. Poli, Riccardo, and William Langdon (1998). "Schema Theory for Genetic Programming with One-Point Crossover and Point Mutation". *Evolutionary Computation* 6(3), pp. 231-252. MIT Press, Cambridge, MA.

The Royal Road Not Taken: A Re-examination of the Reasons for GA Failure on R1

Brian Howard[1] and John Sheppard[1,2]

[1] The Johns Hopkins University, 3400 N. Charles Street, Baltimore, MD 21218
itsbehoward@hotmail.com
[2] ARINC Engineering Services, LLC, 2551 Riva Road, Annapolis, MD 21401
jsheppar@arinc.com
jsheppa2@jhu.edu

Abstract. Previous work investigating the performance of genetic algorithms (GAs) has attempted to develop a set of fitness landscapes, called "Royal Roads" functions, which should be ideally suited for search with GAs. Surprisingly, many studies have shown that genetic algorithms actually perform worse than random mutation hill-climbing on these landscapes, and several different explanations have been offered to account for these observations. Using a detailed stochastic model of genetic search on R1, we attempt to determine a lower bound for the required number of function evaluations, and then use it to evaluate the performance of an actual genetic algorithm on R1.

1 Introduction

Many theoretical frameworks for understanding the overall performance of genetic algorithms assume implicitly that the observed evolutionary dynamics of finite populations should follow those of a hypothetical infinite population. For example, the "implicit parallelism" predicted by the schema theorem assumes that schema sampling is sufficiently unbiased to provide a reasonably accurate estimate of each observed schema's average fitness. However, in reality, sampling anomalies can profoundly influence the way that a population evolves in the short term, especially when the size of that population is small.

For example, Mitchell, *et al.* present an apparent conundrum: on a fitness landscape designed specifically to be amenable for search with a genetic algorithm, the simpler Random Mutation Hill-Climbing (RMHC) approach defeats the genetic algorithm by an order of magnitude [4]. Specifically, their paper demonstrated that the genetic algorithm requires approximately 61,000 fitness evaluations to find the optimal solution, versus only about 6,200 for the hill-climber. This observation has been widely interpreted as evidence against the relevance of the building block hypothesis. These tests used the Royal Roads fitness function R1, which is defined as follows and displayed graphically in Figure 1:

$$R1(x) = \sum_i c_i \sigma_i(x), \quad \text{where } \sigma_i(x) = \begin{cases} 1 & \text{if } x \in s_i \\ 0 & \text{otherwise} \end{cases} \tag{1}$$

K. Deb et al. (Eds.): GECCO 2004, LNCS 3102, pp. 1208–1219, 2004.
© Springer-Verlag Berlin Heidelberg 2004

```
S₁   = 11111111*******************************************************; c1=8
S₂   = ********11111111***********************************************; c2=8
S₃   = ****************11111111***************************************; c3=8
S₄   = ************************11111111*******************************; c4=8
S₅   = ********************************11111111***********************; c5=8
S₆   = ****************************************11111111***************; c6=8
S₇   = ************************************************11111111*******; c7=8
S₈   = ********************************************************11111111; c8=8
Sₒₚₜ = 1111111111111111111111111111111111111111111111111111111111111111
```

Fig. 1. Royal Roads landscape R1

Mitchell, *et al.* attributed the poor performance of the genetic algorithm primarily to the phenomenon of genetic hitchhiking: once an instance of a highly fit schema is discovered, its high fitness allows that schema to spread quickly throughout the population, with zeros in other positions in the string "hitchhiking" along with the building blocks. To reduce the influence of hitchhiking on the GA's performance, they introduced "introns" between the adjacent schemata. However, their experiments showed that, in the case of R1, introns do not improve the GA's performance significantly, highlighting the fact that the mechanism of hitchhiking in genetic algorithms is not clearly understood.

In the following section we will attempt to create a model that defines a reasonable lower bound on the expected performance of a genetic algorithm on landscape R1. We will then use this model to investigate the hypothesis that the poor performance of the genetic algorithm can be partially explained by hitchhiking, but that sampling error and population cost also contribute substantially.

2 A Stochastic Model of Genetic Search on Landscape R1

In the original Royal Roads work [4], Mitchell *et al.*, introduced the "Ideal Genetic Algorithm", or IGA, to determine an approximate lower bound on the number of fitness evaluations required by a genetic algorithm searching landscape R1. According to this model, the researchers assert that an ideal genetic algorithm should be able to find the optimum on R1 in approximately 696 evaluations on average. However, the following points illustrate the need for a more realistic lower bound:

1. The IGA algorithm requires perfect knowledge of all eight schemata, *a priori*. Normally, fitness functions are designed to compute the relative fitness of an entire string as a whole, without any facility to explicitly recognize partial solutions.

2. The algorithm does not use a population and, thus, the cost of maintaining a population is not considered in the model.

3. In general, the mutation and crossover mechanisms proposed do not provide a realistic analogy to what might actually be expected from a real genetic algorithm.

Perhaps the most "exact" method of modeling search using a genetic algorithm is described in Vose [9,10]. Here the genetic algorithm is described as a specific instance of the more general Random Heuristic Search process (RHS). Genetic algorithms, simulated annealing [2], PBIL [1], and hill-climbing can all be modeled under this framework. In general, RHS begins with a randomly initialized population, represented as a frequency vector that describes the relative proportions of each possible member of the search space. At each step, the current population is passed to a heuristic function, G, that returns a multinomial distribution from which the next generation can be generated via stochastic sampling. The transition from one generation to the next can be viewed equivalently as the application of a transition function, τ. Although the function τ is an induced mapping, the process is Markovian because it produces each new population contingent only on the state of the previous population.

Vose describes a precise model of a heuristic function that can be used to model a binary encoding of the simple genetic algorithm [10]. Other authors have extended this model to cover, for example, general cardinality representations [3] and alternate forms of selection [8]. In theory, a variant of Vose's simple genetic algorithm could be employed to model a GA's search of R1; however, in practice, the state space for such a model would require a transition matrix with 2^{64} states, making this approach intractable.

In an attempt to construct a more manageable model of the R1 landscape, Suzuki and Sawai reduce the search space by describing each set of eight bits in terms of the number of ones present, without tracking the exact positions of those ones within the schemata [6]. This reduces the search space somewhat, but still required Suzuki and Sawai to limit their analysis to smaller versions of Royal Roads functions having only a fraction of the total number of bits in R1. Nevertheless, using such a model, they were able to show that crossover accelerates GA search when compared to an equivalent GA without crossover.

Another attempt to model the search of R1 with a genetic algorithm was introduced by van Nimwegen, Crutchfield, and Mitchell [7]. This model groups together all strings that have the same fitness value. Using this model, van Nimwegen *et al.* were primarily interested in exploring the high-level dynamics of the search process, in particular, the occurrence of punctuated equilibria and fitness epochs. In their model, they excluded crossover from the analysis, acknowledging that such an approach may leave out many key details of search necessary to explore other problems.

In this paper, we introduce a new model of genetic search on R1. Using Vose's general RHS framework, our model groups population members by the schemata that they contain. Furthermore, when any one of the eight schemata is missing from a chromosome, our model assumes that the eight bits at the corresponding locations are true "don't cares," freshly sampling, at each generation, from the set of all binary strings of length eight. In this manner, our model engages in an "implicitly parallel search" and eliminates the possibility of bit-wise hitchhiking. By comparing the performance of such a model to the performance of an actual genetic algorithm, we should, therefore, be able to determine the approximate degree to which hitchhiking is actually a problem on R1. The following sections define this model more precisely.

2.1 Representation of the Population

In our model, each of the 256 possible combinations of the schemata s_1-s_8 is represented as a unique 8-bit string c, in which each bit, c_i, indicates the presence or absence of schema s_i. A one in a particular position indicates that the corresponding schema is present, while a zero signifies that the schema is absent. For example, the bit string, 10010001, describes all strings matching the pattern:

`11111111 ******* ******** 11111111 ******** ******* ******* 11111111`

Throughout this paper we will refer to such 8-bit strings as *schema configurations,* or simply as *configurations.* Given this representation, a particular population can be characterized by a real vector $\vec{p}$ of length 256, which indicates the frequencies of each of the possible R1 schema configurations in that population. We will index these configurations with the integers 0–255, such that each index maps to the integer representation of the corresponding schema configuration, interpreted as a binary number. Hence, the configuration 00000000 is given index 0, while configuration 00001000 is given index 8. Using this indexing scheme, the ith position in a particular population vector represents the frequency with which this schema configuration occurs in the underlying population. For example, the following population vector:

$$\vec{p} = [0.25 \quad 0.05 \quad \cdots \quad 0.01] \tag{2}$$

indicates that in 25% of the underlying population, none of the schemata s_1-s_8 are present (configuration 00000000). In 5% of the population, only schema s_8 is present (configuration 00000001), while in 1% of the population all of the schemata s_1-s_8 are present (configuration 11111111).

By grouping together all bit strings with the same arrangement of schemata, this representation reduces the state space from the set of 2^{64} 64-bit binary strings to the set of 2^8 8-bit binary strings. Furthermore, this model forces us to assume that when a schema is not present in a string, then nothing else is known about the eight bits in that schema's partition. If we build our model's operators such that the distribution of these unknown "don't care" bits is assumed to be uniform, then we have a model that implicitly removes the possibility of bit-level hitchhiking.

2.2 Mutation

If the population is represented by the vector, $\vec{p}$, then the mutation operation can be described as a 256 × 256 matrix, $\mathbf{M} = [m_{ij}]$, where each element m_{ij} represents the probability of transitioning from a population member with the schema configuration i to a population member with schema configuration j, in one generation, due to mutation. With this representation, the mutation operator can be applied to a population using matrix multiplication:

$$\vec{p}_m = \vec{p}\mathbf{M} \tag{3}$$

Given any particular eight-bit parent configuration, c, we can calculate the probability that mutation transforms this schema configuration into child configuration c' by considering each of the eight schemata independently. For each of the eight schemata, $s_i \in \{ s_1, \ldots, s_8 \}$, there are four distinct cases corresponding to whether the child string has the schema given the parent does/does not have that same schema.

The individual likelihoods of these four cases can be calculated with the application of basic probability theory, assuming that if a particular population member lacks schema s_i, then the actual bits within partition i are equally likely to be zeroes as they are ones. Starting with the child having the schema given the parent does not, we need to compute:

$$\Pr(c_i' = 1 \mid c_i = 0) \tag{4}$$

Equation (4) can be decomposed as follows. Let "$i \mapsto m$" denote the occurrence of at least one mutation in partition i.

$$\Pr(c_i' = 1 \mid c_i = 0) = \Pr(c_i' = 1 \mid c_i = 0, i \mapsto m)\Pr(i \mapsto m \mid c_i = 0) \tag{5}$$
$$+ \Pr(c_i' = 1 \mid c_i = 0, i \not\mapsto m)\Pr(i \not\mapsto m \mid c_i = 0)$$
$$= \Pr(c_i' = 1 \mid c_i = 0, i \mapsto m)\Pr(i \mapsto m)$$

Here, the probability of mutating partition i is independent of the state of the parent, so $\Pr(i \mapsto m \mid c_i = 0) = \Pr(i \mapsto m)$. Also, note that the second term in the sum drops out because $\Pr(c_i' = 1 \mid c_i = 0, i \not\mapsto m) = 0$. So, in other words, the probability of creating schema s_i in an offspring from a parent that lacks s_i is equal to the probability of creating this schema from such a parent given that at least one mutation occurs in this schema partition, times the probability that such a mutation occurs.

Let μ be the bitwise mutation probability. If the parent string lacks a particular schema, s_i, and the child gains this schema after applying the mutation operator, then we know that a mutation must have occurred somewhere within the eight bit partition attributed to this schema. The probability that a mutation occurs somewhere within a particular 8-bit schema is:

$$\Pr(i \mapsto m) = 1 - (1 - \mu)^8 \tag{6}$$

Assuming a uniform probability distribution for bits in the parent that are not part of a previously discovered schema, the probability that any one of these bits is a one is 0.5, and the probability that any one of these bits is a zero is also 0.5[†]. Let μ' be the bitwise mutation rate within a schema, *given that at least one such mutation has occurred.* Then the first half of the right side of equation (5) reduces to the following:

$$\Pr(c_i' = 1 \mid c_i = 0, i \mapsto m) = (0.5(1 - \mu') + 0.5(\mu'))^8 = (0.5^8) \tag{7}$$

Substituting the above result, along with the probability that a mutation occurs in a given schema, back into equation (5), we get the following result:

$$\Pr(c_i' = 1 \mid c_i = 0) = \Pr(c_i' = 1 \mid c_i = 0, i \mapsto m)\Pr(i \mapsto m) \tag{8}$$
$$= (0.5)^8(1 - (1 - \mu)^8)$$

Note further that

[†] Actually, the probability that a bit is 0 is slightly greater than 0.5, since we *know* that the parent's bit sequence is not 11111111; that is, at least one of the bits must be zero.

$$\Pr(c_i' = 0 \mid c_i = 0) = 1 - \Pr(c_i' = 1 \mid c_i = 0) \tag{9}$$

The remaining cases are straightforward. If the parent contains a particular schema, s_i, then the probability that the child does not contain this schema after mutation is simply the probability that a mutation occurs anywhere within this schema partition:

$$\Pr(c_i' = 0 \mid c_i = 1) = \Pr(i \mapsto m) \tag{10}$$

Likewise,

$$\Pr(c_i' = 1 \mid c_i = 1) = 1 - \Pr(i \mapsto m) \tag{11}$$

Given these four cases, it is now possible, for any particular parent configuration c, and child configuration c' to calculate the probability that c' is the offspring of c after applying the mutation matrix, $\mathbf{M}$, where:

$$m_{ij} = \prod_{k=1}^{8} \Pr(c_k' \mid c_k) \tag{12}$$

2.3 Crossover

Crossover is a bit more complex because we need an operator that takes into account the interaction between *two* parents to produce each offspring. For our model, we represent one-point crossover with a $256 \times 256 \times 256$ hypermatrix $\mathbf{C}$, such that when this hypermatrix is divided into planes along its third dimension, each plane with index $\breve{c}$ represents a transition matrix, where each entry $m_{\hat{c}c'}$ describes the probability that a parent with a particular schema configuration, $\hat{c}$, produces a child c', given that the second parent is $\breve{c}$. At each generation, $\mathbf{C}$ can be applied to the next generation as follows. Let $\Pr(\breve{c})$ be the probability of selecting configuration $\breve{c}$ as the second parent in crossover, given the distribution of configurations resulting from mutation, $\vec{p}_m$, and $\mathbf{C}_{\breve{c}}$ be the slice of hypermatrix $\mathbf{C}$ corresponding to the second parent $\breve{c}$. Then, the new distribution of configurations, $\vec{p}_{m\chi}$, after mutation and crossover is calculated as follows:

$$\vec{p}_{m\chi} = \sum_{\breve{c}} \Pr(\breve{c})[\vec{p}_m \mathbf{C}_{\breve{c}}] \tag{13}$$

Next we will describe a procedure for building the crossover hypermatrix, $\mathbf{C}$. Note that crossover can occur in the middle of a particular schema partition or between schema partitions, resulting in 15 distinct crossover points. Assuming that crossover is equally likely to occur at any of the 64 bit positions in the R1 chromosome, the probability that crossover occurs within a particular partition, given that crossover occurs, is 7/63, while the probability of crossover occurring between two particular adjacent partitions is 1/63.

Given parent strings, $\hat{c}$ and $\breve{c}$, a crossover event at a particular point can be described with a bit mask. We will describe building the crossover hypermatrix, $\mathbf{C}$ with an example. Suppose $\hat{c} = 26$, $\breve{c} = 52$, and the crossover point is 10. Using the indexing scheme described in Section 3.1, $\hat{c} = 26$ maps to chromosome 00011010 and $\breve{c} =$ 52 maps to chromosome 00110100. With crossover point 10, the masks are $\mathbf{A} =$ 11111000 and $\mathbf{B} = 00000111$ respectively. The children are determined by applying the following logical operators:

$$c' = (\hat{c} \wedge \mathbf{A}) \vee (\breve{c} \wedge \mathbf{B})$$
$$c'' = (\breve{c} \wedge \mathbf{A}) \vee (\hat{c} \wedge \mathbf{B})$$

where $\mathbf{A}$ is Mask A and $\mathbf{B}$ is Mask B. Given this information, there are two possible children, $c' = 00011100$ and $c'' = 00110010$.

Given this, we can determine the probability either child will result via crossover, given parents $\hat{c}$ and $\breve{c}$. Let χ be the probability crossover occurs. Let ψ_i indicate crossover point i was selected (e.g., ψ_{10} corresponds to crossover point 10). Finally, let n = the number of children generated by crossover. Then

$$\Pr(c') = \Pr(c'') = \left(\frac{1}{n}\right)\Pr(\psi_i)\chi \qquad (14)$$

The procedure described above works perfectly when crossover occurs between schemata. When crossover occurs within a partition, the situation is a bit more complicated. For example, at crossover point 11, $\mathbf{A} = 11111000$ and $\mathbf{B} = 00000011$. Because bit six in both of the masks is a zero, simply applying the masks in the logical operations would prevent either child from ever having schema six. In reality, the offspring may indeed have the schema, either because both parents had the schema, or because the schema is created in the child after mixing bits from both parents. Thus in the event a schema partition is disrupted by crossover, we need to explicitly calculate the probability that the child inherits the disrupted schema and modify our list of children and their probabilities accordingly. There are 3 distinct possibilities.

If both parents have the schema, then the probability that the child will have the schema after crossover, even if it occurs in the middle of this schema, is 1. However, because the crossover masks turn off the schema bit in the offspring, we must modify the children produced by the procedure outlined above by turning this bit back on. The probabilities are calculated as shown in equation (14), using the appropriate $\Pr(\psi_i)$.

If exactly one of the parents has the schema that is disrupted, then there are actually *four* possible children. First, it is possible that both children lose the schema. At the same time, based on the actual configuration of the parents, it is possible that either child could "regain" the schema. Thus, the probability that the schema in question is "preserved" in the offspring is calculated as follows:

$$\Pr(s_i \mid \psi_{2i-1}) = \frac{1}{7}\sum_{j=1}^{7}(0.5)^{8-j}(1-(0.5)^j) \qquad (15)$$

where i is the index of the schema partition considered, and the sum is taken over the seven possible crossover positions within the schema partition. That is, if that crossover occurs in some schema, it is equally likely to occur in each of seven distinct places, between each of the eight bits comprising the partition.

Thus the probability of producing children c''' and c'''', in which schema i is preserved after crossover is calculated as follows:

$$\Pr(c''') = \Pr(c'''') = \left(\frac{1}{4}\right)\Pr(s_i \mid \psi_{2i-1})\Pr(\psi_i)\chi \tag{16}$$

While for children c' and c'', in which schema i is lost, it is:

$$\Pr(c') = \Pr(c'') = 1 - \Pr(c''') \tag{17}$$

If neither parent has the schema that is disrupted, it is still possible that the schema could be created in the child simply due to the chance mixing of bits from the two parents. Once again, there are four possible children, which can be determined in exactly the same manner as shown above. In this case, we calculate the probability of generating a new schema in the child, assuming that both parents lack the schema, and that the bits in both parents are uniformly distributed. This can be computed as:

$$\Pr(s_i \mid \psi_{2i-1}) = \frac{(0.5)^8}{7}\sum_{j=1}^{7}(1-(0.5)^j)(1-(0.5)^{8-j}) \tag{18}$$

Again, there are seven distinct places where crossover can occur. If crossover occurs at bit j, then parent $\hat{c}$ must have only ones in positions 1 through j, and also must *not* have all ones in positions $j + 1$ through 8 (since we know that the other parent does not have the schema). Similarly, parent $\check{c}$ must have all ones in positions $j + 1$ through 8, and also must not have all ones in positions 1 through j.

These calculations lead directly to Equation (18) for computing the probability that two parents that lack a particular schema create the schema in an offspring, given that a crossover occurs within this schema's bit locations. Given these three distinct cases, it is now possible to compute the crossover hypermatrix, C, by enumerating all possible combinations of parents $\hat{c}$ and $\check{c}$ to produce various children.

2.4 Sigma Truncation Selection

Before applying crossover and mutation, the population is first subject to sigma truncation selection (Mitchell, 1998) and frequencies are updated accordingly, where

$$\text{Expected Offspring for } \vec{p}_i = \begin{cases} 1 + \dfrac{f(\vec{p}_i) - \bar{f}}{2\sigma}, & \text{if } \sigma \neq 0 \\ 1.0, & \text{if } \sigma = 0 \end{cases} \tag{19}$$

3 Experiments

3.1 Quantifying Hitchhiking

To simulate a search using our model, we begin by creating an initial population probability distribution that describes the relative likelihood of each possible configuration of schemata, given a uniform distribution of bits in the underlying population. Then we sample stochastically from this population probability distribution vector to create a new population vector for some particular finite population size. Next, the new (sampled) population vector is used as the input to the model outlined in Section 2 to generate the sampling distribution for the subsequent generation. These two steps are then repeated until a population member with the optimum fitness is selected during the sampling phase of the search process, at which point the search terminates and the number of generations required to find the optimum is recorded.

By performing 100 replications of search using this model, we found that, on average, the hitch-hiking free genetic algorithm required approximately 18,432 fitness evaluations to find the optimum on landscape R1, using a population size of 128, a bitwise mutation rate of .005, and a crossover rate of 0.7. In contrast, when we tested an implementation of a standard simple genetic algorithm using the same parameters as our model, the average number of evaluations required to achieve the optimum was approximately 64,490. This result agrees with the original Royal Roads research [5]. Thus, removing the effect of bitwise hitchhiking in our model has a major impact on the performance of the GA. However, given that RMHC requires only 6,200 evaluations, there must be factors in addition to hitchhiking that influence the GA's performance relative to the hill-climbing algorithm such as sampling errors.

3.2 Quantifying Sampling Errors

Genetic drift and sampling errors are artifacts of a finite population size. As population size increases, these effects should diminish. For Random Heuristic Search, it can be shown that in the limit of an infinite population, the heuristic function, G, converges to the transition function, τ. In other words, for an infinite population, the expected distribution at time t, $\vec{p}_t$, can be calculated by iterating the heuristic function on the initial population vector, bypassing the sampling phase altogether.

$$\vec{p}_t = \vec{p}_0(G^t) \tag{20}$$

To assess the performance of the infinite population model, we need to calculate the expected number of steps required for an arbitrary string selected from an initially random population to achieve the optimum state, via the transition function G. One way to compute this is to iterate G on an initially uniformly distributed population and record the frequency with which strings enter the optimum state after each time step:

$$\text{Avg\,\#\,Steps to Optimum} = \sum_{t=0}^{t \le \text{maxsteps}} \eta_t * t \tag{21}$$

where η_t represents the percentage of strings that first visit the optimum after t steps, and "maxsteps" is chosen large enough such that $\Sigma \eta_i > 0.99999$.

To ensure that we don't recount strings that visit the optimum state more than once during the time interval, we need to make a minor modification to our model. By eliminating all rows and columns that directly transition into or out of the optimum state from the population vector, $\vec{p}$, the mutation matrix, $\mathbf{M}$, and the crossover matrix, $\mathbf{C}$, we transform the optimum state into an absorbing state. At each generation, the proportion of population members in the optimum state can be calculated by subtracting from 1.0 the total proportion of members remaining in the abbreviated population.

This technique was used to compute η_t, the percent of the population newly arriving at the goal state for each generation, t. Given this information, the average number of steps required for an arbitrary population member to reach the goal state, assuming the transition function of an infinite population, can be calculated as described in equation (21). In this manner we calculated that the approximate number of generations required for an initially random string to first visit the optimum state should be about 38 generations, assuming an infinite population.

The preceding calculation quantifies the number of generations required for a *single* string to arrive at the optimum. To determine the role sampling error plays in limiting search efficiency, we want to compare the empirical, finite population model of Section 3.1 to a theoretical, finite population model that has the *dynamics* of an infinite population. For such a model we need to calculate, at each generation, the probability that *at least one population member* out of a population of some particular size has reached the optimum. At each generation this is:

$$\text{Pr(Optimum visited by at least 1 member by time } t) = 1 - \left(1 - \sum_{i=0}^{i \le t} \eta_i \right)^{PopSize} \tag{22}$$

Using this derived distribution we again applied equation (21) to calculate the average number of generations for a "drift-less" population of size 128 to discover the optimum on R1. The resulting calculation yields an estimate of approximately 30.44 generations, or about 3,896 function evaluations.

3.3 Population Cost

In Section 3.2, we calculated that if a population of size 128 could evolve free from the effects of sampling error and hitchhiking, then the number of generations required to discover the optimum would be approximately 30.4, a performance level that actually surpasses that of RMHC. Unfortunately, the limited population size ensures that drift and hitchhiking will remain a problem. If our model is accurate, however, we should observe that, as the population size is increased, the performance of a real genetic algorithm should approximate our finite population model from Section 3.2.

Fig. 2 shows the relationship between population size and search efficiency for an implementation of a genetic algorithm using the same parameters as in the original Royal Roads research. The performance predicted by the model described in Section 3.2 is also displayed as a function of population size. When population size increases, the number of generations required by the actual genetic algorithm converges to the number predicted by the model. In fact, when the population size increases to about 8,000, the performance of the model and the actual GA are approximately the same.

Fig. 3 shows the relationship between population size and the total number of evaluations required to find the optimum, for the actual genetic algorithm. Note that increasing the population size to 8,000 drastically impacts the performance of the genetic algorithm in terms of the total number of fitness evaluations. For the settings used in the original Royal Roads work, the optimum performance occurs at a population size of around 500, at which the average number of evaluations required was 33,041. Thus, it appears the primary shortcomings of the genetic algorithm are related to the cost of maintaining a population, given a serial implementation, rather than hitchhiking, which is, after all, a consequence of utilizing a population size that is too small.

4 Conclusions

In this paper we have re-examined the performance of genetic algorithms on Royal Roads fitness landscapes in comparison to the performance of simpler hill-climbing algorithms. By building a model of genetic search on landscape R1, we have attempted to show that the poor performance of the standard GA can be explained in terms of the trade-off between the cost required to maintain a large population and the consequences due to hitchhiking and drift when utilizing a population that is too small.

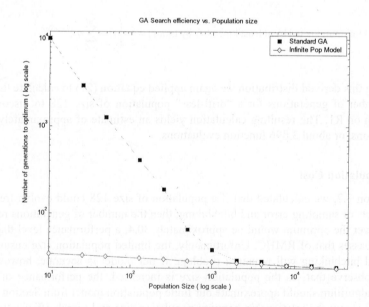

Fig. 2. GA search efficiency vs. population size

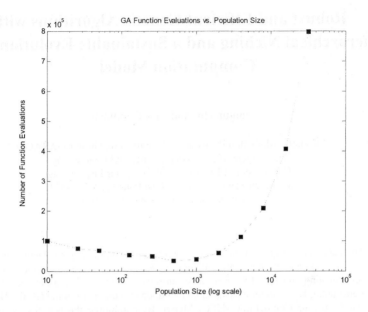

Fig. 3. GA function evaluations vs. population size

References

1. **Baluja, S. & Caruana, R.** (1995). "Removing the Genetics from the Standard Genetic Algorithm." In Prieditis, A., & Russell, S. (Eds.), *The Proceedings of the 12th Annual Conference on Machine Learning*, (pp.38-46). Morgan Kauffman.
2. **Kirkpatrick, S., Gelatt Jr., C. D., & Vecchi, M. P.** (1983). "Optimization by Simulated Annealing." *Science*, vol. 220, pp 671-680.
3. **Koehler, G., Bhattacharyya, S., & Vose, M. D.** (1997). "General Cardinality Genetic Algorithms." *Evolutionary Computation*, vol. 5, no. 4, pp. 439-459.
4. **Mitchell, M., Holland, J. H., & Forrest, S.** (1994). "When Will a Genetic Algorithm Outperform Hill Climbing?" In J. D. Cowan, G. Tesauro, & J. Alspector (Eds.), *Advances In Neural Information Processing Systems 6*, San Mateo, CA: Morgan Kaufmann.
5. **Mitchell, M.** (1998). *Introduction To Genetic Algorithms*. Cambridge, MA: MIT Press.
6. **Suzuki, H., & Sawai, H.** (2001). "Crossover Accelerates Evolution In GA with a Royal Road Function." *2001 Genetic and Evolutionary Computation Conference Late Breaking Papers*, pp. 401-412.
7. **van Nimwegen, E., Crutchfield, J. P., & Mitchell, M.** (1999). "Statistical Dynamics of the Royal Road Genetic Algorithm." *Theoretical Computer Science,* vol. 229, pp. 41-102.
8. **Vose, M. D.** (1995). "Modeling Alternate Selection Schemes For Genetic Algorithms." In Koppel, M. & Shamir, E. (Eds.), *Proceedings of BISFAI '95*, (pp. 166-178). Ramat Gan and Jerusalem, Israel: AAAI Press.
9. **Vose, M. D.** (1999a). "Random Heuristic Search." *Theoretical Computer Science,* vol 229, pp. 103-142.
10. **Vose, M. D.** (1999b). The Simple Genetic Algorithm: Foundations and Theory. Cambridge, MA: MIT Press.

Robust and Efficient Genetic Algorithms with Hierarchical Niching and a Sustainable Evolutionary Computation Model

Jianjun Hu[1] and Erik Goodman[2]

[1,2]Genetic Algorithm Research and Application Group(GARAGe)
[1]Department of Computer Science and Engineering
[2]Department of Electrical and Computer Engineering
Michigan State University, East Lansing, MI, 48823
{Hujianju, Goodman}@egr.msu.edu

Abstract. This paper proposes a new niching method named hierarchical niching, which combines spatial niching in search space and a continuous temporal niching concept. The method is naturally implemented as a new genetic algorithm, QHFC, under a sustainable evolutionary computation model: the Hierarchical Fair Competition (HFC) Model. By combining the benefits of the temporally continuing search capability of HFC and this spatial niching capability, QHFC is able to achieve much better performance than deterministic crowding and restricted tournament selection in terms of robustness, efficiency, and scalability, simultaneously, as demonstrated using three massively multi-modal benchmark problems. HFC-based genetic algorithms with hierarchical niching seem to be very promising for solving difficult real-world problems.

1 Introduction

Genetic algorithms are widely applied to challenging engineering problems today. However, there are still several undesirable properties with current genetic algorithms. The first one is the lack of a quality guarantee of genetic search. For example, genetic algorithms are usually sensitive to the population size in terms of their search capability. Unfortunately, it is difficult to estimate the required population size, despite the extant population sizing theory [1,2]. Too large a population size leads to low efficiency, and one that is too small may simply fail to achieve satisfactory results. The second undesirable property is that once a genetic algorithm stagnates during a search, it usually loses most of its search capability, and there is no good way to rejuvenate the run in an efficient manner. Simple restart or strong mutations may waste the computations spent before by destroying the building blocks in the population. Weak mutations may perturb the solutions a little bit, but they cannot incur significant move in search space once the framework of the individual is established. The third problem of current genetic algorithms is the lack of robustness such as large variation of the performance of several runs due to the opportunistic and convergent nature of current genetic algorithms.

K. Deb et al. (Eds.): GECCO 2004, LNCS 3102, pp. 1220–1232, 2004.

In the past three decades, many niching techniques have been proposed, which have greatly improved the scalability and robustness of genetic search for difficult multi-modal problems [3]. However, due to the convergent nature of the current genetic algorithm framework, these niching approaches still meet difficulty in many hard problems. Based on a sustainable evolutionary computation framework and a hierarchical niching mechanism, this paper proposes a new genetic algorithm, named QHFC, which can significantly improve robustness, efficiency and scalability over that of a representative modern niching approach.

The rest of the paper is organized as follows. In section 2, existing commonly used niching techniques including temporal niching and spatial niching are surveyed, and their three inherent difficulties are outlined. Section 3 then presents the ideas of the sustainable evolutionary computation framework of HFC [4,5], which underlies the design of a new genetic algorithm with hierarchical niching, QHFC to be described in Section 4. A set of three well-known genetic algorithm benchmark problems are used to evaluate QHFC in section 5 and the results are compared to genetic algorithms with deterministic crowding and restricted tournament selection in terms of scalability, efficiency, and robustness. A conclusion is then drawn in Section 6 along with future work to be done.

2   Related Work

The basic framework of genetic algorithms was laid down by John Holland in the 1960s, as summarized in his book [6], following the Darwinian evolution theory of natural selection. Most of the early formulations of evolutionary computation employed the principle of survival of the fittest. But it turned out that incautious keeping of the best individuals leads to bad performance, as population diversity is critical to good evolutionary search. The most widely used techniques to maintain diversity today are niching techniques, including many well-known methods—for example, De Jong crowding [7], deterministic crowding [8], fitness sharing [9], sequential niching [10] and restricted tournament selection [11]. Niching is useful for many application cases of genetic algorithms. It can be used to maintain interim sub-solutions to find a single final solution or to find multiple final solutions. It is also widely used as an effective mechanism to form and maintain diversity in genetic algorithms to solve hard problems. Other methods like reducing selection pressure, selection noise and operator disruption do not typically result in a GA with strong niching behavior. Readers are referred to Mahfoud [3] for an excellent and almost exhaustive review of niching methods.

Niching methods can be classified by their underlying mechanisms [3]. According to the fitness functions employed, they can be categorized as single-environment approaches (such as crowding and sharing) and multiple-environment approaches (such as implicit fitness sharing [12], and multi-objective function optimization). Since multi-environment approaches are usually specific to special types of problems, we are only interested in single-environment niching approaches in this paper. According to whether niching is achieved across space or over time, we have spatial niching and temporal niching. The former includes the widely used crowding and sharing, which form and maintain multiple niches within the space of a single population. The latter form and maintain multiple niches over time. Only one temporal niching approach, called sequential niching [10], has received attention in the literature to date.

Many experimental comparisons and analytical analyses have been conducted to evaluate the advantages and disadvantages of existing niching methods [13, 14, 15]. Mahfoud [13] showed that sequential niching is weak on easy problems and also incapable of solving hard problems due to its lack of cooperation of individuals in niches and the increasing difficulty to find remaining optima. Fitness sharing is a widely used approach and is very strong if used with intelligent scaling and appropriate setting of the sharing radius parameters, both of which, however, are difficult to achieve; bad results have therefore been reported [14,15]. An undesirable property of both sequential niching and fitness sharing is that they modify the search landscape and thus may incur false optima and other unexpected search behaviors. It turns out that deterministic crowding is one of the best spatial niching approaches. It is capable and easy-to-use and its performance has been confirmed by several comparative studies [13, 14]. Compared to fitness sharing, deterministic crowding succeeds with smaller subpopulations and can often find global optima for hard problems [13]. Assuming the selection pressure for high-fitness leads to premature convergence, Hutter [16] proposed a Fitness Uniform Selection Scheme (FUSS) to preserve genetic diversity. However, this approach suffers from insufficient selective pressure for exploitation and unbalanced fitness distribution of the search space. More detailed comparison of FUSS and other diversity maintaining mechanisms with HFC framework [4] is described in [5].

However, there are several difficulties in applying genetic algorithms to practical real-world problems, which lead to situations in which current spatial niching approaches tend to fail miserably. The first constraint of using a genetic algorithm is that we can often use only a very limited population size, at least relative to the size that various sizing methods indicate is needed. However, as spatial niching methods work by spreading the population out across much of the search space, and there are a huge number of local optima, an enormous population size is usually needed to achieve a satisfactory search solution. This has been proved by the population sizing theory associated with deterministic crowding [3]. However, too large a population size leads to a large number of evaluations, which is usually undesirable. This dependence on population size is even made worse by the fact that each niche has to be supported by multiple individuals to search effectively around it.

As a result of the limited population size, spatial niching methods normally fail to maintain a stable subpopulation at low-fitness area of the search space. For example, fitness sharing tends to focus on several high-fitness niches during the later stages of search. The consequence of the loss of low-fitness-level search is that the genetic algorithm may lose the chance of discovering some essential building blocks or other beneficial genetic material in later search stages, focusing instead on building blocks discovered during the very limited sampling experiments in the early search stage. The reason is that the increasingly high average fitness of the population makes it almost impossible to maintain effective search niches at very low fitness levels. This principle is can be interpreted in biological terms as the cost of specialization, or adaptation limiting diversification: adaptation to a specific niche (corresponding to high fitness in a genetic algorithm) theoretically constrains a population's ability to subsequently diversify into other niches [17]. It is in this sense that the ordinary genetic algorithm model is convergent. The progress of fitness corresponds to an entrenching process; the more progress a genetic algorithm makes, the less opportunity it has to find radically new, beneficial structures and then possibly better solutions.

Another difficulty of current spatial niching methods is the uneven pace of progress in the various niches in the early stages. It is often the case that some early-discovered niches tend to attract most of the individuals of the population, while other niches with higher fitness do not attract enough individuals to explore their search domains and expose their potential.

To handle the three difficulties mentioned above—the limited population size, loss of exportation capability, and unbalanced pace of progress of different niches—a new niching approach is needed, based on a new evolutionary algorithm model. In the following section, a new niching method, called hierarchical niching, is proposed. It combines the benefits of both spatial niching and temporal niching, and is implemented in a new sustainable evolutionary search model called the Hierarchical Fair Competition (HFC) model.

3   Hierarchical Niching and the HFC Sustainable Evolutionary Search Model

The basic idea of hierarchical niching is to introduce a continuous version of temporal niching together with spatial niching to address the three difficulties outlined in the previous section. Hierarchical niching here refers to a type of niching technique that maintains continuing search at all (absolute) fitness levels, each of which is subject to a spatial niching technique. It is naturally implemented under a sustainable continuing evolutionary computation model, Hierarchical Fair Competition [4,5,18].

HFC employs an assembly-line structure in which subpopulations are hierarchically organized into different fitness levels [4]. Offspring of a given level are exported to higher levels if their fitness qualifies them for migration. The openings that create are filled by individuals imported from lower levels or generated by mutating other individuals of the same level. The bottom level continuously generates raw genetic material to explore for new building blocks, which are eventually exported to higher fitness levels. The motivation of HFC is to maintain effective search at all fitness levels to sustain the search process indefinitely and thus remove the problem of insufficient sampling and limited population size. The continuing search capability of HFC is achieved by ensuring a continuous supply and incorporation of genetic material in a hierarchical manner, and by culturing and maintaining, but continually renewing, populations of individuals of intermediate fitness levels. It also has the effect of reducing the selection pressure within each subpopulation while maintaining the global selection pressure to help ensure exploitation of good genetic material found. When each subpopulation (level) in an HFC algorithm is updated by application of a spatial niching technique, the hierarchical niching is established.

Hierarchical niching handles the three difficulties mentioned in Section 2 as follows. Since the available population size is too limited to accommodate all local optima simultaneously, hierarchical niching resorts to the continuing search at lower fitness levels to ensure sequential identification of useful building blocks. This is different from sequential niching in the fact that hierarchical niching only allows partial import of recently discovered building blocks from lower levels, which promotes recombination of building blocks discovered early and later. This is in sharp contrast to sequential niching. The issue of loss of explorative capability is handled by the HFC model. In HFC, the lowest fitness level can continuingly generate genetic diversity and export good building blocks to upper levels, so the search power of the genetic algorithm is sustained, and it exhibits no tendency to converge. And because of the mixing of late-discovered building blocks and early-discovered building blocks, HFC works much better than other naïve sustainable search strategies like restarting or multiple runs, in which random genetic material essentially just perturbs current individuals by destroying its building blocks rather than discovering new building blocks. The insufficient sampling and unbalanced pace of progress problems are all handled by the continuing search capability of the HFC model, since lower-level search may go on indefinitely if needed.

Based on hierarchical niching and the HFC model, we have developed a genetic algorithm named QHFC (Q means "quick"), which can achieve significant performance improvement compared to a GA employing another state-of-the-art niching technique, deterministic crowding and restricted tournament selection. The spatial niching used in the current version of QHFC is deterministic crowding, so the demonstration illustrates that QHFC can improve significantly on deterministic crowding alone.

4   The QHFC Algorithm with Hierarchical Niching

QHFC algorithm is designed based on the HFC sustainable evolutionary computation model, the hierarchical niching concept, and the adaptive breeding strategy. Like the multi-population implementation of HFC, the whole population is divided into several levels, each accommodating individuals with fitness within a certain fitness range, except in special situations (to be explained in Table 1 at the end). The QHFC algorithm can be viewed as a set of cooperating GA agents, each searching at a different fitness level, from the lowest (base) level to the top level. Hierarchical niching is implemented as follows: the top level works as a generational GA with deterministic crowding; all other levels update as steady-state GAs with deterministic crowding.

Compared to previous HFC genetic algorithms, one of the most important innovations of QHFC is the *adaptive breeding strategy* implemented using potency testing (discussed next). It provides a generic mechanism to maintain automatically the balance of exploration and exploitation. More specifically, it allows the algorithm to search as greedily as possible, so long as the greedy strategy is sustainable. For easy problems, the top level automatically gets more breeding opportunities and the search is very aggressive. For hard problems where sustained diversity is a necessity, lower levels are automatically bred more frequently to provide the needed influx of diverse individuals for higher levels.

Potency here is defined as the capability of a fitness level in HFC to produce offspring with fitness high enough for export to higher HFC levels. This mechanism for maintaining the potency of all but the top level works as follows: starting from the level just below the top level, breeding is conducted successively in each level, moving toward the lower levels, using steady-state breeding methods, while tracking the number of offspring produced that are eligible for *promotion* (migration to the next-higher fitness level). If a given number of promotable offspring are not produced within a specified number of evaluations at a given level, then a "catch-up" procedure is conducted: a specified fraction of that level's individuals is replaced by individuals taken from (and removed from) the next lower level, and *popsize* genetic operations and evaluations are performed. (*popsize* is the size of the population at the receiving fitness level.) Then, in turn, the openings created at the next lower level are immediately filled with individuals removed from the level below that, etc., until, at the lowest level, the openings are filled by new randomly generated individuals. However, except for the further genetic operations and evaluations performed at the level where the "catch-up" procedure was initiated, further genetic operations and evaluations are not performed as part of this "ripple down" filling of openings. This "double loop" procedure assures that each level, before it next breeds, has either recently produced individuals worthy of promotion to the next level or has received new individuals from the next lower level, thus ensuring its potency to export higher-level individuals. This mechanism for sustaining the potency of search does not require evaluating any measure of the distance among genotypes or phenotypes, and could also be applied to GP and other sorts of problems.

The QHFC algorithm is summarized in Table 1 at the end. Compared with HFC-GP [4] and AHFC-GP [5], QHFC has many fewer parameters to specify, and the admission thresholds are automatically adjusted.

5   Experiments

As discussed in Section 2, we are interested in hard problems with a large number of local optima, typically massively multimodal, with deception. These factors can often expose the limitation of current niching methods if used with a conventional evolutionary computation model. Here, three widely-used massively multimodal and/or deceptive GA test problems are used to evaluate the performance of QHFC with hierarchical niching, and the performance is compared to the modern niching methods deterministic crowding [8] and restricted tournament selection [11], whose performances have been deemed excellent by several other researchers [13,14].

The three benchmark problems used here include:

1) f3deceptive: order-3 deceptive problem [19], with problem sizes n=60, 90, 120, 150, 180, 240, 300

This deceptive function is composed of separable building blocks of order 3 and has one global optimum at 111...1 and a deceptive attractor at 000...0. There are many local optima in the landscape of this function.

2) 6bipolar: order-6 bipolar deceptive problem [19], with problem sizes n=60, 90, 120, 150, 180, 240, 300

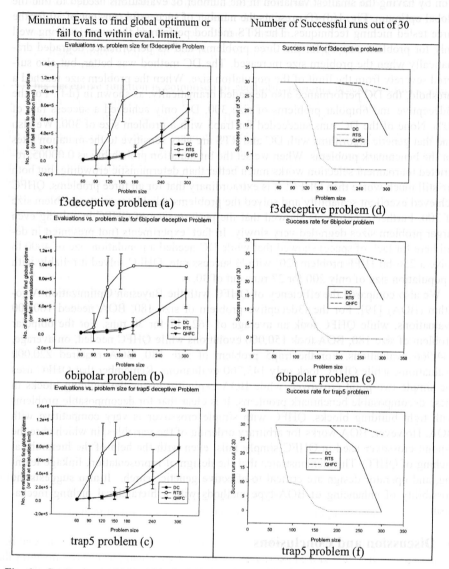

Fig. 1. Comparison of hierarchical niching (QHFC), deterministic crowding (DC), and restricted tournament selection (RTS) in terms of scalability, robustness and efficiency. It is clear that for simple problems or when the problem size is small enough for a population size of 500 is sufficient, DC and RTS work as well as QHFC. However, both DC and RTS suffer from the limited population size and fail for more difficult problems. QHFC clearly has better scalability, robustness, and efficiency.

individuals in all levels, while in the latter method, individuals in the previous run stage cannot help and usually hinder the discovery of later solutions. Compared to spatial niching, hierarchical niching here does not lose the search capability at low fitness levels, while spatial niching methods such as fitness sharing and deterministic crowding are strongly limited by population size and eventually lose search capability at low fitness levels. Another feature of hierarchical niching is that the niching technique used within each level could easily be some method other than the deterministic crowding used in this paper.

Table 1. QHFC Genetic Algorithm with Hierarchical Niching

Procedure do_potency_testing (l)

l is the level for potency testing
catchup_evaluation ? 0
exportedIndividual ? 0

while *catchup_evaluation* $<$ **catchupGen*** $\|P_l$ and *exportedIndividual* $<$ *detectExportNo*

randomly pick two individuals from level l
crossover, mutate, and evaluate

if fitness of offspring $>$ f_{adm}^{l+1},

    promote it (them) to level $l+1$ (replacing randomly any but the best individual or other
    individuals just promoted) and call *import_from_below* to replace its (their) closest
    parent(s)
        exportedIndividual ? *exportedIndividual* +1
else
        do deterministic crowding with the 4 -member family
endif
end while
if fail to promote *detectExportNo* individuals
return not success
else
return success
Procedure end

Procedure import_from_below *(l, nImport, victimList)*

l : the level into which to import new individuals from next lower level
nImport: the number of individuals to import from next lower level
victimList : a list of indices of individuals which will be replaced by the imported new individuals

if l =0

    randomly generate *nImport* new individuals and import into (lowest) level l
else

    randomly choose *nImport* individuals from level $l-1$ to replace individuals in *victimList*.
    If *victimList* is empty, randomly choose victim individual from current level. Put the
    indices of the new immigrant individuals from level $l-1$ into the level $l-1$
    newVictimList, whose openings will eventually be filled with individuals from level
    $l-2$ (this assures the replacement of individuals removed from level $l-1$)
call *import_from_below* (l -1, nImport , newVictimList)
Procedure end

Parameters:

Total population size $|P_t|$ L: number of subpopulations (levels) of QHFC

γ: size factor parameter, the ratio of higher level archive size w.r.t next lower level archive

size $|P_{k-1}| = |P_k| \cdot \gamma$

breedTopFreq: number of generations to breed top level between potency testing of lower levels (via breeding)

detectExportNo: number of individuals from a level that must be promoted for the level to be considered potent

catchupGen: maximum evaluations in any but top level, normalized by level's popsize, for potency test

percentRefill: percentage of this level's popsize to import from next lower level when there is no progress in the top level, or when lower levels fail potency test (do not furnish *detectExportNo* qualified immigrants within specified number of evaluations)

noprogressGen: maximum number of generations without any fitness progress in top level before triggering importing of *percentRefill* individuals from next lower level

QHFC Main procedure

1. initialization

    rancomly initialize and evaluate the HFC subpopulations

    calculate the average fitness of the whole population and set it as the admission fitness of

    the bottom level, f_{min}, which is fixed thereafter

    remove individuals with fitness less than f_{min}, and equally distribute the rest of the individuals among the levels, according to fitness, thereby determining the admission threshold of each level

    generate random individuals to fill the openings in each archive

2. while termination_condition is false

    breed the top level for **breedTopFreq** generations using generational deterministic crowding and applying mutation after each crossover

    if no progress on best fitness of the whole population for **noprogressGen** generations, call **import_from_below**, but ensuring the best individual is not replaced if average fitness of top level $> 2 f_{adm}^{L-1} - f_{adm}^{L-2}$, adjust admission thresholds by evenly allocating fitness range to each level:

$$f_{adm}^k = f_{min} + k(f_{max} - f_{min})/L \quad \text{for k=0 to } L\text{-1}$$

    where f_{adm}^k is the admission fitness of level k, f_{max} is the maximum fitness of the

    whole population

//potency testing

for each level from L-2 to 0

call **do_potency_testing**

    if not succeed

        call **import_from_below** to replace (at random) **percentRefill** percentage of the current level, breed one generation at this level

    endif

end for

end while

End Main

The significant performance gain in terms of search sustainability, efficiency, and robustness of QHFC again demonstrates the usefulness of hierarchical niching and of the hierarchical fair competition (HFC) model for sustainable evolutionary search. These algorithms seem to be especially useful for large-scale long-term artificial evolution experiments such as topologically opened synthesis of electric circuits, mechatronic systems, etc.

Our future work will include an experimental comparison study of QHFC with FUSS [16] and fitness sharing with different parameter configurations such as the population sizes. Although our previous work [21] shows that depending on large population size to maintain diversity is not a scalable solution to premature convergence problem, more experiments with more test problems would be helpful to further justify this hypothesis.

References

1   Goldberg, D.E: Sizing Populations for Serial and Parallel Genetic Algorithms, in J.D. Schaffer (ed.), *Proceedings of the Third International Conference on Genetic Algorithms*, Kaufmann, San Mateo, Calif. (1989)

2   Harik, G. R. and Lobo, F.G.: A parameter-less genetic algorithm. In Proceedings of the Genetic and Evolutionary Computation Conference (1999).

3   Mahfoud, S.W.: Niching Methods for Genetic Algorithms, Ph.D. Thesis, University of Illinois at Urbana-Champaign. (1995).

4   Hu, J., Goodman, E.D.: Hierarchical Fair Competition Model for Parallel Evolutionary Algorithms. In *Proceedings, Congress on Evolutionary Computation, CEC 2002*, IEEE World Congress on Computational Intelligence, Honolulu, Hawaii, May. (2002).

5   Hu, J., Goodman, E. D. and Seo, K.: Continuous Hierarchical Fair Competition Model for Sustainable Innovation in Genetic Programming. In *Genetic Programming Theory and Practice*, Kluwer, (2003), pp. 81-98.

6   Holland, J.H.: *Adaptation in natural and artificial systems*. Ann Arbor, MI: University of Michigan Press. (1975).

7   De Jong, K.A.: An analysis of the behavior of a class of genetic adaptive systems. (Doctoral dissertation, University of Michigan). *Dissertation Abstracts International*, 36(10),514B. (University Microfilms No. 76-9381). (1975).

8   Mahfoud, S. W. Crowding and preselection revisited. In *Proc. Parallel problem Solving from Nature, PPSN '92*, Brussels, (1992).

9   Goldberg, D. E. and Richardson, J.: Genetic algorithms with sharing for multimodal function optimization. In *Proceedings of the 2nd International Conference on Genetic Algorithms*, J. J. Grefenstette, Ed. Hillsdale, NJ: Lawrence Erlbaum, (1987). pp. 41--49.

10   Beasley, D., Bull, D. R. and R. R. Martin: A sequential niche technique for multimodal function optimization. *Evolutionary Computation*, 1(2): (1993) pp. 101--125

11   Harik, G.: Finding multimodal solutions using restricted tournament selection. In *Proceedings of Sixth International Conference on Genetic Algorithms*, (1995).

12   Darwen, P. and Yao, X.: Every niching method has its niche: Fitness sharing and implicit sharing compared. In H.-M. Voigt, W. Ebeling, I. Rechenberg, and H.-P. Schwefel, editors, *Parallel Problem Solving from Nature-- PPSN IV*, pages 398--407, Berlin, Springer. (1996).

13   Mahfoud, S.M. (1995): A Comparison of Parallel and Sequential Niching Methods. In *Proceedings of Genetic and Evolutionary Computation Conference*, (1995) pp136--143.

14   Sareni, B., Krahenbuhl, L.: Fitness Sharing and Niching Methods revisited. *IEEE Trans. on Evolutionary Computation*, 2(3), September (1998) pp. 97-106

15   Ursem, R.K: When Sharing Fails. In *Proceedings of the Third Congress on Evolutionary Computation* (CEC-2001), (2001)

16   Hutter, M. Fitness Uniform Selection to Preserve Genetic Diversity. In Proceedings of the 2002 Congress on Evolutionary Computation: 783—788 (CEC-2002), Hawaii, (2002)

17   Buckling, A. et al.: Adaptation limits diversification of experimental bacterial populations. *Science*, December 19, 302, (2003) pp.2107-2109.

18   Hu, J., E. D. Goodman, K. Seo, Z. Fan, R. C. Rosenberg: HFC: a Continuing EA Framework for Scalable Evolutionary Synthesis. In *Proceedings of the 2003 AAAI Spring Symposium - Computational Synthesis: From Basic Building Blocks to High Level Functionality*, Stanford, California, March, 24-26, (2003) pp. 106-113.

19   Pelikan, M, Goldberg, D.E. & Erick Cantú-Paz, E. BOA: The Bayesian optimization algorithm. *Proceedings of the Genetic and Evolutionary Computation Conference* (GECCO-99), I, 525-532.

20   Watson, R. Analysis of recombinative algorithms on a nonseparable building block problem. In *Foundations of Genetic Algorithms 6*, W. Martin and W. Spears, Eds., San Mateo, CA: Morgan Kaufmann, (2001), pp. 69-89.

21   Hu, J., Goodman, E., Seo, K., Fan, Z., Rosenberg, R. The Hierarchical Fair Competition (HFC) Framework for Sustainable Evolutionary Algorithms. *Evolutionary Computation*, 13(1), 2005 (To appear)

A Systematic Study of Genetic Algorithms with Genotype Editing

Chien-Feng Huang and Luis M. Rocha

Modeling, Algorithms, and Informatics Group (CCS-3),
Computer and Computational Sciences,
Los Alamos National Laboratory, MS B256,
Los Alamos, NM 87545, USA
{cfhuang, rocha}@lanl.gov

Abstract. This paper continues our systematic study of an RNA-editing computational model of Genetic Algorithms (GA). This model is constructed based on several genetic editing characteristics that are gleaned from the RNA editing system as observed in several organisms. We have expanded the traditional Genetic Algorithm with artificial editing mechanisms as proposed in [11] and [12]. The incorporation of editing mechanisms, which stochastically alter the information encoded in the genotype, provides a means for artificial agents with genetic descriptions to gain greater phenotypic plasticity, which may be environmentally regulated. The systematic study of this artificial genotype editing model has shed some light into the evolutionary implications of RNA editing and how to select proper genotype editors to design more robust GAs. Our results also show promising applications to complex real-world problems. We expect that the framework here developed will both facilitate determining the evolutionary role of RNA editing in biology, and advance the current state of research in Evolutionary Computation.

1 Introduction

Evidence for the important role of non-protein coding RNA (ncRNA) in complex organisms (higher eukaryotes) has accumulated in recent years. "ncRNA dominates the genomic output of the higher organisms and has been shown to control chromosome architecture, mRNA turnover and the developmental timing of protein expression, and may also regulate transcription and alternative splicing." ([9], p 930).

RNA Editing ([2]; [1]), a process of post-transcriptional alteration of genetic information, can be performed by ncRNA structures (though it can also be performed by proteins). The term initially referred to the insertion or deletion of particular bases (e.g. uridine), or some sort of base conversion. Basically, RNA Editing instantiates a non-inheritable stochastic alteration of genes, which is typically developmentally and/or environmentally regulated to produce appropriate phenotypical responses to different stages of development or states of the environment.

K. Deb et al. (Eds.): GECCO 2004, LNCS 3102, pp. 1233–1245, 2004.

The most famous RNA editing system is that of the African Trypanosomes [2]. Its genetic material was found to possess strange sequence features such as genes without translational initiation and termination codons, frame shifted genes, etc. Furthermore, observation of mRNA's showed that many of them were significantly different from the genetic material from which they had been transcribed. These facts suggested that mRNA's were edited post-transcriptionally. It was later recognized that this editing was performed by guide RNA's (gRNA's) coded mostly by what was previously thought of as non-functional genetic material [13]. In this particular genetic system, gRNA's operate by inserting, and sometimes deleting, uridines. To appreciate the effect of this edition let us consider Fig. 1. The first example (p. 14 in [2]) shows a massive uridine insertion (lowercase u's); the amino acid sequence that would be obtained prior to any edition is shown on top of the base sequence, and the amino acid sequence obtained after edition is shown in the gray box. The second example shows how, potentially, the insertion of a single uridine can change dramatically the amino acid sequence obtained; in this case, a termination codon is introduced. It is important to retain that a mRNA molecule can be more or less edited according to the concentrations of the editing operators it encounters. Thus, several different proteins coded by the same gene may coexist in an organism or even a cell, if all (or some) of the mRNA's obtained from the same gene, but edited differently, are meaningful to the translation mechanism.

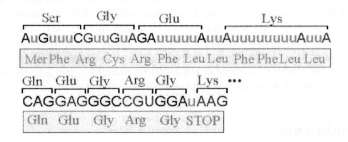

Fig. 1. U-insertion in Trypanosomes' RNA

The role of RNA editing in the development of more complex organisms has also been shown to be important. Lomeli et al. [8] discovered that the extent of RNA editing affecting a type of receptor channels responsible for the mediation of excitatory postsynaptic currents in the central nervous system, increases in rat brain development. As a consequence, the kinetic aspects of these channels differ according to the time of their creation in the brain's developmental process. Another example is that the development of rats without a gene (ADAR1) known to be involved in RNA editing, terminates midterm [14]. This showed that RNA Editing is more prevalent and important than previously thought. More recently, Hoopengardner et al. [5] found that RNA editing plays a central role in nervous system function. Indeed, many edited sites recode conserved and functionally important amino acids, some of which may play a role in nervous system disorders such as epilepsy and Parkinson Disease.

Although RNA editing seems to play an essential role in the development of some genetic systems and more and more editing mechanisms have been identified, not much has been advanced to understand the potential evolutionary advantages, if any, that RNA editing processes may have provided. To acquire insights for answering this question, we started the systematic study of a Genetic Algorithm with Edition (GAE) initially proposed by Rocha [11], [12]. Specifically, we reported in [7] some preliminary results on how Genotype Editing may provide evolutionary advantages. Here, we continue this study by presenting results based on simulations with much larger numbers of runs with randomized parameters, yielding a more statistically significant treatment of the conclusions reached in [7] from individual examples of genotype editing. Our goal is to gain a deeper understanding of the nature of RNA editing and exploit its insights to improve evolutionary computation tools and their applications to complex problems. Before delving fully into this paper, the next section summarizes our prior work in Genetic Algorithms with Genotype Edition in [7].

2   Prior Work on Genetic Algorithms with Edition

In science and technology Genetic Algorithms (GA) [4] have been used as computational models of natural evolutionary systems and as adaptive algorithms for solving optimization problems. Table 1 depicts the process of a simple genetic algorithm.

Table 1. Mechanism of a simple GA.

1. Randomly generate an initial population of l n-bit chromosomes.
2. Evaluate each individual's fitness.
3. Repeat until l offspring have been created.
a. select a pair of parents for mating;
b. apply crossover operator;
c. apply mutation operator.
4. Replace the current population with the new population.
5. Go to Step 2 until terminating condition.

GAs operate on an evolving population of artificial organisms, or agents. Each agent is comprised of a genotype and a phenotype. Evolution occurs by iterated stochastic variation of genotypes, and selection of the best phenotypes in an environment according to a fitness function. In machine learning, the phenotype is a candidate solution to some optimization problem, while the genotype is an encoding, or description, of that solution by means of a domain independent representation, namely, binary symbol strings (or chromosomes). In traditional GAs, this code between genotype and phenotype is a direct and unique mapping.

In biological genetic systems, however, there exists a multitude of processes, taking place between the transcription of genes and their expression, responsible for the establishment of a one-to-many relation between genotype and phenotype. For instance, it was shown that RNA editing has the power to dramatically alter gene expression [10] (p. 78): "cells with different mixes of (editing mechanisms) may edit a transcript from the same gene differently, thereby making different proteins from the same opened gene."

In a genetic system with RNA editing, in other words, before a gene is translated into the space of proteins it may be altered through interactions with other types of molecules, namely RNA editors such as gRNA's. Based upon this analogy, Rocha [11], [12] proposed an expanded framework of GA with a process of stochastic edition of the genetic descriptions (chromosomes) of agents, prior to being translated into solutions. The editing process is implemented by a set of editors with different editing functions, such as insertion or deletion of symbols in the original chromosomes. Before these descriptions can be translated into the space of solutions, they must "pass" through successive layers of editors, present in different concentrations. In each generation, each chromosome has a certain probability (given by the concentrations) of encountering an editor in its layer. If an editor matches some subsequence of the chromosome when they encounter each other, the editor's function is applied and the chromosome is altered. The implementation of a GA with Edition (GAE) is described in the following:

The GAE model consists of a family of r m-bit strings, denoted as $(E_1, E_2, \ldots, E_r)$, that is used as the set of editors for the chromosomes of the agents in a GA population. The length of the editor strings is assumed much smaller than that of the chromosomes: $m << n$, usually an order of magnitude. An editor E_j is said to match a substring, of size m, of a chromosome, S, at position k if $e_i = s_{k+i}, i = 1, 2, \ldots, m, 1 \leq k \leq n-m$, where e_i and s_i denote the ith bit value of E_j and S, respectively. For each editor, E_j, there exists an associated editing function, F_j, that specifies how a particular editor edits the chromosomes: when the editor matches a portion of a chromosome, a number of bits are inserted into or deleted from the chromosome.

For instance, if the editing function of editor E_j is to add one randomly generated allele at s_{k+m+1} when E_j matches S at position k, then all alleles of S from position $k + m + 1$ to $n - 1$ are shifted one position to the right (the allele at position n is removed). Analogously, if the editing function of editor E_j is to delete an allele, this editor will instead delete the allele at s_{k+m+1} when E_j matches S at position k. All the alleles after position $k + m + 1$ are shifted in the inverse direction (one randomly generated allele is assigned at position n).

Finally, let the concentrations of the editor family be defined by $(v_1, v_2, \ldots, v_r)$; i.e., the concentration of editor E_j is denoted as v_j. Then the probability that S encounters E_j is given by v_j. With these settings, the algorithm for the GA with genotype editing is essentially the same as the regular GA, except that step 2 in Table 1 is now redefined as:

"For each individual in the GA population, apply each editor E_j with probability v_j (i.e., concentration). If E_j matches the individual's chromosome S, then edit S with the editing function associated with E_j and evaluate the resulting individual's fitness."

It is important to notice that the "post-transcriptional" edition of genotypes is not a process akin to mutation, because editions are not inheritable. Just like in biological systems, it is the unedited genotype that is reproduced. One can also note that Genotype Editing is not a process akin to the Baldwin effect as studied by, e.g., Hinton and Nowlan [3]. The phenotypes of our agents with genotype edition, do not change (or learn) ontogenetically. In Hinton and Nowlan's experiments, the environment is defined by a very difficult ("needle in a haystack") fitness function, which can be made more amenable to evolutionary search by endowing the phenotypes to "learn" ontogenetically. Eventually, they observed, this learning allows genetic variation to discover, and genetically encode fit individuals. In contrast, genotype edition does not grant agents more "ontogenetic learning time", it simply changes inherited genetic information ontogenetically but the phenotype, once produced, is fixed. Also, as we show below, it is advantageous in environments very amenable to evolution, such as Royal Road functions (the opposite of "needle in a haystack") [7].

In [7], based on specific examples, we have demonstrated how the editing mechanism can improve the GA's search performance by suppressing the effects of hitchhiking. We have also showed that editing frequency plays a critical role in the evolutionary advantage provided by the editors – only a moderate degree of editing processes would facilitate organisms' exploration of the search space. Therefore, one needs to choose proper editor parameters to avoid over or under-editions in order to develop more robust GAs. In this paper, we conduct a larger statistical exploration using numerous sets of families of editors to elaborate on the conditions where genotype editing truly enhances the GA's search power.

3   Effects of Genotype Editing

How rapid is evolutionary change, and what determines the rates, patterns, and causes of change, or lack thereof? Answers to these questions can tell us much about the evolutionary process. The study of evolutionary rate in the context of GA usually involves defining a performance measure that captures the idea of rate of improvement, so that its change over time can be monitored for investigation. In many practical problems, a traditional performance metric is the "best-so-far" curve that plots the fitness of the best individual that has been seen thus far by generation n. As a step towards a deeper understanding of how Genotype Editing works, we employ a testbed, the small Royal Road S_1, which is a miniature of the class of the "Royal Road" functions [7].

Table 2 illustrates the schematic of the small Royal Road function S_1. This function involves a set of schemata $S = \{s_1, \ldots, s_8\}$ and the fitness of a bit string (chromosome) x is defined as $F(x) = \sum_{s \in S} c_s \sigma_s(x)$, where each c_s is a value assigned to the schema s as defined in the table; $\sigma_s(x)$ is defined as 1 if

Table 2. Small royal road function S_1

$$
\begin{aligned}
s_1 &= 11111\text{*********************************}; \; c_1 = 10 \\
s_2 &= \text{*****}11111\text{****************************}; \; c_2 = 10 \\
s_3 &= \text{**********}11111\text{***********************}; \; c_3 = 10 \\
s_4 &= \text{***************}11111\text{******************}; \; c_4 = 10 \\
s_5 &= \text{********************}11111\text{**************}; \; c_5 = 10 \\
s_6 &= \text{*************************}11111\text{**********}; \; c_6 = 10 \\
s_7 &= \text{******************************}11111\text{*****}; \; c_7 = 10 \\
s_8 &= \text{***********************************}11111; \; c_8 = 10
\end{aligned}
$$

x is an instance of s and 0 otherwise. In this function, the fitness of the global optimum string (40 1's) is $10 \times 8 = 80$.

There are several factors that play a role in the GAE's search power – e.g., *size of the family of editors, editor length, editor concentration* and *editor function* [7]. Our aim here is to investigate these four parameters. Since a multitude of parameter combinations are possible, we conduct numerous GAE runs and focus on a single parameter while other parameters are randomly generated in the beginning of each GAE run and then fixed until the end of that run. The results are then averaged over the number of the GAE runs so that we may zero in on the effects of that parameter.

3.1   Effects of Size of the Family of Editors

To study the effect of the size of editor family parameter r, two sets of values, $r \in 1,2,3,4,5$ and $r \in \{6,7,8,9,10\}$, are tested. The GAE was run 100 times for each set and in each GAE run the value of r is randomly chosen from the respective sets.[1] Figure 2.a and 2.b display the results on averaged best-so-far performance and averaged editing frequency (the total number of times all editors edited chromosomes in a generation) over 100 runs, respectively.[2] One can see that the GAEs with less editors (i.e., 1 to 5) clearly outperform the GAEs with more editors (i.e., 6 to 10).[3] The results also show that the editing frequency for the GAEs with less editors is substantially smaller than that of the GAEs with more

---

[1] The settings of other editor parameters are: each editor is a randomized bit-string of a randomly chosen number of bits from $\{1,10\}$; the editor concentration is randomly generated from $[0,1]$; and the editor function inserts or deletes a randomly chosen number of bits from $\{1,10\}$, as well. For the GA part, throughout this section, we use a population of 40 chromosomes, a binary tournament selection, and crossover and mutation rates of 0.7 and 0.005, respectively.

[2] The value of the averaged best-so-far performance is calculated by averaging the best-so-fars obtained at each generation for all 100 runs; and so is the averaged editing frequency, where the vertical bars overlaying the metric curves represent the 95-percent confidence intervals. This applies to all the results presented in this paper.

[3] We do not contrast the performance of traditional GAs with that of the GAE here, since the purpose in this section is to study the effects of the editor parameters per se. Please see [7] for specific examples of the GAE outperforming GA, as well as guidelines for choosing proper editors so that the GAEs can outperform the GAs.

editors. These results are intuitive, since more editors naturally tend to incur more editing processes.

To further elucidate the effects of this parameter, Figure 2.c displays the results of editing frequency in individual runs for r: 2, 5 and 10 editors. The corresponding maximal fitness reached by the GAE with 2, 5, and 10 editors is 70, 80 and 50, respectively (the detailed results are not displayed here due to the limit of the paper length). One can notice that in the run of the GAE with 10 editors, where the maximal fitness attained is far from the optimum, the editing frequency does not significantly drop to zero near the end of the experiments. It appears that the GAE's population continues utilizing the editors to explore the search space. This is the reason why the corresponding population diversity displayed in Figure 2.d is far from zero in the case of the GAE with 10 editors.[4] For the GAE with 2 editors, the best-so-far fitness located is close to the optimum – the results in Figure 2.c and 2.d show that the degree of editing is then reduced and the population is not as diverse as that of the GAE with 10 editors. All this indicates that the system settles into a dynamic equilibrium in which the exploratory power of the editing process is balanced by the exploitative pressure of selection.

In the case of the GAE with 5 editors, whose best-so-far fitness reaches the optimum, the striking difference is that the corresponding editing frequency declines dramatically as the GAE's population evolves, and tends to drop to zero at the end of the experiments. This shows that the editing process ultimately comes to an end and the population diversity is lost (as shown in Figure 2.d). Based on the effects of editor length and concentration, in the next two subsections we will present more results to support this observation.

3.2   Effects of Editor Length

To test the effect of the editor length parameter m, two sets of values, $m \in \{1,2\}$ and $m \in \{3,4,5\}$, are investigated. The GAE was run 100 times for each set and in each GAE run the value of m is randomly chosen from respective sets.[5] Figure 3.a illustrates the results for these two sets of GAEs, in which the GAEs with longer editor length (3 to 5 bits) outperform the other. This is

---

[4] To measure diversity at the ith locus of a GA string, a simple bitwise diversity metric is defined as [7]: $D_i = 1 - 2|0.5 - p_i|$, where p_i is the proportion of 1s at locus i in the current generation. Averaging the bitwise diversity metric over all loci offers a combined allelic diversity measure for the population: $D = \frac{\sum_{i=1}^{l} D_i}{l}$. D has a value of 1 when the proportion of 1s at each locus is 0.5 and 0 when all of the loci are fixed to either 0 or 1. Effectively it measures how close the allele frequency is to a random population (1 being closest).

[5] The size of the editor family in this subsection is randomly chosen from $\{1,10\}$. The other two parameters (concentration and function) of an editor are generated by the same way as in the preceding subsection.

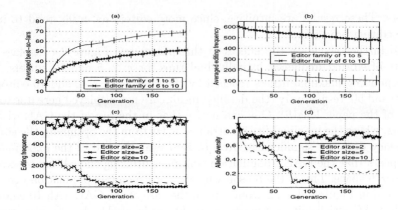

Fig. 2. Effects of size of the family of editors

also an intuitive result, since when the length of editors is too short, numerous matchings occur and the GAEs' population undergos too many editing processes. This typically results in serious disruptive effect on fit individuals.

In other words, the performance discrepancy of the GAEs with different editor length again depends on editing frequency. The empirical results for editing frequency shown in Figure 3.b confirm our assertion. The editing frequency for the GAEs with 1 to 2-bit editors is much higher than that of the GAEs with 3 to 5-bit editors. Therefore, beneficial genotype editing requires moderate editing frequency.

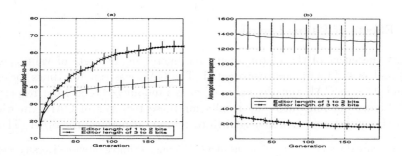

Fig. 3. Effects of editor length

3.3   Effects of Editor Concentration

To test the effect of the editor concentration parameter v_j, we again ran the GAE 100 times for two different sets of values of this parameter for each editor

E_j: $v_j \in [0,0.5]$ and $v_j \in$ in $[0.5,1]$.[6] Thus, the probability that chromosomes encounter editors in the second set of GAEs is higher than in the first set of GAEs. Figure 4.a and 4.b display the results. Since the probability of the chromosomes meeting with editors is higher in the second set of GAEs, the population naturally undergos more editions than in the first set of GAEs.

These results again indicate that the performance difference lies in the number of the performed editions. When the GAE's population is considerably edited by the editors, too much exploration of the search space generates deleterious effects on performance advancement. Appropriate editor concentration is thus essential for the GAE, since beneficial genotype edition requires a moderate quantity of editions.

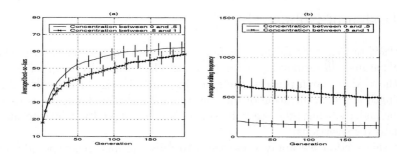

Fig. 4. Effects of editor concentration

3.4   Effects of Editor Function

To test the effect of the editor functions F_j, we again ran the GAE 100 times for two different sets of functions for each editor E_j. The scope of possible functions is open-ended, but here we contrast moderate edition with massive edition. The first set of functions F_j insert or delete a randomly chosen small number of bits in $\{1,2,3\}$. The second set of functions F_j insert or delete a randomly chosen larger number of bits in $\{10,11,12,13,14,15\}$.[7] Figure 5.a and 5.b display the corresponding results. Since the gene deletion or insertion frequency in chromosomes is now much higher in the second set of GAEs, the population naturally undergos more disruptive processes than in the first set of GAEs.

These results demonstrate that the performance difference lies in the degree of gene deletion (or insertion) in chromosomes. Appropriate editor function is thus also very important for the GAE to gain substantial search progress.

---

[6] The size of the editor family in this subsection is randomly chosen from $\{1,10\}$. The other two parameters (length and function) of an editor are generated by the same way as in Section 3.1.

[7] The size of the editor family in this subsection is randomly chosen from $\{1,5\}$; an editor's concentration is randomly generated from $[0,1]$ and its length is a randomly chosen number of bits from $\{1,10\}$.

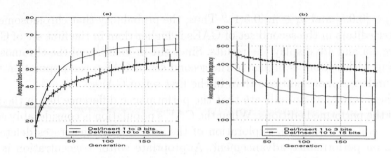

Fig. 5. Effects of editor function

4   Design of Robust GA

The study of Genotype Editing has provided us with insights into how to choose editor parameters for developing more robust GAs. Basically, in order to facilitate the GAE's search process, the guidelines are: the size of the editor family, the length and concentration of the editors need to be moderate so as to avoid over or under-editing processes; the editor function should not lead to massive deletions (or insertions).

In this section we apply these rules to select proper genotype editors for the design of more robust GAEs, and test them on a multimodal, non-building-block-based test function – the modified Schaffer's function F_7 [6]:

$$f(\overline{x}) = 2.5 - (x_1^2 + x_2^2)^{0.25}[sin^2(50(x_1^2 + x_2^2)^{0.1}) + 1],$$

where $-1 \leq x_i \leq 1$ for $1 \leq i \leq 2$. A sketch of this function is displayed in Figure 6. To attain the global optimum at the center of the search space, the population has to cross over many deep wells and high barriers. Since there are many local optima in the search space, a traditional GA's population can easily converge on any of them. The multimodality of this testbed is hence expected to present substantial difficulty to the GA's search.

Each of the two variables is encoded by 30 bits, and thus each individual is a binary string of length 60. We use a GA of population size 50, a binary tournament selection, and crossover and mutation rates of 0.7 and 0.005, respectively. We contrast the traditional GA with a GAE with the same parameters, but with genotype edition performed by a family of five editors as shown in Table 3. The experiments are conducted for 100 runs, each run with 200 generations.

Figure 7.a displays the averaged best-so-far performance, where one can see that the search performance of the GAE is better than that of the traditional GA. We also record the value of best-so-far attained at the end of each run and generate histograms as illustrated in Figure 7.b (for the GA) and 7.c (for the GAE). The results show that the GAE tends to locate more best-so-fars that are close to the optimum. One can also notice that there are several runs in which the traditional GA does not even locate best-so-fars of more than 2.3, meaning

Table 3. Parameters of the five editors

	editor 1	editor 2	editor 3	editor 4	editor 5
length	5	4	5	3	6
alleles	{0,0,1,1,0}	{1,0,0,1}	{0,1,1,0,1}	{0,1,1}	{1,1,1,1,0,0}
concentration	0.1410	0.7936	0.2524	0.5885	0.0871
function	delete 2 bits	delete 1 bit	add 3 bits	add 2 bits	add 5 bits

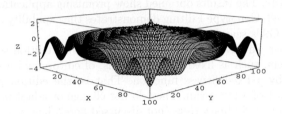

Fig. 6. Modified Schaffer function F_7

that the population in these runs prematurely converge on these "lower" local optima.

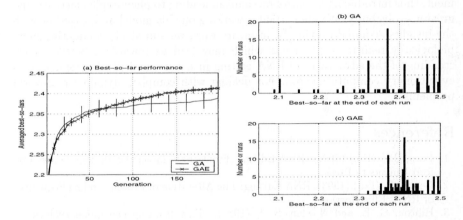

Fig. 7. Best-so-far performance and distribution on the modified Schaffer function F_7

5   Conclusion and Future Work

We have continued our systematic investigation of Genotype Editing in GA and tested several evolutionary scenarios. The results obtained have provided the following insights:

Editing frequency plays a critical role in the evolutionary advantage provided by the editors – only a moderate degree of editing processes can facilitate organisms' exploration of the search space. Our results also indicate that editing frequency declines dramatically as the population diversity is lost, indicating that the editing process ultimately comes to an end. If the editing frequency does not substantially decrease, the system settles into a dynamic equilibrium where the exploratory power of the editing process is balanced by the exploitative pressure of selection.

We have also learned some rules for setting up editors' parameters to develop robust GAs. The results obtained show promising applications to practical problems. Indeed, Genotype Editing demonstrates the capability of substantially improving the GA's search power.

In this paper we have thus far discussed GAs with edition solely with constant parameters, such as fixed concentrations, of editors and a stable environment defined by a fixed fitness function. That is, the edition parameters are fixed at the start of a given run. They do not change or adapt in the evolutionary process. Our preliminary tests (not discussed here), however, also show that constant concentrations of editors may not grant the system any evolutionary advantage when the environment changes. In order to simulate a genetic system in which the linking of editors' concentrations with environmental states may be advantageous in time-varying environments, Rocha [11], [12] proposed a new type of GA known as Contextual Genetic Algorithms (CGA). In this class of algorithms, the concentrations of editors change with the states of the environment, thus introducing a control mechanism leading to phenotypic plasticity and greater evolvability. We are currently working on this model and, together with the insights acquired previously, in future work we aim at (1) conducting more biologically realistic experiments which may lead us towards a better understanding of the advantages of RNA editing in nature, and (2) developing novel evolutionary computation tools for dealing with complex, dynamic real-world problems.

References

1. Bass, B.L. (Ed.) (2001). RNA Editing. Frontiers in Molecular Biology Series. Oxford University Press.
2. Benne, R. (Ed.) (1993). RNA Editing: The Alteration of Protein Coding Sequences of RNA. Ellis Horwood.
3. Hinton, G. E. and Nowlan, S. J. (1987). "How learning can guide evolution." Complex Systems. Vol. 1, pp. 495-502.
4. Holland, J. H. (1975). Adaptation in Natural and Artificial Systems. University of Michigan Press.
5. Hoopengardner, B., Bhalla, T., Staber, C., and Reenan, R. (2003). "Nervous System Targets of RNA Editing Identified by Comparative Genomics." Science 301: 832-836.
6. Huang, C-F. (2002). A Study of Mate Selection in Genetic Algorithms. Doctoral dissertation. Ann Arbor, MI: University of Michigan, Electrical Engineering and Computer Science.
7. Huang, C-F. and Rocha, L. M. (2003). "Exploration of RNA Editing and Design of Robust Genetic Algorithms." Proceedings of the 2003 IEEE Congress on Evolutionary Computation, Canberra, Australia, December 2003. R. Sarker et al (Eds). IEEE Press, pp. 2799-2806.

8. Lomeli, H. et al. (1994). "Control of Kinetic Properties of AMPA Receptor Channels by RNA Editing." Science, 266: 1709-1713.

9. Mattick, J. S. (2003). "Challenging the Dogma: the Hidden Layer of Non-protein-coding RNAs in Complex Organisms." BioEssays. 25: 930-939.
    and Scott, H. S. (1997). "Localization of a Novel Human RNA-editing Deaminase (hRED2 or ADARB2) to Chromosome 10p15." Human Genetics, 100: 398-400.

10. Pollack, R. (1994). Signs of Life: The Language and Meanings of DNA. Houghton Mifflin.

11. Rocha, Luis M. (1995). "Contextual Genetic Algorithms: Evolving Developmental Rules." Advances in Artificial Life. Moran, J., Moreno, A., Merelo, J. J., and Chacon, P. (Eds.). Springer Verlag, pp. 368-382.

12. Rocha, Luis M. (1997). Evidence Sets and Contextual Genetic Algorithms: Exploring Uncertainty, Context and Embodiment in Cognitive and biological Systems. PhD. Dissertation. State University of New York at Binghamton. Science.
    "RNA Editing." Annual Review of Neuroscience, 19: 27-52.

13. Sturn, N. R. and Simpson, L. (1990). "Kinetoplast Dna Minicircles Encode Guide Rna'S for Editing of Cytochrome Oxidase Subunit Iii Mrna." Cell, 61: 879-884.

14. Wang, Q., Khillan, J., Gadue, P., and Nishikura, K. (2000). "Requirement of the RNA Editing Deaminase Adar1 Gene for Embryonic Erythropoiesis." Science, 290 (5497): 1765-1768.

Some Issues on the Implementation of Local Search in Evolutionary Multiobjective Optimization

Hisao Ishibuchi and Kaname Narukawa

Department of Industrial Engineering, Osaka Prefecture University,
1-1 Gakuen-cho, Sakai, Osaka 599-8531, Japan
{hisaoi, kaname}@ie.osakafu-u.ac.jp

Abstract. This paper discusses the implementation of local search in evolutionary multiobjective optimization (EMO) algorithms for the design of a simple but powerful memetic EMO algorithm. First we propose a basic framework of our memetic EMO algorithm, which is a hybrid algorithm of the NSGA-II and local search. In the generation update procedure of our memetic EMO algorithm, the next population is constructed from three populations: the current population, its offspring population generated by genetic operations, and an improved population obtained from the offspring population by local search. We use Pareto ranking and the concept of crowding in the same manner as in the NSGA-II for choosing good solutions to construct the next population from these three populations. For implementing local search in our memetic EMO algorithm, we examine two approaches, which have been often used in the literature: One is based on Pareto ranking, and the other is based on a weighted scalar fitness function. The main difficulty of the Pareto ranking approach is that the movable area of the current solution by local search is very small. On the other hand, the main difficulty of the weighted scalar approach is that the offspring population can be degraded by local search. These difficulties are clearly demonstrated through computational experiments on multiobjective knapsack problems using our memetic EMO algorithm. Our experimental results show that better results are obtained from the weighted scalar approach than the Pareto ranking approach. For further improving the weighted scalar approach, we examine some tricks that can be used for overcoming its difficulty.

1 Introduction

Evolutionary multiobjective optimization (EMO) algorithms have been applied to various problems for efficiently finding their Pareto-optimal or near Pareto-optimal solutions. Recent EMO algorithms usually share some common ideas such as elitism, fitness sharing and Pareto ranking for improving both the diversity of solutions and the convergence to the Pareto front (e.g., see Coello et al. [1] and Deb [2]). In some studies, local search was combined for further improving the search ability of EMO algorithms [4]-[12].

K. Deb et al. (Eds.): GECCO 2004, LNCS 3102, pp. 1246–1258, 2004.
© Springer-Verlag Berlin Heidelberg 2004

Hybridization of EMO algorithms with local search is often referred to as MOGLS (multiobjective genetic local search) algorithms. Such a hybrid algorithm is also called a memetic EMO algorithm. It is clearly shown by Jaszkiewicz [9] that memetic EMO algorithms have higher search ability than pure EMO algorithms. Memetic EMO algorithms can be roughly classified into two categories according to their solution evaluation mechanisms in local search: One uses a weighted scalar fitness function, and the other uses Pareto ranking. A memetic EMO algorithm based on the weighted scalar fitness function with random weights was proposed by Ishibuchi & Murata [6], improved by Jaszkiewicz [10] and Ishibuchi et al. [7], and simplified by Ishibuchi & Kaige [5]. On the other hand, Knowles & Corne [11] proposed a memetic EMO algorithm called M-PAES (memetic Pareto archived evolution strategy) where each solution was evaluated based on Pareto ranking. Some Pareto ranking-based acceptance rules of local search moves were examined in Ishibuchi et al. [7] and Murata et al. [13]. The MOGLS of Jaszkiewicz [10] and the M-PAES of Knowles & Corne [11], which are well-known memetic EMO algorithms with high search ability, have been compared with each other in some comparative studies [4], [8], [9], [12].

Our goal in this paper is to design a simple but powerful memetic EMO algorithm. For achieving this goal, we discuss some issues related to the implementation of local search in EMO algorithms through computational experiments on multiobjective knapsack problems in Zitzler & Thiele [14]. Our computational experiments are performed using a framework of a simple MOGLS algorithm, which is proposed in this paper by combining the NSGA-II [3] with local search. In order to emphasize its simplicity, we refer to our MOGLS algorithm as the simple MOGLS (i.e., S-MOGLS) algorithm in this paper. As in the NSGA-II, we use Pareto ranking and the concept of crowding for generation update in our S-MOGLS algorithm. One characteristic feature of our generation update procedure is the use of three populations for generating the next population: the current population, its offspring population generated by genetic operations, and an improved population obtained from the offspring population by local search. In the existing MOGLS algorithms [5]-[10], the offspring population that had been improved by local search was not used in their generation update procedures. In this sense, our S-MOGLS can be viewed as an improved version of the S-MOGLS of Ishibuchi & Kaige [5].

Through computational experiments on multiobjective knapsack problems using our S-MOGLS algorithm, we compare the above-mentioned two approaches to the implementation of local search. We show that better results are obtained by the weighted scalar approach than the Pareto ranking approach. We also demonstrate a serious difficulty of the weighted scalar approach: Local search often degrades the offspring population generated by genetic operations. For overcoming this difficulty, we examine the effectiveness of the following tricks:

(1) The use of the three populations in the generation update procedure in our S-MOGLS algorithm. Generation update procedures with/without the offspring population are compared with each other.

(2) The choice of good initial solutions for local search from the offspring population. The tournament selection of initial solutions based on the weighted scalar fitness function is examined using various specifications of the tournament size.

(3) The modification of the acceptance rule of local search moves. We examine a modified acceptance rule in the weighted scalar approach.

This paper is organized as follows. In Section 2, we propose a basic framework of our S-MOGLS algorithm. While we try to maximize the search ability of our S-MOGLS algorithm, we also try to minimize its algorithmic complexity so that it can be easily understood, easily implemented and efficiently executed using small memory storage within short CPU time. In Section 3, we show various variants of our S-MOGLS algorithm. In one variant, the weighted scalar fitness function is used in the selection of parent solutions and the local search for their offspring. In another variant, Pareto ranking instead of the weighted scalar fitness function is used for both the parent selection and the local search. Of course, there exist many intermediate variants between these two extremes. In Section 4, we show experimental results on multiobjective knapsack problems using some variants of our S-MOGLS algorithm. Finally Section 5 concludes this paper.

2 Basic Framework of Our S-MOGLS Algorithm

Let us consider the following k-objective maximization problem:

$$\text{Maximize } \mathbf{f}(\mathbf{x}) = (f_1(\mathbf{x}), f_2(\mathbf{x}), ..., f_k(\mathbf{x})), \tag{1}$$
$$\text{subject to } \mathbf{x} \in \mathbf{X}, \tag{2}$$

where $\mathbf{f}(\mathbf{x})$ is the objective vector, $f_i(\mathbf{x})$ is the i-th objective to be maximized, $\mathbf{x}$ is the decision vector, and $\mathbf{X}$ is the feasible region in the decision space. When the following two conditions are satisfied, a feasible solution $\mathbf{x} \in \mathbf{X}$ is said to be dominated by another feasible solution $\mathbf{y} \in \mathbf{X}$ (i.e., $\mathbf{y}$ dominates $\mathbf{x}$: $\mathbf{y}$ is better than $\mathbf{x}$):

$$\forall i, \ f_i(\mathbf{x}) \le f_i(\mathbf{y}) \text{ and } \exists j, \ f_j(\mathbf{x}) < f_j(\mathbf{y}). \tag{3}$$

If there is no feasible solution $\mathbf{y}$ that dominates $\mathbf{x}$, $\mathbf{x}$ is said to be a Pareto-optimal solution of the multiobjective optimization problem. The task of EMO algorithms is to find Pareto-optimal or near Pareto-optimal solutions as many as possible.

The following weighted scalar fitness function was used in the MOGLS algorithms of Ishibuchi et al. [5]-[7] and Jaszkiewicz [8]-[10]:

$$f(\mathbf{x}, \lambda) = \sum_{i=1}^{k} \lambda_i f_i(\mathbf{x}), \tag{4}$$

where

$$\forall i, \ \lambda_i \geq 0 \quad \text{and} \quad \sum_{i=1}^{k} \lambda_i = 1. \tag{5}$$

In those MOGLS algorithms, the weight vector $\lambda = (\lambda_1, ..., \lambda_k)$ was randomly specified whenever a pair of parent solutions was to be selected. The roulette wheel selection was used in the original MOGLS algorithm [6]. In Ishibuchi et al. [7], better results were obtained by the tournament selection than the roulette wheel selection. On the other hand, a pair of parent solutions was randomly chosen from the best K solutions in the current population with respect to the weighted scalar fitness function in Jaszkiewicz [8]-[10]. The same weighted scalar fitness function with the current weight vector, which had been used for choosing a pair of parent solutions, was also used in local search for their offspring generated by genetic operations.

In our S-MOGLS algorithm, we use the weighted scalar fitness function in (4) for choosing a pair of parent solutions from the current population. Since the tournament selection needs less CPU time than the random selection from the best K solutions, we use the tournament selection in our S-MOGLS algorithm (more specifically the binary tournament slection). The selected parents are recombined to generate their offspring to which a mutation operation is applied. The selection, recombination and mutation are iterated to generate the offspring population (see Fig. 1). Local search is applied only to good solutions in the offspring population. For choosing initial solutions for local search from the offspring population, we use the tournament selection based on the weighed scalar fitness function in (4). Whenever an initial solution is chosen, the weight vector is randomly updated. Local search is probabilistically applied to the selected initial solution. The idea of choosing good initial solutions was proposed in Ishibuchi et al. [7]. Only when the initial solution is updated (i.e., improved) by local search, the improved solution is added to the improved population in Fig. 1.

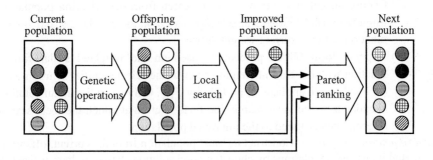

Fig. 1. Generation update mechanism in our S-MOGLS algorithm.

Recently it has been widely recognized in the EMO community that some form of elitism is necessary for designing EMO and memetic EMO algorithms with high search ability (e.g., see Deb [2]). One straightforward implementation of elitism is to store non-dominated solutions in the secondary population separately from the main (i.e., current) population. For example, the secondary population was used in the MOGLS algorithms of Ishibuchi et al. [6]-[7] and Jaszkiewicz [8]-[10], the M-PAES [11] and the SPEA [14]. While the use of the secondary population significantly improves the search ability of EMO and memetic EMO algorithms, it also increases algorithmic complexity, memory storage and CPU time. Thus we do not use the secondary population in our S-MOGLS algorithm. Instead of the secondary population, we use the generation update scheme of the NSGA-II [3] for choosing good solutions from the three populations in Fig. 1 (i.e., current, offspring and improved populations). That is, each solution in the three populations is evaluated by Pareto ranking and the concept of crowding in the same manner as in the NSGA-II [3].

The outline of our S-MOGLS algorithm is written as follows:

Basic Framework of Our S-MOGLS Algorithm:

Step 1 (Initialization): Generate an initial population with N_{pop} solutions where N_{pop} is the population size.

Step 2 (Genetic operations): Generate an offspring population by iterating the following procedures N_{pop} times:
    (1) Randomly specify the weight vector.
    (2) Choose a pair of parent solutions from the current population using the binary tournament selection based on the weighted scalar fitness function with the current weight vector.
    (3) Generate an offspring from the selected parents by crossover and mutation.

Step 3 (Local search): Generate an improved population by iterating the following procedures N_{pop} times:
    (1) Randomly specify the weight vector.
    (2) Choose an initial solution for local search from the offspring population using the binary tournament selection based on the weighted scalar fitness function with the current weight vector.
    (3) Apply a local search procedure based on the weighted scalar fitness function with the current weight vector to the selected initial solution with the local search application probability P_{LS}. Only when the initial solution is updated by the local search, the final solution at which the local search is terminated is added to the improved population.

Step 4 (Generation update): Construct the next population from the current, offspring and improved populations by choosing good solutions based on Pareto ranking and the concept of crowding in the same manner as in the NSGA-II.

Step 5 (Termination test): If the pre-specified stopping condition is not satisfied, return to Step 2. Otherwise terminate the execution of the algorithm.

3 Several Variants of Our S-MOGLS Algorithm

Various variants of our S-MOGLS algorithm can be implemented. In this section, we briefly describe those variants of our S-MOGLS algorithm. Some of them are used in computational experiments in the next section for discussing the implementation of local search in memetic EMO algorithms.

Selection of Parent Solutions: In our S-MOGLS algorithm, we use the weighted scalar fitness function for parent selection. This is to choose similar parents in the objective space, from which good offspring are likely to be generated. Of course, we can use other selection schemes. For example, we can use the parent selection scheme of the NSGA-II [3] (i.e., Pareto ranking and the concept of crowding).

Selection of Initial Solutions for Local Search: In our S-MOGLS algorithm, we choose good initial solutions from the offspring population for local search. It is also possible to probabilistically apply local search to offspring solutions independent of their performance. Moreover it is possible to apply local search to all offspring as in some memetic EMO algorithms (e.g., the S-MOGLS algorithms in [6]-[10]).

Local Search: In our S-MOGLS algorithm, we use the weighted scalar fitness function in local search as well as parent selection. We can also use Pareto ranking in local search. When we use Pareto ranking, the current solution **x** is replaced with its neighboring solution **y** (i.e., the local search move from **x** to **y** is accepted) only when **y** dominates **x** (i.e., **y** is better than **x**: see (3)). That is, the local search move is rejected when **x** and **y** are non-dominated with each other. In the M-PAES [11], a more sophisticated acceptance rule was used for handling the situation where **y** and **x** are incomparable with each other. The acceptance rule in [11] involves not only the comparison between the current solution **x** and the candidate solution **y**, but also the comparison with other solutions. This may somewhat degrade the inherent advantage of local search: simplicity. Thus we do not use the local search procedure in [11].

Let us further discuss the weighted scalar approach and the Pareto ranking approach to the implementation of local search. In Fig. 2, we show the movable area of the current solution **x** by each approach in the case of a two-objective maximization problem. In Fig. 2 (a), the weight vector was specified as $\lambda = (\lambda_1, \lambda_2) = (0.5, 0.5)$. As shown in Fig. 2, we can see that the movable area in the Pareto ranking approach is much smaller than the weighted scalar approach. This difference exponentially increases with the number of objectives because the movable area in the Pareto ranking approach is $1/2^k$ of the k-dimensional objective space while it is always 1/2 in the case of the weighted scalar approach.

As shown by computational experiments in the next section (and other studies [7], [13]), too small movable areas in the Pareto ranking approach prevent local search from efficiently searching for good solutions. On the other hand, too large movable areas in the weighted scalar approach sometimes lead to the deterioration of the offspring population. Thus we examine a simple modification of the weighted scalar approach for decreasing the movable areas as shown in Fig. 3 (a). More specifically, the local search move from the current solution **x** to the candidate solution **y** is accepted when the inequality relation $d/r < a$ holds in Fig. 3 (b) where a is a user-definable parameter. Of course, the candidate solution **y** should be better than the current solution **x** with respect to the weighted scalar fitness function. It should be noted that the weighted scalar approach corresponds to the case of $a = \infty$.

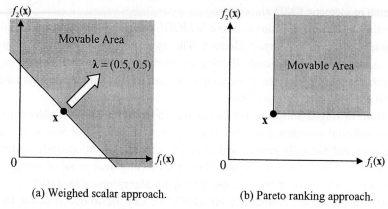

(a) Weighed scalar approach.   (b) Pareto ranking approach.

Fig. 2. Movable area of the current solution **x** in the two-dimensional objective space.

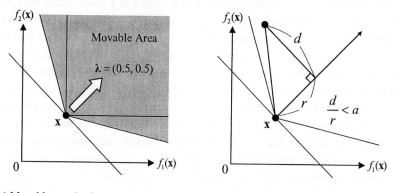

(a) Movable area in the modified approach.   (b) Definition of the acceptance rule.

Fig. 3. Modification of the weighted scalar approach.

4   Computational Experiments

In our computational experiments, we used nine knapsack problems in Zitzler & Thiele [14]: 2-250, 2-500, 2-750, 3-250, 3-500, 3-750, 4-250, 4-500, 4-750 where "k-m" means a k-objective m-item problem. Our computational experiments were performed in the same manner as in other comparative studies (e.g., Ishibuchi & Kaige [4], [5], Jaszkiewicz [9], and Knowles & Corne [12], and Zitzler & Thiele

[14]). We used the same parameter specifications as the NSGA-II in those comparative studies for the EMO part of our S-MOGLS algorithm and as the M-PAES and the MOGLS in those studies for the local search part.

For examining the performance of obtained non-dominated solution sets by our S-MOGLS algorithm, we use the generational distance (GD) and the $D1_R$ measure (see Coello [1] and Deb [2] for various performance measures). These measures evaluate the quality of a non-dominated solution set using a reference solution set. The reference solution set is a set of Pareto-optimal or near Pareto-optimal solutions. In our computational experiments, the reference solution set for each test problem was constructed by choosing non-dominated solutions among all solutions obtained in our previous computational experiments. The GD measure is the average distance from each solution in the obtained solution set to its nearest reference solution. This measure evaluates the convergence to the Pareto front. On the other hand, the $D1_R$ measure is the average distance from each reference solution to its nearest solution in the obtained solution set. This measure evaluates both the convergence and the diversity of obtained solutions. In our computational experiments, the average values of these measures were calculated over 30 runs for each test problem.

Comparison between two approaches: We first compared the Pareto ranking approach to the weighted scalar approach using the S-MOGLS algorithm in Section 2. The local search application probability P_{LS} was specified as $P_{LS} = 0.1$. The relative performance of the Pareto ranking approach with respect to the weighted scalar approach was calculated for each test problem. Experimental results are summarized in Fig. 4. In these figures, the horizontal axis shows the nine test problems. From Fig. 4, we can see that the relative performance of the Pareto ranking approach is larger than 1.00 (which is the relative performance of the weighted scalar approach) for many test problems. This means that the Pareto ranking approach is inferior to the weighted scalar approach because the GD and $D1_R$ measures should be minimized.

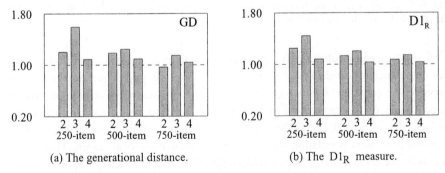

(a) The generational distance.   (b) The $D1_R$ measure.

Fig. 4. Relative performance of the Pareto ranking approach with respect to the weighted scalar approach for each test problem.

While better results were obtained from the weighted scalar approach, it has a serious difficulty: The offspring population can be degraded by the local search. This difficulty is illustrated in Fig. 5. In Fig. 5 (a), we show an offspring population and its improved population. These populations are intermediate results during a computational experiment using our S-MOGLS algorithm on the 2-500 problem. Special parameter values were used in this computational experiment for illustration purpose (e.g., a small population size and a large local search application probability). Fig. 5 (b) shows why the offspring population can be degraded by the local search. As shown in Fig. 5 (b), the offspring population is likely to be degraded when the local search direction for each solution is not appropriate. Ideally the local search direction should be vertical to the non-dominated front of the offspring population. In such an ideal case, the offspring population will not be degraded (i.e., the local search moves in Fig. 5 (b) won't happen). In the remaining of this section, we examine some tricks for overcoming the above-mentioned difficulty of the weighted scalar approach.

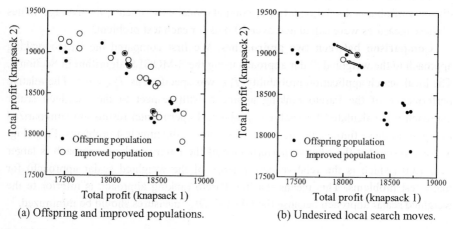

(a) Offspring and improved populations.   (b) Undesired local search moves.

Fig. 5. Illustration of the deterioration of the offspring population by the local search based on the weighted scalar fitness function.

Use of three populations for generation update: A straightforward remedy for the above-mentioned difficulty is to use the offspring population as well as the improved population in the generation update procedure. For examining the effectiveness of this generation update scheme, we examined the performance of a variant of our S-MOGLS algorithm where offspring solutions were not used for generating the next population when they were updated by local search. This variant is exactly the same as the S-MOGLS algorithm of Ishibuchi & Kaige [5]. The relative performance of this variant with respect to our S-MOGLS algorithm in Section 2 is summarized in Fig. 6 in the same manner as Fig. 4. From this figure, we can see that the performance of the S-MOGLS algorithm was degraded by modifying its generation update scheme with respect to the GD measure.

Increase in the selection pressure of initial solutions: As we have already discussed, the inappropriate specification of the local search direction causes the deterioration of offspring populations by local search. Since we use the tournament selection based on the weighted scalar fitness function for choosing initial solutions for local search, the increase in the selection pressure of initial solutions (i.e., the increase in the tournament size) may lead to the selection of an appropriate initial solution for the current weight vector. That is, the tournament selection with a large tournament size is likely to choose a very good offspring solution with respect to the current weight vector. Such an offspring solution is likely to locate near the non-dominated front of the offspring population. Moreover the current weight vector is likely to be vertical to the non-dominated front around the selected initial solution.

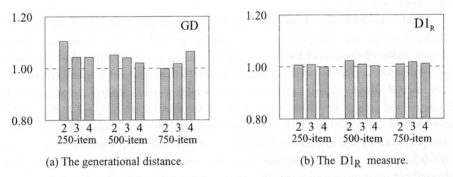

(a) The generational distance.   (b) The $D1_R$ measure.

Fig. 6. Relative performance of a variant of our S-MOGLS algorithm where intermediate offspring solutions were not used for generating the next population.

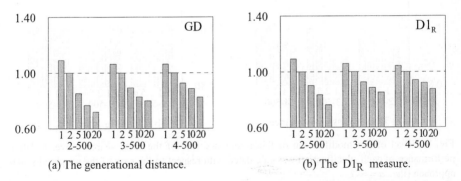

(a) The generational distance.   (b) The $D1_R$ measure.

Fig. 7. Effect of the specification of the tournament size. Relative performance of each specification was calculated with respect to the case of the binary tournament selection.

We examined various specifications of the tournament size for the selection of initial solutions in our S-MOGLS algorithm. Experimental results are summarized in Fig. 7 where horizontal axis shows the tournament size (i.e., 1, 2, 5, 10, 20). In this figure, the relative performance was calculated with respect to the binary tournament selection. Thus the relative performance is always 1.00 in this figure when the tournament size is 2. From this figure, we can see that the performance of our S-MOGLS algorithm was improved by increasing the selection pressure of initial solutions for local search. When the tournament size was specified as 1, initial solutions were randomly chosen from the offspring population. In this case, the performance of our S-MOGLS algorithm was degraded by the following two reasons: One is the inappropriate specification of the local search direction for each initial solution, and the other is the selection of poor offspring solutions as initial solutions. On the other hand, good solutions together with appropriate local search directions are chosen as initial solutions for local search when the selection pressure is high. As a result, the performance of our S-MOGLS algorithm was improved by increasing the tournament size in Fig. 7.

Modification of the acceptance rule: We also examined the effectiveness of the modification of the acceptance rule illustrated in Fig. 3. We examined various specifications of the user-definable parameter a using our S-MOGLS algorithm with the binary tournament selection. The relative performance of each specification of a was calculated with respect to the case of the weighted scalar approach (i.e., $a = \infty$). Experimental results are summarized in Fig. 8 for $a = 0.5, 1, 2, 5, \infty$. In Fig. 8, we can see that the convergence to the Pareto front (i.e., the GD measure in Fig. 8 (a)) was improved by the modification of the acceptance rule from the case of $a = \infty$ to $\alpha = 0.5, 1, 2, 5$. This modification, however, may have a negative effect on the diversity of solutions because the Dl_R measure was not improved in Fig. 8 (b).

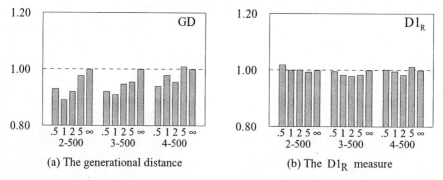

(a) The generational distance   (b) The Dl_R measure

Fig. 8. Effect of the modification of the acceptance rule of the local search move. Relative performance of each specification was calculated with respect to the case of the weighted scalar approach (i.e., $a = \infty$).

5 Concluding Remarks

We proposed a new memetic EMO algorithm called S-MOGLS by combining local search with the NSGA-II [3]. Our intention is to implement a simple but powerful memetic EMO algorithm, which can be easily understood, easily implemented, and efficiently executed using small memory storage within short CPU time. Using our S-MOGLS algorithm, we examined several issues related to the implementation of local search in memetic EMO algorithms. Through computational experiments, we showed that the weighted scalar approach outperforms the Pareto ranking approach. We also showed that the weighted scalar approach has a serious difficulty: The offspring population can be degraded by local search. We implemented three remedies for this difficulty in our S-MOGLS algorithm: the use of three populations (i.e., parent, offspring and improved populations) in generation update, the choice of good initial solutions for local search, and the modification of the acceptance rule. The effectiveness of these tricks was demonstrated by our experimental results.

The authors would like to thank the financial support from Kayamori Foundation of Information Science Advancement, and Japan Society for the Promotion of Science (JSPS) through Grand-in-Aid for Scientific Research (B): KAKENHI (14380194).

References

1. Coello, C. A. C., Van Veldhuizen, D. A., and Lamont, G. B.: *Evolutionary Algorithms for Solving Multi-Objective Problems*, Kluwer Academic Publishers, Boston (2002).
2. Deb, K.: *Multi-Objective Optimization Using Evolutionary Algorithms*, John Wiley & Sons, Chichester (2001).
3. Deb, K., Pratap, A., Agarwal, S., and Meyarivan, T.: A Fast and Elitist Multiobjective Genetic Algorithm: NSGA-II, *IEEE Trans. on Evolutionary Computation* 6 (2002) 182 – 197.
4. Ishibuchi, H., and Kaige, S.: Effects of Repair Procedures on the Performance of EMO Algorithms for Multiobjective 0/1 Knapsack Problems, *Proc. of 2003 Congress on Evolutionary Computation* (2003) 2254-2261.
5. Ishibuchi, H., and Kaige, S.: Implementation of Simple Multiobjective Memetic Algorithms and Its Application to Knapsack Problems, *International Journal of Hybrid Intelligent System* 1 (2004) 22-35.
6. Ishibuchi, H., and Murata, T.: A Multi-Objective Genetic Local Search Algorithm and Its Application to Flowshop Scheduling, *IEEE Trans. on Systems, Man, and Cybernetics - Part C: Applications and Reviews* 28 (1998) 392-403.
7. Ishibuchi, H., Yoshida, T., and Murata, T.: Balance between Genetic Search and Local Search in Memetic Algorithms for Multiobjective Permutation Flowshop Scheduling, *IEEE Trans. on Evolutionary Computation* 7 (2003) 204-223.
8. Jaszkiewicz, A.: Comparison of Local Search-Based Metaheuristics on the Multiple Objective Knapsack Problem, *Foundations of Computing and Decision Sciences* 26 (2001) 99-120.
9. Jaszkiewicz, A.: On the Performance of Multiple-Objective Genetic Local Search on the 0/1 Knapsack Problem - A Comparative Experiment, *IEEE Trans. on Evolutionary Computation* 6 (2002) 402-412.
10. Jaszkiewicz, A.: Genetic Local Search for Multi-Objective Combinatorial Optimization, *European Journal of Operational Research* 137 (2002) 50-71.
11. Knowles, J. D., and Corne, D. W.: M-PAES: A Memetic Algorithm for Multiobjective Optimization, *Proc. of 2000 Congress on Evolutionary Computation* (2000) 325-332.

12. Knowles, J. D., and Corne, D. W.: A Comparison of Diverse Approaches to Memetic Multiobjective Combinatorial Optimization, *Proc. of 2000 Genetic and Evolutionary Computation Conference Workshop Program: WOMA I* (2000) 103-108.
13. Murata, T., Kaige, S., and Ishibuchi, H.: Generalization of Dominance Relation-Based Replacement Rules for Memetic EMO Algorithms, *Lecture Notes in Computer Sciences* 2723 (2003) 1234-1245. *Proc. of 2003 Genetic and Evolutionary Computation Conference.*
14. Zitzler, E., Thiele, L.: Multiobjective Evolutionary Algorithms: A Comparative Case Study and the Strength Pareto Approach, *IEEE Transactions on Evolutionary Computation* 3 (1999) 257-271.

Mating Scheme for Controlling the Diversity-Convergence Balance for Multiobjective Optimization

Hisao Ishibuchi and Youhei Shibata

Department of Industrial Engineering, Osaka Prefecture University,
1-1 Gakuen-cho, Sakai, Osaka 599-8531, Japan
{hisaoi, shibata}@ie.osakafu-u.ac.jp

Abstract. The aim of this paper is to clearly demonstrate the potential ability of a similarity-based mating scheme to dynamically control the balance between the diversity of solutions and the convergence to the Pareto front in evolutionary multiobjective optimization. The similarity-based mating scheme chooses two parents in the following manner. For choosing one parent (say Parent A), first a pre-specified number of candidates (say α candidates) are selected by iterating the standard fitness-based binary tournament selection. Then the average solution of those candidates is calculated in the objective space. The most similar or dissimilar candidate to the average solution is chosen as Parent A. When we want to increase the diversity of solutions, the selection probability of Parent A is biased toward extreme solutions by choosing the most dissimilar candidate. The strength of this diversity-preserving effort is adjusted by the parameter α. We can also bias the selection probability toward center solutions by choosing the most similar candidate when we want to decrease the diversity. The selection probability of the other parent (i.e., the mate of Parent A) is biased toward similar solutions to Parent A for increasing the convergence speed to the Pareto front. This is implemented by choosing the most similar one to Parent A among a pre-specified number of candidates (say β candidates). The strength of this convergence speed-up effort is adjusted by the parameter β. When we want to increase the diversity of solutions, the most dissimilar candidate to Parent A is chosen as its mate. Our idea is to dynamically control the diversity-convergence balance by changing the values of two control parameters α and β during the execution of evolutionary multiobjective optimization algorithms. We examine the effectiveness of our idea through computational experiments on multiobjective knapsack problems.

1 Introduction

The goal of evolutionary multiobjective optimization (EMO) is to efficiently find Pareto-optimal (or near Pareto-optimal) solutions of multiobjective optimization problems as many as possible. There are two sub-goals for achieving this goal: maintaining the diversity of solutions in each population and increasing the

K. Deb et al. (Eds.): GECCO 2004, LNCS 3102, pp. 1259–1271, 2004.
© Springer-Verlag Berlin Heidelberg 2004

convergence speed of solutions to the Pareto front. In this paper, we propose an idea of using a similarity-based mating scheme for controlling the balance between these two sub-goals during the execution of EMO algorithms.

EMO algorithms have been applied to various problems for efficiently finding their Pareto-optimal or near Pareto-optimal solutions. Recent EMO algorithms usually share some common ideas such as elitism, fitness sharing and Pareto ranking for improving both the diversity of solutions and the convergence speed to the Pareto front (e.g., see Coello et al. [1] and Deb [2]). In some studies, local search was combined with EMO algorithms for further improving the convergence speed to the Pareto front [10], [13], [14]. While mating restriction has been often discussed in the literature, its effect has not been clearly demonstrated. As a result, it is not used in many EMO algorithms as pointed out in some reviews on EMO algorithms [5], [20], [22]. Mating restriction was suggested by Goldberg [6] and used in EMO algorithms by Hajela & Lin [7] and Fonseca & Fleming [4]. The basic idea of mating restriction is to ban the crossover of dissimilar parents from which good offspring are not likely to be generated. The necessity of mating restriction in EMO algorithms was stressed in Jaszkiewicz [14]. On the other hand, Zitzler & Thiele [21] reported that no improvement was achieved by mating restriction in their computational experiments. Moreover, there was also an argument for the selection of dissimilar parents. Horn et al. [8] argued that information from very different types of tradeoffs could be combined to yield other kinds of good tradeoffs. Schaffer [19] examined the selection of dissimilar parents but observed no improvement.

A similarity-based mating scheme was proposed in Ishibuchi & Shibata [11] for examining positive and negative effects of mating restriction on the search ability of EMO algorithms. In their mating scheme, one parent (say Parent A) was chosen by the standard fitness-based binary tournament scheme while its mate (say Parent B) was chosen among a pre-specified number of candidates (say β candidates) based on their similarity or dissimilarity to Parent A. Those candidates were selected by iterating the standard fitness-based binary tournament selection β times. Almost the same idea was independently proposed in Huang [9] where Parent B was chosen from two candidates (i.e., the value of β was fixed as $\beta = 2$). Ishibuchi & Shibata [12] extended their similarity-based mating scheme as shown in Fig. 1. That is, first a pre-specified number of candidates (say α candidates) were selected by iterating the standard fitness-based binary tournament selection α times. Next the average solution of those candidates was calculated in the objective space. The most dissimilar candidate to the average solution was chosen as Parent A. On the other hand, the most similar one to Parent A among β candidates was chosen as Parent B.

In this paper, we further extend this mating scheme in the following manner:
(1) Selection of Parent A: We examine the choice of the most similar candidate to the average solution as well as the choice of the most dissimilar one.
(2) Selection of Parent B: We examine the choice of the most dissimilar candidate to Parent A as well as the choice of the most similar one.

(3) Parameters α and β: We do not assume that these parameters are fixed during the execution of EMO algorithms. That is, they are handled as control parameters.

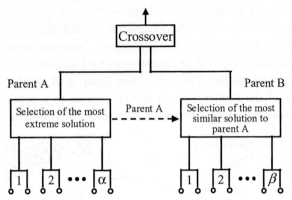

Fig. 1. Mating scheme in Ishibuchi & Shibata [12].

This paper is organized as follows. In Section 2, we describe the motivation behind the above-mentioned extensions of the mating scheme. The extended mating scheme is described in Section 3. In Section 4, we perform computational experiments on multiobjective knapsack problems by combining the extended mating scheme with the NSGA-II algorithm [3]. Finally we conclude this paper in Section 5.

2 Motivation

In the implementation of EMO algorithms, it is very important to find an appropriate balance between the diversity of solutions and the convergence to the Pareto front. Let us consider a two-objective maximization problem in Fig. 2. For increasing the diversity, the current population should be expanded in the direction of the two arrows labeled as D in Fig. 2 (a). On the other hand, the current population should be driven in the direction of the other arrow C in Fig. 2 (a) for the convergence to the Pareto front. When the diversity-convergence balance is appropriate, good solution sets may be obtained as shown in Fig. 2 (b). It is, however, not easy to find an appropriate balance because each problem and each EMO algorithm may require its own specification. Moreover, different specifications may be required in different stages of evolution. For example, it may be a good idea to emphasize the diversity in the early state of evolution and the convergence in the later stage of evolution.

This paper is motivated by Knowles & Corne [16], [17] where they proposed an idea of designing a memetic EMO algorithm based on the landscape analysis of multiobjective optimization problems. For example, if the search along the Pareto front is much easier than the convergence to the Pareto front, it seems to be a good strategy to first drive the population to the Pareto front as close as possible (see Fig. 3

(a)). On the other hand, if the search along the Pareto front is much more difficult than the convergence to the Pareto front, it seems to be a good strategy to first increase the diversity of solutions as large as possible (see Fig. 3 (b)). These discussions suggest the necessity to dynamically control the diversity-convergence balance during the execution of EMO algorithms. Thus we extend the similarity-based mating scheme in Fig. 1 in the above-mentioned manner in Section 1, which is explained in the next section. The aim of this paper is to clearly demonstrate the potential ability of the similarity-based mating scheme to dynamically control the diversity-convergence balance during the execution of EMO algorithms.

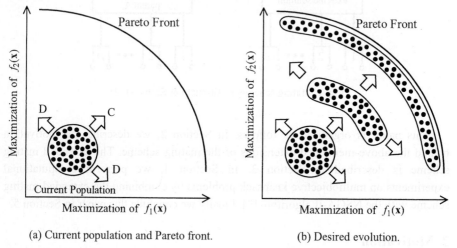

(a) Current population and Pareto front.   (b) Desired evolution.

Fig. 2. Search directions for increasing the diversity of solutions (i.e., Arrows D) and for improving the convergence speed to the Pareto front (i.e., Arrow C).

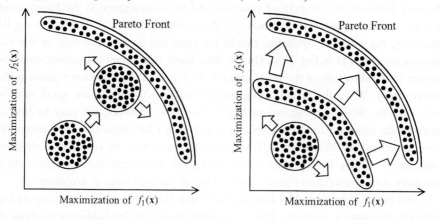

(a) First-convergence-then-diversity strategy.   (b) First-diversity-then-convergence strategy.

Fig. 3. Two search strategies. Emphasis is first placed on the convergence then on the diversity in (a) while emphasis is first placed on the diversity then on the convergence in (b).

3 Similarity-Based Mating Scheme

We explain the extended similarity-based mating scheme using the following n-objective optimization problem:

$$\text{Optimize } \mathbf{f}(\mathbf{x}) = (f_1(\mathbf{x}), f_2(\mathbf{x}), ..., f_n(\mathbf{x})) \text{ subject to } \mathbf{x} \in \mathbf{X}, \tag{1}$$

where $\mathbf{f}(\mathbf{x})$ is the objective vector, $f_i(\mathbf{x})$ is the i-th objective to be minimized or maximized, $\mathbf{x}$ is the decision vector, and $\mathbf{X}$ is the feasible region in the decision space. The distance between two solutions $\mathbf{x}$ and $\mathbf{y}$ is measured in the objective space by the Euclidean distance between $\mathbf{f}(\mathbf{x})$ and $\mathbf{f}(\mathbf{y})$.

As shown in the left-hand side of Fig. 1, the standard fitness-based binary tournament selection with replacement is iterated α times for choosing α candidates (say $\mathbf{x}_1$, $\mathbf{x}_2$, ..., $\mathbf{x}_\alpha$) among which the first parent (i.e., Parent A in Fig. 1) is selected. Then the average solution of the α candidates is calculated in the objective space as

$$\bar{\mathbf{f}}(\mathbf{x}) = (\bar{f}_1(\mathbf{x}), \bar{f}_2(\mathbf{x}), ..., \bar{f}_n(\mathbf{x})), \tag{2}$$

where

$$\bar{f}_i(\mathbf{x}) = \frac{1}{\alpha} \sum_{j=1}^{\alpha} f_i(\mathbf{x}_j) \text{ for } i = 1, 2, ..., n. \tag{3}$$

Finally the most similar or dissimilar candidate to the average solution is selected as Parent A. The similarity is measured by the Euclidean distance to the average solution $\bar{\mathbf{f}}(\mathbf{x})$ in the objective space. When the value of α is specified as $\alpha = 1$, the choice of the first parent is the same as the standard fitness-based binary tournament selection. The case of $\alpha = 2$ is also actually the same as the case of $\alpha = 1$ because two candidates always have the same distance from their average solution. In this case, one candidate is randomly chosen (i.e., random tiebreak). The selection probability of the first parent is biased toward extreme solutions or center solutions only when $\alpha \geq 3$.

On the other hand, the standard fitness-based binary tournament selection with replacement is iterated β times for choosing β candidates of the second parent (i.e., Parent B) as shown in the right-hand side of Fig. 1. Then the most similar or dissimilar candidate to the first parent (i.e., Parent A in Fig. 1) is chosen as Parent B. In this manner, similar or dissimilar parents are recombined in our mating scheme.

As shown in Table 1, our mating scheme is divided into nine (i.e., 3×3) operation modes depending on the strategies to choose Parent A and Parent B. The first operation mode with $\alpha = 1$ and $\beta = 1$ is the same as the standard fitness-based binary tournament selection. The first three operation modes were examined in [9] and [11] while the eighth operation mode was recommended in [12]. In those studies, the values of α and β were fixed throughout the execution of EMO algorithms. In this paper, we examine all the nine operation modes where the values of α and β are not fixed but variable. Moreover our mating scheme can move from one operation mode to another during the execution of EMO algorithms.

Table 1. Nine operation modes of our mating scheme.

Operation mode	Selection of Parent A	Selection of Parent B
1	No bias: $\alpha = 1$	No bias: $\beta = 1$
2	No bias: $\alpha = 1$	Similar solution ($\beta \geq 2$)
3	No bias: $\alpha = 1$	Dissimilar solution ($\beta \geq 2$)
4	Center solution: Similar ($\alpha \geq 3$)	No bias: $\beta = 1$
5	Center solution: Similar ($\alpha \geq 3$)	Similar solution ($\beta \geq 2$)
6	Center solution: Similar ($\alpha \geq 3$)	Dissimilar solution ($\beta \geq 2$)
7	Extreme solution: Dissimilar ($\alpha \geq 3$)	No bias: $\beta = 1$
8	Extreme solution: Dissimilar ($\alpha \geq 3$)	Similar solution ($\beta \geq 2$)
9	Extreme solution: Dissimilar ($\alpha \geq 3$)	Dissimilar solution ($\beta \geq 2$)

4 Computational Experiments

4.1 Conditions of Computational Experiments

We combined our mating scheme with the NSGA-II [3], which was used for choosing candidate parents in our mating scheme. First we examined $7 \times 7 = 49$ combinations of constant values of α and β: $\alpha = 1$, $\alpha = 3, 5, 9$ (similar), $\alpha = 3, 5, 9$ (dissimilar) and $\beta = 1$, $\beta = 3, 5, 9$ (similar), $\beta = 3, 5, 9$ (dissimilar). Then we examined the effect of changing the values of α and β during the execution of the NSGA-II.

As test problems, we used four knapsack problems in Zitzler & Thiele [22]: 2-250 (i.e., two-objective 250-item), 2-500, 3-250 and 3-500 problems. Each solution for an m-item problem was coded as a binary string of the length m. Each string was evaluated in the same manner as in Zitzler & Thiele [22]. The NSGA-II algorithm with our mating scheme was applied to the four knapsack problems under the following parameter specifications: Crossover probability: 0.8, mutation probability: $1/m$, population size: 200, and stopping condition: 2000 generations.

Various performance measures have been proposed for evaluating a set of non-dominated solutions in the literature. As explained in [15], [18], [23], no performance measure can simultaneously evaluate various aspects of a solution set. In this paper, we use three performance measures: the generational distance (GD), the $D1_R$ measure, and a spread measure. The GD measure is the average distance from each solution in the solution set to its nearest Pareto-optimal solution. This measure has been often used for evaluating the convergence to the Pareto front. On the other hand, the $D1_R$ measure is the average distance from each Pareto-optimal solution to its nearest solution in the solution set. This measure has been used in some studies for simultaneously evaluating both the convergence and the diversity. As a spread measure of the solution set S, we use the sum of the width for each objective:

$$Spread = \sum_{i=1}^{n} [\max_{x \in S}\{f_i(x)\} - \min_{x \in S}\{f_i(x)\}], \tag{4}$$

where n is the number of objectives.

The GD and $D1_R$ measures need all Pareto-optimal solutions of each test problem. For the 2-250 and 2-500 test problems, the Pareto-optimal solutions are available from the homepage of the first author of [22]. For the 3-250 and 3-500 test problems, we found near Pareto-optimal solutions using the SPEA [22], the NSGA-II [3] and a single objective genetic algorithm using much longer CPU time and larger memory storage (e.g., 30000 generations with the population size 400 for the NSGA-II) than the other computational experiments.

4.2 Experimental Results with Constant Parameters

The NSGA-II algorithm with our mating scheme was applied to the four test problems using the 7×7 combinations of α and β. For each combination, we performed ten runs from different initial populations for each test problem.

Average results over ten runs for the 2-500 and 3-500 test problems are summarized in Figs. 4-6 where smaller values (i.e., shorter bars) of the GD and $D1_R$ measures mean better results in Fig. 4 and Fig. 5 while larger values (i.e., longer bars) of the spread measure mean better results in Fig. 6. In each figure, the original NSGA-II algorithm corresponds to the bar at the center with $\alpha = 1$ and $\beta = 1$, which is depicted by a white bar. The effect of the selection of the first parent (i.e., the specification of α) is clear in Fig. 6. That is, the bias toward extreme solutions (i.e., the use of a large value of α (dissimilar)) increased the spread of solution sets. On the other hand, the recombination of similar parents (i.e., the use of a large value of β (similar)) improved the convergence to the Pareto front as shown in Fig. 4 with some exceptions in Fig. 4 (a). In the case of α (similar) and β (similar) in Fig. 4 (a), too small diversity of solutions had a bad effect on the evolution of solutions (see Fig. 6 (a)). Experimental results on the 2-250 and 3-250 test problems were very similar to those on the 2-500 and 3-500 test problems, respectively.

For visually demonstrating the effect of our mating scheme on the evolution of solutions, we show some intermediate and final solution sets in Fig. 7 where each solution set is a result of a single run on the 2-500 test problem using each combination of α and β. In Fig. 7, we also depict the true Pareto front for comparison. From Fig. 7, we can see that the search ability of the original NSGA-II algorithm in Fig. 7 (a) was significantly improved by our mating scheme in terms of the diversity of solutions. More specifically, we can see that the recombination of dissimilar parents (i.e., the use of $\beta = 9$ (dissimilar)) increased the diversity of solutions in Fig. 7 (b) if compared with Fig. 7 (a) obtained by the original NSGA-II algorithm. The diversity of solutions was further increased in Fig. 7 (d) by the choice of extreme solutions (i.e., the use of $\alpha = 9$ (dissimilar)). It should be noted that the convergence was not seriously deteriorated in Fig. 7 (d). This is because the recombination of similar solutions (i.e., the use of $\beta = 9$ (similar)) improved the convergence as shown in Fig. 4 when the selection probability of the first parent was biased toward extreme solutions (i.e., when α was specified as $\alpha \geq 3$ (dissimilar)).

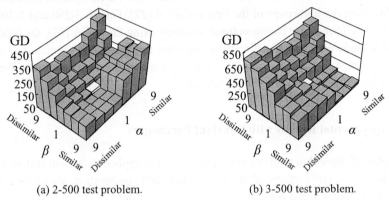

(a) 2-500 test problem.   (b) 3-500 test problem.

Fig. 4. Average values of the GD measure.

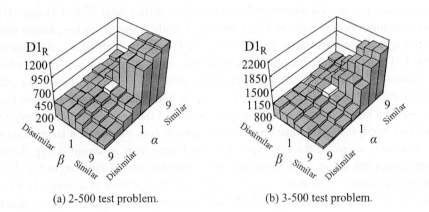

(a) 2-500 test problem.   (b) 3-500 test problem.

Fig. 5. Average values of the $D1_R$ measure.

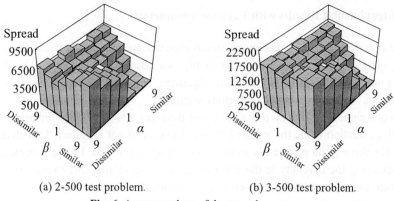

(a) 2-500 test problem.   (b) 3-500 test problem.

Fig. 6. Average values of the spread measure.

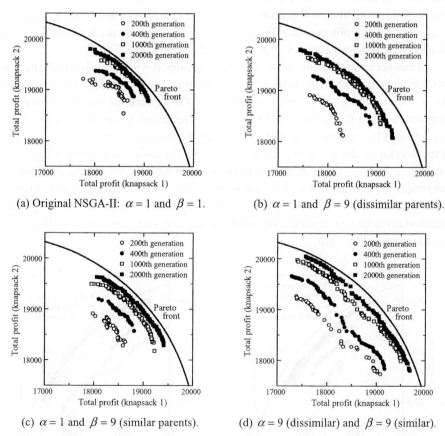

(a) Original NSGA-II: $\alpha = 1$ and $\beta = 1$.   (b) $\alpha = 1$ and $\beta = 9$ (dissimilar parents).

(c) $\alpha = 1$ and $\beta = 9$ (similar parents).   (d) $\alpha = 9$ (dissimilar) and $\beta = 9$ (similar).

Fig. 7. Solution set at each generation in a single run with each combination of α and β.

4.3 Experimental Results with Variable Parameters

We also performed the same computational experiments as in the previous subsection using variable parameters. More specifically, we changed the values of α and β during the execution of the NSGA-II algorithm with our mating scheme. In this subsection, we only report some illustrative cases.

In one case, the value of α was changed from $\alpha = 1$ to $\alpha = 9$ (dissimilar) after the 1000th generation while the value of β was fixed as $\beta = 9$ (similar). This change of α is for driving the population to the Pareto front in the first 1000 generations and for increasing the diversity in the last 1000 generations. Using this change of α, we intended to implement the first-convergence-then-diversity strategy in Fig. 3 (a). Experimental results are shown in Fig. 8 (a). From this figure, we can see that our intention was successfully implemented. That is, the diversity was increased in the last 1000 generations while the population with small diversity was driven to the Pareto front in the first 1000 generations. From the comparison between Fig. 7 (c) and Fig. 8 (a), we can observe the effect of changing the value of α. It should be noted that the first 1000 generations in Fig. 8 (a) are the same as Fig. 7 (c).

In another case, the value of α was changed from $\alpha = 9$ (dissimilar) to 1 after the 1000th generation while the value of β was fixed as $\beta = 9$ (similar). This change of α is for increasing the diversity in the first 1000 generations and for driving the population to the Pareto front in the last 1000 generations. Using this change of α, we intended to implement the first-diversity-then-convergence strategy in Fig. 3 (b). Experimental results are shown in Fig. 8 (b). From the comparison between Fig. 7 (d) and Fig. 8 (b), we can see that the convergence around the center of the Pareto front was improved by the change of the value of α from $\alpha = 9$ (dissimilar) to 1. That is, our intention to implement the first-diversity-then-convergence strategy was successfully achieved. It should be noted that the first 1000 generations in Fig. 8 (b) are the same as Fig. 7 (d).

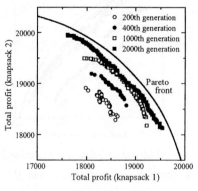
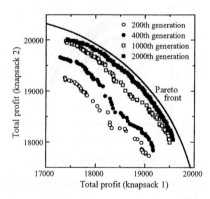

(a) α : 1 to 9 (dissimilar), $\beta = 9$ (similar).   (b) α : 9 (dissimilar) to 1, $\beta = 9$ (similar).

Fig. 8. Experimental results using variable parameters.

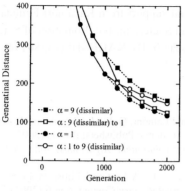
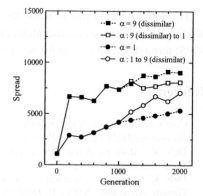

(a) Average values of the GD measure.   (b) Average values of the spread measure.

Fig. 9. Average results over ten runs for each specification of α and β ($\beta = 9$ (similar)).

We further examined the above-mentioned four cases of α and β: those in Fig. 7 (c), (d) and Fig. 8 (a), (b). Using each specification, we applied the NSGA-II with our mating scheme to the 2-500 test problem ten times. Then we calculated the average values of the GD and spread measures over ten runs. Experimental results are shown in Fig. 9. From the results depicted by white squares in Fig. 9 (a), we can see that the change of α from $\alpha = 9$ (dissimilar) to $\alpha = 1$ improved the convergence, which corresponds to the improvement from Fig. 7 (d) to Fig. 8 (b). This improvement was achieved at the cost of the decrease in the diversity as shown in Fig. 9 (b). We can also see from the results depicted by closed circles in Fig. 9 (b) that the change of α from $\alpha = 1$ to $\alpha = 9$ (dissimilar) improved the diversity.

5 Concluding Remarks

We proposed an idea of dynamically controlling the balance between the diversity of solutions and the convergence to the Pareto front using a similarity-based mating scheme. Through computational experiments on multiobjective knapsack problems, we demonstrated that our mating scheme has the potential ability to dynamically control the diversity-convergence balance. We also visually demonstrated that the performance of the NSGA-II algorithm was improved in terms of the diversity of solutions. Our mating scheme has two parameters α and β, which are the number of candidates from which two parents are selected. Since the interpretation of these parameters is very easy, we can intuitively control the values of these parameters.

As we described using several figures, different specifications of the diversity-convergence balance may be required in different stages of generation for each multiobjective optimization problem. Thus an automated adaptation method of α and β may be required in our mating scheme. This is left for future research.

Encoding Bounded-Diameter Spanning Trees with Permutations and with Random Keys

Bryant A. Julstrom

Department of Computer Science
St. Cloud State University, St. Cloud, MN, 56301 USA
julstrom@eeyore.stcloudstate.edu

Abstract. Permutations of vertices can represent constrained spanning trees for evolutionary search via a decoder based on Prim's algorithm, and random keys can represent permutations. Though we might expect that random keys, with an additional level of indirection, would provide inferior performance compared with permutations, a genetic algorithm that encodes spanning trees with random keys is as effective as one whose genotypes are permutations of vertices in comparisons on a variety of instances of the bounded-diameter minimum spanning tree problem. These results suggest that either coding may be used, at the programmer's convenience, in evolutionary algorithms for problems involving constrained spanning trees.

1   Introduction

Evolutionary algorithms often search spaces of permutations. However, when simple positional evolutionary operators like k-point crossover and position-by-position mutation are applied to permutations, the offspring they build are generally not permutations, so researchers must apply specialized operators. A large number of these, particularly crossovers, have been developed [1].

Bean [2] proposed an indirect representation of permutations for evolutionary search to which positional operators can be applied. He called this representation random keys. In it, a genotype is a sequence of floating-point values, called keys and usually falling between 0.0 and 1.0, associated with the items to be ordered. Sorting the keys yields a permutation of the items; this is the permutation the genotype represents. For example, if the integers 0 through 9 label ten items, then the key sequence

$$(0.48, 0.66, 0.07, 0.33, 0.38, 0.72, 0.88, 0.54, 0.25, 0.42)$$

represents this permutation of those items: (2 8 3 4 9 0 7 1 5 6). Every sequence of keys represents a valid permutation, so simple operators like k-point crossover and position-by-position mutation always yield valid genotypes.

Rothlauf [3, pp.180-182] has argued that random keys have good properties for evolutionary search. Researchers have used them to represent permutations, in turn representing candidate solutions, in evolutionary algorithms for a variety

K. Deb et al. (Eds.): GECCO 2004, LNCS 3102, pp. 1272–1281, 2004.

of problems, including machine scheduling, vehicle routing, and the quadratic assignment problem [2], constrained facility layout problems [4], deceptive ordering problems [5] [6], cellular manufacturing [7], and job shop scheduling [8].

Rothlauf, Goldberg, and Heinzl [9] [3, pp.178–190] described a random-key representation of spanning trees called Network Random Keys, or NetKeys. Given a connected, undirected graph G, NetKeys associates keys with G's edges. Sorting the keys yields a permutation of the edges, and a decoding algorithm based on Kruskal's algorithm [10] examines the edges in their permuted order to build a spanning tree on G. The decoder can enforce constraints on the spanning tree, such as a bound on its vertices' degrees. If G has n vertices, it may have up to $\binom{n}{2}$ edges, so the size of each NetKeys genotype is $O(n^2)$, and the time required to sort it—and impose an ordering on G's edges—is $O(n^2 \log n)$.

Permutations of a graph's vertices can also represent spanning trees, via a decoder based on Prim's algorithm [11], and these permutations can in turn be represented by random keys. Each of these codings offers advantages and disadvantages of implementation; does either provide better performance?

Permutations of vertices and random keys, with operators appropriate to them, were implemented in generational genetic algorithms and compared on 25 instances of the bounded-diameter minimum spanning tree problem, which Section 2 describes. These trials revealed no significant or consistent differences in the performance of the two codings. This suggests that either coding may be used, at the programmer's convenience, in evolutionary algorithms for problems involving constrained spanning trees.

Following the description of the bounded-diameter minimum spanning tree problem, this paper presents permutation and random-key codings of bounded-diameter spanning trees, genetic algorithms that use them, and the comparisons of the two genetic algorithms.

2   The Bounded-Diameter MST Problem

In a tree, the eccentricity of a vertex v is the maximum number of edges on any path from v to another vertex. The diameter of a tree is the maximum eccentricity of its vertices, thus the maximum number of edges along any path in it. The center of a tree is the single vertex (if its diameter is even) or the two vertices (if odd) of minimum eccentricity, and a vertex's depth is the number of edges on the path from it to the tree's center.

Given a connected, undirected graph G on n vertices and a bound D, a bounded-diameter spanning tree (BDST) is a spanning tree on G with diameter no greater than D. Figure 1 shows two BDSTs on $n = 20$ vertices. One has even diameter; the vertex v is its center. The other has odd diameter; the vertices v and w together form its center.

If weights are associated with G's edges, then a bounded-diameter minimum spanning tree (BDMST) is a BDST on G of minimum total weight. The search for such a tree is the bounded-diameter minimum spanning tree problem. This problem is NP-hard for $4 \leq D < n - 1$ [12, p.206], though it is solvable in

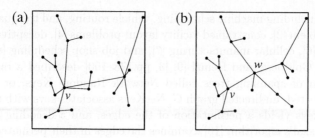

Fig. 1. (a) A tree on 20 vertices with diameter 4; v is its center. (b) A tree on the same vertices with diameter 5; v and w together form its center

polynomial time if $D \leq 3$ or if all the edge weights in G are equal. The bounded-diameter minimum spanning tree problem has applications in a variety of areas, including communications network design [13], mutual exclusion in distributed systems [14], and data compression [15],

Abdalla, Deo, and Gupta [16] [17] described two heuristics for this problem, one of which imitates Prim's algorithm [11] but eschews edges whose inclusion would violate the diameter bound. It maintains the lengths of all paths and the eccentricities of all vertices in the growing BDST, and it requires time that is $O(n^3)$.

More recently, Raidl and Julstrom [18] described a Prim-based, greedy heuristic that builds a BDST beginning at its center. It avoids edges that would place vertices at depths greater than $\lfloor D/2 \rfloor$—that is, more than $\lfloor D/2 \rfloor$ edges from the center—so that no two vertices are more than D edges apart. This algorithm requires time that is $O(n^2)$, and it underlies the decoding algorithms of the two codings of BDSTs that the next section describes.

3 Encoding Bounded-Diameter Spanning Trees

Spanning trees can be encoded for evolutionary search in a variety of ways [19] [3, pp.128–197], most of which can be adapted to represent bounded-diameter spanning trees. Our focus, however, is particularly on permutations and random keys as codings of BDSTs. This section describes these codings and operators appropriate to them.

3.1   With Permutations

Julstrom and Raidl [20] encoded bounded-diameter spanning trees on a graph G as permutations of G's vertices. Evaluating a permutation requires making explicit the tree it represents; the following decoder is based on Prim's algorithm.

Let $c[\cdot]$ be a permutation of G's vertices. The first vertex (if the diameter bound D is even) or the first two vertices (if D is odd) that $c[\cdot]$ lists form the center of $c[\cdot]$'s tree. The decoder appends the remaining vertices to the tree in their order in $c[\cdot]$. The depth of each new vertex is one greater than the depth of the vertex in the tree to which it is joined, so each new vertex is attached by its lowest-weight edge to a vertex of degree less than $\lfloor D/2 \rfloor$. Then no vertex has depth greater than $\lfloor D/2 \rfloor$, and the tree's diameter does not exceed D.

Figure 2 summarizes the permutation decoding algorithm. A vertex's depth is fixed when it joins the tree and does not change thereafter, but as in Prim's algorithm, the nearest-connection information for the remaining unconnected vertices must be updated after each vertex joins the tree, so the decoder's time is $O(n^2)$. The space a permutation requires is $O(n)$.

$$T \leftarrow \emptyset$$
$$v_o \leftarrow c[0]$$
$$U \leftarrow V - \{v_o\}$$
$$C \leftarrow \{v_o\}$$
$$\text{depth}[v_o] \leftarrow 0$$
if D is odd
$$\quad v_1 \leftarrow c[1]$$
$$\quad T \leftarrow \{(v_o, v_1)\}$$
$$\quad U \leftarrow U - \{v_1\}$$
$$\quad C \leftarrow C \cup \{v_1\}$$
$$\quad \text{depth}[v_1] \leftarrow 0$$
while $U \neq \emptyset$ do
$$\quad u \leftarrow \text{the next vertex in } c[\cdot]$$
$$\quad v \leftarrow \text{the vertex in } C \text{ nearest } u$$
$$\quad T \leftarrow T \cup \{(u, v)\}$$
$$\quad U \leftarrow U - \{u\}$$
$$\quad \text{depth}[u] \leftarrow \text{depth}[v] + 1$$
$$\quad \text{if depth}[u] < \lfloor D/2 \rfloor$$
$$\quad\quad C \leftarrow C \cup \{u\}$$
return T

Fig. 2. The decoding algorithm for the permutation coding of bounded-diameter spanning trees. $c[\cdot]$ is the permutation, T is the tree's edge set, V is the graph's vertex set, U is the set of unconnected vertices, and C is the set of connected vertices to which a new edge may be connected without violating the diameter bound

A random permutation can be generated in time that is $O(n)$. An appropriate crossover operator is Reeves' C1 [21] (also described, with different names, by Smith [22] and Prosser [23]). C1 chooses a crossover point at random and copies one parent into the offspring up to that point. It then copies the remaining values in order from the second parent into the offspring's remaining positions. This operator usually preserves the center of the first parent. Its time is $O(n)$.

A mutation operator swaps the vertices at two random positions in the parent genotype, thus exchanging the times at which the decoder includes those vertices in the tree. Identifying the two positions and exchanging their contents require only constant time, but copying the parent genotype into the offspring is $O(n)$.

3.2   With Random Keys

To use random keys to represent bounded-diameter spanning trees, concatenate the discussions of random keys and permutations of vertices. In particular, let G be a graph with n vertices and let D be the diameter bound. A genotype is a sequence of n random keys in $[0.0, 1.0]$, one for each vertex in G. To identify the BDST a genotype represents, sort the keys to obtain a permutation of G's vertices. Then, as in Section 3.1, the Prim-based decoder builds the tree with diameter no more than D corresponding to the permutation.

The time required to sort a random-key genotype is $O(n \log n)$ and the time to identify the tree from the resulting permutation is $O(n^2)$. Thus the time of the entire decoding process is $O(n^2)$. The space a genotype requires is again $O(n)$.

With random keys, appropriate operators are two-point crossover and position-by-position mutation: with a small probability, each value from the parent genotype is replaced by a new random value in $[0.0, 1.0]$. These operators' times are linear in n.

3.3   Two Genetic Algorithms

The permutation and random-key codings of bounded-diameter spanning trees were implemented in two genetic algorithms for the BDMST problem. The GAs are generational and initialize their populations with random genotypes, and they select parents in tournaments. Both generate new genotypes from those parents using either crossover or mutation, never both; each offspring is generated by exactly one operator. Both GAs run through fixed numbers of generations.

4   Comparisons

The permutation-coded GA and the random-key-coded GA were compared on 25 Euclidean instances of the bounded-diameter minimum spanning tree problem, five instances each of $n = 50$, 70, 100, 250, and 500 points. The instances are found in Beasley's OR-Library[1] [24], where they are listed as instances of the Euclidean Steiner problem. The tests used the first five instances of each size; the library contains fifteen instances of each size (and others).

Each instance consists of random points in the unit square. The points are treated as the vertices of complete graphs whose edge weights are the Euclidean distances between the points. When $n = 50$, the diameter bound D was set to 5; when $n = 70$, $D = 7$; when $n = 100$, $D = 10$; when $n = 250$, $D = 15$, and when $n = 500$, $D = 20$.

---

[1] mscmga.ms.ic.ac.uk/info.html

The two EAs' common parameters were set identically. For a BDMST problem instance with n points, their population sizes were $20\sqrt{n}$ and their numbers of generations in each run were $50\sqrt{n}$, so that a run always generated $1000n$ new genotypes. Both algorithms selected parent genotypes in tournaments of size two, without replacement, and a tournament's winner always became a parent. Both applied crossover with probability 60%, and mutation, therefore, with probability 40%. The random-key-coded GA mutated each position in its genotypes with probability $3/n$. Table 1 lists the population sizes and run lengths for the two GAs and the mutation probabilities in the random-key-coded GA.

Table 1. Population sizes and numbers of generations per run of the two genetic algorithms, and the probability of mutating one symbol in the random-key-coded GA. On a problem instance of n vertices, the total number of new genotypes is always $1000n$, with the algorithms' population sizes set to $20\sqrt{n}$ and their numbers of generations to $50\sqrt{n}$. In the random-key-coded GA, $P[mu] = 3/n$

n	PopSize	Generations	P[mu]
50	141	353	0.060
70	167	418	0.043
100	200	500	0.030
250	316	790	0.012
500	447	1118	0.006

On each instance, both EAs were run 50 independent times. Table 2 summarizes the results of these trials. For each instance, the table lists the number of points n and the diameter bound D. For each GA and each instance, it lists the length of the shortest bounded-diameter tree found and the mean and standard deviation of the 50 trials' tree lengths. The smaller best values and the smaller mean values for each instance are bold.

Because of the additional level of indirection imposed by random keys, we might expect the GA using them to be less effective than the permutation-coded GA, but this is not the case. There are no significant or consistent differences between the performance of the permutation-coded GA and that of the random-key-coded GA on these BDMST problem instances. For example, the two GAs return equal best tree lengths on four of the five instances with $n = 50$ vertices, and on the remaining instance the values differ by less than 0.15%. The permutation-coded GA returns the shorter mean tree lengths on three instances, and the random-key-coded GA on two.

Similarly, on the instances with $n = 100$ and $D = 10$, the permutation-coded GA identifies the shortest tree twice and returns the shorter mean tree length twice; the random-key-coded GA wins the remaining three contests of each kind. The results on the instances of other sizes are similar. No consistent winner emerges, and the differences between the best and mean tree lengths are always small compared to the respective standard deviations.

References

1. Whitley, D.: Permutations. In Bäck, T., Fogel, D.B., Michalewicz, Z., eds.: Evolutionary Computation 1: Basic Algorithms and Operators. Institute of Physics Publishing, Philadelphia (2000) 274–284 In Chapter 33, Recombination.
2. Bean, J.C.: Genetic algorithms and random keys for sequencing and optimization. ORSA Journal on Computing **6** (1994) 154–160
3. Rothlauf, F.: Representations for Genetic and Evolutionary Algorithms. Volume 104 of Studies in Fuzziness and Soft Computing. Physica-Verlag, Heidelberg (2002)
4. Norman, B.A., Smith, A.E.: Random keys genetic algorithm with adaptive penalty function for optimization of constrained facility layout problems. In: Proceedings of 1997 IEEE International Conference on Evolutionary Computation, IEEE, IEEE Neural Network Council, Evolutionary Programming Society, IEEE (1997) 407–411
5. Knjazew, D., Goldberg, D.E.: OMEGA - Ordering messy GA: Solving permutation problems with the fast messy genetic algorithm and random keys. Technical Report 2000004, Illinois Genetic Algorithms Laboratory, University of Illinois at Urbana-Champaign (2000)
6. Bosman, P.A.N., Thierens, D.: Permutation optimization by iterated estimation of random keys marginal product factorizations. In Guervós, J.J.M. et al., eds: Parallel Problem Solving from Nature – PPSN VII. Volume 2439 of LNCS., Berlin, Springer-Verlag (2002) 331–340
7. Gonçalves, J.F., Resende, M.G.C.: A hybrid genetic algorithm for manufacturing cell formation. Technical report, AT&T Labs (2002) TD-5FE6RN.
8. Gonçalves, J.F., Mendes, J.J.M., Resende, M.G.C.: A hybrid genetic algorithm for the job shop scheduling problem. Technical report, AT&T Labs (2002) TD-5EAL6J.
9. Rothlauf, F., Goldberg, D., Heinzl, A.: Network random keys – a tree network representation scheme for genetic and evolutionary algorithms. Technical Report 8/2000, University of Bayreuth (2000)
10. Kruskal, J.B.: On the shortest spanning subtree of a graph and the traveling salesman problem. Proceedings of the American Mathematics Society **7** (1956) 48–50
11. Prim, R.C.: Shortest connection networks and some generalizations. Bell System Technical Journal **36** (1957) 1389–1401
12. Garey, M.R., Johnson, D.S.: Computers and Intractibility: A Guide to the Theory of NP-Completeness. W. H. Freeman, New York (1979)
13. Bala, K., Petropoulos, K., Stern, T.E.: Multicasting in a linear lightwave network. In: IEEE INFOCOM'93. (1993) 1350–1358
14. Raymond, K.: A tree-based algorithm for distributed mutual exclusion. ACM Transactions on Computer Systems **7** (1989) 61–77
15. Bookstein, A., Klein, S.T.: Compression of correlated bit-vectors. Information Systems **16** (1991) 110–118
16. Abdalla, A., Deo, N., Gupta, P.: Random-tree diameter and the diameter constrained MST. Congressus Numerantium **144** (2000) 161–182
17. Deo, N., Abdalla, A.: Computing a diameter-constrained minimum spanning tree in parallel. In Bongiovanni, G., Gambosi, G., Petreschi, R., eds.: Algorithms and Complexity. Number 1767 in LNCS. Springer, Berlin (2000) 17–31
18. Raidl, G.R., Julstrom, B.A.: Greedy heuristics and an evolutionary algorithm for the bounded-diameter minimum spanning tree problem. In Lamont, G. et al., eds: Proceedings of the 2003 ACM Symposium on Applied Computing, ACM Press (2003) 747–752

19. Raidl, G.R., Julstrom, B.A.: Edge-sets: An effective evolutionary coding of spanning trees. IEEE Transactions on Evolutionary Computation **7** (2003) 225–239
20. Julstrom, B.A., Raidl, G.R.: A permutation-coded evolutionary algorithm for the bounded-diameter minimum spanning tree problem. In Barry, A., ed.: 2003 Genetic and Evolutionary Computation Conference Workshop Program, Chicago, IL (2003) 2–7
21. Reeves, C.R.: A genetic algorithm for flowshop sequencing. Computers and Operations Research **22** (1995) 5–13
22. Smith, D.: Bin packing with adaptive search. In Greffenstette, J.J., ed.: Proceedings of the First International Conference on Genetic Algorithms, Hillsdale, NJ, Lawrence Erlbaum (1985) 202–207
23. Prosser, P.: A hybrid genetic algorithm for pallet loading. In: Proceedings of the 8th European Conference on Artificial Intelligence, London, Pitman (1988)
24. Beasley, J.E.: OR-library: Distributing test problems by electronic mail. Journal of the Operational Research Society **41** (1990) 1069–1072
25. Schindler, B., Rothlauf, F., Pesch, H.J.: Evolution strategies, network random keys, and the one-max-tree problem. In Cagnoni, S., Gottlieb, J., Hart, E., Middendorf, M., Raidl, G.R., eds.: Applications of Evolutionary Computing. Volume 2279 of LNCS., Berlin, Springer-Verlag (2002) 143–152

Three Evolutionary Codings of Rectilinear Steiner Arborescences

Bryant A. Julstrom[1] and Athos Antoniades[2]

[1] St. Cloud State University, St. Cloud, MN 56301 USA
julstrom@eeyore.stcloudstate.edu
[2] University of Nicosia, Nicosia, Cyprus
athos@athosonline.com

Abstract. A rectilinear Steiner arborescence connects points in the Euclidean plane's first quadrant and the origin with directed rectilinear edges from the origin up and to the right. The search for arborescences of minimum total length is NP-hard and finds applications in VLSI design. A greedy heuristic for this problem often returns near-optimum arborescences. Three genetic algorithms encode candidate arborescences as permutations of the points, as perturbations of the points' locations, and as perturbations of the points' rectilinear distances from the origin. In comparisons on twenty collections of 50 to 250 points in the first quadrant, the permutation-coded GA returns arborescences that are longer than those of the greedy heuristic. The two perturbation-coded GAs return nearly identical results, their arborescences are consistently shorter than those of the heuristic, and they preserve their advantage as the number of points grows. These results support the usefulness of perturbation codings in evolutionary algorithms for geometric problems like the search for short rectilinear Steiner arborescences.

1 Introduction

An arborescence is a connected, directed graph that contains no cycles and in which each vertex has at most one entering edge. Because an arborescence is acyclic, one of its vertices has no entering edge. This vertex is the root, and there is a directed path in the arborescence from it to every other vertex.

In a rectilinear Steiner arborescence (RSA), the vertices are points in the first quadrant of the Euclidean plane and the origin. The origin is an RSA's root, and vertical and horizontal edges lead from it to all the points. On a path from the origin, every edge leads either up or to the right; no edge ever turns back toward either coordinate axis. Figure 1 shows a rectilinear Steiner arborescence on twenty points in the first quadrant and the origin. Note that in an RSA, the path connecting the origin to any point p has minimum length: the rectilinear distance from the origin to p.

A minimum rectilinear Steiner arborescence (MRSA) is a RSA of minimum total length. The minimum rectilinear Steiner arborescence problem, which seeks a MRSA, has applications in VLSI design, where signals from a source must

K. Deb et al. (Eds.): GECCO 2004, LNCS 3102, pp. 1282–1291, 2004.

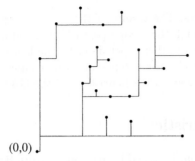

Fig. 1. A rectilinear Steiner arborescence on 20 points in the first quadrant and the origin

be delivered to terminals via rectilinear paths as quickly and economically as possible [1]. Shi and Su [2] established that the MRSA problem is NP-hard.

Rao *et al.* [3] presented an elegant greedy heuristic, described in Section 2 below, for the problem. Córdova and Lee [4] extended this heuristic to points in all four quadrants and observed that it often returns solutions that are very nearly optimal. Leung and Cong [5] based a branch-and-bound exact algorithm on the heuristic, and described another dynamic programming solution. Ramnath [6] described a new heuristic for the problem that he suggested extends more effectively to three dimensions.

Given a collection of points in the plane, a rectilinear Steiner tree (RStT) is a tree made up of vertical and horizontal line segments that connects them all; rectilinear Steiner arborescences are a special case of RStTs. Several evolutionary algorithms that seek short RStTs have been described [7] [8], but their codings of RStTs are based on representations of simple spanning trees and do not adapt well to RSAs. We describe and compare three codings designed to represent RSAs. In the first coding, permutations of points dictate the order in which they are greedily attached to growing arborescences.

The second and third codings are closely related and based on the heuristic of Rao *et al.* Both of them encode RSAs as strings of real values that perturb the points' locations or distances. Such perturbations have been used in EAs for other geometric problems, including the traveling salesman problem (Valenzuela and Williams, 1997; Cohoon et al., 1998).

In the second coding, floating-point perturbations jiggle point locations, and the heuristic of Rao *et al.* builds an arborescence based on the perturbed locations. In the third coding, perturbations are applied directly to the rectilinear distances between points and the origin, and the heuristic of Rao *et al.* builds an arborescence by examining the perturbed distances.

Genetic algorithms using the codings are compared to each other and to the heuristic of Rao *et al.* on twenty instances of the minimum rectilinear Steiner arborescence problem of 50 to 250 points. The GA that uses the permutation coding cannot compete with the greedy heuristic, but the perturbation-coded GAs, which take advantage of the heuristic, almost always improve on its already short arborescences. The results of the two perturbation-coded GAs are essentially identical, though the second perturbation coding uses less space.

The following sections of this paper describe the heuristic of Rao *et al.*; the permutation coding of RSAs; the two perturbation codings of RSAs; the genetic algorithms; and comparisons of the heuristic and the GAs.

2   A Greedy Heuristic

The heuristic of Rao *et al.* [3] greedily develops a short RSA on a collection of points in the first quadrant and the origin. This presentation follows theirs, and begins with some definitions.

For a point $p = (p_x, p_y)$ in the first quadrant, let $\|p\| = p_x + p_y$; that is, $\|p\|$ is the rectilinear distance from p to the origin. For two points $p = (p_x, p_y)$ and $q = (q_x, q_y)$, let $\min(p, q)$ be the point $(\min\{p_x, q_x\}, \min\{p_y, q_y\})$. Figure 2 shows $\min(p, q)$ for two arrangements of p and q.

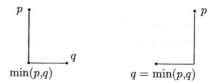

Fig. 2. $\min(p, q)$ for two arrangements of p and q

One iteration of the heuristic identifies the points p and q for which $\|\min(p, q)\|$ is largest, replaces p and q by $\min(p, q)$, and appends the horizontal and vertical segments connecting p and q to $\min(p, q)$ to the developing arborescence. (One of these segments is degenerate when p and q share a horizontal or vertical line.) The heuristic repeats this step until only the origin, which is one of the points, remains.

An efficient implementation of the heuristic maintains a priority queue of the points eligible to be merged and requires time that is $O(n \log n)$. An arborescence that the heuristic finds is never more than twice as long as a shortest RSA [3].

3   A Permutation Coding

Prim's algorithm [11] builds a minimum spanning tree in a weighted, connected, undirected graph from an arbitrary vertex by repeatedly extending the tree with the lowest-weight edge between a vertex in the tree and one not yet in it. In the permutation coding, permutations of point labels represent RSAs via a decoding algorithm based on Prim's algorithm.

The decoder identifies the arborescence a permutation represents by beginning with an arborescence consisting only of the origin. It attaches the points to the arborescence in their listed order, always increasing the arborescence's length as little as possible. The length of the resulting structure is the permutation's fitness, which we seek to minimize.

As in Prim's algorithm, the decoder keeps track of the shortest legal connection from the growing arborescence to each point not yet in it. There are three cases: an unconnected point's nearest connection may be a point below it and to its left, a horizontal segment below it, or a vertical segment to its left. When a point joins the arborescence, the decoder uses this shortest connection.

In the first case, the decoder uses the horizontal and vertical segments whose intersection lies nearer the line $y = x$, to facilitate shorter subsequent connections. The decoder records the segment or segments, accumulates their lengths, and updates the shortest-connection information for the remaining unconnected points. Because updating occurs after each point joins the arborescence, the decoder's time, like that of Prim's algorithm, is $O(n^2)$.

Figure 3 illustrates the operation of the decoder as it builds the arborescence represented by the permutation (3 8 1 2 0 7 6 5 9 4). The origin is at the figure's lower left. The solid lines are the segments the decoder identified as it attached the points 3, 8, 1, 2, and 0. The dotted lines indicate the segments added to the arborescence as the decoder attaches the next three points. These points illustrate the three cases listed above, in order.

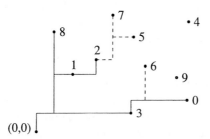

Fig. 3. A snapshot as the decoding algorithm identifies the arborescence that the permutation (3 8 1 2 0 7 6 5 9 4) represents. The solid segments attach the first five points to the arborescence. The dotted segments will connect the next three

In particular, point 7 is joined to point 2, which is below point 7 and to the left; one vertical and one horizontal segment make this connection. Next, point 6 is above a horizontal segment, to which a vertical segment joins it. Then point 5 is to the right of the recent vertical segment, to which the decoder connects it with a horizontal segment. The remaining two steps will join point 9 to the segment descending from point 6, and point 4 to point 5.

Random permutations can be generated in time that is $O(n)$. Though an earlier permutation-coded GA for this problem [12] used Goldberg and Lingle's Partially Mapped Crossover [13], the crossover operator here is Reeves' C1 [14], which has also been described, with different names, by Smith [15] and Prosser [16]. C1 chooses a crossover point at random and copies one parent into the offspring up to that point. It then copies the remaining values in order from the second parent into the offspring's remaining positions. C1's time is $O(n)$.

Mutation chooses two positions at random in its one parent permutation and exchanges the points at those positions, thus exchanging the times at which the decoder appends the points to the arborescence. Copying the parent into the offspring requires linear time.

4   The Long Perturbation Coding

In this coding, a sequence of $2n$ floating-point perturbations, two for each point, represents a rectilinear Steiner arborescence. Identifying a genotype's RSA requires two steps. The first perturbs the points' locations: if $c[\cdot]$ is a genotype and $p = (p_x, p_y)$ is the ith point, p is temporarily replaced by $(p_x + c[2i], p_y + c[2i+1])$. The second applies the heuristic of Rao $et\ al.$ to the perturbed point locations, but the segments it records and their lengths (thus the length of the RSA) are based on the original coordinates. The length of the RSA is the genotype's fitness. Because the number of values in a genotype is twice the number of points, we call this the long perturbation coding.

Random genotypes consist of values from a normal distribution $N(0, \sigma_1)$. k-point or uniform crossover can be used with these genotypes. Mutation perturbs the values in the parent genotype with values from a normal distribution $N(0, \sigma_2)$. The standard deviations σ_1 and σ_2 of the two normal distributions depend on the magnitudes and ranges of the points' coordinates.

5   The Short Perturbation Coding

In the long perturbation coding, the only effect of the perturbations a genotype lists is to modify the rectilinear distances from the points to the origin. This modifies the pairs of points the decoder chooses to merge and the order in which it merges them. The same effect can be had more economically by coding the changes in the points' rectilinear magnitudes directly.

In the second perturbation coding, then, a genotype is a sequence of n floating-point values, one for each point. The rectilinear Steiner arborescence a genotype represents is identified by adding each point's value to its rectilinear distance from the origin. For a merged point $\min(p, q)$, the change in distance is the sum of the changes associated with p and q. The heuristic of Rao $et\ al.$ uses the modified rectilinear distances to construct the RSA. Because the number of values in a genotype is n rather than $2n$, we call this the short perturbation coding.

Again, random genotypes are built of values from a normal distribution $N(0, \sigma_1)$. Positional crossover operators like k-point crossover are appropriate, and mutation modifies each value in a parent genotype with values from a normal distribution $N(0, \sigma_2)$. The values of the standard deviations again depend on the magnitudes and ranges of the points' coordinates.

6   Three Genetic Algorithms

Three generational genetic algorithms seek short rectilinear Steiner arborescences using the permutation, long perturbation, and short perturbation codings and their respective operators. All the GAs initialize their populations with random genotypes. They choose parents in tournaments without replacement and generate each offspring by applying either crossover or mutation, never both. The GAs are 1-elitist: they preserve the one best genotype of the current generation unchanged into the next. They run through a preset number of generations, then report the shortest RSA represented in their populations.

When applied to a problem instance with n points, each GA's population contained n genotypes. Each selection tournament compared two random contestants, and a tournament's winner always became a parent. The probability that crossover generated each offspring was 70%, and the probability of mutation therefore 30%. The GAs ran through $3n$ generations.

7   Comparisons

The heuristic of Rao *et al.*, the permutation-coded GA, the long-perturbation-coded GA, and the short-perturbation-coded GA were compared on twenty instances of the rectilinear Steiner arborescence problem. The instances are found in Beasley's [17] OR-Library[1], where they appear as instances of the Euclidean Steiner problem. The instances consist of random points in the unit square. The library lists fifteen instances of each of many sizes; the present algorithms were exercised on five instances each of $n = 50$, 70, 100, and 250 points.

The perturbation-coded GAs applied two-point crossover. With the long perturbation coding, the standard deviations of the initializing and mutating normal distributions were $\sigma_1 = 0.020$ and $\sigma_2 = 0.010$ when $n = 50$ and 70, $\sigma_1 = 0.010$ and $\sigma_2 = 0.005$ when $n = 100$, and $\sigma_1 = 0.004$ and $\sigma_2 = 0.002$ when $n = 250$. With the short perturbation coding, the two standard deviations were always half of these values. The standard deviations are small because the ranges of the points' coordinates are small and, as the results below show, the points need not "move" far to obtain shorter RSAs.

The heuristic of Rao *et al.* was run once on each instance, and the GAs were each run 40 independent times. Table 1 summarizes the results of these trials. For each instance, the table lists the size and number of the instance and the length of the RSA identified on it by the heuristic of Rao *et al.* For each genetic algorithm applied to each instance, the table lists the length of the shortest RSA the GA discovered, the mean of the 40 RSA lengths from its trials, and the standard deviation of those lengths.

Based on these results we make five observations. First, the permutation-coded GA cannot compete effectively with the heuristic of Rao *et al.*, particularly on the larger instances. On the 50-point instances, the GA's shortest arborescences are sometimes slightly shorter than those returned by the heuristic, but on

---

[1]   mcsmga.ms.ic.ac.uk/info.html

Table 1. Results of the trials of the heuristic of Rao *et al.* and the three genetic algorithms. For each instance, the table lists its number of points, number, and the length of the heuristic's RSA. For each GA on each instance, it lists the length of its best RSA and the mean $\overline{X}$ and standard deviation s of the GA's 40 RSA lengths

Instance		Rao	Permutations			Long Perturbations			Short Perturbations		
n	num	*et al.*	best	$\overline{X}$	s	best	$\overline{X}$	s	best	$\overline{X}$	s
50	1	7.163	7.115	7.294	0.144	7.072	7.075	0.006	7.072	7.074	0.004
	2	6.576	6.569	6.855	0.171	6.524	6.534	0.009	6.524	6.525	0.002
	3	6.589	6.665	6.970	0.187	6.590	6.590	0.002	6.590	6.590	0.000
	4	6.509	6.384	6.657	0.159	6.272	6.281	0.010	6.271	6.273	0.002
	5	6.771	6.800	7.064	0.140	6.687	6.687	0.002	6.687	6.687	0.002
70	1	7.971	8.045	8.328	0.151	7.836	7.853	0.017	7.837	7.851	0.016
	2	7.594	7.789	8.015	0.174	7.501	7.505	0.007	7.501	7.502	0.003
	3	7.483	7.610	8.115	0.215	7.462	7.470	0.010	7.462	7.462	0.001
	4	7.835	7.894	8.165	0.194	7.643	7.661	0.011	7.643	7.655	0.009
	5	7.121	7.231	7.587	0.193	7.114	7.125	0.009	7.114	7.118	0.003
100	1	8.870	9.127	9.418	0.155	8.869	8.869	0.002	8.869	8.869	0.000
	2	9.161	9.400	9.753	0.191	9.003	9.006	0.003	9.003	9.004	0.001
	3	9.039	9.324	9.715	0.212	9.027	9.029	0.002	9.027	9.029	0.001
	4	9.408	9.272	9.658	0.156	9.046	9.050	0.006	9.049	9.055	0.009
	5	8.840	9.141	9.570	0.198	8.810	8.810	0.001	8.810	8.810	0.000
250	1	14.158	14.811	15.373	0.247	13.993	13.998	0.006	13.995	14.010	0.005
	2	14.358	14.880	15.346	0.189	13.977	13.987	0.005	13.984	13.998	0.008
	3	13.953	14.635	15.100	0.241	13.802	13.808	0.006	13.807	13.822	0.009
	4	14.277	14.881	15.296	0.187	14.017	14.024	0.006	14.023	14.050	0.016
	5	14.442	15.031	15.455	0.179	14.222	14.231	0.006	14.227	14.243	0.009

average its RSAs range from 1.8% to 5.8% longer, and the GA's relative performance deteriorates as the instances grow. On the 250-point instances, its shortest RSAs average 4.3% longer than the heuristic's, and its average RSA lengths range from 6.9% to 8.6% longer. Though the permutation coding's Prim-based decoding algorithm greedily joins points to arborescences with short connections, it is not as effective as the heuristic of Rao *et al.* The permutation-coded GA carries out a lot of computation for no discernible benefit.

Second, the results of the two perturbation-coded GAs are very similar, particularly on the smaller instances. Their mean results differ more than their best results, but those differences are still small relative to the standard deviations of the sets of RSA lengths. On the 250-point instances, the long-perturbation-coded GA enjoys a small but consistent advantage over the short-perturbation-coded GA, but in general it makes little difference to the perturbation decoding algorithm whether the choices of points to merge are modified by jiggling the points' positions or their distances from the origin.

Third, both perturbation-coded GAs almost always return shorter arborescences than those discovered by the heuristic of Rao *et al.* Their shortest arborescences are shorter than the heuristic's arborescences on every instance except the third with 50 points, and the average lengths of their trials' arborescences are shorter than the heuristic's results on every instance except that one and, for the long-perturbation-coded GA only, one more. This illustrates the effectiveness of the heuristic of Rao *et al.* and suggests that perturbation codings should be effective in GAs for other geometric problems for which good heuristics exist.

Fourth, the advantage of the perturbation-coded GAs over the heuristic is small. On none of the instances does it exceed 3.9% (the long-perturbation-coded GA on the fourth 100-point instance), and it is usually less. This supports Córdova and Lee's observation [4] that the heuristic of Rao *et al.* often identifies short RSAs. It also explains the small standard deviations of the initializing and mutating distributions of both perturbation-coded GAs. The test instances' points lie in the unit square, and the heuristic is effective, so only small perturbations of points' locations or distances from the origin are necessary to find slightly shorter arborescences.

Last, the performance of the perturbation-coded GAs does not deteriorate as the instances get larger. The lengths of the GAs' arborescences are on average 1.0% to 1.6% shorter than the heuristic's RSAs, regardless of the size of the instances.

The genetic algorithms are, of course, much slower than the heuristic of Rao *et al.* The perturbation-coded GAs execute the heuristic on every evaluation; with population size n and running through $3n$ generations, they perform $3n^2$ such evaluations.

Figure 4 shows, for the first 250-point instance, the rectilinear Steiner arborescence found by the heuristic of Rao *et al.* and the best RSAs found by the three GAs. As reported in Table 1, the heuristic's RSA has length 14.158. The permutation-coded GA's RSA has length 14.811, 4.6% longer than the heuristic's RSA. The long-perturbation-coded and short-perturbation-coded GAs trees have lengths 13.993 and 13.995, about 1.2% shorter than the heuristic's RSA.

8   Conclusion

A rectilinear Steiner arborescence connects the plane's origin to points in its first quadrant with horizontal and vertical directed edges that lead up and to the right. The search for RSAs of minimum total length, an NP-hard problem, has applications to VLSI design.

The greedy heuristic of Rao *et al.* identifies arborescences that are often near-optimum. Three genetic algorithms encode RSAs as permutations of their points, as perturbations of the points' locations, and as perturbations of the points' rectilinear distances from the origin. The permutation-coded GA uses a Prim-like decoder to identify the RSAs that permutations represent. The two perturbation-coded GAs decode sequences of perturbations via the greedy heuristic.

In tests on twenty sets of 50 to 250 points, the permutation-coded GA identified arborescences longer than those returned by the greedy heuristic, and its

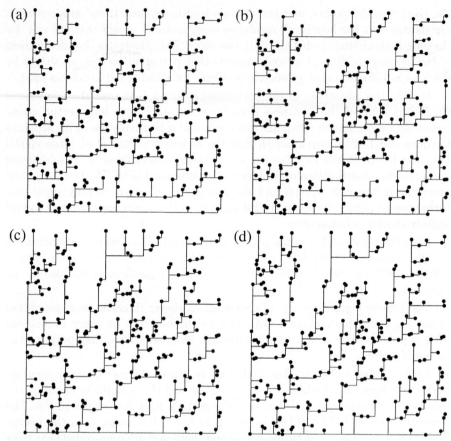

Fig. 4. On the first instance with $n = 250$ points: The RSA identified by the heuristic of Rao *et al.* (a); it has length 14.158, and the shortest RSAs found by the permutation-coded GA (b) (14.811), the long-perturbation-coded GA (c) (13.993), and the short-perturbation-coded GA (d) (13.995)

disadvantage grew with the number of points. Changing this GA's parameter values or crossover operator might improve its performance, but it is unlikely that it can ever compete effectively with the greedy heuristic. The two perturbation-coded GAs returned nearly identical results that almost always improved the heuristic's, and they maintained this advantage on the larger instances. These results illustrate both the effectiveness of the greedy heuristic and the usefulness of perturbation codings in evolutionary algorithms for geometric problems like the search for short rectilinear Steiner arborescences.

References

1. Cong, J., Khang, A.B., Leung, K.S.: Efficient algorithms for the minimum shortest path Steiner arborescence problem with applications to VLSI physical design. IEEE Transactions on Computer-Aided Design of Integrated Circuits and Systems **17** (1998) 24–39
2. Shi, W., Su, C.: The rectilinear Steiner arborescence problem is NP-complete. In: Proceedings of the ACM-SIAM Symposium on Discrete Algorithms. (2000) 780–786
3. Rao, S.K., Sadayappan, P., Hwang, F.K., Shor, P.W.: The rectilinear Steiner arborescence problem. Algorithmica **7** (1992) 277–288
4. Córdova, J., Lee, Y.H.: A heuristic algorithm for the rectilinear Steiner arborescence problem. Technical Report TR-94-025, Department of Computer Science, University of Florida (1994)
5. Leung, K.S., Cong, J.: Fast optimal algorithms for the minimum rectilinear Steiner arborescence problem. In: Proceedings of the International Symposium on Circuits and Systems. (1997) 1568–1571
6. Ramnath, S.: New approximations for the rectilinear Steiner arborescence problem. IEEE Transactions on Computer-Aided Design **22** (2003) 859–869
7. Julstrom, B.A.: Encoding rectilinear Steiner trees as lists of edges. In Lamont, G.B., Yfantis, E.A., Haddad, H., Papadopoulos, G.A., Carroll, J., eds.: Proceedings of the 16th ACM Symposium on Applied Computing, New York, ACM Press (2001) 356–360
8. Julstrom, B.A.: A hybrid evolutionary algorithm for the rectilinear Steiner problem. In Barry, A., ed.: 2003 Genetic and Evolutionary Computation Workshop Program, Chicago, IL (2003) 49–55
9. Valenzuela, C.L., Williams, L.P.: Improving simple heuristic algorithms for the traveling salesman problem using a genetic algorithm. In Bäck, T., ed.: Proceedings of the Seventh International Conference on Genetic Algorithms, San Francisco, CA, Morgan Kaufmann Publishers (1997) 458–464
10. Cohoon, J.P., Karro, J.E., Martin, W.N., Niebel, W.D.: Perturbation method for probabilistic search for the traveling salesperson problem. In: Applications and Science of Neural Networks, Fuzzy Systems, and Evolutionary Computation. Volume 3455 of Proceedings of SPIE. SPIE Press (1998) 118–127
11. Prim, R.C.: Shortest connection networks and some generalizations. Bell System Technical Journal **36** (1957) 1389–1401
12. Julstrom, B.A., Antoniades, A.: Two hybrid evolutionary algorithms for the rectilinear Steiner arborescence problem. In: Proceedings of the 2004 ACM Symposium on Applied Computing, Nicosia, Cyprus (2004)
13. Goldberg, D.E., Robert Lingle, J.: Alleles, loci, and the traveling salesman problem. [18] 154–159
14. Reeves, C.R.: A genetic algorithm for flowshop sequencing. Computers and Operations Research **22** (1995) 5–13
15. Smith, D.: Bin packing with adaptive search. [18] 202–207
16. Prosser, P.: A hybrid genetic algorithm for pallet loading. In: Proceedings of the 8th European Conference on Artificial Intelligence, London, Pitman (1988)
17. Beasley, J.E.: OR-library: Distributing test problems by electronic mail. Journal of the Operational Research Society **41** (1990) 1069–1072
18. Greffenstette, J.J., ed.: Proceedings of the First International Conference on Genetic Algorithms. In Greffenstette, J.J., ed.: Proceedings of the First International Conference on Genetic Algorithms, Hillsdale, NJ, Lawrence Erlbaum (1985)

Central Point Crossover for Neuro-genetic Hybrids

Soonchul Jung and Byung-Ro Moon

School of Computer Science and Engineering, Seoul National University
Shillim-dong, Kwanak-gu, Seoul, 151-742 Korea
{samuel, moon}@soar.snu.ac.kr

Abstract. In this paper, we consider each neural network as a point in a multi-dimensional problem space and suggest a crossover that locates the central point of a number of neural networks. By this, genetic algorithms can spend more time around attractive areas. We also apply representational normalization to neural networks to maintain genotype consistency in crossover. For the normalization, we utilize the Hungarian method of matching problems. The experimental results of our neuro-genetic algorithm overall showed better performance over the traditional multi-start heuristic and the genetic algorithm with a traditional crossover. These results are evidence that it is attractive to exploit central areas of local optima.

1 Introduction

Artificial neural networks (ANNs) have been used to address a variety of problems in pattern recognition, pattern classification, optimization, associative memory, control mechanism, prediction, function approximation, etc. [10] [11]. Even when such a problem addressed by ANNs is fairly small, the problem space is often intractably large that it is almost impossible to find an optimal solution by exhaustive or simple search methods. Thus, heuristic algorithms have been used as alternatives. They provide reasonable solutions — local optima — in acceptable computing time.

A lot of studies have been conducted on the ruggedness and the properties of problem search spaces. A good insight into problem spaces can provide a motivation for a good search algorithm. Kauffman [13] proposed the NK-landscape model that can control the ruggedness of a problem space. Sorkin [28] defined the fractalness of a solution space and proposed that simulated annealing is efficient when the space is fractal. Manderick *et al.* [20] measured the ruggedness of a problem space by autocorrelation function and correlation length obtained from a time series of solutions. Weinberger [30] conjectured that, if all points on a fitness landscape are correlated relatively highly, the landscape is bowl-shaped. Boese *et al.* [3] suggested that, through measuring cost-distance correlation for the traveling salesman and graph bisection problems, the cost surfaces are globally convex. Jones and Forrest [12] introduced fitness-distance correlation as a measure of search difficulty.

K. Deb et al. (Eds.): GECCO 2004, LNCS 3102, pp. 1292–1303, 2004.

ANN is an information processing paradigm that was inspired by the way that biological nervous systems process information. An ANN learns from its environment through an iterative process of adjusting its weights. Although ANNs have been widely applied to a variety of problems, their quality is limited. They often get trapped in local optima due to the limit of their learning algorithms [10]; the back-propagation algorithm, the most popular learning method for ANNs, is basically a hill-climbing technique.

One way to overcome the shortcomings of hill-climbing techniques is to adopt a global search algorithm like evolutionary algorithms. Evolving neural networks has shown successful results [25] [18] [24]. According to Yao [32], evolution of ANNs can be classified into three representative approaches: i) finding a near-optimal set of connection weights globally for an ANN with a fixed architecture [31] [23] [27], ii) evolving architectures of an ANN (i.e., scale and connectivity) in order to adapt to different tasks [16] [9] [19], and iii) evolving learning rules to improve an ANN's learning ability [5] [7] [14].

Kim and Moon [15] extensively investigated the properties around central areas of local optima for the graph bipartitioning problem. Based on their observation, they suggested a new multi-parent crossover that exploits central areas of local optima. Their genetic algorithm (GA) was peculiar enough to select the total population as the parent set and perform the crossover on them. In this paper, we modify their multi-parent crossover to suit neural networks and incorporate the modified crossover into the conventional genetic algorithm. We name it *central point crossover*. We consider each neural network as a point in a multi-dimensional problem space, and compute the central point of multiple neural networks during crossover.

An ANN has a lot of other ANNs whose functionality is equivalent to it, which implies the many-to-one mapping between genotypes and phenotypes of ANNs. The existence of redundancy in encodings is known to be harmful in producing good offspring [6]. Thierens [29] transformed a neural network into its canonical form by flipping the signs of connection weights and reordering hidden neurons. Although the transformation to the canonical form reduced the redundancy, the "similarities" of neural networks were not well considered. We resolve this problem by defining a distance measure for neural networks and normalizing parental neural networks on the basis of the distance measure.

This paper is structured as follows. In the next section, we describe the architecture, isomorphism, and distance measure of neural networks. In Section 3 and 4, we explain the central point crossover and the experimental results, respectively. Finally, we make conclusions in Section 5.

2   Preliminaries

2.1   Architecture

We use a typical multilayer feed-forward network (Fig. 1). A network consists of a set of sensory units that constitutes the input layer, one or more hidden layers of computation neurons, and an output layer of computation neurons. In this paper, we focus on neural networks with one hidden layer. A neuron in any

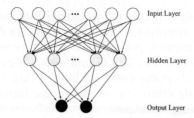

Fig. 1. Architectural graph of a multilayer feed-forward network

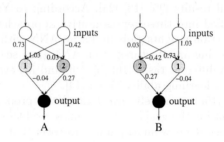

Fig. 2. Two isomorphic neural networks

layer of the network is connected to all the neurons in the previous layer. Each associated edge has a weight. Since we consider neural networks with only one hidden layer, the kth output, y_k^n, of a neural network n, is given simply by

$$y_k^n = \varphi\left\{\sum_{j=1}^{J} w_{kj}^n \cdot \varphi(\sum_{i=1}^{I} w_{ji}^n \cdot x_i)\right\}$$

where I and J are the numbers of input and hidden neurons, respectively, x_i is the value of the ith input neuron, and $\varphi(x)$[1] is a sigmoidal transfer function; w_{kj}^n is the connection weight from j^{th} hidden neuron to k^{th} output neuron and w_{ji}^n is the weight from i^{th} input neuron to j^{th} hidden neuron.

2.2   Isomorphism of Neural Networks

In Fig. 2, two neural networks A and B look different from each other. In other words, they have different representations. However, they are isomorphic because of their equivalent functionality; they output the exactly same value with respect to the same input vector. In fact, any permutation of the hidden neurons including incoming and outgoing weights, produces the same neural network with a different chromosomal representation. Therefore, there are $n!$ representations with respect to a neural network with n hidden neurons. This results in the many-to-one mapping between genotypes and phenotypes. Each genotype corresponds to a permutation from the base neural network. This problem, known as

---

[1] $\varphi(x) = \frac{1}{1+\exp(-x)}$

the permutation problem [1], makes crossover hard to consistently inherit traits of the parents.

2.3   Distance Measure

It is useful in GAs to define a measure of distance between neural networks. It can be used in selection, crossover, replacement operators, etc. Each neuron in the hidden layer can be represented by a weight vector of edges incident to input and output neurons. Thus, a neural network can be represented by a 2D matrix of weights. The distance between two neural networks can be defined as follows:

Definition 1. *Let I, J, and K be the numbers of input neurons, hidden neurons, and output neurons, respectively. Consider two neural networks $\mathfrak{n} = \{\mathfrak{h}_1, \mathfrak{h}_2, \ldots, \mathfrak{h}_J\}$ and $\mathfrak{n}' = \{\mathfrak{h}'_1, \mathfrak{h}'_2, \ldots, \mathfrak{h}'_J\}$ where $\mathfrak{h}_j{}^2$, $\mathfrak{h}'_j \in \mathcal{R}^{I+K}$. Given a metric $\mathfrak{d} : \mathcal{R}^{I+K} \times \mathcal{R}^{I+K} \to \mathcal{R}$ in Euclidean space[3], the distance between the two networks is defined as*

$$\Delta_J(\mathfrak{n}, \mathfrak{n}') = \sum_{j=1}^{J} \mathfrak{d}(\mathfrak{h}_j, \mathfrak{h}'_j).$$

The measure Δ_J is computed easily and fast, but it is not very useful due to the permutation problem. For example, consider a neural network A and its isomorphic neural network A'. The distance between them has diverse values according to the representations of A'; the distance little reflects the functional similarity of the two neural networks.

We define a better distance measure D_J as follows:

Definition 2. *Given the metric $\mathfrak{d}$, and the two neural networks $\mathfrak{n}, \mathfrak{n}'$ as in Definition 1, we define the distance D_J between the two networks as*

$$D_J(\mathfrak{n}, \mathfrak{n}') = min_{\sigma \in \Sigma_J} \left(\sum_{j=1}^{J} \mathfrak{d}(\mathfrak{h}_j, \mathfrak{h}'_{\sigma(j)}) \right)$$

where σ denotes a permutation.

Since D_J is computed according to such a mapping that each hidden neuron of $\mathfrak{n}'$ is mapped to a similar one to $\mathfrak{n}$, it better reflects the functional similarity of the two neural networks than does Δ_J.

D_J is a *metric* in the neural network space and thus satisfies the triangle inequality $(D_J(x, z) \le D_J(x, y) + D_J(y, z))$. The proof is omitted by space limitation. A brute-force algorithm to calculate D_J consumes $O(J!)$ time if it enumerates all the $J!$ permutations to find the optimal one. Fortunately, there is an efficient way to compute D_J; we can formulate the problem of computing D_J as the optimal assignment problem [8]. In other words, this is exactly the

---

[2] $\mathfrak{h}_j = [w_{j1}, \ldots, w_{ji}, \ldots, w_{jI}, w_{1j}, \ldots, w_{kj}, \ldots, w_{Kj}]^T$

[3] $\mathfrak{d}(\mathfrak{h}_a, \mathfrak{h}_b) = \|\mathfrak{h}_a - \mathfrak{h}_b\| = \sqrt{\sum_{i=1}^{I}(w_{ai} - w_{bi})^2 + \sum_{k=1}^{K}(w_{ka} - w_{kb})^2}$

problem of finding an assignment (permutation) with the minimum summation of Definition 2. It can be computed efficiently in $O(J^3)$ by the Hungarian method [17]. Fig. 3 shows the assignment weight matrix $M = (m_{ij})$ between two neural networks $\mathfrak{n}$ and $\mathfrak{n}'$. Each element m_{ij} means $\mathfrak{d}(\mathfrak{h}_i, \mathfrak{h}'_j)$. We reorder the hidden neurons of $\mathfrak{n}'$ by the Hungarian method and pursue maximal genotypical consistency between neural networks.

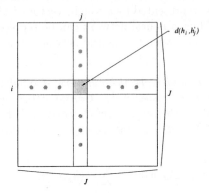

Fig. 3. The assignment weight matrix between two neural networks

3   Central Point Crossover

3.1   Normalization of Neural Networks

As mentioned before, a neural network has a number of isomorphic ones. It causes the existence of redundant encodings in a GA where neural networks are represented by chromosomes. Redundant encodings are known to be harmful in keeping the respectfulness and combination power that crossovers have to possess [26] [6]. The permutation problem is highly related to the redundant encoding in GAs. From the viewpoint of GAs, normalization is an approach that transforms the genotype of one parent to be consistent with that of the other one [6]. It alleviates inconsistency caused by redundant encodings in GAs. We perform normalization by the Hungarian method.

Since our central point crossover manipulates a set of parents, the normalization routine has to process more than two neural networks unlike usual GAs. The algorithm is as follows:

1. Find out the fittest one $\mathfrak{n}_b$ among a set of N parent neural networks.
2. For each neural network $\mathfrak{n}$ in the set, a) use the Hungarian method to find the optimal assignment between $\mathfrak{n}$ and $\mathfrak{n}_b$, and b) transform $\mathfrak{n}$ into the isomorphic one by relocating its hidden neurons and connection weights according to the assignment of (a).

3.2   Central Point Crossover for Neural Networks

Boese *et al.* [3] analyzed relationships among local minima for the traveling salesman problem and the graph bisection problem. They conjectured that cost surfaces of both problems are globally convex, which implies that good solutions are highly probable to be located near other good solutions. Kim and Moon [15] conjectured from their experimental results that, given a subspace of local optima, the "central point" of the subspace is near the optimal solution. They performed experiments on approximate central points[4] for the graph bipartitioning problem, and showed that it was attractive to exploit central areas of multiple solutions.

Fig. 4 shows the template of the central point crossover for neural networks. The offspring of the central point crossover is exactly the central point of the parents. Fig. 5 shows an example crossover with three parents. The central point crossover on a real-number vector like a neural network is technically a generalization of the arithmetic crossover [22]. Because the use of the arithmetic mean tends to lead to premature convergence of a GA, we need to use a rather strong mutation to slow down the convergence speed.

---

central_point_crossover(P, n)
// P: a set of parents, n: the number of parents
// I, J, K: the numbers of neurons in the input, hidden, and output layers,
//   respectively.
{
$\quad P \leftarrow$ normalize(P);
$\quad$**for each** j **in** $\{1, 2, ..., J\}$
$\quad\quad$**for each** i **in** $\{1, 2, ..., I\}$
$\quad\quad${
$$w_{ji}^{off} \leftarrow \frac{\sum_{p \in P} w_{ji}^{p}}{n};$$
$\quad\quad$}
$\quad$**for each** k **in** $\{1, 2, ..., K\}$
$\quad\quad$**for each** j **in** $\{1, 2, ..., J\}$
$\quad\quad${
$$w_{kj}^{off} \leftarrow \frac{\sum_{p \in P} w_{kj}^{p}}{n};$$
$\quad\quad$}
$\quad$**return** the offspring *off*;
}

Fig. 4. The template of central point crossover

---

[4] In the graph bipartitioning problem, it is not easy to calculate the exact central point.

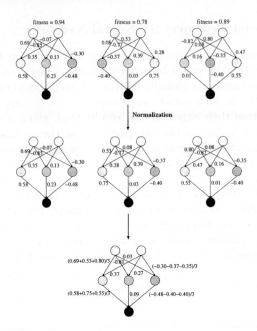

Fig. 5. An example of the central point crossover using three parents

4   Experimental Results

4.1   GA Framework

Fig. 6 shows the steady-state hybrid GA using neural networks as chromosomes. This GA allows more than two parents. Note that the parents are normalized during crossover. We denote the GA by CXGA.

- Initialization of population — Each weight of neural networks in the population ($n_{pop} = 200$) is initialized to a random number ranging from -0.5 to 0.5. After that, neural networks are local-optimized by the back-propagation algorithm.
- Selection — The selection operator here is based on the binary tournament selection. Two parent candidates are selected at random in the population, and the fitter is chosen as a parent. This process is repeated until n_{parent} distinct parents are selected.
- Mutation — The mutation operator is applied to the offspring with a specific probability ($p_{mut} = 0.1$). Each weight of the offspring is set to a random number between -0.5 and 0.5 with probability $p_{mut2} = 0.5$. Consequently, about 50% of weights are changed with new values by mutation in one of every ten offspring. Because the central point crossover may decrease population diversity quickly, we use a rather strong random mutation not to lose the diversity. Another reason of this high mutation rate is that local optimization follows the mutation.

```
GA( n_pop, n_parent, p_mut )
// n_pop: population size, n_parent: the number of parents
// p_mut: probability to apply the mutation operator
{
    // Generate the local-optimized population P = {n_1, n_2, ...}.
    for each i in {1, 2, ..., n_pop}
    {
        initialize( n_i );
        n_i ← back_propagate( n_i );
    }
    B ← the best among {n_1, n_2, ..., n_pop};

    // Start the main loop.
    do
    {
        Parent ← select( P, n_parent );
        off ← central_point_crossover( Parent, n_parent );
        if ( random(0.0, 1.0) < p_mut ) off ← mutate( off );
        off ← back_propagate( off );
        P ← replace( P, off );
        B ← the best between B and off;
    } while ( check_stop_condition() );
    return B;
}
```

Fig. 6. The steady-state hybrid genetic algorithm for neural networks

- Local optimization — We use the back-propagation learning algorithm [10] as the local optimizer. It is the most popular algorithm for the supervised training of multilayer feed-forward networks due to its simple implementation and fast computational speed. We use the cross-validation method [10] to prevent over-fitting and identify when to stop training. After each epoch of training, the network is tested on the validation set and then its learning degree $l(n)$[5] is computed. When its current learning degree does not show improvement during $n_{epoch}(=12)$ consecutive epochs, the learning stops.
- Replacement — We use an extended version of the genitor-style replacement operator [4]. Among parents whose fitnesses are worse than that of the offspring, it replaces the most similar one to the offspring (using distance measure D_J). If none of the parents are worse, it replaces the worst in the population with the offspring.
- Stop condition — The GA stops after a fixed number (=1500) of generations.

The fitness of a neural network is defined as $f(n) = \mu_t - \epsilon_t$ where μ_t and ϵ_t are the accuracy[6] and mean-square error on the training set, respectively.

---

[5] $l(n) = \mu_v - \epsilon_v$ where μ_v and ϵ_v are the accuracy and mean-square error on the validation set, respectively.

[6] $\mu_t = \dfrac{\text{\# of correctly classified examples}}{\text{\# of all the examples}}$

Table 1. The Accuracies of MS, TGA, and CXGA on the Test Sets

	CHDD ave[†]	CHDD best[‡]	WBCD ave	WBCD best	ACCAD ave	ACCAD best
MS	82.77	85.91	96.20	96.78	85.48	87.46
TGA	80.84	82.55	96.29	96.78	83.88	85.63
CXGA	83.26	85.91	96.20	96.20	86.42	87.16

† Average accuracy in percent over 100 runs on the test set.
‡ Best accuracy in percent over 100 runs on the test set.

4.2 Problem Instances

We selected three well-known problem instances from the UCI machine learning repository [2] to measure the performance of our GA. We removed the ones with missing attributes from the original examples.

Cleveland heart disease database (CHDD) contains 297 examples that consist of 160 healthy and 137 unhealthy cases. Each example has 13 attributes. We used as chromosomes neural networks with 13 input, 10 hidden, and 2 output neurons for CHDD. The examples were divided into 111 training, 37 validation, and 149 test examples.

Wisconsin breast cancer database (WBCD) [21] contains 683 examples with 444 benign and 239 malignant cases. Each example is described by a case number, nine attributes, and a binary class label. We used the 9-16-2 network structure for WBCD. The examples were divided into 256 training, 85 validation, and 342 test examples.

Finally, Australian credit card assessment database (ACCAD) has 653 examples with 296 granted and 357 ungranted credits. The output has two classes. Each example is comprised of six continuous attributes and nine discrete ones. We used the 15-14-2 network structure for ACCAD. The 653 examples were partitioned into 245 training, 81 validation, and 327 test examples.

4.3 Comparison with Other Heuristics

We compare CXGA with the traditional multi-start heuristic (MS) and a traditional GA without normalization using linear strings as chromosomes (TGA). CXGA here used 20 parents for crossover.

MS first generates a specific number of neural networks ($n_{ms} = 1800$) obtained with the back-propagation, and then returns the best one among them. In order to make a fair comparison, the value of n_{ms} was determined for MS to spend the same time as CXGA. TGA used the two-point crossover on neural networks represented by one-dimensional arrays. The other settings were the same as CXGA.

We tried 100 runs for each of MS, TGA, and CXGA. Table 1 shows the accuracies on the test sets. CXGA overall outperformed MS and TGA. It is extraordinary that MS worked better than TGA for two problems. We suspect that the problem of redundant encoding caused some negative effect on TGA.

Table 2. The Accuracies of CXGAs on the Test Sets

	CHDD		WBCD		ACCAD	
	ave[†]	best[‡]	ave	best	ave	best
$CXGA_2$	82.15	83.22	96.35	96.78	85.68	86.24
$CXGA_{10}$	81.56	83.22	96.20	96.20	85.38	86.24
$CXGA_{20}$	83.26	85.91	96.20	96.20	**86.42**	87.16
$CXGA_{30}$	78.40	82.55	96.19	96.20	86.35	86.85
$CXGA_{40}$	81.44	83.22	**96.77**	96.78	86.36	86.85
$CXGA_{50}$	82.29	84.56	96.13	96.49	86.28	87.16
$CXGA_{60}$	82.17	83.89	96.19	96.49	85.62	86.85
$CXGA_{70}$	81.29	82.55	95.88	96.20	85.68	86.85
$CXGA_{80}$	**84.03**	85.91	95.88	96.20	85.19	86.54
$CXGA_{90}$	83.56	84.56	95.91	96.20	85.22	86.85
$CXGA_{100}$	82.54	85.23	95.93	96.20	85.30	86.85

† Average accuracy in percent over 100 runs on the test set.
‡ Best accuracy in percent over 100 runs on the test set.

4.4 Comparison of CXGAs with Different Parent Sizes

In this section, we examine the performance of CXGAs on a spectrum of different degrees of central area exploitation. We denote by $CXGA_{n_{parent}}$ the CXGA with n_{parent} parents. We set the number of parents n_{parent} to be 2 through 100 in the experiments. One hundred trials were conducted for each CXGA.

Table 2 shows the accuracies of CXGAs on the test sets in percent. $CXGA_{80}$, $CXGA_{40}$, and $CXGA_{20}$ were the best for CHDD, WBCD, and ACCAD, respectively. The best numbers of parents were different depending on problems. Overall, the results show that too low or too high exploitation of the central areas is not desirable.

5 Conclusion

Kim and Moon [15] examined central areas of local optima for the graph bipartitioning problem, and showed that it was attractive to exploit central areas. We extended their study to the field of neural networks. We represented a neural network as a weight matrix, and regarded it as a point in a multi-dimensional problem space. The proposed crossover computes the exact central point of parents. Overall, our GA outperformed the multi-start heuristic and a traditional neuro-genetic hybrid. The experimental results showed that it was attractive to exploit central areas of local optima, and that too low or too high exploitation was not desirable.

We also defined a new distance measure for neural networks. The semantic distance between two networks could be computed fast by the Hungarian method. By normalizing neural networks on the basis of the distance measure, we avoided the permutation problem that occurs in evolving neural networks.

Acknowledgments. This study was supported by a grant of the International Mobile Telecommunications 2000 R&D Project, Ministry of Information & Communication, Republic of Korea. This study was also supported by Brain Korea 21 Project. The Institute of Computer Technology at Seoul National University provided research facilities for this study. Y.-H. Kim contributed to the definition of the distance measure for neural networks.

References

1. R. K. Belew, J. McInerney, and N. N. Schraudolph. Evolving networks: Using the genetic algorithm with connectionist learning. In C. G. Langton, C. Taylor, J. D. Farmer, and S. Rasmussen, editors, *Artificial Life II*, pages 511–547. Addison-Wesley, Redwood City, CA, 1992.
2. C. L. Blake and C. J. Merz. UCI repository of machine learning databases, 1998.
3. K. D. Boese, A. B. Kahng, and S. Muddu. A new adaptive multi-start technique for combinatorial global optimizations. *Operations Research Letters*, 15:101–113, 1994.
4. T. N. Bui and B. R. Moon. A new genetic approach for the traveling salesman problem. In *IEEE Conference on Evolutionary Computation*, pages 7–12, 1994.
5. D. J. Chalmers. The evolution of learning: An experiment in genetic connectionism. In *Proceedings of the 1990 Connectionist Models Summer School*, pages 81–90, 1990.
6. S. S. Choi and B. R. Moon. Normalization in genetic algorithms. In *Genetic and Evolutionary Computation Conference*, pages 862–873, 2003.
7. D. Crosher. The artificial evolution of a generalized class of adaptive processes. In *Preprints of AI'93 Workshop on Evolutionary Computation*, pages 18–36, 1993.
8. D. Gale. *The Theory of Linear Economic Models*. McGraw-Hill Book Company, Inc., 1960.
9. P. J. B. Hancock. Genetic algorithms and permutation problems: a comparison of recombination operators for neural net structure specification. In *Proc. Int. Workshop Combinations of Genetic Algorithms and Neural Networks*, pages 108–122, 1992.
10. S. Haykin. *Neural Networks, A Comprehensive Foundation*. Prentice Hall, 1999.
11. A. James and M. David. *Neural Networks, Algorithms, Applications, and Programming Techniques*. Addison Wesley, 1994.
12. T. Jones and S. Forrest. Fitness distance correlation as a measure of problem difficulty for genetic algorithms. 1995. To appear in *Sixth International Conference on Genetic Algorithms*.
13. S. Kauffman. Adaptation on rugged fitness landscapes. *Lectures in the Science of Complexity*, pages 527–618, 1989.
14. H. B. Kim, S. H. Jung, T. G. Kim, and K. H. Park. Fast learning method for back-propagation neural network by evolutionary adaptation of learning rates. *Neurocomputating*.
15. Y. H. Kim and B. R. Moon. Investigation of the fitness landscapes and multi-parent crossover for graph bipartitioning. In *Genetic and Evolutionary Computation Conference*, pages 1123–1134, 2003.
16. J. R. Koza and J. P. Rice. Genetic generation of both the weights and architecture for a neural network. In *IEEE Int. Joint Conf. Neural Networks*, pages 71–76, 1991.

17. H. W. Kuhn. The Hungarian method for the assignment problem. *Naval Res. Logist. Quart.*, 2:83–97, 1955.
18. C. T. Lin and C. P. Jou. Controlling chaos by GA-based reinforcement learning neural network. *IEEE Trans. on Neural Networks*, 10(4):846–869, 1999.
19. Y. Liu and X. Yao. Evolutionary design of artificial neural networks with different nodes. In *IEEE Conference on Evolutionary Computation*, pages 670–675, 1996.
20. B. Manderick, M. de Weger, and P. Spiessens. The genetic algorithm and the structure of the fitness landscape. In *International Conference on Genetic Algorithms*, pages 143–150, 1991.
21. O. L. Mangasarian and W. H. Wolberg. Cancer diagnosis via linear programming. *SIAM News*, 23(5):1–18, 1990.
22. Z. Michalewicz. *Genetic Algorithms + Data Structure = Evolution Programs*. Springer-Verlag, Berlin, 1992.
23. D. Montana and L. Davis. Training feedforward neural network using genetic algorithms. In *11th International Joint Conference on Artificial Intelligence*, pages 762–767, 1989.
24. V. Petridis, E. Paterakis, and A. Kehagias. A hybrid neural-genetic multimodel parameter estimation algorithm. *IEEE Trans. on Neural Networks*, 9(5):862–876, 1998.
25. J. Pujol and R. Poli. Evolving neural networks using a dual representation with a combined crossover operator. In *IEEE Conference on Evolutionary Computation*, pages 416–421, 1998.
26. N. J. Radcliffe. Forma analysis and random respectful recombination. In *International Conference on Genetic Algorithms*, pages 222–229, 1991.
27. R. S. Sexton, R. E. Dorsey, and J. D. Johnson. Toward global optimization of neural networks: A comparison of the genetic algorithm and backpropagation. *Decision Support Systems*, 22(2):171–185, 1998.
28. G. B. Sorkin. Efficient simulated annealing on fractal landscapes. *Algorithmica*, 6:367–418, 1991.
29. D. Thierens. Non-redundant genetic coding of neural networks. In *IEEE Conference on Evolutionary Computation*, pages 571–575, 1996.
30. E. D. Weinberger. Fourier and Taylor series on fitness landscapes. *Biological Cybernetics*, 65:321–330, 1991.
31. D. Whitley, T. Starkweather, and C. Bogart. Genetic algorithms and neural networks: Optimizing connections and connectivity. *Parallel Computing*, 14(3):347–361, 1990.
32. X. Yao. Evolving artificial neural networks. *Proceedings of the IEEE*, 87(9):1423–1447, 1999.

Combining a Memetic Algorithm with Integer Programming to Solve the Prize-Collecting Steiner Tree Problem*

Gunnar W. Klau[1], Ivana Ljubić[1], Andreas Moser[1], Petra Mutzel[1],
Philipp Neuner[1], Ulrich Pferschy[2], Günther Raidl[1], and René Weiskircher[1]

[1] Institute of Computer Graphics and Algorithms, Vienna University of Technology,
Favoritenstraße 9–11/186, 1040 Vienna, Austria
{klau,ljubic,moser,mutzel,neuner,raidl,weiskircher}@ads.tuwien.ac.at
[2] Department of Statistics and Operations Research
University of Graz, Austria
pferschy@uni-graz.at

Abstract. The prize-collecting Steiner tree problem on a graph with
edge costs and vertex profits asks for a subtree minimizing the sum of
the total cost of all edges in the subtree plus the total profit of all vertices
not contained in the subtree. For this well-known problem we develop a
new algorithmic framework consisting of three main parts:
(1) An extensive preprocessing phase reduces the given graph without
changing the structure of the optimal solution. (2) The central part of
our approach is a memetic algorithm (MA) based on a steady-state evolutionary algorithm and an exact subroutine for the problem on trees. (3)
The solution population of the memetic algorithm provides an excellent
starting point for post-optimization by solving a relaxation of an integer
linear programming (ILP) model constructed from a model for finding
the minimum Steiner arborescence in a directed graph.
Extensive experiments on benchmark instances from the literature show
that our combination of an MA with ILP-based post-optimization compares favorably with previously published results. While our solution
values are almost always the same (not surprisingly, since an extension
of our ILP approach shows the optimality of these values), we obtain a
significant reduction of running time for medium and large instances.

1 Introduction

We consider the prize-collecting Steiner tree problem, an extension of the well-
known *Steiner problem*, where the input is a graph whose vertices are associated
with profits and edges with costs. Our goal is to find a connected subgraph
that minimizes the sum of the profits of the vertices that are **not** contained in
the subgraph plus the costs of the edges in the subgraph. The problem finds

* Partly supported by the Doctoral Scholarship Program of the Austrian Academy of
Sciences (DOC) and by the Austrian Science Fund (FWF), grant P16263-N04.

K. Deb et al. (Eds.): GECCO 2004, LNCS 3102, pp. 1304–1315, 2004.

its application in the design of networks for communication or distribution of utilities such as district heating or water.

Let $G = (V, E, c, p)$ be an undirected connected graph with $p : V \to \mathbb{R}^{\geq 0}$ a profit function on the vertices and $c : E \to \mathbb{R}^{\geq 0}$ a cost function on the edges. The *prize-collecting Steiner tree problem* (PCSTP) is to find a connected subgraph $T = (V_T, E_T)$ of G, that minimizes

$$c(T) = \sum_{v \notin V_T} p_v + \sum_{e \in E_T} c_e. \tag{1}$$

Note that if the goal is to find a subgraph T that **maximizes** the sum of the profits of the vertices in T minus the cost of the edges in T, every optimal solution is an optimal solution for our minimization problem and vice versa. Furthermore, it is easy to see that every optimal solution T is a tree. Throughout this paper we will distinguish between *positive vertices*, defined as $R = \{v \in V \mid p_v > 0\}$, and *non-positive vertices*. An example of a PCSTP instance and its feasible solution are shown in Figure 1(a) and 1(b), respectively.

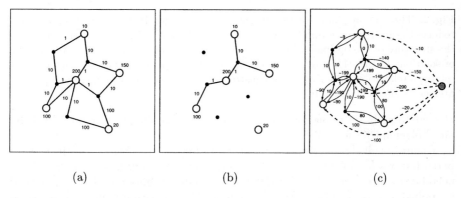

(a)   (b)   (c)

Fig. 1. Example of a PCSTP instance. Each connection has fixed costs, hollow circles and filled circles represent positive and non-positive vertices, respectively (Fig. 1(a)). Figure 1(b) shows a feasible solution and Figure 1(c) the transformation into the Steiner arborescence problem.

Previous Work. The PCSTP has been introduced by Bienstock et al. [1], where a factor 3 approximation algorithm has been proposed. Several other approximation algorithms have been developed (see [7,8]). Segev [16] defined the *node weighted Steiner tree problem* (NWSTP) – another extension of the Steiner problem in graphs, where, in contrast to PCSTP, some vertices must be contained in every solution. Polyhedral studies of this problem can be found in [5,6]. Engevall et al. [4] proposed a Lagrangean relaxation approach based on the *shortest spanning tree* integer linear programming (ILP) formulation for NWSTP.

Lucena and Resende [11] presented a cutting plane algorithm for solving PCSTP based on generalized subtour elimination constraints. The algorithm

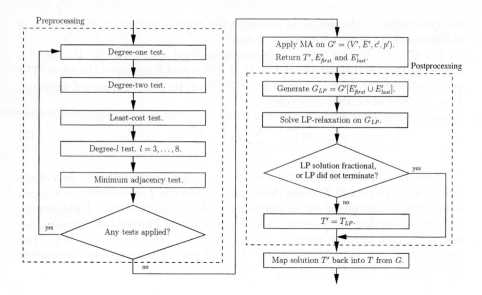

Fig. 2. Three main phases of the proposed approach for PCSTP: (1) Preprocessing reduces the given input graph $G = (V, E, c, p)$ into $G' = (V', E', c', p')$ without changing the structure of the optimal solution. (2) A memetic algorithm (MA). (3) A collection of solutions of the MA provides an excellent starting point for post-optimization by solving a relaxation of an ILP model constructed from a model for finding the minimum Steiner arborescence in a directed graph.

also contains basic reduction steps similar to those already proposed by Duin and Volgenant [3] for NWSTP.

Canuto et al. [2] developed a multi-start local-search-based algorithm with perturbations for PCSTP. It comprises Goemans-Williamson's algorithm, 1-flip neighborhood search and path relinking. A variable neighborhood search method is applied as a post-optimization procedure. The algorithm found optimal solutions on nearly all instances from [11] for which the optima were known.

Our Contribution. A new algorithmic framework is developed as outlined in Figure 2. The computational results given in Section 3 show that our new approach is significantly faster than the previous approach by Canuto et al. [2] while the solutions have the same quality. For a number of instances we manage to find new best solutions, while on the majority of instances our solution values are identical, which is not surprising: Extending our ILP approach shows that these values are indeed optimal. The progress we obtain with respect to running time gives rise to the possibility of solving much larger instances in the future.

2   Combining the Memetic Algorithm with an ILP Model

Within this section, we propose basic ideas of our new algorithmic framework for the PCSTP whose outline is given in Fig. 2. After the input graph G has been reduced into a graph $G' = (V', E', c', p')$, we apply a memetic algorithm that uses problem-dependent operators and strongly interacts with an exact subroutine for the PCSTP problem on trees.

Our ILP-based post-optimization procedure utilizes the combined context of the MA-solutions to produce a final tree that is superior to any single one in the population. Furthermore, the post-optimization algorithm benefits from the fact that solving the PCSTP restricted to a sparse edge set can be much simpler than solving the original problem.

As input for the ILP algorithm, we take a subgraph G_{LP} of G' induced by $E_{LP} = E'_{first} \cup E'_{last}$, the sets of edges that appear in any single solution of the first, respectively, last population. Note that taking the edges from the first generation enables us to escape local optima found by MA.

The best-found subtree T of the original graph G is finally determined by mapping back the solution T' found by the ILP-relaxation.

2.1 Preprocessing

In this section, we briefly describe reduction techniques adopted from the work of Duin and Volgenant [3] for the NWSTP, which have been partially used also in [11]. From the implementation point of view, we transform the graph $G = (V, E, c, p)$ into a reduced graph $G' = (V', E', c', p')$ by applying the steps described below and maintain a *backmapping* function to transform each feasible solution T' of G' into a feasible solution T of G.

Least-Cost Test. Let d_{ij} represent the shortest path length between any two vertices i and j from V (considering only edge-costs). If $\exists e = (i, j)$ such that $d_{ij} < c_{ij}$ then edge e can simply be discarded from G. The procedure's time complexity is dominated by the computation of all-pair shortest paths, which is $O(|E||V| + |V|^2 \log |V|)$ in the worst case.

Degree-l Test. Consider a vertex $v \notin R$ of degree $l \geq 3$, connected to vertices from $Adj(v) = \{v_1, v_2, \ldots, v_l\}$. For any subset $K \subset V$, denote with $\mathrm{MST}_d(K)$, the minimum spanning tree of K with distances d_{ij}. If

$$\mathrm{MST}_d(K) \leq \sum_{w \in K} c_{vw}, \quad \forall K \subseteq Adj(v), \quad |K| \geq 3, \tag{2}$$

then v's degree in an optimal solution must be zero or two. Hence, we can remove v from G by replacing each pair (v_i, v), (v, v_j) with (v_i, v_j) either by adding a new edge $e = (v_i, v_j)$ of cost $c_e = c_{v_i v} + c_{v v_j} - p_v$ or in case e already exists, by defining $c_e = \min\{c_e, c_{v_i v} + c_{v v_j} - p_v\}$.

The procedure's worst case running time is dominated by the computation of all-pair shortest paths, which is $O(|E||V| + |V|^2 \log |V|)$. It is straightforward to apply a simplified version of this test to all vertices $v \in V$ with $l = 1$ and $l = 2$.

Minimum Adjacency Test. This test is also known as $V \setminus K$ *reduction test* from [3]. If there are adjacent vertices $i, j \in R$ such that:

$$\min\{p_i, p_j\} - c_{ij} > 0 \text{ and } c_{ij} = \min_{it \in E} c_{it},$$

then i and j can be fused into one vertex of weight $p_i + p_j - c_{ij}$.

Summary of the Preprocessing Procedure. We apply the steps described above iteratively, as long as any of them changes the input graph (see Fig. 2). The total number of iterations is bounded by the number of edges in G. Each iteration is dominated by the time complexity of the least-cost test. Thus, the preprocessing procedure requires $O(|E|^2|V| + |E||V|^2 \log |V|)$ time in the worst case, in which the input graph would be reduced to a single vertex. However, in practice, the running time is much lower, as documented in Section 3. The space complexity of preprocessing does not exceed $O(|E|^2)$.

2.2   A Memetic Algorithm for the PCSTP

For many hard combinatorial optimization problems, combinations of evolutionary algorithms and problem-dependent heuristics, approximation algorithms or local improvement techniques have been applied with great success. In a memetic algorithm (MA), candidate solutions created by an evolutionary algorithm framework are fine-tuned by some of these procedures [13].

We propose an MA based on a straight-forward steady-state evolutionary algorithm combined with an exact algorithm for solving the PCSTP on trees. In each iteration, we apply k-ary tournament selection with replacement in order to select two parental solutions for mating. A new candidate solution is always created by recombining these parents, mutating it with probability $p_{\text{mut}} \in [0, 1]$, and pruning the obtained tree to optimality. Such a solution replaces always the worst solution in the population with one exception: To guarantee a minimum diversity, a new candidate whose set of edges $E_{T'}$ is identical to that of a solution already contained in the population is discarded [14].

Each randomly created initial solution and each solution derived by recombination and possibly mutation is optimally pruned with respect to its subtrees, using the local improvement algorithm described below.

Local Improvement. The algorithm we use here solves tree instances of the PCSTP to optimality and runs in $O(|V'|)$ time (see also [8,10]).

Given a tree instance $T' = (V_{T'}, E_{T'}, p', c')$ created by an MA, a subtree of T' is *optimal*, if there is no subtree of T' with costs lower than $c(T')$. The algorithm we use here maximizes the sum of the profits of the vertices in T' minus the sum of the edge-costs in T'. We label the vertices $v \in V_{T'}$ and traverse them in bottom-up order, until we end-up with a single vertex. Finally, the optimal solution corresponds to the subtree shrunk within the vertex v^* such that $v^* = \arg\max_{v \in V_{T'}} l_v$. The algorithm is as follows:

1. Set $l_v = p'_v$, for all $v \in V_{T'}$;
2. For all leaves $u \in V_{T'}$: (a) if $c'_{uv} \leq l_u$, shrink u and v into one vertex and set $l_v = l_v + l_u - c'_{uv}$; (b) Delete u;
3. Goto 2. until a single vertex is left;

Clustering. Employing *clustering* as a grouping procedure within variation operators, we can group the subsets of vertices and insert or delete them at once. For each positive vertex $z \in R'$, we define a cluster set $N(z)$ [12]:

$$N(z) := \{v \in V' \setminus R' \mid \forall c \in R' : d'_{vz} \leq d'_{vc}\} \cup \{z\},$$

where d'_{vz} denotes the shortest path length between v and z. Hence, each non-positive vertex v is assigned to the cluster set of its nearest positive vertex $z = base(v)$. Note that the sets $N(z)$ are analogous to Voronoi regions in the Euclidean plane.

Mehlhorn [12] proposed an efficient implementation of the clustering algorithm which runs in $O(|V'|\log|V'| + |E'|)$ time.

Edge-Set Encoding. From spanning tree problems, we know that a direct representation of spanning trees as sets of their edges exhibits significant advantages over indirect encodings [15]. In our approach, the PCSTP solution edges are stored in hash-tables, requiring only $O(|V'|)$ space. Thus, insertion and deletion of edges, as well as checking for existence of an edge, can be done in expected constant time.

Initialization. Given an input graph $G' = (V', E', c', p')$ and its set of positive vertices R', the *distance network* $G_D(R', E_D, c_D)$ is an undirected complete graph whose edge costs $c_D(u, v)$ are given by the shortest path lengths between u and v in G'. For generating initial solutions we use the following modification of the *distance network heuristic* for the Steiner tree problem [12]:

1. Randomly select a subset $V'_{init} \subset R'$ of size $\lceil p_{init} \cdot |R'| \rceil$, $p_{init} \in (0, 1)$;
2. Construct the minimum spanning tree (MST) T'_{init} on the subgraph of G_D induced by V'_{init};
3. Replace each edge of T'_{init} by its corresponding shortest path in G' to obtain $G'_r = (V'_r, E'_r)$;
4. Find the MST T'_r on the subgraph of G' induced by V'_r;
5. Apply the exact algorithm for trees to solve T'_r to optimality;

Recombination. The recombination operator is designed with strong inheritance in mind; we try to adopt the structural properties of two parental solutions. If the two solutions to be combined share at least one vertex, we just construct the spanning tree over the union of their edge sets. Due to the deterministic nature of our local improvement subroutine, we build a random spanning tree on the union of parental edges to avoid premature convergence.

When the parent solutions are disjoint, we randomly choose a vertex out of each solution, look up the shortest path between these two vertices and add for each vertex v along the path all the edges that belong to cluster $N(base(v))$. Finally, we build a random spanning tree over all these edges and apply local improvement.

Mutation. The aim of the mutation operator is to make small changes in the current solution which we achieve by connecting one cluster to the solution. To find an appropriate cluster to add, the algorithm randomly chooses a *border vertex* v which is a vertex adjacent to at least one vertex outside our current solution. We incorporate the vertices of cluster $N(base(v))$ into our solution and search for a neighboring cluster whose base vertex v' is preferably not yet an element of the current solution; the vertices of $N(base(v'))$ will be added to our solution. Finally we construct a minimum spanning tree and apply local improvement.

Assuming the complete distance network is determined once in the preprocessing phase and its edges are pre-sorted in non-increasing order, as well as the edges of E', the running time complexity of initialization and variation operators is $O(|E'| \cdot \alpha(|E'|, |V'|))$.

2.3   ILP Formulation

Our ILP formulation relies on a transformation of the PCSTP to the problem of finding a minimum subgraph in a related, directed graph as proposed by Fischetti [5]. We transform the graph $G_{\text{ILP}} = (V_{\text{ILP}}, E_{\text{ILP}}, c', p')$ that results from the application of the memetic algorithm as described in Section 2.2 into the directed graph $G'_{\text{ILP}} = (V_{\text{ILP}} \cup \{r\}, A_{\text{ILP}}, c'')$ (see Figure 1(c) for an example).

In addition to the vertices of the input graph G_{ILP}, the vertex set of the transformed graph contains an artificial root r. The arc set A_{ILP} contains two directed edges (v, w) and (w, v) for each edge $(v, w) \in E_{\text{ILP}}$ plus a set of arcs from the root r to the positive vertices $\{v \in V_{\text{ILP}} \mid p_v > 0\}$. We define the cost vector c'' as follows:

$$c''_{vw} = c'_{vw} - p'_w \quad \forall (v, w) \in A_{\text{ILP}}, v \neq r \quad \text{and} \quad c''_{rv} = -p'_v \quad \forall (r, v) \in A_{\text{ILP}} \ .$$

A subgraph T_{ILP} of G'_{ILP} that forms a directed tree rooted at r is called a *Steiner arborescence*. It is easy to see that such a subgraph corresponds to a solution of the PCSTP if r has degree 1 in G'_{ILP} (*feasible arborescence*). In particular, a feasible arborescence with minimal total edge cost corresponds to an optimal prize-collecting Steiner tree.

We model the problem of finding a minimum Steiner arborescence T_{ILP} by means of an integer linear program. Therefore, we introduce a variable vector $x \in \{0, 1\}^{|A_{\text{ILP}}| + |\bar{V}_{\text{ILP}}|}$ with the following interpretation:

$$x_{vw} = \begin{cases} 1 & (v, w) \in T_{\text{ILP}} \\ 0 & \text{otherwise} \end{cases} \forall (v, w) \in A_{\text{ILP}}, \quad x_{vv} = \begin{cases} 1 & v \notin T_{\text{ILP}} \\ 0 & \text{otherwise} \end{cases} \forall v \in V_{\text{ILP}} \setminus \{r\}$$

The ILP is then as follows:

$$\min \quad \sum_{a \in A_{\text{ILP}}} c''_a x_a \tag{3}$$

$$\text{subject to} \quad x(\delta^-(\{v\})) + x_{vv} = 1 \qquad\qquad \forall v \in V_{\text{ILP}} \setminus \{r\} \tag{4}$$

$$x(\delta^-(S)) \geq 1 - x_{vv} \qquad\qquad v \in S, r \notin S, \forall S \subset V_{\text{ILP}} \tag{5}$$

$$\sum_{(r,v) \in A_{\text{ILP}}} x_{rv} \leq 1 \tag{6}$$

$$x_{vw}, x_{vv} \in \{0,1\} \qquad\qquad \forall(v,w) \in A_{\text{ILP}}, \forall v \in V_{\text{ILP}}, \tag{7}$$

where $\delta^-(S) = \{(u,v) \in A_{\text{ILP}} \mid u \notin S, v \in S\}$.

Constraint (4) states that every vertex that is part of the solution must have at least one incoming edge while (5) states that for each vertex v in the solution, there must be a directed path from r to v. Constraint (6) ensures that at most one of the edges starting at the artificial root is chosen. We use CPLEX as linear program solver to solve the *ILP-relaxation* of the problem obtained by replacing constraints (7) with $0 \leq x_{vw}, x_{vv} \leq 1$, $(v,w) \in A_{\text{ILP}}, v \in V_{\text{ILP}}$.

There are exponentially many constraints of type (5), so we do not insert them at the beginning but rather *separate* them during the optimization process; that is, we only add constraints violated by the current solution of the ILP-relaxation. These violated constraints can be found efficiently using a maximum flow algorithm on the graph with arc-capacities given by the current solution. We also use *pricing* which means that we do not start with all the variables but rather add them only if needed to prove optimality. A detailed description of this approach that also includes *flow-balance* and *asymmetry constraints* can be found in [9].

3   Computational Results

We tested our new approach extensively on 114 benchmark instances[1] described in [2,11]. The instances range in size from 100 vertices and 284 edges to 1000 vertices and 25 000 edges. Because of space limitations, we present detailed results for the 60 most challenging instances from Steiner series C and D. Graphs from series C have 500, and graphs from series D 1000 vertices. Table 1 lists the instance name, its number of edges $|E|$, the size of the graph after the reductions described in Section 2.1 ($|V'|$, $|E'|$) and the time spent on preprocessing (t_p [s]).

The following setup was used for the memetic algorithm as it proved to be robust in preliminary tests: Population size $|P| = 800$; group size for tournament selection $k = 5$; parameter for initializing solutions $p_{init} = 0.9$; mutation probability $p_{\text{mut}} = 0.3$. Each run was terminated when no new best solution could be identified during the last $\Omega = 10\,000$ iterations.

Because of its stochastic nature, the MA was performed 30 times on each instance and the average results are presented in Table 1 which also contains the

---

[1] Benchmark instances are available from http://research.att.com/~mgcr/data/.

average costs $c(T)_{avg}$ and their standard deviation $\sigma(c)$. Furthermore, we show the average CPU-time and the average number of evaluated solutions until the best solution was found (t, respectively *evals*), and the success rates (*sr* [%]), i.e. the percentage of instances for which optimal solutions could be found.

We also list the results of our combined approach, MA+ILP, where one MA run (with a fixed seed-value) was post-optimized with the ILP method. The value of the obtained solution and *only* the post-optimization CPU-time in seconds are given in columns $c(T)$ and t [s], respectively. Note that the time presented for MA excludes preprocessing times.

We compared the results of our new approach (MA+ILP) to those of Canuto et al. (CRR) obtained using multi-start local search with perturbations and variable neighborhood search [2]. Table 1 provides the solution values of CRR ($c(T)$) and the total running time in seconds (t). In most cases our solution values are identical to CRR. The cases where one of the two is superior are marked by a box.

Finally, to see if we can obtain provably optimal solutions using the ILP approach, we continued the optimization: starting from the ILP-solution of the restricted MA+ILP problem, the rest of variables from G' was considered within pricing of the ILP-relaxation. In column OPT, we show the values of the obtained integer solutions. If we did not obtain an integer solution, or if our ILP-based algorithm terminated abnormally (because of memory consumption) we show the values obtained by Lucena & Resende [11], denoting it with [+], respectively [*]. Note that all values given in OPT are optimal except for D14-B where the best-known lower bound is printed [11]. The last column t [s] lists the *additional* CPU-time needed to compute a provably optimal solution.

When comparing our running time data (achieved on a Pentium IV with 2.8 GHz, 2 GB RAM, SPECint2000=1204) with the results of Canuto et al. [2] (Pentium II with 400 MHz, 64 MB RAM), the widely used SPEC[©] performance evaluation (www.spec.org) does not provide a direct scaling factor. However, taking a comparison to the respective benchmark machines both for SPEC 95 and SPEC 2000 into account, we can argue by a conservative estimate that dividing the Canuto et al. running times by a factor of 10 gives a very reasonable basis of comparison to our data.

Table 2 summarizes our results over all benchmark instances used in [2]. The second and third column show that using sophisticated preprocessing reduces the number of nodes and edges in the problem graph by 30-45% on average. We also provide the average quality (%-*gap*) and the average *total* running time for the approach of Canuto et al. (CRR), our memetic algorithm (MA) and the MA combined with linear programming post-processing (MA+ILP), respectively. The last column gives the average running time for computing a provably optimal solution with our ILP-based approach or a question mark where we could not find an optimal solution for all instances.

The summarized results indicate that MA alone is substantially faster than CRR (by an order of magnitude for the largest group D), but the average solution quality is slightly worse. Solutions of MA+ILP are not significantly worse than CRR solutions, but MA+ILP is much faster than CRR, even when we take the difference in hardware into account.

Table 1. Results obtained by Canuto et al. (CRR), the memetic algorithm (MA) and the combination of MA with ILP (MA+ILP) on selected instances from Steiner series C and D. Running times in (CRR) to be divided by 10 for comparison (cf. SPEC comparison).

Instance	Orig. $|E|$	Preprocessing $|V'|$	$|E'|$	t_p [s]	MA $c(T)_{avg}$	$\sigma(c)$	t [s]	evals	sr [%]	MA+ILP $c(T)$	t [s]	CRR $c(T)$	t [s]	OPT-ILP OPT	t [s]
C11-A	2500	489	2143	9.4	18.0	0.0	6.1	500	100.0	18	0.4	18	128	18	0.2
C11-B	2500	489	2143	9.5	32.0	0.0	9.1	1103	100.0	32	0.4	32	140	32	4.7
C12-A	2500	484	2186	6.8	38.7	0.5	9.0	2456	33.3	38	0.4	38	162	38	0.3
C12-B	2500	484	2186	6.8	46.0	0.0	8.7	590	100.0	46	0.5	46	156	46	0.8
C13-A	2500	472	2113	9.8	237.0	0.2	17.9	5326	0.0	**236**	0.6	237	1050	236	0.5
C13-B	2500	471	2112	9.8	258.5	0.7	35.9	15455	60.0	258	18.5	258	733	258	52.5
C14-A	2500	466	2081	7.5	293.0	0.0	21.0	3163	100.0	293	1.7	293	829	293	0.4
C14-B	2500	459	2048	7.5	318.6	0.5	29.8	9211	43.3	318	1.0	318	766	318	0.4
C15-A	2500	406	1871	6.5	502.2	0.8	45.4	14727	20.0	501	4.7	501	957	501	0.5
C15-B	2500	370	1753	6.0	551.8	0.9	45.7	15607	46.7	551	0.8	551	837	551	0.4
C16-A	12500	500	4740	2.4	12.0	0.0	10.6	500	0.0	12	1.9	**11**	1920	11	0.9
C16-B	12500	500	4740	2.4	12.0	0.0	11.5	503	0.0	12	3.5	**11**	1758	11	13.8
C17-A	12500	498	4694	2.4	19.0	0.0	11.2	620	0.0	19	2.9	**18**	549	18	1.9
C17-B	12500	498	4694	2.3	18.2	0.4	12.7	1951	76.7	18	2.1	18	434	18	1.4
C18-A	12500	469	4569	2.6	112.4	0.7	24.1	7446	6.7	112	2.1	**111**	3990	111+	—
C18-B	12500	465	4538	2.9	115.0	0.7	26.2	8361	6.7	116	219.5	**113**	3262	113+	—
C19-A	12500	430	3982	2.9	146.2	0.4	17.9	5402	80.0	146	2.3	146	3928	146	0.6
C19-B	12500	416	3867	2.8	149.0	0.6	15.8	4035	0.0	147	3.0	**146**	3390	146	0.6
C20-A	12500	241	1222	6.1	266.0	0.0	7.3	598	100.0	266	0.2	266	4311	266	0.0
C20-B	12500	133	563	5.0	267.0	0.0	5.2	500	100.0	267	0.1	267	3800	267	0.1
D1-A	1250	231	440	4.9	18.0	0.0	3.1	500	100.0	18	0.0	18	6	18	0.0
D1-B	1250	233	443	4.9	106.0	0.0	3.8	1950	100.0	106	0.1	106	257	106	0.0
D2-A	1250	257	481	4.9	50.0	0.0	3.5	500	100.0	50	0.1	50	7	50	0.0
D2-B	1250	264	488	4.9	218.3	1.0	7.3	4157	93.3	**218**	0.1	228	486	218	0.0
D3-A	1250	301	529	5.5	807.0	0.0	7.4	500	100.0	807	0.1	807	734	807	0.1
D3-B	1250	372	606	6.3	1516.2	1.3	51.0	15976	0.0	**1509**	0.6	1510	2184	1509	0.3
D4-A	1250	311	541	5.6	1203.8	0.4	10.4	974	16.7	1203	0.3	1203	1263	1203	0.3
D4-B	1250	387	621	7.2	1885.2	2.0	49.6	9671	0.0	1881	11.0	1881	2233	1881	1.3
D5-A	1250	348	588	7.6	2157.0	0.0	29.1	1963	100.0	2157	3.1	2157	3352	2157	8.8
D5-B	1250	411	649	11.5	3137.7	0.9	65.1	7316	0.0	3135	2.2	3135	2555	3135	0.4
D6-A	2000	740	1707	14.4	18.0	0.0	7.7	500	100.0	18	0.3	18	20	18	0.1
D6-B	2000	741	1708	14.7	72.6	0.8	10.5	1192	0.0	71	0.5	**70**	702	67	0.9
D7-A	2000	734	1705	11.3	50.0	0.0	8.2	500	100.0	50	0.3	50	195	50	0.1
D7-B	2000	736	1707	11.4	105.0	0.0	9.5	520	0.0	105	0.3	105	711	103	0.1
D8-A	2000	764	1738	11.7	755.5	0.5	19.1	2788	50.0	755	1.7	755	1727	755	41.8
D8-B	2000	778	1757	12.3	1045.7	3.9	123.8	36313	0.0	**1037**	1013.4	1038	3175	1036	2.8
D9-A	2000	752	1716	17.9	1074.7	1.0	52.1	13718	0.0	1075	354.5	**1072**	4109	1070+	—
D9-B	2000	761	1724	20.9	1436.4	3.0	151.2	31361	0.0	1420	1769.6	1420	2754	1420	4539.6
D10-A	2000	694	1661	14.6	1674.4	1.4	122.2	21289	0.0	1671	9.0	1671	4193	1671	2.2
D10-B	2000	629	1586	18.5	2089.8	2.1	107.3	14598	0.0	2079	44.1	2079	2644	2079	4.1
D11-A	5000	986	4658	27.7	18.0	0.0	15.4	500	100.0	18	1.8	18	540	18	0.5
D11-B	5000	986	4658	23.6	29.0	0.0	17.4	814	100.0	**29**	2.0	30	1280	29	4.7
D12-A	5000	991	4639	23.1	42.0	0.0	13.9	500	100.0	42	2.3	42	844	42	13.2
D12-B	5000	991	4639	22.3	42.0	0.0	15.1	620	100.0	42	2.3	42	687	42	0.4
D13-A	5000	966	4572	27.7	446.7	0.5	58.7	14308	0.0	445	1126.4	445	5047	445	5643.4
D13-B	5000	961	4566	28.0	491.7	1.9	97.2	22843	0.0	486	15.9	486	4288	486	2.6
D14-A	5000	946	4500	35.5	605.6	1.2	102.3	21486	0.0	602	34.2	602	6388	602*	—
D14-B	5000	931	4469	37.2	674.2	1.4	102.8	17746	0.0	665	3409.5	665	6178	664*	—
D15-A	5000	832	4175	47.1	1048.7	1.3	145.7	18343	0.0	1042	185.8	1042	7840	1042	12.8
D15-B	5000	747	3896	49.2	1114.7	0.8	95.6	11026	0.0	1108	117.0	1108	5220	1108	4.8
D16-A	25000	1000	10595	10.8	14.0	0.0	23.1	500	0.0	14	8.9	**13**	1397	13	24.8
D16-B	25000	1000	10595	10.8	13.3	0.4	26.4	1313	73.3	13	9.3	13	1043	13	42.0
D17-A	25000	999	10534	10.8	23.0	0.0	24.8	1983	100.0	23	9.5	23	3506	23	167.1
D17-B	25000	999	10534	10.7	23.0	0.0	23.7	948	100.0	23	10.2	23	2089	23	60.1
D18-A	25000	944	9949	11.7	220.8	0.7	81.4	19864	0.0	218	197.0	218	30044	218+	—
D18-B	25000	929	9816	12.0	230.2	1.3	98.7	25585	0.0	224	25.2	224	36643	223	34.9
D19-A	25000	897	9532	12.4	317.7	2.7	87.6	18480	0.0	308	151.9	308	40955	306	1446.5
D19-B	25000	862	9131	13.1	317.8	2.2	81.9	17912	0.0	311	13.6	311	38600	310	62.8
D20-A	25000	488	2511	37.3	537.0	0.0	18.4	1036	0.0	536	1.0	536	28139	536	0.5
D20-B	25000	307	1383	32.9	537.0	0.0	12.7	1587	100.0	537	0.5	537	22104	537	0.1

Table 3 further illustrates the importance of using both, recombination and mutation, and that it is necessary to apply local improvement immediately after each variation operator. Shown are average results of 30 runs for the following three variants of the MA: In C+LI, new candidate solutions are created only by recombination followed by local improvement. M+LI applies always only mutation followed by local improvement. In C+M+LI, recombination and mutation are used, and local improvement is performed before a solution is inserted into the population. All strategy parameters were set identical as in the previous experiments with the only exception that in M+LI, the probability of applying mutation was $p_{mut} = 1$. The performance values of these variants can therefore directly be compared to those of the original MA in Table 2.

C+M+LI converged fastest, but the obtained solutions were in nearly all cases substantially poorer (1.7% of average gap over all instances) than those of the original MA (0.6% of average gap). This points out the particular importance of applying local improvement after *both* variation operators. C+LI, on the other side, generally needed much more evaluations and also more time to converge. Although its total running time hardly deviates form our original MA, the average gap obtained over all instances was 1.2 %. Finally, the worst results were obtained by running M+LI, with 2% of average gap, which clearly indicates the crossover's importance.

4   Conclusions and Future Research

Our results show that exact algorithms used as local improvement or post-optimization procedures can improve the performance of memetic algorithms. We conjecture that combining linear programming or integer linear programming methods with evolutionary algorithms as described in this paper can yield high quality solutions in short computation time also for other hard optimization problems.

In our future research, we want to combine memetic algorithms with a Branch & Cut approach for solving integer linear programs to obtain even better solutions. Since almost all the currently available benchmark instances are now solved to optimality within a rather short time, the frontier of tractable instances can be pushed further. Based on a real-world utility network design problem we plan to establish new sets of difficult benchmark instances to give new challenges to the community.

Table 2. Summarized results. Running times from Canuto et al. should be divided by 10 for comparison (cf. SPEC comparison). $\%\text{-}gap = (c(T) - OPT)/OPT \cdot 100\%$.

	Preprocessing			MA		MA+ILP		CRR		ILP
Group	$|V'|/|V|$ [%]	$|E'|/|E|$ [%]	t_{prep} [s]	%-gap	t [s]	%-gap	t [s]	%-gap	t [s]	t_{OPT} [s]
K	42.8	46.4	1.6	0.17	4.4	0.13	5.5	0.03	74.5	139.3
P	80.9	74.7	1.0	0.06	12.0	0.01	12.3	0.00	215.1	12.6
C	69.7	59.9	3.8	1.01	20.0	0.70	27.3	0.04	956.2	?
D	70.5	62.9	16.9	0.98	62.7	0.44	232.2	0.41	6834.6	?

Table 3. Average performance over 30 runs of different MA-variants, for K, P, C and D groups of PCSTP instances.

Grp.	C+LI					M+LI					C+M+LI				
	%-gap	σ	t [s]	evals	sr [%]	%-gap	σ	t [s]	evals	sr [%]	%-gap	σ	t [s]	evals	sr [%]
K	0.2	< 0.1	4.2	592	69.1	0.2	< 0.1	4.3	907	70.1	0.3	< 0.1	3.7	727	70.3
P	0.3	< 0.1	10.1	5076	46.1	0.3	0.1	11.6	7478	27.3	0.6	0.1	5.8	3040	19.1
C	2.2	0.1	17.4	6222	41.7	3.9	0.2	18.4	4264	24.6	2.4	0.2	11.0	1313	28.8
D	1.9	0.3	60.5	11582	27.4	3.7	0.9	64.7	9479	20.2	3.5	0.2	37.2	1697	18.2

References

1. D. Bienstock, M. X. Goemans, D. Simchi-Levi, and D. Williamson. A note on the prize collecting traveling salesman problem. *Math. Prog.*, 59:413–420, 1993.
2. S. A. Canuto, M. G. C. Resende, and C. C. Ribeiro. Local search with perturbations for the prize-collecting Steiner tree problem in graphs. *Networks*, 38:50–58, 2001.
3. C. W. Duin and A. Volgenant. Some generalizations of the Steiner problem in graphs. *Networks*, 17(2):353–364, 1987.
4. S. Engevall, M. Göthe-Lundgren, and P. Värbrand. A strong lower bound for the node weighted Steiner tree problem. *Networks*, 31(1):11–17, 1998.
5. M. Fischetti. Facets of two Steiner arborescence polyhedra. *Mathematical Programming*, 51:401–419, 1991.
6. M. X. Goemans. The Steiner tree polytope and related polyhedra. *Mathematical Programming*, 63:157–182, 1994.
7. M. X. Goemans and D. P. Williamson. The primal-dual method for approximation algorithms and its application to network design problems. In D. S. Hochbaum, editor, *Approximation algorithms for NP-hard problems*, pages 144–191. P. W. S. Publishing Co., 1996.
8. D. S. Johnson, M. Minkoff, and S. Phillips. The prize-collecting Steiner tree problem: Theory and practice. In *Proceedings of 11th ACM-SIAM Symposium on Discrete Algorithms*, pages 760–769, San Francisco, CA, 2000.
9. G. Klau, I. Ljubić, A. Moser, P. Mutzel, P. Neuner, U. Pferschy, and R. Weiskircher. A new lower bounding procedure for the prize-collecting Steiner tree problem. Technical Report TR-186-1-04-01, Vienna University of Technology, 2004.
10. G. Klau, I. Ljubić, P. Mutzel, U. Pferschy, and R. Weiskircher. The fractional prize-collecting Steiner tree problem on trees. In G. D. Battista and U. Zwick, editors, *ESA 2003*, volume 2832 of *LNCS*, pages 691–702. Springer-Verlag, 2003.
11. A. Lucena and M. Resende. Strong lower bounds for the prize-collecting Steiner problem in graphs. Technical Report 00.3.1, AT&T Labs Research, 2000.
12. K. Mehlhorn. A faster approximation for the Steiner problem in graphs. *Information Processing Letters*, 27:125–128, 1988.
13. P. Moscato. Memetic algorithms: A short introduction. In D. Corne and et al., editors, *New Ideas in Optimization*, pages 219–234. McGraw Hill, England, 1999.
14. G. R. Raidl and J. Gottlieb. On the importance of phenotypic duplicate elimination in decoder-based evolutionary algorithms. In S. Brave and A. S. Wu, editors, *Late Breaking Papers at the 1999 Genetic and Evolutionary Computation Conference*, pages 204–211, Orlando, FL, 1999.
15. G. R. Raidl and B. A. Julstrom. Edge-sets: An effective evolutionary coding of spanning trees. *IEEE Trans. on Evolutionary Computation*, 7(3):225–239, 2003.
16. A. Segev. The node-weighted Steiner tree problem. *Networks*, 17:1–17, 1987.

On the Evolution of Analog Electronic Circuits Using Building Blocks on a CMOS FPTA

Jörg Langeheine, Martin Trefzer, Daniel Brüderle,
Karlheinz Meier, and Johannes Schemmel

University of Heidelberg, Kirchhoff-Institute for Physics,
INF 227, D-69120 Heidelberg, Germany,
ph.: ++49 6221 54 9838 langehei@kip.uni-heidelberg.de
http://www.kip.uni-heidelberg.de/vision/projects/eh/index.html

Abstract. This article summarizes two experiments utilizing building blocks to find analog electronic circuits on a CMOS Field Programmable Transistor Array (FPTA). The FPTA features 256 programmable transistors whose channel geometry and routing can be configured to form a large variety of transistor level analog circuits. The transistor cells are either of type PMOS or NMOS and are arranged in a checkerboard pattern. Two case studies focus on improving artificial evolution by using a building block library of four digital gates consisting of a NOR, a NAND, a buffer and an inverter. The methodology is applied to the design of the more complex logic gates XOR and XNOR as well as to the evolution of circuits discriminating between square waves of different frequencies.

1 Introduction

The design of complex competitive analog electronics is a difficult task. In fact, to date existing technologies fail to automatically synthesize new transistor level circuit topologies for problems of medium or high complexity ([1] contains an overview of recent efforts). In engineering science, if a problem is hard to solve, usually a divide and conquer approach is used to simplify it. This leads to hierarchical approaches as e.g. described in [2]. Unfortunately, the division into subproblems often is a nontrivial task itself. Another approach, corresponding to the bottom up design principle, is to use functional subunits (building blocks) and assemble them to form solutions to more complex problems as for example done in [3].

The Field Programmable Transistor Array (FPTA) utilized in this work is a fine grained analog substrate dedicated to hardware evolution that offers a fairly high degree of complexity (cf. [4] for an overview of existing hardware). Hence, it is well suited to host hardware-in-the-loop experiments that can take advantage of both worlds: Find new circuit solutions exploiting transistor physics as well as accelerate the evolution process by using (predefined) building blocks and thereby relieving the evolutionary algorithm of reinventing substructures that have been proven useful in analog circuit design.

K. Deb et al. (Eds.): GECCO 2004, LNCS 3102, pp. 1316–1327, 2004.

In order to test the proposed building block concept, a small library of well known building blocks that are well suited to the posed problem is sought. In this regard, the evolution of the analog dc behavior of the more complex gates XOR and XNOR by means of a building block library comprising the four simple logic gates NOR, NAND, inverter and buffer is considered a good test case. On one hand, the used building blocks are known to be useful for the design of the XOR/XNOR gates. On the other hand, the evolution of the analog behavior of XOR/XNOR gates can be easily stated in terms of the fitness function and is considered to be nontrivial, because it is not linearly separable. In a second case study, the same building block library is used to enhance the evolution of circuits that distinguish between square waves of different frequencies referred to as tone discriminators (TDs) (cf. [5]). In contrast to the first test problem, the usefulness of the used building blocks is not obvious at all in this case, because tone discrimination (in the absence of an external clock) is an inherently analog problem.

2 Evolution System

The used evolution system can be divided into three main parts: The actual FPTA chip serving as the silicon substrate to host the candidate circuits, the software that contains the search algorithm running on a standard PC and a PCI interface card that connects the PC to the FPTA chip. The software uploads the configuration bit strings to be tested to the FPTA chip via the PCI card. In order to generate an analog test pattern at the inputs of the FPTA chip, the input data is written to the FPGA on the PCI interface card. There it is converted into an analog signal by a 16 bit DAC. After applying the analog signal to the FPTA, the output of the FPTA is sampled and converted into a digital signal by means of a 12 bit ADC. The digital output is then fed back to the search algorithm, which in turn generates the new individuals for the next generation.

2.1 FPTA Chip

The FPTA consists of 16×16 programmable transistor cells. As CMOS transistors come in two flavors, namely N- and PMOS transistors, half of the transistor cells are designed as programmable NMOS transistors and half as programmable PMOS transistors. P- and NMOS transistor cells are arranged in a checkerboard pattern as depicted on the left hand side of Fig. 1.

Each cell contains the programmable transistor itself, three decoders that allow to connect the three transistor terminals to one of the four cell borders, *vdd* or *gnd*, and six routing switches. A simplified block diagram of the transistor cell is shown on the right hand side of Fig. 1. Width W and Length L of the programmable transistor can be chosen to be $1, 2, \ldots, 15\,\mu m$ and $0.6, 1, 2, 4, 8\,\mu m$ respectively. The three terminals *drain, gate* and *source* of the programmable transistor can be connected to either of the four cell borders named after the four cardinal points, *vdd* or *gnd*. The only means of routing signals through the

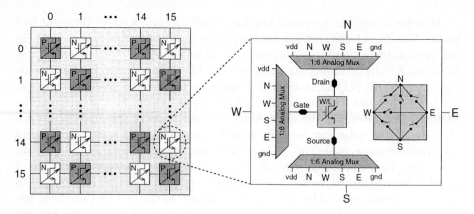

Fig. 1. Left: Schematic diagram of the 16 x 16 programmable transistor cell array. **Right:** Close-up on one NMOS transistor cell.

chip is given by the six routing switches that connect the four cell borders with each other. Thus, in some cases it is not possible to use a transistor cell for routing *and* as a transistor. More details on the FPTA can be found in [6].

2.2   Evolutionary Algorithm

The experiments of all three case studies were performed employing a straight forward genetic algorithm implementation in conjunction with a truncation selection scheme. In order to keep the algorithm stable in case of noisy and/or unreliable fitness measurements, relatively large values for the *reproduction fraction* (the fraction of the population that is moved to the new generation unchanged) are used, as can be seen from Table 1.

Table 1. Genetic algorithm parameters used throughout the presented experiments.

GA Parameter	TD:BB	TD: Cell	X(N)OR: BB	X(N)OR: Cell
generation size	50	50	50	50
reprod. fraction	0.3	0.2	0.3	0.2
mutation fraction	0.4	0.4	0.4	0.4
crossover fraction	–	0.5	–	0.5
crossover rate	0	1 %	0	1 %
mut. rate routing	4 %	1 %	4 %	1 %
mut. rate W/L	3 %	1 %	3 %	1 %
mut. rate term. con.	–	1 %	–	1 %
mut. rate BB	2 %	–	1 %	–
no. of used blocks	16	–	16	–
no. of used cells	112	64	112	64
no. of generations	10000	10000	5000	5000

3   Building Block Concept

The standard genotype representation reflects the structure of the FPTA chip: For each cell the transistor geometry, its terminal connections as well as the state of the routing switches can be mutated individually. The crossover operation is cell based: A rectangular array of cells is copied from one individual to the same location of another individual producing one offspring. A more detailed description of the underlying representation can be found in [7].

This representation is extended for the usage of building blocks (BBs) by introducing genetic access rights for any of the transistor cells and by new crossover and mutation operators: The access rights define the genetic operations the EA is allowed to apply to the according cell. The new crossover operation preserves the building block structure, i.e., the chosen crossover blocks are extended such that they embed all partially covered BBs. The new mutation operator replaces a randomly selected BB of the genotype with one randomly chosen from the building block library. The first generation is initialized with BBs chosen randomly from the used library and randomly configured transistor cells (according to the used genetic access rights). As a result, the genotype can be freely divided into BB sites and simple transistor cells that can be altered in only exactly the ways defined in the particular experimental setup.

For the case studies a library of four simple logic building blocks implemented using 3×3 transistor cells is utilized. Both experiments use the complete chip and a total of 16 building block sites, as depicted in Fig. 2. While genetic operations

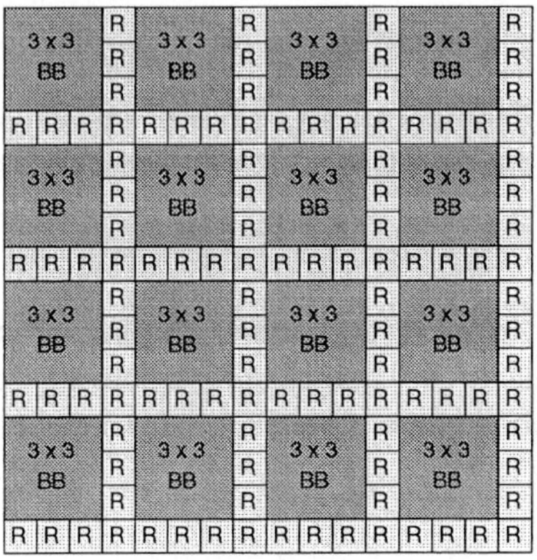

Fig. 2. Geometrical setup for case studies I,II. The R denotes routing cells.

for the cells denoted with an R are restricted to changes of their routing, the GA is allowed to change the channel dimensions W and L for the cells reserved for the BBs. On insertion, all transistors of all BBs possess an aspect ration of $W/L = 4/2$. The input signals are applied to the left hand side, the circuit's output is measured on its right hand side. The crossover operation was omitted in the experiments using BBs. However, the reference experiments using the pure transistor cell implementation did use crossover, but were restricted to 8×8 cells to constrain the design space to a size comparable to that of the building block experiments.

3.1 Logic Gate Library

Fig. 3 illustrates the four logic gates making up the used building block library. Each block possesses two inputs A and B at its western edge, which are short

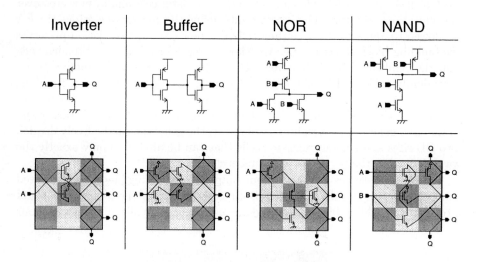

Fig. 3. Building block library used for case studies I and II. The second row shows the schematics of the used circuits and the third one displays their implementation as a block of 3×3 transistor cells. PMOS transistor cells are shaded in darker gray than their NMOS counterparts.

circuited in case of the inverter and buffer implementations. The output Q is available at five terminals at the eastern side. Thus, the proposed building blocks support the aforementioned signal flow from left to right that is used throughout all experiments.

4   Case Study I: XOR and XNOR

As a first test of the building block concept the BB library shown in Fig. 3 is used to evolve the more complex logic gates XOR and XNOR. A total of four experiments each featuring 30 runs were carried out, two using the described building block setup of Fig. 2 and two using the plain cell genotype respectively. For the experiments using plain transistor cells the array provided to the GA was restricted to 8×8 cells. Both input voltages, V_{in1} and V_{in2}, are applied to the western side of the array while the output is measured at the opposite side.

4.1 Experimental Setup

All experiments are run at a fixed generation size of 50 individuals and a number of generations of 5000. During evolution, the used test pattern consists of a set of eight curves with $V_{in1} = 0 \ldots 2\,V, 3 \ldots 5\,V$ each in 4 steps and $V_{in2} = 0 \ldots 2\,V, 3 \ldots 5\,V$ each in 16 steps. A target voltage of $V_{tar} = 0\,V$ corresponds to the logic zero and $V_{tar} = 5\,V$ to the logic one. The input voltage range between 2 and 3 V, where the gate switches its output, is not of interest for the application of logic gates and therefore not covered in the test pattern. Moreover, it would constrain any possible solution more severely than necessary. The sample voltages are applied in randomly chosen random orders with a sample frequency of 244 kHz. Hence, the settling time must be less than 4.1 μs. For measuring the voltage characteristics of the evolved logic gates, a modified test pattern is used that covers the full range of $V_{in2} = 0 \ldots 5\,V$, thus including the transition region.

4.2 Fitness Calculation

Throughout all experiments the sum of squared errors is used as the fitness criterion:

$$\text{Fitness} = \sum_{i=1}^{256} (V_{tar}(i) - V_{out}(i))^2 \; . \tag{1}$$

Hence, the GA has to minimize this fitness. However, in order to add a physical meaning to the fitness measure, the fitness is converted to the root mean square error per data point in mV for all results presented in the remainder of this section:

$$\text{RMS Error [mV]} = \sqrt{\frac{\text{Fitness}}{256}} \cdot 1000 \; . \tag{2}$$

4.3 Evolution Results

The fitness values cover a theoretical range of $0 \ldots 5000\,\text{mV}$. Practically, typical random individuals that are used for the initialization of the population obtain a fitness of about $2500 \pm 500\,\text{mV}$; a circuit that exhibits the exact inverse of the target behavior is as improbable as the desired one.

Comparison of the Results of the Different Experiments. The *RMS error* values of all experiments are shown in the histograms in Fig. 4. The results confirm that, as expected, the building blocks extensively help the GA in finding good solutions for more complex logic gates. While the results of the experiments using the standard representation are comparable to those presented in [7], the use of building blocks boosts the rate of runs finishing with the desired output behavior from 0 to more than 80%.

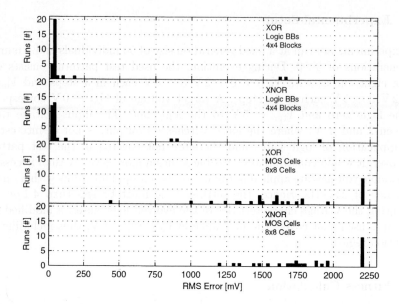

Fig. 4. Comparison of the achieved fitness values of 30 runs per XOR/XNOR experiment.

Voltage Characteristics of the Evolved Logic Gates. While the transition region was not considered during evolution, it is measured and plotted for the best circuits of each experiment in Fig. 5 to obtain information about the complete voltage characteristic of the evolved gates. Both, the best XOR as well as the best XNOR gate evolved with building blocks exhibit an output voltage characteristic that perfectly matches the fitness criterion. Conversely, the XOR and XNOR evolved using the plain transistor cell genotype both fail to reach the voltage rails for at least one of the four logic input combinations. However, assuming a threshold of 2.5 V, they would manage to produce the correct logical result.

The fact that none of the circuits evolved without building blocks perfectly meets the target specifications can be explained as follows: For one, the fitness criterion may be too ambitious in that the region the gate is allowed to switch in is very narrow. In fact, an XOR/XNOR circuit from the standard cell library provided by the manufacturer of the used process technology would be evaluated with a non-zero fitness as shown in [7],[1]. For the other, the difficulty of the task is increased by the used representation that is closely related to the structure of the FPTA chip. Therefore, the design of the circuit and its physical layout on the FPTA chip have to be processed in one single step.

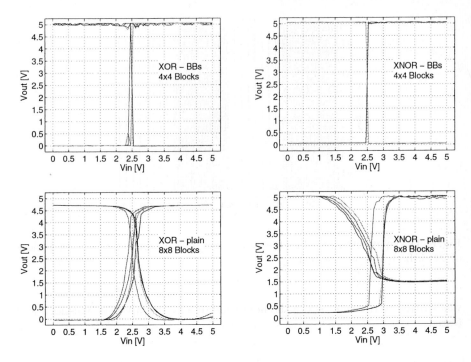

Fig. 5. Measured voltage characteristics of the evolved XORs and XNORs using Building Blocks and standard cells.

5   Case Study II: Tone Discrimination

The problem of discriminating square waves of different frequencies suits hardware evolution well: On one hand, the problem definition in terms of test patterns and fitness function is relatively simple; on the other hand, the design of an analog tone discriminator is a nontrivial task. Within the field of evolvable hardware the problem was first tackled by Adrian Thompson (see e.g. [5]) who used an FPGA to discriminate tones of 1 and 10 kHz.

5.1   Problem Definition

Test Pattern. In contrast to the original experiment, the frequencies to be distinguished were shifted to 40 and 200 kHz, in order to decrease the time necessary for one fitness evaluation. As can be seen from Fig. 6 the test pattern consists of 20 periods of the 200 kHz square wave followed by 4 periods of the 40 kHz one. This pattern is applied twice for each fitness test. The output is sampled with a frequency of 2 MHz, resulting in 800 test points and a total time of 400 µs. In order to prevent successful candidate solutions from exploiting the charge distribution left from the test of its predecessor, a randomly created gene

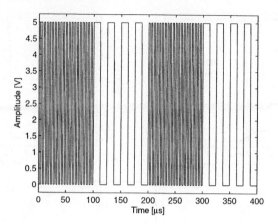

Fig. 6. The input pattern used for the evolution of tone discriminators. 2 × (20 periods of 200 kHz + 4 periods of 40 kHz).

was written to the chip before the next candidate solution was downloaded and tested.

Fitness Function. During the evolution process the fitness is evaluated by

$$\text{Fitness} = \sum_{i=1}^{800}(V_{\text{tar}}(i) - V_{\text{out}}(i))^2 + 3\sum_{i=2}^{800}(V_{\text{out}}(i) - V_{\text{out}}(i-1))^2 , \qquad (3)$$

with the target voltage defined as

$$V_{\text{tar}} = \begin{cases} 0,5 & \text{for} \quad f = 200\,\text{kHz} \\ 5,0 & \text{for} \quad f = 40\,\text{kHz} \end{cases} . \qquad (4)$$

The actual $V_{\text{tar}}(f)$ is chosen to minimize the fitness value; thereby the GA is relieved of the constraint of finding a solution with a prescribed output polarity. While the left term of (3) yields the sum of squared deviations from the target voltage (4), the right sum penalizes unwanted glitches and oscillations of the output. The weighting factor of 3 was chosen based on the experience gathered in preliminary studies.

However, for the analysis within this section the fitness is calculated as the root mean square error per data point given in mV,

$$\text{RMS Error [mV]} = \sqrt{\frac{\sum_{i=1}^{800}(V_{\text{tar}}(i) - V_{\text{out}}(i))^2}{800}} \cdot 1000 , \qquad (5)$$

which adds a physical meaning to the fitness measure.

5.2   Results

The geometrical setup is identical to the one described in case study I and the structure of the two experiments – one using the building block representation described in Fig. 3, the other one the plain cell genotype – is similar to those of section 4. In order to acquire information about the reliability of the evolved circuits, the best individuals of all evolution runs were tested 100 times. Fig. 7 compares the results of both series using the worst fitness values measured during the verification tests. During the course of the experiments it was observed that

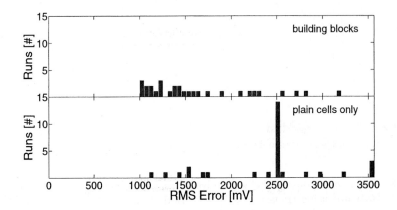

Fig. 7. Comparison of the worst fitness from 100 verification tests for the experiment with and without building blocks respectively.

the algorithm frequently chooses to clamp the output to 2.5 V. On one hand, this realizes the minimum RMS error without having to discriminate between the two frequencies. On the other hand, a circuit producing such a constant output voltage can easily be realized on the FPTA. Accordingly, all runs should finish with a fitness smaller than or equal to 2500 mV – the value resulting from applying (5) to the situation described above. In the histograms of Fig. 7 however, some circuits manage to behave even worse, which indicates that these solutions were performing better during evolution, but fail to work reliably under the verification test conditions.

The results suggest that the GA was more successful in finding tone discriminators of moderate quality when it was allowed to use BBs. The large peak in the histogram for the runs utilizing only plain transistor cells indicates that a large fraction of them got stuck in the local minimum described above.

The circuit responses of the best individual with and without BB usage are plotted in Fig. 8. The left half of the figure captures the circuit response to the test pattern used during evolution. In the right half, the output is plotted versus frequency, where output is defined as the output voltage averaged over one period of a square wave. As can be seen from Fig. 8, the best solutions found

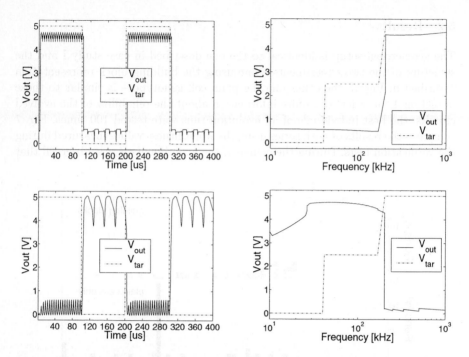

Fig. 8. Measured response of the best evolved tone discriminators: **Top**: using building blocks, **Bottom**: using transistor cells only, **Left**: Original fitness criterion. **Right**: Frequency Sweep. The outputs have different polarities with reference to the input frequency, which is allowed by the fitness criterion described above.

with and without BB usage do not differ significantly. Both solutions clearly distinguish between the two input frequencies but fail to reach the rails of the power supply range and carry a considerable amount of ripple. Considering the frequency sweep tests, it can be observed that both tone discriminators correctly distinguish between frequencies lower than approximately 200 kHz and those above 200 kHz in the measured frequency range of 10 kHz to 1 MHz. Since the best circuits obtained in this work are not as good as the results achieved in the original experiments documented in [5], it should be noted that both experiments do differ in a variety of ways. Most prominently, the FPGA used by Thompson was able to use a larger amount of resources to fulfill the task.

6   Discussion

The use of building blocks was introduced and tested in two case studies, namely the evolution of XOR/XNOR gates and the evolution of tone discriminating circuits. While the success rate as well as the performance of the best evolved circuits could be greatly enhanced in case of the gates, the building blocks mainly support the GA in finding solutions of moderate quality more frequently for the evolution of tone discriminators. The latter result is remarkable insofar as the used building block library is far from being especially devised to the task of tone discrimination.

The proposed building block library of simple logic gates is not expected to be a particularly good choice to solve analog problems, but on the contrary, the choice of good building blocks is a key to efficiently solve a particular problem. Besides the actual functionality of the blocks the geometry of their in- and outputs as well as the geometrical setup they are embedded in are expected to play an important role. To find answers to these questions, future experiments will have to apply different building block libraries and topologies to a wider range of test problems. From the resulting data valuable information can be gathered about better FPTA cells and architectures that will eventually lead to a second generation FPTA.

Acknowledgment. This work is supported by the Ministerium für Wissenschaft, Forschung und Kunst, Baden-Württemberg, Stuttgart, Germany.

References

1. Langeheine, J., Meier, K., Schemmel, J.: Intrinsic evolution of analog electronic circuits using a CMOS FPTA chip. In: Proc. of the 5th Conf. on Evolutionary Methods for Design, Optimization and Control (EUROGEN 2003), Barcelona, Spain, IEEE Press (2003) 87–88 Published on CD: ISBN: 84-95999-33-1.
2. Zebulum, R.S., Stoica, A., Keymeulen, D.: Experiments on the evolution of digital to analog converters. In: Proc. of the IEEE Aerospace Conference, Montana, USA (2001) ISBN: 0-78-3-6600-X (Published on CD).
3. Shibata, H.: Computer-Aided Design of analog Circuits Based on Genetic Algorithm. PhD thesis, Tokyo Institute of Technology (2001)
4. Zebulum, R.S., Stoica, A., Keymeulen, D.: A flexible model of a CMOS field programmable transistor array targeted for hardware evolution. In Miller, J., Thompson, A., Thomson, P., Fogarty, T.C., eds.: Proc. of the Third Int. Conference on Evolvable Systems: From Biology to Hardware (ICES2000), Edinburgh, UK, Springer (2000) 274–283 LNCS 1801.
5. Thompson, A., Layzell, P., Zebulum, R.S.: Explorations in design space: Unconventional electronics design through artificial evolution. IEEE Trans. Evol. Comp. **3** (1999) 167–196
6. Langeheine, J., Becker, J., Fölling, S., Meier, K., Schemmel, J.: Initial studies of a new VLSI field programmable transistor array. In: Proc. 4th Int. Conf. on Evolvable Systems: From Biology to Hardware, Tokio, Japan, Springer Verlag (2001) 62–73
7. Langeheine, J., Meier, K., Schemmel, J.: Intrinsic evolution of quasi dc solutions for transistor level analog electronic circuits using a CMOS FPTA chip. In: Proc. of the Fourth NASA/DOD Workshop on Evolvable Hardware, Alexandria, VA, USA, IEEE Press (2002) 76–85

Parameter-Less Optimization with the Extended Compact Genetic Algorithm and Iterated Local Search

Cláudio F. Lima and Fernando G. Lobo

ADEEC-FCT, Universidade do Algarve
Campus de Gambelas, 8000 Faro, Portugal
{clima,flobo}@ualg.pt

Abstract. This paper presents a parameter-less optimization framework that uses the extended compact genetic algorithm (ECGA) and iterated local search (ILS), but is not restricted to these algorithms. The presented optimization algorithm (ILS+ECGA) comes as an extension of the parameter-less genetic algorithm (GA), where the parameters of a selecto-recombinative GA are eliminated. The approach that we propose is tested on several well known problems. In the absence of domain knowledge, it is shown that ILS+ECGA is a robust and easy-to-use optimization method.

1 Introduction

One of the major topics of discussion within the evolutionary computation community has been the parameter specification of the evolutionary algorithms (EAs). After choosing the encoding and the operators to use, the EA user needs to specify a number of parameters that have little to do with the problem (from the user perspective), but more with the EA mechanics itself. In order to release the user from the task of setting and tuning the EA parameters, several techniques have been proposed. One of these techniques is the parameter-less GA, which controls the parameters of a selecto-recombinative GA. This technique can be applied to various types of (selecto-recombinative) GAs, and in conjunction with a high-order estimation of distribution algorithm (EDA), such as the extended compact GA (ECGA) [1] or the Bayesian optimization algorithm (BOA) [2], results in a powerful and easy-to-use search algorithm. Multivariate EDAs have shown to outperform the simple GA (SGA) by several orders of magnitude, especially on very difficult problems. However, these advanced search algorithms don't come for free, requiring more computational effort than the SGA when moving from population to population. In many problems this extra effort is well worth it, but for other (less complex) problems, a simpler algorithm can easily outperform a multivariate EDA.

Typical EAs are based on two variation operators: recombination and mutation. Recombination and mutation search the solution space in different ways and with different resources. While recombination needs large populations to

K. Deb et al. (Eds.): GECCO 2004, LNCS 3102, pp. 1328–1339, 2004.

combine effectively the necessary information, mutation works best when applied to small populations during a large number of generations. Spears [3] did a comparative study between crossover and mutation operators, and theoretically demonstrates that there were some important characteristics of each operator that were not captured by the other.

Based on these observations, we propose a new parameter-less optimization framework, that consists of running two different search models simultaneously. The idea is to use the best of both search strategies in order to obtain an algorithm that works reasonably well in a large class of problems. The first method can be a parameter-less ECGA, based on selection and wise recombination to improve a population of solutions. As a second method we can use an iterated local search (ILS) algorithm with adaptive perturbation strength. Instead of working with a population of solutions, the ILS iterates a single solution by means of selection and mutation. We called optimization framework, instead of optimization algorithm, since what we propose here is not tied up with the ECGA or our ILS implementation. Other algorithms, such as BOA or other ILS implementations, can be considered with similar or better results. However, in this paper we restrict ourselves to the concrete implementation of the ILS+ECGA algorithm and the discussion of the corresponding results.

The next section reviews some of the work done in the topic of EA parameter tuning/control, then describes the parameter-less GA technique, the ECGA, and the ILS framework. Then, Section 3 describes the basic principles of the parameter-less optimization framework and our ILS+ECGA implementation. In Section 4, computational experiments are done to validate the proposed approach. Section 5 highlights some extensions of this work. Finally, in Section 6, a summary and conclusions are presented.

2   Related Work

This section reviews some of the research efforts done in setting and adapting the EAs parameters, describes the parameter-less GA technique and the mechanics of the ECGA and ILS.

2.1   Parameter Tuning and Parameter Control in EAs

Parameter tuning in EAs involves the empirical and theoretical studies done to find optimal settings and understand the interactions between the various parameters. An example of that was the work of De Jong [4], where various combinations of parameters were tested on a set of five functions. On those experiments, De Jong verified that the parameters that gave better overall performance were: population size in the range 50-100, crossover probability of 0.6, mutation probability of 0.001, and generation gap of 1.0 (full replacement of the population in each generation). Some other empirical studies have been conducted on a larger set of problems yielding somewhat similar results [5,6]. Almost 30 years later, these parameters are still known as the "standard" parameters, being sometimes incorrectly applied to many problems. Besides these empirical studies, some work was done to analyze the effect of one or two parameters in isolation, ignoring the others. Among the most relevant studies, are the ones done on selection [7], population sizing [8,9], mutation [10,11], and control maps [12,13]. The work on population sizing is of special relevance, showing that setting the population size to 50-100 for all problems is a mistake. The control maps study gave regions of the parameter space (selection and crossover values) where the GA was expected to work well, under the assumption of proper linkage identification.

In parameter control we are interested in adapting the parameters during the EA run. Parameter control techniques can be subdivided in three types: deterministic, adaptive, and self-adaptive [14]. In deterministic control, the parameters are changed according to deterministic rules without using any feedback from the search. The adaptive control takes place when there is some form of feedback that influences the parameter specification. Examples of adaptive control are the works of Davis [15], Julstrom [16], and Smith & Smuda [17]. The parameter-less GA technique is a mix of deterministic and adaptive rules of control, as we will see in the next section. Finally, self-adaptive control is based on the idea that evolution can be also applied in the search for good parameter values. In this type of control, the operator probabilities are encoded together with the corresponding solution, and undergo recombination and mutation. This way, the best parameter values will tend to survive because they originate better solutions. Self-adaptive evolution strategies (ESs) [18] are an example of the application of this type of parameter control.

2.2   Parameter-Less Genetic Algorithm

The parameter-less genetic algorithm [19] is a technique that eliminates the parameters of a selecto-recombinative GA. Based on the schema theorem [20] and various facet-wise theoretical studies of GAs [9,12], Harik & Lobo automated the specification of the selection pressure (s), crossover rate (p_c), and population size (N) parameters.

The selection pressure and crossover rate are set to fixed values, according to a simplification of the schema theorem in order to ensure the growth of promising building blocks. Simplifying the schema theorem, and under the conservative hypothesis that a schema is destroyed during the crossover operation, the growth ratio of a schema can be expressed by $s\,(1-p_c)$. Thus, setting $s = 4$ and $p_c = 0.5$, gives a net growth factor of 2, ensuring that the necessary building blocks will grow. If these building blocks will be able to mix in a single individual or not is now a matter of having the right population size.

In order to achieve the right population size, multiple populations with different sizes are run in a concurrent way. The GA starts by firing the first population, with size $N_1 = 4$, and whenever a new population is created its size is doubled. The parameter-less GA gives an advantage to smaller populations by giving them more function evaluations. Consequently, the smaller populations have the chance to converge faster than the large ones. The reader is referred to Harik & Lobo [19] for details on this approach.

Extended Compact Genetic Algorithm (ECGA)

---

(1) Create a random population of N individuals.
(2) Apply selection.
(3) Model the population using a greedy MPM search.
(4) Generate a new population according to the MPM found in step 3.
(5) If stopping criteria is not satisfied, return to step 2.

Fig. 1. Steps of the extended compact genetic algorithm (ECGA).

2.3   Extended Compact Genetic Algorithm

The extended compact genetic algorithm (ECGA) [1] is based on the idea that the choice of a good probability distribution is equivalent to linkage learning. The ECGA uses a product of marginal distributions on a partition of the decision variables. These kind of probability distributions are a class of probability models known as marginal product models (MPMs). The measure of a good MPM is quantified based on the minimum description length (MDL) principle. According to Harik, good distributions are those under which the representation of the distribution using the current encoding, along with the representation of the population compressed under that distribution, is minimal. Formally, the MPM complexity is given by the sum $C_m + C_p$. The model complexity C_m is given by

$$C_m = \log_2(N+1) \sum_i (2^{S_i} - 1), \tag{1}$$

where N is the population size and S_i is the length of the ith subset of genes. The compressed population complexity C_p is quantified by

$$C_p = N \sum_i E(M_i), \tag{2}$$

where $E(M_i)$ is the entropy of the marginal distribution of subset i. Entropy is a measure of the dispersion (or randomness) of a distribution, and is defined as $E = \sum_{j=1}^{n} -p_j \log_2(p_j)$, where p_j is the probability of observing the outcome j in a total of n possible outcomes.

As we can see in Figure 1, steps 3 and 4 of the ECGA differ from the simple GA operation. Instead of applying crossover and mutation, the ECGA searches for a MPM that better represents the current population and then generates a new population sampling from the MPM found in step 3. This way, new individuals are generated without destroying the building blocks.

2.4   Iterated Local Search

The iterated local search (ILS) [21] is a simple and general purpose metaheuristic that iteratively builds a sequence of solutions generated by an embedded heuristic, leading to better solutions than repeated random trials of that

---

Iterated Local Search (ILS)

$s_0 = \texttt{GenerateInitialSolution}(seed)$
$s^* = \texttt{LocalSearch}(s_0)$
repeat
    $s' = \texttt{Perturbation}(s^*, history)$
    $s^{*\prime} = \texttt{LocalSearch}(s')$
    $s^* = \texttt{AcceptanceCriterion}(s^*, s^{*\prime}, history)$
until termination condition met

Fig. 2. Pseudo-code of Iterated Local Search (ILS).

heuristic. This simple idea is not new, but Lourenço et al. formulated as a general framework. The key idea of ILS is to build a biased randomized walk in the space of local optima, defined by some local search algorithm. This walk is done by iteratively perturbing a locally optimal solution, next applying a local search algorithm to obtain a new locally optimal solution, and finally using an acceptance criterion for deciding from which of these two solutions to continue the search. The perturbation must be strong enough to allow the local search to escape from local optima and explore different areas of the search space, but also weak enough to avoid that the algorithm degenerates into a simple random restart algorithm (that typically performs poorly).

Figure 2 depicts the four components that have to be specified to apply an ILS algorithm. The first one is the procedure `GenerateInitialSolution` that generates an initial solution s_0. The second one is the procedure `LocalSearch` that implements the local search algorithm, giving the mapping from a solution s to a local optimal solution s^*. Any local search algorithm can be used, however, the performance of the ILS algorithm depends strongly on the one chosen. The `Perturbation` is responsible for perturbing the local optima s^*, returning a perturbed solution s'. Finally, the procedure `AcceptanceCriterion` decides which solution (s^* or $s^{*\prime}$) will be perturbed in the next iteration. An important aspect in the perturbation and the acceptance criterion is to introduce a bias between intensification and diversification of the search. Intensification in the search can be reached by applying the perturbation always to the best solution found and using small perturbations. On the other hand, diversification is achieved by accepting every new solution $s^{*\prime}$ and applying large perturbations.

3   Two Search Models, Two Tracks, One Objective

Different approaches have been proposed to combine global search with local search. A common practice is to combine GAs with local search heuristics. It has been used so often that originated a new class of search methods called memetic algorithms [22]. In this work we propose something different, the combination of two global search methods based on distinct principles. By principles we mean variation operators, selection methods, and population management policies. The ECGA is a powerful search algorithm based on recombination to improve solutions, however at the cost of extra computation time (needed to search for a good MPM) in each generation. For hard problems this effort is well worth it, but for other problems, less complex search algorithms may do. This is where ILS comes in. As a light mutation-based algorithm, ILS can quickly and reliably find good solutions for simpler or mutation-tailed problems. What we propose is to run ILS and ECGA simultaneously. This "pseudo-parallelism" is done by giving an equal number of function evaluations to each search method alternately. ILS and ECGA will have their own track in the exploration of the search space, without influencing each other. In the resulting optimization algorithm, that we call ILS+ECGA, the search will be done by alternating between ILS and ECGA.

3.1   Parameter-Less ECGA

The parameter-less GA technique is coupled together with the ECGA. An important aspect of our implementation is the saving of function evaluations. Since the crossover probability is always equal to 0.5, there is no need of reevaluating the individuals that are not sampled from the model. This way, half of the total number of function evaluations are saved.

3.2   ILS with Adaptive Perturbation

In this section we describe the ILS implementation used for this work, and present a simple but effective way to eliminate the need of specifying its parameters. The four components chosen for the ILS algorithm are:

Local Search: next ascent hill climber (NAHC). NAHC consists in having one individual and keep mutating each gene, one at a time, in a predefined random sequence, until the resulting individual is fitter than the original. In that case, the new individual replaces the original and the procedure is repeated until no improvement can be made further.

Initial Solution: randomly generated. Since the NAHC is fast in getting local optima solutions, there is no need to use a special greedy algorithm.

Acceptance Criterion: accept always the last local optima obtained ($s^{*\prime}$) as the solution from where the search will continue. In a way, this is done to compensate the intensive selection criterion from NAHC, where just better solutions are accepted. On the other side, with this kind of acceptance criterion we promote a stochastic search in the space of local optima.

Perturbation: probabilistic and greater than the mutation rate of the NAHC (equal to $1/l$). The perturbation strength is proportional to the problem size (number of genes l). This way, the perturbation is always strong enough, whatever the problem size. Each allele is perturbed with probability $p_p = 0.05l/l = 0.05$. This means that on average 5% of the genes are perturbed. However, if the problem length is too small (for example, $l \leq 60$), then the perturbation becomes of the same order of magnitude than the mutation done by NAHC. To avoid this, we fix the perturbation probability to $3/l$ for problems where $l \leq 60$. This way, we ensure that on average the perturbation strength is at least 3 times greater than the mutation strength of NAHC. This is done to prevent perturbation from being easily cancelled by the local search algorithm. Nevertheless, the perturbation strength may not be strong enough if the attraction area of a specific local optima is too big, leading to a situation where frequently $s^{*\prime} = s^*$. In that case, we need to increase the perturbation strength until we get out from the attraction area of the local optima. Therefore, the perturbation strength α is updated as follows:

$$\alpha_{new} = \begin{cases} \alpha_{current} + 0.02l, & \text{if } s^{*\prime} = s^* \\ 0.05l, & \text{if } s^{*\prime} \neq s^* \end{cases} \tag{3}$$

This way, the updated perturbation probability is equal to α_{new}/l.

3.3   ILS+ECGA

The parameter-less optimization framework proposed consists of running the two different search models more or less simultaneously. This is accomplished by switching back and forth between one method and the other after a predefined number of function evaluations have elapsed ($fe_{elapsed}$). Notice however that there are minimum execution units that must be completed. For example, a generation of the parameter-less ECGA cannot be left half done. Likewise, a NAHC search cannot be interrupted in the middle. Therefore, care must be taken to ensure that both methods receive approximately the same number of evaluations and closest as possible from the defined value. For our experiments we used $fe_{elapsed} = 500$. The ideal $fe_{elapsed}$ would be equal to one, since the computational cost of changing between methods is minimal. However, in practice, it will never happen because of the minimal execution units of the ILS and ECGA. Since the main objective of this work is to propose a parameter-less search method, $fe_{elapsed}$ was fixed to a reasonable value.

Initially, ILS with adaptive perturbation runs during 500 function evaluations, plus the ones needed to finish the current NAHC search. Then, the parameter-less ECGA will run during another 500 evaluations, plus the ones needed to complete the current generation. And the process repeats *ad eternum* until the user is satisfied with the solution quality obtained or run out of time. This approach supplies robustness, small intervention from the user (just the fitness function needs to be specified), and good results in a broad class of problems.

4   Experiments

This section presents computer simulations on five test problems. These problems were carefully chosen to represent different types of problem difficulty. For each problem, the performance of ILS+ECGA algorithm is compared with other four search algorithms: the simple GA with "standard" parameters (SGA1), the simple GA with tuned parameters (SGA2), the ILS with adaptive perturbation alone (ILS), and the parameter-less ECGA alone (ECGA). For the GAs, we use binary encoding, tournament selection, and uniform crossover (except for ECGA). SGA1 represents a typical GA parameter configuration: population size $N = 100$, crossover probability $p_c = 0.6$, mutation probability $p_m = 0.001$, and selection pressure $s = 2$. SGA2 represents a tuned GA parameter configuration. For each problem, the GA parameters were tuned to obtain the best performance. Note that they aren't optimal parameters, but the best parameters found after a period of wise trials[1]. ILS and ECGA are tested alone to compare with ILS+ECGA and understand the advantages of running the two search models simultaneously.

For each problem, 20 independent runs were performed in order to get results with statistical significance. For each run, 2,000,000 function evaluations were allowed to be spent. For each algorithm, the mean and standard deviation of the number of function evaluations spent to find the target solution were calculated. For function optimization testing, each run was considered well succeeded if it found a solution with a function value $f(x_1, \ldots, x_n)$ in a given neighborhood of the optimal function value $f(x_{1opt}, \ldots, x_{nopt})$. The number of runs ($R_{ts}$) in which a target solution was found was also recorded. For each problem, all algorithms started with the same 20 seed numbers in order to avoid initialization (dis)advantages among algorithms.

4.1   Test Functions

The first problem is the onemax function, that simply returns the number of ones in a string. A string length of 100 bits is used. The optimal solution is the string with all ones. After some tuning, SGA2 was set with $N = 30$, $p_c = 0.9$, $p_m = 0.005$, and $s = 2$.

The second test function is the unimodal Himmelblau's function, defined as $f(x_1, x_2) = (x_1^2 + x_2 - 11)^2 + (x_1 + x_2^2 - 7)^2$. The search space considered is in the range $0 \leq x_1, x_2 \leq 6$, in which the function has a single minimum at $(3,2)$ with a function value equal to zero. Each variable x_i is encoded with 12 bits, totalizing a 24-bit chromosome. For a successful run, a solution with a function value smaller or equal to 0.001 must be found. After some tuning, SGA2 was set with the parameters $N = 100$, $p_c = 0.9$, $p_m = 0.01$, and $s = 2$.

The third function is the four-peaked Himmelblau's function, defined as $f(x_1, x_2) = (x_1^2 + x_2 - 11)^2 + (x_1 + x_2^2 - 7)^2 + 0.1(x_1 - 3)^2(x_2 - 2)^2$. This function is similar to the previous one, but the range is extended to $-6 \leq x_1, x_2 \leq 6$. Since the original Himmelblau's function has four minima in this range (one in each quadrant), the added term causes the point $(3,2)$ to be global minimum. Each variable x_i is encoded with 13 bits, giving a chromosome with 26 bits. Once more, a run is considered successful if the function value is within 0.001 of the global optima. The SGA2 uses $N = 200$, $p_c = 0.5$, $p_m = 1/l$, and $s = 4$.

---

[1] These trials were based on the work of Deb & Agrawal [23], since they used the same test functions. For each trial, 5 runs were performed to get some statistical significance.

Table 1. Mean and standard deviation of the number of function evaluations spent to find the target solution for the tested problems. The number of runs (R_{ts}) in which a target solution was found was also recorded.

		SGA1	SGA2	ECGA	ILS	ILS+ECGA
Onemax	mean	2,990	1,256	13,735	451	451
	std. dev.	±189	±258	±5,371	±65	±65
	R_{ts}	20	20	20	20	20+0
Unimodal	mean	2,019	1,750	3,731	1,400	3,174
Himmelblau	std. dev.	±790	±497	±2,290	±1,385	±2,766
	R_{ts}	16	20	20	20	14+6
Four-peaked	mean	2,414	2,850	5,205	2,593	4,990
Himmelblau	std. dev.	±750	±668	±2,725	±3,002	±3,432
	R_{ts}	14	20	20	20	12+8
10-variable	mean	1,555,300	570,000	149,635	>2,000,000	275,170
Rastrigin	std. dev.	±306,600	±87,240	±85,608	—	±87,472
	R_{ts}	3	20	20	0	0+20
Bounded	mean	—	741,000	15,388	>2,000,000	31,870
Deceptive	std. dev.	—	±95,416	±3,417	—	±15,306
	R_{ts}	0	20	20	0	0+20

The fourth function tested is the 10-variable Rastrigin's function. This is a massively multimodal function, known to be difficult to any search algorithm. It is defined as $f(x_1, \ldots, x_{10}) = 100 + \sum_{i=1}^{10} x_i^2 - 10 \cos(2\pi x_i)$, being each variable defined in the range $-6 \leq x_i \leq 6$. This function has a global minimum at $(0,0,\ldots,0)$ with a function value equal to zero. There are a total of 13^{10} minima, of which 2^{10} are close to the global minimum. A solution with a function value smaller or equal to 0.01 is considered a target solution. For best performance, SGA2 was set to $N = 10,000$, $p_c = 0.9$, $p_m = 1/l$, and $s = 8$.

The fifth and last problem is a bounded deceptive function, that results from the concatenation of 10 copies of a 4-bit trap function. In a 4-bit trap function the fitness value depends on the number of ones (u) in a 4-bit string. If $u \leq 3$, the fitness is $3 - u$, if $u = 4$, the fitness is equal to 4. The overall fitness is the sum of the 10 independent sub-function values. For such a problem, the SGA is only able to mix the building blocks with very large population sizes. To assure that we find the optimal solution in all 20 runs, the SGA2 was set with $N = 60,000$, $p_c = 0.5$, $p_m = 0$, and $s = 4$.

4.2   Results

The results obtained can be seen in Table 1. The growing difficulty of the five tested problems can be verified by the number of runs (R_{ts}) in which algorithms found a target solution, and by the number of function evaluations needed to do so. For the onemax problem, all the algorithms found the target solution in the 20 runs. Both ILS and ILS+ECGA got the same (and the best) results. This can be explained because the ILS is the first search method to run (500 function evaluations) in the ILS+ECGA framework. Taking into account that both algorithms used the same seed numbers, it was expected that they did similar since the NAHC always returned the optimal solution in the first time that it was solicited. This eventually happens because the problem is linear in Hamming space. In fact, that's the reason why mutation-based algorithms outperformed the rest of the algorithms for this problem. For the remaining problems, the SGA1 (with "standard" parameters) couldn't find a satisfiable solution in all runs. Although SGA1 performed well for the Himmelblau's functions, it wasn't robust enough to achieve a target solution in all runs. For the 10-variable Rastrigin's function, SGA1 found only 3 good solutions, and for the deceptive function, the "standard" parameter configuration failed completely, converging always to sub-optimal solutions. These results confirmed that setting these parameters to all kind of problems is a mistake.

For the Himmelblau's functions, SGA2 and ILS obtained the best results, taking half of the evaluations spent by ECGA. The ILS+ECGA algorithm, mostly due to the ILS performance, obtained a reasonable performance. Note that ILS was the algorithm responsible for getting a good solution in 14 and 12 (in a total of 20) runs, for unimodal and four-peaked Himmelblau's functions, respectively.

For the 10-variable Rastrigin's function, a different scenario occurred. ILS failed all the attempts to find a satisfiable solution. This is not a surprising result, since search algorithms based on local search don't do well in massively multimodal functions. Remember that some of the components (NAHC and adaptive perturbation scheme) of our ILS implementation were chosen in order to solve linear, non-correlated, or mutation-tailed problems in a quick and reliable way. For other kind of problems, parameter-less ECGA performance is quite good, making ILS+ECGA a robust and easy-to-use search algorithm. The ECGA was the best algorithm for this problem, and because of it, ILS+ECGA got the second best result, taking half of the evaluations of the SGA2.

For the bounded deceptive problem, SGA1 (converged to sub-optimal solutions) and ILS (spent all of the 2,000,000 evaluations available) didn't find the optimal solution. For this problem, the real power of ECGA could be verified. SGA2 took almost 50 more times function evaluations than ECGA to find the best solution in all runs. Taking advantage of the ECGA performance, ILS+ECGA was the second best algorithm, finding the target solution in 2 times more evaluations than the ECGA alone (as expected).

5   Extensions

There are a number of extensions that can be done based on this work:

- investigate other workload strategies;
- investigate interactions between the two search methods;
- investigate how other algorithms perform in the framework.

For many problems, the internal mechanisms needed by the ECGA to build the MPM may contribute to a significant fraction of the total execution time. Therefore, it makes sense (and it's more fair) to divide the workload between the two methods based on total execution time rather than on fitness function evaluations. Another aspect is to investigate interactions between the two methods. How much beneficial is it to insert one (or more) ILS local optimal solution(s) in one (or more) population(s) of the parameter-less ECGA? What about the reverse situation? Finally, other algorithm instances such as BOA could be used instead of the ECGA, as well as other concrete ILS implementation.

We are currently exploring some of these extensions.

6 Summary and Conclusions

This paper presented a concrete implementation of the proposed parameter-less optimization framework that eliminates the need of specifying the configuration parameters, and combines population-based search with iterated local search in a novel way. The user just needs to specify the fitness function in order to achieve good solutions for the optimization problem.

Although the combination might not perform as well as the best algorithm for a specific problem, it is more robust than either method alone, working reasonably well on problems with different characteristics.

Acknowledgments. This work was sponsored by FCT/MCES under grants POSI/SRI/42065/2001 and POCTI/MGS/37970/2001.

References

1. Harik, G.R.: Linkage learning via probabilistic modeling in the ECGA. IlliGAL Report No. 99010, Illinois Genetic Algorithms Laboratory, University of Illinois at Urbana-Champaign, Urbana, IL (1999)
2. Pelikan, M., Goldberg, D.E., Cant Paz, E.: BOA: The Bayesian Optimization Algorithm. In Banzhaf, W., et al., eds.: Proceedings of the Genetic and Evolutionary Computation Conference GECCO-99, San Francisco, CA, Morgan Kaufmann (1999) 525–532
3. Spears, W.M.: Crossover or mutation? In Whitley, L.D., ed.: Foundations of Genetic Algorithms 2. Morgan Kaufmann, San Mateo, CA (1993) 221–237
4. De Jong, K.A.: An analysis of the behavior of a class of genetic adaptive systems. PhD thesis, University of Michigan, Ann Arbor (1975)
5. Grefenstette, J.J.: Optimization of control parameters for genetic algorithms. In Sage, A.P., ed.: IEEE Transactions on Systems, Man, and Cybernetics. Volume SMC–16(1),. IEEE, New York (1986) 122–128
6. Schaffer, J.D., Caruana, R.A., Eshelman, L.J., Das, R.: A study of control parameters affecting online performance of genetic algorithms for function optimization. In Schaffer, J.D., ed.: Proceedings of the Third International Conference on Genetic Algorithms, San Mateo, CA, Morgan Kaufman (1989) 51–60
7. Goldberg, D.E., Deb, K.: A comparative analysis of selection schemes used in genetic algorithms. Proceedings of the First Workshop on Foundations of Genetic Algorithms 1 (1991) 69–93 (Also TCGA Report 90007).

8. Goldberg, D.E., Deb, K., Clark, J.H.: Genetic algorithms, noise, and the sizing of populations. Complex Systems **6** (1992) 333–362

9. Harik, G., Cant Paz, E., Goldberg, D.E., Miller, B.L.: The gambler's ruin problem, genetic algorithms, and the sizing of populations. In: Proceedings of the International Conference on Evolutionary Computation 1997 (ICEC '97), Piscataway, NJ, IEEE Press (1997) 7–12

10. Mühlenbein, H.: How genetic algorithms really work: I.Mutation and Hillclimbing. In Männer, R., Manderick, B., eds.: Parallel Problem Solving from Nature 2, Amsterdam, The Netherlands, Elsevier Science (1992) 15–25

11. Bäck, T.: Optimal mutation rates in genetic search. In: Proceedings of the Fifth International Conference on Genetic Algorithms. (1993) 2–8

12. Goldberg, D.E., Deb, K., Thierens, D.: Toward a better understanding of mixing in genetic algorithms. Journal of the Society of Instrument and Control Engineers **32** (1993) 10–16

13. Thierens, D., Goldberg, D.E.: Mixing in genetic algorithms. In: Proceedings of the Fifth International Conference on Genetic Algorithms. (1993) 38–45

14. Eiben, A.E., Hintering, R., Michalewicz, Z.: Parameter Control in Evolutionary Algorithms. IEEE Transactions on Evolutionary Computation **3** (1999) 124–141

15. Davis, L.: Adapting operator probabilities in genetic algorithms. In Schaffer, J.D., ed.: Proceedings of the Third International Conference on Genetic Algorithms, San Mateo, CA, Morgan Kaufman (1989) 61–69

16. Julstrom, B.A.: What have you done for me lately? Adapting operator probabilities in a steady-state genetic algorithm. In Eshelman, L., ed.: Proceedings of the Sixth International Conference on Genetic Algorithms, San Francisco, CA, Morgan Kaufmann (1995) 81–87

17. Smith, R.E., Smuda, E.: Adaptively resizing populations: Algorithm, analysis, and first results. Complex Systems **9** (1995) 47–72

18. Bäck, T., Schwefel, H.P.: Evolution strategies I: Variants and their computational implementation. In Winter, et al., eds.: Genetic Algorithms in Engineering and Computer Science. John Wiley and Sons, Chichester (1995) 111–126

19. Harik, G.R., Lobo, F.G.: A parameter-less genetic algorithm. In Banzhaf, W., et al., eds.: Proceedings of the Genetic and Evolutionary Computation Conference GECCO-99, San Francisco, CA, Morgan Kaufmann (1999) 258–265

20. Holland, J.H.: Adaptation in Natural and Artificial Systems. University of Michigan Press, Ann Arbor, MI (1975)

21. Lourenço, H.R., Martin, O., Stützle, T.: Iterated local search. In Glover, F., Kochenberger, G., eds.: Handbook of Metaheuristics, Norwell, MA, Kluwer Academic Publishers (2002) 321–353

22. Moscato, P.: On evolution, search, optimization, genetic algorithms and martial arts: Towards memetic algorithms. Technical Report C3P 826, Caltech Concurrent Computation Program, California Institute of Technology, Pasadena, CA (1989)

23. Deb, K., Agrawal, S.: Understanding interactions among genetic algorithm parameters. In Banzhaf, W., Reeves, C., eds.: Foundations of Genetic Algorithms 5 (FOGA'98), Amsterdam, Morgan Kaufmann, San Mateo CA, 1999 (1998) 265–286

Comparing Search Algorithms for the Temperature Inversion Problem

Monte Lunacek, Darrell Whitley, Philip Gabriel, and Graeme Stephens

Colorado State University
Fort Collins, Colorado 80523 USA

Abstract. Several inverse problems exist in the atmospheric sciences that are computationally costly when using traditional gradient based methods. Unfortunately, many standard evolutionary algorithms do not perform well on these problems. This paper investigates why the temperature inversion problem is so difficult for heuristic search. We show that algorithms imposing smoothness constraints find more competitive solutions. Additionally, a new algorithm is presented that rapidly finds approximate solutions.

1 Introduction

There are a number of problems in the atmospheric sciences where forward models are used to map a set of atmospheric properties to a set of observations.

$$\textbf{MODEL}(\textbf{Atmospheric.properties}) \longrightarrow \textbf{Observations}$$

What is actually needed is the inverse: given the observed data, what atmospheric properties produced those observations? Typically, observations are noisy. In many cases, it is necessary to solve these inverse problems in real-time. For example, in several satellite missions, it is necessary to solve these inverse problems several times a second in order to keep up with data collection.

Traditional gradient based methods can be used, but such methods are computationally costly [1]. It would seem that these problems are perfect candidates for heuristic search methods. However, we have found that well-known, well-tested evolutionary algorithms and local search methods applied to inversion problems do not always yield acceptable solutions.

This paper describes the temperature inversion problem that is central to the retrieval of water vapor profiles. These profiles are used in global atmospheric circulation and weather prediction models. Every set of observations that is collected results in a new temperature inversion problem that must be solved. Results are formally presented for evolution strategies, the CHC algorithm, and a local search bit climber. A number of algorithms have been been applied to the temperature inversion problem on a more limited basis, including Population-Based Incremental Learning, or PBIL [2], and Differential Evolution [3]. All of these algorithms fail in similar ways.

K. Deb et al. (Eds.): GECCO 2004, LNCS 3102, pp. 1340–1351, 2004.

This paper also looks at why the temperature inversion problem is difficult for evolutionary algorithms and local search methods. While the problem is nonlinear, the 2-D slices are smooth and uniformly unimodal. However, there are ridges in the fitness landscape that can induce false local minima. There are also biases in the evaluation functions so that some parameters (i.e., estimated temperatures) exert a larger effect on the evaluation function than others.

An algorithm proposed by Salomon is tested that exploits known properties of temperature profiles and produces useful results [4]. Finally a new algorithm called "Tube Search" is developed and tested. It ignores bias in the evaluation function, and uses smoothness constraints to avoid ridge problems. It quickly produces good approximate solutions.

2   Background

Atmospheric sciences researchers use a *forward model* that relates vertical temperature profiles to observed measurements. The forward model, as described in this paper, generates 2000 radiance measurements (observations) given a 43 dimensional temperature profile. The parameter indexed by $(44 - k)$ in the profile is the estimated temperature at an altitude of approximately k kilometers in the atmosphere: the parameters are enumerated in reverse order, and the spacing is somewhat greater at higher altitudes. We actually want to solve the inverse problem: given a set of observations, what is the corresponding temperature profile? In practice, radiance measurements from a constellation of satellites are used in an inverse radiative transfer model. Examples of extant observing systems are: Operational Vertical Sounder (TOVS), the Special Sensor Microwave Imager (SSM/I), and the Advanced Microwave Sounder Unit (AMSU). The inverse solution must be accurate and *fast*; measurements are often collected at a high spatial resolution from satellites whose orbital period is about 90 minutes (or moving at about 8 km/sec).

The forward model is the simplified form of the equation of radiative transfer that does not account for the presence of clouds. The equation of transfer is solved for the radiances at different wavelengths observed at the top of the atmosphere. This model is "plane parallel" (e.g., with no horizontal variations in its properties). Radiances are calculated at a viewing angle θ as:

$$I_{(\tau,\mu)} = B_\nu(T_s)e^{-\tau_s/\mu} + \int_0^\nu B_\nu(T)e^{-\tau/\mu}\mu^{-1}d\tau$$

$$\text{where } I_{(\tau,\mu)} = \text{radiance}$$

$$\mu = cos(\theta)$$

$$\tau = \text{optical depth}$$

$$s = \text{surface}$$

$$B_\nu(T) = \text{Planck radiance for temperature T.}$$

An analytical inversion of this model is impossible because radiances are non-linearly related to the temperature profiles. Alternatively, the inverse temperature model can be formulated as an optimization problem, where the target temperature profile is the global optimum of the search space. Specifically, the objective function is the root mean squared error between the observable measurements, and the output of the forward model at any point in the search space.

First order derivatives can be calculated analytically for the temperature inversion problem [1]. When clouds or aerosols are present, the analytical calculation of these derivatives is impossible. In the simple model where only blue sky exists, success has been achieved using *Newtonian iteration* and a good starting guess. However, achieving a quadratic convergence rate for solutions near the optimum is highly dependent on a good a priori guess of the temperature profile, and search is still very costly. We attempt to improve computational efficiency by solving the inverse problem using non-derivative search methods.

3   Evolutionary Algorithms and Local Search

One of the more successful variants of genetic algorithms is CHC [5]. CHC uses a bit representation. In this study, the standard binary reflected Gray code is used. CHC uses *cross generational* selection: newly created offspring must compete with the parent population for survival. Parents are not allowed to cross unless they are sufficiently different. CHC uses a modified version of uniform crossover, where half of the non-matching bits are exchanged. No mutation is used (note that uniform crossover already randomly assigns non-matching bits). CHC also includes a restart mechanism that reinitializes the population by randomly flipping 35% of the bits of the best individual.

Evolution strategies emphasize mutations over recombination. Individuals are represented as real valued vectors. Each individual modifies its parameters to produce offspring. Depending on the implementation, there can be several parents in the population, and each can generate one or more offspring. If many offspring are generated, selection is used to keep each generation the same size. In a (μ, λ) selection strategy, the new population is chosen only from the offspring. An elitist strategy, on the other hand, selects the next generation from both the parents and the offspring. This is known as a $(\mu + \lambda)$ selection strategy. Mutation is usually performed based on a distribution around the individual undergoing mutation. A global distribution can be used for all individuals, or each individual may maintain its own distribution, σ, often interpreted as a step size. Self-adaptive strategies allow the angle of mutation to change. Correlated mutations attempt to estimate the covariance for each pair of object parameters. In other words, an n dimensional problem requires $n(n-1)/2$ rotation parameters, in addition to the n object parameters, and n step size multipliers, σ_i.

Local search encompasses a broad range of algorithms that search from a current state, moving only if new states improve objective fitness. This has proven to be a simple, yet often effective search method. In this paper, local search refers to a Gray coded *steepest ascent bit climber*. Each parameter is encoded as a Gray bit string and, by flipping one bit at a time, a neighborhood pattern forms around the current best solution. Local search evaluates all these neighborhood points before taking the best, or *steepest*, step. Because each neighbor

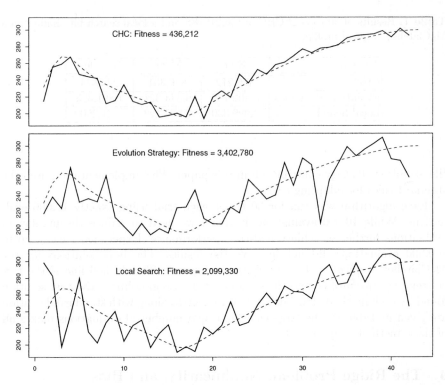

Fig. 1. The best solutions in 30 trials on a McClatchey *tropical* profile. The dashed line indicates the target *tropical* profile, and the solid black line is the best solution found by each algorithm. Even CHC's dominating performance finds a disappointing "zig-zag" solution. None of the solutions finds a useful temperature profile.

differs from the current best by only one dimension, the neighborhood forms a coordinate pattern. Local search terminates when no improving move is found.

Empirical Results

In order to evaluate the various search algorithms on the temperature inversion problem, we used five, well-known, McClatchey temperature profiles [6], which represent conditions ranging from subarctic winter to tropical summer.

The range of the temperatures is $(190, 310)$ Kelvin, a difference of 120. In order to represent this with integer precision, seven bits are needed ($2^7 = 128$). CHC and local search used a Gray encoding scheme. A population size of 50 was used for CHC. For evolution strategies, the high dimension space means that using the correlated mutations model would require $43(42)/2 = 903$ rotation parameters. This much additional overhead is impractical; rotations were not used. Bäck and Schwefel [7] recommend a (μ, λ) selection strategy and indicate that the ideal ratio of parents and offspring is $\mu/\lambda = 1/7$. The (30,210)ES we tested on the temperature problem outperformed the (30+210)ES and is reported as

Table 1. Results of 30 runs of CHC, a (30,210)ES and a local search bit climber on a McClatchey *tropical* profile.

Algorithm	Best	Mean	Std Dev
CHC	436,212	850,381	226,674
Evolution Strategies	3,402,780	6,344,321	1,891,605
Local Search	2,099,330	2,886,128	621,260

the evolution strategy contribution in this paper. This implementation used the standard rules for adapting σ [7].

Each algorithm was run for 30 trials, each trial using exactly 10,000 evaluations. While 10,000 evaluations is small, we need to reduce the number of evaluations further to achieve real-time performance. Experiments using up to 100,000 evaluations did not improve the results. The best solutions for the McClatchey *tropical* profile are shown in figure 1. The dashed line is the target temperature profile, and the solid black, zig-zagging line is the best solution found by each method. The best and average error along with standard deviation are given in Table 1. The large error and high number of evaluations makes all of these methods impractical.

4   The Ridge Problem, Nonlinearity, and Bias

What makes the temperature inversion problem hard? Without question, the nonlinearity of the problem plays a major role. Specifically, changing parameter k in the temperature profile changes the error surface almost everywhere in the space. An incorrect temperature at location k makes it impossible to correctly assign temperature at other locations. In future work, we may be able to modify the evaluation function to localize the nonlinear effects, since the atmosphere should display physical locality.

Additionally, there seems to be two other major factors. One problem is bias in the evaluation function. The other problem is ridges in the landscape.

Starting from a globally optimal solution, we varied each parameter by $+/-$ 2.0. Every move increases the objective error, which is zero when no change is applied. Figure 2 shows the average of the two numbers over the range of the temperature problem. The upper dimensions have greater influence on the error value returned by the evaluation function. The parameters that offer the greatest opportunity to reduce the error will be in the upper dimensions. The bias can cause search algorithms to fit the upper dimensions of the temperature profile first–and to potentially assign incorrect values to the temperature parameters in the lower atmosphere.

Perhaps the most serious problem is that there are *ridges* in the search space. Figure 3 shows several representative 2-D slices of the search space. Although each slice is smooth and unimodal, the curved ridge that cuts through each slice can cause search to become stuck.

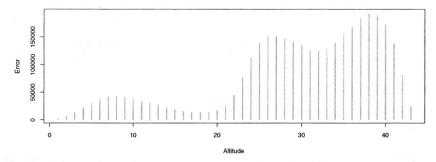

Fig. 2. The average error profile near the optimal solution. Higher dimension parameters contribute more to the error profile.

Rosenbrock was among the first to notice that search methods, including derivative-based methods such as *steepest descent*, are crippled by ridge features [8]. Winston also notes that ridges cause problems for simple hill climbers [9]. A ridge can cause a search algorithm to believe it has found a local optima, when, in reality, the algorithm is simply stuck on the ridge. Even when an algorithm is not stuck, convergence can be slowed dramatically.

The ridge problem involves two factors: precision and search direction. If an algorithm looks for improving moves by changing only one dimension at a time (in a coordinate pattern), it will not see better points that fall between the neighborhood axis. This is the *direction* problem. Instead, the search will find improvements close to the current best solution that lie on or near the ridge. Precision dictates how close an algorithm looks for improving neighbors. If the ridge is very steep and narrow, higher precision will be needed to find an improving move.

Increasing the precision generally decreases the number of false optima. A lower precision search will get stuck on a ridge, blindly assuming it has found a local optima. Increasing precision allows more improving moves to be found, but it forces search algorithms to take smaller steps and move very slowly through the landscape. This causes an increase in evaluations and a much slower convergence.

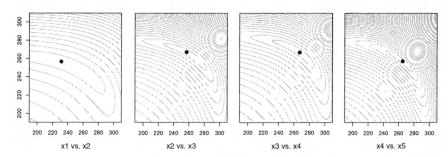

Fig. 3. Two dimensional contours of the first five parameters in the temperature problem. The black dots represent the optimal solution.

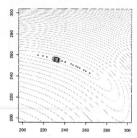 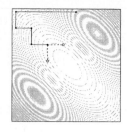 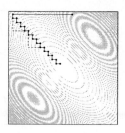

Fig. 4. In the leftmost graph, a high precision local search (large circles) finds the global optima, whereas the low precision search gets stuck in local optima (black dots). In the middle graph, a low precision search induces local optima on a simple parabolic ridge because all the neighbors (dashed lines) have poorer evaluation. The higher precision search (rightmost) is able to make more progress, but at the expense of significantly more evaluations.

This phenomena is called *creeping*. Figure 4 graphically explains this problem on a simple parabolic ridge and also documents the existence of this problem on the first two dimensions of the temperature problem. The higher precision search is able to move along the ridge and find a better solution. Low precision induces false optima.

Local search uses a coordinate pattern to search for a globally competitive solution. Therefore, local search performs poorly in the presence of ridges.

Salomon [10] showed that ridges can be created by rotating common benchmark problems. Salomon also points out that the performance of evolution strategies are invariant with respect to a rotation of the coordinate systems. Mutations can move in any direction, and multiple parameters normally change. This implies that offspring will not be reproduced on the coordinate axes.

Salomon contrasts this with the Breeder Genetic Algorithm (BGA). On common benchmarks, if the coordinate system is rotated in the n–dimensional space, the breeder genetic algorithm often fails. The reason for this failure is largely due to the low probability that a parameter is modified under mutation (commonly $1/l$, where l is the chromosome length). More specifically, the probability that two or more parameters change simultaneously is small. When a ridge runs through a space that is offset from the coordinate axis, it is necessary for all the parameters that align with the ridge to change. The conclusions drawn by Salomon indicate that "crossover's niche" is quite small, and not suitable for problems that have ridges.

The limitations of the Breeder Genetic Algorithm do not extend to all genetic algorithms that use crossover. CHC uses a variation of uniform crossover that changes many parameters at once. Nevertheless, CHC does use a fixed coordinate system. Our results indicate that, in fact, CHC performs better than evolution strategies on the temperature inversion problem.

Salomon suggests that evolution strategies are impervious to the ridge problem because they are invariant to rotations of the search space. However, Oyman et al. [11] define conditions on a simple parabolic ridge where the elitist ES *limps*, or *creeps*. The problem occurred using a $(1 + 10)$ES where a single parent produces ten offspring; the best offspring replaces the parent if an improvement is found. The "1/5 rule" was used, which means that the step size is adjusted to produce an improving move one out of every five tries. When the parent encounters a ridge, the step size will decrease because of this rule. After reaching the ridge, it is difficult for the evolution strategy to re-adapt its step size and follow the ridge. Thus, evolution strategies can also *creep*.

5   Optimize and Refine

Salomon [4] also notes that some search algorithms produce results that "zig-zag" the actual solution when the desired solutions displays physical smoothness. Salomon suggests an *optimize and refine* evolution strategy.

The *optimize and refine* technique was inspired by manufacturing methods: many products start with a rough approximation that is refined to be more smooth. The smooth target profile of the temperature inversion problem may be tackled in the same way. The procedure starts by approximating the target with a linear fit. The endpoints, x_1 and x_{43}, are searched for the position where linear interpolation minimizes the objective error. Refinement reduces the regions by half, and the solution becomes a piecewise linear approximation. For example, the next iteration would increase the dimensionality from two to three by adding the point x_{20}. This two piece linear approximation is optimized before more points are added in the next refinement phase.

This method is efficient in several ways. First, a close approximation to the target is found by searching small landscapes. In the temperature inversion problem, a linear approximation reduces the dimensionality of the search space from 43 to only two. This gives higher dimensional searches a good place to start. Second, it forces a smoothness constraint on the problem. Neighboring points in the domain are forced to be relatively close in the range.

Salomon used a $(1,6)$ evolution strategy, where a single parent produces six offspring, and uses a non-elitist selection strategy. Instead, in the optimize procedure, we implemented a simple *binary search* to locate the minimum at each inflection point of the piecewise linear solution. The search started at the endpoints, x_1 and x_{43}. The *binary search* moved to the optimum in each dimension for several iterations until no improvement could be found. Then, the point x_{20} was added to break the linear region in two, and the optimize procedure was repeated, this time with three points instead of two. At each step, the regions, defined by the current set of points, were cut in half after they had been fully optimized. Figure 5 shows this procedure for the McClatchey *subarctic summer* profile. Although this method shows promise, it is able to fit some data examples better than others. Sometimes the solution still "zig-zags" the target. This method also struggles to fit the ends of the profile.

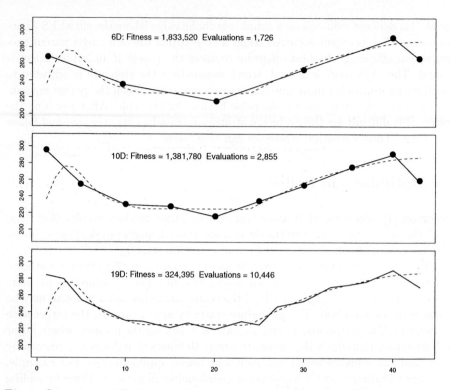

Fig. 5. Optimize and Refine. A fully convergent *subarctic summer* solution required an average of 10,446 evaluations. Although this solution fits the profile better than previous methods, it still wanders the target solution.

6   A New Algorithm: Tube Search

We know that the target temperature profile we are trying to retrieve is relatively smooth, a constraint that is exploited by the *optimize and refine* algorithm.

We implemented a new algorithm called *tube search*. Like optimize and refine, tube search starts with a linear fit. This provides a consistent starting point that is smooth, a quality we hope to retain throughout the search. Once the linear fit has been determined, tube search begins. A fixed step is taken on either side of the linear fit – in effect defining a tube about that solution – and the change in evaluation is recorded and stored in a vector. Some moves will offer improvement, while others will not. Once improving moves have been determined, a step of the same magnitude is taken in each improving dimension simultaneously. A three-parameter moving average is run on the solution every five iterations to maintain a smoothness. Each parameter, except the first and last two end points, is replaced by the average of itself and its two neighbors.

$$\text{temp}[i] = \frac{\text{temp}[i-1] + \text{temp}[i] + \text{temp}[i+1]}{3}$$

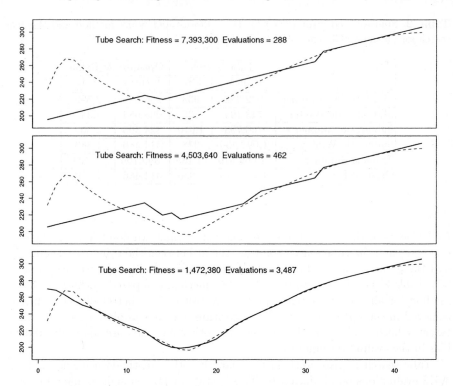

Fig. 6. Tube Search: The top two graphs show select iterations of the tube search. The bottom graph shows the final solution after 3,487 evaluations. Of all the profiles tested, this was the worst fit. The step distance for each parameter is exactly the same, so bias has no impact on tube search.

Figure 6 graphically explains the tube search and shows the final solution generated by searching the temperature problem. Note that 43*2 evaluations are needed to evaluate the moves defined by the tube. Given the small number of moves used by the tube search, the total number of evaluations is less than half of that used by the the optimize and refine algorithm.

The error values associated with the move forming the *tube* around the current best solution will drive the search toward better points while maintaining smoothness. Because all of the parameters change at once, tube search is not a simple coordinate search scheme. Additionally, when each step is taken the magnitude of the step is the same independent of the magnitude of the error. In this way, tube search ignores the bias in the evaluation function. Lower dimension parameters can change just as much as higher dimension parameters, even when they have a smaller contribution to the error.

Tube search works surprisingly well on all temperature profiles we have optimized. Oddly enough, the errors associated with the tube search solutions are not particularly low: the errors are generally much lower for optimize and refine. Even CHC achieves lower errors. However, if we compute a sum-squared error

Table 2. Sum-squared error (SSE) for the *optimize and refine* method and the *tube* search for all the McClatchey profiles we tested.

Profile	Tube Search		Optimize & Refine	
	Fitness	SSE	Fitness	SSE
Mid-latitude Summer	932,322	933	256,605	2,592
Mid-latitude Winter	743,194	738	298,684	3,342
Sub-arctic Summer	760,703	1,610	324,395	3,502
Sub-arctic Winter	1,092,430	348	314,486	1,383
Tropical Summer	1,664,570	1,189	399,106	1,423
Original Profile	1,472,380	1,950	314,486	1,370

(SSE) between the actual target temperature (which we don't have in the general case) and the tube search solution, the fit between the tube search solutions and the actual profile is better, on average, than is achieved with other methods. Table 2 shows the *optimize and refine* method compared to the *tube* search method for all the McClatchey profiles we tested. The better objective fitness achieved in the *optimize and refine* algorithm does not imply a closer fit to the target solution. This may be because the other methods are more affected by bias in the evaluation function.

Tube search is also much faster than the other methods using fewer than 3,612 evaluations on all data sets. This is still not fast enough to allow for real-time evaluation. However, tube search has another attractive feature. Each of the 86 evaluations required to evaluate the moves defined by the tube around the current best solution are independent and can be done in parallel. This would allow us to use parallelism to speed up Tube Search by a factor of 86. Parallel tube search could obtain a solution in the amount of time taken to do $3,612/86 = 42$ sequence evaluations. This is a major advantage given the goal of doing real-time temperature inversion.

7   Conclusions

Temperature inversion is a practical example of an optimization problem that has not been efficiently solved using derivative-based search methods. Attempts to solve this problem using widely used evolutionary algorithms and local search methods produce poor results. Three algorithms were formally evaluated in this study, including CHC, a (30,210)ES and local search. We also applied PBIL and Differential Evolution to the temperature problem using 100,000 evaluations and the results were similarly poor. Methods that exploit the smoothness of the temperature profile are more effective and, in the case of the tube search, more efficient. Other types of smoothing, such as splines, may be a useful addition to the *tube search*, as well as other evolutionary algorithms.

The temperature inversion application highlights two difficulties that can cause a problem for optimization algorithms: bias and ridges. The *ridge problem* is relatively well documented in the mathematical literature on derivative-free minimization algorithms [8] [12]. The ridge problem seems to be largely unexplored in the genetic algorithm community, but has received attention in the evolution strategies community [10] [11] [13]. Recently, we have begun looking at the Covariance Matrix Adaptation method [14] [15] for rotating the representation space; on test functions it is highly effective, but it has not been tested on the temperature inversion problem.

Acknowledgments. This work was supported by National Science Foundation grant IIS-0117209.

References

1. R.J. Englen, A.S. Denning, K. Gurney, and G.L. Stephens. Global observations of the carbon budget: I. expected satellite capabilities for emission spectroscopy in the eos and npoess eras. *J. of Geophysical Research*, 106:20,055–20,068, 2001.
2. S. Baluja and R. Caruana. Removing the genetics from the standard genetic algorithm. *The Int. Conf. on Machine Learning*, 38–46, 1995. Morgan Kaufmann.
3. R. Storn and K. Price. Differential evolution - a simple and efficient adaptive scheme for global optimization over continuous spaces. *Journal of Global Optimization*, 11:341 – 359, 1997.
4. R. Salomon. Applying evolutionary algorithms to real-world-inspired problems with physical smoothness constraints. *Proceedings Congress on Evolutionary Computation*, 2:921–928, 1999. IEEE Press.
5. L. J. Eshelman. The CHC adaptive search algorithm. *Foundations of Genetic Algorithms*, 265–283. Morgan Kaufmann, 1991.
6. R.A. McClatchey, R.W. Senn, J.E.A. Feldy, S.E. Voltz, and J.S. Garing. Optical properties of the atmosphere. Technical Report TR-354, AFCRL, 1971.
7. T. Back and H.-P. Schwefel. An overview of evolutionary algorithms for parameter optimization. *Evolutionary Computation*, 1(1):1–23, 1993.
8. H.H. Rosenbrock. An automatic method for finding the greatest or least value of a function. *Computer Journal*, 3:175–184, 1960.
9. P. Winston. *Artifical Intelligence (2nd ed.)*. Addison-Wesley, 1984.
10. R. Salomon. Re-evaluating genetic algorithm performance under coordinate rotation of benchmark functions. *BioSystems*, 39:263–278, 1996.
11. A. I. Oyman, H. Beyer, and H. Schwefel. Where elitists start limping evolution strategies at ridge functions. *Parallel Problem Solving from Nature – PPSN V*, 34–43, 1998. Springer.
12. R. P. Brent. *Algorithms for Minimization Without Derivatives*. Prentice-Hall, New Jersey, 2002.
13. D. Yuret and M. Maza. Dynamic hillclimbing. *Second Turkish Symposium on Artificial Intelligence and Neural Networks*, 208–212, 1993.
14. Nikolaus Hansen and Andreas Ostermeier. Completely derandomized self-adaptation in evolution strategies. *Evolutionary Computation*, 9(2):159–195, 2001.
15. N. Hansen, S. Müller and P. Koumoutsakos. Reducing the Time Complexity of the Derandomized Evolution Strategy with Covariance Matrix Adaptation (CMA-ES). *Evolutionary Computation*, 11(1):1–18, 2003.

Inequality's Arrow: The Role of Greed and Order in Genetic Algorithms

Anil Menon

ProductSoft, Inc.
10707 Bailey Drive
Cheltenham, MD 20623
anilm@acm.org

Abstract. Moderated greedy search is based on the idea that it is helpful for greedy search algorithms to make non-optimal choices "once in a while." This notion can be made precise by using the majorization-theoretic approach to greedy algorithms. Majorization is the study of pre-orderings induced by doubly stochastic matrices. A majorization operator when applied to a distribution makes it "less unequal," where inequality is defined with respect to a very wide class of measures known as Schur-convex functions. It is shown that proportional selection, point crossover and point mutations are all majorization operators. It is also shown that with respect to the majorization-theoretic definition, the standard GA is a moderated greedy algorithm. Some consequences of this result are discussed.

1 Introduction

Gordon Gecko, in his paean to greed in the movie *Wall Street*, makes several bold claims:

"Greed is good. Greed is right. Greed works. Greed clarifies, cuts through and captures the essence of the evolutionary spirit."

The questions as to whether greed is any good (efficiency questions), or whether it is right (sufficiency questions), or whether it works (optimality questions) are important topics in the theory of algorithms. But the focus of this paper is on Gecko's last claim, namely, to show that greed does indeed clarify and capture the essence of the evolutionary process.

The concept of a greedy algorithm can be studied in several different (but equivalent) formalisms: decision theory, greedoids, submodular functions and majorization theory [3,4,10]. In the first part of the paper, the majorization approach is briefly reviewed and then used to define the concept of moderated greed. In the second part of the paper, proportional selection, point crossover and point mutation are shown to be majorization operators, and this result used to demonstrate that the simple GA is a moderated greedy algorithm.

K. Deb et al. (Eds.): GECCO 2004, LNCS 3102, pp. 1352–1364, 2004.

2   Preliminaries

A square matrix is said to be column (row) stochastic if it is non-negative and its column (row) sums are unity. For any $x = (x_1, x_2, \ldots, x_n) \in R^n$ let $x_{[1]} \geq x_{[2]} \geq \ldots \geq x_{[n]}$ denote the components of x sorted in non-increasing order, and let $x_\downarrow \equiv (x_{[1]}, \ldots, x_{[n]})$. The following definition is central to this paper.

Definition 1 *(Lorenz Majorization)* [5, pp. 7] If $x, y \in R^n$ then, y is said to *majorize* x, denoted $x \preceq y$ (equivalently, $y \succeq x$) if the following conditions are satisfied:

$$\sum_{i=1}^{k} x_{[i]} \leq \sum_{i=1}^{k} y_{[i]} \quad \forall k = 1, \ldots, n-1, \quad \text{and} \quad \sum_{i=1}^{n} x_{[i]} = \sum_{i=1}^{n} y_{[i]}.$$

If at least one of the above inequalities is strict, then y is said to *strictly* majorize x, that is, $x \prec y$. ∎

A *pre-order* on a set is a binary relation that is reflexive and transitive. A *partial order* is a pre-order that is also anti-symmetric (if aRb and bRa then $a = b$). The '$\preceq$' relation is a pre-order on the set of real vectors. Finally, results marked "Proposition" are results cited from the works of other authors.

3   Greed and Inequality

The behavior of a search algorithm in real-valued, multivariable optimization problems may be visualized as movements in state space. This state space is essentially defined by the domain of $F(x)$, the function to be optimized; the algorithm's behavior is described by the sequence of real vectors $x(0), x(1), x(2), \ldots$ it generates in search of the optimal solution.

In greedy search, the state transition is always toward that state which provides the largest, most immediate gain. Specifically, at time t, the algorithm applies a scoring function to a list of candidate states $x_1(t+1), x_2(t+1), \ldots$ and selects ("moves to") that state with the largest score amongst the candidates. Quite commonly, the scoring function is nothing more than the values of $F(\cdot)$ on these candidate states $x_k(t+1)$. A scoring function represents a value judgement on what is considered preferable (desirable); in unmoderated greed, these preferences are typically held as fixed.

It can be shown that the state selection problem in greedy algorithms can be converted into a state construction problem; the new state $x(t+1)$ is obtained from a specific manipulation of $x(t)$'s components in what is known as an exchange transformation [10].

It is here that majorization theory enters the picture; the field originated more than a hundred years ago in the study of exchange transformations [5]. The following is an informal review of the key concepts.

It is useful to interpret the components of a vector $x(t) \in R^n$ as indicating the amounts "possessed" of some commodity (income, energy, proportion, scores, weights etc.) by n entities at time t. The exchanges that are of interest are those that transfer an amount ϵ *from* entity j *to* entity i such that three constraints are satisfied:

$$j \xrightarrow{\epsilon} i: \quad \epsilon > 0, \quad x_j(t) > x_i(t), \quad x_j(t+1) \geq x_i(t+1). \tag{1}$$

In short, non-zero amounts have to be transferred, the "richer" entity is the source of the amount, and the transfer cannot be so large that it reverses the original inequality between the two entities. For example, $(2, -1, 3) \to (1.5, -0.5, 3)$ is such a transformation because it can be interpreted as the transfer of an amount of $\epsilon = 0.5$ units from entity 1 to entity 2.

Depending on how $x_j(t+1)$ and $x_i(t+1)$ are related to $x_j(t), x_i(t)$ and ϵ, there are (at least) two ways in which the conditions in (1) can be satisfied:

$$x_j(t+1) = x_j(t) - \epsilon, \quad x_i(t+1) = x_i(t) + \epsilon. \tag{2}$$

Such exchanges were first studied by the economist Hugh Dalton in connection with income inequality distributions, and have come to be called Dalton exchanges [5, pp. 6].

Alternatively, proportional exchanges may be considered [10]:

$$x_j(t+1) = x_j(t) - \epsilon x_j(t), \quad x_i(t+1) = x_i(t) + \epsilon x_j(t). \tag{3}$$

We will focus on Dalton exchanges, because their relationship with evolutionary operators is particularly simple to demonstrate. The results obtained herein have analogues in the Parker-Ram exchange system as well.

Dalton exchanges are best expressed in terms of matrix transformations. Define the non-negative fraction $\lambda = \epsilon/(x_j - x_i)$. Then,

$$x_i(t+1) = x_i + \epsilon = (1 - \lambda)x_i + \lambda x_j \tag{4}$$
$$x_j(t+1) = x_j - \epsilon = \lambda x_i + (1 - \lambda)x_j. \tag{5}$$

Define the $n \times n$ matrix $T_\lambda(i,j)$, $0 \leq \lambda \leq 1$ by,

$$\begin{pmatrix}
1 \cdots & 0 & \cdots & 0 & \cdots 0 \\
\vdots & \vdots & & \vdots & \vdots \\
0 \cdots & 1 - \lambda \cdots & & \lambda & \cdots 0 \\
\vdots & \vdots & & \vdots & \vdots \\
0 \cdots & \lambda & \cdots & 1 - \lambda \cdots & 0 \\
\vdots & \vdots & & \vdots & \vdots \\
0 \cdots & 0 & \cdots & 0 & \cdots 1
\end{pmatrix}$$

Then the *T-transform* of a vector $\boldsymbol{x}$, defined for some $1 \leq i, j \leq n$ and $0 < \lambda < 1$ by $\boldsymbol{y} = T_\lambda(i,j)(\boldsymbol{x})$. A T-transform represents a *single* Dalton transfer between a pair of entities in the population. To extend the matrix formalism to multiple exchanges between different pairs of individuals, the expression $x' = T_\lambda(i,j)x$ has to be replaced by,

$$x' = Mx, \tag{6}$$

where the matrix M is a *doubly stochastic* matrix (that is, both column-stochastic and row-stochastic). To see this, it suffices to note that any doubly stochastic matrix can be written as a product of at most $(n-1)$ T-transforms [5], and that a T matrix is, by definition, a doubly stochastic matrix.

The following proposition [Hardy-Littlewood-Polya theorem] relates T-transforms, doubly stochastic matrices and Lorenz majorization (see Definition 1).

Proposition 1 [5, pp. 107] For two vectors $x, y \in R^n$ the following statements are equivalent:

1. $y \preceq x$.
2. There exists a doubly stochastic matrix M such that $y = Mx$.
3. y can be obtained from x by a finite number of T-transforms (Dalton exchanges).

It can be shown that the matrix M can always be chosen to be non-negative definite. ∎

Lorenz majorization is related to optimization problems through the concept of Schur-convex functions [5].

Definition 2 (Schur-Convexity) A function $F : R^n \rightarrow R$ is said to be Schur-convex, if $x, y \in R^n$ and $y \preceq x$ implies that $F(y) \leq F(x)$. If the inequalities listed above are strict then F is said to be *strictly* Schur-convex. A function F is said to be Schur-concave if $-F$ is Schur-convex. ∎

Schur-convex functions occupy a great deal of mathematical real estate; almost all diversity measures and many statistical functionals belong to this class of functions [5, pp. 115-128,139-168]. Their importance is also based on the fact that an ordering relation '$\preceq$' on vectors imposes an ordering on the values that a function takes at these vectors, that is, '$\preceq$' is *order-preserving*. It is for this reason that the study of inequalities was transformed by majorization theory.

Definition 2 in conjunction with Proposition 1 suggests that one way to obtain the maximum of a Schur-convex function is to find a vector x' that strictly majorizes the current state vector x, that is, $x \prec x'$. This implies $F(x) < F(x')$ and the process can be repeated till a boundary point of the domain is reached. Proposition 2 is the simplest example of the kind of optimality results achievable with the machinery of majorization and Schur functions.

Proposition 2 (Greedy Optimization) [11, Thm. 5.1] Let $C \subseteq R^n$, $G : C \rightarrow R$ be a Schur-convex function on C, and a, b be constant vectors. Then the optimization problem,

$$\text{Maximize} \quad G(x), \quad a \preceq x \preceq b, \ x \in C, \tag{7}$$

is *greedy-solvable*. In particular, there exists a vector $a \preceq x_o \preceq b$ such that a global optimum of $G(x)$ can be found by a finite number of iterative T-transforms on x_o. ∎

Lorenz majorization is the pre-order associated with the semigroup of doubly stochastic matrices (see Proposition 1). By considering the majorization pre-orders defined by other matrix semigroups (for example, lower triangular stochastic matrices, orthostochastic matrices etc.), it is possible to significantly generalize Proposition 2 [9,10].

4   Defining Moderated Greed

If the objective function is Schur-convex, then optimization is a relatively simple task (at least, in principle). If the objective function is to be maximized, the basic strategy would be to generate a sequence of feasible solutions $x(0), x(1), \ldots$ such that $x(t-1) \preceq x(t)$ (or $x(t) \preceq x(t-1)$ if the objective function is to be minimized).

If the function is *not* Schur-convex, then other majorization pre-orders may prove to be useful. But failing that, it is clear that an alternate approach is needed. One solution to making a greedy algorithm less myopic is to use greed in a more moderate manner. For example, since monotonicity in objective functions values is a hallmark of the greedy strategy, one relaxation could be to allow movements to states that could potentially *decrease* the value of the objective function.

Unfortunately, there is no universally accepted notion of moderated greed; techniques like simulated annealing and randomized gradient descent are suggestive of what is meant by the concept, but they are not definitive examples. The main difficulty in defining moderated greed lies in restricting the scope of the definition. For example, if moderated greed is defined as an algorithm that is "occasionally" greedy, then any random search algorithm which makes at least one greedy step is eligible as a candidate. At the other extreme, a purely greedy algorithm would also be eligible. There are other problems. Is being "less greedy" to be interpreted as investing in gathering more landscape information, forecasting and better scoring functions? Is there a continuum of greed based on how far ahead an algorithm looks into the consequences of its choices?

Finally, is the concept to be defined probabilistically? If an algorithm undertakes a greedy move based on the toss of a (biased) die, then does that constitute being greedy in a moderate way, or does it merely compound two vices — greed and gambling — into one?

It is helpful to consider the problem from a slightly more general angle. In Rational Choice theory, the *choice set* $C(S)$ is a subset of a set of alternatives S such that $a_i \in C(S)$ implies that there is no other alternative $a_j \in S$ *strictly preferred* to a_i. Here, "preference" is a pre-order on the set of alternatives. Abstract rationality is rooted in the idea that an agent has to choose the most preferred option from a list of options. A greedy algorithm is abstractly rational in that it invariably picks the highest scored alternative from a list of alternatives (scores are assumed to reflect preferences). The problem, however, is that it never varies its preferences.

Optimization is the art of the possible in that it requires the balancing of tradeoffs: quality with running times, storage requirements with CPU cycles, exploration with exploitation et cetera. The trouble with (unmoderated) greedy approaches is that they are myopic; they are heavily biased towards a particular aspect of each one of these tradeoffs. Ideally, moderated greed should be based on the idea that no single handle of a tradeoff dominates the algorithm's behavior. The choices a moderated greedy algorithm makes is always greedy, but not necessarily with respect to the same preference orderings.

This need to balance tradeoffs meshes nicely with the fact that there are two types of majorization processes. Let $\mathcal{F}$ be a vector valued operator such that $x(t+1) = \mathcal{F}(x(t))$. If $x(t) \preceq x(t+1)$, then $\mathcal{F}$ is said to be a *contractive* majorization operator, and the sequence, a contractive sequence. On the other hand, if $x(t+1) \preceq x(t)$, then $\mathcal{F}$ is said to be an *expansive* majorization operator, and the sequence, an expansive sequence. In a contractive (expansive) process Schur-convex (Schur-concave) functions increase over time. The contractive/expansive nature of a majorization process gives it a direction; inequality's arrow, as it were.

It seems reasonable that a moderated greedy algorithm should be defined as one which consists of contractive and expansive phases. The contractive phase optimizes one handle of a given tradeoff, while in the expansive phase, the other handle is worked on. For example, if it desired to balance exploration with exploitation, and Shannon entropy — a Schur-concave function — is used as a criterion measure, then in the contractive phase, entropy will decrease (exploitation), and in the expansive phase, entropy will increase (exploration). Also needed is a schedule (protocol) which specifies when each phase is to start and end.

If these ideas are put together, a moderated greedy algorithm is a triple $(\{\mathcal{F}, \mathcal{G}\}, S)$ where,

1. $\mathcal{F}, \mathcal{G} : R^n \to R^n$, $\mathcal{F}$ ($\mathcal{G}$) is a contractive (expansive) majorization operator.
2. $S : N \to \{\mathcal{F}, \mathcal{G}\}$, is the *schedule* , a computable procedure rule that determines which operator is to be applied at time instant t. N is the set of natural numbers.

Then, given a state vector $x \in R^n$, the action of the moderated greedy algorithm is given by the sequence $x(0), x(1), \ldots$, where,

$$x(t+1) = \begin{cases} \mathcal{F}(x(t)) & if \quad S(t) = \mathcal{F}, \\ \mathcal{G}(x(t)) & if \quad S(t) = \mathcal{G}. \end{cases} \tag{8}$$

Note that the operators are stochastic if they are implemented through random, doubly stochastic matrices. A slightly more sophisticated definition would make $\mathcal{F}, \mathcal{G}$ into functionals (so as to model adaptive modifications of parameters), and also permit a family of operators $(\mathcal{F}_1, \mathcal{F}_2, \ldots, \mathcal{F}_k)$, rather than just two operators. Another line of generalization is to define moderated greed in terms of pre-orders defined over more abstract settings. Any serious consideration of these possibilities, however, are outside the scope of this paper.

In the next section, it will be shown that with respect to the above definition, the overall effect of the evolutionary operators is to make the simple GA a moderated greedy algorithm.

5   Abstract Evolutionary Process

Consider a GA sample of size N, consisting of replicators (chromosomes) drawn from a set of n possible types[1]. The ith replicator type is characterized by a non-negative proportion $(p_i(t))$ and a non-negative fitness $(f_i(t))$. Here, $p_i(t) = n_i(t)/N$ where $n_i(t)$ is the number of replicators of type i in the sample. The state of the system is completely characterized by $\boldsymbol{p}(t)$ and $\boldsymbol{f}(t)$, the n-dimensional proportion and fitness vectors.

The sample is subject to the action of three operators: proportional selection, point crossover and point mutation. There are a variety of mathematical models for each of these operators; we use the discrete models derived from the replicator framework [2]. Our results hold for the continuous replicator models as well, but space limitations precludes the consideration of both types of models.

5.1   Proportional Selection

The effect of proportional selection is modeled by the replicator selection equation,

$$p_i(t+1) = \frac{p_i(t)f_i(t)}{\sum_j p_j(t)f_j(t)} = \frac{p_i(t)f_i(t)}{f_{\text{avg}}(t)}, \quad i = 1, \ldots, n. \tag{9}$$

Proportional selection attempts to respect the principle that replicators with above average-fitness should gain at the expense of the those with below-average fitness. This suggests that changes in the replicator proportions during proportional selection may be modeled as arising from a series of Dalton exchanges between these two sub-groups of replicators.

There is, however, a complication. In a Dalton exchange, if entity i gains at the expense of entity j, then it must have been the case that x_i was less than x_j ($\boldsymbol{x}$ represents the "incomes" possessed by the entities before the exchange took place). In proportional selection, if replicator i gains in proportion at the expense of replicator j (at time t), then it does not necessarily imply that $p_i(t-1)$ was less than $p_j(t-1)$; this is because the updates to replicator proportions are mediated by relative *fitness* ratios and not relative proportions. Unlike in a Dalton exchange, the item of exchange (proportion) differs from the item used to measure "income" (fitness).

---

[1] The standard example of replicators are binary chromosomes of constant length l (which implies $n = 2^l$). However, each schema in a schema partition (schemas that partition all possible chromosomes) can also act as a replicator.

There are two ways to address this complication. The first approach is stated in Theorem 1, which states that it suffices for fitness vector $f(t)$ to be similarly ordered[2] as the proportion vector $p(t)$ for proportional selection to be a (contractive) majorization operator.

Theorem 1 [7] Let $\{p(t), t \geq 0\}$ be a sequence of proportions such that $p_i(t + 1) = p_i(t)f_i/\bar{f}(t)$, where $f_i(t)$ are the fitnesses of the n possible replicator types in the sample at time t. Assume, without loss of generality, that $p(t)$ is strictly positive. If $p(t)$ is similarly ordered as $f(t)$, then $p(t) \preceq p(t + 1)$, that is, proportional selection is a contractive majorization operator. ∎

The assumption that $p(t)$ is strictly positive is only made to simplify the statement of the theorem. The proof of Theorem 1 only requires that the vector of *sample proportions* (by definition, non-zero) be similarly ordered as the vector of corresponding fitnesses. Also, note that the theorem does *not* require that the fitnesses be constant. This is significant for two reasons. First, some non-proportional selection operators (like ranking selection) can be modeled as proportion selection on non-constant fitnesses. Second, the proportional selection equations are self-similar under aggregation of chromosome types into schemas, provided the schemas define a schema partition. Even if chromosome fitnesses are constant, schema fitnesses are not. Theorem 1 applies to both replicators-as-chromosomes as well as replicators-as-schemas (in a schema partition).

For constant fitness functions, the situation is particularly simple. If at a time instant τ, $f \sim p(\tau)$, then for all $t > \tau$, $f \sim p(t)$. In other words, once similarity ordering is achieved in the sample, it is preserved under proportional selection. Equivalently, once for some $\tau > 0$, $p(\tau) \preceq p(\tau+1)$, then for all $t > \tau$, $p(t) \preceq p(t + 1)$.

Theorem 2 uses a different approach; it uses a scaling technique to show that proportional selection over constant fitnesses induces a Lorenz-majorization ordering.

Theorem 2 Let $\{p(t), t \geq 0\}$ be a sequence of proportions such that $p_i(t + 1) = p_i(t)f_i/\bar{f}(t)$, where f_i are the constant fitnesses of the n replicators in the sample. Let $w^t = (w_1, \ldots, w_n)$ be a set of weights such that, $f_i \geq f_j \Rightarrow w_i\,p_i(0) \geq w_j\,p_j(0)$. Define the *scaled* proportion vector $r(t) = (r_1, \ldots, r_n)^t$ by, $r_i(t) = p_i(t)w_i/\sum_{j=1}^{n} p_j(t)w_j$. Then, for all $t \geq 0$, $r(t)$ satisfies the same discrete dynamics as $p(t)$, namely, $r_i(t + 1) = r_i(t)f_i/f_{\text{avg}}(t)$. Furthermore, $r(t) \preceq r(t + 1)$. ∎

Proof: See Appendix I.

The above theorems can be stated in more general contexts (matrix semiorders) but the main point should be clear. Proportional selection, subject to some mild assumptions, induces a Lorenz majorization on sample proportions (or functions of sample proportions). The case of the point crossover operator is considered next.

---

[2] Two vectors x and y are *similarly ordered* and denoted $x \sim y$, if for all i, j, $(x_i - x_j)(y_i - y_j) \geq 0$.

5.2   Point Crossover

The quadratic dynamical system characterizing multiplicative recombination processes was worked out by Moran in 1961 [8]. Perhaps because of its simplicity, the resulting system of equations has been discovered and re-discovered several times [7].

By considering the change in the proportion of the ith replicator in terms of collision arguments, Moran obtained the following discrete model.

$$p_i(t+1) = \sum_{j,k,l=1}^{n} \pi(i,j|k,l)\, p_k(t)p_l(t) = \boldsymbol{p}^t A^{(i)} \boldsymbol{p}. \tag{10}$$

The interaction term $\pi(i,j|k,l)$ is a *non-negative* (possibly time-dependent) factor measuring the probability that replicators i and j are produced as offspring in a mating between replicator types k and l. Moran's model was based on the assumption that the interaction coefficients satisfied three conditions:

$$\sum_{i,j} \pi(i,j|k,l) = 1, \tag{11}$$

$$\pi(i,j|k,l) = \pi(i,j|l,k) = \pi(j,i|k,l), \tag{12}$$

$$\pi(i,j|k,l) = \pi(k,l|i,j). \tag{13}$$

The first condition (normalization)is necessarily true, and merely says that any mating must have a definite outcome. The second condition (symmetry) is reasonable under the assumption of the random mating of replicators. The third condition (bi-exchangeability) implies that any crossover operation can be "reversed," so that if replicators k and l mate to produce i and j with a certain probability, then replicators i and j mate to produce k and l with the same probability. An argument derived from Feller can be used to show that for unbiased point crossover operators, bi-exchangeability is always satisfied [6]. Theorem 3 shows that these properties imply that dynamics of the quadratic dynamical system is an expansive majorization process in a space of dimensionality n^2.

Theorem 3 If $p_i(t+1) = \sum_{j,k,l=1}^{n} \pi(i,j|k,l)p_k(t)p_l(t)$, and the transition probabilities satisfy the conditions of Moran's model, then $\boldsymbol{p}(t+1) \otimes \boldsymbol{p}(t+1) \preceq \boldsymbol{p}(t) \otimes \boldsymbol{p}(t)$. Here, $\boldsymbol{p}(t) \otimes \boldsymbol{p}(t)$ denotes the Kronecker product of $\boldsymbol{p}(t)$ with itself. Thus, point crossover is an expansive majorization operator for the sequence $\{\boldsymbol{p}(t) \otimes \boldsymbol{p}(t) | t \geq 0\}$.

Proof: See Appendix I.

The Kronecker $\boldsymbol{p}(t) \otimes \boldsymbol{p}(t)$ has n^2 components; it consists of terms of the form $p_i(t)p_j(t)$ for where i and j range from 1 through n.

The connection between his model and double stochasticity was known to Moran (at least, implicitly), but its significance appears to have been neglected. For point crossover with more than two parents, an approach similar to that used for discrete Boltzmann maps in Quantum Mechanics can be used to extend Theorem 3; essentially, majorization shifts to even higher dimensional spaces.

Majorization induced by point crossover differs from that induced by proportional selection in two important ways:

1. The *direction* is different. In proportion selection, the process is contractive. That is, $p(t) \preceq p(t+1)$. In point crossover, the process is expansive.
2. The *dimension* is different. Point crossover is a *quadratic transformation*; pairs producing pairs. Here, majorization occurs, not in an n-dimensional setting, but in an n^2-dimensional one.

The above consideration raises the question whether there exists an operator, that like proportional selection, also operates in an n-dimensional space, but like point crossover, is an expansive majorization operator. It turns out that point mutation is just such an operator.

5.3 Point Mutation

Point mutation is a unary operator that transforms one replicator to another. In replicator theory, point mutation effects are usually modeled as a set of master equation equations [2, pp. 249-256]. Ruch and Mead have shown that a master equation system (with symmetric mutation rates[3]), imply a expansive process in the proportions vector [12]. That is, $p(t+1) \preceq p(t)$. The symmetry of the transition matrix is responsible for this; any stochastic matrix that is symmetric is automatically a doubly stochastic matrix. Hence a Markov process defined by symmetric mutation matrices (not necessarily homogeneous) induces a majorization process. Since mutation "expands out" a distribution, it is natural that it be described by a expansive process on $p(t)$.

6   Genetic Algorithms and Moderated Greed

The results of the last section show that the three basic evolutionary operators of a simple GA are majorization operators; they differ in direction (expansive/contractive) and dimension (n-dimensional, n^2-dimensional). In the simple GA, the three operators are applied in phases as per a schedule; each phase consists of multiple application of the same operator. In the terminology of Section 4, the simple GA is a moderated greedy algorithm.

This identification is not to be seen as a negative result on the capabilities of genetic algorithms. Moderated greedy algorithms are not minor variants on greedy algorithms; they are capable of optimality results that are far beyond the reach of (unmoderated) greedy algorithms. A case in point would be simulated annealing. It is based on the successive transformations of a state vector by means of time-dependent, symmetric, stochastic matrices, that is, inhomogeneous doubly stochastic matrices. It is not hard to show that simulated annealing is an expansive majorization process. Similarly, the annealed GA (moderated greed with a particular schedule) also has global optimization capabilities [13].

---

[3] The transition matrix consists of elements ϵ_{ij}, defined as the proportion of replicators of type j undergoing mutation and producing replicators of type i.

Perhaps the most significant aspect of the analysis is its emphasis on the concept of inequality rather than diversity. The importance of diversity as both cause and consequence of evolution has been stressed so many times that any further emphasis is to flog a cliché. Yet, diversity has proved to be a very hard concept to pin down [1, pp. 1-7]. One problem is that most diversity measures are really relative abundance measures, and so a habitat consisting of one mosquito and hundred pandas is just as diverse as that consisting of hundred mosquitoes and one panda. There are ways to incorporate preference criteria [14], but such efforts also serve as demonstrations that diversity is a highly value-laden, observer-dependent concept. In contrast, inequality is at heart a binary *relation* between the cardinal attributes of entities. Yet it can not only be reified into a property of statistical distributions (inequality measures), but it can also be generalized to order collections (majorization pre-orders). Inequality, to use a classification from elementary logic, is both a collective term as well as a distributive one. The importance of "population thinking" has often been stressed in evolutionary theory. But perhaps "relation thinking" is equally important for understanding evolutionary processes, be they real or artificial.

7   Conclusion

In the majorization-theoretic interpretation, greedy algorithms apply exchange transformations on vectors to generate optimal solutions. The net result is to either increase the inequality amongst the components (contractive transforms) or reduce it (expansive transforms). A moderated greedy algorithm is one where contractive and expansive operators are alternatively applied as per a schedule. It was shown that proportional selection is a contractive majorization operator, while point crossover and point mutation are expansive operators. On the other hand, both selection and mutation majorize in an n-dimensional space, while point crossover majorizes in an n^2-dimensional space. The majorization pre-order delineates the role of (moderated) greed in genetic algorithms.

"Inequality," Leonardo da Vinci is reputed to have said, "is the cause of all local motion." The Renaissance genius found dozens of practical uses for this idea. Whether the "da Vinci principle" will be likewise useful for the *design* of genetic algorithms is a subject for future investigations.

Acknowledgements. I would like to thank the reviewers for their constructive and thought-provoking comments.

References

1. K. J. Gaston, editor. *Biodiversity: A Biology of Numbers and Difference.* Blackwell, Oxford, 1996.
2. J. Hofbauer and K. Sigmund. *The Theory of Evolution and Dynamical Systems.* Cambridge University Press, Cambridge, 1988.

3. R. M. Karp and M. Held. Finite-state processes and dynamic programming. *SIAM J. of Applied Mathematics*, 15:693–718, 1967.
4. B. Korte, L. Lovász, and R. Schrader. *Greedoids*. Springer-Verlag, 1991.
5. A. W. Marshall and I. Olkin. *Inequalities: Theory of Majorization and its Applications*. Academic Press, New York, 1979.
6. A. Menon. The point of point crossover: Shuffling to randomness. In W. B. Langdon et. al., editor, *Proceedings of the Genetic and Evolutionary Computation Conference, GECCO'2002*, pages 463–471, San Francisco, CA, 2002. Morgan Kaufmann.
7. A. Menon, K. Mehrotra, C. Mohan, and S. Ranka. Replicators, majorization and genetic algorithms: New models, connections and analytical tools. In R. Belew and M. Vose, editors, *Foundations of Genetic Algorithms*, volume 4, pages 155–180. Morgan Kaufman, 1997.
8. P. A. P. Moran. Entropy, Markov processes and Boltzmann's H-theorem. *Proc. Cambridge Phil. Soc.*, 57:833–842, 1961.
9. D. Stott Parker and P. Ram. The construction of Huffman codes is a submodular ("convex") optimization problem over a lattice of binary trees. *SIAM J. of Computation*, 28(5):1875–1905, 1999.
10. P. Ram. *A New Understanding Of Greed*. PhD thesis, Dept. of Computer Science, University of California, Los Angeles, 1993.
11. P. Ram and D. Stott Parker. Greed and majorization. Technical Report CSD-960003, UCLA Computer Science Dept., 1997.
12. E. Ruch and A. Mead. The principal of mixing character and some of its consequences. *Theoretica Chimica Acta*, 41:95–117, 1976.
13. L. M. Schmitt. Asymptotic convergence of scaled genetic algorithms to global optima. In *Frontiers of Evolutionary Computation*, volume 11, pages 157–192. Kluwer Academic Publishers, 2004.
14. M. L. Weitzman. The Noah's Ark problem. *Econometrica*, 66(6):1279–1298, 1998.

Appendix I

Proof of Theorem 2:   First, it will be shown that the vector $r(t)$ satisfies the same discrete replicator equations as $p(t)$. For $i \in \{1, \dots, n\}$,

$$r_i(t+1) = \frac{p_i(t+1)w_i}{\sum_{j=1}^{n} p_j(t+1)w_j} = \frac{\frac{p_i(t)w_i}{\sum_{k=1}^{n} p_k(t)w_k} f_i}{\sum_{j=1}^{n} \frac{p_j(t)w_j}{\sum_{k=1}^{n} p_k(t)w_k} f_j} = \frac{r_i(t)f_i}{\sum_{j=1}^{n} r_j(t)f_j}. \quad (14)$$

The non-negativity of the weights and Equation (14) imply that $r(t)$ is in the unit simplex. From $f_i \geq f_j \Rightarrow p_i(0)w_i \geq p_j(0)w_j$,

$$f_i \geq f_j \quad \Rightarrow \quad r_i(0) \geq r_j(0). \quad (15)$$

If $f_i \geq f_j$, then it can be shown (by induction) that for for all $t \geq 0$, $r_i(t) \geq r_j(t)$. In other words $f \sim r(t)$. Also the dynamics of $r(t)$ is given by the proportional selection equation (Equation (14)). The conditions of Theorem 1 apply, and hence $r(t) \preceq r(t+1)$. ∎

Proof of Theorem 3: Let T be the $n^2 \times n^2$ matrix whose (ij, kl)th element is $\pi(i, j | k, l)$. Normalization implies that T is row-stochastic. Normalization together with bi-exchangeability implies that T is also column stochastic, that it, $\sum_{k,l} \pi(i, j | k, l) = 1$. Let $\hat{p}(t) \equiv p(t) \otimes p(t)$. From the given dynamics, $\hat{p}(t + 1) = T \hat{p}(t)$. The theorem then follows from T's doubly stochasticity, and the definition of Lorenz majorization (Proposition 1). ∎

Trap Avoidance in Strategic Computer Game Playing with Case Injected Genetic Algorithms

Chris Miles, Sushil J. Louis, and Rich Drewes

Evolutionary Computing Systems Lab
Department of Computer Science
University of Nevada
Reno - 89557
{miles,sushil,drewes}@cs.unr.edu

Abstract. We use case injected genetic algorithms to learn to competently play computer strategy games. Such games are characterized by player decision in anticipation of opponent moves and imperfect knowledge of game state. Within the broad goal of developing effective and general methods of anticipatory play, this paper investigates anticipation in the context of trap avoidance in an immersive, 3D strike planning game. Case injection allows acquiring player knowledge from experience and incorporating acquired knowledge into future game play. Results show that with an appropriate representation case injection is effective at biasing the genetic algorithm toward producing plans that both avoid traps and carry out the mission effectively.

1 Introduction

The computer gaming industry is now bigger than the movie industry and both gaming and entertainment drive research in graphics and modeling. Although AI research has in the past been interested in games like checkers and chess, popular computer games like Starcraft and counter-strike are very different from chess and checkers. These games are situated in a virtual world, involve both long-term and reactive planning, and provide an immersive, fun experience. At the same time, we can pose many business, training, planning, and scientific problems as games where player decisions determine the final solution. A decision support system for a player in such games corresponds closely with a decision support system in the "real" world.

This paper applies a case-injected genetic algorithm that combines genetic algorithms with case-based reasoning to provide player decision support in the context of domains modeled by computer games [1]. The genetic algorithm "plays" the game by attempting to solve the underlying decision problem. Specifically, we develop and use a strike force asset allocation game, which maps to a broad category of resource allocation problems in industry, as our test problem. Strike force planning consists of allocating a collection of strike assets on flying platforms to a set of targets and threats on the ground. The problem is dynamic; weather and other environmental factors affect asset performance, unknown threats can

K. Deb et al. (Eds.): GECCO 2004, LNCS 3102, pp. 1365–1376, 2004.

popup, and new targets can be assigned. These complications as well as the varying effectiveness of assets on targets make the problem suitable for genetic and evolutionary computing approaches.

The idea behind a case-injected genetic algorithm is that as the genetic algorithm component iterates over a problem it selects members of its population and caches them (in memory) for future storage into a case base. Cases are therefore members of the genetic algorithm's population and represent an encoded candidate solution to the problem at hand. Periodically, the system injects appropriate cases from the case base, containing cases from previous attempts at *other* problems, into the evolving population replacing low fitness population members. When done with the current problem, the system stores the cached population members into the case base for retrieval and use on new problems.

Case injection is used to handle the dynamic nature of the game which places a premium on re-planning or re-allocation of assets when needed. We have shown that case-injected genetic algorithms learn to increase performance with experience at solving similar problems [1,2,3,4,5]. This implies that a case-injected genetic algorithm should quickly produce new plans (a new allocation) in response to changing game dynamics. Beyond purely responding to immediate scenario changes we use case injection in order to produce plans that anticipate opponent moves in the future. Doing this teaches our Genetic Algorithm Player (GAP) where traps are likely to occur, so that GAP acts in anticipation of changing game states. Specifically we try to influence GAP to produce plans that avoid areas similar to those in which it has encountered traps in the past. Our results show that GAP makes an effective Blue player with the ability to quickly replan to deal with changing game dynamics, and that case-injection can bias GAP to produce solutions that are suboptimal with respect to the game simulation's evaluation function but that avoid potential traps.

In the rest of the paper we define the game, the particular scenario being played, and the trap being encountered. We outline GAP's architecture; detailing the use of genetic algorithms, the encoding of strategy, the routing system, and the incorporation of case injection. Section 7 presents results showing that GAP can effectively play the game in the absence of the trap, and that GAP can quickly re-plan in the face of changing game dynamics. Preliminary results also show the effectiveness of case injection in acquiring and using player knowledge in learning to avoid traps. The last section presents our conclusions and directions for future work.

2   The Strike Planning Game

The strike planning game is based on an underlying resource allocation and routing problem and the genetic algorithm plays by solving the underlying problem. Our game involves two sides: Blue and Red, both seeking to allocate their respective resources to minimize damage received while maximizing the effectiveness of the strike.

Blue plays by allocating its resources, a set of assets on aircraft (platforms), to Red's buildings (targets) and defensive installations (threats). Blue determines which targets to attack, which weapons (assets) to use on them, and how to route each platform to minimize risk and maximize effectiveness.

Red's defensive installations (threats) protect targets by threatening platforms that come within range. Red plays by placing these threats in space and time to best protect targets. Potential threats and targets can also "pop-up" on Red's command in the middle of a mission, allowing a range of strategic game-playing options. By cleverly locating threats Red can feign vulnerability and lure Blue into a deviously located popup trap, or keep Blue from exploiting such a weakness out of fear of a trap. The scenario in this paper involves Red presenting Blue with a corridor of easy access to unprotected targets, a corridor containing a popup threat.

In this paper, a human player scripts Red's play while a Genetic Algorithm Player (GAP) plays Blue. The fitness of an individual in GAP's population solving the underlying allocation problem is evaluated by running the game. We explain our fitness evaluation in more detail in a later section. GAP develops strategies for the attacking strike force, including flight plans and weapon targeting for all available aircraft. When confronted with popups, GAP responds by replanning with the genetic algorithm in order to produce a new plan of action that responds to changes. Beyond purely responding to immediate scenario changes we use case injection in order to produce plans that anticipate opponent moves in the future.

2.1   Previous Work

Previous work in strike force asset allocation has been done in optimizing the allocation of assets to targets, the majority of it focusing on static pre-mission planning. Griggs [6] formulated a mixed-integer problem (MIP) to allocate platforms and assets for each objective. The MIP is augmented with a decision tree that determines the best plan based upon weather data. Li [7] converts a nonlinear programming formulation into a MIP problem. Yost [8] provides a survey of the work that has been conducted to address the optimization of strike allocation assets. Louis [9] applied case injected genetic algorithms to strike force asset allocation. A large body of work exists in which evolutionary methods have been applied to games [10,11,12,13,14]. However the majority of this work has been applied to board, card, and other well defined games. Such games have many differences from popular real time strategy (RTS) games such as Starcraft, Total Annihilation, and Homeworld[15,16,17]. Chess, checkers and many others use entities (pieces) that have a limited space of positions (such as on a board) and restricted sets of actions. Players in these games also have well defined roles and the domain of knowledge available to each player is well identified. These characteristics make the game state easier to specify and analyze.

In contrast, entities in our game exist and interact over time in continuous three dimensional space. Entities are not directly controlled by players but instead sets of parametrized algorithms control them in order to meet goals out-

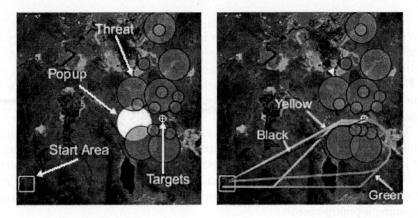

Fig. 1. Left: The Scenario Right: Route Categories

lined by players. This adds a level of abstraction not found in more traditional games. In most such computer games, players have incomplete knowledge of the game state and even the domain of this incomplete knowledge is difficult to determine. John Laird [18,19,20] surveys the state of research in using Artificial Intelligence (AI) techniques in interactive computers games. He describes the importance of such research and provides a taxonomy of games. Several military simulations share some of our game's properties [21,22,23], however these attempt to model reality while ours is designed to provide a platform for research in strategic planning, knowledge acquisition and re-use, and to have fun. The next section describes the scenario (or mission) used in our experiments.

3   The Scenario

Figure 1-Left shows an overview of our test scenario - chosen to be simple, easily analyzable but to still encapsulate the dynamics of traps and anticipation.

The scenario takes place in Northern Nevada and California, Lake Tahoe is visible below (south) of the popup on the bottom of the map. Red has four targets on the right hand side of the map with their locations denoted by the white cross-hair. As the targets represent different buildings comprising a larger facility, they appear as a single cross-hair from our point of view which is at a significant distance. Red has twenty two (22) threats placed to defend the targets and the translucent blue hemispheres show the effective radii of these threats. Red has the potential to play a popup threat to trap platforms venturing into the corridor formed by the threats and this trap is displayed as the solid white circle near the middle.

Blue has eight platforms, all of which start in the lower left hand corner. Each platform has one weapon, with three classes of weapons being distributed among the platforms. A weapon-target effectiveness table determines the effectiveness of each weapon against each target. Each of the eight weapons can be allocated to any of the four targets, giving $4^8 = 2^{16} = 64k$ allocations. This space is exhaustively search-able, but more complex scenarios quickly become intractable.

In this scenario, GAP's router can produce the three broad types of routes shown in Figure 1-Right.

1. Black - Flies inside the perimeter of known threats.
2. Yellow - Flies through the corridor in order to reach the targets.
3. Green - Flies around the threats, attacking the targets from behind.

Black routes expose platforms to unnecessary risk from threats and thus receive low fitness. The naively optimal strategy contains yellow routes which are the most direct routes to the target that still manage to avoid known threats. However in the presence of the popup, Green routes become optimal although they are longer then yellow routes. The evaluator looks only at known threats, so plans containing green routes receive lower fitness then those containing yellow routes. With experience GAP should learn to anticipate traps and to prefer green routes even though green routes have lower fitness then yellow routes.

In order to search for good routes and allocations, GAP must be able to compute and compare their fitnesses. Computing this fitness is dependent on the representation of entities states inside the game, and our way of representing this state is rather unusual so we next detail it.

4   Probabilistic Health Metrics

In many games, entities (platforms, threats and targets in our game) posses hit-points which represents their ability to take damage. Each attack removes a number of hit-points and when reduced to zero hit-points the entity is destroyed and cannot participate further. However, weapons have a more hit or miss effect, entirely destroying an entity or leaving it functional. A single attack may be effective while multiple attacks may have no effect. Although more realistic, this introduces a degree of stochastic error into the game. In the worst case, evaluating a individual plan can result in outcomes ranging from total failure to perfect success making it difficult to compare two plans based on a single evaluation. Lacking a good comparison it is difficult to search for an optimal strategy. By taking a statistical analysis of survival we can achieve better results. Consider the state of each entity at the end of the mission as a random variable. Comparing the expected values for those variables becomes an effective means to judge the effectiveness of a plan. These expected values can then be estimated by executing each plan a number of times and averaging the results. However, doing multiple runs to determine a single evaluation increases the computational expense many-fold.

We use a different approach based on probabilistic health metrics. Instead of monitoring whether or not an object has been destroyed we monitor the probability of its survival. Being attacked no longer destroys objects and removes

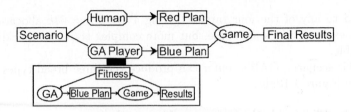

Fig. 2. System Architecture.

them from the game, it just reduces their probability of survival according to Equation 1 below.

$$S(E) = S_{t_0}(E) * (1 - D(E)) \qquad (1)$$

E is the entity being considered, a platform, target, or threat. $S(E)$ is the probability of survival of entity E after the attack. $S_{t_0}(E)$ is probability of survival of E up until the attack and $D(E)$ is the probability of that platform being destroyed by the attack and is given by equation 2 below.

$$D(E) = S(A) * E(W) \qquad (2)$$

Here, $S(A)$ is the attackers probability of survival up until the time of the attack and $E(W)$ is the effectiveness of the attackers weapon as given in the weapon-entity effectiveness matrix. This method gives us the true expected values of survival for all entities in the game within one run of the game, thereby producing a representative evaluation of the value of a plan. As a side effect, we also gain a smoother gradient for the GA to search as well as consistently reproducible evaluations. This technique is impractical when applied to more complicated relationships, but is effective at this stage of research.

The gaming system's architecture reflects the flow of action in the game and is described next.

5   System Architecture

Figure 2 outlines our system's architecture. Starting at the left, Red and Blue, human and GAP, are presented with the scenario and given time to prepare their strategy. GAP works by applying the genetic algorithm to the underlying resource allocation and routing problem. We chose the best plan produced by the GA in the time available to play against Red. These plans then execute and during execution, Red can script the emergence of a popup threat. When the popup is detected by GAP, the genetic algorithm re-plans and begins execution of the new plan.

To play the game GAP must produce routing data for each of Blue's platforms. Figure 3 shows how routes are built using the A* algorithm [24]. A* builds routes from current platform locations to target locations and back and tends to prefer short routes that avoid threats while seeking targets.

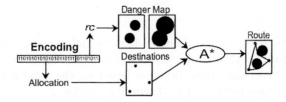

Fig. 3. How Routes are Built From an Encoding.

In order to avoid traps the routing system must be somehow parameterized to avoid areas with particular characteristics. Note that traps are most effective in areas confined by other threats. If we artificially inflate threat radii, threats will expand to fill in potential trap corridors and A* will find routes that go around these expanded threats. We therefore add a multiplier parameter rc that increases threats' effective radii. Larger rc's expand threats and fill in confined areas. A* then routes around those confined areas. Combined with case injection, rc allows GAP to learn coefficients that avoid traps and re-use them in new scenarios. In our scenario $rc < 1.0$ produce black routes, $1.0 < rc < 1.35$ produce yellow routes and $rc > 1.35$ produce green routes. rc is limited to the range $[0, 3]$ and encoded with eight (8) bits at the end of our chromosome. We are encoding a single rc for each plan, future work may include rc's for each section of routing contained in the plan.

5.1   Encoding

Most of the encoding specifies the asset to target allocation with rc encoded at the end as detailed above. Figure 4 shows how we represent the allocation data as an enumeration of assets to targets. The scenario involves two platforms (P1, P2), each with a pair of assets, attacking four targets. The left box illustrates the allocation of asset A1 on platform P1 to target T3, asset A2 to target T1 and so on. Tabulating the asset to target allocation gives the table in the center. Letting the position denote the asset and reducing the target id to binary then produces a binary string representation for the allocation.

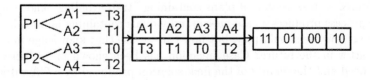

Fig. 4. Allocation Encoding

5.2   Fitness

Blue's goals are to maximize damage done to red targets, while minimizing damage done to its platforms. Shorter simpler routes are also desirable, so we include a penalty in the fitness function based on the total distance traveled. This gives the fitness calculated as shown in Equation 3

$$fit(plan) = TotalDamage(Red) - TotalDamage(Blue) - d * c \qquad (3)$$

d is the total distance traveled by Blue's platforms and c is chosen such that $d*c$ has a 10-20% effect on the fitness ($fit(plan)$). Total damage done is calculated below.

$$TotalDamage(Player) = \sum_{E \in F} E_v * (1 - E_s)$$

E is an entity in the game and F is the set of all forces belonging to that side. E_v is the value of E, while E_s is the probability of survival for entity E.

6   Avoiding Traps with Case-Injection

We address the problem of learning from experience to avoid traps with a two part approach. First we learn from experience where traps are likely to be, then we apply that knowledge and avoid potential traps in the future. Case injection provides an implementation of these steps: building a case-base of individuals from past games stores important knowledge, the injection of those individuals applies the knowledge towards future search.

GAP records games played against opponents and runs offline after a game playing episode in order to determine the optimal way to win that game. The simulation now contains knowledge about opponents moves, in our case, the game contains the popup trap. Allowing the search to progress towards the optimal strategy in the presence of the popup, GAP saves individuals from this search into the case-base, building a case-base with routes that go around the popup trap – green routes. When faced with other opponents, GAP then injects individuals from the case-base, biasing the current search towards containing this learned anticipatory knowledge.

In this paper GAP first plays the scenario, likely picking a yellow route and falling into Red's trap. Afterward GAP replays the game, including Red's trap into the evaluator. Yellow routes then receive poor fitness, and GAP searches towards the optimal green route. Saving individuals to the case-base from this search stores a cross-section of plans containing "trap avoiding" knowledge.

The process produces a case-base of individuals containing important knowledge about how we should play, but how can we use that knowledge in order to play smarter in the future? Case Injection has been shown [2] to increase the search speed and the quality of the final solution produced by a GA working on a similar problem. It also tends to produce answers similar to old ones by biasing the search to look in areas that were previously successful – exploiting this effect gives our GA its learning behavior. When playing the game we periodically inject a number of individuals from the case-base into the population, biasing our current search towards information from those individuals. Injection occurs by replacing the worst members of the population with individuals chosen from the case database through a "Probabilistic Closest to the Best" strategy [1]. Those individuals bring their "trap avoiding" knowledge into the population, increasing the likelihood of that knowledge being used in the final solution and therefore increasing GAP's ability to avoid the trap.

7 Results

We present results showing

1. GAP can play the game effectively.
2. Replanning can effectively react to popups.
3. We can use case injection to learn to avoid the trap.

We also analyze the effect of altering the population size and number of generations on the strength of the biasing provided by case injection.

We first show that GAP can form efficient strategies. GAP is run against our test scenario 50 times, and we graph the min, max, and average population fitness against generation in Figure 5-Left. The graph shows a strong approach toward the optimum and in more the 95% of runs it gets within 5% of the optimum. This indicates that GAP can form effective strategies for playing the game.

To deal with opponent moves and the dynamic nature of the game we look at the effects of re-planning. Figure 5-Right illustrates the effect of replanning by showing the final route followed inside a game. A yellow route was chosen, and when the popup occurred, trapping the platforms, GAP redirected the strike

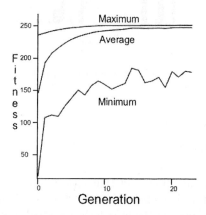
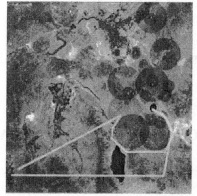

Fig. 5. Left: Best/Worst/Average Individual Fitness as a function of Generation - Averaged over 50 runs. Right: Final routes used during a mission involving replanning.

force to retreat and attack from the rear. Replanning allows GAP to rebuild its routing information as well as modify its allocation to compensate for damaged platforms.

GAP's ability to learn to avoid the trap is shown in Figure 6. The figure compares the histograms of rc values produced by GAP with and without case injection. Case injection leads to a strong shift in the kinds of rc's produced, biasing the population towards using green routes. The effect of this bias being a large and statistically significant increase in the frequency at which strategies containing green routes were produced $(2\% - > 42\%)$. These results were based on 50 independent runs of the system and show that case injection does bias the search toward avoiding the trap.

Figure 7-left compares the fitnesses with and without case injection. Without case injection the search shows a strong approach toward the optimal yellow plan; with injection the population quickly converges toward the optimal green plan. Case injection applies a bias towards green routes, however the GA has a tendency to act in opposition of this bias, trying to search towards ever shorter routes. The ability of the GA to overcome the bias through manipulation of injected material is dependent on the size of the population and the number of generations it runs. Figure 7-Right illustrates this effect. As the number of evaluations alloted to the GA is increased, the frequency of green routes being produced as a final solution decrease. Counteracting this tendency requires a careful balance of GA and case-injection parameters.

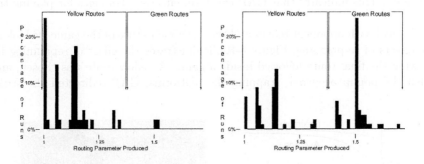

Fig. 6. Histogram of Routing Parameters produced without Case Injection.

8   Conclusions and Future Work

Results show that GAP is able to play the game, and that case injection can be used to to bias the search to incorporate knowledge from past game playing experience. We had expected difficulty in biasing the search, but we had under-estimated the GA's resilience towards searching away from the optimum. We expected a stronger bias from case-injection - while 50% green is a significant

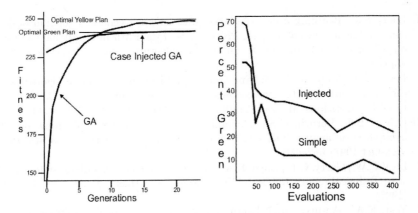

Fig. 7. Left: Effect of Case Injection on Fitness Inside the GA over time Right: Effect of Population Size and the Number of Generations on Percentage Green routes Produced

improvement on 2% we had hoped for numbers in the range of 80 to 90%. Even after extensive testing with the parameters involved we were unable to bias the search towards consistently producing plans containing green routes. However, preliminary results from new work show that artificially inflating the fitness of individuals in the population that contain injected material is an effective way of maintaining the preferred bias. This method appears to consistently produce green routes while maintaining an effective search across a range of problems without the need for parameter tuning.

There are a large number of interesting avenues in which to continue this research. Fitness inflation appears to solve one of our major problem in using case injection, further exploration of this technique is underway. We are also interested in capturing information from human players in order to better emulate their style of play. The game itself is also under major expansion, the next phase of research should involve a symmetric game involving aspects of resource management and much deeper strategies than those seen at the current level.

Acknowledgment. This material is based upon work supported by the Office of Naval Research under contract number N00014-03-1-0104.

References

1. Louis, S.J., McDonnell, J.: Learning with case injected genetic algorithms. IEEE Transactions on Evolutionary Computation (To Appear in 2004)
2. Louis, S.J.: Evolutionary learning from experience. Journal of Engineering Optimization (To Appear in 2004)
3. Louis, S.J.: Genetic learning for combinational logic design. Journal of Soft Computing (To Appear in 2004)

4. Louis, S.J.: Learning from experience: Case injected genetic algorithm design of combinational logic circuits. In: Proceedings of the Fifth International Conference on Adaptive Computing in Design and Manufacturing, Springer-Verlag (2002) to appear

5. Louis, S.J., Johnson, J.: Solving similar problems using genetic algorithms and case-based memory. In: Proceedings of the Seventh International Conference on Genetic Algorithms, Morgan Kauffman, San Mateo, CA (1997) 283–290

6. Griggs, B.J., Parnell, G.S., Lemkuhl, L.J.: An air mission planning algorithm using decision analysis and mixed integer programming. Operations Research **45** (Sep-Oct 1997) 662–676

7. Li, V.C.W., Curry, G.L., Boyd, E.A.: Strike force allocation with defender suppression. Technical report, Industrial Engineering Department, Texas A&M University (1997)

8. Yost, K.A.: A survey and description of usaf conventional munitions allocation models. Technical report, Office of Aerospace Studies, Kirtland AFB (Feb 1995)

9. Louis, S.J., McDonnell, J., Gizzi, N.: Dynamic strike force asset allocation using genetic algorithms and case-based reasoning. In: Proceedings of the Sixth Conference on Systemics, Cybernetics, and Informatics. Orlando. (2002) 855–861

10. Fogel, D.B.: Blondie24: Playing at the Edge of AI. Morgan Kauffman (2001)

11. Rosin, C.D., Belew, R.K.: Methods for competitive co-evolution: Finding opponents worth beating. In Eshelman, L., ed.: Proceedings of the Sixth International Conference on Genetic Algorithms, San Francisco, CA, Morgan Kaufmann (1995) 373–380

12. Pollack, J.B., Blair, A.D., Land, M.: Coevolution of a backgammon player. In Langton, C.G., Shimohara, K., eds.: Artificial Life V: Proc. of the Fifth Int. Workshop on the Synthesis and Simulation of Living Systems, Cambridge, MA, The MIT Press (1997) 92–98

13. Kendall, G., Willdig, M.: An investigation of an adaptive poker player. In: Australian Joint Conference on Artificial Intelligence. (2001) 189–200

14. Samuel, A.L.: Some studies in machine learning using the game of checkers. IBM Journal of Research and Development **3** (1959) 210–229

15. Blizzard: Starcraft (1998, www.blizzard.com/starcraft)

16. Cavedog: Total annihilation (1997, www.cavedog.com/totala)

17. Inc., R.E.: Homeworld (1999, homeworld.sierra.com/hw)

18. Laird, J.E.: Research in human-level ai using computer games. Communications of the ACM **45** (2002) 32–35

19. Laird, J.E., van Lent, M.: The role of ai in computer game genres (2000)

20. Laird, J.E., van Lent, M.: Human-level ai's killer application: Interactive computer games (2000)

21. Tidhar, G., Heinze, C., Selvestrel, M.C.: Flying together: Modelling air mission teams. Applied Intelligence **8** (1998) 195–218

22. Serena, G.M.: The challenge of whole air mission modeling (1995)

23. McIlroy, D., Heinze, C.: Air combat tactics implementation in the smart whole air mission model. In: Proceedings of the First Internation SimTecT Conference, Melbourne, Australia, 1996. (1996)

24. Stout, B.: The basics of a* for path planning. In: Game Programming Gems, Charles River media (2000) 254–262

Topological Interpretation of Crossover

Alberto Moraglio and Riccardo Poli

Department of Computer Science, University of Essex,
Wivenhoe Park, Colchester, CO4 3SQ, UK
{amoragn,rpoli}@essex.ac.uk

Abstract. In this paper we give a representation-independent topological definition of crossover that links it tightly to the notion of fitness landscape. Building around this definition, a geometric/topological framework for evolutionary algorithms is introduced that clarifies the connection between representation, genetic operators, neighbourhood structure and distance in the landscape. Traditional genetic operators for binary strings are shown to fit the framework. The advantages of this interpretation are discussed.

1 Introduction

Fitness landscapes and genetic operators have been studied for considerable time in connection with evolutionary algorithms. However, a unifying theory of the two is missing and many questions about their relationship remain unanswered. Below we will briefly analyze the current situation in this respect.

Fitness landscapes and genetic operators are undoubtedly connected. Mutation is intuitively associated with the neighbourhood structure of the search space. However, the connection between landscape and crossover is less clear. Complicated topological structures, hyper-neighbourhoods, have been proposed [Culberson, 1995; Jones, 1995; Gitchoff & Wagner, 1996; Reidys & Stadler, 2002] to formally link crossover to fitness landscapes. However, even using these ideas, effectively one is left with a different landscape for each operator [Culberson, 1995], which is deeply unsatisfactory. Important questions then are: is there an easier way of interpreting crossover in connection to fitness landscapes? Are crossover and mutation really unrelated?

An established way of defining a fitness landscape for search spaces where a natural notion of distance exists is to imagine that the neighbourhood of each point includes the points that are at minimum distance from that point [Back et al, 1997]. Once a landscape is defined, typically the notion of distance is not used further. Couldn't distance play a much more important role in explaining the relationship between landscapes and crossover?

Local search and many other meta-heuristics are naturally defined over the neighbourhood structure of the search space [Glover, 2002]. However, a peculiarity of evolutionary algorithms (seen as meta-heuristics) is that the neighbourhood structure over the search space is specified by the way genetic operators act on the representa-

K. Deb et al. (Eds.): GECCO 2004, LNCS 3102, pp. 1377–1388, 2004.

tion for solutions. One may wonder whether it is possible to naturally reconcile these two ways of defining structure over the search space.

Yet in another sense, solution representation and neighbourhood structure are just two different perspectives on the solution space. An example is the classical binary string representation and its geometric dual, a hypercube, which has been extremely useful in explaining genetic algorithms [Whitley, 1994]. Can solution representation and neighbourhood structure be two sides of the same coin for other representations, like permutation lists or syntax trees?

The traditional mutation and crossover operators defined for binary strings have been *extended* to other representations [Langdon & Poli, 2002]. Also, there are general guidelines for the *design* of such operators for representations other than binary [Radcliffe, 1994; Surry, 1998]. Is there a way to rigorously *define*, rather than *design* or *extend*, mutation and crossover in general, independently of the representation adopted?

Except for solution representations, many evolutionary algorithms are very similar which suggests that unification might be possible [Stephens & Poli, 2004]. Are all evolutionary algorithms really the same algorithm in some sense?

In this paper we clarify the connection between representation, genetic operators, neighbourhood structure and distance and we propose a new answer to the previous questions. The results of our work are surprising: all the previous questions are connected, and that the central question to address is really only one: what is crossover?

The paper is organized as follows. In section 2, we introduce some necessary definitions. Geometric/topological definitions of crossover and mutation are given in section 3, where we also prove some properties of these operators. As an example, in section 4, we show how traditional mutation and crossover defined over binary strings, fit our general topological definitions for mutation and crossover. In section 5, we discuss some implications of our topological interpretation of crossover. Finally, in section 6, we draw some conclusions and we indicate our future research directions.

2 Preliminary Definitions

2.1 Search Problem

Let S denote the *solution set[1]* comprising all the candidate solutions to a given *search problem* P. The members of this set must be seen as *formal solutions* not relaying on any specific underlying representation.

The goal of a search problem P is to find specific solution/s in the search space that maximize (minimize) an *objective function*:

$$g : S \to R$$

---

[1] We distinguish between *solution set* and *solution space*. The first refers to a collection of elements, while the second implies a structure over the elements.

Let us assume, without loss of generality, that the goal of the search problem P is to maximize g. The *global optima* x^* are points in S for which g is a maximum:

$$x^* \in S^* \Leftrightarrow g(x^*) = \max_{x \in S} g(x)$$

Notice that global optima are well defined when the objective function is well defined and are independent of any structure defined on S. On the contrary, *local optima* are definable only when a structure over S is defined. A search problem in itself does not come with any predefined structure over the solution set.

2.2 Fitness Landscape

A *configuration space* C is a pair (G, Nhd) where G is a set of syntactic configurations (syntactic objects or genotypes) and $Nhd : G \rightarrow 2^G$ is a syntactic neighbourhood function which maps every configuration in C to the set of all its neighbour configurations in C which can be obtained by applying a unitary syntactic modification operator. The unitary syntactic modification operator must be reversible (i.e. $y \in Nhd(x) \Leftrightarrow x \in Nhd(y)$) and connected (any configuration can be transformed into any other by applying the operator a finite number of times). Notice that a configuration set may lead to more than one configuration space if multiple syntactic neighbourhood functions are available.

A configuration space $C=(G, Nhd)$ is said to be a space endowed with a *neighbourhood structure*. This is induced by the syntax of the configurations and the particular notion of syntactic neighbourhood function adopted. Such a neighbourhood structure can be associated with an undirected *neighbourhood graph* $W= (V, E)$, where V is the set of vertices representing configurations and E is the set of edges representing the relationship of neighbourhood between configurations.

Since the neighbourhood is symmetric ($y \in Nhd(x) \Leftrightarrow x \in Nhd(y)$) and the neighbourhood structure is connected, this space is also a *metric space* provided with a *distance function* d induced by the neighbourhood function (see formal definition below) [Back et al, 1997]. So, we can equivalently write $C=(G, Nhd)$ or $C=(G, d)$. However, we must keep in mind that the notion of distance in the metric space of syntactic configurations has a syntactic nature (and, therefore, may have special features other than respecting distance axioms). Distances arising from graphs are known as *graphic distances* [Van der Vel, 1993].

Formally, a *metric space* (M, d) is a set M provided with a metric or distance d that is a real-valued map on $M \times M$ which fulfils the following axioms for all $s_1, s_2 \in M$:

1. $d(s_1, s_2) \geq 0$ and $d(s_1, s_2) = 0$ if and only if $s_1 = s_2$;
2. $d(s_1, s_2) = d(s_2, s_1)$, i.e. d is symmetric; and
3. $d(s_1, s_3) \leq d(s_1, s_2) + d(s_2, s_3)$, i.e. d satisfies the triangle inequality.

A *representation mapping* is a function $r : G \rightarrow S$ associating any syntactic configuration in G with a formal solution in S. Ideally this mapping would be bijective. However, there are cases where the sizes of G and S differ.

A *fitness landscape* F is a pair *(C, f)* where $C=(G, d)$ is a configuration space and $f : G \to R$ is a *fitness function* mapping a syntactic configuration to its *fitness value*. The fitness value is a positive real number. It may or may not coincide with the objective function value of the solution represented by the input genotype. For the sake of simplicity, we assume that it is. Therefore, the fitness function is the composition of the representation mapping r with the objective function g: $f = g \circ r$.

2.3 Topological and Geometric Preliminaries: Balls and Segments

In a metric space (S, d) a *closed ball* is the set of the form $B(x; y) = \{y \in S \mid d(x, y) \leq r\}$ where $x \in S$ and r is a positive real number called the radius of the ball. A *line segment* (or closed interval) is the set of the form $[x; y] = \{z \in S \mid d(x, z) + d(z, y) = d(x, y)\}$ where $x, y \in S$ are called extremes of the segment. Note that $[x; y] = [y; x]$. The length l of the segment $[x; y]$ is the distance between a pair of extremes $l([x; y]) = d(x, y)$. Let H be a segment and $x \in H$ is an extreme of H, there exists only one point $y \in H$, its conjugate extreme, such as $[x; y] = H$. Examples of balls and segments for different spaces are shown in Figure 1. Note how the same set can have different geometries (see Euclidean and Manhattan spaces) and how segments can have more than a pair of extremes. E.g. in the Hamming space, a segment coincides with a hypercube and the number of ex-

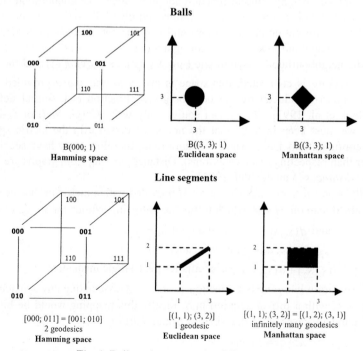

Fig. 1. Balls and segments for different spaces

tremes varies with the length of the segment, while in the Manhattan space, a segment is a rectangle and it has two pairs of extremes. Also, a segment is not necessarily "slim", it may include points that are not on the boundaries. Finally, a segment does not coincide with a shortest path connecting its extremes (*geodesic*). In general, there may be more than one geodesic connecting two extremes.

3   Topological Genetic Operators

We *define*, postponing the justifications of these definitions to the discussion, two *classes* of operators in the landscape (i.e. using the notion of *distance* coming with the landscape): topological mutation and topological crossover. Within these classes, we identify two *specific* operators: topological uniform mutation and topological uniform crossover. *These definitions are representation-independent and therefore the operators are well-defined for any representation.*

A *g*-ary genetic operator *OP* takes *g* parents $p_1, p_2, \ldots p_g$ and produces one offspring *c* according to a given conditional probability distribution:

$$\Pr\{OP(p_1, p_2, \ldots p_g) = c\} = \Pr\{OP = c \mid P_1 = p_1, P_2 = p_2, \ldots, P_g = p_g\} = f_{OP}(c \mid p_1, p_2, \ldots p_g)$$

Mutation is a unary operator while *crossover* is typically a binary operator.

Definition 1 *The image set of a genetic operator OP is the set of all possible offspring produced by OP when the parents are $p_1, p_2, \ldots p_g$ with non-zero probability:*

$$\mathrm{Im}[OP(p_1, p_2, \ldots p_g)] = \{c \in S \mid f_{OP}(c \mid p_1, p_2, \ldots p_g) > 0\}$$

Notice that the image set is a *mapping* from a vector of parents to a set of offspring.

Definition 2 *A unary operator M is a topological ε-mutation operator if* $\mathrm{Im}[M(p)] \subseteq B(p; \varepsilon)$ *where ε is the smallest real for which this condition holds true.*

In other words, in *a topological ε-mutation* all offspring are at most *ε away* from their parent.

Definition 3 *A binary operator CX is a topological crossover if* $\mathrm{Im}[CX(p_1, p_2)] \subseteq [p_1; p_2]$.

This simply means that in a topological crossover offspring lay *between* parents. We use the term *recombination* as a synonym of any binary genetic operator.

We now introduce two *specific* operators belonging to the *families* defined above.

Definition 4 *Topological uniform ε-mutation UM is a topological ε-mutation where all z at most ε away from parent x have the same probability of being the offspring:*

$$f_{UM\varepsilon}(z \mid x) = \Pr\{UM = z \mid P = x\} = \frac{\delta(z \in B(x, \varepsilon))}{\mid B(x, \varepsilon) \mid}$$

$$\mathrm{Im}[UM_\varepsilon(x)] = \{z \in S \mid f_{M\varepsilon}(z \mid x) > 0\} = B(x, \varepsilon)$$

where δ is a function which returns 1 if the argument is true, 0 otherwise.

When ε is not specified, we mean ε = 1.

Definition 5 *Topological uniform crossover UX is a topological crossover where all z laying between parents x and y have the same probability of being the offspring:*

$$f_{UX}(z \mid x, y) = \Pr\{UX = z \mid P1 = x, P2 = y\} = \frac{\delta(z \in [x, y])}{|[x, y]|}$$

$$\mathrm{Im}[UX(x, y)] = \{z \in S \mid f_{UX}(z \mid x, y) > 0\} = [x, y].$$

Theorem 1 *The structure over the configuration space C can equivalently be defined by the set G of the syntactic configurations and one of the following objects: 1. The neighborhood function Nhd, 2. The neighborhood graph W= (V, E), 3. The graphic distance function d, 4. Uniform topological mutation UM, 5. Uniform topological crossover UX, 6. The set of all balls **B**, 7. The set of all segments **H**.*

Proof.

Equivalences 1, 2 and 3 are trivial consequences of the fitness landscape definition.

Equivalence 4: given *UM* one has its conditional density function $f_{UM}(z \mid x)$ and, consequently, the image set mapping $\mathrm{Im}[UM(x)]$, i.e. the mapping $x \mapsto B(x,1)$. The structure of the space is therefore given by $Nhd : x \mapsto (B(x,1) \setminus \{x\})$.

Equivalence 5: analogously, given *UX* one has the mapping $(x, y) \mapsto [x; y]$. By restricting this mapping through its co-domain considering only segments of size 2, the corresponding restricted domain coincides with the set of edges *E* of the neighborhood graph, hence the structure of the space is determined.

Equivalence 6: the relation of inclusion between sets $\subseteq$ induces a partial order in **B**. The set of all balls of radius 1 $\mathbf{B}_1$ can be determined by considering all those balls in **B** that have, as only predecessors, balls of size 1 (i.e. balls of radius zero). Given a ball $b \in \mathbf{B}_1$ a point $x \in b$ is the center of the ball if and only if $\forall x' \in (b \setminus \{x\}) \exists b' \in \mathbf{B}_1 : b \neq b' \wedge x, x' \in b'$.[2] Knowing the center of each ball of radius 1, it is possible to form the map $x \mapsto B(x,1)$ and proceed as in equivalence 4.

Equivalence 7: by considering only segments in **H** of size 2, one can form the set *E* of the edges of the neighborhood graph; hence, the structure of the space is determined.■

Corollary 1 Uniform topological mutation *UM* and uniform topological crossover *UX* are isomorphic.

Proof.

Since both *UM* and *UX* identify the structure of the configuration space univocally and also the configuration space structure identify both operators univocally then they are isomorphic.■

Corollary 2 *Given a structure of the configuration search space in terms of neighborhood function or graphic distance function, UM and UX are unique.*

---

[2] Given two different points in the same ball of radius 1 $x, x' \in b$, they are either at distance 1 or distance 2. If they are at a distance 2, *b* is the only ball in $\mathbf{B}_1$ satisfying this condition since the two points are extremes of a diameter of the ball *b* and identify the ball univocally. If they are at a distance 1, there must exist at least two balls in $\mathbf{B}_1$ containing x, x' one in which one is the center and the other is not, and another one in which the roles are reversed; this symmetry holds because the neighborhood is symmetric.

Proof.
This follows trivially from the definition of *UM* and *UX* over the space structure. ∎

Corollary 3 *Given a representation, there are as many UM and UX operators as notions of graphic/syntactic distance for the representation.*

Proof.
Given a representation, one has a configuration set for which the structure is not specified. A specific notion of graphic distance transforms the set into a space with a structure. Given such a structure, UM and UX are unique (corollary 2). ∎

4   Generalization of Binary String Crossover

Given two binary strings $s_1 = (x_1,..., x_n)$ and $s_2 = (y_1,..., y_n)$ of length n, the *Hamming distance* $d_H(s_1, s_2)$ is the number of places in which the two strings differ, i.e.

$$d_H(s_1, s_2) = \sum_{i=1}^{n} \delta(x_i \neq y_i)$$

A *property* of the Hamming distance is that a binary string $s_3 = (z_1,..., z_n)$ lays between s_1 and s_2 if and only if every bit of s_3 equals al least one of the corresponding bits of s_1 and s_2, i.e. $\forall i : z_i \in \{x_i, y_i\} \Leftrightarrow s_3 \in [s_1, s_2]$.

Traditional (one-point, two-point, uniform, etc.) crossovers for binary strings belong to the class of *mask-based crossover operators* [Syswerda, 1989]. A crossover operator is a probabilistic mapping $cx_m : S \times S \xrightarrow{m} S$ where the mask m is a random variable with different probability distributions for different crossover operators. The mask m takes the form of a binary string of length n that specifies for each position from which parent to copy the corresponding bit to the offspring, i.e. $cx_m(s_1, s_2) = s_3$ and $m = (m_1,..., m_n)$ then $z_i = x_i \cdot \delta(m_i = 0) + y_i \cdot \delta(m_i = 1)$.

Theorem 2 *All mask-based crossover operators for binary strings are topological crossovers. All mutations for binary strings are topological ε-mutations.*

Proof.
We need to show that for any probability distribution over m it holds $\mathrm{Im}[cx_m(s_1, s_2)] \subseteq [s_1, s_2]$. Out of all possible mask-based crossovers, those with a non-zero probability of using all the 2^n masks produce the biggest image set for any given pair of parents. Formally, this is given by $\mathrm{Im}[cx(s_1, s_2)] = \{cx_m(s_1, s_2) \mid m \in B^n\}$. So, it is sufficient to prove that $\mathrm{Im}[cx(s_1, s_2)] \subseteq [s_1, s_2]$ for this image set. This is equivalent to prove that $\forall m \in B^n : s_3 = cx_m(s_1, s_2) \rightarrow s_3 \in [s_1, s_2]$.

Given $s_1 = (x_1,..., x_n)$, $s_2 = (y_1,..., y_n)$ and any mask m there exists a unique $s_3 = (z_1,..., z_n)$. From the definition of mask-based crossover it follows that $\forall i : z_i \in \{x_i, y_i\}$. Then, from the Hamming distance property mentioned above, it follows that $\forall m : s_3 \in [s_1, s_2]$, which completes the proof of the first part of the theorem.

Let $s_2 = \mu(s_1)$ be the result of mutating s_1, that is $s_2 \in \mathrm{Im}[\mu(s_1)]$, then $\exists \varepsilon : \forall s_2 : d_H(s_1, s_2) \leq \varepsilon$ whereby $s_2 \in B(s_1, \varepsilon)$ with ε being the smallest possible. ∎

Theorem 3. *The topological uniform crossover for the configuration space of binary strings endowed with Hamming distance is the traditional uniform crossover. The topological uniform 1-mutation for the configuration space of binary strings endowed with Hamming distance is equivalent to a zero-or-one-bit mutation.*

Proof.

Let us start by proving that the image sets of traditional uniform crossover and topological uniform crossover coincide. We need to show that $\mathrm{Im}[cx(s_1, s_2)] = [s_1, s_2]$, where $\mathrm{Im}[cx(s_1, s_2)]$ was defined in the proof of theorem 2, from which we know that $\mathrm{Im}[cx(s_1, s_2)] \subseteq [s_1, s_2]$. Consequently, all we need to prove is that $\forall s_3 \in [s_1, s_2] \rightarrow \exists m \in B^n : cx_m(s_1, s_2) = s_3$. For the Hamming distance property this is equivalent to say $\forall s_3 \forall i : z_i \in \{x_i, y_i\} \rightarrow \exists m \in B^n : cx_m(s_1, s_2) = s_3$, where z_i are the bits of s_3. From the definition of crossover this is equivalent to proving that $\forall s_3 \forall i : z_i \in \{x_i, y_i\} \rightarrow \exists m \in B^n : z_i = x_i \cdot \delta(m_i = 0) + y_i \cdot \delta(m_i = 1)$. This is true because it always exists at least a mask for which when the bits in the parents differ, it specifies the parent for which the bit equals the offspring bit. If the bits do not differ, the mask indifferently specifies one parent or the other for that bit. This shows that the image sets of traditional uniform crossover and topological uniform crossover coincide.

Every element of the image set of the traditional uniform crossover has identical probability of being the offspring [Whitley, 1994] and the same is true for the elements of the image set of the topological uniform crossover (by definition). This completes the proof of the first part of this theorem.

Let us now consider the zero-or-one-bit mutation. This is an operator where a string is either mutated by flipping one bit or is not mutated with equal probability. The image sets of this mutation and topological 1-mutation coincide as it is trivial to see by noting that the Hamming ball of radius 1, which is the image set of topological 1-mutation, coincides with the image set of the zero-or-one-bit mutation. Every element of the image set of zero-or-one-bit mutation has identical probability of being the offspring and the same is true for the elements of the image set of the topological uniform 1-mutation (by definition). ∎

5   Discussion

In the introduction, we raised various questions, claiming that this way of interpreting crossover lays a foundation to connect all these questions. In the following, we show how our framework answers those questions by highlighting the properties of the class of topological crossovers.

1. *Generalization*: topological crossover is a generalization of the family of crossovers based on masks for binary representation in that it captures and generalizes the distinction between crossover and recombination for binary representation.

2. *Unification*: from preliminary research, we believe that a variety of operators developed for other important representations, such as real-valued vectors, permutations and syntax trees, fit our topological definitions given suitable notions of distance (naturally not *all* pre-existing operators do this, but many do). Hence, topological crossover has the potential to lead to a unification of the different evolutionary algorithms.

3. *Representation independence*: evolutionary computation theory is fragmented. One of the reasons is that there is not a unified way to deal with different solution representation (although steps forward in this direction have recently been made [Langdon & Poli 2002; Stephens & Poli 2004]), which has led to the development of significantly different theories for different representations. In this context, one important theoretical implication of our topological definitions is that the genetic operators are fully defined without any reference to the representation. This may pave the route to a unified treatment of evolutionary theory.

4. *Clarification*: the connections between operators, representation, distance and neighborhood are completely clear when using topological operators.

5. *Analysis*: given a certain representation with pre-existing genetic operators, it is easy to check whether they fit our topological definitions. If they do, their properties are unveiled.

6. *Geometric interpretation*: an evolutionary algorithm using topological operators does a form of geometric search based on segments (crossover) and balls (mutation). This suggests looking at the solution space not as a graph or hyper-graph, as normally done, but rather as a geometric discrete space. The notion of distance arising from the syntax of configurations reveals therefore a natural *dual* interpretation:[3] (i) it is a measure of similarity (or dissimilarity) between two syntactic objects; (ii) and it is a measure of spatial remoteness between points in a geometric space.

7. *Principled design*: one important practical implication of the topological definition of crossover is the possibility of doing crossover principled design. When applying evolutionary algorithms to optimization problems, a domain specific solution representation is often the most effective [Davis, 1991; Radcliffe, 1992]. However, for any non-standard representation, it is not always clear how to define a good crossover operator. Given a good neighborhood structure for a problem, all meta-heuristics defined over such a structure tend to be good. Indeed, the most important step in using a meta-heuristic is the definition of good neighborhood structure for the problem at hand [Glover, 2002]. With topological crossover, given a good neighborhood structure or a good mutation operator, a crossover operator that respects such a structure is *automatically defined*. This has good chances of performing well, being effectively a composition of unitary moves on

---

[3] Any mathematical object/property that admits a definition only based on the concept of distance possesses a dual nature: a syntactic one and a geometric one.

the landscape. An example is shown in Figure 2, where we assume that we want to evolve graphs with four nodes and we are given a mutation operator for such graphs that either adds or removes exactly one edge. We want to define a good crossover operator that would, for example, produce meaningful offspring when applied to the parent graphs in Figure 2(a). The configuration space for this problem is shown in Figure 2(b). The parent graphs are boxed while the graphs belonging to the segment defined by the parents are encircled. With our definition of topological crossover these are all possible successors, as shown in Figure 2(c).

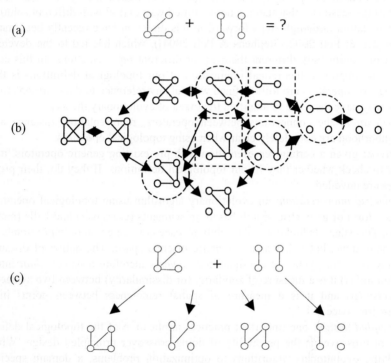

Fig. 2. Inducing crossover from mutation (see text).

8.   *Landscape and knowledge*: the landscape structure is relevant to a search method only when the move operators used in that search method are strongly related to those which induce the neighborhood structure used to define the landscape [Back et al, 1997]. This is certainly the case for the topological operators. The problem knowledge used by an evolutionary algorithm that uses topological operators is embedded in the connectivity structure of the landscape. The landscape is therefore a *knowledge interface* between a *formal problem* and a *formal search algorithm* that has no knowledge of the problem whatsoever. In order for the knowledge to be transmissible from the problem to the search algorithm *through the landscape*, there are two requirements: (i) the search operators have to be defined over the connectivity structure of the landscape (i.e. using a distance function);

(ii) the landscape has to be *designed* around the specific definitions of the operators employed in such a way to bias the search towards good areas of the search space so as to perform better than random search.

9. *Landscape conditions*: for the no free lunch theorem [Wolpert & Macready, 1996], over all the problems, on average any search algorithm performs the same as random search. So in itself a given search algorithm (any meta-heuristics) is not inherently superior to any other. A search algorithm therefore, to be of use, has to specify the class of problems for which it works better than random search. The geometric definition of mutation (connected with the concept of ball) and the geometric definition of crossover (connected with the concept of segment) suggest, respectively, conditions over the landscape in terms of *continuity* and *convexity*. These conditions, in various guises, are important to guarantee good performance in optimisation [Pardalos & Resende, 2002] and ensuring them should guide the landscape design for the topological operators.

6 Conclusions

In this paper, we have introduced a geometric/topological framework for evolutionary algorithms that clarifies the connections between representation, genetic operators, neighbourhood structure and distance in the landscape. Thanks to this framework a novel and general way of looking at crossover (and mutation) that is based on landscape topology and geometry has been put forward. Traditional crossover and mutation for binary strings have been shown to fit our topological framework, which, from preliminary investigations, appears to also encompass a variety of other representations and associated operators.

This framework presents a number of additional advantages. The theory is representation independent, and therefore it offers a unique opportunity for generality and unification. The theory provides a natural, direct and automatic way of deriving (designing) both mutation *and* crossover from the neighbourhood structure of a landscape. Conversely, if one adopts our topological operators, one and only one fitness landscape is induced: that is we *do not* have a different landscape for each operator, but a common one for both.

In future work we expect to further extend the applications of our framework to other representations and operators, to study the connections between this theory and other evolutionary computation theories (including those based on the notions of schema) and to investigate the links with generalized notions of convexity and continuity for the landscape.

References

[Back et al, 1997] Fogel, T. Back, D. B. Fogel, Z. Michalewicz (eds). *Handbook of Evolutionary Computation*. Oxford press, 1997.

[Culberson, 1995] J.C.Culberson. *Mutation-crossover isomorphism and the construction of discriminating functions*. Evol. Comp., 2:279-311, 1995.

[Davis, 1991] L. Davis, *Handbook of Genetic Algorithms*, van Nostrand Reinhold, New York, 1991.

[Gitchoff & Wagner, 1996] P. Gitchoff, G. P. Wagner. *Recombination Induced HyperGraphs: A New Approach to Mutation- Recombination Isomorphism*. Journal of Complexity, 37-43, 1996.

[Glover, 2002] F. W. Glover (ed). *Handbook of Metaheuristics*. Kluwer, 2002.

[Jones, 1995] T. Jones. *Evolutionary Algorithms, Fitness Landscapes and Search*. PhD dissertation, University of New Mexico, 1995.

[Langdon & Poli, 2002] W. B. Langdon, R. Poli. *Foundations of Genetic Programming*. Springer, 2002.

[Pardalos & Resende, 2002] P. M. Pardalos, M. G. C. Resende (eds) *Handbook of Applied Optimization*. Oxford University Press, 2002

[Radcliffe, 1992] N. J. Radcliffe. *Nonlinear genetic representations*. In R. Manner and B. Manderick, editors, Proceedings of the 2 nd Conference on Parallel Problems Solving from Nature, pages 259-268. Morgan Kaufman, 1992.

[Radcliffe, 1994] N. J. Radcliffe, 1994. *The Algebra of Genetic Algorithms*, Annals of Maths and Artificial Intelligence 10, 339 --384.

[Reidys & Stadler, 2002] C. M. Reidys, P. F. Stadler. *Combinatorial Landscapes*. SIAM Review 44, 3-54, 2002.

[Stephens & Poli, 2004] R. Stephens and R. Poli. *EC Theory "in Theory": Towards a Unification of Evolutionary Computation Theory*. In A. Menon (ed), *Frontiers of Evolutionary Computation*, pp. 129-156, Kluwer, Boston, 2004.

[Surry, 1998] P. D. Surry. *A Prescriptive Formalism for Constructing Domain-specific Evolutionary Algorithms*. PhD dissertation, University of Edinburgh, 1998.

[Syswerda, 1989] G. Syswerda. *Uniform crossover in genetic algorithms*. In J. D. Scha er, editor, Proceedings of the International Conference on Genetic Algorithms, San Mateo (CA), 1989. Morgan Kaufmann Publishers.

[Van der Vel, 1993] M. van de Vel. *Theory of Convex Structures*, Elsevier, Amsterdam, 1993.

[Whitley, 1994] D. Whitley. *A Genetic Algorithm Tutorial*. Statistics and Computing (4):65-85, 1994.

[Wolpert & Macready, 1996] D. H. Wolpert, W. G. Macready. *No Free Lunch Theorems for Optimization*. IEEE Transaction on Evolutionary Computation, April 1996.

Simple Population Replacement Strategies for a Steady-State Multi-objective Evolutionary Algorithm

Christine L. Mumford

School of Computer Science, Cardiff University
PO Box 916, Cardiff CF24 3XF, United Kingdom
christine@cs.cardiff.ac.uk

Abstract. This paper explores some simple evolutionary strategies for an elitist, steady-state Pareto-based multi-objective evolutionary algorithm. The experimental framework is based on the SEAMO algorithm which differs from other approaches in its reliance on simple population replacement strategies, rather than sophisticated selection mechanisms. The paper demonstrates that excellent results can be obtained without the need for dominance rankings or global fitness calculations. Furthermore, the experimental results clearly indicate which of the population replacement techniques are the most effective, and these are then combined to produce an improved version of the SEAMO algorithm. Further experiments indicate the approach is competitive with other state-of-the-art multi-objective evolutionary algorithms.

1 Introduction

Multi-objective optimization problems are common in the real world and involve the simultaneous optimization of several (often competing) objectives. Such problems are characterized by optimum sets of alternative solutions, known as *Pareto sets*, rather than by a single optimum. Pareto-optimal solutions are *non-dominated solutions* in the sense that it is not possible to improve the value of any one of the objectives, in such a solution, without simultaneously degrading the quality of one or more of the other objectives in the vector.

Evolutionary algorithms (EAs) are ideally suited to multi-objective optimization problems because they produce many solutions in parallel. However, traditional approaches to EAs require scalar fitness information and converge on a single compromise solution, so need to be adapted if a set of viable alternatives is required for multi-objective optimization. Like their single objective counterparts however, most multi-objective EAs focus the genetic search on the selection stage, and use a fitness function to bias the choice of parents for breeding, favoring the 'better individuals'. In a multi-objective context, fitness functions are usually based either on a count of how many contemporaries in the population are dominated by a particular individual, or alternatively, on a count of by how many contemporaries the individual is itself dominated. This technique, known as Pareto-based selection, was first proposed by Goldberg [3], and is favored, in

K. Deb et al. (Eds.): GECCO 2004, LNCS 3102, pp. 1389–1400, 2004.

one form or another, by most researchers (for example see [1,2,10,11]). In contrast, SEAMO (a Simple Evolutionary Algorithm for Multi-objective Optimization, [7,9]) uses uniform selection and thus does not need any fitness functions to bias the selection of parents. Instead progression of the genetic search relies entirely on a few simple rules for replacing individuals with newly generated offspring in a steady-state environment. The implementation of these rules usually requires nothing more complicated than a simple 'who shall live and who shall die' decision, based on the outcome of a straight comparison between the solution generated by an offspring with those produced by its parents (or other population members). Despite its simplicity, SEAMO has produced some very good results in earlier studies [7,9].

The present study explores a range of simple population replacement strategies for a steady-state multi-objective EA, based on the SEAMO framework. Its purpose is twofold:

− to discover the best strategies
− and use them to improve the original SEAMO algorithm.

The evolutionary strategies are developed and compared using the multiple knapsack problem (MKP) as a testbed. The instances chosen are kn500.2 and kn750.2 of [10], consisting of 500 and 750 items, respectively, in two knapsacks. The best strategies are finally combined to produce an improved version of the SEAMO algorithm, and its performance is compared to other state-of-the-art multi-objective EAs, on various multi-objective functions.

2   A Basic Framework for the SEAMO Algorithm

The SEAMO framework, outlined in Figure 1, illustrates a simple steady-state approach, which sequentially selects every individual in the population to serve as the first parent once, and pairs it with a second parent that is selected at random (uniformly). A single crossover is then applied to produce one offspring, and this is followed by a single mutation. Each new offspring will either replace an existing population member, or it will die, depending on the outcome of the chosen replacement strategy. This paper will investigate different replacement strategies for lines 10 − 13 in Figure 1.

2.1   The Original SEAMO Algorithm

In the original SEAMO algorithm, an offspring is evaluated using the following criteria:

1. Does offspring dominate either parent?
2. Does offspring produce any global improvements on any Pareto components?

On the basis of this 'superiority test', the offspring will replace one or other of its parents, if it is deemed to be better.

Procedure *SEAMO*
1. **Begin**
2.   Generate N random individuals {N is the population size}
3.   Evaluate the objective vector for each population member and store it
4.   **Repeat**
5.   **For** each member of the population
6.   This individual becomes the first parent
7.   Select a second parent at random
8.   Apply crossover to produce single offspring
9.   Apply a single mutation to the offspring
10.   Evaluate the objective vector produced by the offspring
11.   **if** offspring qualifies
12.   **Then** the offspring replaces a member of the population
13.   **else** it dies
14.   **Endfor**
15.   **Until** stopping condition satisfied
16.   **Print** all non-dominated solutions in the final population
17. **End**

Fig. 1. Algorithm 1 A basic framework for SEAMO

On average an offspring will have 50 % genetic material in common with each parent, and, for this reason, parental replacement is favored in SEAMO in the hope that it will encourage the maintenance of genetic diversity within the population and thus help avoid premature convergence. One purpose of the current study is to put assumptions like this to the test, and also try some alternative strategies.

In more detail, the superiority test applied in the original SEAMO algorithm progresses as follow. To start with, a new offspring is compared with its first parent, and replaces that parent in the population if it dominates it, provided that the offspring is not a duplicate, in which case it dies immediately (the deletion of duplicates is explained see later in the present section). Any offspring that fails the first test, and thus does not dominate its first parent, is next compared with its second parent. Similar to before, a non-duplicate, dominating offspring will replace its second parent in this situation. If an offspring fails to dominate either parent, however, it will usually die at this stage. The replacement of population members by dominating offspring ensures that the solution vectors move closer to the Pareto front as the search progresses. To additionally ensure an improved range of coverage, the dominance condition is relaxed whenever a new global best value is discovered for any of the individual components of the solution vector (i.e. for improved maximum profits in individual knapsacks). Care has to be taken, however, to ensure that global best values for other components (i.e. maximum profits in other knapsacks) are not lost when a dominance condition is relaxed. Ensuring that global best components are not lost is straightforward if multi-objective optimization is restricted to two components in the solution vector, as is the case in this paper: whenever an offspring produces an improved global best for either of the components, if the global best for the second component happens to occur in one of the parents, the offspring will simply replace the other parent. One weakness with the replacement strategies applied in the original SEAMO algorithm is that offspring that neither dominate nor are dominated by their parents will usually die immediately and their potential is wasted.

To complete the description of the original SEAMO algorithm, an explanation of the 'deletion of duplicates' policy is now given. A simple way to help promote genetic diversity is avoid the propagation of genetic duplicates through the population. Thus, before a final decision is made on replacement of a parent, a dominating offspring is compared with every individual in the current population, and if the offspring is duplicated elsewhere in the population, the offspring dies and does not replace its parent. For speed and simplicity it is the phenotypic values of the offspring that are compared to those of other population members (i.e. the values of the Pareto vectors) rather than the genotypic values (i.e. the permutation lists of items). Ideally, the genotypes should be compared, but due to the lengths of the permutation lists, this would be very time consuming.

3   Experimental Design

An order-based representation with a first fit decoder is used for the MKP, and Cycle Crossover (CX) [8] is used as the recombination operator. A simple mutation operator swaps two arbitrarily selected objects within a single permutation list. The representational scheme was chosen because it produced the best results in a recent comparative study, [5]. The reader is referred to the earlier work for full details.

In all the experiments that follow, each strategy is tested by 30 replicate runs, initialized with different random seeds. 2D plots are obtained by combining all 30 results files, for each experiment, and extracting the non-dominated solutions from the combined results. 2D plots give a good fast visual indication the solution quality, spread and range of the approximate Pareto sets produced by the competing strategies. Additionally, some performance metrics are used to compare the improved SEAMO approach with other state-of-the-art EAs.

4   Simple Strategies: Replacing a Population Member with a Dominating Offspring

When using a steady-state evolutionary algorithm, a decision has to be made each time that a new offspring is created, whether that offspring will live or die. If it is allowed to live, one has to determine, which population member to replace. In the SEAMO framework no selective pressure is applied when choosing parents, so if the population is to improve over time, new individuals entering the population need to be generally superior to those individuals that they are replacing. In the first set of experiments we shall compare three simple strategies

Table 1. Average run times of experiments in seconds

Problem	1a	1b	2a	2b	3a	3b
kn500.2	19	19	9	9	19	19
kn750.2	31	32	15	15	31	32

that replace a current population member with an offspring that dominates that individual:

1. offspring replaces a population member that it dominates at random
2. offspring replaces a parent that it dominates
3. offspring replaces a parent if it dominates either parent, otherwise it replaces a population member that it dominates at random.

To implement the first strategy, and part of the third, the population is sampled without replacement until a suitable candidate for replacement is found, or until the whole population has been exhausted. In the latter case the offspring will be allowed to die. The pseudocode is given below.

11.	**Repeat**
12a.	Select population member at random without replacement
12b.	**If** offspring dominates selected individual
12c.	**Then** offspring replaces it in the population; ****quitloop****
12d.	**Until** all members of population are tried
13.	{offspring dies if it does not replace any member of the population}

The second strategy is implemented by testing the offspring with the first parent and then the second parent, in the way described in the earlier section for the original SEAMO algorithm. An offspring will replace a parent that it dominates.

The third strategy is a combination of the first two. A new offspring will replace a parent if it dominates either of them. When this is not the case the offspring will replace a population member that it dominates at random. If it fails to dominate any individual, it dies. For each strategy, we assess the effect that deleting duplicates has on the results. We use population sizes of 200 and 250 for kn500.2 and kn750.2 respectively, and stop the runs after 500 generations have elapsed.

4.1   Results for the Simple Strategies

Figure 2 summarizes the results for replacement strategies 1, 2 and 3 on kn500.2 and kn750.2. For each trace the non-dominated solutions are extracted from the combined results of 30 replicated experiments. Clearly strategy 3 appears to be the most successful. Figure 3 indicates that failing to delete duplicates has a serious deleterious effect on the results for strategy 3. (A similar pattern was observed for strategies 1 and 2.)

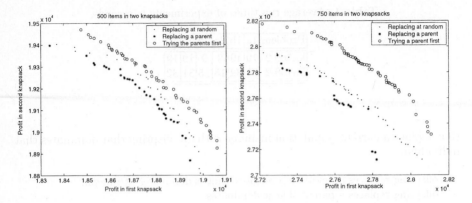

Fig. 2. Comparing replacement strategies with duplicates deleted

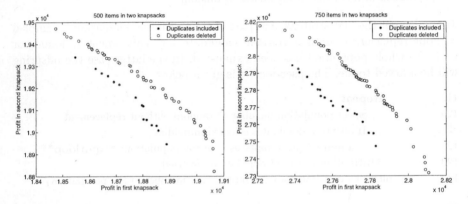

Fig. 3. Examining the effect the deleting duplicates has on the results produced by strategy 3

Table 1 compares the average run times, on a 1.5 GHz PC laptop for the three strategies. For experiments 1a, 2a and 3a phenotypic duplicates are allowed, but in 1b, 2b and 3b, the duplicates are deleted. From table 1 it would appear that including a routine to test and exclude phenotypic duplicates, does not add to the run time of the EA. Although this may seem counter-intuitive, closer examination reveals that, as a direct result of deleting the duplicates, fewer new offspring genotypes are copied into the population, and copying permutation lists 500 or 750 item long is indeed a lengthy business. In the next section we will try improving on strategy 3. Phenotypic duplicates will be deleted in all future experiments.

5   Further Strategies

As discussed in Section 2.1, replacing parents with their offspring is likely to more successfully preserve genetic diversity than replacing arbitrary members of the population with the offspring of other individuals. Nevertheless, replacement strategy 3 will frequently maintain offspring that are dominated by both of their parents. Perhaps it would make better sense if such individuals were allowed to die? Strategy 4 will investigate the following:

Replacement Strategy 4

1. **if** offspring dominates either parent it replaces it
2. **else if** offspring is neither dominated by nor dominates either parent it replaces another individual that it dominates at random
3. **otherwise** it dies

Strategy 4 differs from strategy 3 by killing off offspring that are dominated by both parents. Unlike the simpler strategy 2, though, strategy 4 will maintain offspring that are neither dominated by nor dominate their parents, provided a weaker candidate can be found elsewhere in the population. The loss of such offspring is a weakness of the original SEAMO algorithm. Unfortunately, occasional loss of a non-dominated individual will occur, even applying stategy 4, if a weaker individual cannot be found. This is inevitable when maintaining a constant population size.

In the original SEAMO algorithm, dominance rules are relaxed when new global best components appear in the objective vectors, and an offspring is then allowed to replace one of its parents (or occasionally another individual) whether it dominates that parent or not. This approach tends to increase the range of values in the solution set. Strategy 5 extends strategy 4 to incorporate new global best components. The precise mechanism is outlined below:

Replacement Strategy 5

1. **if** offspring harbors a new best-so-far Pareto component
   a) it replaces a parent, if possible
   b) **else** it replaces another individual at random
2. **else if** offspring dominates either parent it replaces it
3. **else if** offspring is neither dominated by nor dominates either parent it replaces another individual that it dominates at random
4. **otherwise** it dies

(Note: Condition 1 (b) in strategy 5 is not needed for problems with only two objectives, but is required for three or more.) The parameters for population sizes and the number of generations are the same as set previously in section 4

5.1   Results for Strategies 4 and 5

Strategies 3, 4 and 5 are compared in Figure 4. Clearly, strategy 5 produces the best results, as they are much more widely spread. An additional set of experiments confirmed that SEAMO using strategy 5 (SEAMO2) is able to produce better results than the original SEAMO algorithm (see Figure 5).

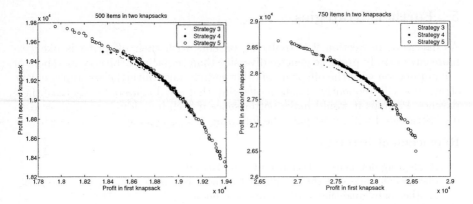

Fig. 4. Comparing strategies 3, 4 and 5

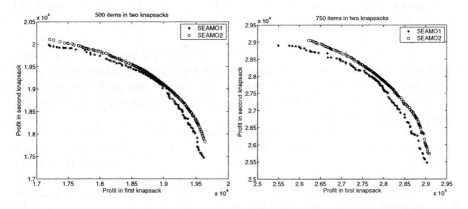

Fig. 5. Comparing SEAMO with strategy 5 (SEAMO2) with the original SEAMO (SEAMO1)

6   Comparing SEAMO Using Strategy 5 with Other State-of-the-Art EAs

A final set of experiments compares the performance of SEAMO using strategy 5 (i.e.SEAMO2) with NGSA2 (a fast elitist non-dominated sorting genetic algorithm) [2], PESA (the Pareto envelope-based seletion algorithm) [1], and SPEA2 an improved version of SPEA (the strength Pareto evolutionary algorithm) [10]. The test problems used are kn750.2, plus four continuous functions, SPH-2, ZDT6, QV, and KUR, [11]. The results for PESA, NSGA2 and SPEA2, were obtained from [11].

For kn750.2 a population of 250 is used and 30 replicate runs collected for SEAMO2, as previously. However, the experiments in this section are allowed to run for 1920 generations, to make results comparable with those in [11]. The parameters for the continuous function experiments (domain size, population

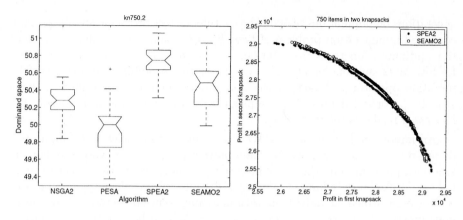

Fig. 6. Comparing SEAMO2 with SPEA2

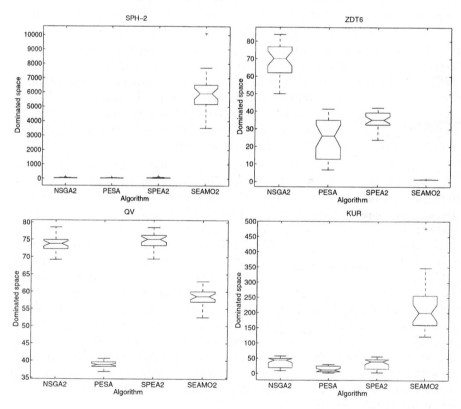

Fig. 7. Comparing SEAMO2 with NSGA2, PESA, and SPEA2

size, and number of evaluations etc.) are as given in [11]. For each algorithm on each function, 30 replicate runs are collected, each run consisting of 10,000 generations on populations of 100. The continuous test problems are all specially contrived minimization problem of two objectives and 100 variables. For all of the continuous functions the solutions are coded as real vectors of length 100 for SEAMO2, and one-point crossover acts as the recombination operator. The mutation operator is based on the non-uniform mutation described on page 111 of [4]. For full details of the implementation of non-uniform mutation, the interested reader is referred to [7].

An important feature of SEAMO algorithms is their deletion of duplicates, designed to help maintain diversity and prevent premature convergence. For the knapsack problem and other combinatorial problems, where the objective functions can take on only limited number of discrete values, phenotypic duplicates are easily identified as individuals with matching solution vectors. With continuous functions, however, exact duplicates are likely to be rare. For this reason, values for component objective functions x_i and x'_i of $\mathbf{x}$ and $\mathbf{x}'$, respectively, are deemed to be equal if and only if $x_i - \epsilon \leq x'_i \leq x_i + \epsilon$, where ϵ is an error term, which is set at $0.00001 \times x_i$ for the purpose of these experiments.

SEAMO2 is compared with its competitors using the two metrics, S and C, described in [10]. For the purpose of the S metric, the minimization problems, SPH-2, ZDT6, QV and KUR, have been transformed into maximization problems by replacing the Pareto values with their reciprocals. Furthermore, all the S hypervolumes have been scaled as percentages of suitable reference values for ease of tabulation. The reference values are 1.649e+009, 40, 500, 1 and 0.002 for kn750.2, SPH-2, ZDT6, QV and KUR respectively.

Figure 6 compares the performance of the various algorithms on kn750.2 The boxplots on the left show the spread of dominated space produced by the 30 replicate runs collected for each algorithm, and the 2D plot on the right compares the non-dominated solutions produced by SEAMO2 and SPEA2 directly. From the boxplots it is clear that SPEA2 and SEAMO2 are the leaders with SPEA2 performing a little better than SEAMO2. However, the 2D plots suggest that the solution quality produced by SEAMO2 is slightly better than that of SPEA2. (Further evidence for is provided by the coverage metric). Figure 7 gives the boxplots showing the dominated space obtained from the experiments with the continuous functions. Clearly SEAMO2 performs extremely well, with respect to this metric, on SPH-2 and KUR, not so well in QV and very poorly indeed on ZDT6. (Note: a high average of 5907 was obtained for SEAMO2 on SPH-2, distorting the plots for SPH-2 in Figure 7. This distortion seems to be an unfortunate feature of the transformation process used to convert functions from a minimization to maximization, and does not reflect superiority on a scale suggested by this result.)

Table 2 gives the average values for $C =$ Coverage $(A \succeq B)$ (the number of points in set B that are weakly dominated by points in set A). The standard deviations are given in brackets. Table 2 shows a very strong performance for SEAMO2 on kn750.2, SPH-2, and KUR, and a performance comparable with NSGA2 and SPEA2 on QV. Notably, SEAMO2 performs very poorly on ZDT6 for coverage as well as for hypervolume.

Table 2. Average values (and standard deviations) for Coverage ($A \succeq B$)

Coverage ($A \succeq B$)						
Algorithm		Test problems				
A	B	kn750.2	SPH-2	ZDT6	QV	KUR
SEAMO2	NSGA2	73.5 (20.0)	85.5 (14.1	0 (0)	36.9 (11.8)	93.1 (8.9)
	PESA	69.4 (19.4)	88.0 (9.5)	0 (0)	52.1 (11.5)	89.6 (16.8)
	SPEA2	72.5 (13.1)	81.4 (13.4)	0 (0)	35.0 (11.7)	93.4 (7.4)
NSGA2	SEAMO2	11.7 (15.5)	0 (0)	97.7 (0.3)	35.5 (15.7)	0.2 (0.8)
PESA		10.8 (11.8)	0 (0)	96.9 (1.4)	0.23 (0.6)	0.15 (0.8)
SPEA2		9.7 (9.4)	0 (0)	97.7 (0.3)	33.6 (19.7)	0(0)

To summarize, Figures 6 and 7 and Table 2 show that SEAMO2 outperforms its competitors on SPH-2 and KUR for both metrics and additionally outperforms the other EAs on kn750.2 and QV (marginally) for Coverage ($A \succeq B$). SEAMO2 performs very poorly on ZDT6, however.

Some caution is required in interpreting the results in this section. For the knapsack problems SEAMO2 uses a different representation scheme to that of its competitors, and slightly different mutation and recombination operators are used by SEAMO2 on the continuous problems. The results for SEAMO2 are, nevertheless, encouraging. Perhaps the performance of the other EAs could be improved with some changes to the representations and operators.

7   Conclusions and Future Work

This paper explores some simple evolutionary strategies for an elitist, steady-state Pareto-based multi-objective evolutionary algorithm. It validates the approach developed earlier for the SEAMO algorithm and also produces some improvements. The paper demonstrates experimentally that simple population replacement strategies coupled with the deletion of duplicates can produce excellent results, without the need for dominance ranking or global fitness calculations. Furthermore, the results clearly indicate that, despite its simplicity, the SEAMO approach is competitive with other state-of-the-art multi-objective evolutionary algorithms. Since the original submission of the present paper, further work has produced encouraging results for some hierarchical versions of the SEAMO2 algorithm, [6]. However, even these improvements have failed to lift performance on the ZDT6 continuous function. Work in progress is focussed on parallel implementations and also on improving the performance of the algorithm on non-uniformly spread functions, such as ZDT6.

References

1. Corne D W, Knowles J D, and Oates M J: The Pareto envelope-based selection algorithm for multiobjective optimization. *Parallel Problem Solving from Nature – PPSN VI*, Lecture Notes in Computer Science 1917 (2000) 839–848, Springer.
2. Deb K, Agrawal S, Pratap A, and Meyarivan T: A fast elitist non-dominated sorting genetic algorithm for mult-objective optimization: NSGA-II, *Parallel Problem Solving from Nature – PPSN VI*, Lecture Notes in Computer Science 1917 (2000) 849–858, Springer.
3. Goldberg D E: *Genetic Algorithms in Search, Optimization, and Machine Learning*, Addison-Wesley (1989).
4. Michalewicz, Z.: Genetic Algorithms + Data Structures = Evolution Programs. 3rd edn. Springer-Verlag, Berlin Heidelberg New York (1996).
5. Mumford C L (Valenzuela): Comparing representations and recombination operators for the multi-objective 0/1 knapsack problem, *Congress on Evolutionary Computation (CEC)* Canberra Australia (2003) 854–861.
6. Mumford C L (Valenzuela): A hierarchical approach to multi-objective optimization, *Congress on Evolutionary Computation (CEC)* Portland, Oregon (2004) (to appear).
7. Mumford-Valenzuela C L: A Simple Approach to Evolutionary Multi-Objective Optimization, In *Evolutionary Computation Based Multi-Criteria Optimization: Theoretical Advances and Applications*, edited by Ajith Abraham, Lakhmi Jain and Robert Goldberg. Springer Verlag (2004) London.
8. Oliver I M, Smith D J, and Holland J R C: A study of permutation crossover operators on the traveling salesman problem, *Genetic Algorithms and their Applications:Proceedings of the Second International Conference on Genetic Algorithms* (1987) 224–230.
9. Valenzuela C L: A simple evolutionary algorithm for multi-objective optimization (SEAMO), *Congress on Evolutionary Computation (CEC)*, Honolulu, Hawaii (2002) 717–722.
10. Zitzler E and Thiele L: Multiobjective evolutionary algorithms: a comparative case study and the strength pareto approach, *IEEE Transactions on Evolutionary Computation*, 3(4) (1999) 257–271.
11. Zitzler E, Laumanns M, and Thiele L: SPEA2: Improving the strength Pareto evolutionary algorithm, TIK-Report 103, Department of Electrical Engineering, Swiss Federal Institute of Technology (ETH), Zurich, Switzerland, {zitzler, laumanns, thiele}@tik.ee.ethz.ch.(2001) (Data and results downloaded from: http://www.tik.ee.ethz.ch/zitzler/testdata.html)

Dynamic and Scalable Evolutionary Data Mining: An Approach Based on a Self-Adaptive Multiple Expression Mechanism

Olfa Nasraoui, Carlos Rojas, and Cesar Cardona

Department of Electrical and Computer Engineering, The University of Memphis
Memphis, TN 38152
{onasraou,crojas,ccardona}@memphis.edu

Abstract. Data mining has recently attracted attention as a set of efficient techniques that can discover patterns from huge data. More recent advancements in collecting massive evolving data streams created a crucial need for dynamic data mining. In this paper, we present a genetic algorithm based on a new representation mechanism, that allows several phenotypes to be simultaneously expressed to different degrees in the same chromosome. This *gradual multiple expression* mechanism can offer a simple model for a *multiploid* representation with self-adaptive dominance, including *co-dominance* and *incomplete dominance*. Based on this model, we also propose a data mining approach that considers the data as a reflection of a dynamic environment, and investigate a new evolutionary approach based on continuously mining non-stationary data sources that do not fit in main memory. Preliminary experiments are performed on real Web clickstream data

1 Introduction and Motivation

1.1 The Need for "Adaptive Representation and Dynamic Learning" in Data Mining

Data mining has recently attracted attention as a set of efficient techniques that can discover patterns from huge data sets, and thus alleviate the information overload problem. The further advancement in data collection and measurements led to an even more drastic proliferation of data, such as sensor data streams, web clickstreams, network security data, news and intelligence feeds in form of speech, video and text, which in addition to scalability challenges, further stressed the fact that the environment in which we live is constantly changing. Thus, there is a crucial need for dynamic data mining, Specifically, within the context of data mining, there are two scenarios that call on dynamic learning:

(i) Scenario 1: The data supporting the learning task (including its nature, structure, and distribution), the goals of the learning task, or the constraints governing the feasible solutions for this task may be changing. A typical example today lies in mining sensor and data streams.

(ii) Scenario 2: The mechanism that is used to process the data for data mining may mimic the previous dynamic learning scenario. For instance, the size of the data may be huge, and thus it cannot fit in main memory, and we opt to process it incrementally, one

K. Deb et al. (Eds.): GECCO 2004, LNCS 3102, pp. 1401–1413, 2004.

sample at a time, or in chunks of data. In this case, there is no warranty that the different increments of data will reflect the same distribution. Hence this can be mapped to the previous dynamic learning scenario.

The type of flexibility and adaptation that is called for when learning in dynamic environments is nowhere to be found more than in nature itself. For instance, the way that DNA gets transcribed and synthesized into elaborate protein structures is dynamic. Genes get promoted and suppressed with varying degrees and in a dynamic way that adapts to the environment even within a single lifetime.

1.2   Contributions and Organization of This Paper

In this paper, we present the *Soft Structured Genetic Algorithm* (s^2GA) algorithm, and illustrate its use for non-stationary objective function optimization. We also adapt this approach to *evolutionary data mining* in non-stationary environments. s^2GA uses a *gradual multiple expression* mechanism that offers a simple model for a *multiploid* representation with self-adaptive dominance, including *co-dominance*, where both haploid phenotypes are expressed at the same time, as well as *incomplete dominance*, where a phenoptypical trait is expressed only to a certain degree (such as in certain flowers' colors).

Justifying the Choice of Multiploidy as the Underlying Adaptation Mechanism. Some work on dynamic optimization has solely relied on hypermutation to recover from environmental changes [1]. Furthermore, Lewis et al. [2] have empirically shown that high mutation rates, applied when an enviroment change is detected, can outperform a simple diploid representation scheme. However, in many data mining problems, the dimensionality is extremely high, ranging in the millions in the case of web usage and gene sequence data. For example, each URL on a website can be mapped to a different attribute. This will lead to an excessive devotion of the computing resources just for the bit mutations, and slow the search process. Moreover, the comparative results in [2] were based on diploidy with a simple adaptive dominance mechanism and uniform crossover that does not take into account the arbitrary permutations of the subchromosomes within the diploid chromosome. In fact, most existing multiploidy schemes perform the crossover in a blind way between two parent chromosomes without any consideration to the important information that differentiates each subchromosome from the others. When the dominance genes are evolved together with the structural information genes, this blind crossover can be shown to cause all the chromosomes and even their subchromosomes to converge to an identical copy in the long term. This in turn defeats the purpose of multiploidy which serves primarily as a memory bank and a source of diversity. For these reasons, we present a new *specialized* crossover that avoids this problem by encouraging crossover between only the most *similar* subchromosomes, hence preserving the diversity *within each* chromosome.

Problems with the Current State of the Art in Web Usage Mining and New Contributions. The majority of web mining techniques (see Section 2.2) assume that the entire Web usage data can reside in main memory. This can be a disadvantage for systems with limited main memory, since the I/O operations would have to be extensive to shuffle chunks of data in and out, and thus compromise scalability. Today's web sites are a source of an exploding amount of clickstream data that can put the scalability of any data mining technique into question. Moreover, the Web access patterns on a web site are very dynamic in nature, due not only to the dynamics of Web site content and structure, but also to changes in the user's interests, and thus their navigation patterns. The access patterns can be observed to change depending on the time of day, day of week, and according to seasonal and external events. As an alternative to locking the state of the Web access patterns in a frozen state depending on when the Web log data was collected and preprocessed, we propose an approach that considers the Web usage data as a reflection of a dynamic environment, and investigate a new evolutionary approach, based on a self-adaptive multiploidy representation, that continuously learns dynamic Web access patterns from non-stationary Web usage environments. This approach can be generalized to fit the needs of mining dynamic data or huge data sets that do not fit in main memory.

Organization of this Paper. The remainder of this paper is organized as follows. We start with a background overview in Section 2. Then, in Section 3, we present a modification to the GA, based on a soft multiple Expression mechanism, for non-stationary function optimization. Based on the soft multiple Expression GA model, we present in Section 4, an evolutionary approach, called *DynaWeb*, for mining dynamic Web profiles automatically from changing clickstream environments. In Section 5, we present simulation results for synthetic non-stationary fitness functions. Then, in Section 6, we present experimental results that illustrate the performance of *DynaWeb* in mining profiles from dynamic environments on a real website. Finally, we present our conclusions in Section 7.

2   Background

2.1   Genetic Optimization in Dynamic Environments

Dynamic objective functions can make the evolutionary search extremely difficult. Some work has focused on altering the evolutionary process, including the selection strategy, genetic operators, replacement strategy, or fitness modification [3,2,1], while other work focused on the concept of genotype to phenotype mapping or gene expression. This line of work includes models based on diploidy and dominance [4], messy GAs [5], Gene Expression Messy GA [6], overlapping genes such as in DNA coding methods [7,8, 9], the floating point representation [10], and the structured GA [11]. In particular, the structured GA (sGA) uses a structured hierarchical chromosome representation, where lower level genes are collectively switched on or off by specific higher level genes. Genes that are switched on are expressed into the final phenotype, while genes that are switched off do not contribute to coding the phenotype. A modification of the sGA based on the concept of soft activation mechanism was recently proposed with some preliminary results in [12]. This approach is detailed in Section 3.

2.2 Mining the Web for User Profiles

The World Wide Web is a hypertext body of close to 10 Billion pages (not including dynamic pages, crucial for interaction with Web Databases and Web services) that continues to grow at a roughly exponential rate in terms of not only content (total number of Web pages), but also reach (accessibility) and usage (user activity). Data on the Web exceeds 30 Terabytes on roughly three million servers. Almost 1 million pages get added daily, and typically, several hundred Gigabytes are changed every month. Hence, the Web constitutes one of the largest dynamic data repositories. In addition to its ever-expanding size and lack of structure, the World Wide Web has not been responsive to user preferences and interests. *Personalization* deals with tailoring a user's interaction with the Web information space based on information about him/her, in the same way that a reference librarian uses background knowledge about a person or *context* in order to help them better. The concept of *contexts* can be mapped to distinct user *profiles*. Mass profiling is based on general trends of usage patterns (thus protecting privacy) compiled from all users on a site, and can be achieved by mining user profiles from the historical *web clickstream* data stored in server access logs. A *web clickstream* is a virtual trail that a user leaves behind while surfing the Internet, such as a record of every page of a Web site that the user visits. Recently, data mining techniques have been applied to discover mass usage patterns or profiles from Web log data [13,14,15,16,17]. In [17], a *linear* complexity Evolutionary Computation technique, called Hierarchical Unsupervised Niche Clustering (H-UNC), was presented for mining both user profile clusters and URL associations in a *single* step. The evolutionary search allowed HUNC to exploit a subjective domain specific similarity measure, but it was limited to a stationary environment.

3 The Soft Multiple Expression Genetic Algorithm (s²GA)

In the *Soft Structured Genetic Algorithm* (s²GA), the lower level or structural information genes are no longer limited to total expression or to none. Instead, they can be expressed to different continuous degrees. Hence, several phenotypes can be simultaneously expressed in the same chromosome, but to different degrees. This *gradual multiple expression* mechanism can offer a simple model for a *multiploid* representation with self-adaptive dominance, including *co-dominance*, where both haploid phenotypes are expressed at the same time, as well as *incomplete dominance*, where a phenoptypical trait is expressed only to a certain degree (such as in the color of some flowers). Compared to the structured GA, in the soft activation mechanism, the activation of the subchromosomes in the lower levels is not a crisp value (active or not). Instead, every subchromosome has a soft activation/expression value in the interval $[0, 1]$. This allows the expression of multiple subchromosomes. To get this soft activation, the number of redundant subchromosomes is fixed to N_A. The dominance mechanism, traditionally used to decide the final phenotype that gets expressed is not fixed a priori, but rather adapts by evolution to express the best-fit subchromosomes depending on the current environment. The *dominance* or *activation* value for each subchromosome is controlled by a soft activation gene, A_i, a real number in the interval $[0, 1]$. The values for the soft activations are obtained as follows. In general, if there are N_A soft activation genes

A_i, $i \in 1, 2, \cdots, N_A$, each encoded on l_a bits, the value a_i for the soft activation gene A_i is:

$$a_i = \begin{cases} \frac{D_i}{\sum_{j=1}^{N_A} D_j}, & \text{if } \sum_{j=1}^{N_A} D_j \neq 0 \\ \frac{1}{N_A}, & \text{if } \sum_{j=1}^{N_A} D_j = 0 \end{cases} \qquad (1)$$

Where D_j is the decimal value of the l_a bits coding the A_j soft activation gene. Therefore $a_i \in [0, 1]$, and $\sum_{i=1}^{N_A} a_i = 1$. This has the advantage of keeping a chromosome with the same data encoding (binary) for both the activation and the information genes. The activation genes are constrained to sum to 1 in the preliminary model, but this constraint is not required. Hence, $\sum_{i=1}^{N_A} a_i = 1$. But they can be nonzero simultaneously. This means that *several* different expressions can co-exist in the same population, same generation, and same chromosome. It is this feature that is expected to allow for gradual adaptations of the genome to dynamic environments. The fitness computation of this genetic algorithm can consider all the subchromosome expressions in order to compute an aggregate fitness for the entire chromosome. This is accomplished by a weighted fitness. However, other aggregation mechanisms, such as the fitness of the *maximally activated* subchromosome, or the maximum of the fitnesses among the *sufficiently activated* subchromosomes, are possible. The weighted fitness is given by

$$f = \sum_{i=1}^{N_A} a_i f_i. \qquad (2)$$

Modified Two Point Crossover. In this modification, first, a usual two point crossover is made on the structural genes. The crossover points are selected such that an offspring inherits the same proportion of activation bits from the parent, as the proportion of structural bits, that is inherited. Then, a usual two point crossover is performed on the activation genes.

A New Specialized Crossover for Multiploid Chromosomes. This specialization performs an independent crossover for each information subchromosome. First, a measure of the distance (the phenotypical distance) between the subchromosomes of the parents is computed, and each subchromosome from one parent is paired with the most similar unpaired subchromosome from the other parent. Next, a one point crossover between the paired subchromosomes is done (some care is taken to guarantee that all the subchromosomes participate in the crossover). Finally, the activation genes are crossed, by performing a one point crossover between each pair of corresponding activation strings (the correspondence is obtained from the matching between the paired subchromosomes).

Advantages of the soft activation mechanism. The soft multiple expression and activation mechanism is expected to have the following advantages:

1. All the genotype data in the chromosome can be expressed to some degree. However, this level of expression can depend on the goodness and activation of *all* the subchromosomes.
2. The inherently redundant information, and the soft activation mechanism provide a *robust* chromosome. In order to damage the quality of the chromosome, a significant change must *concurrently* disrupt the data in the *activation and information* genes.
3. Depending on the activation values, and on how they are interpreted, more than one soft genotype can map to a single phenotype. Similarly, a single soft genotype can map to several phenotypes. This property has been lately recognized as very desirable to solve highly complex optimization problems [6].

4 DynaWeb: Mining Web Usage Data in Dynamic Environments

4.1 Extracting Web User Sessions

The access log for a given Web server consists of a record of all files accessed by users. Each log entry consists of: (i) User's IP address, (ii) Access time, (iii) URL of the page accessed, $\cdots$, etc. A user session consists of accesses originating from the same IP address within a predefined time period. Each URL in the site is assigned a unique number $j \in \{1, \ldots, N_U\}$, where N_U is the total number of valid URLs. Thus, the ith user session is encoded as an N_U-dimensional binary attribute vector $\mathbf{s}^{(i)}$ with the property

$$s_j^{(i)} = \begin{cases} 1 \text{ if the user accessed the } j^{th} \text{ URL during the } i^{th} \text{ session} \\ 0 \text{ otherwise} \end{cases}$$

4.2 Assessing Web User Session Similarity

Due to the asymmetric binary nature of the URL attributes, in this paper, we use the cosine similarity measure between two user-sessions, $\mathbf{s}^{(k)}$ and $\mathbf{s}^{(l)}$, given by $S_{kl} = \frac{\sum_{i=1}^{N_u} s_i^{(k)} s_i^{(l)}}{\sqrt{\sum_{i=1}^{N_u} s_i^{(k)}} \sqrt{\sum_{i=1}^{N_u} s_i^{(l)}}}$. Finally, this similarity is mapped to the dissimilarity measure $d_s^2(k, l) = (1 - S_{kl})^2$.

4.3 Mining Web User Profiles by Clustering Web Sessions

The proposed dynamic evolutionary Web mining algorithm, *DynaWeb* uses the s^2GA algorithm in representing and evolving the population. It uses the following representation: Each chromosome consists of N_A subchromosomes. Each subchromosome encodes a possible session prototype or profile that consists of a binary string of length N_U URLs, with same format as the binary session attribute vectors s_i defined in Section 4.1. Hence, each chromosome may encode different profiles, where each profile can be expressed to a certain degree in $[0, 1]$. The cosine based dissimilarity measure, defined in Section 4.2, is used to compute the distance between session data and candidate profiles.

The fitness value, f_i, for the ith candidate profile, $\mathbf{P}_i$, is defined as the density of a hypothetical cluster of Web sessions with $\mathbf{P}_i$ as a summarizing prototype or medoid. It is defined as $f_i = \frac{\sum_{j=1}^{N} w_{ij}}{\sigma_i^2}$, where w_{ij} is a robust weight that measures how typical a session s_j is in the ith profile, and is given by

$$w_{ij} = \exp -\frac{d_{ij}^2}{2\sigma_i^2}. \tag{3}$$

σ_i^2 is a robust measure of scale (dispersion) for the ith profile, d_{ij}^2 is a distance measure from session s_j to profile $\mathbf{P}_i$, and N is the number of data points. Note that the robust weights w_{ij} will be small for outliers, hence offering a means of distinguishing between good data and noise. The scale parameter that maximizes the fitness value for the ith profile can be found by setting $\frac{\partial f_i}{\partial \sigma_i^2} = 0$ to obtain $\sigma_i^2 = \frac{\sum_{j=1}^{N} w_{ij} d_{ij}^2}{2\sum_{j=1}^{N} w_{ij}}$. To get unbiased scale estimates, the above scale measure should be compensated by a factor of 2, which results in

$$\sigma_i^2 = \frac{\sum_{j=1}^{N} w_{ij} d_{ij}^2}{\sum_{j=1}^{N} w_{ij}}. \tag{4}$$

Therefore, w_{ij} and σ_i^2 will be alternatively updated using (3) and (4) respectively, for 3 iterations for each individual, starting with an initial value of $\sigma_{initial}^2$, and using the previous values of σ_i^2 to compute the weights w_{ij}. This *hybrid* genetic optimization converges much faster than a purely genetic search. More details about the underlying mechanism for stationary environments can be found in [18] and [17].

5   Simulation Results for Synthetic Non-stationary Fitness Functions

The s^{2}GA was applied to the alternating optimization of two non-overlapping objective functions, $F1$ and $F2$, defined in the interval $[0, 1]$, and each having a single peak with height $= 1$. These functions are translations of the function $F(x) = \left(\frac{27(-x^3+x^2)}{4}\right)^{10}$, and given by $F1(x) = F(0.8 - x)$ and $F2(x) = F(x - 0.2)$. The non-stationary optimization was based on periodical swappings between $F1$ and $F2$, as fitness functions, every $n = 15$ generations for a total of 300 generations. In all experiments, the population size was 200, the crossover rate was 0.9, and the mutation rates were 0.01 and 0.05 for the structural and activation bits, respectively. First, we plot the proportion of Good chromosomes (individuals that accomplish more than 80% of the optimal fitness value) for each one of the evaluated functions versus the generation number. Next, we plot the average and best chromosome performance (defined below) against the generation number. The entire procedure was repeated 30 times and average results are reported in the plots. The s^{2}GA representation consisted of 2 binary subchromosomes, each consisting of 10 structural information bits encoding a real number in $[0, 1]$. Each subchromosome was expressed by a 3-bit activation gene, resulting in a total chromosome length of 26.

The fitness function of the chromosome was defined as the weighted (by the activation values) aggregation of the fitnesses of all their subchromosomes. However, a single chromosome truly expresses different phenotypes. This led us to define the following measures: **(i) Activation threshold,** α: Sufficient activation value for considering a sub-chromosome as "activated". In our experimentats, we used $\alpha = 0.4$, i.e., 80% of the expected activation per gene, (i.e., $0.8(1/N_A)$ given a uniform activation distribution on N_A subchromosomes). **(ii) Subchromosome fitness:** subchromosome fitness evaluated using the current objective function. **(iii) Best Expressed Subchromosome:** subchromosome with highest subchromosome fitness among the ones with activation exceeding α. **(iv) Chromosome performance:** Fitness of the Best Expressed Subchromosome. In the new *specialized crossover*, special care is taken so that only similar subchromosomes are combined, regardless of their order inside the chromosome. From the point of view of exploitation, this recombination operator performs very well, contributing to the fast adaptation of the population to each new environment (see Figs. 1(c) and (d)).

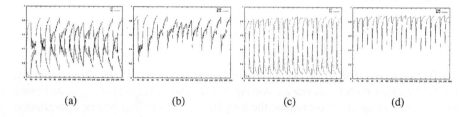

(a)   (b)   (c)   (d)

Fig. 1. Results for non-stationary function optimization, averaged over 30 runs (a,b) with *modified two point crossover* versus (c,d) with *specialized crossover*. (a,c) show Proportion of Good subchromosomes, while (b,d) show Average and Best Chromosome Performance.

6   Dynamic Web Usage Mining Experimental Results

The real clickstream data used in this section consists of 1703 sessions and 369 URLs extracted from Web logs of a department's website. The following experiment was performed to illustrate how an evolutionary algorithm can be used for mining dynamic data to discover Web user profiles. In order to simulate a non-stationary environment for Web mining in a controlled experiment, we used a coarse partition previously obtained and validated using H-UNC [17], and partially listed in Table 1, in order to consider the sessions that were assigned to each cluster as representing a different environment. Thus, each environment corresponds to a different Web usage trend. The sessions from these clusters were split into 20 different clickstream data sets, each one consisting of the sessions that are closest to one of the 20 profiles. The Genetic algorithm tried to evolve profiles, while facing a changing data set obtained by alternating the data from each of the 20 usage trends. The process was repeated for several epochs, each time presenting the succession of different data sets in alternation, simulating non-stationary observed usage trends.

Table 1. Summary of some usage trends previously discovered using Hierarchical Unsupervised Niche Clustering (only URLs with top 3 to 4 relevance weights shown in each profile)

i	$\|P_{T_i}\|$	P_{T_i}
0	106	{0.99 - /people_index.html}, {0.98 - /people.html}, {0.97 - /faculty.html}
1	104	{0.99 - /}, {1.00 - /cecs_computer.class}
2	177	{0.90 - /courses_index.html}, {0.88 - /courses100.html}, {0.87 - /courses.html}, {0.81 - /}
3	61	{0.80 - /}, {0.48 - /degrees.html}, {0.23 - /degrees_grad.html}
4	58	{0.97 - /degrees_undergrad.html}, {0.97 - /bsce.html}, {0.95 - /degrees_index.html}
5	50	{0.56 - /faculty/springer.html}, {0.38 - /faculty/palani.html}
6	116	{0.91 - /~saab/cecs333/private}, {0.78 - /~saab/cecs333}
12	74	{0.57 - /~shi/cecs345}, {0.45 - /~shi/cecs345/java_examples}, {0.46 - /~shi/cecs345/Lectures/07.html}
13	38	{0.82 - /~shi/cecs345}, {0.47 - /~shi}, {0.34 - /~shi/cecs345/references.html}
14	33	{0.55 - /~shi/cecs345}, {0.55 - /~shi/cecs345/java_examples}, {0.33 - /~shi/cecs345/Projects/1.html}
15	51	{0.92 - /courses_index.html}, {0.90 - /courses100.html}, {0.86 - /courses.html}, {0.78 - /courses200.html}
16	77	{0.78 - /~yshang/CECS341.html}, {0.56 - /~yshang/W98CECS341}, {0.29 - /~yshang}
19	120	{0.27 - /access}, {0.23 - /access/details.html}

We simulated the following *dynamic* scenarios:

scenario 1 (straight): We presented the sessions to *DynaWeb* one profile at a time for 50 generations each: sessions assigned to trend 0, then sessions assigned to trend 1, $\cdots$, until trend 19.

scenario 2 (reverse): We presented the sessions to *DynaWeb* one profile at a time for 50 generations each, but in *reverse* order: sessions assigned to trend 19, $\cdots$, until sessions assigned to trend 0.

scenario 3 (multi-trend): The sessions are presented in bursts of simultaneous multiple usage trends for 200 generation per multi-trend: First the sessions in profiles 7 and 8 are presented together for 200 generations, followed by the sessions in profiles 9 and 14, and finally by profiles 15 and 16, to test diversity as well as dynamic adaptation.

The proposed algorithm, *DynaWeb*, was applied with *specialized crossover*, a population of $N_P = 50$ individuals, initialized by selecting sessions randomly from the input data set, and with chromosome encoding based on 5 subchromosomes, each activated by one of $N_A = 5$ continuous valued activation genes. Each activation gene is encoded on 3 bits. The crossover probability was 0.9 per subchromosome, and the mutation probability was 0.01 per bit for the structural genes, and 0.05 per bit for the activation genes. The fitness of a chromosome was computed as the fitness of the subchromosome with maximum activation value in the case of scenarios 1 and 2, and as the combined fitness for scenario 3 to encourage diversity in this multimodal scenario. The ability of the population to evolve in a dynamic way when facing each new environment was evaluated in each generation by comparing the *good* individuals in the population to the ground-truth profiles, P_{T_i}, $(i = 0, \cdots, 19)$. To do this, we defined as *good* individuals, those individuals that have a combined fitness exceeding $(f_{max} + f_{avg})/2$, where f_{max} and f_{avg} are the maximal and average fitness in the current generation, respectively. Before comparing an individual to the ground truth profiles, an expressed phenotype must first be extracted. In our case, the active (i.e., with activation gene value $> \alpha$)

subchromosome with highest fitness, was used to yield the final expressed phenotype. It is this phenotype that is compared with each of the ground-truth profiles in each generation. We do this by computing the cosine similarity between the phenotype expressed by each *good* chromosome and each of the ground-truth profiles, P_{Ti}, $i = 0, \cdots, 19$. The similarities computed using all the good chromosomes are averaged in each generation, to yield measures $\hat{S}_i$ for each ground-truth profile, P_{Ti}, $i = 0, \cdots, 20$. These measures are used to assess whether the evolution is able to adapt to each change in the environment. Ideally, adaptation to the ith environment is quantified by the fact that $\hat{S}_i$ gradually becomes higher than all other $\hat{S}_j$, $j \neq i$.

The above procedure was repeated 20 times and the results are averaged. Stochastic *Viral injection/replacement* was used. This phenomenon is different from traditional evolutionary techniques, in that genetic material from an external organism gets injected into the host organism's DNA. It is common with viruses such as the AIDS virus. Given the nature of our *data driven* approach, it is expected that this operation will refresh the current genome with vital and current information from the new environment. This step stochastically replaced with a 0.3 injection rate per generation the most active subchromosome from the worst individual of the current population (based on their combined chromosome fitness) with data randomly selected from the data set being presented in the current generation. The results for *scenario 1: straight order* are shown in Fig. 2 and Fig. 3, for *DynaWeb* and the *Simple GA*, respectively. Fig. 2, *which is better viewed in color*, shows that as each environment comes into context, the genomes in the current population gradually evolve to yield candidate profiles that match the new environment. That is, whenever the environment changes from j to i, the similarity measure that is the highest gradually switches from being $\hat{S}_j$ to becoming $\hat{S}_i$. Hence, the genome succeeds in tracking the dynamic web usage trends, which is the desired goal. We have also observed a successful adaptation of the expression/activation genes, switching between different parts of the chromosome to track the changing environments. We note that the average similarity, $\hat{S}_i$, achieved for certain usage environments (such as profile 19) are relatively low. This is because the sessions in these environments have more variability, contain more noise, and thus form a less compact cluster, as can be judged by their lower URL relevance weights in Table 1. Fig. 2 also shows a desired property in the cross-reaction between overlapping usage trends. For example the first 5 usage trends overlap significantly since they represent outside visitors to the website, mostly prospective students, with slightly different interests. Fig. 3 shows that the simple GA yields a population that is too slow to adapt, and with lower quality.

The results using *DynaWeb* for *scenario 2: reverse order* and for *scenario 3: multi-trend* are shown in Figure 4 and Figure 5, respectively. Figure 4 shows that the order of presentation of the environments is not important, since it is merely a vertical reflection of the evolution for scenario 1. Figure 5 shows the ability of *DynaWeb* to track multiple profiles simultaneously, even as they change. Except for the first epoch, the remaining epochs show a consistent adaptation to the presented usage trends, since the population achieves highest similarity to the two current usage trends, as compared to the remaining 4 trends. The improvement in adaptation starting from the second cycle shows the presence of a good *memory* mechanism that is distributed over the different subchromosomes of

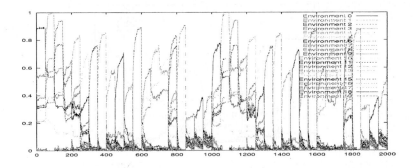

Fig. 2. Average similarity to ground-truth profiles among good individuals averaged for 20 runs, for scenario 1 with DynaWeb, $N_A = 5$ subchromosomes, 0.3 injection, for scenario 1 (Straight order of usage trends)

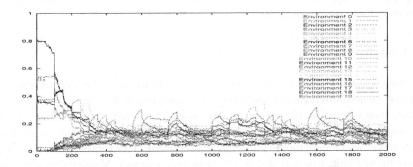

Fig. 3. Average similarity to ground-truth profiles among good individuals averaged for 20 runs, for scenario 1 with the Simple GA

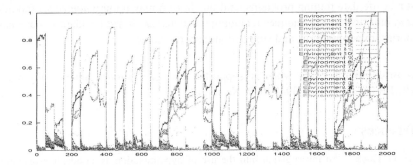

Fig. 4. Average similarity to ground-truth profiles among good individuals averaged for 20 runs, for DynaWeb with $N_A = 5$ subchromosomes, 0.3 injection, for scenario 2 (Reverse order of usage trends)

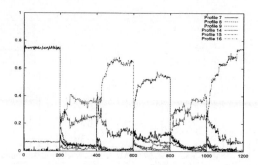

Fig. 5. Average similarity to ground-truth profiles among good individuals averaged for 30 runs, for DynaWeb with $N_A = 5$ subchromosomes, 0.3 injection, for scenario 3 (alternating multi-usage trends)

the population, a memory that comes into context, i.e. becomes expressed when it is relevant in the current context, and goes dormant in other contexts.

7   Conclusion

For many data mining tasks, the subjective objective functions and/or dissimilarity measure may be non-differentiable. Evolutionary techniques can handle a vast array of subjective, even non-metric dissimilarities. We proposed a new framework that considers evolving data, such as in the context of mining stream data, as a reflection of a dynamic environment which therefore requires dynamic learning. This approach can be generalized to mining huge data sets that do not fit in main memory. Massive data sets can be mined in parts that can fit in the memory buffer, while the evolutionary search adapts to the changing trends automatically. While it is interesting to compare the proposed approach against other standard dynamic optimization strategies, one must keep in mind that *domain knowledge*, *scalability*, and a *data-driven* learning framework are crucial to most real life data mining problems, and this in turn may require *nontrivial* modifications to most existing techniques including those that are based on adaptive case-based memories, hypermutation, and simple dominance schemes.

Acknowledgment. This work is supported by the National Science Foundation (CAREER Award IIS-0133948 to O. Nasraoui).

References

1. H. G. Cobb, "An investigation into the use of hypermutation as an adaptive operator in genetic algorithms having continuous, time-dependent nonstationary environments," Tech. Rep. AIC-90-001, Naval Research Laboratory, Washington, 1990.
2. J. Lewis, E. Hart, and R. A.Graeme, "A comparison of dominance mechanisms and simple mutation on non-stationary problems," in *5th International Conference on Parallel Problem Solving from Nature*, 1998, pp. 139–148.

3. J. Branke, "Evolutionary approaches to dynamic optimization problems: A survey," *Evolutionary Algorithms for Dynamic Optimization*, pp. 134–137, 1999.
4. D. Goldberg and R. E. Smith, "Nonstationary function optimization using genetic algorithms with diloidy and dominance," in *2nd International Conference on Genetic Algorithms*, J. J. Grefensette, Ed., Lawrence, 1987, pp. 59–68.
5. D. Goldberg, K. Deb, and B. Korb, "Nonstationary messy genetic algorithms: motivation, analysis, and first results," *Complex Systems*, vol. 3, pp. 493–530, 1987.
6. H. Kargupta, "The gene expression messy genetic algorithm," in *International Conference on Evolutionary Computation*, 1996.
7. W. Wienholt, "A refined genetic algorithm for parameter optimization problems," in *5th International Conference on Genetic Algorithms*, 1993.
8. D. K.Burke, J. DeJong, C. Grefensette, and A. Wu, "Putting more genetics into genetic algorithms," *Evolutionary Computation*, vol. 6, no. 4, 1998.
9. A. Wu and R. K. Lindsay, "Empirical studies of the genetic algorithm with non-coding segments," *Evolutionary Computation*, vol. 3, no. 2, pp. 121–147, 1995.
10. A. Wu and R. K. Lindsay, "A comparison of the fixed and floating building block representation in the genetic algorithm," *Evolutionary Computation*, vol. 4, no. 2, pp. 169–193, 1996.
11. D. Dasgupta and D. McGregor, "Nonstationary function optimization using structured genetic algorithm," in *Parallel Problem Solving For Nature Conference*, Belgium, 1992.
12. O. Nasraoui, C. Rojas, C. Cardona, and D. Dasgupta, "Soft adaptive multiple expression mechanism for structured and multiploid chromosome representations," in *Genetic and Evolutionary Computation Conference, late breaking papers*, Chicago, July 2003.
13. O. Zaiane, M. Xin, and J. Han, "Discovering web access patterns and trends by applying olap and data mining technology on web logs," in *Advances in Digital Libraries*, Santa Barbara, CA, 1998, pp. 19–29.
14. M. Perkowitz and O. Etzioni, "Adaptive web sites: Automatically synthesizing web pages," in *AAAI 98*, 1998.
15. R. Cooley, B. Mobasher, and J. Srivastava, "Data preparation for mining world wide web browsing patterns," *Knowledge and Information Systems*, vol. 1, no. 1, 1999.
16. O. Nasraoui, H. Frigui, R. Krishnapuram, and A. Joshi, "Mining web access logs using relational competitive fuzzy clustering," in *8th International World Wide Web Conference*, Toronto, Canada, 1999.
17. O. Nasraoui and R. Krishnapuram, "A new evolutionary approach to web usage and context sensitive associations mining," *International Journal on Computational Intelligence and Applications - Special Issue on Internet Intelligent Systems*, vol. 2, no. 3, pp. 339–348, 2002.
18. O. Nasraoui and R. Krishnapuram, "A novel approach to unsupervised robust clustering using genetic niching," in *Ninth IEEE International Conference on Fuzzy Systems*, San Antonio, TX, May 2000, pp. 170–175.

Crossover, Population Dynamics, and Convergence in the GAuGE System

Miguel Nicolau and Conor Ryan

Biocomputing and Developmental Systems Group
Computer Science and Information Systems Department
University of Limerick, Ireland
{Miguel.Nicolau, Conor.Ryan}@ul.ie

Abstract. This paper presents a study of the effectiveness of a recently presented crossover operator for the GAuGE system. This crossover, unlike the traditional crossover employed previously, preserves the association of positions and values which exists in GAuGE genotype strings, and as such is more adequate for problems where the meaning of an allele is dependent on its placement in the phenotype string. Results obtained show that the new operator improves the performance of the GAuGE system on simple binary problems, both when position-sensitive data is manipulated and not.

1 Introduction

The GAuGE (Genetic Algorithms using Grammatical Evolution) [13,9] system is a recent approach to position-independence in the field of genetic algorithms. Each individual in a GAuGE genotype string is composed of a sequence of position and value specifications which, through a genotype-to-phenotype mapping process similar to that of the GE (Grammatical Evolution) [12] system, ensures that each position in the resulting phenotype string is always specified, but only once. This mapping process produces a very compact and efficient representation, with neither under- nor over-specification of phenotypic information.

Until recently, a simple genetic algorithm [6] was used to select, combine and mutate genotype strings, which are then mapped using the GAuGE system to produce phenotypic information, to be evaluated. This approach, simple and elegant as it may be, presents some drawbacks. By allowing genetic operators which do not respect the representation of GAuGE genotype strings to manipulate those strings, associations between positions and values, which were previously discovered, may be lost in future generations.

To prevent this scenario from happening, a set of new crossover operators has been introduced [10], which, when applied, do not disrupt the associations between positions and values. Three different problem domains were tackled on that study, and some of the new operators showed a significant improvement of performance, when compared to the original GAuGE approach.

Of those operators, the *pure* crossover showed the most significant improvement in performance, and therefore a more detailed analysis of that operator is

K. Deb et al. (Eds.): GECCO 2004, LNCS 3102, pp. 1414–1425, 2004.
© Springer-Verlag Berlin Heidelberg 2004

done in the current paper. By restricting the problem domain to that of binary pattern matching, and by turning off the mutation operator on all systems, the advantages of the new crossover become clearer and easier to analyse. The results obtained show that, by respecting the underlying GAuGE representation of the genotype strings, the new crossover operator significantly improves the performance of the GAuGE system, even on irregular binary pattern problems, where using the traditional crossover resulted in a heavy loss of performance.

This paper is structured as follows. The next section presents the GAuGE system, its mapping process, and the crossover operators tested. Section 3 presents the experiments conducted and their results, while Section 4 analyses those results. Finally, Section 5 draws some conclusions on this work, and highlights future work directions.

2   GAuGE

The main principle behind the GAuGE system is the separate encoding of the position and value of each phenotypic variable. Its mapping process interprets each (fixed-length) genotype binary strings as a sequence of (*position,value*) pairs, which are used to build a fixed-length phenotype string. This mapping process ensures that each position of the phenotype string is specified exactly once, in much the same way that the GE system, an automatic programming system using grammars, ensures that each codon from the genotype strings chooses an existing production from a specified grammar.

Another feature that GAuGE shares with GE is functional dependency between genes, which is a direct result from the mapping process used. In GE, when a production has been chosen by a codon, the set of available choices available to the next codons changes; in other words, previous choices of grammar productions affect the available choices for the current codon. A similar effect occurs in GAuGE; in its mapping process, the set of free positions left in the phenotype string for the current gene to choose from depends on the choices of previous genes. In short, there is a functional dependency across the genotype strings in both GE and GAuGE, as the function of previous genes dictates the function of the following ones.

Finally, as GE uses the *mod* operator to map gene values to a choice of productions from a grammar rule, this creates a many-to-one mapping from the genotype strings onto the phenotype programs, leading to the occurrence of neutral mutations [7], which in turn introduce variety at the genotypic level. This also occurs in GAuGE, as each position specified at the genotypic level is mapped onto a set of available positions in the phenotype string. It has also been shown that the explicit introduction of degeneracy can reduce structural bias at the genotypic level [9].

2.1   Previous Work

Previous work has used similar approaches and techniques as the ones employed in GAuGE. Some of Bagley's [1] computer simulations used an extended string representation to encode both the position and the value of each allele, and used an inversion operator to affect the ordering of genes. Holland [6] later presented modifications to the schema theorem, to include the approximate effect of the inversion operator. To tackle the problems associated with the combination of the inversion and crossover operators, these were later combined into a single operation, and a series of reordering operators were created [11].

The so-called messy genetic algorithms applied the principle of separating the *gene* and *locus* specifications with considerable success [4], and have since been followed by many competent GAs.

Work by Bean [2] with the Random Keys Genetic Algorithm (RKGA) hinted that a tight linkage between genes would result in both a smoother transition between parents and offspring when genetic operators are applied, and an error-free mapping to a sequence of ordinal numbers. More recently, Harik [5] has applied the principles of functional dependency in the Linkage Learning Genetic Algorithm (LLGA), in which a chromosome is expressed as a circular list of genes, with the functionality of a gene being dependent on a chosen interpretation point, and the genes between that point and itself.

2.2   GAuGE Mapping

A full description and analysis of the GAuGE mapping process can be found elsewhere [9]. As an example of this process, consider a simple problem composed of four phenotypic variables ($\ell = 4$), ranging between the values 0 and 7 (*range* = 8). The evolutionary algorithm maintains a genotype population G, of N individuals.

The length of each individual depends on a chosen position field size (pfs) and a value field size (vfs). As this problem is composed of four variables, $pfs = 2$ has been chosen, as that is the minimum number of bits required to encode four positions; for the value fields, a value of $vfs = 4$ has been chosen, to introduce degeneracy in the coding of values (the minimum number of bits required for the range specified is three). The required length of each string G_i, of the genotypic space G, is therefore $L = (pfs + vfs) \times \ell = (2 + 4) \times 4 = 24$.

For example, take the following individual as an example genotype string:

$$G_i = 000101111101111001010010$$

The mapping process will proceed to create a phenotype string P_i. It consists in four steps[1]:

$$\Phi : G \xrightarrow{\Phi_1} X \xrightarrow{\Phi_2} D \xrightarrow{\Phi_3} R \xrightarrow{\Phi_4} P$$

The first mapping process (Φ_1) consists in creating an integer string, using the chosen pfs and vfs values:

$$X_i = \left((X_i^j, \tilde{X}_i^j) \right)_{0 \le j \le \ell - 1} = ((0, 5), (3, 13), (3, 9), (1, 2))$$

---

[1] In the actual implementation of GAuGE, some of these steps can be reduced.

The second mapping process (Φ_2) consists in interpreting this string as a sequence of four (*position,value*) pairs, to create a string of *desired* positions D_i and a string of *desired* values $\tilde{D}_i$.

These are created by mapping each position field onto the number of positions left in the phenotype string. For the first position field, $X_i^0 = 0$, the desired position specified is calculated by $(X_i^0 \bmod \ell) = (0 \bmod 4) = 0$, as at this stage no positions have been specified yet. The value field is calculated using the range of phenotypic values, giving $(\tilde{X}_i^0 \bmod range) = (0 \bmod 8) = 5$.

The second set of specifications is calculated in a similar way. For the position field, the desired position specified is calculated by $(X_i^1 \bmod (\ell - 1)) = (3 \bmod 3) = 0$, as only three positions remain unspecified in the phenotype string. The value field is calculated as before, giving $(\tilde{X}_i^1 \bmod range) = (13 \bmod 8) = 6$.

After processing all four pairs, the string of desired specifications are:

$$D_i = (0, 0, 1, 0) \qquad \tilde{D}_i = (5, 6, 1, 2)$$

At this stage, it can be seen that there are some conflicts in the position specifications (position 0 is specified three times, and positions 2 and 3 are still unspecified). The third mapping process (Φ_3) consists in removing these conflicts, creating a string of *real* positions R_i and a string of *real* values $\tilde{R}_i$.

These are created as follows. The first position specified, 0, is kept, as there are no conflicts at this stage, so $R_i^0 = 0$ (i.e. the first position on the phenotype string). The desired value specified, 5, is mapped to the range of the first phenotypic variable; as all variables share the same range in this problem, the *real* value specification is the same as before, $(5 \bmod 8) = 5$. An X sign is used to signal positions already taken in the phenotype string:

$$R_i = (0, ?, ?, ?) \qquad \tilde{R}_i = (5, ?, ?, ?) \qquad P_i = (X, ?, ?, ?)$$

We then take the second desired position, 0, and perform a similar mapping. As the value specified is 0, it is interpreted as being the first **available** position of the phenotype string; as the position 0 has already been taken, the first available position is 1. The value specification is calculated as before, giving:

$$R_i = (0, 1, ?, ?) \qquad \tilde{R}_i = (5, 6, ?, ?) \qquad P_i = (X, X, ?, ?)$$

The third set of specifications is calculated in the same fashion. Its position specification is calculated by $(1 \bmod 2) = 1$, that is, the second available position in the phenotype string, while the value specification remains unchanged, giving:

$$R_i = (0, 1, 3, ?) \qquad \tilde{R}_i = (5, 6, 1, ?) \qquad P_i = (X, X, ?, X)$$

Finally, the fourth pair is handled in the same fashion, giving the final *real* specification strings:

$$R_i = (0, 1, 3, 2) \qquad \tilde{R}_i = (5, 6, 1, 2) \qquad P_i = (X, X, X, X)$$

The fourth and final mapping step (Φ_4) simply consists in interpreting these specifications, creating a phenotype string by using the formula:

$$P_i^{R_i^j} = \tilde{R}_i^j \tag{1}$$

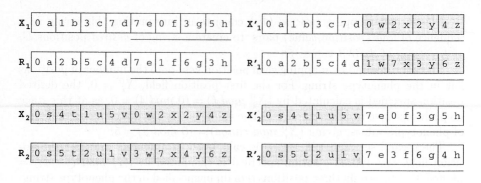

Fig. 1. Standard crossover operator for the GAuGE system. Two individuals, X_1 and X_2 exchange information after the fourth pair, generating the offspring X'_1 and X'_2.

In other words, through a permutation defined by R_i, the elements of $\tilde{R}_i$ are placed in their final positions. The phenotype string, ready for evaluation, is:

$$P_i = (5, 6, 2, 1)$$

2.3   Crossover Operators

Standard Crossover. This crossover operator has been used with GAuGE in all experiments up to now. It is a one-point crossover, operating at the genotype level, but with crossover points limited to pair boundaries; that means that there are $\ell - 1$ possible crossover points between each individual (every $pfs + vfs$ bits).

An example of how this operator works is shown in Figure 1. Two individuals, randomly generated using a problem of size $\ell = 8$, are shown, already expressed as X_i strings and their corresponding R_i strings[2]. By choosing to crossover these individuals after the fourth pair, two offspring are generated, X'_1 and X'_2.

As can be seen, each child keeps the information from the first half of one parent, and uses the second half of the other parent to fill in the remaining unspecified positions. This has the side effect that the values specified in the second half of each parent do not necessarily stay in their original positions. In the example, the first parent specified that values (e,f,g,h) should be located at positions (7,0,3,5), respectively, which correspond to the *real* positions $R_1 =$ (... ,7,1,6,3). However, when those specifications are interpreted within the context of the second child, they now correspond to the *real* positions $R'_2 =$ (... ,7,3,6,4), as the *real* position 1 was already specified in that child's left side, creating a chain of changes.

This change (or adaptation) of the second half specifications to the new context upon which they are now interpreted is known as the *ripple effect* [12]. Although the way those specifications are interpreted can be quite different when in a new context, it is not random; indeed, the ordering relationship between those specifications is kept. In the example provided, this means that since the

---

[2] With values a...h for the first individual, and s...z for the second individual.

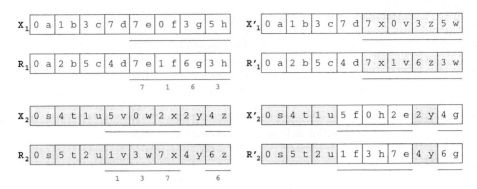

Fig. 2. Pure crossover for the GAuGE system. A crossover point is chosen on the first parent, and the corresponding value specifications from the second parent are used to create the first offspring; the complementary operation is used to create the second offspring. Both offspring keep the structure of their corresponding parent.

values (e,f,g,h) appeared in the order (g,h,f,e) in the phenotype string, then this ordering will be kept in the second child's phenotype.

Pure Crossover. This is a new crossover operator, designed to respect the *(position,value)* associations of GAuGE strings. It works by maintaining the structure of each parent on the offspring strings, but exchanging value specifications, corresponding to the positions specified after the crossover point.

An example of how this operator works is shown in Figure 2. The first offspring (X'_1) keeps the position specifications of the first parent (X_1), and the value specifications of the first half of that parent, up to the crossover point. After that point, the values specified by the second parent, corresponding to the same *real* positions, are used instead.

The second offspring (X'_2) is produced in a similar fashion. It keeps the position specifications of the second parent (X_2), and the value specifications *which are not required by the first offspring*; all other value specifications are taken from the first parent, corresponding to the same positions.

In the example provided, it can be seen that the *real* positions (1,3,7,6), in the 4th, 5th, 6th and 8th pairs of the second offspring, receive the corresponding values (f,h,e,g) from the first parent, as these are the *real* positions specified in the second half of that parent.

3   Experiments and Results

To test how effectively the new operator maintains the association between positions and values, a set of four binary problems was used. These problems share the common feature that the fitness contribution of each variable is the same, regardless of its location (i.e. no salience). In the two first problems, onemax and zeromax, all alleles have the same value on the global optimum, regardless of

Table 1. Experimental setup, used on all experiments

Problem length (ℓ):	128
Population size (N):	100
Number of generations:	100
Position field size (pfs):	7 bits
Value field size (vfs):	1 bit
Crossover probability:	1.0
Position field mutation probability:	0.0
Value field mutation probability:	0.0

their position, whereas on the other problems, the association between position and value is important and must be kept.

These problems were used as it is easy to demonstrate and visualise the effects on the population of the genetic operators used. By understanding how these operators affect population dynamics and performance in these simple binary problem domains, important information is gathered which can be used on the design and refinement of these operators.

The GAuGE system using the two crossover operators was compared to a simple GA. The experimental setup used on all experiments is shown in Table 1. In these experiments, the mutation operator was turned off, to test how effectively the crossover operators combine the information that is currently on the population.

3.1   Onemax

The onemax problem is a well-known problem in the GA community. It is defined by the following formula:

$$f(x) = \sum_{i=0}^{\ell-1} x_i \qquad x_i \in \{0, 1\}$$

where ℓ is the phenotype length, and x_i the allele at position i within that string (with positions ranging from 0 to $\ell - 1$). The best individual is a binary string composed of all 1s.

This problem has been used before to demonstrate that the GAuGE mapping process does not impair its performance on simple binary maximisation problems.

3.2   Zeromax

This problem is the opposite of the onemax problem. It is defined by the formula:

$$f(x) = \sum_{i=0}^{\ell-1} 1 - x_i \qquad x_i \in \{0, 1\}$$

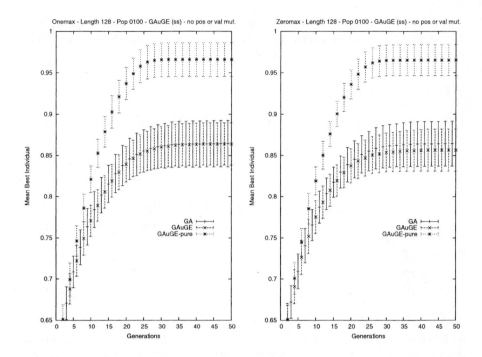

Fig. 3. Results obtained for the onemax (left) and zeromax (right) problems with length 128. The *x-axis* shows the generation number, and the *y-axis* the mean best individual (from 100 independent runs). The vertical error bars plot the standard deviation for all runs, for each system.

where ℓ is the phenotype length, and x_i the allele at position i within that string (with positions ranging from 0 to $\ell - 1$). In this case, the best individual is a binary string composed of all 0s; as with the onemax problem, the fitness contribution of each variable is the same.

This experiment was chosen, along with the onemax problem, to test the performance of the systems being compared under easy maximisation problems, where the location of an allele in the genotype string is unimportant. The results obtain for the onemax and zeromax problems are shown in Figure 3.

3.3   Zero-Onemax

This is yet another binary matching problem. It is defined by the formula:

$$f(x) = \sum_{i=0}^{\ell-1} |(i+1) \bmod 2 - x_i| \qquad x_i \in \{0, 1\}$$

where ℓ is the phenotype length, and x_i the allele at position i within that string (with positions ranging from 0 to $\ell - 1$). For this problem, the best individual is a binary string composed of 0s and 1s constantly alternated.

This problem was used as the location of an allele on the phenotype string is important. An operator which does not respect the association between alleles and their positions should perform badly on this kind of problem.

3.4 Binary Matching

This is the last problem analysed. It is defined by the formula:

$$f(x) = \sum_{i=0}^{\ell-1} |y_i - x_i| \qquad x_i, y_i \in \{0, 1\}$$

where ℓ is the phenotype length, x_i is the allele at position i within that string (with positions ranging from 0 to $\ell - 1$), and y_i is the element at position i of a randomly created binary string. The best individual in this case is a binary string equal to the randomly created one. The following string was used:

001110110000110110111101011010110011000010000111000101100011 0010
1001011101011011000100001100010111000111110001000100110010101011

As with the zero-onemax problem, in this problem the associations between positions and values are important, as a value 1 will only contribute to the fitness of an individual if placed on the second half of its phenotype string. The results obtain for this and the zero-onemax problems are shown in Figure 4.

4 Analysis

The results obtained for the first two problems show both the simple GA and the original GAuGE system have a similar performance, whereas GAuGE with the new crossover operator shows a significantly better performance than those two systems. All three systems have a similar behaviour for these two problems, which was to be expected.

For the remaining problems, however, there is a significant drop in performance for the original GAuGE system, whereas GAuGE with the new crossover and the simple GA have a similar performance as on the previous problems (Figure 4). The reason for this difference in performance between the standard and pure crossover operators is explained by representation convergence: while the standard crossover requires the population to converge in its representation to keep the association between positions and values [14], the pure crossover always respects those associations even when individuals do not share the same representation. As a result, from the first generation, the pure crossover exchanges valuable information between individuals, while the population slowly converges in its representation [8] (due to selection pressure); the standard crossover is however actively working in the population to achieve a convergence in the representation of individuals, and only then is it capable of exchanging sensible information, as at that stage a crossover operation between two individuals will not break position-value associations.

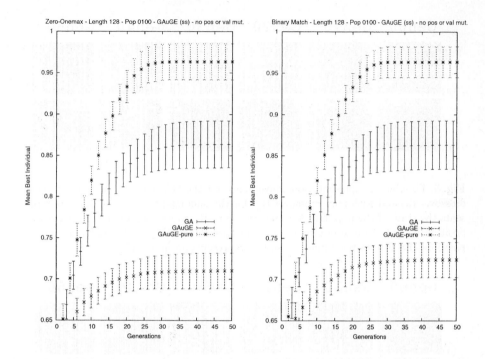

Fig. 4. Results obtained for the zero-onemax and binary-matching problems with length 128. The *x-axis* shows the generation number, and the *y-axis* the mean best individual (from 100 independent runs). The vertical error bars plot the standard deviation for all runs, for each system.

This behaviour can be seen by observing Figures 5 and 6, for the onemax and zero-onemax experiments (respectively). These figures plot a typical run of the GAuGE system, with both crossovers. Each square represents the state of the population at the specified generation; each horizontal line inside the square represents an individual (run settings are the same as on the previous experiments). The information plotted for each individual is its representation, that is, the R_i string: position 0 is represented by a black dot, position 127 is represented by a white dot, and all interim positions are represented with grey levels in between. Individuals sharing the same representation cause the vertical lines observed. It can be seen in both figures that representation converges faster with the standard crossover, as this is required for sensible information to be exchanged between individuals; on the contrary, the representation is slower to converge with the pure crossover[3], and even at generation 50, when all individuals share the same fitness, different representations co-exist in the population.

---

[3] It does converge, although at a slower rate, as a result of selection pressure.

Gen00   Gen10   Gen20   Gen30   Gen40   Gen50

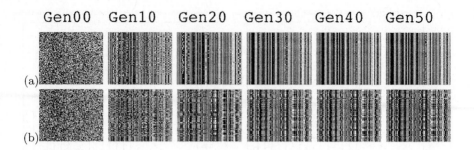

Fig. 5. Population representation convergence for the GAuGE system, with standard crossover (a) and with pure crossover (b), for the onemax problem. Each square represents the state of the population at generation 0, 10, and so on; a black dot represents position 0, and a white dot represents position 127, with grey levels for all interim positions. Each horizontal line in a square represents an individual.

Gen00   Gen10   Gen20   Gen30   Gen40   Gen50

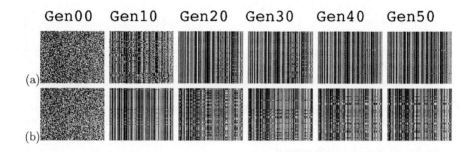

Fig. 6. Population representation convergence for the GAuGE system, with standard crossover (a) and with pure crossover (b), for the zero-onemax problem. Each square represents the state of the population at generation 0, 10, and so on; a black dot represents position 0, and a white dot represents position 127, with grey levels for all interim positions. Each horizontal line in a square represents an individual.

5   Conclusions and Future Work

The performance of the pure crossover for the GAuGE system has been analysed in this paper. By adapting to the representation of GAuGE genotype strings, this crossover does not depend on representation convergence to exchange context-sensitive data, and as such is fast an effective in combining information present in the population.

Future work will continue the analysis and possible enhancement to the presented operator, and the design of a reordering genetic operator [3], to maintain diversity at representation level, but without breaking the association between values and their positions. This operator should also allow for the discovery and maintenance of linkages between genotypic locations.

Acknowledgments. The authors would like to thank an anonymous reviewer of a previous paper, whose comments and suggestions lead to the investigation presented in this work.

References

1. Bagley, J. D.: The behaviour of adaptive systems which employ genetic and correlation algorithms. Doctoral Dissertation, University of Michigan (1967)
2. Bean, J.: Genetic Algorithms and Random Keys for Sequencing and Optimization. ORSA Journal on Computing, Vol. **6**, No. 2. (1994) 154-160
3. Chen, Y. and Goldberg, D. E.: An Analysis of a Reordering Operator with Tournament Selection on a GA-Hard Problem. In: Cantu-Paz et al., (eds.): Genetic and Evolutionary Computation - GECCO 2003. Springer. (July 2003) 825-836
4. Goldberg, D. E., Korb, B., and Deb, K.: Messy genetic algorithms: Motivation, analysis, and first results. Complex Systems, Vol. **3**. (1989) 493-530
5. Harik, G.: Learning Gene Linkage to Efficiently Solve Problems of Bounded Difficulty Using Genetic Algorithms. Doctoral Dissertation, University of Illinois (1997)
6. Holland, J. H.: Adaptation in Natural and Artificial Systems. Ann Arbor, MI: University of Michigan Press. (1975)
7. Kimura, M.: The Neutral Theory of Molecular Evolution. Cambridge University Press. (1983)
8. Nicolau, M. and Ryan, C.: How Functional Dependency Adapts to Salience Hierarchy in the GAuGE System. In: Ryan et al, (eds.): Proceedings of EuroGP-2003. Lecture Notes in Computer Science, Vol. 2610. Springer-Verlag. (2003) 153-163
9. Nicolau, M., Auger, A., and Ryan, C.: Functional Dependency and Degeneracy: Detailed Analysis of the GAuGE System. In: Liardet et al, (eds.): Proceedings of Évolution Artificielle 2003. Lecture Notes in Computer Science (to be published). Springer-Verlag. (2003)
10. Nicolau, M. and Ryan, C.: Efficient Crossover in the GAuGE system. In: Keijzer et al, (eds.): Proceedings of EuroGP-2004. Lecture Notes in Computer Science (to be published). Springer-Verlag. (2004)
11. Oliver, I. M., Smith, D. J., and Holland, J. R. C.: A Study of Permutation Crossover Operators on the Traveling Salesman Problem. In: Proceedings of the Second International Conference on Genetic Algorithms. (1987) 224-230
12. O'Neill, M. and Ryan, C.: Grammatical Evolution - Evolving programs in an arbitrary language. Kluwer Academic Publishers. (2003)
13. Ryan, C., Nicolau, M., and O'Neill, M.: Genetic Algorithms using Grammatical Evolution. In: Foster et al, (eds.): Proceedings of EuroGP-2002. Lecture Notes in Computer Science, Vol. 2278. Springer-Verlag. (2002) 278-287
14. Ryan, C. and Nicolau, M.: Doing Genetic Algorithms the Genetic Programming Way. In: Riolo, R., and Worzel, B. (eds.): Genetic Programming Theory and Practice. Kluwer Publishers, Boston, MA. (2003) 189-204

Inducing Sequentiality Using Grammatical Genetic Codes

Kei Ohnishi, Kumara Sastry, Ying-Ping Chen, and David E. Goldberg

Illinois Genetic Algorithms Laboratory (IlliGAL)
University of Illinois at Urbana-Champaign
104 S. Mathews Ave, Urbana, IL 61801, USA
{kei,kumara,ypchen,deg}@illigal.ge.uiuc.edu

Abstract. This paper studies the inducement of *sequentiality* in genetic algorithms (GAs) for uniformly-scaled problems. Sequentiality is a phenomenon in which sub-solutions converge sequentially in time in contrast to uniform convergence observed for uniformly-scaled problems. This study uses three different grammatical genetic codes to induce sequentiality. Genotypic genes in the grammatical codes are interpreted as phenotypes according to the grammar, and the grammar induces sequential interactions among phenotypic genes. The experimental results show that the grammatical codes can indeed induce sequentiality, but the GAs using them need exponential population sizes for a reliable search.

1 Introduction

Identification and exchange of important building blocks (BBs) is one of the key challenges in the design of genetic algorithms (GAs). Fixed recombination operators that do not adapt linkage of BBs have been shown to be inadequate and scale-up exponentially with the problem size [1]. Furthermore, GAs that adaptively identify and efficiently exchange BBs successfully solve boundedly difficult problems, usually requiring only polynomial number of function evaluations [2]. GAs that identify and exchange BBs and thereby solve difficult problems quickly, reliably, and accurately are called *competent* GAs [3].

One of the approaches to achieve competence is by means of linkage learning GA (LLGA) [4]. The LLGA takes the position that tightly linked BBs are evolutionarily advantageous. The LLGA is designed to achieve tight linkage between interacting variables. While the LLGA has been successful in solving non-uniformly scaled problems, it can only solve uniformly scaled problems of limited size [5,6]. In non-uniformly-scaled problems, since a selection operator identifies BBs sequentially, it helps the LLGA achieve tight linkage. However, in uniformly-scaled problems, a selection operator identifies BBs simultaneously. Therefore, it is difficult for the LLGA to achieve tight linkage for all BBs in parallel [7].

Recently, a genetic algorithm using grammatical evolution (GAuGE) [8,9], which was inspired by grammatical evolution [10,11], has been proposed to solve problems through a process of getting salient phenotypic genes clustered in a genotypic chromosome. The GAuGE relies on a grammatical genetic code in

K. Deb et al. (Eds.): GECCO 2004, LNCS 3102, pp. 1426–1437, 2004.

which genes are in a certain order interpreted according to the grammar, and the grammar induces sequential interactions among phenotypic genes corresponding to their determined order. In addition, the grammatical genetic code allows phenotypic genes to locate at any positions in a genotypic chromosome. If salient phenotypic genes get clustered on a specific part of a genotypic chromosome, they can be kept from their disruption due to a specific crossover operator as well as grammatical decoding.

We hypothesized that sequential interactions among phenotypic genes induced by grammar could induce prioritized phenotypic convergence for search problems including uniformly-scaled problems. Therefore, the objective of this paper is to investigate whether or not *sequentiality* can be induced in uniformly-scaled problem using grammatical genetic codes. Sequentiality is a phenomenon in which sub-solutions converge sequentially in time. The grammatical genetic codes used in this paper are based on similar principal as in the GAuGE.

This paper is organized as follows. Section 2 briefly describes studies on grammatical genetic codes and sequentiality. In section 3, three grammatical genetic codes used in this paper are explained. We empirically examine if GAs using the grammatical codes can induce sequentiality in section 4. Finally, we summarize our results and draw our conclusions.

2   Related Studies

Representation of the variables of a search problem play an important role in genetic and evolutionary search, and effects of a variety of genetic codes on the performance of GAs have been extensively studied. An exhaustive review of studies on genetic representations is beyond the scope of this paper and the reader is referred elsewhere [12,13,14] and to the references therein.

One of the motivations for this study came from [15,16], in which GAs with seemingly disruptive and highly epistatic genetic codes were successful in solving difficult combinatorial problems. Some researchers have also used grammar-based genetic codes, which are also highly epistatic, with reasonable GA success [17,8]. In [17], the genes encode production rules, which are in turn used in a fixed manner to generate a structured phenotype.

The grammar-based genetic code used in the GAuGE [8,9] allows phenotypic genes to locate at any positions in a genotype chromosome similar to the representation used by Goldberg and Lingle [18] and the representations used in messy GAs [14] and the LLGA [4]. The grammar in GAuGE also induces sequential interactions among phenotypic loci, which is determined by the genotype-to-phenotype decoding procedure. LINKGAUGE [19], which is a variant of GAuGE uses grammars that induce sequential interactions not only among phenotypic alleles, but also among phenotypic loci.

At the first glance, it looks like such highly epistatic genetic codes should yield poor results. However, based on their empirical success, we wondered if such genetic codes might be simplifying the search problem by implicitly focusing on a single or few subproblems at a time. That is, we hypothesized that the genetic codes with high epistasis might be inducing sequentiality into search problems, which we investigate in this paper.

3 Grammatical Codes

Since we would like to verify if the strength of sequential interactions among phenotypic genes is directly related to inducing sequentiality, we employ three kinds of grammatical genetic codes which induce sequential interactions among phenotypic genes with different strength. The three codes are : (1) GAuGE code which is slight variant of [8], (2) complex grammatical code, and (3) cellular grammatical code. The codes (2) and (3) are meant to induce stronger interactions among phenotypic genes than the GAuGE code.

All the grammatical codes use integers as the genotypic genes, and all the genotypes are decoded from left to right. The grammatical codes (1) and (2) determine both the phenotypic loci and their alleles by applying modulus operation (%) to integers which are obtained in the decoding process. Interactions among phenotypic genes which are common to all the codes comes from relative phenotypic loci. All the phenotypic loci are labeled as integers, and they are relabeled every time one phenotypic locus is occupied. Those grammatical codes are in detail explained below, where a ℓ-bit optimization problem is assumed.

(1) GAuGE Code (Base 10)

The difference between the original GAuGE code proposed in [8] and the one used here is that in the original GAuGE, every integer, which is 0 to 255, is encoded into an eight-bit binary number. Here we directly use a base 10 integer from 1 to ℓ. That is, a GAuGE genotype used here can be written as $(p_1, v_1, p_2, v_2, \cdots, p_\ell, v_\ell)$, where $p_q, v_q \in [1, \ell]$ and $q = 1, 2, \cdots, \ell$. In this code, there are sequential interactions only among the phenotypic loci. The decoding procedure is as follows.

1. Let q be 1.
2. When $1 \leq q \leq \ell$, the unoccupied phenotypic loci are labeled as integers in $[1, \ell - q - 1]$ from left to right, which is as $(1, 2, \cdots, \ell - q - 1)$. The locus and its allele are determined as $p_q \% (\ell - q - 1) \in [1, \ell - q - 1]$, which represents one of the labels of the unoccupied loci, and $v_q \% 2 \in \{0, 1\}$, respectively.
3. In the case of $q = \ell$, the whole decoding process ends. Otherwise q increases by one, and return to procedure 2.

(2) Complex Grammatical Code

A genotype in the complex grammatical code consists of $\ell + 1$ real and imaginary parts in complex numbers and ℓ operations applied to two complex numbers. The genotypic genes are arranged as $(r_1, i_1, o_1, \cdots, r_\ell, i_\ell, o_\ell, r_{\ell+1}, i_{\ell+1})$, where $r_* \in [1, 3]$ represents the real part, $i_* \in [1, \ell]$ is the imaginary part, and $o_* \in \{\times, \times_t\}$ is the operation. In this code, there are sequential interactions among both the phenotypic loci and their alleles. This decoding procedure is described below. Since the decoding is done through ℓ iterations, the iteration number is denoted by $q = 1, 2, \cdots, \ell$.

1. Let q be 1.
2. In the case of $q = 1$, a new complex number is calculated as $R_1 + I_1 j = (r_1 + i_1 j) \times (r_2 + i_2 j)$ no matter what o_1 is, where j is a imaginary number. A phenotypic locus is obtained as $P_1 = |I_1|\%\ell + 1 \in [1, \ell]$, which points out one of the phenotypic loci labeled as 1 to ℓ from left to right. An allele at the locus is obtained as $V_1 = |R_1|\%2 \in \{0, 1\}$. If $o_1 = \times$, a new complex number is defined as $rr_2 + ii_2 j = (|R_1|\%3 + 1) + (|I_1|\%\ell + 1)j$. If $o_1 = \times_t$, a new complex number is defined as $rr_2 + ii_2 j = r_2 + i_2 j$.
3. In the case of $2 \leq q \leq \ell$, a new complex number is calculated as $R_q + I_q j = (rr_q + ii_q j) \times (r_{q+1} + i_{q+1} j)$. A phenotypic locus is obtained as $P_q = |I_q|\%(\ell - q - 1) + 1 \in [1, \ell - q]$, which points out one of the unoccupied loci relabeled as 1 to $(\ell - q)$ from left to right. An allele at the locus is obtained as $V_q = |R_q|\%2 \in \{0, 1\}$. If $o_q = \times$, a new complex number is defined as $rr_{q+1} + ii_{q+1} j = (|R_q|\%3 + 1) + (|I_q|\%\ell + 1)j$. If $o_q = \times_t$, a new complex number is defined as $rr_{q+1} + ii_{q+1} j = r_q + i_q j$.
4. In the case of $q = \ell$, the whole decoding process ends. Otherwise q increases by one, and return to procedure 3.

(3) Cellular Grammatical Code

A genotype in this code is interpreted as a system which is composed of a series connection of simple cellular automata. Each cellular automaton, $C_q (q = 1, 2, \cdots, \ell)$, is composed of four transition rules and an output timing. The inputs to the cellular automata, the outputs from them, and their inside states are represented by integers in a range of $[1, 4]$. The transition rules convert one integer ($\in [1, 4]$) into another one ($\in [1, 4]$). Therefore, integers ($\in [1, 4]$) are propagated among the cellular automata. The transition rules have not only their outputs but also information on a phenotypic locus and its allele, so that each cellular automaton can determine a phenotypic locus and its allele at its output timing. The output timing is also an integer ($\in [1, 8]$), which represents the number of the transitions. In this code, there are sequential interactions among both the phenotypic loci and their alleles. This decoding procedure is described below. Since the decoding is done through ℓ iterations, the iteration number is denoted by $q = 1, 2, \cdots \ell$.

1. In the case of $q = 1$, the initial input is given to the first cellular automaton C_1. In the case of $2 \leq q \leq \ell$, the output of the $(q - 1)$-th cellular automaton is give to the q-th one as its input.
2. When the input value to the cellular automaton is $i_q \in [1, 4]$, the i_q-th transition rule is activated. The state of the cell moves from i_q to $s_1 \in [1, 4]$ according to the i_q-th transition rule. This state transition is repeated until the number of times of the state transitions reaches the output timing o_t. After reaching o_t, the current state of the cell $s_{o_t} \in [1, 4]$ becomes the input value $i_{q+1} = s_{o_t}$ to the next cell C_{q+1}. Finally, one more the state transition is done according to the s_{o_t}-th transition rule, and the phenotypic locus and its allele are determined as $p_{o_t+1} \in [1, \ell - q]$, which represents one of the labels of the unoccupied loci labeled as 1 to $\ell - q$, and its allele $v_{o_t+1} \in [0, 1]$ that the s_{o_t+1}-th transition rule has, respectively.
3. In the case of $q = \ell$, the whole decoding process ends. Otherwise q increases by one, and return to procedure 1.

4   Experiments

4.1   Test Problems

We use three types of uniformly-scaled problems for investigating the GAs using the grammatical codes. Those are (1) OneMax problem with ℓ bits, (2) 4-bit trap deceptive function with tightly linked m BBs [20], and (3) 4-bit trap deceptive function with loosely linked m BBs. They are thereafter called OneMax-ℓ, $(m, 4)$-Trap-T, and $(m, 4)$-Trap-L, respectively.

(1) OneMax Problem with ℓ Bits (**OneMax-ℓ**)

This problem gives the number of ones in the phenotypes to their corresponding genotypes as their fitness values.

(2) 4-bit Trap Deceptive Function with Tightly Linked m BBs ($(m, 4)$-**Trap-T**)

A BB in the phenotype consists of four bits, and each BB is close to one another like $(B_1, B_2, \cdots, B_m)$, where B_q is the q-th BB. A fitness value of a genotype is the sum of fitness values that m BBs give. A fitness value of each BB is calculated in the same way. When the number of ones in a BB is u, the fitness value of the BB, $f_{BB}(u)$, is given by

$$f_{BB}(u) = \begin{cases} 4 & u = 4, \\ 3 - u & otherwise. \end{cases}$$

(3) 4-bit Trap Deceptive Function with Loosely Linked m BBs ($(m, 4)$-**Trap-L**)

A BB in the phenotype consists of four bits, and each BB is distant from one another. Concretely, the q-th BB is denoted by $(q, q+\ell/4, q+\ell/2, q+3\ell/4)$, where $q = 1, 2, \cdots, \ell/4$, $\ell(= 4m)$ is the length of the phenotype, and each element in that vector notation of the BB represents a phenotypic locus. A fitness value of a BB is calculated in the same way as done in $(m, 4)$-Trap-T.

4.2   Genotype-Phenotype-Mapping Characteristics

First of all the experiments, the characteristics of genotype-phenotype mappings of the three grammatical codes are examined. We observe two things: (1) how many small perturbations in the genotypes change their phenotypes, which was called *locality* in [21], and (2) how many small perturbations in the genotypes change their fitness values when concrete optimization problems are assumed. We use OneMax-32, $(8, 4)$-Trap-T, and $(8, 4)$-Trap-L as test problems. The experimental procedure is as follows:

1. A genotype is randomly generated, and then its phenotype is obtained by the genotype-phenotype-mapping. The genotype and phenotype generated are called *original genotype* and *original phenotype*, respectively. Also, the fitness value of the original phenotype, which is called *original fitness value*, is calculated.

2. A new genotype is obtained by modifying an allele at a certain locus in the original genotype, and then its phenotype is obtained. Also, the fitness value of the new phenotypes is calculated. The difference between the original and the new genotypes is just one allele at the chosen locus.

3. A Hamming distance between the original and the new generated phenotypes is calculated. Absolute value of the difference between two fitness values that the original and the new phenotypes have is calculated.
4. Iterating the procedure 2 to 3 until all the genotypes that are adjacent to the original genotype are generated and compared with the original genotype.
5. Iterating the procedure 1 to 4 until 100 original genotypes are generated and compared with all the genotypes adjacent to them.

The experimental results are shown in Fig. 1, which represents two things: (1) the averaged Hamming distance between the original phenotype and each of the other phenotypes corresponding to all the genotypes adjacent to the original one, and (2) the averaged difference between the original fitness value and each of the other fitness ones that all the genotypes adjacent to the original one have. The two kinds of averaged values were calculated for each genotypic locus.

Figure 1(a)-1(c) show that small perturbations on the left parts of the genotypes in the three grammatical codes caused bigger changes in their phenotypes than small perturbation on the right parts of them.

However, the changes in the fitness values that result from the changes in the phenotypes were not always like the ones in the phenotypes. When we assumed OneMax-32 and $(8, 4)$-Trap-T, the changes in the fitness values were almost flat over all the genotypic loci for almost all the grammatical codes used (Figure 1(d)-1(i)). When we assumed $(8, 4)$-Trap-L, the more left the genotypic loci were, the bigger the changes in the fitness values became (Figure 1(j)-1(h)). In this case, we could say that uniformly-scaled problems became non-uniformly-scaled ones at least in the local regions of their genotype spaces.

4.3   Genetic Algorithm

The results shown in the previous section suggest that the three grammatical codes have genotype-phenotype-mappings that give low correspondences between their genotypic and phenotypic neighborhoods. In addition, fitness landscapes on their genotypic spaces have multi-modalities because the three codes are redundant genetic representations. As a result, the fitness landscapes on their genotypic spaces should be highly rugged ones. We now briefly describe the GA used in this paper to investigate the grammatical genetic codes to induce sequentiality.

Minimal generation gap model (MGG) [22] is used as a generation gap model. This model literally minimizes a generation gap. Since it seems that fitness landscapes on the grammatical genetic codes are highly rugged, this generation gap model should be better than ones which change GA populations drastically at generation gaps. We will thereafter regard generating genotypes amounting a population size as one generation.

We use a one-point crossover operator from the viewpoint of not exploiting genotypes but minimizing the disruption of good genetic materials in the left part of the genotype. Since there are sequential interactions among the genotypic genes from left to right, the genes in the left part of the genotype should be kept from their disruption. A one-point crossover operator should be suitable from this point. Mutation operator is not used.

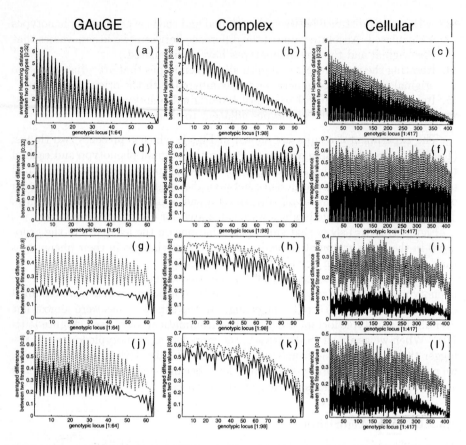

Fig. 1. Genotype-phenotype-mapping characteristics of the three grammatical codes. Labels of *GAuGE*, *Complex*, and *Cellular* represent the results for the GAuGE, the complex, and the cellular grammatical codes, respectively. The sub-figures (a)(b)(c) represent the averaged Hamming distance between the original phenotype and each of the other phenotypes corresponding to all the genotypes adjacent to the 100 original ones. The other sub-figures from (d) to (l) represent the averaged difference between the original fitness value and each of the other fitness ones that all the genotypes adjacent to the 100 original ones have. The sub-figures (d)(e)(f), (g)(h)(i), and (j)(k)(l) are for OneMax-32, $(8, 4)$-Trap-T8, and $(8, 4)$-Trap-L, respectively. Those two kinds of averaged values (**solid lines**) were calculated for each genotypic locus. When there are K genotypic alleles at a certain locus, $K - 1$ genotypes adjacent to an original genotype are obtained by modifying the allele at the locus in the original one. Then averaged value over $K - 1$ comparisons is obtained for the locus. The same procedure is applied to 100 original genotypes, so that the 100 averaged values are obtained. The final averaged value for the locus is obtained by averaging the 100 values. Standard deviations of the observed values are also plotted in all the sub-figures (**dash lines**).

4.4 Inducing Sequentiality

We examine if sequentiality is actually induced in uniformly-scaled problems by the GAs using the grammatical genetic codes. A GA using an identical map between a genotype and a phenotype spaces, which is called *standard GA* thereafter, is also used to compare with them. The population size is appropriately sized so that 95 out of 100 independent runs converge to the optimum. We observe convergence of both phenotypic alleles and loci. As for phenotypic alleles, we obtain averaged generations at which proportion of correct BBs or bits in the GA population are over 0.9. Since all the BBs or bits do not always converge in a fixed order, the generations at which the BBs or bits converged are sorted in ascending order and the sorted generations at the same order are averaged. As for phenotypic loci, we obtain averaged proportion of loci into which a set of genotypic genes at some order from the most left (1st) set in the GA population are mapped the most frequently. Those two averaged values are calculated using data of success runs out of 100. OneMax-80, $(8, 4)$-Trap-T, and $(8, 4)$-Trap-L are used here. The experimental results for the convergence of the phenotypic alleles and loci are shown in Table 1 and Fig. 2, respectively.

Table 1. Averaged generations over success runs out of 100 at which proportion of correct BBs or bits in the GA population were over 0.9. As for OneMax-80, the generations for 1st, 10th, 20th, 30th, 40th, 55th, 70th, and 80th converged bits are shown. As for $(8, 4)$-Trap-*, the generations for all the converged BBs are shown.

OneMax-80	pop. size	1st	10th	20th	30th	40th	55th	70th	80th
Standard GA	500	10.53	13.64	15.41	17.26	19.36	23.45	29.66	39.33
GAuGE	500	14.44	18.03	19.78	21.10	22.46	24.61	27.67	35.02
Complex	500	14.61	18.39	20.17	21.49	22.89	25.13	28.61	36.65
Cellular	2000	22.20	28.30	31.68	34.44	37.34	42.29	49.75	66.54
(8, 4)-Trap-T	pop. size	1st	2nd	3rd	4th	5th	6th	7th	8th
Standard GA	500	15.49	16.86	18.22	19.23	20.47	21.74	23.30	25.36
GAuGE	22000	33.70	35.44	36.54	38.09	39.29	42.08	44.65	48.94
Complex	22000	32.52	34.26	35.57	37.15	38.89	40.61	43.37	47.65
Cellular	50000	38.64	40.96	43.13	45.24	48.09	51.85	56.31	70.09
(8, 4)-Trap-L	pop. size	1st	2nd	3rd	4th	5th	6th	7th	8th
Standard GA	540000	39.26	40.98	42.65	43.94	45.59	48.03	51.41	57.94
GAuGE	32000	34.74	36.26	37.48	38.58	39.88	41.64	44.25	47.76
Complex	32000	34.08	35.89	37.26	38.30	39.46	41.15	43.58	46.90
Cellular	80000	38.84	41.08	43.29	45.03	46.97	50.18	54.52	63.73

Table 1 shows that the GAs using the grammatical code can induce sequentiality. However, since the standard GA also induced sequentiality, we can not conclude that the grammatical genetic code is the only factor to induce sequentiality. It is suggested that the genetic operators used, especially the generation gap model, could also be a possible factor.

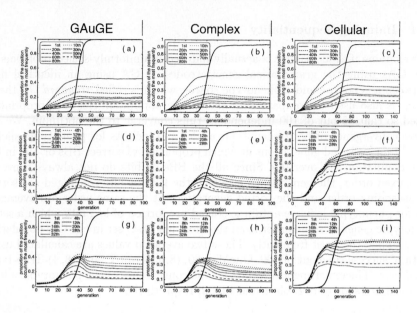

Fig. 2. Averaged proportion of loci into which the q-th set of genotypic genes from the most left (1st) set are mapped the most frequently over success runs out of 100. Labels of "GAuGE", "Complex", and "Cellular" mean the results for the GAuGE, the complex, and the cellular grammatical codes, respectively. As for OneMax-80 (the sub-figures (a)(b)(c)), the proportions for the 1st, 10th, 20th, 30th, 40th, 50th, 60th, 70th, and 80th sets of genotypic genes are plotted. As for $(8, 4)$-Trap-T $((d)(e)(f))$ and $(8, 4)$-Trap-L $((g)(h)(i))$, the proportions for the 1st, 4th, 8th, 12th, 16th, 20th, 24th, 28th, and 32th sets of genotypic genes are plotted. Also, the proportion of the genotypes with the optimal fitness value in the GA population is plotted in every sub-figure (**the thickest solid line**).

Figure 2 shows that the more left the genotypic genes are located in the genotypes, the more frequently they are mapped into the same locus. It is suggested that this fixations of the loci should be essential to induce sequentiality. However, as especially in the GAs using the GAuGE and complex codes for $(8, 4)$-Trap-T and $(8, 4)$-Trap-L, the degree of the fixations of the loci was low, and a variety of genotypes resided together in the GA populations even when the fitness values of all the genotypes almost converged. Considering the low degree of the loci fixations, it could be thought that the big reliable population sizes for $(8, 4)$-Trap-T and $(8, 4)$-Trap-L should result from the fact that the crossover operator used was not able to mix the genotypes effectively due to the lack of the mechanism to fix the loci properly.

4.5   Reliable Population Size

In the previous section, we verified that the GAs using the grammatical codes can induce sequentiality. However, the scalability of the GAs has not been revealed. Therefore, we examine population sizes with which the GAs using grammatical codes can reliably find global optima for given optimization problems. The reliable population sizes are determined as minimal population ones with which the GAs succeed in finding global optima for given optimization problems over 95 times out of 100 runs. The experimental results are shown in Table 2.

Table 2. The reliable population sizes for OneMax-40,60,80, $(m, 4)$-Trap-T $(m = 4, 6, 8)$, and $(m, 4)$-Trap-L $(m = 4, 6, 8)$. The reliable population size is determined as a minimal population size with which each GA can find global optima for the given optimization problems over 95 times out of 100 runs.

	OneMax-ℓ			$(m, 4)$-Trap-T			$(m, 4)$-Trap-L		
	40 bits	60 bits	80 bits	4 BBs	6 BBs	8 BBs	4 BBs	6 BBs	8 BBs
Standard GA	150	280	460	120	250	400	6000	50000	540000
GAuGE	140	280	420	1300	6000	20000	1400	7000	30000
Complex	140	300	460	1400	5000	18000	1000	5000	26000
Cellular	380	900	1700	1800	11000	48000	1400	12000	76000

From Table 2, we can predict that the reliable population sizes of the GAs using the three grammatical genetic codes for $(m, 4)$-Trap-T and $(m, 4)$-Trap-L exponentially increase with problem size. The function evaluations that need to find the global optima can be predicted to increase exponentially as well, though those data are not shown in this paper. In terms of scalability, the GAs using the three grammatical codes are impractical for GA-hard uniformly-scaled problems. However, we should examine the performances of them when smaller cardinal numbers are used for representing their genotypic genes in the further work.

5   Summary and Conclusion

We empirically examined grammatical genetic codes as one of the factors that induce sequentiality in uniformly-scaled problems. The factors to induce sequentiality are manifold, such as optimization problems, genotype-phenotype-mapping, population size, and genetic operators. This work focused on genotype-phenotype-mapping, and empirically observed their effects on sequentiality. The observed effects are: (1) the grammatical codes get uniformly-scaled problems to be non-uniformly-scaled ones, and help GAs induce sequentiality, (2) the genetic operators used help GAs induce sequentiality, and (3) impractical population sizes are needed for a successful search with sequentiality.

The results suggest that while the grammatical codes help GAs induce sequentiality together with the genetic operators, they are not enough to cause strong fixations of the genotypic genes for a recombination operator to mix the genotypes effectively, so that the GAs using the grammatical codes scale-up exponentially with problem size. On the other hand, selectomutative GAs might be more economical for grammatical codes in which genotypic genes are represented by integers than selectorecombinative GAs. If we are waiting for discovery of good genes block one after another by mutation, large population size might not really be needed. Therefore, by using mutation, we can do away with smaller populations, but might require longer time—in terms of number of generations—than in the case of crossover.

Our results are also useful in isolating some of the features of grammatical evolution (GE). One of the attributes for the success of GE might be a balanced mixture of inherent interactions among components of a program and interactions induced by grammar. Furthermore, unlike integer codes, the use of binary-coded genotypic genes in GE likely bring diversity and flexibility into search. Finally, the selectomutative part of GE might also be playing a more important role than it appears on a first glance.

Acknowledgments. This work was sponsored by the Air Force Office of Scientific Research, Air Force Materiel Command, USAF, under grant F49620-03-1-0129, and by the Technology Research Center (TRECC), a program of the University of Illinois at Urbana-Champaign, administered by the National Center for Supercomputing Applications (NCSA) and funded by the Office of Naval Research under grant N00014-01-1-0175. The US Government is authorized to reproduce and distribute reprints for Government purposes notwithstanding any copyright notation thereon.

The views and conclusions contained herein are those of the authors and should not be interpreted as necessarily representing the official policies or endorsements, either expressed or implied, of the Air Force Office of Scientific Research, or the U.S. Government.

References

1. Thierens, D., Goldberg, D.E.: Mixing in genetic algorithms. In: Proceedings of the 5th International Conference on Genetic Algorithms (ICGA-93). (1993) 38–45
2. Goldberg, D.E.: The race, the hurdle, and the sweet spot: Lessons from genetic algorithms for the automation of design innovation and creativity. Evolutionary Design by Computers (1999) 105–118
3. Goldberg, D.E.: The Design of Innovation: Lessons from and for Competent Genetic Algorithms. Kluwer Academic Publishers, Norwell, MA (2002)
4. Harik, G.R., Goldberg, D.E.: Learning linkage. Foundations of Genetic Algorithms 4 (1996) 247–262
5. Harik, G.R.: Learning gene linkage to efficiently solve problems of bounded difficulty using genetic algorithms. PhD thesis, University of Michigan, Ann Arbor (1997) Also IlliGAL Report No. 97005.
6. Chen, Y.P., Goldberg, D.E.: Introducing start expression genes to the linkage learning genetic algorithm. In: Proceedings of Parallel Problem Solving from Nature VII. (2002) 351–360
7. Chen, Y.P., Goldberg, D.E.: Convergence time for the linkage learning genetic algorithm. IlliGAL Report No. 2003025, Illinois Genetic Algorithms Lab., Univ. of Illinois, Urbana, IL (2003)
8. Ryan, C., Nicolau, M., O'Neill, M.: Genetic algorithms using grammatical evolution. In: Proceedings of the Fifth European Conference on Genetic Programming (EuroGP 2002). (2002) 278–287

9. Nicolau, M., Ryan, C.: How functional dependency adapts to salience hierarchy in the GAuGE system. In: Proceedings of the Sixth European Conference on Genetic Programming (EuroGP 2003). (2003) 153–163

10. Ryan, C., Collins, J., O'Neill, M.: Grammatical evolution: Evolving programs for an arbitrary language. In: Proceedings of the First European Conference on Genetic Programming. (1998) 83–96

11. O'Neill, M., Ryan, C.: Grammatical evolution. IEEE Transactions on Evolutionary Computation 5 (2001) 349–358

12. Rothlauf, F., Goldberg, D.E.: Representations for Genetic and Evolutionary Algorithms. Physica-Verg, Heidelberg, New York (2002)

13. Whitley, D., Rana, S., Heckendorn, R.: Representation issues in neighborhood search and evolutionary algorithms. In: Genetic Algorithms and Evolution Strategy in Engineering and Computer Science. John Wiley & Sons Ltd, West Sussex, England (1997) 39–58

14. Goldberg, D.E., Korb, B., Deb, K.: Messy genetic algorithms: Motivation, analysis, and first results. Complex Systems 3 (1989) 493–530

15. Anderson, P.G.: Ordered greed. In: Proceedings of Third International ICSC Symposium on Soft Computing. (1999)

16. Anderson, P.G.: Ordered greed, ii: Graph coloring. In: Proceedings of the Internatinal NAISO Congress on Information science innovations (ISI2001). (2001)

17. Kitano, H.: Designing neural networks using genetic algorithms with graph generation system. Complex Systems 4 (1990) 461–476

18. Goldberg, D.E., Lingle, Jr., R.: Alleles, loci, and the traveling salesman problem. In: Proceedings of an International Conference on Genetic Algorithms and Their Applications. (1985) 154–159

19. Nicolau, M., Ryan, C.: LINKGAUGE: Tackling hard deceptive problems with a new linkage learning genetic algorithm. In: Proceedings of the Genetic and Evolutionary Computation Conference 2002 (GECCO 2002). (2002) 488–494

20. Deb, K., Goldberg, D.E.: Analyzing deception in trap functions. Foundations of Genetic Algorithms 2 (1993) 93–108

21. Rothlauf, F.: Towards a Theory of Representations for Genetic and Evolutionary Algorithms— Development of Basic Concepts and their Application to Binary and Tree Representations. Unpublished doctoral dissertation, University of Illinois at Urbana-Champaign, Urbana, IL (2001)

22. Satoh, H., Yamamura, M., Kobayashi, S.: Minimal generation gap model for GAs considering both exploration and expolation. In: Proceedings of the International Conference on Fuzzy Systems, Neural Networks and Soft Computing (Iizuka'96). (1996) 494–497

Author Index

Lecture Notes in Computer Science

For information about Vols. 1–3010

please contact your bookseller or Springer-Verlag

Vol. 3056: H. Dai, R. Srikant, Knowledge Discovery and L ... 2004. (Subseries LNAI).

Vol. 3055: H. Christiansen, M.-S. Hacid, T. Andreasen, H.L. Larsen (Eds.), Flexible Query Answering Systems. X, 500 pages. 2004. (Subseries LNAI).

Vol. 3054: I. Crnkovic, J.A. Stafford, H.W. Schmidt, K. Wallnau (Eds.), Component-Based Software Engineering. XI, 311 pages. 2004.

Vol. 3053: C. Bussler, J. Davies, D. Fensel, R. Studer (Eds.), The Semantic Web: Research and Applications. XIII, 490 pages. 2004.

Vol. 3052: W. Zimmermann, B. Thalheim (Eds.), Abstract State Machines 2004. Advances in Theory and Practice. XII, 235 pages. 2004.

Vol. 3051: R. Berghammer, B. Möller, G. Struth (Eds.), Relational and Kleene-Algebraic Methods in Computer Science. X, 279 pages. 2004.

Vol. 3050: J. Domingo-Ferrer, V. Torra (Eds.), Privacy in Statistical Databases. IX, 367 pages. 2004.

Vol. 3049: M. Bruynooghe, K.-K. Lau (Eds.), Program Development in Computational Logic. VIII, 539 pages. 2004.

Vol. 3047: F. Oquendo, B. Warboys, R. Morrison (Eds.), Software Architecture. X, 279 pages. 2004.

Vol. 3046: A. Laganà, M.L. Gavrilova, V. Kumar, Y. Mun, C.J.K. Tan, O. Gervasi (Eds.), Computational Science and Its Applications – ICCSA 2004. LIII, 1016 pages. 2004.

Vol. 3045: A. Laganà, M.L. Gavrilova, V. Kumar, Y. Mun, C.J.K. Tan, O. Gervasi (Eds.), Computational Science and Its Applications – ICCSA 2004. LIII, 1040 pages. 2004.

Vol. 3044: A. Laganà, M.L. Gavrilova, V. Kumar, Y. Mun, C.J.K. Tan, O. Gervasi (Eds.), Computational Science and Its Applications – ICCSA 2004. LIII, 1140 pages. 2004.

Vol. 3043: A. Laganà, M.L. Gavrilova, V. Kumar, Y. Mun, C.J.K. Tan, O. Gervasi (Eds.), Computational Science and Its Applications – ICCSA 2004. LIII, 1180 pages. 2004.

Vol. 3042: N. Mitrou, K. Kontovasilis, G.N. Rouskas, I. Iliadis, L. Merakos (Eds.), NETWORKING 2004, Networking Technologies, Services, and Protocols; Performance of Computer and Communication Networks; Mobile and Wireless Communications. XXXIII, 1519 pages. 2004.

Vol. 3040: R. Conejo, M. Urretavizcaya, J.-L. Pérez-de-la-Cruz (Eds.), Current Topics in Artificial Intelligence. XIV, 689 pages. 2004. (Subseries LNAI).

Vol. 3039: M. Bubak, G.D.v. Albada, P.M.A. Sloot, J.J. Dongarra (Eds.), Computational Science - ICCS 2004. LXVI, 1271 pages. 2004.

Vol. 3038: M. Bubak, G.D.v. Albada, P.M.A. Sloot, J.J. Dongarra (Eds.), Computational Science - ICCS 2004. LXVI, 1311 pages. 2004.

Vol. 3037: M. Bubak, G.D.v. Albada, P.M.A. Sloot, J.J. Dongarra (Eds.), Computational Science - ICCS 2004. LXVI, 745 pages. 2004.

Vol. 3036: M. Bubak, G.D.v. Albada, P.M.A. Sloot, J.J. Dongarra (Eds.), Computational Science - ICCS 2004. LXVI, 713 pages. 2004.

Vol. ... Ed.), Knowledge Management ... XII, 326 pages. 2004. (Subseries LNAI).

Vol. 3034: J. Favela, E. Menasalvas, E. Chávez (Eds.), Advances in Web Intelligence. XIII, 227 pages. 2004. (Subseries LNAI).

Vol. 3033: M. Li, X.-H. Sun, Q. Deng, J. Ni (Eds.), Grid and Cooperative Computing. XXXVIII, 1076 pages. 2004.

Vol. 3032: M. Li, X.-H. Sun, Q. Deng, J. Ni (Eds.), Grid and Cooperative Computing. XXXVII, 1112 pages. 2004.

Vol. 3031: A. Butz, A. Krüger, P. Olivier (Eds.), Smart Graphics. X, 165 pages. 2004.

Vol. 3030: P. Giorgini, B. Henderson-Sellers, M. Winikoff (Eds.), Agent-Oriented Information Systems. XIV, 207 pages. 2004. (Subseries LNAI).

Vol. 3029: B. Orchard, C. Yang, M. Ali (Eds.), Innovations in Applied Artificial Intelligence. XXI, 1272 pages. 2004. (Subseries LNAI).

Vol. 3028: D. Neuenschwander, Probabilistic and Statistical Methods in Cryptology. X, 158 pages. 2004.

Vol. 3027: C. Cachin, J. Camenisch (Eds.), Advances in Cryptology - EUROCRYPT 2004. XI, 628 pages. 2004.

Vol. 3026: C.V. Ramamoorthy, R. Lee, K.W. Lee (Eds.), Software Engineering Research and Applications. XV, 377 pages. 2004.

Vol. 3025: G.A. Vouros, T. Panayiotopoulos (Eds.), Methods and Applications of Artificial Intelligence. XV, 546 pages. 2004. (Subseries LNAI).

Vol. 3024: T. Pajdla, J. Matas (Eds.), Computer Vision - ECCV 2004. XXVIII, 621 pages. 2004.

Vol. 3023: T. Pajdla, J. Matas (Eds.), Computer Vision - ECCV 2004. XXVIII, 611 pages. 2004.

Vol. 3022: T. Pajdla, J. Matas (Eds.), Computer Vision - ECCV 2004. XXVIII, 621 pages. 2004.

Vol. 3021: T. Pajdla, J. Matas (Eds.), Computer Vision - ECCV 2004. XXVIII, 633 pages. 2004.

Vol. 3019: R. Wyrzykowski, J.J. Dongarra, M. Paprzycki, J. Wasniewski (Eds.), Parallel Processing and Applied Mathematics. XIX, 1174 pages. 2004.

Vol. 3018: M. Bruynooghe (Ed.), Logic Based Program Synthesis and Transformation. X, 233 pages. 2004.

Vol. 3017: B. Roy, W. Meier (Eds.), Fast Software Encryption. XI, 485 pages. 2004.

Vol. 3016: C. Lengauer, D. Batory, C. Consel, M. Odersky (Eds.), Domain-Specific Program Generation. XII, 325 pages. 2004.

Vol. 3015: C. Barakat, I. Pratt (Eds.), Passive and Active Network Measurement. XI, 300 pages. 2004.

Vol. 3014: F. van der Linden (Ed.), Software Product-Family Engineering. IX, 486 pages. 2004.

Vol. 3012: K. Kurumatani, S.-H. Chen, A. Ohuchi (Eds.), Multi-Agnets for Mass User Support. X, 217 pages. 2004. (Subseries LNAI).

Vol. 3011: J.-C. Régin, M. Rueher (Eds.), Integration of AI and OR Techniques in Constraint Programming for Combinatorial Optimization Problems. XI, 415 pages. 2004.